McGraw–Hill Dictionary of PHYSICS and MATHEMATICS

Daniel N. Lapedes

Editor in Chief

McGraw-Hill Book Company

New York   St. Louis   San Francisco

Auckland
Bogotá
Düsseldorf
Johannesburg
London
Madrid
Mexico
Montreal

New Delhi
Panama
Paris
São Paulo
Singapore
Sydney
Tokyo
Toronto

Library of Congress Cataloging in Publication Data

McGraw-Hill dictionary of physics and mathematics.

   1. Physics—Dictionaries. 2. Mathematics—
Dictionaries. 3. Science—Dictionaries. I. Lapedes,
Daniel N. II. Title: Dictionary of physics and mathe-
matics.
QC5.M23   530′.03   78-8983
ISBN 0-07-045480-9

Contents

Preface

The *McGraw-Hill Dictionary of Physics and Mathematics* is intended to provide the high school and college student, librarian, teacher, engineer, researcher, and the general public with the vocabulary of physics, mathematics, and related disciplines such as statistics, astronomy, electronics, and geophysics. Care has been taken to include not only the basic vocabulary but also the most current and specialized terminology.

The more than 20,000 terms and definitions in the *McGraw-Hill Dictionary of Physics and Mathematics* are, in the opinion of the Board of Consulting Editors, fundamental to understanding physics and mathematics. The definitions either were written especially for this work or were drawn from the broader *McGraw-Hill Dictionary of Scientific and Technical Terms* (2d Ed., 1978). The present dictionary is a product of data-base operations, for the terms and definitions selected from the larger work were extracted from a master file stored on magnetic tape. As additional terms were written and reviewed by the consulting editors, they were alphabetically collated with the extracted terms on a new tape. The present dictionary was generated from this tape utilizing computer composition.

Each definition is preceded by an abbreviation identifying the field in which it is primarily used. Some of the fields covered are atomic physics, quantum mechanics, optics, electronics, crystallography, spectroscopy, fluid mechanics, acoustics, physical chemistry, geophysics, and astronomy. When a definition applies to more than one field, it is identified by a more general field. For example, a definition that applies to both crystallography and solid-state physics is assigned to physics.

The usefulness of this dictionary is enhanced by illustrations, cross-references, and the Appendix. There are approximately 700 illustrations to amplify the definitions. Synonyms are given in the alphabetical sequence and are cross-referenced to the term where the definition appears. The Appendix has an explanation of the International System of Units, with conversion tables; tables giving properties of the chemical elements, elementary particles, and of planets and stars; diagrams of crystal lattices; and mathematical and special constants.

An explanation of the alphabetization, cross-referencing, format, field abbreviations, and other information on how to use the dictionary begins on page ix.

The *McGraw-Hill Dictionary of Physics and Mathematics* is the result of the ideas and efforts of the editorial staff and the consulting editors. It is a reference tool which the editors hope will serve the information and communication needs of the community in the fields of physics and mathematics.

Daniel N. Lapedes
Editor in Chief

Editorial Staff

Consulting Editors

How to Use the Dictionary

I. ALPHABETIZATION

The terms in the *McGraw-Hill Dictionary of Physics and Mathematics* are alphabetized on a letter-by-letter basis; word spacing, hyphen, comma, solidus, and apostrophe in a term are ignored in the sequencing. For example, an ordering of terms would be:

Ah
A/in.²
air-core coil
air current
airflow

II. FORMAT

The basic format for a defining entry provides the term in boldface, the field in small capitals, and the single definition in lightface:

term [FIELD] Definition.

A field may be followed by multiple definitions, each introduced by a boldface number:

term [FIELD] **1.** Definition. **2.** Definition. **3.** Definition.

A term may have definitions in two or more fields:

term [ASTROPHYS] Definition. [MATH] Definition.

A simple cross-reference entry appears as:

term *See* another term.

A cross-reference may also appear in combination with definitions:

term [ASTROPHYS] Definition. [MATH] *See* another term.

III. CROSS-REFERENCING

A cross-reference entry directs the user to the defining entry. For example, the user looking up "accumulative error" finds:

accumulative error *See* cumulative error.

The user then turns to the "C" terms for the definition. Cross-references are also made from variant spellings, acronyms, abbreviations, and symbols.

ATR *See* attenuated total reflectance.
at. wt *See* atomic weight.
Au *See* gold.
Tchebycheff *See* Chebyshev.

IV. ALSO KNOWN AS . . . , etc.

A definition may conclude with a mention of a synonym of the term, a variant spelling, an abbreviation for the term, or other such information, introduced by "Also known as . . .," "Also spelled . . .," "Abbreviated . . .," "Symbolized . . .," "Derived from. . . ." When a term has more than one definition, the positioning of any of these phrases conveys the extent of applicability. For example:

> **term** [ELECTROMAG] **1.** Definition. Also known as synonym. **2.** Definition. Symbolized T.

In the above arrangement, "Also known as . . ." applies only to the first definition: "Symbolized . . ." applies only to the second definition.

> **term** [ELECTROMAG] **1.** Definition. **2.** Definition. [RELAT] Definition. Also known as synonym.

In the above arrangement "Also known as . . ." applies only to the second field.

> **term** [ELECTROMAG] Also known as synonym. **1.** Definition. **2.** Definition. [RELAT] Definition.

In the above arrangement, "Also known as . . ." applies to both definitions in the first field.

> **term** Also known as synonym. [ELECTROMAG] **1.** Definition. **2.** Definition. [RELAT] Definition.

In the above arrangement, "Also known as . . ." applies to all definitions in both fields.

Field Abbreviations

ACOUS	acoustics	HOROL	horology
ADP	automatic data processing	IND ENG	industrial engineering
AERO ENG	aerospace engineering	INORG CHEM	inorganic chemistry
ANALY CHEM	analytical chemistry	MATER	materials
ASTRON	astronomy	MATH	mathematics
ASTROPHYS	astrophysics	MECH	mechanics
ATOM PHYS	atomic physics	MED	medicine
BIOCHEM	biochemistry	MICROBIO	microbiology
BIOL	biology	MINERAL	mineralogy
BIOPHYS	biophysics	NUCLEO	nucleonics
CHEM	chemistry	NUC PHYS	nuclear physics
CHEM ENG	chemical engineering	OPTICS	optics
COMMUN	communications	ORD	ordnance
CONT SYS	control systems	PARTIC PHYS	particle physics
CRYO	cryogenics	PHYS	physics
CRYSTAL	crystallography	PHYS CHEM	physical chemistry
CYTOL	cytology	PHYSIO	physiology
ELEC	electricity	PL PHYS	plasma physics
ELECTR	electronics	QUANT MECH	quantum mechanics
ELECTROMAG	electromagnetism	RELAT	relativity
ENG	engineering	SCI TECH	science and technology
ENG ACOUS	engineering acoustics	SOLID STATE	solid-state physics
FL MECH	fluid mechanics	SPECT	spectroscopy
GEOCHEM	geochemistry	STAT	statistics
GEOD	geodesy	STAT MECH	statistical mechanics
GEOL	geology	SYS ENG	systems engineering
GEOPHYS	geophysics	THERMO	thermodynamics
GRAPHICS	graphic arts		

Scope of Fields

acoustics — The science of the production, transmission, and effects of sound.

aerospace engineering — Engineering pertaining to the design and construction of aircraft and space vehicles and of power units, and dealing with the special problems of flight in both the earth's atmosphere and space, such as in the flight of air vehicles and the launching, guidance, and control of missiles, earth satellites, and space vehicles and probes.

analytical chemistry — Science and art of determining composition of materials in terms of elements and compounds which they contain.

astronomy — The science concerned with celestial bodies and with the observation and interpretation of radiation received in the vicinity of earth from the component parts of the universe.

astrophysics — A branch of astronomy that treats of the physical properties of celestial bodies, such as luminosity, size, mass, density, temperature, and chemical composition, and their origin and evolution.

atomic physics — A branch of physics concerned with the structures of the atom, the characteristics of the electrons and other elementary particles of which the atom is composed, the arrangement of the atom's energy states, and the processes involved in the radiation of light and x-rays.

automatic data processing — The machine performance, with little or no human assistance, of any of a variety of tasks involving informational data; examples include automatic and responsive reading, computation, writing, speaking, directing artillery, and running of an entire factory.

biochemistry — The study of the chemical substances that occur in living organisms, the processes by which these substances enter into or are formed in the organisms and react with each other and the environment, and the methods by which the substances and processes are identified, characterized, and measured.

biology — The science of living organisms, concerned with the study of embryology, anatomy, physiology, cytology, morphology, taxonomy, genetics, evolution, and ecology.

biophysics — The hybrid science involving the methods and ideas of physics and chemistry to study and explain the structures of living organisms and the mechanics of life processes.

chemical engineering — A branch of engineering that deals with the development and application of manufacturing processes, such as refinery processes, which chemically convert raw materials into a variety of products, and that deals with the design and operation of plants and equipment to perform such work.

chemistry — The scientific study of the properties, composition, and structure of matter, the changes in structure and composition of matter, and accompanying energy changes.

communications — The science and technology by which information is collected from an originating source, transformed into electric currents or fields, transmitted over electrical networks or space to another point, and reconverted into a form suitable for interpretation by a receiver.

control systems — The study of those systems in which one or more outputs are forced to change in a desired manner as time progresses.

cryogenics — The science of producing and maintaining very low temperatures, of phenomena at those temperatures, and of technical operations performed at very low temperatures.

crystallography — The branch of science that deals with the geometric description of crystals, their internal arrangement, and their properties.

cytology — The branch of biological science which deals with the structure, behavior, growth, and reproduction of cells and the function and chemistry of cells and cell components.

electricity — The science of physical phenomena involving electric charges and their effects when at rest and when in motion.

electromagnetism — The branch of physics dealing with the observations and laws relating electricity to magnetism, and with magnetism produced by an electric current.

electronics — The branch of science and technology relating to the conduction of electricity through gases or vacuum or through semiconducting materials; concerned with the design, manufacture, and application of electron tubes.

engineering — The science by which the properties of matter and the sources of power in nature are made useful to humans in structures, machines, and products.

engineering acoustics — A field of acoustics that deals with the production, detection, and control of sound by electrical devices, including the study, design, and construction of such things as microphones, loudspeakers, sound recorders and reproducers, and public address systems.

fluid mechanics — The science concerned with fluids, either at rest or in motion, and dealing with pressures, velocities, and accelerations in the fluid, including fluid deformation and compression or expansion.

geochemistry — The study of the chemical composition of the various phases of the earth and the physical and chemical processes which have produced the observed distribution of the elements and nuclides in these phases.

geodesy — A subdivision of geophysics which includes determinations of the size and shape of the earth, the earth's gravitational field, and the location of points fixed to the earth's crust in an earth-referred coordinate system.

geology — The study or science of earth, its history, and its life as recorded in the rocks; includes the study of the geologic features of an area, such as the geometry of rock formations, weathering and erosion, and sedimentation.

geophysics — A branch of geology in which the principles and practices of physics are used to study the earth and its environment, that is, earth, air, and (by extension) space.

graphic arts — The fine and applied arts of representation, decoration, and writing or printing on flat surfaces together with the techniques and crafts associated with each; includes painting, drawing, engraving, etching, lithography, photography, and printing arts.

horology — Science of time measurement and the principles and technology of constructing time-measuring instruments.

industrial engineering — The application of engineering principles and training and the techniques of scientific management to the maintenance of a high level of productivity at optimum cost in industrial enterprises, as by analytical study, improvement, and installation of methods and systems, operating procedures, quantity and quality measurements and controls, safety measures, and personnel administration.

inorganic chemistry — A branch of chemistry that deals with reactions and properties of all chemical elements and their compounds, excluding hydrocarbons but usually including carbides and other simple carbon compounds (such as CO_2, CO, and HCN).

materials — The study of admixtures of matter or the basic matter from which products are made; includes adhesives, building materials, fuels, paints, leathers, and so on.

mathematics — The deductive study of shape, quantity, and dependence; the two main areas are applied mathematics and pure mathematics, the former arising from the study of physical phenomena, the latter involving the intrinsic study of mathematical structures.

mechanics — The branch of physics which seeks to formulate general rules for predicting the behavior of a physical system under the influence of any type of interaction with its environment.

medicine — The study of cause and treatment of human disease, including the healing arts dealing with diseases which are treated by a physician or a surgeon.

microbiology — The science and study of microorganisms, especially bacteria and rickettsiae, and of antibiotic substances.

mineralogy — The science concerning the study of natural inorganic substances called minerals, including origin, description, and classification.

nuclear physics — The study of the characteristics, behavior, and internal structure of the atomic nucleus.

nucleonics — The technology based on phenomena of the atomic nucleus such as radioactivity, fission, and fusion; includes nuclear reactors, various applications of radioisotopes and radiation, particle accelerators, and radiation detection devices.

optics — The study of phenomena associated with the generation, transmission, and detection of electromagnetic radiation in the spectral range extending from the long-wave edge of the x-ray region to the short-wave edge of the radio region; and the science of light.

ordnance — That military area concerned with supplies, including weapons, ammunition, combat vehicles, and the necessary repair equipment; and with heavy firearms discharged from mounts, including cannons and artillery.

particle physics — The branch of physics concerned with understanding the properties, behavior, and structure of elementary particles, especially through study of collisions or decays involving energies of hundreds of MeV or more.

physical chemistry — The description and prediction of chemical behavior by means of physical theory, with extensive use of graphs and mathematical formulas; main subject areas are structure, thermodynamics, and kinetics.

physics — The science concerned with those aspects of nature which can be understood in terms of elementary principles and laws.

physiology — The branch of biological science concerned with the basic activities that occur in cells and tissues of living organisms and involving physical and chemical studies of these organisms.

plasma physics — The study of highly ionized gases.

quantum mechanics — The modern theory of matter, of electromagnetic radiation, and of the interaction between matter and radiation; it differs from classical physics, which it generalizes and supersedes, mainly in the realm of atomic and subatomic phenomena.

relativity — The study of physics theory which recognizes the universal character of the propagation speed of light and the consequent dependence of space, time, and other mechanical measurements on the motion of the observer performing the measurements; the two main divisions are special theory and general theory.

science and technology — The study of the natural sciences and the application of this knowledge for practical purposes.

solid-state physics — The branch of physics centering on the physical properties of solid materials; it is usually concerned with the properties of crystalline materials only, but it is sometimes extended to include the properties of glasses or polymers.

spectroscopy — The branch of physics concerned with the production, measurement, and interpretation of electromagnetic spectra arising from either emission or absorption of radiant energy by various substances.

statistical mechanics — That branch of physics which endeavors to explain and predict the macroscopic properties and behavior of a system on the basis of the known characteristics and interactions of the microscopic constituents of the system, usually when the number of such constituents is very large.

statistics — The science dealing with the collection, analysis, interpretation, and presentation of masses of numerical data.

systems engineering — The branch of engineering dealing with the design of a complex interconnection of many elements (a system) to maximize an agreed-upon measure of system performance.

thermodynamics — The branch of physics which seeks to derive, from a few basic postulates, relations between properties of substances, especially those which are affected by changes in temperature, and a description of the conversion of energy from one form to another.

a *See* ampere; atto-.

aΩ *See* abohm.

(aΩ)⁻¹ *See* abmho.

A *See* ampere; angstrom.

Å *See* angstrom.

A+ *See* A positive.

aA *See* abampere.

aAcm² *See* abampere centimeter squared.

aA/cm² *See* abampere per square centimeter.

A AND NOT B gate *See* AND NOT gate.

ab- [ELECTROMAG] A prefix used to identify centimeter-gram-second electromagnetic units, as in abampere, abcoulomb, abfarad, abhenry, abmho, abohm, and abvolt.

abac *See* nomograph.

abacus [MATH] An instrument for performing arithmetical calculations manually by sliding markers on rods or in grooves.

abampere [ELEC] The unit of electric current in the electromagnetic centimeter-gram-second system; 1 abampere equals 10 amperes in the absolute meter-kilogram-second-ampere system. Abbreviated aA. Also known as Bi; biot.

abampere centimeter squared [ELECTROMAG] The unit of magnetic moment in the electromagnetic centimeter-gram-second system. Abbreviated aAcm².

abampere per square centimeter [ELEC] The unit of current density in the electromagnetic centimeter-gram-second system. Abbreviated aA/cm².

Abbe condenser [OPTICS] A variable large-aperture lens system arranged substage to image a light source into the focal plane of a microscope objective.

Abbe number [OPTICS] A number which expresses the deviating effect of an optical glass on light of different wavelengths.

Abbe prism [OPTICS] A system used for image erection which is composed of two double right-angle prisms and involves four reflections.

Abbe refractometer [OPTICS] An optical instrument for the measurement of the refractive index of liquids.

Abbe's sine condition [OPTICS] A relationship which must hold to prevent aberration of a mirror or lens from producing a coma.

Abbe's theory [OPTICS] The theory that for a lens to produce a true image, it must be large enough to transmit the entire diffraction pattern of the object.

abcoulomb [ELEC] The unit of electric charge in the electromagnetic centimeter-gram-second system, equal to 10 coulombs. Abbreviated aC.

abcoulomb centimeter [ELEC] The electromagnetic centimeter-gram-second unit of electric dipole moment. Abbreviated aCcm.

ABACUS

Drawing of an abacus.

abcoulomb per cubic centimeter [ELEC] The electromagnetic centimeter-gram-second unit of volume density of charge. Abbreviated aC/cm³.

abcoulomb per square centimeter [ELEC] The electromagnetic centimeter-gram-second unit of surface density of charge, electric polarization, and displacement. Abbreviated aC/cm².

Abel-Gonchorov interpolation problem [MATH] The problem of finding a polynominal whose ith derivative is equal to the ith derivative of a given n-times differentiable function at x_i for $i = 0, 1, \ldots, n$, where $x_0, \ldots, x_n$ are $n + 1$ points of interpolation.

Abelian domain *See* Abelian field.

Abelian extension [MATH] A Galois extension whose Galois group is Abelian.

Abelian field [MATH] A set of elements $a, b, c, \ldots$ forming Abelian groups with addition and multiplication as group operations where $a(b + c) = ab + ac$. Also known as Abelian domain; domain.

Abelian group [MATH] A group whose binary operation is commutative; that is, $ab = ba$ for each a and b in the group.

Abelian operation *See* commutative operation.

Abelian theorems [MATH] A class of theorems which assert that if a sequence or function behaves regularly, then some average of the sequence or function behaves regularly; examples include the Abel theorem (second definition) and the statement that if a sequence converges to s, then its Cesaro summation exists and is equal to s.

Abelian tower [MATH] A normal tower in which each quotient group G_i/G_{i+1} is an Abelian group.

Abel's formula [MATH] A formula in the theory of differential equations which states that the Wronskian of any two solutions of the self-adjoint linear differential equation $(d/dx)[p(x)y'] + q(x)y = 0$ is equal to a constant divided by $p(x)$.

Abel's identities [MATH] Formulas satisfied by sums of products of any two infinite series, a_i and b_i; they may be written

$$\sum_{i=r}^{s} a_i b_i = b_{s+1}A_s - b_r A_{r-1} + \sum_{i=r}^{s} (b_i - b_{i+1})A_i$$

$$\sum_{i=r}^{s} a_i b_1 = b_r A'_r - b_s A'_{s+1} + \sum_{i=r}^{s-1} (b_{i+1} - b_i)A'_{i+1}$$

where

$$A_r = \sum_{i=1}^{r} a_i, \ A'_r = \sum_{i=r}^{\infty} a_i.$$

Abel's inequality [MATH] An inequality which states that the absolute value of the sum of n terms, each in the form ab, where the bs are positive numbers, is not greater than the product of the largest b with the largest absolute value of a partial sum of the as.

Abel's integral equation [MATH] The equation

$$f(x) = \int_a^x u(z)(x - z)^{-a} \, dz \ (0 < a < 1, x \geq a)$$

where $f(x)$ is a known function and $u(z)$ is the function to be determined; when $a = \frac{1}{2}$, this equation has application to Abel's problem.

Abel's matrix [MATH] An infinite matrix $[a_{mn}]$ whose elements are defined by the equation $a_{mn} = m^n/(m + 1)^{n+1}$.

Abel's problem [MATH] The problem which asks what path a particle will follow if it moves under the influence of gravity alone and its altitude-time function is to follow a specific law.

Abel-summable [MATH] The series

$$\sum_{n=1}^{\infty} a_n$$

is said to be Abel-summable to s if the function

$$f(x) = \sum_{n=1}^{\infty} a_n x^n, \, 0 \le x < 1,$$

has s as a limit on the left at $x = 1$. Also known as A-summable.

Abel theorem [MATH] **1.** A theorem stating that if a power series in z converges for $z = a$, it converges absolutely for $|z| < |a|$. **2.** A theorem stating that if a power series in z converges to $f(z)$ for $|z| < 1$ and to a for $z = 1$, then the limit of $f(z)$, as z approaches 1, is equal to a. **3.** A theorem stating that if the three series with nth term a_n, b_n, and $c_n = a_0 b_n + a_1 b_{n-1} + \cdots + a_n b_0$, respectively, converge, then the third series equals the product of the first two series.

aberration [ASTRON] The apparent angular displacement of the position of a celestial body in the direction of motion of the observer, caused by the combination of the velocity of the observer and the velocity of light. [OPTICS] *See* optical aberration.

abfarad [ELEC] A unit of capacitance in the electromagnetic centimeter-gram-second system equal to 10^9 farads. Abbreviated aF.

abhenry [ELEC] A unit of inductance in the electromagnetic centimeter-gram-second system of units equal to 10^{-9} henry. Abbreviated aH. Also known as centimeter.

abmho [ELEC] A unit of conductance in the electromagnetic centimeter-gram-second system of units equal to 10^9 mhos. Abbreviated $(a\Omega)^{-1}$. Also known as absiemens (aS).

Abney effect [OPTICS] A shift in the apparent hue of a light which occurs as colored light is desaturated by the addition of white light.

Abney law [OPTICS] The shift in apparent hue of spectral color that is desaturated by addition of white light is toward the red end of the spectrum if the wavelength is below 570 nanometers and toward the blue if it is above.

Abney mounting [SPECT] A modification of the Rowland mounting in which only the slit is moved to observe different parts of the spectrum.

abnormal glow discharge [ELECTR] A discharge of electricity in a gas tube at currents somewhat higher than those of an ordinary glow discharge, at which point the glow covers the entire cathode and the voltage drop decreases with increasing current.

abnormal reflections [ELECTROMAG] Sharply defined reflections of substantial intensity at frequencies greater than the critical frequency of the ionized layer of the ionosphere.

abnormal series *See* anomalous series.

abohm [ELEC] The unit of electrical resistance in the centimeter-gram-second system; 1 abohm equals 10^{-9} ohm in the meter-kilogram-second system. Abbreviated $a\Omega$.

abohm centimeter [ELEC] The centimeter-gram-second unit of resistivity. Abbreviated $a\Omega$cm.

A bomb *See* atomic bomb.

abrupt junction [ELECTR] A pn junction in which the concentration of impurities changes suddenly from acceptors to donors.

abscissa [MATH] One of the coordinates of a two-dimensional coordinate system, usually the horizontal coordinate, denoted by x.

absiemens *See* abmho.

absolute convergence [MATH] That property of an infinite series (or infinite product) of real or complex numbers if the

ABERRATION

The aberration of light as seen in astronomy. Starlight arriving along AB and seen in this direction by a stationary observer (left) appears to the observer in transverse motion AA′ (right) to come from the direction AA′ (or A′B). *(From G. de Vaucouleurs, Discovery of the Universe, 1957; reprinted by permission of Faber and Faber Ltd.)*

series (product) of absolute values converges; absolute convergence implies convergence.

absolute density *See* absolute gravity.

absolute deviation [STAT] The difference, without regard to sign, between a variate value and a given value.

absolute efficiency [ENG ACOUS] The ratio of the power output of an electroacoustic transducer, under specified conditions, to the power output of an ideal electroacoustic transducer.

absolute electrometer [ELEC] A very precise type of attracted disk electrometer in which the attraction between two disks is balanced against the force of gravity.

absolute error [MATH] In an approximate number, the numerical difference between the number and a number considered exact.

absolute expansion [THERMO] The true expansion of a liquid with temperature, as calculated when the expansion of the container in which the volume of the liquid is measured is taken into account; in contrast with apparent expansion.

absolute gain of an antenna [ELECTROMAG] Gain in a given direction when the reference antenna is an isotropic antenna isolated in space. Also known as isotropic gain of an antenna.

absolute gravity [CHEM] Density or specific gravity of a fluid reduced to standard conditions; for example, with gases, to 760 mmHg (101,325 newtons per square meter) pressure and 0°C temperature. Also known as absolute density.

absolute humidity [PHYS] The ratio of the mass of water vapor in a sample of air to the volume of the sample.

absolute inequality *See* unconditional inequality.

absolute linear momentum *See* absolute momentum.

absolute luminosity [OPTICS] The luminosity of an object expressed in units of fundamental quantities.

absolute magnitude [ASTROPHYS] **1.** A measure of the brightness of a star equal to the magnitude the star would have at a distance of 10 parsecs from the observer. **2.** The stellar magnitude any meteor would have if placed in the observer's zenith at a height of 100 kilometers. [MATH] The absolute value of a number or quantity.

absolute momentum [METEOROL] The sum of the (vector) momentum of a particle relative to the earth and the (vector) momentum of the particle due to the earth's rotation. Also known as absolute linear momentum.

absolute motion [PHYS] The motion of an object described by its measurement in a frame of reference that is preferred over all other frames.

absolute permeability [ELECTROMAG] The ratio of the magnetic flux density to the intensity of the magnetic field in a medium; measurement is in webers per square meter in the meter-kilogram-second system. Also known as induced capacity.

absolute pitch [ACOUS] The pitch of a musical tone expressed as the frequency of the sound wave of that tone.

absolute potential vorticity *See* potential vorticity.

absolute pressure [PHYS] The pressure above the absolute zero value of pressure that theoretically obtains in empty space or at the absolute zero of temperature, as distinguished from gage pressure.

absolute scale *See* absolute temperature scale.

absolute space-time [PHYS] A concept underlying Newtonian mechanics which postulates the existence of a preferred reference system of time and spatial coordinates; replaced in relativistic mechanics by Einstein's equivalency principle. Also known as absolute time.

absolute specific gravity [MECH] The ratio of the weight of a given volume of a substance in a vacuum at a given temperature to the weight of an equal volume of water in a vacuum at a given temperature.

absolute standard [PHYS] A particle or object designated as a standard by assigning to it a mass of one unit; used in defining quantities in Newton's second law of motion.

absolute system of units [PHYS] A set of units for measuring physical quantities, defined by interrelated equations in terms of arbitrary fundamental quantities of length, mass, time, and charge or current.

absolute temperature [THERMO] 1. The temperature measurable in theory on the thermodynamic temperature scale. 2. The temperature in Celsius degrees relative to the absolute zero at $-273.16°C$ (the Kelvin scale) or in Fahrenheit degrees relative to the absolute zero at $-459.69°F$ (the Rankine scale).

absolute temperature scale [THERMO] A scale with which temperatures are measured relative to absolute zero. Also known as absolute scale.

absolute time *See* absolute space-time.

absolute unit [PHYS] A unit defined in terms of units of fundamental quantities such as length, time, mass, and charge or current.

absolute vacuum [PHYS] A void completely empty of matter. Also known as perfect vacuum.

absolute value of a complex number [MATH] The modulus of a complex number; the square root of the sum of the squares of its real and imaginary part. Also known as magnitude of a complex number; modulus of a complex number.

absolute value of a real number [MATH] The number if it is nonnegative, and the negative of the number if it is negative. Also known as magnitude of a real number; numerical value of a real number.

absolute value of a vector [MATH] The length of a vector, disregarding its direction; the square root of the sum of the squares of its orthogonal components. Also known as magnitude of a vector.

absolute velocity [PHYS] The vector sum of the velocity of a fluid parcel relative to the earth and the velocity of the parcel due to the earth's rotation; the east-west component is the only one affected.

absolute viscosity [FL MECH] The tangential force per unit area of two parallel planes at unit distance apart when the space between them is filled with a fluid and one plane moves with unit velocity in its own plane relative to the other.

absolute vorticity [FL MECH] The vorticity of a fluid relative to an absolute coordinate system; especially, the vorticity of the atmosphere relative to axes not rotating with the earth.

absolute wavemeter [ELECTROMAG] A type of wavemeter in which the frequency of an injected radio-frequency voltage is determined by measuring the length of a resonant line.

absolute zero [THERMO] The temperature of $-273.16°C$, or $-459.69°F$, or 0 K, thought to be the temperature at which molecular motion vanishes and a body would have no heat energy.

absorbance [PHYS CHEM] The common logarithm of the reciprocal of the transmittance of a pure solvent. Also known as absorbancy; extinction.

absorbancy *See* absorbance.

absorbed charge [ELEC] Charge on a capacitor which arises only gradually when the potential difference across the capacitor is maintained, due to gradual orientation of permanent dipolar molecules.

ABSOLUTE TEMPERATURE

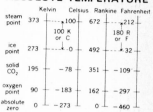

Comparisons of Kelvin, Celsius, Rankine, and Fahrenheit temperature scales. Temperatures are rounded off to nearest degree. *(From M. W. Zemansky, Temperatures Very Low and Very High, Van Nostrand, 1964)*

absorbed dose [NUCLEO] The amount of energy imparted by ionizing particles to a unit mass of irradiated material at a place of interest. Also known as dosage; dose.

absorbency index *See* absorptivity.

absorber [ELECTR] A material or device that takes up and dissipates radiated energy; may be used to shield an object from the energy, prevent reflection of the energy, determine the nature of the radiation, or selectively transmit one or more components of the radiation. [NUCLEO] A material that absorbs neutrons or other ionizing radiation.

absorber control *See* absorption control.

absorbing rod *See* control rod.

absorbing set [MATH] A set S in a real or complex vector space X such that for any x in X there exists a positive number ε such that ax belongs to S whenever $0 < |a| \leq \varepsilon$.

absorbing state [MATH] A special case of recurrent state in a Markov process in which the transition probability, P_{ii}, equals 1; a process will never leave an absorbing state once it enters.

absorptance [PHYS] The ratio of the total unabsorbed radiation to the total incident radiation; equal to 1 (unity) minus the transmittance.

absorption [CHEM] The taking up of matter in bulk by other matter, as in dissolving of a gas by a liquid. [ELEC] The property of a dielectric in a capacitor which causes a small charging current to flow after the plates have been brought up to the final potential, and a small discharging current to flow after the plates have been short-circuited, allowed to stand for a few minutes, and short-circuited again. Also known as dielectric soak. [ELECTROMAG] The taking up of energy from radiation by the medium through which the radiation is passing.

absorption band [PHYS] A range of wavelengths or frequencies in the electromagnetic spectrum within which radiant energy is absorbed by a substance.

absorption cell [OPTICS] A vessel with transparent walls for holding a gas or liquid whose absorptivity or absorption spectrum is to be measured.

absorption coefficient Also known as absorption factor; absorption ratio; coefficient of absorption. [ACOUS] The ratio of the sound energy absorbed by a surface of a medium or material to the sound energy incident on the surface. [PHYS] If a flux through a material decreases with distance x in proportion to e^{-ax}, then a is called the absorption coefficient.

absorption constant *See* absorptivity.

absorption control [NUCLEO] Control of a nuclear reactor by a material that absorbs neutrons, such as cadmium or boron steel. Also known as absorber control.

absorption cross section [ELECTROMAG] In radar, the ratio of the amount of power removed from a beam by absorption of radio energy by a target to the power in the beam incident upon the target.

absorption current [ELEC] The component of a dielectric current that is proportional to the rate of accumulation of electric charges within the dielectric.

absorption curve [PHYS] A graph showing the curvilinear relationship of the variation in absorbed radiation as a function of wavelength.

absorption edge [SPECT] The wavelength corresponding to a discontinuity in the variation of the absorption coefficient of a substance with the wavelength of the radiation. Also known as absorption limit.

absorption-emission pyrometer [ENG] A thermometer for

determining gas temperature from measurement of the radiation emitted by a calibrated reference source before and after this radiation has passed through and been partially absorbed by the gas.

absorption factor See absorption coefficient.

absorption index [OPTICS] The complex index of refraction may be written as $n(1 + ik)$; the coefficient k is the absorption index. Also known as index of absorption.

absorption lens [OPTICS] Glass which prevents selected wavelengths from passing through it; used in eyeglasses.

absorption limit See absorption edge.

absorption line [SPECT] A minute range of wavelength or frequency in the electromagnetic spectrum within which radiant energy is absorbed by the medium through which it is passing.

absorption meter [ENG] An instrument designed to measure the amount of light transmitted through a transparent substance, using a photocell or other light detector.

absorption nebula See dark nebula.

absorption peak [SPECT] A wavelength of maximum electromagnetic absorption by a chemical sample; used to identify specific elements, radicals, or compounds.

absorption ratio See absorption coefficient.

absorption spectrophotometer [SPECT] An instrument used to measure the relative intensity of absorption spectral lines and bands. Also known as difference spectrophotometer.

absorption spectroscopy [SPECT] The study of spectra obtained by the passage of radiant energy from a continuous source through a cooler, selectively absorbing medium.

absorption spectrum [SPECT] The array of absorption lines and absorption bands which results from the passage of radiant energy from a continuous source through a cooler, selectively absorbing medium.

absorption unit See sabin.

absorption wavemeter [ELECTR] A frequency- or wavelength-measuring instrument consisting of a calibrated tunable circuit and a resonance indicator.

absorptive laws [MATH] Either of two laws satisfied by the operations, usually denoted ∪ and ∩, on a Boolean algebra, namely $a \cup (a \cap b) = a$ and $a \cap (a \cup b) = a$, where a and b are any two elements of the algebra; if the elements of the algebra are sets, then ∪ and ∩ represent union and intersection of sets.

absorptive power See absorptivity.

absorptivity [ANALY CHEM] The constant a in the Beer's law relation $A = abc$, where A is the absorbance, b the path length, and c the concentration of solution. Also known as absorptive power. Formerly known as absorbency index; absorption constant; extinction coefficient. [THERMO] The ratio of the radiation absorbed by a surface to the total radiation incident on the surface.

absorptivity-emissivity ratio [ASTROPHYS] In space applications, the ratio of absorptivity for solar radiation of a material to its infrared emissivity. Also known as A/E ratio.

abstract algebra [MATH] The study of mathematical systems consisting of a set of elements, one or more binary operations by which two elements may be combined to yield a third, and several rules (axioms) for the interaction of the elements and the operations; includes group theory, ring theory, and number theory.

abstract Cauchy problem [MATH] The initial value problem for the evolution equation, that is, the problem of finding a solution u to this equation when $u(0)$ is given.

abstract Lie algebra [MATH] A finite-dimensional vector space with a multiplication operation such that xy is linear in x and y, $xx = 0$ for all vectors x in the space, and $x(yz) + y(zx) + z(xy) = 0$ for any three vectors x, y, and z in the space.

abstract theory [SCI TECH] A theory in which a system is described without specifying a structure.

abT *See* gauss.

abtesla *See* gauss.

abundance *See* abundance ratio.

abundance ratio [NUCLEO] The ratio of the number of atoms of one isotope to the total number of atoms in a mixture of isotopes. Also known as abundance.

abundant number [MATH] A positive integer which is greater than the sum of all its divisors, including unity.

abvolt [ELEC] The unit of electromotive force in the electromagnetic centimeter-gram-second system; 1 abvolt equals 10^{-8} volt in the absolute meter-kilogram-second system. Abbreviated aV.

abvolt per centimeter [ELEC] The electromagnetic centimeter-gram-second unit of electric field strength. Abbreviated aV/cm.

abwatt [ELEC] The unit of electrical power in the centimeter-gram-second system; 1 abwatt equals 1 watt in the absolute meter-kilogram-second system.

abWb *See* maxwell.

abweber *See* maxwell.

ac *See* alternating current.

aC *See* abcoulomb.

Ac *See* actinium.

accelerating electrode [ELECTR] An electrode used in cathode-ray tubes and other electron tubes to increase the velocity of the electrons that contribute the space current or form a beam.

acceleration [MECH] The rate of change of velocity with respect to time.

acceleration error constant [CONT SYS] The ratio of the acceleration of a controlled variable of a servomechanism to the actuating error when the actuating error is constant.

acceleration measurement [MECH] The technique of determining the magnitude and direction of acceleration, including translational and angular acceleration.

acceleration mechanisms [ASTROPHYS] The ways in which cosmic-ray and solar-flare particles may have acquired their high energies.

acceleration of free fall *See* acceleration of gravity.

acceleration of gravity [MECH] The acceleration imparted to bodies by the attractive force of the earth; has an international standard value of 980.665 cm/sec² but varies with latitude and elevation. Also known as acceleration of free fall; apparent gravity.

acceleration potential [FL MECH] The sum of the potential of the force field acting on a fluid and the ratio of the pressure to the fluid density; the negative of its gradient gives the acceleration of a point in the fluid.

acceleration voltage [ELECTR] The voltage between a cathode and accelerating electrode of an electron tube.

accelerator *See* particle accelerator.

accelerometer [ENG] An instrument which measures acceleration or gravitational force capable of imparting acceleration.

acceptor [SOLID STATE] An impurity element that increases the number of holes in a semiconductor crystal such as germanium or silicon; aluminum, gallium, and indium are examples. Also known as acceptor impurity; acceptor material; electron acceptor.

acceptor atom [SOLID STATE] An atom of a substance added to a semiconductor crystal to increase the number of holes in the conduction band.

acceptor impurity *See* acceptor.

acceptor level [SOLID STATE] An energy level in a semiconductor that results from the presence of acceptor atoms.

acceptor material *See* acceptor.

accessibility condition [MATH] The condition that any state of a finite Markov chain can be reached from any other state.

accessory plate [OPTICS] Thin plate of quartz, gypsum, or mica used with a petrological microscope to modify the effects of polarized light and intensify qualities in translucent minerals.

aCcm *See* abcoulomb centimeter.

aC/cm² *See* abcoulomb per square centimeter.

aC/cm³ *See* abcoulomb per cubic centimeter.

accommodation coefficient [STAT MECH] The ratio of the average energy actually transferred between a surface and impinging gas molecules scattered by the surface, to the average energy which would theoretically be transferred if the impinging molecules reached complete thermal equilibrium with the surface.

accommodation time [ELECTR] The time from the production of the first electron to the production of a steady electric discharge in a gas.

accretion hypothesis [ASTRON] Any hypothesis which assumes that the earth originated by the gradual addition of solid bodies, such as meteorites, that were formerly revolving about the sun but were drawn by gravitation to the earth.

accretion theory [ASTRON] A theory that the solar system originated from vortices in a disk-shaped mass.

accretive operator [MATH] A linear operator T defined on a subspace D of a Hilbert space which satisfies the following condition: the real part of the inner product of Tu with u is nonnegative for all u belonging to D.

accumulation factor [MATH] The quantity $(1+r)$ in the formula for compound interest, where r is the rate of interest; measures the rate at which the principal grows.

accumulation point *See* cluster point.

accumulative error *See* cumulative error.

accumulator *See* storage battery.

accumulator battery *See* storage battery.

Ac-Em *See* actinon.

a-c fracture [CRYSTAL] A type of tension fracture lying parallel to the a-c fabric plane and normal to plane b in a crystal.

achromat *See* achromatic lens.

achromatic [OPTICS] Capable of transmitting light without decomposing it into its constituent colors.

achromatic color [OPTICS] A color that has no hue or saturation but only brightness, such as white, black, and various shades of gray.

achromatic condenser [OPTICS] A condenser designed to eliminate chromatic and spherical aberrations, usually through the use of four elements, two of which are achromatic lenses; used in microscopes having high magnification.

achromatic fringe [OPTICS] An interference fringe of light whose position is independent of the wavelength of the light used; the first fringe of a Lloyd's mirror system and the central fringe of a Fresnel biprism system are examples.

achromatic lens [OPTICS] A combination of two or more lenses having a focal length that is the same for two quite different wavelengths, thereby removing a major portion of chromatic aberration. Also known as achromat.

achromatic prism [OPTICS] A prism consisting of two or more prisms with different refractive indices combined so that light passing through the device is deviated but not dispersed.

achronal set [RELAT] A set of points in a space-time with no two points of the set having timelike separation.

ACI *See* acoustic comfort index.

acid solution [CHEM] An aqueous solution containing more hydrogen ions than hydroxyl ions.

Ackeret method [FL MECH] A method of studying the behavior of an airfoil in a supersonic airstream based on the hypothesis that the disturbance caused by the airfoil consists of two plane waves, at the leading and trailing edges, which propagate outward like sound waves and each makes an angle equal to the Mach angle with the direction of flow.

aclastic [OPTICS] Having the property of not refracting light.

aclinic [GEOPHYS] Referring to a situation where a freely suspended magnetic needle remains in a horizontal position.

aclinic line *See* magnetic equator.

aΩcm *See* abohm centimeter.

acnode *See* isolated point.

acorn tube [ELECTR] An ultra-high-frequency electron tube resembling an acorn in shape and size.

acoustic [ACOUS] Relating to, containing, producing, arising from, actuated by, or carrying sound.

acoustic absorption *See* sound absorption.

acoustic absorption coefficient *See* sound absorption coefficient.

acoustic absorptivity *See* sound absorption coefficient.

acoustical [ACOUS] Having a characteristic concerning sound, of an object or quantity that in and of itself does not have properties associated with sound, such as a device, measurement, or symbol.

acoustical Doppler effect [ACOUS] The change in pitch of a sound observed when there is relative motion between source and observer.

acoustical holography [PHYS] A technique for using sound to form visible images, in which acoustic beams form an interference pattern of an object and a beam of light interacts with this pattern and is focused to form an optical image.

acoustical stiffness [ACOUS] The acoustic reactance associated with the potential energy of a medium multiplied by 2π times the sound frequency.

acoustic approximation [FL MECH] The approximation that leads from the nonlinear hydrodynamic equations of a gas to the linear wave equation for sound wave propagation.

acoustic branch [SOLID STATE] One of the parts of the dispersion relation, frequency as a function of wave number, for crystal lattice vibrations, representing vibration at low (acoustic) frequencies.

acoustic capacitance *See* acoustic compliance.

acoustic center [ENG ACOUS] The center of the spherical sound waves radiating outward from an acoustic transducer.

acoustic comfort index [ACOUS] An arbitrarily designed scale to indicate the noise inside the passenger cabin of an aircraft; on this scale $+100$ represents ideal conditions or zero noise, 0 represents barely tolerable conditions, and -100 represents intolerable conditions. Abbreviated ACI.

acoustic compliance [ACOUS] The reciprocal of acoustic stiffness. Also known as acoustic capacitance.

acoustic delay line [ELECTR] A device in which acoustic signals are propagated in a medium to make use of the sonic propagation time to obtain a time delay for the signals. Also known as sonic delay line.

acoustic dispersion [ACOUS] A complex sound wave's separation into its frequency components as it passes through a medium; usually measured by the rate of change of velocity with frequency.

acoustic domain [ACOUS] A concentration of crystal lattice vibrations traveling at the speed of sound; used to generate light from an array of *pn* junctions.

acoustic energy *See* sound energy.

acoustic fatigue [MECH] The tendency of a material, such as a metal, to lose strength after acoustic stress.

acoustic grating [ACOUS] A series of rods or other suitable objects of equal size placed in a row a fixed distance apart; causes sounds with different wavelengths to be diffracted in different directions.

acoustic hologram [ENG] The phase interference pattern, formed by acoustic beams, that is used in acoustical holography; when light is made to interact with this pattern, it forms an image of an object placed in one of the beams.

acoustic horn *See* horn.

acoustic image [ACOUS] The geometric space figure that is made up of the acoustic foci of an acoustic lens, mirror, or other acoustic optical system and is the acoustic counterpart of an extended source of sound. Also known as image.

acoustic imaging [ACOUS] The production of real-time images of the internal structure of a metallic or biological object that is opaque to light. Also known as ultrasonic imaging.

acoustic impedance [ACOUS] The complex ratio of the sound pressure on a given surface to the sound flux through that surface, expressed in acoustic ohms.

acoustic inertance *See* acoustic mass.

acoustic interferometer [ACOUS] A device for measuring the velocity and attenuation of sound waves in a gas or liquid by an interference method.

acoustic lens [MATER] Selected materials shaped to refract sound waves in accordance with the principles of geometrical optics, as is done for light. Also known as lens.

acoustic mass [ACOUS] The quantity which, after multiplication by 2π times the frequency, results in the acoustic reactance associated with the kinetic energy of the sound medium. Also known as acoustic inertance.

acoustic mass reactance [ACOUS] The part of the acoustic reactance associated with the kinetic energy of a medium. Also known as mass reactance.

acoustic measurement [ACOUS] The process of quantitatively determining one or more properties of sound.

acoustic microscope [OPTICS] An instrument which employs acoustic radiation at microwave frequencies to allow visualization of the microscopic detail exhibited in elastic properties of an object.

acoustic mode [SOLID STATE] The type of crystal lattice vibrations which for long wavelengths act like an acoustic wave in a continuous medium, but which for shorter wavelengths approach the Debye frequency, showing a dispersive decrease in phase velocity.

acoustic noise [ACOUS] Noise in the acoustic spectrum; usually measured in decibels.

acoustic ohm [ACOUS] The unit of acoustic impedance. Also known as acoustic reactance unit; acoustic resistance unit.

acoustic phonon [SOLID STATE] A quantum of excitation of an acoustic mode of vibration.

acoustic power *See* sound power.

ACOUSTIC INTERFEROMETER

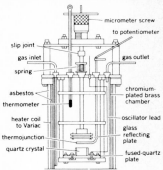

micrometer screw
to potentiometer
slip joint
gas inlet
gas outlet
spring
asbestos
chromium-plated brass chamber
thermometer
heater coil to Variac
oscillator lead
glass reflecting plate
thermojunction
quartz crystal
fused-quartz plate

The interferometric chamber in which measurements of change in velocity and attenuation of sound waves are made.

acoustic radiation [ACOUS] Infrasonic, sonic, or ultrasonic waves propagating through a solid, liquid, or gaseous medium.

acoustic radiation pressure [ACOUS] A unidirectional, steady-state pressure exerted upon a surface exposed to a sound wave.

acoustic radiometer [ENG] An instrument for measuring sound intensity by determining the unidirectional steady-state pressure caused by the reflection or absorption of a sound wave at a boundary.

acoustic ratio [ENG ACOUS] The ratio of the intensity of sound radiated directly from a source to the intensity of sound reverberating from the walls of an enclosure, at a given point in the enclosure.

acoustic reactance [ACOUS] The imaginary component of the acoustic impedance.

acoustic reactance unit *See* acoustic ohm.

acoustic reciprocity theorem [ACOUS] A theorem which states that in the acoustic field due to a sound source at point A, the sound pressure received at any other point B is the same as that which would be produced at A if the source were placed at B, and that this can be generalized for multiple sources and receivers.

acoustic reflection coefficient *See* acoustic reflectivity.

acoustic reflectivity [ACOUS] Ratio of the rate of flow of sound energy reflected from a surface, on the side of incidence, to the incident rate of flow. Also known as acoustic reflection coefficient; sound reflection coefficient.

acoustic refraction [ACOUS] Variation of the direction of sound transmission due to spatial variation of the wave velocity in the medium.

acoustic resistance [ACOUS] The real component of the acoustic impedance.

acoustic resistance unit *See* acoustic ohm.

acoustic resonance [ACOUS] A phenomenon exhibited by an acoustic system, such as an organ pipe or Helmholtz resonator, in which the response of the system to sound waves becomes very large when the frequency of the sound approaches a natural vibration frequency of the air in the system.

acoustic resonator [ACOUS] An enclosure that produces sound-wave resonance at a particular frequency.

acoustics [PHYS] **1.** The science of the production, transmission, and effects of sound. **2.** The characteristics of a room that determine the qualities of sound in it relevant to hearing.

acoustic scattering [ACOUS] The irregular reflection, refraction, and diffraction of sound in many directions.

acoustic shadow [ACOUS] A region immediately behind an object placed in the path of a sound wave whose wavelength is much smaller than the object, in which the initial sound wave is cut off by the object and the sound intensity is determined by the diffraction and interference of sound waves bent around the obstacle.

acoustic shielding [ACOUS] A sound barrier that prevents the transmission of acoustic energy.

acoustic spectrometer [ENG ACOUS] An instrument that measures the intensities of the various frequency components of a complex sound wave. Also known as audio spectrometer.

acoustic spectrum [ACOUS] The range of acoustic frequencies, extending from subsonic to ultrasonic frequencies, that is, approximately from zero to at least 1 megahertz.

acoustic stiffness [ACOUS] The product of the angular frequency and the acoustic stiffness reactance.

acoustic stiffness reactance [ACOUS] The part of acoustic reactance associated with the potential energy of a medium or its boundaries. Also known as stiffness reactance.

acoustic streaming [FL MECH] Unidirectional flow currents in a fluid that are due to the presence of sound waves.

acoustic theory [AERO ENG] The linearized small-disturbance theory used to predict the approximate airflow past an airfoil when the disturbance velocities caused by the flow are small compared to the flight speed and to the speed of sound.

acoustic transmission coefficient *See* sound transmission coefficient.

acoustic transmissivity *See* sound transmission coefficient.

acoustic wave [ACOUS] **1.** An elastic nonelectromagnetic wave that has a frequency which may extend into the gigahertz range; one type is a surface acoustic wave, and the other type is a bulk or volume acoustic wave. Also known as elastic wave. **2.** *See* sound.

acoustic-wave amplifier [ELECTR] An amplifier in which the charge carriers in a semiconductor are coupled to an acoustic wave that is propagated in a piezoelectric material, to produce amplification.

acoustoelectric effect [ELECTR] The development of a direct-current voltage in a semiconductor or metal by an acoustic wave traveling parallel to the surface of the material. Also known as electroacoustic effect.

acoustoelectronics [ENG ACOUS] The branch of electronics that involves use of acoustic waves at microwave frequencies (above 500 megahertz), traveling on or in piezoelectric or other solid substrates. Also known as pretersonics.

acoustooptical cell [ELEC] An electric-to-optical transducer in which an acoustic or ultrasonic electric input signal modulates or otherwise acts on a beam of light.

acoustooptical filter [OPTICS] An optical filter that is tuned across the visible spectrum by acoustic waves in the frequency range of 40 to 68 megahertz.

acoustooptic interaction [OPTICS] A way to influence the propagation characteristics of an optical wave by applying a low-frequency acoustical field to the medium through which the wave passes.

acoustooptic modulator [OPTICS] A device utilizing acoustooptic interaction ultrasonically to vary the amplitude or the phase of a light beam. Also known as Bragg cell.

acoustooptics [OPTICS] The science that deals with interactions between acoustic waves and light.

a-c plane [CRYSTAL] A plane at right angles to the surface of movement in a crystal.

acre [MECH] A unit of area, equal to 43,560 square feet, or to 4046.8564224 square meters.

actinic [PHYS] Pertaining to electromagnetic radiation capable of initiating photochemical reactions, as in photography or the fading of pigments.

actinic achromatism [OPTICS] **1.** The design of a photographic lens system so that light sources at the wavelength of the Fraunhofer D line near 589 nanometers and the G line at 430.8 nanometers are focused at the same point and produce images of the same size. **2.** The design of an astronomical lens system so that light sources at the wavelength of the Fraunhofer F line at 486.1 nanometers and the G line at 430.8 nanometers are focused at the same point and produce images of the same size. Also known as FG achromatism.

actinic focus [OPTICS] The point in an optical system at which the chemically most effective rays (usually those in the ultraviolet) converge. Also known as chemical focus.

actinic glass [OPTICS] Glass that transmits more of the visible components of incident radiation and less of the infrared and ultraviolet components.

actinide series [CHEM] The group of elements of atomic number 89 through 103. Also known as actinoid elements.

actinism [CHEM] The production of chemical changes in a substance upon which electromagnetic radiation is incident.

actinium [CHEM] A radioactive element, symbol Ac, atomic number 89; its longest-lived isotope is Ac^{227} with a half-life of 21.7 years; the element is trivalent; chief use is, in equilibrium with its decay products, as a source of alpha rays.

actinium decay series [NUCLEO] A series of radioactive disintegration products derived from uranium-235.

actinium emanation *See* actinon.

actinochemistry [CHEM] A branch of chemistry concerned with chemical reactions produced by light or other radiation.

actinodielectric [ELEC] Of a substance, exhibiting an increase in electrical conductivity when electromagnetic radiation is incident upon it.

actinoelectricity [ELEC] The electromotive force produced in a substance by electromagnetic radiation incident upon it.

actinogram [ENG] The record of heat from a source, such as the sun, as detected by a recording actinometer.

actinoid elements *See* actinide series.

actinology [PHYS] The branch of physics dealing with electromagnetic radiation and its chemical effects.

actinometry [ASTROPHYS] The science of measurement of radiant energy, particularly that of the sun, in its thermal, chemical, and luminous aspects.

actinon [NUC PHYS] A radioactive isotope of radon, symbol An, atomic number 86, atomic weight 219, belonging to the actinium series. Also known as actinium emanation (Ac-Em).

actinotherapy *See* radiation therapy.

actinouranium [NUC PHYS] A naturally occurring radioactive isotope of the actinium series, emitting only alpha decay; symbol AcU; atomic number 92; mass number 235; half-life 7.1×10^8 years; isotopic symbol U^{235}.

action [MECH] An integral associated with the trajectory of a system in configuration space, equal to the sum of the integrals of the generalized momenta of the system over their canonically conjugate coordinates. Also known as phase integral.

action at a distance theory [PHYS] A theory of the interaction of two bodies separated in space, without concern for a detailed mechanism of the propagation of effects between bodies.

action integral *See* action variable.

action-reaction law [PHYS] The law that when one body exerts force on another, the second body exerts a collinear force on the first equal in magnitude but oppositely directed.

action variable [PHYS] The integral $\int p\,dq$ over a cycle of a dynamical system; q is some coordinate, and p the conjugate momentum. Also known as action integral.

activate [PHYS] To start activity or motion in a device or material.

activated cathode [ELECTR] A thermionic cathode consisting of a tungsten filament to which thorium has been added, and then brought to the surface, by a process such as heating in the absence of an electric field in order to increase thermionic emission.

activated complex [PHYS CHEM] An energetically excited state which is intermediate between reactants and products in a chemical reaction.

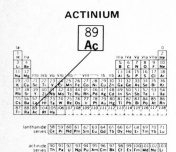

ACTINIUM

89
Ac

Periodic table of the chemical elements showing the position of actinium.

activated diffusion [SOLID STATE] Movement of atoms, ions, or lattice defects across a potential barrier in a solid.

activation [NUCLEO] The process of inducing radioactivity by bombardment with neutrons or other types of radiation.

activation analysis [NUCLEO] A method of chemical analysis based on the detection of characteristic radionuclides following a nuclear bombardment. Also known as radioactivity analysis.

activation cross section [NUC PHYS] The cross section for formation of a radionuclide by a particular interaction.

activation energy [PHYS CHEM] The energy, in excess over the ground state, which must be added to an atomic or molecular system to allow a particular process to take place.

active area [ELECTR] The area of a metallic rectifier that acts as the rectifying junction and conducts current in the forward direction.

active center [ASTRON] A localized, transient region of the solar atmosphere in which sunspots, faculae, plages, prominences, solar flares, and so forth are observed.

active component [ELEC] In the phasor representation of quantities in an alternating-current circuit, the component of current, voltage, or apparent power which contributes power, namely, the active current, active voltage, or active power. Also known as power component. [ELECTR] *See* active element.

active element [ELECTR] Any generator of voltage or current in an impedance network. Also known as active component. [NUC PHYS] A chemical element which has one or more radioactive isotopes.

active filter [ELECTR] A filter that uses an amplifier with conventional passive filter elements to provide a desired fixed or tunable pass or rejection characteristic.

active leg [ELECTR] An electrical element within a transducer which changes its electrical characteristics as a function of the application of a stimulus.

active material [ELEC] **1.** A fluorescent material used in screens for cathode-ray tubes. **2.** An energy-storing material, such as lead oxide, used in the plates of a storage battery. **3.** A material, such as the iron of a core or the copper of a winding, that is involved in energy conversion in a circuit. [ELECTR] The material of the cathode of an electron tube that emits electrons when heated. [NUCLEO] A material capable of releasing substantial quantities of nuclear energy during fission.

active power [ELEC] The product of the voltage across a branch of an alternating-current circuit and the component of the electric current that is in phase with the voltage.

active prominence region [ASTRON] Portions of the solar limb that display active prominences, characterized by downflowing knots and streamers, sprays, frequent surges, and curved loops. Abbreviated APR.

active region [ASTRON] A localized, transient, nonuniform region on the sun's surface, penetrating well down into the lower chromosphere. [ELECTR] The region in which amplifying, rectifying, light emitting, or other dynamic action occurs in a semiconductor device.

active substrate [SOLID STATE] A semiconductor or ferrite material in which active elements are formed; also a mechanical support for the other elements of a semiconductor device or integrated circuit.

active transducer [ELECTR] A transducer whose output is dependent upon sources of power, apart from that supplied by any of the actuating signals, which power is controlled by one or more of these signals.

active voltage [ELEC] In an alternating-current circuit, the component of voltage which is in phase with the current.

activity [NUC PHYS] The intensity of a radioactive source. Also known as radioactivity. [PHYS CHEM] A thermodynamic function that correlates changes in the chemical potential with changes in experimentally measurable quantities, such as concentrations or partial pressures, through relations formally equivalent to those for ideal systems.

activity coefficient [PHYS CHEM] A characteristic of a quantity expressing the deviation of a solution from ideal thermodynamic behavior; often used in connection with electrolytes.

actual height [ELECTROMAG] Highest altitude at which refraction of radio waves actually occurs.

actuator [CONT SYS] A mechanism to activate process control equipment by use of pneumatic, hydraulic, or electronic signals; for example, a valve actuator for opening or closing a valve to control the rate of fluid flow.

AcU *See* actinouranium.

acutance [OPTICS] An objective measure of the ability of a photographic system to show a sharp edge between contiguous areas of low and high illuminance.

acute angle [MATH] An angle of less than 90°.

acute triangle [MATH] A triangle each of whose angles is less than 90°.

acyclic [MATH] **1.** A transformation on a set to itself for which no nonzero power leaves an element fixed. **2.** A chain complex all of whose homology groups are trivial. [PHYS] Continually varying without a regularly repeated pattern.

acyclic motion *See* irrotational flow.

AD *See* average deviation.

Adams-Bashforth process [MATH] A method of numerically integrating a differential equation of the form $(dy/dx) = f(x,y)$ that uses one of Gregory's interpolation formulas to expand f.

adaptation brightness *See* adaptation luminance.

adaptation illuminance *See* adaptation luminance.

adaptation level *See* adaptation luminance.

adaptation luminance [OPTICS] The average luminance, or brightness, of objects and surfaces in the immediate vicinity of an observer estimating the visual range. Also known as adaptation brightness; adaptation illuminance; adaptation level; brightness level; field brightness; field luminance.

adaptive control [CONT SYS] A control method in which one or more parameters are sensed and used to vary the feedback control signals in order to satisfy the performance criteria.

adaptometer [ENG] An instrument that measures the lowest brightness of an extended area that can barely be detected by the eye.

adatom [PHYS CHEM] An atom adsorbed on a surface so that it will migrate over the surface.

Adcock antenna [ELECTROMAG] A pair of vertical antennas separated by a distance of one-half wavelength or less and connected in phase opposition to produce a radiation pattern having the shape of a figure eight.

adconductor cathode [ELECTR] A cathode in which adsorbed alkali metal atoms provide electron emission in a glow or arc discharge.

addend [MATH] One of a collection of numbers to be added.

adder [ELECTR] A circuit in which two or more signals are combined to give an output-signal amplitude that is proportional to the sum of the input-signal amplitudes. Also known as adder circuit.

adder circuit *See* adder.

adding circuit [ELECTR] A circuit that performs the mathematical operation of addition.

addition [MATH] An operation by which two elements of a set are combined to yield a third; denoted +; usually reserved for the operation in an Abelian group or the group operation in a ring or vector space.

addition of complex quantities [MATH] The combining of complex quantities in which the individual real parts and the individual imaginary parts are separately added.

addition of vectors [MATH] The combining of vectors in a prescribed way; for example, by algebraically adding corresponding components of vectors or by forming the third side of the triangle whose other sides each represent a vector. Also known as composition of vectors.

addition solid solution [CRYSTAL] Random addition of atoms or ions in the interstices within a crystal structure.

additive [MATH] Pertaining to addition. [STAT] That property of a process in which increments of the dependent variable are independent for nonoverlapping intervals of the independent variable.

additive function [MATH] Any function which preserves addition, such as $f(x+y) = f(x) + f(y)$.

additive primary colors [OPTICS] The three colors, usually red, green, and blue, which are mixed together in an additive process.

additive process [OPTICS] The process of producing colors by mixing lights of additive primary colors in various proportions.

additive set function [MATH] A set function with the property that the value of the function at a finite union of disjoint sets is equal to the sum of the values at each set in the union.

adhesion [ELECTROMAG] Any mutually attractive force holding together two magnetic bodies, or two oppositely charged nonconducting bodies. [MECH] The force of static friction between two bodies, or the effects of this force. [PHYS] The tendency, due to intermolecular forces, for matter to cling to other matter.

adhesional work [THERMO] The work required to separate a unit area of a surface at which two substances are in contact. Also known as work of adhesion.

adiabatic [THERMO] Referring to any change in which there is no gain or loss of heat.

adiabatic approximation [ASTROPHYS] The approximation that the pressure and density of gas in a star are related by the adiabatic law. [PHYS CHEM] *See* Born-Oppenheimer approximation.

adiabatic calorimeter [PHYS CHEM] An instrument used to study chemical reactions which have a minimum loss of heat.

adiabatic compression [THERMO] A reduction in volume of a substance without heat flow, in or out.

adiabatic cooling [THERMO] A process in which the temperature of a system is reduced without any heat being exchanged between the system and its surroundings.

adiabatic demagnetization [CRYO] A method of cooling paramagnetic salts to temperatures of 10^{-3} K; the sample is cooled to boiling point of helium in a strong magnetic field, thermally isolated, and then removed from the field to demagnetize it. Also known as Giaque-Debye method; magnetic cooling; paramagnetic cooling.

adiabatic ellipse [FL MECH] A plot of the speed of sound as a function of the speed of flow for the adiabatic flow of a gas, which forms one quadrant of an ellipse.

adiabatic envelope [THERMO] A surface enclosing a thermodynamic system in an equilibrium which can be disturbed only by long-range forces or by motion of part of the envelope; intuitively, this means that no heat can flow through the surface.

adiabatic expansion [THERMO] Increase in volume without heat flow, in or out.

adiabatic flame temperature [PHYS CHEM] The highest possible temperature of combustion obtained under the conditions that the burning occurs in an adiabatic vessel, that it is complete, and that dissociation does not occur.

adiabatic flow [FL MECH] Movement of a fluid without heat transfer.

adiabatic invariant [PHYS] A physical quantity which may be quantized and which, to a certain degree of approximation, remains unchanged under the slow variation of any parameter.

adiabatic law [PHYS] The relationship which states that, for adiabatic expansion of gases, $P\rho^{-\gamma}$ = constant, where P = pressure, ρ = density, and γ = ratio of specific heats C_P/C_V.

Adiabatic Low-energy Injection and Capture Experiment [NUCLEO] An experimental apparatus for research on controlled fusion that uses a beam of neutral deuterium atoms. Also known as Alice.

adiabatic process [THERMO] Any thermodynamic procedure which takes place in a system without the exchange of heat with the surroundings.

adiabatic recovery temperature [FL MECH] **1.** The temperature reached by a moving fluid when brought to rest through an adiabatic process. Also known as recovery temperature; stagnation temperature. **2.** The final and initial temperature in an adiabatic, Carnot cycle.

adiabatic system [SCI TECH] A body or system whose condition is altered without gaining heat from or losing heat to the surroundings.

adiabatic wall temperature [FL MECH] The temperature assumed by a wall in a moving fluid stream when there is no heat transfer between the wall and the stream.

adiathermanous [PHYS] Not capable of transmitting radiant heat. Also known as adiathermic.

adiathermic *See* adiathermanous.

adjacency matrix of a graph [MATH] For a graph with p vertices, this is a p by p matrix (a_{ij}) for which $a_{ij} = 1$ if the ith and jth vertices are joined by a common edge, and $a_{ij} = 0$ otherwise.

adjacent angle [MATH] One of a pair of angles with common side formed by two intersecting straight lines.

adjoint equation [MATH] The adjoint equation of the linear differential equation

$$L(y) \equiv \sum_{i=0}^{n} p_i(x)\frac{d^i y}{dx^i} = f(x)$$

is the equation

$$\bar{L}(y) \equiv \sum_{i=0}^{n} (-1)^{n-i}\frac{d^i}{dx^i}[p^i(x)y] = f(x).$$

adjoint of a matrix *See* adjugate of a matrix; Hermitian conjugate of a matrix.

adjoint operator [MATH] An operator B such that the inner products (Ax,y) and (x,By) are equal for a given operator A and for all elements x and y of a Hilbert space. Also known as associate operator; Hermitian conjugate operator.

adjoint variable [PHYS] In classical dynamics, the canonically conjugate p_i interpreted as generalized momenta.

adjoint vector space [MATH] The complete normed vector

space constituted by a class of bounded, linear, homogeneous scalar functions defined on a normed vector space.

adjoint wave functions [QUANT MECH] Functions in the Dirac electron theory which are formed by applying the Dirac matrix B to the Hermitian conjugates of the original wave functions.

adjugate of a matrix [MATH] The matrix obtained by replacing each element with the cofactor of the transposed element. Also known as adjoint of a matrix.

adjustable transformer *See* variable transformer.

adjusted decibel [ELECTR] A unit used to show the relationship between the interfering effect of a noise frequency, or band of noise frequencies, and a reference noise power level of −85 dBm. Abbreviated dBa. Also known as decibel adjusted.

adjusted value [SCI TECH] A value of a quantity derived from observed data by some orderly process which eliminates discrepancies arising from errors in those data.

admissible arc [MATH] In the calculus of variations, a set of state variables, $x_j(t)$, $j = 1, \ldots, m$, and a set of control variables, $u_j(t)$, $j = 1, \ldots, n$, where t is an independent variable which ranges from an initial value t_i to a final value t_f, which satisfy a set of differential constraints of the form $dx_j/dt - f_j(t, x_1, \ldots, x_m, u_1, \ldots, u_n) = 0$ where $j = 1, \ldots, m$; a set of initial conditions of the form $g_j(t_i, x_1(t_i), \ldots, x_m(t_i)) = 0$, where $j = 1, \ldots, p$; and a set of final conditions of the form $h_j(t_f, x_1(t_f), \ldots, x_m(t_f)) = 0$, where $j = 1, \ldots, q$, with $n \geq 1$, $p \leq m + 1$, $q \leq m + 1$, and $p + q \leq 2m + 1$; this concept is used in formulating the Mayer problem.

admittance [ELEC] A measure of how readily alternating current will flow in a circuit; the reciprocal of impedance, it is expressed in mhos.

ADP *See* automatic data processing.

adsorption [CHEM] The surface retention of solid, liquid, or gas molecules, atoms, or ions by a solid or liquid, as opposed to absorption, the penetration of substances into the bulk of the solid or liquid.

adsorption isobar [PHYS CHEM] A graph showing how adsorption varies with some parameter, such as temperature, while holding pressure constant.

adsorption isotherm [PHYS CHEM] The relationship between the gas pressure p and the amount w, in grams, of a gas or vapor taken up per gram of solid at a constant temperature.

advanced gas-cooled reactor [NUCLEO] A power-generating nuclear reactor which has steel-clad uranium dioxide fuel elements and is cooled by carbon dioxide gas.

advanced potential [ELECTROMAG] Any electromagnetic potential arising as a solution of the classical Maxwell field equations, analogous to a retarded potential solution, but lying on the future light cone of space-time; the potential appears, at present, to have no physical interpretation.

advantage factor [NUCLEO] The ratio of the radiation dose received in a specified time interval at a position in a nuclear reactor where some enhanced effect is produced, to the radiation dose in the same time interval at a reference position in the reactor.

aeolotropic *See* anisotropic.

aeon [ASTRON] A billion (10^9) years.

aerated flow [ENG] Flowing liquid in which gas is dispersed as fine bubbles throughout the liquid.

A/E ratio *See* absorptivity-emissivity ratio.

aeration cell [PHYS CHEM] An electrolytic cell whose electromotive force is due to electrodes of the same material located

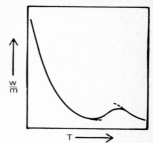

ADSORPTION ISOBAR

A typical adsorption isobar; w/m is weight of material adsorbed per unit weight of adsorbent, and T is absolute temperature.

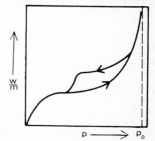

ADSORPTION ISOTHERM

A typical adsorption isotherm; w/m is weight of material adsorbed per weight of adsorbent, and p is pressure.

in different concentrations of dissolved air. Also known as oxygen cell.

aerial *See* antenna.

aerial perspective [OPTICS] The effect produced by diffusion of light in the atmosphere whereby more distant objects have less clarity of outline and are lighter in tone.

aeriform [PHYS] Having the form or nature of air.

aeroballistics [MECH] The study of the interaction of projectiles or high-speed vehicles with the atmosphere.

aerodynamic [FL MECH] Pertaining to forces acting upon any solid or liquid body moving relative to a gas (especially air).

aerodynamically rough surface [FL MECH] A surface whose irregularities are sufficiently high that the turbulent boundary layer reaches right down to the surface.

aerodynamically smooth surface [FL MECH] A surface whose irregularities are sufficiently small to be entirely embedded in the laminar sublayer.

aerodynamic coefficient [FL MECH] Any nondimensional coefficient relating to aerodynamic forces or moments, such as a coefficient of drag or a coefficient of lift.

aerodynamic drag [FL MECH] A retarding force that acts upon a body moving through a gaseous fluid and that is parallel to the direction of motion of the body; it is a component of the total fluid forces acting on the body. Also known as aerodynamic resistance.

aerodynamic force [FL MECH] The force between a body and a gaseous fluid caused by their relative motion. Also known as aerodynamic load.

aerodynamic heating [FL MECH] The heating of a body produced by passage of air or other gases over its surface; caused by friction and by compression processes and significant chiefly at high speeds.

aerodynamic lift [FL MECH] That component of the total aerodynamic force acting on a body perpendicular to the undisturbed airflow relative to the body. Also known as lift.

aerodynamic load *See* aerodynamic force.

aerodynamic moment [AERO ENG] The torque about the center of gravity of a missile or projectile moving through the atmosphere, produced by any aerodynamic force which does not act through the center of gravity.

aerodynamic noise [ACOUS] Acoustic noise caused by turbulent airflow over the surface of a body.

aerodynamic phenomena [FL MECH] Acoustic, thermal, electrical, and mechanical effects, among others, that result from the flow of air over a body.

aerodynamic resistance *See* aerodynamic drag.

aerodynamics [FL MECH] The science that deals with the motion of air and other gaseous fluids and with the forces acting on bodies when they move through such fluids or when such fluids move against or around the bodies.

aerodynamic size [PHYS] Particle size determined from inertia or settling velocity, assuming Stokes' law for the resistance to a sphere moving through a fluid. Also known as inertial size.

aerodynamic time [AERO ENG] A characteristic time equal to the mass of an aircraft divided by the product of the gross wing area, the density of air, and the air speed.

aerodynamic trail [FL MECH] A condensation trail formed by adiabatic cooling to saturation (or slight supersaturation) of air passing over the surfaces of high-speed aircraft.

aerodynamic trajectory [MECH] A trajectory or part of a trajectory in which the missile or vehicle encounters sufficient air resistance to stabilize its flight or to modify its course significantly.

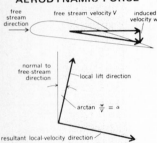

AERODYNAMIC FORCE

free stream direction

free stream velocity V

induced velocity w

normal to free-stream direction

local lift direction

arctan $\frac{w}{V} = \alpha$

resultant local-velocity direction

Effect of the induced velocity on the direction of local velocity and lift.

aerodynamic turbulence [FL MECH] A state of fluid flow in which the instantaneous velocities exhibit irregular and apparently random fluctuations.

aerodynamic wave drag [FL MECH] The force retarding an airplane, especially in supersonic flight, as a consequence of the formation of shock waves ahead of it.

aeroelasticity [MECH] The deformation of structurally elastic bodies in response to aerodynamic loads.

aeromechanics [FL MECH] The science of air and other gases in motion or equilibrium; has two branches, aerostatics and aerodynamics.

aeronautical flutter [FL MECH] An aeroelastic, self-excited vibration in which the external source of energy is the airstream and which depends on the elastic, inertial, and dissipative forces of the system in addition to the aerodynamic forces.

aeronautics [FL MECH] The science that deals with flight through the air.

aeronomy [GEOPHYS] The study of the atmosphere of the earth or other bodies, particularly in relation to composition, properties, relative motion, and radiation from outer space or other bodies.

aerophysics [AERO ENG] The physics dealing with the design, construction, and operation of aerodynamic devices.

aerostatic balance [ENG] An instrument for weighing air.

aerostatics [FL MECH] The science of the equilibrium of gases and of solid bodies immersed in them when under the influence only of natural gravitational forces.

aerothermochemistry [FL MECH] The study of gases which takes into account the effect of motion, heat, and chemical changes.

aerothermodynamics [FL MECH] The study of aerodynamic phenomena at sufficiently high gas velocities that thermodynamic properties of the gas are important.

aerothermoelasticity [FL MECH] The study of the response of elastic structures to the combined effects of aerodynamic heating and loading.

AES *See* Auger electron spectroscopy.

af *See* audio frequency.

aF *See* abfarad.

AFC *See* automatic frequency control.

affine connection [MATH] A structure on an n-dimensional space that, for any pair of neighboring points P and Q, specifies a rule whereby a definite vector at Q is associated with each vector at P; the two vectors are said to be parallel.

affine geometry [MATH] The study of geometry using the methods of linear algebra.

affine Hjelmslev plane [MATH] A generalization of an affine plane in which more than one line may pass through two distinct points. Also known as Hjelmslev plane.

affine plane [MATH] In projective geometry, a plane in which (1) every two points lie on exactly one line, (2) if p and L are a given point and line such that p is not on L, then there exists exactly one line that passes through p and does not intersect L, and (3) there exist three noncollinear points.

affine space [MATH] An n-dimensional vector space which has an affine connection defined on it.

affine strain [GEOPHYS] A strain in the earth that does not differ from place to place.

affine transformation [MATH] A function on a linear space to itself, which is the sum of a linear transformation and a fixed vector.

afocal lens [OPTICS] A lens of zero convergent power, whose focal points are infinitely distant.

AERODYNAMIC WAVE DRAG

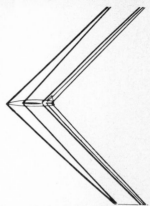

Shock waves generated by an airplane at supersonic speeds act to retard the plane.

afocal system [OPTICS] An optical system of zero convergent power, for example, a telescope.

aftercooling [NUCLEO] The cooling of a reactor after it has been shut down.

afterglow [ATOM PHYS] *See* phosphorescence. [PL PHYS] The transient decay of a plasma after the power has been turned off.

afterheat [NUCLEO] Heat derived from residual radioactivity after a reactor has been shut down.

afterwind [NUCLEO] A wind produced by the updraft accompanying the rise of the fireball of a nuclear explosion and directed toward the burst center.

Ag *See* silver.

AGC *See* automatic gain control.

agglomeration [SCI TECH] An indiscriminately formed cluster of particles.

aggregate recoil [NUC PHYS] The ejection of atoms from the surface of a sample as a result of their being attached to one atom that is recoiling as the result of α-particle emission.

aging [ELEC] Allowing a permanent magnet, capacitor, meter, or other device to remain in storage for a period of time, sometimes with a voltage applied, until the characteristics of the device become essentially constant. [ELECTROMAG] Change in the magnetic properties of iron with passage of time, for example, increase in the hysteresis. [NUCLEO] The slowing down of neutrons.

agonic line [GEOPHYS] The imaginary line through all points on the earth's surface at which the magnetic declination is zero; that is, the locus of all points at which magnetic north and true north coincide.

agravic [GEOPHYS] Of or pertaining to a condition of no gravitation.

agreement residual [SOLID STATE] The sum of the differences between the observed and calculated structure amplitudes of a crystal, for all observed reflections, divided by the sum of the observed amplitudes.

aH *See* abhenry.

Ah *See* ampere-hour.

A/in.² *See* ampere per square inch.

airblast circuit breaker [ELEC] An electric switch which, on opening, utilizes a high-pressure gas blast (air or sulfur hexafluoride) to break the arc.

air capacitor [ELEC] A capacitor having only air as the dielectric material between its plates. Also known as air condenser.

air cell [ELECTR] A cell in which depolarization at the positive electrode is accomplished chemically by reduction of the oxygen in the air.

air compression [PHYS] The decrease of volume of a quantity of air as a result of an increase in pressure, as is accomplished by a piston moving in a cylinder.

air condenser *See* air capacitor.

air-core coil [ELECTR] An inductor without a magnetic core.

air-core transformer [ELECTROMAG] Transformer (usually radio-frequency) having a nonmetallic core.

air current [FL MECH] Very generally, any moving stream of air. [GEOPHYS] *See* air-earth conduction current.

air depolarized battery [ELEC] A primary battery which is kept depolarized by atmospheric oxygen rather than chemical compounds. Also known as metal-air battery.

air-earth conduction current [GEOPHYS] That part of the air-earth current contributed by the electrical conduction of the atmosphere itself; represented as a downward movement of

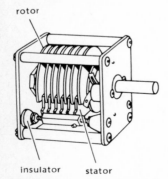

AIR CAPACITOR

rotor

insulator   stator

A drawing of a variable air capacitor showing the flat, parallel metallic plates separated by air.

positive space charge in storm-free regions all over the world. Also known as air current.

air-earth current [GEOPHYS] The transfer of electric charge from the positively charged atmosphere to the negatively charged earth; made up of the air-earth conduction current, a precipitation current, a convection current, and miscellaneous smaller contributions.

air equivalent [NUCLEO] A measure of the effectiveness of an absorber of nuclear radiation, equal to the thickness of a layer of air at standard pressure and temperature that absorbs the same fraction of radiation or results in the same energy loss as does the absorber.

airflow [FL MECH] 1. A flow or stream of air which may take place in a wind tunnel or, as a relative airflow, past the wing or other parts of a moving craft. Also known as airstream. 2. A rate of flow, measured by mass or volume per unit of time.

airflow stack effect [FL MECH] The variation of pressure with height in air flowing in a vertical duct due to a difference in temperature between the flowing air and the air outside the duct.

airglow [GEOPHYS] The quasi-steady radiant emission from the upper atmosphere over middle and low latitudes, as distinguished from the sporadic emission of auroras which occur over high latitudes. Also known as light-of-the-night-sky; night-sky light; night-sky luminescence; permanent aurora.

airlight [METEOROL] In determinations of visual range, light from sun and sky which is scattered into the eyes of an observer by atmospheric suspensoids (and, to slight extent, by air molecules) lying in the observer's cone of vision.

airlight formula [OPTICS] A fundamental equation of visual-range theory, relating the apparent luminance of a distant black object, the apparent luminance of the background sky above the horizon, and the extinction coefficient of the air layer near the ground.

air line [SPECT] Lines in a spectrum due to the excitation of air molecules by spark discharges, and not ordinarily present in arc discharges.

air pressure [PHYS] The force per unit area that the air exerts on any surface in contact with it, arising from the collisions of the air molecules with the surface.

air properties [PHYS] Characteristics of air as a gas, such as density, molecular weight, specific heats, boiling point, critical temperature, and critical pressure.

air resistance [MECH] Wind drag giving rise to forces and wear on buildings and other structures.

air shower *See* cosmic-ray shower.

air-spaced coax [ELECTROMAG] Coaxial cable in which air is basically the dielectric material; the conductor may be centered by means of a spirally wound synthetic filament, beads, or braided filaments.

air-standard cycle [THERMO] A thermodynamic cycle in which the working fluid is considered to be a perfect gas with such properties of air as a volume of 12.4 ft^3/lb at 14.7 psi (approximately 0.7756 m^3/kg at 101.36 kPa) and 492°R and a ratio of specific heats of 1:4.

airstream *See* airflow.

air terminal [ELEC] A structure, such as a tower, that serves as a lightning arrester.

air-velocity measurement [FL MECH] The measurement of the rate of displacement of air or gas at a specific location, as when ascertaining wind speed or airspeed of an aircraft.

air wall [NUCLEO] A wall of an ionization chamber designed so that its effect on ionizing radiation approximates that of air.

air–water vapor mixture [PHYS] A mixture of dry air and water vapor, such as the atmosphere.

Airy differential equation [MATH] The differential equation $(d^2f/dz^2) - zf = 0$, where z is the independent variable and f is the value of the function; used in studying the diffraction of light near caustic surface.

Airy disk [OPTICS] The bright, diffuse central spot of light formed by an optical system imaging a point source of light.

Airy function [MATH] Either of the solutions of the Airy differential equation.

Airy isostasy [GEOPHYS] A theory of hydrostatic equilibrium of the earth's surface which contends that mountains are floating on a fluid lava of higher density, and that higher mountains have a greater mass and deeper roots.

Airy phase [ACOUS] An acoustic wave formed by an explosion in shallow water over a flat bottom.

Airy spirals [OPTICS] Spiral interference patterns formed by quartz cut perpendicularly to the axis in convergent circularly polarized light.

Airy stress function [MECH] A biharmonic function of two variables whose second partial derivatives give the stress components of a body subject to a plane strain.

Aitken's formula [ASTRON] The expression used to determine the separation limit for true binary stars: $\log p'' = 2.5 - 0.2m$, where $p'' = $ limit, $m = $ magnitude.

Al *See* aluminum.

Albada finder [OPTICS] A viewfinder used with a camera held at eye level; the field of view is enclosed by a white frame that is made to appear very distant by reflection from the rear surface of the objective lens.

albedo [NUCLEO] The reflection factor a surface, such as paraffin, has for neutrons. [OPTICS] That fraction of the total light incident on a reflecting surface, especially a celestial body, which is reflected back in all directions.

albedo neutrons *See* albedo particles.

albedo particles [GEOPHYS] Neutrons or other particles, such as electrons or protons, which leave the earth's atmosphere, having been produced by nuclear interactions of energetic particles within the atmosphere. Also known as albedo neutrons.

aleph null [MATH] The cardinal number of any set which can be put in one-to-one correspondence with the set of positive integers.

Alexandroff compactification *See* one-point compactification.

Alford loop [ELECTROMAG] An antenna utilizing multielements which usually are contained in the same horizontal plane and adjusted so that the antenna has approximately equal and in-phase currents uniformly distributed along each of its peripheral elements and produces a substantially circular radiation pattern in the plane of polarization; it is known for its purity of polarization.

Alfvén number [PHYS] The ratio of the speed of the Alfvén wave to the speed of the fluid at a point in the fluid.

Alfvén speed [PHYS] The speed of motion of the Alfvén wave, which is $v_a = B_0/\sqrt{\rho\mu}$, where B_0 is the magnetic field strength, ρ the fluid density, and μ the magnetic permeability (in meter-kilogram-second units).

Alfvén wave [PHYS] A hydromagnetic shear wave which moves along magnetic field lines; a major accelerative mechanism of charged particles in plasma physics and astrophysics.

algebra [MATH] **1.** A method of solving practical problems

AIRY DISK

Picture of an Airy disk, one type of diffraction pattern. *(R. W. Ditchburn)*

by using symbols, usually letters, for unknown quantities. **2.** The study of the formal manipulations of equations involving symbols and numbers. **3.** An abstract mathematical system consisting of a vector space together with a multiplication by which two vectors may be combined to yield a third, and some axioms relating this multiplication to vector addition and scalar multiplication. Also known as hypercomplex system.

algebraic addition [MATH] The addition of algebraic quantities in the sense that adding a negative quantity is the same as subtracting a positive one.

algebraically closed field [MATH] **1.** A field F such that every polynomial of degree equal to or greater than 1 with coefficients in F has a root in F. **2.** A field F is said to be algebraically closed in an extension field K if any root in K of a polynominal with coefficients in F also lies in F.

algebraically independent [MATH] A subset S of a commutative ring B is said to be algebraically independent over a subring A of B (or the elements of S are said to be algebraically independent over A) if, whenever a polynominal in elements of S, with coefficients in A, is equal to 0, then all the coefficients in the polynomial equal 0.

algebraic closure of a field [MATH] An algebraic extension field which has no algebraic extensions but itself.

algebraic curve [MATH] **1.** The set of points in the plane satisfying a polynomial equation in two variables. **2.** More generally, the set of points in n-space satisfying a polynomial equation in n variables.

algebraic element [MATH] An element x of a field E is said to be algebraic over a subfield F if it is the root of a polynomial with coefficients in F; that is, it satisfies an equation of the form $a_0 + a_1 x + \cdots + a_n x^n = 0$, where $a_1, \ldots, a_n$ are elements of F.

algebraic equation [MATH] An equation in which zero is set equal to an algebraic expression.

algebraic expression [MATH] An expression which is obtained by performing a finite number of the following operations on symbols representing numbers: addition, subtraction, multiplication, division, raising to a power.

algebraic extension of a field [MATH] An extension field E of a given field F such that every element of E is algebraic over F.

algebraic function [MATH] A function whose value is obtained by performing only the following operations to its argument: addition, subtraction, multiplication, division, raising to a rational power.

algebraic geometry [MATH] The study of geometric properties of figures using methods of abstract algebra.

algebraic identity [MATH] A relation which holds true for all possible values of the literal symbols occurring in it; for example, $(x + y)(x - y) = x^2 - y^2$.

algebraic integer [MATH] The root of a polynomial whose coefficients are integers and whose leading coefficient is equal to 1.

algebraic invariant [MATH] A polynomial in coefficients of a quadratic or higher form in a collection of variables whose value is unchanged by a specified class of linear transformations of the variables.

algebraic language [MATH] The conventional method of writing the symbols, parentheses, and other signs of formulas and mathematical expressions.

algebraic number [MATH] Any root of a polynomial with rational coefficients.

algebraic number field   [MATH] A finite extension field of the field of rational numbers.

algebraic number theory   [MATH] The study of properties of real numbers, especially integers, using the methods of abstract algebra.

algebraic set   [MATH] **1.** A set made up of all zeros of some specified set of polynomials in n variables with coefficients in a specified field F, in a specified extension field of F. **2.** *See* algebraic variety.

algebraic sum   [MATH] The result of the addition of two or more quantities, with the addition of a negative quantity equivalent to subtraction of the corresponding positive quantity.

algebraic topology   [MATH] The study of topological properties of figures using the methods of abstract algebra; includes homotopy theory, homology theory, and cohomology theory.

algebraic variety   [MATH] **1.** An algebraic variety of an n-dimensional affine space $A_n(K)$ over a field K is a subset of $A_n(K)$ consisting of points $(x_1, \ldots, x_n)$ which are zeros of a finite set of polynomials in $x_1, \ldots, x_n$ with coefficients in K. Also known as algebraic set. **2.** An algebraic variety of an n-dimensional projective space $P_n(K)$ over a field K is a subset of $P_n(K)$ consisting of points $(x_0, \ldots, x_n)$ which are zeros of a finite set of polynomials in $x_0, \ldots, x_n$ with coefficients in K.

algebra of subsets   [MATH] An algebra of subsets of a set S is a family of subsets of S that contains the null set, the complement (relative to S) of each of its members, and the union of any two of its members. Also known as Boolean algebra of subsets.

algebra with identity   [MATH] An algebra which has an element, not equal to 0 and denoted by 1, such that, for any element x in the algebra, $x1 = 1x = x$.

algorithm   [MATH] A set of well-defined rules for the solution of a problem in a finite number of steps.

alias   [STAT] Either of two effects in a factorial experiment which cannot be differentiated from each other on the basis of the experiment.

aliasing   [MATH] Introduction of error into the computed amplitudes of the lower frequencies in a Fourier analysis of a function carried out using discrete time samplings whose interval does not allow the proper analysis of the higher frequencies present in the analyzed function.

Alice   *See* Adiabatic Low-energy Injection and Capture Experiment.

alignment   [ELECTR] The process of adjusting components of a system for proper interrelationship, including the adjustment of tuned circuits for proper frequency response and the time synchronization of the components of a system.   [NUC PHYS] A population $p(m)$ of the $2I + 1$ orientational substates of a nucleus, $m = -I$ to $+I$, such that $p(m) = p(-m)$.

alignment chart   *See* nomograph.

alive   *See* energized.

alkali   [CHEM] Any compound having highly basic qualities.

alkali emission   [GEOPHYS] Light from free lithium, potassium, and especially sodium in the atmosphere, observed in twilight.

alkali metal   [CHEM] Any of the elements of group Ia in the periodic table: lithium, sodium, potassium, rubidium, cesium, and francium.

alkaline   [CHEM] **1.** Having properties of an alkali. **2.** Having a pH greater than 7.

alkaline cell   [ELEC] A primary cell that uses an alkaline electrolyte, usually potassium hydroxide, and delivers about 1.5 volts at much higher current rates than the common carbon-zinc cell. Also known as alkaline-manganese cell.

alkaline earth [INORG CHEM] An oxide of an element of group IIa in the periodic table, such as barium, calcium, and strontium. Also known as alkaline-earth oxide.

alkaline-earth metals [CHEM] The heaviest members of group IIa in the periodic table; usually calcium, strontium, magnesium, and barium.

alkaline-earth oxide *See* alkaline earth.

alkaline-manganese cell *See* alkaline cell.

alkaline storage battery [ELEC] A storage battery in which the electrolyte consists of an alkaline solution, usually potassium hydroxide.

alkalinity [CHEM] The property of having excess hydroxide ions in solution.

Allard's law [OPTICS] A mathematical formula defining the relationship between the intensity of a light, atmospheric conditions, and the amount of light received at any given distance.

all-diffused monolithic integrated circuit [ELECTR] Microcircuit consisting of a silicon substrate into which all of the circuit parts (both active and passive elements) are fabricated by diffusion and related processes.

allobar [NUC PHYS] A form of an element differing in its atomic weight from the naturally occurring form and hence being of different isotopic composition. [PHYS] A barometric pressure change.

allochromy [PHYS] Emission of electromagnetic radiation that results from incident radiation at a different wavelength, as occurs in fluorescence or the Raman effect.

allogyric birefringence [OPTICS] The phenomenon in active optical media whereby circularly polarized light is transmitted unchanged but the velocity of right-handed circularly polarized light is different from that of left-handed.

allomerism [CRYSTAL] A constancy in crystal form in spite of a variation in chemical composition.

allometry [MATH] A relation between two variables x and y that can be written in the form $y = ax^n$, where a and n are constants.

allowable load [MECH] The maximum force that may be safely applied to a solid, or is permitted by applicable regulators.

allowable stress [MECH] The maximum force per unit area that may be safely applied to a solid.

allowed energy bands [SOLID STATE] The restricted regions of possible electron energy levels in a solid.

allowed transition [QUANT MECH] A transition between two states which is permitted by the selection rules and which consequently has a relatively high priority.

alloy junction [ELECTR] A junction produced by alloying one or more impurity metals to a semiconductor to form a p or n region, depending on the impurity used. Also known as fused junction.

alloy junction diode [ELECTR] A junction diode made by placing a pill of doped alloying material on a semiconductor material and heating until the molten alloy melts a portion of the semiconductor, resulting in a pn junction when the dissolved semiconductor recrystallizes. Also known as fused-junction diode.

alloy junction transistor [ELECTR] A junction transistor made by placing pellets of a p-type impurity such as indium above and below an n-type wafer of germanium, then heating until the impurity alloys with the germanium to give a pnp transistor. Also known as fused-junction transistor.

alloy nuclear fuel [NUCLEO] A material used in nuclear

ALLOY JUNCTION TRANSISTOR

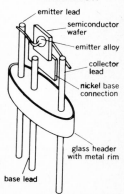

emitter lead

semiconductor wafer

emitter alloy

collector lead

nickel base connection

glass header with metal rim

base lead

View of an alloy junction transistor.

reactors that is an alloy of a fissionable substance and a nonfissionable metal or metals.

almost every [MATH] A proposition concerning the points of a measure space is said to be true at almost every point, or to be true almost everywhere, if it is true for every point in the space, with the exception at most of a set of points which form a measurable set of measure zero.

almost periodic function [MATH] A continuous function $f(x)$ such that for any positive number ε there is a number M so that for any real number x, any interval of length M contains a nonzero number t such that $|f(x+t) - f(x)| < \varepsilon$.

almucantar *See* parallel of altitude.

alpha [SCI TECH] The first letter in the Greek alphabet; α, A.

alpha cross section [NUCLEO] Total cross section for interaction with α-particles.

alpha cutoff frequency [ELECTR] The frequency at the high end of a transistor's range at which current amplification drops 3 decibels below its low-frequency value.

alpha decay [NUC PHYS] A radioactive transformation in which an alpha particle is emitted by a nuclide.

alpha emission [NUC PHYS] Ejection of alpha particles from the atom's nucleus.

alpha irradiation [NUCLEO] Subjection of a substance to a flux of alpha particles.

alpha-limit point [MATH] For the vector differential equation $dx/dt = f(x)$, where f is continuously differentiable, describing the motion of points in euclidean n-space, an α-limit point of x is a point y for which there exists a sequence $t_n \to -\infty$ such that $x(t_n) \to y$ as $n \to \infty$. Also known as limit point.

alphameric characters *See* alphanumeric characters.

alphanumeric characters [ADP] All characters used by a computer, including letters, numerals, punctuation marks, and such signs as \$, @, and #. Also known as alphameric characters.

alphanumeric display device [ELECTR] A device which visibly represents alphanumeric output information from some signal source.

alpha particle [ATOM PHYS] A positively charged particle consisting of two protons and two neutrons, identical with the nucleus of the helium atom; emitted by several radioactive substances.

alpha-particle detector [NUCLEO] A device used to indicate the presence of alpha particles.

alpha-particle scattering [ATOM PHYS] Deviation at various angles of a stream of alpha particles passing through a foil of material.

alpha ray [NUCLEO] A stream of alpha particles.

alpha-ray vacuum gage [ENG] An ionization gage in which the ionization is produced by alpha particles emitted by a radioactive source, instead of by electrons emitted from a hot filament; used chiefly for pressures from 10^{-3} to 10 torrs. Also known as alphatron.

alphatopic [NUCLEO] Pertaining to the relationship between two nuclides that differ in composition or in mass by an alpha particle.

alphatron *See* alpha-ray vacuum gage.

alt *See* altitude.

alternant [MATH] A determinant for which the element in the ith column and jth row is of the form $f_i(x_j)$, where $f_1, f_2, \ldots, f_n$ are functions, $x_1, x_2, \ldots, x_n$ are quantities, and n is the order of the determinant.

alternate angles [MATH] A pair of nonadjacent angles that a transversal forms with each of two lines; they lie on opposite

ALPHA–PARTICLE SCATTERING

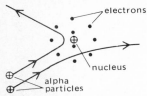

Scattering by an atom of alpha particles.

sides of the transversal, and are both interior, or both exterior, to the two lines.

alternating current [ELEC] Electric current that reverses direction periodically, usually many times per second. Abbreviated ac.

alternating-current circuit theory [ELEC] The mathematical description of conditions in an electric circuit driven by an alternating source or sources.

alternating-current generator [ELEC] A machine, usually rotary, which converts mechanical power into alternating-current electric power.

alternating-current motor [ELEC] A machine that converts alternating-current electrical energy into mechanical energy by utilizing forces exerted by magnetic fields produced by the current flow through conductors.

alternating-current power supply [ELEC] A power supply that provides one or more alternating-current output voltages, such as an ac generator, dynamotor, inverter, or transformer.

alternating-current resistance *See* high-frequency resistance.

alternating form [MATH] A bilinear form f which changes sign under interchange of its independent variables; that is, $f(x,y) = -f(y,x)$ for all values of the independent variables x and y.

alternating function [MATH] A function in which the interchange of two independent variables causes the dependent variable to change sign.

alternating gradient [ELECTROMAG] A magnetic field in which successive magnets have gradients of opposite sign, so that the field increases with radius in one magnet and decreases with radius in the next; used in synchrotrons and cyclotrons.

alternating-gradient synchrotron [NUCLEO] A proton synchrotron using an alternating magnetic-field gradient for focusing; beams of protons having extremely high energy (above 25 GeV) are produced.

alternating group [MATH] A group made up of all the even permutations of n objects.

alternating series [MATH] Any series of real numbers in which consecutive terms have opposite signs.

alternating stress [MECH] A stress produced in a material by forces which are such that each force alternately acts in opposite directions.

alternating voltage [ELEC] Periodic voltage, the average value of which over a period is zero.

alternation [PHYS] Variation, either positive or negative, of a waveform from zero to maximum and back to zero, equaling one-half of a cycle.

alternation of multiplicities law [CHEM] The law that the periodic table arranges the elements in such a sequence that their number of orbital electrons, and hence their multiplicities, alternates between even and odd numbers.

alternative algebra [MATH] A nonassociative algebra which is an alternative ring.

alternative hypothesis [STAT] Value of the parameter of a population other than the value hypothesized or believed to be true by the investigator.

alternative ring [MATH] A nonassociative ring R such that $(xx)y = x(xy)$ and $(yx)x = y(xx)$ for all x and y in R.

alternator [ELEC] A mechanical, electrical, or electromechanical device which supplies alternating current.

altitude [MATH] The perpendicular distance from the base to the top (a vertex or parallel line) of a geometric figure such as a triangle or parallelogram. Abbreviated alt.

altitude circle   *See* parallel of altitude.

aluminium   *See* aluminum.

aluminum   [CHEM] A chemical element, symbol Al, atomic number 13, atomic weight 26.9815. Also spelled aluminium.

aluminum arrester   *See* aluminum-cell arrester.

aluminum-cell arrester   [ELEC] A lightning arrester consisting of a number of electrolytic cells in series formed from aluminum trays containing electrolyte. Also known as aluminum arrester; electrolytic arrester.

Am   *See* americium.

AM   *See* amplitude modulation.

A/m   *See* ampere per meter.

Am²   *See* ampere meter squared.

A/m²   *See* ampere per square meter.

amacratic lens   *See* amasthenic lens.

Amagat density unit   [PHYS] A unit of density in the Amagat system, used in the study of the behavior of gases under pressure; it is equal to the density of a gas at a pressure of 1 atmosphere (101,325 newtons per square meter) and a temperature of 0°C; for an ideal gas this is 44.616 ± 0.006 moles per cubic meter.

Amagat diagram   [PHYS] A diagram that plots a series of isothermal curves for a gas pressure versus the gas pressure-volume product.

Amagat law   *See* Amagat-Leduc rule.

Amagat-Leduc rule   [PHYS] The rule which states that the volume taken up by a gas mixture equals the sum of the volumes each gas would occupy at the temperature and pressure of the mixture. Also known as Amagat law; Leduc law.

Amagat system   [PHYS] A system of units in which the unit of pressure is the atmosphere and the unit of volume is the gram-molecular volume (22.4 liters at standard conditions).

Amagat volume unit   [PHYS] A unit of volume in the Amagat system, used in the study of the behavior of gases under pressure; it is equal to the volume occupied by 1 mole of a gas at a pressure of 1 atmosphere (101,325 newtons per square meter) and a temperature of 0°C; for an ideal gas this is 0.022413 ± 0.000003 cubic meter.

amasthenic lens   [OPTICS] A lens that refracts the rays of light into one focus. Also known as amacratic lens.

ambient light   [OPTICS] The surrounding light, such as that reaching a television picture-tube screen from light sources in a room.

ambient noise   [ACOUS] The pervasive noise associated with a given environment, being usually a composite of sounds from sources both near and distant.

ambient pressure   [FL MECH] The pressure of the surrounding medium, such as a gas or liquid, which comes into contact with an apparatus or with a reaction.

ambient temperature   [PHYS] The temperature of the surrounding medium, such as gas or liquid, which comes into contact with the apparatus.

ambiguity   [ELECTR] The condition in which a synchro system or servosystem seeks more than one null position.

ambipolar diffusion   [PHYS] The diffusion in a plasma of charged particles, such as electrons or ions, as a result of the almost exact local charge neutrality required.

americium   [CHEM] A chemical element, symbol Am, atomic number 95; the mass number of the isotope with the longest half-life is 243.

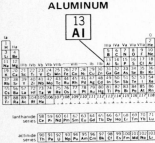

ALUMINUM

Periodic table of the chemical elements showing the position of aluminum.

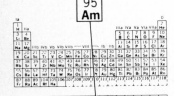

AMERICIUM

Periodic table of the chemical elements showing the position of americium.

amicable numbers [MATH] Two numbers such that the exact divisors of each number (except the number itself) add up to the other number.

Amici prism [OPTICS] A compound prism, used in direct-vision spectroscopes, that disperses a beam of light into a spectrum without causing the beam as a whole to undergo any net deviation; it is made up of alternate crown and flint glass components, refracting in opposite directions. Also known as direct-vision prism.

A min *See* ampere-minute.

Am²/Js *See* ampere square meter per joule second.

ammeter [ENG] An instrument for measuring the magnitude of electric current flow.

ammonia-beam maser [PHYS] A gas maser using ammonia as the paramagnetic material.

ammonia clock [HOROL] A time-measuring device dependent on the pyramidal ammonia molecule's property of turning inside out readily and oscillating between the two extreme positions at the precise frequency of 2.387013×10^{10} hertz.

ammonia maser clock [HOROL] A gas maser that utilizes the transition of high-energy ammonia molecules to generate a stable microwave output signal for use as a time standard.

amorphous [PHYS] Pertaining to a solid which is noncrystalline, having neither definite form nor structure.

amorphous laser *See* glass laser.

amorphous semiconductor [SOLID STATE] A semiconductor material which is not entirely crystalline, having only short-range order in its structure.

amp *See* amperage; ampere.

ampacity [ELEC] Current-carrying capacity in amperes; used as a rating for power cables.

amperage [ELEC] The amount of electric current in amperes. Abbreviated amp.

ampere [ELEC] The unit of electric current in the rationalized meter-kilogram-second system of units; defined in terms of the force of attraction between two parallel current-carrying conductors. Abbreviated a; A; amp.

Ampère currents [ELECTROMAG] Postulated "molecular-ring" currents to explain the phenomena of magnetism as well as the apparent nonexistence of isolated magnetic poles.

ampere-hour [ELEC] A unit for the quantity of electricity, obtained by integrating current flow in amperes over the time in hours for its flow; used as a measure of battery capacity. Abbreviated Ah; amp-hr.

ampere-hour capacity [ELEC] The charge, measured in ampere-hours, that can be delivered by a storage battery up to the limit to which the battery may be safely discharged.

Ampère law [ELECTROMAG] **1.** A law giving the magnetic induction at a point due to given currents in terms of the current elements and their positions relative to the point. Also known as Laplace law. **2.** A law giving the line integral over a closed path of the magnetic induction due to given currents in terms of the total current linking the path.

ampere meter squared [ELECTROMAG] The SI unit of electromagnetic moment. Abbreviated Am².

ampere-minute [ELEC] A unit of electrical charge, equal to the charge transported in 1 minute by a current of 1 ampere, or to 60 coulombs. Abbreviated A min.

ampere per meter [ELECTROMAG] The SI unit of magnetic field strength and magnetization. Abbreviated A/m.

ampere per square inch [ELEC] A unit of current density, equal to the uniform current density of a current of 1 ampere

AMPERE LAW

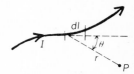

A graphic representation of the Ampère law (def. 1). Contribution of current element to magnetic induction at P is proportional to $I\,dl\sin\theta/r^2$. I = current dl = length of current element, r = distance of point P from current elements, angle θ is between current element and the line forming the element to point P. The contribution points perpendicularly into the page.

flowing through an area of 1 square inch. Abbreviated A/in^2.

ampere per square meter [ELEC] The SI unit of current density. Abbreviated A/m^2.

Ampère rule [ELECTROMAG] The rule which states that the direction of the magnetic field surrounding a conductor will be clockwise when viewed from the conductor if the direction of current flow is away from the observer.

ampere square meter per joule second [ELECTROMAG] The SI unit of gyromagnetic ratio. Abbreviated Am^2/Js.

Ampère theorem [ELECTROMAG] The theorem which states that an electric current flowing in a circuit produces a magnetic field at external points equivalent to that due to a magnetic shell whose bounding edge is the conductor and whose strength is equal to the strength of the current.

ampere-turn [ELECTROMAG] A unit of magnetomotive force in the meter-kilogram-second system defined as the force of a closed loop of one turn when there is a current of 1 ampere flowing in the loop. Abbreviated amp-turn.

amperometric titration [PHYS CHEM] A titration that involves measuring an electric current or changes in current during the course of the titration.

amp-hr *See* ampere-hour.

amplification factor [ELECTR] In a vacuum tube, the ratio of the incremental change in plate voltage to a given small change in grid voltage, under the conditions that the plate current and all other electrode voltages are held constant.

amplified back bias [ELECTR] Degenerative voltage developed across a fast time-constant circuit within a stage of an amplifier and fed back into a preceding stage.

amplifier [ENG] A device capable of increasing the magnitude or power level of a physical quantity, such as an electric current or a hydraulic mechanical force, that is varying with time, without distorting the wave shape of the quantity.

amplitron [ELECTR] Crossed-field continuous cathode reentrant beam backward-wave amplifier for microwave frequencies.

amplitude [MATH] The angle between a vector representing a specified complex number on an Argand diagram and the positive real axis. Also known as argument. [PHYS] The maximum absolute value attained by the disturbance of a wave or by any quantity that varies periodically.

amplitude discriminator *See* pulse-height discriminator.

amplitude factor *See* crest factor.

amplitude-frequency response *See* frequency response.

amplitude gate [ELECTR] A circuit which transmits only those portions of an input signal which lie between two amplitude boundary level values. Also known as slicer; slicer amplifier.

amplitude level [PHYS] The natural logarithm of the ratio of two amplitudes, each measured in the same units.

amplitude limiter *See* limiter.

amplitude-limiting circuit *See* limiter.

amplitude modulation [ELECTR] Abbreviated AM. **1.** Modulation in which the aplitude of a wave is the characteristic varied in accordance with the intelligence to be transmitted. **2.** In telemetry, those systems of modulation in which each component frequency f of the transmitted intelligence produces a pair of sideband frequencies at carrier frequency plus f and carrier minus f.

amplitude modulator [PHYS] Any device which imposes amplitude modulation upon a carrier wave in accordance with a desired program.

amplitude resonance [PHYS] The frequency at which a given sinusoidal excitation produces the maximum amplitude of oscillation in a resonant system.

amplitude response [ELECTR] The maximum output amplitude obtainable at various points over the frequency range of an instrument operating under rated conditions.

amplitude selector *See* pulse-height selector.

amplitude separator [ELECTR] A circuit used to isolate the portion of a waveform with amplitudes above or below a given value or between two given values.

amplitude splitting [OPTICS] A technique in which light falls on a partially reflecting surface; part of the light is transmitted, part reflected, and after further manipulation, these parts are recombined to give interference.

amplitude versus frequency distortion [ELECTR] Distortion caused by the nonuniform attenuation or gain of the system, with respect to frequency under specified terminal conditions.

amp-turn *See* ampere-turn.

amu *See* atomic mass unit.

amyriotic field [QUANT MECH] A quantized field that has creation and annihilation operators satisfying specified commutation rules and a vacuum state.

An *See* actinon.

anacoustic zone [GEOPHYS] The zone of silence in space, starting at about 100 miles (160 kilometers) altitude, where the distance between air molecules is greater than the wavelength of sound, and sound waves can no longer be propagated.

analog [ELECTR] A physical variable which remains similar to another variable insofar as the proportional relationships are the same over some specified range; for example, a temperature may be represented by a voltage which is its analog.

analog comparator [ELECTR] **1.** A comparator that checks digital values to determine whether they are within predetermined upper and lower limits. **2.** A comparator that produces high and low digital output signals when the sum of two analog voltages is positive and negative, respectively.

analog computer [ADP] A computer is which quantities are represented by physical variables; problem parameters are translated into equivalent mechanical or electrical circuits as an analog for the physical phenomenon being investigated.

analog-digital computer *See* hybrid computer.

analogous pole [SOLID STATE] The pole of a crystal that acquires a positive charge when the crystal is heated.

analog output [CONT SYS] Transducer output in which the amplitude is continuously proportional to a function of the stimulus.

analog-to-digital converter [ELECTR] A device which translates continuous analog signals into proportional discrete digital signals.

analysis [MATH] The branch of mathematics most explicitly concerned with the limit process or the concept of convergence; includes the theories of differentiation, integration and measure, infinite series, and analytic functions. Also known as mathematical analysis.

analysis line [SPECT] The spectral line used in determining the concentration of an element in spectrographic analysis.

analysis of variance [STAT] A method for partitioning the total variance in experimental data into components assignable to specific sources.

analytical chemistry [CHEM] The branch of chemistry dealing with techniques which yield any type of information about chemical systems.

analytic continuation [MATH] The process of extending an analytic function to a domain larger than the one on which it was originally defined.

analytic curve [MATH] A curve whose parametric equations are real analytic functions of the same real variable.

analytic extension [RELAT] An extension, in a real analytic manner, past a coordinate singularity of a solution to Einstein's equations of general relativity.

analytic function [MATH] A function which can be represented by a convergent Taylor series. Also known as holomorphic function.

analytic geometry [MATH] The study of geometric figures and curves using a coordinate system and the methods of algebra. Also known as cartesian geometry.

analytic mechanics [MECH] The application of differential and integral calculus to classical (nonquantum) mechanics.

analytic number theory [MATH] The study of problems concerning the discrete domain of integers by means of the mathematics of continuity.

analytic regularization [QUANT MECH] A method of extracting a finite piece from an infinite result in quantum field theory, based on analytically continuing the propagators that appear in typically divergent integrals.

analytic trigonometry [MATH] The study of the properties and relations of the trigonometric functions.

analyzer [OPTICS] A device, such as a Nicol prism, which passes only plane polarized light; used in the eyepiece of instruments such as the polariscope.

anamorphic lens [OPTICS] A lens that produces different magnifications along lines in different directions in the image plane.

anamorphic system [OPTICS] An optical system incorporating a cylindrical surface in which the image is distorted so that the angle of coverage in a direction perpendicular to the cylinder is different for the image than for the object.

anamorphoscope [OPTICS] An optical instrument, usually consisting of a cylindrical lens or mirror, that restores an image distorted by anamorphosis to its normal proportions.

anamorphosis [OPTICS] The production of a distorted image by an optical system.

anamorphote lens [OPTICS] A lens designed to produce anamorphosis.

anastigmat *See* anastigmatic lens.

anastigmatic lens [OPTICS] A compound lens corrected for astigmatism and curvature of field. Also known as anastigmat.

anchylosis *See* ankylosis.

AND circuit *See* AND gate.

Anderson bridge [ELECTR] A six-branch modification of the Maxwell-Wien bridge, used to measure self-inductance in terms of capacitance and resistance; bridge balance is independent of frequency.

AND function [MATH] An operation in logical algebra on statements *P, Q, R,* such that the operation is true if all the statements *P, Q, R,* . . . are true, and the operation is false if at least one statement is false.

AND gate [ELECTR] A circuit which has two or more input-signal ports and which delivers an output only if and when every input signal port is simultaneously energized. Also known as AND circuit; passive AND gate.

AND NOT gate [ELECTR] A coincidence circuit that performs the logic operation AND NOT, under which a result is true only if statement A is true and statement B is not. Also known as A AND NOT B gate.

AND-OR circuit [ELECTR] Gating circuit that produces a prescribed output condition when several possible combined input signals are applied; exhibits the characteristics of the AND gate and the OR gate.

Andrade's creep law [MECH] A law which states that creep exhibits a transient state in which strain is proportional to the cube root of time and then a steady state in which strain is proportional to time.

Andrews's curves [THERMO] A series of isotherms for carbon dioxide, showing the dependence of pressure on volume at various temperatures.

Andronikashvili experiment [CRYO] An experiment to determine the fractional densities of the superfluid and normal fluid components of liquid helium by measuring the period and decrement of a torsional pendulum immersed in the helium.

anechoic chamber [ENG] **1.** A test room in which all surfaces are lined with a sound-absorbing material to reduce reflections of sound to a minimum. Also known as dead room; free-field room. **2.** A room completely lined with a material that absorbs radio waves at a particular frequency or over a range of frequencies; used principally at microwave frequencies, such as for measuring radar beam cross sections.

anelasticity [MECH] Deviation from a proportional relationship between stress and strain.

anelectric [PHYS] Not becoming charged by friction.

aneroid [ENG] **1.** Containing no liquid or using no liquid. **2.** *See* aneroid barometer.

aneroid barometer [ENG] A barometer which utilizes an aneroid capsule. Also known as aneroid.

aneroid calorimeter [ENG] A calorimeter that uses a metal of high thermal conductivity as a heat reservoir.

aneroid capsule [ENG] A thin, disk-shaped box or capsule, usually metallic, partially evacuated and sealed, held extended by a spring, which expands and contracts with changes in atmospheric or gas pressure. Also known as bellows.

Anger function [MATH] A solution $J_\nu(x)$ of a generalization of the Bessel equation, $x^2y'' + xy' + (x^2 - \nu^2)y = [(x - \nu)$ $\sin \nu\pi]/\pi$, where ν is a parameter; it can be written as the integral from 0 to π of $\pi^{-1} \cos (\nu\theta - x \sin \theta) \, d\theta$. Also known as complete Anger function.

angle [MATH] The geometric figure, arithmetic quantity, or algebraic signed quantity determined by two rays emanating from a common point or by two planes emanating from a common line.

angle bisection [MATH] The division of an angle by a line or plane into two equal angles.

angle modulation [ELECTR] The variation in the angle of a sine-wave carrier; particular forms are phase modulation and frequency modulation. Also known as sinusoidal angular modulation.

angle of arrival [ELECTROMAG] A measure of the direction of propagation of electromagnetic radiation upon arrival at a receiver (the term is most commonly used in radio); it is the angle between the plane of the phase front and some plane of reference, usually the horizontal, at the receiving antenna.

angle of contact [FL MECH] The angle between the surface of a liquid and the surface of a partially submerged object or of the container at the line of contact. Also known as contact angle.

angle of deviation *See* deviation.

angle of fall [MECH] The vertical angle at the level point, between the line of fall and the base of the trajectory.

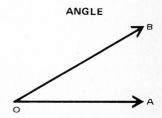

ANGLE

Angle formed by rays *OA* and *OB* emanating from common point *O*. *O* is vertex of angle; *OA* and *OB* are sides of angle.

angle of friction *See* angle of repose.

angle of impact [MECH] The acute angle between the tangent to the trajectory at the point of impact of a projectile and the plane tangent to the surface of the ground or target at the point of impact.

angle of incidence [OPTICS] The angle formed by a ray arriving at a surface and the perpendicular to that surface at the point of arrival. Also known as incidence angle.

angle of lag *See* lag angle.

angle of lead *See* lead angle.

angle of orientation [MECH] Of a projectile in flight, the angle between the plane determined by the axis of the projectile and the tangent to the trajectory (direction of motion), and the vertical plane including the tangent to the trajectory.

angle of radiation [ELECTROMAG] Angle between the surface of the earth and the center of the beam of energy radiated upward into the sky from a transmitting antenna.

angle of reflection [PHYS] The angle between the direction of propagation of a wave reflected by a surface and the line perpendicular to the surface at the point of reflection. Also known as reflection angle.

angle of refraction [PHYS] The angle between the direction of propagation of a wave that is refracted by a surface and the line that is perpendicular to the surface at the point of refraction.

angle of repose [MECH] The angle between the horizontal and the plane of contact between two bodies when the upper body is just about to slide over the lower. Also known as angle of friction.

angle of slide [MIN ENG] The slope, measured in degrees of deviation from the horizontal, on which loose or fragmented solid materials will start to slide.

angle of torsion [MECH] The angle through which a part of an object such as a shaft or wire is rotated from its normal position when a torque is applied. Also known as angle of twist.

angle of twist *See* angle of torsion.

angle of view [OPTICS] The angle subtended by an image at the second nodal point of a lens.

angle-resolved photoelectron spectroscopy [SPECT] A type of photoelectron spectroscopy which measures the kinetic energies of photoelectrons emitted from a solid surface and the angles at which they are emitted relative to the surface. Abbreviated ARPES.

angle variable [MECH] The dynamical variable w conjugate to the action variable J, defined only for periodic motion.

angstrom [MECH] A unit of length, 10^{-10} meter, used primarily to express wavelengths of optical spectra. Abbreviated A; Å. Also known as tenthmeter.

Ångström coefficient [PHYS] The multiplying amplitude parameter inserted in Ångström's formula for the scattering of electromagnetic radiation by atmospheric dust.

Ångström compensation pyrheliometer [ENG] A pyrheliometer consisting of two identical Manganin strips, one shaded, the other exposed to sunlight; an electrical current is passed through the shaded strip to raise its temperature to that of the exposed strip, and the electric power required to accomplish this is a measure of the solar radiation.

Ångström's formula [PHYS] A formula stating that the scattering coefficient for dust in the atmosphere is inversely proportional to a positive power of the wavelength of the radiation, with the power depending on the size of the dust particles.

angular acceleration [MECH] The time rate of change of angular velocity.

angular displacement [PHYS] A vector measure of the rotation of an object about an axis; the vector points along the axis according to the right-hand rule; the length of the vector is the rotation angle, in degrees or radians.

angular distribution [NUCLEO] The distribution in angle, relative to an experimentally specified direction, of the intensity of photons or particles resulting from a nuclear or extranuclear process.

angular frequency [PHYS] For any oscillation, the number of vibrations per unit time, multiplied by 2π. Also known as angular velocity; radian frequency.

angular impulse [MECH] The integral of the torque applied to a body over time.

angular length [MECH] A length expressed in the unit of the length per radian or degree of a specified wave.

angular magnification [OPTICS] For an optical system, the ratio of the angle subtended by the image at the eye to the angle subtended by the object at the eye.

angular momentum [MECH] **1.** The cross product of a vector from a specified reference point to a particle, with the particle's linear momentum. Also known as moment of momentum. **2.** For a system of particles, the vector sum of the angular momenta (first definition) of the particles.

angular momentum operator [QUANT MECH] Any vector operator satisfying communication rules of the type $[J_x, J_y] = iJ_z$.

angular radius [MATH] For a circle drawn on a sphere, the smaller of the angular distances from one of the two poles of the circle to any point on the circle.

angular rate *See* angular speed.

angular speed [MECH] Change of direction per unit time, as of a target on a radar screen, without regard to the direction of the rotation axis; in other words, the magnitude of the angular velocity vector. Also known as angular rate.

angular travel error [MECH] The error which is introduced into a predicted angle obtained by multiplying an instantaneous angular velocity by a time of flight.

angular velocity [MECH] The time rate of change of angular displacement. [PHYS] *See* angular frequency.

anharmonicity [PHYS] **1.** Mechanical vibration where the restoring force acting on a system does not vary linearly with displacement from equilibrium position. **2.** Variation from a linear relationship of dipole moment with internuclear distance in the infrared portion of the electromagnetic spectrum.

anharmonic oscillator [PHYS] An oscillating system in which the restoring force opposing a displacement from the position of equilibrium is a nonlinear function of the displacement.

anharmonic oscillator spectrum [SPECT] A molecular spectrum which is significantly affected by anharmonicity of the forces between atoms in the molecule.

anion [CHEM] An ion that is negatively charged.

anisotropic [PHYS] Showing different properties as to velocity of light transmission, conductivity of heat or electricity, compressibility, and so on, in different directions. Also known as aeolotropic.

anisotropy constant [ELECTROMAG] In a ferromagnetic material, temperature-dependent parameters relating the magnetization in various directions to the anisotropy energy.

anisotropy energy [ELECTROMAG] Energy stored in a ferromagnetic crystal by virtue of the work done in rotating the magnetization of a domain away from the direction of easy magnetization.

anisotropy factor *See* dissymmetry factor.

ANGULAR VELOCITY

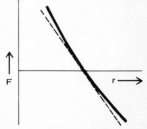

Illustration of angular velocity. Particle p moves on circular path with radius r. θ is angular displacement of p. Angular velocity is rate of change of θ with respect to time.

ANHARMONIC OSCILLATOR

Interatomic force F as a function of the atomic separation r. Anharmonicity is produced by the departure of the actual force (solid curve) from the dashed line.

anker [MECH] A unit of capacity equal to 10 U.S. gallons (37.854 liters); used to measure liquids, especially honey, oil, vinegar, spirits, and wine.

ankylosis [PHYS] The loss by a system of one or more degrees of freedom through development of one or more frictional constraints. Also spelled anchylosis.

annealing point [THERMO] The temperature at which the viscosity of a glass is $10^{13.0}$ poises. Also known as annealing temperature; 13.0 temperature.

annealing temperature *See* annealing point.

annihilation [PARTIC PHYS] A process in which an antiparticle and a particle combine and release their rest energies in other particles.

annihilation operator [QUANT MECH] An operator which reduces the occupation number of a single state by unity; for example, an annihilation operator applied to a state of one particle yields the vacuum.

annihilation radiation [PARTIC PHYS] Electromagnetic radiation arising from the collision, and resulting annihilation, of an electron and a positron, or of any particle and its antiparticle.

annihilator [MATH] The annihilator of an element x in a module over a commutative ring R is the ideal in R consisting of all elements a such that $ax = 0$.

annual aberration [ASTRON] Aberration caused by the velocity of the earth's revolution about the sun.

annual magnetic change *See* magnetic annual change.

annual magnetic variation *See* magnetic annual variation.

annual parallax [ASTRON] The apparent displacement of a celestial body viewed from two separated observation points whose base line is the radius of the earth's orbit.

annular eclipse [ASTRON] An eclipse in which a thin ring of the source of light appears around the obscuring body.

annular effect [FL MECH] A phenomenon observed in the flow of fluid in a tube when its motion is alternating rapidly, as in the propagation of sound waves, in which the mean velocity rises progressing from the center of the tube toward the walls and then falls within a thin laminar boundary layer to zero at the wall itself.

annular solid [MATH] A solid generated by rotating a closed plane curve about a line which lies in the plane of the curve and does not intersect the curve.

annular transistor [ELECTR] Mesa transistor in which the semiconductor regions are arranged in concentric circles about the emitter.

annulus [MATH] The ringlike figure that lies between two concentric circles.

annulus conjecture [MATH] For dimension n, the assertion that if f and g are locally flat embeddings of the $(n - 1)$ sphere, S^{n-1}, in real n space, R^n, with $f(S^{n-1})$ in the bounded component of $R^n - g(S^{n-1})$, then the closed region in R^n bounded by $f(S^{n-1})$ and $g(S^{n-1})$ is homeomorphic to the direct product of S^{n-1} and the closed interval $[0,1]$; it is established for $n \neq 4$.

anode [ELEC] The positive terminal of a primary cell or of a storage battery. [ELECTR] **1.** The collector of electrons in an electron tube. Also known as plate; positive electrode. **2.** In a semiconductor diode, the terminal toward which forward current flows from the external circuit. [PHYS CHEM] The positive terminal of an electrolytic cell.

anode balancing coil [ELEC] Set of mutually coupled windings used to maintain approximately equal currents in anodes operating in parallel from the same transformer terminal.

anode characteristic [ELECTR] Relationship of anode current to anode voltage in a vacuum tube.

anode circuit [ELECTR] Complete external electrical circuit connected between the anode and the cathode of an electron tube. Also known as plate circuit.

anode current [ELECTR] The electron current flowing through an electron tube from the cathode to the anode. Also known as plate current.

anode dark space [ELECTR] A thin, dark region next to the anode glow in a glow-discharge tube.

anode dissipation [ELECTR] Power dissipated as heat in the anode of an electron tube because of bombardment by electrons and ions.

anode drop *See* anode fall.

anode efficiency [ELECTR] The ratio of the ac load circuit power to the dc anode power input for an electron tube. Also known as plate efficiency.

anode fall [ELECTR] **1.** A very thin space-charge region in front of an anode surface, characterized by a steep potential gradient through the region. **2.** The voltage across this region. Also known as anode drop.

anode glow [ELECTR] A thin, luminous layer on the surface of the anode in a glow-discharge tube.

anode impedance [ELECTR] Total impedance between anode and cathode exclusive of the electron stream. Also known as plate impedance; plate-load impedance.

anode input power [ELECTR] Direct-current power delivered to the plate (anode) of a vacuum tube by the source of supply. Also known as plate input power.

anode rays [ELECTR] Positive ions coming from the anode of an electron tube; generally due to impurities in the metal of the anode.

anode resistance [ELECTR] The resistance value obtained when a small change in the anode voltage of an electron tube is divided by the resulting small change in anode current. Also known as plate resistance.

anode saturation [ELECTR] The condition in which the anode current of an electron tube cannot be further increased by increasing the anode voltage; the electrons are then being drawn to the anode at the same rate as they are emitted from the cathode. Also known as current saturation; plate saturation; saturation; voltage saturation.

anode sheath [ELECTR] The electron boundary which exists in a gas-discharge tube between the plasma and the anode when the current demanded by the anode circuit exceeds the random electron current at the anode surface.

anodic [PHYS] Pertaining to the anode.

anomalistic month [ASTRON] The average period of revolution of the moon from perigee to perigee, a period of 27 days 13 hours 18 minutes 33.2 seconds.

anomalistic period [ASTRON] The interval between two successive perigee passages of a satellite in orbit about a primary. Also known as perigee-to-perigee period.

anomalistic year [ASTRON] The period of one revolution of the earth about the sun from perihelion to perihelion; 365 days 6 hours 13 minutes 53.0 seconds in 1900 and increasing at the rate of 0.26 second per century.

anomaloscope [OPTICS] An optical instrument for testing color vision, in which a yellow light whose intensity may be varied is matched against red and green lights whose intensity is fixed.

anomalous dispersion [OPTICS] Extraordinary behavior in the curve of refractive index versus wavelength which occurs in the vicinity of absorption lines or bands in the absorption spectrum of a medium.

anomalous magnetic moment [PARTIC PHYS] The difference between the observed magnetic moment and the value predicted by Dirac's theory.

anomalous series [ATOM PHYS] A series of spectral lines associated with atomic energy levels whose Rydberg corrections do not vary smoothly with total quantum number, generally because they involve excitation of two electrons. Also known as abnormal series.

anomalous viscosity *See* non-Newtonian viscosity.

anomalous Zeeman effect [SPECT] A type of splitting of spectral lines of a light source in a magnetic field which occurs for any line arising from a combination of terms of multiplicity greater than one; due to a nonclassical magnetic behavior of the electron spin.

anomaly [ASTRON] In celestial mechanics, the angle between the radius vector to an orbiting body from its primary (the focus of the orbital ellipse) and the line of apsides of the orbit, measured in the direction of travel, from the point of closest approach to the primary (perifocus). Also known as true anomaly.

anorthic crystal *See* triclinic crystal.

Anosov diffeomorphism [MATH] A diffeomorphism which has a hyperbolic structure.

Anosov flow [MATH] A differentiable flow on a manifold of dimension greater than 1 which has a hyperbolic structure.

Anosov's theorem [MATH] Every Anosov flow or diffeomorphism is structurally stable.

anotron [ELECTR] A cold-cathode glow-discharge diode having a copper anode and a large cathode of sodium or other material.

ante meridian [ASTRON] **1.** A section of the celestial meridian; it lies below the horizon, and the nadir is included. **2.** Before noon, or the period of time between midnight (0000) and noon (1200).

antenna [ELECTROMAG] A device used for radiating or receiving radio waves. Also known as aerial; radio antenna.

antenna amplifier [ELECTROMAG] One or more stages of wide-band electronic amplification placed within or physically close to a receiving antenna to improve signal-to-noise ratio and mutually isolate various devices receiving their feed from the antenna.

antenna circuit [ELECTR] A complete electric circuit which includes an antenna.

antenna coil [ELECTROMAG] Coil through which antenna current flows.

antenna counterpoise *See* counterpoise.

antenna coupler [ELECTROMAG] A radio-frequency transformer, tuned line, or other device used to transfer energy efficiently from a transmitter to a transmission line or from a transmission line to a receiver.

antenna crosstalk [ELECTROMAG] The ratio or the logarithm of the ratio of the undesired power received by one antenna from another to the power transmitted by the other.

antenna directivity diagram [ELECTROMAG] Curve representing, in polar or cartesian coordinates, a quantity proportional to the gain of an antenna in the various directions in a particular plane or cone.

antenna effect [ELECTROMAG] A distortion of the directional properties of a loop antenna caused by an input to the direction-finding receiver which is generated between the loop and ground, in contrast to that which is generated between the two terminals of the loop. Also known as electrostatic error; vertical component effect.

antenna effective area [ELECTROMAG] In any specified direction, the square of the wavelength multiplied by the power gain (or directive gain) in that direction, and divided by 4π.

antenna field [ELECTROMAG] A group of antennas placed in a geometric configuration.

antenna gain [ELECTROMAG] A measure of the effectiveness of a directional antenna as compared to a standard nondirectional antenna. Also known as gain.

antenna loading [ELECTR] **1.** The amount of inductance or capacitance in series with an antenna, which determines the antenna's electrical length. **2.** The practice of loading an antenna in order to increase its electrical length.

antenna matching [ELECTROMAG] Process of adjusting impedances so that the impedance of an antenna equals the characteristic impedance of its transmission line.

antenna pattern *See* radiation pattern.

antenna power [ELECTROMAG] Radio-frequency power delivered to an antenna.

antenna power gain [ELECTROMAG] The power gain of an antenna in a given direction is 4π times the ratio of the radiation intensity in that direction to the total power delivered to the antenna.

antenna resistance [ELECTROMAG] The power supplied to an entire antenna divided by the square of the effective antenna current measured at the point where power is supplied to the antenna.

antenna scanner [ELECTROMAG] A microwave feed horn which moves in such a way as to illuminate sequentially different reflecting elements of an antenna array and thus produce the desired field pattern.

antenna temperature [ELECTROMAG] The temperature of a blackbody enclosure which would produce the same amount of noise as the antenna if it completely surrounded the antenna and was in thermal equilibrium with it.

antiatom [ATOM PHYS] An atom made up of antiprotons, antineutrons, and positrons in the same way that an ordinary atom is made up of protons, neutrons, and electrons.

antibaryon [ATOM PHYS] One of a class of antiparticles, including the antinucleons and the antihyperons, with strong interactions, baryon number -1, and hypercharge and charge opposite to those for the particles.

antibonding orbital [PHYS] An atomic or molecular orbital whose energy increases as atoms are brought closer together, indicating a net repulsion rather than a net attraction and chemical bonding.

anticathode [ELECTR] The anode or target of an x-ray tube, on which the stream of electrons from the cathode is focused and from which x-rays are emitted.

anticlastic [MATH] Having the property of a surface or portion of a surface whose two principal curvatures at each point have opposite signs, so that one normal section is concave and the other convex.

anticoincidence [NUC PHYS] The occurrence of an event at one place without a simultaneous event at another place.

anticoincidence circuit [ELECTR] Circuit that produces a specified output pulse when one (frequently predesignated) of two inputs receives a pulse and the other receives no pulse within an assigned time interval.

anticommutator [MATH] The anticommutator of two operators, A and B, is the operator $AB + BA$.

anticommute [MATH] Two operators anticommute if their anticommutator is equal to zero.

anticorona [OPTICS] A diffraction phenomenon appearing at a point before an observer with the sun or moon directly behind him; consists of rings of colored lights complementary to the coronal rings. Also known as Brocken bow.

antiderivative *See* indefinite integral.

antideuteron [ATOM PHYS] The antiparticle to the deuteron, composed of an antineutron and an antiproton.

antiferroelectric crystal [SOLID STATE] A crystalline substance characterized by a state of lower symmetry consisting of two interpenetrating sublattices with equal but opposite electric polarization, and a state of higher symmetry in which the sublattices are unpolarized and indistinguishable.

antiferromagnetic domain [SOLID STATE] A region in a solid within which equal groups of elementary atomic or molecular magnetic moments are aligned antiparallel.

antiferromagnetic resonance [ELECTROMAG] Magnetic resonance in antiferromagnetic materials which may be observed by rotating magnetic fields in either of two opposite directions.

antiferromagnetic substance [ELECTROMAG] A substance composed of antiferromagnetic domains.

antiferromagnetic susceptibility [ELECTROMAG] The magnetic response to an applied magnetic field of a substance. whose atomic magnetic moments are aligned in antiparallel fashion.

antiferromagnetism [SOLID STATE] A property possessed by some metals, alloys, and salts of transition elements by which the atomic magnetic moments form an ordered array which alternates or spirals so as to give no net total moment in zero applied magnetic field.

antifriction [MECH] Making friction smaller in magnitude.

antigravity [PHYS] The repulsion of one body by another by means of a gravitational type of force; this has never been observed.

antihyperon [PARTIC PHYS] An antiparticle to a hyperon, having the same mass, lifetime, and spin as the hyperon, but with charge and magnetic moment reversed in sign.

antilinear map [MATH] A function f from a module E over a ring R to another module over R such that if x and y are in E and c is in R, $f(x + y) = f(x) + f(y)$, and $f(cx) = \bar{c}f(x)$, where $\bar{c}$ is the image of c under an automorphism of R with period 2 (such as complex conjugation over the complex numbers). Also known as semilinear map.

antilogarithm of a number [MATH] A second number whose logarithm is the first number. Abbreviated antilog. Also known as inverse logarithm of a number.

antilogous pole [SOLID STATE] That crystal pole which becomes electrically negative when the crystal is heated or is expanded by decompression.

antimagnetic [ENG] Constructed so as to avoid the influence of magnetic fields, usually by the use of nonmagnetic materials and by magnetic shielding.

antimatter [PHYS] Material consisting of atoms which are composed of positrons, antiprotons, and antineutrons.

antimolecule [ATOM PHYS] A molecule made up of antiprotons, antineutrons, and positrons in the same way that an ordinary molecule is made up of protons, neutrons, and electrons.

antimony [CHEM] A chemical element, symbol Sb, atomic number 51, atomic weight 121.75.

antimony-124 [NUC PHYS] Radioactive antimony with mass number of 124; 60-day half-life; used as tracer in solid-state and pipeline flow studies.

antineutrino [PARTIC PHYS] The antiparticle to the neutrino;

ANTIMONY

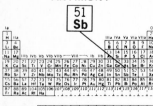

Periodic table of the chemical elements showing the position of antimony.

it has zero mass, spin $\frac{1}{2}$, and positive helicity; there are two antineutrinos, one associated with electrons and one with muons.

antineutron [PARTIC PHYS] The antiparticle to the neutron; a strongly interacting baryon which has no charge, mass of 939.6 MeV, spin $\frac{1}{2}$, and mean life of about 10^3 seconds.

antinode [ASTRON] Either of the two points on an orbit where a line in the orbit plane, perpendicular to the line of nodes and passing through the focus, intersects the orbit. [PHYS] A point, line, or surface in a standing-wave system at which some characteristic of the wave has maximum amplitude. Also known as loop.

antinucleon [PARTIC PHYS] An antineutron or antiproton, that is, particles having the same mass as their nucleon counterparts but opposite charge or opposite magnetic moment.

antinucleus [NUC PHYS] A nucleus made up of antineutrons and antiprotons in the same way that an ordinary nucleus is made up of neutrons and protons.

antiparallel [PHYS] Property of two displacements or other vectors which lie along parallel lines but point in opposite directions.

antiparticle [PARTIC PHYS] A counterpart to a particle having mass lifetime and spin identical to the particle but with charge and magnetic moment reversed in sign.

antipodal points [MATH] The points at opposite ends of a diameter of a sphere.

antiproton [PARTIC PHYS] The antiparticle to the proton; a strongly interacting baryon which is stable, carries unit negative charge, has the same mass as the proton (938.3 MeV), and has spin $\frac{1}{2}$.

antiprotonic atom [ATOM PHYS] An atom consisting of an ordinary nucleus with an orbiting antiproton.

antiquark [PARTIC PHYS] The hypothetical antiparticle of a quark, having electric charge, baryon number, and strangeness opposite in sign to that of the corresponding quark.

antiresonance [ELEC] *See* parallel resonance. [ENG] The condition for which the impedance of a given electric, acoustic, or dynamic system is very high, approaching infinity.

antiresonant circuit *See* parallel resonant circuit.

antisolar point [ASTRON] The point on the celestial sphere which lies directly opposite the sun from the observer, that is, on the line from the sun through the observer.

anti-Stokes lines [MATH] Lines in the complex plane at which changes in the asymptotic behavior of the Airy function $Ai(z)$ occur, located at $\arg z = \frac{1}{3}\pi$, $\arg z = -\frac{1}{3}\pi$, and $\arg z = \pi$. [SPECT] Lines of radiated frequencies which are higher than the frequency of the exciting incident light.

antisymmetric dyadic [MATH] A dyadic equal to the negative of its conjugate.

antisymmetric matrix [MATH] A matrix which is equal to the negative of its transpose. Also known as skew-symmetric matrix.

antisymmetric relation [MATH] A relation, which may be denoted $\geq$, among the elements of a set such that if $a \geq b$ and $b \geq a$ then $a = b$.

antisymmetric tensor [MATH] A tensor in which interchanging two indices of an element changes the sign of the element.

antisymmetric wave function [PHYS] A many-particle wave function which changes its sign when the coordinates of two of the particles are interchanged.

antithetic variable [STAT] One of two random variables having high negative correlation, used in the antithetic variate

ANTINEUTRON

Propane bubble-chamber photograph taken in proton and meson beam of the bevatron. Large arrow **indicates** where antineutron was formed. Five-pronged star indicated by small arrow is annihilation star of antineutron. Actual distance between formation and annihilation is about 12 centimeters. Density of propane is 0.42 g/cm³. *(University of California Lawrence Berkeley Laboratory)*

method of estimating the mean of a series of observations.

anti-transmit-receive tube [ELECTR] A switching tube that prevents the received echo signal from being dissipated in the transmitter.

Antoine equation [PHYS] The empirical relationship between temperature and vapor pressure of liquids; $\log P = B - A/(C + T)$, where A, B, C are experimental constants, T is absolute temperature, and P is vapor pressure.

Antonoff's rule [PHYS] The rule which states that the surface tension at the interface between two saturated liquid layers in equilibrium is equal to the difference between the individual surface tensions of similar layers when exposed to air.

AOQ *See* average outgoing quality.

apastron [ASTRON] That point of the orbit of one member of a binary star system at which the stars are farthest apart.

aperiodic antenna [ELECTROMAG] Antenna designed to have constant impedance over a wide range of frequencies because of the suppression of reflections within the antenna system; includes terminated wave and rhombic antennas.

aperiodic damping [PHYS] Condition of a system in which the amount of damping is so large that, when the system is subjected to a single disturbance, either constant or instantaneous, the system comes to a position of rest without passing through that position; while an aperiodically damped system is not strictly an oscillating system, it has such properties that it should become an oscillating system if the damping were sufficiently reduced.

aperiodic waves [ELEC] The transient current wave in a series circuit with resistance R, inductance L, and capacitance C when $R^2C = 4L$. [PHYS] Waves without a definite repetitive pattern; for example, transient waves.

apertometer [OPTICS] An instrument designed to measure the numerical aperture of microscope objectives.

aperture [ELECTR] An opening through which electrons, light, radio waves, or other radiation can pass. [OPTICS] The diameter of the objective of a telescope or other optical instrument, usually expressed in inches, but sometimes as the angle between lines from the principal focus to opposite ends of a diameter of the objective.

aperture aberration [OPTICS] Errors in optical imaging which occur because rays of different distances from the axis do not come to the same focus.

aperture angle [OPTICS] The angle subtended by the radius of the entrance pupil of an optical instrument at the object.

aperture conductivity [ACOUS] The ratio of the density of a medium to the acoustic mass at an aperture.

aperture illumination [ELECTROMAG] Field distribution in amplitude and phase over an aperture.

aperture ratio [OPTICS] The ratio of the effective diameter of a lens to its focal length.

aperture splitting [OPTICS] A technique in which light from a single slit is divided by passing it through two other slits and is combined by a lens.

aperture stop [OPTICS] That opening in an optical system that determines the size of the bundle of rays which traverse the system from a given point of the object to the corresponding point of the image.

apex [MATH] **1.** The vertex of a triangle opposite the side which is regarded as the base. **2.** The vertex of a cone or pyramid.

aphelion [ASTRON] The point on a planetary orbit farthest from the sun.

apical angle [MECH] The angle between the tangents to the curve outlining the contour of a projectile at its tip. [OPTICS]

The dihedral angle between the refracting faces of a prism. Also known as refracting angle.

aplanatic lens [OPTICS] A lens corrected for spherical abberation.

aplanatic points [OPTICS] Two points on the axis of an optical system which are located so that all the rays emanating from one converge to, or appear to diverge from, the other.

apoapsis [ASTRON] The point in an orbit farthest from the center of attraction.

apocenter *See* apofocus.

apochromat *See* apochromatic lens.

apochromatic lens [OPTICS] A lens with corrections for chromatic and spherical aberration. Also known as apochromat.

apochromatic system [OPTICS] An optical system which is free from both spherical and chromatic aberration for two or more colors.

apodization [ELECTR] A technique for modifying the response of a surface acoustic wave filter by varying the overlap between adjacent electrodes of the interdigital transducer. [OPTICS] The modification of the amplitude transmittance of the aperture of an optical system so as to reduce or suppress the energy of the diffraction rings relative to that of the central Airy disk.

apofocus [ASTRON] The point on an elliptic orbit at the greatest distance from the principal focus. Also known as apocenter.

apogee [ASTRON] That point in an orbit at which the moon or an artificial satellite is most distant from the earth; the term is sometimes loosely applied to positions of satellites of other planets.

Apollonius' circle *See* circle of Apollonius.

Apollonius' theorem [MATH] The locus of a point P such that $(AP)^2 + (BP)^2$ is a constant, where A and B are fixed points, forms a circle whose center is at the midpoint of AB.

A positive [ELEC] Also known as A+. **1.** Positive terminal of an A battery or positive polarity of other sources of filament voltage. **2.** Denoting the terminal to which the positive side of the filament voltage source should be connected.

a posteriori probability *See* empirical probability.

apostilb [OPTICS] A luminance unit equal to one ten-thousandth of a lambert. Also known as blandel.

apothecaries' dram *See* dram.

apothecaries' ounce *See* ounce.

apothecaries' pound *See* pound.

apothem [MATH] The perpendicular distance from the center of a regular polygon to one of its sides.

A power supply *See* A supply.

apparent [ASTRON] A term used to designate certain measured or measurable astronomic quantities to refer them to real or visible objects, such as the sun or a star.

apparent additional mass [FL MECH] A fictitious mass of fluid added to the mass of the body to represent the force required to accelerate the body through the fluid.

apparent candlepower [OPTICS] For an extended source of light, at a specified distance, the candlepower of a point source that would produce the same illumination as the extended source at the same distance.

apparent expansion [THERMO] The expansion of a liquid with temperature, as measured in a graduated container without taking into account the container's expansion.

apparent force [MECH] A force introduced in a relative coordinate system in order that Newton's laws be satisfied in the system; examples are the Coriolis force and the centrifugal force incorporated in gravity.

apparent gravity *See* acceleration of gravity.

apparent horizon [RELAT] The boundary of a region in space-time in which the gravitational field is so strong that the cross-sectional area of an outgoing light pulse decreases.

apparent libration in longitude *See* lunar libration.

apparent luminance [OPTICS] Luminance, created by air light, of that portion of the visual field subtended by a dark, distant object; that is, the light scattered into the eye by particles, including air molecules, lying along the optic path from eye to object.

apparent magnitude [ASTRON] An index of a star's brightness relative to that of the other stars; it does not take into account the difference in distance between the stars and is not an indication of the star's true luminosity.

apparent motion *See* relative motion.

apparent noon [ASTRON] Twelve o'clock apparent time, or the instant the apparent sun is over the upper branch of the meridian.

apparent place *See* apparent position.

apparent position [ASTRON] The position on the celestial sphere at which a heavenly body (or a space vehicle) would be seen from the center of the earth at a particular time. Also known as apparent place.

apparent power [ELEC] The product of the root-mean-square voltage and the root-mean-square current delivered in an alternating-current circuit, no account being taken of the phase difference between voltage and current.

apparent precession *See* apparent wander.

apparent solar day [ASTRON] The duration of one rotation of the earth on its axis with respect to the apparent sun. Also known as true solar day.

apparent solar time [ASTRON] Time measured by the apparent diurnal motion of the sun. Also known as apparent time; true solar time.

apparent sun [ASTRON] The sun as it appears to an observer. Also known as true sun.

apparent time *See* apparent solar time.

apparent vertical [GEOPHYS] The direction of the resultant of gravitational and all other accelerations. Also known as dynamic vertical.

apparent viscosity [FL MECH] The value obtained by applying the instrumental equations used in obtaining the viscosity of a Newtonian fluid to viscometer measurements of a non-Newtonian fluid.

apparent volume [PHYS] The difference between the volume of a binary solution and the volume of the pure solvent at the same temperature.

apparent wander [GEOPHYS] Apparent change in the direction of the axis of rotation of a spinning body, such as a gyroscope, due to rotation of the earth. Also known as apparent precession; wander.

apparent weight [MECH] For a body immersed in a fluid (such as air), the resultant of the gravitational force and the buoyant force of the fluid acting on the body; equal in magnitude to the true weight minus the weight of the displaced fluid.

appearance potential [PHYS] The minimal potential which the electron beam in the ion source of a mass spectrometer must traverse in order to acquire enough energy to produce ions of a specified nuclide or molecular fragment.

Appleton layer *See* F_2 layer.

applied research [ENG] Research directed toward using

knowledge gained by basic research to make things or to create situations that will serve a practical or utilitarian purpose.

appliqué [OPTICS] A combination of lenses that provides for the same focal length at three or more wavelengths.

approximate [MATH] **1.** To obtain a result that is not exact but is near enough to the correct result for some specified purpose. **2.** To obtain a series of results approaching the correct result.

approximate absolute temperature [PHYS] A temperature scale with the ice point at 273° and boiling point of water at 373°; it is intended to approximate the Kelvin temperature scale with sufficient accuracy for many sciences, notably meteorology, and is widely used in the meteorological literature. Also known as tercentesimal thermometric scale.

approximate continuity [MATH] A measurable function of a real variable, $f(x)$, is said to be approximately continuous at c and to have approximate continuity at c if, for every pair of real numbers a_1 and a_2 such that $a_1 < f(c) < a_2$, the set of values of x such that $a_1 < f(x) < a_2$ has metric density 1 at c.

approximate derivative [MATH] The number D is called the approximate derivative of the function $f(x)$ at the point x_0 if there exists a set E, having x_0 as density point, such that the limit, as x approaches x_0, of $[f(x) - f(x_0)]/(x - x_0)$ restricted to E is equal to D.

approximation [MATH] **1.** A result that is not exact but is near enough to the correct result for some specified purpose. **2.** A procedure for obtaining such a result.

appulse [ASTRON] **1.** The near approach of one celestial body to another on the celestial sphere, as in occultation or conjunction. **2.** A penumbral eclipse of the moon.

APR *See* active prominence region.

a priori [MATH] Pertaining to deductive reasoning from assumed axioms or supposedly self-evident principles, supposedly without reference to experience.

a priori probability *See* mathematical probability.

apse *See* apsis.

apsis [ASTRON] In celestial mechanics, either of the two orbital points nearest or farthest from the center of attraction. Also known as apse.

APW method *See* augmented plane-wave method.

aquadag [ELECTR] Graphite coating on the inside of certain cathode-ray tubes for collecting secondary electrons emitted by the face of the tube.

Ar *See* argon.

arabic numerals [MATH] The numerals 0, 1, 2, 3, 4, 5, 6, 7, 8, and 9.

Arago distance [ASTRON] The angular distance from the antisolar point to the Arago point.

Arago point [OPTICS] A neutral point located about 20° directly above the antisolar point in relatively clear air and at higher elevations in turbid air.

arc [ELEC] *See* electric arc. [MATH] **1.** A continuous piece of the circumference of a circle. **2.** *See* edge.

arcback [ELECTR] The flow of a principal electron stream in the reverse direction in a mercury-vapor rectifier tube because of formation of a cathode spot on an anode; this results in failure of the rectifying action. Also known as backfire.

arc converter [ELECTR] A form of oscillator using an electric arc as the generator of alternating or pulsating current.

arc cosecant of a number [MATH] **1.** Any angle whose cosecant is that number. **2.** The angle between $-\pi/2$ radians and $\pi/2$ radians whose cosecant is that number; it is the value of the inverse of the restriction of the cosecant function to the interval between $-\pi/2$ and $\pi/2$, at this number.

arc cosine of a number [MATH] **1.** Any angle whose cosine is that number. **2.** The angle between 0 radians and π radians whose sine is that number; it is the value of the inverse of the restriction of the cosine function to the interval between 0 and π, at this number.

arc cotangent of a number [MATH] **1.** Any angle whose cotangent is that number. **2.** The angle between 0 radians and π radians whose cotangent is that number; it is the value of the inverse of the restriction of the cotangent function to the interval between 0 and π, at this number.

arc discharge [ELEC] A direct-current electrical current between electrodes in a gas or vapor, having high current density and relatively low voltage drop.

arc excitation [ATOM PHYS] Use of electric-arc energy to move electrons into higher energy orbits.

Archimedean ordered field [MATH] A field with a linear order that satisfies the axiom of Archimedes.

Archimedean principle [PHYS] The principle that a body immersed in a fluid undergoes an apparent loss in weight equal to the weight of the fluid it displaces.

Archimedean solid [MATH] One of 13 possible solids whose faces are all regular polygons, though not necessarily all of the same type, and whose polyhedral angles are all equal.

Archimedes' axiom *See* axiom of Archimedes.

Archimedes number [FL MECH] One of a dimensionless group of numbers denoting the ratio of gravitational force to viscous force.

Archimedes' problem [MATH] The problem of dividing a hemisphere into two parts of equal volume with a plane parallel to the base of the hemisphere; it cannot be solved by euclidean methods.

Archimedes' spiral *See* spiral of Archimedes.

arc hyperbolic function [MATH] An inverse function of one of the hyperbolic functions.

arc lamp [ELEC] An electric lamp in which the light is produced by an arc made when current flows through ionized gas between two electrodes. Also known as electric-arc lamp.

arcmin *See* minute.

arcometer [ENG] A device for determining the density of a liquid by measuring the apparent weight loss of a solid of known mass and volume when it is immersed in the liquid.

arc-over [ELEC] An unwanted arc resulting from the opening of a switch or the breakdown of insulation.

arc secant of a number [MATH] **1.** Any angle whose secant is that number. **2.** The angle between 0 radians and π radians whose secant is that number; it is the value of the inverse of the restriction of the secant function to the interval between 0 and π, at this number.

arc sine of a number [MATH] **1.** Any angle whose sine is that number. **2.** The angle between $-\pi/2$ radians and $\pi/2$ radians whose sine is that number; it is the value of the inverse of the restriction of the sine function to the interval between $-\pi/2$ and $\pi/2$, at this number.

arc sine transformation [STAT] A technique used to convert data made up of frequencies or proportions into a form that can be analyzed by analysis of variance or by regression analysis.

arc spectrum [SPECT] The spectrum of a neutral atom, as opposed to that of a molecule or an ion; it is usually produced by vaporizing the substance in an electric arc; designated by the roman numeral I following the symbol for the element, for example, HeI.

arc tangent of a number [MATH] **1.** Any angle whose tangent is that number. **2.** The angle between $-\pi/2$ radians and $\pi/2$ radians whose tangent is that number; it is the value of the inverse of the restriction of the tangent function to the interval between $-\pi/2$ and $\pi/2$, at this number.

are [MECH] A unit of area, used mainly in agriculture, equal to 100 square meters.

area [MATH] A measure of the size of a two-dimensional surface, or of a region on such a surface.

areal velocity [ASTROPHYS] In celestial mechanics, the area swept out by the radius vector per unit time.

area sampling [STAT] The area to be sampled is subdivided into smaller blocks which are selected at random and then subsampled or fully surveyed; method is used when a complete frame of reference is not available.

Aren's product [MATH] Let S be a semigroup, $m(S)$ the space of all bounded real-valued functions on S with supremum norm, and $m(S)^*$ the space of bounded real linear functionals on $m(S)$; then the Aren's product is a multiplication on $m(S)^*$, denoted $\odot$, given by $(F \odot G)(f) = F(h)$ for all F, G in $m(S)^*$ and f in $m(S)$, where $h(x) = G(f_x)$ for all x in S, f_x being defined by the equation $f_x(y) = f(xy)$.

Argand diagram [MATH] A two-dimensional cartesian coordinate system for representing the complex numbers, the number $x + iy$ being represented by the point whose coordinates are x and y.

Argelander method [ASTRON] A technique to estimate the brightness of variable stars; it involves estimating the difference in magnitude between the variable stars as compared to one or more stars that are invariable.

argon [CHEM] A chemical element, symbol Ar, atomic number 18, atomic weight 39.998.

argon ionization detector [NUCLEO] An ionization chamber that is filled with argon gas.

argon laser [OPTICS] A gas laser using ionized argon; emits a 4880-angstrom line as well as infrared radiation.

Arguesian plane *See* Desarguesian plane.

argument [ASTRON] An angle or arc, as in argument of perigee. [MATH] *See* amplitude; independent variable.

argument of latitude [ASTRON] The angular distance measured in the orbit plane from the ascending node to the orbiting object; the sum of the argument of perigee and the true anomaly.

argument of perigee [ASTRON] The angle or arc, as seen from a focus of an elliptical orbit, from the ascending node to the closest approach of the orbiting body to the focus; the angle is measured in the orbital plane in the direction of motion of the orbiting body.

arithlog paper [MATH] Graph paper marked with a semi-logarithmic coordinate system.

arithmetic [MATH] Addition, subtraction, multiplication, and division, usually of integers, rational numbers, real numbers, or complex numbers.

arithmetical addition [MATH] The addition of positive numbers or of the absolute values of signed numbers.

arithmetic mean [MATH] The average of a collection of numbers obtained by dividing the sum of the numbers by the quantity of numbers. Also known as average (av).

arithmetic progression [MATH] A sequence of numbers for which there is a constant d such that the difference between any two successive terms is equal to d.

arithmetic series [MATH] A series whose terms form an arithmetic progression.

arithmetic sum [MATH] **1.** The result of the addition of two or

ARGON

18
Ar

Periodic table of the chemical elements showing the position of argon.

more positive quantities. **2.** The result of the addition of the absolute values of two or more quantities.

arithmetization [MATH] The study of various branches of higher mathematics by methods that make use of only the basic concepts and operations of arithmetic.

arm [ELEC] *See* branch. [MATH] A side of an angle. [PHYS] The perpendicular distance from the line along which a force is applied to a reference point.

armature [ELECTROMAG] **1.** That part of an electric rotating machine that includes the main current-carrying winding in which the electromotive force produced by magnetic flux rotation is induced; it may be rotating or stationary. **2.** The movable part of an electromagnetic device, such as the movable iron part of a relay, or the spring-mounted iron part of a vibrator or buzzer.

armature chatter [ELECTROMAG] Vibration of the armature of a relay caused by pulsating coil current or by marginally low coil current.

armature reactance [ELECTROMAG] The inductive reactance due to the flux produced by the armature current and enclosed by the conductors in the armature slots and the end connections.

armature reaction [ELECTROMAG] Interaction between the magnetic flux produced by armature current and that of the main magnetic field in an electric motor or generator.

armature resistance [ELEC] The ohmic resistance in the main current-carrying windings of an electric generator or motor.

Armstrong oscillator [ELECTR] Inductive feedback oscillator that consists of a tuned-grid circuit and an untuned-tickler coil in the plate circuit; control of feedback is accomplished by varying the coupling between the tickler and the grid circuit.

ARPES *See* angle-resolved photoelectron spectroscopy.

array [ELECTR] A group of components such as antennas, reflectors, or directors arranged to provide a desired variation of radiation transmission or reception with direction. [STAT] The arrangement of a sequence of items in statistics according to their values, such as from largest to smallest.

Arrhenius equation [PHYS CHEM] The relationship that the specific reaction rate constant k equals the frequency factor constant s times exp $(-\Delta H_{act}/RT)$, where ΔH_{act} is the heat of activation, R the gas constant, and T the absolute temperature.

Arrhenius-Guzman equation [PHYS] The relation between the viscosity η of a liquid and the Kelvin temperature T at constant pressure: $\eta = A$ exp (B/RT), where A and B are constants and R is the gas constant.

Arrhenius viscosity formulas [PHYS] A series of three equations which relate the viscosity of a liquid to the temperature, the viscosity of a solution to its concentration and to the viscosity of the solvent, and the viscosity of a sol to the viscosity of the medium.

arsenic [CHEM] A chemical element, symbol As, atomic number 33, atomic weight 74.9216.

artificial delay line *See* delay line.

artificial echo [ELECTROMAG] **1.** Received reflections of a transmitted pulse from an artificial target, such as an echo box, corner reflector, or other metallic reflecting surface. **2.** Delayed signal from a pulsed radio-frequency signal generator.

artificial line [ELEC] A circuit made up of lumped constants, which is used to simulate various characteristics of a transmission line.

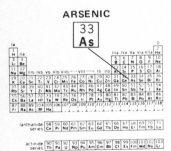

Periodic table of the chemical elements showing the position of arsenic.

artificial line duct [ELEC] A balancing network simulating the impedance of the real line and distant terminal apparatus, which is employed in a duplex circuit to make the receiving device unresponsive to outgoing signal currents.

artificial load [ELEC] Dissipative but essentially nonradiating device having the impedance characteristics of an antenna, transmission line, or other practical utilization circuit.

artificial radiation belt [GEOPHYS] High-energy electrons trapped in the earth's geomagnetic field as a result of high-altitude nuclear explosions.

artificial radioactivity *See* induced radioactivity.

artificial voice [ENG ACOUS] **1.** Small loudspeaker mounted in a shaped baffle which is proportioned to simulate the acoustical constants of the human head; used for calibrating and testing close-talking microphones. **2.** Synthetic speech produced by a multiple tone generator; used to produce a voice reply in some real-time computer applications.

Artinian ring [MATH] A ring in which every sequence of left ideals, $a_1, a_2, \ldots$, with a_{i+1} a proper subset of a_i, is finite.

Artin-Rees theorem [MATH] If E is a finitely generated module over a commutative Noetherian ring R containing an ideal I, and F is a submodule of E, then there exists a positive integer s such that for all positive integers n the equation $I^n E \cap F = I^{n-s}(I^s E \cap F)$ is satisfied, where $\cap$ denotes intersection, and $I^n E$ denotes the set of all elements in E that can be written in the form $a_1 \ldots a^n x$, where $a_1, \ldots, a_n$ are in I and x is in E.

aS *See* abmho.

As *See* arsenic.

A scale [ACOUS] A system used to filter out sound below 55 decibels; its characteristics are equal to those of the human ear.

ascending branch [MECH] The portion of the trajectory between the origin and the summit on which a projectile climbs and its altitude constantly increases.

ascending node [ASTRON] Also known as northbound node. **1.** The point at which a planet, planetoid, or comet crosses to the north side of the ecliptic. **2.** The point at which a satellite crosses to the north side of the equatorial plane of its primary.

ascending series [MATH] **1.** A series each of whose terms is greater than the preceding term. **2.** *See* power series.

Ascoli's theorem [MATH] A set of uniformly bounded, equi-continuous, real-valued functions on a closed set of a real euclidean n-dimensional space contains a sequence of functions which converges uniformly on compact subsets.

aspect [ASTRON] The apparent position of a celestial body relative to another; particularly, the apparent position of the moon or a planet relative to the sun.

aspheric surface [OPTICS] A lens or mirror surface which is altered slightly from a spherical surface in order to reduce aberrations.

aspiration condenser [NUCLEO] An ion-counter collecting element consisting of a cylindrical condenser which when charged produces a radial field that collects ions from the aspirated air.

associate curve *See* Bertrand curve.

associated corpuscular emission [GEOPHYS] The full complement of secondary charged particles associated with the passage of an x-ray or gamma-ray beam through air.

associated Laguerre polynomials [MATH] Polynomials which result from repeated differentiation of Laguerre polynomials and satisfy the differential equation $xy'' + (m + 1 - x)y' + (n - m)y = 0$, where n is a positive integer, m is a nonnegative integer, and $m \leq n$.

associated Legendre function [MATH] Any solution of Legendre's associated equation.

associated Legendre polynomials [MATH] Polynomials which satisfy Legendre's associated equation with integral values of the parameters, and which appear in spherical harmonics.

associated prime ideal [MATH] A prime ideal I in a commutative ring R is said to be associated with a module M over R if there exists an element x in M such that I is the annihilator of x.

associated production [PARTIC PHYS] Production of strange particles invariably in twos, never one particle alone.

associate matrix *See* Hermitian conjugate of a matrix.

associate operator *See* adjoint operator.

associates [MATH] Two elements x and y in a commutative ring with identity such that $x = ay$, where a is a unit. Also known as equivalent elements.

associative algebra [MATH] An algebra in which the vector multiplication obeys the associative law.

associative law [MATH] For a binary operation designated ∘, the relationship expressed by $a \circ (b \circ c) = (a \circ b) \circ c$.

astable multivibrator [ELECTR] A multivibrator in which each active device alternately conducts and is cut off for intervals of time determined by circuit constants, without use of external triggers. Also known as free-running multivibrator.

astatic [PHYS] Without orientation or directional characteristics; having no tendency to change position.

astatic galvanometer [ENG] A sensitive galvanometer designed to be independent of the earth's magnetic field.

astatic pair [ELECTROMAG] A pair of parallel magnets, equal in strength and having polarities in opposite directions, and perpendicular to an axis which bisects both of them; there is no net force or torque on the pair in a uniform field.

astatic system [ELECTROMAG] A system of magnets arranged so that the net force and torque exerted on the system by a uniform magnetic field equals 0.

astatic wattmeter [ENG] An electrodynamic wattmeter designed to be insensitive to uniform external magnetic fields.

astatine [CHEM] A radioactive chemical element, symbol At, atomic number 85, the heaviest of the halogen elements.

astatized gravimeter [ENG] A gravimeter, sometimes referred to as unstable, where the force of gravity is maintained in an unstable equilibrium with the restoring force.

asterism [ASTRON] A constellation or small group of stars. [OPTICS] A starlike optical phenomenon seen in gemstones called star stones; due to reflection of light by lustrous inclusions reduced to sharp lines of light by a domed cabochon style of cutting. [SPECT] A star-shaped pattern sometimes seen in x-ray spectrophotographs.

asteroid [ASTRON] One of the many small celestial bodies revolving around the sun, most of the orbits being between those of Mars and Jupiter. Also known as minor planet; planetoid.

astigmat *See* astigmatic lens.

astigmatic difference [OPTICS] **1.** The distance between the primary and secondary foci of an astigmatic optical system. **2.** The difference between the reciprocals of the distances of the primary and secondary foci from an astigmatic thin lens or mirror.

astigmatic foci [OPTICS] The two lines on which rays emanating from a point are focused by an astigmatic optical system. Also known as focal lines.

astigmatic interval [OPTICS] The portion of a pencil of rays

ASTATINE

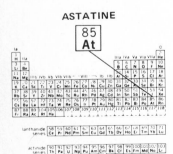

Periodic table of the chemical elements showing the position of astatine.

in an astigmatic optical system that lies between the primary and secondary foci. Also known as conoid of Sturm; interval of Sturm.

astigmatic lens [OPTICS] A planocylindrical, spherocylindrical, or spherotoric lens used in eyeglasses to correct astigmatism. Also known as astigmat.

astigmatic mounting [SPECT] A mounting designed to minimize the astigmatism of a concave diffraction grating.

astigmatic surfaces [OPTICS] Two surfaces containing the astigmatic foci of points in a plane perpendicular to the optical axis of an astigmatic system.

astigmatism [ELECTR] In an electron-beam tube, a focus defect in which electrons in different axial planes come to focus at different points. [OPTICS] The failure of an optical system, such as a lens or a mirror, to image a point as a single point; the system images the point on two line segments separated by an interval.

astigmatizer [OPTICS] A device, as attached to a rangefinder, for drawing out a point of light into a line or band.

astigmometer [OPTICS] An instrument which measures the amount of astigmatism in an optical system.

Aston dark space [ELECTR] A dark region in a glow-discharge tube which extends for a few millimeters from the cathode up to the cathode glow.

Aston whole-number rule [PHYS] The rule which states that when expressed in atomic weight units, the atomic weights of isotopes are very nearly whole numbers, and the deviations found in samples of elements are due to the presence of several isotopes with different weights.

astre fictif [ASTRON] Any of several fictitious stars assumed to move along the celestial equator at uniform rates corresponding to the speeds of the several harmonic constituents of the tide-producing force.

astrionics [ELECTR] The science of adapting electronics to aerospace flight.

astro- [ASTRON] A prefix meaning star or stars and, by extension, sometimes used as the equivalent of celestial, as in astronautics.

astroballistics [MECH] The study of phenomena arising out of the motion of a solid through a gas at speeds high enough to cause ablation; for example, the interaction of a meteoroid with the atmosphere.

astrodynamics [ASTROPHYS] The dynamics of celestial objects.

astrogeology [ASTRON] The science that applies the principles of geology, geochemistry, and geophysics to the moon and planets other than the earth.

astrographic position *See* astrometric position.

astrograph mean time [ASTRON] A form of mean time, used in setting an astrograph; mean-time setting of 1200 occurs when the local hour angle of Aries is 0°.

astroid [MATH] The locus of points satisfying the equation $x^{2/3} + y^{2/3} = a^{2/3}$, where x and y are cartesian coordinates; it has the shape of a four-pointed star with points formed from cusps on the x and y axes.

astrometric position [ASTRON] The position of a heavenly body or space vehicle on the celestial sphere corrected for aberration but not for planetary aberration. Also known as astrographic position.

astrometry [ASTRON] The branch of astronomy dealing with the geometrical relations of the celestial bodies and their real and apparent motions.

astron [NUC PHYS] A proposed thermonuclear device in

which a deuterium plasma is confined by an axial magnetic field produced by a shell of relativistic electrons.

astronomic   *See* astronomical.

astronomical   [ASTRON] Of or pertaining to astronomy or to observations of the celestial bodies. Also known as astronomic.

astronomical almanac   [ASTRON] A publication giving the tables of coordinates of a number of celestial bodies at a number of specific times during a given period.

astronomical camera   [OPTICS] A camera designed to record either point sources (stars), extended sources (nebulae, galaxies, planets, or the sun and moon), or the spectra of celestial bodies.

astronomical constants   [ASTROPHYS] The elements of the orbits of the bodies of the solar system, their masses relative to the sun, their size, shape, orientation, rotation, and inner constitution, and the velocity of light.

astronomical coordinate system   [ASTRON] Any system of spherical coordinates serving to locate astronomical objects on the celestial sphere.

astronomical date   [ASTRON] Designation of epoch by year, month, day, and decimal fraction.

astronomical day   [ASTRON] A mean solar day beginning at mean noon, 12 hours later than the beginning of the civil day of the same date; astronomers now generally use the civil day.

astronomical distance   [ASTRON] The distance of a celestial body expressed in units such as the light-year, astronomical unit, and parsec.

astronomical eclipse   *See* eclipse.

astronomical ephemeris   *See* ephemeris.

astronomical nutation   [ASTRON] A small periodic motion of the celestial pole of celestial bodies, including the earth, with respect to the pole of the ecliptic.

astronomical photography   [OPTICS] The use of the photographic process to record surface features of celestial objects, their positions and motions (for measurement), and their radiation (photometry) and spectra (spectroscopy).

astronomical refraction   [GEOPHYS] The bending of a ray of celestial radiation as it passes through atmospheric layers of increasing density.

astronomical scintillation   [ASTROPHYS] Any scintillation phenomena, such as irregular oscillatory motion, variation of intensity, and color fluctuation, observed in the light emanating from an extraterrestrial source. Also known as stellar scintillation.

astronomical spectrograph   [SPECT] An instrument used to photograph spectra of stars.

astronomical spectroscopy   [SPECT] The use of spectrographs in conjunction with telescopes to obtain observational data on the velocities and physical conditions of astronomical objects.

astronomical telescope   [OPTICS] A telescope designed for viewing astronomical objects.

astronomical time   [ASTRON] Solar time in an astronomical day that begins at noon.

astronomical twilight   [ASTRON] The period of incomplete darkness when the center of the sun is more than 6° but not more than 18° below the celestial horizon.

astronomical unit   [ASTRON] Abbreviated AU. **1.** A measure for distance within the solar system equal to the mean distance between earth and sun, that is, 149,599,000 kilometers. **2.** The semimajor axis of the elliptical orbit of earth.

astronomy   [SCI TECH] The science concerned with celestial bodies and the observation and interpretation of the radiation

ASTRONOMICAL COORDINATE SYSTEM

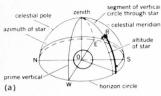

(a)

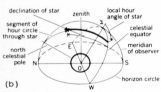

(b)

Two types of astronomical coordinate systems. *(a)* Horizon system. *(b)* Equatorial system.

received in the vicinity of the earth from the component parts of the universe.

astrophysics [ASTRON] A branch of astronomy that treats of the physical properties of celestial bodies, such as luminosity, size, mass, density, temperature, and chemical composition, and with their origin and evaluation.

A-summable *See* Abel-summable.

A supply [ELECTR] Battery, transformer filament winding, or other voltage source that supplies power for heating filaments of vacuum tubes. Also known as A power supply.

asymmetrical deflection [ELECTR] A type of electrostatic deflection in which one deflector plate is maintained at a fixed potential and the deflecting voltage is supplied to the other plate.

asymmetric top [MECH] A system that has all three principal moments of inertia different.

asymmetry [PHYS CHEM] The geometrical design of a molecule, atom, or ion that cannot be divided into like portions by one or more hypothetical planes. Also known as molecular asymmetry.

asymptote [MATH] **1.** A line approached by a curve in the limit as the curve approaches infinity. **2.** The limit of the tangents to a curve as the point of contact approaches infinity.

asymptotically flat [RELAT] A space-time is asymptotically flat if it approaches Minkowski space-time at a prescribed rate at large spatial distances.

asymptotically simple [RELAT] A space-time is asymptotically simple if it satisfies certain mathematical requirements on the conformal structure of null infinity; these requirements are a definition of a type of asymptotic flatness.

asymptotic cone of acceptance [GEOPHYS] The solid angle in the celestial sphere from which particles have to come in order to contribute significantly to the counting rate of a given neutron monitor on the surface of the earth.

asymptotic curve [MATH] A curve on a surface whose osculating plane at each point is the same as the tangent plane to the surface.

asymptotic direction of arrival [GEOPHYS] The direction at infinity of a positively charged particle, with given rigidity, which impinges in a given direction at a given point on the surface of the earth, after passing through the geomagnetic field.

asymptotic directions [MATH] For a hyperbolic point on a surface, the two directions in which the normal curvature vanishes; equivalently, the directions of the asymptotic curves passing through the point.

asymptotic efficiency [STAT] The efficiency of an estimator within the limiting value as the size of the sample increases.

asymptotic expansion [MATH] A series of the form $a_0 + (a_1/x) + (a_2/x^2) + \cdot \cdot \cdot + (a_n/x^n) + \cdot \cdot \cdot$ is an asymptotic expansion of the function $f(x)$ if there exists a number N such that for all $n > N$ the quantity $x^n [f(x) - S_n(x)]$ approaches zero as x approaches infinity, where $S_n(x)$ is the sum of the first n terms in the series. Also known as asymptotic series.

asymptotic formula [MATH] A statement of equality between two functions which is not a true equality but which means the ratio of the two functions approaches 1 as the variable approaches some value, usually infinity.

asymptotic freedom [PARTIC PHYS] In some gage theories, the property of the strong interactions of growing steadily weaker at high energies.

asymptotic series *See* asymptotic expansion.

asynchronous [PHYS] Not synchronous.

asynchronous control [CONT SYS] A method of control in

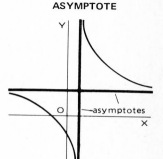

ASYMPTOTE

Asymptotes of a hyperbola.

which the time allotted for performing an operation depends on the time actually required for the operation, rather than on a predetermined fraction of a fixed machine cycle.

asynchronous device [CONT SYS] A device in which the speed of operation is not related to any frequency in the system to which it is connected.

asynchronous machine [ELEC] An ac machine whose speed is not proportional to the frequency of the power line.

at *See* technical atmosphere.

At *See* astatine.

ata [MECH] A unit of absolute pressure in the metric technical system equal to 1 technical atmosphere.

athermal transformation [PHYS] A chemical or physical change not requiring a change in the temperature of the substance, as in the formation of martensite.

athermancy [ELECTROMAG] Property of a substance which cannot transmit infrared radiation.

A1 time [ASTRON] A particular atomic time scale, established by the U.S. Naval Observatory, with the origin on January 1, 1958, at zero hours Universal Time and with the unit (second) equal to 9,192,631,770 cycles of cesium at zero field.

atlas [MATH] An atlas for a manifold is a collection of coordinate patches that covers the manifold.

atm *See* atmosphere.

atmidometer *See* atmometer.

atmolysis [FL MECH] The separation of gas mixtures by using their relative diffusibility through a porous partition.

atmo-meter *See* meter-atmosphere.

atmometer [ENG] The general name for an instrument which measures the evaporation rate of water into the atmosphere. Also known as atmidometer; evaporation gage; evaporimeter.

atmosphere [MECH] A unit of pressure equal to 1.013250 × 10^6 dynes/cm^2, which is the air pressure at mean sea level. Abbreviated atm. Also known as standard atmosphere.

atmospheric acoustics [ACOUS] The science of sound waves in the open air.

atmospheric attenuation [GEOPHYS] A process in which the flux density of a parallel beam of energy decreases with increasing distance from the source as a result of absorption or scattering by the atmosphere.

atmospheric boil *See* terrestrial scintillation.

atmospheric drag [FL MECH] A major perturbation of close artificial satellite orbits caused by the resistance of the atmosphere; the secular effects are decreasing eccentricity, semidiameter, and period.

atmospheric duct [GEOPHYS] A stratum of the troposphere within which the refractive index varies so as to confine within the limits of the stratum the propagation of an abnormally large proportion of any radiation of sufficiently high frequency, as in a mirage.

atmospheric electric field [GEOPHYS] The atmosphere's electric field strength in volts per meter at any specified point in time and space; near the earth's surface, in fair-weather areas, a typical datum is about 100 and the field is directed vertically in such a way as to drive positive charges downward.

atmospheric electricity [GEOPHYS] The electrical processes occurring in the lower atmosphere, including both the intense local electrification accompanying storms and the much weaker fair-weather electrical activity over the entire globe produced by the electrified storms continuously in progress.

atmospheric ionization [GEOPHYS] The process by which neutral atmospheric molecules are rendered electrically charged chiefly by collisions with high-energy particles.

atmospheric optics *See* meteorological optics.

atmospheric physics [GEOPHYS] The study of the physical phenomena of the atmosphere.

atmospheric pressure [PHYS] The pressure at any point in an atmosphere due solely to the weight of the atmospheric gases above the point concerned. Also known as barometric pressure.

atmospheric radiation [GEOPHYS] Infrared radiation emitted by or being propagated through the atmosphere.

atmospheric radio wave [ELECTROMAG] Radio wave that is propagated by reflection in the atmosphere; may include either the ionospheric wave or the tropospheric wave, or both.

atmospheric refraction [GEOPHYS] **1.** The angular difference between the apparent zenith distance of a celestial body and its true zenith distance, produced by refraction effects as the light from the body penetrates the atmosphere. **2.** Any refraction caused by the atmosphere's normal decrease in density with height.

atmospheric scattering [GEOPHYS] A change in the direction of propagation, frequency, or polarization of electromagnetic radiation caused by interaction with the atoms of the atmosphere.

atmospheric shimmer *See* terrestrial scintillation.

atmospheric tide [GEOPHYS] Periodic global motions of the earth's atmosphere, produced by gravitational action of the sun and moon; amplitudes are minute except in the upper atmosphere.

atom [CHEM] The individual structure which constitutes the basic unit of any chemical element. [MATH] An element, A, of a measure algebra, other than the zero element, which has the property that any element which is equal to or less than A is either equal to A or equal to the zero element.

atomic absorption coefficient [PHYS] The linear absorption coefficient divided by the number of atoms per unit volume.

atomic battery *See* nuclear battery.

atomic beam [PHYS] A stream of atoms, which may or may not be ionized.

atomic-beam frequency standard [PHYS] A source of precisely timed signals which are derived from an atomic-beam resonance, such as a cesium-beam cell or a hydrogen maser.

atomic-beam resonance [PHYS] Phenomenon in which an oscillating magnetic field, superimposed on a uniform magnetic field at right angles to it, causes transitions between states with different magnetic quantum numbers of the nuclei of atoms in a beam passing through the field; the transitions occur only when the frequency of the oscillating field assumes certain characteristic values.

atomic bomb [ORD] Also known as A bomb. **1.** A device for suddenly producing an explosively rapid neutron chain reaction in a fissile material such as uranium-235 or plutonium-239. Also known as fission bomb. **2.** Any explosive device which derives its energy from nuclear reactions, including a fusion bomb. Also known as nuclear bomb.

atomic charge [ATOM PHYS] The electric charge of an ion, equal to the number of electrons the atom has gained or lost in its ionization multiplied by the charge on one electron.

atomic clock [HOROL] An electronic clock whose frequency is supplied or governed by the natural resonance frequencies of atoms or molecules of suitable substances.

atomic cloud [NUCLEO] The cloud of hot gases, smoke, dust, and other matter that is carried aloft after the explosion of a nuclear weapon in the air or near the surface; frequently has a mushroom shape.

atomic constants [PHYS] The physical constants which play a fundamental role in atomic physics, including the electronic charge, electronic mass, speed of light, Avogadro number, and Planck's constant.

atomic core [ATOM PHYS] An atom stripped of its valence electrons, so that its remaining electrons are all in closed shells.

atomic diamagnetism [ATOM PHYS] Diamagnetic ionic susceptibility, important in providing correction factors for measured magnetic susceptibilities; calculated theoretically by considering electron density distributions summed for each electron shell.

atomic emission spectroscopy [SPECT] A form of atomic spectroscopy in which one observes the emission of light at discrete wavelengths by atoms which have been electronically excited by collisions with other atoms and molecules in a hot gas.

atomic energy *See* nuclear energy.

atomic energy level [ATOM PHYS] A definite value of energy possible for an atom, either in the ground state or an excited condition.

atomic fallout *See* fallout.

atomic F curve [PHYS] A graph plotting the atomic scattering factor F as a function of $\sin\theta/\lambda$, where θ is the scattering angle and λ is the wavelength. Also known as F curve.

atomic fission *See* fission.

atomic fluorescence spectroscopy [SPECT] A form of atomic spectroscopy in which the sample atoms are first excited by absorbing radiation from an external source containing the element to be detected, and the intensity of radiation emitted at characteristic wavelengths during transitions of these atoms back to the ground state is observed.

atomic form factor *See* atomic scattering factor.

atomic frequency [SOLID STATE] One of the vibrational frequencies of an atom in a crystal lattice.

atomic fusion *See* fusion.

atomic gas laser [OPTICS] A gas laser, such as the helium-neon laser, in which electrons and ions accelerated between electrodes by an electric field collide and excite atoms and ions to higher energy levels; laser action occurs during subsequent decay back to lower energy levels.

atomic ground state [ATOM PHYS] The state of lowest energy in which an atom can exist. Also known as atomic unexcited state.

atomic heat capacity [PHYS CHEM] The heat capacity of a gram-atomic weight of an element.

atomic hydrogen [CHEM] Gaseous hydrogen whose molecules are dissociated into atoms.

atomic hydrogen maser [PHYS] A maser in which dissociated hydrogen atoms from an electric discharge source are formed into a beam that undergoes selective magnetic processing; can be used as an atomic clock.

atomic magnet [ATOM PHYS] An atom which possesses a magnetic moment either in the ground state or in an excited state.

atomic magnetic moment [ATOM PHYS] A magnetic moment, permanent or temporary, associated with an atom, measured in magnetons.

atomic mass [PHYS] The mass of a neutral atom usually expressed in atomic mass units.

atomic mass unit [PHYS] An arbitrarily defined unit in terms of which the masses of individual atoms are expressed; the standard is the unit of mass equal to $\frac{1}{12}$ the mass of the carbon atom, having as nucleus the isotope with mass number 12. Abbreviated amu. Also known as dalton.

atomic nucleus *See* nucleus.

atomic number [NUC PHYS] The number of protons in an atomic nucleus. Also known as proton number.

atomic orbital [ATOM PHYS] The space-dependent part of a wave function describing an electron in an atom.

atomic paramagnetism [ELECTROMAG] The result of a permanent magnetic moment in an atom.

atomic particle [ATOM PHYS] One of the particles of which an atom is constituted, as an electron, neutron, or proton.

atomic photoelectric effect *See* photoionization.

atomic physics [PHYS] The science concerned with the structure of the atom, the characteristics of the elementary particles of which the atom is composed, and the processes involved in the interactions of radiant energy with matter.

atomic pile *See* nuclear reactor.

atomic polarization [PHYS CHEM] Polarization of a material arising from the change in dipole moment accompanying the stretching of chemical bonds between unlike atoms in molecules.

atomic power plant [NUCLEO] A means for converting stored nuclear energy into work, such as a nuclear electric power generating station.

atomic radius [PHYS CHEM] **1.** Half the distance of closest approach of two like atoms not united by a bond. **2.** The experimentally determined radius of an atom in a covalently bonded compound.

atomic ratio [CHEM] The ratio of the number of atoms of an element in a given sample to the total number of atoms in the sample.

atomic reactor *See* nuclear reactor.

atomic scattering factor [PHYS] A quantity which expresses the efficiency with which x-rays of a stated wavelength are scattered into a given direction by a particular atom, measured in terms of the corresponding scattering by a point electron. Also known as atomic form factor.

atomic second [PHYS] As defined in 1967, the duration of 9,192,631,770 periods of the radiation corresponding to the two hyperfine levels of the fundamental state of the atom of cesium-133.

atomic spectroscopy [SPECT] The branch of physics concerned with the production, measurement, and interpretation of spectra arising from either emission or absorption of electromagnetic radiation by atoms.

atomic spectrum [SPECT] The spectrum of radiations due to transitions between energy levels in an atom, either absorption or emission.

atomic standard [PHYS] Any supposedly immutable property of an atom, such as the wavelength or frequency of a characteristic spectral line, in terms of which a unit of a physical quantity is defined.

atomic stopping power [NUCLEO] For an ionizing particle passing through an element, the particle's energy loss per atom within a unit area normal to the particle's path; equal to the linear energy transfer (energy loss per unit path length) divided by the number of atoms per unit volume.

atomic structure [ATOM PHYS] The arrangement of the parts of an atom, which consists of a massive, positively charged nucleus surrounded by a cloud of electrons arranged in orbits describable in terms of quantum mechanics.

atomic susceptibility [ELECTROMAG] The magnetization of a material per atom per unit of applied field; measured in ergs/oersted/atom.

atomic theory [CHEM] The assumption that matter is composed of particles called atoms and that these are the limit to which matter can be subdivided.

atomic time [HOROL] Any time system standardized with reference to an atomic resonance, such as the international standard cesium-133 transition.

atomic unexcited state *See* atomic ground state.

atomic units *See* Hartree units.

atomic vibration [ATOM PHYS] Periodic, nearly harmonic changes in position of the atoms in a molecule giving rise to many properties of matter, including molecular spectra, heat capacity, and heat conduction.

atomic volume [PHYS CHEM] The volume occupied by 1 gram-atom of an element in the solid state.

atomic weight [CHEM] The relative mass of an atom based on a scale in which a specific carbon atom (carbon-12) is assigned a mass value of 12. Abbreviated at. wt.

atoms-in-molecules method [PHYS CHEM] The description of the electronic structure of a molecule as a perturbation of the isolated states of its constituent atoms.

atom smasher *See* particle accelerator.

ATR *See* attenuated total reflectance.

attached shock *See* attached shock wave.

attached shock wave [FL MECH] An oblique or conical shock wave that appears to be in contact with the leading edge of an airfoil or the nose of a body in a supersonic flow field. Also known as attached shock.

attaching gas [ELECTR] A gas in which electron attachment takes place.

attachment coefficient [ELECTR] The probability that an electron drifting through a gas under the influence of a uniform electric field will undergo electron attachment in a unit distance of drift.

attempt frequency [NUC PHYS] The frequency with which an alpha particle attempts to cross the Gamow barrier in the Gamow-Condon-Gurney theory.

attenuated total reflectance [SPECT] A method of spectrophotometric analysis based on the reflection of energy at the interface of two media which have different refractive indices and are in optical contact with each other. Abbreviated ATR. Also known as frustrated internal reflectance; internal reflectance spectroscopy.

attenuation [PHYS] The reduction in level of a quantity, such as the intensity of a wave, over an interval of a variable, such as the distance from a source.

attenuation coefficient [ELECTROMAG] The space rate of attenuation of any transmitted electromagnetic radiation.

attenuation constant [PHYS] A wave with space-time dependence $\exp[i(kx - wt)]$, where $k = k_r + ik_i$, is attenuated as it propagates according to the factor $\exp(=k_i x)$ and k_i is called the attenuation constant. Also known as attenuation factor.

attenuation factor *See* attenuation constant.

attenuation length [PHYS] The reciprocal of the attenuation coefficient.

attenuation network [ELECTR] Arrangement of circuit elements, usually impedance elements, inserted in circuitry to introduce a known loss or to reduce the impedance level without reflections.

attenuation ratio [PHYS] The magnitude of the propagation ratio.

attenuator [ELECTR] An adjustable or fixed transducer for reducing the amplitude of a wave without introducing appreciable distortion.

atto- [PHYS] A prefix representing 10^{-18}, which is 0.000 000 000 000 000 000 001, or one-millionth of a millionth of a millionth. Abbreviated a.

attracted-disk electrometer [ELEC] A type of electrometer in which the attraction between two oppositely charged disks is measured.

atu [PHYS] A unit of underpressure or pressure below atmospheric pressure in the metric technical system; equal to 1 technical atmosphere.

atü [PHYS] A unit of overpressure or gage pressure in the metric technical system; equal to 1 technical atmosphere.

at. wt *See* atomic weight.

A-type star [ASTRON] In star classification based on spectral characteristics, the type of star in whose spectrum the hydrogen absorption lines are at a maximum.

Au *See* gold.

AU *See* astronomical unit.

audibility [ACOUS] **1.** The state or quality of being heard. **2.** The intensity of a received audio signal, usually expressed in decibels above or below 1 milliwatt using a stated single-frequency sine wave.

audibility curve [ACOUS] **1.** The limits of hearing represented graphically as an area by plotting the minimum audible intensity of a sine wave sound versus frequency. **2.** *See* equal loudness contour.

audibility threshold [ACOUS] The sound intensity at a given frequency which is the minimum perceptible by a normal human ear under specified standard conditions.

audio [ACOUS] **1.** Of or pertaining to sound in the range of frequencies considered audible at reasonable listening intensities to the average young adult listener, approximately 15 to 20,000 hertz. **2.** Pertaining to equipment for the recording, transmission, reproduction, or amplification of such sound.

audio amplifier *See* audio-frequency amplifier.

audio frequency [ACOUS] A frequency that can be detected as a sound by the average young adult, approximately 15 to 20,000 hertz. Abbreviated af. Also known as sonic frequency; sound frequency.

audio-frequency amplifier [ELECTR] An electronic circuit for amplification of signals within, and in some cases above, the audible range of frequencies in equipment used to record and reproduce sound. Also known as audio amplifier.

audio-frequency choke [ELECTROMAG] Choke used to impede the flow of audio-frequency currents; generally a coil wound on an iron core.

audio-frequency meter [ENG] One of a number of types of frequency meters usable in the audio range; for example, a resonant-reed frequency meter.

audio-frequency oscillator [ELECTR] An oscillator circuit using an electron tube, transistor, or other nonrotating device to produce an audio-frequency alternating current. Also known as audio oscillator.

audio-frequency range [ACOUS] The range of frequencies to which the human ear is sensitive, approximately 15 to 20,000 hertz. Also known as audio range.

audio-frequency transformer [ELEC] An iron-core transformer used for coupling audio-frequency circuits. Also known as audio transformer.

audiogram [ACOUS] A graph showing hearing loss, percent hearing loss, or percent hearing as a function of frequency.

audioimpedance measurement [ACOUS] The measurement of acoustic impedance, as in the direct assessment of the dynamic motor control of sound feedback of different parts of the ear.

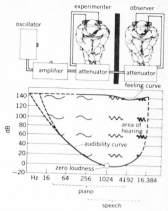

The audibility curve plotted by measuring range of sound intensities in decibels (dB) against the sound frequencies in hertz (Hz).

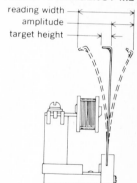

Reed-type frequency meter.

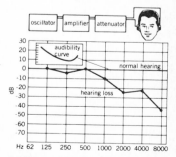

Audiogram for determining the audibility curve for pure-tone hearing loss at various frequency levels.

audiology [ACOUS] The science of hearing.

audio masking *See* masking.

audiometer [ENG] An instrument composed of an oscillator, amplifier, and attenuator and used to measure hearing acuity for pure tones, speech, and bone conduction.

audiometry [ACOUS] The study of hearing ability by means of audiometers.

audio oscillator *See* audio-frequency oscillator.

audio range *See* audio-frequency range.

audio signal [ACOUS] An electric signal having the frequency of a mechanical wave that can be detected as a sound by the human ear.

audio spectrometer *See* acoustic spectrometer.

audio transformer *See* audio-frequency transformer.

auditory perspective [ACOUS] Three-dimensional realism of sound, as produced by an actual orchestra or by a stereophonic sound system.

augend [MATH] A quantity to which another quantity is added.

Auger coefficient [ATOM PHYS] The ratio of the number of Auger electrons to the number of ejected x-ray photons.

Auger effect [ATOM PHYS] The radiationless transition of an electron in an atom from a discrete electronic level to an ionized continuous level of the same energy. Also known as autoionization.

Auger electron [ATOM PHYS] An electron that is expelled from an atom in the Auger effect.

Auger electron spectroscopy [SPECT] The energy analysis of Auger electrons produced when an excited atom relaxes by a radiationless process after ionization by a high-energy electron, ion, or x-ray beam. Abbreviated AES.

Auger recombination [ATOM PHYS] Recombination of an electron and a hole in which no electromagnetic radiation is emitted, and the excess energy and momentum of the recombining electron and hole are given up to another electron or hole.

Auger shower [ASTRON] A very large cosmic-ray shower. Also known as extensive air shower.

augmentation [ASTRON] The apparent increase in the semidiameter of a celestial body, as observed from the earth, as the body's altitude (angular distance above the horizon) increases, due to the reduced distance from the observer. The term is used principally in reference to the moon.

augmentation distance [NUC PHYS] The extrapolation distance, which is the distance between the time boundary of a nuclear reactor and its boundary calculated by extrapolation.

augmented matrix [MATH] The matrix of the coefficients, together with the constant terms, in a system of linear equations.

augmented plane-wave method [SOLID STATE] A method of approximating the energy states of electrons in a crystal lattice; the potential is assumed to be spherically symmetrical within spheres centered at each atomic nucleus and constant in the interstitial region, wave functions (the augmented plane waves) are constructed by matching solutions of the Schrödinger equation within each sphere with plane-wave solutions in the interstitial region, and linear combinations of these wave functions are then determined by the variational method. Abbreviated APW method.

A* unit [PHYS] An atomic standard unit of length, based on the tungsten $K\alpha_1$ line, approximately 10^{-11} centimeter; used for measurements of x-ray wavelengths and of crystal dimensions.

aural masking *See* masking.

aural signal [ACOUS] **1.** A signal that can be heard. **2.** The sound portion of a television signal.

aurora [ELEC] *See* corona discharge. [GEOPHYS] The most intense of the several lights emitted by the earth's upper atmosphere, seen most often along the outer realms of the Arctic and Antarctic, where it is called the aurora borealis and aurora australis, respectively; excited by charged particles from space.

aurora australis [GEOPHYS] The aurora of southern latitudes. Also known as southern lights.

aurora borealis [GEOPHYS] The aurora of northern latitudes. Also known as northern lights.

auroral absorption event [GEOPHYS] A large increase in D-region electron density and associated radio-signal absorption, caused by electron-bombardment of the atmosphere during an aurora or a geomagnetic storm.

auroral caps [GEOPHYS] The regions surrounding the auroral poles, lying between the poles and the auroral zones.

auroral electrojet [GEOPHYS] An intense electric current in the magnetosphere, flowing along the auroral zones during a polar substorm.

auroral frequency [GEOPHYS] The percentage of nights on which an aurora is seen at a particular place, or on which one would be seen if clouds did not interfere.

auroral isochasm [GEOPHYS] A line connecting places of equal auroral frequency, averaged over a number of years.

auroral line [SPECT] A prominent green line in the spectrum of the aurora at a wavelength of 5577 angstroms (557.7 nanometers), resulting from a certain forbidden transition of oxygen.

auroral oval [GEOPHYS] An oval-shaped region centered on the earth's magnetic pole in which auroral emissions occur.

auroral poles [GEOPHYS] The points on the earth's surface on which the auroral isochasms are centered; coincide approximately with the magnetic-axis poles of the geomagnetic field.

auroral region [GEOPHYS] The region within 30° geomagnetic latitude of each auroral pole.

auroral zone [GEOPHYS] A roughly circular band around either geomagnetic pole within which there is a maximum of auroral activity; lies about 10–15° geomagnetic latitude from the geomagnetic poles.

auroral zone blackout [GEOPHYS] An increase of ionization in the lower ionosphere near the auroral zone.

aurora polaris [GEOPHYS] A high-altitude aurora borealis or aurora australis.

austausch coefficient *See* exchange coefficient.

austral axis pole [GEOPHYS] The southern intersection of the geomagnetic axis with the earth's surface.

autocollimation [OPTICS] A procedure for collimating a telescope or other optical instrument with objective and crosshairs, in which the instrument is directed toward a plane mirror and the crosshairs and lens are adjusted so that the crosshairs coincide with their reflected image.

autocollimator [OPTICS] **1.** A device by which a single lens collimates diverging light from a slit, and then focuses the light on an exit slit after it has passed through a prism to a mirror and been reflected back through the prism. **2.** A telescope which has a graduated reticle, enabling an observer to read off the angles subtended by distant objects. **3.** A convex mirror at the focus of the principal mirror of a reflecting telescope, which causes light to leave the telescope in a parallel beam. **4.** A telescope equipped with an eyepiece designed for autocollimation.

autocorrelation [STAT] In a time series, to determine the

AURORA

Photograph of aurora borealis taken at College, Alaska. *(V. P. Hessler, University of Alaska)*

relationship between values of a variable taken at certain times in the series and values of a variable taken at other, usually earlier times.

autocorrelation function [MATH] For a function $f(t)$, the limit as T approaches infinity of $1/2T$ times the integral from $-T$ to T of $f(t)f(t-\tau)\,dt$, where τ is a time-delay parameter.

autocorrelator [ELECTR] A correlator in which the input signal is delayed and multiplied by the undelayed signal, the product of which is then smoothed in a low-pass filter to give an approximate computation of the autocorrelation function; used to detect a nonperiodic signal or a weak periodic signal hidden in noise.

autodyne circuit [ELECTR] A circuit in which the same tube elements serve as oscillator and detector simultaneously.

autofocus rectifier [OPTICS] A precise, vertical photoenlarger which permits the correction of distortion in an aerial negative caused by tilt.

autogenous electrification [PHYS] The process by which net charge is built up on an object, such as an airplane, moving relative to air containing dust or ice crystals; produced by frictional effects (triboelectrification) accompanying contact between the object and the particulate matter.

autogenous ignition temperature *See* ignition temperature.

autoignition *See* spontaneous combustion.

autoionization *See* Auger effect.

autoluminescence [ATOM PHYS] Luminescence of a material (such as a radioactive substance) resulting from energy originating within the material itself.

automatic C bias *See* self-bias.

automatic check-out system [CONT SYS] A system utilizing test equipment capable of automatically and simultaneously providing actions and information which will ultimately result in the efficient operation of tested equipment while keeping time to a minimum.

automatic control [CONT SYS] Control in which regulating and switching operations are performed automatically in response to predetermined conditions. Also known as automatic regulation.

automatic-control block diagram [CONT SYS] A diagrammatic representation of the mathematical relationships defining the flow of information and energy through the automatic control system, in which the components of the control system are represented as functional blocks in series and parallel arrangements according to their position in the actual control system.

automatic-control error coefficient [CONT SYS] Three numerical quantities that are used as a measure of the steady-state errors of an automatic control system when the system is subjected to constant, ramp, or parabolic inputs.

automatic-control frequency response [CONT SYS] The steady-state output of an automatic control system for sinusoidal inputs of varying frequency.

automatic controller [CONT SYS] An instrument that continuously measures the value of a variable quantity or condition and then automatically acts on the controlled equipment to correct any deviation from a desired preset value. Also known as automatic regulator; controller.

automatic-control servo valve [CONT SYS] A mechanically or electrically actuated servo valve controlling the direction and volume of fluid flow in a hydraulic automatic control system.

automatic-control stability [CONT SYS] The property of an automatic control system whose performance is such that the amplitude of transient oscillations decreases with time and the system reaches a steady state.

automatic-control system [CONT SYS] A control system having one or more automatic controllers connected in closed loops with one or more processes. Also known as regulating system.

automatic-control transient analysis [CONT SYS] The analysis of the behavior of the output variable of an automatic control system as the system changes from one steady-state condition to another in terms of such quantities as maximum overshoot, rise time, and response time.

automatic cutout [ELEC] A device, usually operated by centrifugal force or by an electromagnet, that automatically shorts part of a circuit at a particular time.

automatic data processing [ENG] The machine performance, with little or no human assistance, of any of a variety of tasks involving informational data; examples include automatic and responsive reading, computation, writing, speaking, directing artillery, and the running of an entire factory. Abbreviated ADP.

automatic focus [OPTICS] A device in a camera or enlarger which automatically keeps the objective lens in focus through a range of magnification.

automatic frequency control [ELECTR] A circuit used to maintain the frequency of an oscillator within specified limits, as in a transmitter. Abbreviated AFC.

automatic gain control [ELECTR] A control circuit that automatically changes the gain (amplification) of a receiver or other piece of equipment so that the desired output signal remains essentially constant despite variations in input signal strength. Abbreviated AGC.

automatic grid bias *See* self-bias.

automatic peak limiter *See* limiter.

automatic regulation *See* automatic control.

automatic regulator *See* automatic controller.

automatic scanning receiver [ELECTR] A receiver which can automatically and continuously sweep across a preselected frequency, either to stop when a signal is found or to plot signal occupancy within the frequency spectrum being swept.

automatic sensitivity control [ELECTR] Circuit used for automatically maintaining receiver sensitivity at a predetermined level; it is similar to automatic gain control, but it affects the receiver constantly rather than during the brief interval selected by the range gate.

automatic tuning system [CONT SYS] An electrical, mechanical, or electromechanical system that tunes a radio receiver or transmitter automatically to a predetermined frequency when a button or lever is pressed, a knob turned, or a telephone-type dial operated.

automatic voltage regulator *See* voltage regulator.

automorphic function [MATH] A single-valued, meromorphic function $f(z)$, defined on a domain D, with the property that, for each transformation T of a specified group of linear fractional transformations and each point z in D, $T(z)$ is also in D and $f[T(z)] = f(z)$.

automorphic number [MATH] A natural number a is said to be an automorphic number in the scale n with index k if $a^2 - a$ is an integral multiple of n^k.

automorphism [MATH] An isomorphism of an algebraic structure with itself.

autonomous system of differential equations [MATH] A system of first-order differential equations having the form $dx_i/dt = X_i(x_1, \ldots, x_n)$, $i = 1, \ldots, n$; the functions X_i do not depend on the independent variable t.

autoradiography [ENG] A technique for detecting radioactiv-

AUTORADIOGRAPHY

A reproduction of a photograph produced by autoradiography of metatarsus of a calf. The dense areas indicate highest concentration of the radioisotope. *(From C. L. Comar, Radioisotopes in Biology and Agriculture, McGraw-Hill, 1955)*

ity in a specimen by producing an image on a photographic film or plate. Also known as radioautography.

autoregressive series [MATH] A function of the form $f(t) = a_1 f(t-1) + a_2 f(t-2) + \cdots + a_m f(t-m) + k$, where k is any constant.

autorotation [MECH] **1.** Rotation about any axis of a body that is symmetrical and exposed to a uniform airstream and maintained only by aerodynamic moments. **2.** Rotation of a stalled symmetrical airfoil parallel to the direction of the wind.

autostability [CONT SYS] The ability of a device (such as a servomechanism) to hold a steady position, either by virtue of its shape and proportions, or by control by a servomechanism.

autostarter [ELEC] **1.** Automatic starting and switchover generating system consisting of a standby generator coupled to the station load through an automatic power transfer control unit. **2.** *See* autotransformer starter.

autotransformer [ELEC] A power transformer having one continuous winding that is tapped; part of the winding serves as the primary and all of it serves as the secondary, or vice versa; small autotransformers are used to start motors.

autotransformer starter [ELEC] Motor starter having an autotransformer to furnish a reduced voltage for starting; includes the necessary switching mechanism. Also known as autostarter; compensator.

autumn [ASTRON] The season of the year which is the transition period from summer to winter, occurring as the sun approaches the winter solstice; beginning is marked by the autumnal equinox. Also known as fall.

autumnal equinox [ASTRON] The point on the celestial sphere at which the sun's rays at noon are 90° above the horizon at the Equator, or at an angle of 90° with the earth's axis, and neither North nor South Pole is inclined to the sun; occurs in the Northern Hemisphere on approximately September 23 and marks the beginning of autumn.

auxiliary circle [ASTRON] In celestial mechanics, a circumscribing circle to an orbital ellipse with radius a, the semimajor axis.

auxiliary relay [ELEC] Relay that operates in response to the opening or closing of its operating circuit to assist another relay or device in performing a function.

auxiliary switch [ELEC] A switch actuated by the main device (such as a circuit breaker) for signaling, interlocking, or other purposes.

av *See* arithmetic mean.

aV *See* abvolt.

availability [PHYS] The difference between the enthalpy per unit mass of substance and the product of entropy per unit mass multiplied by the lowest temperature available to the substance for heat discard; used in determining the ratio of actual work performed during a process by a working substance to that which theoretically should have been performed.

available power gain [ELECTR] Ratio, in an electronic transducer, of the available power from the output terminals of the transducer, under specified input termination conditions, to the available power from the driving generator.

avalanche [ELECTR] **1.** The cumulative process in which an electron or other charged particle accelerated by a strong electric field collides with and ionizes gas molecules, thereby releasing new electrons which in turn have more collisions, so that the discharge is thus self-maintained. Also known as avalanche effect; cascade; cumulative ionization; Townsend avalanche; Townsend ionization. **2.** Cumulative multiplication of carriers in a semiconductor as a result of avalanche breakdown. Also known as avalanche effect.

avalanche breakdown [ELECTR] Nondestructive breakdown in a semiconductor diode when the electric field across the barrier region is strong enough so that current carriers collide with valence electrons to produce ionization and cumulative multiplication of carriers.

avalanche diode [ELECTR] A semiconductor breakdown diode, usually made of silicon, in which avalanche breakdown occurs across the entire *pn* junction and voltage drop is then essentially constant and independent of current; the two most important types are IMPATT and TRAPATT diodes.

avalanche effect *See* avalanche.

avalanche-induced migration [ELECTR] A technique of forming interconnections in a field-programmable logic array by applying appropriate voltages for shorting selected base-emitter junctions.

avalanche oscillator [ELECTR] An oscillator that uses an avalanche diode as a negative resistance to achieve one-step conversion from direct-current to microwave outputs in the gigahertz range.

avalanche photodiode [ELECTR] A photodiode operated in the avalanche breakdown region to achieve internal photocurrent multiplication, thereby providing rapid light-controlled switching operation.

avalanche transistor [ELECTR] A transistor that utilizes avalanche breakdown to produce chain generation of charge-carrying hole-electron pairs.

aV/cm *See* abvolt per centimeter.

average *See* arithmetic mean.

average deviation [MATH] In statistics, the average or arithmetic mean of the deviation, taken without regard to sign, from some fixed value, usually the arithmetic mean of the data. Abbreviated AD. Also known as mean deviation.

average life *See* mean life.

average noise figure [ELECTR] Ratio in a transducer of total output noise power to the portion thereof attributable to thermal noise in the input termination, the total noise being summed over frequencies from zero to infinity, and the noise temperature of the input termination being standard (290 K).

average outgoing quality [IND ENG] A quality-control concept which specifies the percent defective in lots of products which are finally passed. Abbreviated AOQ.

average outgoing quality limit [IND ENG] In acceptance sampling in quality control, the maximum possible value of the average outgoing quality.

averaging operator *See* central mean operator.

Avogadro's number [PHYS] The number (6.02×10^{23}) of molecules in a gram-molecular weight of a substance.

Avogadro's hypothesis *See* Avogadro's law.

Avogadro's law [PHYS] The law which states that under the same conditions of pressure and temperature, equal volumes of all gases contain equal numbers of molecules; for example, 359 cubic feet (10.167 cubic meters) at 32°F (0°C) and 1 atmosphere (101,325 newtons per square meter) for a perfect gas. Also known as Avogadro's hypothesis.

avogram [MECH] A unit of mass, equal to 1 gram divided by the Avogadro number.

avoirdupois pound *See* pound.

avoirdupois weight [MECH] The system of units which has been commonly used in English-speaking countries for measurement of the mass of any substance except precious stones, precious metals, and drugs; it is based on the pound (approximately 453.6 grams) and includes the short ton (2000 pounds), long ton (2240 pounds), ounce (one-sixteenth pound), and dram (one-sixteenth ounce).

axes of inertia [PHYS] The three principal axes of inertia, namely, one about which the moment of inertia is a maximum, one about which the moment of inertia is a minimum, and one perpendicular to both.

axial [SCI TECH] Of, pertaining to, or along an axis.

axial angle [CRYSTAL] **1.** The acute angle between the two optic axes of a biaxial crystal. Also known as optic angle; optic-axial angle. **2.** In air, the larger angle between the optic axes after refraction on leaving the crystal.

axial dipole field [GEOPHYS] A postulated magnetic field for the earth, consisting of a dipolar field centered at the earth's center, with its axis coincident with the earth's rotational axis.

axial element [CRYSTAL] The lengths, length ratios, and angles which define a crystal's unit cell.

axial flow [FL MECH] Flow of fluid through an axially symmetric device such that the direction of the flow is along the axis of symmetry. Also known as axisymmetric flow.

axial jet [FL MECH] A flowing, turbulent stream which mixes with standing water in three dimensions.

axial lead [ELEC] A wire lead extending from the end along the axis of a resistor, capacitor, or other component.

axial length [CRYSTAL] The length of one of the edges of a unit cell.

axial load [MECH] A force with its resultant passing through the centroid of a particular section and being perpendicular to the plane of the section.

axial modulus [MECH] The ratio of a simple tension stress applied to a material to the resulting strain parallel to the tension when the sides of the sample are restricted so that there is no lateral deformation. Also known as modulus of simple longitudinal extension.

axial moment of inertia [MECH] For any object rotating about an axis, the sum of its component masses times the square of the distance to the axis.

axial plane [CRYSTAL] **1.** A plane that includes two of the crystallographic axes. **2.** The plane of the optic axis of an optically biaxial crystal.

axial quadrupole *See* longitudinal quadrupole.

axial ratio [CRYSTAL] The ratio obtained by comparing the length of a crystallographic axis with one of the lateral axes taken as unity. [ELECTR] The ratio of the major axis to the minor axis of the polarization ellipse of a waveguide. Also known as ellipticity.

axial symmetry [MATH] Property of a geometric configuration which is unchanged when rotated about a given line.

axial vector *See* pseudovector.

axiom [MATH] Any of the assumptions upon which a mathematical theory (such as geometry, ring theory, and the real numbers) is based. Also known as postulate.

axiomatic S-matrix theory [PARTIC PHYS] An approach to the study of elementary particles that seeks to formulate S-matrix theory in a rigorous manner based on a few fundamental axioms that include Lorentz invariance, unitarity, analyticity near the physical values of the energy and momentum variables, and singularities in the physical region that correspond to known particles and scattering thresholds.

axiom of Archimedes [MATH] If x is any real number, there exists an integer n such that n is greater than x. Also known as Archimedes' axiom.

axis [MATH] **1.** In a coordinate system, the line determining one of the coordinates, obtained by setting all other coordinates to zero. **2.** A line of symmetry for a geometric figure. [MECH] A line about which a body rotates.

axis of abscissas [MATH] The horizontal or x axis of a two-dimensional cartesian coordinate system, parallel to which abscissas are measured.

axis of circulation [ELECTROMAG] The axis where the equiphase surfaces of a circulating electromagnetic wave converge.

axis of ordinates [MATH] The vertical or y axis of a two-dimensional cartesian coordinate system, parallel to which ordinates are measured.

axis of perspective [MATH] The line specified by Desargues' theorem, on which are located the intersections of corresponding sides of two triangles.

axis of rotation [MECH] A straight line passing through the points of a rotating rigid body that remain stationary, while the other points of the body move in circles about the axis.

axis of symmetry [MECH] An imaginary line about which a geometrical figure is symmetric. Also known as symmetry axis.

axis of torsion [MECH] An axis parallel to the generators of a cylinder undergoing torsion, located so that the displacement of any point on the axis lies along the axis. Also known as axis of twist.

axis of twist *See* axis of torsion.

axisymmetric flow *See* axial flow.

Ayrton-Jones balance [ELEC] A type of balance with which force between current-carrying conductors is measured; uses single-layer solenoids as the fixed and movable coils.

Ayrton-Perry winding [ELEC] Winding of two wires in parallel but opposite directions to give better cancellation of magnetic fields than is obtained with a single winding.

Ayrton shunt [ELEC] A shunt used to increase the range of a galvanometer without changing the damping. Also known as universal shunt.

azimuth [ASTRON] The horizontal direction of a celestial point from a terrestrial point, expressed as the angular distance from a reference direction, usually measured from 0° at the reference direction clockwise through 360°.

azimuthal quantum number [ATOM PHYS] The orbital angular momentum quantum number l, such that the eigenvalue of L^2 is $l(l + 1)$.

azimuth tables [ASTRON] Publications providing tabulated azimuths or azimuth angles of celestial bodies for various combinations of declination, altitude, and hour angle; great-circle course angles can also be obtained by substitution of values.

b *See* barn; bel.

B *See* bel; boron; brewster.

Ba *See* barium.

Babcock coefficient of friction [FL MECH] An approximation to the coefficient of friction for steam flowing in a circular pipe of diameter d inches, given by $0.0027(1 + 3.6/d)$.

Babcock magnetograph [ASTRON] An instrument used to measure weak magnetic fields on the sun.

Babinet compensator [OPTICS] A device for working with polarized light, made of two quartz prisms, assembled in a rhomb, to enable the optical retardation to be adjusted to positive or negative values.

Babinet point [OPTICS] A neutral point located 15 to 20° directly above the sun.

Babinet's principle [OPTICS] The principle that the diffraction patterns produced by complementary screens are identical; two screens are said to be complementary when the opaque parts of one correspond to the transparent parts of the other.

Babo's law [PHYS CHEM] A law stating that the relative lowering of a solvent's vapor pressure by a solute is the same at all temperatures.

backbending [NUC PHYS] A discontinuity in the rotational levels of some rare-earth nuclei around spin 20 $\hbar$ (where $\hbar$ is Planck's constant divided by 2π), which appears as a backbend on a graph that plots the moment of inertia versus the square of the rotational frequency.

back bond [SOLID STATE] A chemical bond between an atom in the surface layer of a solid and an atom in the second layer.

back-coated mirror [OPTICS] Glass with a reflective coating applied against the rear surface.

back contact [ELEC] Normally closed stationary contact on a relay that is opened when the relay is energized.

back echo [ELECTROMAG] An echo signal produced on a radar screen by one of the minor back lobes of a search radar beam.

back electromotive force *See* counterelectromotive force.

back-emission electron radiography [ELECTR] A technique used in microradiography to visualize, among other things, the presence of material of different atomic numbers in the surface of the specimen being observed; the polished side of the specimen is facing and in close contact with the emulsion side of a fine-grain photographic plate; a light-tight cover holds the specimen and plate in place to be subjected to hardened x-rays.

backfire *See* arcback.

backfire antenna [ELECTROMAG] An antenna which exhibits significant gain in a direction 180° from its principal lobe.

back focal length [OPTICS] The distance from the rear surface of a lens to its focal plane.

Back-Goudsmit effect [ATOM PHYS] Breakdown of the coupling between the nuclear-spin angular momentum and the total angular momentum of the electrons in an atom at relatively small magnetic fields.

background count [PHYS] Responses of the radiation counting system to radiation coming from sources other than the source to be measured.

background discrimination [ENG] The ability of a measuring instrument, circuit, or other device to distinguish signal from background noise.

background luminance [OPTICS] In visual-range theory, the brightness of the background against which a target is viewed.

background noise [ACOUS] The unwanted residual sound that is present whether or not the sound source being studied is in operation. [ENG] The undesired signals that are always present in an electronic or other system, independent of whether or not the desired signal is present.

background radiation [NUCLEO] The radiation in humans' natural environment, including cosmic rays and radiation from the naturally radioactive elements. Also known as natural radiation. [PHYS] Radiation which is due to sources other than the source of interest in a measurement of radiation and which is detected by the measuring apparatus.

backlash [ELECTR] A small reverse current in a rectifier tube caused by the motion of positive ions produced in the gas by the impact of thermoelectrons. [ENG] **1.** Relative motion of mechanical parts caused by looseness. **2.** The difference between the actual values of a quantity when a dial controlling this quantity is brought to a given position by a clockwise rotation and when it is brought to the same position by a counterclockwise rotation.

back lobe [ELECTROMAG] The three-dimensional portion of the radiation pattern of a directional antenna that is directed away from the intended direction.

back pressure [MECH] Pressure due to a force that is operating in a direction opposite to that being considered, such as that of a fluid flow.

back radiation *See* backscattering; counterradiation.

back-reflection photography [CRYSTAL] A method of studying crystalline structure by x-ray diffraction in which the photographic film is placed between the source of x-rays and the crystal specimen.

back resistance [ELECTR] The resistance between the contacts opposing the inverse current of a metallic rectifier.

backscattering Also known as back radiation; backward scattering. [ELECTROMAG] **1.** Radar echos from a target. **2.** Undesired radiation of energy to the rear by a directional antenna. [PHYS] The deflection of radiation or nuclear particles by scattering processes through angles greater than 90° with respect to the original direction of travel.

back-to-front ratio [ELECTROMAG] Ratio used in connection with an antenna, metal rectifier, or any device in which signal strength or resistance in one direction is compared with that in the opposite direction.

backward-acting regulator [ELECTR] Transmission regulator in which the adjustment made by the regulator affects the quantity which caused the adjustment.

backward difference [MATH] One of a series of quantities obtained from a function whose values are known at a series of equally spaced points by repeatedly applying the backward difference operator to these values; used in interpolation and numerical calculation and integration of functions.

backward difference operator [MATH] A difference operator, denoted ∇, defined by the equation $\nabla f(x) = f(x) - f(x - h)$, where h is a constant denoting the difference between successive points of interpolation or calculation.

backward diode [ELECTR] A semiconductor diode similar to a tunnel diode except that it has no forward tunnel current; used as a low-voltage rectifier.

backward scattering *See* backscattering.

backward wave [ELECTROMAG] An electromagnetic wave traveling opposite to the direction of motion of some other physical quantity in an electronic device such as a traveling-wave tube or mismatched transmission line.

backward-wave oscillator [ELECTR] An electronic device which amplifies microwave signals simultaneously over a wide band of frequencies and in which the traveling wave produced is reflected backward so as to sustain the wave oscillations. Abbreviated BWO. Also known as carcinotron.

backward-wave tube [ELECTR] A type of microwave traveling-wave electron tube in which electromagnetic energy on a slow-wave circuit flows opposite in direction to the travel of electrons in a beam.

badge meter *See* film badge.

Badger's rule [PHYS CHEM] An empirical relationship between the stretching force constant for a molecular bond and the bond length.

baffle [ELECTR] An auxiliary member in a gas tube used, for example, to control the flow of mercury particles or deionize the mercury following conduction.

baffle plate [ELECTROMAG] Metal plate inserted in a waveguide to reduce the cross-sectional area for wave conversion purposes.

Baily's beads [ASTRON] Bright points of sunlight appearing around the edge of the moon just before and after the central phase of a total solar eclipse.

Baire function [MATH] The smallest class of functions on a topological space which contains the continuous functions and is closed under pointwise limits.

Baire measure [MATH] A measure defined on the class of all Baire sets such that the measure of any closed, compact set is finite.

Baire set [MATH] A member of the smallest sigma algebra containing all closed, compact subsets of a topological space.

Bairstow number [FL MECH] A term previously used for Mach number.

Baker-Nunn camera [OPTICS] A large camera with a Schmidt-type lens system used to track earth satellites.

Baker-Schmidt telescope [OPTICS] A type of Schmidt telescope in which the light reflected from the near-spheroidal primary mirror is again reflected from a smaller, near-spheroidal secondary mirror, producing an image that is free of astigmatism and distortion.

balance [CHEM] To bring a chemical equation into balance so that reaction substances and reaction products obey the laws of conservation of mass and charge. [ELEC] The state of an electrical network when it is adjusted so that voltage in one branch induces or causes no current in another branch. [ENG] An instrument for measuring mass or weight.

balance coil [ELEC] An iron-core solenoid with adjustable taps near the center; used to convert a two-wire circuit to a three-wire circuit, the taps furnishing a neutral terminal for the latter.

balanced amplifier [ELECTR] An electronic amplifier in which there are two identical signal branches connected so as to operate with the inputs in phase opposition and with the output connections in phase, each balanced to ground.

balanced bridge [ELEC] Wheatstone bridge circuit which,

**BACKWARD-WAVE
OSCILLATOR**

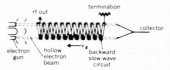

O-type backward-wave oscillator (or O-carcinotron), which uses a helix as the slow-wave circuit and has a hollow cylindrical electron beam; v_g is the velocity at which microwave energy travels along the helix toward the gun.

when in a quiescent state, has an output voltage of zero.

balanced circuit [ELEC] A circuit whose two sides are electrically alike and symmetrical with respect to a common reference point, usually ground.

balanced converter *See* balun.

balanced currents [ELEC] Currents flowing in the two conductors of a balanced line which, at every point along the line, are equal in magnitude and opposite in direction. Also known as push-pull currents.

balanced incomplete block design [MATH] For positive integers b, v, r, k, and λ, this is an arrangement of v elements into b subsets or blocks so that each block contains exactly k distinct elements, each element occurs in r blocks, and every combination of two elements occurs together in exactly λ blocks. Also known as (b,v,r,k,λ)-design.

balanced line [ELEC] A transmission line consisting of two conductors capable of being operated so that the voltages of the two conductors at any transverse plane are equal in magnitude and opposite in polarity with respect to ground.

balanced network [ELEC] Hybrid network in which the impedances of the opposite branches are equal.

balanced oscillator [ELECTR] Any oscillator in which, at the oscillator frequency, the impedance centers of the tank circuits are at ground potential, and the voltages between either end and their centers are equal in magnitude and opposite in phase.

balanced range of error [STAT] A range of error in which the maximum and minimum possible errors are opposite in sign and equal in magnitude.

balanced set [MATH] A set S in a real or complex vector space X such that if x is in S and $|a| \leq 1$, then ax is in S.

balanced voltages [ELEC] Voltages that are equal in magnitude and opposite in polarity with respect to ground. Also known as push-pull voltages.

balanced wire circuit [ELEC] Circuit wherein the two sides are electrically alike and symmetrical with respect to ground and other conductors.

balance equation [MATH] An equation expressing a balance of quantities in the sense that the local or individual rates of change are zero.

balance method *See* null method.

balancer [ELEC] A mechanism for equalizing the loads on the outer lines of a three-wire system for electric power distribution, consisting of two similar shunt or compound machines coupled together with the armatures connected in series across the outer lines.

balance-to-unbalance transformer [ELEC] Device for matching a pair of lines, balanced with respect to earth, to a pair of lines not balanced with respect to earth.

balancing unit [ELEC] **1.** Antenna-matching device used to permit efficient coupling of a transmitter or receiver having an unbalanced output circuit to an antenna having a balanced transmission line. **2.** Device for converting balanced to unbalanced transmission lines, and vice versa, by placing suitable discontinuities at the junction between the lines instead of using lumped components.

B* algebra *See* C* algebra.

ballast lamp [ELEC] A light-producing electrical resistance device which maintains nearly constant current by increasing in resistance as the current increases.

ballast resistor [ELEC] A resistor that increases in resistance as current through it increases, and decreases in resistance as current decreases. Also known as barretter (British usage).

ballast tube [ELEC] A ballast resistor mounted in an evacuated glass or metal envelope, like that of a vacuum tube, to reduce radiation of heat from the resistance element and thereby improve the voltage-regulating action.

ballistic camera [OPTICS] A ground-based camera using multiple exposures on the same plate to record the trajectory of a rocket.

ballistic coefficient [MECH] The numerical measure of the ability of a missile to overcome air resistance; dependent upon the mass, diameter, and form factor.

ballistic conditions [MECH] Conditions which affect the motion of a projectile in the bore and through the atmosphere, including muzzle velocity, weight of projectile, size and shape of projectile, rotation of the earth, density of the air, temperature or elasticity of the air, and the wind.

ballistic curve [MECH] The curve described by the path of a bullet, a bomb, or other projectile as determined by the ballistic conditions, by the propulsive force, and by gravity.

ballistic deflection [MECH] The deflection of a missile due to its ballistic characteristics.

ballistic density [MECH] A representation of the atmospheric density encountered by a projectile in flight, expressed as a percentage of the density according to the standard artillery atmosphere.

ballistic efficiency [MECH] **1.** The ability of a projectile to overcome the resistance of the air; depends chiefly on the weight, diameter, and shape of the projectile. **2.** The external efficiency of a rocket or other jet engine of a missile.

ballistic entry [MECH] Movement of a ballistic body from without to within a planetary atmosphere.

ballistic galvanometer [ELEC] A galvanometer having a long period of swing so that the deflection may measure the electric charge in a current pulse or the time integral of a voltage pulse.

ballistic instrument [ENG] Any instrument, such as a ballistic galvanometer or a ballistic pendulum, that measures an impact or sudden pulse of energy.

ballistic limit [MECH] The minimum velocity at which a particular armor-piercing projectile is expected to consistently and completely penetrate armor plate of given thickness and physical properties at a specified angle of obliquity.

ballistic measurement [MECH] Any measurement in which an impulse is applied to a device such as the bob of a ballistic pendulum, or the moving part of a ballistic galvanometer, and the subsequent motion of the device is used to determine the magnitude of the impulse, and, from this magnitude, the quantity to be measured.

ballistic pendulum [ENG] A device which uses the deflection of a suspended weight to determine the momentum of a projectile.

ballistics [MECH] Branch of applied mechanics which deals with the motion and behavior characteristics of missiles, that is, projectiles, bombs, rockets, guided missiles, and so forth, and of accompanying phenomena.

ballistics of penetration [MECH] That part of terminal ballistics which treats of the motion of a projectile as it forces its way into targets of solid or semisolid substances, such as earth, concrete, or steel.

ballistic table [MECH] Compilation of ballistic data from which trajectory elements such as angle of fall, range to summit, time of flight, and ordinate at any time, can be obtained.

ballistic temperature [MECH] That temperature (in °F) which, when regarded as a surface temperature and used in conjunction with the lapse rate of the standard artillery

BALLISTIC PENDULUM

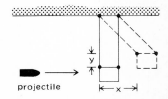

Ballistic pendulum before (solid lines) and after (broken lines) projectile impact. Measurements of x or y and knowledge of mass and length of the pendulum determine the initial momentum of the projectile.

atmosphere, would produce the same effect on a projectile as the actual temperature distribution encountered by the projectile in flight.

ballistic trajectory [MECH] The trajectory followed by a body being acted upon only by gravitational forces and resistance of the medium through which it passes.

ballistic uniformity [MECH] The capability of a propellant, when fired under identical conditions from round to round, to impart uniform muzzle velocity and produce similar interior ballistic results.

ballistic wave [MECH] An audible disturbance caused by compression of air ahead of a missile in flight.

ballistic wind [MECH] That constant wind which would produce the same effect upon the trajectory of a projectile as the actual wind encountered in flight.

ball lightning [GEOPHYS] A relatively rare form of lightning, consisting of a reddish, luminous ball, of the order of 1 foot (30 centimeters) in diameter, which may move rapidly along solid objects or remain floating in midair. Also known as globe lightning.

balloting [MECH] A tossing or bounding movement of a projectile, within the limits of the bore diameter, while moving through the bore under the influence of the propellant gases.

Balmer continuum [SPECT] A continuous range of wavelengths (or wave numbers or frequencies) in the spectrum of hydrogen at wavelengths less than the Balmer limit, resulting from transitions between states with principal quantum number $n = 2$ and states in which the single electron is freed from the atom.

Balmer discontinuity [SPECT] A discontinuity in the intensity of the hydrogen spectrum at the Balmer limit.

Balmer limit [SPECT] The lower limit of wavelengths of Balmer lines (365 nanometers), or the corresponding upper limit in frequency, energy of quanta, or wave number.

Balmer lines [SPECT] Lines in the hydrogen spectrum, produced by transitions between $n=2$ and $n>2$ levels either in emission or absorption; here n is the principal quantum number.

Balmer series [SPECT] The set of Balmer lines.

balun [ELEC] A device used for matching an unbalanced coaxial transmission line or system to a balanced two-wire line or system. Also known as balanced converter; bazooka; line-balance converter.

Banach algebra [MATH] An algebra which is a Banach space satisfying the property that for every pair of vectors, the norm of the product of those vectors does not exceed the product of their norms.

Banach's fixed-point theorem [MATH] If a mapping f of a metric space E into itself is a contraction, then there exists a unique element x of E such that $fx = x$. Also known as Caccioppoli-Banach principle.

Banach space [MATH] A real or complex vector space in which each vector has a nonnegative length, or norm, and in which every Cauchy sequence converges to a point of the space. Also known as complete normed linear space.

Banach-Steinhaus theorem [MATH] If a sequence of bounded linear transformations of a Banach space is pointwise bounded, then it is uniformly bounded.

band [COMMUN] A range of electromagnetic-wave frequencies between definite limits, such as that assigned to a particular type of radio service. [SOLID STATE] A restricted range in which the energies of electrons in solids lie, or from which they are excluded, as understood in quantum-mechanical terms. [SPECT] *See* band spectrum.

band-elimination filter *See* band-stop filter.

band gap [SOLID STATE] An energy difference between two allowed bands of electron energy in a metal.

band head [SPECT] A location on the spectrogram of a molecule at which the lines of a band pile up.

band lightning *See* ribbon lightning.

band-pass amplifier [ELECTR] An amplifier designed to pass a definite band of frequencies with essentially uniform response.

band-pass filter [ELECTR] An electric filter which transmits more or less uniformly in a certain band, outside of which frequency components are attenuated. [OPTICS] *See* Christiansen filter.

band-pass response [ELECTR] Response characteristics in which a definite band of frequencies is transmitted uniformly. Also known as flat top response.

band-rejection filter *See* band-stop filter.

band scheme [SOLID STATE] The identification of energy bands of a solid with the levels of independent atoms from which they arise as the atoms are brought together to form the solid, together with the width and spacing of the bands.

band spectrum [SPECT] A spectrum consisting of groups or bands of closely spaced lines in emission or absorption, characteristic of molecular gases and chemical compounds. Also known as band.

band-stop filter [ELECTR] An electric filter which transmits more or less uniformly at all frequencies of interest except for a band within which frequency components are largely attenuated. Also known as band-elimination filter; band-rejection filter.

band theory of ferromagnetism [SOLID STATE] A theory according to which ferromagnetism is caused by electrons in the unfilled energy bands of a crystal.

band theory of solids [SOLID STATE] A quantum-mechanical theory of the motion of electrons in solids that predicts certain restricted ranges or bands for the energies of these electrons.

bang-bang circuit [ELECTR] An operational amplifier with double feedback limiters that drive a high-speed relay (1–2 milliseconds) in an analog computer; involved in signal-controlled programming.

bang-bang control [CONT SYS] A type of automatic control system in which the applied control signals assume either their maximum or minimum values.

bank-and-wiper switch [ELEC] Switch in which electromagnetic ratchets or other mechanisms are used, first, to move the wipers to a desired group of terminals, and second, to move the wipers over the terminals of the group to the desired bank contacts.

banked winding [ELECTR] A radio-frequency coil winding which proceeds from one end of the coil to the other without return by having, side by side, many flat spirals formed by winding single turns one over the other, thereby reducing the distributed capacitance of the coil.

bantam tube [ELECTR] Vacuum tube having a standard octal base, but a considerably smaller glass tube than a standard glass tube.

bar [MECH] A unit of pressure equal to 10^5 pascals, or 10^5 newtons per square meter, or 10^6 dynes per square centimeter.

bar chart *See* bar graph.

Bardeen-Cooper-Schrieffer theory [SOLID STATE] A theory of superconductivity that describes quantum-mechanically

BAND-STOP FILTER

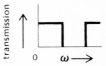

band-stop

Transmission function of a band-stop filter. Frequency (ω) components are largely attenuated at the stop band.

those states of the system in which conduction electrons cooperate in their motion so as to reduce the total energy appreciably below that of other states by exploiting their effective mutual attraction; these states predominate in a superconducting material. Abbreviated BCS theory.

bare charm [PARTIC PHYS] Charm that is carried by a quark and is not canceled by the charm of the corresponding antiquark, so that the hadron of which the quark is a constituent has net charm different from 0.

bare value [QUANT MECH] The value which some physical property of a particle, such as its mass or charge, is supposed to have in the absence of any interactions with fields.

bar graph [STAT] A diagram of frequency-table data in which a rectangle with height proportional to the frequency is located at each value of a variate that takes only certain discrete values. Also known as bar chart.

BARITT diode *See* barrier injection transit-time diode.

barium [CHEM] A chemical element, symbol Ba, with atomic number 56 and atomic weight of 137.34.

barium-140 [NUC PHYS] A radioactive isotope of barium with atomic mass 140; the half-life is 12.8 days, and the decay is by negative beta-particle emission.

barium fuel cell [ELEC] A fuel cell in which barium is used with either oxygen or chlorine to convert chemical energy into electrical energy.

Barkhausen effect [ELECTROMAG] The succession of abrupt changes in magnetization occurring when the magnetizing force acting on a piece of iron or other magnetic material is varied.

Barkhausen-Kurz oscillator [ELECTR] An oscillator of the retarding-field type in which the frequency of oscillation depends solely on the transit time of electrons oscillating about a highly positive grid before reaching the less positive anode. Also known as Barkhausen oscillator; positive-grid oscillator.

Barkhausen oscillator *See* Barkhausen-Kurz oscillator.

Barlow lens [OPTICS] A lens with one plane surface and one concave surface that is placed between the objective and eyepiece of a telescope to decrease the convergence of the beam from the objective and thereby increase the effective focal length.

Barlow's equation [MECH] A formula, $t = DP/2S$, used in computing the strength of cylinders subject to internal pressures, where t is the thickness of the cylinder in inches, D the outside diameter in inches, P the pressure in pounds per square inch, and S the allowable tensile strength in pounds per square inch.

Barlow's rule [PHYS CHEM] The rule that the volume occupied by the atoms in a given molecule is proportional to the valences of the atoms, using the lowest valency values.

bar magnet [ELECTROMAG] A bar of hard steel that has been strongly magnetized and holds its magnetism, thereby serving as a permanent magnet.

barn [NUC PHYS] A unit of area equal to 10^{-24} square centimeter; used in specifying nuclear cross sections. Symbolized b.

Barnes' integral [MATH] **1.** The contour integral of $(1/2\pi i)$ $[\Gamma(a + s)\Gamma(b + s)\Gamma(-s)/\Gamma(c - s)](-z)^s \, ds$, taken along the entire imaginary axis of the complex plane, with loops, if necessary, to ensure that the origin lies on the right of the contour and $-a$ and $-b$ lie on the left of the contour; used to represent $[\Gamma(a)\Gamma(b)/\Gamma(c)]F(a,b;c;z)$, where F is the hypergeometric function. **2.** One of several related integrals used to

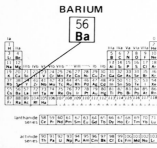

BARIUM

56
Ba

Periodic table of the chemical elements showing the position of barium.

represent confluent hypergeometric functions and generalized hypergeometric functions.

Barnett effect [ELECTROMAG] The development of a slight magnetization in an initially unmagnetized iron rod when it is rotated at high speed about its axis.

Barnett method [ELETROMAG] Use of the Barnett effect to determine the gyromagnetic moment of ferromagnetic material.

baroclinicity *See* baroclinity.

baroclinity [PHYS] The state of stratification in a fluid in which surfaces of constant pressure (isobaric surfaces) intersect surfaces of constant density (isosteric surfaces). Also known as baroclinicity; barocliny.

barocliny *See* baroclinity.

barodynamics [MECH] The mechanics of heavy structures which may collapse under their own weight.

barogram [ENG] The record of an aneroid barograph.

barograph [ENG] A recording barometer.

barometer [ENG] An absolute pressure gage specifically designed to measure atmospheric pressure.

barometric [ENG] Pertaining to a barometer or to the results obtained by using a barometer. [PHYS] Loosely, pertaining to atmospheric pressure; for example, barometric gradient (meaning pressure gradient).

barometric corrections [PHYS] The corrections which must be applied to the reading of a mercury barometer in order that the observed value may be rendered accurate. Also known as barometric errors.

barometric errors *See* barometric corrections.

barometric gradient *See* pressure gradient.

barometric pressure *See* atmospheric pressure.

barometric surface [PHYS] A surface at each point of which the barometric pressure is the same.

barometric tide [GEOPHYS] A daily variation in atmospheric pressure due to the gravitational attraction of the sun and moon.

barometry [ENG] The study of the measurement of atmospheric pressure, with particular reference to ascertaining and correcting the errors of the different types of barometer.

baromil [MECH] The unit of length used in graduating a mercury barometer in the centimeter-gram-second system.

baroscope [ENG] An apparatus which demonstrates the equality of the weight of air displaced by an object and its loss of weight in air.

barostat [ENG] A mechanism which maintains constant pressure inside a chamber.

barotropic [PHYS] Of, pertaining to, or characterized by a condition of barotropy.

barotropy [PHYS] The state of a fluid in which surfaces of constant density (or temperature) are coincident with surfaces of constant pressure; it is the state of zero baroclinity.

barrel [MECH] Abbreviated bbl. 1. The unit of liquid volume equal to 31.5 gallons (approximately 119 liters). 2. The unit of liquid volume for petroleum equal to 42 gallons (approximately 158 liters). 3. The unit of dry volume equal to 105 quarts (approximately 116 liters). 4. A unit of weight that varies in sizes according to the commodity being weighed. [OPTICS] A tapering cylindrical housing containing the lenses of a camera and the iris diaphragm. [ORD] The cylindrical metallic part of a gun which controls the initial direction of a projectile.

barrel distortion [OPTICS] A defect in an optical system whereby lateral magnification decreases with object size; the image of a square then appears barrel-shaped.

BARRETTER

Cutaway view of commercial barretter about 1 inch (2.5 centimeters) in length. *(Sperry Division, Sperry Rand Corp.)*

barretter [ELEC] **1.** Bolometer that consists of a fine wire or metal film having a positive temperature coefficient of resistivity, so that resistance increases with temperature; used for making power measurements in microwave devices. **2.** *See* ballast resistor.

barrier injection transit-time diode [ELECTR] A microwave diode in which the carriers that traverse the drift region are generated by minority carrier injection from a forward-biased junction instead of being extracted from the plasma of an avalanche region. Abbreviated BARITT diode.

barrier layer *See* depletion layer.

barrier-layer cell *See* photovoltaic cell.

barrier-layer rectification *See* depletion-layer rectification.

barrier penetration [QUANT MECH] The passage of a particle through a potential barrier, that is, through a region of finite extent in which the particle's potential energy is greater than its total energy.

Bartlett force [NUC PHYS] A force between nucleons in which spin is exchanged.

Bartlett's test [STAT] A method to test for the equalities of variances from a number of independent normal samples by testing the hypothesis.

bar winding [ELEC] An armature winding made up of a series of metallic bars connected at their ends.

barycenter [MATH] The center of mass of a system of finitely many equal point masses distributed in euclidean space in such a way that their position vectors are linearly independent.

barycentric coordinates [MATH] The coefficients in the representation of a point in a simplex as a linear combination of the vertices of the simplex.

barycentric element [ASTROPHYS] An orbital element referred to the center of mass of the solar system.

barycentric energy [MECH] The energy of a system in its center-of-mass frame.

barye [MECH] The pressure unit of the centimeter-gram-second system of physical units; equal to 1 dyne/cm^2 (0.001 millibar). Also known as microbar.

baryon [PARTIC PHYS] Any elementary particle which can be transformed into a nucleon and some number of mesons and lighter particles. Also known as heavy particle.

baryon number [PARTIC PHYS] A conserved quantum number, equal to the number of baryons minus the number of antibaryons in a system; neutrons and protons have baryon number one; mesons and leptons have baryon number zero.

baryon octet [PARTIC PHYS] The group of one lambda, three sigma, and two xi hyperons and two nucleons, all having spin ½ and positive parity, and forming a symmetrical pattern as suggested by SU$_3$ symmetry.

baryon resonance [PARTIC PHYS] A cross section anomaly indicating the existence of an unstable baryon.

baryon spectroscopy [PARTIC PHYS] The science of the energy levels and changes of state occurring among baryon particles.

basal cleavage [CRYSTAL] Cleavage parallel to the base of the crystal structure or to the lattice plane which is normal to one of the lattice axes.

basal orientation [CRYSTAL] A crystal orientation in which the surface is parallel to the base of the lattice or to the lattice plane which is normal to one of the lattice axes.

basal plane [CRYSTAL] The plane perpendicular to the long, or c, axis in all crystals except those of the isometric system.

base [ELECTR] **1.** The region that lies between an emitter and a collector of a transistor and into which minority carriers are

BARYON OCTET

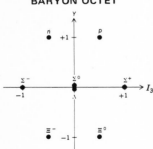

The baryon octet, arrayed with respect to I_3 as abscissa and Y as ordinate. The charge number Q is given by $Q = I_3 + Y/2$.

injected. **2.** The part of an electron tube that has the pins, leads, or other terminals to which external connections are made either directly or through a socket. **3.** The plastic, ceramic, or other insulating board that supports a printed wiring pattern. **4.** A plastic film that supports the magnetic powder of magnetic tape or the emulsion of photographic film.

baseball [PL PHYS] A machine used in controlled fusion research to confine a plasma; consists of a linear magnetic bottle sealed by magnetic mirrors at both ends, and has current-carrying structures, which resemble the seams of a baseball in shape, to stabilize the plasma.

base bias [ELECTR] The direct voltage that is applied to the majority-carrier contact (base) of a transistor.

base-centered lattice [CRYSTAL] A space lattice in which each unit cell has lattice points at the centers of each of two opposite faces as well as at the vertices; in a monoclinic crystal, they are the faces normal to one of the lattice axes.

base curve [OPTICS] The reciprocal of the largest focal length of an astigmatic lens.

base drag [FL MECH] Drag owing to a base pressure lower than the ambient pressure; it is a part of the pressure drag.

base electrode [ELECTR] An ohmic or majority carrier contact to the base region of a transistor.

base for the neighborhood system *See* local base.

base insulator [ELEC] Heavy-duty insulator used to support the weight of an antenna mast and insulate the mast from the ground or some other surface.

base line [SCI TECH] A line drawn in the graphical representation of a varying physical quantity, such as a voltage or current, to indicate a reference value, such as the voltage value of a bias. Abbreviated BL.

baseload [ELEC] Minimum load of a power generator over a given period of time.

base-loaded antenna [ELECTROMAG] Vertical antenna having an impedance in series at the base for loading the antenna to secure a desired electrical length.

base magnification [OPTICS] The ratio of the distance between the centers of the objectives of a pair of binoculars to the distance between the centers of the eyepieces.

base modulation [ELECTR] Amplitude modulation produced by applying the modulating voltage to the base of a transistor amplifier.

base notation *See* radix notation.

base of a logarithm [MATH] The number of which the logarithm is the exponent.

base of a number system [MATH] The number whose powers determine place value.

base of a topological space [MATH] A collection of sets, unions of which form all open sets.

base period [STAT] The period of a year, or other unit of time, used as a reference in constructing an index number. Also known as base year.

base pin *See* pin.

base pressure [FL MECH] The pressure exerted on the base or extreme aft end of a body, as of a cylindrical or boat-tailed body or of a blunt-trailing-edge wing in fluid flow. [MECH] A pressure used as a reference base, for example, atmospheric pressure.

base quantity [PHYS] One of a small number of physical quantities in a system of measurement that are defined, independent of other physical quantities, by means of a physical standard and by procedures for comparing the

BASE-CENTERED LATTICE

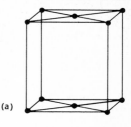

(a)

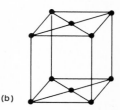

(b)

Two of fourteen Bravais lattices.
(*a*) Base-centered orthorhombic.
(*b*) Base-centered monoclinic.

quantity to be measured with the standard. Also known as fundamental quantity.

base space of a bundle [MATH] The topological space B in the bundle (E,p,B).

base unit [PHYS] One of a small number of units in a system of measurement that are defined, independent of other units, by means of a physical standard; equivalently, a unit of a base quantity. Also known as fundamental unit.

base vector [MATH] One of a set of linearly independent vectors in a vector space such that each vector in the space is a linear combination of vectors from the set; that is, a member of a basis.

base year *See* base period.

base-year method *See* Laspeyre's index.

basic frequency [PHYS] Frequency, in any wave, which is considered the most important; in a driven system, it would generally be the driving frequency, while in most periodic waves it would correspond to the fundamental frequency.

basic hypergeometric series [MATH] The series

$$_r\Phi_s\left(\begin{matrix} a_1, a_2, \ldots, a_r; q; z \\ b_1, b_2, \ldots, b_s \end{matrix}\right) = \sum_{n=0}^{\infty} \frac{(a_1)_{q,n}(a_2)_{q,n} \ldots (a_r)_{q,n}}{(q)_{q,n}(b_1)_{q,n} \ldots (b_s)_{q,n}} z^n$$

where $a_{q,n} = (1-a)(1-aq)(1-aq^2) \ldots (1-aq^{n-1})$, $a_{q,0} = 1$, and $|q| < 1$, $|z| < 1$.

basic Q *See* nonloaded Q.

basic solution [MATH] In bifurcation theory, a simple, explicitly known solution of a nonlinear equation, in whose neighborhood other solutions are studied.

basic truss [MECH] A framework of bars arranged so that for any given loading of the bars the forces on the bars are uniquely determined by the laws of statics.

basis [MATH] A set of linearly independent vectors in a vector space such that each vector in the space is a linear combination of vectors from the set.

basket coil *See* basket winding.

basket winding [ELECTR] A crisscross coil winding in which successive turns are far apart except at points of crossing, giving low distributed capacitance. Also known as basket coil.

bass [ACOUS] Sounds having frequencies at the lower end of the audio range, below about 250 hertz.

bass response [ELECTR] A measure of the output of an electronic device or system as a function of an input of low audio frequencies.

Batchinsky relation [FL MECH] The relation stating that the fluidity of a liquid is proportional to the difference between the specific volume and a characteristic specific volume, approximately equal to the specific volume appearing in the van der Waals equation.

Bateman equations [NUC PHYS] A set of equations that give the number of atoms of each nuclide of a radioactive decay chain produced after a specified time, when a specified number of atoms of the parent nuclide are initially present.

Bateman function [MATH] A function of the form

$$k_\nu(x) = (2/\pi)\int_0^{\pi/2} \cos(x \tan\theta - \nu\theta)\, d\theta.$$

Bateman-Pasternack polynomial [MATH] A polynomial defined by either of the generalized hypergeometric functions

$$F_n(z) = {}_3F_2(-n, n+1, \tfrac{1}{2} + \tfrac{1}{2}z; 1,1; 1)$$

$$Z_n(z) = {}_2F_2(-n, n+1; 1,1; z)$$

where n is a nonnegative integer.

bat-handle switch [ELEC] A toggle switch having an actuating lever shaped like a baseball bat.

bathochromatic shift [PHYS CHEM] The shift of the fluorescence of a compound toward the red part of the spectrum due to the presence of a bathochrome radical in the molecule.

bathtub capacitor [ELEC] A capacitor enclosed in a metal housing having broadly rounded corners like those on a bathtub.

battery charger [ELEC] A rectifier unit used to change alternating to direct power for charging a storage battery. Also known as charger.

battery clip [ELEC] A terminal of a connecting wire having spring jaws that can be quickly snapped on a terminal of a device, such as a battery, to which a temporary wire connection is desired.

battery command periscope [OPTICS] An optical instrument consisting of dual telescope tubes positioned vertically on a common mounting; it provides periscopic vision for the observer, and may be used to observe artillery fire.

battery eliminator [ELECTR] A device which supplies electron tubes with voltage from electric power supply mains.

battery separator [ELEC] An insulating plate inserted between the positive and negative plates of a battery to prevent them from touching.

battle short [ELEC] Switch for short-circuiting safety interlocks and lighting a red warning light.

Baumé hydrometer scale [PHYS CHEM] A calibration scale for liquids that is reducible to specific gravity by the following formulas: for liquids heavier than water, specific gravity $= 145 \div (145 - n)$ (at 60°F); for liquids lighter than water, specific gravity $= 140 \div (130 + n)$ (at 60°F); n is the reading on the Baumé scale, in degrees Baumé. Baumé is abbreviated Bé.

b axis [CRYSTAL] A crystallographic axis that is oriented horizontally, right to left.

bay [ELECTROMAG] One segment of an antenna array.

Bayard-Alpert ionization gage [ELECTR] A type of ionization vacuum gage using a tube with an electrode structure designed to minimize x-ray-induced electron emission from the ion collector.

Bayes' decision rule [STAT] A decision rule under which the strategy chosen from among several available ones is the one for which the expected value of payoff is the greatest.

Bayes' theorem [MATH] The probability of a hypothesis, given the original data and some new data, is proportional to the probability of the hypothesis, given the original data only, and the probability of the new data, given the original data and the hypothesis. Also known as inverse probability principle.

bazooka *See* balun.

B battery [ELECTR] The battery that furnishes required direct-current voltages to the plate and screen-grid electrodes of the electron tubes in a battery-operated circuit.

BBD *See* bucket brigade device.

bbl *See* barrel.

b-boundary [RELAT] A means of attaching a boundary to singular space-times; points on the b-boundary represent end points of equivalence classes of b-incomplete curves.

BCD system *See* binary coded decimal system.

b-complete [RELAT] A criterion determining whether a space-time is free of singularities based on whether curves of finite length have an end point, where length is defined by a generalized affine parameter along the curve.

BCS theory *See* Bardeen-Cooper-Schrieffer theory.

Be *See* beryllium.

Bé *See* Baumé hydrometer scale.

BE *See* binding energy.

bead [ELECTROMAG] A glass, ceramic, or plastic insulator through which passes the inner conductor of a coaxial transmission line and by means of which the inner conductor is supported in a position coaxial with the outer conductor.

beaded transmission line [ELECTROMAG] Line using beads to support the inner conductor in coaxial transmission lines.

bead thermistor [ELEC] A thermistor made by applying the semiconducting material to two wire leads as a viscous droplet, which cements the leads upon firing.

beam [PHYS] A concentrated, nearly unidirectional flow of particles, or a like propagation of electromagnetic or acoustic waves.

beam angle *See* beam width.

beam antenna [ELECTROMAG] An antenna that concentrates its radiation into a narrow beam in a definite direction.

beam attenuator [SPECT] An attachment to the spectrophotometer that reduces reference to beam energy to accommodate undersized chemical samples.

beam-condensing unit [SPECT] An attachment to the spectrophotometer that condenses and remagnifies the beam to provide reduced radiation at the sample.

beam current [ELECTR] The electric current determined by the number and velocity of electrons in an electron beam.

beam-deflection tube [ELECTR] An electron-beam tube in which current to an output electrode is controlled by transversely moving the electron beam.

beam divergence [PHYS] The angular spread in the directions of the components of a beam of particles or radiation.

beam edge [PHYS] The locus of positions at which the intensity of a beam of particles or radiation is 10% of that along the axis of the beam.

beam equation [MECH] The equation of motion for the transverse displacement y of a beam as a function of distance x along the beam and time t, when subjected to a force of magnitude $F(x,t)$ per unit length; the equation is

$$(\partial^2/\partial x^2)\,(EI\ \partial^2 y/\partial x^2) + \rho A\ \partial^2 y/\partial t^2 = F(x,t)$$

where ρ is the beam's mass density, A is its cross-sectional area, E is Young's modulus, and I is the moment of inertia about the central axis.

beam extractor [NUCLEO] A magnetic or electrostatic device for removing charged particles from a circular particle accelerator when they have been accelerated to the desired energy.

beam-foil spectroscopy [ATOM PHYS] A method of studying the structure of atoms and ions in which a beam of ions energized in a particle accelerator passes through a thin carbon foil from which the ions emerge with various numbers of electrons removed and in various excited energy levels; the light or Auger electrons emitted in the deexcitation of these levels are then observed by various spectroscopic techniques. Abbreviated BFS.

beam-forming electrode [ELECTR] Electron-beam focusing elements in power tetrodes and cathode-ray tubes.

beam hole [NUCLEO] A hole through the shield, and usually the reflector, of a nuclear reactor which allows a beam of radiation, especially fast neutrons, to escape for experimental purposes. Also known as glory hole.

beam parametric amplifier [ELECTR] Parametric amplifier that uses a modulated electron beam to provide a variable reactance.

beam pattern *See* directivity pattern.

beam power tube [ELECTR] An electron-beam tube which uses directed electron beams to provide most of its power-handling capability and in which the control grid and screen grid are essentially aligned. Also known as beam tetrode.

beam splitter [OPTICS] A mirror that reflects part of a beam of light falling on it and transmits part.

beam splitting [OPTICS] The division of a beam of light into two beams by placing a special type of mirror in the path of the beam that reflects part of the light falling on it and transmits part.

beam-switching tube [ELECTR] An electron tube which has a series of electrodes arranged around a central cathode and in which an electron beam is switched from one electrode to another. Also known as cyclophon.

beam tetrode *See* beam power tube.

beam width [ELECTROMAG] The angle, measured in a horizontal plane, between the directions at which the intensity of an electromagnetic beam, such as a radar or radio beam, is one-half its maximum value. Also known as beam angle.

bearing capacity [MECH] Load per unit area which can be safely supported by the ground.

bearing pressure [MECH] Load on a bearing surface divided by its area. Also known as bearing stress.

bearing strain [MECH] The deformation of bearing parts subjected to a load.

bearing strength [MECH] The maximum load that a column, wall, footing, or joint will sustain at failure, divided by the effective bearing area.

bearing stress *See* bearing pressure.

beat [PHYS] The periodic variation in amplitude of a wave that is the superposition of two simple harmonic waves of different frequencies.

beat-frequency oscillator [ELECTR] An oscillator in which a desired signal frequency, such as an audio frequency, is obtained as the beat frequency produced by combining two different signal frequencies, such as two different radio frequencies. Abbreviated BFO. Also known as heterodyne oscillator.

beating-in [ELECTR] Interconnecting two transmitter oscillators and adjusting one until no beat frequency is heard in a connected receiver; the oscillators are then at the same frequency.

beat reception *See* heterodyne reception.

Beattie and Bridgman equation [THERMO] An equation that relates the pressure, volume, and temperature of a real gas to the gas constant.

Becker and Kornetzki effect [PHYS] A reduction in the internal friction of a ferromagnetic substance when it is subjected to a magnetic field that is large enough to produce magnetic saturation.

becquerel [NUCLEO] The International System unit of activity of a radionuclide, equal to the activity of a quantity of a radionuclide having one spontaneous nuclear transition per second. Symbolized Bq.

Becquerel effect [ELEC] The phenomenon of a current flowing between two unequally illuminated electrodes of a certain type when they are immersed in an electrolyte.

Becquerel rays [NUC PHYS] Formerly, radiation emitted by radioactive substances; later renamed alpha, beta, and gamma rays.

bedspring array *See* billboard array.

BEND

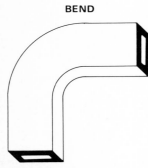

H-plane 90° bend for a rectangular waveguide.

BENDING MOMENT

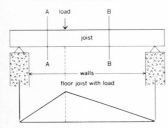

Schematic of bending moment on an end-supported joist with a concentrated load. Bending moment at section A-A is upward reactive force of left wall times distance from wall to A-A. Bending moment at section B-B is upward reactive force of left wall times the distance from wall to B-B plus the moment produced by the downward load acting at its distance from B-B. The resulting bending moment at each cross section is shown on the lower portion of the diagram.

Beer-Lambert-Bouguer law *See* Bouguer-Lambert-Beer law.

Beer's law [PHYS CHEM] The law which states that the absorption of light by a solution changes exponentially with the concentration, all else remaining the same.

Behrens-Fisher problem [STAT] The problem of calculating the probability of drawing two random samples whose means differ by some specified value (which may be zero) from normal populations, when one knows the difference of the means of these populations but not the ratio of their variances.

bei function [MATH] One of the functions defined by $ber_n(z) \pm i\ bei_n(z) = J_n(ze^{\pm 3\pi i/4})$, where J_n is the nth Bessel function.

Békésy audiometry [ACOUS] A subject-controlled auditory threshold testing procedure.

bel [PHYS] A dimensionless unit expressing the ratio of two powers or intensities, or the ratio of a power to a reference power, such that the number of bels is the common logarithm of this ratio. Symbolized b; B.

B eliminator [ELECTR] Power pack that changes the alternating-current powerline voltage to the direct-current source required by plant circuits of vacuum tubes or semiconductor devices.

Bellatrix [ASTRON] A bluish-white star of stellar magnitude 1.7, spectral classification B2-III, in the constellation Orion; the star γ Orionis.

bell glass *See* bell jar.

bell jar [ENG] A bell-shaped vessel, usually made of glass, which is used for enclosing a vacuum, holding gases, or covering objects. Also known as bell glass.

Bell numbers [MATH] The numbers

$$B_n = \sum_{k=1}^{n} S(n,k)$$

where $S(n,k)$ is the Stirling number of the second kind.

bellows [OPTICS] An accordionlike component of a camera which forms a passage between the lens and the film and allows one to vary the distance between them.

bell-shaped curve [STAT] The curve representing a continuous frequency distribution with a shape having the overall curvature of the vertical cross section of a bell; usually applied to the normal distribution.

Bénard convection cells [PHYS] A regular array of hexagonal cells which sometimes appear in convection in a layer of liquid heated from below.

bench photometer [ENG] A device which uses an optical bench with the two light sources to be compared mounted one at each end; the comparison between the two illuminations is made by a device moved along the bench until matching brightnesses appear.

bend [ELECTROMAG] A smooth change in the direction of the longitudinal axis of a waveguide.

bender element [ELECTR] A combination of two thin strips of different piezoelectric materials bonded together so that when a voltage is applied, one strip increases in length and the other becomes shorter, causing the combination to bend.

bending moment [MECH] The algebraic sum of all moments located between a cross section and one end of a structural member; a bending moment that bends the beam convex downward is positive, and one that bends it convex upward is negative.

bending moment diagram [MECH] A diagram showing the bending moment at every point along the length of a beam plotted as an ordinate.

bending stress [MECH] An internal tensile or compressive longitudinal stress developed in a beam in response to curvature induced by an external load.

bend plane   *See* tilt boundary.

Benedicks effect   [PHYS] An electromotive force produced in a circuit containing one metal only, but having impurities or internal strains, in the presence of an asymmetrical temperature distribution.

Benedict equation of state   [PHYS CHEM] An empirical equation relating pressures, temperatures, and volumes for gases and gas mixtures; superseded by the Benedict-Webb-Rubin equation of state.

Benham top   [OPTICS] A disk whose surface has black and white portions and which, when rotated at certain speeds and subjected to certain lighting, produces sensations of color.

Beranek scale   [ACOUS] A scale which measures the subjective loudness of a noise; noises are arranged into six arbitrary categories: very quiet, quiet, moderately quiet, noisy, very noisy, and intolerably noisy.

ber function   [MATH] One of the functions defined by $\text{ber}_n(z) \pm i\,\text{bei}_n(z) = J_n(ze^{\pm 3\pi i/4})$, where J_n is the nth Bessel function.

berkelium   [CHEM] A radioactive element, symbol Bk, atomic number 97, the eighth member of the actinide series; properties resemble those of the rare-earth cerium.

Bernal chart   [CRYSTAL] A chart used to determine the coordinates in reciprocal space of x-ray reflections that produce the spots on an x-ray diffraction photograph of a single crystal.

Bernoulli differential equation   *See* Bernoulli equation.

Bernoulli distribution   *See* binomial distribution.

Bernoulli effect   [FL MECH] As a consequence of the Bernoulli theorem, the pressure of a stream of fluid is reduced as its speed of flow is increased.

Bernoulli equation   [FL MECH] *See* Bernoulli theorem. [MATH] A nonlinear first-order differential equation of the form $(dy/dx) + yf(x) = y^n g(x)$, where n is a number different from unity and f and g are given functions. Also known as Bernoulli differential equation.

Bernoulli-Euler law   [MECH] A law stating that the curvature of a beam is proportional to the bending moment.

Bernoulli law   *See* Bernoulli theorem.

Bernoulli number   [MATH] Numerical value of the coefficient of $x^{2n}/(2n)!$ is the expansion of $xe^x/(e^x - 1)$.

Bernoulli polynomial   [MATH] The nth one is

$$\sum_{k=0}^{n} \binom{n}{k} B_k Z^{n-k}$$

where $\binom{n}{k}$ is a binomial coefficient, and B_k is a Bernoulli number.

Bernoulli's lemniscate   [MATH] A curve shaped like a figure eight whose equation in rectangular coordinates is expressed as $(x^2 + y^2)^2 = a^2(x^2 - y^2)$.

Bernoulli theorem   [FL MECH] An expression of the conservation of energy in the steady flow of an incompressible, inviscid fluid; it states that the quantity $(p/\rho) + gz + (v^2/2)$ is constant along any streamline, where p is the fluid pressure, v is the fluid velocity, ρ is the mass density of the fluid, g is the acceleration due to gravity, and z is the vertical height. Also known as Bernoulli equation; Bernoulli law. [STAT] *See* law of large numbers.

Bernstein polynomials   [MATH] Given a function $f(x)$, defined on the interval [0,1], these are the polynomials

$$B_n f(x) = \sum_{k=0}^{n} \binom{n}{k} x^k (1-x)^{n-k} f(k/n)$$

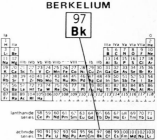

BERKELIUM

Periodic table of the chemical elements showing the position of berkelium.

BERNOULLI'S LEMNISCATE

Curve known as Bernoulli's lemniscate.

where n is a positive integer; used to approximate $f(x)$.

Berthelot equation [PHYS CHEM] A form of the equation of state which relates the temperature, pressure, and volume of a gas with the gas constant.

Berthelot method [THERMO] A method of measuring the latent heat of vaporization of a liquid that involves determining the temperature rise of a water bath that encloses a tube in which a given amount of vapor is condensed.

Berthelot relation [PHYS] A relationship between molecular attraction constants of like and unlike species.

Berthelot-Thomsen principle [PHYS CHEM] The principle that of all chemical reactions possible, the one developing the greatest amount of heat will take place, with certain obvious exceptions such as changes of state.

Bertrand curve [MATH] One of a pair of curves having the same principal normals. Also known as associate curve; conjugate curve.

Bertrand lens [OPTICS] An auxiliary lens that can be inserted in the tube of a polarizing microscope to obtain interference figures.

Bertrand's postulate [MATH] The proposition that there exists at least one prime number between any integer greater than three and twice the integer minus two.

Bertrand's test [MATH] A test for the convergence of an infinite integral; it states that the integral from some number $a>0$ to infinity of $f(x)\ dx$ converges if there exist numbers $\alpha<0$, A, and N such that the absolute value of $f(x)$ is bounded by $Ax^{\alpha-1}$ or by $Ax^{-1}[\log x]^{\alpha-1}$ for $x>N$.

beryllium [CHEM] A chemical element, symbol Be, atomic number 4, atomic weight 9.0122.

Bessel-Clifford equation [MATH] The differential equation $x(d^2y/dx^2) + (n+1)(dy/dx) + y = 0$.

Bessel equation [MATH] The differential equation $z^2f''(z) + zf'(z) + (z^2 - n^2)f(z) = 0$.

Bessel function [MATH] A solution of the Bessel equation. Symbolized $J_n(z)$.

Besselian elements [ASTRON] Data on a solar eclipse, giving, for selected times, the coordinates of the axis of the moon's shadow with respect to the fundamental plane, and the radii of umbra and penumbra in that plane; the data allow one to derive local circumstances of the eclipse at any point on the earth's surface.

Besselian star numbers [ASTRON] Constants used in the reduction of a mean position of a star to an apparent position; used to account for short-term variations in precession, nutation, aberration, and parallax.

Besselian year *See* fictitious year.

Bessel inequality [MATH] The statement that the sum of the squares of the inner product of a vector with the members of an orthonormal set is no larger than the square of the norm of the vector.

Bessel's interpolation formula [MATH] A formula for estimating the value of a function at an intermediate value of the independent variable, when its value is known at a series of equally spaced points (such as those that appear in a table), in terms of the central differences of the function; if $y_i = f(x_0 + ih)$, with i running over the integers, are the known values of the function, the formula states that $f(x_0 + uh)$ is approximated by a series whose kth term is $B_k(\delta^k y_0 + \delta^k y_1)$ for even k, and $B_k\ \delta^k y_{1/2}$ for odd k, where the δ is the central difference operator and the B_k are polynomial functions of u.

Bessel transform *See* Hankel transform.

Be star [ASTRON] A star of spectral type B in the Draper

BERYLLIUM

Periodic table of the chemical elements showing the position of beryllium.

catalog that has emission lines indicating mass loss and a surrounding gaseous shell.

best estimate [STAT] A term applied to unbiased estimates which have a minimum variance.

best fit *See* goodness of fit.

beta [ELECTR] The current gain of a transistor that is connected as a grounded-emitter amplifier, expressed as the ratio of change in collector current to resulting change in base current, the collector voltage being constant. [SCI TECH] The second letter of the Greek alphabet; β, B.

beta-absorption gage *See* beta gage.

beta coefficient [STAT] Also known as beta weight. **1.** One of the coefficients in a regression equation. **2.** A moment ratio, especially one used to describe skewness and kurtosis.

beta decay [NUC PHYS] Radioactive transformation of a nuclide in which the atomic number increases or decreases by unity with no change in mass number; the nucleus emits or absorbs a beta particle (electron or positron). Also known as beta disintegration.

beta decay spectrum [NUC PHYS] The distribution in energy or momentum of the beta particles arising from a nuclear disintegration process.

beta disintegration *See* beta decay.

beta distribution [STAT] The probability distribution of a random variable with density function $f(x) = [x^{\alpha-1}(1-x)^{\beta-1}]/B(\alpha,\beta)$, where B represents the beta function, α and β are positive real numbers, and $0<x<1$. Also known as Pearson Type I distribution.

beta emitter [NUC PHYS] A radionuclide that disintegrates by emission of a negative or positive electron.

beta factor [PL PHYS] In plasma physics, the ratio of the plasma kinetic pressure to the magnetic pressure.

beta filter [PHYS] A filter used in x-ray diffraction analysis to remove $K\beta$ radiation from a beam of characteristic K x-rays, allowing only $K\alpha$ radiation to pass. Also known as K-beta filter.

beta function [MATH] A function of two positive variables, defined by

$$B\,(m,n) \;=\; \int_0^1 x^{m-1}\,(1-x)^{n-1}dx.$$

beta gage [NUCLEO] A penetration-type thickness gage that measures the absorption of beta rays in the sample. Also known as beta-absorption gage.

beta-gamma survey meter [NUCLEO] An ionization-chamber type of monitor that is sensitive primarily to beta particles and gamma rays.

beta interaction *See* weak interaction.

beta particle [NUC PHYS] An electron or positron emitted from a nucleus during beta decay.

beta ray [NUC PHYS] A stream of beta particles.

beta-ray spectrometer [SPECT] An instrument used to determine the energy distribution of beta particles and secondary electrons. Also known as beta spectrometer.

beta spectrometer *See* beta-ray spectrometer.

betatron [NUCLEO] A device for accelerating electrons in an evacuated ring by means of a time-varying magnetic flux encircled by the ring. Also known as induction accelerator; rheotron.

betatron oscillations [NUCLEO] Oscillations of particles about an equilibrium orbit in a particle accelerator.

beta weight *See* beta coefficient.

BET equation *See* Brunauer-Emmett-Teller equation.

Bethe-Bloch formula [NUCLEO] The linear stopping power of a material for a fast charged particle is equal to $4\pi e^4 z^2 nB/mv^2$, where e is the electron charge in electrostatic units, z the charge number of the incident particle, n the number of atoms per unit volume of material, m the electron mass, v the particle velocity, and B the stopping number of the material.

Bethe-Heitler theory [NUCLEO] A theory for the energy loss of charged particles passing through matter, based on the Dirac equation and the Born approximation for the interaction of the particle with the field of a nucleus.

Bethe-Salpeter equation [PARTIC PHYS] The relativistic analog of the integral form of the two-body Schrödinger equation, the two-particle interaction kernel being the analog of the potential.

Bethe-Slater curve [SOLID STATE] A graph of the exchange energy for the transition elements versus the ratio of the interatomic distance to the radius of the $3d$ shell.

Betti group *See* homology group.

Betti number *See* connectivity number.

Betti reciprocal theorem [MECH] A theorem in the mathematical theory of elasticity which states that if an elastic body is subjected to two systems of surface and body forces, then the work that would be done by the first system acting through the displacements resulting from the second system equals the work that would be done by the second system acting through the displacements resulting from the first system.

Betti's method [MECH] A method of finding the solution of the equations of equilibrium of an elastic body whose surface displacements are specified; it uses the fact that the dilatation is a harmonic function to reduce the problem to the Dirichlet problem.

BeV [PHYS] A billion (10^9) electron volts, a unit used in the United States; international unit is GeV (gigaelectronvolts), which has the same value.

Bevatron [NUCLEO] Name of the 6-GeV proton synchrotron at the University of California at Berkeley.

beyond-the-horizon communication *See* scatter propagation.

Bézout domain [MATH] An integral domain in which all finitely generated ideals are principal.

BFO *See* beat-frequency oscillator.

BFS *See* beam-foil spectroscopy.

Bhabha scattering [PARTIC PHYS] The scattering of positrons by electrons.

B-H curve [ELECTROMAG] A graphical curve showing the relation between magnetic induction B and magnetizing force H for a magnetic material. Also known as magnetization curve.

B-H meter [ENG] A device used to measure the intrinsic hysteresis loop of a sample of magnetic material.

Bi *See* abampere; bismuth.

Bianchi classification [RELAT] A classification of possible types of spatially homogeneous space-times.

Bianchi identity [MATH] A differential identity satisfied by the Riemann curvature first tensor: the antisymmetric first covariant derivative of the Riemann tensor vanishes identically.

bias [STAT] In estimating the value of a parameter of a probability distribution, the difference between the expected value of the estimator and the true value of the parameter.

bias cell [ELECTR] A small dry cell used singly or in series to provide the required negative bias for the grid circuit of an electron tube. Also known as grid-bias cell.

bias current [ELECTR] **1.** An alternating electric current above about 40,000 hertz added to the audio current being recorded on magnetic tape to reduce distortion. **2.** An electric current flowing through the base-emitter junction of a transistor and adjusted to set the operating point of the transistor.

bias distortion [ELECTR] Distortion resulting from the operation on a nonlinear portion of the characteristic curve of a vacuum tube or other device, due to improper biasing.

biased sample [STAT] A sample obtained by a procedure that incorporates a systematic error introduced by taking items from a wrong population or by favoring some elements of a population.

biased statistic [STAT] A statistic whose expected value, as obtained from a random sampling, does not equal the parameter or quantity being estimated.

bias error [STAT] A measurement error that remains constant in magnitude for all observations; a kind of systematic error.

bias oscillator [ELECTR] An oscillator used in a magnetic recorder to generate the alternating-current signal that is added to the audio current being recorded on magnetic tape to reduce distortion.

bias resistor [ELECTR] A resistor used in the cathode or grid circuit of an electron tube to provide a voltage drop that serves as the bias.

bias voltage [ELECTR] A voltage applied or developed between two electrodes as a bias.

biaxial crystal [CRYSTAL] A crystal of low symmetry in which the index ellipsoid has three unequal axes.

biaxial indicatrix [CRYSTAL] An ellipsoid whose three axes at right angles to each other are proportional to the refractive indices of a biaxial crystal.

biaxial stress [MECH] The condition in which there are three mutually perpendicular principal stresses; two act in the same plane and one is zero.

bicompact set *See* compact set.

biconcave lens *See* double-concave lens.

biconditional operation [MATH] A logic operator on two statements P and Q whose result is true if P and Q are both true or both false, and whose result is false otherwise. Also known as if and only if operation; match.

biconical antenna [ELECTROMAG] An antenna consisting of two metal cones having a common axis with their vertices coinciding or adjacent and with coaxial-cable or waveguide feed to the vertices.

bicontinuous function *See* homeomorphism.

biconvex lens *See* double-convex lens.

bicuspidal [MATH] A plane curve whose equation in cartesian coordinates x and y is $(x^2 - a^2)(x - a)^2 + (y^2 - a^2)^2 = 0$, where a is a constant.

bidirectional antenna [ELECTROMAG] An antenna that radiates or receives most of its energy in only two directions.

bidirectional clamping circuit [ELECTR] A clamping circuit that functions at the prescribed time irrespective of the polarity of the signal source at the time the pulses used to actuate the clamping action are applied.

bidirectional clipping circuit [ELECTR] An electronic circuit that prevents transmission of the portion of an electrical signal that exceeds a prescribed maximum or minimum voltage value.

bidirectional transducer [ELECTR] A transducer capable of measuring in both positive and negative directions from a reference position. Also known as bilateral transducer.

bidirectional transistor [ELECTR] A transistor that provides switching action in either direction of signal flow through a circuit; widely used in telephone switching circuits.

Biedenharn identity [NUC PHYS] A relationship among the six-j symbols of Wigner.

BIDIRECTIONAL CLIPPING CIRCUIT

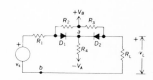

Circuit diagram of bidirectional clipping obtained by connecting two diodes.

BIFILAR ELECTROMAGNETIC OSCILLOGRAPH

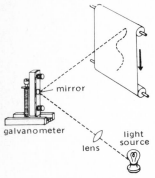

Bifilar electromagnetic oscillograph with light-beam, photographic-film recorder. (*General Electric Co.*)

Bienayme-Chebyshev inequality [STAT] The probability that the magnitude of the difference between the mean of the sample values of a random variable and the mean of the variable is less than st, where s is the standard deviation and t is any number greater than 1, is equal to or greater than $1 - (1/t^2)$.

bifilar electromagnetic oscillograph [ELECTROMAG] A writing low-frequency light-beam oscillograph usually using a moving coil with a single U-shaped turn (bifilar type).

bifilar resistor [ELECT] A resistor wound with a wire doubled back on itself to reduce the inductance.

bifilar suspension [ENG] The suspension of a body from two parallel threads, wires, or strips.

bifilar transformer [ELEC] A transformer in which wires for the two windings are wound side by side to give extremely tight coupling.

bifilar winding [ELEC] A winding consisting of two insulated wires, side by side, with currents traveling through them in opposite directions.

bifocal lens [OPTICS] **1.** A lens with two parts having different focal lengths. **2.** In particular, an eyeglass lens having one part that corrects for distant vision and one part for near vision.

bifurcated contact [ELEC] A contact having a forked shape such that it can slide over and interlock with an identical mating contact.

bifurcation theory [MATH] The study of the local behavior of solutions of a nonlinear equation in the neighborhood of a known solution of the equation; in particular, the study of solutions which appear as a parameter in the equation is varied and at first approximate the known solution, thus seeming to branch off from it. Also known as branching theory.

big bang theory [ASTRON] A theory of the origin and evolution of the universe which holds that approximately 2×10^{10} years ago all the matter in the universe was packed into a small agglomeration of extremely high density and temperature which exploded, sending matter in all directions and giving rise to the expanding universe. Also known as superdense theory.

bigit *See* binary digit.

bigraded module [MATH] A collection of modules $E_{s,t}$, indexed by pairs of integers s and t, with each module over a fixed principal ideal domain.

biharmonic function [MATH] A solution to the partial differential equation $\Delta^2 u(x,y,z) = 0$, where Δ is the Laplacian operator; occurs frequently in problems in electrostatics.

bijection [MATH] A mapping f from a set A onto a set B which is both an injection and a surjection; that is, for every element b of B there is a unique element a of A for which $f(a) = b$. Also known as bijective mapping.

bijective mapping *See* bijection.

bilateral antenna [ELECTROMAG] An antenna having maximum response in exactly opposite directions, 180° apart, such as a loop.

bilateral circuit [ELEC] Circuit wherein equipment at opposite ends is managed, operated, and maintained by different services.

bilateral element [ELECTR] A two-terminal circuit element in which a given current flow in either direction results in the same voltage drop.

bilateral Laplace transform [MATH] A generalization of the Laplace transform in which the integration is done over the negative real numbers as well as the positive ones.

bilateral slit [SPECT] A slit for spectrometers and spectrographs that is bounded by two metal strips which can be moved symmetrically, allowing the distance between them to be adjusted with great precision.

bilateral transducer *See* bidirectional transducer.

bilinear concomitant [MATH] An expression $B(u,v)$, where u, v are functions of x, satisfying $vL(u) - u\overline{L}(v) = (d/dx)B(u,v)$, where L, $\overline{L}$ are given adjoint differential equations.

bilinear expression [MATH] An expression which is linear in each of two variables separately.

bilinear form [MATH] **1.** A polynomial of the second degree which is homogeneous of the first degree in each of two sets of variables; thus, it is a sum of terms of the form $a_{ij}x_iy_j$, where $x_1, \ldots, x_m$ and $y_1, \ldots, y_n$ are two sets of variables and the a_{ij} are constants. **2.** More generally, a mapping $f(x,y)$ from $E \times F$ into R, where R is a commutative ring and $E \times F$ is the cartesian product of two modules E and F over R, such that for each x in E the function which takes y into $f(x,y)$ is linear, and for each y in F the function which takes x into $f(x,y)$ is linear.

bilinear transformations *See* Möbius transformations.

billboard array [ELECTROMAG] A broadside antenna array consisting of stacked dipoles spaced 1/4 to 3/4 wavelength apart in front of a large sheet-metal reflector. Also known as bedspring array; mattress array.

Billet split lens [OPTICS] A lens cut into two halves, along the optic axis; used in interferometry.

billion [MATH] **1.** The number 10^9. **2.** In British usage, the number 10^{12}.

bimetal [MATER] A laminate of two dissimilar metals, with different coefficients of thermal expansion, bonded together.

binary [ADP] Possessing a property for which there exists two choices or conditions, one choice excluding the other. [SCI TECH] Composed of or characterized by two parts or elements.

binary code [ADP] A code in which each allowable position has one of two possible states, commonly 0 and 1; the binary number system is one of many binary codes.

binary coded decimal system [ADP] A system of number representation in which each digit of a decimal number is represented by a binary number. Abbreviated BCD system.

binary counter *See* binary scaler.

binary digit [MATH] A digit in a binary number, equal to either 0 or 1. Also known as bigit.

binary encounter approximation [ATOM PHYS] An approximation for predicting the probability that an incident proton will eject an inner shell electron from an atom; it uses a semiclassical treatment of momentum transfer from the incident proton to the ejected electron.

binary magnetic core [SOLID STATE] A ferromagnetic core that can be made to take either of two stable magnetic states.

binary notation *See* binary number system.

binary number [MATH] A number expressed in the binary number system of positional notation.

binary number system [MATH] A representation for numbers using only the digits 0 and 1 in which successive digits are interpreted as coefficients of successive powers of the base 2. Also known as binary notation; binary system.

binary numeral [MATH] One of the two digits 0 and 1 used in writing a number in binary notation.

binary operation [MATH] A rule for combining two elements of a set to obtain a third element of that set, for example, addition and multiplication.

binary scaler [ELECTR] A scaler that produces one output

BILLET SPLIT LENS

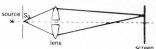

Billet split-lens interference.

pulse for every two input pulses. Also known as binary counter; scale-of-two circuit.

binary star [ASTRON] A pair of stars located sufficiently near each other in space to be connected by the bond of mutual gravitational attraction, compelling them to describe an orbit around their common center of gravity. Also known as binary system.

binary system [ASTRON] *See* binary star. [MATH] *See* binary number system.

binary to decimal conversion [MATH] The process of converting a number written in binary notation to the equivalent number written in ordinary decimal notation.

binaural intensity effect [ACOUS] The relationship wherein, if sound of the same frequency and phase is incident at both ears, the angle between the apparent direction of the sound and the median plane of the line joining the ears is proportional to the logarithm of the ratio of the intensities of sound received at the left and right ears.

binaural phase effect [ACOUS] A displacement in the apparent direction of a sound that results when a difference in phase is introduced between otherwise identical sound signals applied to the two ears; the angular displacement from the median plane is proportional to the phase difference.

binaural sound [ACOUS] The sound resulting from a reproduction system which has two channels, each fed into a different earphone or loudspeaker, so that a listener hears sounds coming from their original directions (with reference to the separated microphones used in recording the original sounds).

b-incomplete curve [RELAT] A b-incomplete curve in a space is a curve of finite length, where length is defined by a generalized affine parameter, and has an end point at a space-time singularity.

binding energy [PHYS] Abbreviated BE. Also known as total binding energy (TBE). **1.** The net energy required to remove a particle from a system. **2.** The net energy required to decompose a system into its constituent particles.

binding fraction [NUC PHYS] The ratio of the binding energy of a nucleus to the atomic mass number.

binding post [ELEC] A manually turned screw terminal used for making electrical connections.

Binet's formula [MATH] Either of two formulas for the logarithm of the gamma function $\Gamma(z)$, namely:

$$\log \Gamma(z) = (z - \tfrac{1}{2}) \log z - z + \tfrac{1}{2} \log (2\pi)$$
$$+ \int_0^\infty \left(\frac{1}{e^t - 1} - \frac{1}{t} + \frac{1}{2} \right) \frac{e^{-tz}}{t} \, dt$$

$$\log \Gamma(z) = (z - \tfrac{1}{2}) \log z - z + \tfrac{1}{2} \log (2\pi)$$
$$+ 2 \int_0^\infty \frac{\tan^{-1}(u/z)}{e^{2\pi u} - 1} \, du,$$

both valid when the real part of z is greater than 0.

Bingham number [FL MECH] A dimensionless number used to study the flow of Bingham plastics.

Bingham plastic [FL MECH] A non-Newtonian fluid exhibiting a yield stress which must be exceeded before flow starts; thereafter the rate-of-shear versus shear stress curve is linear.

binocular [OPTICS] Any optical instrument designed for use with both eyes to give enhanced views of distant objects, whose distinguishing performance feature is the depth perception obtainable.

binocular microscope [OPTICS] A microscope having two oculars, allowing the use of both eyes at once.

BINOCULAR

Modern prism binocular. (*Bausch and Lomb Optical Co.*)

binode [ELECTR] An electron tube with two anodes and one cathode used as a full-wave rectifier. Also known as double diode.

binomial [MATH] A polynomial with only two terms.

binomial array *See* Pascal triangle.

binomial array antenna [ELECTROMAG] Directional antenna array for reducing minor lobes and providing maximum response in two opposite directions.

binomial coefficient [MATH] A coefficient in the expansion of $(x+y)^n$, where n is a positive integer; the $(k+1)$st coefficient is equal to the number of ways of choosing k objects out of n without regard for order. Symbolized $\binom{n}{k}$; $_nC_k$; $C(n,k)$; C_k^n.

binomial differential [MATH] A differential of the form $x^p(a + bx^q)^r dx$, where p, q, r are integers.

binomial distribution [STAT] The distribution of a binomial random variable; the distribution (n,p) is given by $P(B = r) = \binom{n}{r} p^r q^{n-r}$, $p + q = 1$. Also known as Bernoulli distribution.

binomial equation [MATH] An equation of the form $x^n - a = 0$.

binomial expansion *See* binomial series.

binomial law [MATH] The probability of an event occurring r times in n Bernoulli trials is equal to $\binom{n}{r} p^r (1-p)^{n-r}$, where p is the probability of the event.

binomial probability paper [STAT] Graph paper designed to aid in the analysis of data from a binomial population, that is, data in the form of proportions or as percentages; both axes are marked so that the graduations are square roots of the variable.

binomial series [MATH] The expansion of $(x + y)^n$ when n is neither a positive integer nor zero. Also known as binomial expansion.

binomial surd [MATH] A polynomial having two terms, at least one of which is a surd.

binomial theorem [MATH] The rule for expanding $(x+y)^n$.

binomial trials [STAT] A sequence of trials, on each trial of which a certain result may or may not happen.

binomial trials model [STAT] A product model in which each factor has two simple events with probabilities p and $q = 1 - p$.

binormal [MATH] A vector on a curve at a point so that, together with the positive tangent and principal normal, it forms a system of right-handed rectangular cartesian axes.

biological shield [NUCLEO] A radiation-absorbing shield used to protect personnel from the effects of nuclear particles or radiation in the vicinity of a nuclear reactor.

biometrics [STAT] The use of statistics to analyze observations of biological phenomena.

biophysics [SCI TECH] The hybrid science involving the application of physical principles and methods to study and explain the structures of living organisms and the mechanics of life processes.

biot [ELEC] *See* abampere. [OPTICS] A unit of rotational strength in substances exhibiting circular dichroism, equal to 10^{-40} times the corresponding centimeter-gram-second unit.

biotar lens [OPTICS] A modern camera lens which is a modified Gauss objective with a large aperture and a field of about 24°.

Biot-Fourier equation [THERMO] An equation for heat conduction which states that the rate of change of temperature at any point divided by the thermal diffusivity equals the Laplacian of the temperature.

Biot number [FL MECH] A dimensionless group, used in the study of mass transfer between a fluid and a solid, which gives the ratio of the mass-transfer rate at the interface to the mass-

BIOTAR LENS

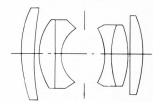

Biotar camera lens.

transfer rate in the interior of a solid wall of specified thickness.

Biot-Savart law [ELECTROMAG] A law that gives the intensity of the magnetic field due to a wire carrying a constant electric current.

Biot's law [OPTICS] The law that an optically active substance rotates plane-polarized light through an angle inversely proportional in its wavelength.

bipartite cubic [MATH] The points satisfying the equation $y^2 = x(x - a)(x - b)$.

bipartite graph [MATH] A linear graph (network) in which the nodes can be partitioned into two groups G_1 and G_2 such that for every arc (i,j) node i is in G_1 and node j in G_2.

biplate [OPTICS] **1.** Two plates of glass cemented together with a small angle between them, for producing a double image of a slit in interference experiments. **2.** Two half-wave plates of doubly refracting material, each cut parallel to its optical axis and cemented together with axes perpendicular; used to detect optical polarization. Also known as Bravais biplate.

bipolar [SCI TECH] **1.** Having two poles. **2.** Capable of assuming positive or negative values, such as an electric charge, or pertaining to a quantity with this property, such as a bipolar transistor.

bipolar amplifier [ELECTR] An amplifier capable of supplying a pair of output signals corresponding to the positive or negative polarity of the input signal.

bipolar circuit [ELECTR] A logic circuit in which zeros and ones are treated in a symmetric or bipolar manner, rather than by the presence or absence of a signal; for example, a balanced arrangement in a square-loop-ferrite magnetic circuit.

bipolar coordinate system [MATH] **1.** A two-dimensional coordinate system defined by the family of circles that pass through two common points, and the family of circles that cut the circles of the first family at right angles. **2.** A three-dimensional coordinate system in which two of the coordinates depend on the x and y coordinates in the same manner as in a two-dimensional bipolar coordinate system and are independent of the z coordinate, while the third coordinate is proportional to the z coordinate.

bipolar electrode [ELEC] Electrode, without metallic connection with the current supply, one face of which acts as anode surface and the opposite face as a cathode surface when an electric current is passed through a cell.

bipolar power supply [ELEC] A high-precision, regulated, direct-current power supply that can be set to provide any desired voltage between positive and negative design limits, with a smooth transition from one polarity to the other.

bipotential electrostatic lens [ELECTR] An electron lens in which image and object space are field-free, but at different potentials; examples are the lenses formed between apertures of cylinders at different potentials. Also known as immersion electrostatic lens.

biprism [OPTICS] A prism with apex angle only a little less than 180°, which produces a double image of a point source, giving rise to interference fringes on a nearby screen.

biprism interference [OPTICS] Light interference fringes seen on a screen near a biprism.

biquadratic [MATH] Any fourth-degree algebraic expression. Also known as quartic.

biquadratic equation *See* quartic equation.

biquartic filter [ELECTR] An active filter that uses operational amplifiers in combination with resistors and capacitors to provide infinite values of Q and simple adjustments for band-pass and center frequency.

biquartz [OPTICS] A device consisting of two adjoining pieces of quartz of equal thickness that rotate the plane of polarization of light in opposite directions; used with a Nicol prism or other analyzer to increase the accuracy of the latter in determining the properties of polarized light.

biquinary abacus [MATH] An abacus in which the frame is divided into two parts by a bar which separates each wire into two- and five-counter segments.

biquinary notation [MATH] A mixed-base notation system in which the first of each pair of digits counts 0 or 1 unit of five, and the second counts 0, 1, 2, 3, or 4 units. Also known as biquinary number system.

biquinary number system   *See* biquinary notation.

birectangular [MATH] Property of a geometrical object that has two right angles.

birefringence [OPTICS] Splitting of a light beam into two components, which travel at different velocities, by a material. Also known as double refraction.

birefringent filter [OPTICS] A filter consisting of alternate layers of polarizing films and plates cut from a birefringent crystal; transmits light in a series of sharp, widely spaced wavelength bands. Also known as Lyot filter; monochromatic filter.

birefringent plate [OPTICS] A piece of birefringent optical material with parallel plane surfaces.

Birge-Mieck rule [ATOM PHYS] The product of the equilibrium vibrational frequency and the square of the internuclear distance is a constant for various electronic states of a diatomic molecule.

Birge-Sponer extrapolation [SPECT] A method of calculating the dissociation limit of a diatomic molecule when the convergence limit cannot be observed directly, based on the assumption that vibrational energy levels converge to a limit for a finite value of the vibrational quantum number.

Birkhoff ergodic theorem [MATH] A generalization of the mean ergodic theorem which states that under the same hypotheses the same conclusion is true with convergence in the mean replaced by point-wise convergence almost everywhere.

Birkhoff's theorem [RELAT] The general relativistic theorem proving that a spherically symmetric gravitational field in empty space must be static and locally isometric to the Schwarzschild solution.

birth-death process [STAT] A method for describing the size of a population in which the population increases or decreases by one unit or remains constant over short time periods.

birth process [STAT] A stochastic process that defines a population whose members may have offspring; usually applied to the case where the population increases by one.

bisector [MATH] The ray dividing an angle into two equal angles.

bisectrix [CRYSTAL] A line that is the bisector of the angle between the optic axes of a biaxial crystal.

biserial correlation coefficient [STAT] A measure of the relationship between two qualities, one of which is a measurable random variable and the other a variable which is dichotomous, classified according to the presence or absence of an attribute; not a product moment correlation coefficient.

bismuth [CHEM] A metallic element, symbol Bi, of atomic number 83 and atomic weight 208.980.

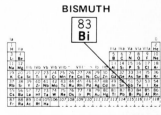

Periodic table of the chemical elements showing the position of bismuth.

bisphenoid [CRYSTAL] A form apparently consisting of two sphenoids placed together symmetrically.

bispherical lens [OPTICS] A lens one of whose surfaces consists of portions of two spheres of different radius, one near the center and the other near the edge.

bistable circuit [ELECTR] A circuit with two stable states such that the transition between the states cannot be accomplished by self-triggering.

bistable multivibrator [ELECTR] A multivibrator in which either of the two active devices may remain conducting, with the other nonconducting, until the application of an external pulse. Also known as Eccles-Jordan circuit; Eccles-Jordan multivibrator; flip-flop circuit; trigger circuit.

bistable unit [ENG] A physical element that can be made to assume either of two stable states; a binary cell is an example.

bistatic reflectivity [OPTICS] The characteristic of a reflector which reflects energy along a line, or lines, different from or in addition to that of the incident ray.

bit [ADP] **1.** A unit of information content equal to one binary decision or the designation of one of two possible and equally likely values or states of anything used to store or convey information. **2.** A dimensionless unit of storage capacity specifying that the capacity of a storage device is expressed by the logarithm to the base 2 of the number of possible states of the device.

bitangent *See* double tangent.

Bitter pattern [SOLID STATE] A pattern produced when a drop of a colloidal suspension of ferromagnetic particles is placed on the surface of a ferromagnetic crystal; the particles collect along domain boundaries at the surface.

bivariate distribution [STAT] The joint distribution of a pair of variates for continuous or discontinuous data.

Bk *See* berkelium.

BL *See* base line.

black [OPTICS] Quality of an object which uniformly absorbs large percentages of light of all visible wavelengths.

black-and-white groups *See* Shubnikov groups.

blackbody [THERMO] An ideal body which would absorb all incident radiation and reflect none. Also known as hohlraum; ideal radiator.

blackbody radiation [THERMO] The emission of radiant energy which would take place from a blackbody at a fixed temperature; it takes place at a rate expressed by the Stefan-Boltzmann law, with a spectral energy distribution described by Planck's equation.

blackbody temperature [THERMO] The temperature of a blackbody that emits the same amount of heat radiation per unit area as a given object; measured by a total radiation pyrometer. Also known as brightness temperature.

black-bulb thermometer [ENG] A thermometer whose sensitive element has been made to approximate a blackbody by covering it with lampblack.

black carbon counter [NUCLEO] The original type of radiation counter used in radiocarbon dating, in which the sample, whose carbon has first been converted to carbon black, is mounted on the inside of a steel cylinder which is inserted into a sensitive Geiger counter. Also known as Libby counter.

black hole [RELAT] A region of space-time from which nothing can escape, according to classical physics; quantum corrections indicate a black hole radiates particles with a temperature inversely proportional to the mass and directly proportional to Planck's constant.

black light [OPTICS] Invisible light, such as ultraviolet rays

Bitter powder patterns on a (100) surface of silicon-iron. *(Photograph by H. J. Williams)*

which fall on fluorescent materials and cause them to emit visible light.

blackout [COMMUN] *See* radio blackout. [ELEC] Shutting off of power in an electrical power transmission system, either deliberately or through failure of the system.

black-surface enclosure [THERMO] An enclosure for which the interior surfaces of the walls possess the radiation characteristics of a blackbody.

blade [ELEC] A flat moving conductor in a switch.

Blagden's law [PHYS CHEM] The law that the lowering of a solution's freezing point is proportional to the amount of dissolved substance.

Blake number [FL MECH] A dimensionless number used in the study of beds of particles.

blandel *See* apostilb.

blanket [NUCLEO] A layer of fertile uranium-238 or thorium-232 material placed around or within the core of a nuclear reactor to breed new fuel.

Blaschke product [MATH] A function of the form

$$z^k \prod_{n=1}^{\infty} \frac{a_n - z}{1 - \bar{a}_n z} \frac{|a_n|}{a_n},$$

defined for $|z| < 1$, where $\{a_n\}$ is a sequence of complex numbers with $0 < |a_n| < 1$, and

$$\sum_{n=1}^{\infty} (1 - |a_n|) < \infty;$$

$\bar{a}_n$ denotes the complex conjugate of a_n.

Blasius equation [FL MECH] An empirical formula relating the pressure loss coefficient λ for fully developed turbulent flow in a smooth pipe to the Reynolds number Re (where the characteristic length is the pipe diameter): $\lambda = 0.3164 (Re)^{-0.25}$. [MATH] A differential equation $2 d^3 y/dx^3 + y d^2 y/dx^2 = 0$, used in the study of boundary-layer flow.

Blasius function [MATH] The solution $y(x)$ of the Blasius equation that satisfies the boundary conditions $y(0) = 0$ and $y(\infty) = 1$.

blast effect [PHYS] Violent air movements and pressure changes and the destruction or damage resulting therefrom, generally caused by an explosion on or above the surface of the earth.

blast pressure [PHYS] The impact pressure of the air set in motion by an explosion.

blast wave [PHYS] The air wave set in motion by an explosion.

blaze-of-grating technique [OPTICS] A technique whereby the ruled grooves of a diffraction grating are given a controlled shape so that they reflect as much as 80% of the incoming light into one particular order for a given wavelength.

BLC *See* boundary-layer control.

bleeder current [ELEC] Current drawn continuously from a voltage source to lessen the effect of load changes or to provide a voltage drop across a resistor.

bleeder resistor [ELEC] A resistor connected across a power pack or other voltage source to improve voltage regulation by drawing a fixed current value continuously; also used to dissipate the charge remaining in filter capacitors when equipment is turned off.

blind controller system [CONT SYS] A process control arrangement that separates the in-plant measuring points (for example, pressure, temperature, and flow rate) and control points (for example, a valve actuator) from the recorder or indicator at the central control panel.

blink [MECH] A unit of time equal to 10^{-5} day or to 0.864 second.

blink comparator [OPTICS] An optical instrument used to alternately view two pictures in the same visual field in rapid succession, to detect small differences in similar images.

blink microscope [OPTICS] A blink comparator which magnifies the compared pictures.

blip [ELECTR] **1.** The display of a received pulse on the screen of a cathode-ray tube. Also known as pip. **2.** An ideal infrared radiation detector that detects with unit quantum efficiency all of the radiation in the signal for which the detector was designed, and responds only to the background radiation noise that comes from the field of view of the detector.

blister [NUCLEO] A protuberance that sometimes develops on the surface of a nuclear-reactor fuel element during use, generally because of entrapped gases.

BL Lacertae objects [ASTRON] A class of extragalactic sources of extremely intense, highly variable electromagnetic radiation which are related to quasars but have a featureless optical spectrum, and display strong optical polarization and a radio spectrum that increases in intensity at shorter wavelengths.

Bloch equations [SOLID STATE] Approximate equations for the rate of change of magnetization of a solid in a magnetic field due to spin relaxation and gyroscopic precession.

Bloch function [SOLID STATE] A wave function for an electron in a periodic lattice, of the form $u(\mathbf{r}) \exp [i\mathbf{k}\cdot\mathbf{r}]$ where $u(\mathbf{r})$ has the periodicity of the lattice.

Bloch theorem [QUANT MECH] The theorem that the lowest state of a quantum-mechanical system without a magnetic field can carry no current. [SOLID STATE] The theorem that, in a periodic structure, every electronic wave function can be represented by a Bloch function.

Bloch wall [SOLID STATE] A transition layer, with a finite thickness of a few hundred lattice constants, between adjacent ferromagnetic domains. Also known as domain wall.

block [STAT] In experimental design, a homogeneous aggregation of items under observation, such as a group of contiguous plots of land or all animals in a litter; allows the experimenter to isolate sources of heterogeneity.

blocked impedance [ELEC] The impedance at the input of a transducer when the impedance of the output system is made infinite, as by blocking or clamping the mechanical system.

blocking [ELECTR] **1.** Applying a high negative bias to the grid of an electron tube to reduce its anode current to zero. **2.** Overloading a receiver by an unwanted signal so that the automatic gain control reduces the response to a desired signal. **3.** Distortion occurring in a resistance-capacitance-coupled electron tube amplifier stage when grid current flows in the following tube. [STAT] The grouping of sample data into subgroups with similar characteristics.

blocking capacitor *See* coupling capacitor.

blocking layer *See* depletion layer.

blocking oscillator [ELECTR] A relaxation oscillator that generates a short-time-duration pulse by using a single transistor or electron tube and associated circuitry. Also known as squegger; squegging oscillator.

block protector [ELEC] Rectangular piece of carbon, bakelite with a metal insert, or porcelain with a carbon insert which, in combination with each other, make one element of a protector; they form a gap which will break down and provide a path to ground for excessive voltages.

blondel *See* apostilb.

Blondel-Rey law [OPTICS] A law utilized to determine the apparent point brilliance of a flashing light.

blow [ELEC] Opening of a circuit because of excess current, particularly when the current is heavy and a melting or breakdown point is reached.

blown-fuse indicator [ELEC] A neon warning light connected across a fuse so that it lights when the fuse is blown.

blowout [ELEC] The melting of an electric fuse because of excessive current. [ELECTROMAG] The extinguishing of an electric arc by deflection in a magnetic field. Also known as magnetic blowout.

blowout coil [ELECTROMAG] A coil that produces a magnetic field in an electrical switching device for the purpose of lengthening and extinguishing an electric arc formed as the contacts of the switching device part to interrupt the current.

blowout magnet [ELECTROMAG] An electromagnet or permanent magnet used to deflect and extinguish the arc formed when a high-current circuit breaker or switch is opened.

blue [OPTICS] The hue evoked in an average observer by monochromatic radiation having a wavelength in the approximate range from 455 to 492 nanometers; however, the same sensation can be produced in a variety of other ways.

blue flash *See* green flash.

blue glow [ELECTR] A glow normally seen in electron tubes containing mercury vapor, due to ionization of the mercury molecules.

blue-green flame *See* green flash.

blue magnetism [GEOPHYS] The magnetism displayed by the south-seeking end of a freely suspended magnet; this is the magnetism of the earth's north magnetic pole.

blue shift [RELAT] A systematic shift of all wavelengths toward the blue end of the spectrum, owing to relativistic effects.

blue star [ASTRON] A star of spectral type O, B, A, or F according to the Draper catalog.

blur circle [OPTICS] The patch of light produced on a screen by rays from a point source reflected or refracted through an optical system, if the screen is not at the correct focus or if the system suffers from aberration.

B meson [PARTIC PHYS] An elementary particle with strong nuclear interactions, baryon number $B = 0$, and mass 1237 MeV.

Board of Trade unit *See* kilowatt-hour.

bobbin [ELECTROMAG] An insulated spool serving as a support for a coil.

bobbin core [ELECTROMAG] A magnetic core having a form or bobbin on which the ferromagnetic tape is wrapped for support of the tape.

Bobillier's law [MECH] The law that, in general plane rigid motion, when a and b are the respective centers of curvature of points A and B, the angle between Aa and the tangent to the centrode of rotation (pole tangent) and the angle between Bb and a line from the centrode to the intersection of AB and ab (collineation axis) are equal and opposite.

Bode diagram [ELECTR] A diagram in which the phase shift or the gain of an amplifier, a servomechanism, or other device is plotted against frequency to show frequency response; logarithmic scales are customarily used for gain and frequency.

Bodenstein number [FL MECH] A numberless group used in the study of diffusion in reactors.

Bode's law [ASTRON] An empirical law giving mean distances of planets to the sun by the formula $a = 0.4 + 0.3 \times 2^n$, where a is in astronomical units and n equals $-\infty$ for Mercury, 0 for Venus, 1 for Earth, and so on; the asteroids are included as planets. Also known as Titius-Bode law.

BLOWOUT COIL

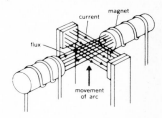

The relation of direction of current, magnetic flux, and movement of arc in a blowout coil.

BOBILLIER'S LAW

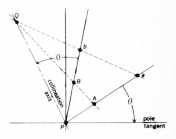

Application of Bobillier's law of four-bar linkage consisting of fixed member ab, cranks aA and bB and connecting rod AB. Law states that angle θ between Aa and pole tangent equals angle θ between Bb and collineation axis joining centrode P and intersection Q of AB and ab.

BODY-CENTERED LATTICE

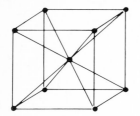

Drawing of a body-centered cubic lattice.

BOHR-SOMMERFELD
THEORY

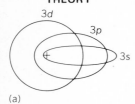

(a)

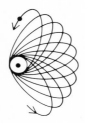

(b)

Possible elliptical orbits, according to the Bohr-Sommerfeld theory. (a) The three permitted orbits for $n=3$. (b) Precession of the 3s orbit caused by the relativistic variation of mass.

bodily tide   *See* earth tide.

body burden   [NUCLEO] The amount of radioactive material present in the body of a human or animal.

body capacitance   [ELEC] Capacitance existing between the human hand or body and a circuit.

body-centered lattice   [CRYSTAL] A space lattice in which the point at the intersection of the body diagonals is identical to the points at the corners of the unit cell.

body centrode   [MECH] The path traced by the instantaneous center of a rotating body relative to the body.

body cone   [MECH] The cone in a rigid body that is swept out by the body's instantaneous axis during Poinsot motion. Also known as Polhode cone.

body force   [MECH] An external force, such as gravity, which acts on all parts of a body.

body of revolution   [MATH] A symmetrical body having the form described by rotating a plane curve about an axis in its plane.

Bohr atom   [ATOM PHYS] An atomic model having the structure postulated in the Bohr theory.

Bohr-Breit-Wigner theory   *See* Breit-Wigner theory.

Bohr frequency condition   [ATOM PHYS] The frequency of the radiation emitted or absorbed during the transition of an atomic system between two stationary states equals the difference in the energies of the states divided by Planck's constant.

Bohr magneton   [ATOM PHYS] The amount $he/4\pi mc$ of magnetic moment, where h is Planck's constant, e and m are the charge and mass of the electron, and c is the speed of light.

Bohr-Mollerup theorem   [MATH] The gamma function Γ is the only function whose value is defined and is positive for all positive, real values of the independent variable, and which satisfies the equation $\Gamma(x+1)=x\Gamma(x)$, has a logarithm which is a convex function, and satisfies the condition $\Gamma(1)=1$.

Bohr orbit   [ATOM PHYS] One of the electron paths about the nucleus in Bohr's model of the hydrogen atom.

Bohr radius   [ATOM PHYS] The radius of the ground-state orbit of the hydrogen atom in the Bohr theory.

Bohr's correspondence principle   *See* correspondence principle.

Bohr-Sommerfeld theory   [ATOM PHYS] A modification of the Bohr theory in which elliptical as well as circular orbits are allowed.

Bohr's theorem   [MATH] The basic result in the theory of almost-periodic functions, which states that any almost-periodic function $f(t)$ can be uniformly approximated by a sequence of functions each of which is a finite sum of terms of the form $a_k \exp(i\omega_k t)$, where the a_k are constants and the ω_k are real constants.

Bohr theory   [ATOM PHYS] A theory of atomic structure postulating an electron moving in one of certain discrete circular orbits about a nucleus with emission or absorption of electromagnetic radiation necessarily accompanied by transitions of the electron between the allowed orbits.

Bohr–van Leeuwen theorem   [QUANT MECH] The theorem that magnetism is inexplicable in classical physics and is a quantum phenomenon.

Bohr-Wheeler theory of fission   [NUC PHYS] A theory accounting for the stability of a nucleus against fission by treating it as a droplet of incompressible and uniformly charged liquid endowed with surface tension.

boiling point   [PHYS CHEM] Abbreviated bp. **1.** The temperature at which the transition from the liquid to the gaseous phase occurs in a pure substance at fixed pressure. **2.** *See* bubble point.

boiling-water reactor [NUCLEO] A nuclear reactor in which the coolant is water, maintained at such a pressure as to allow it to boil and form steam. Abbreviated BWR.

boil-off [THERMO] The vaporization of a liquid, such as liquid oxygen or liquid hydrogen, as its temperature reaches its boiling point under conditions of exposure, as in the tank of a rocket being readied for launch.

bolide [ASTRON] A brilliant meteor, especially one which explodes; a detonating fireball meteor.

bolograph [ENG] Any graphical record made by a bolometer; in particular, a graph formed by directing a pencil of light reflected from the galvanometer of the bolometer at a moving photographic film.

bolometer [ENG] An instrument that measures the energy of electromagnetic radiation in certain wavelength regions by utilizing the change in resistance of a thin conductor caused by the heating effect of the radiation. Also known as thermal detector.

Boltzmann constant [STAT MECH] The ratio of the universal gas constant to the Avogadro number.

Boltzmann distribution [STAT MECH] A function giving the probability that a molecule of a gas in thermal equilibrium will have generalized position and momentum coordinates within given infinitesimal ranges of values, assuming that the molecules obey classical mechanics.

Boltzmann engine [THERMO] An ideal thermodynamic engine that utilizes blackbody radiation; used to derive the Stefan-Boltzmann law.

Boltzmann entropy hypothesis [STAT MECH] The hypothesis that the entropy of a system in a given state is directly proportional to the logarithm of the probability of finding it in that state.

Boltzmann factor [STAT MECH] The quantity, $\exp(-E/kT)$, that appears in the Boltzmann distribution, where E is the total energy of the particle in question, T is the temperature, and k is the Boltzmann constant.

Boltzmann H theorem [STAT MECH] The theorem that the entropy of a system never decreases; Boltzmann proved this for a classical gas of colliding particles. Also known as H theorem of Boltzmann.

Boltzmann statistics *See* Maxwell-Boltzmann statistics.

Boltzmann transport equation [STAT MECH] An equation used to study the nonequilibrium behavior of a collection of particles; it states that the rate of change of a function which specifies the probability of finding a particle in a unit volume of phase space is equal to the sum of terms arising from external forces, diffusion of particles, and collisions of the particles. Also known as Maxwell-Boltzmann equation.

Boltzmann-Vlasov equations [PL PHYS] The equations that govern a high-temperature plasma in which the collisional mean free path is much larger than all the characteristic lengths of the system.

Bolyai geometry *See* Lobachevski geometry.

Bolzano's theorem [MATH] The theorem that a single-valued, real-valued, continuous function of a real variable is equal to zero at some point in an interval if its values at the end points of the interval have opposite sign.

Bolzano-Weierstrass theorem [MATH] The theorem that every bounded, infinite set in finite dimensional euclidean space has a cluster point.

bombard [NUCLEO] To direct a stream of particles or photons against a target.

bombardment [ELECTR] Induction heating of electrodes of electron tubes to drive out gases during evacuation.

bond [CHEM] The strong attractive force that holds together atoms in molecules and crystalline salts. Also known as chemical bond. [ELEC] The connection made by bonding electrically.

Bond albedo [OPTICS] The ratio of the total light reflected in all directions from a sphere illuminated by parallel rays, to the light incident on the sphere.

bond angle [PHYS CHEM] The angle between bonds sharing a common atom. Also known as valence angle.

bond distance [PHYS CHEM] The distance separating the two nuclei of two atoms bonded to each other in a molecule. Also known as bond length.

bond energy [PHYS CHEM] The heat of formation of a molecule from its constituent atoms.

bonding [CHEM] The joining together of atoms to form molecules or crystalline salts. [ELEC] The use of low-resistance material to connect electrically a chassis, metal shield cans, cable shielding braid, and other supposedly equipotential points to eliminate undesirable electrical interaction resulting from high-impedance paths between them.

bonding electron [PHYS CHEM] An electron whose orbit spans the entire molecule and so assists in holding it together.

bonding strength [MECH] Structural effectiveness of adhesives, welds, solders, glues, or of the chemical bond formed between the metallic and ceramic components of a cermet, when subjected to stress loading, for example, shear, tension, or compression.

bonding wire [ELEC] Wire used to connect metal objects so they have the same potential (usually ground potential).

bond length *See* bond distance.

Bond number [FL MECH] A dimensionless number used in the study of atomization and the study of bubbles and drops, equal to $(\rho - \rho')L^2g/\sigma$, where ρ is the density of a bubble or drop, ρ' is the density of the surrounding medium, L is a characteristic dimension, g is the acceleration of gravity, and σ is the surface tension of the bubble or drop. Also known as Eötvös number.

bone seeker [NUCLEO] A radioisotope that tends to accumulate in the bones when it is introduced into the body; an example is strontium-90, which behaves chemically like calcium.

Bonnet's mean value theorem *See* second law of the mean for integrals.

book capacitor [ELEC] A trimmer capacitor consisting of two plates which are hinged at one end; capacitance is varied by changing the angle between them.

Boolean algebra [MATH] An algebraic system with two binary operations and one unary operation important in representing a two-valued logic.

Boolean algebra of subsets *See* algebra of subsets.

Boolean calculus [MATH] Boolean algebra modified to include the element of time.

Boolean determinant [MATH] A function defined on Boolean matrices which depends on the elements of the matrix in a manner analogous to the manner in which an ordinary determinant depends on the elements of an ordinary matrix, with the operation of multiplication replaced by intersection and the operation of addition replaced by union.

Boolean function [MATH] A function $f(x, y, \ldots, z)$ assembled by the application of the operations AND, OR, NOT on the variables $x, y, \ldots, z$ and elements whose common domain is a Boolean algebra.

Boolean matrix [MATH] A rectangular array of elements each of which is a member of a Boolean algebra.

Boolean operation table [MATH] A table which indicates, for a particular operation on a Boolean algebra, the values that result for all possible combinations of values of the operands; used particularly with Boolean algebras of two elements which may be interpreted as "true" and "false."

Boolean ring [MATH] A commutative ring with the property that for every element a of the ring, $a \times a = a$ and $a + a = 0$; it can be shown to be equivalent to a Boolean algebra.

bootstrapping [ELECTR] A technique for lifting a generator circuit above ground by a voltage value derived from its own output signal.

bootstrap scheme [PARTIC PHYS] A theory of elementary particles in which the existence of each particle contributes to forces between it and other particles; these forces lead to bound systems which are the particles themselves.

Borda mouthpiece [FL MECH] A reentrant tube in a hydraulic reservoir, whose contraction coefficient (the ratio of the cross section of the issuing jet of liquid to that of the opening) can be calculated more simply than for other discharge openings.

Borel measurable function [MATH] **1.** A real-valued function such that the inverse image of the set of real numbers greater than any given real number is a Borel set. **2.** More generally, a function to a topological space such that the inverse image of any open set is a Borel set.

Borel measure [MATH] A measure defined on the class of all Borel sets of a topological space such that the measure of any compact set is finite.

Borel set [MATH] A member of the smallest sigma algebra containing the compact subsets of a topological space.

Born approximation [QUANT MECH] A method used for the computation of cross sections in scattering problems; the interactions are treated as perturbations of free-particle systems.

Born equation [PHYS CHEM] An equation for determining the free energy of solvation of an ion in terms of the Avogadro number, the ionic valency, the ion's electronic charge, the dielectric constant of the electrolytic, and the ionic radius.

Born-Haber cycle [SOLID STATE] A sequence of chemical and physical processes by means of which the cohesive energy of an ionic crystal can be deduced from experimental quantities; it leads from an initial state in which a crystal is at zero pressure and 0 K to a final state which is an infinitely dilute gas of its constituent ions, also at zero pressure and 0 K.

Born-Madelung model [SOLID STATE] A classical theory of cohesive energy, lattice spacing, and compressibility of ionic crystals.

Born-Mayer equation [SOLID STATE] An equation for the cohesive energy of an ionic crystal which is deduced by assuming that this energy is the sum of terms arising from the Coulomb interaction and a repulsive interaction between nearest neighbors.

Born-Oppenheimer approximation [PHYS CHEM] The approximation, used in the Born-Oppenheimer method, that the electronic wave functions and energy levels at any instant depend only on the positions of the nuclei at that instant and not on the motions of the nuclei. Also known as adiabatic approximation.

Born-Oppenheimer method [PHYS CHEM] A method for calculating the force constants between atoms by assuming that the electron motion is so fast compared with the nuclear motions that the electrons follow the motions of the nuclei adiabatically.

Born-von Kármán theory [SOLID STATE] A theory of specific heat which considers an acoustical spectrum for the vibra-

BORON

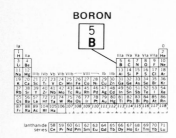

Periodic table of the chemical elements showing the position of boron.

tions of a system of point particles distributed like the atoms in a crystal lattice.

boron [CHEM] A chemical element, symbol B, atomic number 5, atomic weight 10.811; it has three valence electrons and is nonmetallic.

boron-10 [NUC PHYS] A nonradioactive isotope of boron with a mass number of 10; it is a good absorber for slow neutrons, simultaneously emitting high-energy alpha particles, and is used as a radiation shield in Geiger counters.

boron chamber [NUCLEO] An ionization chamber that is lined with boron or boron compounds or filled with a gaseous boron compound.

boron counter tube [NUCLEO] A counter tube filled with boron fluoride or having electrodes coated with boron or boron compounds; used for detecting slow neutrons.

boron thermopile [NUCLEO] A thermopile in which alternate thermocouple junctions are coated with boron; exposure to a flux of slow neutrons generates heat in these junctions, producing an output voltage proportional to neutron flux.

Borrman effect [PHYS] The irregular transmission of x-rays when a single crystal of high perfection is placed in a monochromatic x-ray beam in a reflecting position.

borrow [MATH] An arithmetically negative carry; it occurs in direct subtraction by raising the low-order digit of the minuend by one unit of the next-higher-order digit; for example, when subtracting 67 from 92, a tens digit is borrowed from the 9, to raise the 2 to a factor of 12; the 7 of 67 is then subtracted from the 12 to yield 5 as the units digit of the difference; the 6 is then subtracted from 8, or $9-1$, yielding 2 as the tens digit of the difference.

Bosanquet's law [ELECTROMAG] The statement that, in analogy to Ohm's law for the resistance of an electric circuit, in a magnetic circuit the ratio of the magnetomotive force to the magnetic flux is a constant known as the reluctance.

Bose distribution *See* Bose-Einstein distribution.

Bose-Einstein condensation [CRYO] A phenomenon that occurs in the study of systems of bosons; there is a critical temperature below which the ground state is highly populated. Also known as condensation; Einstein condensation.

Bose-Einstein distribution [STAT MECH] For an assembly of independent bosons, such as photons or helium atoms of mass number 4, a function that specifies the number of particles in each of the allowed energy states. Also known as Bose distribution.

Bose-Einstein statistics [STAT MECH] The statistical mechanics of a system of indistinguishable particles for which there is no restriction on the number of particles that may exist in the same state simultaneously. Also known as Einstein-Bose statistics.

Bose gas [STAT MECH] An assemblage of noninteracting or weakly interacting bosons.

boson [STAT MECH] A particle that obeys Bose-Einstein statistics; includes photons, pi mesons, and all nuclei having an even number of particles and all particles with integer spin.

boson commutation relations [QUANT MECH] The antisymmetric operator relations, $a_r a_s - a_s a_r = 0$ and $a_r a_s^+ - a_s^+ a_r = \delta_{rs}$, between variables and their Hermitian conjugates that are central to the canonical formalism for bosons in quantum theory.

bougie decimale [OPTICS] Formerly, a unit of luminous intensity equal to 0.96 international standard candle.

Bouguer-Lambert-Beer law [ANALY CHEM] The intensity of a beam of monochromatic radiation in an absorbing medium decreases exponentially with penetration distance. Also known as Beer-Lambert-Bouguer law; Lambert-Beer law.

Bouguer-Lambert law [ANALY CHEM] The law that the change in intensity of light transmitted through an absorbing substance is related exponentially to the thickness of the absorbing medium and a constant which depends on the sample and the wavelength of the light. Also known as Lambert's law.

boule [CRYSTAL] A pure crystal, such as silicon, having the atomic structure of a single crystal, formed synthetically by rotating a small seed crystal while pulling it slowly out of molten material in a special furnace.

boundary [ELECTR] An interface between p- and n-type semiconductor materials, at which donor and acceptor concentrations are equal.

boundary condition [MATH] A requirement to be met by a solution to a set of differential equations on a specified set of values of the independent variables.

boundary friction [MECH] Friction between surfaces that are neither completely dry nor completely separated by a lubricant.

boundary-layer control [FL MECH] Control over the development of a boundary layer by reduction of surface roughness and choice of surface contours. Abbreviated BLC.

boundary-layer flow [FL MECH] The flow of that portion of a viscous fluid which is in the neighborhood of a body in contact with the fluid and in motion relative to the fluid.

boundary-layer photocell *See* photovoltaic cell.

boundary-layer separation [FL MECH] That point where the boundary layer no longer continues to follow the contour of the boundary because the residual momentum of the fluid (left after overcoming viscous forces) may be insufficient to allow the flow to proceed into regions of increasing pressure. Also known as flow separation.

boundary-layer theory *See* film theory.

boundary of a set *See* frontier of a set in a topological space.

boundary operator [MATH] If $\{C_n\}$ is a sequence of Abelian groups, boundary operators are homomorphisms $\{d_n\}$ such that $d_n:C_n \rightarrow C_{n-1}$ and $d_{n-1} \circ d_n = 0$.

boundary point [MATH] In a topological space, a point of a set with the property that every neighborhood of the point contains points of both the set and its complement.

boundary value problem [MATH] A problem, such as the Dirichlet or Neumann problem, which involves finding the solution of a differential equation or system of differential equations which meets certain specified requirements, usually connected with physical conditions, for certain values of the independent variable.

boundary wavelength *See* quantum limit.

bound charge [ELEC] Electric charge which is confined to atoms or molecules, in contrast to free charge, such as metallic conduction electrons, which is not. Also known as polarization charge.

bounded function [MATH] **1.** A function whose image is a bounded set. **2.** A function of a metric space to itself which moves each point no more than some constant distance.

bounded growth [MATH] The property of a function f defined on the positive real numbers which requires that there exist numbers M and a such that the absolute value of $f(t)$ is less than Ma^t for all positive values of t.

bounded linear transformation [MATH] A linear transformation T for which there is some positive number A such that the norm of $T(x)$ is equal to or less than A times the norm of x for each x.

BOUNDARY-LAYER CONTROL

In this airfoil, retarded flow in the boundary layer is reenergized by supplying high-velocity flow through a slot in the surface.

bounded set [MATH] **1.** A collection of numbers whose absolute values are all smaller than some constant. **2.** A set of points, the distance between any two of which is smaller than some constant.

bounded variation [MATH] A real-valued function is of bounded variation on an interval if its total variation there is bounded.

bound electron [ATOM PHYS] An electron whose wave function is negligible except in the vicinity of an atom.

bound level [NUC PHYS] An energy level in a nucleus so close to the ground state that it can only decay by gamma emission.

bound particle [PHYS] A particle which is confined to some finite region.

bound variable [MATH] In logic, a variable that occurs within the scope of a quantifier, and cannot be replaced by a constant.

bound vector [MECH] A vector whose line of application and point of application are both prescribed, in addition to its direction.

Bourdon pressure gage [ENG] A mechanical pressure-measuring instrument employing as its sensing element a curved or twisted metal tube, flattened in cross section and closed. Also known as Bourdon tube.

Bourdon tube *See* Bourdon pressure gage.

Boussinesq approximation [FL MECH] The assumption (frequently used in the theory of convection) that the fluid is incompressible except insofar as the thermal expansion produces a buoyancy, represented by a term $g\alpha T$, where g is the acceleration of gravity, α is the coefficient of thermal expansion, and T is the perturbation temperature.

Boussinesq number [FL MECH] A dimensionless number used to study wave behavior in open channels.

Boussinesq-Papkovich solution *See* Papkovich-Neuber solution.

Boussinesq's problem [MECH] The problem of determining the stresses and strains in an infinite elastic body, initially occupying all the space on one side of an infinite plane, and indented by a rigid punch having the form of a surface of revolution with axis of revolution perpendicular to the plane. Also known as Cerruti's problem.

bow [MATH] A plane curve whose equation in cartesian coordinates x and y is $x^4 = x^2y - y^3$.

Bowie formula [GEOPHYS] A correction used for calculation of the local gravity anomaly on earth.

bowshock [ASTROPHYS] The shock wave set up by the interaction of the supersonic solar wind with a planet's magnetic field.

Bow's notation [MECH] A graphical method of representing coplanar forces and stresses, using alphabetical letters, in the solution of stresses or in determining the resultant of a system of concurrent forces.

bow wave [FL MECH] A shock wave occurring in front of a body, such as an airfoil, or apparently attached to the forward tip of the body.

boxcar function [MATH] A function whose value is zero except for a finite interval of its argument, for which it has a constant nonzero value.

Boyle's law [PHYS] The law that the product of the volume of a gas times its pressure is a constant at fixed temperature. Also known as Mariotte's law.

Boyle's temperature [THERMO] For a given gas, the temperature at which the virial coefficient B in the equation of state $Pv = RT\,[1 + (B/v) + (C/v^2) + \ldots]$ vanishes.

Boys camera [OPTICS] A type of camera used for the observation of lightning flashes.

Boys' method [OPTICS] A method of measuring the refractive index of a lens, in which the curvatures of the lens surfaces are determined by positioning a light source so that reflection from a surface gives an image coincident with the object; these curvatures and the focal length are used to calculate the refractive index.

bp *See* boiling point.

Bq *See* becquerel.

Br *See* bromine.

brachistochrone [MECH] The curve along which a smooth-sliding particle, under the influence of gravity alone, will fall from one point to another in the minimum time.

brachyaxis [CRYSTAL] The shorter lateral axis, usually the *a* axis, of an orthorhombic or triclinic crystal. Also known as brachydiagonal.

brachydiagonal *See* brachyaxis.

Brackett series [SPECT] A series of lines in the infrared spectrum of atomic hydrogen whose wave numbers are given by $R_H[(1/16) - (1/n^2)]$, where R_H is the Rydberg constant for hydrogen and n is any integer greater than 4.

Bradley aberration [ASTRON] Stellar aberration with a maximum of 20.5 seconds of arc; can be used to compute an approximate velocity for light.

Bragg angle [SOLID STATE] One of the characteristic angles at which x-rays reflect specularly from planes of atoms in a crystal.

Bragg cell *See* acoustooptic modulator.

Bragg curve [ATOM PHYS] **1.** A curve showing the average number of ions per unit distance along a beam of initially monoenergetic ionizing particles, usually alpha particles, passing through a gas. Also known as Bragg ionization curve. **2.** A curve showing the average specific ionization of an ionizing particle of a particular kind as a function of its kinetic energy, velocity, or residual range.

Bragg diffraction *See* Bragg scattering.

Bragg ionization curve *See* Bragg curve.

Bragg-Kleeman rule *See* Bragg rule.

Bragg-Pierce law [PHYS] A relationship for determining an element's atomic absorption coefficient for x-rays when the atomic number of the element and the wavelength of the x-rays are known.

Bragg reflection *See* Bragg scattering.

Bragg rule [ATOM PHYS] An empirical rule according to which the mass stopping power of an element for alpha particles is inversely proportional to the square root of the atomic weight. Also known as Bragg-Kleeman rule.

Bragg scattering [SOLID STATE] Scattering of x-rays or neutrons by the regularly spaced atoms in a crystal, for which constructive interference occurs only at definite angles called Bragg angles. Also known as Bragg diffraction; Bragg reflection.

Bragg's equation *See* Bragg's law.

Bragg's law [SOLID STATE] A statement of the conditions under which a crystal will reflect a beam of x-rays with maximum intensity. Also known as Bragg's equation; Bravais' law.

Bragg spectrometer [ENG] An instrument for x-ray analysis of crystal structure and measuring wavelengths of x-rays and gamma rays, in which a homogeneous beam of x-rays is directed on the known face of a crystal and the reflected beam is detected in a suitably placed ionization chamber. Also

BRAGG SPECTROMETER

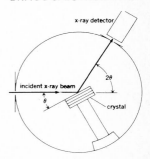

Schematic of Bragg spectrometer. θ is angle between incident beam and crystallographic planes; 2θ is angle between incident and diffracted beams.

known as crystal diffraction spectrometer; crystal spectrometer; ionization spectrometer.

branch [ELEC] A portion of a network consisting of one or more two-terminal elements in series. Also known as arm. [MATH] A complex function which is analytic in some domain and which takes on one of the values of a multiple-valued function in that domain. [NUC PHYS] A product resulting from one mode of decay of a radioactive nuclide that has two or more modes of decay. [SCI TECH] An area of study representing an independent offshoot of a related basic discipline.

branch-and-bound technique [MATH] A technique in nonlinear programming in which all sets of feasible solutions are divided into subsets, and those having bounds inferior to others are rejected.

branch circuit [ELEC] A portion of a wiring system in the interior of a structure that extends from a final overload protective device to a plug receptable or a load such as a lighting fixture, motor, or heater.

branch cut [MATH] A line or curve of singular points used in defining a branch of a multiple-valued complex function.

branch gain *See* branch transmittance.

branching [NUC PHYS] The occurrence of two or more modes by which a radionuclide can undergo radioactive decay. Also known as multiple decay; multiple disintegration.

branching diagram [MATH] In bifurcation theory, a graph in which a parameter characterizing solutions of a nonlinear equation is plotted against a parameter that appears in the equation itself.

branching fraction [NUC PHYS] That fraction of the total number of atoms involved which follows a particular branch of the disintegration scheme; usually expressed as a percentage.

branching process [STAT] A stochastic process in which the members of a population may have offspring and the lines of descent branch out as the new members are born.

branching ratio [NUC PHYS] The ratio of the number of parent atoms or particles decaying by one mode to the number decaying by another mode; the ratio of two specified branching fractions.

branching theory *See* bifurcation theory.

branch joint [ELEC] Joint used for connecting a branch conductor or cable, where the latter continues beyond the branch.

branch point [ELEC] A terminal in an electrical network that is common to more than two elements or parts of the network. Also known as junction point; node. [MATH] **1.** A point at which two or more sheets of a Riemann surface join together. **2.** In bifurcation theory, a value of a parameter in a nonlinear equation at which solutions branch off from the basic solution.

branch transmittance [CONT SYS] The amplification of current or voltage in a branch of an electrical network; used in the representation of such a network by a signal-flow graph. Also known as branch gain.

Braun tube *See* cathode-ray tube.

Bravais biplate *See* biplate.

Bravais indices [CRYSTAL] A modification of the Miller indices; frequently used for hexagonal and trigonal crystalline systems; they refer to four axes: the c axis and three others at $120°$ angles in the basal plane.

Bravais lattice [CRYSTAL] One of the 14 possible arrangements of lattice points in space such that the arrangement of points about any chosen point is identical with that about any other point.

Bravais' law *See* Bragg's law.

bra vector [QUANT MECH] A vector describing the state of a dynamic system in Hilbert space; the dual of a ket vector.

Brayton cycle [THERMO] A thermodynamic cycle consisting of two constant-pressure processes interspersed with two constant-entropy processes. Also known as complete-expansion diesel cycle; Joule cycle.

breakaway [FL MECH] Boundary-layer separation in which the boundary layer does not become reattached to the surface.

break-before-make contact [ELEC] One of a pair of contacts that interrupt one circuit before establishing another.

break contact [ELEC] The contact of a switching device which opens a circuit upon the operation of the device.

breakdown [ELEC] A large, usually abrupt rise in electric current in the presence of a small increase in voltage; can occur in a confined gas between two electrodes, a gas tube, the atmosphere (as lightning), an electrical insulator, and a reverse-biased semiconductor diode.

breakdown impedance [ELECTR] Of a semiconductor, the small-signal impedance at a specified direct current in the breakdown region.

breakdown law [STAT] The law that if the event E is broken down into the exclusive events $E_1, E_2, \ldots$ so that E is the event E_1 or E_2 or $\ldots$, then if F is any event, the probability of F is the sum of the products of the probabilities of E_i and the conditional probability of F given E_i.

breakdown potential *See* breakdown voltage.

breakdown voltage [ELEC] **1.** The voltage measured at a specified current in the electrical breakdown region of a semiconductor diode. Also known as Zener voltage. **2.** The voltage at which an electrical breakdown occurs in a dielectric. **3.** The voltage at which an electrical breakdown occurs in a gas. Also known as breakdown potential; sparking potential; sparking voltage.

break frequency [CONT SYS] The frequency at which a graph of the logarithm of the amplitude of the frequency response versus the logarithm of the frequency has an abrupt change in slope. Also known as corner frequency; knee frequency.

breaking-drop theory [GEOPHYS] A theory of thunderstorm charge separation based upon the suggested occurrence of the Lenard effect in thunderclouds, that is, the separation of electric charge due to the breakup of water drops.

breaking load [MECH] The stress which, when steadily applied to a structural member, is just sufficient to break or rupture it.

breaking strength [MECH] The ability of a material to resist breaking or rupture from a tension force.

breaking stress [MECH] The stress required to fracture a material whether by compression, tension, or shear.

breeder reactor [NUCLEO] A nuclear reactor that produces more fissionable material that it consumes.

breeding factor *See* breeding ratio.

breeding gain [NUCLEO] The excess of fissionable atoms produced per fissionable atom consumed in a breeder reactor.

breeding ratio [NUCLEO] The ratio of the number of fissionable atoms produced in a breeder reactor to the number of fissionable atoms consumed in the reactor; breeding gain is the breeding ratio minus 1. Also known as breeding factor.

Breit-Wigner formula [NUC PHYS] A formula which relates the cross section of a particular nuclear reaction with the energy of the incident particle, when the energy is near that required to form a discrete resonance level of the component nucleus.

Breit-Wigner theory [NUC PHYS] A theory of nuclear reactions from which the Breit-Wigner formula is derived. Also known as Bohr-Breit-Wigner theory.

bremsstrahlung [ELECTROMAG] Radiation that is emitted by an electron accelerated in its collision with the nucleus of an atom.

brewster [OPTICS] A unit of stress optical coefficient of a material; it is equal to the stress optical coefficient of a material in which a stress of 1 bar (10^5 newtons per square meter) produces a relative retardation between the components of a linearly polarized light beam of 1 angstrom (0.1 nanometer) when the light passes through a thickness of 1 millimeter in a direction perpendicular to the stress. Abbreviated B.

Brewster fringes [OPTICS] Interference fringes observed when white light is viewed through two plane parallel plates of nearly equal thickness.

Brewster point [OPTICS] A neutral point located 15 to 20° directly below the sun.

Brewster's angle [OPTICS] The angle of incidence of light reflected from a dielectric surface at which the reflectivity for light whose electrical vector is in the plane of incidence becomes zero; given by Brewster's law. Also known as polarizing angle.

Brewster's law [OPTICS] The law that the index of refraction for a material is equal to the tangent of the polarizing angle for the material.

Brewster stereoscope [OPTICS] A type of stereoscope that uses prisms to enable the eyes to form a fused image of two pictures whose separation is greater than the interocular distance.

Brewster window [OPTICS] A special glass window used at opposite ends of some gas lasers to transmit one polarization of the laser output beam without loss.

Brianchon's theorem [MATH] The theorem that if a hexagon circumscribes a conic section, the three lines joining three pairs of opposite vertices are concurrent (or are parallel).

bridge circuit [ELEC] An electrical network consisting basically of four impedances connected in series to form a rectangle, with one pair of diagonally opposite corners connected to an input device and the other pair to an output device.

bridge hybrid *See* hybrid junction.

bridge magnetic amplifier [ELECTR] A magnetic amplifier in which each of the gate windings is connected in series with an arm of a bridge rectifier; the rectifiers provide self-saturation and direct-current output.

bridging [MATH] The operation of carrying in addition or multiplication.

bridging amplifier [ELECTR] Amplifier with an input impedance sufficiently high so that its input may be bridged across a circuit without substantially affecting the signal level of the circuit across which it is bridged.

Bridgman anvil [PHYS] A device for producing high static pressures using two large massive opposed pistons bearing on a small thin sample confined by a gasket material.

Bridgman effect [SOLID STATE] The phenomenon that when an electric current passes through an anisotropic crystal, there is an absorption or liberation of heat due to the nonuniformity in current distribution.

Bridgman relation [SOLID STATE] $P = QT\sigma$ in a metal or semiconductor, where P is the Ettingshausen coefficient, Q

BRIDGMAN ANVIL

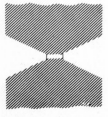

Configuration of Bridgman anvil, a basic type of static high-pressure equipment.

the Nernst-Ettingshausen coefficient, T the temperature, and σ the thermal conductivity in a transverse magnetic field.

brig [PHYS] A unit to express the ratio of two quantities, as a logarithm to the base 10; that is, a ratio of 10^x is equal to x brig; it is analogous to the bel, but the latter is restricted to power ratios. Also known as dex.

Briggs' logarithm *See* common logarithm.

bright-field [OPTICS] Having a brightly lighted background.

bright-line spectrum [SPECT] An emission spectrum made up of bright lines on a dark background.

brightness [OPTICS] **1.** The characteristic of light that gives a visual sensation of more or less light. **2.** *See* luminance.

brightness control [ELECTR] A control that varies the luminance of the fluorescent screen of a cathode-ray tube, for a given input signal, by changing the grid bias of the tube and hence the beam current. Also known as brilliance control; intensity control.

brightness level *See* adaptation luminance.

brightness temperature *See* blackbody temperature.

bril [OPTICS] A unit of subjective luminance; 100 brils is the luminance level that corresponds to a luminance of 1 millilambert, and a doubling of luminance level corresponds to an increase of 1 bril.

brilliance control *See* brightness control.

Brillouin function [SOLID STATE] A function of x with index (or parameter) n that appears in the quantum-mechanical theories of paramagnetism and ferromagnetism and is expressed as $[(2n+1)/2n]$ coth $[(2n+1)x/2n]$ $-$ $(1/2n)\cdot$ coth $(x/2n)$.

Brillouin scattering [SOLID STATE] Light scattering by acoustic phonons.

Brillouin zone [SOLID STATE] A fundamental region of wave vectors in the theory of the propagation of waves through a crystal lattice; any wave vector outside this region is equivalent to some vector inside it.

Brinkmann number [FL MECH] A dimensionless number used to study viscous flow.

British absolute system of units [PHYS] A measurement system based on the foot, the second, and the pound mass; force unit is the poundal. Also known as foot-pound-second system of units (fps system of units).

British engineering system of units *See* British gravitational system of units.

British gravitational system of units [PHYS] A measurement system based on the foot, the second, and the slug mass; 1 slug weighs 32.174 pounds at sea level and 45° latitude, and equals 14.594 kilograms. Also known as British engineering system of units; engineer's system of units.

British imperial pound [MECH] The British standard of mass, of which a standard is preserved by the government.

British thermal unit [THERMO] Abbreviated Btu. **1.** A unit of heat energy equal to the heat needed to raise the temperature of 1 pound of air-free water from 60° to 61°F at a constant pressure of 1 standard atmosphere; it is found experimentally to be equal to 1054.5 joules. Also known as sixty degrees Fahrenheit British thermal unit ($Btu_{60/61}$). **2.** A unit of heat energy equal to 1/180 of the heat needed to raise 1 pound of air-free water from 32°F (0°C) to 212°F (100°C) at a constant pressure of 1 of 1 standard atmosphere; it is found experimentally to be equal to 1055.79 joules. Also known as mean British thermal unit (Btu_{mean}). **3.** A unit of heat energy whose magnitude is such that 1 British thermal unit per pound equals 2326 joules per kilogram; it is equal to exactly

BRILLOUIN ZONE

Brillouin zone for the body-centered cubic lattice.

1055.05585262 joules. Also known as international table British thermal unit (Btu$_{IT}$).

brittleness [MECH] That property of a material manifested by fracture without appreciable prior plastic deformation.

brittle temperature [THERMO] The temperature point below which a material, especially metal, is brittle; that is, the critical normal stress for fracture is reached before the critical shear stress for plastic deformation.

broad-band amplifier [ELECTR] An amplifier having essentially flat response over a wide range of frequencies.

broad-band antenna [ELECTROMAG] An antenna that functions satisfactorily over a wide range of frequencies, such as for all 12 very-high-frequency television channels.

broad-band klystron [ELECTR] Klystron having three or more resonant cavities that are externally loaded and stagger-tuned to broaden the bandwidth.

broad beam [PHYS] In measurements of the attenuation of a beam of ionizing radiation, a beam in which much of the scattered radiation reaches the detector, along with the unscattered radiation.

broadening of spectral line [SPECT] A widening of spectral lines by collision or pressure broadening, or possibly by Doppler effect.

broadside [ELECTROMAG] Perpendicular to an axis or plane.

broadside array [ELECTROMAG] An antenna array whose direction of maximum radiation is perpendicular to the line or plane of the array.

broadside-on position [ELECTROMAG] The position of a point which lies on a line through the center of a magnet, perpendicular to the magnetic axis. Also known as Gauss B position.

Broca galvanometer [ELECTROMAG] A type of astatic galvanometer in which a current-carrying coil encloses consequent poles at the centers of two parallel magnets with opposite polarities.

Brocken bow *See* anticorona.

broken line [MATH] A line which is composed of a series of line segments lying end to end, and which does not form a continuous line.

bromine [CHEM] A chemical element, symbol Br, atomic number 35, atomic weight 79.904; used to make dibromide ethylene and in organic synthesis and plastics.

Bromwich contour [MATH] A path of integration in the complex plane running from $c - i\infty$ to $c + i\infty$, where c is a real, positive number chosen so that the path lies to the right of all singularities of the analytic function under consideration.

Brooks variable inductometer [ELEC] An inductometer providing a nearly linear scale and consisting of two movable coils, side by side in a plane, sandwiched between two pairs of fixed coils.

broomy flow [FL MECH] A swirling flow of a fluid in a pipe after passing through a constricted section or after a sudden change of direction.

Brouncker's continued fraction [MATH] An expression for π having the form

$$\frac{4}{\pi} = 1 + \frac{1}{2+}\ \frac{9}{2+}\ \frac{25}{2+}\ \frac{49}{2+}\cdots$$

Brouwer's theorem [MATH] A fixed-point theorem which states that for any continuous mapping f of the solid n-sphere into itself there is a point x such that $f(x) = x$.

Brownian movement [STAT MECH] Random movements of

BROMINE

35
Br

Periodic table of the chemical elements showing the position of bromine.

BROOKS VARIABLE INDUCTOMETER

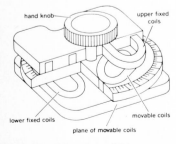

Drawing of Brooks variable inductometer.

small particles suspended in a fluid, caused by the statistical pressure fluctuations over the particle.

Brunauer-Emmett-Teller equation [PHYS CHEM] An extension of the Langmuir isotherm equation in the study of sorption; used for surface area determinations by computing the monolayer area. Abbreviated BET equation.

brush discharge [ELEC] A luminous electric discharge that starts from a conductor when its potential exceeds a certain value but remains too low for the formation of an actual spark.

brush-shifting motor [ENG] A category of alternating-current motor in which the brush contacts shift to modify operating speed and power factor.

brute-force filter [ELEC] Type of powerpack filter depending on large values of capacitance and inductance to smooth out pulsations rather than on resonant effects of tuned filters.

BSR *See* bulk shielding reactor.

Btu *See* British thermal unit.

B-type star [ASTRON] A type in a classification based on stellar spectral characteristics; has strong He I absorption.

bu *See* bushel.

bubble [PHYS] **1.** A small, approximately spherical body of fluid within another fluid or solid. **2.** A thin, approximately spherical film of liquid inflated with air or other gas. [SOLID STATE] *See* magnetic bubble.

bubble cavitation [FL MECH] **1.** Formation of vapor- or gas-filled cavities in liquids by mechanical forces. **2.** The formation of vapor-filled cavities in the interior of liquids in motion when the pressure is reduced without change in ambient temperature.

bubble chamber [NUCLEO] A chamber in which the movements and interactions of charged particles can be observed as visible tracks in a superheated liquid, the tracks being gas bubbles that form along the paths of the moving particles.

bubble point [PHYS CHEM] In a solution of two or more components, the temperature at which the first bubbles of gas appear. Also known as boiling point.

bubble raft [SOLID STATE] A visual demonstration for the structure of dislocations in metal lattices, showing slip propagation; it consists of many identical bubbles floating on a liquid surface in something like a crystalline array.

bucket brigade device [ELECTR] A semiconductor device in which majority carriers store charges that represent information, and minority carriers transfer charges from point to point in sequence. Abbreviated BBD.

bucking coil [ELECTROMAG] A coil connected and positioned in such a way that its magnetic field opposes the magnetic field of another coil; for example, the hum-bucking coil of an excited-field loudspeaker.

Buckingham's π theorem [PHYS] The theorem that if there are n physical quantities, $x_1, x_2, \ldots, x_n$, which can be expressed in terms of m fundamental quantities and if there exists one and only one mathematical expression connecting them which remains formally true no matter how the units of the fundamental quantities are changed, namely $\phi(x_1, x_2, \ldots, x_n) = 0$, then the relation ϕ can be expressed by a relation of the form $F(\pi_1, \pi_2, \ldots, \pi_{n-m}) = 0$, where the πs are $n - m$ independent dimensionless products of $x_1, x_2, \ldots, x_n$. Also known as pi theorem.

bucking transformer [ELEC] A transformer whose voltage opposes that of a second transformer.

bucking voltage [ELEC] A voltage having a polarity opposite to that of another voltage against which it acts.

buckling [MECH] Bending of a sheet, plate, or column supporting a compressive load. [NUCLEO] The size-shape factor that appears in the general nuclear reactor equation and is a measure of the curvature of the neutron density distribution in the reactor.

buckling stress [MECH] Force exerted by the crippling load.

Budan's theorem [MATH] The theorem that the number of roots of an nth-degree polynomial lying in an open interval equals the difference in the number of sign changes induced by n differentiations at the two ends of the interval.

buffer [ELEC] An electric circuit or component that prevents undesirable electrical interaction between two circuits or components. [ELECTR] *See* buffer amplifier.

buffer amplifier [ELECTR] An amplifier used after an oscillator or other critical stage to isolate it from the effects of load impedance variations in subsequent stages. Also known as buffer; buffer stage.

buffer stage *See* buffer amplifier.

Buffon's problem [STAT] The problem of calculating the probability that a needle of specified length, dropped at random on a plane ruled with a series of straight lines a specified distance apart, will interesect one of the lines.

bulk acoustic wave [ACOUS] An acoustic wave that travels through a piezoelectric material, as in a quartz delay line. Also known as volume acoustic wave.

bulk diode [ELECTR] A semiconductor microwave diode that uses the bulk effect, such as Gunn diodes and diodes operating in limited space-charge-accumulation modes.

bulk effect [ELECTR] An effect that occurs within the entire bulk of a semiconductor material rather than in a localized region or junction.

bulk-effect device [ELECTR] A semiconductor device that depends on a bulk effect, as in Gunn and avalanche devices.

bulk flow *See* convection.

bulk lifetime [SOLID STATE] The average time that elapses between the formation and recombination of minority charge carriers in the bulk material of a semiconductor.

bulk modulus *See* bulk modulus of elasticity.

bulk modulus of elasticity [MECH] The ratio of the compressive or tensile force applied to a substance per unit surface area to the change in volume of the substance per unit volume. Also known as bulk modulus; compression modulus; hydrostatic modulus; modulus of compression; modulus of volume elasticity.

bulk photoconductor [ELECTR] A photoconductor having high power-handling capability and other unique properties that depend on the semiconductor and doping materials used.

bulk resistor [ELECTR] An integrated-circuit resistor in which the n-type epitaxial layer of a semiconducting substrate is used as a noncritical high-value resistor; the spacing between the attached terminals and the sheet resistivity of the material together determine the resistance value.

bulk shielding reactor [NUCLEO] The prototype swimming-pool reactor located at Oak Ridge, Tennessee; it uses heterogeneous enriched fuel and provides a combination of high thermal-neutron flux, ready accessibility, and versatility. Abbreviated BSR.

bulk strength [MECH] The strength per unit volume of a solid.

Bulygen number [THERMO] A dimensionless number used in the study of heat transfer during evaporation.

bunched pair [ELEC] Group of pairs tied together or otherwise associated for identification.

buncher *See* buncher resonator.

buncher resonator [ELECTR] The first or input cavity resonator in a velocity-modulated tube, next to the cathode; here the faster electrons catch up with the slower ones to produce bunches of electrons. Also known as buncher; input resonator.

bunching [ELECTR] The flow of electrons from cathode to anode of a velocity-modulated tube as a succession of electron groups rather than as a continuous stream.

bunching voltage [ELECTR] Radio-frequency voltage between the grids of the buncher resonator in a velocity-modulated tube such as a klystron; generally, the term implies the peak value of this oscillating voltage.

bunch-map analysis [STAT] A graphic technique in confluence analysis; all subsets of regression coefficients in a complete set are drawn on standard diagrams, and the representation of any set of regression coefficients produces a "bunch" of lines; allows the observer to determine the effect of introducing a new variate on a set of variates.

bundle [MATH] A triple (E,p,B), where E and B are topological spaces and p is a continuous map of E onto B; intuitively E is the collection of inverse images under p of points from B glued together by the topology of X.

bundle of planes [MATH] The set of all planes which pass through a given point.

Buniakowski's inequality *See* Cauchy-Schwarz inequality.

Bunn chart [CRYSTAL] A chart for classifying x-ray diffraction powder photographs of substances whose crystals have tetragonal or hexagonal symmetry.

Bunsen disk [OPTICS] The screen generally used in a grease-spot photometer, with a circular translucent spot at the center.

Bunsen-Kirchhoff law [SPECT] The law that every element has a characteristic emission spectrum of bright lines and absorption spectrum of dark lines.

buoyancy [FL MECH] The resultant vertical force exerted on a body by a static fluid in which it is submerged or floating.

buoyancy parameter [FL MECH] The Grashof number divided by the square of the Reynolds number.

buoyancy-type density transmitter [ENG] An instrument which records the specific gravity of a flowing stream of a liquid or gas, using the principle of hydrostatic weighing.

buoyant density [PHYS] A technique that uses the sedimentation equilibrium in a density gradient to characterize a solute.

buoyant force [FL MECH] The force exerted vertically upward by a fluid on a body wholly or partly immersed in it; its magnitude is equal to the weight of the fluid displaced by the body.

Burali-Forti paradox [MATH] The order-type of the set of all ordinals is the largest ordinal, but that ordinal plus one is larger.

burble [FL MECH] **1.** A separation or breakdown of the laminar flow past a body. **2.** The eddying or turbulent flow resulting from this occurrence.

burble angle *See* burble point.

burble point [FL MECH] A point reached in an increasing angle of attack at which burble begins. Also known as burble angle.

Burgers vector [CRYSTAL] A translation vector of a crystal lattice representing the displacement of the material to create a dislocation.

burnable poison [NUCLEO] A neutron absorber that is incorporated in the fuel or fuel cladding of a nuclear reactor and gradually burns up under neutron irradiation. Also known as burnout poison.

BUOYANCY-TYPE DENSITY TRANSMITTER

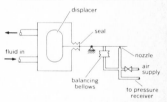

Buoyancy-type density transmitter, used to obtain a continuous record of liquid density. *(From D. M. Considine, ed., Process Instruments and Controls Handbook, McGraw-Hill, 1957)*

BURGERS VECTOR

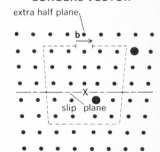

An arrangement of atoms around an edge dislocation. The dashed lines represent Burgers closure circuits which define Burgers vector **b**. Dislocation is perpendicular to the paper at ×. The large dots represent impurity atoms which may be present.

burning glass [OPTICS] A converging lens used to produce intense heat by converging the rays of the sun on a small area.

burnout [ELEC] Failure of a device due to excessive heat produced by excessive current. [NUCLEO] **1.** To receive the greatest amount of radiation permissible during a given time. **2.** The point at which the heat flux across a surface causes film-blanketing of the surface, resulting in a drop in the film heat-transfer coefficient, overheating, and possible surface failure.

burnout poison *See* burnable poison.

burnup [NUCLEO] A measure of nuclear-reactor fuel consumption, expressed either as the percentage of fuel atoms that have undergone fission or as the amount of energy produced per unit weight of fuel.

burst [ELECTR] An exceptionally large electric pulse in the circuit of an ionization chamber due to the simultaneous arrival of several ionizing particles.

bursting strength [MECH] A measure of the ability of a material to withstand pressure without rupture; it is the hydraulic pressure required to burst a vessel of given thickness.

burst pressure [MECH] The maximum inside pressure that a process vessel can safely withstand.

burst slug detector [NUCLEO] A radiation detector used for detecting small leaks in a fuel element of a nuclear reactor by measuring the radiation from short-lived fission products that escape into the coolant.

burst wave [FL MECH] Wave of compressed air caused by a bursting projectile or bomb; a detonation wave; it may produce extensive local damage.

bus [ELEC] A set of two or more electric conductors that serve as common connections between load circuits and each of the polarities (in direct-current systems) or phases (in alternating-current systems) of the source of electric power.

bushel [MECH] Abbreviated bu. **1.** A unit of volume (dry measure) used in the United States, equal to 2150.42 cubic inches or approximately 35.239 liters. **2.** A unit of volume (liquid and dry measure) used in Britain, equal to 2219.36 cubic inches or 8 imperial gallons (approximately 36.369 liters).

busway [ELEC] A prefabricated assembly of standard lengths of busbars rigidly supported by solid insulation and enclosed in a sheet-metal housing.

Butler oscillator [ELEC] Oscillator in which a piezoelectric crystal is connected between the cathode of two tubes, one functioning as a cathode follower, and the other as a grounded-grid amplifier.

butt contact [ELEC] A hemispherically shaped contact designed to mate against a similarly shaped contact.

butterfly capacitor [ELEC] A variable capacitor having stator and rotor plates shaped like butterfly wings, with the stator plates having an outer ring to provide an inductance so that both capacitance and inductance may be varied, thereby giving a wide tuning range.

butterfly lemma [MATH] If U and V are subgroups of a group, and u and v are normal subgroups of U and V, then $u(U \cap v)$ is normal in $u(U \cap V)$, the group $(u \cap V)v$ is normal in $(U \cap V)v$, and the quotient group $u(U \cap V)/u(U \cap v)$ is isomorphic to $(U \cap V)v/(u \cap V)v$; here $\cap$ denotes intersection.

Butterworth filter [ELECTR] An electric filter whose passband (graph of transmission versus frequency) has a maximally flat shape.

butt joint [ELEC] A connection formed by placing the ends of two conductors together and joining them by welding, braz-

BUTTERWORTH FILTER

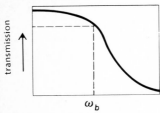

Plot of the passband of the Butterworth filter; ω_b is the point where the transmission function falls below the passband tolerance.

ing, or soldering. [ELECTROMAG] A connection giving physical contact between the ends of two waveguides to maintain electrical continuity.

buttonhook contact [ELEC] A curved, hooklike contact often used on feed-through terminals of headers to facilitate soldering or unsoldering of leads.

buzz [CONT SYS] *See* dither. [ELECTR] The condition of a combinatorial circuit with feedback that has undergone a transition, caused by the inputs, from an unstable state to a new state that is also unstable. [FL MECH] In supersonic diffuser aerodynamics, a nonsteady shock motion and airflow associated with the shock system ahead of the inlet.

(b,v,r,k,λ)-design *See* balanced incomplete block design.

BWO *See* backward-wave oscillator.

BWR *See* boiling-water reactor.

bypass [ELEC] A shunt path around some element or elements of a circuit.

byte [ADP] A sequence of adjacent binary digits operated upon as a unit in a computer and usually shorter than a word.

c *See* calorie; centi-; charmed quark; curie.

C *See* capacitance; capacitor; carbon; coulomb.

Ca *See* calcium.

Ca⁴⁵ *See* calcium-45.

Cabibbo theory [PARTIC PHYS] A theory describing baryon beta-decay processes, according to which the amplitude for such processes is given by $G \{\cos \Theta \, [V(\Delta s = 0) + A \, (\Delta s = 0)] + \sin \Theta \, [V(\Delta s = +1) + A \, (\Delta s = +1)]\}$, where Θ is the Cabibbo angle, Δs is the change in strangeness for the baryon, G is a universal beta-decay amplitude, and V and A are vector and axial vector amplitudes, respectively; it is experimentally determined that $\sin \Theta \approx 0.25$, so that $\cos \Theta = 0.97$.

Caccioppoli-Banach principle *See* Banach's fixed-point theorem.

cadmium [CHEM] A chemical element, symbol Cd, atomic number 48, atomic weight 112.40.

cadmium cell [ELEC] A standard cell used as a voltage reference; at 20°C its voltage is 1.0186 volts.

cadmium cutoff [NUCLEO] The neutron energy, approximately 0.3 electron volt, below which cadmium has a high neutron absorption cross section but above which this cross section falls off sharply.

cadmium difference [NUCLEO] The difference between the response of an uncovered neutron detector to a neutron beam and the response of the same detector under identical conditions when it is covered with a thin layer of cadmium.

cadmium neutron [NUCLEO] A neutron whose energy lies below the cadmium cutoff.

cadmium lamp [ELEC] A lamp containing cadmium vapor; wavelength (6438.4696 international angstroms, or 643.84696 nanometers) of light emitted is a standard of length.

cadmium-nickel storage cell *See* nickel-cadmium battery.

cadmium ratio [NUCLEO] The ratio of the response of an uncovered neutron detector to that of the same detector under identical conditions when it is covered with cadmium of a specified thickness.

cadmium silver oxide cell [ELEC] An alkaline-electrolyte cell that may be used without recharging in primary batteries or that may be recharged for secondary-battery use.

cadmium sulfide cell [ELECTR] A photoconductive cell in which a small wafer of cadmium sulfide provides an extremely high dark-light resistance ratio.

cadmium telluride detector [ELECTR] A photoconductive cell capable of operating continuously at ambient temperatures up to 750°F (400°C); used in solar cells and infrared, nuclear-radiation, and gamma-ray detectors.

cage antenna [ELECTROMAG] Broad-band dipole antenna in which each pole consists of a cage of wires whose overall shape resembles that of a cylinder or a cone.

cal *See* calorie.

CADMIUM

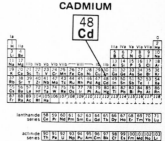

Periodic table of the chemical elements showing the position of cadmium.

CALCIUM

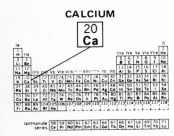

Periodic table of the chemical elements showing the position of calcium.

Cal *See* kilocalorie.

Calabi conjecture [MATH] If the volume of a certain type of surface, defined in a higher dimensional space in terms of complex numbers, is known, then a particular kind of metric can be defined on it; the conjecture was subsequently proved to be correct.

calcium [CHEM] A chemical element, symbol Ca, atomic number 20, atomic weight 40.08; used in metallurgy as an alloying agent for aluminum-bearing metal, as an aid in removing bismuth from lead, and as a deoxidizer in steel manufacture, and also used as a cathode coating in some types of photo tubes.

calcium-45 [NUC PHYS] A radioisotope of calcium having a mass number of 45, often used as a radioactive tracer in studying calcium metabolism in humans and other organisms; half-life is 165 days. Designated Ca^{45}.

calculating machine *See* calculator.

calculator [ADP] A device that performs logic and arithmetic digital operations based on numerical data which are entered by pressing numerical and control keys. Also known as calculating machine.

calculus [MATH] The branch of mathematics dealing with differentiation and integration and related topics.

calculus of enlargement *See* calculus of finite differences.

calculus of finite differences [MATH] A method of interpolation that makes use of formal relations between difference operators which are, in turn, defined in terms of the values of a function on a set of equally spaced points. Also known as calculus of enlargement.

calculus of residues [MATH] The application of the Cauchy residue theorem and related theorems to compute the residues of a meromorphic function at simple poles, evaluate contour integrals, expand meromorphic functions in series, and carry out related calculations.

calculus of tensors [MATH] The branch of mathematics dealing with the differentiation of tensors.

calculus of variations [MATH] The study of problems concerning maximizing or minimizing a given definite integral relative to the dependent variables of the integrand function.

calculus of vectors [MATH] That branch of calculus concerned with differentiation and integration of vector-valued functions.

C* algebra [MATH] An involutive Banach algebra satisfying $\|x\|^2 = \|x^*x\|$ for each element x in the algebra. Also known as B* algebra; completely regular algebra.

californium [CHEM] A chemical element, symbol Cf, atomic number 98; all isotopes are radioactive.

Callendar and Barnes' continuous flow calorimeter [ENG] A calorimeter in which the heat to be measured is absorbed by water flowing through a tube at a constant rate, and the quantity of heat is determined by the rate of flow and the temperature difference between water at ends of the tube.

Callendar's compensated air thermometer [ENG] A type of constant-pressure gas thermometer in which errors resulting from temperature differences between the thermometer bulb and the connecting tubes and manometer used to maintain constant pressure are eliminated by the configuration of the connecting tubes.

Callendar's equation [THERMO] 1. An equation of state for steam whose temperature is well above the boiling point at the existing pressure, but less than the critical temperature: $(V - b) = (RT/p) - (a/T^n)$, where V is the volume, R is the gas constant, T is the temperature, p is the pressure, n equals 10/3, and a and b are constants. 2. A very accurate equation

CALIFORNIUM

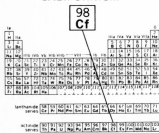

Periodic table of the chemical elements showing the position of californium.

relating temperature and resistance of platinum, according to which the temperature is the sum of a linear function of the resistance of platinum and a small correction term, which is a quadratic function of temperature.

Callendar's thermometer *See* platinum resistance thermometer.

calomel electrode [PHYS CHEM] A reference electrode of known potential consisting of mercury, mercury chloride (calomel), and potassium chloride solution; used to measure pH and electromotive force. Also known as calomel half-cell; calomel reference electrode.

calomel half-cell *See* calomel electrode.

calomel reference electrode *See* calomel electrode.

calorescence [PHYS] The production of visible light by infrared radiation; the transformation is indirect, the light being produced by heat and not by any direct change of wavelength.

calorie [THERMO] Abbreviated cal; often designated c. **1.** A unit of heat energy, equal to 4.1868 joules. Also known as International Table calorie (IT calorie). **2.** A unit of energy, equal to the heat required to raise the temperature of 1 gram of water from 14.5° to 15.5°C at a constant pressure of 1 standard atmosphere (101,325 newtons per square meter); equal to 4.1855 ± 0.0005 joules. Also known as fifteen-degrees calorie; gram-calorie (g-cal); small calorie. **3.** A unit of heat energy equal to 4.184 joules; used in thermochemistry. Also known as thermochemical calorie.

calorific value [ENG] Quantity of heat liberated on the complete combustion of a unit weight or unit volume of fuel.

calorimetry [ENG] The measurement of the quantity of heat involved in various processes, such as chemical reactions, changes of state, and formations of solutions, or in the determination of the heat capacities of substances; fundamental unit of measurement is the joule or the calorie (4.184 joules).

Calutron [NUCLEO] An electromagnetic apparatus for separating isotopes of uranium and other elements according to their masses, using the principle of the mass spectrograph.

Campbell bridge [ELEC] **1.** A bridge designed for comparison of mutual inductances. **2.** A circuit for measuring frequencies by adjusting a mutual inductance, until the current across a detector is zero.

Campbell's formula [ELECTROMAG] A formula which relates the propagation constant of a loaded transmission line to the propagation constant and characteristic impedance of an unloaded line and the impedance of each loading coil.

Campbell-Stokes recorder [ENG] A sunshine recorder in which the time scale is supplied by the motion of the sun and which has a spherical lens that burns an image of the sun upon a specially prepared card.

Camp-Meidell condition [STAT] For determining the distribution of a set of numbers, the guideline stating that if the distribution has only one mode, if the mode is the same as the arithmetic mean, and if the frequencies decline continuously on both sides of the mode, then more than $1 - (1/2.25t^2)$ of any distribution will fall within the closed range $\bar{X} \pm t\sigma$, where t = number of items in a set, $\bar{X}$ = average, and σ = standard deviation.

can *See* jacket.

Canada balsam [MATER] A transparent balsam useful for cementing together lenses and other optical elements because its index of refraction is in the same range as that of glass.

canal ray [ATOM PHYS] The name given in early gaseous discharge experiments to the particles passing through a hole

CALOMEL ELECTRODE

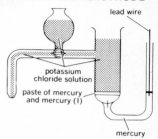

Diagram of calomel electrode.

CAMPBELL BRIDGE

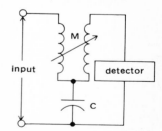

Circuit diagram of Campbell bridge (def. 2).

or canal in the cathode; the ray comprises positive ions of the gas being used in the discharge.

cancellation law [MATH] A rule which allows formal division by common factors in equal products, even in systems which have no division, as integral domains; $ab = ac$ implies $b = c$.

candela [OPTICS] A unit of luminous intensity, defined as $\frac{1}{60}$ of the luminous intensity per square centimeter of a blackbody radiator operating at the temperature of freezing platinum. Formerly known as candle. Also known as new candle.

candle *See* candela.

candlepower [OPTICS] Luminous intensity expressed in candelas. Abbreviated cp.

canning [NUCLEO] Placing a jacket around a slug of uranium before inserting the slug in a nuclear reactor.

canonical [SCI TECH] Relating to the simplest or most significant form of a general function, equation, statement, rule, or expression.

canonical coordinates [MATH] Any set of generalized coordinates of a system together with their conjugate momenta.

canonical correlation [STAT] The maximum correlation between linear functions of two sets of random variables when specific restrictions are imposed upon the coefficients of the linear functions of the two sets.

canonical distribution [STAT MECH] The density of members of the canonical ensemble in phase space.

canonical ensemble [STAT MECH] A hypothetical collection of systems of particles used to describe an actual individual system which is in thermal contact with a heat reservoir but is not allowed to exchange particles with its environment.

canonical equations of motion *See* Hamilton's equations of motion.

canonically conjugate variable [MECH] A generalized coordinate and its conjugate momentum.

canonical matrix [MATH] A member of an equivalence class of matrices that has a particularly simple form, where the equivalence classes are determined by one of the relations defining equivalent, similar, or congruent matrices.

canonical momentum *See* conjugate momentum.

canonical time unit [ASTRON] For geocentric orbits, the time required by a hypothetical satellite to move one radian in a circular orbit of the earth's equatorial radius, that is, 13.447052 minutes.

canonical transformation [MATH] Any function which has a standard form, depending on the context. [MECH] A transformation which occurs among the coordinates and momenta describing the state of a classical dynamical system and which leaves the form of Hamilton's equations of motion unchanged. Also known as contact transformation.

cantilever vibration [MECH] Transverse oscillatory motion of a body fixed at one end.

canting [MECH] Displacing the free end of a beam which is fixed at one end by subjecting it to a sideways force which is just short of that required to cause fracture.

Cantor diagonal process [MATH] A technique of proving statements about infinite sequences, each of whose terms is an infinite sequence by operation on the nth term of the nth sequence for each n; used to prove the uncountability of the real numbers.

Cantor function [MATH] A real-valued nondecreasing continuous function defined on the closed interval [0,1] which maps the Cantor ternary set onto the interval [0,1].

Cantor's axiom [MATH] The postulate that there exists a one-to-one correspondence between the points of a line extending indefinitely in both directions and the set of real numbers.

Cantor ternary set [MATH] A perfect, uncountable, totally disconnected subset of the real numbers having Lebesgue measure zero; it consists of all numbers between 0 and 1 (inclusive) with ternary representations containing no ones.

Cantor theorem [MATH] A theorem that there is no one-to-one correspondence between a set and the collection of its subsets.

capacitance [ELEC] The ratio of the charge on one of the conductors of a capacitor (there being an equal and opposite charge on the other conductor) to the potential difference between the conductors. Symbolized C. Formerly known as capacity.

capacitance box [ELEC] An assembly of capacitors and switches which permits adjustment of the capacitance existing at the terminals in nominally uniform steps, from a minimum value near zero to the maximum which exists when all the capacitors are connected in parallel.

capacitance bridge [ELEC] A bridge for comparing two capacitances, such as a Schering bridge.

capacitance standard *See* standard capacitor.

capacitive coupling [ELEC] Use of a capacitor to transfer energy from one circuit to another.

capacitive divider [ELEC] Two or more capacitors placed in series across a source, making available a portion of the source voltage across each capacitor; the voltage across each capacitor will be inversely proportional to its capacitance.

capacitive electrometer [ENG] An instrument for measuring small voltages; the voltage is applied to the plates of a capacitor when they are close together, then the voltage source is removed and the plates are separated, increasing the potential difference between them to a measurable value. Also known as condensing electrometer.

capacitive load [ELECTROMAG] A load in which the capacitive reactance exceeds the inductive reactance; the load draws a leading current.

capacitive post [ELECTROMAG] Metal post or screw extending across a waveguide at right angles to the E field, to provide capacitive susceptance in parallel with the waveguide for tuning or matching purposes.

capacitive reactance [ELECTROMAG] Reactance due to the capacitance of a capacitor or circuit, equal to the inverse of the product of the capacitance and the angular frequency.

capacitive window [ELECTROMAG] Conducting diaphragm extending into a waveguide from one or both sidewalls, producing the effect of a capacitive susceptance in parallel with the waveguide.

capacitor [ELEC] A device which consists essentially of two conductors (such as parallel metal plates) insulated from each other by a dielectric and which introduces capacitance into a circuit, stores electrical energy, blocks the flow of direct current, and permits the flow of alternating current to a degree dependent on the capacitor's capacitance and the current frequency. Symbolized C. Also known as condenser; electrical condenser.

capacitor antenna [ELECTROMAG] Antenna consisting of two conductors or systems of conductors, the essential characteristic of which is its capacitance. Also known as condenser antenna.

capacitor bank [ELEC] A number of capacitors connected in series or in parallel.

capacitor box [ELECTR] A box-shaped structure in which a

CAPACITOR MOTOR

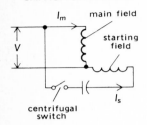

Winding connections of the starting winding connected to the supply through a capacitor. I_m = main winding; I_s = starting winding; V = common voltage.

CAPILLARY TUBE

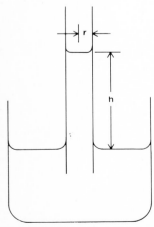

Rise of liquid to a height h in a capillary tube whose radius is r.

capacitor is submerged in a heat-absorbing medium, usually water. Also known as condenser box.

capacitor color code [ELEC] A method of marking the value on a capacitor by means of dots or bands of colors as specified in the Electronic Industry Association color code.

capacitor motor [ELEC] A single-phase induction motor having a main winding connected directly to a source of alternating-current power and an auxiliary winding connected in series with a capacitor to the source of ac power.

capacity [ELEC] *See* capacitance. [SCI TECH] Volume, especially in reference to merchandise or containers thereof.

capacity cell [ELEC] **1.** Capacitance-type device used to measure the dielectric constants of gases, liquids, or solids. **2.** Capacitance-type device used to monitor certain composition changes in flowing streams.

capacity correction [ENG] The correction applied to a mercury barometer with a nonadjustable cistern in order to compensate for the change in the level of the cistern as the atmospheric pressure changes.

cape foot [MECH] A unit of length equal to 1.033 feet or to 0.3148584 meter.

capillarity [FL MECH] The action by which the surface of a liquid where it contacts a solid is elevated or depressed, because of the relative attraction of the molecules of the liquid for each other and for those of the solid.

capillarity correction [ENG] As applied to a mercury barometer, that part of the instrument correction which is required by the shape of the meniscus of the mercury.

capillary attraction [FL MECH] The force of adhesion existing between a solid and a liquid in capillarity.

capillary condensation [PHYS CHEM] Condensation of an adsorbed vapor within the pores of the adsorbate.

capillary curve [FL MECH] The curve along which the surface of a liquid intersects a vertical plane perpendicular to a vertical glass plane surface.

capillary depression [FL MECH] The depression of the meniscus of a liquid contained in a tube where the liquid does not wet the walls of the container, as in a mercury barometer; the meniscus has a convex shape, resulting in a depression.

capillary electrometer [ENG] An electrometer designed to measure a small potential difference between mercury and an electrolytic solution in a capillary tube by measuring the effect of this potential difference on the surface tension between the liquids. Also known as Lippmann electrometer.

capillary ripple *See* capillary wave.

capillary rise [FL MECH] The rise of a liquid in a capillary tube times the radius of the tube.

capillary tube [ENG] A tube sufficiently fine so that capillary attraction of a liquid into the tube is significant.

capillary viscometer [ENG] A long, narrow tube used to measure the laminar flow of fluids.

capillary wave [FL MECH] **1.** A wave occurring at the interface between two fluids, such as the interface between air and water on oceans and lakes, in which the principal restoring force is controlled by surface tension. **2.** A water wave of less than 1.7 centimeters. Also known as capillary ripple; ripple.

capture [ASTROPHYS] Of a central force field, as of a planet, to overcome by gravitational force the velocity of a passing body and bring the body under the control of the central force field, in some cases absorbing its mass. [PHYS] A process in which an atomic or nuclear system acquires an additional particle; for example, the capture of electrons by positive ions, or capture of neutrons by nuclei.

capture cross section [NUC PHYS] The cross section that is effective for radiative capture.

capture gamma rays [NUC PHYS] The gamma rays emitted in radiative capture.

Carathéodory outer measure [MATH] A positive, countably subadditive set function defined on the class of all subsets of a given set; used for defining measures.

Carathéodory's principle [THERMO] An expression of the second law of thermodynamics which says that in the neighborhood of any equilibrium state of a system, there are states which are not accessible by a reversible or irreversible adiabatic process. Also known as principle of inaccessibility.

carbide nuclear fuel [NUCLEO] A nuclear reactor fuel mixed with carbon compounds and a metal to give structural strength and oxidation resistance.

carbon [CHEM] A nonmetallic chemical element, symbol C, atomic number 6, atomic weight 12.01115; occurs freely as diamond, graphite, and coal.

carbon-12 [NUC PHYS] A stable isotope of carbon with mass number of 12, forming about 98.9% of natural carbon; used as the basis of the newer scale of atomic masses, having an atomic mass of exactly 12 u (relative nuclidic mass unit) by definition.

carbon-13 [NUC PHYS] A heavy isotope of carbon having a mass number of 13.

carbon-14 [NUC PHYS] A naturally occurring radioisotope of carbon having a mass number of 14 and half-life of 5780 years; used in radiocarbon dating and in the elucidation of the metabolic path of carbon in photosynthesis. Also known as radiocarbon.

carbon cycle *See* carbon-nitrogen cycle.

carbon-14 dating [NUCLEO] Determining the approximate age of organic material associated with archeological or fossil artifacts by measuring the rate of radiation of the carbon-14 isotope. Also known as radioactive carbon dating; radiocarbon dating.

carbon dioxide gas laser [PHYS] A powerful, continuously operating laser in the infrared that can emit several hundred watts of power at a wavelength of 10.6 micrometers.

carbon isotope ratio [GEOL] Ratio of carbon-12 to either of the less common isotopes, carbon-13 or carbon-14, or the reciprocal of one of these ratios; if not specified, the ratio refers to C^{12}/C^{13}. Also known as carbon ratio.

carbon lamp [ELEC] An arc lamp with carbon electrodes.

carbon monoxide laser [OPTICS] A molecular gas laser in which the active laser molecule is carbon monoxide, and the strongest wavelengths are 4.9 to 5.7 micrometers. Also known as CO laser.

carbon-nitrogen cycle [NUC PHYS] A series of thermonuclear reactions, with release of energy, which presumably occurs in the sun and other stars; the net accomplishment is the synthesis of four hydrogen atoms into a helium atom, the emission of two positrons and much energy, and restoration of a carbon-12 atom with which the cycle began. Also known as carbon cycle; nitrogen cycle.

carbon pile [ELEC] A variable resistor consisting of a stack of carbon disks mounted between a fixed metal plate and a movable one that serve as the terminals of the resistor; the resistance value is reduced by applying pressure to the movable plate.

carbon ratio *See* carbon isotope ratio.

carbon resistance thermometer [ENG] A highly sensitive resistance thermometer for measuring temperatures in the

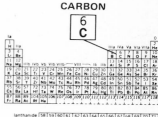

CARBON

Periodic table of the chemical elements showing the position of carbon.

CARDIOID

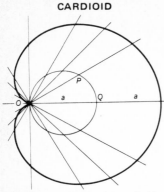

Point-construction of a cardioid. The fixed circle is OPQ which has a diameter a. The cardioid is constructed by laying off, along every secant OP passing through the fixed point O, the distance a in both directions from P.

CARDIOID CONDENSER

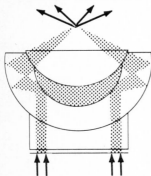

Cardioid condenser, a dark-field device. Shaded meniscus area is air space; unshaded areas are portions of condenser through which passes the light, indicated by arrows and shaded paths. (*Photographic Service Department, Kodak Research Laboratory*)

CARNOT CYCLE

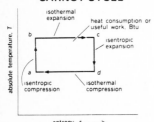

Carnot cycle, comprising the four processes b–c, c–d, d–a, and a–b.

range 0.05–20 K; capable of measuring temperature changes of the order 10^{-5} degree.

carbon star [ASTRON] Any of a class of stars with an apparently high abundance ratio of carbon to hydrogen; a majority of these are low-temperature red giants of the C class.

carcinotron *See* backward-wave oscillator.

cardinal measurement *See* interval measurement.

cardinal number [MATH] The number of members of a set; usually taken as a particular well-ordered set representative of the class of all sets which are in one-to-one correspondence with one another.

cardinal point [OPTICS] Any one of six points in an optical system, namely, the two principal points, two nodal points, and two focal points. Also known as Gauss point.

cardioid [MATH] A heart-shaped curve generated by a point of a circle that rolls without slipping on a fixed circle of the same diameter.

cardioid condenser [OPTICS] A substage condenser that cuts off the direct light and allows only the light diffracted or dispersed from the object to enter the microscope; used in dark-field microscopes.

cardioid pattern [ENG] Heart-shaped pattern obtained as the response or radiation characteristic of certain directional antennas, or as the response characteristic of certain types of microphones.

Carina Nebula [ASTRON] A gaseous nebula near the star η Carinae in the Milky Way.

Carlsbad law [CRYSTAL] A feldspar twin law in which the twinning axis is the c axis, the operation is rotation of 180°, and the contact surface is parallel to the side pinacoid.

Carlsbad turn [CRYSTAL] A twin crystal in the monoclinic system with the vertical axis as the turning axis.

Carnot-Clausius equation [THERMO] For any system executing a closed cycle of reversible changes, the integral over the cycle of the infinitesimal amount of heat transferred to the system divided by its temperature equals 0. Also known as Clausius theorem.

Carnot cycle [THERMO] A hypothetical cycle consisting of four reversible processes in succession: an isothermal expansion and heat addition, an isentropic expansion, an isothermal compression and heat rejection process, and an isentropic compression.

Carnot efficiency [THERMO] The efficiency of a Carnot engine receiving heat at a temperature absolute T_1 and giving it up at a lower temperature absolute T_2; equal to $(T_1 - T_2)/T_1$.

Carnot engine [MECH ENG] An ideal, frictionless engine which operates in a Carnot cycle.

Carnot number [THERMO] A property of two heat sinks, equal to the Carnot efficiency of an engine operating between them.

Carnot's theorem [THERMO] **1.** The theorem that all Carnot engines operating between two given temperatures have the same efficiency, and no cyclic heat engine operating between two given temperatures is more efficient than a Carnot engine. **2.** The theorem that any system has two properties, the thermodynamic temperature T and the entropy S, such that the amount of heat exchanged in an infinitesimal reversible process is given by $dQ = TdS$; the thermodynamic temperature is a strictly increasing function of the empirical temperature measured on an arbitrary scale.

carrier [COMMUN] The radio wave produced by a transmitter when there is no modulating signal, or any other wave, recurring series of pulses, or direct current capable of being modulated. Also known as carrier wave; signal carrier.

[NUCLEO] A substance that, when associated with a radioactive trace of another substance, will carry the trace with it through a chemical or physical process; an isotope is often used for this purpose. Also known as isotopic carrier. [SOLID STATE] *See* charge carrier.

carrier density [SOLID STATE] The density of electrons and holes in a semiconductor.

carrier mobility [SOLID STATE] The average drift velocity of carriers per unit electric field in a homogeneous semiconductor; the mobility of electrons is usually different from that of holes.

carrier system [COMMUN] A system permitting a number of simultaneous, independent communications over the same circuit. Also known as carrier.

carry [MATH] An arithmetic operation that occurs in the course of addition when the sum of the digits in a given position equals or exceeds the base of the number system; a multiple m of the base is subtracted from this sum so that the remainder is less than the base, and the number m is then added to the next-higher-order digit.

Carson transform *See* Laplace-Carson transform.

Carter's theorem [RELAT] Theorem proving that the only stationary, charged black hole solutions to the equations of general relativity are the Kerr-Newman solutions.

cartesian axis [MATH] One of a set of mutually perpendicular lines which all pass through a single point, used to define a cartesian coordinate system; the value of one of the coordinates on the axis is equal to the directed distance from the intersection of axes, while the values of the other coordinates vanish.

cartesian coordinates [MATH] The set of numbers which locate a point in space with respect to a collection of mutually perpendicular axes.

cartesian coordinate system [MATH] A coordinate system in n dimensions where n is any integer made by using n number axes which intersect each other at right angles at an origin, enabling any point within that rectangular space to be identified by the distances from the n lines. Also known as rectangular cartesian coordinate system.

cartesian geometry *See* analytic geometry.

cartesian oval [MATH] A curve consisting of all the points P such that $aFP + bF'P = c$, where F and F' are fixed points and a, b, and c are constants which are not necessarily positive.

cartesian product [MATH] In reference to the product of P and Q, the set $P \times Q$ of all pairs (p,q), where p belongs to P and q belongs to Q.

cartesian space [MATH] A vector space consisting of all ordered sets $(a_1, a_2, \ldots, a_n)$ of n elements of a given field, together with the addition law, $(a_1, a_2, \ldots, a_n) + (b_1, b_2, \ldots, b_n) = (a_1 + b_1, a_2 + b_2, \ldots, a_n b_n)$, and the scalar multiplication law, $c(a_1, a_2, \ldots, a_n) = (ca_1, ca_2, \ldots, ca_n)$, where c is an element of the field.

cartesian surface [MATH] A surface obtained by rotating the curve $n_0(x^2 + y^2)^{1/2} \pm n_1[(x - a)^2 + y^2]^{1/2} = c$ about the x axis.

cartesian tensor [MATH] The aggregate of the functions of position in a tensor field in an n-dimensional cartesian coordinate system.

cartridge *See* jacket.

Cartwright-Littlewood equation [MATH] The second-order nonlinear differential equation

$$d^2y/dx^2 - \lambda f(x)\, dy/dx + g(x) = \lambda a p(t)$$

where f, g, and p are independent of λ and λ is large.

CARTESIAN COORDINATE SYSTEM

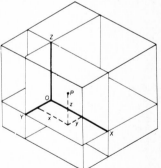

Cartesian coordinate system showing the coordinates x, y, and z of point P from the origin O.

Carvallo paradox [OPTICS] The absurdity that since light is composed from infinitely long wave trains of various frequencies, a spectrograph should show the spectrum of a source both before and after it is illuminated.

cascade [ELECTR] *See* avalanche. [ENG] An arrangement of separation devices, such as isotope separators, connected in series so that they multiply the effect of each individual device. [PHYS] The emission of a series of photons by a quantum system, such as an atomic nucleus or a laser, in an excited state, accompanying transitions of the system to successively lower excited states, until the system reaches the ground state.

cascade amplifier [ELECTR] A vacuum-tube amplifier containing two or more stages arranged in the conventional series manner. Also known as multistage amplifier.

cascade-amplifier klystron A klystron having three resonant cavities to provide increased power amplification and output; the extra resonator, located between the input and output resonators, is excited by the bunched beam emerging from the first resonator gap and produces further bunching of the beam.

cascade compensation [CONT SYS] Compensation in which the compensator is placed in series with the forward transfer function. Also known as series compensation; tandem compensation.

cascade connection [ELECTR] A series connection of amplifier stages, networks, or tuning circuits in which the output of one feeds the input of the next. Also known as tandem connection.

cascade control [CONT SYS] An automatic control system in which various control units are linked in sequence, each control unit regulating the operation of the next control unit in line.

cascade gamma emission [NUC PHYS] The emission by a nucleus of two or more gamma rays in succession.

cascade hyperon *See* xi hyperon.

cascade image tube [ELECTR] An image tube having a number of sections stacked together, the output image of one section serving as the input for the next section; used for light detection at very low levels.

cascade liquefaction [CRYO] A method of liquefying gases in which a gas with a high critical temperature is liquefied by increasing its pressure; evaporation of this liquid cools a second liquid so that it can also be liquefied by compression, and so on.

cascade networks [ELEC] Two networks in tandem such that the output of the first feeds the input of the second.

cascade particle *See* xi hyperon; xi-minus particle.

cascade shower [PARTIC PHYS] A cosmic-ray shower of electrons, positrons, and gamma rays which grows by pair production and bremsstralung events.

cascade transformer [ELEC] A source of high voltage that is made up of a collection of step-up transformers; secondary windings are in series, and primary windings, except the first, are supplied from a pair of taps on the secondary winding of the preceding transformer.

cascade unit *See* radiation length.

Casimir–du Pré theory [SOLID STATE] A theory of spin-lattice relaxation which treats the lattice and spin systems as distinct thermodynamic systems in thermal contact with one another.

cask *See* coffin.

casket *See* coffin.

Cassegrain antenna [ELECTROMAG] A microwave antenna in which the feed radiator is mounted at or near the surface

CASCADE IMAGE TUBE

Two-stage electrostatically focused cascade-type image tube. *(U.S. Army Engineer Development Corps)*

of the main reflector and aimed at a mirror at the focus; energy from the feed first illuminates the mirror, then spreads outward to illuminate the main reflector.

Cassegrain focus [OPTICS] The principal focus of a Cassegrain telescope, located just behind the primary mirror.

Cassegrain telescope [OPTICS] A reflecting telescope in which a small hyperboloidal mirror reflects the convergent beam from the paraboloidal primary mirror through a hole in the primary mirror to an eyepiece in back of the primary mirror.

Cassini's division [ASTRON] The dark ring, 2500 miles (4000 kilometers) wide, that separates ring A from ring B of the planet Saturn.

cast [OPTICS] A change in a color because of the adding of a different hue.

Castigliano's principle *See* Castigliano's theorem.

Castigliano's theorem [MECH] The theorem that the component in a given direction of the deflection of the point of application of an external force on an elastic body is equal to the partial derivative of the work of deformation with respect to the component of the force in that direction. Also known as Castigliano's principle.

casting-out nines [MATH] A method of checking the correctness of elementary arithmetical operations, based on the fact that an integer yields the same remainder as the sum of its decimal digits, when divided by 9.

cataclysmic variable [ASTRON] A star showing a sudden increase in the magnitude of light, followed by a slow fading of light; examples are novae and supernovae. Also known as explosive variable.

catadioptric [OPTICS] Involving both reflection and refraction of light.

Catalan numbers [MATH] The numbers $c_n = (2n - 2)!/n!(n-1)!$, $n = 1, 2, 3, \ldots$, which count the ways to insert parentheses in a string of n terms so that their product may be unambiguously carried out by multiplying two quantities at a time.

catalog number [ASTRON] The designation of a star composed of the name of a particular star catalog and the number of the star as listed there.

catastrophe theory [MATH] The study of certain mathematical models for discontinuous events in nature.

category [MATH] A class of objects together with a set of morphisms for each pair of objects and a law of composition for morphisms; sets and functions form an important category, as do groups and homomorphisms.

catenary [MATH] The curve obtained by suspending a uniform chain by its two ends; the graph of the hyperbolic cosine function.

catenoid [MATH] The surface formed by rotating a catenary with equation $y = a \cosh x/a$ about the x axis.

caterer problem [MATH] A linear programming problem in which it is required to find the optimal policy for a caterer who must choose between buying new napkins and sending them to either a fast or a slow laundry service.

cathode [ELEC] The positively charged pole of a primary cell or a storage battery. [ELECTR] **1.** The primary source of electrons in an electron tube; in directly heated tubes the filament is the cathode, and in indirectly heated tubes a coated metal cathode surrounds a heater. Designated K. Also known as negative electrode. **2.** The terminal of a semiconductor diode that is negative with respect to the other terminal when the diode is biased in the forward direction. [PHYS CHEM] The electrode at which reduction takes place in an electrochemical cell, that is, a cell through which electrons are being forced.

cathode bias [ELECTR] Bias obtained by placing a resistor in the common cathode return circuit, between cathode and ground; flow of electrode currents through this resistor produces a voltage drop that serves to make the control grid negative with respect to the cathode.

cathode-coupled amplifier [ELECTR] A cascade amplifier in which the coupling between two stages is provided by a common cathode resistor.

cathode coupling [ELECTR] Use of an input or output element in the cathode circuit for coupling energy to another stage.

cathode crater [ELECTR] A depression formed in the surface of a cathode by sputtering.

cathode dark space [ELECTR] The relatively nonluminous region between the cathode glow and the negative flow in a glow-discharge cold-cathode tube. Also known as Crookes dark space; Hittorf dark space.

cathode disintegration [ELECTR] The destruction of the active area of a cathode by positive-ion bombardment.

cathode drop [ELECTR] The voltage between the arc stream and the cathode of a glow-discharge tube. Also known as cathode fall.

cathode emission [ELECTR] A process whereby electrons are emitted from the cathode structure.

cathode fall *See* cathode drop.

cathode follower [ELECTR] A vacuum-tube circuit in which the input signal is applied between the control grid and ground, and the load is connected between the cathode and ground. Also known as grounded-anode amplifier; grounded-plate amplifier.

cathode glow [ELECTR] The luminous glow that covers all or part of the cathode in a glow-discharge cold-cathode tube.

cathode interface capacitance [ELECTR] A capacitance which, when connected in parallel with an appropriate resistance, forms an impedance approximately equal to the cathode interface impedance. Also known as layer capacitance.

cathode interface impedance [ELECTR] The impedance between the cathode base and coating in an electron tube, due to a high-resistivity layer or a poor mechanical bond. Also known as layer impedance.

cathode layers [ELECTR] One or more faint layers next to, and on the anode side of, the Aston dark space in a glow-discharge tube.

cathode ray [ELECTR] A stream of electrons, such as that emitted by a heated filament in a tube, or that emitted by the cathode of a gas-discharge tube when the cathode is bombarded by positive ions.

cathode-ray oscillograph [ELECTR] A cathode-ray oscilloscope in which a photographic or other permanent record is produced by the electron beam of the cathode-ray tube.

cathode-ray oscilloscope [ELECTR] A test instrument that uses a cathode-ray tube to make visible on a fluorescent screen the instantaneous values and waveforms of electrical quantities that are rapidly varying as a function of time or another quantity. Abbreviated CRO. Also known as oscilloscope; scope.

cathode-ray tube [ELECTR] An electron tube in which a beam of electrons can be focused to a small area and varied in position and intensity on a surface. Abbreviated CRT. Originally known as Braun tube; also known as electron-ray tube.

CATHODE FOLLOWER

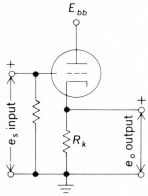

Circuit diagram of a cathode-follower amplifier.

cathode-ray voltmeter [ELEC] An instrument consisting of a cathode-ray tube of known sensitivity, whose deflection can be used to measure voltages.

cathode spot [ELECTR] The small cathode area from which an arc appears to originate in a discharge tube.

cathodic polarization [PHYS CHEM] Portion of electric cell polarization occurring at the cathode.

cathodoluminescence [ELECTR] Luminescence produced when high-velocity electrons bombard a metal in vacuum, thus vaporizing small amounts of the metal in an excited state, which amounts emit radiation characteristic of the metal. Also known as electronoluminescence.

cathodophosphorescence [ELECTR] Phosphorescence produced when high-velocity electrons bombard a metal in a vacuum.

cation [CHEM] A positively charged atom or group of atoms, or a radical which moves to the negative pole (cathode) during electrolysis.

catoptric light [OPTICS] Light reflected from a mirror, for example, light from a filament, concentrated into a parallel beam by means of a reflector.

CATT *See* controlled avalanche transit-time triode.

catwhisker [ELECTR] A sharply pointed, flexible wire used to make contact with the surface of a semiconductor crystal at a point that provides rectification. Also known as whisker.

Cauchy-Abel equation [MATH] The functional equation $f(x)f(y) = f(x + y)$. Also known as Cauchy's functional equation.

Cauchy boundary conditions [MATH] The conditions imposed on a surface in euclidean space which are to be satisfied by a solution to a partial differential equation.

Cauchy condensation test [MATH] A monotone decreasing series of positive terms Σa_n converges or diverges, as does $\Sigma p^n a_p{}^n$, for any positive integer p.

Cauchy data [RELAT] The Cauchy data for a hyperbolic partial differential equation consist of the value of the field and its time derivative on some spacelike surface.

Cauchy dispersion formula [OPTICS] A semiempirical formula for the index of refraction n of a medium as a function of wavelength λ, according to which $n = A + (B/\lambda^2)$, where A and B are constants.

Cauchy distribution [STAT] A distribution function having the form $M/[\pi M^2 + (x - a)^2]$, where x is the variable and M and a are constants. Also known as Cauchy frequency distribution.

Cauchy formula [MATH] An expression for the value of an analytic function f at a point z in terms of a line integral

$$f(z) = \frac{1}{2\pi i} \int_c \frac{f(\zeta)}{\zeta - z}\, d\zeta$$

where C is a simple closed curve containing z. Also known as Cauchy integral formula.

Cauchy frequency distribution *See* Cauchy distribution.

Cauchy horizon [RELAT] Boundary of the region that can be predicted by Cauchy data set on a spacelike surface (partial Cauchy surface).

Cauchy inequality [MATH] The square of the sum of the products of two variables for a range of values is less than or equal to the product of the sums of the squares of these two variables for the same range of values.

Cauchy integral [MATH] An integral of the form

$$\frac{1}{2\pi i} \int_L \frac{f(\zeta)}{\zeta - z}\, d\zeta$$

where L is an arc in the complex plane and $f(\zeta)$ is a function that is defined and satisfies a Hölder condition on L.

Cauchy integral formula   *See* Cauchy formula.

Cauchy integral test   *See* Cauchy's test for convergence.

Cauchy integral theorem   [MATH] The theorem that if γ is a closed path in a region R satisfying certain topological properties, then the integral around γ of any function analytic in R is zero.

Cauchy mean   [MATH] The Cauchy mean value theorem for the ratio of two continuous functions.

Cauchy mean value theorem   [MATH] If f and g are functions satisfying certain conditions on an interval $[a,b]$, then there is a point x in the interval at which the ratio of the derivatives $f'(x)/g'(x)$ equals the ratio of the net change in f, $f(b) - f(a)$, to that of g.

Cauchy number   [FL MECH] A dimensionless number used in the study of compressible flow, equal to the density of a fluid times the square of its velocity divided by its bulk modulus. Also known as Hooke number.

Cauchy principal value   [MATH] Also known as principal value. **1.** The Cauchy principal value of

$$\int_{-\infty}^{\infty} f(x)\ dx \text{ is } \lim_{s\to\infty} \int_{-s}^{s} f(x)\ dx$$

provided the limit exists. **2.** If a function f is bounded on an interval (a,b) except in the neighborhood of a point c, the Cauchy principal value of

$$\int_{a}^{b} f(x)\,dx$$

is
$$\lim_{\delta\to 0} \left[\int_{a}^{c-\delta} f(x)\,dx + \int_{c+\delta}^{b} f(x)\,dx \right]$$

provided the limit exists.

Cauchy problem   [MATH] The problem of determining the solution of a system of partial differential equations of order m from the prescribed values of the solution and of its derivatives of order less than m on a given surface.

Cauchy product   [MATH] A method of multiplying two absolutely convergent series to obtain a series which converges absolutely to the product of the limits of the original series:

$$\left(\sum_{n=0}^{\infty} a_n\right)\left(\sum_{n=0}^{\infty} b_n\right) = \sum_{n=0}^{\infty} c_n \text{ where } c_n = \sum_{k=0}^{n} a_k b_{n-k}.$$

Cauchy radical test   [MATH] A test for convergence of series of positive terms: if the nth root of the nth term is less than some number less than unity, the series converges; if it remains equal to or greater than unity, the series diverges.

Cauchy ratio test   [MATH] A series of nonnegative terms converges if the limit as n approaches infinity of the ratio of the $(n + 1)$st to nth term is smaller than 1, and diverges if it is greater than 1; the test fails if this limit is 1.

Cauchy relations   [SOLID STATE] A set of six relations between the compliance constants of a solid which should be satisfied provided the forces between atoms in the solid depend only on the distances between them and act along the lines joining them, and provided that each atom is a center of symmetry in the lattice.

Cauchy residue theorem   [MATH] The theorem expressing a line integral around a closed curve of a function which is analytic in a simply connected domain containing the curve, except at a finite number of poles interior to the curve, as a sum of residues of the function at these poles.

Cauchy-Riemann equations   [MATH] A pair of partial differential equations satisfied by the real and imaginary parts of a complex function $f(z)$ if and only if the function is analytic:

$\partial u/\partial x = \partial v/\partial y$ and $\partial u/\partial y = -\partial v/\partial x$, where $f(z) = u + iv$ and $z = x + iy$.

Cauchy-Schwarz inequality [MATH] The square of the inner-product of two vectors does not exceed the product of the squares of their norms. Also known as Buniakowski's inequality; Schwarz's inequality.

Cauchy sequence [MATH] A sequence with the property that the difference between any two terms is arbitrarily small provided they are both sufficiently far out in the sequence; more precisely stated: a sequence $\{a_n\}$ such that for every $\varepsilon > 0$ there is an integer N with the property that if n and m are both greater than N, then $|a_n - a_m| < \varepsilon$. Also known as fundamental sequence.

Cauchy's form of remainder [MATH] An expression for the difference R_n between the value of a function $f(x)$ and the sum of the first $n + 1$ terms of its Taylor series about a point a; it may be written $R_n = h^{n+1}(1-\theta)^n f^{(n+1)}(a+\theta h)/n!$, where $h = x - a$ and θ is some number between 0 and 1.

Cauchy's functional equation [MATH] **1.** The functional equation $f(x) + f(y) = f(x + y)$. **2.** The functional equation $f(x) + f(y) = f(xy)$. **3.** The functional equation $f(x)f(y) = f(xy)$. **4.** *See* Cauchy-Abel equation.

Cauchy's test for convergence [MATH] **1.** A series is absolutely convergent if the limit as n approaches infinity of its nth term raised to the $1/n$ power is less than unity. **2.** A series a_n is convergent if there exists a monotonically decreasing function f such that $f(n) = a_n$ for n greater than some fixed number N, and if the integral of $f(x)\ dx$ from N to ∞ converges. Also known as Cauchy integral test; Maclaurin-Cauchy test.

Cauchy surface [RELAT] A surface S in a space-time M is a (global) Cauchy surface if every nonspacelike curve in M intersects S exactly once; that is, the Cauchy development of S equals M.

Cauchy transcendental equation [MATH] An equation whose roots are characteristic values of a certain type of Sturm-Liouville problem: $\tan \sigma\pi = (k + K)/(\sigma^2 - kK)$, where k and K are given, and σ is to be determined.

Cauer form [ELEC] A continued fraction expansion of the impedance used in the network synthesis for a driving point function resulting in a ladder network.

causal boundary [RELAT] A boundary attached to a space-time that depends only on the causal structure; it does not distinguish between boundary points at finite distances (singularities) or those at infinity. Also known as C boundary.

causal curve [RELAT] A curve in space-time that is nowhere spacelike.

causal future [RELAT] The causal future relative to a set of points S in a space-time M is the set of points in M which can be reached from S by future-directed timelike or null curves.

causality [MECH] In classical mechanics, the principle that the specification of the dynamical variables of a system at a given time, and of the external forces acting on the system, completely determines the values of dynamical variables at later times. Also known as determinism. [PHYS] **1.** The principle that an event cannot precede its cause; in a relativistic theory, an event cannot have an effect outside its future light cone. **2.** In relativistic quantum field theory, the principle that the field operators at different space-time points commute (for boson fields; anticommute in the case of fermion fields) if the separation of the points is spacelike. [QUANT MECH] The principle that the specification of the dynamical state of a system at a given time, and of the interaction of the system with its environment, determines the dynamical state of the system at later times, from which a probability distribution for the observation of any dynamical variable may be determined. Also known as determinism. [SCI TECH] The existence of regularities which control natural phenomena.

causality condition [RELAT] The condition of a space-time requiring there be no closed nonspacelike curves.

causally simple [RELAT] A set of points U in a space-time is said to be causally simple if the causal past and causal future of every compact subset of U is closed in U.

causal past [RELAT] The causal past relative to a set of points S in a space-time M is the set of points in M which can be reached from S by past-directed timelike or null curves.

causal system [CONT SYS] A system whose response to an input does not depend on values of the input at later times. Also known as nonanticipatory system; physical system.

cave [NUCLEO] A heavily shielded compartment in which highly radioactive material can be handled, generally by remote control. Also known as hot cell.

Cavendish balance [ENG] An instrument for determining the constant of gravitation, in which one measures the displacement of two small spheres of mass m, which are connected by a light rod suspended in the middle by a thin wire, caused by bringing two large spheres of mass M near them.

cavitation [FL MECH] Formation of gas- or vapor-filled cavities within liquids by mechanical forces; broadly includes bubble formation when water is brought to a boil and effervescence of carbonated drinks; specifically, the formation of vapor-filled cavities in the interior or on the solid boundaries of vaporized liquids in motion where the pressure is reduced to a critical value without a change in ambient temperature.

cavitation noise [ACOUS] Noise resulting from the formation of vapor- or gas-filled cavities in liquids by mechanical forces, as occurs near a propeller.

cavity *See* cavity resonator.

cavity absorbent [ENG ACOUS] A Helmholtz or other acoustic resonator used to absorb and attenuate sound.

cavity coupling [ELECTROMAG] The extraction of electromagnetic energy from a resonant cavity, either waveguide or coaxial, using loops, probes, or apertures.

cavity filter [ELECTROMAG] A microwave filter that uses quarter-wavelength-coupled cavities inserted in waveguides or coaxial lines to provide band-pass or other response characteristics at frequencies in the gigahertz range.

cavity impedance [ELECTR] The impedance of the cavity of a microwave tube which appears across the gap between the cathode and the anode.

cavity magnetron [ELECTR] A magnetron having a number of resonant cavities forming the anode; used as a microwave oscillator.

cavity oscillator [ELECTR] An ultra-high-frequency oscillator whose frequency is controlled by a cavity resonator.

cavity radiator [THERMO] A heated enclosure with a small opening which allows some radiation to escape or enter; the escaping radiation approximates that of a blackbody.

cavity resonator [ELECTROMAG] A space totally enclosed by a metallic conductor and excited in such a way that it becomes a source of electromagnetic oscillations. Also known as cavity; microwave cavity; microwave resonance cavity; resonant cavity; resonant chamber; resonant element; rhumbatron; tuned cavity; waveguide resonator.

cavity tuning [ELECTROMAG] Use of an adjustable cavity

CAVENDISH BALANCE

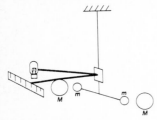

The torsion balance used by H. Cavendish to determine the gravitational constant G.

CAVITY RADIATOR

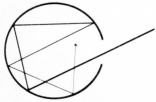

A drawing of a cavity radiator showing the almost complete absorption of entering radiation energy because of the multiple reflections it encounters.

resonator as a tuned circuit in an oscillator or amplifier, with tuning usually achieved by moving a metal plunger in or out of the cavity to change the volume, and hence the resonant frequency of the cavity.

c axis [CRYSTAL] A vertically oriented crystal axis, usually the principal axis; the unique symmetry axis in tetragonal and hexagonal crystals.

Cayley algebra [MATH] The nonassociative division algebra consisting of pairs of quaternions; it may be identified with an eight-dimensional vector space over the real numbers.

Cayley-Dickson algebra [MATH] Any eight-dimensional nonassociative algebra over a field F obtained by the Cayley-Dickson process from a quaternion algebra over F; this generalizes the concept of a Cayley algebra.

Cayley-Dickson process [MATH] A process for constructing a nonassociative algebra B with dimension $2n$ over a field F from a nonassociative algebra A, with unit element 1, of dimension n over F; specifically, B consists of all ordered pairs of elements in A, with addition and multiplication by scalars defined component-wise, and multiplication defined by the equation

$$(x_1, x_2)(x_3, x_4) = (x_1 x_3 + a x_4 x_2^*, x_1^* x_4 + x_3 x_2),$$

for x_1, x_2, x_3, x_4 in A, where a is a fixed nonzero element of F, and the mapping $x \rightarrow x^*$ is a linear function from A into itself satisfying $x^* y^* = (yx)^*$, $(x^*)^* = x$, $x + x^* = t(x)1$, and $xx^* = n(x)1$, with $t(x)$ and $n(x)$ in F, for all x and y in A.

Cayley-Dickson ring [MATH] A torsion-free ring R with non-zero center Z that can be embedded in a Cayley-Dickson algebra over the quotient field of Z.

Cayley-Hamilton theorem [MATH] The theorem that a linear transformation or matrix is a root of its own characteristic polynomial. Also known as Hamilton-Cayley theorem.

Cayley-Klein parameters [MATH] A set of four complex numbers used to describe the orientation of a rigid body in space, or equivalently, the rotation which produces that orientation, starting from some reference orientation.

Cayley numbers [MATH] The elements of a Cayley algebra.

Cayley's formula [MATH] The representation of an orthogonal matrix R, for which the determinant of $R + 1$ does not vanish, in the form $R = (1 - T)(1 + T)^{-1}$, where T is an antisymmetric matrix.

Cayley's sextic [MATH] A plane curve with the equation $r = 4a \cos^3 \frac{1}{3}\theta$, where r and θ are radial and angular polar coordinates and a is a constant.

C-band waveguide [ELECTROMAG] A rectangular waveguide, with dimensions 3.48 by 1.58 centimeters, which is used to excite only the dominant mode (TE_{01}) for wavelengths in the range 3.7–5.1 centimeters.

C battery [ELEC] The battery that supplies the steady bias voltage required by the control-grid electrodes of electron tubes in battery-operated equipment. Also known as grid battery.

C bias *See* grid bias.

C boundary *See* causal boundary.

C core [ELECTROMAG] A spirally wound magnetic core that is formed to a desired rectangular shape before being cut into two C-shaped pieces and placed around a transformer or magnetic amplifier coil.

Cd *See* cadmium.

CD *See* circular dichroism.

Ce *See* cerium.

celerity *See* phase velocity.

celestial coordinates [ASTRON] Any set of coordinates, such as zenithal distance, altitude, celestial latitude, celestial longitude, local hour angle, azimuth and declination, used to define a point on the celestial sphere.

celestial equator [ASTRON] The primary great circle of the celestial sphere in the equatorial system, everywhere 90° from the celestial poles; the intersection of the extended plane of the equator and the celestial sphere. Also known as equinoctial.

celestial equator system of coordinates *See* equatorial system.

celestial geodesy [GEOD] The branch of geodesy which utilizes observations of near celestial bodies and earth satellites to determine the size and shape of the earth.

celestial horizon [ASTRON] That great circle of the celestial sphere which is formed by the intersection of the celestial sphere and a plane through the center of the earth and is perpendicular to the zenith-nadir line. Also known as rational horizon.

celestial latitude [ASTRON] Angular distance north or south of the ecliptic; the arc of a circle of latitude between the ecliptic and a point on the celestial sphere, measured northward or southward from the ecliptic through 90°, and labeled N or S to indicate the direction of measurement. Also known as ecliptic latitude.

celestial longitude [ASTRON] Angular distance east of the vernal equinox, along the ecliptic; the arc of the ecliptic or the angle at the ecliptic pole between the circle of latitude of the vernal equinox and the circle of latitude of a point on the celestial sphere, measured eastward from the circle of latitude of the vernal equinox, through 360°. Also known as ecliptic longitude.

celestial mechanics [ASTROPHYS] The calculation of motions of celestial bodies under the action of their mutual gravitational attractions. Also known as gravitational astronomy.

celestial meridian [ASTRON] A great circle on the celestial sphere, passing through the two celestial poles and the observer's zenith.

celestial parallel *See* parallel of declination.

celestial pole [ASTRON] Either of the two points of intersection of the celestial sphere and the extended axis of the earth, labeled N or S to indicate the north celestial pole or the south celestial pole.

celestial sphere [ASTRON] An imaginary sphere of indefinitely large radius, which is described about an assumed center, and upon which positions of celestial bodies are projected along radii passing through the bodies.

cell [MATH] The homeomorphic image of the unit ball. [NUCLEO] One of a set of elementary regions in a heterogeneous reactor, all of which have the same geometrical form and the same neutron characteristics.

cell complex [MATH] A topological space which is the last term of a finite sequence of spaces, each obtained from the previous by sewing on a cell along its boundary.

cell frequency [STAT] The number of observations of specified conditional constraints on one or more variables; used mainly in the analysis of data obtained by performing actual counts.

cellular horn *See* multicellular horn.

celo [MECH] A unit of acceleration equal to the acceleration of a body whose velocity changes uniformly by 1 foot per second (0.3048 meter per second) in 1 second.

Celor lens system [OPTICS] An anastigmatic lens system consisting of two air-spaced achromatic doublet lenses, one

CELOR LENS SYSTEM

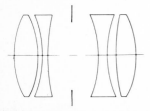

Celor lens system, an anastigmatic photographic objective.

on each side of the stop. Also known as Gauss lens system; Gauss objective lens.

Celsius degree [THERMO] Unit of temperature interval or difference equal to the kelvin.

Celsius temperature scale [THERMO] Temperature scale in which the temperature Θ_c in degrees Celsius (°C) is related to the temperature T_k in kelvins by the formula $\Theta_c = T_k - 273.15$; the freezing point of water at standard atmospheric pressure is very nearly 0°C and the corresponding boiling point is very nearly 100°C. Formerly known as centigrade temperature scale.

cemented lens *See* compound lens.

censored data [STAT] Observations collected by determining in advance whether to record only a specified number of the smallest or largest values, or of the remaining values in a sample of a particular size.

census [STAT] A complete counting of a population, as opposed to a partial counting or sampling.

cent [ACOUS] The interval between two sounds whose basic frequency ratio is the twelve-hundredth root of 2; the interval, in cents, between any two frequencies is 1200 times the logarithm to the base 2 of the frequency ratio. [NUCLEO] A unit of nuclear reactivity equal to one-hundredth of a dollar.

cental *See* hundredweight.

centare *See* centiare.

center [MATH] **1.** The point which is equidistant from all the points on a circle or sphere. **2.** For an ellipsoid or hyperboloid, the point about which the surface is symmetrical. [OPTICS] To adjust the components of an optical system so that their centers of curvature lie on a common optical axis. Also known as square-on.

center of a group [MATH] The subgroup consisting of all elements that commute with all other elements in the given group.

center of area [MATH] For a plane figure, the center of mass of a thin uniform plate having the same boundaries as the plane figure. Also known as center of figure; centroid.

center of a ring [MATH] The subring consisting of all elements a such that $ax = xa$ for all x in the given ring.

center of attraction [MECH] A point toward which a force on a body or particle (such as gravitational or electrostatic force) is always directed; the magnitude of the force depends only on the distance of the body or particle from this point.

center of buoyancy [MECH] The point through which acts the resultant force exerted on a body by a static fluid in which it is submerged or floating; located at the centroid of displaced volume.

center of curvature [MATH] At a given point on a curve, the center of the osculating circle of the curve at that point.

center of figure *See* center of area; center of volume.

center of force [MECH] The point toward or from which a central force acts.

center of gravity [MECH] A fixed point in a material body through which the resultant force of gravitational attraction acts.

center of inertia *See* center of mass.

center of inversion [CRYSTAL] A point in a crystal lattice such that the lattice is left invariant by an inversion in the point. [MATH] The point O with respect to which an inversion is defined, so that every point P is mapped by the inversion into a point Q that is collinear with O and P.

center of mass [MECH] That point of a material body or system of bodies which moves as though the system's total mass existed at the point and all external forces were applied at the point. Also known as center of inertia; centroid.

center of mass coordinate system [MECH] A reference frame which moves with the velocity of the center of mass, so that the center of mass is at rest in this system, and the total momentum of the system is zero. Also known as center of momentum coordinate system.

center of momentum coordinate system *See* center of mass coordinate system.

center of oscillation [MECH] Point in a physical pendulum, on the line through the point of suspension and the center of mass, which moves as if all the mass of the pendulum were concentrated there.

center of percussion [MECH] If a rigid body, free to move in a plane, is struck a blow at a point O, and the line of force is perpendicular to the line from O to the center of mass, then the initial motion of the body is a rotation about the center of percussion relative to O; it can be shown to coincide with the center of oscillation relative to O.

center of perspective [MATH] The point specified by Desargues' theorem, at which lines passing through corresponding vertices of two triangles are concurrent.

center of similitude [MATH] A point of intersection of lines that join the ends of parallel radii of coplanar circles.

center of spherical curvature [MATH] The center of the osculating sphere at a specified point on a space curve.

center of suspension [MECH] The intersection of the axis of rotation of a pendulum with a plane perpendicular to the axis that passes through the center of mass.

center of symmetry [SCI TECH] A point in an object through which any straight line encounters exactly similar points on opposite sides. Also known as symmetry center.

center of twist [MECH] A point on a line parallel to the axis of a beam through which any transverse force must be applied to avoid twisting of the section. Also known as shear center.

center of volume [MATH] For a three-dimensional figure, the center of mass of a homogeneous solid having the same boundaries as the figures. Also known as center of figure; centroid.

center tap [ELEC] A terminal at the electrical midpoint of a resistor, coil, or other device. Abbreviated CT.

centi- [SCI TECH] A prefix representing 10^{-2}, which is 0.01 or one-hundredth. Abbreviated c.

centiare [MECH] Unit of area equal to 1 square meter. Also spelled centare.

centibar [MECH] A unit of pressure equal to 0.01 bar or to 1000 pascals.

centigrade heat unit [THERMO] A unit of heat energy, equal to 0.01 of the quantity of heat needed to raise 1 pound of air-free water from 0 to 100°C at a constant pressure of 1 standard atmosphere; equal to 1900.44 joules. Symbolized CHU; (more correctly) CHU_{mean}.

centigrade temperature scale *See* Celsius temperature scale.

centigram [MECH] Unit of mass equal to 0.01 gram or 10^{-5} kilogram. Abbreviated cg.

centihg *See* centimeter of mercury.

centiliter [MECH] A unit of volume equal to 0.01 liter or to 10^{-5} cubic meter.

centimeter [ELEC] *See* abhenry; statfarad. [MECH] A unit of length equal to 0.01 meter. Abbreviated cm.

centimeter-candle *See* phot.

centimeter-gram-second system [PHYS] An absolute system of metric units in which the centimeter, gram mass, and the second are the basic units. Abbreviated cgs system.

centimeter of mercury [MECH] A unit of pressure equal to the pressure that would support a column of mercury 1 centimeter high, having a density of 13.5951 grams per cubic centimeter, when the acceleration of gravity is equal to its standard value (980.665 centimeters per second per second); equal to 1333. 22387415 pascals; it differs from the decatorr by less than 1 part in 7,000,000. Abbreviated cmHg. Also known as centihg.

centipoise [FL MECH] A unit of viscosity equal to 0.01 poise. Abbreviated cp.

centistoke [FL MECH] A cgs unit of kinematic viscosity in customary use, equal to the kinematic viscosity of a fluid having a dynamic viscosity of 1 centipoise and a density of 1 gram per cubic centimeter. Abbreviated cs.

centner *See* hundredweight.

centrad [MATH] A unit of plane angle equal to 0.01 radian or to about 0.573 degree.

central difference [MATH] One of a series of quantities obtained from a function whose values are known at a series of equally spaced points by repeatedly applying the central difference operator to these values; used in interpolation or numerical calculation and integration of functions.

central difference operator [MATH] A difference operator, denoted ∂, defined by the equation $\partial f(x) = f(x + h/2) - f(x - h/2)$, where h is a constant denoting the difference between successive points of interpolation or calculation.

central field approximation [PHYS] The approximation that the electrons in an atom or the nucleons in a nucleus move in the potential of a central force which is the same for all the particles.

central force [MECH] A force whose line of action is always directed toward a fixed point; the force may attract or repel.

centralizer [MATH] The subgroup consisting of all elements which commute with a given element of a group.

central-limit theorem [STAT] The theorem that the distribution of sample means taken from a large population approaches a normal (Gaussian) curve.

central mean operator [MATH] A difference operator, denoted μ, defined by the equation $\mu f(x) = [f(x + h/2) + f(x - h/2)]/2$, where h is a constant denoting the difference between successive points of interpolation or calculation. Also known as averaging operator.

central orbit [MECH] The path followed by a body moving under the action of a central force.

central quadric [MATH] A quadric surface that has a point about which the surface is symmetrical; namely, a sphere, ellipsoid, or hyperboloid.

central-slice theorem [MATH] If $g(x,y)$ is a function that is continuous except on a finite number of simple arcs and vanishes outside a finite domain, and if $f_\phi(x')$ is the projection given by

$$f_\phi(x') = \int_{-\infty}^{\infty} g(x',y')\,dy'$$

where the x'-y' axes are obtained from the x-y axis by rotation through an angle ϕ, then the Fourier transform of $f_\phi(x')$ is equal to the two-dimensional Fourier transform of $g(x,y)$,

$$F_2(p,q) = \int_{-\infty}^{\infty} dx \int_{+\infty}^{\infty} dy\; g(x,y)e^{2\pi i(px\,+\,qy)},$$

evaluated along a line (slice) passing through the origin of the p-q plane and making an angle ϕ with the p axis. Also known as projection-slice theorem.

centrifugal [MECH] Acting or moving in a direction away from the axis of rotation or the center of a circle along which a body is moving.

centrifugal barrier [MECH] A steep rise, located around the center of force, in the effective potential governing the radial motion of a particle of nonvanishing angular momentum in a central force field, which results from the centrifugal force and prevents the particle from reaching the center of force, or causes its Schrödinger wave function to vanish there in a quantum-mechanical system.

centrifugal distortion [PHYS] Tendency of a molecule to stretch slightly as its speed of rotation increases.

centrifugal force [MECH] **1.** An outward pseudo-force, in a reference frame that is rotating with respect to an inertial reference frame, which is equal and opposite to the centripetal force that must act on a particle stationary in the rotating frame. **2.** The reaction force to a centripetal force.

centrifugal moment [MECH] The product of the magnitude of centrifugal force acting on a body and the distance to the center of rotation.

centrifugal stretching [PHYS] Stretching of the bonds of a rotating molecule caused by centrifugal force, resulting in an increase in the molecule's moment of inertia and a modification of its energy levels.

centrifugation potentials [PHYS CHEM] Electric potential differences between points at different distances from the axis of rotation of a colloidal solution that is being rapidly rotated in a centrifuge.

centrifuge microscope [OPTICS] An instrument which permits magnification and observation of living cells being centrifuged; image of the material magnified by the objective which rotates near the periphery of the centrifuge head is brought to the axis of rotation where it is observed in a stationary ocular.

centripetal [MECH] Acting or moving in a direction toward the axis of rotation or the center of a circle along which a body is moving.

centripetal acceleration [MECH] The radial component of the acceleration of a particle or object moving around a circle, which can be shown to be directed toward the center of the circle.

centripetal force [MECH] The radial force required to keep a particle or object moving in a circular path, which can be shown to be directed toward the center of the circle.

centrobaric [MECH] **1.** Pertaining to the center of gravity, or to some method of locating it. **2.** Possessing a center of gravity.

centrode [MECH] The path traced by the instantaneous center of a plane figure when it undergoes plane motion.

centroid *See* center of area; center of mass; center of volume.

centroid of a probability measure [MATH] A centroid of a probability measure m on a compact convex subset K of a topological vector space is a point x_0 satisfying

$$f(x_0) = \int_K f(x)m(dx)$$

for every continuous linear functional f.

centroid of asymptotes [CONT SYS] The intersection of asymptotes in a root-locus diagram.

centroids of areas and lines [MATH] Points positioned identically with the centers of gravity of corresponding thin homogeneous plates or thin homogeneous wires; involved in the analysis of certain problems of mechanics such as the phenomenon of bending.

centrosymmetry [PHYS] Property of a body or system which is unchanged under space inversion through a specified point.

cepheid [ASTRON] One of a subgroup of periodic variable stars whose brightness does not remain constant with time and whose period of variation is a function of intrinsic mean brightness.

cepstrum [ACOUS] The Fourier transform of the logarithm of a speech power spectrum; used to separate vocal tract information from pitch excitation in voiced speech.

ceramagnet [ELECTROMAG] A ferrimagnet composed of the hard magnetic material $BaO \cdot 6Fe_2O_3$.

ceramic amplifier [ELECTR] An amplifier that utilizes the piezoelectric properties of semiconductors such as silicon.

ceramic capacitor [ELEC] A capacitor whose dielectric is a ceramic material such as steatite or barium titanate, the composition of which can be varied to give a wide range of temperature coefficients.

ceramic magnet [ELECTROMAG] A permanent magnet made from pressed and sintered mixtures of ceramic and magnetic powders. Also known as ferromagnetic ceramic.

ceramic reactor [NUCLEO] A nuclear reactor in which the fuel and moderator assemblies are made from high-temperature-resistant ceramic materials such as metal oxides, carbides, or nitrides.

ceraunograph [ENG] An instrument that detects radio waves generated by lightning discharges and records their occurrence.

Cerenkov counter [NUCLEO] An apparatus for detecting high-energy charged particles by observation of the Cerenkov radiation produced.

Cerenkov radiation [ELECTROMAG] Light emitted by a high-speed charged particle when the particle passes through a transparent, nonconducting material at a speed greater than the speed of light in the material.

Cerenkov rebatron radiator [ELECTR] Device in which a tightly bunched, velocity-modulated electron beam is passed through a hole in a dielectric; the reaction between the higher velocity of the electrons passing through the hole and the slower velocity of the electromagnetic energy passing through the dielectric results in radiation at some frequency higher than the frequency of modulation of the electron beam.

cerium [CHEM] A chemical element, symbol Ce, atomic number 58, atomic weight 140.12; a rare-earth metal, used as a getter in the metal industry, as an opacifier and polisher in the glass industry, in Welsbach gas mantles, in cored carbon arcs, and as a liquid-liquid extraction agent to remove fission products from spent uranium fuel.

cerium-140 [NUC PHYS] An isotope of cerium with atomic mass number of 140, 88.48% of the known amount of the naturally occurring element.

cerium-142 [NUC PHYS] A radioactive isotope of cerium with atomic mass number of 142; emits α-particles and has a half-life of 5×10^{15} years.

cerium-144 [NUC PHYS] A radioactive isotope of the element cerium with atomic mass number of 144; a beta emitter with a half-life of 285 days.

cermet nuclear fuel [NUCLEO] A nuclear reactor fuel mixed with a heat-resistant ceramic and a metal to give it both refractory and damage-resistant properties.

Cerruti's problem *See* Boussinesq's problem.

Cesaro summation [MATH] A method of attaching sums to certain divergent sequences and series by taking averages of the first *n*-terms and passing to the limit.

cesium [CHEM] A chemical element, symbol Cs, atomic number 55, atomic weight 132.905.

cesium-134 [NUC PHYS] An isotope of cesium, atomic mass number of 134; emits negative beta particles and has a half-life

CERAMIC CAPACITOR

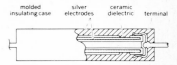

molded insulating case   silver electrodes   ceramic dielectric   terminal

Cutaway drawing of typical molded-case ceramic capacitor. *(From K. Henney and C. Walsh, eds., Electronic Components Handbook, McGraw-Hill, 1957)*

CERIUM

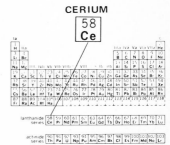

Periodic table of the chemical elements showing the position of cerium.

CESIUM

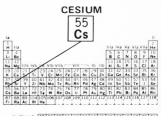

Periodic table of the chemical elements showing the position of cesium.

of 2.19 years; used in photoelectric cells and in ion propulsion systems under development.

cesium-137 [NUC PHYS] An isotope of cesium with atomic mass number of 137; emits negative beta particles and has a half-life of 30 years; offers promise as an encapsulated radiation source for therapeutic and other purposes. Also known as radiocesium.

cesium-antimonide photocathode [ELECTR] A photocathode obtained by exposing a thin layer of antimony to cesium vapor at elevated temperatures; has a maximum sensitivity in the blue and ultraviolet regions of the spectrum.

cesium-beam atomic clock [NUCLEO] An instrument, used as the primary standard of frequency and time, in which a microwave oscillator, which generates radiation in a microwave cavity, is maintained at a frequency such that a hyperfine transition is induced in cesium atoms in a beam passing through the cavity. Also known as cesium-beam atomic oscillator.

cesium-beam atomic oscillator *See* cesium-beam atomic clock.

cesium-beam tube *See* cesium electron tube.

cesium electron tube [ELECTR] An electronic device used as an atomic clock, producing electromagnetic energy that is accurate and stable in frequency. Also known as cesium-beam tube.

cesium magnetometer [ENG] A magnetometer that uses a cesium atomic-beam resonator as a frequency standard in a circuit that detects very small variations in magnetic fields.

cesium phototube [ELECTR] A phototube having a cesium-coated cathode; maximum sensitivity in the infrared portion of the spectrum.

cesium-vapor lamp [ELECTR] A lamp in which light is produced by the passage of current between two electrodes in ionized cesium vapor.

Ceva's theorem [MATH] The theorem that if three concurrent straight lines pass through the vertices A, B, and C of a triangle and intersect the opposite sides, produced if necessary at D, E, and F, then the product $\overline{AF} \cdot \overline{BD} \cdot \overline{CE}$ of the lengths of three alternate segments equals the product $\overline{FB} \cdot \overline{DC} \cdot \overline{EA}$ of the other three.

cevian [MATH] A straight line that passes through a vertex of a triangle or tetrahedron and intersects the opposite side or face.

Cf *See* californium.

C figure *See* C index.

cg *See* centigram.

cgs system *See* centimeter-gram-second system.

chad [NUCLEO] **1.** A unit of neutron flux equal to 1 neutron per square centimeter per second. **2.** A unit of neutron flux equal to 10^{12} neutrons per square centimeter per second.

Chadwick-Goldhaber effect *See* photodisintegration.

chain *See* linearly ordered set.

chain complex [MATH] A sequence $\{C_n\}, -\infty < n < \infty$, of Abelian groups together with a sequence of boundary homomorphisms $d_n: C_n \rightarrow C_{n-1}$ such that $d_{n-1} \circ d_n = 0$ for each n.

chain decay *See* series disintegration.

chain disintegration *See* series disintegration.

chain fission yield [NUC PHYS] The sum of the independent fission yields for all isobars of a particular mass number.

chain homomorphism [MATH] A sequence of homomorphisms $f_n: C_n \rightarrow D_n$ between the groups of two chain complexes such that $f_{n-1} d_n = \overline{d}_n f_n$ where d_n and $\overline{d}_n$ are the boundary homomorphisms of $\{C_n\}$ and $\{D_n\}$ respectively.

chain index [STAT] An index number derived by relating the value at any given period to the value in the previous period rather than to a fixed base.

chain of simplices   [MATH] A member of the free Abelian group generated by the simplices of a given dimension of a simplicial complex.

chain reaction   *See* nuclear chain reaction.

chain rule   [MATH] A rule for differentiating a composition of functions: $(d/dx) f(g(x)) = f'(g(x)) \cdot g'(x)$.

chain structure   [SOLID STATE] A crystalline structure in which forces between atoms in one direction are greater than those in other directions, so that the atoms are concentrated in chains.

chaldron   [MECH] **1.** A unit of volume in common use in the United Kingdom, equal to 36 bushels, or 288 gallons, or approximately 1.30927 cubic meters. **2.** A unit of volume, formerly used for measuring solid substances in the United States, equal to 36 bushels, or approximately 1.26861 cubic meters.

chance variable   *See* random variable.

changeover switch   [ELEC] A means of moving a circuit from one set of connections to another.

channel   [ELECTR] **1.** A path for a signal, as an audio amplifier may have several input channels. **2.** The main current path between the source and drain electrodes in a field-effect transistor or other semiconductor device.   [NUCLEO] A passage for fuel slugs or heat-transfer fluid in a reactor.

channeling   [NUCLEO] The transmission of extra particles through a medium in a nuclear reactor due to the presence of voids in the medium.

channel spin   [NUC PHYS] The vector sum of the spins of the particles involved in a nuclear reaction, either before or after the reaction takes place.

channel width   [NUC PHYS] The part of the total energy width of a nuclear energy level that corresponds to a particular mode of decay.

Chaplygin-Kármán-Tsien relation   [FL MECH] The relation that in the case of isentropic flow of an ideal gas with negligible viscosity and thermal conductivity, the sum of the pressure and a constant times the reciprocal of the density of the fluid is constant along a streamline; a useful, although physically impossible, approximation.

Chapman-Enskog approximations   [STAT MECH] Approximations to a solution of the Boltzmann transport equation in the Chapman-Enskog theory.

Chapman-Enskog solution   [STAT MECH] The solution of the Boltzmann transport equation according to the Chapman-Enskog theory.

Chapman-Enskog theory   [STAT MECH] A method of solving the Boltzmann transport equation by successive approximations, essentially in powers of the mean free path. Also known as Enskog theory.

Chapman equation   [GEOPHYS] A theoretical relation describing the distribution of electron density with height in the upper atmosphere.   [STAT MECH] The relationship that the viscosity of a gas equals $(0.499)mv/[\sqrt{2}\,\pi\sigma^2\,(1 + C/T)]$, where m is the mass of a molecule, v its average speed, σ its collision diameter, C the Sutherland constant, and T the absolute temperature (Kelvin scale).

Chapman-Jouguet plane   [MECH] A hypothetical, infinite plane, behind the initial shock front, in which it is variously assumed that reaction (and energy release) has effectively been completed, that reaction product gases have reached thermodynamic equilibrium, and that reaction gases, streaming backward out of the detonation, have reached such a condition that a forward-moving sound wave located at this precise plane would remain a fixed distance behind the initial shock.

Chapman region [GEOPHYS] A hypothetical region in the upper atmosphere in which the distribution of electron density with height can be described by Chapman's theoretical equation.

character [MATH] **1.** A character of a monoid in a field is a homomorphism of the monoid into the multiplicative group of the field. **2.** In particular, a complex-valued function on a locally compact Abelian group G such that $|f(x)| = 1$ for all x in G and $f(x + y) = f(x)f(y)$ for all x and y in G. **3.** The character of a representation of a group is a function that maps each element of the group into the trace of the associated matrix or linear transformation.

character group [MATH] The set of all continuous homomorphisms of a topological group onto the group of all complex numbers with unit norm.

characteristic [MATH] **1.** That part of the logarithm of a number which is the integral (the whole number) to the left of the decimal point in the logarithm. **2.** For a function u of the variables x and y satisfying a partial differential equation in x and y, a curve in the x-y plane along which certain information about u and its first derivatives is insufficient to determine all the higher derivatives of u, even though such information is sufficient to determine the higher derivatives along other curves. **3.** For a partial differential equation of the form $f(x,y)\ \partial u/\partial x + g(x,y)\ \partial u/\partial y = h(x,y)$, a curve in x,y,u-space, with x, y, and u depending on a parameter t, such that $dx/dt = f(x,y)$, $dy/dt = g(x,y)$, and $du/dt = h(x,y)$; every surface representing a solution of the equation is generated by a family of such curves. **4.** The characteristic of an integral domain D is the smallest positive integer p such that $px = 0$ for some nonzero x in D; if no such p exists, then D is said to be of characteristic 0.

characteristic acoustic impedance [ACOUS] The product of the density and the speed of sound in a medium; it is analogous to the characteristic impedance of an infinitely long transmission line. Also known as intrinsic impedance.

characteristic cone [MATH] A conelike region important in the study of initial value problems in partial differential equations.

characteristic curve [MATH] **1.** One of a pair of conjugate curves in a surface with the property that the directions of the tangents through any point of the curve are the characteristic directions of the surface. **2.** A curve plotted on graph paper to show the relation between two changing values. **3.** A characteristic curve of a one-parameter family of surfaces is the limit of the curve of intersection of two neighboring surfaces of the family as those surfaces approach coincidence.

characteristic directions [MATH] For a point P on a surface S, the pair of conjugate directions which are symmetric with respect to the directions of the lines of curvature on S through P.

characteristic equation [MATH] **1.** Any equation which has a solution, subject to specified boundary conditions, only when a parameter occurring in it has certain values. **2.** Specifically, the equation $A\mathbf{u} = \lambda\ \mathbf{u}$, which can have a solution only when the parameter λ has certain values, where A can be a square matrix which multiplies the vector $\mathbf{u}$, or a linear differential or integral operator which operates on the function $\mathbf{u}$, or in general, any linear operator operating on the vector $\mathbf{u}$ in a finite or infinite dimensional vector space. Also known as eigenvalue equation. **3.** An equation which sets the characteristic polynomial of a given linear transformation on a finite dimensional vector space, or of its matrix representation, equal to zero. **4.** The number preceding the decimal of a common logarithm. [PHYS] An equation relating a set of variables, such as pressure, volume, and temperature, whose values determine a substance's physical condition. [PL PHYS] An equation whose solutions give the frequencies and modes of those perturbations of a hydromagnetic system which decay or grow exponentially in time, and indicate regions of stability of such a system.

characteristic form [MATH] A means of classifying partial differential equations.

characteristic function [MATH] **1.** The function χ_A defined for any subset A of a set by setting $\chi_A(x) = 1$ if x is in A and $\chi_A(x) = 0$ if x is not in A. Also known as indicator function. **2.** *See* eigenfunction. [PHYS] A function, such as the point characteristic function or the principal function, which is the integral of some property of an optical or mechanical system over time or over the path followed by the system, and whose value for a path actually followed by a system is a maximum or a minimum with respect to nearby paths with the same end points. [STAT] A function that uniquely defines a probability distribution; it is equal to $\sqrt{2\pi}$ times the Fourier transform of the frequency function of the distribution.

characteristic impedance [COMMUN] The impedance that, when connected to the output terminals of a transmission line of any length, makes the line appear to be infinitely long, for there are then no standing waves on the line, and the ratio of voltage to current is the same for each point on the line. Also known as surge impedance.

characteristic length [MECH] A convenient reference length (usually constant) of a given configuration, such as overall length of an aircraft, the maximum diameter or radius of a body of revolution, or a chord or span of a lifting surface.

characteristic loss spectroscopy [SPECT] A branch of electron spectroscopy in which a solid surface is bombarded with monochromatic electrons, and backscattered particles which have lost an amount of energy equal to the core-level binding energy are detected. Abbreviated CLS.

characteristic manifold [MATH] A surface used to study the problem of existence of solutions to partial differential equations.

characteristic number *See* eigenvalue.

characteristic point [MATH] The characteristic point of a one-parameter family of surfaces corresponding to the value u_0 of the parameter is the limit of the point of intersection of the surfaces corresponding to the values u_0, u_1, and u_2 of the parameter as u_1 and u_2 approach u_0 independently.

characteristic polynomial [MATH] The polynomial whose roots are the eigenvalues of a given linear transformation on a finite dimensional vector space.

characteristic radiation [ATOM PHYS] Radiation originating in an atom following removal of an electron, whose wavelength depends only on the element concerned and the energy levels involved.

characteristic ray [MATH] For a differential equation, an integral curve which generates all the others.

characteristic root *See* eigenvalue.

characteristic temperature *See* Debye temperature.

characteristic value *See* eigenvalue.

characteristic vector *See* eigenvector.

characteristic x-rays [ATOM PHYS] Electromagnetic radiation emitted as a result of rearrangements of the electrons in the inner shells of atoms; the spectrum consists of lines whose wavelengths depend only on the element concerned and the energy levels involved.

charge [ELEC] **1.** A basic property of elementary particles of matter; the charge of an object may be a positive or negative number or zero; only integral multiples of the proton charge occur, and the charge of a body is the algebraic sum of the charges of its constituents; the value of the charge may be inferred from the Coulomb force between charged objects. Also known as electric charge. **2.** To convert electrical energy to chemical energy in a secondary battery. **3.** To feed electrical energy to a capacitor or other device that can store it. [NUCLEO] The fissionable material or fuel placed in a reactor to produce a chain reaction.

charge carrier [SOLID STATE] A mobile conduction electron or mobile hole in a semiconductor. Also known as carrier.

charge conjugation conservation [PARTIC PHYS] The principle that the laws of motion are left unchanged by the charge conjugation operation; it is violated by the weak interactions, but no other violations have as yet been established.

charge conjugation operation [PARTIC PHYS] The operation of changing every particle into its antiparticle.

charge conjugation parity *See* charge parity.

charge conservation *See* conservation of charge.

charge-coupled devices [ELECTR] Semiconductor devices arrayed so that the electric charge at the output of one provides the input stimulus to the next.

charged-current interaction [PARTIC PHYS] A weak interaction in which the charges of the interacting fermions are changed; easily observed processes such as beta decay are of this type.

charge density [ELEC] The charge per unit area on a surface or per unit volume in space.

charged particle [PARTIC PHYS] A particle whose charge is not zero; the charge of a particle is added to its designation as a superscript, with particles of charge $+1$ and -1 (in terms of the charge of the proton) denoted by $+$ and $-$ respectively; for example, π^+, Σ^-.

charge exchange [PHYS] The transfer of electric charge from one particle to another during a collision between the two particles.

charge independence [NUC PHYS] The principle that the nuclear (strong) force between a neutron and a proton is identical to the force between two protons or two neutrons in the same orbital and spin state. [PARTIC PHYS] As a generalization of the nuclear physics definition, the principle that the strong interactions of particles are unchanged if a particle is replaced by another particle of the same isotopic spin multiplet.

charge invariance [NUC PHYS] The principle that interactions between nucleons are left unchanged by rotations in isotopic spin space.

charge multiplet *See* isospin multiplet.

charge neutrality [PL PHYS] The near equality in the density of positive and negative charges throughout a volume, which is characteristic of a plasma. [SOLID STATE] The condition in which electrons and holes are present in equal numbers in a semiconductor.

charge parity [PARTIC PHYS] The eigenvalue of the charge conjugation operation; it exists only for a system which goes into itself under this operation. Also known as charge conjugation parity.

charge quantization [ELEC] The principle that the electric charge of an object must equal an integral multiple of a universal basic charge.

charger *See* battery charger.

charger-reader [NUCLEO] An auxiliary device used to charge and read small, portable ionization chambers.

charge-storage transistor [ELECTR] A transistor in which the collector-base junction will charge when forward bias is applied with the base at a high level and the collector at a low level.

charge transfer [PHYS CHEM] The process in which an ion takes an electron from a neutral atom, with a resultant transfer of charge.

charge-transfer device [ELECTR] A semiconductor device that depends upon movements of stored charges between predetermined locations, as in charge-coupled and charge-injection devices.

charging current [ELEC] The current that flows into a capacitor when a voltage is first applied.

Charles' law [PHYS] The law that at constant pressure the volume of a fixed mass or quantity of gas varies directly with the absolute temperature; a close approximation. Also known as Gay-Lussac law.

Charlier polynomials [MATH] Families of polynomials which are orthogonal with respect to Poisson distributions.

charm [PARTIC PHYS] A quantum number which has been proposed to account for an apparent lack of symmetry in the behavior of hadrons relative to that of leptons, to explain why certain reactions of elementary particles do not occur, and to account for the longevity of the J particle.

charmed particle [PARTIC PHYS] A particle whose total charm is not equal to zero.

charmed quark [PARTIC PHYS] A quark with an electric charge of $+\frac{2}{3}$, baryon number of $\frac{1}{3}$, zero strangeness, and charm of $+1$. Symbolized c.

charmonium [PARTIC PHYS] An elementary particle, such as the J particle, that is believed to be a bound state of the charmed quark c and its antiquark $\bar{c}$.

Charpak-Massonet current distribution system [PARTIC PHYS] An electronic data readout method used in spark chambers to locate a single spark, as determined by observing how the spark current divides between the two available paths to the ground.

Charpit's method [MATH] A method for finding a complete integral of the general first-order partial differential equation in two independent variables; it involves solving a set of five ordinary differential equations.

chart [MATH] An n-chart is a pair (U,h), where U is an open set of a topological space and h is a homeomorphism of U onto an open subset of n-dimensional euclidean space. [SCI TECH] A form, such as a graph, table, or diagram, which gives information about some variable quantity.

Chattock gage [ENG] A form of micromanometer in which observation of the interface between two immiscible liquids is used to determine when the pressure to be measured has been balanced by the pressure head resulting from tilting of the entire apparatus.

Chebyshev approximation *See* min-max technique.

Chebyshev filter [ELECTR] A filter in which the transmission frequency curve has an equal-ripple shape, with very small peaks and valleys.

Chebyshev polynomials [MATH] A family of orthogonal polynomials which solve Chebyshev's differential equation.

Chebyshev quadrature formula [MATH] A formula of the form

$$\int_a^b G(x)f(x)\ dx = \gamma \sum_0^m f(x_i) + R_m$$

CHEBYSHEV FILTER

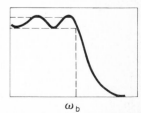

ω_b

Equal ripple shape characteristic in the transmission band of the Chebyshev filter. Frequency ω_b is lower limit of transition region from pass band to stop band.

where the points x_i are chosen so that the integration is exact when $f(x)$ is a polynomial of degree as high as possible.

Chebyshev's differential equation [MATH] A special case of Gauss' hypergeometric second-order differential equation: $(1 - x^2)f''(x) - xf'(x) + n^2 f(x) = 0$.

Chebyshev's inequality [STAT] Given a nonnegative random variable $f(x)$, and $k > 0$, the probability that $f(x) \geq k$ is less than or equal to the expected value of f divided by k.

Chebyshev system [MATH] A set of continuous functions $f_1(x), f_2(x), \ldots, f_n(x)$ is said to be a Chebyshev system on the open interval (a,b) if, for any set of real constants $\{c_k\}$ not all 0, the function

$$\sum_{k=1}^{n} c_k f_k(x)$$

does not vanish more than $n - 1$ times on the interval (a,b). Also known as T system.

cheese antenna [ELECTROMAG] An antenna having a parabolic reflector between two metal plates, dimensioned to permit propagation of more than one mode in the desired direction of polarization.

chelate laser [OPTICS] A liquid laser that uses a rare-earth chelate (a metalloorganic compound), with initial excitation taking place within the organic part of the liquid molecule and then transferring to the metallic ions to give lasing action. Also known as rare-earth chelate laser.

chemical bond *See* bond.

chemical compound *See* compound.

chemical dosimeter [NUCLEO] A dosimeter in which the accumulated radiation-exposure dose is indicated by color changes accompanying chemical reactions induced by the radiation.

chemical element *See* element.

chemical energy [PHYS CHEM] Energy of a chemical compound which, by the law of conservation of energy, must undergo a change equal and opposite to the change of heat energy in a reaction; the rearrangement of the atoms in reacting compounds to produce new compounds causes a change in chemical energy.

chemical film dielectric [ELEC] An extremely thin layer of material on one or both electrodes of an electrolytic capacitor, which conducts electricity in only one direction and thereby constitutes the insulating element of the capacitor.

chemical focus *See* actinic focus.

chemical laser *See* chemically pumped laser.

chemically pumped laser [OPTICS] A laser in which pumping is achieved by using a chemical action rather than electrical energy to produce the required pulses of light. Also known as chemical laser.

chemical polarity [PHYS CHEM] Tendency of a molecule, or compound, to be attracted or repelled by electrical charges because of an asymmetrical arrangement of atoms around the nucleus.

chemical potential [PHYS CHEM] In a thermodynamic system of several constituents, the rate of change of the Gibbs function of the system with respect to the change in the number of moles of a particular constituent.

chemical remanent magnetization [GEOPHYS] Permanent magnetization of rocks acquired when a magnetic material, such as hematite, is grown at low temperature through the oxidation of some other iron mineral, such as magnetite or goethite; the growing mineral becomes magnetized in the direction of any field which is present. Abbreviated CRM.

chemical shift [PHYS CHEM] Shift in a nuclear magnetic-

CHEMICAL SHIFT

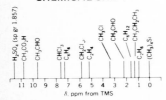

Chemical shifts for representative compounds. Decreasing values of δ correspond to increasing magnetic field in a constant-frequency spectrometer. The scale calibration is obtained from the resonance signal of a small amount of tetramethylsilane (TMS) placed in the sample tube to provide a zero reference point.

resonance spectrum resulting from diamagnetic shielding of the nuclei by the surrounding electrons.

chemical shim [NUCLEO] A chemical, usually boric acid, that is placed in the coolant system of a nuclear reactor to serve as a neutron absorber and that compensates for fuel burnup during normal operation.

chemical shutdown [NUCLEO] Addition of a dissolved poison to the coolant of a nuclear reactor to achieve shutdown.

chemical thermodynamics [PHYS CHEM] The application of thermodynamic principles to problems of chemical interest.

chemical tracer [NUCLEO] A tracer having chemical properties similar to those of the substance with which it is mixed.

chemisorption [CHEM] The process of chemical adsorption.

chemistry [SCI TECH] The scientific study of the properties, composition, and structure of matter, the changes in structure and composition of matter, and accompanying energy changes.

Chevalier lens [OPTICS] A type of magnifying lens composed of an achromatic negative lens combined with a distant collecting front lens; a magnifying power up to 10X with an object distance up to 3 inches (7.62 centimeters) can be obtained.

Chevalley's theorem [MATH] Let $f(x_1, \ldots, x_n)$ be a polynomial of degree less than n with coefficients in a finite field F, and let $f(0, \ldots, 0) = 0$; then there is at least one other set of elements $a_1, \ldots, a_n$ in F such that $f(a_1, \ldots, a_n) = 0$.

Chézy formula [FL MECH] For the velocity V of open-channel flow which is steady and uniform, $V = \sqrt{8g/f} \cdot \sqrt{mS}$, where f is the Darcy-Weisbach friction coefficient, m the hydraulic radius, S the energy dissipation per unit length, and g the acceleration of gravity.

chief ray [OPTICS] A ray in a pencil which passes through the intersection of the axis of an optical system with the plane of the aperture stop.

Child-Langmuir equation *See* Child's law.

Child-Langmuir-Schottky equation *See* Child's law.

Child's law [ELECTR] A law stating that the current in a thermionic diode varies directly with the three-halves power of anode voltage and inversely with the square of the distance between the electrodes, provided the operating conditions are such that the current is limited only by the space charge. Also known as Child-Langmuir equation; Child-Langmuir-Schottky equation; Langmuir-Child equation.

chi meson [PARTIC PHYS] A meson resonance of mass 958 MeV/c^2, designated χ_0, which has 0 isospin and charge, negative parity, positive G parity, and spin probably equal to 0. Also known as eta-prime meson (η'). Also denoted η'_A (958).

Chinese remainder theorem [MATH] **1.** If $m_1, m_2, \ldots, m_n$ is a set of integers which are relatively prime in pairs and $b_1, \ldots, b_n$ is a set of integers, then there exists an integer x such that $x \equiv b_i \pmod{m_i}$ for $i = 1, \ldots, n$. **2.** More generally, if R is a commutative ring with identity, $b_1, \ldots, b_n$ are elements of R, and $I_1, \ldots, I_n$ are ideals in R, such that for all $i \neq j$ any element in R can be written as the sum of an element in I_i and an element in I_j, then there exists an element x in R such that $x \equiv b_i \pmod{I_i}$ for $i = 1, \ldots, n$.

chip [ELECTR] **1.** The shaped and processed semiconductor die that is mounted on a substrate to form a transistor, diode, or other semiconductor device. **2.** An integrated microcircuit performing a significant number of functions and constituting a subsystem.

chip capacitor [ELECTR] A single-layer or multilayer monolithic capacitor constructed in chip form, with metallized

CHEVALIER LENS

Chevalier magnifying lens.

terminations to facilitate direct bonding on hybrid integrated circuits.

chip circuit *See* large-scale integrated circuit.

chip resistor [ELECTR] A thick-film resistor constructed in chip form, with metallized terminations to facilitate direct bonding on hybrid integrated circuits.

chirality [CHEM] The handedness of an asymmetric molecule. [PARTIC PHYS] The characteristic of particles of spin $\frac{1}{2} \hbar$ that are allowed to have only one spin state with respect to an axis of quantization parallel to the particle's momentum; if the particle's spin is always parallel to its momentum, it has positive chirality; antiparallel, negative chirality. [PHYS] The characteristic of an object that cannot be superimposed upon its mirror image.

chiral symmetry group [PARTIC PHYS] A group of symmetry transformations that act differently on the left- and right-handed parts of fermion fields.

chiral twinning *See* optical twinning.

chi-square distribution [STAT] The distribution of the sample variances indicated by

$$S^2 = \sum_{i=1}^{n} (x_i - \bar{x})^2 / (n-1),$$

where $x_1, x_2, \ldots, x_n$ are observations of a random sample n drawn from a normal population.

chi-square statistic [STAT] A statistic which is distributed approximately in the form of a chi-square distribution; used in goodness-of-fit.

chi-square test [STAT] A generalization, and an extension, of a test for significant differences between a binomial population and a multinomial population, wherein each observation may fall into one of several classes and furnishes a comparison among several samples instead of just two.

Chladni's figures [MECH] Figures produced by sprinkling sand or similar material on a horizontal plate and then vibrating the plate while holding it rigid at its center or along its periphery; indicate the nodal lines of vibration.

chlorine [CHEM] A chemical element, symbol Cl, atomic number 17, atomic weight 35.453; used in manufacture of solvents, insecticides, and many non-chlorine-containing compounds, and to bleach paper and pulp.

chlorine-36 [NUCLEO] A radioactive isotope of chlorine with atomic mass number of 36; a beta emitter with a half-life of 3×10^5 years.

choke [ELEC] An inductance used in a circuit to present a high impedance to frequencies above a specified frequency range without appreciably limiting the flow of direct current. Also known as choke coil. [ELECTROMAG] A groove or other discontinuity in a waveguide surface so shaped and dimensioned as to impede the passage of guided waves within a limited frequency range.

choke coil *See* choke.

choke coupling [ELECTROMAG] Coupling between two parts of a waveguide system that are not in direct mechanical contact with each other.

choked flow [FL MECH] Flow in a duct or passage such that the flow upstream of a certain critical section cannot be increased by a reduction of downstream pressure.

choke flange [ELECTROMAG] A waveguide flange having in its mating surface a slot (choke) so shaped and dimensioned as to restrict leakage of microwave energy within a limited frequency range.

choke input filter [ELEC] A power-supply filter in which the first filter element is a series choke. Also known as choke filter.

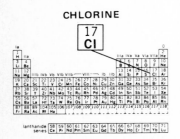

CHLORINE

Periodic table of the chemical elements showing the position of chlorine.

choke joint [ELECTROMAG] A connection between two waveguides that uses two mating choke flanges to provide effective electrical continuity without metallic continuity at the inner walls of the waveguide.

choke piston [ELECTROMAG] A piston in which there is no metallic contact with the walls of the waveguide at the edges of the reflecting surface; the short circuit for high-frequency currents is achieved by a choke system. Also known as noncontacting piston; noncontacting plunger.

choking [FL MECH] The condition which prevails in compressible fluid flow when the upper limit of mass flow is reached, or when the speed of sound is reached in a duct.

choking Mach number [FL MECH] The Mach number at some reference point in a duct or passage (for example, at the inlet) at which the flow in the passage becomes choked.

cholesteric material [PHYS CHEM] A liquid crystal material in which the elongated molecules are parallel to each other within the plane of a layer, but the direction of orientation is twisted slightly from layer to layer to form a helix through the layers.

cholesteric phase [PHYS CHEM] A form of the nematic phase of a liquid crystal in which the molecules are spiral.

chopper [PHYS] A device for interrupting an electric current, beam of light, or beam of infrared radiation at regular intervals, to permit amplification of the associated electrical quantity or signal by an alternating-current amplifier; also used to interrupt a continuous stream of neutrons to measure velocity.

chopping [ELECTR] Removal, by electronic means, of one or both extremities of a wave at a predetermined level. [PHYS] The act of interrupting an electric current, beam of light, beam of infrared radiation, or stream of neutrons at regular intervals.

Choquet theorem [MATH] Let K be a compact convex set in a locally convex Hausdorff real vector space and assume that either (1) the set of extreme points of K is closed or (2) K is metrizable; then for every point x in K there is at least one Radon probability measure m on X, concentrated on the set of extreme points of K, such that x is the centroid of m.

chord [ACOUS] A combination of two or more tones. [MATH] A line segment which intersects a curve or surface only at the end points of the segment.

Christiansen filter [OPTICS] A type of color filter, a solid-in-liquid suspension, which scatters all incident energy except that of a narrow frequency range out of the direct beam. Also known as band-pass filter.

Christoffel-Darboux formula [MATH] For a set of orthogonal polynomials, $f_n(x)$, $n = 0, 1, 2, \ldots$, satisfying

$$\int_a^b w(x) f_n(x) f_m(x) \, dx = 0$$

for m not equal to n, this is the formula

$$\sum_{m=0}^{n} \frac{1}{h_m} f_m(x) f_m(y) = \frac{k_n}{k_{n+1} h_n} \frac{f_{n+1}(x) f_n(y) - f_n(x) f_{n+1}(y)}{x - y}$$

where

$$h_n = \int_a^b w(x) f_n^2(x) \, dx$$

and k_n is the coefficient of x_n in f_n.

Christoffel symbols [MATH] Symbols which represent particular functions of the coefficients and their first-order derivatives of a quadratic form.

chroma [OPTICS] **1.** The dimension of the Munsell system of color that corresponds most closely to saturation, which is the

CHROMATICITY DIAGRAM

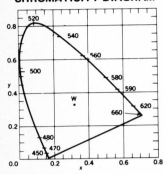

International Committee on Illumination chromaticity diagram. The wavelengths of the visible spectrum in units of 10^{-9} meter are indicated along the curve. W represents a white composed of equal amounts of the three primaries. *(Adapted from A. C. Hardy, ed., Handbook of Colorimetry, copyright 1936 by The MIT Press)*

CHROMIUM

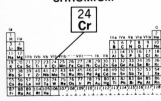

Periodic table of the chemical elements showing the position of chromium.

degree of vividness of a hue. Also known as Munsell chroma. **2.** *See* color saturation.

chromascope [OPTICS] An instrument used to determine the optical effects of color.

chromatic [OPTICS] Relating to color.

chromatic aberration [OPTICS] An optical lens defect causing color fringes, because the lens material brings different colors of light to focus at different points. Also known as color aberration.

chromatic color [OPTICS] A color which has hue and saturation; that is, a color other than white, black, or a shade of gray.

chromatic diagram *See* chromaticity diagram.

chromatic difference of magnification [OPTICS] Variation in the size of the image produced by an optical system with the wavelength (or, equivalently, color) of light. Also known as lateral chromatic aberration.

chromaticity [OPTICS] The color quality of light that can be defined by its chromaticity coordinates; depends only on hue and saturation of a color, and not on its luminance (brightness).

chromaticity coordinates [OPTICS] The fractional amounts of the x, y, and z primary colors, specified by the International Committee on Illumination, in a color sample; more precisely, $x = X/(X + Y + Z)$, $y = Y/(X + Y + Z)$, $z = Z/(X + Y + Z)$, where X, Y, and Z are the integrals over wavelength λ of the product of the amount of light emerging from the sample per unit wavelength, and the tristimulus values, $\bar{x}(\lambda)$, $\bar{y}(\lambda)$, and $\bar{z}(\lambda)$ respectively.

chromaticity diagram [OPTICS] A triangular graph for specifying colors, whose ordinate is the y chromaticity coordinate and whose abscissa is the x chromaticity coordinate; the apexes of the triangle represent primary colors. Also known as chromatic diagram.

chromatic parallax [OPTICS] The change of the apparent position of a line image in an optical instrument, relative to a graticule, when the wavelength of the incident radiation is varied.

chromatic resolving power [OPTICS] The difference between two equally strong spectral lines that can barely be separated by a spectroscopic instrument, divided into the average wavelength of these two lines; for prisms and gratings Rayleigh's criteria are used, and the term is defined as the width of the emergent beam times the angular dispersion.

chromatics [OPTICS] **1.** The branch of optics concerned with the properties of colors. **2.** The part of colorimetry concerned with hue and saturation.

chromatic sensitivity [OPTICS] The smallest change in wavelength of light that produces a change in hue which is just large enough to be detected by human vision.

chromatic vision [PHYSIO] Vision pertaining to the color sense, that is, the perception and evaluation of the colors of the spectrum.

chromatoscope [OPTICS] An instrument in which light beams are used to mix color stimuli.

chrominance [OPTICS] The difference between any color and a specified reference color of equal brightness; in color television, this reference color is white having coordinates $x = 0.310$ and $y = 0.316$ on the chromaticity diagram.

chromium [CHEM] A metallic chemical element, symbol Cr, atomic number 24, atomic weight 51.996.

chromium-51 [NUC PHYS] A radioactive isotope with atomic mass 51 made by neutron bombardment of chromium; radiates gamma rays.

chromodynamics [PARTIC PHYS] A theory of the interaction between quarks carrying color in which the quarks exchange gluons in a manner analogous to the exchange of photons between charged particles in electrodynamics.

chromoradiometer [ENG] A radiation meter that uses a substance whose color changes with x-ray dosage.

chromosphere [ASTRON] A transparent, tenuous layer of gas that rests on the photosphere in the atmosphere of the sun.

chronological future [RELAT] The chronological future relative to a set of points S in a space-time M is the set of points in M which can be reached from S by future-directed timelike curves.

chronological past [RELAT] The chronological past relative to a set of points S in a space-time M is the set of points in M which can be reached from S by past-directed timelike curves.

chronon [PHYS] A hypothetical quantum of time, given approximately by the time taken for light to traverse the classical electron radius, on the order of 10^{-23} second.

Chrystal's equation [MATH] The partial differential equation $(1 + \sqrt{z - x - y})(\partial z/\partial x) + (\partial z/\partial y) = 2$.

CHU *See* centigrade heat unit.

CHU_mean *See* centigrade heat unit.

chugging [NUCLEO] An instability in a water-moderated reactor in which the formation of steam bubbles in the core and their subsequent collapse cause oscillations in the reactivity.

CI *See* color index.

C index [GEOPHYS] A subjectively obtained daily index of geomagnetic activity, in which each day's record is evaluated on the basis of 0 for quiet, 1 for moderately disturbed, and 2 for very disturbed. Also known as C figure; magnetic character figure.

cineradiography [GRAPHICS] A version of flash radiography in which a succession of flashes is used to form a moving picture of an object.

circle [MATH] **1.** The set of all points in the plane at a given distance from a fixed point. **2.** A unit of angular measure, equal to one complete revolution, that is, to 2π radians or $360°$. Also known as turn.

circle diagram [ELEC] A diagram which gives a graphical solution of equations for a transmission line, giving the input impedance of the line as a function of load impedance and electrical length of the line.

circle of Apollonius [MATH] For any two fixed points A and B and positive number r, the locus of points P such that $(AP)/(BP) = r$. Also known as Apollonius' circle.

circle of confusion [OPTICS] The blurred circular image of a point object which is formed by a camera lens, even with the best focusing. Also known as circle of least confusion.

circle of convergence [MATH] The region in which a power series possesses a limit.

circle of curvature [MATH] The circle tangent to a curve on the concave side and having the same curvature at the point of tangency as does the curve.

circle of declination *See* hour circle.

circle of equal declination *See* parallel of declination.

circle of latitude [ASTRON] A great circle of the celestial sphere through the ecliptic poles, and hence perpendicular to the plane of the ecliptic. Also known as parallel of latitude. [GEOD] A meridian of the terrestrial sphere along which latitude is measured.

circle of least confusion *See* circle of confusion.

circle of longitude [ASTRON] A circle of the celestial sphere, parallel to the ecliptic.

circle of perpetual apparition [ASTRON] That circle of the celestial sphere, centered on the polar axis and having a polar distance from the elevated pole approximately equal to the latitude of the observer, within which celestial bodies do not set.

circle of perpetual occultation [ASTRON] That circle of the celestial sphere, centered on the polar axis and having a polar distance from the depressed pole approximately equal to the latitude of the observer, within which celestial bodies do not rise.

circle of right ascension *See* hour circle.

circle polynomials [MATH] A set of polynomials V_n^m *(x,y)*, in two real variables x and y, which are orthogonal over the unit disk and which, when expressed in polar coordinates r and θ, have the form $R_n^m(r)e^{im\theta}$, where $R_n^m(r)$ is a polynomial in r; used in the theory of the phase-contrast microscope. Also known as Zernike polynomials.

circle theorem [FL MECH] If $f(z)$ represents the complex potential of the two-dimensional flow of an incompressible inviscid fluid in the complex z-plane, and if there are no rigid boundaries and all singularities of $f(z)$ are at a distance greater than a from the origin, then if a circular cylinder with equation $|z| = a$ is introduced into the flow, the new complex potential is $f(z) + f^*(a^2/z)$, where f^* represents the complex conjugate of f.

circuit *See* electric circuit.

circuital field *See* rotational field.

circuit analyzer *See* volt-ohm-milliammeter.

circuit element *See* component.

circuit theory [ELEC] The mathematical analysis of conditions and relationships in an electric circuit. Also known as electric circuit theory.

circuit determinant [MATH] A determinant in which the elements of each row are the same as those of the previous row moved one place to the right, with the last element put first.

circulant matrix [MATH] A matrix in which the elements of each row are those of the previous row moved one place to the right.

circular accelerator *See* circular particle accelerator.

circular antenna [ELECTROMAG] A folded dipole that is bent into a circle, so the transmission line and the abutting folded ends are at opposite ends of a diameter.

circular birefringence [OPTICS] The phenomenon in which an optically active substance transmits right circularly polarized light with a different velocity from left circularly polarized light.

circular cylinder [MATH] A solid bounded by two parallel planes and a cylindrical surface whose intersections with planes perpendicular to the straight lines forming the surface are circles.

circular dichroism [OPTICS] A change from planar to elliptic polarization when an initially plane-polarized light wave traverses an optically active medium. Abbreviated CD.

circular electric wave [ELECTROMAG] A transverse electric wave for which the lines of electric force form concentric circles.

circular flow method [FL MECH] A method to determine viscosities of Newtonian fluids by measuring the torque from viscous drag of sample material between a closely spaced rotating plate–stationary cone assembly.

circular functions *See* trigonometric functions.

circular horn [ELECTROMAG] A circular-waveguide section that flares outward into the shape of a horn, to serve as a feed for a microwave reflector or lens.

circular inch [MECH] The area of a circle 1 inch (25.4 millimeters) in diameter.

circular magnetic wave [ELECTROMAG] A transverse magnetic wave for which the lines of magnetic force form concentric circles.

circular mil [MECH] A unit equal to the area of a circle whose diameter is 1 mil (0.001 inch); used chiefly in specifying cross-sectional areas of round conductors. Abbreviated cir mil.

circular motion [MECH] **1.** Motion of a particle in a circular path. **2.** Motion of a rigid body in which all its particles move in circles about a common axis, fixed with respect to the body, with a common angular velocity.

circular nomograph [MATH] A chart with concentric circular scales for three variables, laid out so that any straight line passes through values of the variables satisfying a given equation.

circular orbit [ASTRON] An orbit comprising a complete constant-altitude revolution around the earth.

circular particle accelerator [NUCLEO] A particle accelerator which utilizes a magnetic field to bend charged-particle orbits and confine the extent of particle motion. Also known as circular accelerator.

circular point at infinity [MATH] In projective geometry, one of two points at which every circle intersects the ideal line.

circular polarization [PHYS] Attribute of a transverse wave (either of electromagnetic radiation, or in an elastic medium) whose electric or displacement vector is of constant amplitude and, at a fixed point in space, rotates in a plane perpendicular to the propagation direction with constant angular velocity.

circular segment [MATH] Portion of circle cut off from the main body of the circle by a straight line (chord) through the circle.

circular slide rule [MATH] A slide rule in a circular form whose advantages over a straight slide rule are its precision, because it is equivalent to a straight slide rule many times longer than the circular slide rule's diameter, and ease of multiplication, because the scale is continuous.

circular velocity [MECH] At any specific distance from the primary, the orbital velocity required to maintain a constant-radius orbit.

circular waveguide [ELECTROMAG] A waveguide whose cross-sectional area is circular.

circulating electromagnetic wave [ELECTROMAG] An electromagnetic wave whose equiphase surfaces are half-planes originating at a common axis.

circulating reactor [NUCLEO] A nuclear reactor in which the fissionable material circulates through the core in fluid form or as small particles suspended in a fluid.

circulation [FL MECH] The flow or motion of fluid in or through a given area or volume. [MATH] For the circulation of a vector field around a closed path, the line integral of the field vector around the path.

circulator [ELECTROMAG] A waveguide component having a number of terminals so arranged that energy entering one terminal is transmitted to the next adjacent terminal in a particular direction. Also known as microwave circulator.

circumcenter [MATH] For a triangle or a regular polygon, the center of the circle that is circumscribed about the triangle or polygon.

circumference [MATH] **1.** The length of a circle. **2.** For a sphere, the length of any great circle on the sphere.

circumhorizontal arc [OPTICS] A halo phenomenon consisting of a colored arc, red on its upper margin; it extends for

CIRCULAR NOMOGRAPH

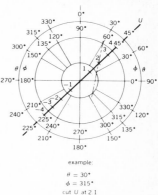

example:

$\theta = 30°$
$\phi = 315°$
cut U at 2 1

Circular nomograph which results from a trigonometric equation expressed in the form of a determinant.

about 90° parallel to the horizon and lies about 46° below the sun.

circummeridian altitude *See* exmeridian altitude.

circumpolar star [ASTRON] A star with its polar distance approximately equal to or less than the latitude of the observer.

circumradius [MATH] The radius of a circle that is circumscribed about a polygon.

circumscribed [MATH] **1.** A closed curve (or surface) is circumscribed about a polygon (or polyhedron) if every vertex of the polygon (or polyhedron) is incident upon the curve (or surface) and the polygon (or polyhedron) is contained in the curve (or surface). **2.** A polygon (or polyhedron) is circumscribed about a closed curve (or surface) if every side of the polygon (or face of the polyhedron) is tangent to the curve (or surface) and the curve (or surface) is contained within the polygon (or polyhedron).

circumzenithal arc [OPTICS] A brilliant rainbow-colored arc of about a quarter of a circle with its center at the zenith and about 46° above the sun, produced by refraction and dispersion of the sun's light striking the top of prismatic ice crystals in the atmosphere, and usually lasting only a few minutes.

cir mil *See* circular mil.

cislunar [ASTRON] Of or pertaining to phenomena, projects, or activity in the space between the earth and moon, or between the earth and the moon's orbit.

cissoid [MATH] A plane curve consisting of all points which lie on a variable line passing through a fixed point on a circle, and whose distance from the fixed point is equal to the distance from the line's intersection with the circle to its intersection with the tangent to the circle at the point diametrically opposite the fixed point; in cartesian coordinates the equation is $y^2(2a - x) = x^3$.

cissoid of Diocles [MATH] A plane curve consisting of all points which lie at the foot of a perpendicular from a fixed point to a variable line that remains tangent to a parabola.

civil day [ASTRON] A mean solar day beginning at midnight instead of at noon; may be based on either apparent solar time or mean solar time.

civil time [ASTRON] Solar time in a day (civil day) that begins at midnight; may be either apparent solar time or mean solar time.

civil twilight [ASTRON] The interval of incomplete darkness between sunrise (or sunset) and the time when the center of the sun's disk is 6° below the horizon.

Cl *See* chlorine.

cladding [NUCLEO] An outer jacket, usually metallic, for a nuclear fuel element; prevents corrosion of fuel and release of fission products into the coolant.

Clairaut equation [MATH] A first-order differential equation that can be written in the form $y - y'x = f(y')$, where $y' = dy/dx$.

Clairaut's formula [GEOD] An approximate formula for gravity at the earth's surface, assuming that the earth is an ellipsoid; states that the gravity is equal to $g_e[1 + (5/2\,m' - f)\sin^2\theta]$, where θ is the latitude, g_e is the gravity at the equator, m' is the ratio of centrifugal acceleration to gravity at the equatorial surface, and f is the earth's flattening, equal to $(a - b)/a$, where a is the semimajor axis and b is the semiminor axis.

clamp *See* clamping circuit.

clamping [ELECTR] The introduction of a reference level that has some desired relation to a pulsed waveform, as at the negative or positive peaks. Also known as direct-current reinsertion; direct-current restoration.

clamping circuit [ELECTR] A circuit that reestablishes the direct-current level of a waveform; used in the dc restorer stage of a television receiver to restore the dc component to the video signal after its loss in capacitance-coupled alternating-current amplifiers, to reestablish the average light value of the reproduced image. Also known as clamp.

clamping diode [ELECTR] A diode used to clamp a voltage at some point in a circuit.

Clapeyron-Clausius equation *See* Clausius-Clapeyron equation.

Clapeyron equation *See* Clausius-Clapeyron equation.

Clapeyron's theorem [MECH] The theorem that the strain energy of a deformed body is equal to one-half the sum over three perpendicular directions of the displacement component times the corresponding force component, including deforming loads and body forces, but not the six constraining forces required to hold the body in equilibrium.

Clapp oscillator [ELECTR] A series-tuned Colpitts oscillator, having low drift.

class [MATH] The class of an algebraic plane curve is the number of tangents to the curve which pass through a general point in the plane.

class A amplifier [ELECTR] 1. An amplifier in which the grid bias and alternating grid voltages are such that anode current in a specific tube flows at all times. 2. A transistor amplifier in which each transistor is in its active region for the entire signal cycle.

class AB amplifier [ELECTR] 1. An amplifier in which the grid bias and alternating grid voltages are such that anode current in a specific tube flows for appreciably more than half but less than the entire electric cycle. 2. A transistor amplifier whose operation is class A for small signals and class B for large signals.

class B amplifier [ELECTR] 1. An amplifier in which the grid bias is approximately equal to the cutoff value, so that anode current is approximately zero when no exciting grid voltage is applied, and flows for approximately half of each cycle when an alternating grid voltage is applied. 2. A transistor amplifier in which each transistor is in its active region for approximately half the signal cycle.

class C amplifier [ELECTR] 1. An amplifier in which the bias on the control element is appreciably greater than the cutoff valve, so that the output current in each device is zero when no alternating control signal is applied, and flows for appreciably less than half of each cycle when an alternating control signal is applied. 2. A transistor amplifier in which each transistor is in its active region for significantly less than half the signal cycle.

class formula [MATH] The order of a finite group G is equal to the sum, over a set of representatives x_i of the distinct conjugacy classes of G, of the index of the normalizer of x_i in G.

class function [MATH] A class function of a group G over a field F is a function f from G to F such that $f(sts^{-1}) = f(t)$ for all s and t in G.

classical approximation [QUANT MECH] The approximation that Planck's constant may be considered infinitely small; the laws of quantum mechanics must then reduce to those of classical mechanics.

classical conductivity theory [STAT MECH] A theory which treats the system of electrons in a metal as a gas and uses the Boltzmann transport equation to calculate conductivity.

classical electron radius [ELECTROMAG] The quantity $e^2/m_e c^2$, where e is the electron's charge in electrostatic units, m_e its mass, and c the speed of light; equal to approximately 2.82×10^{-13} centimeter.

classical field theory [PHYS] The study of distributions of energy, matter, and other physical quantities under circumstances where their discrete nature is unimportant, and they may be regarded as (in general, complex) continuous functions of position. Also known as c-number theory; continuum mechanics; continuum physics.

classical mechanics [MECH] Mechanics based on Newton's laws of motion.

classical wave equation *See* wave equation.

class interval [STAT] One of several convenient intervals into which the values of the variate of a frequency distribution may be grouped.

class mark [STAT] The mid-value of a class interval, or the integral value nearest the midpoint of the interval.

classons [PARTIC PHYS] Massless bosons which are quanta of the two classical fields, gravitational and electromagnetic.

clausius [THERMO] A unit of entropy equal to the increase in entropy associated with the absorption of 1000 international table calories of heat at a temperature of 1 K, or to 4186.8 joules per kelvin.

Clausius-Clapeyron equation [THERMO] An equation governing phase transitions of a substance, $dp/dT = \Delta H/(T\Delta V)$, in which p is the pressure, T is the temperature at which the phase transition occurs, ΔH is the change in heat content (enthalpy), and ΔV is the change in volume during the transition. Also known as Clapeyron-Clausius equation; Clapeyron equation.

Clausius equation [THERMO] **1.** An equation of state for gases which applies a correction to the van der Waals equation: $\{P + (n^2a/[T(V + c)^2])\} (V - nb) = nRT$, where P is the pressure, T the temperature, V the volume of the gas, n the number of moles in the gas, R the gas constant, a depends only on temperature, b is a constant, and c is a function of a and b. **2.** The equation $c_2 - c_1 = T(d/dT)(\Delta H/T)$, where c_1 and c_2 are the specific heat capacities of a liquid and its vapor, and ΔH is the heat of vaporization at absolute temperature T.

Clausius inequality [THERMO] The principle that for any system executing a cyclical process, the integral over the cycle of the infinitesimal amount of heat transferred to the system divided by its temperature is equal to or less than zero. Also known as Clausius theorem; inequality of Clausius.

Clausius law [THERMO] The law that an ideal gas's specific heat at constant volume does not depend on the temperature.

Clausius-Mosotti equation [ELEC] An expression for the polarizability γ of an individual molecule in a medium which has the relative dielectric constant ε and has N molecules per unit volume: $\gamma = (3/4\pi N) [(\varepsilon - 1)/(\varepsilon + 2)]$ (Gaussian units).

Clausius-Mosotti-Lorentz-Lorenz equation [ELECTROMAG] The equation that results from replacing the real relative dielectric constant in the Clausius-Mosotti equation, or the real index of refraction in the Lorentz-Lorenz equation, with its complex counterpart.

Clausius number [THERMO] A dimensionless number used in the study of heat conduction in forced fluid flow, equal to $V^3 L\rho/k\Delta T$, where V is the fluid velocity, ρ is its density, L is a characteristic dimension, k is the thermal conductivity, and ΔT is the temperature difference.

Clausius range [STAT MECH] The condition in which the

mean free path of molecules in a gas is much smaller than the dimensions of the container.

Clausius' statement [THERMO] A formulation of the second law of thermodynamics, stating it is not possible that, at the end of a cycle of changes, heat has been transferred from a colder to a hotter body without producing some other effect.

Clausius theorem *See* Carnot-Clausius equation; Clausius inequality.

Clausius virial theorem [STAT MECH] The theorem that in a system of particles whose positions and velocities are bounded, the total kinetic energy of the system averaged over a long period of time equals the virial of the system. Also known as virial theorem.

cleavage plane [CRYSTAL] Plane along which a crystalline substance may be split.

Clebsch-Gordan coefficient *See* vector coupling coefficient.

Clifford algebra [MATH] An algebra generated by a basis consisting of elements $a_1, \ldots, a_n$, with $a_i^2 = 1, i = 1, \ldots, n$, and $a_i a_j + a_j a_i = 0$ for $i \neq j$.

clinoaxis [CRYSTAL] The inclined lateral axis that makes an oblique angle with the vertical axis in the monoclinic system. Also known as clinodiagonal.

clinodiagonal *See* clinoaxis.

clinographic projection [GRAPHICS] A method of representing objects, especially crystals, in which each point P of the object to be represented is projected onto the foot of a perpendicular from P to a plane which is located so that no place surface of the object is represented by a line.

clinohedral class [CRYSTAL] A rare class of crystals in the monoclinic system having a plane of symmetry but no axis of symmetry. Also known as domatic class.

clinopinacoid [CRYSTAL] A form of monoclinic crystal whose faces are parallel to the inclined and vertical axes.

clipper *See* limiter.

clipper diode [ELECTR] A bidirectional breakdown diode that clips signal voltage peaks of either polarity when they exceed a predetermined amplitude.

clipper-limiter [ELECTR] A device whose output is a function of the instantaneous input amplitude for a range of values lying between two predetermined limits but is approximately constant, at another level, for input values above the range.

clipping *See* limiting.

clipping circuit *See* limiter.

clipping level [ELECTR] The level at which a clipping circuit is adjusted; for example, the magnitude of the clipped wave shape.

clock paradox [RELAT] The apparent contradiction between the principle of relativity, which asserts the equivalence of different observers, and the prediction, also part of the theory of relativity, that the clock of an observer who passes back and forth will be slower than the clock of an observer at rest. Also known as twin paradox.

close coupling [ELEC] **1.** The coupling obtained when the primary and secondary windings of a radio-frequency or intermediate-frequency transformer are close together. **2.** A degree of coupling that is greater than critical coupling. Also known as tight coupling.

closed ball [MATH] The set of points in a metric space whose distance from a specified point is equal to or less than a specified number.

closed circuit [ELEC] A complete path for current.

closed cycle [THERMO] A thermodynamic cycle in which the thermodynamic fluid does not enter or leave the system, but is used over and over again.

closed-cycle reactor [NUCLEO] A nuclear reactor in which the primary coolant flows to a heat exchanger and then recirculates through the core in a completely closed circuit.

closed graph theorem [MATH] If T is a linear transformation on a Banach space X to a Banach space Y whose domain $D(T)$ is closed and whose graph, that is, the set of pairs (x, Tx) for x in $D(T)$, is closed in $X \times Y$, then T is bounded (and hence continuous).

closed intervals [MATH] A closed interval of real numbers, denoted by $[a,b]$, consists of all numbers equal to or greater than a and equal to or less than b.

closed linear manifold [MATH] A topologically closed vector subspace of a topological vector space.

closed loop [CONT SYS] A family of automatic control units linked together with a process to form an endless chain; the effects of control action are constantly measured so that if the controlled quantity departs from the norm, the control units act to bring it back.

closed-loop control system *See* feedback control system.

closed magnetic circuit [ELECTROMAG] A complete circulating path for magnetic flux around a core of ferromagnetic material.

closed map [MATH] A function between two topological spaces which sends each closed set of one into a closed set of the other.

closed operator [MATH] A linear transformation f whose domain A is contained in a normed vector space X satisfying the condition that if $\lim x_n = x$ for a sequence x_n in A, and $\lim f(x_n) = y$, then x is in A and $f(x) = y$.

closed orthonormal set *See* complete orthonormal set.

closed pair [MECH] A pair of bodies that are subject to constraints which prevent any relative motion between them.

closed set [MATH] A set of points which contains all its cluster points. Also known as topologically closed set.

closed shell [PHYS] An atomic or nuclear shell containing the maximum number of electrons or nucleons allowed by the Pauli exclusion principle.

closed system [THERMO] A system which is isolated so that it cannot exchange matter or energy with its surroundings and can therefore attain a state of thermodynamic equilibrium.

closed trapped surface [RELAT] A compact spacelike two-surface in space-time such that outgoing null rays perpendicular to the surface are not expanding.

close-packed crystal [CRYSTAL] A crystal structure in which the lattice points are centers of spheres of equal radius arranged so that the volume of the interstices between the spheres is minimal.

closing line [MECH] The vector required to complete a polygon consisting of a set of vectors whose sum is zero (such as the forces acting on a body in equilibrium).

closure [MATH] The union of a set and its cluster points; the smallest closed set containing the set.

closure domain [SOLID STATE] A small ferromagnetic domain whose position and orientation ensure that the flux lines between adjacent larger domains close on themselves. Also known as flux-closure domain.

clothing monitor [NUCLEO] An instrument designed for monitoring radioactive contamination on clothing.

clothoid *See* Cornu's spiral.

cloud [NUC PHYS] The nucleons that are in the nucleus of an atom but not in closed shells.

cloud attenuation [ELECTROMAG] The attenuation of microwave radiation by clouds (for the centimeter-wavelength band, clouds produce Rayleigh scattering); due largely to scattering, rather than absorption, for both ice and water clouds.

cloud chamber [NUCLEO] A particle detector in which the path of a charged particle is made visible by the formation of liquid droplets along the trail of ions left by the particle as it passes through the gas of the chamber. Also known as expansion chamber; fog chamber.

cloud column [NUCLEO] The column of smoke extending upward from the point of burst of an atomic weapon.

cloud discharge [GEOPHYS] A lightning discharge occurring between a positive charge center and a negative charge center, both of which lie in the same cloud. Also known as cloud flash; intracloud discharge.

cloud flash *See* cloud discharge.

cloud-ion chamber [NUCLEO] An instrument combining the functions of a Wilson cloud chamber with those of an ionization chamber.

cloud-to-cloud discharge [GEOPHYS] A lightning discharge occurring between a positive charge center of one cloud and a negative charge center of a second cloud. Also known as intercloud discharge.

cloud-to-ground discharge [GEOPHYS] A lightning discharge occurring between a charge center (usually negative) in the cloud and a center of opposite charge at the ground. Also known as ground discharge.

cloud track [NUCLEO] The string of minute water droplets that forms along the path of an ionizing particle in the supersaturated vapor of a cloud chamber.

cloudy crystal-ball model [NUC PHYS] An optical analogy used in explaining scattering of nucleons by nuclei, in which the nucleus is thought of as a sphere of nuclear matter which partially refracts and partially absorbs the incident nucleon (de Broglie) wave. Also known as optical model.

cloverleaf antenna [ELECTROMAG] Antenna having radiating units shaped like a four-leaf clover.

CLS *See* characteristic loss spectroscopy.

clusec [MECH ENG] A unit of power used to measure the power of evacuation of a vacuum pump, equal to the power associated with a leak rate of 1 centiliter per second at a pressure of 1 millitorr, or to approximately 1.33322×10^{-6} watt.

Clusins column [NUCLEO] A device for separating isotopes by thermal diffusion, consisting of a long vertical tube with an electrically heated wire along its axis that produces a temperature gradient, causing the lighter isotope to concentrate around the wire and the heavier isotope to concentrate near the walls.

cluster cepheids *See* RR Lyrae stars.

cluster expansion [STAT MECH] A virial expansion in which the virial coefficients (of inverse powers of the volume of the gas in question) are obtained from integrals, over positions of a small number of molecules, of functions involving intermolecular potentials.

cluster point [MATH] A cluster point of a set in a topological space is a point p whose neighborhoods all contain at least one point of the set other than p. Also known as accumulation point; limit point.

cluster sampling [STAT] A random sampling plan in which the population is subdivided into groups called clusters so that there is small variability within clusters and large variability between clusters.

cluster set [MATH] Let f be a function from a domain D into the set R consisting of either the sphere in euclidean n-space

CLOUD CHAMBER

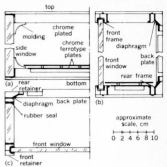

Cloud chamber designed for use in a magnetic field. *(a)* Vertical section parallel to front. *(b)* Vertical section parallel to side. *(c)* Horizontal section. The back plate moves to produce the expansion. The mechanism for compressing the chamber and producing the expansion is not shown.

or the real numbers, and let z_0 be a point in the closure of D; the cluster set $C(f, z_0)$ is the set of points a in R such that there exists a sequence z_n in $D - \{z_0\}$ such that

$$\lim_{n \to \infty} z_n = z_0 \text{ and } \lim_{n \to \infty} f(z_n) = a.$$

cluster variables *See* RR Lyrae stars.

cm *See* centimeter.

Cm *See* curium.

cmHg *See* centimeter of mercury.

CMOS device [ELECTR] A device formed by the combination of a PMOS (p-type-channel metal oxide semiconductor device) with an NMOS (n-type-channel metal oxide semiconductor device). Derived from complementary metal oxide semiconductor device.

C network [ELECTR] Network composed of three impedance branches in series, the free ends being connected to one pair of terminals, and the junction points being connected to another pair of terminals.

C neutron [NUCLEO] A neutron of such energy, up to about 0.3 electron volt, that it is strongly absorbable in cadmium.

c-number theory *See* classical field theory.

Co *See* cobalt.

Co60 *See* cobalt-60.

coalesce [SCI TECH] To come together to form a whole.

Coanda effect [FL MECH] The tendency of a gas or liquid coming out of a jet to travel close to the wall contour even if the wall's direction of curvature is away from the jet's axis; a factor in the operation of a fluidic element.

coated cathode [ELECTR] A cathode that has been coated with compounds to increase electron emission.

coated filament [ELECTR] A vacuum-tube filament coated with metal oxides to provide increased electron emission.

coated lens [OPTICS] A lens whose surfaces have been coated with a thin, transparent film having an index of refraction that minimizes light loss by reflection.

coax *See* coaxial cable.

coaxial [MECH] Sharing the same axes.

coaxial antenna [ELECTROMAG] An antenna consisting of a quarter-wave extension of the inner conductor of a coaxial line and a radiating sleeve that is in effect formed by folding back the outer conductor of the coaxial line for a length of approximately a quarter wavelength.

coaxial attenuator [ELECTROMAG] An attenuator that has a coaxial construction and terminations suitable for use with coaxial cable.

coaxial bolometer [ELECTR] A bolometer in which the desired square-law detection characteristic is provided by a fine Wollaston wire element that has been thoroughly cleaned before being axially located and soldered in position in its cylinder.

coaxial cable [ELECTROMAG] A transmission line in which one conductor is centered inside and insulated from an outer metal tube that serves as the second conductor. Also known as coax; coaxial line; coaxial transmission line; concentric cable; concentric line; concentric transmission line.

coaxial capacitor *See* cylindrical capacitor.

coaxial cavity [ELECTROMAG] A cylindrical resonating cavity having a central conductor in contact with its pistons or other reflecting devices.

coaxial circles [MATH] Family of circles such that any pair have the same radical axis.

coaxial connector [ELECTROMAG] An electric connector between a coaxial cable and an equipment circuit, so con-

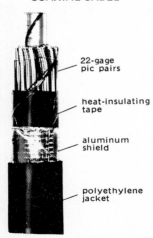

COAXIAL CABLE

22-gage pic pairs

heat-insulating tape

aluminum shield

polyethylene jacket

Cutaway view of coaxial transmission line.

structed as to maintain the conductor configuration, through the separable connection, and the characteristic impedance of the coaxial cable.

coaxial cylinder magnetron [ELECTR] A magnetron in which the cathode and anode consist of coaxial cylinders.

coaxial cylinders [MATH] Two cylinders whose cylindrical surfaces consist of the lines that pass through concentric circles in a given plane and are perpendicular to this plane.

coaxial diode [ELECTR] A diode having the same outer diameter and terminations as a coaxial cable, or otherwise designed to be inserted in a coaxial cable.

coaxial filter [ELECTROMAG] A section of coaxial line having reentrant elements that provide the inductance and capacitance of a filter section.

coaxial isolator [ELECTROMAG] An isolator used in a coaxial cable to provide a higher loss for energy flow in one direction than in the opposite direction; all types use a permanent magnetic field in combination with ferrite and dielectric materials.

coaxial line *See* coaxial cable.

coaxially fed linear array [ELECTROMAG] A beacon antenna having a uniform azimuth pattern.

coaxial stub [ELECTROMAG] A length of nondissipative cylindrical waveguide or coaxial cable branched from the side of a waveguide to produce some desired change in its characteristics.

coaxial transistor [ELECTR] A point-contact transistor in which the emitter and collector are point electrodes making pressure contact at the centers of opposite sides of a thin disk of semiconductor material serving as base.

coaxial transmission line *See* coaxial cable.

cobalt [CHEM] A metallic element, symbol Co, atomic number 27, atomic weight 58.93; used chiefly in alloys.

cobalt-60 [NUC PHYS] A radioisotope of cobalt, symbol Co^{60}, having a mass number of 60; emits gamma rays and has many medical and industrial uses; the most commonly used isotope for encapsulated radiation sources.

cobalt bomb [NUCLEO] A quantity of cobalt-60 mounted in a housing with walls having up to 8 inches (20 centimeters) of lead for protection, and with means for removing a lead plug to release a beam of gamma rays for use in cobalt-beam therapy.

cobinotron [ELECTROMAG] The combination of a corbino disk and a coil arranged to produce a magnetic field perpendicular to the disk.

coboundary [MATH] An image under the coboundary operator.

coboundary operator [MATH] If $\{C^n\}$ is a sequence of Abelian groups, coboundary operators are homomorphisms $\{\delta^n\}$ such that $\delta^n: C^n \rightarrow C^{n+1}$ and $\delta^{n+1} \circ \delta^n = 0$.

cochain complex [MATH] A sequence of Abelian groups C^n, $-\infty < n < \infty$, together with coboundary homomorphisms $\delta^n: C^n \rightarrow C^{n+1}$ such that $\delta^{n+1} \circ \delta^n = 0$.

Cochran's test [STAT] A test used when one estimated variance appears to be very much larger than the remainder of the estimated variances; based on the ratio of the largest estimate of the variance to the total of all the estimates.

Cockcroft-Walton accelerator [NUCLEO] An electrostatic particle accelerator utilizing as a source of high voltage a transformer and an array of rectifiers and condensers, giving voltage multiplication.

cocked hat [MATH] A plane curve whose equation in cartesian coordinates x and y is $(x^2 + 2ay - a^2)^2 = y^2(a^2 - x^2)$, where a is a constant.

COBALT

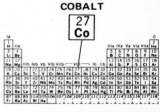

Periodic table of the chemical elements showing the position of cobalt.

cocycle [MATH] A chain of simplices whose coboundary is 0.

Coddington lens [OPTICS] A magnifier consisting of a glass sphere with a deep groove cut around a great circle to serve as a stop.

codistor [ELECTR] A multijunction semiconductor device which provides noise rejection and voltage regulation functions.

coefficient [MATH] A factor in a product.

coefficient of absorption *See* absorption coefficient.

coefficient of alienation [STAT] A statistic that measures the lack of linear association between two variables; computed by taking the square root of the difference between 1 and the square of the correlation coefficient.

coefficient of association [STAT] A statistic used as a measure of the association of data grouped in a 2×2 table; the value of the statistic ranges from -1 to $+1$, with the former indicating perfect negative association and the latter perfect positive association. Usually designated as Q.

coefficient of capacitance [ELEC] One of the coefficients which appears in the linear equations giving the charges on a set of conductors in terms of the potentials of the conductors; a coefficient is equal to the ratio of the charge on a given conductor to the potential of the same conductor when the potentials of all the other conductors are 0.

coefficient of compressibility [MECH] The decrease in volume per unit volume of a substance resulting from a unit increase in pressure; it is the reciprocal of the bulk modulus.

coefficient of concordance [STAT] A statistic that measures the agreement among sets of rankings by two or more judges.

coefficient of condensation [STAT MECH] The ratio of the number of molecules condensed on the surface of a solid or liquid in equilibrium with its vapor phase to the total number of vapor molecules striking the surface.

coefficient of contingency [STAT] A measure of the strength of dependence between two statistical variables, based on a contingency table.

coefficient of contraction [FL MECH] The ratio of the minimum cross-sectional area of a jet of liquid discharging from an orifice to the area of the orifice. Also known as contraction coefficient.

coefficient of cubical expansion [THERMO] The increment in volume of a unit volume of solid, liquid, or gas for a rise of temperature of 1° at constant pressure. Also known as coefficient of expansion; coefficient of thermal expansion; expansion coefficient; expansivity.

coefficient of determination [STAT] A statistic which indicates the strength of fit between two variables implied by a particular value of the sample correlation coefficient r. Designated by r^2.

coefficient of discharge *See* discharge coefficient.

coefficient of eddy viscosity [FL MECH] The portion of the kinematic viscosity of a turbulent fluid that is associated with its eddy viscosity. Also known as coefficient of turbulence.

coefficient of elasticity *See* modulus of elasticity.

coefficient of expansion *See* coefficient of cubical expansion.

coefficient of friction [MECH] The ratio of the frictional force between two bodies in contact, parallel to the surface of contact, to the force, normal to the surface of contact, with which the bodies press against each other. Also known as friction coefficient.

coefficient of friction of rest *See* coefficient of static friction.

coefficient of induction [ELEC] One of the coefficients which appears in the linear equations giving the charges on a set of conductors in terms of the potentials of the conductors;

a coefficient is equal to the ratio of the charge on a given conductor to the potential on another conductor, when the potentials of all the other conductors equal 0.

coefficient of kinematic viscosity *See* kinematic viscosity.

coefficient of kinetic friction [MECH] The ratio of the frictional force, parallel to the surface of contact, that opposes the motion of a body which is sliding or rolling over another, to the force, normal to the surface of contact, with which the bodies press against each other.

coefficient of linear expansion [THERMO] The increment of length of a solid in a unit of length for a rise in temperature of 1° at constant pressure.

coefficient of multiple correlation [STAT] A measure used as an index of the strength of a relationship between a variable y and a set of one or more variables x_i; computed by deriving the square root of the ratio of the explained variation to the total variation.

coefficient of nondetermination [STAT] The coefficient of alienation squared; represents that part of the dependent variable's total variation not accounted for by linear association with the independent variable.

coefficient of performance [THERMO] In a refrigeration cycle, the ratio of the heat energy extracted by the heat engine at the low temperature to the work supplied to operate the cycle; when used as a heating device, it is the ratio of the heat delivered in the high-temperature coils to the work supplied.

coefficient of permeability *See* permeability coefficient.

coefficient of potential [ELEC] One of the coefficients which appears in the linear equations giving the potentials of a set of conductors in terms of the charges on the conductors.

coefficient of reflection *See* reflection coefficient.

coefficient of resistance [FL MECH] The ratio of the loss of head of fluid, issuing from an orifice or passing over a weir, to the remaining head.

coefficient of restitution [MECH] The constant e, which is the ratio of the relative velocity of two elastic spheres after direct impact to that before impact; e can vary from 0 to 1, with 1 equivalent to an elastic collision and 0 equivalent to a perfectly elastic collision.

coefficient of rolling friction [MECH] The ratio of the frictional force, parallel to the surface of contact, opposing the motion of a body rolling over another, to the force, normal to the surface of contact, with which the bodies press against each other.

coefficient of sliding friction [MECH] The ratio of the frictional force, parallel to the surface of contact, opposing the motion of a body sliding over another, to the force, normal to the surface of contact, with which the bodies press against each other.

coefficient of static friction [MECH] The ratio of the maximum possible frictional force, parallel to the surface of contact, which acts to prevent two bodies in contact, and at rest with respect to each other, from sliding or rolling over each other, to the force, normal to the surface of contact, with which the bodies press against each other. Also known as coefficient of friction of rest.

coefficient of strain [MATH] Multiplier used in transformations to elongate or compress configurations in a direction parallel to an axis. [MECH] For a substance undergoing a one-dimensional strain, the ratio of the distance along the strain axis between two points in the body, to the distance between the same points when the body is undeformed.

coefficient of thermal expansion *See* coefficient of cubical expansion.

coefficient of turbulence *See* coefficient of eddy viscosity.

coefficient of variation [STAT] The ratio of the standard deviation of a distribution to its arithmetic mean.

coefficient of velocity *See* velocity coefficient.

coercimeter [ENG] An instrument that measures the magnetic intensity of a natural magnet or electromagnet.

coercive force [ELECTROMAG] The magnetic field H which must be applied to a magnetic material in a symmetrical, cyclicly magnetized fashion, to make the magnetic induction B vanish. Also known as magnetic coercive force.

coercivity [ELECTROMAG] The coercive force of a magnetic material in a hysteresis loop whose maximum induction approximates the saturation induction.

cofactor *See* minor.

coffin [NUCLEO] A box of heavy shielding material, usually lead, used for transporting radioactive objects and having walls thick enough to attenuate radiation from the contents to an allowable level. Also known as cask; casket.

cofinal [MATH] A subset C of a directed set D is cofinal if for each element of D there is a larger element in C.

coherence [PHYS] **1.** The existence of a correlation between the phases of two or more waves, so that interference effects may be produced between them. **2.** Property of moving in unison, such as is characteristic of the particles in a synchrotron.

coherence distance *See* coherence length.

coherence length [PHYS] For a beam of particles, the typical length of a wave packet along the beam; the more monochromatic the beam, the greater its coherence length. [SOLID STATE] A measure of the distance through which the effect of any local disturbance is spread out in a superconducting material. Also known as coherence distance.

coherent light [OPTICS] Radiant electromagnetic energy of the same, or almost the same, wavelength, and with definite phase relationships between different points in the field.

coherent radiation [PHYS] Radiation in which there are definite phase relationships between different points in a cross section of the beam.

coherent scattering [PHYS] Scattering in which there is a definite phase relationship between incoming and scattered particles or photons.

coherent source [PHYS] A source in which there is a constant phase difference between waves emitted from different parts of the source.

coherent units [PHYS] A system of units, such as the International System, in which the units of derived quantities are formed as products or quotients of units of the base quantities according to the algebraic relations linking these quantities.

cohesional work [PHYS] The work per unit area required to separate a column of liquid into two parts.

cohesive energy [SOLID STATE] The difference between the energy per atom of a system of free atoms at rest far apart from each other, and the energy of the solid.

cohesive strength [MECH] **1.** Strength corresponding to cohesive forces between atoms. **2.** Hypothetically, the stress causing tensile fracture without plastic deformation.

cohomology group [MATH] One of a series of Abelian groups $H^n(K)$ that are used in the study of a simplicial complex K and are closely related to homology groups, being associated with cocycles and coboundaries in the same manner as homology groups are associated with cycles and boundaries.

cohomology theory [MATH] A theory which uses algebraic groups to study the geometric properties of topological spaces; closely related to homology theory.

coil [ELECTROMAG] A number of turns of wire used to introduce inductance into an electric circuit, to produce magnetic flux, or to react mechanically to a changing magnetic flux; in high-frequency circuits a coil may be only a fraction of a turn. Also known as electric coil; inductance; inductance coil; inductor. [SCI TECH] An arrangement of flexible material into a spiral or helix.

coil antenna [ELECTROMAG] An antenna that consists of one or more complete turns of wire.

coincidence amplifier [ELECTR] An electronic circuit that amplifies only that portion of a signal present when an enabling or controlling signal is simultaneously applied.

coincidence circuit [ELECTR] A circuit that produces a specified output pulse only when a specified number or combination of two or more input terminals receives pulses within an assigned time interval. Also known as coincidence counter; coincidence gate.

coincidence counter *See* coincidence circuit.

coincidence counting [NUCLEO] A method of distinguishing particular types of events from background events and of measuring the velocities or directions of particles, by registering the occurrence of counts in two or more particle detectors within a given time interval by means of coincidence circuits.

coincidence gate *See* coincidence circuit.

coincidence magnet [NUCLEO] An electromagnet used in one type of scram mechanism for a nuclear reactor; has three electrically independent coils and releases when any two coils are deenergized.

coincidence rangefinder [OPTICS] An optical rangefinder in which one-eyed viewing through a single eyepiece provides the basis for manipulation of the rangefinder adjustment to cause two images of the target or parts of each, viewed over different paths, to match or coincide.

CO laser *See* carbon monoxide laser.

Colburn analogy [FL MECH] Dimensionless Reynolds equation for fluid-flow resistance modified to be analogous to the Colburn *j* factor heat-transfer equation.

Colburn j factor equation [THERMO] Dimensionless heat-transfer equation to calculate the natural convection movement of heat from vertical surfaces or horizontal cylinders to fluids (gases or liquids) flowing past these surfaces.

cold [ELEC] Pertaining to electrical circuits that are disconnected from voltage supplies and at ground potential; opposed to hot, pertaining to carrying an electrical charge.

cold cathode [ELECTR] A cathode whose operation does not depend on its temperature being above the ambient temperature.

cold-cathode counter tube [ELECTR] A counter tube having one anode and three sets of 10 cathodes; two sets of cathodes serve as guides that direct the flow discharge to each of the 10 output cathodes in correct sequence in response to driving pulses.

cold-cathode discharge *See* glow discharge.

cold-cathode ionization gage *See* Philips ionization gage.

cold-cathode rectifier [ELECTR] A cold-cathode gas tube in which the electrodes differ greatly in size so electron flow is much greater in one direction than in the other. Also known as gas-filled rectifier.

cold-cathode tube [ELECTR] An electron tube containing a cold cathode, such as a cold-cathode rectifier, mercury-pool rectifier, neon tube, phototube, or voltage regulator.

cold emission *See* field emission.

cold junction [ELECTR] The reference junction of thermo-

COINCIDENCE AMPLIFIER

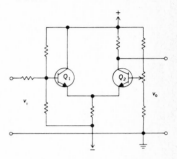

Simple coincidence amplifier.

couple wires leading to the measuring instrument; normally at room temperature.

cold light [PHYS] **1.** Light emitted in luminescence. **2.** Visible light which is accompanied by little or no infrared radiation, and therefore has little heating effect.

cold neutron [SOLID STATE] A very-low-energy neutron in a reactor, used for research into solid-state physics because it has a wavelength of the order of crystal lattice spacings and can therefore be diffracted by crystals.

cold stress [MECH] Forces tending to deform steel, cement, and other materials, resulting from low temperatures.

Colebrook equation [FL MECH] An empirical equation for the flow of liquids in ducts, relating the friction factor to the Reynolds number and the relative roughness of the duct.

colidar *See* ladar.

collapsar *See* black hole.

collapse properties [MECH] Strength and dimensional attributes of piping, tubing, or process vessels, related to the ability to resist collapse from exterior pressure or internal vacuum.

collapsing pressure [MECH] The external pressure which causes a thin-walled body or structure to collapse.

collar vortex *See* vortex ring.

collateral series [NUC PHYS] A radioactive decay series, initiated by transmutation, that eventually joins into one of the four radioactive decay series encountered in natural radioactivity.

collecting power [OPTICS] The power of a lens to make parallel rays converge or reduce the divergence of divergent rays.

collective electron theory [SOLID STATE] A theory of ferromagnetism in which electrons responsible for ferromagnetism are supposed to move more or less freely throughout a crystal, and to align with one another as the result of an exchange interaction.

collective motion [NUC PHYS] Motion of nucleons in a nucleus correlated so that their overall space pattern is essentially constant or undergoes changes which are slow compared to the motions of individual nucleons.

collective paramagnetism [ELECTROMAG] Magnetization of a collection of extremely small ferromagnetic particles, each containing only one magnetic domain, that resembles paramagnetism of a collection of atoms or molecules. Also known as superparamagnetism.

collective transition [NUC PHYS] A nuclear transition from one state of collective motion to another.

collector [ELECTR] **1.** A semiconductive region through which a primary flow of charge carriers leaves the base of a transistor; the electrode or terminal connected to this region is also called the collector. **2.** An electrode that collects electrons or ions which have completed their functions within an electron tube; a collector receives electrons after they have done useful work, whereas an anode receives electrons whose useful work is to be done outside the tube. Also known as electron collector.

collector capacitance [ELECTR] The depletion-layer capacitance associated with the collector junction of a transistor.

collector cutoff [ELECTR] The reverse saturation current of the collector-base junction.

collector junction [ELECTR] A semiconductor junction located between the base and collector electrodes of a transistor.

collector plate [ELEC] One of several metal inserts that are sometimes embedded in the lining of an electrolyte cell to make the resistance between the cell lining and the current leads as small as possible.

collector resistance [ELECTR] The back resistance of the collector-base diode of a transistor.

collector voltage [ELECTR] The direct-current voltage, obtained from a power supply, that is applied between the base and collector of a transistor.

colligative properties [PHYS CHEM] Properties dependent on the number of molecules but not their nature.

collimate [PHYS] To render parallel to a certain line or direction; paths of electrons in a flooding beam, or paths of various rays of a scanning beam are collimated to cause them to become more nearly parallel as they approach the storage assembly of a storage tube.

collimating lens [OPTICS] A lens on a collimator used to focus light from a source near one of its focal points into a parallel beam.

collimator [OPTICS] An instrument which produces parallel rays of light. [PHYS] A device for confining the elements of a beam within an assigned solid angle.

collinear array *See* linear array.

collinear vectors [MATH] Two vectors one of which is a nonzero scaler multiple of the other.

collineation [MATH] A mapping which transforms points into points, lines into lines, and planes into planes. Also known as collineatory transformation.

collineatory transformation *See* collineation.

Collins helium liquefier [CRYO] A machine which uses the Joule-Thomson effect and work done by helium gas in expansion against a movable piston to liquefy helium.

collision [PHYS] An interaction resulting from the close approach of two or more bodies, particles, or systems of particles, and confined to a relatively short time interval during which the motion of at least one of the particles or systems changes abruptly.

collision broadening *See* collision line-broadening.

collision cross section *See* cross section.

collision density [PHYS] The number of collisions of a specified type per unit volume per unit time.

collision diameter [PHYS CHEM] The distance between the centers of two molecules taking part in a collision at the time of their closest approach.

collision excitation [ATOM PHYS] The excitation of a gas by collisions of moving charged particles.

collision frequency [PHYS] The average number of collisions undergone by a particle traveling through a material, such as an electron traveling through a gas, in a unit time.

collision ionization [ATOM PHYS] The ionization of atoms or molecules of a gas or vapor by collision with other particles.

collisionless Boltzmann equation *See* Vlasov equation.

collision line-broadening [SPECT] Spreading of a spectral line due to interruption of the radiation process when the radiator collides with another particle. Also known as collision broadening.

collision matrix *See* scattering matrix.

collision of the first kind [PHYS] An inelastic collision in which some of the kinetic energy of translational motion is converted to internal energy of the colliding systems. Also known as endoergic collision.

collision of the second kind [PHYS] An inelastic collision in which some of the internal energy of the colliding systems is converted to kinetic energy of translation. Also known as exoergic collision.

collision probability [PHYS] The ratio of the cross section for a given type of collision between two particles to the total cross section for all types of collision between the particles.

collision-radiative recombination [ATOM PHYS] The capture of an electron by an ion in a gas, accompanied by the emission of one or more photons.

collision theory [PHYS CHEM] Theory of chemical reaction proposing that the rate of product formation is equal to the number of reactant-molecule collisions multiplied by a factor that corrects for low-energy-level collisions. [QUANT MECH] Theory to describe collisions of simple or complex particles, the derivation of collision cross sections from postulated interactions and the study of properties of collision amplitudes which follow from invariance principles such as conservation of probability and time-reversal invariance.

colog *See* cologarithm.

cologarithm [MATH] The cologarithm of a number is the logarithm of the reciprocal of that number. Abbreviated colog.

color [OPTICS] A general term that refers to the wavelength composition of light, with particular reference to its visual appearance. [PARTIC PHYS] A hypothetical quantum number carried by quarks, so that each type of quark comes in three varieties which are identical in all measurable qualities but which differ in this additional property; this quantity determines the coupling of quarks to the gluon field.

color aberration *See* chromatic aberration.

color center [SOLID STATE] A point lattice defect which produces optical absorption bands in an otherwise transparent crystal.

color circle [OPTICS] An arrangement of hues about the circumference of a circle in the order in which they appear in the electromagnetic spectrum, with pairs of complementary colors at opposite ends of diameters.

color class [MATH] In a given coloring of a graph, the set of vertices which are assigned the same color.

color code [ELEC] A system of colors used to indicate the electrical value of a component or to identify terminals and leads.

color comparator [ANALY CHEM] A photoelectric instrument that compares an unknown color with that of a standard color sample for matching purposes. Also known as photoelectric color comparator.

color correction [OPTICS] The construction of an optical system so that the image positions of an object are the same for two or more wavelengths, and chromatic aberration is thus minimized.

color disk [OPTICS] A rotating circular disk having three filter sections to produce the individual red, green, and blue pictures in a field-sequential color television system.

color emissivity *See* monochromatic emissivity.

color equation [ASTRON] In astronomy, a measure of the color sensitivity and response of a method of observation; photographic, visual, or photoelectric techniques may be employed. [OPTICS] An algebraic equation that expresses a specified color as an additive mixture of primary colors.

color filter [OPTICS] An optical element that partially absorbs incident light, consisting of a pane of glass or other partially transparent material, or of films separated by narrow layers; the absorption may be either selective or nonselective with respect to wavelength. Also known as light filter.

colorimeter [OPTICS] An instrument that measures color by determining the intensities of the three primary colors that will give that color.

colorimetric photometer [OPTICS] A photometer that can

measure light intensities in several spectral regions, using color filters placed in the path of the light.

colorimetry [OPTICS] Any technique by which an unknown color is evaluated in terms of standard colors; the technique may be visual, photoelectric, or indirect by means of spectrophotometry; used in chemistry and physics.

color index [ASTRON] Of a star, the numerical difference between the apparent photographic magnitude and the apparent photovisual magnitude. Abbreviated CI.

coloring of a graph [MATH] An assignment of colors to the vertices of a graph so that adjacent vertices are assigned different colors.

color medium [OPTICS] Any colored, transparent material that is placed in front of a lighting unit to color the light transmitted.

color radiography [GRAPHICS] A radiographic technique in which various intensities are displayed as different colors.

color rendering [OPTICS] For a light source, the extent of the agreement between the perceived color of a surface illuminated by the source and that of the same surface illuminated by a reference source under specified viewing conditions, measured and expressed in terms of the chromaticity coordinates of the source and the luminance of the source in agreed spectral bands.

color saturation [OPTICS] The degree to which a color is mixed with white; high saturation means little white, low saturation means much white. Also known as chroma; saturation.

color solid [OPTICS] A three-dimensional diagram which represents the relationship of three attributes of surface color: hue, saturation, and brightness.

color system [OPTICS] Any three-component coordinate system used to represent the attributes of colors.

color temperature [STAT MECH] Of a solid surface, that temperature of a blackbody from which the radiant energy has essentially the same spectral distribution as that from the surface.

color-translating microscope [OPTICS] A type of compound microscope that employs three different wavelengths of light to reveal details produced by ultraviolet or other nonvisible radiation.

color triangle [OPTICS] A triangle on a chromaticity diagram representing the range of colors that can be obtained from three specified primary colors by an additive process.

column [MATH] *See* place. [NUCLEO] A hollow cylinder of water and spray thrown up from an underwater burst of an atomic weapon, through which hot, high-pressure gases are vented to the atmosphere; a somewhat similar column of dirt is formed in an underground explosion. Also known as plume.

columnar ionization [PHYS] Ionization of atoms in a region confined to one or more paths of very small cross-sectional area.

columnar resistance [GEOPHYS] The electrical resistance of a column of air 1 centimeter square, extending from the earth's surface to some specified altitude.

column operations [MATH] A set of rules for manipulating the columns of a matrix so that the image of the corresponding linear transformation remains unchanged.

column rank [MATH] The number of linearly independent columns of a matrix; the dimension of the image of the corresponding linear transformation.

column space [MATH] The vector space spanned by the columns of a matrix.

column vector [MATH] A matrix consisting of only one column.

colure [ASTRON] A great circle of the celestial sphere through the celestial poles and either the equinoxes or solstices, called respectively the equinoctial colure or the solstitial colure.

coma [ASTRON] The gaseous envelope that surrounds the nucleus of a comet. Also known as head. [ELECTR] A cathode-ray tube image defect that makes the spot on the screen appear comet-shaped when away from the center of the screen. [OPTICS] A manifestation of errors in an optical system, so that a point has an asymmetrical image (that is, appears as a pear-shaped spot).

coma lobe [ELECTROMAG] Side lobe that occurs in the radiation pattern of a microwave antenna when the reflector alone is tilted back and forth to sweep the beam through space because the feed is no longer always at the center of the reflector; used to eliminate the need for a rotary joint in the feed waveguide.

comatic circle [OPTICS] A circle formed in the focal plane by rays from an off-axis point passing through a given zone of a lens that displays coma.

comb antenna [ELECTROMAG] A broad-band antenna for vertically polarized signals, in which half of a fishbone antenna is erected vertically and fed against ground by a coaxial line.

combescure transformation [MATH] A one-to-one continuous mapping of one space curve onto another space curve so that tangents to corresponding points are parallel.

combination [MATH] A selection of one or more of the elements of a given set without regard to order.

combinational circuit [ELECTR] A switching circuit whose outputs are determined only by the concurrent inputs.

combination principle *See* Ritz's combination principle.

combination tone [ACOUS] A subjective tone produced by simultaneously sounding two pure tones whose frequencies differ by a large amount.

combination vibration [SPECT] A vibration of a polyatomic molecule involving the simultaneous excitation of two or more normal vibrations.

combinatorial analysis [MATH] **1.** The determination of the number of possible outcomes in ideal games of chance by using formulas for computing numbers of combinations and permutations. **2.** The study of large finite problems.

combinatorial theory [MATH] The branch of mathematics which studies the arrangements of elements into sets.

combinatorial topology [MATH] The study of polyhedrons, simplicial complexes, and generalizations of these. Also known as piecewise-linear topology.

combinatorics [MATH] Combinatorial topology which studies geometric forms by breaking them into simple geometric figures.

combined flexure [MECH] The flexure of a beam under a combination of transverse and longitudinal loads.

combined stresses [MECH] Bending or twisting stresses in a structural member combined with direct tension or compression.

combining-volumes principle [CHEM] The principle that when gases take part in chemical reactions the volumes of the reacting gases and those of the products (if gaseous) are in the ratio of small whole numbers, provided that all measurements are made at the same temperature and pressure. Also known as Gay-Lussac law.

combining weight [CHEM] The weight of an element that

chemically combines with 8 grams of oxygen or its equivalent.

combustion [CHEM] The burning of gas, liquid, or solid, in which the fuel is oxidized, evolving heat and often light.

comes [ASTRON] The smaller star in a binary system. Also known as companion.

comet [ASTRON] A nebulous celestial body having a fuzzy head surrounding a bright nucleus, one of two major types of bodies moving in closed orbits about the sun; in comparison with the planets, the comets are characterized by their more eccentric orbits and greater range of inclination to the ecliptic.

comma [ACOUS] The difference between the larger and smaller whole tones in the just scale, corresponding to a frequency ratio of 81/80.

common-base connection See grounded-base connection.

common branch [ELEC] A branch of an electrical network which is common to two or more meshes. Also known as mutual branch.

common-collector connection See grounded-collector connection.

common denominator [MATH] Any common multiple of the denominators of a collection of fractions.

common-drain amplifier [ELECTR] An amplifier using a field-effect transistor so that the input signal is injected between gate and drain, while the output is taken between the source and drain. Also known as source-follower amplifier.

common-emitter connection See grounded-emitter connection.

common fraction [MATH] A fraction whose numerator and denominator are both integers.

common-gate amplifier [ELECTR] An amplifier using a field-effect transistor in which the gate is common to both the input circuit and the output circuit.

common impedance coupling [ELECTROMAG] The interaction of two circuits by means of an inductance or capacitance in a branch which is common to both circuits.

common logarithm [MATH] The exponent in the representation of a number as a power of 10. Also known as Briggs' logarithm.

common mode [ELECTR] Having signals that are identical in amplitude and phase at both inputs, as in a differential operational amplifier.

common-mode error [ELECTR] The error voltage that exists at the output terminals of an operational amplifier due to the common-mode voltage at the input.

common-mode voltage [ELECTR] A voltage that appears in common at both input terminals of a device with respect to the output reference (usually ground).

common multiple [MATH] A quantity (polynomial number) divisible by all quantities in a given set.

common-source amplifier [ELECTR] An amplifier stage using a field-effect transistor in which the input signal is applied between gate and source and the output signal is taken between drain and source.

commutating capacitor [ELECTR] A capacitor used in gas-tube rectifier circuits to prevent the anode from going highly negative immediately after extinction.

commutation [ELECTR] The transfer of current from one channel to another in a gas tube. [ELECTROMAG] The process of current reversal in the armature windings of a direct-current rotating machine to provide direct current at the brushes.

commutation rules [QUANT MECH] The specification of the

commutators of operators corresponding to the dynamical variables of a system, which are equal to $i\hbar$ times the Poisson brackets of the classical variables to which the operators correspond.

commutative algebra [MATH] An algebra in which the multiplication operation obeys the commutative law.

commutative diagram [MATH] A diagram in which any two mappings between the same pair of sets, formed by composition of mappings represented by arrows in the diagram, are equal.

commutative law [MATH] A rule which requires that the result of a binary operation be independent of order; that is, $ab = ba$.

commutative operation [MATH] A binary operation that obeys a commutative law, such as addition and multiplication on the real or complex numbers. Also known as Abelian operation.

commutative ring [MATH] A ring in which the multiplication obeys the commutative law.

commutator [ELECTROMAG] That part of a direct-current motor or generator which serves the dual function, in combination with brushes, of providing an electrical connection between the rotating armature winding and the stationary terminals, and of permitting reversal of the current in the armature windings. [MATH] The commutator of a and b is the element c of a group such that $bac = ab$. [QUANT MECH] The commutator of a and b is $[a,b] = ab - ba$.

compact-open topology [MATH] A topology on the space of all continuous functions from one topological space into another; a subbase for this topology is given by the sets $W(K,U) = \{ f : f(K) \subset U \}$, where K is compact and U is open.

compact set [MATH] A set in a topological space with the property that every open cover has a finite subset which is also a cover. Also known as bicompact set.

compact space [MATH] A topological space which is a compact set.

companion *See* comes.

comparative experiments [STAT] Experiments conducted to determine statistically whether one procedure is better than another.

comparative lifetime [NUC PHYS] The product of the mean life of a nucleus that undergoes beta decay and the probability per unit time that beta decay would occur if the matrix element between the initial and final states of this transition were unity.

comparator [CONT SYS] A device which detects the value of the quantity to be controlled by a feedback control system and compares it continuously with the desired value of that quantity.

comparator circuit [ELECTR] An electronic circuit that produces an output voltage or current whenever two input levels simultaneously satisfy predetermined amplitude requirements; may be linear (continuous) or digital (discrete).

comparator method [THERMO] A method of determining the coefficient of linear expansion of a substance in which one measures the distance that each of two traveling microscopes must be moved in order to remain centered on scratches on a rod-shaped specimen when the temperature of the specimen is raised by a measured amount.

comparing unit [ELECTR] An electromechanical device which compares two groups of timed pulses and signals to establish either identity or nonidentity.

comparison bridge [ELECTR] A bridge circuit in which any change in the output voltage with respect to a reference

COMMUTATOR

Direct-current generator showing the typical commutator and brush assembly. *(Allis-Chalmers)*

voltage creates a corresponding error signal, which, by means of negative feedback, is used to correct the output voltage and thereby restore bridge balance.

comparison lamp [OPTICS] An incandescent lamp whose luminous intensity is constant (although not necessarily known), and which is compared against other lamps in a photometer.

comparison microscope [OPTICS] **1.** An arrangement of two microscopes connected by a special receiving ocular so that the field of one microscope is seen at one side of a dividing line and the field of the other microscope at the opposite side. **2.** A projection type of microscope in which the image is compared with a template or known pattern.

comparison spectrum [SPECT] A line spectrum whose wavelengths are accurately known, and which is matched with another spectrum to determine the wavelengths of the latter.

comparison test [MATH] A simple test for the convergence of an infinite series, according to which a series converges if the absolute values of each of its terms are equal to or less than the corresponding term of a series that is known to converge, and diverges if each of its terms is equal to or greater than the absolute value of the corresponding term of a series that is known to diverge.

compatibility conditions [MECH] A set of six differential relations between the strain components of an elastic solid which must be satisfied in order for these components to correspond to a continuous and single-valued displacement of the solid.

compatible charts [MATH] Two charts (U_1, h_1) and (U_2, h_2) such that the mapping $h_2 \circ h_1^{-1}$ from the open set $h_1(U_1 \cap U_2)$ in euclidean space to $h_2(U_1 \cap U_2)$ is infinitely differentiable and has an infinitely differentiable inverse function; here $\cap$ denotes intersection.

compensated ionization chamber [NUCLEO] An arrangement of two ionization chambers in parallel, with potentials reversed, used as a radiation null indicator.

compensated pendulum [DES ENG] A pendulum made of two materials with different coefficients of expansion so that the distance between the point of suspension and center of oscillation remains nearly constant when the temperature changes.

compensated semiconductor [ELECTR] Semiconductor in which one type of impurity or imperfection (for example, donor) partially cancels the electrical effects on the other type of impurity or imperfection (for example, acceptor).

compensating eyepiece [OPTICS] A type of Huygens eyepiece in which the eye lens is achromatized to compensate for the color errors of the objective.

compensating leads [ENG] A pair of wires, similar to the working leads of a resistance thermometer or thermocouple, which are run alongside the working leads and are connected in such a way that they balance the effects of temperature changes in the working leads.

compensating network [CONT SYS] A network used in a low-energy-level method for suppression of excessive oscillations in a control system.

compensating plate [OPTICS] The first of two plates in a Brace compensator, which covers the entire field of view.

compensation [CONT SYS] Introduction of additional equipment into a control system in order to reshape its root locus so as to improve system performance. Also known as stabilization. [ELECTR] Modification of the amplitude-frequency response of an amplifier to broaden the bandwidth or to make the response more nearly uniform over the existing bandwidth. Also known as frequency compensation.

compensator [CONT SYS] A device introduced into a feedback control system to improve performance and achieve stability. Also known as filter. [ELECTR] A component that offsets an error or other undesired effect. [OPTICS] A device, usually consisting of two quartz wedges, for determining the phase difference between the two components of elliptically polarized light.

complement [MATH] **1.** The complement of a number A is another number B such that the sum $A + B$ will produce a specified result. **2.** See radix complement. **3.** For a subset of a set, the collection of all members of the set which are not in the given subset.

complementarity [QUANT MECH] The principle that nature has complementary aspects, particle and wave; the two aspects are related by $p = h/\lambda$ and $E = h\nu$, where p and E are the momentum and energy of the particle, λ and ν are the wavelength and frequency of the wave, and h is Planck's constant.

complementary [ELECTR] Having pnp and npn or p- and n-channel semiconductor elements on or within the same integrated-circuit substrate or working together in the same functional amplifier state.

complementary angle [MATH] One of a pair of angles whose sum is 90°.

complementary colors [OPTICS] Two colors which lie on opposite sides of the white point in the chromaticity diagram so that an additive mixture of the two, in appropriate proportions, can be made to yield an achromatic mixture.

complementary function [MATH] Any solution of the equation obtained from a given linear differential equation by replacing the inhomogeneous term with zero.

complementary logic switch [ELECTR] A complementary transistor pair which has a common input and interconnections such that one transistor is on when the other is off, and vice versa.

complementary metal oxide semiconductor device See CMOS device.

complementary minor See minor.

complementary operation [MATH] An operation on a Boolean algebra of two elements (labeled "true" and "false") whose result is the negation of a given operation; for example, NAND is complementary to the AND function.

complementary variables See conjugate variables.

complementary wave [ELECTROMAG] Wave brought into existence at the ends of a coaxial cable, or two-conductor transmission line, or any discontinuity along the line.

complementary wavelength [OPTICS] The wavelength of light that, when combined with a sample color in suitable proportions, matches a reference standard light.

complementation [MATH] The act of replacing a set by its complement.

complementation law [STAT] The law that the probability of an event E is 1 minus the probability of the event not E.

complemented lattice [MATH] A lattice with distinguished elements a and b, and with the property that corresponding to each point x of the lattice, there is a y such that the greatest lower bound of x and y is a, and the least upper bound of x and y is b.

complete Anger function See Anger function.

complete class of decision functions [STAT] A concept in decision theory which states that for a class of decision functions to be complete it must include a uniformly better decision function, which is a decision function that is sometimes better but never worse (according to some criterion) than each decision function not in the class.

complete degeneracy [QUANT MECH] The condition in which all the states of interest have the same energy.

complete elliptic integrals [MATH] Legendre normal elliptic integrals in which the upper limit of integration, sin ϕ, is equal to 1; that is, $\phi = \pi/2$; designated $F(\pi/2, k) = K(k)$, $E(\pi/2, k) = E(k)$, and $\Pi(\pi/2, n, k) = \Pi(n,k)$.

complete-expansion diesel cycle *See* Brayton cycle.

complete four-point *See* four-point.

complete graph [MATH] A graph with exactly one edge connecting each pair of distinct ve..ices and no loops.

complete induction *See* mathematical induction.

complete integral [MATH] **1.** A solution of an nth order ordinary differential equation which depends on n arbitrary constants as well as the independent variable. Also known as complete primitive. **2.** A solution of a first-order partial differential equation with n independent variables which depends upon n arbitrary parameters as well as the independent variables.

complete lattice [MATH] A partially ordered set in which every subset has both a supremum and an infimum.

completely inelastic collision *See* perfectly inelastic collision.

completely normal space [MATH] A topological space with the property that any pair of sets with disjoint closures can be separated by open sets.

completely reducible representation of a group [MATH] A representation of a group as a family of linear operators of a vector space V such that V is the direct sum of subspaces V_1, ..., V_n which are invariant under these operators, but $V_1, \ldots,$ V_n do not have any proper closed subspaces which are also invariant under these operators. Also known as semisimple representation of a group.

completely regular algebra *See* C* algebra.

completely regular space [MATH] A topological space X where for every point x and neighborhood U of x there is a continuous function from X to [0,1] with $f(x) = 1$ and $f(y) = 0$, if y is not in U. Also known as Tychonoff space.

complete metric space [MATH] A metric space in which every Cauchy sequence converges to a point of the space. Also known as complete space.

complete normed linear space *See* Banach space.

complete orthonormal set [MATH] A set of mutually orthogonal unit vectors in a (possibly infinite dimensional) vector space which is contained in no larger such set, that is no nonzero vector is perpendicular to all the vectors in the set. Also known as closed orthonormal set.

complete primitive *See* complete integral.

complete space *See* complete metric space.

complete T system *See* Haar system.

completion of a metric space [MATH] A complete metric space obtained from another metric space by formally adding limits to Cauchy sequences.

complex [MATH] A space which is represented as a union of simplices which intersect only on their faces.

complex conjugate [MATH] One of a pair of complex numbers with identical real parts and with imaginary parts differing only in sign. Also known as conjugate.

complex fraction [MATH] A fraction whose numerator or denominator is a fraction.

complex frequency [ENG] A complex number used to characterize exponential and damped sinusoidal motion in the same way that an ordinary frequency characterizes simple harmonic motion; designated by the constant s corresponding to a motion whose amplitude is given by Ae^{st}, where A is a constant and t is time.

complex impedance *See* electrical impedance; impedance.

complex notation [PHYS] The representation of a physical quantity by a complex number whose real component equals the instantaneous value of the physical quantity, a sinusoidally varying quantity thus being represented by a point rotating in a circle centered at the origin of the complex plane with uniform speed.

complex number [MATH] Any number of the form $a + bi$, where a and b are real numbers, and $i^2 = -1$.

complex permeability [ELECTROMAG] A property, designated by μ^*, of a magnetic material, equal to $\mu_0 (L/L_0)$, where L is the complex inductance of an inductance coil in which the magnetic material forms the core when the coil is connected to a sinusoidal voltage source, and L_0 is the vacuum inductance of the coil.

complex permittivity [ELEC] A property of a dielectric, equal to $\varepsilon_0(C/C_0)$, where C is the complex capacitance of a capacitor in which the dielectric is the insulating material when the capacitor is connected to a sinusoidal voltage source, and C_0 is the vacuum capacitance of the capacitor.

complex potential [FL MECH] An analytic function in ideal aerodynamics whose real part is the velocity potential and whose imaginary part is the stream function. [NUC PHYS] Generalization of the potential in the Schrödinger equation describing the scattering of a nucleon by a nucleus in the cloudy crystal-ball model.

complex sphere *See* Riemann sphere.

complex tone [ACOUS] A sound wave produced by the combination of simple sinusoidal components of different frequencies.

complex unit [MATH] Any complex number, $x + iy$, whose absolute value, $\sqrt{x^2 + y^2}$, equals 1.

complex variable [MATH] A variable which assumes complex numbers for values.

complex velocity [FL MECH] In ideal aerodynamic flow, the derivative of the complex potential with respect to $z = x + iy$, where x and y are the chosen coordinates.

compliance [MECH] The displacement of a linear mechanical system under a unit force.

compliance constant [MECH] Any one of the coefficients of the relations in the generalized Hooke's law used to express strain components as linear functions of the stress components. Also known as elastic constant.

component [ELEC] Any electric device, such as a coil, resistor, capacitor, generator, line, or electron tube, having distinct electrical characteristics and having terminals at which it may be connected to other components to form a circuit. Also known as circuit element; element.

component bar chart [STAT] A bar chart which shows within each bar the components that make up the bar; each component is represented by a section proportional in size to its representation in the total of each bar.

composite balance [ELEC] An electric balance made by modifying the Kelvin balance to measure amperage, voltage, or wattage.

composite cable [ELEC] Cable in which conductors of different gages or types are combined under one sheath.

composite flash [GEOPHYS] A lightning discharge which is made up of a series of distinct lightning strokes with all strokes following the same or nearly the same channel, and with successive strokes occurring at intervals of about 0.05 second. Also known as multiple discharge.

composite function [MATH] A function of a variable which is itself a function of a second variable; for example, $y = f(x)$ where $x = g(t)$.

composite hypothesis [STAT] In hypothesis testing, a hypothesis composed of a group of simple hypotheses, that is, the mean and standard deviation of a distribution are both equal to specific values; as opposed to a simple hypothesis, in which only the mean is equal to a specified value.

composite number [MATH] Any positive integer which is not prime. Also known as composite quantity.

composite quantity *See* composite number.

composite wave filter [ELECTR] A combination of two or more low-pass, high-pass, band-pass, or band-elimination filters.

composition [SCI TECH] The elements or compounds making up a material or produced from it by analysis.

composition face *See* composition surface.

composition of forces [MECH] The determination of a force whose effect is the same as that of two or more given forces acting simultaneously; all forces are considered acting at the same point.

composition of mappings [MATH] The composition of two mappings, f and g, denoted $g \circ f$, where the domain of g includes the range of f, is the mapping which assigns to each element x in the domain of f the element $g(y)$, where $y = f(x)$.

composition of vectors *See* addition of vectors.

composition of velocities law [MECH] A law relating the velocities of an object in two references frames which are moving relative to each other with a specified velocity.

composition plane [CRYSTAL] A planar composition surface in a crystal uniting two individuals of a contact twin.

composition series [MATH] A normal series $G_1, G_2, \ldots,$ of a group, where each G_i is a proper normal subgroup of G_{i-1} and no further normal subgroups both contain G_i and are contained in G_{i-1}.

composition surface [CRYSTAL] The surface uniting individuals of a crystal twin; may or may not be planar. Also known as composition face.

compositum of fields [MATH] Let E and F be fields, both contained in some field L; the compositum of E and F, denoted EF, is the smallest subfield of L containing E and F.

compound cryosar [ELECTR] A cryosar consisting of two normal cryosars with different electrical characteristics in series.

compound curve [MATH] A curve made up of two arcs of differing radii whose centers are on the same side, connected by a common tangent; used to lay out railroad curves because curvature goes from nothing to a maximum gradually, and vice versa.

compound distribution [STAT] A frequency distribution resulting from the combining of two or more separate distributions that have the same general type of distribution.

compound elastic scattering [NUC PHYS] Scattering in which the final state is the same as the initial state, but there is an intermediate state with the colliding systems amalgamating to form a compound system.

compound lens [OPTICS] **1.** A combination of two or more lenses in which the second surface of one lens has the same radius as the first surface of the following lens, and the two lenses are cemented together. Also known as cemented lens. **2.** Any optical system consisting of more than one element, even when they are not in contact.

COMPOUND MICROSCOPE

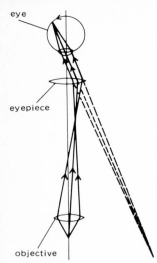

eye

eyepiece

objective

Compound microscope diagram.
*(From F. A. Jenkins and H. E.
White, Fundamentals of Optics,
3d ed., McGraw-Hill, 1957)*

compound microscope [OPTICS] A microscope which utilizes two lenses or lens systems; one lens forms an enlarged image of the object, and the second magnifies the image formed by the first.

compound nucleus [NUC PHYS] An intermediate state in a nuclear reaction in which the incident particle combines with the target nucleus and its energy is shared among all the nucleons of the system.

compound number [MATH] A quantity which is expressed as the sum of two or more quantities in terms of different units, for example, 3 feet 10 inches, or 2 pounds 5 ounces.

compound twins [CRYSTAL] Individuals of one mineral group united in accordance with two or more different twin laws.

compound wave [FL MECH] A plane wave of finite amplitude in which neither the sum of the velocity potential and the component of velocity in the direction of wave motion, nor the difference of these two quantities, is constant.

compound winding [ELEC] A winding that is a combination of series and shunt winding.

compressadensity function [MECH] A function used in the acoustic levitation technique to determine either the density or the adiabatic compressibility of a submicroliter droplet suspended in another liquid, if the other property is known.

compressed air [MECH] Air whose density is increased by subjecting it to a pressure greater than atmospheric pressure.

compressibility [MECH] The property of a substance capable of being reduced in volume by application of pressure; quantitively, the reciprocal of the bulk modulus.

compressibility burble [FL MECH] A region of disturbed flow, produced by and rearward of a shock wave.

compressibility correction [FL MECH] The correction of the calibrated airspeed caused by compressibility error.

compressibility error [FL MECH] The error in the readings of a differential-pressure-type airspeed indicator due to compression of the air on the forward part of the pitot tube component moving at high speeds.

compressibility factor [THERMO] The product of the pressure and the volume of a gas, divided by the product of the temperature of the gas and the gas constant; this factor may be inserted in the ideal gas law to take into account the departure of true gases from ideal gas behavior. Also known as deviation factor; gas-deviation factor; supercompressibility factor.

compressible flow [FL MECH] Flow in which the fluid density varies.

compressible-flow principle [FL MECH] The principle that when flow velocity is large, it is necessary to consider that the fluid is compressible rather than to assume that it has a constant density.

compression [ELECTR] **1.** Reduction of the effective gain of a device at one level of signal with respect to the gain at a lower level of signal, so that weak signal components will not be lost in background and strong signals will not overload the system. **2.** *See* compression ratio. [MECH] Reduction in the volume of a substance due to pressure; for example in building, the type of stress which causes shortening of the fibers of a wooden member.

compressional wave [PHYS] A disturbance traveling in an elastic medium; characterized by changes in volume and by particle motion parallel with the direction of wave movement. Also known as dilatational wave; irrotational wave; pressure wave; P wave.

compression modulus *See* bulk modulus of elasticity.

compression ratio [ELECTR] The ratio of the gain of a device at a low power level to the gain at some higher level, usually expressed in decibels. Also known as compression.

compression wave [FL MECH] A wave in a fluid in which a compression is propagated.

compressive strength [MECH] The maximum compressive stress a material can withstand without failure.

compressive stress [MECH] A stress which causes an elastic body to shorten in the direction of the applied force.

Compton absorption [QUANT MECH] The absorption of an x-ray or gamma-ray photon in Compton scattering, accompanied by the emission of another photon of lower energy.

Compton cross section [QUANT MECH] The differential cross section for the elastic scattering of photons by electrons.

Compton-Debye effect *See* Compton effect.

Compton effect [QUANT MECH] The increase in wavelength of electromagnetic radiation in the x-ray and gamma-ray region on being scattered by material objects; the scattering is due to the interaction of the photons with electrons that are effectively free. Also known as Compton-Debye effect.

Compton electron *See* Compton recoil electron.

Compton equation [QUANT MECH] The equation for the change in wavelength $\Delta\lambda$ of radiation scattered by electrons in the Compton effect, $\Delta\lambda = \lambda_c(1 - \cos\theta)$, where λ_c is the Compton wavelength of the electron, and θ is the angle between the directions of incident and scattered radiation.

Compton-Getting effect [ASTROPHYS] The sidereal diurnal variation of the intensity of cosmic rays which would be expected from the rotation of the galaxy if cosmic radiation originated in extragalactic regions and was isotropic in intergalactic space, and if this radiation was unaffected at entry to and passage through the galaxy.

Compton incoherent scattering [NUC PHYS] Scattering of gamma rays by individual nucleons in a nucleus or electrons in an atom when the energy of the gamma rays is large enough so that binding effects may be neglected.

Compton meter [NUCLEO] An ionization chamber having a balance chamber with a uranium source that is adjusted until it balances out normal cosmic radiation; variations in cosmic radiation are then shown on an electrometer.

Compton process *See* Compton scattering.

Compton recoil electron [QUANT MECH] An electron set in motion by its interaction with a photon in Compton scattering. Also known as Compton electron.

Compton recoil particle [QUANT MECH] Any particle that has acquired its momentum in a scattering process similar to Compton scattering.

Compton rule [PHYS CHEM] An empirical law stating that the heat of fusion of an element times its atomic weight divided by its melting point in degrees Kelvin equals approximately 2.

Compton scattering [QUANT MECH] The elastic scattering of photons by electrons. Also known as Compton process; gamma-ray scattering.

Compton shift [QUANT MECH] The change in wavelength of scattered radiation due to the Compton effect.

Compton wavelength [QUANT MECH] A convenient unit of length that is characteristic of an elementary particle, equal to Planck's constant divided by the product of the particle's mass and the speed of light.

computable function [MATH] A function whose value can be calculated by some Turing machine in a finite number of steps. Also known as effectively computable function.

computation [MATH] **1.** The act or process of calculating. **2.** The result so obtained.

computer [ADP] A device that receives, processes, and presents data; the two types are analog and digital. Also known as computing machine.

computer-controlled system [CONT SYS] A feedback control system in which a computer operates on both the input signal and the feedback signal to effect control.

computing machine *See* computer.

Comstock refraction formula [ASTROPHYS] A formula for the apparent angular displacement of an object outside the earth's atmosphere due to refraction, in terms of the barometric pressure, the temperature of the atmosphere, and the observed zenith distance.

concave function [MATH] A function $f(x)$ is said to be concave over the interval a,b if for any three points x_1, x_2, x_3 such that $a < x_1 < x_2 < x_3 < b$, $f(x_2) \geq L(x_2)$, where $L(x)$ is the equation of the straight line passing through the points $[x_1, f(x_1)]$ and $[x_3, f(x_3)]$.

concave grating [SPECT] A reflection grating which both collimates and focuses the light falling upon it, made by spacing straight grooves equally along the chord of a concave spherical or paraboloid mirror surface. Also known as Rowland grating.

concave polygon [MATH] A polygon at least one of whose angles is greater than 180°.

concave spherical mirror [OPTICS] A round mirror having a concavely curved surface, in the form of a portion of a sphere.

concentrated [MATH] A measure (or signed measure) m is concentrated on a measurable set A if any measurable set B with nonzero measure has a nonnull intersection with A.

concentrated load [MECH] A force that is negligible because of a small contact area; a beam supported on a girder represents a concentrated load on the girder.

concentration cell [PHYS CHEM] **1.** Electrochemical cell for potentiometric measurement of ionic concentrations where the electrode potential electromotive force produced is determined as the difference in emf between a known cell (concentration) and the unknown cell. **2.** An electrolytic cell in which the electromotive force is due to a difference in electrolyte concentrations at the anode and the cathode.

concentration polarization [PHYS CHEM] That part of the polarization of an electrolytic cell resulting from changes in the electrolyte concentration due to the passage of current through the solution.

concentric cable *See* coaxial cable.

concentric circles [MATH] A family of coplanar circles with the same center.

concentric lens [OPTICS] A lens whose two spherical surfaces have the same center.

concentric line *See* coaxial cable.

concentric slip ring [ELEC] A large slip-ring assembly consisting of concentrically arranged insulators and conducting materials.

concentric transmission line *See* coaxial cable.

concentric windings [ELEC] Transformer windings in which the low-voltage winding is in the form of a cylinder next to the core, and the high-voltage winding, also cylindrical, surrounds the low-voltage winding.

conchoid [MATH] A plane curve consisting of the locus of both ends of a line segment of constant length on a line which rotates about a fixed point, while the midpoint of the segment remains on a fixed line which does not contain the fixed point.

condensation [ACOUS] A measure of the increase in the instantaneous density at a given point resulting from a sound wave, namely $(\rho - \rho_0)/\rho_0$, where ρ is the density and ρ_0 is the

CONCENTRIC SLIP RING

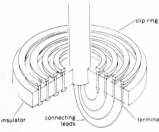

Concentric slip-ring assembly configuration.

CONCENTRIC WINDINGS

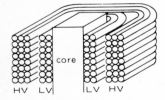

Section through the core and concentric winding of a transformer. LV = low-voltage winding; HV = high-voltage winding.

constant mean density at the point. [CRYO] *See* Bose-Einstein condensation. [ELEC] An increase of electric charge on a capacitor conductor. [MECH] An increase in density. [OPTICS] Focusing or collimation of light.

condensation number [THERMO] The ratio of the number of molecules that condense on a solid surface to the total number of molecules incident on the surface.

condensation shock wave [FL MECH] A sheet of discontinuity associated with a sudden condensation and fog formation in a field of flow; it occurs, for example, on a wing where a rapid drop in pressure causes the temperature to drop considerably below the dew point.

condensation temperature [ANALY CHEM] In boiling-point determination, the temperature established on the bulb of a thermometer on which a thin moving film of liquid coexists with vapor from which the liquid has condensed, the vapor phase being replenished at the moment of measurement from a boiling-liquid phase.

condensed matter [PHYS] Matter in the liquid or solid state.

condenser [ELEC] *See* capacitor. [OPTICS] A system of lenses or mirrors in an optical projection system, which gathers as much of the light from the source as possible and directs it through the projection lens.

condenser antenna *See* capacitor antenna.

condenser box *See* capacitor box.

condenser ionization chamber [NUCLEO] An ionization chamber which is charged before irradiation, and in which the charge remaining after irradiation indicates the dose received.

condensing electrometer *See* capacitive electrometer.

condensing flow [FL MECH] The flow and simultaneous condensation (partial or complete) of vapor through a cooled pipe or other closed conduit or container.

conditional convergence [MATH] The property of a series that is convergent but not absolutely convergent.

conditional distribution [STAT] If W and Z are random variables with discrete values w_1, w_2, . . . , and z_1, z_2, . . . , the conditional distribution of W given $Z = z$ is the distribution which assigns to w_i, $i = 1, 2, . . .$, the conditional probability of $W = w_i$ given $Z = z$.

conditional expectation [MATH] If X is a random variable on a probability space (Ω, F, P), the conditional expectation of X with respect to a given sub σ-field F' of F is an F'-measurable random variable whose expected value over any set in F' is equal to the expected value of X over this set. [STAT] The expected value of a conditional distribution.

conditional frequency [STAT] If r and s are possible outcomes of an experiment which is performed n times, the conditional frequency of s given that r has occurred is the ratio of the number of times both r and s have occurred to the number of times r has occurred.

conditional implication *See* implication.

conditionally compact set [MATH] A set whose closure is compact. Also known as relatively compact set.

conditionally periodic function *See* quasi-periodic function.

conditionally periodic motion [MECH] Motion of a system in which each of the coordinates undergoes simple periodic motion, but the associated frequencies are not all rational fractions of each other so that the complete motion is not simply periodic.

conditionally stable circuit [ELECTR] A circuit which is stable for certain values of input signal and gain, and unstable for other values.

conditional probability [STAT] The probability that a second

CONDENSER

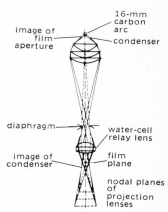

A relay condenser system having a water cell incorporated in second stage.

event will be B if the first event is A, expressed as $P(B/A)$.

Condon-Shortley-Wigner phase convention [QUANT MECH] Convention relating the phases of states having the same eigenvalue of $J^2 = J_x^2 + J_y^2 + J_z^2$, and different eigenvalues of J_z, where **J** is the total angular momentum, according to which the matrix elements of $\mathbf{J}_+ = J_x + iJ_y$ and $\mathbf{J}_- = J_x - iJ_y$ between such states are real.

conductance [ELEC] The real part of the admittance of a circuit; when the impedance contains no reactance, as in a direct-current circuit, it is the reciprocal of resistance, and is thus a measure of the ability of the circuit to conduct electricity. Also known as electrical conductance. Designated G. [FL MECH] For a component of a vacuum system, the amount of a gas that flows through divided by the pressure difference across the component. [THERMO] *See* thermal conductance.

conductance coefficient [PHYS CHEM] The ratio of the equivalent conductance of an electrolyte, at a given concentration of solute, to the limiting equivalent conductance of the electrolyte as the concentration of the electrolyte approaches 0.

conductance-variation method [ELEC] A technique for measuring low admittances; measurements in a parallel-resonance circuit with the terminals open-circuited, with the unknown admittance connected, and then with the unknown admittance replaced by a known conductance standard are made; from them the unknown can be calculated.

conduction [ELEC] The passage of electric charge, which can occur by a variety of processes, such as passage of electrons or ionized atoms. Also known as electrical conduction. [PHYS] Transmission of energy by a medium which does not involve movement of the medium itself.

conduction band [SOLID STATE] An energy band in which electrons can move freely in a solid, producing net transport of charge.

conduction current [SOLID STATE] A current due to a flow of conduction electrons through a body.

conduction electron [SOLID STATE] An electron in the conduction band of a solid, where it is free to move under the influence of an electric field. Also known as outer-shell electron; valence electron.

conduction field [ELECTROMAG] Energy surrounding a conductor when an electric current is passed through the conductor, which, because of the difference in phase between the electrical field and magnetic field set up in the conductor, cannot be detached from the conductor.

conductive coupling [ELEC] Electric connection of two electric circuits by their sharing the same resistor.

conductive gasket [ELEC] A flexible metallic gasket used to reduce radio-frequency leakage at joints in shielding.

conductivity [ELEC] The ratio of the electric current density to the electric field in a material. Also known as electrical conductivity; specific conductance.

conductivity bridge [ELEC] A modified Kelvin bridge for measuring very low resistances.

conductivity cell [ELEC] A glass vessel with two electrodes at a definite distance apart and filled with a solution whose conductivity is to be measured.

conductivity ellipsoid [ELEC] For an anisotropic material, an ellipsoid whose axes are the eigenvectors of the conductivity tensor.

conductivity modulation [ELECTR] Of a semiconductor, the variation of the conductivity of a semiconductor through variation of the charge carrier density.

conductivity modulation transistor [ELECTR] Transistor in which the active properties are derived from minority carrier modulation of the bulk resistivity of the semiconductor.

conductivity tensor [ELEC] For an anisotropic material, a tensor whose product with the electric field vector gives the current density.

conductivity theory [STAT MECH] Theory which treats the system of electrons in a metal as a gas and uses the Boltzmann transport equation to calculate conductivity.

conductometer [ENG] An instrument designed to measure thermal conductivity; in particular, one that compares the rates at which different rods transmit heat.

conductor [ELEC] A wire, cable, or other body or medium that is suitable for carrying electric current.

conductor skin effect *See* skin effect.

conduit [ELEC] Solid or flexible metal or other tubing through which insulated electric wires are run.

cone [MATH] A solid bounded by a region enclosed in a closed curve on a plane and a surface formed by the segments joining each point of the closed curve to a point not in the plane.

cone antenna *See* conical antenna.

cone flow *See* conical flow.

cone of escape [GEOPHYS] A hypothetical cone in the exosphere, directed vertically upward, through which an atom or molecule would theoretically be able to pass to outer space without a collision.

cone of friction [MECH] A cone in which the resultant force exerted by one flat horizontal surface on another must be located when both surfaces are at rest, as determined by the coefficient of static friction.

cone of nulls [ELECTROMAG] In antenna practice, a conical surface formed by directions of negligible radiation.

cone of revolution [MATH] The surface obtained by rotating a line around another line which it intersects, using the intersection point as a pivot.

confidence [STAT] The degree of assurance that a specified failure rate is not exceeded.

confidence coefficient [STAT] The probability associated with a confidence interval; that is, the probability that the interval contains a given parameter or characteristic. Also known as confidence level.

confidence interval [STAT] An interval which has a specified probability of containing a given parameter or characteristic.

confidence level *See* confidence coefficient.

confidence limit [STAT] One of the end points of a confidence interval.

configuration [ELEC] A group of components interconnected to perform a desired circuit function. [MATH] An arrangement of geometric objects. [MECH] The positions of all the particles in a system.

configurational free energy [STAT MECH] The free energy of a solid lattice associated with the interaction between neighboring atoms, and with external electric and magnetic fields.

configuration interaction [PHYS CHEM] Interaction between two different possible arrangements of the electrons in an atom (or molecule); the resulting electron distribution, energy levels, and transitions differ from what would occur in the absence of the interaction.

confinement of plasma [PL PHYS] Restriction of a hot plasma to a given volume as long as possible, by such means as magnetic mirrors and pinch effect.

confluent hypergeometric function [MATH] A solution to the differential equation $z(d^2 w/dz^2) + (\rho - z)(dw/dz) - \alpha w = 0$.

CONDUCTOR

19-strand

7-strand

37-strand

End views of stranded round electric conductors.

confocal conics [MATH] **1.** A system of ellipses and hyperbolas which have the same pair of foci. **2.** A system of parabolas which have the same focus and the same axis of symmetry.

confocal coordinates [MATH] Coordinates of a point in the plane with norm greater than 1 in terms of the system of ellipses and hyperbolas whose foci are at $(1,0)$ and $(-1,0)$.

confocal resonator [ELECTROMAG] A wavemeter for millimeter wavelengths, consisting of two spherical mirrors facing each other; changing the spacing between the mirrors affects propagation of electromagnetic energy between them, permitting direct measurement of free-space wavelength.

conformable matrices [MATH] Two matrices which can be multiplied together; this is possible if and only if the number of columns in the first matrix equals the number of rows in the second.

conformal diagram *See* Penrose diagram.

conformal mapping [MATH] An angle-preserving analytic function of a complex variable.

conformal reflection chart [ELECTROMAG] An Argand diagram for plotting the complex reflection coefficient of a waveguide junction and its image, the two being related by a conformal transformation.

conformation [PHYS CHEM] The spatial arrangement of the atoms in a molecule; mainly employed when a given molecule has two or more stable arrangements with the same set of chemical bonds.

conformational analysis [PHYS CHEM] The determination of the arrangement in space of the constituent atoms of a molecule that may rotate about a single bond.

confounding [STAT] Method used in design of factorial experiments in which some information about higher-order interactions is sacrificed so that estimates of main effects in lower-order interactions can be more precise.

congruence [MATH] **1.** The property of geometric figures that can be made to coincide by a rigid transformation. **2.** The property of two integers having the same remainder on division by another integer.

congruence transformation [MATH] Also known as transformation. **1.** A mapping which associates with each real quadratic form on a set of coordinates the quadratic form that results when the coordinates are subjected to a linear transformation. **2.** A mapping which associates with each square matrix A the matrix $B = SAT$, where S and T are nonsingular matrices, and T is the transpose of S; if A represents the coefficients of a quadratic form, then this definition is equivalent to definition 1.

congruent matrices [MATH] Two matrices A and B related by the transformation $B = SAT$, where S and T are nonsingular matrices and T is the transpose of S.

congruent melting point [THERMO] A point on a temperature composition plot of a nonstoichiometric compound at which the one solid phase and one liquid phase are adjacent.

congruent numbers [MATH] Two numbers having the same remainder when divided by a given quantity called the modulus.

conic [MATH] A curve which may be represented as the intersection of a cone with a plane; the four types of conics are circle, ellipse, parabola, and hyperbola. Also known as conic section.

conical antenna [ELECTROMAG] A wide-band antenna in which the driven element is conical in shape. Also known as cone antenna.

conical flow [FL MECH] Steady supersonic flow of a perfect, inviscid gas past a conical solid body in a region of the flow field where the principal physical quantities such as velocity, pressure, and density are constant on rays passing through a fixed point. Also known as cone flow.

conical function [MATH] Functions of the form

$$P^{\mu}_{-\frac{1}{2}+i\lambda} (\cos \theta) \text{ and } Q^{\mu}_{-\frac{1}{2}+i\lambda} (\cos \theta)$$

where $P^{\mu}_{-\frac{1}{2}+i\lambda}$ and $Q^{\mu}_{-\frac{1}{2}+i\lambda}$ are associated Legendre functions, satisfying Legendre's associated equation with parameters μ and $\nu = -\frac{1}{2} + i\lambda$.

conical helimagnet [SOLID STATE] A helimagnet in which the directions of atomic magnetic moments all make the same angle with a specified axis of the crystal, this angle is greater than 0° and less than 90°, moments of atoms in successive basal planes are separated by equal azimuthal angles, and all moments have the same magnitude.

conical horn [ACOUS] A horn having a circular cross section and straight sides.

conical-horn antenna [ELECTROMAG] A horn antenna having a circular cross section and straight sides.

conical pendulum [MECH] A weight suspended from a cord or light rod and made to rotate in a horizontal circle about a vertical axis with a constant angular velocity.

conical projection [MATH] A projection which associates with each point P in a plane Q the point p in a second plane q which is collinear with O and P, where O is a fixed point lying outside Q.

conical refraction [OPTICS] Phenomenon in which a ray incident on the surface of a biaxial crystal at a certain direction splits into a family of rays which lie along a cone.

conical surface [MATH] A surface formed by the lines which pass through each of the points of a closed plane curve and a fixed point which is not in the plane of the curve.

conic section *See* conic.

conjugate *See* complex conjugate.

conjugate binomial surds *See* conjugate radicals.

conjugate branches [ELEC] Any two branches of an electrical network such that a change in the electromotive force in either branch does not result in a change in current in the other. Also known as conjugate conductors.

conjugate conductors *See* conjugate branches.

conjugate convex functions [MATH] Two functions $f(x)$ and $g(y)$ are conjugate convex functions if the derivative of $f(x)$ is 0 for $x = 0$ and constantly increasing for $x > 0$, and the derivative of $g(y)$ is the inverse of the derivative of $f(x)$.

conjugate curve **1.** A member of one of two families of curves on a surface such that exactly one member of each family passes through each point P on the surface, and the directions of the tangents to these two curves at P are conjugate directions. **2.** *See* Bertrand curve.

conjugate diameters [MATH] **1.** For a conic section, any pair of straight lines either of which bisects all the chords that are parallel to the other. **2.** For an ellipsoid or hyperboloid, any three lines passing through the point of symmetry of the surface such that the plane containing the conjugate diameters (first definition) of one of the lines also contains the other two lines.

conjugate directions [MATH] For a point on a surface, a pair of directions, one of which is the direction of a curve on the surface through the point, while the other is the direction of the characteristic of the planes tangent to the surface at points on the curve.

conjugate elements [MATH] Two elements a and b in a group G for which there is an element x in G such that $ax = xb$.

conjugate foci *See* conjugate points.

conjugate functions [MATH] A pair of functions of two variables, $u(x,y)$ and $v(x,y)$, such that $u + iv$ is an analytic function of $x + iy$.

conjugate hyperbolas [MATH] Two hyperbolas having the same asymptotes with semiaxes interchanged.

conjugate impedances [ELEC] Impedances having resistance components that are equal, and reactance components that are equal in magnitude but opposite in sign.

conjugate lines [MATH] **1.** For a conic section, two lines each of which passes through the intersection of the tangents to the conic at its points of intersection with the other line. **2.** For a quadric surface, two lines each of which intersects the polar line of the other.

conjugate momentum [MECH] If q_j $(j = 1, 2, \ldots)$ are generalized coordinates of a classical dynamical system, and L is its Lagrangian, the momentum conjugate to q_j is $p_j = \partial L / \partial q_j$. Also known as canonical momentum; generalized momentum.

conjugate particles [PARTIC PHYS] A particle and its antiparticle.

conjugate planes [MATH] For a quadric surface, two planes each of which contains the pole of the other.

conjugate points [MATH] For a conic section, two points either of which lies on the line that passes through the points of contact of the two tangents drawn to the conic from the other. [OPTICS] Any pair of points such that all rays from one are imaged on the other within the limits of validity of Gaussian optics. Also known as conjugate foci.

conjugate quaternion [MATH] For a quaternion $q = s + ia + jb + kc$, this is the quaternion $q' = s - (ia + jb + kc)$.

conjugate radicals [MATH] Binomial surds that are of the type $a\sqrt{b} + c\sqrt{d}$ and $a\sqrt{b} - c\sqrt{d}$, where a, b, c, d are rational but $\sqrt{b}$ and $\sqrt{d}$ are not both rational. Also known as conjugate binomial surds.

conjugate roots [MATH] Conjugate complex numbers which are roots of a given equation.

conjugate space [MATH] The set of all continuous linear functionals defined on a normed linear space.

conjugate subgroups [MATH] Two subgroups A and B of a group G for which there exists an element x in G such that B consists of the elements of the form xax^{-1}, where a is in A.

conjugate torsion function [MECH] The conjugate function ψ to the torsion function ϕ of a cylinder undergoing torsion; that is, $\phi + i\psi$ is an analytic function of $x + iy$, where x and y are coordinates of a plane perpendicular to the axis of torsion.

conjugate triangles [MATH] Two triangles in which the poles of the sides of each with respect to a given curve are the vertices of the other.

conjugate variables [QUANT MECH] A pair of physical variables describing a quantum-mechanical system such that their commutator is a nonzero constant; either of them, but not both, can be precisely specified at the same time. Also known as complementary variables.

conjugation [MATH] An operation of a group G on itself which associates with each ordered pair (x,y) of elements in the group the element xyx^{-1}.

conjunction [MATH] The connection of two statements by the word "and."

conjunctive matrices [MATH] Two matrices A and B related by the transformation $B = SAT$, where S and T are nonsingular matrices and S is the Hermitian conjugate of I.

conjunctive transformation [MATH] The transformation $B =$

SAT, where S is the Hermitian conjugate of T, and matrices A and B are equivalent.

connected load [ELEC] The sum of the continuous power ratings of all load-consuming apparatus connected to an electric power distribution system or any part thereof.

connected set [MATH] A set in a topological space which is not the union of two nonempty sets A and B for which both the intersection of the closure of A with B and the intersection of the closure of B with A are empty; intuitively, a set with only one piece.

connected space [MATH] A topological space which cannot be written as the union of two nonempty disjoint open subsets.

connectivity number [MATH] **1.** The number of points plus 1 which can be removed from a curve without separating the curve into more than one piece. **2.** The number of closed cuts or cuts joining points of previous cuts (or joining points on the boundary) plus 1 which can be made on a surface without separating the surface. **3.** In general, the n-dimensional connectivity number of a topological space X is the number of infinite cyclic groups whose direct sum with the torsion group $G_n(X)$ forms the homology group $H_n(X)$. Also known as Betti number.

conode *See* tie line.

conoid of Sturm *See* astigmatic interval.

conoscope [OPTICS] An instrument, essentially a wide-angle microscope, used for study and observation of interference figures and related phenomena of specially cut crystal plates, especially for measuring the axial angle. Also known as hodoscope.

Conrad discontinuity [GEOPHYS] A relatively abrupt discontinuity in the velocity of elastic waves in the earth, increasing from 6.1 to 6.4–6.7 kilometers per second; occurs at various depths and marks contact of granitic and basaltic layers.

consequent poles [ELECTROMAG] Pairs of magnetic poles in a magnetized body that are in excess of the usual single pair.

conservation of angular momentum [MECH] The principle that, when a physical system is subject only to internal forces that bodies in the system exert on each other, the total angular momentum of the system remains constant, provided that both spin and orbital angular momentum are taken into account.

conservation of areas [MECH] A principle governing the motion of a body moving under the action of a central force, according to which a line joining the body with the center of force sweeps out equal areas in equal times.

conservation of charge [ELEC] A law which states that the total charge of an isolated system is constant; no violation of this law has been discovered. Also known as charge conservation.

conservation of condensation [FL MECH] The principle that the rapid rise of pressure associated with the spherical wave propagating outward from an explosion must be followed by a region of diminished pressure.

conservation of energy [PHYS] The principle that energy cannot be created or destroyed, although it can be changed from one form to another; no violation of this principle has been found. Also known as energy conservation.

conservation of mass [PHYS] The notion that mass can neither be created nor destroyed; it is violated by many microscopic phenomena.

conservation of matter [PHYS] The notion that matter can be neither created nor destroyed; it is violated by microscopic phenomena.

conservation of momentum [MECH] The principle that, when a system of masses is subject only to internal forces that masses of the system exert on one another, the total vector momentum of the system is constant; no violation of this principle has been found. Also known as momentum conservation.

conservation of parity [QUANT MECH] The law that, if the wave function describing the initial state of a system has even (odd) parity, the wave function describing the final state has even (odd) parity; it is violated by the weak interactions. Also known as parity conservation.

conservation of probability [QUANT MECH] The requirement that the sum of the probabilities of finding a system in each of its possible states is constant.

conservation of vorticity [FL MECH] **1.** The principle that the vertical component of the absolute vorticity of each particle in an inviscid, autobarotropic fluid flowing horizontally remains constant. **2.** The hypothesis that the vorticity of fluid particles remains constant during the turbulent mixing of the fluid.

conservative force field [MECH] A field of force in which the work done on a particle in moving it from one point to another depends only on the particle's initial and final positions.

conservative property [THERMO] A property of a system whose value remains constant during a series of events.

conservative system [MECH] A physical system in which the work done on the particles in moving from one configuration to another depends only on the initial and final configurations of the particles.

conserved vector current [PARTIC PHYS] The hypothesis that the weak hadronic vector current is identical to the conserved isotopic-spin current. Abbreviated CVC.

consistency condition [MATH] The requirement that a mathematical theory be free from contradiction.

consistent estimate [STAT] A method of estimation which has the property that the estimate is practically certain to fall very close to a parameter being estimated, provided there are sufficient observations.

consolute temperature [THERMO] The upper temperature of immiscibility for a two-component liquid system. Also known as upper consolute temperature; upper critical solution temperature.

consonance [ACOUS] The interval between two tones whose frequencies are in a ratio approximately equal to the quotient of two whole numbers, each equal to or less than 6, or to such a quotient multiplied or divided by some power of 2.

constant-angle fringes *See* Haidinger fringes.

constant-bandwidth analyzer [ACOUS] A tunable sound analyzer which has a fixed pass band that is swept through the frequency range of interest. Also known as constant-bandwidth filter.

constant-bandwidth filter *See* constant-bandwidth analyzer.

constant-conductance network *See* constant-resistance network.

constant-current dc potentiometer [ELEC] A potentiometer in which the unknown electromotive force is balanced by a constant current times the resistance of a calibrated resistor or slide-wire. Also known as Poggendorff's first method.

constant-current transformer [ELEC] A transformer that automatically maintains a constant current in its secondary circuit under varying loads, when supplied from a constant-voltage source.

constant-deviation fringes *See* Haidinger fringes.

constant-deviation prism [OPTICS] A prism whose deviation is constant and does not depend on the index of refraction or wavelength.

constant-deviation spectrometer [SPECT] A spectrometer in which the collimator and telescope are held fixed and the observed wavelength is varied by rotating the prism or diffraction grating.

constant-effect model [STAT] A model of a test in which the effect of a treatment is the same for all subjects.

constant field *See* stationary field.

constant-gradient synchrotron [NUCLEO] A synchrotron in which the radial gradient of the magnetic field is constant as a function of angle around the orbit.

constant-k filter [ELECTR] A filter in which the product of the series and shunt impedances is a constant that is independent of frequency.

constant-k lens [ELECTROMAG] A microwave lens that is constructed as a solid dielectric sphere; a plane electromagnetic wave brought to a focus at one point on the sphere emerges from the opposite side of the sphere as a parallel beam.

constant-k network [ELECTR] A ladder network in which the product of the series and shunt impedances is independent of frequency within the operating frequency range.

constant of aberration [ASTRON] The maximum aberration of a star observed from the surface of the earth, equal to 20.49 seconds of arc.

constant of gravitation *See* gravitational constant.

constant of motion [MECH] A dynamical variable of a system which remains constant in time.

constant-potential accelerator [NUCLEO] An accelerator in which constant direct-current voltage is applied to an accelerating tube to produce high-energy ions or electrons.

constant-resistance dc potentiometer [ELEC] A potentiometer in which the ratio of an unknown and a known potential are set equal to the ratio of two known constant resistances. Also known as Poggendorff's second method.

constant-resistance network [ELECTR] A network having at least one driving-point impedance that is a positive constant. Also known as constant-conductance network.

constant-volume gas thermometer *See* gas thermometer.

constellation [ASTRON] **1.** Any one of the star groups interpreted as forming configurations in the sky; examples are Orion and Leo. **2.** Any one of the definite areas of the sky.

constituent day [ASTRON] The duration of one rotation of the earth on its axis with respect to an astre fictif, that is, a fictitious star representing one of the periodic elements in the tidal forces; approximates the length of a lunar or solar day.

constitutive equations [ELECTROMAG] The equations $D = \varepsilon E$ and $B = \mu H$, which relate the electric displacement D with the electric field intensity E, and the magnetic induction B with the magnetic field intensity H.

constitutive property [CHEM] Any physical or chemical property that depends on the constitution or structure of the molecule.

constraint [MECH] A restriction on the natural degrees of freedom of a system; the number of constraints is the difference between the number of natural degrees of freedom and the number of actual degrees of freedom.

constringence *See* nu value.

constructive interference [PHYS] Phenomenon in which the phases of waves arriving at a specified point over two or more paths of different lengths are such that the square of the

CONSTANT-K NETWORK

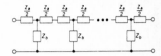

T sections, one form of constant-*k* recurrent ladder network.

resultant amplitude is greater than the sum of the squares of the component amplitudes.

contact [ELEC] *See* electric contact. [FL MECH] The surface between two immiscible fluids in a reservoir.

contact anemometer [ENG] An anemometer which actuates an electrical contact at a rate dependent upon the wind speed. Also known as contact-cup anemometer.

contact angle *See* angle of contact.

contact clip [ELEC] The clip which the blade of a knife switch is clamped to in the closed condition.

contact-cup anemometer *See* contact anemometer.

contact electricity [ELEC] An electric charge at the surface of contact of two different materials.

contact electromotive force *See* contact potential difference.

contact lens [OPTICS] **1.** A thin lens fitted over the cornea to correct defects of vision. **2.** A similar lens or prism used with a gonioscope in eye examinations.

contactor control system [CONT SYS] A feedback control system in which the control signal is a discontinuous function of the sensed error and may therefore assume one of a limited number of discrete values.

contact piston [ELECTROMAG] A waveguide piston that makes contact with the walls of the waveguide. Also known as contact plunger.

contact plunger *See* contact piston.

contact potential *See* contact potential difference.

contact potential difference [ELEC] The potential difference that exists across the space between two electrically connected materials. Also known as contact electromotive force; contact potential; Volta effect.

contact protection [ELEC] Any method for suppressing the surge which results when an inductive circuit is suddenly interrupted; the break would otherwise produce arcing at the contacts, leading to their deterioration.

contact rectifier *See* metallic rectifier.

contact transformation *See* canonical transformation.

contact twin [CRYSTAL] Twinned crystals whose members are symmetrically arranged about a twin plane.

contagious distribution [STAT] A probability distribution which is dependent on a parameter that itself has a probability distribution.

containment [NUCLEO] **1.** Provision of a gastight enclosure around the highly radioactive components of a nuclear power plant, to contain the radioactivity released by a possible major accident. **2.** The use of remote-control devices (slave apparatus) to remove spent cores from nuclear power plants or, in shielded laboratory hoods, to perform chemical studies of dangerous radioactive materials.

containment vessel [NUCLEO] A gas-tight shell or other enclosure around a reactor.

contamination [NUCLEO] The deposit of radioactive materials, such as fission fragments or radiological warfare agents, on any objective or surface or in the atmosphere.

contamination monitor [NUCLEO] A radiation counter used to detect radioactive contamination of surface areas or of the atmosphere.

content [MATH] The content of a polynomial f with coefficients in the quotient field of a unique factorization domain R is the product

$$\prod p^{\mathrm{ord}_p f}$$

(where ord $_p f$ denotes the order of p at f), the product being taken over all primes p in R for which $\mathrm{ord}_p f \neq 0$, or any multiple of this product by a unit of R. Denoted cont (f).

contiguous functions [MATH] Any pair of hypergeometric functions in which one of the parameters differs by unity and the other two are equal.

contingency table [STAT] A table for classifying elements of a population according to two variables, the rows corresponding to one variable and the columns to the other.

continuant [MATH] The determinant of a continuant matrix.

continuant matrix [MATH] A square matrix all of whose nonzero elements lie on the principal diagonal or the diagonals immediately above and below the principal diagonal. Also known as triple-diagonal matrix.

continued fraction [CONT SYS] The sum of a number and a fraction whose denominator is the sum of a number and a fraction, and so forth; it may have either a finite or an infinite number of terms.

continued-fraction expansion [MATH] **1.** An expansion of a driving-point function about infinity (or zero) in a continued fraction, in which the terms are alternately constants and multiples of the complex frequency (or multiples of the reciprocal of the complex frequency). **2.** A representation of a real number by a continued fraction, in a manner similar to the representation of real numbers by a decimal expansion.

continued product [MATH] A product of three or more factors, or of an infinite number of factors.

continuity [ELEC] Continuous effective contact of all components of an electric circuit to give it high conductance by providing low resistance.

continuity equation [PHYS] An equation obeyed by any conserved, indestructible quantity such as mass, electric charge, thermal energy, electrical energy, or quantum-mechanical probability, which is essentially a statement that the rate of increase of the quantity in any region equals the total current flowing into the region. Also known as equation of continuity.

continuity of state [THERMO] Property of a transition between two states of matter, as between gas and liquid, during which there are no abrupt changes in physical properties.

continuous at a point [MATH] A function f is continuous at a point x if for every sequence $\{x_n\}$ whose limit is x, the sequence $f(x_n)$ converges to $f(x)$; in a general topological space, for every neighborhood W of $f(x)$, there is a neighborhood N of x such that $f^{-1}(W)$ is contained in N.

continuous control [CONT SYS] Automatic control in which the controlled quantity is measured continuously and corrections are a continuous function of the deviation.

continuous distribution [STAT] Distribution of a continuous population, which is a class of pairs such that the second member of each pair is a value, and the first member of the pair is a proportion density for that value.

continuous extension [MATH] A continuous function which is equal to another continuous function defined on a smaller domain.

continuous function [MATH] A function which is continuous at each point of its domain. Also known as continuous transformation.

continuous geometry [MATH] A generalization of projective geometry.

continuous image [MATH] The image of a set under a continuous function.

continuous leader *See* dart leader.

continuously adjustable transformer *See* variable transformer.

continuous operator [MATH] A linear transformation of Banach spaces which is continuous with respect to their topologies.

continuous population [STAT] A population in which a random variable is measuring a continuous characteristic.

continuous radiation [ELECTROMAG] Electromagnetic radiation that includes all the wavelengths in some interval. Also known as white radiation.

continuous set [STAT] In an infinite number of outcomes of an experiment, those outcomes in which any value in a given interval can occur.

continuous spectrum [MATH] The portion of the spectrum of a linear operator which is a continuum. [SPECT] A radiation spectrum which is continuously distributed over a frequency region without being broken up into lines or bands.

continuous system [CONT SYS] A system whose inputs and outputs are capable of changing at any instant of time. Also known as continuous-time signal system.

continuous-time signal system *See* continuous system.

continuous transformation *See* continuous function.

continuous-wave gas laser [OPTICS] A laser having a quartz envelope filled with a mixture of helium and neon at low pressure, with Brewster-angle mirrors at opposite ends and an external optical system.

continuous-wave laser [OPTICS] A laser in which the beam of coherent light is generated continuously, as required for communication and certain other applications. Abbreviated CW laser.

continuous x-rays [ELECTROMAG] The electromagnetic radiation, having a continuous spectral distribution, that is produced when high-velocity electrons strike a target.

continuum [MATH] A compact, connected set.

continuum mechanics *See* classical field theory.

continuum physics *See* classical field theory.

contour [PHYS] A curve drawn on a two-dimensional diagram through points satisfying $f(x,y) = c$, where c is a constant and f is some function, such as the field strength for a transmitter.

contour integral [MATH] A line integral of a complex function, usually over a simple closed curve.

contracted curvature tensor [MATH] A symmetric tensor of second order, obtained by summation on two indices of the Riemann curvature tensor which are not antisymmetric. Also known as contracted Riemann-Christoffel tensor; Ricci tensor.

contracted Riemann-Christoffel tensor *See* contracted curvature tensor.

contraction [MATH] **1.** A continuous function of a metric space to itself which moves each pair of points closer together. **2.** The operation of setting one of the contravariant indices of a tensor equal to one of the covariant indices and summing over this index, yielding a tensor of order two less than that of the original tensor. [MECH] The action or process of becoming smaller or pressed together, as a gas on cooling.

contraction coefficient *See* coefficient of contraction.

contraction loss [FL MECH] In fluid flow, the loss in mechanical energy in a stream flowing through a closed duct or pipe when there is a sudden contraction of the cross-sectional area of the passage.

contraction semigroup [MATH] A strongly continuous semigroup all of whose elements have norms which are equal to or less than a constant which is, in turn, less than 1.

contracurrent system *See* katoptric system.

contrapositive [MATH] The contrapositive of the statement "if p, then q" is the equivalent statement "if not q, then not p."

contrast sensitivity *See* threshold contrast.

contrast threshold *See* threshold contrast.

contravariant functor [MATH] A functor which reverses the sense of morphisms.

contravariant index [MATH] A tensor index such that, under a transformation of coordinates, the procedure for obtaining a component of the transformed tensor for which this index has the value p involves taking a sum over q of the product of a component of the original tensor for which the index has the value q times the partial derivative of the pth transformed coordinate with respect to the qth original coordinate; it is written as a superscript.

contravariant tensor [MATH] A tensor with only contravariant indices.

contravariant vector [MATH] A contravariant tensor of degree 1, such as the tensor whose components are differentials of the coordinates.

control [CONT SYS] A means or device to direct and regulate a process or sequence of events. [ELECTR] An input element of a cryotron. [STAT] **1.** A test made to determine the extent of error in experimental observations or measurements. **2.** A procedure carried out to give a standard of comparison in an experiment. **3.** Observations made on subjects which have not undergone treatment, to use in comparison with observations made on subjects which have undergone treatment.

control characteristic [ELECTR] **1.** The relation, usually shown by a graph, between critical grid voltage and anode voltage of a gas tube. **2.** The relation between control ampere-turns and output current of a magnetic amplifier.

control drive [NUCLEO] The system of control rods which regulate the reaction rate of a nuclear reactor.

control electrode [ELECTR] An electrode used to initiate or vary the current between two or more electrodes in an electron tube.

control grid [ELECTR] A grid, ordinarily placed between the cathode and an anode, that serves to control the anode current of an electron tube.

control-grid bias [ELECTR] Average direct-current voltage between the control grid and cathode of a vacuum tube.

control-grid plate transconductance [ELECTR] Ratio of the amplification factor of a vacuum tube to its plate resistance, combining the effects of both into one term.

controllability [CONT SYS] Property of a system for which, given any initial state and any desired state, there exists a time interval and an input signal which brings the system from the initial state to the desired state during the time interval.

controlled atmosphere [SCI TECH] A specified gas or mixture of gases at a predetermined temperature, and sometimes humidity, in which selected processes take place.

controlled avalanche device [ELECTR] A semiconductor device that has rigidly specified maximum and minimum avalanche voltage characteristics and is able to operate and absorb momentary power surges in this avalanche region indefinitely without damage.

controlled avalanche rectifier [ELECTR] A silicon rectifier in which carefully controlled, nondestructive internal avalanche breakdown across the entire junction area protects the junction surface, thereby eliminating local heating that would impair or destroy the reverse blocking ability of the rectifier.

controlled avalanche transit-time triode [ELECTR] A solid-state microwave device that uses a combination of IMPATT diode and *npn* bipolar transistor technologies; avalanche and drift zones are located between the base and collector regions. Abbreviated CATT.

controlled fusion [NUCLEO] The use of thermonuclear fusion reactions in a controlled manner to generate power.

controlled parameter [ENG] In the formulation of an optimization problem, one of the parameters whose values determine the value of the criterion parameter.

controlled thermonuclear reaction [NUCLEO] A fusion reaction generated in a controlled manner for research purposes or for production of useful power.

controlled thermonuclear reactor [NUCLEO] The heart of a fusion spacecraft propulsion system, based on the thermonuclear reaction of deuterium with a helium-3 isotope to produce helium-4 and protons. Abbreviated CTR.

controller *See* automatic controller.

controlling magnet [ENG] An auxiliary magnet used with a galvanometer to cancel the effect of the earth's magnetic field.

control rod [NUCLEO] Any rod used to control the reactivity of a nuclear reactor; may be a fuel rod or part of the moderator; in a thermal reactor, commonly a neutron absorber. Also known as absorbing rod.

control system [ENG] A system in which one or more outputs are forced to change in a desired manner as time progresses.

control system feedback [CONT SYS] A signal obtained by comparing the output of a control system with the input, which is used to diminish the difference between them.

control variable [CONT SYS] One of the input variables of a control system, such as motor torque or the opening of a valve, which can be varied directly by the operator to maximize some measure of performance of the system.

convection [FL MECH] Diffusion in which the fluid as a whole is moving in the direction of diffusion. Also known as bulk flow. [PHYS] Transmission of energy or mass by a medium involving movement of the medium itself.

convection cell [GEOPHYS] A concept in plate tectonics that accounts for the lateral or the upward and downward movement of subcrustal mantle material as due to heat variation in the earth.

convection coefficient *See* film coefficient.

convection current [ELECTR] The time rate at which the electric charges of an electron stream are transported through a given surface. [GEOPHYS] Mass movement of subcrustal or mantle material as a result of temperature variations.

convection modulus [FL MECH] An intrinsic property of a fluid which is important in determining the Nusselt number, equal to the acceleration of gravity times the volume coefficient of thermal expansion divided by the product of the kinematic viscosity and the thermal diffusivity.

convective discharge [ELECTR] The movement of a visible or invisible stream of charged particles away from a body that has been charged to a sufficiently high voltage. Also known as electric wind; static breeze.

conventional current [ELEC] The concept of current as the transfer of positive charge, so that its direction of flow is opposite to that of electrons which are negatively charged.

converge in measure [MATH] A sequence of functions $f_n(x)$ converges in measure to $f(x)$ if given any $\varepsilon > 0$, the measure of the set of points at which $|f_n(x) - f(x)| > \varepsilon$ is less than ε, provided n is sufficiently large.

converge in the mean [MATH] A sequence of functions f_n is said to converge in the mean of order p to a function f on a measure space (X,μ) if the integral over X of $|f(x) - f_n(x)|^p \, d\mu$ approaches 0 as n approaches infinity.

convergence [ELECTR] A condition in which the electron beams of a multibeam cathode-ray tube intersect at a specified point, such as at an opening in the shadow mask of a three-gun color television picture tube; both static and dynamic convergence are required. [MATH] The property of having a limit for infinite series, sequences, products, and so on.

convergence in mean [MATH] A sequence of functions $f_n(x)$ converges in mean to $f(x)$ if the integral of $|f_n(x) - f(x)|^2 \, dx$ over the domain of definition of the functions approaches 0 as n approaches infinity.

convergence limit [SPECT] **1.** The short-wavelength limit of a set of spectral lines that obey a Rydberg series formula; equivalently, the long-wavelength limit of the continuous spectrum corresponding to ionization from or recombination to a given state. **2.** The wavelength at which the difference between successive vibrational bands in a molecular spectrum decreases to 0.

convergence pressure [PHYS CHEM] The pressure at which the different constant-temperature K (liquid-vapor equilibrium) factors for each member of a two-component system converge to unity.

convergence ratio [OPTICS] The ratio of the tangent of the angle between a meridional ray and the optical axis after it passes through an optical system to the tangent of the angle between the ray and the axis before it passes through the system.

convergence zone [ACOUS] A sound transmission channel produced in sea water by a combination of pressure and temperature changes in the depth range between 2500 and 15,000 feet (750 and 4500 meters); utilized by sonar systems.

convergent integral [MATH] An improper integral which has a finite value.

convergent sequence [MATH] A sequence which has a limit.

convergent series [MATH] A series whose sequence of partial sums has a limit.

converging lens [OPTICS] A lens that has a positive focal length, and therefore causes rays of light parallel to its axis to converge.

converse [MATH] The converse of the statement "if p, then q" is the statement "if q, then p."

conversion [ADP] *See* data conversion. [NUC PHYS] Nuclear transformation of a fertile substance into a fissile substance. [PHYS] Change in a quantity's numerical value as a result of using a different unit of measurement.

conversion coefficient Also known as conversion fraction; internal conversion coefficient. [NUC PHYS] **1.** The ratio of the number of conversion electrons emitted per unit time to the number of photons emitted per unit time in the de-excitation of a nucleus between two given states. **2.** In older literature, the ratio of the number of conversion electrons emitted per unit time to the number of conversion electrons plus the number of photons emitted per unit time in the de-excitation of a nucleus between two given states.

conversion electron [NUC PHYS] An electron which receives energy directly from a nucleus in an internal conversion process and is thereby expelled from the atom.

conversion factor [MATH] The numerical factor by which one must multiply (or divide) a quantity expressed in terms of one unit to express the quantity in terms of another unit. [NUCLEO] *See* conversion ratio.

conversion fraction *See* conversion coefficient.

conversion gain [ELECTR] **1.** Ratio of the intermediate-frequency output voltage to the input signal voltage of the first detector of a superheterodyne receiver. **2.** Ratio of the available intermediate-frequency power output of a converter or mixer to the available radio-frequency power input. [NU-

CLEO] The conversion ratio minus one in a nuclear reactor.

conversion length [PHYS] The average distance traveled by an energetic photon in a given medium before it is converted into an electron and a positron through pair production.

conversion ratio [MATH] *See* conversion factor. [NUCLEO] The number of fissionable atoms produced per fissionable atom fissioned in a converter type of nuclear reactor. Also known as conversion factor.

converter [NUCLEO] Also known as nuclear converter. **1.** A nuclear reactor that converts fertile atoms into fuel by neutron capture, using one kind of fuel and producing another. **2.** A nuclear reactor that produces some fissionable fuel, but less than it consumes; the fuel produced may be the same as that consumed or different.

convex angle [MATH] A polyhedral angle that lies entirely on one side of each of its faces.

convex combination [MATH] A linear combination of vectors in which the sum of the coefficients is 1.

convex curve [MATH] A plane curve for which any straight line that crosses the curve crosses it at just two points.

convex function [MATH] A function $f(x)$ is said to be convex over the interval a,b if for any three points x_1, x_2, x_3 such that $a < x_1 < x_2 < x_3 < b$, $f(x_2) \leq L(x_2)$, where $L(x)$ is the equation of the straight line passing through the points $[x_1, f(x_1)]$ and $[x_3, f(x_3)]$.

convex hull [MATH] The smallest convex set containing a given collection of points in a real linear space. Also known as convex linear hull.

convex linear hull *See* convex hull.

convex polyhedron [MATH] A polyhedron in the plane which is a convex set, for example, any regular polyhedron.

convex set [MATH] A set which contains the entire line segment joining any pair of its points.

convolution [STAT] A method for finding the distribution of the sum of two or more random variables; computed by direct integration or summation as contrasted with, for example, the method of characteristic functions.

convolution family *See* faltung.

convolution of two functions [MATH] The convolution of the functions f and g is the function F, defined by

$$F(x) = \int_0^x f(t) g(x-t)\, dt.$$

Conwell-Weisskopf equation [SOLID STATE] An equation for the mobility of electrons in a semiconductor in the presence of donor or acceptor impurities, in terms of the dielectric constant of the medium, the temperature, the concentration of ionized donors (or acceptors), and the average distance between them.

Cooke objective [OPTICS] A three-lens objective consisting of one biconcave lens, the dispersive component, between two biconvex lens, the collective components; used in astronomical cameras.

cooled infrared detector [ELECTR] An infrared detector that must be operated at cryogenic temperatures, such as at the temperature of liquid nitrogen, to obtain the desired infrared sensitivity.

Coolidge tube [ELECTROMAG] An x-ray tube in which the needed electrons are produced by a hot cathode.

cooling [NUCLEO] Setting aside a highly radioactive material until the radioactivity has diminished to a desired level.

cooling correction [THERMO] A correction that must be employed in calorimetry to allow for heat transfer between a body and its surroundings. Also known as radiation correction.

cooling curve [THERMO] A curve obtained by plotting time

COOKE OBJECTIVE

Cooke objective, a type of wide-field lens objective for astronomical cameras.

against temperature for a solid-liquid mixture cooling under constant conditions.

cooling method [THERMO] A method of determining the specific heat of a liquid in which the times taken by the liquid and an equal volume of water in an identical vessel to cool through the same range of temperature are compared.

cooling-power anemometer [ENG] Any anemometer operating on the principle that the heat transfer to air from an object at an elevated temperature is a function of airspeed.

cool star [ASTROPHYS] A low-temperature star, generally visible in the infrared range of the electromagnetic spectrum.

cooperative phenomenon [SOLID STATE] A process involving a simultaneous collective interaction among many atoms or electrons in a crystal, such as ferromagnetism, superconductivity, and order-disorder transformations.

Cooper pairs [SOLID STATE] Pairs of bound electrons which occur in a superconducting medium according to the Bardeen-Cooper-Schrieffer theory.

coordinate axes [MATH] One of a set of lines or curves used to define a coordinate system; the value of one of the coordinates uniquely determines the location of a point on the axis, while the values of the other coordinates vanish on the axis.

coordinate basis [MATH] A basis for tensors on a manifold induced by a set of local coordinates.

coordinate bond *See* dative bond.

coordinated complex *See* coordination compound.

coordinates [MATH] A set of numbers which locate a point in space.

coordinate singularity [MATH] A singularity in the coordinate system describing a geometry; no intrinsic property of the geometry itself is singular.

coordinate systems [MATH] A rule for designating each point in space by a set of numbers.

coordinate transformation [MATH] A mathematical or graphic process of obtaining a modified set of coordinates by performing some nonsingular operation on the coordinate axes, such as rotating or translating them.

coordination compound [CHEM] A compound with a central atom or ion and a group of ions or molecules surrounding it. Also known as coordinated complex; Werner complex.

coordination lattice [CRYSTAL] The crystal structure of a coordination compound.

coordination number [PHYS] The number of nearest neighbors of a point in a space lattice, of an atom or an ion in a solid, or of an anion or cation in a solution.

Copernican system [ASTRON] The system of planetary motions according to Copernicus, who maintained that the earth revolves about an axis once every day and revolves around the sun once every year while the other planets also move in orbits centered near the sun.

coplanar electrodes [ELECTR] Electrodes mounted in the same plane.

coplanar forces [MECH] Forces that act in a single plane; thus the forces are parallel to the plane and their points of application are in the plane.

copper [CHEM] A chemical element, symbol Cu, atomic number 29, atomic weight 63.546.

copper-64 [NUC PHYS] Radioactive isotope of copper with mass number of 64; derived from pile-irradiation of metallic copper; used as a research aid to study diffusion, corrosion, and friction wear in metals and alloys.

copper oxide photovoltaic cell [ELECTR] A photovoltaic cell in which light acting on the surface of contact between layers of copper and cuprous oxide causes a voltage to be produced.

COPPER

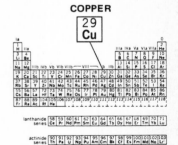

Periodic table of the chemical elements showing the position of copper.

COPPER OXIDE RECTIFIER

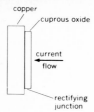

Cross section of a copper oxide rectifier.

copper oxide rectifier [ELECTR] A metallic rectifier in which the rectifying barrier is the junction between metallic copper and cuprous oxide.

copper sulfide rectifier [ELECTR] A semiconductor rectifier in which the rectifying barrier is the junction between magnesium and copper sulfide.

Corbino disk [ELECTROMAG] A variable-resistance device utilizing the effect of a magnetic field on the flow of carriers from the center to the circumference of a disk made of semiconducting or conducting material.

Corbino effect [ELECTROMAG] The production of an electric current around the circumference of a disk when a magnetic field perpendicular to the disk acts on a radial current in the disk.

cord [ELEC] A small, very flexible insulated cable.

cordwood module [ELECTR] High-density circuit module in which discrete components are mounted between and perpendicular to two small, parallel printed circuit boards to which their terminals are attached.

core [ELECTR] See magnetic core. [ELECTROMAG] See magnetic core. [NUCLEO] The active portion of a nuclear reactor, containing the fissionable material.

core logic [ELECTR] Logic performed in ferrite cores that serve as inputs to diode and transistor circuits.

core loss [ELECTROMAG] The rate of energy conversion into heat in a magnetic material due to the presence of an alternating or pulsating magnetic field. Also known as excitation loss; iron loss.

Coriolis acceleration [MECH] **1.** An acceleration which, when added to the acceleration of an object relative to a rotating coordinate system and to its centripetal acceleration, gives the acceleration of the object relative to a fixed coordinate system. **2.** A vector which is equal in magnitude and opposite in direction to that of the first definition.

Coriolis deflection See Coriolis effect.

Coriolis effect [MECH] Also known as Coriolis deflection. **1.** The deflection relative to the earth's surface of any object moving above the earth, caused by the Coriolis force; an object moving horizontally is deflected to the right in the Northern Hemisphere, to the left in the Southern. **2.** The effect of the Coriolis force in any rotating system.

Coriolis force [MECH] A velocity-dependent pseudoforce in a reference frame which is rotating with respect to an inertial reference frame; it is equal and opposite to the product of the mass of the particle on which the force acts and its Coriolis acceleration.

Coriolis operator [SPECT] An operator which gives a large contribution to the energy of an axially symmetric molecule arising from the interaction between vibration and rotation when two vibrations have equal or nearly equal frequencies.

Coriolis parameter [GEOPHYS] Twice the component of the earth's angular velocity about the local vertical $2\Omega \sin \phi$, where Ω is the angular speed of the earth and ϕ is the latitude; the magnitude of the Coriolis force per unit mass on a horizontally moving fluid parcel is equal to the product of the Coriolis parameter and the speed of the parcel.

Coriolis resonance interactions [SPECT] Perturbation of two vibrations of a polyatomic molecule, having nearly equal frequencies, on each other, due to the energy contribution of the Coriolis operator.

Coriolis theorem [MECH] The formula $\mathbf{a} = \mathbf{a}' + \omega \times (\omega \times \mathbf{r}) + 2\omega \times \mathbf{v}' + (d\omega/dt) \times \mathbf{r}$, giving the acceleration $\mathbf{a}$ of a particle in an inertial reference frame in terms of its acceleration **a'**, velocity **v'**, and position vector **r** in a coordinate system rotating with angular velocity ω.

corkscrew rule [ELECTROMAG] The rule that the direction of the current and that of the resulting magnetic field are related to each other as the forward travel of a corkscrew and the direction in which it is rotated.

corner conditions [MATH] In a problem in which an integral over x of a function of the form $f(y_1, \ldots, y'_1, \ldots, y'_n, x)$ with $y'_i = dy_i/dx$, is to be minimized, with the y_i continuous and the y'_i piecewise-continuous, corner condtions are equations which must be satified at values of x where one of the y_i has a discontinuity.

corner frequency *See* break frequency.

corner reflector [ELECTROMAG] An antenna consisting of two conducting surfaces intersecting at an angle that is usually 90°, with a dipole or other antenna located on the bisector of the angle. [OPTICS] A reflector which returns a laser beam in the direction of its source, consisting of perpendicular reflecting surfaces; used to make precise determinations of distances in surveying.

Cornu-Hartmann formula *See* Hartmann dispersion formula.

Cornu quartz prism [OPTICS] A prism constructed of two 30° quartz prisms, left- and right-handed, used in conjunction with left- and right-handed lenses, so that the rotation of polarization occurring in one half of the optical path is exactly compensated by the reverse rotation in the other; used in a quartz spectrograph.

Cornu's spiral [MATH] The graph of the function

$$f(x,y) = \int_{-\infty}^{+\infty} \exp\left[-y^2z^2 - z - xe^{-z}\right]\, dz.$$

Also known as clothoid.

corona [ASTRON] *See* solar corona. [ELEC] *See* corona discharge.

corona current [ELEC] The current of electricity equivalent to the rate of charge transferred to the air from an object experiencing corona discharge.

corona discharge [ELEC] A discharge of electricity appearing as a bluish-purple glow on the surface of and adjacent to a conductor when the voltage gradient exceeds a certain critical value; due to ionization of the surrounding air by the high voltage. Also known as aurora; corona; electric corona.

coronagraph [ASTRON] An instrument for photographing the corona and prominences of the sun at times other than at solar eclipse.

corona resistance [ELEC] Ability of a conductor to resist destruction when a high-voltage electrostatic field ionizes within insulation voids.

corona shield [ELEC] A shield placed about a point of high potential to redistribute electrostatic lines of force.

corona tube [ELEC] A gas-discharge voltage-reference tube employing a corona discharge.

corona voltmeter [ELEC] A voltmeter in which the crest value of a voltage is indicated by the inception of corona at a known electrode spacing.

corpuscle [OPTICS] A particle of light in the corpuscular theory, corresponding to the photon in the quantum theory.

corpuscular radiation [PHYS] Radiation consisting of subatomic particles, such as electrons, protons, deuterons, and neutrons, as distinguished from electromagnetic radiation.

corpuscular theory of light [OPTICS] Theory that light consists of a stream of particles; now considered a limiting case of the quantum theory. Also known as Newton's theory of light.

correcting plate *See* corrector plate.

CORNER REFLECTOR

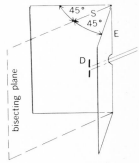

Configuration of a 90° corner reflector. Distance S from the driven radiator D to edge E need not be critically chosen with respect to wavelength; reflector D may lie between 0.25 and 0.7 wavelength.

correction for attenuation [STAT] A method used to adjust correlation coefficients upward because of errors of measurement when two measured variables are correlated; the errors always serve to lower the correlation coefficient as compared with what it would have been if the measurement of the two variables had been perfectly reliable.

correction time [CONT SYS] The time required for the controlled variable to reach and stay within a predetermined band about the control point following any change of the independent variable or operating condition in a control system. Also known as settling time.

corrective action [CONT SYS] The act of varying the manipulated process variable by the controlling means in order to modify overall process operating conditions.

corrective network [ELEC] An electric network inserted in a circuit to improve its transmission properties, impedance properties, or both. Also known as shaping circuit; shaping network.

corrector plate [OPTICS] A thin lens or system of lenses used to correct the spherical aberration of a spherical lens or the coma of a parabolic lens; used particularly in telescopes such as the Schmidt telescope. Also known as correcting plate.

correlation [STAT] The interdependence or association between two variables that are quantitative or qualitative in nature.

correlation array See multiplicative acoustic array.

correlation coefficient [STAT] A measurement, which is unchanged by both addition and multiplication of the random variable by positive constants, of the tendency of two random variables X and Y to vary together; it is given by the ratio of the covariance of X and Y to the square root of the product of the variance of X and the variance of Y.

correlation curve See correlogram.

correlation energy [SOLID STATE] The modification of the Coulomb energy of a crystal that results from the tendency of electrons to stay apart from each other.

correlation ratio [STAT] A measure of the nonlinear relationship between two variables; in a two-way frequency table it may be regarded as the ratio of the variance between arrays to the total variance.

correlation table [STAT] A table designed to categorize paired quantitative data; used to calculate correlation coefficients.

correlation-type receiver See correlator.

correlator [ELECTR] A device that detects weak signals in noise by performing an electronic operation approximating the computation of a correlation function. Also known as correlation-type receiver.

correlogram [MATH] A curve showing the assumed correlation between two mathematical variables. Also known as correlation curve.

correspondence principle [QUANT MECH] The principle that quantum mechanics has a classical limit in which it is equivalent to classical mechanics. Also known as Bohr's correspondence principle.

corresponding states [PHYS CHEM] The condition when two or more substances are at the same reduced pressures, the same reduced temperatures, and the same reduced volumes.

corrugated lens [OPTICS] A lens having circular sections cut out from the surface to reduce its weight without lowering its focal power.

cos See cosine function.

cosecant [MATH] The reciprocal of the sine. Denoted csc.

cosh See hyperbolic cosine.

cosine emission law [OPTICS] The law that the energy emitted by a radiating surface in any direction is proportional to the cosine of the angle which that direction makes with the normal.

cosine function [MATH] In a right triangle having an angle θ, the cosine function gives the ratio of adjacent side to hypotenuse; more generally, it is the function which assigns to any real number θ the abscissa of the point on the unit circle obtained by moving from (1,0) counterclockwise θ units along the circle, or clockwise $|\theta|$ units if θ is less than 0. Denoted cos.

cosine pulse [PHYS] A pulse whose amplitude varies during some time interval in proportion to the cosine function over the range from $-\pi/2$ to $\pi/2$, and vanishes outside this time interval.

cosine-squared pulse [PHYS] A pulse whose amplitude varies during some time interval in proportion to the square of the cosine function over the range from $-\pi/2$ to $\pi/2$, and vanishes outside this time interval.

cosmic electrodynamics [ASTROPHYS] The science concerned with electromagnetic phenomena in ionized media encountered in interstellar space, in stars, and above the atmosphere.

cosmic expansion [ASTRON] The recession of all distant galaxies from each other, as manifested in the red shift of their spectral lines.

cosmic radiation *See* cosmic rays.

cosmic radio waves [ASTRON] Radio waves reaching the earth from interstellar or intergalactic sources.

cosmic rays [NUC PHYS] Electrons and the nuclei of atoms, largely hydrogen, that impinge upon the earth from all directions of space with nearly the speed of light. Also known as cosmic radiation; primary cosmic rays.

cosmic-ray shower [NUC PHYS] The simultaneous appearance of a number of downward-directed ionizing particles, with or without accompanying photons, caused by a single cosmic ray. Also known as air shower; shower.

cosmic-ray telescope [NUCLEO] An array of counters, sensitive to the direction of the rays detected.

cosmic year [ASTRON] The period of rotation of the Milky Way Galaxy, about 220 million years.

cosmochemistry [ASTROPHYS] The branch of science which treats of the chemical composition of the universe and its origin.

cosmogony [ASTROPHYS] Study of the origin and evolution of specific astronomical systems and of the universe as a whole.

cosmological constant [RELAT] The multiplicative constant for a term proportional to the metric in Einstein's equation relating the curvature of space to the energy-momentum tensor.

cosmological term [RELAT] A term proportional to the metric tensor in Einstein's field equations for special relativity.

cosmology [ASTRON] The study of the overall structure of the physical universe.

cospectrum [PHYS] **1.** The spectral decomposition of the in-phase components of the covariance of two functions of time. **2.** The real part of the cross spectrum of two functions.

cot *See* cotangent.

cotangent [MATH] The reciprocal of the tangent. Denoted cot; ctn.

Cotes formulas [MATH] Formulas, such as the trapezoidal rule and Simpson's rule, for calculating approximate values of definite integrals, having the general form

$$\int_a^b f(x)\, dx \cong \sum_{i=0}^{n} a_i f(a + ih)$$

where $h = (b - a)/n$.

cotidal hour [ASTRON] The average interval expressed in solar or lunar hours between the moon's passage over the meridian of Greenwich and the following high water at a specified place.

Cotton balance [ENG] A device which employs a current-carrying conductor of special shape to determine the strength of a magnetic field.

Cotton effect [ANALY CHEM] The characteristic wavelength dependence of the optical rotatory dispersion curve or the circular dichroism curve or both in the vicinity of an absorption band.

Cotton-Mouton birefringence *See* Cotton-Mouton effect.

Cotton-Mouton constant [OPTICS] A constant giving the strength of the Cotton-Mouton effect in a liquid; when multiplied by the path length and the square of the magnetic field, it gives the phase difference between the components of light parallel and perpendicular to the field.

Cotton-Mouton effect [OPTICS] Double refraction (birefringence) of light in a liquid in a magnetic field at right angles to the direction of light propagation. Also known as Cotton-Mouton birefringence.

Cottrell atmosphere [SOLID STATE] A cluster of impurity atoms surrounding a dislocation in a crystal.

Cottrell hardening [SOLID STATE] Hardening of a material caused by locking of its dislocations when impurity atoms whose size differs from that of the solvent cluster around them.

coudé focus [OPTICS] Focus achieved with a coudé telescope.

coudé-Newtonian-Cassegrain telescope [OPTICS] A reflecting telescope designed so that observations can be made at the coudé, Newtonian, or Cassegrain focus.

coudé spectrograph [SPECT] A stationary spectrograph that is attached to the tube of a coudé telescope.

coudé spectroscopy [SPECT] The production and investigation of astronomical spectra using a coudé spectrograph.

coudé telescope [OPTICS] An instrument in which light is reflected along the polar axis to come to focus at a fixed place where it is viewed through a fixed eyepiece or where a spectrograph can be mounted.

Couette flow [FL MECH] Low-speed, steady motion of a viscous fluid between two infinite plates moving parallel to each other.

Couette viscometer [ENG] A viscometer in which the liquid whose viscosity is to be measured fills the space between two vertical coaxial cylinders, the inner one suspended by a torsion wire; the outer cylinder is rotated at a constant rate, and the resulting torque on the inner cylinder is measured by the twist of the wire. Also known as rotational viscometer.

coul *See* coulomb.

coulomb [ELEC] A unit of electric charge, defined as the amount of electric charge that crosses a surface in 1 second when a steady current of 1 absolute ampere is flowing across the surface; this is the absolute coulomb and has been the legal standard of quantity of electricity since 1950; the previous standard was the international coulomb, equal to 0.999835 absolute coulomb. Abbreviated coul. Symbolized C.

Coulomb attraction [ELEC] The electrostatic force of attraction exerted by one charged particle on another charged

COTTON EFFECT

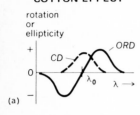

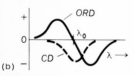

Behavior of optical rotatory dispersion (ORD) and circular dichroism (CD) curves in the vicinity of an absorption band at wavelength λ_0 (idealized).
(a) Positive Cotton effect.
(b) Negative Cotton effect.

particle of opposite sign. Also known as electrostatic attraction.

Coulomb barrier [NUC PHYS] **1.** The Coulomb repulsion which tends to keep positively charged bombarding particles out of the nucleus. **2.** Specifically, the Coulomb potential associated with this force.

Coulomb energy [PHYS] The part of the binding energy of a system of particles, such as an atomic nucleus of a solid, which is associated with electrostatic forces between the particles. [PHYS CHEM] The energy associated with the electrostatic interaction between two or more electron distributions in terms of which the actual electron distribution of a covalent bond is described.

Coulomb excitation [NUC PHYS] Inelastic scattering of a positively charged particle by a nucleus and excitation of the nucleus, caused by the interaction of the nucleus with the rapidly changing electric field of the bombarding particle.

Coulomb field [ELEC] The electric field created by a stationary charged particle.

Coulomb force [ELEC] The electrostatic force of attraction or repulsion exerted by one charged particle on another, in accordance with Coulomb's law.

Coulomb friction [MECH] Friction occurring between dry surfaces.

Coulomb gage [ELECTOMAG] A gage in which the divergence of the magnetic vector potential is equal to 0.

Coulomb interactions [ELEC] Interactions of charged particles associated with the Coulomb forces they exert on one another. Also known as electrostatic interactions.

coulombmeter [ENG] A measuring instrument that measures quantity of electricity in coulombs by integrating a stored charge in a circuit which has very high input impedance.

Coulomb potential [ELEC] A scalar point function equal to the work per unit charge done against the Coulomb force in transferring a particle bearing an infinitesimal positive charge from infinity to a point in the field of a specific charge distribution.

Coulomb repulsion [ELEC] The electrostatic force of repulsion exerted by one charged particle on another charged particle of the same sign. Also known as electrostatic repulsion.

Coulomb scattering [PHYS] A collision of two charged particles in which the Coulomb force is the dominant interaction.

Coulomb's law [ELEC] The law that the attraction or repulsion between two electric charges acts along the line between them, is proportional to the product of their magnitudes, and is inversely proportional to the square of the distance between them. Also known as law of electrostatic attraction.

Coulomb's theorem [ELEC] The intensity of an electric field near the surface of a conductor is equal to the surface charge density on the nearby conductor surface divided by the absolute permittivity of the surrounding medium.

count [NUCLEO] **1.** A single response of the counting system in a radiation counter. **2.** The total number of events indicated by a counter.

countability axioms [MATH] Two conditions which are satisfied by a euclidean space and one or the other of which is often assumed in the study of a general topological space; the first states that any point in the topological space has a countable local base, while the second states that the topological space has a countable base.

countable [MATH] Either finite or denumerable. Also known as enumerable.

countably additive [MATH] Given a measure m, and a sequence of pairwise disjoint measurable sets, the property that the measure of the union is equal to the sum of the measures of the sets.

countably additive set function [MATH] A real-valued function defined on a class of sets such that the value of the function on the union of any pairwise disjoint sequence of sets is equal to the sum of the sequence of the values of the function on the sets.

countably compact set [MATH] A set with the property that every cover with countably many open sets contains a finite number of sets which is also a cover.

countably subadditive [MATH] A set function m is countably subadditive if, given any sequence of sets, the measure of the union is less than or equal to the sum of the measures of the sets.

countably subadditive set function [MATH] A real-valued function defined on a class of sets such that the value of the function on the union of any sequence of sets is equal to or less than the sum of the sequence of the values of the function on the sets.

counter *See* radiation counter; scaler.

counter circuit *See* counting circuit.

counter dead time [NUCLEO] The time interval between the start of a counted event and the earliest instant at which a new event can be counted by a radiation counter.

counter decade *See* decade scaler.

counterelectromotive cell [ELEC] Cell of practically no ampere-hour capacity, used to oppose the line voltage.

counterelectromotive force [ELECTROMAG] The voltage developed in an inductive circuit by a changing current; the polarity of the induced voltage is at each instant opposite that of the applied voltage. Also known as back electromotive force.

counterflow [ENG] Fluid flow in opposite directions in adjacent parts of an apparatus, as in a heat exchanger.

counter/frequency meter [ENG] An instrument that contains a frequency standard and can be used to measure the number of events or the number of cycles of a periodic quantity that occurs in a specified time, or the time between two events.

counterglow *See* gegenschein.

counterions [PHYS CHEM] The mobile ions which are distributed through a liquid in the neighborhood of an electric double layer, and which are of opposite charge to the ions bound to the solid surface. Also known as gegenions.

counterpoise [ELEC] A system of wires or other conductors that is elevated above and insulated from the ground to form a lower system of conductors for an antenna. Also known as antenna counterpoise.

counterpoise method *See* substitution weighing.

counterradiation [GEOPHYS] The downward flux of atmospheric radiation passing through a given level surface, usually taken as the earth's surface. Also known as back radiation.

counter terms [QUANT MECH] Additional terms added to a Lagrangian in quantum field theory in order to absorb the typical divergences that occur in a perturbation expansion of the theory.

counter tube [ELECTR] An electron tube having one signal-input electrode and 10 or more output electrodes, with each input pulse serving to transfer conduction sequentially to the next output electrode; beam-switching tubes and cold-cathode counter tubes are examples. [NUCLEO] An electron tube that converts an incident particle or burst of incident radiation into a discrete electric pulse, generally by utilizing the current flow through a gas that is ionized by the radiation;

used in radiation counters. Also known as radiation counter tube.

counting circuit [ELECTR] A circuit that counts pulses by frequency-dividing techniques, by charging a capacitor in such a way as to produce a voltage proportional to the pulse count, or by other means. Also known as counter circuit.

counting ionization chamber *See* pulse ionization chamber.

counting rate [PHYS] The average rate of occurrence of events as observed by means of a counting system.

counting rate meter [NUCLEO] An instrument that indicates the time rate of occurrence of input pulses to a radiation counter, averaged over a time interval. Also known as rate meter.

counting rate–voltage characteristic *See* plateau characteristic.

couple [MECH] A system of two parallel forces of equal magnitude and opposite sense.

coupled antenna [ELECTROMAG] An antenna electromagnetically coupled to another.

coupled circuits [ELEC] Two or more electric circuits so arranged that energy can transfer electrically or magnetically from one to another.

coupled field vectors [ELECTROMAG] The electric- and magnetic-field vectors, which depend upon each other according to Maxwell's field equations.

coupled harmonic oscillators [PHYS] Linear oscillators with an interaction, often also linear or weak.

coupled modes [ACOUS] Modes of acoustic transmission along a duct having a discontinuity, so that the reflected and transmitted waves contain modes other than the incident ones.

coupled oscillators [ELECTROMAG] A set of alternating-current circuits which interact with each other, for example, through mutual inductances or capacitances. [MECH] A set of particles subject to elastic restoring forces and also to elastic interactions with each other.

coupled systems [PHYS] Mechanical, electrical, or other systems which are connected in such a way that they interact and exchange energy with each other.

coupler [ELEC] A component used to transfer energy from one circuit to another. [ELECTROMAG] **1.** A passage joining two cavities or waveguides, allowing them to exchange energy. **2.** A passage joining the ends of two waveguides, whose cross section changes continuously from that of one to that of the other.

coupling [ELEC] **1.** A mutual relation between two circuits that permits energy transfer from one to another, through a wire, resistor, transformer, capacitor, or other device. **2.** A hardware device used to make a temporary connection between two wires.

coupling aperture [ELECTROMAG] An aperture in the wall of a waveguide or cavity resonator, designed to transfer energy to or from an external circuit. Also known as coupling hole; coupling slot.

coupling capacitor [ELECTR] A capacitor used to block the flow of direct current while allowing alternating or signal current to pass; widely used for joining two circuits or stages. Also known as blocking capacitor; stopping capacitor.

coupling coefficient [ELECTR] The ratio of the maximum change in energy of an electron traversing an interaction space to the product of the peak alternating gap voltage and the electronic charge. [PHYS] *See* coupling constant.

coupling constant [PARTIC PHYS] A measure of the strength of a type of interaction between particles, such as the strong

COUPLED CIRCUITS

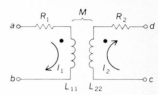

A pair of coupled circuits.

interaction between mesons and nucleons, and the weak interaction between four fermions; analogous to the electric charge, which is the coupling constant between charged particles and electromagnetic radiation. [PHYS] **1.** A measure of the strength of the coupling between two systems, especially electric circuits; maximum coupling is 1 and no coupling is 0. Also known as coefficient of coupling; coupling coefficient. **2.** A measure of the dependence of one physical quantity on another.

coupling hole *See* coupling aperture.

coupling loop [ELECTROMAG] A conducting loop projecting into a waveguide or cavity resonator, designed to transfer energy to or from an external circuit.

coupling probe [ELECTROMAG] A probe projecting into a waveguide or cavity resonator, designed to transfer energy to or from an external circuit.

coupling slot *See* coupling aperture.

course programmer [CONT SYS] An item which initiates and processes signals in a manner to establish a vehicle in which it is installed along one or more projected courses.

covalence [CHEM] The number of covalent bonds which an atom can form.

covalent bond [CHEM] A bond in which each atom of a bound pair contributes one electron to form a pair of electrons. Also known as electron pair bond.

covariance [STAT] A measurement of the tendency of two random variables, X and Y, to vary together, given by the expected value of the variable $(X - \bar{X})(Y - \bar{Y})$, where $\bar{X}$ and $\bar{Y}$ are the expected values of the variables X and Y respectively.

covariance analysis [STAT] An extension of the analysis of variance which combines linear regression with analysis of variance; used when members falling into classes have values of more than one variable.

covariant [RELAT] A scalar, vector, or higher-order tensor.

covariant components [MATH] Vector or tensor components which, in a transformation from one set of basis vectors to another, transform in the same manner as the basis vectors.

covariant derivative [MATH] For a tensor field at a point P of an affine space, a new tensor field equal to the difference between the derivative of the original field defined in the ordinary manner and the derivative of a field whose value at points close to P are parallel to the value of the original field at P as specified by the affine connection.

covariant equation [PHYS] An equation which has the same form in all inertial frames of reference; that is, its form is unchanged by Lorentz transformations.

covariant functor [MATH] A functor which does not change the sense of morphisms.

covariant index [MATH] A tensor index such that, under a transformation of coordinates, the procedure for obtaining a component of the transformed tensor for which this index has value p involves taking a sum over q of the product of a component of the original tensor for which the index has the value q times the partial derivative of the qth original coordinate with respect to the pth transformed coordinate; it is written as a subscript.

covariant tensor [MATH] A tensor with only covariant indices.

covariant theory [PHYS] A theory in which the equations have the same form in any inertial reference frame, the frames being related to each other by Lorentz transformations.

covariant vector [MATH] A covariant tensor of degree 1, such as the gradient of a function.

covering of a set [MATH] A collection of sets whose union contains the given set. Also known as cover of a set.

covering power [OPTICS] The field of view over which a camera lens can produce a sharp image, frequently expressed as an angle.

cover of a set *See* covering of a set.

covers *See* coversed sine.

coversed sine [MATH] The coversed sine of A is $1 - \sin A$. Denoted covers. Also known as coversine; versed cosine.

coversine *See* coversed sine.

cp *See* candlepower; centipoise.

CP invariance [PARTIC PHYS] The principle that the laws of physics are left unchanged by a combination of the operations of charge conjugation C and space inversion P; a small violation of this principle has been observed in the decay of neutral K mesons.

cpm *See* cycles per minute.

cps *See* hertz.

CPT theorem [PARTIC PHYS] A theorem which states that a Lorentz invariant field theory is invariant to the product of charge conjugation C, space inversion P, and time reversal T.

Cr *See* chromium.

Crab Nebula [ASTRON] A gaseous nebula in the constellation Taurus; an amorphous mass which radiates a continuous spectrum involved in a mesh of filaments that radiate a bright-line spectrum.

cracovian [MATH] An object which is the same as a matrix except that the product of cracovians A and B is equal to the matrix product $A'B$, where A' is the transpose of A.

Cramér-Rao inequality [STAT] A method for determining a lower bound to the variance of an estimator of a parameter.

Cramer's rule [MATH] The method of solving a system of linear equations by means of determinants.

creation operator [QUANT MECH] An operator which increases the occupation number of a single state by unity and leaves all the other occupation numbers unchanged.

creep [ELECTR] A slow change in a characteristic with time or usage. [MECH] A time-dependent strain of solids caused by stress.

creepage [ELEC] The conduction of electricity across the surface of a dielectric.

creep buckling [MECH] Buckling that may occur when a compressive load is maintained on a member over a long period, leading to creep which eventually reduces the member's bending stiffness.

creep limit [MECH] The maximum stress a given material can withstand in a given time without exceeding a specified quantity of creep.

creep recovery [MECH] Strain developed in a period of time after release of load in a creep test.

creep rupture strength [MECH] The stress which, at a given temperature, will cause a material to rupture in a given time.

creep strength [MECH] The stress which, at a given temperature, will result in a creep rate of 1% deformation in 100,000 hours.

creep test [ENG] Any one of a number of methods of measuring creep, for example, by subjecting a material to a constant stress or deforming it at a constant rate.

crest factor [PHYS] The ratio of the peak value to the effective value of any periodic quantity such as a sinusoidal alternating current. Also known as amplitude factor; peak factor.

crest value *See* peak value.

CRAB NEBULA

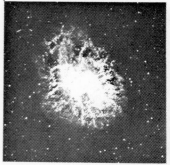

Crab Nebula, in the constellation Taurus, emitter of strong radio waves and of x-rays.

crest voltmeter [ELEC] A voltmeter reading the peak value of the voltage applied to its terminals.

cricondenbar [PHYS CHEM] Maximum pressure at which two phases (for example, liquid and vapor) can coexist.

cricondentherm [PHYS CHEM] Maximum temperature at which two phases (for example, liquid and vapor) can coexist.

crinal [MECH] A unit of force equal to 0.1 newton.

crispation number [PHYS] A dimensionless number used in the study of convection currents, equal to the product of a fluid's dynamic viscosity and its thermal diffusivity, divided by the product of its undisturbed surface tension and a layer thickness.

crit [NUCLEO] The mass of fissionable material that is critical under a given set of conditions; sometimes applied to the mass of an untamped critical sphere of fissionable material.

crith [MECH] A unit of mass, used for gases, equal to the mass of 1 liter of hydrogen at standard pressure and temperature; it is found experimentally to equal 8.9885×10^{-5} kilogram.

critical [NUCLEO] Capable of sustaining a chain reaction at a constant level.

critical absorption wavelength [SPECT] The wavelength, characteristic of a given electron energy level in an atom of a specified element, at which an absorption discontinuity occurs.

critical angle [PHYS] An angle associated with total reflection of electromagnetic or acoustic radiation back into a medium from the boundary with another medium in which the radiation has a higher phase velocity; it is the smallest angle with the normal to the boundary at which total reflection occurs.

critical angle refractometer [OPTICS] A refractometer, such as the Abbe or Pulfrich refractometer, in which the index of refraction of a medium A is measured by observing its critical angle with respect to another medium B with a known index of refraction, or by measuring the critical angle of B with respect to A.

critical assembly [NUCLEO] An assembly of sufficient fissionable and moderator material to sustain a fission chain reaction at a low power level.

critical condensation temperature [PHYS CHEM] The temperature at which the sublimand of a sublimed solid recondenses; used to analyze solid mixtures, analogous to liquid distillation. Also known as true condensing point.

critical constant [PHYS CHEM] A characteristic temperature, pressure, and specific volume of a gas above which it cannot be liquefied.

critical coupling [ELEC] The degree of coupling that provides maximum transfer of signal energy from one radio-frequency resonant circuit to another when both are tuned to the same frequency. Also known as optimum coupling.

critical current [SOLID STATE] The current in a superconductive material above which the material is normal and below which the material is superconducting, at a specified temperature and in the absence of external magnetic fields.

critical damping [PHYS] Damping in a linear system on the threshold between oscillatory and exponential behavior.

critical density [THERMO] The density of a substance at the liquid-vapor critical point.

critical equation [NUCLEO] Any equation relating parameters of a reactor that must be satisfied for the reactor to be critical.

critical experiment [NUCLEO] An experiment in which fissionable material is assembled gradually until the arrangement will support a self-sustaining chain reaction.

critical exponent [THERMO] A parameter n that characterizes the temperature dependence of a thermodynamic property of a substance near its critical point; the temperature dependence has the form $|T - T_c|^n$, where T is the temperature and T_c is the critical temperature.

critical facility [NUCLEO] A facility where critical experiments are conducted.

critical field [ELECTR] The smallest theoretical value of steady magnetic flux density that would prevent an electron emitted from the cathode of a magnetron at zero velocity from reaching the anode. Also known as cutoff field.

critical flicker frequency [OPTICS] That frequency of an intermittent light source at which the light appears half the time as flickering and half the time as continuous.

critical flow [FL MECH] The rate of flow of a fluid equivalent to the speed of sound in that fluid.

critical frequency [ELECTR] *See* cutoff frequency. [ELECTROMAG] The limiting frequency below which a radio wave will be reflected by an ionospheric layer at vertical incidence at a given time. [GEOPHYS] The minimum frequency of a vertically directed radio wave which will penetrate a particular layer in the ionosphere; for example, all vertical radio waves with frequencies greater than the E-layer critical frequency will pass through the E layer. Also known as penetration frequency.

critical function [MATH] A function satisfying the Euler equations in the calculus of variations.

critical grid current [ELECTR] Instantaneous value of grid current when the anode current starts to flow in a gas-filled vacuum tube.

critical grid voltage [ELECTR] The grid voltage at which anode current starts to flow in a gas tube. Also known as firing point.

critical isotherm [THERMO] A curve showing the relationship between the pressure and volume of a gas at its critical temperature.

criticality [NUCLEO] The condition in which a nuclear reactor is just self-sustaining.

critical level of escape [GEOPHYS] **1.** That level, in the atmosphere, at which a particle moving rapidly upward will have a probability of $1/e$ (e is base of natural logarithm) of colliding with another particle on its way out of the atmosphere. **2.** The level at which the horizontal mean free path of an atmospheric particle equals the scale height of the atmosphere.

critical locus [PHYS CHEM] The line connecting the critical points of a series of liquid-gas phase-boundary loops for multicomponent mixtures plotted on a pressure versus temperature graph.

critical magnetic field [SOLID STATE] The field below which a superconductive material is superconducting and above which the material is normal, at a specified temperature and in the absence of current.

critical magnetic scattering [SOLID STATE] Intense scattering of low-energy neutrons by a ferromagnetic crystal at temperatures near the Curie point.

critical mass [NUCLEO] The mass of fissionable material of a particular shape that is just sufficient to sustain a nuclear chain reaction.

critical opalescence [OPTICS] Extreme opalescence resulting from strong density fluctuations in a medium near a critical point.

critical phenomena [PHYS CHEM] Physical properties of liquids and gases at the critical point (conditions at which two phases are just about to become one); for example, critical pressure is that needed to condense a gas at the critical temperature, and above the critical temperature the gas cannot be liquefied at any pressure.

critical point [MATH] **1.** A point at which the first derivative of a function is either 0 or does not exist. **2.** A critical point of an autonomous system of differential equations is a point where the given functions vanish simultaneously. [PHYS CHEM] **1.** The temperature and pressure at which two phases of a substance in equilibrium with each other become identical, forming one phase. **2.** The temperature and pressure at which two ordinarily partially miscible liquids are consolute.

critical potential [ATOM PHYS] The energy needed to raise an electron to a higher energy level in an atom (resonance potential) or to remove it from the atom (ionization potential). [ELEC] A potential which results in sudden change in magnitude of the current.

critical pressure [FL MECH] For a nozzle whose cross section at each point is such that a fluid in isentropic flow just fills it, the pressure at the section of minimum area of the nozzle; if the nozzle is cut off at this point with no diverging section, decrease in discharge pressure below the critical pressure (at constant admission pressure) does not result in increased flow. [THERMO] The pressure of the liquid-vapor critical point.

critical pressure ratio [FL MECH] The ratio of the critical pressure of a nozzle to the admission pressure of the nozzle (equals 0.53 for gases).

critical ratio [STAT] The ratio of a particular deviation from the mean value to the standard deviation.

critical reactor [NUCLEO] A nuclear reactor in which the ratio of moderator to fuel is either subcritical or just critical; used to study the properties of the system and determine critical size.

critical region [STAT] In hypothesis testing, that region of the sampling space in which the hypothesis tested is rejected if the space contains the sample point.

critical Reynolds number [FL MECH] The Reynolds number at which there is a transition from laminar to turbulent flow.

critical scattering [PHYS] Intense scattering of some form of radiation by a substance at a temperature near a second-order transition, as in critical opalescence or critical magnetic scattering.

critical shear stress [SOLID STATE] The shear stress needed to cause slip in a given direction along a given crystallographic plane of a single crystal.

critical size [NUCLEO] A set of physical dimensions for the core and reflector of a nuclear reactor at which a critical chain reaction is maintained.

critical speed *See* critical velocity.

critical state [PHYS CHEM] Unique condition of pressure, temperature, and composition wherein all properties of coexisting vapor and liquid become identical.

critical table [MATH] A table, usually for a function that varies slowly, which gives only values of the argument near which changes in the value of the function, as rounded to the number of decimal places displayed in the table, occur.

critical temperature [PHYS CHEM] The temperature of the liquid-vapor critical point, that is, the temperature above which the substance has no liquid-vapor transition.

critical value [MATH] The value of the independent variable at a critical point of a function. [STAT] A number which causes rejection of the null hypothesis if a given test statistic is this number or more, and acceptance of the null hypothesis if the test statistic is smaller than this number.

critical velocity [CRYO] The velocity of a superfluid in very narrow channels (on the order of 10^{-5} centimeter), which is nearly constant. Also known as critical speed. [FL MECH] **1.** The speed of flow equal to the local speed of sound. Also known as critical speed. **2.** The speed of fluid flow through a given conduit above which it becomes turbulent.

critical voltage [ELECTR] The highest theoretical value of steady anode voltage, at a given steady magnetic flux density, at which electrons emitted from the cathode of a magnetron at zero velocity would fail to reach the anode. Also known as cutoff voltage.

critical volume [PHYS] The volume occupied by one mole of a substance at the liquid-vapor critical point, that is, at the critical temperature and pressure.

critical zone [FL MECH] In fluid flow, the area on a graph of the Reynolds number versus friction factor indicating unstable flow (Reynolds number 2000 to 4000) between laminar flow and the transition to turbulent flow.

CR law [ELEC] A law which states that when a constant electromotive force is applied to a circuit consisting of a resistor and capacitor connected in series, the time taken for the potential on the plates of the capacitor to rise to any given fraction of its final value depends only on the product of capacitance and resistance.

CRM *See* chemical remanent magnetization.

CRO *See* cathode-ray oscilloscope.

Crocco's equation [FL MECH] A relationship, expressed as $\mathbf{v} \times \omega = -T\,\mathrm{grad}\,S$, between vorticity and entropy gradient for the steady flow of an inviscid compressible fluid; $\mathbf{v}$ is the fluid velocity vector, ω ($= \mathrm{curl}\,\mathbf{v}$) is the vorticity vector, T is the fluid temperature, and S is the entropy per unit mass of the fluid.

crocodile [ELEC] A unit of potential difference or electromotive force, equal to 10^6 volts; used informally at some nuclear physics laboratories.

Crookes dark space *See* cathode dark space.

Crookes glass [MATER] A type of glass that contains cerium and other rare earths and has a high absorption of ultraviolet radiation; used in sunglasses.

Crookes radiometer [PHYS] A radiometer used to demonstrate that radiant energy from the sun can produce motion; a miniature four-vane windmill is mounted in a glass-envelope vacuum tube, with each vane polished on one side and black on the other.

Crookes tube [ELECTR] An early form of low-pressure discharge tube whose cathode was a flat aluminum disk at one end of the tube, and whose anode was a wire at one side of the tube, outside the electron stream; used to study cathode rays.

cross-correlation [STAT] **1.** Correlation between corresponding members of two or more series: if $q_1, \ldots, q_n$ and $r_1, \ldots, r_n$ are two series, correlation between q_i and r_i, or between q_i and r_{i+j} (for fixed j), is a cross correlation. **2.** Correlation between or expectation of the inner product of two series of random variables, where the difference in indices between the corresponding values of the two series is fixed.

cross-correlation function [COMMUN] A function, $\phi_{12}(\tau)$, where τ is a time-delay parameter, equal to the limit, as T approaches infinity, of the reciprocal of $2T$ times the integral over t from $-T$ to T of $f_1(t)f_2(t-\tau)$, where f_1 and f_2 are functions of time, such as the input and output of a communication system.

cross-correlator [ELECTR] A correlator in which a locally generated reference signal is multiplied by the incoming signal and the result is smoothed in a low-pass filter to give an approximate computation of the cross-correlation function. Also known as synchronous detector.

crosscurrent [FL MECH] A current that flows across or opposite to another current.

crossed cylinder [OPTICS] **1.** A thin lens whose surfaces are portions of circular cylinders whose axes cross at right angles or obliquely. **2.** A weak lens whose effect is equivalent to that of lenses with convex and concave cylindrical surfaces of equal curvature crossed at right angles.

crossed-field multiplier phototube [ELECTR] A multiplier phototube in which repeated secondary emission is obtained from a single active electrode by the combined effects of a strong radio-frequency electric field and a perpendicular direct-current magnetic field.

crossed lens [OPTICS] A lens designed with radii of curvature which give minimum spherical aberration for parallel incident rays.

crossed prisms [OPTICS] A pair of Nicol prisms whose principal planes are perpendicular to each other, so that light passing through one is extinguished by the other.

cross flux [ELECTROMAG] A component of magnetic flux perpendicular to that produced by the field magnets in an electrical rotating machine.

crossing symmetry [PARTIC PHYS] The amplitude for a process that involves creation of a particle with four-momentum P_μ is equal to the amplitude for a process which is the same except it involves destruction of the antiparticle with four-momentum $-P_\mu$.

cross-magnetizing effect [ELECTROMAG] The distortion in the flux-density distribution in the air gap of an electric rotating machine caused by armature reaction.

cross multiplication [MATH] Multiplication of the numerator of each of two fractions by the denominator of the other, as when eliminating fractions from an equation.

cross-over [ELECTR] The plane at which the cross section of a beam of electrons in an electron gun is a minimum.

crossover frequency [ENG ACOUS] **1.** The frequency at which a dividing network delivers equal power to the upper and lower frequency channels when both are terminated in specified loads. **2.** See transition frequency.

cross-polarization [ELECTROMAG] The component of the electric field vector normal to the desired polarization component.

cross product [MATH] An anticommutative multiplication on the vectors of euclidean three-dimensional space. Also known as vector product.

cross ratio [MATH] For four collinear points, A, B, C, and D, the ratio $(AB)(CD)/(AD)(CB)$, or one of the ratios obtained from this quantity by a permutation of A, B, C, and D.

cross section [MATH] **1.** The intersection of an n-dimensional geometric figure in some euclidean space with a lower dimensional hyperplane. **2.** A right inverse for the projection of a fiber bundle. [PHYS] An area characteristic of a collision reaction between atomic or nuclear particles or systems, such that the number of reactions which occur equals the product of the number of target particles or systems and the number of incident particles or systems which would pass through this area if their velocities were perpendicular to it. Also known as collision cross section.

cross section per atom [NUC PHYS] The microscopic cross section for a given nuclear reaction referred to the natural element, even though the reaction involves only one of the natural isotopes.

cross spectrum [PHYS] The complex vector sum of the cospectrum and quadrature spectrum.

Crout reduction [MATH] Modification of the Gauss procedure for numerical solution of simultaneous linear equations; adapted for use on desk calculators and digital computers.

Crova wavelength [STAT MECH] The wavelength in the spectrum of a radiator whose intensity divided by the intensity of the total radiation equals the derivative of the intensity of the wavelength with respect to temperature divided by the derivative of the total intensity with respect to temperature.

crown cell [ELEC] Generic name for alkaline zinc–manganese dioxide dry-cell battery; manganese dioxide–graphite cathode mix is pressed into a steel can onto which a steel cap is spot-welded to contain the amalgamated powdered-zinc anode.

CRT *See* cathode-ray tube.

cruciform core [ELEC] A transformer core in which all windings are on one center leg, and four additional legs arranged in the form of a cross serve as return paths for magnetic flux.

crunode [MATH] A point on a curve through which pass two branches of the curve with different tangents. Also known as node.

crushing strength [MECH] The compressive stress required to cause a solid to fail by fracture; in essence, it is the resistance of the solid to vertical pressure placed upon it.

cry-, cryo- [SCI TECH] Combining form meaning cold, freezing.

cryoelectronics [ELECTR] A branch of electronics concerned with the study and application of superconductivity and other low-temperature phenomena to electronic devices and systems. Also known as cryolectronics.

cryogen *See* cryogenic fluid.

cryogenic coil [CRYO] A high-purity coil refrigerated to very low temperatures to reduce effective coil resistivity.

cryogenic conductor *See* superconductor.

cryogenic device [CRYO] A device whose operation depends on superconductivity as produced by temperatures near absolute zero. Also known as superconducting device.

cryogenic fluid [CRYO] A liquid which boils at temperatures of less than about 110 K at atmospheric pressure, such as hydrogen, helium, nitrogen, oxygen, air, or methane. Also known as cryogen.

cryogenic pump [CRYO] A high-speed vacuum pump that can produce an extremely low vacuum and has a low power consumption; to reduce the pressure, gases are condensed on surfaces within an enclosure at extremely low temperatures, usually attained by using liquid helium or liquid or gaseous hydrogen. Also known as cryopump.

cryogenics [PHYS] The production and maintenance of very low temperatures, and the study of phenomena at these temperatures.

cryogenic temperature [CRYO] A temperature within a few degrees of absolute zero.

cryolectronics See cryoelectronics.

cryomagnetic [CRYO] Pertaining to production of very low temperatures by adiabatic demagnetization of paramagnetic salts.

cryometer [ENG] A thermometer for measuring low temperatures.

cryophysics [CRYO] Physics as restricted to phenomena occurring at very low temperatures, approaching absolute zero.

cryopump *See* cryogenic pump.

CRYOTRON

gate wire (Ta)

control
coil (Nb)

The wire-wound cryotron operated at temperatures close to absolute zero.

CRYSTAL COUNTER

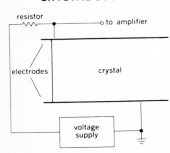

resistor — o to amplifier

electrodes — crystal

voltage supply

Circuit diagram of a crystal counter.

cryosar [ELECTR] A cryogenic, two-terminal, negative-resistance semiconductor device, consisting essentially of two contacts on a germanium wafer operating in liquid helium.

cryoscope [ENG] A device to determine the freezing point of a liquid.

cryosistor [ELECTR] A cryogenic semiconductor device in which a reverse-biased *pn* junction is used to control the ionization between two ohmic contacts.

cryostat [ENG] An apparatus used to provide low-temperature environments in which operations may be carried out under controlled conditions.

cryotron [ELECTR] A switch that operates at very low temperatures at which its components are superconducting; when current is sent through a control element to produce a magnetic field, a gate element changes from a superconductive zero-resistance state to its normal resistive state.

crystal [CRYSTAL] A homogeneous solid made up of an element, chemical compound or isomorphous mixture throughout which the atoms or molecules are arranged in a regularly repeating pattern. [ELECTR] A natural or synthetic piezoelectric or semiconductor material whose atoms are arranged with some degree of geometric regularity.

crystal axis [CRYSTAL] A reference axis used for the vectoral properties of a crystal.

crystal base [CRYSTAL] The contents of a primitive cell of a crystal.

crystal blank [ELECTR] The result of the final cutting operation on a piezoelectric or semiconductor crystal.

crystal calibrator [ELECTR] A crystal-controlled oscillator used as a reference standard to check frequencies.

crystal class [CRYSTAL] One of 32 categories of crystals according to the inversions, rotations about an axis, reflections, and combinations of these which leaves the crystal invariant. Also known as symmetry class.

crystal control [ELECTR] Control of the frequency of an oscillator by means of a quartz crystal unit.

crystal-controlled oscillator [ELECTR] An oscillator whose frequency of operation is controlled by a crystal unit.

crystal counter [NUCLEO] A particle detector in which the sensitive material is a dielectric (nonconducting) crystal mounted between two metallic electrodes.

crystal defect [CRYSTAL] Any departure from crystal symmetry caused by free surfaces, disorder, impurities, vacancies and interstitials, dislocations, lattice vibrations, and grain boundaries. Also known as lattice defect.

crystal diffraction [SOLID STATE] Diffraction by a crystal of beams of x-rays, neutrons, or electrons whose wavelengths (or de Broglie wavelengths) are comparable with the interatomic spacing of the crystal.

crystal diffraction spectrometer *See* Bragg spectrometer.

crystal diode *See* semiconductor diode.

crystal face [CRYSTAL] One of the outward planar surfaces which define a crystal and reflect its internal structure. Also known as face.

crystal field theory [PHYS CHEM] The theory which assumes that the ligands of a coordination compound are the sources of negative charge which perturb the energy levels of the central metal ion and thus subject the metal ion to an electric field analogous to that within an ionic crystalline lattice.

crystal form [CRYSTAL] A collection of crystal faces generated by operating on a single face with a subgroup of the symmetry elements of the crystal class.

crystal gliding [CRYSTAL] Slip along a crystal plane due to plastic deformation; often produces crystal twins. Also known as translation gliding.

crystal grating [SPECT] A diffraction grating for gamma rays or x-rays which uses the equally spaced lattice planes of a crystal.

crystal growth [CRYSTAL] The growth of a crystal, which involves diffusion of the molecules of the crystallizing substance to the surface of the crystal, diffusion of these molecules over the crystal surface to special sites on the surface, incorporation of molecules into the surface at these sites, and diffusion of heat away from the surface.

crystal habit [CRYSTAL] The size and shape of the crystals in a crystalline solid. Also known as habit.

crystal indices *See* Miller indices.

crystal laser [OPTICS] A laser that uses a pure crystal of ruby or other material for generating a coherent beam of output light.

crystal lattice [CRYSTAL] A lattice from which the structure of a crystal may be obtained by associating with every lattice point an assembly of atoms identical in composition, arrangement, and orientation.

crystalline [CRYSTAL] Of, pertaining to, resembling, or composed of crystals.

crystalline anisotropy [SOLID STATE] The tendency of crystals to have different properties in different directions; for example, a ferromagnet will spontaneously magnetize along certain crystallographic axes.

crystalline double refraction [OPTICS] The splitting which a wavefront experiences when a wave disturbance propagates through an anisotropic crystal.

crystalline field [SOLID STATE] The internal electric field in a solid due to localized charges, especially ions, inside.

crystalline laser [OPTICS] A solid laser in which the lasing material is a pure crystal like ruby or a doped crystal like neodymium-doped ruby or neodymium-doped yttrium aluminum garnet.

crystallogram [CRYSTAL] A photograph of the x-ray diffraction pattern of a crystal.

crystallographic axis [CRYSTAL] One of three lines (sometimes four, in the case of a hexagonal crystal), passing through a common point, that are chosen to have definite relation to the symmetry properties of a crystal, and are used as a reference in describing crystal symmetry and structure.

crystallography [PHYS] The branch of science that deals with the geometric description of crystals and their internal arrangement.

crystallomagnetic [SOLID STATE] Pertaining to magnetic properties of crystals.

crystal momentum [SOLID STATE] The product of Planck's constant and the wave vector associated with an elementary excitation in a crystal (the magnitude of the wave vector being taken as the reciprocal of the wavelength).

crystal monochromator [SPECT] A spectrometer in which a collimated beam of slow neutrons from a reactor is incident on a single crystal of copper, lead, or other element mounted on a divided circle.

crystal optics [OPTICS] The study of the propagation of light, and associated phenomena, in crystalline solids.

crystal oscillator [ELECTR] An oscillator in which the frequency of the alternating-current output is determined by the mechanical properties of a piezoelectric crystal. Also known as piezoelectric oscillator.

crystal plane [CRYSTAL] One of a set of parallel, equally

CRYSTAL GROWTH

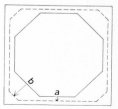

Schematic representation of cross section of crystal at three stages of growth; a represents slower growing faces, b faster growing faces.

CRYSTAL OSCILLATOR

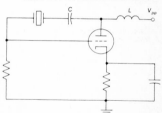

Circuit diagram of Pierce crystal oscillator; C is capacitor, L is inductor, and V_{pp} is plate voltage.

spaced planes in a crystal structure, each of which contains an infinite periodic array of lattice points.

crystal plate [ELECTR] A precisely cut slab of quartz crystal that has been lapped to final dimensions, etched to improve stability and efficiency, and coated with metal on its major surfaces for connecting purposes. Also known as quartz plate.

crystal projection [CRYSTAL] Any method of displaying the positions of the poles of a crystal by projecting them on a plane.

crystal rectifier *See* semiconductor diode.

crystal resonator [ELECTR] A precisely cut piezoelectric crystal whose natural frequency of vibration is used to control or stabilize the frequency of an oscillator. Also known as piezoelectric resonator.

crystal spectrometer *See* Bragg spectrometer.

crystal structure [CRYSTAL] The arrangement of atoms or ions in a crystalline solid.

crystal symmetry [CRYSTAL] The existence of nontrivial operations, consisting of inversions, rotations around an axis, reflections, and combinations of these, which bring a crystal into a position indistinguishable from its original position.

crystal system [CRYSTAL] One of seven categories (cubic, hexagonal, tetragonal, trigonal, orthorhombic, monoclinic, and triclinic) into which a crystal may be classified according to the shape of the unit cell of its Bravais lattice, or according to the dominant symmetry elements of its crystal class.

crystal whiskers [CRYSTAL] Single crystals that have grown in a filamentary form.

cs *See* centistoke.

Cs *See* cesium.

csc *See* cosecant.

csch *See* hyperbolic cosecant.

CT *See* center tap.

ctn *See* cotangent.

CTR *See* controlled thermonuclear reactor.

cu *See* cubic.

Cu *See* copper.

cubature [MATH] The numerical integration of a function of two variables.

cube [MATH] Regular polyhedron whose faces are all square.

cube of a number [MATH] The new number obtained by taking the three-fold product of the given number a with itself: $a \times a \times a$.

cube root [MATH] Another number whose cube is the original number.

cube-surface coil [ELECTROMAG] A system of five equally spaced square coils that produces a region of uniform magnetic field over a large volume which is easily accessible from outside the coils.

cubic [MECH] Denoting a unit of volume, so that if x is a unit of length, a cubic x is the volume of a cube whose sides have length $1x$; for example, a cubic meter, or a meter cubed, is the volume of a cube whose sides have a length of 1 meter. Abbreviated cu.

cubical antenna [ELECTROMAG] An antenna array, the elements of which are positioned to form a cube.

cubical dilation [MECH] The isotropic part of the strain tensor describing the deformation of an elastic solid, equal to the fractional increase in volume.

cubical expansion [PHYS] The increase in volume of a substance with a change in temperature or pressure.

cubical parabola [MATH] A plane curve whose equation in cartesian coordinates x and y is $y = x^3$.

cubic crystal [CRYSTAL] A crystal whose lattice has a unit cell with perpendicular axes of equal length.

cubic curve [MATH] A plane curve which has an equation of the form $f(x,y) = 0$, where $f(x,y)$ is a polynomial of degree three in x and y.

cubic determinant [MATH] A mathematical form analogous to an ordinary determinant, with the elements forming a cube instead of a square.

cubic equation [MATH] A polynomial equation with no exponent larger than 3.

cubic foot per second *See* cusec.

cubic measure [MECH] A unit or set of units to measure volume.

cubic packing [CRYSTAL] The spacing pattern of uniform solid spheres in a clastic sediment or crystal lattice in which the unit cell is a cube.

cubic plane [CRYSTAL] A plane that is at right angles to any one of the three crystallographic axes of the cubic system.

cubic polynomial [MATH] A polynomial in which all exponents are no greater than 3.

cubic spline [MATH] One of a collection of cubic polynomials used in interpolating a function whose value is specified at each of a collection of distinct ordered values, X_i ($i = 1, \ldots, n$), and whose slope is specified at X_1 and X_n; one cubic polynomial is found for each interval, such that the interpolating system has the prescribed values at each of the X_i, the prescribed slope at X_1 and X_n, and a continuous slope at each of the X_i.

cuboctahedron [MATH] A polyhedron whose faces consist of six equal squares and eight equal equilateral triangles, and which can be formed by cutting the corners off a cube; it is one of the 13 Archimedean solids. Also spelled cubooctahedron.

cubooctahedron *See* cuboctahedron.

cumec [MECH] A unit of volume flow rate equal to 1 cubic meter per second.

cumulants [STAT] A set of parameters k_h ($h = 1, \ldots, r$) of a one-dimensional probability distribution defined by

$$\ln \chi_x(q) = \sum_{h=1}^{r} k_h[(iq)^h / h!] + o(q^r)$$

where $\chi_x(q)$ is the characteristic function of the probability distribution of x. Also known as semi-invariants.

cumulative compound generator [ELEC] A compound generator in which the series field is connected to aid the shunt field magnetomotive force.

cumulative dose [NUCLEO] The total dose resulting from repeated exposures to radiation.

cumulative error [STAT] An error whose magnitude does not approach zero as the number of observations increases. Also known as accumulative error.

cumulative excitation [ATOM PHYS] Process by which the atom is raised from one excited state to a higher state by collision, for example, with an electron.

cumulative ionization [ATOM PHYS] Ionization of an excited atom in the metastable state by cumulative excitation. [ELECTR] *See* avalanche.

cup core [ELECTROMAG] A core that encloses a coil to provide magnetic shielding; usually has a powdered iron center post through the coil.

cup electrometer [ENG] An electrometer that has a metal cup attached to its plate so that a charged body touching the inside of the cup gives up its entire charge to the instrument.

cup product [MATH] A multiplication defined on cohomology classes; it gives cohomology a ring structure.

curie [NUCLEO] A unit of radioactivity, defined as that quantity of any radioactive nuclide which has 3.700×10^{10} disintegrations per second. Abbreviated c; Ci.

Curie balance [ENG] An instrument for determining the susceptibility of weakly magnetic materials, in which one measures the deflection produced by a strong permanent magnet on a suspended tube containing the specimen.

Curie constant [ELECTROMAG] The electric or magnetic susceptibility at some temperature times the difference of the temperature and the Curie temperature, which is a constant at temperatures above the Curie temperature according to the Curie-Weiss law.

Curie point *See* Curie temperature.

Curie principle [THERMO] The principle that a macroscopic cause never has more elements of symmetry than the effect it produces; for example, a scalar cause cannot produce a vectorial effect.

Curie scale of temperature [THERMO] A temperature scale based on the susceptibility of a paramagnetic substance, assuming that it obeys Curie's law; used at temperatures below about 1 kelvin.

Curie's law [ELECTROMAG] The law that the magnetic susceptibilities of most paramagnetic substances are inversely proportional to their absolute temperatures.

Curie temperature [ELECTROMAG] The temperature marking the transition between ferromagnetism and paramagnetism, or between the ferroelectric phase and the paraelectric phase. Also known as Curie point.

Curie-Weiss law [ELECTROMAG] A relation between magnetic or electric susceptibilities and the absolute temperatures which is followed by ferromagnets, antiferromagnets, nonpolar ferroelectrics, antiferroelectrics, and some paramagnets.

curium [CHEM] An element, symbol Cm, atomic number 96; the isotope of mass 244 is the principal source of this artificially produced element.

curium-242 [NUC PHYS] An isotope of curium, mass number 242; half-life is 165.5 days for α-particle emission; 7.2×10^6 years for spontaneous fission.

curium-244 [NUC PHYS] An isotope of curium, mass number 244; half-life is 16.6 years for α-particle emission; 1.4×10^7 years for spontaneous fission; potential use as compact thermoelectric power source.

curl [MATH] The curl of a vector function is a vector which is formally the cross product of the del operator and the vector. Also known as rotation (rot).

current [ELEC] The net transfer of electric charge per unit time; a specialization of the physics definition. Also known as electric current. [PHYS] **1.** The rate of flow of any conserved, indestructible quantity across a surface per unit time. **2.** *See* current density.

current algebra [PARTIC PHYS] The application of algebraic relationships among currents derived from approximate symmetries, such as broken SU_3 symmetry, to the study of hadrons.

current antinode [ELEC] A point at which current is a maximum along a transmission line, antenna, or other circuit element having standing waves. Also known as current loop.

current balance [ELEC] An apparatus with which force is measured between current-carrying conductors, with the purpose of assigning the value of the ampere.

current-carrying capacity [ELEC] The maximum current that can be continuously carried without causing permanent deterioration of electrical or mechanical properties of a device or conductor.

current-controlled switch [ELECTR] A semiconductor device

CURIUM

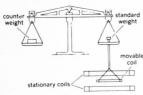

Periodic table of the chemical elements showing the position of curium.

CURRENT BALANCE

Rayleigh current balance.

in which the controlling bias sets the resistance at either a very high or very low value, corresponding to the "off" and "on" conditions of a switch.

current density [ELEC] The current per unit cross-sectional area of a conductor; a specialization of the physics definition. Also known as electric current density. [PHYS] A vector quantity whose component perpendicular to any surface equals the rate of flow of some conserved, indestructible quantity across that surface per unit area per unit time. Also known as current.

current efficiency [PHYS CHEM] The ratio of the amount of electricity, in coulombs, theoretically required to yield a given quantity of material in an electrochemical process, to the amount actually consumed.

current feedback [ELECTR] Feedback introduced in series with the input circuit of an amplifier.

current feedback circuit [ELECTR] A circuit used to eliminate effects of amplifier gain instability in an indirect-acting recording instrument, in which the voltage input (error signal) to an amplifier is the difference between the measured quantity and the voltage drop across a resistor.

current function *See* Lagrange stream function.

current intensity [ELEC] The magnitude of an electric current. Also known as current strength.

current loop *See* current antinode.

current measurement [ELEC] The measurement of the flow of electric current.

current node [ELEC] A point at which current is zero along a transmission line, antenna, or other circuit element having standing waves.

current noise [ELECTR] Electrical noise of uncertain origin which is observed in certain resistances when a direct current is present, and which increases with the square of this current.

current phasor [ELEC] A line referenced to a point, whose length and angle represent the magnitude and phase of a current.

current ratio [ELECTROMAG] In a waveguide, the ratio of maximum to minimum current.

current saturation *See* anode saturation.

current strength *See* current intensity.

current transformer [ELEC] An instrument transformer intended to have its primary winding connected in series with a circuit carrying the current to be measured or controlled; the current is measured across the secondary winding.

current transformer phase angle [ELEC] Angle between the primary current vector and the secondary current vector reversed; it is conveniently considered as positive when the reversed secondary current vector leads the primary current vector.

current-voltage dual [ELEC] A circuit which is equivalent to a specified circuit when one replaces quantities with dual quantities; current and voltage impedance and admittance, and meshes and nodes are examples of dual quantities.

curtain [NUCLEO] A thin shield, usually cadmium, used in a nuclear reactor to shut off a flow of slow neutrons.

curtain array [ELECTROMAG] An antenna array consisting of vertical wire elements stretched between two suspension cables.

curtain rhombic antenna [ELECTROMAG] A multiple-wire rhombic antenna having a constant input impedance over a wide frequency range; two or more conductors join at the feed and terminating ends but are spaced apart vertically from 1 to 5 feet (30 to 150 centimeters) at the side poles.

curvature [MATH] The reciprocal of the radius of the circle

CURRENT FEEDBACK CIRCUIT

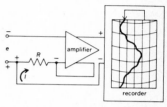

Diagram of a current feedback circuit.

which most nearly approximates a curve at a given point; the rate of change of the unit tangent vector to a curve with respect to arc length of the curve.

curvature correction [ASTRON] A correction applied to the mean of a series of observations on a star or planet to take account of the divergence of the apparent path of the star or planet from a straight line. [GEOD] The correction applied in some geodetic work to take account of the divergence of the surface of the earth (spheroid) from a plane.

curvature of field [OPTICS] Error in the image of a plane object formed on a flat screen by an optical system when the best image lies on a curved surface.

curvature of space [RELAT] **1.** The deviation of a spacelike three-dimensional subspace of curved space-time from euclidean geometry. **2.** The Gaussian curvature of a spacelike three-dimensional subspace of curved space-time.

curvature scalar *See* scalar curvature.

curvature tensor *See* Riemann-Christoffel tensor.

curve [MATH] The continuous image of the unit interval.

curved space-time [RELAT] A four-dimensional Riemannian space, in which there are no straight lines but only curves, which is a generalization of the Minkowski universe in the general theory of relativity.

curve fitting [STAT] The calculation of a curve of some particular character (as a logarithmic curve) that most closely approaches a number of points in a plane.

curve tracing [MATH] The method of graphing a function by plotting points and analyzing symmetries, derivatives, and so on.

curvilinear correlation [STAT] In regression analysis, a nonlinear relationship between two or more variables.

curvilinear motion [MECH] Motion along a curved path.

curvilinear regression [STAT] Regression study of jointly distributed random variables where the function measuring their statistical dependence is analyzed in terms of curvilinear coordinates.

curvilinear solid [MATH] A solid whose surfaces are not planes.

curvilinear transformation [MATH] A transformation from one coordinate system to another in which the coordinates in the new system are arbitrary twice-differentiable functions of the coordinates in the old system.

curvilinear trend [STAT] A nonlinear trend which may be expressed as a polynomial or a smooth curve.

cusec [MECH] A unit of volume flow rate, used primarily to describe pumps, equal to a uniform flow of 1 cubic foot (approximately 0.0283 cubic meter) in 1 second. Also known as cubic foot per second (cfs).

cusp [MATH] A singular point of a curve at which the limits of the tangents of the portions of the curve on either side of the point coincide. Also known as spinode.

cusped magnetic field [ELECTROMAG] A magnetic field created by adjacent parallel coils that carry current in opposite directions; used in fusion research, to contain a plasma of high-energy deuterium ions.

cuspidal cubic [MATH] A cubic curve that has one cusp, one point of inflection, and no node.

cuspidal locus [MATH] A curve consisting of the cusps of some family of curves.

cut [CRYSTAL] A section of a crystal having two parallel major surfaces; cuts are specified by their orientation with respect to the axes of the natural crystal, such as X cut, Y cut, BT cut, and AT cut. [NUCLEO] The fraction that is removed as product or advanced to the next separative element in an isotope separation process.

cut capacity [MATH] For a network whose points have been partitioned into two specified classes, C_1 and C_2, the sum of the capacities of all the segments directed from a point in C_1 to a point in C_2. Also known as cut value.

cutie pie [NUCLEO] A radiation dose-rate meter having a pistol grip, a plastic cylinder or barrel containing an ionization chamber, and an indicating meter mounted above the grip.

cutoff [ELECTR] **1.** The minimum value of negative grid bias that will prevent the flow of anode current in an electron tube. **2.** *See* cutoff frequency. [PHYS] Technique used when the contribution to the value of a physical quantity given by integration over a certain variable is absurd (in particular, when the contribution is infinite); involves cutting off the integral at some limit.

cutoff attenuator [ELECTROMAG] Variable length of waveguide used below its cutoff frequency to introduce variable nondissipative attenuation.

cutoff bias [ELECTR] The direct-current bias voltage that must be applied to the grid of an electron tube to stop the flow of anode current.

cutoff field *See* critical field.

cutoff frequency [ELECTR] A frequency at which the attenuation of a device begins to increase sharply, such as the limiting frequency below which a traveling wave in a given mode cannot be maintained in a waveguide, or the frequency above which an electron tube loses efficiency rapidly. Also known as critical frequency; cutoff.

cutoff voltage [ELECTR] **1.** The electrode voltage value that reduces the dependent variable of an electron-tube characteristic to a specified low value. **2.** *See* critical voltage.

cutoff wavelength [ELECTROMAG] **1.** The ratio of the velocity of electromagnetic waves in free space to the cutoff frequency in a uniconductor waveguide. **2.** The wavelength corresponding to the cutoff frequency.

cut-set [ELEC] A set of branches of a network such that the cutting of all the branches of the set increases the number of separate parts of the network, but the cutting of all the branches except one does not.

cut value *See* cut capacity.

CVC *See* conserved vector current.

CW laser *See* continuous-wave laser.

cwt *See* hundredweight.

cyanogen absorption [ASTROPHYS] Bands in the absorption spectra of stars at wavelengths near 418 nanometers, caused by atmospheric cyanogen; used as a measure of absolute stellar magnitude.

cyanometer [OPTICS] An instrument designed to measure or estimate the degree of blueness of light, as of the sky.

cyanometry [OPTICS] The study and measurement of the blueness of light.

cybernetics [SCI TECH] **1.** The science of control and communication in all of their manifestations within and between machines, animals, and organizations. **2.** Specifically, the interaction between automatic control and living organisms, especially humans and animals.

cybotaxis [PHYS] A transient molecular orientation in a liquid evidenced by x-ray diffraction effects.

cycle [FL MECH] A system of phases through which the working substance passes in an engine, compressor, pump, turbine, power plant, or refrigeration system. [MATH] **1.** A member of the kernel of a boundary homomorphism. **2.** A closed path in a graph that does not pass through any vertex more than once and passes through at least three vertices. [SCI TECH] **1.** One complete sequence of values of an alternating quantity, or of a sequence of process operations. **2.** A set of operations that is repeated as a unit. [STAT] A periodic movement in a time series.

cycle per minute [PHYS] A unit of frequency of action, equal to 1/60 hertz. Abbreviated cpm.

cycle per second *See* hertz.

cyclic [SCI TECH] **1.** Pertaining to some cycle. **2.** Repeating itself in some manner in space or time.

cyclic coordinate [MECH] A generalized coordinate on which the Lagrangian of a system does not depend explicitly. Also known as ignorable coordinate.

cyclic currents *See* mesh currents

cyclic curve [MATH] **1.** A curve (such as a cycloid, cardioid, or epicycloid) generated by a point of a circle that rolls (without slipping) on a given curve. **2.** The intersection of a quadric surface with a sphere. Also known as spherical cyclic curve. **3.** The stereographic projection of a spherical cyclic curve. Also known as plane cyclic curve.

cyclic extension [MATH] A Galois extension whose Galois group is cyclic.

cyclic functional equation [MATH] An equation of the form $f(x_1,x_2,\ldots,x_m) + f(x_2,x_3,\ldots,x_{m+1}) + \cdots + f(x_{n-m+1}, \ldots,x_n) + f(x_{n-m+2},\ldots,x_n,x_1) + \cdots + f(x_n,x_1,\ldots,x_{m-1}) = 0$.

cyclic group [MATH] A finite group that consists of the powers of a single element.

cyclic identity [MATH] The principle that the sum of any component of the Riemann-Christoffel tensor and two other components obtained from it by cyclic permutation of any three indices, while the fourth is held fixed, is zero.

cyclic magnetization [ELECTROMAG] A magnetizing force varying between two specific limits long enough so that the magnetic induction has the same value for corresponding points in successive cycles.

cyclic permeability *See* normal permeability.

cyclic permutation [MATH] A permutation of an ordered set of symbols which sends the first to the second, the second to the third, . . . , the last to the first.

cyclic tower [MATH] A normal tower in which each quotient group G_i / G_{i+1} is a cyclic group.

cyclic twinning [CRYSTAL] Repeated twinning of three or more individuals in accordance with the same twinning law but without parallel twinning axes.

cyclic voltammetry [PHYS CHEM] An electrochemical technique for studying variable potential at an electrode involving application of a triangular potential sweep, allowing one to sweep back through the potential region just covered.

cycling [CONT SYST] A periodic change of the controlled variable from one value to another in an automatic control system. Also known as oscillation.

cycloconverter [ELEC] A device that produces an alternating current of constant or precisely controllable frequency from a variable-frequency alternating-current input, with the output frequency usually one-third or less of the input frequency.

cycloid [MATH] The curve traced by a point on the circumference of a circle as the circle rolls along a straight line.

cycloidal mass spectrometer [SPECT] Small mass spectrometer of limited mass range fitted with a special-type analyzer that generates a cycloidal-path beam of the sample mass.

cycloidal pendulum [MECH] A modification of a simple pendulum in which a weight is suspended from a cord which is slung between two pieces of metal shaped in the form of cycloids; as the bob swings, the cord wraps and unwraps on the cycloids; the pendulum has a period that is independent of the amplitude of the swing.

cyclonic [GEOPHYS] Having a sense of rotation about the local vertical that is the same as that of the earth's rotation: as viewed from above, counterclockwise in the Northern Hemisphere, clockwise in the Southern Hemisphere, undefined at the Equator.

cyclophon See beam-switching tube.

cyclostrophic flow [FL MECH] A form of gradient flow in which the centripetal acceleration exactly balances the horizontal pressure force.

cyclotomic equation [MATH] An equation which has the form $x^{n-1} + x^{n-2} + \cdots + x + 1 = 0$, where n is a prime number.

cyclotomic field [MATH] The extension field of a given field K which is the smallest extension field of K that includes the pth roots of unity for some integer n.

cyclotomic polynomial [MATH] A polynomial of the form $x^{n-1} + x^{n-2} + \cdots + x + 1$, where n is a prime number.

cyclotomy [MATH] Theory of dividing the circle into equal parts or constructing regular polygons or, analytically, of finding the nth roots of unity.

cyclotron [NUCLEO] An accelerator in which charged particles are successively accelerated by a constant-frequency alternating electric field that is synchronized with movement of the particles on spiral paths in a constant magnetic field normal to their path. Also known as phasotron.

cyclotron D See dee.

cyclotron emission See cyclotron radiation.

cyclotron frequency [ELECTROMAG] The frequency at which an electron traverses an orbit when moving subject to a uniform magnetic field, at right angles to the field. Also known as gyrofrequency.

cyclotron magnets [NUCLEO] The magnets which bend charged-particle orbits and confine the extent of particle motion in a cyclotron.

cyclotron radiation [ELECTROMAG] The electromagnetic radiation emitted by charged particles as they orbit in a magnetic field, at a speed which is not close to the speed of light. Also known as cyclotron emission.

cyclotron resonance [PHYS] Resonance absorption of energy from an alternating-current electric field by electrons in a uniform magnetic field when the frequency of the electric field equals the cyclotron frequency, or the cyclotron frequency corresponding to the electron's effective mass if the electrons are in a solid. Also known as diamagnetic resonance.

cyclotron resonance heating [PL PHYS] A modification of magnetic pumping that involves compressing and expanding plasma at a frequency approximating the cyclotron frequency of the ions in the plasma; the goal is temperatures above several million degrees.

cylinder [MATH] **1.** A solid bounded by a cylindrical surface and two parallel planes, or the surface of such a solid. **2.** See cylindrical surface.

cylinder function [MATH] Any solution of the Bessel equation, including Bessel functions, Neumann functions, and Hankel functions.

cylindrical antenna [ELECTROMAG] An antenna in which hollow cylinders serve as radiating elements.

cylindrical capacitor [ELEC] A capacitor made of two concentric metal cylinders of the same length, with dielectric

CYCLOTRON

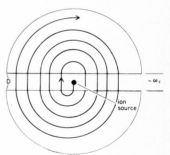

Principle of the cyclotron. The ions are formed in the ion source and are drawn out by one of the D's (electrodes). Arrows show circular orbit path of the ions. Period of orbit equals period $(2\pi/\omega_c)$ of applied D voltage.

filling the space between the cylinders. Also known as coaxial capacitor.

cylindrical cavity [ELECTROMAG] A cavity resonator in the shape of a right circular cylinder.

cylindrical coordinates [MATH] A system of curvilinear coordinates in which the position of a point in space is determined by its perpendicular distance from a given line, its distance from a selected reference plane perpendicular to this line, and its angular distance from a selected reference line when projected onto this plane.

cylindrical helix [MATH] A curve lying on a cylinder which intersects the elements of the cylinder at a constant angle.

cylindrical lens [OPTICS] A lens one or both of whose surfaces are a portion of a circular cylinder.

cylindrical pinch *See* pinch effect.

cylindrical reflector [ELECTROMAG] A reflector that is a portion of a cylinder; this cylinder is usually parabolic.

cylindrical surface [MATH] A surface consisting of each of the straight lines which are parallel to a given straight line and pass through a given curve. Also known as cylinder.

cylindrical wave [ELECTROMAG] A wave whose equiphase surfaces form a family of coaxial cylinders.

cylindrical winding [ELEC] The current-carrying element of a core-type transformer, consisting of a single coil of one or more layers wound concentrically with the iron core.

D *See* dee; diopter.

Da I *See* Damköhler number I.

Da II *See* Damköhler number II.

Da III *See* Damköhler number III.

Da IV *See* Damköhler number IV.

dac *See* digital-to-analog converter.

daily aberration *See* diurnal aberration.

d'Alembertian [MATH] A differential operator in four-dimensional space,

$$\frac{\partial^2}{\partial x^2} + \frac{\partial^2}{\partial y^2} + \frac{\partial^2}{\partial z^2} - \frac{1}{c^2}\frac{\partial^2}{\partial t^2},$$

which is used in the study of relativistic mechanics.

d'Alembert's equation [MATH] The functional equation $2f(x)f(y) = f(x + y) + f(x - y)$.

d'Alembert's paradox [FL MECH] The paradox that no forces act on a body moving at constant velocity in a straight line through a large mass of incompressible, inviscid fluid which was initially at rest, or in uniform motion.

d'Alembert's principle [MECH] The principle that the resultant of the external forces and the kinetic reaction acting on a body equals zero.

d'Alembert's system of equations [MATH] The set of ordinary linear differential equations

$$\frac{dy_i}{dt} = \sum_{j=1}^{n} a_{ij}y_j, \ i = 1, 2, \ldots, n,$$

where the a_{ij} are constants.

d'Alembert's test for convergence [MATH] A series Σa_n converges if there is an N such that the absolute value of the ratio a_n/a_{n-1} is always less than some fixed number smaller than 1, provided n is at least N, and diverges if the ratio is always greater than 1.

d'Alembert's wave equation *See* wave equation.

Dalitz pair [PARTIC PHYS] The electron and positron resulting from the decay of a neutral pion to these particles and a photon.

Dalitz plot [PARTIC PHYS] Pictorial representation for data on the distribution of certain three-particle configurations that result from elementary-particle decay processes or high-energy nuclear reactions.

dalton *See* atomic mass unit.

Dalton's atomic theory [CHEM] Theory forming the basis of accepted modern atomic theory, according to which matter is made of particles called atoms, reactions must take place between atoms or groups of atoms, and atoms of the same element are all alike but differ from atoms of another element.

Dalton's law [PHYS] The law that the pressure of a gas mixture is equal to the sum of the partial pressures of the gases composing it. Also known as law of partial pressures.

DALITZ PLOT

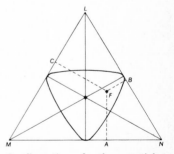

Configuration of a three-particle system *(abc)* in its barycentric frame is specified by a point F such that the three perpendiculars FA, FB, and FC to the sides of an equilateral triangle LMN (of height Q) are equal in magnitude to the kinetic energies T_a, T_b, T_c, where Q denotes the sum. Heavy curve encloses points which correspond to physically allowed configurations.

Dalton's temperature scale [THERMO] A scale for measuring temperature such that the absolute temperature T is given in terms of the temperature on the Dalton scale τ by $T = 273.15(373.15/273.15)^{\tau/100}$.

damaging stress [MECH] The minimum unit stress for a given material and use that will cause damage to the member and make it unfit for its expected length of service.

Damköhler number I [PHYS] A dimensionless number, equal to the ratio of the time it takes a fluid to flow some characteristic distance, to the time it takes some chemical reaction or other physical process to be completed. Symbolized Da I. Also known as Damköhler's ratio.

Damköhler number II [PHYS] A measure of the ratio of the rate of a chemical reaction to the rate of molecular diffusion, equal to the square of a characteristic length divided by the product of the diffusivity and the time it takes for a chemical reaction or other physical process to be completed. Symbolized Da II.

Damköhler number III [PHYS] A measure of the ratio of the heat liberated by a chemical reaction to the bulk transport of heat in a fluid, equal to the time it takes the fluid to travel a characteristic length, divided by the product of the fluid temperature and the time it would take for the chemical reaction to raise this temperature one unit if all the heat liberated by it were immediately absorbed by the fluid. Symbolized Da III.

Damköhler number IV [PHYS] A measure of the ratio of the heat liberated by a chemical reaction to the conductive heat transfer, equal to a characteristic length times the heat liberated per unit volume per unit time divided by the product of the thermal conductivity and the temperature. Symbolized Da IV.

Damköhler number V *See* Reynolds number.

Damköhler's ratio *See* Damköhler number I.

damped harmonic motion [PHYS] Also known as damped oscillation; damped vibration. **1.** The linear motion of a particle subject both to an elastic restoring force proportional to its displacement and to a frictional force in the direction opposite to its motion and proportional to its speed. **2.** A similar variation in a quantity analogous to the displacement of a particle, such as the charge on a capacitor in a simple series circuit containing a resistance.

damped oscillation [PHYS] **1.** Any oscillation in which the amplitude of the oscillating quantity decreases with time. Also known as damped vibration. **2.** *See* damped harmonic motion.

damped vibration *See* damped harmonic motion; damped oscillation.

damped wave [PHYS] **1.** A wave whose amplitude drops exponentially with distance because of energy losses which are proportional to the square of the amplitude. **2.** A wave in which the amplitudes of successive cycles progressively diminish at the source.

damping [PHYS] **1.** The dissipation of energy in motion of any type, especially oscillatory motion and the consequent reduction or decay of the motion. **2.** The extent of such dissipation and decay.

damping capacity [MECH] A material's capability in absorbing vibrations.

damping coefficient *See* damping factor; resistance.

damping constant *See* resistance.

damping factor [PHYS] **1.** The ratio of the logarithmic decrement of any underdamped harmonic motion to its period. Also known as damping coefficient. **2.** *See* decrement.

damping magnet [ELECTROMAG] A permanent magnet used

in conjunction with a disk or other moving conductor to produce a force that opposes motion of the conductor and thereby provides damping.

damping ratio [PHYS] The ratio of the actual resistance in damped harmonic motion to that necessary to produce critical damping. Also known as relative damping ratio.

danger coefficient [NUCLEO] The change in reactivity per unit mass of a substance resulting from inserting the substance in a particular nuclear reactor.

Daniell hygrometer [ENG] An instrument for measuring dew point; dew forms on the surface of a bulb containing ether which is cooled by evaporation into another bulb, the second bulb being cooled by the evaporation of ether on its outer surface.

daraf [ELEC] The unit of elastance, equal to the reciprocal of 1 farad.

Darboux's equation [MATH] The differential equation

$$\left[f(x,y) + xg(x,y) \right] \frac{dy}{dx} = h(x,y) + yg(x,y)$$

where f, g, and h are homogeneous in x and y of the same degree.

Darboux's monodromy theorem [MATH] If the function $f(z)$ of the complex variable z is analytic in a domain D bounded by a simple closed curve C, and $f(z)$ is continuous in the union of D and C and is injective for z on C, then $f(z)$ is injective for z in D.

Darboux's theorem [MATH] If the function $f(x)$ is bounded in the interval $[a,b]$, and if M_k and m_k are the least upper and greatest lower bounds respectively of $f(x)$ on (x_{k-1}, x_k), where $a = x_0 < x_1 < \cdots < x_{k-1} < x_k < \cdots < x_n = b$, then the sums

$$S = \sum_{k=1}^{n} M_k \, \delta_k \quad \text{and} \quad s = \sum_{k=1}^{n} m_k \, \delta_k,$$

where $\delta_k = x_k - x_{k-1}$, approach definite limits as n approaches infinity and the maximum of the δ_k approaches 0.

darcy [PHYS] A unit of permeability, equivalent to the passage of 1 cubic centimeter of fluid of 1 centipoise viscosity flowing in 1 second under a pressure of 1 atmosphere (101,325 newtons per square meter) through a porous medium having a cross-sectional area of 1 square centimeter and a length of 1 centimeter.

Darcy number 1 [FL MECH] A dimensionless group, equal to four times the Fanning friction factor. Symbolized Da_1. Also known as Darcy-Weisbach coefficient; resistance coefficient 2.

Darcy number 2 [FL MECH] A dimensionless group used in the study of the flow of fluids in porous media, equal to the fluid velocity times the flow path divided by the permeability of the medium. Symbolized Da_2.

Darcy's law [FL MECH] The law that the rate at which a fluid flows through a permeable substance per unit area is equal to the permeability, which is a property only of the substance through which the fluid is flowing, times the pressure drop per unit length of flow, divided by the viscosity of the fluid.

Darcy-Weisbach coefficient *See* Darcy number 1.

Darcy-Weisbach equation [FL MECH] An equation for the loss of head due to friction h_f during turbulent flow of a fluid through a duct of any shape; for a circular pipe, it is $h_f = f(L/d)(V^2/2g)$, where L and d are the length and diameter of the pipe, V is the fluid velocity, g the acceleration of gravity, and f a dimensionless number called Darcy number 1.

dark conduction [ELECTR] Residual conduction in a photosensitive substance that is not illuminated.

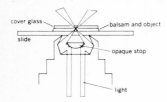

Diagram of dark-field illumination. *(American Optical Corp.)*

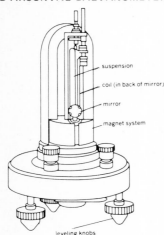

Drawing of d'Arsonval galvanometer. *(From D. M. Considine, ed., Process Instruments and Control Handbook, McGraw-Hill, 1957)*

dark current *See* electrode dark current.

dark-current pulse [ELECTR] A phototube dark-current excursion that can be resolved by the system employing the phototube.

dark discharge [ELECTR] An invisible electrical discharge in a gas.

dark-eclipsing variables [ASTRON] A binary star system, comprising a bright star and an almost dark companion that revolve about each other.

dark-field illumination [OPTICS] A method of microscope illumination in which the illuminating beam is a hollow cone of light formed by an opaque stop at the center of the condenser large enough to prevent direct light from entering the objective; the specimen is placed at the concentration of the light cone, and is seen with light scattered or diffracted by it.

dark-line spectrum [SPECT] The absorption spectrum that results when white light passes through a substance, consisting of dark lines against a bright background.

dark nebula [ASTRON] A cloud of solid particles which absorbs or scatters away radiation directed toward an observer and becomes apparent when silhouetted against a bright nebula or rich star field. Also known as absorption nebula.

dark resistance [ELECTR] The resistance of a selenium cell or other photoelectric device in total darkness.

dark space [ELECTR] A region in a glow discharge that produces little or no light.

dark star [ASTRON] A star that is not visible but is a part of a binary star system; in particular, a star which causes, in an eclipsing variable, a primary eclipse.

d'Arsonval galvanometer [ENG] A galvanometer in which a light coil of wire, suspended from thin copper or gold ribbons, rotates in the field of a permanent magnet when current is carried to it through the ribbons; the position of the coil is indicated by a mirror carried on it, which reflects a light beam onto a fixed scale. Also known as light-beam galvanometer.

dart leader [GEOPHYS] The leader which, after the first stroke, initiates each succeeding stroke of a composite flash of lightning. Also known as continuous leader.

Darwin curve [CRYSTAL] A plot of the intensity of diffracted x-rays from a perfect crystal as a function of angle.

dasymeter [PHYS] A thin glass globe used to measure the density of gas by weighing the globe in the gas.

data [ADP] **1.** General term for numbers, letters, symbols, and analog quantities that serve as input for computer processing. **2.** Any representations of characters or analog quantities to which meaning, if not information, may be assigned. [SCI TECH] Numerical or qualitative values derived from scientific experiments.

data conversion [ADP] The changing of the representation of data from one form to another, as from binary to decimal, or from one physical recording medium to another, as from card to disk. Also known as conversion.

data processing [ADP] Any operation or combination of operations on data, including everything that happens to data from the time they are observed or collected to the time they are destroyed. Also known as information processing.

data reduction [STAT] The conversion of all information in a data set into fewer dimensions for a particular purpose, as, for example, a single measure such as a reliability measure.

daughter [NUC PHYS] The immediate product of radioactive decay of an element, such as uranium. Also known as decay product; radioactive decay product.

Dauphine law [CRYSTAL] A twin law in which the twinned parts are related by a rotation of 180° around the *c* axis.

Davis correction [FL MECH] Empirical relation of flow-line diameters used to correct data calculated from the Atherton equation (friction loss in annular passages).

Davis-Gibson color filter [OPTICS] A two-component filter for converting the spectral energy distribution of an incandescent light source to that of white light.

Davisson-Calbick formula [ELECTR] The focal length of a simple electrostatic lens consisting of a circular hole in a conducting plate is equal to four times the potential of the plate divided by the difference in the potential gradients on either side of the plate.

Davisson-Germer experiment [QUANT MECH] The first experiment to demonstrate electron diffraction, in which a beam of electrons was directed at the surface of a nickel crystal, and the distribution of electrons scattered back from the crystal was measured by a Faraday cylinder.

Dawes' limit [OPTICS] The resolving power of a telescope, limited by diffraction effects, is $4.5/a$ seconds of arc, where a is the aperture in inches.

day [ASTRON] One of various units of time equal to the period of rotation of the earth with respect to one or another direction in space; specific examples are the mean solar day and the sidereal day.

dB _See_ decibel.

dBa _See_ adjusted decibel.

dBf _See_ decibels above 1 femtowatt.

dBk _See_ decibels above 1 kilowatt.

dBm _See_ decibels above 1 milliwatt.

dBp _See_ decibels above 1 picowatt.

dBrn _See_ decibels above reference noise.

dBV _See_ decibels above 1 volt.

dBW _See_ decibels above 1 watt.

dBx _See_ decibels above reference coupling.

dc _See_ direct current.

D cable [ELEC] Two-conductor cable, each conductor having the shape of the letter D, with insulation between the conductors and between the conductors and the sheath.

D center _See_ R center.

d constant [SOLID STATE] The ratio of the induced strain in a piezoelectric material to the applied electric field that produces this strain.

DCTL _See_ direct-coupled transistor logic.

dc-to-dc converter [ELEC] An electronic circuit which converts one direct-current voltage into another, consisting of an inverter followed by a step-up or step-down transformer and rectifier.

dead [ELEC] Free from any electric connection to a source of potential difference and from electric charge; not having a potential different from that of earth; the term is used only with reference to current-carrying parts which are sometimes alive or charged.

dead band [ELEC] The portion of a potentiometer element that is shortened by a tap; when the wiper traverses this area, there is no change in output.

deadbeat [MECH] Coming to rest without vibration or oscillation, as when the pointer of a meter moves to a new position without overshooting. Also known as deadbeat response.

deadbeat response _See_ deadbeat.

dead load _See_ static load.

dead room _See_ anechoic chamber.

dead space [THERMO] A space filled with gas whose temperature differs from that of the main body of gas, such as the gas in the capillary tube of a constant-volume gas thermometer.

dead time [CONT SYS] The time interval between a change in the input signal to a process control system and the response to the signal.

deadweight gage [ENG] An instrument used as a standard for calibrating pressure gages in which known hydraulic pressures are generated by means of freely balanced (dead) weights loaded on a calibrated piston.

Dean number [FL MECH] A dimensionless number giving the ratio of the viscous force acting on a fluid flowing in a curved pipe to the centrifugal force; equal to the Reynolds number times the square root of the ratio of the radius of the pipe to its radius of curvature. Symbolized N_D.

Deborah number [MECH] A dimensionless number used in rheology, equal to the relaxation time for some process divided by the time it is observed. Symbolized D.

de Broglie equation *See* de Broglie relation.

de Broglie relation [QUANT MECH] The relation in which the de Broglie wave associated with a free particle of matter, and the electromagnetic wave in a vacuum associated with a photon, has a wavelength equal to Planck's constant divided by the particle's momentum and a frequency equal to the particle's energy divided by Planck's constant. Also known as de Broglie equation.

de Broglie's theory [QUANT MECH] The theory that particles of matter have wavelike properties which can give rise to interference effects, and electrons in an atom are associated with standing waves on a Bohr orbit.

de Broglie wave [QUANT MECH] The quantum-mechanical wave associated with a particle of matter. Also known as matter wave.

de Broglie wavelength [QUANT MECH] The wavelength of the wave associated with a particle as given by the de Broglie relation.

debunching [ELECTR] A tendency for electrons in a beam to spread out both longitudinally and transversely due to mutual repulsion; the effect is a drawback in velocity modulation tubes.

debye [ELEC] A unit of electric dipole moment, equal to 10^{-18} Franklin centimeter.

Debye effect [ELECTROMAG] Selective absorption of electromagnetic waves by a dielectric, due to molecular dipoles.

Debye equation [SOLID STATE] The equation for the Debye specific heat, which satisfies the Dulong and Petit law at high temperatures and the Debye T^3 law at low temperatures.

Debye equation for polarization [STAT MECH] The Langevin-Debye formula for the polarization of a dielectric material, relating the total polarization for n molecules to the permanent moment of the specific molecule and its polarizability.

Debye-Falkenhagen effect [PHYS CHEM] The increase in the conductance of an electrolytic solution when the applied voltage has a very high frequency.

Debye force *See* induction force.

Debye frequency [SOLID STATE] The maximum allowable frequency in the computation of the Debye specific heat.

Debye-Hückel screening radius *See* Debye shielding length.

Debye-Hückel theory [PHYS CHEM] A theory of the behavior of strong electrolytes, according to which each ion is surrounded by an ionic atmosphere of charges of the opposite sign whose behavior retards the movement of ions when a current is passed through the medium.

Debye-Jauncey scattering [SOLID STATE] Incoherent background scattering of x-rays from a crystal in directions between those of the Bragg reflections.

Debye length *See* Debye shielding length.

Debye potentials [ELECTROMAG] Two scalar potentials, designated Π_e and Π_m, in terms of which one can express the electric and magnetic fields resulting from radiation or scattering of electromagnetic waves by a distribution of localized sources in a homogeneous isotropic medium.

Debye relaxation time [PHYS CHEM] According to the Debye-Hückel theory, the time required for the ionic atmosphere of a charge to reach equilibrium in a current-carrying electrolyte, during which time the motion of the charge is retarded.

Debye-Scherrer method [SOLID STATE] An x-ray diffraction method in which the sample, consisting of a powder stuck to a thin fiber or contained in a thin-walled silica tube, is rotated in a monochromatic beam of x-rays, and the diffraction pattern is recorded on a cylindrical film whose axis is parallel to the axis of rotation of the sample.

Debye-Sears ultrasonic cell [ACOUS] A process in ultrasonic imaging for which the acoustic wavefronts act as optical gratings to diffract the light on either side of the central spot.

Debye shielding length [PL PHYS] A characteristic distance in a plasma beyond which the electric field of a charged particle is shielded by particles having charges of the opposite sign. Also known as Debye-Hückel screening radius; Debye length; shielding distance.

Debye specific heat [SOLID STATE] The specific heat of a solid under the assumption that the energy of the lattice arises entirely from acoustic lattice vibration modes which all have the same sound velocity, and that frequencies are cut off at a maximum such that the total number of modes equals the number of degrees of freedom of the solid.

Debye temperature [SOLID STATE] The temperature Θ arising in the computation of the Debye specific heat, defined by $k\Theta = h\nu$, where k is the Boltzmann constant, h is Planck's constant, and ν is the Debye frequency. Also known as characteristic temperature.

Debye T³ law [SOLID STATE] The law that the specific heat of a solid at constant volume varies as the cube of the absolute temperature T at temperatures which are small with respect to the Debye temperature.

Debye unit [ELEC] A unit of electric moment that is equal to 10^{-18} statcoulomb-centimeter.

Debye-Waller factor [SOLID STATE] A reduction factor for the intensity of coherent (Bragg) scattering of x-rays, neutrons, or electrons by a crystal, arising from thermal motion of the atoms in the lattice.

deca-, deka- [SCI TECH] A prefix denoting 10.

decade [ELEC] A group or assembly of 10 units; for example, a decade counter counts 10 in one column, and a decade box inserts resistance quantities in multiples of powers of 10. [SCI TECH] The interval between any two quantities having the ratio of 10 to 1.

decade box [ELEC] An assembly of precision resistors, coils, or capacitors whose individual values vary in submultiples and multiples of 10; by appropriately setting a 10-position selector switch for each section, the decade box can be set to any desired value within its range.

decade bridge [ELECTR] Electronic apparatus for measurement of unknown values of resistances or capacitances by comparison with known values (bridge); one secondary section of the oscillator-driven transformer is tapped in decade steps, the other in 10 uniform steps.

decade counter *See* decade scaler.

decade scaler [ELECTR] A scaler that produces one output pulse for every 10 input pulses. Also known as counter decade; decade counter; scale-of-ten circuit.

decagon [MATH] A 10-sided polygon.

decahedron [MATH] A polyhedron that has 10 faces.

decaliter [MECH] A unit of volume, equal to 10 liters, or to 0.01 cubic meter. Also spelled dekaliter.

decameter [MECH] A unit of length in the metric system equal to 10 meters. Also spelled dekameter.

decanning [NUCLEO] Removing the outer container of an enriched uranium fuel rod, in preparation for reprocessing of the fuel.

decastere [MECH] A unit of volume, equal to 10 cubic meters.

decay [PHYS] Gradual reduction in the magnitude of a quantity, as of current, magnetic flux, a stored charge, or phosphorescence.

decay chain *See* radioactive series.

decay coefficient *See* decay constant.

decay constant [PHYS] The constant c in the equation $I = I_0 e^{-ct}$, for the time dependence of rate of decay of a radioactive species; here, I is the number of disintegrations per unit time. Also known as decay coefficient; disintegration constant; radioactive decay constant; transformation constant.

decay curve [NUC PHYS] A graph showing how the activity of a radioactive sample varies with time; alternatively, it may show the amount of radioactive material remaining at any time.

decay family *See* radioactive series.

decay gammas [NUC PHYS] The characteristic gamma rays emitted during the decay of most radioisotopes.

decay heat [NUCLEO] Heat produced by the decay of radioactive nuclides.

decay mode [NUC PHYS] A possible type of decay of a radionuclide or elementary particle.

decay product *See* daughter.

decay rate [NUC PHYS] The time rate of disintegration of radioactive material, generally accompanied by emission of particles or gamma radiation.

decay series *See* radioactive series.

decay time [PHYS] The time taken by a quantity to decay to a stated fraction of its initial value; the fraction is commonly $1/e$. Also known as storage time (deprecated).

decelerating electrode [ELECTR] Of an electron-beam tube, an electrode to which a potential is applied to decrease the velocity of the electrons in the beam.

deceleration [MECH] The rate of decrease of speed of a motion.

decentered lens [OPTICS] A lens whose optical center does not coincide with the geometrical center of the rim of the lens; has the effect of a lens combined with a weak prism.

deci- [SCI TECH] A prefix indicating 10^{-1}, 0.1, or a tenth.

deciare [MECH] A unit of area, equal to 0.1 are or 10 square meters.

decibar [MECH] A metric unit of pressure equal to one-tenth bar.

decibel [PHYS] A unit for describing the ratio of two powers or intensities, or the ratio of a power to a reference power; in the measurement of sound intensity, the pressure of the reference sound is usually taken as 2×10^{-4} dyne per square centimeter; equal to one-tenth bel; if P_1 and P_2 are two amounts of power, the first is said to be n decibels greater, where $n = 10 \log_{10}(P_1/P_2)$. Abbreviated dB.

decibel adjusted *See* adjusted decibel.

decibels above 1 femtowatt [ELEC] A power level equal to 10 times the common logarithm of the ratio of the given power in watts to 1 femtowatt (10^{-15} watt). Abbreviated dBf.

decibels above 1 kilowatt [ELEC] A measure of power equal to 10 times the common logarithm of the ratio of a given power to 1000 watts. Abbreviated dBk.

decibels above 1 milliwatt [ELEC] A measure of power equal to 10 times the common logarithm of the ratio of a given power to 0.001 watt; a negative value, such as -2.7 dBm, means decibels below 1 milliwatt. Abbreviated dBm.

decibels above 1 picowatt [ELEC] A measure of power equal to 10 times the common logarithm of the ratio of a given power to 1 picowatt. Abbreviated dBp.

decibels above 1 volt [ELEC] A measure of voltage equal to 20 times the common logarithm of the ratio of a given voltage to 1 volt. Abbreviated dBV.

decibels above 1 watt [ELEC] A measure of power equal to 10 times the common logarithm of the ratio of a given power to 1 watt. Abbreviated dBW.

decibels above reference coupling [ELEC] A measure of the coupling between two circuits, expressed in relation to a reference value of coupling that gives a specified reading on a specified noise-measuring set when a test tone of 90 dBa is impressed on one circuit. Abbreviated dBx.

decibels above reference noise [ELEC] Units used to show the relationship between the interfering effect of a noise frequency, or band of noise frequencies, and a fixed amount of noise power commonly called reference noise; a 1000-hertz tone having a power level of -90 dBm was selected as the reference noise power; superseded by the adjusted decibel unit. Abbreviated dBrn.

decigram [MECH] A unit of mass, equal to 0.1 gram.

decile [STAT] Any of the points which divide the total number of items in a frequency distribution into 10 equal parts.

deciliter [MECH] A unit of volume, equal to 0.1 liter, or 10^{-4} cubic meter.

decimal [MATH] A number expressed in the scale of tens.

decimal-binary switch [ELEC] A switch that connects a single input lead to appropriate combinations of four output leads (representing 1, 2, 4, and 8) for each of the decimal-numbered settings of its control knob; thus, for position 7, output leads 1, 2, and 4 would be connected to the input.

decimal fraction [MATH] Any number written in the form: an integer followed by a decimal point followed by a (possibly infinite) string of digits.

decimal number [MATH] A number signifying a decimal fraction by a decimal point to the left of the numerator with the number of figures to the right of the point equal to the power of 10 of the denominator.

decimal number system [MATH] A representational system for the real numbers in which place values are read in powers of 10.

decimal place [MATH] Reference to one of the digits following the decimal point in a decimal fraction; the kth decimal place registers units of 10^{-k}.

decimal point [MATH] A dot written either on or slightly above the line; used to mark the point at which place values change from positive to negative powers of 10 in the decimal number system.

decimal system [MATH] A number system based on the number 10; in theory, each unit is 10 times the next smaller one.

decimeter [MECH] A metric unit of length equal to one-tenth meter.

decineper [PHYS] One-tenth of a neper.

decision calculus [SYS ENG] A guide to the process of decision-making, often outlined in the following steps: analysis of the decision area to discover applicable elements; location or creation of criteria for evaluation; appraisal of the known information pertinent to the applicable elements and correction for bias; isolation of the unknown factors; weighting of the pertinent elements, known and unknown, as to relative importance; and projection of the relative impacts on the objective, and synthesis into a course of action.

decision-making under uncertainty [STAT] The process of drawing conclusions from limited information or conjecture.

decision rule [SYS ENG] In decision theory, the mathematical representation of a physical system which operates upon the observed data to produce a decision.

decision theory [SYS ENG] A broad spectrum of concepts and techniques which have been developed to both describe and rationalize the process of decision making, that is, making a choice among several possible alternatives.

declination [GEOPHYS] The angle between the magnetic and geographical meridians, expressed in degrees and minutes east or west to indicate the direction of magnetic north from true north. Also known as magnetic declination; variation.

declination axis [ENG] For an equatorial mounting of a telescope, an axis of rotation that is perpendicular to the polar axis and allows the telescope to be pointed at objects of different declinations.

declination circle [ENG] For a telescope with an equatorial mounting, a setting circle attached to the declination axis that shows the declination to which the telescope is pointing.

declination compass *See* declinometer.

declination variometer [ENG] An instrument that measures changes in the declination of the earth's magnetic field, consisting of a permanent bar magnet, usually about 1 centimeter long, suspended with a plane mirror from a fine quartz fiber 5–15 centimeters in length; a lens focuses to a point a beam of light reflected from the mirror to recording paper mounted on a rotating drum. Also known as D variometer.

declinometer [ENG] A magnetic instrument similar to a surveyor's compass, but arranged so that the line of sight can be rotated to conform with the needle or to any desired setting on the horizontal circle; used in determining magnetic declination. Also known as declination compass.

decoder [ELECTR] A matrix of logic elements that selects one or more output channels, depending on the combination of input signals present.

decomposable process [MATH] A process which can be reduced to several basic events.

decomposition potential [PHYS CHEM] The electrode potential at which the electrolysis current begins to increase appreciably. Also known as decomposition voltage.

decomposition voltage *See* decomposition potential.

decontamination factor [NUCLEO] The ratio of initial specific radioactivity to final specific radioactivity resulting from a separation process.

decontamination index [NUCLEO] The logarithm of the ratio of initial specific radioactivity to final specific radioactivity resulting from a separation process.

decreasing function [MATH] **1.** A function of x whose value gets smaller as x gets larger, that is, if $x<y$, then $f(x)>f(y)$. **2.** *See* monotone nonincreasing function.

decrement [MATH] The quantity by which a variable is decreased. [PHYS] The ratio of the amplitudes of an underdamped harmonic motion during two successive oscillations. Also known as damping factor; numerical decrement.

decrement gage [ENG] A type of molecular gage consisting of a vibrating quartz fiber whose damping is used to determine the viscosity and, thereby, the pressure of a gas. Also known as quartz-fiber manometer.

Dedekind cut [MATH] A set of rational numbers satisfying certain properties, with which a unique real number may be

---

DECLINATION VARIOMETER

Declination variometer equipped with Helmholtz coil for calibration. *(U.S. Coast and Geodetic Survey)*

associated; used to define the real numbers as an extension of the rationals.

Dedekind test [MATH] If the series $\sum_i (b_i - b_{i+1})$ converges absolutely, the b_i converge to zero, and the series $\sum_i a_i$ has bounded partial sums, then the series $\sum_i a_i b_i$ converges.

deduction [MATH] The process of deriving a statement from certain assumed statements by applying the rules of logic.

dee [NUCLEO] A hollow accelerating-cyclotron electrode in the shape of the letter D. Also known as cyclotron D; D.

dee line [NUCLEO] A structural member that supports the dee of a cyclotron and acts with the dee to form the resonant circuit.

deenergize [ELEC] To disconnect from the source of power.

deep inelastic transfer See quasifission.

deep space [ASTRON] Space beyond the gravitational influence of the earth.

defect cluster [CRYSTAL] A macroscopic cluster of crystal defects which can arise from attraction among defects.

defect conduction [SOLID STATE] Electric conduction in a semiconductor by holes in the valence band.

defective number See deficient number.

defect motion [CRYSTAL] Movement of a point defect from one lattice point to another.

defect scattering [SOLID STATE] Scattering of particles or electromagnetic radiation by crystal defects.

defect structure [SOLID STATE] A crystal structure in which some atomic positions are occupied by atoms other than those that would be found in a perfect crystal, or are unoccupied.

deficiency index [MATH] For a curve or equation involving two complex variables this is the genus of the Riemann surface associated to the equation.

deficient number [MATH] A positive integer the sum of whose divisors, including 1 but excluding itself, is less than itself. Also known as defective number.

defining contrast [STAT] In the analysis of experimental designs which use one-half of the complete number of treatment combinations in a basic design, the comparison of those treatment combinations which have been used with those which have not been used.

definite proportions law See definite composition law.

definite Riemann integral [MATH] A number associated with a function defined on an interval $[a,b]$ which is
$$\lim_{N \to \infty} \sum_{k=0}^{N-1} f\left(a + \frac{k}{N}\right) \cdot \frac{b-a}{N},$$
if f is bounded and continuous; denoted by
$$\int_a^b f(x)\,dx;$$
if f is a positive function, the definite integral measures the area between the graph of f and the x axis.

definition [OPTICS] Lens image clarity or discernible detail.

deflecting torque [MECH] An instrument's moment, resulting from the quantity measured, that acts to cause the pointer's deflection.

deflection [ELECTR] The displacement of an electron beam from its straight-line path by an electrostatic or electromagnetic field.

deflection circuit [ELECTR] A circuit which controls the deflection of an electron beam in a cathode-ray tube.

deflection defocusing [ELECTR] Defocusing that becomes greater as deflection is increased in a cathode-ray tube, because the beam hits the screen at a greater slant and the beam spot becomes more elliptical as it approaches the edges of the screen.

deflection electrode [ELECTR] An electrode whose potential provides an electric field that deflects an electron beam. Also known as deflection plate.

deflection factor [ELECTR] The reciprocal of the deflection sensitivity in a cathode-ray tube.

deflection plate *See* deflection electrode.

deflection polarity [ELECTR] Relationship between the direction of a displacement of the cathode beam and the polarity of the applied signal wave.

deflection sensitivity [ELECTR] The displacement of the electron beam at the target or screen of a cathode-ray tube per unit of change in the deflection field; usually expressed in inches per volt applied between deflection electrodes or inches per ampere in a deflection coil.

deflection voltage [ELECTR] The voltage applied between a pair of deflection electrodes to produce an electric field.

deflection yoke [ELECTR] An assembly of one or more electromagnets that is placed around the neck of an electron-beam tube to produce a magnetic field for deflection of one or more electron beams. Also known as scanning yoke; yoke.

deformation [MATH] A homotopy of the identity map to some other map. [MECH] Any alteration of shape or dimensions of a body caused by stresses, thermal expansion or contraction, chemical or metallurgical transformations, or shrinkage and expansions due to moisture change.

deformation curve [MECH] A curve showing the relationship between the stress or load on a structure, structural member, or a specimen and the strain or deformation that results. Also known as stress-strain curve.

deformation ellipsoid *See* strain ellipsoid.

deformation energy [NUC PHYS] The energy which must be supplied to an initially spherical nucleus to give it a certain deformation in the Bohr-Wheeler theory.

deformation potential [SOLID STATE] The effective electric potential experienced by free electrons in a semiconductor or metal resulting from a local deformation in the crystal lattice.

degas [ELECTR] To drive out and exhaust the gases occluded in the internal parts of an electron tube or other gastight apparatus, generally by heating during evacuation.

degasser *See* getter.

degauss [ELECTR] To remove, erase, or clear information from a magnetic tape, disk, drum, or core. [ELECTROMAG] To neutralize (demagnetize) a magnetic field of, for example, a ship hull or television tube; a direct current of the correct value is sent through a cable around the ship hull; a current-carrying coil is brought up to and then removed from the television tube. Also known as deperm.

degeneracy [MATH] The condition in which two characteristic functions of an operator have the same characteristic value. [PHYS] The condition in which two or more modes of a vibrating system have the same frequency; a special case of the mathematics definition. [QUANT MECH] The condition in which two or more stationary states of the same system have the same energy even though their wave functions are not the same; a special case of the mathematics definition.

degenerate conduction band [SOLID STATE] A band in which two or more orthogonal quantum states exist that have the same energy, the same spin, and zero mean velocity.

degenerate electron gas [STAT MECH] An electron gas that is far below its Fermi temperature and is therefore described in first approximation by the Fermi distribution; most of the electrons completely fill the lower energy levels and are unable to take part in physical processes until excited out of these levels.

degenerate matter [PHYS] Matter that has been stripped of its orbital electrons, so the nuclei are packed close together.

degenerate semiconductor [SOLID STATE] A semiconductor in which the number of electrons in the conduction band approaches that of a metal.

degeneration [ELECTR] The loss or gain in an amplifier through unintentional negative feedback. [STAT MECH] A phenomenon which occurs in gases at very low temperatures when the molecular heat drops to less than $\frac{3}{2}$ the gas constant.

degradation [PHYS] Loss of energy of a particle, such as a neutron or photon, through a collision. [THERMO] The conversion of energy into forms that are increasingly difficult to convert into work, resulting from the general tendency of entropy to increase.

degree [FL MECH] One of the units in any of various scales of specific gravity, such as the Baumé scale. [MATH] **1.** A unit for measurement of plane angles, equal to $1/360$ of a complete revolution, or $1/90$ of a right angle. Symbolized °. **2.** For a term in one variable, the exponent of that variable. **3.** For a term in several variables, the sum of the exponents of its variables. **4.** For a polynomial, the degree of the highest-degree term. **5.** For a differential equation, the greatest power to which the highest-order derivative occurs. **6.** For an algebraic curve defined by the polynomial equation $f(x,y) = 0$, the degree of the polynomial $f(x,y)$. [THERMO] One of the units of temperature or temperature difference in any of various temperature scales, such as the Celsius, Fahrenheit, and Kelvin temperature scales (the Kelvin degree is now known as the kelvin).

degree Engler [FL MECH] A measure of viscosity; the ratio of the time of flow of 200 milliliters of the liquid through a viscometer devised by Engler, to the time for the flow of the same volume of water.

degree of degeneracy [MATH] The number of characteristic functions of an operator having the same characteristic value. Also known as order of degeneracy.

degree of enrichment [NUCLEO] The enrichment factor minus 1.

degree of freedom [MECH] Of a gyro, the number of orthogonal axes about which the spin axis is free to rotate, the spin axis freedom not being counted; this is not a universal convention; for example, the free gyro is frequently referred to as a three-degree-of-freedom gyro, the spin axis being counted. [PHYS CHEM] Any one of the variables, including pressure, temperature, composition, and specific volume, which must be specified to define the state of a system. [STAT] A number one less than the number of frequencies being tested with a chi-square test.

de Gua's rule [MATH] The rule that if, in a polynomial equation $f(x) = 0$, a group of r consecutive terms is missing, then the equation has at least r imaginary roots if r is even, or the equation has at least $r + 1$ or $r - 1$ imaginary roots if r is odd (depending on whether the terms immediately preceding and following the group have like or unlike signs).

de Haas-Van Alphen effect [SOLID STATE] An effect occurring in many complex metals at low temperatures, consisting of a periodic variation in the diamagnetic susceptibility of conduction electrons with changes in the component of the applied magnetic field at right angles to the principal axis of the crystal.

deionization [ELECTR] The return of an ionized gas to its neutral state after all sources of ionization have been removed, involving diffusion of ions to the container walls and volume recombination of negative and positive ions.

deionization potential   [ELECTR] The potential at which ionization of the gas in a gas-filled tube ceases and conduction stops.

deionization time   [ELECTR] The time required for a gas tube to regain its preconduction characteristics after interruption of anode current, so that the grid regains control. Also called recontrol time.

dekapoise   [FL MECH] A unit of absolute viscosity, equal to 10 poises.

Delaborne prism   [OPTICS] A special compound prism which, when rotated about an axis parallel to the reflecting face and lying in a plane perpendicular to the refracting faces, rotates the image through twice the angle. Also known as Dove prism.

de la Rue and Miller's law   [ELECTR] The law that in a field between two parallel plates, the sparking potential of a gas is a function of the product of gas pressure and sparking distance only.

Delaunay orbit element   [MECH] In the n-body problem, certain functions of variable elements of an ellipse with a fixed focus along which one of the bodies travels; these functions have rates of change satisfying simple equations.

delay circuit   *See* time-delay circuit.

delay Doppler mapping   [MAP] Mapping of a planet by illuminating it with a radar beam and measuring the Doppler shift caused by rotation of the planet.

delayed alpha particle   [NUC PHYS] An alpha particle emitted by an excited nucleus that was formed an appreciable time after a beta disintegration process.

delayed coincidence   [NUCLEO] Occurrence of a count in one detector at a short but measurable time later than a count in another detector, the two counts being due to successive events in the same nucleus.

delayed critical   [NUCLEO] The condition in which a nuclear reactor is critical because of delayed neutrons alone, without requiring the contribution of prompt neutrons.

delayed gamma ray   [NUCLEO] A gamma ray emitted during radioactive decay of a fission product.

delayed neutron   [NUC PHYS] A neutron emitted spontaneously from a nucleus as a consequence of excitation left from a preceding radioactive decay event; in particular, a delayed fission neutron.

delayed neutron fraction   [NUC PHYS] The ratio of the mean number of delayed fission neutrons per fission to the mean total number of neutrons (prompt plus delayed) per fission.

delayed proton   [NUC PHYS] A proton emitted spontaneously from a nucleus as a consequence of excitation left from a previous radioactive decay event.

delay lens   [ELECTROMAG] A regular array of conductors for focusing microwaves of relatively long wavelength.

delay line   [ELECTR] A transmission line (as dissipationless as possible), or an electric network approximation of it, which, if terminated in its characteristic impedance, will reproduce at its output a waveform applied to its input terminals with little distortion, but at a time delayed by an amount dependent upon the electrical length of the line. Also known as artificial delay line.

delay time   [CONT SYS] The amount of time by which the arrival of a signal is retarded after transmission through physical equipment or systems.   [ELECTR] The time taken for collector current to start flowing in a transistor that is being turned on from the cutoff condition.

Delbrück scattering   [NUC PHYS] Elastic scattering of gamma

DELABORNE PRISM

Diagram of Delaborne prism.

DELAY LINE

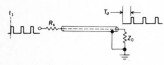

Circuit diagram of a transmission line as a delay line. R_s is series resistance of line; Z_O is its characteristic impedance. Graphs at left and right represent input and output pulses respectively. Series of pulses starting at time t_1 require time T_d to propagate down line.

rays by a nucleus caused by virtual electron-positron pair production.

d electron [ATOM PHYS] An atomic electron that has an orbital angular momentum of 2 in the central field approximation.

del operator [MATH] The rule which replaces the function f of three variables, x, y, z, by the vector valued function whose components in the x, y, z directions are the respective partial derivatives of f. Written ∇f. Also known as nabla.

delta baryon [PARTIC PHYS] **1.** Any excited baryon state belonging to a multiplet having a total isospin of 3/2, a hypercharge of $+1$, a spin of 3/2, positive parity, and an approximate mass of 1236 MeV. Designated $\Delta(1236)$. **2.** Any excited baryon state belonging to any multiplet having a total isospin of 3/2 and a hypercharge of $+1$.

delta E effect [ELECTROMAG] Magnetization of a ferromagnetic substance that is caused by elastic tension.

delta function [MATH] A distribution δ such that

$$\int_{-\infty}^{\infty} f(t) \, \delta(x-t) \, dt$$

is $f(x)$. Also known as Dirac delta function.

delta meson [PARTIC PHYS] Any scalar meson resonance, with positive charge conjugation parity, belonging to a multiplet with a total isospin of 1, a hypercharge of zero, a mass of 962 ± 5 MeV, and a width <5 MeV. Designated $\delta(962)$.

delta ray [ATOM PHYS] An electron or proton ejected by recoil when a rapidly moving alpha particle or other primary ionizing particle passes through matter.

delta-Y transformation *See* Y-delta transformation.

deltohedron [CRYSTAL] A polyhedron which has 12 quadrilateral faces, and is the form of a crystal belonging to the cubic system and having hemihedral symmetry. Also known as deltoid dodecahedron; tetragonal tristetrahedron.

deltoid [MATH] The plane curve traced by a point on a circle while the circle rolls along the inside of another circle whose radius is twice as great. Also known as Steiner's hypocycloid.

deltoid dodecahedron *See* deltohedron.

demagnetization [ELECTROMAG] **1.** The process of reducing or removing the magnetism of a ferromagnetic material. **2.** The reduction of magnetic induction by the internal field of a magnet.

demagnetization coefficient *See* demagnetizing factor.

demagnetization curve [ELECTROMAG] Graph of magnetic induction B versus magnetic field H in a ferromagnetic material, as the magnetic field is reduced to 0 from its saturation value.

demagnetizing factor [ELECTROMAG] The ratio of the negative of the demagnetizing field to the magnetization of a sample. Also known as demagnetization coefficient.

demagnetizing field [ELECTROMAG] An additional magnetic field that is produced in a magnetic material subject to an applied magnetic field, due to the magnetic material itself.

Dember effect [ELECTR] Creation of a voltage in a conductor or semiconductor by illumination of one surface. Also known as photodiffusion effect.

demodulator *See* detector.

De Moivre's theorem [MATH] The nth power of the quantity $\cos \theta + i \sin \theta$ is $\cos n\theta + i \sin n\theta$ for any integer n.

demon of Maxwell [THERMO] Hypothetical creature who controls a trapdoor over a microscopic hole in an adiabatic wall between two vessels filled with gas at the same temperature, so as to supposedly decrease the entropy of the gas as a whole and thus violate the second law of thermodynamics. Also known as Maxwell's demon.

De Morgan's rules [MATH] The complement of the union of two sets equals the intersection of their respective complements; the complement of the intersection of two sets equals the union of their complements.

De Morgan's test [MATH] A series with term u_n, for which $|u_{n+1}/u_n|$ converges to 1, will converge absolutely if there is $c > 0$ such that the limit superior of $n(|u_{n+1}/u_n| - 1)$ equals $-1 - c$.

denaturant [NUCLEO] A nonfissionable isotope that can be added to fissionable material to make it unsuitable for use in atomic weapons without extensive processing.

Deneb [ASTRON] A white star of spectral classification A2-Ia in the constellation Cygnus; the star α Cygni.

Denebola [ASTRON] A white star of stellar magnitude 2.2, spectral classification A2, in the constellation Leo; the star β Leonis.

denominator [MATH] In a fraction, the term that divides the other term (called the numerator), and is written below the line.

dense-air refrigeration cycle *See* reverse Brayton cycle.

dense-in-itself set [MATH] A set every point of which is an accumulation point; a set without any isolated points.

dense subset [MATH] A subset of a topological space whose closure is the entire space.

densimeter [ENG] An instrument which measures the density or specific gravity of a liquid, gas, or solid. Also known as densitometer; density gage; density indicator; gravitometer.

densitometer [ENG] **1.** An instrument which measures optical density by measuring the intensity of transmitted or reflected light; used to measure photographic density. **2.** *See* densimeter.

density [MECH] The mass of a given substance per unit volume. [OPTICS] **1.** The degree of opacity of a translucent material. **2.** The common logarithm of opacity. [PHYS] The total amount of a quantity, such as energy, per unit of space.

density bottle *See* specific gravity bottle.

density effect [NUCLEO] The reduction in the stopping power of dense materials for relativistic particles that is caused by the reduction of the effective electric field of the particles by the polarization of adjacent atoms.

density function [MATH] A density function for a measure m is a function which gives rise to m when it is integrated with respect to some other specified measure. [STAT] *See* probability density function.

density gage *See* densimeter.

density indicator *See* densimeter.

density matrix [QUANT MECH] A matrix ρ_{mn} describing an ensemble of quantum-mechanical systems in a representation based on an orthonormal set of functions ϕ_n; for any operator G with representation G_{mn}, the ensemble average of the expectation value of G is the trace of ρG.

density of states [SOLID STATE] A function of energy E equal to the number of quantum states in the energy range between E and $E + dE$ divided by the product of dE and the volume of the substance.

density point [MATH] A density point of a measurable subset S of the real numbers is a real number at which the metric density of S exists and equals 1.

density-wave theory [ASTROPHYS] A theory explaining the spiral structure of galaxies by a periodic variation in space in the density of matter which rotates with a fixed angular velocity while the angular velocity of the matter itself varies with distance from the galaxy's center.

dependence [STAT] The existence of a relationship between frequencies obtained from two parts of an experiment which does not arise from the direct influence of the result of the first part on the chances of the second part but indirectly from the fact that both parts are subject to influences from a common outside factor.

dependent variable [MATH] If y is a function of x, that is, if the function assigns a single value of y to each value of x, then y is the dependent variable.

deperm *See* degauss.

depleted material [NUCLEO] Material in which the amount of one or more isotopes of a constituent has been reduced by an isotope separation process or by a nuclear reaction.

depleted uranium [NUCLEO] Uranium having a smaller percentage of uranium-235 than the 0.7% found in natural uranium.

depletion [ELECTR] Reduction of the charge-carrier density in a semiconductor below the normal value for a given temperature and doping level. [NUCLEO] The percentage reduction in the quantity of fissionable atoms in the fuel assemblies or fuel mixture that occurs during operation of a nuclear reactor.

depletion layer [ELECTR] An electric double layer formed at the surface of contact between a metal and a semiconductor having different work functions, because the mobile carrier charge density is insufficient to neutralize the fixed charge density of donors and acceptors. Also known as barrier layer (deprecated); blocking layer (deprecated); space-charge layer.

depletion-layer rectification [ELECTR] Rectification at the junction between dissimilar materials, such as a *pn* junction or a junction between a metal and a semiconductor. Also known as barrier-layer rectification.

depletion-layer transistor [ELECTR] A transistor that relies directly on motion of carriers through depletion layers, such as spacistor.

depletion-mode field-effect transistor *See* junction field-effect transistor.

depletion region [ELECTR] The portion of the channel in a metal oxide field-effect transistor in which there are no charge carriers.

depolarization [ELEC] The removal or prevention of polarization in a substance (for example, through the use of a depolarizer in an electric cell) or of polarization arising from the field due to the charges induced on the surface of a dielectric when an external field is applied. [OPTICS] The resolution of polarized light in an optical depolarizer.

depolarization factor [ELEC] The ratio of the internal electric field induced by the charges on the surface of a dielectric when an external field is applied to the polarization of the dielectric.

deposit dose [NUCLEO] The residual radioactivity deposited on the surface after a nuclear explosion, as by water falling as rain from the base surge of an underwater atomic explosion.

depositional remanent magnetization [GEOPHYS] Remanent magnetization occurring in sedimentary rock following the depositional alignment of previously magnetized grains. Abbreviated DRM.

depth dose [NUCLEO] The radiation dose delivered at a particular depth beneath the surface of a body; usually expressed as percent of surface dose or of air dose.

depth magnification [OPTICS] The ratio of the distance between two nearby points of the axis on the image side of an optical system to the distance between their conjugate points on the object side.

depth of compensation [GEOPHYS] That depth at which density differences occurring in the earth's crust are compensated isostatically; calculated to be between 100 and 113–117 kilometers.

depth of field [OPTICS] The range of distances over which a camera gives satisfactory definition, its lens in the best focus for a certain specific distance.

depth of focus [OPTICS] The range of image distances corresponding to the range of object distances included in depth of field.

derangement numbers [MATH] The numbers D_n, $n = 1, 2, 3, \ldots$, giving the number of permutations of a set of n elements that carry no element of the set into itself.

derivation [MATH] **1.** The process of deducing a formula. **2.** A function from a ring into itself which satisfies the equations $D(u+v) = Du + Dv$ and $D(uv) = uD(v) + D(u)v$.

derivative [MATH] The slope of a graph $y = f(x)$ at a given point c; more precisely, it is the limit as h approaches zero of $f(c+h) - f(c)$ divided by h. Also known as differential coefficient; rate of change.

derivative action [CONT SYS] Control action in which the speed at which a correction is made depends on how fast the system error is increasing. Also known as derivative compensation; rate action.

derivative compensation *See* derivative action.

derivative network [CONT SYS] A compensating network whose output is proportional to the sum of the input signal and its derivative. Also known as lead network.

derived curve [MATH] A curve whose ordinate, for each value of the abscissa, is equal to the slope of some given curve. Also known as first derived curve.

derived quantity [PHYS] A physical quantity which, in a specified system of measurement, is defined by operations based on other physical quantities.

derived set [MATH] The set of cluster points of a given set.

derived unit [PHYS] A unit that is formed, in a specified system of measurement, by combining base units and other derived units according to the algebraic relations linking the corresponding quantities.

Deryagin number [PHYS] A dimensionless group equal to the ratio of the thickness of a film coating a liquid to the capillary length of the liquid. Symbolized De.

Desarguesian plane [MATH] Any projective plane in which points and lines satisfy Desargues' theorem. Also known as Arguesian plane.

Desargues' theorem [MATH] If the three lines passing through corresponding vertices of two triangles are concurrent, then the intersections of the three pairs of corresponding sides lie on a straight line, and conversely.

desaturated color [OPTICS] A color that is neither a pure spectral color nor a purple formed from a mixture of deep red and violet.

DeSauty's bridge [ELEC] A four-arm bridge used to compare two capacitances; two adjacent arms contain capacitors in series with resistors, while the other two arms contain resistors only. Also known as Wien–DeSauty bridge.

Descartes laws of refraction *See* Snell laws of refraction.

Descartes ray [OPTICS] A ray of light incident on a sphere of transparent material, such as a water droplet, which after one internal reflection leaves the drop at the smallest possible angle of deviation from the direction of the incident ray; these rays make the primary rainbow.

Descartes' rule of signs [MATH] A polynomial with real coefficients has at most k real positive roots, where k is the number of sign changes in the polynomial.

descending branch [MECH] That portion of a trajectory

which is between the summit and the point where the trajectory terminates, either by impact or air burst, and along which the projectile falls, with altitude constantly decreasing. Also known as descent trajectory.

descending node [ASTRON] The point at which a planet, planetoid, or comet crosses the ecliptic from north to south.

descent trajectory *See* descending branch.

describing function [CONT SYS] A function used to represent a nonlinear transfer function by an approximately equivalent linear transfer function; it is the ratio of the phasor representing the fundamental component of the output of the nonlinearity, determined by Fourier analysis, to the phasor representing a sinusoidal input signal.

descriptive geometry [MATH] The application of graphical methods to the solution of three-dimensional space problems.

descriptive statistics [STAT] Presentation of data in the form of tables and charts or summarization by means of percentiles and standard deviations.

de Sitter space [RELAT] The four-dimensional surface of a sphere in five-dimensional space, used as a model of the universe.

Destriau effect [SOLID STATE] Sustained emission of light by suitable phosphor powders that are embedded in an insulator and subjected only to the action of an alternating electric field.

destructive breakdown [ELECTR] Breakdown of the barrier between the gate and channel of a field-effect transistor, causing failure of the transistor.

destructive interference [OPTICS] The interaction of superimposed light from two different sources when the phase relationship is such as to reduce or cancel the resultant intensity to less than the sum of the individual lights.

detached shock wave [FL MECH] A shock wave not in contact with the body which originates it.

detailed balance [STAT MECH] The hypothesis that when a system is in equilibrium any process occurs with the same frequency as the reverse process.

detector [ELECTR] The stage in a receiver at which demodulation takes place; in a superheterodyne receiver this is called the second detector. Also known as demodulator; envelope detector.

determinant tensor [MATH] A tensor whose components are each equal to the corresponding component of the Levi-Civita tensor density times the square root of the determinant of the metric tensor, and whose contravariant components are each equal to the corresponding component of the Levi-Civita density divided by the square root of the metric tensor. Also known as permutation tensor.

determinate structure [MECH] A structure in which the equations of statics alone are sufficient to determine the stresses and reactions.

determinism *See* causality.

detonation wave [FL MECH] A shock wave that accompanies detonation and has a shock front followed by a region of decreasing pressure in which the reaction occurs.

deuterium [CHEM] The isotope of the element hydrogen with one neutron and one proton in the nucleus; atomic weight 2.0144. Designated D, d, H^2, or ^2H.

deuterium cycle *See* proton-proton chain.

deuterium discharge tube [ELECTR] A tube similar to a hydrogen discharge lamp, but with deuterium replacing the hydrogen; source of high-intensity ultraviolet radiation for spectroscopic microanalysis.

deuterium oxide *See* heavy water.

deuteron [NUC PHYS] The nucleus of a deuterium atom,

consisting of a neutron and a proton. Designated *d*. Also known as deuton.

deuteron accelerator [NUCLEO] An accelerator that produces a flux of slow neutrons by bombarding a metal-tritium target with deuterons.

deuteron capture [NUC PHYS] The absorption of a deuteron by a nucleus, giving rise to a compound nucleus which subsequently decays.

deuton *See* deuteron.

developable surface [MATH] A surface that can be obtained from a plane sheet by deformation, without stretching or shrinking.

deviation [OPTICS] The angle between the incident ray on an object or optical system and the emergent ray, following reflection, refraction, or diffraction. Also known as angle of deviation. [STAT] The difference between any given number in a set and the mean average of those numbers.

deviation factor *See* compressibility factor.

deviatonic stress [MECH] The portion of the total stress that differs from an isostatic hydrostatic pressure; it is equal to the difference between the total stress and the spherical stress.

Dewar calorimeter [ENG] **1.** Any calorimeter in which the sample is placed inside a Dewar flask to minimize heat losses. **2.** A calorimeter for determining the mean specific heat capacity of a solid between the boiling point of a cryogenic liquid, such as liquid oxygen, and room temperature, by measuring the amount of the liquid that evaporates when the specimen is dropped into the liquid.

DEWAR FLASK

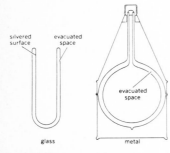

Typical Dewar containers.

Dewar flask [PHYS] A vessel having double walls, the space between being evacuated to prevent the transfer of heat and the surfaces facing the vacuum being heat-reflective; used to hold liquid gases and to study low-temperature phenomena.

dew point [CHEM] The temperature at which water vapor begins to condense.

dex *See* brig.

dextro *See* dextrorotatory.

dextrorotatory [OPTICS] Rotating clockwise the plane of polarization of a wave traveling through a medium in a clockwise direction, as seen by an eye observing the light. Abbreviated dextro.

di- [SCI TECH] Prefix meaning two.

diabatic [THERMO] Involving a thermodynamic change of state of a system in which there is a transfer of heat across the boundaries of the system.

diad axis [CRYSTAL] A rotation axis whose multiplicity is equal to 2.

diafocal point [OPTICS] For a ray of light refracted by a lens, a point on the ray which lies on a plane passing through the axis of the lens which is parallel to the ray on the opposite side of the lens.

diagonal [MATH] **1.** The set of points all of whose coordinates are equal to one another in an n-dimensional coordinate system. **2.** A line joining opposite vertices of a polygon with an even number of sides.

diagonal horn antenna [ELECTROMAG] Horn antenna in which all cross sections are square and the electric vector is parallel to one of the diagonals; the radiation pattern in the far field has almost perfect circular symmetry.

diagonal matrix [MATH] A matrix whose nonzero entries all lie on the principal diagonal.

diagram [MATH] A picture in which sets are represented by symbols and mappings between these sets are represented by arrows.

diamagnet [ELECTROMAG] A substance which is diamagnetic, such as the alkali and alkaline earth metals, the halogens, and the noble gases.

diamagnetic [ELECTROMAG] Having a magnetic permeability less than 1; materials with this property are repelled by a magnet and tend to position themselves at right angles to magnetic lines of force.

diamagnetic Faraday effect [OPTICS] Faraday effect at frequencies near an absorption line which is split due to the splitting of the upper level only.

diamagnetic resonance See cyclotron resonance.

diamagnetic susceptibility [ELECTROMAG] The susceptibility of a diamagnetic material, which is always negative and usually on the order of -10^{-5} cm³/mole.

diamagnetism [ELECTROMAG] The property of a material which is repelled by magnets.

diameter [MATH] **1.** A line segment which passes through the center of a circle, and whose end points lie on the circle. **2.** The length of such a line.

diameter of a conic [MATH] Any straight line that passes through the midpoints of all the chords of the conic that are parallel to a given chord.

diameter of a set [MATH] The smallest number which is greater than or equal to the distance between every pair of points of the set.

diametral curve [MATH] A curve that passes through the midpoints of a family of parallel chords of a given curve.

diametral plane [MATH] **1.** A plane that passes through the center of a sphere. **2.** A plane that passes through the midpoints of a family of parallel chords of a quadric surface that are parallel to a given chord.

diametral surface [MATH] A surface that passes through the midpoints of a family of parallel chords of a given surface that are parallel to a given chord.

diamond antenna See rhombic antenna.

diaphragm [ELECTROMAG] See iris. [OPTICS] Any opening in an optical system which controls the cross section of a beam of light passing through it, to control light intensity, reduce aberration, or increase depth of focus. Also known as lens stop. [PHYS] **1.** A separating wall or membrane, especially one which transmits some substances and forces but not others. **2.** In general, any opening, sometimes adjustable in size, which is used to control the flow of a substance or radiation.

diaphragm setting [OPTICS] The position of a camera's diaphragm after opening or closing it.

diasporometer [OPTICS] Oppositely rotating wedges used in optical rangefinders to obtain deviation of the axis of the image.

diathermanous [PHYS] Capable of transmitting radiant heat. Also known as diathermic.

diathermic See diathermanous.

diathermous envelope [THERMO] A surface enclosing a thermodynamic system in equilibrium that is not an adiabatic envelope; intuitively, this means that heat can flow through the surface.

diatonic scale [ACOUS] A musical scale in which the octave is divided into intervals of two different sizes, five of one and two of the other, with adjustments in tuning systems other than equal temperament.

dichotic listening See dichotic presentation.

dichotic presentation [ACOUS] The simultaneous reception of one message through one ear and another message through the other ear. Also known as dichotic listening.

dichotomic variable [QUANT MECH] A variable with a range

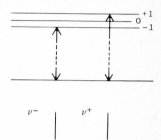

**DIAMAGNETIC
FARADAY EFFECT**

Diamagnetic Faraday effect which is temperature-independent; ν = frequency of light. Splitting of the absorption line is due to the splitting of the upper level only, and the lower level of the line is not split.

consisting of two values; used, for example, to describe a particle with spin ½.

dichotomy [ASTRON] The phase of the moon or an inferior planet at which exactly half of its disk is illuminated and the terminator is a straight line.

dichroic mirror [OPTICS] A glass surface coated with a special metal film that reflects certain colors of light while allowing others to pass through.

dichroism [OPTICS] In certain anisotropic materials, the property of having different absorption coefficients for light polarized in different directions.

Dicke radiometer [ELECTR] A radiometer-type receiver that detects weak signals in noise by modulating or switching the incoming signal before it is processed by conventional receiver circuits.

dielectric absorption [ELEC] The persistence of electric polarization in certain dielectrics after removal of the electric field. [ELECTROMAG] *See* dielectric loss.

dielectric aging [ELEC] The gradual change in the properties of a dielectric with time, usually for the worse.

dielectric antenna [ELECTROMAG] An antenna in which a dielectric is the major component used to produce a desired radiation pattern.

dielectric breakdown [ELECTR] Breakdown which occurs in an alkali halide crystal at field strengths on the order of 10^6 volts per centimeter.

dielectric circuit [ELEC] Any electric circuit which has capacitors.

dielectric constant [ELEC] **1.** For an isotropic medium, the ratio of the capacitance of a capacitor filled with a given dielectric to that of the same capacitor having only a vacuum as dielectric. **2.** More generally, $1 + \gamma\chi$, where γ is 4π in Gaussian and cgs electrostatic units or 1 in rationalized mks units, and χ is the electric susceptibility tensor. Also known as relative dielectric constant; relative permittivity; specific inductive capacity (SIC).

dielectric crystal [ELEC] A crystal which is electrically nonconducting.

dielectric current [ELEC] The current flowing at any instant through a surface of a dielectric that is located in a changing electric field.

dielectric displacement *See* electric displacement.

dielectric ellipsoid [ELEC] For an anisotropic medium in which the dielectric constant is a tensor quantity **K**, the locus of points **r** satisfying $\mathbf{r} \cdot \mathbf{K} \cdot \mathbf{r} = 1$.

dielectric fatigue [ELECTR] The property of some dielectrics in which resistance to breakdown decreases after a voltage has been applied for a considerable time.

dielectric field [ELEC] The average total electric field acting upon a molecule or group of molecules inside a dielectric. Also known as internal dielectric field.

dielectric flux density *See* electric displacement.

dielectric hysteresis *See* ferroelectric hysteresis.

dielectric imperfection levels [SOLID STATE] Energy levels that occur in the forbidden zone between the valence and conduction bands of a dielectric crystal, because of imperfections in the crystal.

dielectric leakage [ELEC] A very small steady current that flows through a dielectric subject to a steady electric field.

dielectric lens [ELECTROMAG] A lens made of dielectric material so that it refracts radio waves in the same manner that an optical lens refracts light waves; used with microwave antennas.

dielectric-lens antenna [ELECTROMAG] An aperture antenna in which the beam width is determined by the dimensions of a dielectric lens through which the beam passes.

dielectric loss [ELECTROMAG] The electric energy that is converted into heat in a dielectric subjected to a varying electric field. Also known as dielectric absorption.

dielectric loss angle [ELEC] Difference between 90° and the dielectric phase angle.

dielectric loss factor [ELEC] Product of the dielectric constant of a material and the tangent of its dielectric loss angle.

dielectric matching plate [ELECTROMAG] In waveguide technique, a dielectric plate used as an impedance transformer for matching purposes.

dielectric phase angle [ELEC] Angular difference in phase between the sinusoidal alternating potential difference applied to a dielectric and the component of the resulting alternating current having the same period as the potential difference.

dielectric polarization *See* polarization.

dielectric power factor [ELEC] Cosine of the dielectric phase angle (or sine of the dielectric loss angle).

dielectric-rod antenna [ELECTROMAG] A surface-wave antenna in which an end-fire radiation pattern is produced by propagation of a surface wave on a tapered dielectric rod.

dielectric shielding [ELEC] The reduction of an electric field in some region by interposing a dielectric substance, such as polystyrene, glass, or mica.

dielectric soak *See* absorption.

dielectric strength [ELEC] The maximum electrical potential gradient that a material can withstand without rupture; usually specified in volts per millimeter of thickness. Also known as electric strength.

dielectric susceptibility *See* electric susceptibility.

dielectric waveguide [ELEC] A waveguide consisting of a dielectric cylinder surrounded by air.

dielectric wedge [ELECTROMAG] A wedge-shaped piece of dielectric used in a waveguide to match its impedance to that of another waveguide.

dielectric wire [ELECTROMAG] A dielectric waveguide used to transmit ultra-high-frequency radio waves short distances between parts of a circuit.

dielectronic recombination [ATOM PHYS] The combination of an electron with a positive-ion in a gas, so that the energy released is taken up by two electrons of the resulting atom.

diesel cycle [THERMO] An internal combustion engine cycle in which the heat of compression ignites the fuel.

Dieterici equation of state [THERMO] An empirical equation of state for gases, $pe^{a/vRT}(v-b) = RT$, where p is the pressure, T is the absolute temperature, v is the molar volume, R is the gas constant, and a and b are constants characteristic of the substance under consideration.

difference [MATH] **1.** The result of subtracting one number from another. **2.** The difference between two sets A and B is the set consisting of all elements of A which do not belong to B; denoted $A - B$.

difference amplifier *See* differential amplifier.

difference equation [MATH] An equation expressing a functional relationship of one or more independent variables, one or more functions dependent on these variables, and successive differences of these functions.

difference methods [MATH] Versions of the predictor-corrector methods of calculating numerical solutions of differential equations in which the prediction and correction formulas express the value of the solution function in terms of finite differences of a derivative of the function.

difference number *See* neutron excess.

difference operator [MATH] One of several operators, such as the displacement operator, forward difference operator, or central mean operator, which can be used to conveniently express formulas for interpolation or numerical calculation or integration of functions and can be manipulated as algebraic quantities.

difference spectrophotometer *See* absorption spectrophotometer.

difference tone [ACOUS] A combination tone whose frequency equals the difference of the frequencies of the pure tones producing it.

differentiable atlas [MATH] A family of embeddings $h_i:E^n\to M$ of euclidean space into a topological space M with the property that $h_i^{-1}h_j:E^n\to E^n$ is a differentiable map for each pair of indices, i, j.

differentiable function [MATH] A function which has a derivative at each point of its domain.

differentiable manifold [MATH] A topological space with a maximal differentiable atlas; roughly speaking, a smooth surface.

differential [MATH] **1.** The differential of a real-valued function $f(x)$, where x is a vector, evaluated at a given vector c, is the linear, real-valued function whose graph is the tangent hyperplane to the graph of $f(x)$ at $x=c$; if x is a real number, the usual notation is $df=f'(c)\ dx$. **2.** The differential of a smooth map h from a smooth manifold M to a smooth manifold N, at a point p of M, is a mapping which carries each tangent vector X at p into the tangent victor L at $h(p)$, where L is defined by the equation $Lf = X(f\circ h)$, where f is any smooth, real-valued function on N. **3.** A differential on a bigraded module E of bidegree $(-r, r-1)$ is a collection of homomorphisms $d:E_{s,t}\to E_{s-r,t+r-1}$ for all s and t such that $d^2 = 0$. **4.** *See* total differential.

differential air thermometer [ENG] A device for detecting radiant heat, consisting of a U-tube manometer with a closed bulb at each end, one clear and the other blackened.

differential amplifier [ELECTR] An amplifier whose output is proportional to the difference between the voltages applied to its two inputs. Also called difference amplifier.

differential calculus [MATH] The study of the manner in which the value of a function changes as one changes the value of the independent variable; includes maximum-minimum problems and expansion of functions into Taylor series.

differential calorimetry [THERMO] Technique for measurement of and comparison (differential) of process heats (reaction, absorption, hydrolysis, and so on) for a specimen and a reference material.

differential capacitance [ELECTR] The derivative with respect to voltage of a charge characteristic, such as an alternating charge characteristic or a mean charge characteristic, at a given point on the characteristic.

differential capacitor [ELEC] A two-section variable capacitor having one rotor and two stators so arranged that as capacitance is reduced in one section it is increased in the other.

differential coefficient *See* derivative.

differential correction [ASTRON] A method for finding from the observed residuals minus the computed residuals $(O - C)$ small corrections which, when applied to the orbital elements or constants, will reduce the deviations from the observed motion to a minimum.

differential cross section [PHYS] The cross section for a collision process resulting in the emission of particles or photons at a specified angle relative to the direction of the incident particles, per unit angle or per unit solid angle.

differential effects [MECH] The effects upon the elements of the trajectory due to variations from standard conditions.

differential equation [MATH] An equation expressing a relationship between functions and their derivatives.

differential form [MATH] A homogeneous polynomial in differentials.

differential galvanometer [ELEC] A galvanometer having a magnetic needle which is free to rotate in the magnetic field produced by currents flowing in opposite directions through two separate identical coils, so that there is no deflection when the currents are equal.

differential game [CONT SYS] A two-sided optimal control problem. [MATH] A game in which the describing equations are differential equations.

differential geometry [MATH] The study of curves and surfaces using the methods of differential calculus.

differential heat of dilution *See* heat of dilution.

differential heat of solution [THERMO] The partial derivative of the total heat of solution with respect to the molal concentration of one component of the solution, when the concentration of the other component or components, the pressure, and the temperature are held constant.

differential instrument [ENG] Galvanometer or other measuring instrument having two circuits or coils, usually identical, through which currents flow in opposite directions; the difference or differential effect of these currents actuates the indicating pointer.

differential ionization chamber [NUCLEO] A two-section ionization chamber in which electrode potentials are such that output current is equal to the difference between the separate ionization currents of the two sections.

differential operational amplifier [ELECTR] An amplifier that has two input terminals, used with additional circuit elements to perform mathematical functions on the difference in voltage between the two input signals.

differential operator [MATH] An operator on a space of functions which maps a function f into a linear combination of higher-order derivatives of f.

differential permeability [ELECTROMAG] The slope of the magnetization curve for a magnetic material.

differential pressure [PHYS] The difference in pressure between two points of a system, such as between the well bottom and wellhead or between the two sides of an orifice.

differential pressure pickup [ELEC] An instrument that measures the difference in pressure between two pressure sources and translates this difference into a change in inductance, resistance, voltage, or some other electrical quality.

differential reaction rate [PHYS CHEM] The order of a chemical reaction expressed as a differential equation with respect to time; for example, $dx/dt = k(a - x)$ for first order, $dx/dt = k(a - x)(b - x)$ for second order, and so on, where k is the specific rate constant, a is the concentration of reactant A, b is the concentration of reactant B, and dx/dt is the rate of change in concentration for time t.

differential selection [STAT] A biased selection of a conditioned sample.

differential spectrophotometry [SPECT] Spectrophotometric analysis of a sample when a solution of the major component of the sample is placed in the reference cell; the recorded spectrum represents the difference between the sample and the reference cell.

differential stream calorimeter [ENG] An instrument for measuring small specific-heat capacities, such as those of gases, in which the amount of steam condensing on a body containing the substance whose heat capacity is to be measured is compared with the amount condensing on a similar body which is evacuated or contains a substance of known heat capacity.

differential topology [MATH] The branch of mathematics dealing with differentiable manifolds.

differential transducer [ELEC] A transducer that simultaneously senses two separate sources and provides an output proportional to the difference between them.

differential transformer [ELEC] A transformer used to join two or more sources of signals to a common transmission line.

differential voltmeter [ELEC] A voltmeter that measures only the difference between a known voltage and an unknown voltage.

differentiating circuit [ELEC] A circuit whose output voltage is proportional to the rate of change of the input voltage. Also known as differentiating network.

differentiating network *See* differentiating circuit.

differentiation [MATH] The act of taking a derivative.

differentiator [ELECTR] A device whose output function is proportional to the derivative, or rate of change, of the input function with respect to one or more variables.

diffracted wave [PHYS] A wave whose front has been changed in direction by an obstacle or other nonhomogeneity in a medium, other than by reflection or refraction.

diffraction [PHYS] Any redistribution in space of the intensity of waves that results from the presence of an object causing variations of either the amplitude or phase of the waves; found in all types of wave phenomena.

diffractional pulse-height discriminator *See* pulse-height selector.

diffraction analysis [PHYS] The study of the atomic structure of solids, liquids, or gases by means of diffraction of x-rays or particles, such as neutrons or electrons.

diffraction grating [SPECT] An optical device which consists of an assembly of narrow slits or grooves which produce a large number of beams that can interfere to produce spectra. Also known as grating.

diffraction instrument *See* diffractometer.

diffraction mottle [GRAPHICS] A mottled appearance of a radiograph caused by superposition of diffraction effects.

diffraction pattern [PHYS] Pattern produced on a screen or plate by waves which have undergone diffraction.

diffraction propagation [ELECTROMAG] Propagation of electromagnetic waves around objects or over the horizon by diffraction.

diffraction ring [OPTICS] Circular light pattern which appears to surround particles in a microscope field.

diffraction scattering [PHYS] Elastic scattering that occurs when inelastic processes remove particles from a beam.

diffraction spectrum [SPECT] Parallel light and dark or colored bands of light produced by diffraction.

diffraction symmetry [CRYSTAL] Any symmetry in a crystal lattice which causes the systematic annihilation of certain beams in x-ray diffraction.

diffraction velocimeter [OPTICS] A velocity-measuring instrument that uses a continuous-wave laser to send a beam of coherent light at objects moving at right angles to the beam; the needlelike diffraction lobes reflected by the moving objects sweep past the optical grating in the receiver, thereby generating in a photomultiplier a series of impulses from which velocity can be determined and read out. Also known as laser velocimeter; optical diffraction velocimeter.

diffractometer [PHYS] An instrument used to study the structure of matter by means of the diffraction of x-rays, electrons, neutrons, or other waves. Also known as diffraction instrument.

diffractometry [CRYSTAL] The science of determining crystal structures by studying the diffraction of beams of x-rays or other waves.

diffuse-cutting filter [OPTICS] A color filter that gradually changes in absorption with wavelength.

diffused-alloy transistor [ELECTR] A transistor in which the semiconductor wafer is subjected to gaseous diffusion to produce a nonuniform base region, after which alloy junctions are formed in the same manner as for an alloy-junction transistor; it may also have an intrinsic region, to give a *pnip* unit. Also known as drift transistor.

diffused-base transistor [ELECTR] A transistor in which a nonuniform base region is produced by gaseous diffusion; the collector-base junction is also formed by gaseous diffusion, while the emitter-base junction is a conventional alloy junction.

diffused emitter-collector transistor [ELECTR] A transistor in which both the emitter and collector are produced by diffusion.

diffused junction [ELECTR] A semiconductor junction that has been formed by the diffusion of an impurity within a semiconductor crystal.

diffused-junction rectifier [ELECTR] A semiconductor diode in which the *pn* junction is produced by diffusion.

diffused-junction transistor [ELECTR] A transistor in which the emitter and collector electrodes have been formed by diffusion by an impurity metal into the semiconductor wafer without heating.

diffused resistor [ELECTR] An integrated-circuit resistor produced by a diffusion process in a semiconductor substrate.

diffuse nebula [ASTRON] A type of nebula ranging from huge masses presenting relatively high surface brightness down to faint, milky structures that are detectable only with long exposures and special filters; may contain both dust and gas or may be purely gaseous.

diffuse radiation [PHYS] Radiant energy propagating in many different directions through a given small volume of space.

diffuse reflection [PHYS] Reflection of light, sound, or radio waves from a surface in all directions according to the cosine law.

diffuse reflection model [PHYS] A model for the behavior of gas molecules striking the surface of a solid body, in which the molecules are absorbed and reemitted with a Maxwellian velocity distribution corresponding to a temperature intermediate between that of the surface and that of the incoming flow of gas.

diffuse reflector [OPTICS] Any surface whose irregularities are so large compared to the wavelength of the incident radiation that the reflected rays are sent back in a multiplicity of directions.

diffuse series [SPECT] A series occurring in the spectra of many atoms having one, two, or three electrons in the outer shell, in which the total orbital angular momentum quantum number changes from 2 to 1.

diffuse skylight *See* diffuse sky radiation.

diffuse sky radiation [ASTROPHYS] Solar radiation reaching the earth's surface after having been scattered from the direct solar beam by molecules or suspensoids in the atmosphere. Also known as diffuse skylight; skylight; sky radiation.

diffuse sound [ACOUS] Sound that has uniform energy density in a given region so that all directions of energy flux at all parts of the region are equally probable.

DIFFUSE REFLECTION

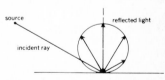

Diffuse reflection, such as from a mat surface of microscopic roughness. *(From W. B. Boast, Illumination Engineering, 2d ed., McGraw-Hill, 1953)*

diffuse transmission [PHYS] Transmission of electromagnetic or acoustic radiation in all directions by a transmitting body.

diffuse transmission density [OPTICS] The value of the photographic transmission density obtained when light flux impinges normally on the sample and all the transmitted flux is collected and measured.

diffusing disk *See* diffusion disk.

diffusing power [OPTICS] The quantity $(L_{20} + L_{70})/2L_5$, where L_5, L_{20}, and L_{70} are the luminances of the medium in question when illuminated by a narrow light beam perpendicular to the surface and observed at angles of 5, 20, and 70°, respectively, from the direction of incidence.

diffusion [ACOUS] The degree of variation in the propagation directions of sound waves over the volume of a sound field. [ELECTR] A method of producing a junction by diffusing an impurity metal into a semiconductor at a high temperature. [OPTICS] **1.** The distribution of incident light by reflection. **2.** Transmission of light through a translucent material. [PHYS] The spontaneous movement and scattering of particles (atoms and molecules), of liquids, gases, and solids. [SOLID STATE] **1.** The actual transport of mass, in the form of discrete atoms, through the lattice of a crystalline solid. **2.** The movement of carriers in a semiconductor.

diffusion area [PHYS] One-sixth of the mean-square displacement between the appearance and disappearance of a subatomic particle of a given type.

diffusion capacitance [ELECTR] The rate of change of stored minority-carrier charge with the voltage across a semiconductor junction.

diffusion cloud chamber [NUCLEO] A cloud chamber in which vapor diffuses from a source near a hot plate and condenses on a cold plate; the resulting layer of supersaturated vapor between the plates is sensitive to the passage of ionizing particles.

diffusion coefficient [PHYS] The weight of a material, in grams, diffusing across an area of 1 square centimeter in 1 second in a unit concentration gradient. Also known as diffusivity.

diffusion constant [SOLID STATE] The diffusion current density in a homogeneous semiconductor divided by the charge carrier concentration gradient.

diffusion diameter [STAT MECH] For a gas, the diameter of identical hard spheres that display the same diffusion as that observed for the molecules of the actual gas when their motion is treated classically.

diffusion disk [OPTICS] A piece of transparent material that is marked or embossed, and is used with a camera lens to give the image a hazy softened quality. Also known as diffusing disk.

diffusion equation [PHYS] **1.** An equation for diffusion which states that the rate of change of the density of the diffusing substance, at a fixed point in space, equals the sum of the diffusion coefficient times the Laplacian of the density, the amount of the quantity generated per unit volume per unit time, and the negative of the quantity absorbed per unit volume per unit time. **2.** More generally, any equation which states that the rate of change of some quantity, at a fixed point in space, equals a positive constant times the Laplacian of that quantity.

diffusion gradient [PHYS] The graphed distance of penetration (diffusion) versus concentration of the material (or effect) diffusing through a second material; applies to heat, liquids, solids, or gases.

diffusion kernel [NUCLEO] The neutron flux resulting from a point source emitting one neutron per second; it is a function of the distance between the source and the point where the flux is measured.

diffusion length [PHYS] The average distance traveled by a particle, such as a minority carrier in a semiconductor or a thermal neutron in a nuclear reactor, from the point at which it is formed to the point at which it is absorbed.

diffusion number [FL MECH] A dimensionless number used in the study of mass transfer, equal to the diffusivity of a solute through a stationary solution contained in the solid, times a characteristic time, divided by the square of the distance from the midpoint of the solid to the surface. Symbolized β.

diffusion plant [NUCLEO] A plant which separates isotopes by isotopic diffusion or thermal diffusion.

diffusion potential [PHYS CHEM] A potential difference across the boundary between electrolytic solutions with different compositions. Also known as liquid junction potential.

diffusion pump [ENG] A vacuum pump in which a stream of heavy molecules, such as mercury vapor, carries gas molecules out of the volume being evacuated; also used for separating isotopes according to weight, the lighter molecules being pumped preferentially by the vapor stream.

diffusion theory [ELEC] The theory that in semiconductors, where there is a variation of carrier concentration, a motion of the carriers is produced by diffusion in addition to the drift determined by the mobility and the electric field. [PHYS] The derivation of rates of diffusion of particles in a substance from the diffusion equation.

diffusion transistor [ELECTR] A transistor in which current flow is a result of diffusion of carriers, donors, or acceptors, as in a junction transistor.

diffusion velocity [FL MECH] **1.** The relative mean molecular velocity of a selected gas undergoing diffusion in a gaseous atmosphere, commonly taken as a nitrogen (N_2) atmosphere; a molecular phenomenon that depends upon the gaseous concentration as well as upon the pressure and temperature gradients present. **2.** The velocity or speed with which a turbulent diffusion process proceeds as evidenced by the motion of individual eddies.

diffusivity [PHYS] *See* diffusion coefficient. [THERMO] The quantity of heat passing normally through a unit area per unit time divided by the product of specific heat, density, and temperature gradient. Also known as thermometric conductivity.

digit [MATH] A character used to represent one of the nonnegative integers smaller than the base of a system of positional notation. Also known as numeric character.

digital computer [ADP] A computer operating on discrete data by performing arithmetic and logic processes on these data.

digital converter [ELECTR] A device that converts voltages to digital form; examples include analog-to-digital converters, pulse-code modulators, encoders, and quantizing encoders.

digital filter [ELECTR] An electrical filter that responds to an input which has been quantified, usually as pulses.

digital-to-analog converter [ELECTR] A converter in which digital input signals are changed to essentially proportional analog signals. Abbreviated dac.

digital transducer [ELECTR] A transducer that measures physical quantities and transmits the information as coded digital signals rather than as continuously varying currents or voltages.

digital voltmeter [ELECTR] A voltmeter in which the unknown voltage is compared with an internally generated

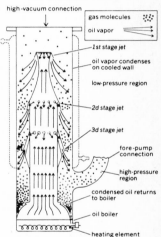

DIFFUSION PUMP

Main operating features of diffusion pump.

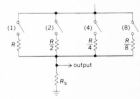

DIGITAL-TO-ANALOG CONVERTER

Circuit diagram for simple digital-to-analog converter. Each of the four switches represents binary 1 when closed and 0 when open. Current flowing through summing resistor R_s and voltage at output are proportional to value of number represented.

analog voltage, the result being indicated in digital form rather than by a pointer moving over a meter scale.

digit place *See* digit position.

digit position [MATH] The position of a particular digit in a number that is expressed in positional notation, usually numbered from the lowest significant digit of the number. Also known as digit place.

dihedral *See* dihedron.

dihedral angle [MATH] The angle between two planes; it is said to be zero if the planes are parallel; if the planes intersect, it is the plane angle between two lines, one in each of the planes, which pass through a point on the line of intersection of the two planes and are perpendicular to it.

dihedral reflector [OPTICS] A corner reflector having two sides meeting at a line.

dihedron [MATH] A geometric figure formed by two half planes that are bounded by the same straight line. Also known as dihedral.

dihexagonal [CRYSTAL] Of crystals, having a symmetrical form with 12 sides.

dihexagonal-dipyramidal [CRYSTAL] Characterized by the class of crystals in the hexagonal system in which any section perpendicular to the sixfold axis is dihexagonal.

dihexahedron [CRYSTAL] A type of crystal that has 12 faces, such as a double six-sided pyramid.

dilatation [PHYS] The increase in volume per unit volume of any continuous substance, caused by deformation.

dilatational wave *See* compressional wave.

dilation [MATH] A transformation which changes the size, and only the size, of a geometric figure.

dilatometer [ENG] An instrument for measuring thermal expansion and dilation of liquids or solids.

dilatometry [PHYS] The measurement of changes in the volume of a liquid or dimensions of a solid which occur in phenomena such as allotropic transformations, thermal expansion, compression, creep, or magnetostriction.

DIM *See* nonthermal decimetric emission.

dimensional analysis [PHYS] A technique that involves the study of dimensions of physical quantities, used primarily as a tool for obtaining information about physical systems too complicated for full mathematical solutions to be feasible.

dimensional constant [PHYS] A physical quantity whose numerical value depends on the units chosen for fundamental quantities but not on the system being considered.

dimensional formula [PHYS] The expression of a derived quantity as a product of powers of the fundamental quantities.

dimensional regularization [QUANT MECH] A method of extracting a finite piece from an infinite result in quantum field theory based on analytically continuing a typically divergent integral in its number of space-time dimensions.

dimensionless group [PHYS] Any combination of dimensional or dimensionless quantities possessing zero overall dimensions; an example is the Reynold's number.

dimensionless number [MATH] A ratio of various physical properties (such as density or heat capacity) and conditions (such as flow rate or weight) of such nature that the resulting number has no defining units of weight, rate, and so on. Also known as nondimensional parameter.

dimension of a simplex [MATH] One less than the number of vertices of the simplex.

dimension of a vector space [MATH] The number of vectors in any basis of the vector space.

dimensions [PHYS] The product of powers of fundamental quantities (or of convenient derived quantities) which are

---

DIHEDRON

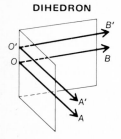

Dihedron is formed by the half-planes bounded by the line passing through O and O'. AOB and $A'O'B'$ are dihedral angles.

used to define a physical quantity; the fundamental quantities are often mass, length, and time.

dimension theory [MATH] The study of abstract notions of dimension, which are topological invariants of a space.

dimuon event [PARTIC PHYS] An inelastic collision of a neutrino or antineutrino with a nucleus in which there are two muons among the products of the collision.

dineric [PHYS CHEM] 1. Having two liquid phases. 2. Pertaining to the interface between two liquids.

dineutron [NUC PHYS] 1. A hypothetical bound state of two neutrons, which probably does not exist. 2. A combination of two neutrons which has a transitory existence in certain nuclear reactions.

Dini derivate *See* Dini derivative.

Dini derivative [MATH] One of four quantities associated with any real-valued function of a real variable $f(x)$, at a point x_0; it is either the limit superior or the limit inferior of $[f(x) - f(x_0)]/(x - x_0)$ as x approaches x_0, where x is restricted either to values greater than x_0 or to values less than x_0. Also known as Dini derivate.

Dini series [MATH] For a function $f(x)$, defined and integrable over the integral $(0,1)$, this is the series

$$a_0(x) + \sum_{n=1}^{\infty} a_n J_\nu(\lambda_n x)$$

where J_ν is a Bessel function, λ_n are the positive roots of the equation $xJ'_\nu(x) + hJ_\nu(x) = 0$, h and ν are real constants, and the coefficients a_n are defined in such a way that the series converges to $f(x)$ provided $f(x)$ has suitable properties.

dioctahedral [CRYSTAL] Pertaining to a crystal structure in which only two of the three available octahedrally coordinated positions are occupied. [MATH] Having 16 faces.

diode [ELECTR] 1. A two-electrode electron tube containing an anode and a cathode. 2. *See* semiconductor diode.

diode amplifier [ELECTR] A microwave amplifier using an IMPATT, TRAPATT, or transferred-electron diode in a cavity, with a microwave circulator providing the input/output isolation required for amplification; center frequencies are in the gigahertz range, from about 1 to 100 gigahertz, and power outputs are up to 20 watts continuous-wave or more than 200 watts pulsed, depending on the diode used.

diode-capacitor transistor logic [ELECTR] A circuit that uses diodes, capacitors, and transistors to provide logic functions.

diode characteristic [ELECTR] The composite electrode characteristic of an electron tube when all electrodes except the cathode are connected together.

diode-connected transistor [ELECTR] A bipolar transistor in which two terminals are shorted to give diode action.

diode function generator [ELECTR] A function generator that uses the transfer characteristics of resistive networks containing biased diodes; the desired function is approximated by linear segments.

diode laser *See* semiconductor laser.

diode theory [ELEC] The theory that in a semiconductor, when the barrier thickness is comparable to or smaller than the mean free path of the carriers, then the carriers cross the barrier without being scattered, much as in a vacuum tube diode.

diode transistor logic [ELECTR] A circuit that uses diodes, transistors, and resistors to provide logic functions. Abbreviated DTL.

diode voltage regulator [ELECTR] A voltage regulator with a Zener diode, making use of its almost constant voltage over a range of currents. Also known as Zener diode voltage regulator.

DIODE VOLTAGE REGULATOR

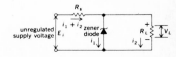

A Zener diode voltage regulator.

diophantine analysis [MATH] A means of determining integer solutions for certain algebraic equations.

diophantine equations [MATH] Equations with more than one independent variable and with integer coefficients for which integer solutions are desired.

diopter [OPTICS] A measure of the power of a lens or a prism, equal to the reciprocal of its focal length in meters. Abbreviated D.

dioptometer [OPTICS] An instrument for determining ocular refraction.

dioptric [OPTICS] **1.** Serving in or effecting refraction. **2.** Produced by means of refraction.

dioptrics [OPTICS] The branch of optics that treats of the refraction of light, especially by the transparent medium of the eye, and by lenses.

dip inductor *See* earth inductor.

diploid [CRYSTAL] A crystal form in the isometric system having 24 similar quadrilateral faces arranged in pairs.

dipolar gas [PHYS CHEM] A gas whose molecules have a permanent electric dipole moment.

dipole [ELECTROMAG] Any object or system that is oppositely charged at two points, or poles, such as a magnet or a polar molecule; more precisely, the limit as either charge goes to infinity, the separation distance to zero, while the product remains constant. Also known as doublet; electric doublet.

dipole antenna [ELECTROMAG] An antenna approximately one-half wavelength long, split at its electrical center for connection to a transmission line whose radiation pattern has a maximum at right angles to the antenna. Also known as doublet antenna; half-wave dipole.

dipole-dipole force *See* orientation force.

dipole disk feed [ELECTROMAG] Antenna, consisting of a dipole near a disk, used to reflect energy to the disk.

dipole moment *See* electric dipole moment; magnetic dipole moment.

dipole polarization *See* orientation polarization.

dipole radiation [ELECTROMAG] The electromagnetic radiation generated by an oscillating electric or magnetic dipole.

dipole relaxation [ELEC] The process, occupying a certain period of time after a change in the applied electric field, in which the orientation polarization of a substance reaches equilibrium.

dipole sound field [ACOUS] A sound field generated by an oscillating dipole source.

dipole transition [ATOM PHYS] A transition of an atom or nucleus from one energy state to another in which dipole radiation is emitted or absorbed.

dipping refractometer *See* immersion refractometer.

dip pole *See* magnetic pole.

diproton [NUC PHYS] A hypothetical bound state of two protons, which probably does not exist.

dipyramid [CRYSTAL] A crystal having the form of two pyramids that melt at a plane of symmetry.

Dirac covariants [QUANT MECH] Quantities which behave as a scalar, a pseudoscalar, a vector, an axial vector, or a second-rank tensor under Lorentz transformations, and whose elements consist of basis elements of the Dirac gamma algebra multiplied by the Dirac wave function on the right and its adjoint on the left.

Dirac delta function *See* delta function.

Dirac electron theory *See* Dirac theory.

Dirac equation [QUANT MECH] A relativistic wave equation for an electron in an electromagnetic field, in which the wave function has four components corresponding to four internal states specified by a two-valued spin coordinate and an

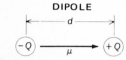

DIPOLE

Electric dipole with moment $\mu = Qd$.

energy coordinate which can have a positive or negative value.

Dirac fields [QUANT MECH] Operators, arising in the second quantization of the Dirac theory, which correspond to the Dirac wave functions in the original theory.

Dirac gamma algebra [QUANT MECH] An algebra whose basis consists of 16 linearly independent 4×4 matrices constructed from products of the four basic Dirac matrices.

Dirac h *See* h-bar.

Dirac matrix [QUANT MECH] Any one of four matrices, designated γ_μ (μ = 1, 2, 3, 4), each having four rows and four columns and satisfying $\gamma_\mu \gamma_\nu + \gamma_\nu \gamma_\mu = \delta_{\mu\nu}$, where $\delta_{\mu\nu}$ is the Kronecker delta function, which matrices operate on the four-component wave function in the Dirac equation. Also known as gamma matrix.

Dirac moment [QUANT MECH] Magnetic moment of the electron according to the Dirac theory, equal to $e\hbar/2mc$, where e and m are the charge and mass of the positron respectively, $\hbar$ is Planck's constant divided by 2π, and c is the speed of light.

Dirac monopole [QUANT MECH] A magnetic monopole whose magnetic charge is an integral multiple of $\hbar c/2e$, where $\hbar$ is Planck's constant divided by 2π, c is the speed of light, and e is the charge of the electron.

Dirac particle [PARTIC PHYS] A particle behaving according to the Dirac theory, which describes the behavior of electrons and muons except for radiative corrections, and is envisaged as describing a central core of a hadron of spin ½$\hbar$ which remains when the effects of nuclear forces are removed.

Dirac quantization [QUANT MECH] The condition, arising from conservation of angular momentum, that for any electric charge q and magnetic monopole with magnetic charge m, one has $2qm = n\hbar c$, where n is an integer, $\hbar$ is Planck's constant divided by 2π, and c is the speed of light (Gaussian units).

Dirac spinor *See* spinor.

Dirac theory [QUANT MECH] Theory of the electron based on the Dirac equation, which accounts for its spin angular momentum and gives its magnetic moment and its behavior in an electromagnetic field (except for higher-order corrections). Also known as Dirac electron theory.

Dirac wave function [QUANT MECH] A function appropriate for describing a spin ½ particle and antiparticle; it is a column matrix with four entries, each of which is a function of the space and time coordinates; the four-components form two first-rank Lorentz spinors.

direct-aperture antenna [ELECTROMAG] An antenna whose conductor or dielectric is a surface or solid, such as a horn, mirror, or lens.

direct-coupled amplifier [ELECTR] A direct-current amplifier in which a resistor or a direct connection provides the coupling between stages, so small changes in direct currents can be amplified.

direct-coupled transistor logic [ELECTR] Integrated-circuit logic using only resistors and transistors, with direct conductive coupling between the transistors; speed can be up to 1 megahertz. Abbreviated DCTL.

direct coupling [ELEC] Coupling of two circuits by means of a non-frequency-sensitive device, such as a wire, resistor, or battery, so both direct and alternating current can flow through the coupling path.

direct current [ELEC] Electric current which flows in one direction only, as opposed to alternating current. Abbreviated dc.

direct-current amplifier [ELECTR] An amplifier that is capable of amplifying dc voltages and slowly varying voltages.

DIRECT-APERTURE ANTENNA

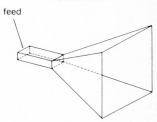

feed

Pyramidal horn, a type of direct-aperture antenna.

DIRECT-CURRENT MOTOR

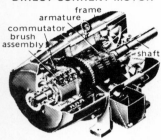

frame
armature
commutator
brush
assembly
shaft

Cutaway view of typical
direct-current motor.
(General Electric)

direct-current circuit [ELEC] Any combination of dc voltage or current sources, such as generators and batteries, in conjunction with transmission lines, resistors, and power converters such as motors.

direct-current circuit theory [ELEC] An analysis of relationships within a dc circuit.

direct-current coupling [ELECTR] That type of coupling in which the zero-frequency term of the Fourier series representing the input signal is transmitted.

direct-current generator [ELEC] A rotating electric machine that converts mechanical power into dc power.

direct-current motor [ELEC] An electric rotating machine energized by direct current and used to convert electric energy to mechanical energy.

direct-current offset [ELECTR] A direct-current level that may be added to the input signal of an amplifier or other circuit.

direct-current plate resistance [ELECTR] Value or characteristic used in vacuum-tube computations; it is equal to the direct-current plate voltage divided by the direct-current plate current.

direct-current reinsertion *See* clamping.

direct-current restoration *See* clamping.

direct-current transducer [ELECTR] A transducer that requires dc excitation and provides a dc output that varies with the parameter being sensed.

direct-current vacuum-tube voltmeter [ELECTR] The amplifying and indicating portions of the diode rectifier-amplifier meter, which are usually designed so that the diode rectifier can be disconnected for dc measurements.

direct-current voltage *See* direct voltage.

direct-cycle reactor [NUCLEO] A nuclear power plant in which the heat-transfer fluid circulates through the reactor and then passes directly to the turbine in a continuous cycle.

directed angle [MATH] An angle for which one side is designated as initial, the other as terminal.

directed graph [MATH] A graph in which a direction is shown for every arc. Also known as digraph.

directed line [MATH] A line on which a positive direction has been specified.

directed number [MATH] A number together with a sign.

directed set [MATH] A partially ordered set with the property that for every pair of elements a,b in the set, there is a third element which is larger than both a and b.

direct electromotive force [ELEC] Unidirectional electromotive force in which the changes in values are either zero or so small that they may be neglected.

direct feedback system [CONT SYS] A system in which electrical feedback is used directly, as in a tachometer.

direct-gap semiconductor [SOLID STATE] A semiconductor in which the minimum of the conduction band occurs at the same wave vector as the maximum of the valence band, and recombination radiation consequently occurs with relatively large intensity.

direct grid bias *See* grid bias.

direct interelectrode capacitance *See* interelectrode capacitance.

directional antenna [ELECTROMAG] An antenna that radiates or receives radio waves more effectively in some directions than others.

directional counter [NUCLEO] A counter that is more sensitive to nuclear radiation from some directions than from others.

directional derivative [MATH] The rate of change of a function in a given direction; more precisely, if f maps an n-

dimensional euclidean space into the real numbers, and $\mathbf{x} = (x_1, \ldots, x_n)$ is a vector in this space, and $\mathbf{u} = (u_1, \ldots, u_n)$ is a unit vector in the space (that is, $u_1{}^2 + \cdots + u_n{}^2 = 1$), then the directional derivative of f at $\mathbf{x}$ in the direction of $\mathbf{u}$ is the limit as h approaches zero of $[f(\mathbf{x} + h\mathbf{u}) - f(\mathbf{x})]/h$.

directional gain   *See* directivity index.

directional gyro   [MECH] A two-degrees-of-freedom gyro with a provision for maintaining its spin axis approximately horizontal.

directional pattern   *See* radiation pattern.

directional response pattern   *See* directivity pattern.

direction angles   [MATH] The three angles which a line in space makes with the positive x, y, and z axes.

direction cosine   [MATH] The cosine of one of the direction angles of a line in space.

direct ionization   *See* extrinsic photoemission.

direction numbers   [MATH] Any three numbers proportional to the direction cosines of a line in space. Also known as direction ratios.

direction of propagation   [PHYS] **1.** The normal to a surface of constant phase, in a propagating wave. **2.** The direction of the group velocity. **3.** The direction of time-average energy flow. (In a homogeneous isotropic medium, these three directions coincide.)

direction ratios   *See* direction numbers.

directive gain   [ELECTROMAG] Of an antenna in a given direction, 4π times the ratio of the radiation intensity in that direction to the total power radiated by the antenna.

directivity   [ELECTROMAG] **1.** The value of the directive gain of an antenna in the direction of its maximum value. **2.** The ratio of the power measured at the forward-wave sampling terminals of a directional coupler, with only a forward wave present in the transmission line, to the power measured at the same terminals when the direction of the forward wave in the line is reversed; the ratio is usually expressed in decibels.

directivity factor   [ENG ACOUS] **1.** The ratio of radiated sound intensity at a remote point on the principal axis of a loudspeaker or other transducer, to the average intensity of the sound transmitted through a sphere passing through the remote point and concentric with the transducer; the frequency must be stated. **2.** The ratio of the square of the voltage produced by sound waves arriving parallel to the principal axis of a microphone or other receiving transducer, to the mean square of the voltage that would be produced if sound waves having the same frequency and mean-square pressure were arriving simultaneously from all directions with random phase; the frequency must be stated.

directivity index   [ENG ACOUS] The directivity factor expressed in decibels; it is 10 times the logarithm to the base 10 of the directivity factor. Also known as directional gain.

directivity pattern   [ENG ACOUS] A graphical or other description of the response of a transducer used for sound emission or reception as a function of the direction of the transmitted or incident sound waves in a specified plane and at a specified frequency. Also known as beam pattern; directional response pattern.

directly heated cathode   *See* filament.

direct nuclear reaction   [NUC PHYS] A nuclear reaction which is completed in the time required for the incident particle to transverse the target nucleus, so that it does not combine with the nucleus as a whole but interacts only with the surface or with some individual constituent.

director   [ELECTR] Telephone switch which translates the digits dialed into the directing digits actually used to switch the call.   [ELECTROMAG] A parasitic element placed a fraction of a wavelength ahead of a dipole receiving antenna to increase the gain of the array in the direction of the major lobe.

director circle [MATH] A circle consisting of the points of intersection of pairs of perpendicular tangents to an ellipse or hyperbola.

direct product of a finite family of sets [MATH] Given sets $A_1, \ldots, A_n$, the direct product is the set of all n-tuples $(a_1, \ldots, a_n)$, where a_i belongs to A_i for $i = 1, \ldots, n$.

direct proportion [MATH] A statement that the ratio of two variable quantities is equal to a constant.

direct reflection *See* specular reflection.

directrix [MATH] A fixed line used in one method of defining a conic; the distance from this line divided by the distance from a fixed point (called the focus) is the same for all points on the conic.

direct solar radiation [ASTROPHYS] That portion of the radiant energy received at the actinometer direct from the sun, as distinguished from diffuse sky radiation, effective terrestrial radiation, or radiation from any other source.

direct sum [MATH] If each of the sets in a finite direct product of sets has a group structure, this structure may be imposed on the direct product by defining the composition "componentwise"; the resulting group is called the direct sum.

direct tide [GEOPHYS] A gravitational solar or lunar tide in the ocean or atmosphere which is in phase with the apparent motions of the attracting body, and consequently has its local maxima directly under the tide-producing body, and on the opposite side of the earth.

direct viewfinder [OPTICS] A viewfinder in which the user views the subject directly through a lens or sight.

direct-vision nephoscope [OPTICS] A type of nephoscope in which the cloud motion is observed by looking directly into the instrument.

direct voltage [ELEC] A voltage that forces electrons to move through a circuit in the same direction continuously, thereby producing a direct current. Also known as direct-current voltage.

direct-writing galvanometer [ENG] A direct-writing recorder in which the stylus or pen is attached to a moving coil positioned in the field of the permanent magnet of a galvanometer.

direct-writing recorder [ENG] A recorder in which the permanent record of varying electrical quantities or signals is made on paper, directly by a pen attached to the moving coil of a galvanometer or indirectly by a pen moved by some form of motor under control of the galvanometer. Also known as mechanical oscillograph.

Dirichlet conditions [MATH] The requirement that a function be bounded, and have finitely many maxima, minima, and discontinuities on the closed interval $[-\pi, \pi]$.

Dirichlet drawer principle *See* pigeonhole principle.

Dirichlet integral [MATH] 1. One of the partial sums in the Fourier series of the function $f(x)$; it may be written

$$s_n(x) = \frac{1}{\pi} \int_0^{\pi/2} \left[f(x + 2t) + f(x - 2t) \right] \frac{\sin (2n + 1) t}{\sin t} \, dt.$$

2. The integral specified in Dirichlet's integral formula. 3. The integral specified in the Dirichlet principle.

Dirichlet principle [MATH] The principle that the solution $f(x, y)$ of Laplace's equation in two variables, satisfying given boundary conditions, is given by the function, among those functions satisfying the boundary conditions, which minimizes the integral

$$\int \int \left[(\partial f / \partial x)^2 + (\partial f / \partial y)^2 \right] dx \, dy.$$

Dirichlet problem [MATH] To determine a solution to Laplace's equation which satisfies certain conditions in a region and on its boundary.

Dirichlet series [MATH] A series whose nth term is a complex number divided by n to the zth power.

Dirichlet's formula [MATH] The formula

$$\int_a^b dy \int_a^y f(x,y) \, dx = \int_a^b dx \int_x^b f(x,y) \, dy$$

for a change of variable in a double integral whose range of integration is the isosceles triangle bounded by the lines $x = a$, $y = b$, and $x = y$.

Dirichlet's integral formula [MATH] **1.** The formula

$$\int \int \cdots \int f(x_1 + x_2 + \cdots + x_n) x_1^{p_1 - 1} x_2^{p_2 - 1} \cdots x_n^{p_n - 1}$$

$$\times \, dx_1 \, dx_2 \ldots dx_n$$

$$= \frac{\Gamma(p_1)\Gamma(p_2) \ldots \Gamma(p_n)}{\Gamma(p_1 + p_2 + \cdots + p_n)} \int_0^1 f(u) u^{p_1 + p_2 + \cdots p_n - 1} \, du$$

where $p_i < 0$ ($i = 1, 2, \ldots, n$) and the integration on the left-hand side of the equation extends over nonnegative values of $x_1, x_2, \ldots, x_n$ satisfying $0 \le x_1 + x_2 + \cdots + x_n \le 1$. **2.** The formula

$$\lim_{\omega \to \infty} \frac{1}{\pi} \int_{-\infty}^{\infty} f(y) \frac{\sin \omega(x - y)}{(x - y)} \, dy = \frac{1}{2}[f(x + 0) + f(x - 0)]$$

where $f(x + 0)$ and $f(x - 0)$ represent the limits on the right and left of $f(x)$.

Dirichlet test for convergence [MATH] If Σb_n is a series whose sequence of partial sums is bounded, and if $\{a_n\}$ is a monotone decreasing null sequence, then the series

$$\sum_{n=1}^{\infty} a_n b_n \text{ converges.}$$

Dirichlet transform [MATH] For a function $f(x)$, this is the integral of $f(x) \cdot \sin (kx)/x$; its convergence determines the convergence of the Fourier series of $f(x)$.

disability glare *See* glare.

disappearing filament pyrometer *See* optical pyrometer.

disc *See* disk.

discharge coefficient [FL MECH] In a nozzle or other constriction, the ratio of the mass flow rate at the discharge end of the nozzle to that of an ideal nozzle which expands an identical working fluid from the same initial conditions to the same exit pressure. Also known as coefficient of discharge.

discharge lamp [ELECTR] A lamp in which light is produced by an electric discharge between electrodes in a gas (or vapor) at low or high pressure. Also known as electric-discharge lamp; gas-discharge lamp; vapor lamp.

discharge tube [ELECTR] An evacuated enclosure containing a gas at low pressure, through which current can flow when sufficient voltage is applied between metal electrodes in the tube. Also known as electric-discharge tube.

discomfort glare *See* glare.

discomposition [NUCLEO] The process in which an atom is knocked out of its position in a crystal lattice by direct nuclear impact, as by fast neutrons or by fast ions that have been previously knocked out of their lattice positions.

discomposition effect [NUCLEO] Changes in physical or chemical properties of a substance caused by discomposition. Also known as Wigner effect.

discone antenna [ELECTROMAG] A biconical antenna in which one of the cones is spread out to 180° to form a disk;

DISCONE ANTENNA

A high-frequency discone antenna.

the center conductor of the coaxial line terminates at the center of the disk, and the cable shield terminates at the vertex of the cone.

discontinuity [ELECTROMAG] An abrupt change in the shape of a waveguide. [GEOPHYS] A boundary at which the velocity of seismic waves changes abruptly. [MATH] A point at which a function is not continuous. [PHYS] A break in the continuity of a medium or material at which a reflection of wave energy can occur.

discontinuous group of conformal mappings [MATH] **1.** A group G of conformal mappings of a Riemann surface onto itself is said to be discontinuous at a point x if G_x, the stabilizer of x, is finite and if there is a neighborhood U of x such that $g(U) = U$ for all g in G_x, and the intersection of $g(U)$ with U is empty for all g in $G - G_x$. **2.** The group G is said to be discontinuous if it is discontinuous at one or more points.

discord *See* dissonance.

discrete [SCI TECH] **1.** Composed of separate and distinct parts. **2.** Having an individually distinct identity.

discrete group [MATH] A totally ordered Abelian group G of finite rank n such that if $G_0 = 0, G_1, \ldots, G_{n-1}$ are its isolated subgroups, with $G_0 \subset G_1 \subset \cdots \subset G_{n-1} \subset G$, then the quotient groups G_{i+1}/G_i, $i = 0, 1, \ldots, n-1$ $(G_n = G)$, are each isomorphic to the group of integers.

discrete radio source [ASTROPHYS] A source of radio waves coming from a small area of the sky.

discrete set [MATH] A set with no cluster points.

discrete spectrum [SPECT] A spectrum in which the component wavelengths constitute a discrete sequence of values rather than a continuum of values.

discrete transfer function *See* pulsed transfer function.

discrete valuation [MATH] A mapping v of the nonzero elements of a field K into an additive, Abelian, totally ordered group such that $v(xy) = v(x) + v(y)$ and $v(x + y) \geq \min [v(x), v(y)]$.

discrete variable [MATH] A variable for which the possible values form a discrete set.

discriminant [MATH] The quantity $b^2 - 4ac$ where a, b, c are the coefficients of a given quadratic polynomial: $ax^2 + bx + c$.

dish *See* parabolic reflector.

disinclination [CRYSTAL] A type of crystal imperfection in which one part of the crystal is rotated and therefore displaced relative to the rest of the crystal; observed in liquid crystals and protein coats of viruses.

disintegration [NUC PHYS] Any transformation of a nucleus, whether spontaneous or induced by irradiation, in which particles or photons are emitted.

disintegration chain *See* radioactive series.

disintegration constant *See* decay constant.

disintegration energy [NUC PHYS] The energy released, or the negative of the energy absorbed, during a nuclear or particle reaction. Designated Q. Also known as Q value; reaction energy.

disintegration family *See* radioactive series.

disintegration of measure [MATH] The representation of a measure as an integral of a family of positive measures.

disintegration rate [NUCLEO] **1.** The absolute rate of decay of a radioactive substance, usually expressed in terms of disintegrations per unit of time. **2.** The absolute rate of transformation of a nuclide under bombardment.

disintegration series *See* radioactive series.

disintegration voltage [ELECTR] The lowest anode voltage at which destructive positive-ion bombardment of the cathode occurs in a hot-cathode gas tube.

disjoint sets [MATH] Sets with no elements in common.

disk [MATH] The region in the plane consisting of all points with norm less than 1 (sometimes less than or equal to 1). Also spelled disc.

disk telescope [OPTICS] A telescope designed for observations of the brilliant solar disk; examples are the tower telescope and the horizontal fixed telescope.

disk thermistor [ELECTR] A thermistor which is produced by pressing and sintering an oxide binder mixture into a disk, 0.2–0.6 inch (5–15 millimeters) in diameter and 0.04–0.5 inch (1.0–13 millimeters) thick, coating the major surfaces with conducting material, and attaching leads.

dislocation [CRYSTAL] A defect occurring along certain lines in the crystal structure and present as a closed ring or a line anchored at its ends to other dislocations, grain boundaries, the surface, or other structural feature. Also known as line defect.

dislocation line [CRYSTAL] A curve running along the center of a dislocation.

dispenser cathode [ELECTR] An electron tube cathode having provisions for continuously replacing evaporated electron-emitting material.

dispersing prism [OPTICS] An optical prism which deviates light of different wavelengths by different amounts and can therefore be used to separate white light into its monochromatic parts.

dispersion [ELECTROMAG] Scattering of microwave radiation by an obstruction. [PHYS] **1.** The separation of a complex of electromagnetic or sound waves into its various frequency components. **2.** Quantitatively, the rate of change of refractive index with wavelength or frequency at a given wavelength or frequency. **3.** The rate of change of deviation with wavelength or frequency. **4.** In general, any process separating radiation into components having different frequencies, energies, velocities, or other characteristics, such as the sorting of electrons according to velocity in a magnetic field. [STAT] The degree of spread shown by observations in a sample or a population.

dispersion equation *See* dispersion formula.

dispersion force [PHYS CHEM] The force of attraction that exists between molecules that have no permanent dipole.

dispersion formula [PHYS] Any formula which gives the refractive index as a function of wavelength of electromagnetic radiation. Also known as dispersion equation.

dispersion index [STAT] Statistics used to determine the homogeneity of a set of samples.

dispersion medium [PHYS CHEM] Any fluid, either gas or liquid, in which colloidal particles are dispersed.

dispersion of a random variable [STAT] The spread of a random variable's distribution about its mean.

dispersion relation [NUC PHYS] A relation between the cross section for a given effect and the de Broglie wavelength of the incident particle, which is similar to a classical dispersion formula. [PHYS] An integral formula relating the real and imaginary parts of some function of frequency or energy, such as a refractive index or scattering amplitude, based on the causality principle and the Cauchy integral formula. [PL PHYS] A relation between the radian frequency and the wave vector of a wave motion or instability in a plasma.

dispersion strengthening [SOLID STATE] The reduction of plastic deformation of a solid by the presence of a uniform dispersion of another substance which inhibits the motion of plastic dislocations.

dispersive line [ELECTROMAG] A delay line that delays each frequency a different length of time.

DISPERSING PRISM

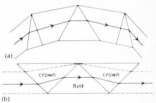

Two types of dispersing prisms. *(a)* Rayleigh prism system. *(b)* Amici direct-vision system consisting of flint-glass prism and two crown-glass prisms.

dispersive medium [ELECTROMAG] A medium in which the phase velocity of an electromagnetic wave is a function of frequency.

dispersive power [OPTICS] A measure of the power of a medium to separate different colors of light, equal to $(n_2 - n_1)/(n - 1)$, where n_1 and n_2 are the indices of refraction at two specified widely differing wavelengths, and n is the index of refraction for the average of these wavelengths, or for the D line of sodium.

disphenoid [CRYSTAL] **1.** A crystal form with four similar triangular faces combined in a wedge shape; can be tetragonal or orthorhombic. **2.** A crystal form with eight scalene triangles combined in pairs.

displacement [ELEC] *See* electric displacement. [FL MECH] **1.** The weight of fluid which is displaced by a floating body, equal to the weight of the body and its contents; the displacement of a ship is generally measured in long tons (1 long ton = 2240 pounds). **2.** The volume of fluid which is displaced by a floating body. [MECH] **1.** The linear distance from the initial to the final position of an object moved from one place to another, regardless of the length of path followed. **2.** The distance of an oscillating particle from its equilibrium position.

displacement angle [ELEC] The change in the phase of an alternator's terminal voltage when a load is applied.

displacement collision [NUCLEO] The collision of an energetic particle with an atom in a solid resulting in the atom being moved permanently from its original site.

displacement current [ELECTROMAG] The rate of change of the electric displacement vector, which must be added to the current density to extend Ampère's law to the case of time-varying fields (meter-kilogram-second units). Also known as Maxwell's displacement current.

displacement kernel [NUCLEO] In nuclear reactor theory, a function of two locations that depends only on the distance between the locations, such as the diffusion kernel or slowing-down kernel.

displacement law *See* radioactive displacement law; Wien's displacement law.

displacement manometer [ENG] A differential manometer which indicates the pressure difference across a solid or liquid partition which can be displaced against a restoring force.

displacement meter [ENG] A water meter that measures water flow quantitatively by recording the number of times a vessel of known capacity is filled and emptied.

displacement operator [MATH] A difference operator, denoted E, defined by the equation $Ef(x) = f(x + h)$, where h is a constant denoting the difference between successive points of interpolation or calculation. Also known as forward shift operator.

displacement spike [NUCLEO] A region in a solid in which atoms have been permanently moved from their original locations as the result of energetic particle bombardment.

displacement thickness [FL MECH] A measure of the thickness of a boundary-layer flow, given by the expression

$$\int_0^\infty (1 - u/V)\, dy$$

where u is the fluid velocity at a distance y from the surface of the body and V is the velocity outside the boundary layer. Symbolized δ^*.

display loss *See* visibility factor.

disruptive discharge [ELEC] A sudden and large increase in current through an insulating medium due to complete failure of the medium under electrostatic stress.

dissecting microscope [OPTICS] Either of two types of optical microscope used to magnify materials undergoing dissection.

dissipation [PHYS] Any loss of energy, generally by conversion into heat; quantitatively, the rate at which this loss occurs. Also known as energy dissipation.

dissipation constant [GEOPHYS] In atmospheric electricity, a measure of the rate at which a given electrically charged object loses its charge to the surrounding air.

dissipation factor [ELEC] The inverse of Q, the storage factor.

dissipation function *See* Rayleigh's dissipation function; viscous dissipation function.

dissipative tunneling [SOLID STATE] Quantum-mechanical tunneling of individual electrons, rather than pairs, across a thin insulating layer separating two superconducting metals when there is a voltage across this layer, resulting in partial disruption of cooperative motion.

dissipator *See* heat sink.

dissociation [PHYS CHEM] Separation of a molecule into two or more fragments (atoms, ions, radicals) by collision with a second body or by the absorption of electromagnetic radiation.

dissociation constant [PHYS CHEM] A constant whose numerical value depends on the equilibrium between the undissociated and dissociated forms of a molecule; a higher value indicates greater dissociation.

dissociation energy [PHYS CHEM] The energy required for complete separation of the atoms of a molecule.

dissociation limit [SPECT] The wavelength, in a series of vibrational bands in a molecular spectrum, corresponding to the point at which the molecule dissociates into its constituent atoms; it corresponds to the convergence limit.

dissociative recombination [ATOM PHYS] The combination of an electron with a positive molecular ion in a gas followed by dissociation of the molecule in which the resulting atoms carry off the excess energy.

dissonance [ACOUS] An unpleasant combination of harmonics heard when certain musical tones are played simultaneously. Also known as discord.

dissymmetry factor [OPTICS] A quantity which expresses the strength of circular dichroism, equal to the difference in the absorption indices for left and right circularly polarized light divided by the absorption index for ordinary light of the same wavelength. Also known as anisotropy factor.

distal flow [MATH] A transformation group on a compact metric space such that every point in the metric space is distal.

distal map [MATH] A continuous function T of a compact metric space, with distance $d(x,y)$, into itself such that if x and y are elements of the metric space with $x \neq y$, then there exists $\varepsilon > 0$ such that $d(T^n x, T^n y) \geq \varepsilon$, for $n = 1, 2, \dots$.

distal point [MATH] A point x in a metric space X with metric d is distal with respect to a transformation group (G, X, π) if, for any point y in X that is distinct from x, there is a number $m > 0$ such that $d[\pi(g,x), \pi(g,y)] \geq m$ for all g in G.

distance [MATH] A nonnegative number associated with pairs of geometric objects. [MECH] The spatial separation of two points, measured by the length of a hypothetical line joining them.

distance-luminosity relation [ASTRON] The relation in which the light intensity from a star is inversely proportional to the square of its distance.

distance modulus *See* modulus of distance.

distance/velocity lag [CONT SYS] The delay caused by the amount of time required to transport material or propagate a signal or condition from one point to another. Also known as transportation lag; transport lag.

distant field [ELECTROMAG] The electromagnetic field at a distance of five wavelengths or more from a transmitter, where the radial electric field becomes negligible.

distortion [ENG] In general, the extent to which a system fails to accurately reproduce the characteristics of an input signal at its output. [ENG ACOUS] Any undesired change in the waveform of a sound wave. [OPTICS] A type of aberration in which there is variation in magnification with the distance from the axis of an optical system, so that images are not geometrically similar to their objects.

distortional wave *See* S wave.

distributed capacitance [ELEC] Capacitance that exists between the turns in a coil or choke, or between adjacent conductors or circuits, as distinguished from the capacitance concentrated in a capacitor.

distributed constant [ELECTROMAG] A circuit parameter that exists along the entire length of a transmission line. Also known as distributed paramater.

distributed inductance [ELECTROMAG] The inductance that exists along the entire length of a conductor, as distinguished from inductance concentrated in a coil.

distributed parameter *See* distributed constant.

distribution [MATH] An abstract object which generalizes the idea of function; used in applied mathematics, quantum theory, and probability theory; the delta function is an example. Also known as generalized function.

distributional derivative *See* weak derivative.

distributional partial derivative *See* weak partial derivative.

distribution coefficient [OPTICS] One of the tristimulus values of monochromatic radiations having equal power, usually denoted by $\bar{x}$, $\bar{y}$, $\bar{z}$. [PHYS CHEM] The ratio of the amounts of solute dissolved in two immiscible liquids at equilibrium.

distribution control *See* linearity control.

distribution curve [STAT] The graph of the distribution function of a random variable.

distribution factor [NUCLEO] A term used to express the modification of the effect of radiation in a biological system attributable to the nonuniform distribution of an internally deposited isotope, such as radium's being concentrated in bones.

distribution-free method [STAT] Any method of inference that does not depend on the characteristics of the population from which the samples are obtained.

distribution function *See* distribution of a random variable.

distribution law [STAT MECH] A law which gives a density function specifying the probability of finding a particle in a unit volume of phase space, or the number of particles in each of the states which a particle may occupy, or the number of particles per unit volume of phase space.

distribution of a random variable [STAT] For a discrete random variable, a function (or table) which assigns to each possible value of the random variable the probability that this value will occur; for a continuous random variable x, the monotone nondecreasing function which assigns to each real t the probability that x is less than or equal to t. Also known as distribution function; probability distribution; statistical distribution.

distribution photometer *See* light-distribution photometer.

distributive lattice [MATH] A lattice in which "greatest lower bound" obeys a distributive law with respect to "least upper bound," and vice versa.

distributive law [MATH] A rule which stipulates how two binary operations on a set shall behave with respect to one another; in particular, if $+$, $\circ$ are two such operations then $\circ$ distributes over $+$ means $a \circ (b + c) = (a \circ b) + (a \circ c)$ for all a, b, c in the set.

disturbed sun noise   [ASTROPHYS] Noise at times of sunspot or solar flare activity.

dither   [CONT SYS] A force having a controlled amplitude and frequency, applied continuously to a device driven by a servomotor so that the device is constantly in small-amplitude motion and cannot stick at its null position. Also known as buzz.

Dittus-Boelter equation   [FL MECH] An equation used to calculate the surface coefficient of heat transfer for fluids in turbulent flow inside clean, round pipes.

diurnal aberration   [ASTRON] Aberration caused by the rotation of the earth; its value varies with the latitude of the observer and ranges from zero at the poles to 0.31 second of arc. Also known as daily aberration.

diurnal arc   [ASTRON] That part of a celestial body's diurnal circle which lies above the horizon of the observer.

diurnal circle   [ASTRON] The apparent daily path of a celestial body, approximating a parallel of declination.

diurnal motion   [ASTRON] The apparent daily motion of a celestial body as observed from a rotating body.

diurnal parallax   *See* geocentric parallax.

diurnal variation   [GEOPHYS] Daily variations of the earth's magnetic field at a given point on the surface, with both solar and lunar periods having their source in the horizontal movements of air in the ionosphere.

divariant system   [THERMO] A system composed of only one phase, so that two variables, such as pressure and temperature, are sufficient to define its thermodynamic state.

divergence   [FL MECH] The ratio of the area of any section of fluid emerging from a nozzle to the area of the throat of the nozzle.   [NUCLEO] In a nuclear reactor, the condition wherein the number of neutrons produced increases in each succeeding generation.

divergence of a vector-valued function   [MATH] The sum of the diagonal entries of the Jacobian matrix; it is the scalar product of the del operator and the vector.

divergence theorem   *See* Gauss' theorem.

divergent beam technique   [PHYS] A method of x-ray diffraction analysis in which a divergent beam of x-rays is used to produce Kossel lines.

divergent integral   [MATH] An improper integral which does not have a finite value.

divergent sequence   [MATH] A sequence which does not converge.

divergent series   [MATH] An infinite series whose sequence of partial sums does not converge.

diverging lens   [OPTICS] A lens whose focal length is negative, so that light incident parallel to its axis diverges after passing through it. Also known as negative lens.

diverging meniscus lens   *See* negative meniscus lens.

divide   [SCI TECH] A point or line of division.

divided differences   [MATH] Quantities which are used in the interpolation or numerical calculation or integration of a function when the function is known at a series of points which are not equally spaced, and which are formed by various operations on the difference between the values of the function at successive points.

dividend   [MATH] A quantity which is divided by another quantity in the operation of division.

division   [MATH] The inverse operation of multiplication; the number a divided by the number b is the number c such that b multiplied by c is equal to a.

division algebra [MATH] An algebra which is also a division ring.

division modulo p [MATH] Division in the finite field with p elements, where p is a prime number.

division ring [MATH] **1.** A ring in which the set of nonzero elements form a group under multiplication. **2.** More generally, a nonassociative ring with nonzero elements in which, for any two elements a and b, there are elements x and y such that $ax = b$ and $ya = b$.

divisor [MATH] **1.** The quantity by which another quantity is divided in the operation of division. **2.** An element b in a commutative ring with identity is a divisor of an element a if there is an element c in the ring such that $a = bc$.

divisor of zero [MATH] A nonzero element x of a commutative ring such that $xy = 0$ for some nonzero element y of the ring. Also known as zero divisor.

Dixon theorem [MATH] The formula

$$_3F_2\,(a,b,c;\ 1 + a - b,\ 1 + a - c;\ 1)$$

$$= \frac{\Gamma(1 + \tfrac{1}{2}a)\ \Gamma(1 + a - b)\ \Gamma(1 + a - c)\ \Gamma(1 - b - c + \tfrac{1}{2}a)}{\Gamma(1 + a)\ \Gamma(1 - b + \tfrac{1}{2}a)\ \Gamma(1 - c + \tfrac{1}{2}a)\ \Gamma(1 + a - b - c)}$$

where $_3F_2$ is a generalized hypergeometric function.

D line [SPECT] The yellow line that is the first line of the major series of the sodium spectrum; the doublet in the Fraunhofer lines whose almost equal components have wavelengths of 5895.93 and 5889.96 angstroms (589.593 and 588.996 nanometers) respectively.

D meson [PARTIC PHYS] **1.** A neutral pseudovector meson resonance having a mass of 1285 ± 4 MeV, width of about 30 MeV, and positive charge conjugation parity; the only singlet state consistent with the $(1^+)^+$ nonet. **2.** Collective name for three charmed mesons that form an isotopic spin triplet, have masses of approximately 1865 MeV, and are pseudoscalar particles.

Dobson spectrophotometer [SPECT] A photoelectric spectrophotometer used in the determination of the ozone content of the atmosphere; compares the solar energy at two wavelengths in the absorption band of ozone by permitting the radiation of each to fall alternately upon a photocell.

dodecagon [MATH] A 12-sided polygon.

dodecahedron [MATH] A polyhedron with 12 faces.

dollar [NUCLEO] A unit of reactivity, equal to the difference between the reactivities for delayed critical and prompt critical conditions in a given nuclear reactor.

domain [MATH] **1.** A nonempty open connected set in euclidean space. Also known as region. **2.** See Abelian field. [SOLID STATE] A region in a solid within which elementary atomic or molecular magnetic or electric moments are uniformly arrayed.

domain growth [SOLID STATE] A stage in the process of magnetization in which there is a growth of those magnetic domains in a ferromagnet oriented most nearly in the direction of an applied magnetic field.

domain of a function [MATH] The set of values of the independent variable.

domain of dependence See future Cauchy development.

domain rotation [SOLID STATE] The stage in the magnetization process in which there is rotation of the direction of magnetization of magnetic domains in a ferromagnet toward the direction of a magnetic applied field and against anisotropy forces.

domain theory [SOLID STATE] A theory of the behavior of ferromagnetic and ferroelectric crystals according to which changes in the bulk magnetization and polarization arise from

DOMAIN GROWTH

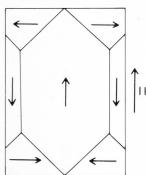

Domain growth, a stage of magnetization in an applied magnetic field. Arrow at right with symbol "∥" represents direction of applied magnetic field. Other arrows represent direction of magnetization of domains.

changes in size and orientation of domains that are each polarized to saturation but which point in different directions.

domain wall *See* Bloch wall.

domatic class *See* clinohedral class.

dome [CRYSTAL] An open crystal form consisting of two faces astride a symmetry plane.

dominant energy condition [RELAT] The condition used in general relativity theory that all observers see a nonnegative energy density and a nonnegative energy flux.

dominant mode *See* fundamental mode.

dominant wave [ELECTROMAG] The electromagnetic wave that has the lowest cutoff frequency in a given uniconductor waveguide.

dominant wavelength [OPTICS] The single wavelength of light that, when combined in suitable proportions with a reference standard light, matches the color of a given sample.

dominated convergence theorem [MATH] If a sequence $\{f_n\}$ of Lebesgue measurable functions converges almost everywhere to f and if the absolute value of each f_n is dominated by the same integrable function, then f is integrable and $\lim \int f_n \, dm = \int f \, dm$.

dominating integral [MATH] An improper integral whose nonnegative, nonincreasing integrand function has the property that its value for all sufficiently large positive integers n is no smaller than the nth term of a given series of positive terms; used in the integral test for convergence.

dominating series [MATH] A series, each term of which is larger than the respective term in some other given series; used in the comparison test for convergence of series.

Donders reduced eye [OPTICS] An optical model used to simplify calculations of the size and position of images produced by the human eye, consisting of a convex refracting surface that separates air in front from water, with a refractive index of 4/3, behind, and having an anterior focal distance of 15 millimeters and a posterior focal distance of 20 millimeters.

donkey power [PHYS] A unit of power equal to 250 watts; it is approximately ⅓ horsepower.

Donnan distribution coefficient [PHYS CHEM] A coefficient in an expression giving the distribution, on two sides of a boundary between electrolyte solutions in Donnan equilibrium, of ions which can diffuse across the boundary.

Donnan equilibrium [PHYS CHEM] The particular equilibrium set up when two coexisting phases are subject to the restriction that one or more of the ionic components cannot pass from one phase into the other; commonly, this restriction is caused by a membrane which is permeable to the solvent and small ions but impermeable to colloidal ions or charged particles of colloidal size. Also known as Gibbs-Donnan equilibrium.

Donnan potential [PHYS CHEM] The potential difference across a boundary between two electrolytic solutions in Donnan equilibrium.

Donohue equation [THERMO] Equation used to determine the heat-transfer film coefficient for a fluid on the outside of a baffled shell-and-tube heat exchanger.

donor [SOLID STATE] An impurity that is added to a pure semiconductor material to increase the number of free electrons. Also known as donor impurity; electron donor.

donor impurity *See* donor.

donor level [SOLID STATE] An intermediate energy level close to the conduction band in the energy diagram of an extrinsic semiconductor.

donut *See* doughnut.

doorknob capacitor [ELEC] A high-voltage, plastic-encased capacitor resembling a doorknob in size and shape.

dopant *See* doping agent.

dope *See* doping agent.

doped junction [ELECTR] A junction produced by adding an impurity to the melt during growing of a semiconductor crystal.

doping [ELECTR] The addition of impurities to a semiconductor to achieve a desired characteristic, as in producing an n-type or p-type material. Also known as semiconductor doping.

doping agent [ELECTR] An impurity element added to semiconductor materials used in crystal diodes and transistors. Also known as dopant; dope.

doping compensation [ELECTR] The addition of donor impurities to a p-type semiconductor or of acceptor impurities to an n-type semiconductor.

Doppler-averaged cross section [PHYS] A cross section averaged over the energy of the incident particles and weighted to take into account the Doppler shifts associated with the thermal motions of the target particles.

Doppler broadening [SPECT] Frequency spreading that occurs in single-frequency radiation when the radiating atoms, molecules, or nuclei do not all have the same velocity and may each give rise to a different Doppler shift.

Doppler effect [PHYS] The change in the observed frequency of an acoustic or electromagnetic wave due to relative motion of source and observer.

Doppler frequency *See* Doppler shift.

Doppler shift [PHYS] The amount of the change in the observed frequency of a wave due to Doppler effect, usually expressed in hertz. Also known as Doppler frequency.

doroid [ELECTROMAG] A coil resembling half a toroid, using a removable core segment to simplify the winding process.

dosage *See* absorbed dose.

dose *See* absorbed dose.

dose equivalent [NUCLEO] The product of absorbed dose, in rads, and a number of modifying factors due to nonuniform distribution of internally deposited isotopes in radiobiology; the unit is the rem.

dosemeter *See* dosimeter.

dose rate [NUCLEO] The rate at which nuclear radiation is delivered.

dose-rate meter [NUCLEO] An instrument that measures radiation dose rate.

dosimeter [NUCLEO] An instrument that measures the total dose of nuclear radiation received in a given period. Also spelled dosemeter.

dosimetry [NUCLEO] The measurement of radiation doses. Also known as radiation dosimetry.

dot product *See* inner product.

double-base diode *See* unijunction transistor.

double-base junction diode *See* unijunction transistor.

double-base junction transistor [ELECTR] A tetrode transistor that is essentially a junction triode transistor having two base connections on opposite sides of the central region of the transistor. Also known as tetrode junction transistor.

double-beam cathode-ray tube [ELECTR] A cathode-ray tube having two beams and capable of producing two independent traces that may overlap; the beams may be produced by splitting the beam of one gun or by using two guns.

double-beam spectrophotometer [SPECT] An instrument that uses a photoelectric circuit to measure the difference in absorption when two closely related wavelengths of light are passed through the same medium.

double beta decay [NUC PHYS] A nuclear transformation in which the atomic number changes by 2 and the mass number does not change; either two electrons are emitted or two orbital electrons are captured.

double bond [PHYS CHEM] A type of linkage between atoms in which two pair of electrons are shared equally.

double-bond isomerism [PHYS CHEM] Isomerism in which two or more substances possess the same elementary composition but differ in having double bonds in different positions.

double-break switch [ELEC] Switch which opens the connected circuit at two points.

double bridge *See* Kelvin bridge.

double Compton scattering [QUANT MECH] A process in which a photon collides with a free electron and two photons are given off.

double-concave lens [OPTICS] A lens having surfaces that are adjacent portions of nonintersecting spheres whose centers lie on opposite sides of the plane of the lens. Also known as biconcave lens.

double-convex lens [OPTICS] A lens having surfaces that are adjacent portions of intersecting spheres whose centers lie on opposite sides of the plane of the lens. Also known as biconvex lens.

double cusp [MATH] A point on a curve through which two branches of the curve with the same tangent pass, and at which each branch extends in both directions of the tangent. Also known as point of osculation; tacnode.

double-diffused transistor [ELECTR] A transistor in which two *pn* junctions are formed in the semiconductor wafer by gaseous diffusion of both *p*-type and *n*-type impurities; an intrinsic region can also be formed.

double diode *See* binode.

double-doped transistor [ELECTR] The original grown-junction transistor, formed by successively adding *p*-type and *n*-type impurities to the melt during growing of the crystal.

double-doublet antenna [ELECTROMAG] Two half-wave doublet antennas criss-crossed at their center, one being shorter than the other to give broader frequency coverage.

double electron excitation [ATOM PHYS] An excited state of an atom in which two electrons are excited rather than one.

double group [QUANT MECH] A type of group useful in studying systems of half-integral spin; it is formed by modifying a finite point group by introducing an element which is a rotation through an angle of 2π about an arbitrary axis and which is not the unit element but gives the unit element when applied twice.

double integral [MATH] The Riemann integral of functions of two variables.

double-integrating gyro [MECH] A single-degree-of-freedom gyro having essentially no restraint of its spin axis about the output axis.

double layer *See* electric double-layer.

double mirror [OPTICS] Two plane mirrors inclined at an angle to each other.

double pendulum [MECH] Two masses, one suspended from a fixed point by a weightless string or rod of fixed length, and the other similarly suspended from the first; often the system is constrained to remain in a vertical plane.

double point [MATH] A point on a curve at which a curve crosses or touches itself, or has a cusp; that is, a point at which the curve has two tangents (which may be coincident).

double-quantum stimulated-emission device [OPTICS] A laser in which the crystal contains two species of fluorescent ions whose fluorescence frequencies are so related that, when the flash lamp coils produce pumping action, the ions of one

DOUBLE-CONCAVE LENS

Double-concave lens, a type of diverging or negative lens. *(From F. A. Jenkins and H. E. White, Fundamentals of Optics, 3d ed., McGraw-Hill, 1957)*

DOUBLE-CONVEX LENS

A double-convex lens, a type of collecting or positive lens. *(From F. A. Jenkins and H. E. White, Fundamentals of Optics, 3d ed., McGraw-Hill, 1957)*

DOUBLE CUSP

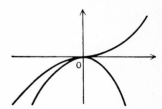

Example of a double cusp.

species contribute photons to the fluorescence of the other species, which is the active ion.

double quantum transition [ATOM PHYS] A radiative transition between atomic or molecular states in which two or more photons are simultaneously emitted or absorbed.

doubler *See* voltage doubler.

double refraction *See* birefringence.

double root [MATH] A number a such that $(x-a)^2 p(x) = 0$ where $p(x)$ is a polynomial of which a is not a root.

double-shield enclosure [ELEC] Type of shielded enclosure or room in which the inner wall is partially isolated electrically from the outer wall.

double star [ASTRON] A star which appears as a single point of light to the eye but which can be resolved into two points by a telescope.

doublet [ATOM PHYS] Two stationary states which have the same orbital and spin angular momentum but which have different total angular momenta, and therefore have slightly different energies due to spin-orbit coupling. [ELECTROMAG] *See* dipole. [FL MECH] A source and a sink separated by an infinitesimal distance, each having an infinitely large strength so that the product of this strength and the separation is finite. [OPTICS] A lens made up of two components, especially an achromat. [PARTIC PHYS] Two elementary particles which have slightly differing masses and the same baryon number, spin, parity, and charge conjugation parity (if self-conjugate), but have different charges. [PHYS CHEM] Two electrons which are shared between two atoms and give rise to a nonpolar valence bond. [SPECT] Two closely separated spectral lines arising from a transition between a single state and a pair of states forming a doublet as described in the atomic physics definition.

double tangent [MATH] **1.** A line which is tangent to a curve at two distinct noncoincident points. Also known as bitangent. **2.** Two coincident tangents to branches of a curve at a given point, such as the tangents to a cusp.

doublet antenna *See* dipole antenna.

doublet flow [FL MECH] The motion of a fluid in the vicinity of a doublet; can be superposed with uniform flow to yield flow around a cylinder or a sphere.

double weighing [MECH] A method of weighing to allow for differences in lengths of the balance arms, in which object and weights are balanced twice, the second time with their positions interchanged. Also known as Gauss method of weighing.

doubling time [NUCLEO] The time required for a breeding reactor to double its fuel inventory.

doubly periodic function [MATH] A periodic function $f(x)$ which is not simply periodic, and for which there exist two numbers a and b such that any period is of the form $ma + nb$, where m and n are integers.

doubly stochastic matrix [MATH] A matrix of nonnegative real numbers such that every row sum and every column sum are equal to 1.

Dougall's first theorem [MATH] The formula

$${}_7F_6(a,\ 1+\tfrac{1}{2}a,\ b,\ c,\ d,\ e,\ -n;\ \tfrac{1}{2}a,\ a-b+1,\ a-c+1,\ a-d+1,$$
$$a-e+1,\ a+n+1;\ 1)$$

$$= \frac{(a+1;n)(a-b-c+1;n)(a-c-d+1;n)(a-d-b+1;n)}{(a-b+1;n)(a-c+1;n)(a-d+1;n)(a-b-c-d+1;n)}$$

where ${}_7F_6$ is a generalized hypergeometric function, $1+2a = b+c+d+e-n$, n is a nonnegative integer, $(a;0) = 1$, and $(a;n) = a(a+1)\cdots(a+n-1)$.

Dougall's second theorem [MATH] The formula

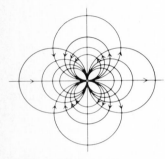

DOUBLET

Equipotential lines and streamlines for the two-dimensional doublet, as used in fluid mechanics.

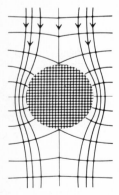

DOUBLET FLOW

Streamlines and equipotential lines for uniform flow about a sphere at rest.

$_5F_4(a,\ 1+\tfrac{1}{2}a,\ b,\ c,\ d;\ \tfrac{1}{2}a,\ a-b+1,\ a-c+1,\ a-d+1;\ 1)$

$$= \frac{\Gamma(a-b+1)\ \Gamma(a-c+1)\ \Gamma(a-d+1)\ \Gamma(a-b-c-d+1)}{\Gamma(a+1)\ \Gamma(a-b-c+1)\ \Gamma(a-c-d+1)\ \Gamma(a-d-b+1)}$$

where $_5F_4$ is a generalized hypergeometric function and the real part of $a-b-c-d$ is greater than -1.

doughnut [NUCLEO] Also spelled donut. **1.** The toroidal vacuum chamber in which electrons are accelerated in a betatron or synchrotron. Also known as toroid. **2.** An assembly of enriched fissionable material, often doughnut-shaped, used in a thermal reactor to provide a local increase in fast neutron flux for experimental purposes.

Dove prism *See* Delaborne prism.

down quark [PARTIC PHYS] A quark with an electric charge of $-\tfrac{1}{3}$, baryon number of $\tfrac{1}{3}$, and 0 strangeness and charm.

dr *See* dram.

drachm *See* dram.

draconitic month *See* nodical month.

draft [FL MECH] **1.** An air current in a confined space, such as that in a cooling tower or chimney. **2.** The difference between atmospheric pressure and some lower pressure in a confined space that causes air to flow, such as exists in the furnace or gas passages of a steam-generating unit or in a chimney. Also spelled draught.

drag [FL MECH] Resistance caused by friction in the direction opposite to that of the motion of the center of gravity of a moving body in a fluid.

drag coefficient [FL MECH] A characteristic of a body in a flowing inviscous fluid, equal to the ratio of twice the force on the body in the direction of flow to the product of the density of the fluid, the square of the flow velocity, and the effective cross-sectional area of the body.

drag force [PL PHYS] A force on an electrically conducting fluid arising from inelastic collisions of electrons and ions and proportional to the fluid velocity.

drain [ELEC] *See* current drain. [ELECTR] One of the electrodes in a thin-film transistor.

dram [MECH] **1.** A unit of mass, used in the apothecaries' system of mass units, equal to $\tfrac{1}{8}$ apothecaries' ounce or 60 grains or 3.8879346 grams. Also known as apothecaries' dram (dram ap); drachm (British). **2.** A unit of mass, formerly used in the United Kingdom, equal to $\tfrac{1}{16}$ ounce (avoirdupois) or approximately 1.77185 grams. Abbreviated dr.

dram ap *See* dram.

Draper catalog [ASTRON] A nine-volume catalog of stars completed in 1924; it gives positions, magnitudes, and spectral classes of 225,300 stars.

draught *See* draft.

D region [GEOPHYS] The region of ionosphere up to about 60 miles (97 kilometers) above the earth, below the E and F regions, in which the D layer forms.

Drew number [PHYS CHEM] A dimensionless group used in the study of diffusion of a solid material A into a stream composed of substance B, equal to

$$\frac{Z_A(M_A-M_B)+M_B}{(Z_A-Y_{AW})(M_B-M_A)}\cdot\ln\frac{M_V}{M_W},$$

where M_A and M_B are the molecular weights of components A and B, M_V and M_W are the molecular weights of the mixture in the vapor and at the wall, and Y_{AW} and Z_A are the mole fractions of A at the wall and in the diffusing stream, respectively. Symbolized N_D.

drift [SOLID STATE] The movement of current carriers in a semiconductor under the influence of an applied voltage.

drift current [PL PHYS] A current of free charged particles in perpendicular electric and magnetic fields that results from an average motion of the particles in a direction perpendicular to both fields.

drift mobility [SOLID STATE] The average drift velocity of carriers per unit electric field in a homogeneous semiconductor. Also known as mobility.

drift space [ELECTR] A space in an electron tube which is substantially free of externally applied alternating fields and in which repositioning of electrons takes place.

drift speed [ELEC] Average speed at which electrons or ions progress through a medium.

drift transistor [ELECTR] **1.** A transistor having two plane parallel junctions, with a resistivity gradient in the base region between the junctions to improve the high-frequency response. **2.** *See* diffused-alloy transistor.

drift tube [NUCLEO] A tubular electrode placed in the vacuum chamber of a circular accelerator, to which radio-frequency voltage is applied to accelerate the particles.

drift velocity [SOLID STATE] The average velocity of a carrier that is moving under the influence of an electric field in a semiconductor, conductor, or electron tube.

drift wave [PL PHYS] An oscillation in a magnetically confined plasma which arises in the presence of density gradients, for example, at the plasma's surface, and which resembles the waves that propagate at the interface of two fluids of different density in a gravity field.

drive *See* excitation.

driving point function [CONT SYS] A special type of transfer function in which the input and output variables are voltages or currents measured between the same pair of terminals in an electrical network.

DRM *See* depositional remanent magnetization.

drop [FL MECH] The quantity of liquid that coalesces into a single globule; sizes vary according to physical conditions and the properties of the fluid itself.

drop model of nucleus *See* liquid-drop model of nucleus.

drop weight [FL MECH] The weight of the largest drop that can hang from the end of a tube of given radius.

drop-weight method [FL MECH] A method of measuring surface tension by measuring the weight of a slowly increasing drop of the liquid hanging from the end of a tube, just before it is detached from the tube.

dropwise condensation [THERMO] Condensation of a vapor on a surface in which the condensate forms into drops.

Drude equation [OPTICS] An equation which states that the rotation of the plane of polarization of plane-polarized light passing through an optically active substance is inversely proportional to the difference between the square of the wavelength of the light and the square of a constant wavelength.

Drude's theory of conduction [SOLID STATE] A theory which treats the electrons in a metal as a gas of classical particles.

dry-bulb temperature [PHYS] The actual air temperature as measured by a dry-bulb thermometer.

dry-bulb thermometer [ENG] An ordinary thermometer, especially one with an unmoistened bulb; not dependent upon atmospheric humidity.

dry criticality [NUCLEO] Reactor criticality achieved without a coolant.

dry-disk rectifier *See* metallic rectifier.

dry electrolytic capacitor [ELEC] An electrolytic capacitor in which the electrolyte is a paste rather than a liquid; the dielectric is a thin film of gas formed on one of the plates by chemical action.

dry flashover voltage [ELECTR] Voltage at which the air surrounding a clean dry insulator or shell completely breaks down between electrodes.

dry friction [MECH] Resistance between two dry solid surfaces, that is, surfaces free from contaminating films or fluids.

dry pint *See* pint.

dry-plate rectifier *See* metallic rectifier.

dry pt *See* pint.

dry well [NUCLEO] The first containment tank surrounding a water-cooled nuclear reactor that uses the pressure-suppressing containment system.

Dst [GEOPHYS] The "storm-time" component of variation of the terrestrial magnetic field, that is, the component which correlates with the interval of time since the onset of a magnetic storm; used as an index of intensity of the ring current.

DTL *See* diode transistor logic.

dual coordinates [MATH] Point coordinates and plane coordinates are dual in geometry since an equation about one determines an equation about the other.

dual-cycle boiling-water reactor [NUCLEO] A boiling-water reactor in which part of the steam used to run the steam turbine is generated in the reactor core and part is generated in an external heat exchanger. Also known as dual-cycle reactor system.

dual-cycle reactor system *See* dual-cycle boiling-water reactor.

dual-emitter transistor [ELECTR] A passivated *pnp* silicon planar epitaxial transistor having two emitters, for use in low-level choppers.

dual graph [MATH] A planar graph corresponding to a planar map obtained by replacing each country with its capital and each common boundary by an arc joining the two countries.

dual group [MATH] **1.** The group of all homomorphisms of an Abelian group G into the cyclic group of order n, where n is the smallest integer such that g^n is the identity element of G. **2.** The dual group of a locally continuous Abelian group G is the group of continuous characters on G, with addition of any two characters f and g defined by $(f + g)(x) = f(x)g(x)$ for all x in G.

dual-gun cathode-ray tube [ELECTR] A dual-trace oscilloscope in which beams from two electron guns are controlled by separate balanced vertical-deflection plates and also have separate brightness and focus controls.

dual integral equations [MATH] A pair of equations of the form

$$\int_0^\infty f(y)K(x,y)\ dy = g_1(x),\ x>1$$

$$\int_0^\infty G(y)f(y)K(x,y)\ dy = g_2(x),\ 0<x<1$$

where $K(x,y)$, $G(y)$, $g_1(x)$, and $g_2(x)$ are known functions, and the function $f(y)$ is to be determined; these equations arise in solving certain boundary value problems in mathematical physics.

duality principle Also known as principle of duality. [ELEC] The principle that for any theorem in electrical circuit analysis there is a dual theorem in which one replaces quantities with dual quantities; current and voltage, impedance and admittance, and meshes and nodes are examples of dual quantities. [ELECTR] The principle that analogies may be drawn between a transistor circuit and the corresponding vacuum tube circuit. [ELECTROMAG] The principle that one can obtain new solutions of Maxwell's equations from known

DUAL GRAPH

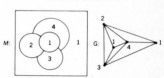

A map requiring four colors, indicated by numbers, and its planar graph.

solutions by replacing **E** with **H**, **H** with $-\mathbf{E}$, ε with μ, and μ with ε. [MATH] A principle that if a theorem is true, it remains true if each object and operation is replaced by its dual; important in projective geometry and Boolean algebra. [QUANT MECH] *See* wave-particle duality.

duality theorem [MATH] **1.** A theorem which asserts that for a given space, the $(n - p)$ dimensional homology group is isomorphic to a p-dimensional cohomology group for each $p = 0, \ldots, n$, provided certain conditions are met. **2.** Let G be either a compact group or a discrete group, let X be its character group, and let G' be the character group of X; then there is an isomorphism of G onto G' so that the groups G and G' may be identified. **3.** If either of two dual linear-programming problems has a solution, then so does the other.

dual laser [OPTICS] A gas laser having Brewster windows and concave mirrors at opposite ends, the mirrors having different reflectivities so as to produce two different visible or infrared wavelengths from a helium-neon laser beam.

dual linear programming [MATH] Linear programming in which the maximum and minimum number are the same number.

dual linear-programming problems [MATH] The problem of finding nonnegative numbers $x_1, \ldots, x_n$ that maximize the quantity $c_1x_1 + \cdots + c_nx_n$ subject to the constraints $a_{i1}x_1 + \cdots + a_{in}x_n \leq b_i$, $i = 1, \ldots, m$, and the problem of finding nonnegative numbers $y_1, \ldots, y_m$ that minimize the quantity $b_1y_1 + \cdots + b_my_m$ subject to the contraints $a_{ij}y_1 + \cdots + a_{mj}y_m \geq c_j$, $j = 1, \ldots, n$.

dual meter [ENG] Meter constructed so that two aspects of an electric circuit may be read simultaneously.

dual network [ELEC] A network which has the same number of terminal pairs as a given network, and whose open-circuit impedance network is the same as the short-circuit admittance matrix of the given network, and vice versa.

dual operation [MATH] An operation on a Boolean algebra of two elements that can be obtained from another such operation by reversing the values of each element in the latter's truth table; for example, the AND function is dual to the OR function.

dual-purpose reactor [NUCLEO] Any nuclear reactor which both acts as a source of heat energy for a power plant and produces fissionable material.

dual radioactive decay [NUC PHYS] Property exhibited by a nucleus which has two or more independent and alternative modes of decay.

dual space [MATH] The vector space consisting of all linear transformations from a given vector space into its scalar field.

dual tensor [MATH] The product of a given tensor, covariant in all its indices, with the contravariant form of the determinant tensor, contracting over the indices of the given tensor.

dual-trace amplifier [ELECTR] An oscilloscope amplifier that switches electronically between two signals under observation in the interval between sweeps, so that waveforms of both signals are displayed on the screen.

dual-trace oscilloscope [ELECTR] An oscilloscope which can compare two waveforms on the face of a single cathode-ray tube, using any one of several methods.

dual variables [MATH] Mutually dependent variables.

Duane-Hunt law [QUANT MECH] The law that the frequency of x-rays resulting from electrons striking a target cannot exceed eV/h, where e is the charge of the electron, V is the exciting voltage, and h is Planck's constant.

Duane-Hunt limit [QUANT MECH] The upper limit on the frequency of radiation from an x-ray tube given by the Duane-Hunt law.

DuBois-Reymond's mean value theorem *See* second law of the mean for integrals.

DuBois-Reymond theorem [MATH] A theorem used in the calculus of variations which states that a continuous function $f(x)$ must be constant on an interval $[a,b]$ if the integral of $f(x)g(x)\,dx$ over $[a,b]$ vanishes for every continuous function $g(x)$ for which the integral of $g(x)\,dx$ over $[a,b]$ vanishes.

Duchemin's formula [PHYS] An expression for normal wind pressure per square foot on an inclined surface, $N = F[(2\sin a)/(1 + \sin^2 a)]$, where F = normal wind force in pounds per square foot on a vertical surface, and a = angle of inclination of inclined surface.

ductile fracture *See* fibrous fracture.

Duddell oscillograph [ELECTROMAG] A moving-coil oscillograph; the current to be observed passes through a coil in a magnetic field and a mirror attached to the coil reveals its movement.

Duddle arc [PHYS] An electric arc with which a coil and capacitor are connected in parallel, causing the arc to oscillate and emit musical tones. Also known as singing arc.

Duffing's equation [MATH] A nonlinear, second-order differential equation, $d^2x/dt^2 + a\,dx/dt + x + bx^3 = 0$, used to describe the oscillations of a nonlinear spring.

Dufour effect [THERMO] Energy flux due to a mass gradient occurring as a coupled effect of irreversible processes.

Dufour number [THERMO] A dimensionless number used in studying thermodiffusion, equal to the increase in enthalpy of a unit mass during isothermal mass transfer divided by the enthalpy of a unit mass of mixture. Symbol Du_2.

Duhamel's theorem [MATH] **1.** If f and g are continuous functions, then

$$\lim_{|\Delta x| \to 0} \sum_{i=1}^{n} f(x_i')g(x_i'')\Delta x_i = \int_a^b f(x)g(x)\,dx$$

where x_i' and x_i'' are between x_{i-1} and x_i, $i = 1, \ldots, n$, and $|\Delta x| = \max\{x_i - x_{i-1}\}$ for a partition $a = x_0 < x_1 < \cdots < x_n = b$. **2.** If $f_i(n)$, $i = 1, 2, \ldots$, are a set of functions defined on the positive integers, such that the sum over i from 1 to n of $f_i(n)$ approaches a limit L as n approaches infinity, and if $g_i(n)$, $i = 1, 2, \ldots$, are a set of functions such that $g_i(n)/f_i(n)$ approaches 0, uniformly over i, as n approaches infinity, then the sum over i from 1 to n of $f_i(n) + g_i(n)$ also approaches L as n approaches infinity.

Duhem-Margules equation [THERMO] An equation showing the relationship between the two constituents of a liquid-vapor system and their partial vapor pressures:

$$\frac{d\ln p_A}{d\ln x_A} = \frac{d\ln p_B}{d\ln x_B}$$

where x_A and x_B are the mole fractions of the two constituents, and p_A and p_B are the partial vapor pressures.

Duhem's equation *See* Gibbs-Duhem equation.

Dühring's rule [PHYS CHEM] The rule that a plot of the temperature at which a liquid exerts a particular vapor pressure against the temperature at which a similar reference liquid exerts the same vapor pressure produces a straight or nearly straight line.

dull emitter [ELECTR] An electron tube whose cathode is a filament that does not glow brightly.

Dulong number *See* Eckert number.

Dulong-Petit law [THERMO] The law that the product of the specific heat per gram and the atomic weight of many solid elements at room temperature has almost the same value, about 6.3 calories (264 joules) per degree Celsius.

dumbbell [MATH] A plane curve whose equation in cartesian coordinates x and y is $y^2 = x^4 - x^6$.

dummy suffix [MATH] A suffix which has no true mathematical significance and is used only to facilitate notation; usually an index which is summed over.

dummy variable [MATH] A variable which has no true mathematical significance and is used only to facilitate notation; usually a variable which is integrated over.

duodecimal number system [MATH] A representation system for real numbers using 12 as the base.

duolateral coil *See* honeycomb coil.

duoplasmatron [ELECTR] An ion-beam source in which electrons from a hot filament are accelerated sufficiently to ionize a gas by impact; the resulting positive ions are drawn out by high-voltage electrons and focused into a beam by electrostatic lens action.

Dupin indicatrix [MATH] For a point P on a surface S, the conic section $x^2/|r_1| + y^2/|r_2| = 1$, or $x^2/r_1 + y^2/r_2 = \pm 1$, or $x^2 = r_1$, according to whether the curvature of S at P is positive, negative, or 0, where the x and y axes are taken along the principal directions, and r_1 and r_2 are the radii of the corresponding principal curvatures.

Dupin's theorem [MATH] Given three families of mutually orthogonal surfaces, the line of intersection of any two surfaces of different families is a line of curvature for both the surfaces.

duplet lens system [OPTICS] A system of lenses in which there are two groups of lenses separated by a space, and successive lens in each group are in contact.

duplication formula [MATH] A formula for the gamma function: $\Gamma(2z) = 2^{2z-1}\pi^{-1/2}\Gamma(z)\,\Gamma(z+\frac{1}{2})$.

Dupré equation [THERMO] The work W_{LS} done by adhesion at a gas-solid-liquid interface, expressed in terms of the surface tensions γ of the three phases, is $W_{LS} = \gamma_{GS} + \gamma_{GL} - \gamma_{LS}$.

duration [MECH] A basic concept of kinetics which is expressed quantitatively by time measured by a clock or comparable mechanism.

Durer's conchoid [MATH] A plane curve whose equation in cartesian coordinates x and y is $2y^2(x^2 + y^2) - 2by^2(x + y) + (b^2 - 3a^2)y^2 - a^2x^2 + 2a^2b(x + y) + a^2(a^2 - b^2) = 0$, where a and b are constants.

Dushman equation *See* Richardson-Dushman equation.

Dushman-Langmuir equation [SOLID STATE] An expression for the constant D_0 which appears in the equation $D = D_0 \exp(-E/RT)$, giving the diffusion coefficient D of a solid in terms of the activation energy E, the gas constant R, and the absolute temperature T; the expression is $D_0 = Ed^2/Nh$, where d is the perpendicular distance between crystal planes in the direction of diffusion, N is Avogadro's number, and h is Planck's constant.

dust [PHYS] A loose term applied to solid particles predominantly larger than colloidal size and capable of temporary gas suspension.

dust extinction [OPTICS] The contribution to total extinction of light made by scattering and absorption by dust particles in the path of a light beam.

duty cycle [ELECTR] *See* duty ratio. [NUCLEO] The fraction of time during which a pulsed accelerator beam is on target, usually expressed as a percent. Also known as duty factor.

duty factor *See* duty cycle.

duty ratio [ELECTR] In a pulse radar or similar system, the ratio of average to peak pulse power. Also known as duty cycle.

D variometer *See* declination variometer.

dwarf star [ASTRON] A star that typically has surface temperature of 5730 K, radius of 690,000 kilometers, mass of 2×10^{33} grams, and luminosity of 4×10^{33} ergs/sec. Also known as main sequence star.

dwt *See* pennyweight.

Dy *See* dysprosium.

dyad [MATH] An abstract object which is a pair of vectors **AB** in a given order on which certain operations are defined.

dyadic expansion [MATH] The representation of a number in the binary number system.

dyadic rational [MATH] A fraction whose denominator is a power of 2.

dye laser [OPTICS] A type of tunable laser in which the active material is a dye such as acridine red or esculin, with very large molecules, and laser action takes place between the first excited and ground electronic states, each of which comprises a broad vibrational-rotational continuum.

dynamical friction [PHYS] **1.** The drag force between electrons and ions drifitng with respect to each other. **2.** Sliding friction, in contrast to static friction.

dynamical parallax [ASTRON] A parallax of binary stars that is computed from the sum of the masses of the binary system.

dynamical similarity [MECH] Two flow fields are dynamically similar if one can be transformed into the other by a change of length and velocity scales. All dimensionless numbers of the flows must be the same.

dynamical system [MATH] An abstraction of the concept of a family of solutions to an ordinary differential equation; namely, an action of the real numbers on a topological space satisfying certain "flow" properties.

dynamical variable [MECH] One of the quantities used to describe a system in classical mechanics, such as the coordinates of a particle, the components of its velocity, the momentum, or functions of these quantities.

dynamic analogies [PHYS] Analogies that make it possible to convert the differential equations for mechanical and acoustical systems to equivalent electrical equations that can be represented by electric networks and solved by circuit theory.

dynamic balance [MECH] The condition which exists in a rotating body when the axis about which it is forced to rotate, or to which reference is made, is parallel with a principal axis of inertia; no products of inertia about the center of gravity of the body exist in relation to the selected rotational axis.

dynamic boundary condition [FL MECH] The condition that the pressure must be continuous across an internal boundary or free surface in a fluid.

dynamic braking [MECH] A technique of electric braking in which the retarding force is supplied by the same machine that originally was the driving motor.

dynamic characteristic *See* load characteristic.

dynamic creep [MECH] Creep resulting from fluctuations in a load or temperature.

dynamic equilibrium Also known as kinetic equilibrium. [MECH] The condition of any mechanical system when the kinetic reaction is regarded as a force, so that the resultant force on the system is zero according to d'Alembert's principle. [PHYS] A condition in which several processes act simultaneously to maintain a system in an overall state that does not change with time.

dynamic fluidity [FL MECH] The reciprocal of the dynamic viscosity.

dynamic height [PHYS] The amount of work done when a water particle of unit mass is moved vertically from one level to another. Also known as geodynamic height.

dynamic impedance [ELEC] The impedance of a circuit having an inductance and a capacitance in parallel at the frequency at which this impedance has a maximum value. Also known as rejector impedance.

dynamic instability *See* inertial instability.

dynamic meter [PHYS] The standard unit of dynamic height expressed as 10 square meters per second per second.

dynamic parallax [ASTRON] A value for the parallax of a binary star computed from the observations of the period and angular dimensions of the orbit by assuming a value for the mass of the binary system. Also known as hypothetical parallax.

dynamic plate impedance [ELECTR] Internal resistance to the flow of alternating current between the cathode and plate of a tube.

dynamic plate resistance [ELECTR] Opposition that the plate circuit of a vacuum tube offers to a small increment of plate voltage; it is the ratio of a small change in plate voltage to the resulting change in the plate current, other tube voltages remaining constant.

dynamic pressure [FL MECH] **1.** The pressure that a moving fluid would have if it were brought to rest by isentropic flow against a pressure gradient. Also known as impact pressure; stagnation pressure; total pressure. **2.** The difference between the quantity in the first definition and the static pressure.

dynamic programming [MATH] A mathematical technique, more sophisticated than linear programming, for solving a multidimensional optimization problem, which transforms the problem into a sequence of single-stage problems having only one variable each.

dynamic resistance [ELEC] A device's electrical resistance when it is in operation.

dynamics [MECH] That branch of mechanics which deals with the motion of a system of material particles under the influence of forces, especially those which originate outside the system under consideration.

dynamic stability [MECH] The characteristic of a body, such as an aircraft, rocket, or ship, that causes it, when disturbed from an original state of steady motion in an upright position, to damp the oscillations set up by restoring moments and gradually return to its original state. Also known as stability.

dynamic temperature difference [PHYS] The difference between the temperature of a static medium and the surface temperature at the stagnation point of a heat-insulated body immersed in a flowing medium of the same composition.

dynamic vertical *See* apparent vertical.

dynamo *See* generator.

dynamo effect [GEOPHYS] A process in the ionosphere in which winds and the resultant movement of ionization in the geomagnetic field give rise to induced current.

dynamoelectric [PHYS] Pertaining to the conversion of mechanical energy to electric energy, or vice versa.

dynamometer [ENG] **1.** An instrument in which current, voltage, or power is measured by the force between a fixed coil and a moving coil. **2.** A special type of electric rotating machine used to measure the output torque or driving torque of rotating machinery by the elastic deformation produced.

dynamometer multiplier [ELEC] A multiplier in which a fixed and a moving coil are arranged so that the deflection of the moving coil is proportional to the product of the currents flowing in the coils.

dynamo theory [GEOPHYS] The hypothesis which explains the regular daily variations in the earth's magnetic field in terms of electrical currents in the lower ionosphere, generated by tidal motions of the ionized air across the earth's magnetic field.

dynatron [ELECTR] A screen-grid tube in which secondary emission of electrons from the anode causes the anode current to decrease as anode voltage increases, resulting in a negative resistance characteristic. Also known as negatron.

dynatron oscillator [ELECTR] An oscillator in which secondary emission of electrons from the anode of a screen-grid tube causes the anode current to decrease as anode voltage is increased, giving the negative resistance characteristic required for oscillation.

dyne [MECH] The unit of force in the centimeter-gram-second system of units, equal to the force which imparts an acceleration of 1 cm/sec^2 to a 1 gram mass.

dynode [ELECTR] An electrode whose primary function is secondary emission of electrons; used in multiplier phototubes and some types of television camera tubes. Also known as electron mirror.

Dyson microscope [OPTICS] A type of interference microscope, now obsolete, in which a light ray is split into two parallel beams and then recombined by reflections from surfaces of parallel plates, and one of the beams passes through the object under observation.

dysprosium [CHEM] A metallic rare-earth element, symbol Dy, atomic number 66, atomic weight 162.50.

DYSPROSIUM

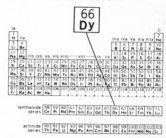

Periodic table of the chemical elements showing the position of dysprosium.

E

e [MATH] The base of the natural logarithms; the number defined by the equation

$$\int_1^e \frac{1}{x} \, dx = 1.$$

E *See* electric field vector.

eagle mounting [SPECT] A mounting for a diffraction grating, based on the principle of the Rowland circle, in which the diffracted ray is returned along nearly the same direction as the incident beam.

Earnshaw's theorem [ELEC] The theorem that a charge cannot be held in stable equilibrium by an electrostatic field.

earth [ASTRON] The third planet in the solar system, lying between Venus and Mars; sometimes capitalized. [ELEC] *See* ground.

earth connection *See* ground.

earth current [ELEC] Return, fault, leakage, or stray current passing through the earth from electrical equipment. Also known as ground current. [GEOPHYS] A current flowing through the ground and due to natural causes, such as the earth's magnetic field or auroral activity. Also known as telluric current.

earth-current storm [GEOPHYS] Irregular fluctuations in an earth current in the earth's crust, often associated with electric field strengths as large as several volts per kilometer, and superimposed on the normal diurnal variation of the earth currents.

earth inductor [ENG] A type of inclinometer that has a coil which rotates in the earth's field and in which a voltage is induced when the rotation axis does not coincide with the field direction; used to measure the dip angle of the earth's magnetic field. Also known as dip inductor; earth inductor compass; induction inclinometer.

earth inductor compass *See* earth inductor.

earth-layer propagation [GEOPHYS] **1.** Propagation of electromagnetic waves through layers of the earth's atmosphere. **2.** Electromagnetic wave propagation through layers below the earth's surface.

earth orbit [ASTRON] The elliptical motion of the earth about the sun (eccentricity 0.01675, average radius 1.496×10^8 kilometers) in a little over a year.

earth oscillations [GEOPHYS] Any rhythmic deformations of the earth as an elastic body; for example, the gravitational attraction of the moon and sun excite the oscillations known as earth tides.

earth radiation *See* terrestrial radiation.

earth rate [ASTRON] The angular velocity or rate of the earth's rotation.

earth rotation [ASTRON] Motion about the earth's axis that occurs 365.2422 times over a year's period.

earth tide [GEOPHYS] The periodic movement of the earth's crust caused by forces of the moon and sun. Also known as bodily tide.

east-west effect [ASTRON] The phenomenon due to the fact that a greater number of cosmic-ray particles approach the earth from a westerly direction than from an easterly.

easy glide [SOLID STATE] A large increase in plastic deformation of a single crystal accompanying a small increase in stress as the result of the passage of many thousands of dislocations through the crystal along a single glide system.

E bend [ELECTROMAG] A smooth change in the direction of the axis of a waveguide, throughout which the axis remains in a plane parallel to the direction of polarization. Also known as E-plane bend.

Ebert ion counter [ENG] An ion counter of the aspiration condenser type, used for the measurement of the concentration and mobility of small ions in the atmosphere.

ebullioscopic constant [PHYS CHEM] The ratio of the elevation of the boiling point of a solvent caused by dissolving a solute to the molality of the solution, taken at extremely low concentrations. Also known as molal elevation of the boiling point.

eccentric angle [MATH] For an ellipse having semimajor and semiminor angles of lengths a and b respectively, lying along the x and y axes of a coordinate system respectively, and for a point (x,y) on the ellipse, the angle arc cos (x/a) = arc sin (y/b).

eccentric anomaly [ASTRON] For a planet in an elliptical orbit, the eccentric angle corresponding to the planet's location.

eccentricity [MATH] The ratio of the distance of a point on a conic from the focus to the distance from the directrix. [MECH] The distance of the geometric center of a revolving body from the axis of rotation.

eccentric orbit [ASTRON] An orbit of a celestial body that deviates markedly from a circle.

Eccles-Jordan circuit *See* bistable multivibrator.

Eccles-Jordan multivibrator *See* bistable multivibrator.

E center *See* R center.

echelette grating [SPECT] A diffraction grating with coarse groove spacing, designed for the infrared region; has grooves with comparatively flat sides and concentrates most of the radiation by reflection into a small angular coverage.

echelle grating [SPECT] A diffraction grating designed for use in high orders and at angles of illumination greater than 45° to obtain high dispersion and resolving power by the use of high orders of interference.

echelon grating [SPECT] A diffraction grating which consists of about 20 plane-parallel plates about 1 centimeter thick, cut from one sheet, each plate extending beyond the next by about 1 millimeter, and which has a resolving power on the order of 10^6.

echo [ELECTR] The signal reflected by a radar target, or the trace produced by this signal on the screen of the cathode-ray tube in a radar receiver. Also known as radar echo; return. [PHYS] A wave packet that has been reflected or otherwise returned with sufficient delay and magnitude to be perceived as a signal distinct from that directly transmitted.

Eckert number [PHYS] A dimensionless group used in the study of compressible flow around a body, equal to the square of the fluid velocity far from the body divided by the product of the specific heat of the fluid at constant temperature and the difference between the temperatures of the fluid and the body. Symbolized N_E. Also known as Dulong number.

ECL *See* emitter-coupled logic.

eclipse [ASTRON] **1.** The reduction in visibility or disappearance of a body by passing into the shadow cast by another body. **2.** The apparent cutting off, wholly or partially, of the light from a luminous body by a dark body coming between it and the observer. Also known as astronomical eclipse.

eclipse seasons [ASTRON] The two times when the sun is near enough to one of the nodes of the moon's orbit for eclipses to occur; this positioning occurs at nearly opposite times of the year, and the eclipse seasons vary yearly because of westward regression of the nodes.

eclipse year [ASTRON] The interval between two successive conjunctions of the sun with the same node of the moon's orbit, equal to 346.62 sidereal days.

eclipsing binary *See* eclipsing variable star.

eclipsing variable star [ASTRON] A binary star whose orbit is such that every time one star passes between the observer and its companion an eclipse results. Also known as eclipsing binary.

ecliptic [ASTRON] The apparent annual path of the sun among the stars; the intersection of the plane of the earth's orbit with the celestial sphere.

ecliptic coordinate system [ASTRON] A celestial coordinate system in which the ecliptic is taken as the primary and the great circles perpendicular to it are then taken as secondaries.

ecliptic diagram [ASTRON] A diagram of the zodiac indicating positions of certain celestial bodies in the ecliptic region.

ecliptic latitude *See* celestial latitude.

ecliptic limits [ASTRON] The distance of the sun from a node of the moon's orbit such that a solar eclipse cannot occur, or the greatest distance of the moon from a node such that an eclipse of the moon cannot occur.

ecliptic longitude *See* celestial longitude.

ecliptic pole [ASTRON] On the celestial sphere, either of two points 90° from the ecliptic.

Eddington transfer equation [ASTROPHYS] A differential equation for the transfer of radiative energy in a stellar atmosphere, taking absorption and scattering of radiation within spectral lines into account:

$$\frac{\cos\theta}{\rho}\frac{dI_\nu(\theta)}{dr} = -(k+l)I_\nu(\theta) + (1-\varepsilon)lJ_\nu + \varepsilon lB + kB$$

where $I_\nu(\theta)$ is the intensity of radiation at angle θ with the outward normal and frequency ν within an absorption line, J_ν is the average of $I_\nu(\theta)$ over all directions, ρ is the mass density of the absorbing material, r is the distance from the center of the star, l is the mass absorption coefficient for selective absorption by the line, k is the continuous mass absorption coefficient at a frequency just outside the line, $(1-\varepsilon)$ is the fraction of the total selectively absorbed energy that is reemitted without frequency change, and B is intensity of blackbody radiation at the frequency of the line and the temperature of the material.

eddy [FL MECH] A vortexlike motion of a fluid running contrary to the main current.

eddy coefficient *See* exchange coefficient.

eddy conduction *See* eddy heat conduction.

eddy conductivity [THERMO] The exchange coefficient for eddy heat conduction.

eddy current [ELECTROMAG] An electric current induced within the body of a conductor when that conductor either moves through a nonuniform magnetic field or is in a region where there is a change in magnetic flux. Also known as Foucault current.

ECLIPSE

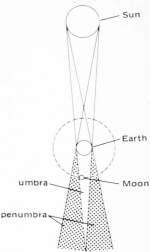

Diagram of a lunar eclipse.

EDDY CURRENT

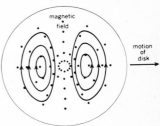

Eddy currents which are induced in a disk moving through a nonuniform magnetic field.

eddy-current loss [ELECTROMAG] Energy loss due to undesired eddy currents circulating in a magnetic core.

eddy diffusion [FL MECH] Diffusion which occurs in turbulent flow, by the rapid process of mixing of the swirling eddies of fluid. Also known as turbulent diffusion.

eddy diffusion coefficient *See* eddy diffusivity.

eddy diffusivity [FL MECH] The exchange coefficient for the diffusion of a conservative property by eddies in a turbulent flow. Also known as coefficient of eddy diffusion; eddy diffusion coefficient.

eddy flux [FL MECH] The rate of transport (or flux) of fluid properties such as momentum, mass heat, or suspended matter by means of eddies in a turbulent motion; the rate of turbulent exchange. Also known as moisture flux; turbulent flux.

eddy heat conduction [THERMO] The transfer of heat by means of eddies in turbulent flow, treated analogously to molecular conduction. Also known as eddy conduction; eddy heat flux.

eddy heat flux *See* eddy heat conduction.

eddy kinetic energy [FL MECH] The kinetic energy of that component of fluid flow which represents a departure from the average kinetic energy of the fluid, the mode of averaging depending on the particular problem. Also known as turbulence energy.

eddy resistance [FL MECH] Resistance or drag of a ship resulting from eddies that are shed from the hull or appendages of the ship and carry away energy.

eddy spectrum [FL MECH] **1.** The distribution of frequencies of rotation of eddies in a turbulent flow, or of those eddies having some range of sizes. **2.** The distribution of kinetic energy among eddies with various frequencies or sizes.

eddy stress *See* Reynolds stress.

eddy velocity [FL MECH] The difference between the mean velocity of fluid flow and the instantaneous velocity at a point. Also known as fluctuation velocity.

eddy viscosity [FL MECH] The turbulent transfer of momentum by eddies giving rise to an internal fluid friction, in a manner analogous to the action of molecular viscosity in laminar flow, but taking place on a much larger scale.

edge [MATH] **1.** A line along which two plane faces of a solid intersect. **2.** A line segment connecting nodes or vertices in a graph (a geometric representation of the relation among situations). Also known as arc.

edge dislocation [CRYSTAL] A dislocation which may be regarded as the result of inserting an extra plane of atoms, terminating along the line of the dislocation. Also known as Taylor-Orowan dislocation.

edge effect [ELEC] An outward-curving distortion of lines of force near the edges of two parallel metal plates that form a capacitor.

edge focusing [ELECTROMAG] Axial focusing of a stream of ions which occurs when it crosses a fringe magnetic field obliquely; used in mass spectrometers and cyclotrons.

edge of regression [MATH] The curve swept out by the characteristic point of a one-parameter family of surfaces.

edge tones [ACOUS] Tones produced when an air jet of sufficient speed is split by a sharp edge.

Edison battery [ELEC] A storage battery composed of cells having nickel and iron in an alkaline solution. Also known as nickel-iron battery.

EDP *See* electronic data processing.

Edser and Butler's bands [OPTICS] Dark bands at intervals of equal frequency in a spectrum of light which has passed through a thin plate of transparent material with parallel sides.

eff *See* efficiency.

effective ampere [ELEC] The amount of alternating current flowing through a resistance that produces heat at the same average rate as 1 ampere of direct current flowing in the same resistance.

effective antenna length [ELECTROMAG] Electrical length of an antenna, as distinguished from its physical length.

effective aperture [OPTICS] The diameter of the image of the aperture stop of an optical system, as viewed from the object.

effective area [ELECTROMAG] Of an antenna in any specified direction, the square of the wavelength multiplied by the power gain (or directive gain) in that direction and divided by 4π (12.57).

effective bandwidth [ELECTR] The bandwidth of an assumed rectangular band-pass having the same transfer ratio at a reference frequency as a given actual band-pass filter, and passing the same mean-square value of a hypothetical current having even distribution of energy throughout that bandwidth.

effective capacitance [ELEC] Total capacitance existing between any two given points of an electric circuit.

effective current [ELEC] The value of alternating current that will give the same heating effect as the corresponding value of direct current. Also known as root-mean-square current.

effective energy [OPTICS] The energy of a quantum of a beam of monochromatic radiation that is absorbed or scattered by a given medium to the same extent as a given beam of polychromatic radiation.

effective force *See* inertial force.

effective half-life [NUCLEO] The half-life of a radioisotope in a biological organism, resulting from a combination of radioactive decay and biological elimination.

effectively computable function *See* computable function.

effective magnetic length [ELECTROMAG] The distance between the effective magnetic poles of a magnet. Also known as equivalent magnetic length.

effective mass [SOLID STATE] A parameter with the dimensions of mass that is assigned to electrons in a solid; in the presence of an external electromagnetic field the electrons behave in many respects as if they were free, but with a mass equal to this parameter rather than the true mass.

effective molecular diameter [PHYS CHEM] The general extent of the electron cloud surrounding a gas molecule as calculated in any of several ways.

effective multiplication factor [NUCLEO] The multiplication factor of an actual reactor, in which there is leakage of neutrons.

effective permeability [PHYS CHEM] The observed permeability exhibited by a porous medium to one fluid phase when there is physical interaction between this phase and other fluid phases present.

effective radiation *See* effective terrestrial radiation.

effective resistance *See* high-frequency resistance.

effective sound pressure [ACOUS] The root-mean-square value of the instantaneous sound pressure at a point during a complete cycle, expressed in dynes per square centimeter. Also known as root-mean-square sound pressure; sound pressure.

effective temperature [ASTROPHYS] A measure of the temperature of a star, deduced by means of the Stefan-Boltzmann law, from the total energy emitted per unit area.

effective terrestrial radiation [GEOPHYS] The amount by which outgoing infrared terrestrial radiation of the earth's surface exceeds downcoming infrared counterradiation from the sky. Also known as effective radiation; nocturnal radiation.

effective thermal resistance [ELECTR] Of a semiconductor device, the effective temperature rise per unit power dissipation of a designated junction above the temperature of a stated external reference point under conditions of thermal equilibrium. Also known as thermal resistance.

effective transformation group [MATH] A transformation group in which the identity element is the only element to leave all points fixed.

effective value *See* root-mean-square value.

effective wavelength [OPTICS] The wavelength of a beam of monochromatic radiation that is absorbed or scattered by a given medium to the same extent as a given beam of polychromatic radiation.

effector [CONT SYS] A motor, solenoid, or hydraulic piston that turns commands to a teleoperator into specific manipulatory actions.

efficiency Abbreviated eff. [NUCLEO] The probability that a count will be produced in a counter tube by a specified particle or quantum incident. [PHYS] The ratio, usually expressed as a percentage, of the useful power output to the power input of a device. [STAT] **1.** An estimator is more efficient than another if it has a smaller variance. **2.** An experimental design is more efficient than another if the same level of precision can be obtained in less time or with less cost. [THERMO] The ratio of the work done by a heat engine to the heat energy absorbed by it. Also known as thermal efficiency.

efficient estimator [STAT] A statistical estimator that has minimum variance.

Egerov's therorem [MATH] If a sequence of measurable functions converges almost everywhere on a set of finite measure to a real-valued function, then given any $\varepsilon > 0$ there is a set of measure smaller than ε on whose complement the sequence converges uniformly.

Egerton's effusion method [THERMO] A method of determining vapor pressures of solids at high temperatures, in which one measures the mass lost by effusion from a sample placed in a tightly sealed silica pot with a small hole; the pot rests at the bottom of a tube that is evacuated for several hours, and is maintained at a high temperature by a heated block of metal surrounding it.

Ehrenfest's adiabatic law [QUANT MECH] The law that, if the Hamiltonian of a system undergoes an infinitely slow change, and if the system is initially in an eigenstate of the Hamiltonian, then at the end of the change it will be in the eigenstate of the new Hamiltonian that derives from the original state by continuity, provided certain conditions are met. Also known as Ehrenfest's theorem.

Ehrenfest's equations [THERMO] Equations which state that for the phase curve $P(T)$ of a second-order phase transition the derivative of pressure P with respect to temperature T is equal to $(C_p{}^f - C_p{}^i)/TV(\gamma^f - \gamma^i) = (\gamma^f - \gamma^i)/(K^f - K^i)$, where i and f refer to the two phases, γ is the coefficient of volume expansion, K is the compressibility, C_p is the specific heat at constant pressure, and V is the volume.

Ehrenfest's theorem [QUANT MECH] **1.** The theorem that a quantum-mechanical wave packet obeys the equations of motion of the corresponding classical particle when the position, momentum, and force acting on the particle are replaced by the expectation values of these quantities. **2.** *See* Ehrenfest's adiabatic law.

Ehrenhaft effect [ELECTROMAG] A helical motion of fine

particles along the lines of force of a magnetic field during exposure to light, resulting from radiometer effects.

E-H T junction [ELECTROMAG] In microwave waveguides, a combination of E- and H-plane T junctions forming a junction at a common point of intersection with the main waveguide.

E-H tuner [ELECTROMAG] Tunable E-H T junction having two arms terminated in adjustable plungers used for impedance transformation.

eigenfrequency [PHYS] One of the frequencies at which an oscillatory system can vibrate.

eigenfunction [MATH] Also known as characteristic function. **1.** An eigenvector for a linear operator on a vector space whose vectors are functions. **2.** A solution to the Sturm-Liouville partial differential equation.

eigenfunction expansion [MATH] By using spectral theory for linear operators defined on spaces composed of functions, in certain cases the operator equals an integral or series involving its eigenvectors; this is known as its eigenfunction expansion and is particularly useful in studying linear partial differential equations.

eigenmatrix [MATH] Corresponding to a diagonalizable matrix or linear transformation, this is the matrix all of whose entries are 0 save those on the principal diagonal where appear the eigenvalues.

eigenstate [QUANT MECH] **1.** A dynamical state whose state vector (or wave function) is an eigenvector (or eigenfunction) of an operator corresponding to a specified physical quantity. **2.** *See* energy state.

eigenvalue [MATH] One of the scalars λ such that $T(v) = \lambda v$, where T is a linear operator on a vector space, and v is an eigenvector. Also known as characteristic number; characteristic root; characteristic value; latent root.

eigenvalue equation *See* characteristic equation.

eigenvalue problem *See* Sturm-Liouville problem.

eigenvector [MATH] A nonzero vector v whose direction is not changed by a given linear transformation T; that is, $T(v) = \lambda v$ for some scalar λ. Also known as characteristic vector.

eikonal equation [PHYS] An equation for propagation of electromagnetic or acoustic waves in a nonhomogeneous medium; it is valid only when the variation of the properties of the medium is small over the distance of a wavelength.

eikonometer [OPTICS] A scale used to measure sizes of objects viewed through a microscope, usually attached to the eyepiece so that it is seen superimposed on the image.

einstein [PHYS] A unit of light energy used in photochemistry, equal to Avogadro's number times the energy of one photon of light of the frequency in question.

Einstein-Bohr equation [QUANT MECH] In a system undergoing a transition between two states so that it emits or absorbs radiation, that equation indicating that the radiation frequency equals the difference in energy between the two states divided by Planck's constant.

Einstein-Bose statistics *See* Bose-Einstein statistics.

Einstein characteristic temperature [SOLID STATE] A temperature, characteristic of a substance, that appears in Einstein's equation for specific heat; it is equal to the product of Planck's constant and the Einstein frequency divided by Boltzmann's constant.

Einstein condensation *See* Bose-Einstein condensation.

Einstein-de Haas effect [ELECTROMAG] A freely suspended body consisting of a ferromagnetic material acquires a rotation when its magnetization changes.

Einstein-de Haas method [ELECTROMAG] Method of measuring the gyromagnetic ratio of a ferromagnetic substance;

one measures the angular displacement induced in a ferromagnetic cylinder suspended from a torsion fiber when magnetization of the object is reversed, and the magnetization change is measured with a magnetometer.

Einstein–de Sitter model [RELAT] A model of the universe in which ordinary euclidean geometry holds good, the distribution of matter extends infinitely at all times, and the universe expands from an infinitely condensed state at such a rate that the density is inversely proportional to the square of the time elapsed since the beginning of the expansion.

Einstein diffusion equation [STAT MECH] An equation which gives the mean square displacement caused by Brownian movement of spherical, colloidal particles in a gas or liquid.

Einstein displacement *See* Einstein shift.

Einstein elevator [RELAT] A windowless elevator freely falling in its shaft, inside of which conditions resemble interstellar space; used to elucidate the principle of equivalence.

Einstein equations [STAT MECH] Equations for the density and pressure of a Bose-Einstein gas in terms of power series in a parameter which appears in the Bose-Einstein distribution law.

Einstein frequency [SOLID STATE] Single frequency with which each atom vibrates independently of other atoms, in a model of lattice vibrations; equal to the frequency observed in infrared absorption studies.

Einstein frequency condition [SOLID STATE] The assumption that all vibrations of a crystal lattice are harmonic with the same characteristic frequency.

einsteinium [CHEM] Synthetic radioactive element, symbol Es, atomic number 99; discovered in debris of 1952 hydrogen bomb explosion; now made in cyclotrons.

Einstein mass-energy relation [RELAT] The relation in which the energy of a system is equivalent to its mass times the square of the speed of light.

Einstein number [PL PHYS] A dimensionless number used in magnetofluid dynamics, equal to the ratio of the velocity of a fluid to the speed of light.

Einstein partition function [STAT MECH] The partition function for a solid, based on the Einstein frequency condition.

Einstein photochemical equivalence law [PHYS CHEM] The law that each molecule taking part in a chemical reaction caused by electromagnetic radiation absorbs one photon of the radiation. Also known as Stark-Einstein law.

Einstein photoelectric law [QUANT MECH] The law that the energy of an electron emitted from a system in the photoelectric effect is $h\nu - W$, where h is Planck's constant, ν is the frequency of the incident radiation, and W is the energy needed to remove the electron from the system; if $h\nu$ is less than W, no electrons are emitted.

Einstein-Planck law [QUANT MECH] The law that the energy of a photon is given by Planck's constant times the frequency. [RELAT] The equation of motion of a charged particle in an electromagnetic field, according to which its rate of change of momentum is equal to the Lorentz force, where the magnitude of the momentum is $mv/(1 - v^2/c^2)^{1/2}$, where m and v are the particle's mass and velocity, and c is the speed of light.

Einstein relation [PHYS] The relation in which the mobility of charges in an ionic solution or semiconductor is equal to the magnitude of the charge times the diffusion coefficient divided by the product of the Boltzmann constant and the absolute temperature.

Einstein-Rosen waves [RELAT] Gravitational waves produced by oscillating ponderable matter, along an infinitely long cylindrical axis, in an exact solution of Einstein's field equations.

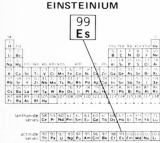

EINSTEINIUM

Periodic table of the chemical elements showing the position of einsteinium.

Einstein's absorption coefficient [ATOM PHYS] Proportionality constant governing the absorption of electromagnetic radiation by atoms, equal to the number of quanta absorbed per second divided by the product of the energy of radiation per unit volume per unit wave number and the number of atoms in the ground state.

Einstein's coefficient of spontaneous emission [ATOM PHYS] Proportionality constant governing the rate at which atoms or molecules pass spontaneously from an upper energy state to a lower one by emission of radiation, equal to the number of such transitions per second divided by the number of atoms in the upper state.

Einstein's coefficient of stimulated emission [ATOM PHYS] Proportionality constant governing the rate at which atoms or molecules pass from an upper energy state to a lower one by stimulated emission of radiation, equal to the number of such transitions per second divided by the product of energy of the radiation inducing the transition per unit volume per unit wave number and the number of atoms in the upper state.

Einstein's equation for specific heat [SOLID STATE] The earliest equation based on quantum mechanics for the specific heat of a solid; uses the assumption that each atom oscillates with the same frequency.

Einstein's equivalency principle *See* equivalence principle.

Einstein's field equations [RELAT] Those equations relevant to the relationship in which the Einstein tensor equals -8π times the energy momentum tensor times the gravitational constant divided by the square of the speed of light. Also known as Einstein's law of gravitation.

Einstein shift [RELAT] A shift toward longer wavelengths of spectral lines emitted by atoms in strong gravitational fields. Also known as Einstein displacement.

Einstein's law of gravitation *See* Einstein's field equations.

Einstein space [MATH] A Riemannian space in which the contracted curvature tensor is proportional to the metric tensor.

Einstein's principle of relativity [RELAT] The principle that all the laws of physics must assume the same mathematical form in any inertial frame of reference; thus, it is impossible to determine the absolute motion of a system by any means.

Einstein's summation convention [MATH] A notational convenience used in tensor analysis whereupon it is agreed that any term in which an index appears twice will stand for the sum of all such terms as the index assumes all of a preassigned range of values.

Einstein static universe [RELAT] A nonvacuum, globally static solution to Einstein's equations of general relativity with cosmological term.

Einstein's unified field theories [RELAT] A series of theories attempting to express a general unifying principle underlying electromagnetism and gravity.

Einstein tensor [RELAT] The tensor expressed as $E_{\mu\nu} = R_{\mu\nu} - \frac{1}{2}(g_{\mu\nu}R - 2\Lambda)$, where $R_{\mu\nu}$ is the contracted curvature tensor, R is the curvature of space-time, $g_{\mu\nu}$ is the metric tensor, and Λ is the cosmological constant.

Einstein universe [RELAT] A model of the universe which is a four-dimensional cylindrical surface in a five-dimensional space.

Einstein viscosity equation [PHYS CHEM] An equation which gives the viscosity of a sol in terms of the volume of dissolved particles divided by the total volume.

Einthoven galvanometer *See* string galvanometer.

Eisenstein's criterion [MATH] A polynomial $a_nX^n + \cdots + a_0$, with coefficients $a_0, \ldots, a_n$ in a unique factorization domain R, is irreducible over the quotient field of R if there

EKMAN SPIRAL

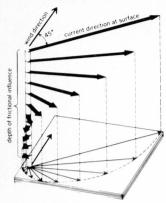

Schematic of pure wind current in deep water, showing decrease in velocity and change of direction at regular intervals of depth (the Ekman spiral).

is a prime p in R such that p does not divide a_n, p divides a_i for all i less than n, and p^2 does not divide a_0.

Ekman spiral [METEOROL] A theoretical representation that a wind blowing steadily over an ocean of unlimited depth and extent and uniform viscosity would cause, in the Northern Hemisphere, the immediate surface water to drift at an angle of 45° to the right of the wind direction, and the water beneath to drift further to the right, and with slower and slower speeds, as one goes to greater depths.

elastance [ELEC] The reciprocal of capacitance.

elastic [MECH] Capable of sustaining deformation without permanent loss of size or shape.

elastica [MECH] The elastic curve formed by a uniform rod that is originally straight, then is bent in a principal plane by applying forces, and couples only at its ends.

elastic aftereffect [MECH] The delay of certain substances in regaining their original shape after being deformed within their elastic limits. Also known as elastic lag.

elastic axis [MECH] The lengthwise line of a beam along which transverse loads must be applied in order to produce bending only, with no torsion of the beam at any section.

elastic body [MECH] A solid body for which the additional deformation produced by an increment of stress completely disappears when the increment is removed. Also known as elastic solid.

elastic buckling [MECH] An abrupt increase in the lateral deflection of a column at a critical load while the stresses acting on the column are wholly elastic.

elastic center [MECH] That point of a beam in the plane of the section lying midway between the flexural center and the center of twist in that section.

elastic collision [MECH] A collision in which the sum of the kinetic energies of translation of the participating systems is the same after the collision as before.

elastic constant *See* compliance constant; stiffness constant.

elastic cross section [PHYS] The cross section for an elastic collision between two particles or systems.

elastic curve [MECH] The curved shape of the longitudinal centroidal surface of a beam when the transverse loads acting on it produced wholly elastic stresses.

elastic deformation [MECH] Reversible alteration of the form or dimensions of a solid body under stress or strain.

elastic equilibrium [MECH] The condition of an elastic body in which each volume element of the body is in equilibrium under the combined effect of elastic stresses and externally applied body forces.

elastic failure [MECH] Failure of a body to recover its original size and shape after a stress is removed.

elastic flow [MECH] Return of a material to its original shape following deformation.

elastic force [MECH] A force arising from the deformation of a solid body which depends only on the body's instantaneous deformation and not on its previous history, and which is conservative.

elastic hysteresis [MECH] Phenomenon exhibited by some solids in which the deformation of the solid depends not only on the stress applied to the solid but also on the previous history of this stress; analogous to magnetic hysteresis, with magnetic field strength and magnetic induction replaced by stress and strain respectively.

elasticity [MECH] 1. The property whereby a solid material changes its shape and size under action of opposing forces, but recovers its original configuration when the forces are removed. 2. The existence of forces which tend to restore to its original position any part of a medium (solid or fluid) which has been displaced.

elasticity modulus *See* modulus of elasticity.

elasticity number 1 [FL MECH] A dimensionless number which is a measure of the ratio of elastic forces to inertial forces on a viscoelastic fluid flowing in a pipe, and is equal to the product of the fluid's relaxation time and its dynamic viscosity, divided by the product of the fluid's density and the square of the radius of the pipe. Symbolized N_{El_1}.

elasticity number 2 [FL MECH] A dimensionless number used in studying the effect of elasticity on a flow process, equal to the fluid's density times its specific heat at constant pressure, divided by the product of its coefficient of bulk expansion and its bulk modulus. Symbolized K_E.

elastic lag *See* elastic aftereffect.

elastic limit [MECH] The maximum stress a solid can sustain without undergoing permanent deformation.

elastic modulus *See* modulus of elasticity.

elasticoviscosity [FL MECH] That property of a fluid whose rate of deformation under stress is the sum of a part corresponding to a viscous Newtonian fluid and a part obeying Hooke's law.

elastic potential energy [MECH] Capacity that a body has to do work by virtue of its deformation.

elastic ratio [MECH] The ratio of the elastic limit to the ultimate strength of a solid.

elastic recovery [MECH] That fraction of a given deformation of a solid which behaves elastically.

elastic scattering [MECH] Scattering due to an elastic collision.

elastic solid *See* elastic body.

elastic strain energy [MECH] The work done in deforming a solid within its elastic limit.

elastic theory [MECH] Theory of the relations between the forces acting on a body and the resulting changes in dimensions.

elastic vibration [MECH] Oscillatory motion of a solid body which is sustained by elastic forces and the inertia of the body.

elastic wave [ACOUS] *See* acoustic wave. [PHYS] A wave propagated by a medium having inertia and elasticity (the existence of forces which tend to restore any part of a medium to its original position), in which displaced particles transfer momentum to adjoining particles, and are themselves restored to their original position.

elastodynamics [MECH] The study of the mechanical properties of elastic waves.

elastoplasticity [MECH] State of a substance subjected to a stress greater than its elastic limit but not so great as to cause it to rupture, in which it exhibits both elastic and plastic properties.

elastoresistance [ELEC] The change in a material's electrical resistance as it undergoes a stress within its elastic limit.

E layer [GEOPHYS] A layer of ionized air occurring at altitudes between 100 and 120 kilometers in the E region of the ionosphere, capable of bending radio waves back to earth. Also known as Heaviside layer; Kennelly-Heaviside layer.

elbow [ELECTROMAG] In a waveguide, a bend of comparatively short radius, normally 90°, and sometimes for acute angles down to 15°.

electret [ELEC] A solid dielectric possessing persistent electric polarization, by virtue of a long time constant for decay of a charge instability.

electric [ELEC] Containing, producing, arising from, or actuated by electricity; often used interchangeably with electrical.

electrical [ELEC] Related to or associated with electricity,

**ELASTIC
POTENTIAL ENERGY**

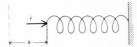

A compressed spring possesses potential energy. Here s is the distance that the spring has been compressed from its normal length, and $\bar{f}$ is the average force of compression. It can be shown from Hooke's law that $\bar{f} = \frac{1}{2}ks$, where k is a constant known as the stiffness coefficient of the spring. It follows that the elastic potential energy of the compressed spring is $\frac{1}{2}ks^2$.

but not containing it or having its properties or characteristics; often used interchangeably with electric.

electrical analog [PHYS] An electric circuit whose behavior may be described by the same mathematical equations as some physical system under study.

electrical angle [ELEC] An angle that specifies a particular instant in an alternating-current cycle or expresses the phase difference between two alternating quantities; usually expressed in electrical degrees.

electrical axis [SOLID STATE] The x axis in a quartz crystal; there are three such axes in a crystal, each parallel to one pair of opposite sides of the hexagon; all pass through and are perpendicular to the optical, or z, axis.

electrical calorimeter [ANALY CHEM] Device to measure heat evolved (from fusion or vaporization, for example); measured quantities of heat are added electrically to the sample, and the temperature rise is noted.

electrical center [ELEC] Point approximately midway between the ends of an inductor or resistor that divides the inductor or resistor into two equal electrical values.

electrical condenser *See* capacitor.

electrical conductance *See* conductance.

electrical conduction *See* conduction.

electrical conductivity *See* conductivity.

electrical degree [ELEC] A unit equal to $1/360$ cycle of an alternating quantity.

electrical distance [ELECTROMAG] The distance between two points, expressed in terms of the duration of travel of an electromagnetic wave in free space between the two points.

electrical equipment [ELEC] Apparatus, appliances, devices, wiring, fixtures, fittings, and material used as a part of or in connection with an electrical installation.

electrical impedance Also known as impedance. [ELEC] **1.** The total opposition that a circuit presents to an alternating current, equal to the complex ratio of the voltage to the current in complex notation. Also known as complex impedance. **2.** The ratio of the maximum voltage in an alternating-current circuit to the maximum current; equal to the magnitude of the quantity in the first definition.

electrical impedance meter [ELEC] An instrument which measures the complex ratio of voltage to current in a given circuit at a given frequency. Also known as impedance meter.

electrical instability [ELEC] A persistent condition of unwanted self-oscillation in an amplifier or other electric circuit.

electrical insulation *See* insulation.

electrical insulator *See* insulator.

electrical interference *See* interference.

electrical length [ELECTROMAG] The length of a conductor expressed in wavelengths, radians, or degrees.

electrical loading *See* loading.

electrically connected [ELEC] Connected by means of a conducting path, or through a capacitor, as distinguished from connection merely through electromagnetic induction.

electrically suspended gyro [ENG] A gyroscope in which the main rotating element is suspended by an electromagnetic or an electrostatic field.

electrical measurement [ELEC] The measurement of any one of the many quantities by which electricity is characterized.

electrical potential energy [ELEC] Energy possessed by electric charges by virtue of their position in an electrostatic field.

electrical pressure transducer *See* pressure transducer.

electrical properties [ELEC] Properties of a substance which

ELECTRICAL IMPEDANCE METER

The Madsen impedance meter uses an acoustic bridge as the basic component for testing deafness.

determine its response to an electric field, such as its dielectric constant or conductivity.

electrical resistance   *See* resistance.

electrical resistance meter   *See* resistance meter.

electrical resistance thermometer   *See* resistance thermometer.

electrical resistivity   [ELEC] The electrical resistance offered by a material to the flow of current, times the cross-sectional area of current flow and per unit length of current path; the reciprocal of the conductivity. Also known as resistivity; specific resistance.

electrical resistor   *See* resistor.

electrical resonator   *See* tank circuit.

electrical unit   [ELEC] A standard in terms of which some electrical quantity is evaluated.

electrical weighing system   [ENG] An instrument which weighs an object by measuring the change in resistance caused by the elastic deformation of a mechanical element loaded with the object.

electric arc   [ELEC] A discharge of electricity through a gas, normally characterized by a voltage drop approximately equal to the ionization potential of the gas. Also known as arc.

electric-arc lamp   *See* arc lamp.

electric charge   *See* charge.

electric chopper   [ELECTROMAG] A chopper in which an electromagnet driven by a source of alternating current sets into vibration a reed carrying a moving contact that alternately touches two fixed contacts in a signal circuit, thus periodically interrupting the signal.

electric circuit   [ELEC] Also known as circuit. **1.** A path or group of interconnected paths capable of carrying electric currents. **2.** An arrangement of one or more complete, closed paths for electron flow.

electric circuit theory   *See* circuit theory.

electric coil   *See* coil.

electric connection   [ELEC] A direct wire path for current between two points in a circuit.

electric constant   [ELEC] The permittivity of empty space, equal to 1 in centimeter-gram-second electrostatic units and to $10^7/4\pi c^2$ farads per meter or, numerically, to 8.854×10^{-12} farads per meter in International System units, where c is the speed of light in meters per second. Symbolized ε_0.

electric contact   [ELEC] A physical contact that permits current flow between conducting parts. Also known as contact.

electric corona   *See* corona discharge.

electric current   *See* current.

electric current density   *See* current density.

electric dipole   [ELEC] A localized distribution of positive and negative electricity, without net charge, whose mean positions of positive and negative charges do not coincide.

electric dipole moment   [ELEC] A quantity characteristic of a charge distribution, equal to the vector sum over the electric charges of the product of the charge and the position vector of the charge. Also known as dipole moment.

electric dipole transition   [ATOM PHYS] A transition of an atom or nucleus from one energy state to another, in which electric dipole radiation is emitted or absorbed.

electric-discharge lamp   *See* discharge lamp.

electric-discharge tube   *See* discharge tube.

electric displacement   [ELEC] The electric field intensity multiplied by the permittivity. Symbolized D. Also known as dielectric displacement; dielectric flux density; displacement; electric displacement density; electric flux density; electric induction.

ELECTRIC CHOPPER

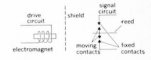

Basic chopper structure.

electric displacement density *See* electric displacement.

electric double layer [PHYS CHEM] A phenomenon found at a solid-liquid interface; it is made up of ions of one charge type which are fixed to the surface of the solid and an equal number of mobile ions of the opposite charge which are distributed through the neighboring region of the liquid; in such a system the movement of liquid causes a displacement of the mobile counterions with respect to the fixed charges on the solid surface. Also known as double layer.

electric doublet *See* dipole.

electric energy [ELECTROMAG] 1. Energy of electric charges by virtue of their position in an electric field. 2. Energy of electric currents by virtue of their position in a magnetic field.

electric energy measurement [ELEC] The measurement of the integral, with respect to time, of the power in an electric circuit.

electric eye *See* cathode-ray tuning indicator; photocell; phototube.

electric field [ELEC] 1. One of the fundamental fields in nature, causing a charged body to be attracted to or repelled by other charged bodies; associated with an electromagnetic wave or a changing magnetic field. 2. Specifically, the electric force per unit test charge.

electric field effect *See* Stark effect.

electric field intensity *See* electric field vector.

electric field strength *See* electric field vector.

electric field vector [ELEC] The force on a stationary positive charge per unit charge at a point in an electric field. Designated E. Also known as electric field intensity; electric field strength; electric vector.

electric filter [ELECTR] 1. A network that transmits alternating currents of desired frequencies while substantially attenuating all other frequencies. Also known as frequency-selective device. 2. *See* filter.

electric flux [ELEC] 1. The integral over a surface of the component of the electric displacement perpendicular to the surface; equal to the number of electric lines of force crossing the surface. 2. The electric lines of force in a region.

electric flux density *See* electric displacement.

electric flux line *See* electric line of force.

electric forming [ELECTR] The process of applying electric energy to a semiconductor or other device to modify permanently its electrical characteristics.

electric fuse *See* fuse.

electric generator *See* generator.

electric hysteresis *See* ferroelectric hysteresis.

electric image [ELEC] A fictitious charge used in finding the electric field set up by fixed electric charges in the neighborhood of a conductor; the conductor, with its distribution of induced surface charges, is replaced by one or more of these fictitious charges. Also known as image.

electric induction *See* electric displacement.

electric instrument [ENG] An electricity-measuring device that indicates, such as an ammeter or voltmeter, in contrast to an electric meter that totalizes or records.

electricity [PHYS] Physical phenomenon involving electric charges and their effects when at rest and when in motion.

electric line of force [ELEC] An imaginary line drawn so that each segment of the line is parallel to the direction of the electric field or of the electric displacement at that point, and the density of the set of lines is proportional to the electric field or electrical displacement. Also known as electric flux line.

electric meter [ENG] An electricity-measuring device that

ELECTRIC IMAGE

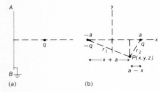

(a)   (b)

Illustration of the method of electric images. *(a)* Point charge q at distance a in front of an infinite grounded conducting plane AB. The problem is to find the potential function and equation for electric field intensity in the region to the right of AB, the field being caused by point charge q, and its equal distributed induced charge on the conducting plane. *(b)* Physically simpler but mathematically equivalent situation for the region of interest. Infinite conducting plane is replaced by point charge $-q$. Potential at P with coordinates (x,y,z) is that resulting from charges $-q$ and q at distances r_1 and r_2 respectively.

totalizes with time, such as a watthour meter or ampere-hour meter, in contrast to an electric instrument.

electric moment [ELEC] One of a series of quantities characterizing an electric charge distribution; an lth moment is given by integrating the product of the charge density, the lth power of the distance from the origin, and a spherical harmonic Y^*_{lm} over the charge distribution.

electric monopole [ELEC] A distribution of electric charge which is concentrated at a point or is spherically symmetric.

electric motor *See* motor.

electric multipole [ELECTROMAG] One of a series of types of static or oscillating charge distributions; the multipole of order 1 is a point charge or a spherically symmetric distribution, and the electric and magnetic fields produced by an electric multipole of order 2^n are equivalent to those of two electric multipoles of order 2^{n-1} of equal strengths, but opposite sign, separated from each other by a short distance.

electric multipole field [ELECTROMAG] The electric and magnetic fields generated by a static or oscillating electric multipole.

electric network *See* network.

electric octupole moment [ELEC] A quantity characterizing an electric charge distribution; obtained by integrating the product of the charge density, the third power of the distance from the origin, and a spherical harmonic Y^*_{3m} over the charge distribution.

electric polarizability [ELEC] Induced dipole moment of an atom or molecule in a unit electric field.

electric polarization *See* polarization.

electric potential [ELEC] The work which must be done against electric forces to bring a unit charge from a reference point to the point in question; the reference point is located at an infinite distance, or, for practical purposes, at the surface of the earth or some other large conductor. Also known as electrostatic potential; potential.

electric power [ELEC] The rate at which electric energy is converted to other forms of energy, equal to the product of the current and the voltage drop.

electric probe [PL PHYS] A device used to measure electron temperatures, electron and ion densities, space and wall potentials, and random electron currents in a plasma; consists substantially of one or two small collecting electrodes to which various potentials are applied, with the corresponding collection currents being measured. Also known as electrostatic probe.

electric quadrupole [ELEC] A charge distribution that produces an electric field equivalent to that produced by two electric dipoles whose dipole moments have the same magnitude but point in opposite directions and which are separated from each other by a small distance.

electric quadrupole lens [ELECTR] A device for focusing beams of charged particles which has four electrodes with alternately positive and negative polarity; used in electron microscopes and particle accelerators.

electric quadrupole moment [ELEC] A quantity characterizing an electric charge distribution, obtained by integrating the product of the charge density, the second power of the distance from the origin, and a spherical harmonic Y^*_{2m} over the charge distribution.

electric quadrupole transition [ATOM PHYS] A transition of an atom or molecule from one energy state to another, in which electric quadrupole radiation is emitted or absorbed.

electric reactor *See* reactor.

electric relay *See* relay.

electric shielding [ELECTROMAG] Any means of avoiding

ELECTRIC PROBE

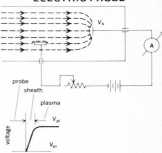

Schematic diagram of a single electric probe.

pickup of undesired signals or noise, suppressing radiation of undesired signals, or confining wanted signals to desired paths or regions, such as electrostatic shielding or electromagnetic shielding. Also known as screening; shielding.

electric shock tube [PL PHYS] A gas-filled tube used in plasma physics to ionize a gas suddenly; a capacitor bank charged to a high voltage is discharged into the gas at one tube end to ionize and heat the gas, producing a shock wave that may be studied as it travels down the tube.

electric solenoid *See* solenoid.

electric spark *See* spark.

electric strength *See* dielectric strength.

electric susceptibility [ELEC] A dimensionless parameter measuring the ease of polarization of a dielectric, equal (in meter-kilogram-second units) to the ratio of the polarization to the product of the electric field strength and the vacuum permittivity. Also known as dielectric susceptibility.

electric switch *See* switch.

electric terminal *See* terminal.

electric transient [ELEC] A temporary component of current and voltage in an electric circuit which has been disturbed.

electric twinning [SOLID STATE] A defect occurring in natural quartz crystals, in which adjacent regions of quartz have their electric axes oppositely poled.

electric vector *See* electric field vector.

electric wave [ELECTROMAG] An electromagnetic wave, especially one whose wavelength is at least a few centimeters. Also known as Hertzian wave.

electric-wave filter *See* filter.

electric wind *See* convective discharge.

electric wire *See* wire.

electric wiring *See* wiring.

electrization [ELEC] The electric polarization divided by the permittivity of empty space.

electroacoustic effect *See* acoustoelectric effect.

electroacoustics [ENG ACOUS] The conversion of acoustic energy and waves into electric energy and waves, or vice versa.

electroacoustic transducer [ENG ACOUS] A transducer that receives waves from an electric system and delivers waves to an acoustic system, or vice versa. Also known as sound transducer.

electrocapillarity [PHYS] A change in the surface tension of a liquid caused by an electric field at the surface.

electrochemical cell [PHYS CHEM] A combination of two electrodes arranged so that an overall oxidation-reduction reaction produces an electromotive force; includes dry cells, wet cells, standard cells, fuel cells, solid-electrolyte cells, and reserve cells.

electrochemical effect [PHYS CHEM] Conversion of chemical to electric energy, as in electrochemical cells; or the reverse process, used to produce elemental aluminum, magnesium, and bromine from compounds of these elements.

electrochemical emf [PHYS CHEM] Electrical force generated by means of chemical action, in manufactured cells (such as dry batteries) or by natural means (galvanic reaction).

electrochemical equivalent [PHYS CHEM] The weight in grams of a substance produced or consumed by electrolysis with 100% current efficiency during the flow of a quantity of electricity equal to 1 faraday ($96,487.0 \pm 1.6$ coulombs).

electrochemical potential [PHYS CHEM] The difference in potential that exists when two dissimilar electrodes are connected through an external conducting circuit and the two electrodes are placed in a conducting solution so that electrochemical reactions occur.

electrochemical reduction cell [PHYS CHEM] The cathode component of an electrochemical cell, at which chemical reduction occurs (while at the anode, chemical oxidation occurs).

electrochemical series [PHYS CHEM] A series in which the metals and other substances are listed in the order of their chemical reactivity or electrode potentials, the most reactive at the top and the less reactive at the bottom. Also known as electromotive series.

electrochemical thermodynamics [THERMO] The application of the laws of thermodynamics to electrochemical systems.

electrochemical valve [ELEC] Electric valve consisting of a metal in contact with a solution or compound, across the boundary of which current flows more readily in one direction than in the other direction, and in which the valve action is accompanied by chemical changes.

electrochemistry [PHYS CHEM] A branch of chemistry dealing with chemical changes accompanying the passage of an electric current; or with the reverse process, in which a chemical reaction is used to produce an electric current.

electrocyclic reaction [PHYS CHEM] The interconversion of a linear π-system containing n π-electrons and a cyclic molecule containing $n - 2$ π-electrons which is formed by joining the ends of the linear molecule.

electrode [ELEC] **1.** An electric conductor through which an electric current enters or leaves a medium, whether it be an electrolytic solution, solid, molten mass, gas, or vacuum. **2.** One of the terminals used in dielectric heating or diathermy for applying the high-frequency electric field to the material being heated.

electrode admittance [ELECTR] Quotient of dividing the alternating component of the electrode current by the alternating component of the electrode voltage, all other electrode voltages being maintained constant.

electrode capacitance [ELECTR] Capacitance between one electrode and all the other electrodes connected together.

electrode characteristic [ELECTR] Relation between the electrode voltage and the current to an electrode, all other electrode voltages being maintained constant.

electrode conductance [ELECTR] Quotient of the inphase component of the electrode alternating current by the electrode alternating voltage, all other electrode voltage being maintained constant; this is a variational and not a total conductance. Also known as grid conductance.

electrode couple [ELEC] The pair of electrodes in an electric cell, between which there is a potential difference.

electrode current [ELECTR] Current passing to or from an electrode, through the interelectrode space within a vacuum tube.

electrode dark current [ELECTR] The electrode current that flows when there is no radiant flux incident on the photocathode in a phototube or camera tube. Also known as dark current.

electrode efficiency [PHYS CHEM] The ratio of the amount of metal actually deposited in an electrolytic cell to the amount that could theoretically be deposited as a result of electricity passing through the cell.

electrode impedance [ELECTR] Reciprocal of the electrode admittance.

electrode inverse current [ELECTR] Current flowing through an electrode in the direction opposite to that for which the tube is designed.

electrodeless discharge [ELECTR] An electric discharge

generated by placing a discharge tube in a strong, high-frequency electromagnetic field.

electrodeless lamp [ELECTR] A lamp based on an electrodeless discharge.

electrode potential Also known as electrode voltage. [ELECTR] The instantaneous voltage of an electrode with respect to the cathode of an electron tube. [PHYS CHEM] The voltage existing between an electrode and the solution or electrolyte in which it is immersed; usually, electrode potentials are referred to a standard electrode, such as the hydrogen electrode.

electrode resistance [ELECTR] Reciprocal of the electrode conductance; this is the effective parallel resistance and is not the real component of the electrode impedance.

electrode voltage *See* electrode potential.

electrodisintegration [NUC PHYS] The breakup of a nucleus into two or more fragments as a result of bombardment by electrons.

electrodynamic drift [GEOPHYS] Motion of ions in the upper atmosphere due to the electric or geomagnetic field.

electrodynamic instrument [ENG] An instrument that depends for its operation on the reaction between the current in one or more movable coils and the current in one or more fixed coils. Also known as electrodynamometer.

electrodynamics [ELECTROMAG] The study of the relations between electrical, magnetic, and mechanical phenomena.

electrodynamic shaker *See* shaker.

electrodynamic wattmeter [ENG] An electrodynamic instrument connected as a wattmeter, with the main current flowing through the fixed coil, and a small current proportional to the voltage flowing through the movable coil. Also known as moving-coil wattmeter.

electrodynamometer *See* electrodynamic instrument.

electrofluid [FL MECH] Newtonian (or shear-thinning) fluid whose rheological or flow properties are changed into those of a viscoplastic type by the addition of electric-field modulation.

electrogasdynamics [PHYS] Conversion of the kinetic energy of a moving gas to electricity, for such applications as high-voltage electric power generation, air-pollution control, and paint spraying.

electrojet [GEOPHYS] A stream of electricity moving in the upper atmosphere around the equator and in polar regions, where it produces auroras.

electrokinetic potential *See* zeta potential.

electrokinetics [ELECTROMAG] The study of the motion of electric charges, especially of steady currents in electric circuits, and of the motion of electrified particles in electric or magnetic fields.

electrokinetic transducer [ELEC] An instrument which converts dynamic physical forces, such as vibration and sound, into corresponding electric signals by measuring the streaming potential generated by passage of a polar fluid through a permeable refractory-ceramic or fritted-glass member between two chambers.

electrokinetograph [ENG] An instrument used to measure ocean current velocities based on their electrical effects in the magnetic field of the earth.

electroluminescence [ELECTR] The emission of light, not due to heating effects alone, resulting from application of an electric field to a material, usually solid.

electrolyte [PHYS CHEM] A chemical compound which when molten or dissolved in certain solvents, usually water, will conduct an electric current.

electrolyte-activated battery [ELEC] A reserve battery in

ELECTRODYNAMIC INSTRUMENT

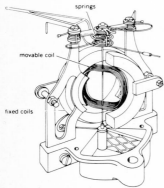

springs

movable coil

fixed coils

Electrodynamic instrument mechanism. *(Weston Instruments, Division of Sangamo Weston, Inc.)*

which an aqueous electrolyte is stored in a separate chamber, and a mechanism, which may be operated from a remote location, drives the electrolyte out of the reservoir and into the cells of the battery for activation.

electrolytic arrester   *See* aluminum-cell arrester.

electrolytic capacitor   [ELEC] A capacitor consisting of two electrodes separated by an electrolyte; a dielectric film, usually a thin layer of gas, is formed on the surface of one electrode. Also known as electrolytic condenser.

electrolytic cell   [PHYS CHEM] A cell consisting of electrodes immersed in an electrolyte solution, for carrying out electrolysis.

electrolytic condenser   *See* electrolytic capacitor.

electrolytic conductance   [PHYS CHEM] The transport of electric charges, under electric potential differences, by charged particles (called ions) of atomic or larger size.

electrolytic conductivity   [PHYS CHEM] The conductivity of a medium in which the transport of electric charges, under electric potential differences, is by particles of atomic or larger size.

electrolytic migration   [PHYS CHEM] The motions of ions in a liquid under the action of an electric field.

electrolytic polarization   [PHYS CHEM] The existence of a minimum potential difference necessary to cause a steady current to flow through an electrolytic cell, resulting from the tendency of the products of electrolysis to recombine.

electrolytic rectifier   [ELEC] A rectifier consisting of metal electrodes in an electrolyte, in which rectification of alternating current is accompanied by electrolytic action; polarizing film formed on one electrode permits current flow in one direction but not the other.

electrolytic rheostat   [ELEC] A rheostat that consists of a tank of conducting liquid in which electrodes are placed, and resistance is varied by changing the distance between the electrodes, the depth of immersion of the electrodes, or the resistivity of the solution. Also known as water rheostat.

electrolytic separation   [PHYS CHEM] Separation of isotopes by electrolysis, based on differing rates of discharge at the electrode of ions of different isotopes.

electrolytic solution   [PHYS CHEM] A solution made up of a solvent and an ionically dissociated solute; it will conduct electricity, and ions can be separated from the solution by deposition on an electrically charged electrode.

electromagnet   [ELECTROMAG] A magnet consisting of a coil wound around a soft iron or steel core; the core is strongly magnetized when current flows through the coil, and is almost completely demagnetized when the current is interrupted.

electromagnetic amplifying lens   [ELECTROMAG] Large numbers of waveguides symmetrically arranged with respect to an excitation medium in order to become excited with equal amplitude and phase to provide a net gain in energy.

electromagnetic cathode-ray tube   [ELECTR] A cathode-ray tube in which electromagnetic deflection is used on the electron beam.

electromagnetic constant   *See* speed of light.

electromagnetic coupling   [ELECTROMAG] Coupling that exists between circuits when they are mutually affected by the same electromagnetic field.

electromagnetic current   [ELECTR] Motion of charged particles (for example, in the ionosphere) giving rise to electric and magnetic fields.

electromagnetic damping   [ELEC] Retardation of motion that results from the reaction between eddy currents in a moving conductor and the magnetic field in which it is moving.

ELECTROMAGNET

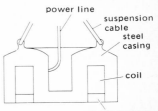

Cross section of circular lifting electromagnet.

electromagnetic deflection [ELECTR] Deflection of an electron stream by means of a magnetic field.

electromagnetic field [ELECTROMAG] An electric or magnetic field, or a combination of the two, as in an electromagnetic wave.

electromagnetic field equations *See* Maxwell field equations.

electromagnetic field tensor [ELECTROMAG] An antisymmetric, second-rank Lorentz tensor, whose elements are proportional to the electric and magnetic fields; the Maxwell field equations can be expressed in a simple form in terms of this tensor.

electromagnetic horn *See* horn antenna.

electromagnetic induction [ELECTROMAG] The production of an electromotive force either by motion of a conductor through a magnetic field so as to cut across the magnetic flux or by a change in the magnetic flux that threads a conductor. Also known as induction.

electromagnetic inertia [ELECTROMAG] **1.** Characteristic delay of a current in an electric circuit in reaching its maximum value, or in returning to zero, after the source voltage has been removed or applied. **2.** The property of a circuit whereby variation of the current in the circuit gives rise to a voltage in the circuit.

electromagnetic interaction [PARTIC PHYS] The interaction of elementary particles that results from the coupling of charge to the electromagnetic field.

electromagnetic lens [ELECTR] An electron lens in which electron beams are focused by an electromagnetic field.

electromagnetic mass [ELECTROMAG] The contribution to the mass of an object from its electric and magnetic field energy.

electromagnetic mirror [ELECTROMAG] Surface or region capable of reflecting radio waves, such as one of the ionized layers in the upper atmosphere.

electromagnetic momentum [ELECTROMAG] The momentum transported by electromagnetic radiation; its volume density equals the Poynting vector divided by the square of the speed of light.

electromagnetic oscillograph [ELECTROMAG] An oscillograph in which the recording mechanism is controlled by a moving-coil galvanometer, such as a direct-writing recorder or a light-beam oscillograph.

electromagnetic potential [ELECTROMAG] Collective name for a scalar potential, which reduces to the electrostatic potential in a time-independent system, and the vector potential for the magnetic field; the electric and magnetic fields can be written in terms of these potentials.

electromagnetic properties [ELECTROMAG] The response of materials or equipment to electromagnetic fields, and their ability to produce such fields.

electromagnetic pulse [ELECTROMAG] The pulse of electromagnetic radiation generated by a large thermonuclear explosion.

electromagnetic radiation [ELECTROMAG] Electromagnetic waves and, especially, the associated electromagnetic energy.

electromagnetic scattering [PHYS] The process in which energy is removed from a beam of electromagnetic radiation and reemitted without appreciable changes in wavelength.

electromagnetic separator [ELECTROMAG] Device in which ions of varying mass are separated by a combination of electric and magnetic fields.

electromagnetic shielding [ELECTROMAG] Means, similar to electrostatic or magnetostatic shielding, for suppressing changing magnetic fields or electromagnetic radiation at a device.

electromagnetic shock wave [ELECTROMAG] Electromagnetic wave of great intensity which results when waves with different intensities propagate with different velocities in a nonlinear optical medium, and faster-traveling waves from a pulse of light catch up with preceding, slower traveling waves.

electromagnetic spectrum [ELECTROMAG] The total range of wavelengths or frequencies of electromagnetic radiation, extending from the longest radio waves to the shortest known cosmic rays.

electromagnetic system of units [ELECTROMAG] A centimeter-gram-second system of electric and magnetic units in which the unit of current is defined as the current which, if maintained in two straight parallel wires having infinite length and being 1 centimeter apart in vacuum, would produce between these conductors a force of 2 dynes per centimeter of length; other units are derived from this definition by assigning unit coefficients in equations relating electric and magnetic quantities. Also known as electromagnetic units (emu).

electromagnetic theory of light [ELECTROMAG] Theory according to which light is an electromagnetic wave whose electric and magnetic fields obey Maxwell's equations.

electromagnetic units *See* electromagnetic system of units.

electromagnetic wave [ELECTROMAG] A disturbance which propagates outward from any electric charge which oscillates or is accelerated; far from the charge it consists of vibrating electric and magnetic fields which move at the speed of light and are at right angles to each other and to the direction of motion.

electromagnetic-wave filter [ELECTROMAG] Any device to transmit electromagnetic waves of desired frequencies while substantially attenuating all other frequencies.

electromagnetism [PHYS] **1.** Branch of physics relating electricity to magnetism. **2.** Magnetism produced by an electric current rather than by a permanent magnet.

electromechanical coupling coefficient [SOLID STATE] The ratio of the mutual elastodielectric energy density in a piezoelectric material to the square root of the product of the stored elastic and dielectric energy densities.

electrometer [ENG] An instrument for measuring voltage without drawing appreciable current.

electrometer amplifier [ELECTR] A low-noise amplifier having sufficiently low current drift and other characteristics required for measuring currents smaller than 10^{-12} ampere.

electrometer tube [ELECTR] A high-vacuum electron tube having a high input impedance (low control-electrode conductance) to facilitate measurement of extremely small direct currents or voltages.

electromodulation [SPECT] Modulation spectroscopy in which changes in transmission or reflection spectra induced by a perturbing electric field are measured.

electromotance *See* electromotive force.

electromotive force [PHYS CHEM] **1.** The difference in electric potential that exists between two dissimilar electrodes immersed in the same electrolyte or otherwise connected by ionic conductors. **2.** The resultant of the relative electrode potential of the two dissimilar electrodes at which electrochemical reactions occur. Abbreviated emf. Also known as electromotance.

electromotive series *See* electrochemical series.

electron [PHYS] **1.** A stable elementary particle which is the negatively charged constituent of ordinary matter, having a mass of about 9.11×10^{-28} gram (equivalent to 0.511 **MeV**), a charge of about -1.602×10^{-19} coulomb, and a spin of $\frac{1}{2}$. Also known as negative electron; negatron. **2.** Collective name for the electron, as in the first definition, and the positron.

ELECTRON ACCELERATOR

A type of electron accelerator, the Stanford Linear Accelerator, installed in underground housing.

electron accelerator [NUCLEO] A device which accelerates electrons to high energies.

electron acceptor [PHYS CHEM] An atom or part of a molecule joined by a covalent bond to an electron donor. [SOLID STATE] *See* acceptor.

electron affinity [ATOM PHYS] The work needed in removing an electron from a negative ion, thus restoring the neutrality of an atom or molecule.

electron attachment [ATOM PHYS] The combination of an electron with a neutral atom or molecule to form a negative ion. [NUC PHYS] *See* electron capture.

electron beam [ELECTR] A narrow stream of electrons moving in the same direction, all having about the same velocity.

electron-beam generator [ELECTR] Velocity-modulated generator, such as a klystron tube, used to generate extremely high frequencies.

electron-beam laser [OPTICS] A semiconductor laser in which the electron beam that provides pumping action in a thin plate of cadmium sulfide or other material is swept electrically in two dimensions by a deflection yoke, much as in a cathode-ray tube.

electron-beam magnetometer [ENG] A magnetometer that depends on the change in intensity or direction of an electron beam that passes through the magnetic field to be measured.

electron-beam parametric amplifier [ELECTR] A parametric amplifier in which energy is pumped from an electrostatic field into a beam of electrons traveling down the length of the tube, and electron couplers impress the input signal at one end of the tube and translate spiraling electron motion into electric output at the other.

electron-beam pumping [ELECTR] The use of an electron beam to produce excitation for population inversion and lasing action in a semiconductor laser.

electron-beam tube [ELECTR] An electron tube whose performance depends on the formation and control of one or more electron beams.

electron capture [ATOM PHYS] The process in which an atom or ion passing through a material medium either loses or gains one or more orbital electrons. [NUC PHYS] A radioactive transformation of nuclide in which a bound electron merges with its nucleus. Also known as electron attachment.

electron charge [PHYS] The charge carried by an electron, equal to about -1.602×10^{-19} coulomb, or -4.803×10^{-10} statcoulomb.

electron cloud [ATOM PHYS] Picture of an electron state in which the charge is thought of as being smeared out, with the resulting charge density distribution corresponding to the probability distribution function associated with the Schrödinger wave function.

electron collector *See* collector.

electron conduction [ELEC] Conduction of electricity resulting from motion of electrons, rather than from ions in a gas or solution, or holes in a solid.

electron configuration [ATOM PHYS] The orbital and spin arrangement of an atom's electrons, specifying the quantum numbers of the atom's electrons in a given state.

electron cyclotron wave [PL PHYS] A wave in a plasma which propagates parallel to the magnetic field produced by currents outside the plasma at frequencies less than that of the electron cyclotron resonance, and which is circularly polarized, rotating in the same sense as electrons in the plasma; responsible for whistlers. Also known as whistler wave.

electron density [PHYS] **1.** The number of electrons in a unit volume. **2.** When quantum-mechanical effects are significant, the total probability of finding an electron in a unit volume.

electron detachment [ATOM PHYS] The separation of an electron from a negative ion to form a neutral atom or molecule.

electron diffraction [PHYS] The phenomenon associated with the interference processes which occur when electrons are scattered by atoms in crystals to form diffraction patterns.

electron diffraction analysis [PHYS] Examination of solid surfaces by observing the diffraction of a stream of electrons by the surface.

electron diffraction camera [OPTICS] A camera used to obtain a photographic record of the position and intensity of the diffracted beams produced when a specimen is irradiated by a beam of electrons.

electron diffractograph [PHYS] A device, allied to the electron microscope, in which a beam of electrons strikes the sample, showing crystal pattern and other physical attributes on the resulting diffraction pattern; used for chemical analysis, atomic structure determination, and so on.

electron dipole moment *See* electron magnetic moment.

electron distribution [PHYS] A function which gives the number of electrons per unit volume of phase space.

electron distribution curve [PHYS CHEM] A curve indicating the electron distribution among the different available energy levels of a solid substance.

electron donor [PHYS CHEM] An atom or part of a molecule which supplies both electrons of a duplet forming a covalent bond. [SOLID STATE] *See* donor.

electronegative [ELEC] **1.** Carrying a negative electric charge. **2.** Capable of acting as the negative electrode in an electric cell. [PHYS CHEM] Pertaining to an atom or group of atoms that has a relatively great tendency to attract electrons to itself.

electronegative potential [PHYS CHEM] Potential of an electrode expressed as negative with respect to the hydrogen electrode.

electron emission [ELECTR] The liberation of electrons from an electrode into the surrounding space, usually under the influence of heat, light, or a high electric field.

electron emitter [ELECTR] The electrode from which electrons are emitted.

electron energy level [ATOM PHYS] A quantum-mechanical concept for energy levels of electrons about the nucleus; electron energies are functions of each particular atomic species.

electron energy loss spectroscopy [SPECT] A technique for studying atoms, molecules, or solids in which a substance is bombarded with monochromatic electrons, and the energies of scattered electrons are measured to determine the distribution of energy loss.

electron flow [ELEC] A current produced by the movement of free electrons toward a positive terminal; the direction of electron flow is opposite to that of current.

electron gas [PHYS] A concentration of electrons whose behavior is, in first approximation, not governed by forces.

electron gun [ELECTR] An electrode structure that produces and may control, focus, deflect, and converge one or more electron beams in an electron tube.

electron-gun density multiplication [ELECTR] Ratio of the average current density at any specified aperture through

ELECTRON GUN

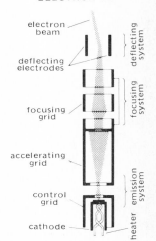

Simplified electron gun employing electrostatic focus and deflection.

which the electron stream passes to the average current density at the cathode surface.

electron hole   *See* hole.

electron hole droplets   [SOLID STATE] A form of electronic excitation observed in germanium and silicon at sufficiently low cryogenic temperatures; it is associated with a liquid-gas phase transition of the charge carriers, and consists of regions of conducting electron-hole Fermi liquid coexisting with regions of insulating exciton gas.

electronic   [ELECTR] Pertaining to electron devices or to circuits or systems utilizing electron devices, including electron tubes, magnetic amplifiers, transistors, and other devices that do the work of electron tubes.

electronic absorption spectrum   [SPECT] Spectrum resulting from absorption of electromagnetic radiation by atoms, ions, and molecules due to excitations of their electrons.

electronic alternating-current voltmeter   [ELECTR] A voltmeter consisting of a direct-current milliammeter calibrated in volts and connected to an amplifier-rectifier circuit.

electronic angular momentum   [ATOM PHYS] The total angular momentum associated with the orbital motion of the spins of all the electrons of an atom.

electronic band spectrum   [SPECT] Bands of spectral lines associated with a change of electronic state of a molecule; each band corresponds to certain vibrational energies in the initial and final states and consists of numerous rotational lines.

electronic circuit   [ELECTR] An electric circuit in which the equilibrium of electrons in some of the components (such as electron tubes, transistors, or magnetic amplifiers) is upset by means other than an applied voltage.

electronic component   [ELECTR] A component which is able to amplify or control voltages or currents without mechanical or other nonelectrical command, or to switch currents or voltages without mechanical switches; examples include electron tubes, transistors, and other solid-state devices.

electronic data processing   [ADP] Processing data by using equipment that is predominantly electronic in nature, such as an electronic digital computer. Abbreviated EDP.

electronic emission spectrum   [SPECT] Spectrum resulting from emission of electromagnetic radiation by atoms, ions, and molecules following excitations of their electrons.

electronic energy curve   [PHYS CHEM] A graph of the energy of a diatomic molecule in a given electronic state as a function of the distance between the nuclei of the atoms.

electronic magnetic moment   [ATOM PHYS] The total magnetic dipole moment associated with the orbital motion of all the electrons of an atom and the electron spins; opposed to nuclear magnetic moment.

electronic microradiography   [ELECTR] Microradiography of very thin specimens in which the emission of electrons from an irradiated object, either the specimen or a lead screen behind it, is used to produce a photographic image of the specimen, which is then enlarged.

electronic multimeter   [ELECTR] A multimeter that uses semiconductor or electron-tube circuits to drive a conventional multiscale meter.

electronic phase-angle meter   [ELECTR] A phasemeter that makes use of electronic devices, such as amplifiers and limiters, that convert the alternating-current voltages being measured to square waves whose spacings are proportional to phase.

electronic photometer   *See* photoelectric photometer.

electronic polarization   [ELEC] Polarization arising from the displacement of electrons with respect to the nuclei with which they are associated, upon application of an external electric field.

electronic power supply *See* power supply.

electronic pumping *See* pumping.

electronic radiography [ELECTR] Radiography in which the image is detached by direct image converter tubes or by the use of television pickup or electronic scanning, and the resultant signals are amplified and presented for viewing on a kinescope.

electronic-raster scanning *See* electronic scanning.

electronic recording [ELECTR] The process of making a graphical record of a varying quantity or signal (or the result of such a process) by electronic means, involving control of an electron beam by electric or magnetic fields, as in a cathode-ray oscillograph, in contrast to light-beam recording.

electronics [PHYS] Study, control, and application of the conduction of electricity through gases or vacuum or through semiconducting or conducting materials.

electronic scanning [ELECTR] Scanning in which an electron beam, controlled by electric or magnetic fields, is swept over the area under examination, in contrast to mechanical or electromechanical scanning. Also known as electronic-raster scanning.

electronic specific heat [SOLID STATE] Contribution to the specific heat of a metal from the motion of conduction electrons.

electronic spectrum [SPECT] Spectrum resulting from emission or absorption of electromagnetic radiation during changes in the electron configuration of atoms, ions, or molecules, as opposed to vibrational, rotational, fine-structure, or hyperfine spectra.

electronic state [QUANT MECH] The physical state of electrons of a system, as specified, for example, by a Schrödinger-Pauli wave function of the positions and spin orientations of all the electrons.

electronic structure [PHYS] The arrangement of electrons in an atom, molecule, or solid, specified by their wave functions, energy levels, or quantum numbers.

electronic switch [ELECTR] **1.** Vacuum tube, crystal diodes, or transistors used as an on and off switching device. **2.** Test instrument used to present two wave shapes on a single gun cathode-ray tube.

electronic voltage regulator [ELECTR] A device which maintains the direct-current power supply voltage for electronic equipment nearly constant in spite of input alternating-current line voltage variations and output load variations.

electronic work function [SOLID STATE] The energy required to raise an electron with the Fermi energy in a solid to the energy level of an electron at rest in vacuum outside the solid.

electron image tube *See* image tube.

electron injection [ELECTR] **1.** The emission of electrons from one solid into another. **2.** The process of injecting a beam of electrons with an electron gun into the vacuum chamber of a mass spectrometer, betatron, or other large electron accelerator.

electron lens [ELECTR] An electric or magnetic field, or a combination thereof, which acts upon an electron beam in a manner analogous to that in which an optical lens acts upon a light beam. Also known as lens.

electron linear accelerator [NUCLEO] A linear accelerator used to accelerate electrons in a straight line, usually by means of radio-frequency fields which are produced in a loaded waveguide and travel with the electrons.

electron magnetic moment [ATOM PHYS] The magnetic dipole moment which an electron possesses by virtue of its spin. Also known as electron dipole moment.

electron mass [PHYS] The mass of an electron, equal to about 9.11×10^{-28} gram, equivalent to 0.511 MeV. Also known as electron rest mass.

electron microprobe [PHYS] An x-ray machine in which electrons emitted from a hot-filament source are accelerated electrostatically, then focused to an extremely small point on the surface of a specimen by an electromagnetic lens; nondestructive analysis of the specimen can then be made by measuring the backscattered electrons, the specimen current, the resulting x-radiation, or any other resulting phenomenon. Also known as electron probe.

electron microscope [ELECTR] A device for forming greatly magnified images of objects by means of electrons, usually focused by electron lenses.

electron mirror [ELECTR] **1.** A device capable of reflecting back an electron beam. **2.** *See* dynode.

electron mobility [SOLID STATE] The drift mobility of electrons in a semiconductor, being the electron velocity divided by the applied electric field.

electron multiplicity [ATOM PHYS] In an atom with Russell-Saunders coupling, the quantity $2S + 1$, where S is the total spin quantum number.

electron multiplier [ELECTR] An electron-tube structure which produces current amplification; an electron beam containing the desired signal is reflected in turn from the surfaces of each of a series of dynodes, and at each reflection an impinging electron releases two or more secondary electrons, so that the beam builds up in strength. Also known as multiplier.

electron-multiplier phototube *See* multiplier phototube.

electron nuclear double resonance [PHYS] A method combining electron paramagnetic resonance and nuclear magnetic resonance, in which the nuclear magnetic resonance is detected through the resulting electron paramagnetic resonance. Abbreviated ENDOR.

electron number [ATOM PHYS] The number of electrons in an ion or atom.

electronographic tube [ELECTR] An image tube used in astronomy in which the electron image formed by the tube is recorded directly upon film or plates.

electronography [ELECTR] The use of image tubes to form intensified electron images of astronomical objects and record them directly on film or plates.

electron optics [ELECTR] The study of the motion of free electrons under the influence of electric and magnetic fields.

electron orbit [PHYS] The path described by an electron.

electron pair [PHYS CHEM] A pair of valence electrons which form a nonpolar bond between two neighboring atoms.

electron pair bond *See* covalent bond.

electron paramagnetic resonance [PHYS] Magnetic resonance arising from the magnetic moment of unpaired electrons in a paramagnetic substance or in a paramagnetic center in a diamagnetic substance. Abbreviated EPR. Also known as electron spin resonance (ESR); paramagnetic resonance.

electron paramagnetism [PHYS] Paramagnetism in a substance whose atoms or molecules possess a net electronic magnetic moment; arises because of the tendency of a magnetic field to orient the electronic magnetic moments parallel to itself.

electron-positron pair [PHYS] An electron and a positron produced at the same time in the interaction of a photon with a high-intensity electric field.

electron-positron storage ring [NUCLEO] An annular vacuum chamber, enclosed by bending and focusing magnets, in which counterrotating beams of electrons and positrons are stored for several hours and can be made to collide with each other.

electron probe *See* electron microprobe.

electron radiography [GRAPHICS] A technique for producing a photographic image of an opaque specimen by transmitting electrons through it onto an adjacent photographic film; the electrons are generated in a metal sheet adjacent to the specimen or in the specimen itself by x-rays.

electron radius [PHYS] The classical value r of 2.81777 $\times$ 10^{-13} centimeter for the radius of an electron; obtained by equating mc^2 for the electron to e^2/r, where e and m are the charge and mass of the electron respectively; any classical model for an electron will have approximately this radius.

electron-ray tube *See* cathode-ray tube.

electron refraction [ELECTR] The bending of an electron beam passing from one region to another of different electric potential.

electron ring accelerator [NUCLEO] Proposed particle accelerator in which protons to be accelerated are trapped by the space charge of a ring of relativisitic electrons which is then accelerated. Abbreviated ERA.

electron shell [ATOM PHYS] **1.** The collection of all the electron states in an atom which have a given principal quantum number. **2.** The collection of all the electron states in an atom which have a given principal quantum number and a given orbital angular momentum quantum number.

electron spectroscopy [SPECT] The study of the energy spectra of photoelectrons or Auger electrons emitted from a substance upon bombardment by electromagnetic radiation, electrons, or ions; used to investigate atomic, molecular, or solid-state structure, and in chemical analysis.

electron spectrum [SPECT] Visual display, photograph, or graphical plot of the intensity of electrons emitted from a substance bombarded by x-rays or other radiation as a function of the kinetic energy of the electrons.

electron spin [QUANT MECH] That property of an electron which gives rise to its angular momentum about an axis within the electron.

electron spin density [PHYS] The vector sum of the spin angular momenta of electrons at each point in a substance per unit volume.

electron spin resonance *See* electron paramagnetic resonance.

electron-stream potential [ELECTR] At any point in an electron stream, the time average of the potential difference between that point and the electron-emitting surface.

electron-stream transmission efficiency [ELECTR] At an electrode through which the electron stream (beam) passes, the ratio of the average stream current through the electrode to the stream current approaching the electrode.

electron synchrotron [NUCLEO] A circular electron accelerator in which the frequency of the accelerating system is constant, the strength of the magnetic guide field increases, and the electrons move in orbits of nearly constant radius.

electron telescope [ELECTR] A telescope in which an infrared image of a distant object is focused on the photosensitive cathode of an image converter tube; the resulting electron image is enlarged by electron lenses and made visible by a fluorescent screen.

electron temperature [PL PHYS] The temperature at which ideal gas molecules would have an average kinetic energy equal to that of electrons in a plasma under consideration.

ELECTRON SPECTROSCOPY

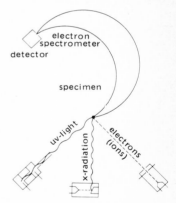

Excitation of electron spectra recorded in high-resolution instruments.

electron transfer [PHYS] The passage of an electron from one constituent of a system to another.

electron transition [QUANT MECH] Change of an electron from one state to another, accompanied by emission or absorption of electromagnetic radiation.

electron trap [SOLID STATE] A defect or chemical impurity in a semiconductor or insulator which captures mobile electrons in a special way.

electron tube [ELECTR] An electron device in which conduction of electricity is provided by electrons moving through a vacuum or gaseous medium within a gastight envelope. Also known as radio tube; tube; valve (British usage).

electron tube static characteristic [ELECTR] Relation between a pair of variables such as electrode voltage and electrode current with all other voltages maintained constant.

electron tunneling [QUANT MECH] The passage of electrons through a potential barrier which they would not be able to cross according to classical mechanics, such as a thin insulating barrier between two superconductors.

electron vacuum gage [ENG] An instrument used to measure vacuum by the ionization effect that an electron flow (from an incandescent filament to a charged grid) has on gas molecules.

electron volt [PHYS] A unit of energy equal to the energy acquired by an electron when it passes through a potential difference of 1 volt in a vacuum; it is equal to $(1.602192 \pm 0.000007) \times 10^{-19}$ volt. Abbreviated eV.

electron voltaic effect [ELECTR] Sensitivity of photovoltaic cells to electron bombardment.

electron wave [QUANT MECH] The de Broglie wave or probability amplitude wave of an electron.

electron wave function [QUANT MECH] Function of the spin orientation and position of one or more electrons, specifying the dynamical state of the electrons; the square of the function's modulus gives the probability per unit volume of finding electrons at a given position.

electron wavelength [QUANT MECH] The de Broglie wavelength of an electron, given by Planck's constant divided by the momentum.

electrooptical birefringence *See* electrooptical Kerr effect.

electrooptical Kerr effect [OPTICS] Birefringence induced by an electric field. Also known as electrooptical birefringence; Kerr effect.

electrooptic material [OPTICS] A material in which the indices of refraction are changed by an applied electric field.

electrooptics [OPTICS] The study of the influence of an electric field on optical phenomena, as in the electrooptical Kerr effect and the Stark effect. Also known as optoelectronics.

electrophonic effect [BIOPHYS] The sensation of hearing produced when an alternating current of suitable frequency and magnitude is passed through a person.

electrophoretic effect [PHYS CHEM] Retarding effect on the characteristic motion of an ion in an electrolytic solution subjected to a potential gradient, which results from motion in the opposite direction by the ion atmosphere.

electrophorus [ELEC] A device used to produce electric charges; it consists of a hard-rubber disk, which is negatively charged by rubbing with fur, and a metal plate, held by an insulating handle, which is placed on the disk; the plate is then touched with a grounded conductor, so that negative charge is removed and the plate has net positive charge.

electrophotoluminescence [ELECTR] Emission of light resulting from application of an electric field to a phosphor which is concurrently, or has been previously, excited by other means.

electrophotophoresis [PHYS] Helical motion of small parti-

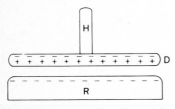

cles suspended in a gas along the direction of an electric field when exposed to a beam of light.

electropositive [ELEC] **1.** Carrying a positive electric charge. **2.** Capable of acting as the positive electrode in an electric cell. [PHYS CHEM] Pertaining to elements, ions, or radicals that tend to give up or lose electrons.

electropositive potential [PHYS CHEM] Potential of an electrode expressed as positive with respect to the hydrogen electrode.

electroresistive effect [ELECTR] The change in the resistivity of certain materials with changes in applied voltage.

electroscope [ENG] An instrument for detecting an electric charge by means of the mechanical forces exerted between electrically charged bodies.

electrostatic [ELEC] Pertaining to electricity at rest, such as an electric charge on an object.

electrostatic accelerator [ELECTR] Any instrument which uses an electrostatic field to accelerate charged particles to high velocities in a vacuum.

electrostatic analyzer [ELECTR] A device which filters an electron beam, permitting only electrons within a very narrow velocity range to pass through.

electrostatic attraction *See* Coulomb attraction.

electrostatic bond [PHYS CHEM] A valence bond in which two atoms are kept together by electrostatic forces caused by transferring one or more electrons from one atom to the other.

electrostatic cathode-ray tube [ELECTR] A cathode-ray tube in which electrostatic deflection is used on the electron beam.

electrostatic deflection [ELECTR] The deflection of an electron beam by means of an electrostatic field produced by electrodes on opposite sides of the beam; used chiefly in cathode-ray tubes for oscilloscopes.

electrostatic energy [ELEC] The potential energy which a collection of electric charges possesses by virtue of their positions relative to each other.

electrostatic error *See* antenna effect.

electrostatic field [ELEC] A time-independent electric field, such as that produced by stationary charges.

electrostatic focus [ELECTR] Production of a focused electron beam in a cathode-ray tube by the application of an electric field.

electrostatic force [ELEC] Force on a charged particle due to an electrostatic field, equal to the electric field vector times the charge of the particle.

electrostatic generator [ELEC] Any machine which produces electric charges by friction or (more commonly) electrostatic induction.

electrostatic gyroscope [ENG] A gyroscope in which a small beryllium ball is electrostatically suspended within an array of six electrodes in a vacuum inside a ceramic envelope.

electrostatic induction [ELEC] The process of charging an object electrically by bringing it near another charged object, then touching it to ground. Also known as induction.

electrostatic instrument [ELEC] A meter that depends for its operation on the forces of attraction and repulsion between electrically charged bodies.

electrostatic interactions *See* Coulomb interactions.

electrostatic lens [ELECTR] An arrangement of electrostatic fields which acts upon beams of charged particles similar to the way a glass lens acts on light beams.

electrostatic octupole lens [ELECTR] A device for controlling beams of electrons or other charged particles, consisting of eight electrodes arranged in a circular pattern with alternating polarities; commonly used to correct aberrations of quadrupole lens systems.

ELECTROSCOPE

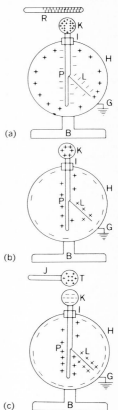

Simple gold-leaf electroscope. *(a)* An electroscope being charged by induction by negative charge on hard-rubber rod R. *(b)* Positive charge left on its leaf after induction process is complete. *(c)* Testing the sign of an unknown charge on test ball T. L = gold leaf, P = metal post, I = insulator, K = metal knob, H = metal housing, B = base, R = rubber rod, J = insulating handle, and G = ground.

ELECTROSTATIC QUADRUPOLE LENS

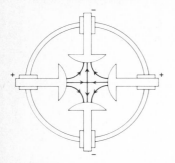

Electrostatic quadrupole lens showing the four electrodes.

electrostatic potential *See* electric potential.

electrostatic probe *See* electric probe.

electrostatic quadrupole lens [ELECTR] A device for focusing beams of electrons or other charged particles, consisting of four electrodes arranged in a circular pattern with alternating polarities.

electrostatic repulsion *See* Coulomb repulsion.

electrostatics [ELEC] The study of electric charges at rest, their electric fields, and potentials.

electrostatic shielding [ELEC] The placing of a grounded metal screen, sheet, or enclosure around a device or between two devices to prevent electric fields from interacting.

electrostatic stress [ELEC] An electrostatic field acting on an insulator, which produces polarization in the insulator and causes electrical breakdown if raised beyond a certain intensity.

electrostatic tape camera [OPTICS] A camera in which images are stored electrostatically on a plastic tape; designed for use in satellites, where the stored image is not damaged by Van Allen or other radiation.

electrostatic units [ELEC] A centimeter-gram-second system of electric and magnetic units in which the unit of charge is that charge which exerts a force of 1 dyne on another unit charge when separated from it by a distance of 1 centimeter in vacuum; other units are derived from this definition by assigning unit coefficients in equations relating electric and magnetic quantities. Abbreviated esu.

electrostatic valence rule [PHYS CHEM] The postulate that in a stable ionic structure the valence of each anion, with changed sign, equals the sum of the strengths of its electrostatic bonds to the adjacent cations.

electrostatic wave [PL PHYS] Wave motion of a plasma whose restoring forces are primarily electrostatic.

electrostriction [MECH] A form of elastic deformation of a dielectric induced by an electric field, associated with those components of strain which are independent of reversal of field direction, in contrast to the piezoelectric effect. Also known as electrostrictive strain.

electrostrictive strain *See* electrostriction.

electrothermal [PHYS] **1.** Pertaining to both heat and electricity. **2.** In particular, pertaining to conversion of electrical energy into heat energy.

electrothermal ammeter *See* thermoammeter.

electrothermal voltmeter [ENG] An electrothermal ammeter employing a series resistor as a multiplier, thus measuring voltage instead of current.

electrovalence [PHYS CHEM] The valence of an atom that has formed an ionic bond.

electrovalent bond *See* ionic bond.

electroviscous effect [FL MECH] Change in a liquid's viscosity induced by a strong electrostatic field.

element [CHEM] A substance made up of atoms with the same atomic number; common examples are hydrogen, gold, and iron. Also known as chemical element. [ELEC] **1.** A part of an electron tube, semiconductor device, or antenna array that contributes directly to the electrical performance. **2.** *See* component. [ELECTROMAG] Radiator, active or parasitic, that is a part of an antenna. [MATH] **1.** In an array such as a matrix or determinant, a quantity identified by the intersection of a given row or column. **2.** In network topology, an edge.

element 104 [CHEM] The first element beyond the actinide series, and the twelfth transuranium element; the atoms of element 104, of mass number 260, were first produced by

ELEMENT 104

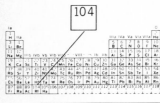

Periodic table of the chemical elements showing the position of element 104.

irradiating plutonium-242 with neon-22 ions in a heavy-ion cyclotron.

element 105 [CHEM] An artificial element whose isotope of mass number 260 was discovered by bombarding californium-249 with nitrogen-15 ions in a heavy-ion linear accelerator.

element 106 [CHEM] An artificial element whose isotope of mass number 263 was discovered by bombarding californium-249 with oxygen-18 ions in a heavy-ion linear accelerator, and whose isotope of mass number 259 was discovered by bombarding lead-207 and lead-208 with chromium-54 ions in a heavy-ion cyclotron.

element 107 [CHEM] An artificial element whose isotope of mass number 261 has been tentatively identified as a reaction product in the bombardment of bismuth-209 with chromium-54 ions and lead-208 with manganese-55 ions in a heavy-ion cyclotron.

elementary charge [PHYS] An electric charge such that the electric charge of any body is an integral multiple of it, equal to the electron charge.

elementary excitation [QUANT MECH] The quantum of energy of some vibration or wave, such as a photon, phonon, plasmon, magnon, polaron, or exciton.

elementary function [MATH] Any function which can be formed from algebraic functions and the exponential, logarithmic, and trigonometric functions by a finite number of operations consisting of addition, subtraction, multiplication, division, and composition of functions.

elementary particle [PARTIC PHYS] A particle which, in the present state of knowledge, cannot be described as compound, and is thus one of the fundamental constituents of all matter. Also known as fundamental particle; particle.

elementary process [PHYS CHEM] In chemical kinetics, the particular events at the atomic or molecular level which make up an overall reaction.

elements of the trajectory [MECH] The various features of the trajectory such as the angle of departure, maximum ordinate, angle of fall, and so on.

elevation head [FL MECH] The energy per unit mass possessed by a fluid as a result of its height above some reference level. Also known as potential head.

elimination [MATH] A process of deriving from a system of equations a new system with fewer variables, but with precisely the same solutions.

eliminator [ELECTR] Device that takes the place of batteries, generally consisting of a rectifier operating from alternating current.

E lines [ELEC] Contour lines of constant electrostatic field strength referred to some reference base.

ellipse [MATH] The locus of all points in the plane at which the sum of the distances from a fixed pair of points, the foci, is a given constant.

ellipsoid [MATH] A surface whose intersection with every plane is an ellipse (or circle).

ellipsoidal coordinates [MATH] Coordinates in space determined by confocal quadrics.

ellipsoidal harmonics [MATH] Lamé functions that play a role in potential problems on an ellipsoid analogous to that played by spherical harmonics in potential problems on a sphere.

ellipsoidal of wave normals *See* index ellipsoid.

ellipsoidal reflector [OPTICS] A concave ellipsoidal surface from which light is specularly reflected; used in a light projector to focus rays from a light source at the near focal point onto the opposite focal point of the ellipse.

ELEMENT 105

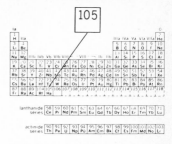

Periodic table of the chemical elements showing the position of element 105.

ellipsoidal wave functions *See* Lamé wave functions.

ellipsoid of revolution [MATH] An ellipsoid generated by rotation of an ellipse about one of its axes. Also known as spheroid.

ellipsoid of wave normals *See* index ellipsoid.

ellipsometer [OPTICS] An instrument for determining the degree of ellipticity of polarized light; used to measure the thickness of very thin transparent films by observing light reflected from the film.

ellipsometry [OPTICS] Techniques for measuring the degree of ellipticity of polarized light.

elliptical galaxy [ASTRON] A galaxy whose overall shape ranges from a spheroid to an ellipsoid, without any noticeable structural features. Also known as spheroidal galaxy.

elliptical orbit [MECH] The path of a body moving along an ellipse, such as that described by either of two bodies revolving under their mutual gravitational attraction but otherwise undisturbed.

elliptical polarization [ELECTROMAG] Polarization of an electromagnetic wave in which the electric field vector at any point in space describes an ellipse in a plane perpendicular to the propagation direction.

elliptic coordinates [MATH] The coordinates of a point in the plane determined by confocal ellipses and hyperbolas.

elliptic cylindrical coordinate system [MATH] A three-dimensional coordinate system in which two of the coordinates depend on the x and y cartesian coordinates in the same manner as elliptic coordinates and are independent of the z coordinate, while the third coordinate is directly proportional to the z coordinate.

elliptic differential equation [MATH] A general type of second-order partial differential equation which includes Laplace's equation, and having the form

$$\sum_{i,j=1}^{N} A_{ij} (\partial^2 u / \partial x_i \, \partial x_j) + B = D$$

where the A_{ij} and B are suitably differentiable real functions of the x_i, u and the $\partial u / \partial x_i$, and there exists at each point $(x_1, \ldots, x_n)$ a real linear transformation on the x_i which reduces the quadratic form

$$\sum_{i,j=1}^{N} A_{ij} x_i x_j$$

to a sum of n squares, all of the same sign. Also known as elliptic partial differential equation.

elliptic function [MATH] An inverse function of an elliptic integral; alternatively, a doubly periodic, meromorphic function of a complex variable.

elliptic geometry [MATH] The geometry obtained from euclidean geometry by replacing the parallel line postulate with the postulate that infinitely many lines may be drawn through a given point, parallel to a given line. Also known as Riemannian geometry.

elliptic integral [MATH] An integral over x whose integrand is a rational function of x and the square root of $p(x)$, where $p(x)$ is a third- or fourth-degree polynomial without multiple roots.

elliptic integral of the first kind [MATH] Any elliptic integral which is finite for all values of the limits of integration and which approaches a finite limit when one of the limits of integration approaches infinity.

elliptic integral of the second kind [MATH] Any elliptic integral which approaches infinity as one of the limits of integration y approaches infinity, or which is infinite for some value of y, but which has no logarithmic singularities in y.

elliptic integral of the third kind [MATH] Any elliptic integral which has logarithmic singularities when considered as a function of one of its limits of integration.

ellipticity [ELECTR] *See* axial ratio. [MATH] Also known as oblateness. **1.** For an ellipse, the difference between the semimajor and semiminor axes of the ellipse, divided by the semimajor axis. **2.** For an oblate spheroid, the difference between the equatorial diameter and the axis of revolution, divided by the equatorial diameter.

elliptic paraboloid [MATH] A surface which can be so situated that sections parallel to one coordinate plane are parabolas while those parallel to the other plane are ellipses.

elliptic partial differential equation *See* elliptic differential equation.

elliptic point [MATH] A point on a surface at which the total curvature is strictly positive.

elliptic system [MATH] A system of equations of the form

$$a_{11} \, \partial u/\partial x + a_{12} \, \partial v/\partial y = F_1(x,y; u,v)$$

$$a_{21} \, \partial u/\partial y + a_{22} \, \partial v/\partial x = F_2(x,y; u,v)$$

where the determinant

$$\begin{vmatrix} a_{11} & a_{12} \\ a_{21} & a_{22} \end{vmatrix}$$

is a positive-definite or negative-definite function of x and y.

elongation [ASTRON] The difference between the celestial longitude of the moon or a planet, as measured from the earth, and that of the sun. [MECH] The fractional increase in a material's length due to stress in tension or to thermal expansion.

ELR scale *See* equal listener response scale.

Elster-Geitel effect [PHYS] The phenomenon in which a heated conductor acquires a positive or negative electric charge in the presence of a gas, while in a vacuum it always acquires a negative charge.

emagram [THERMO] A graph of the logarithm of the pressure of a substance versus its temperature, when it is held at constant volume; in meteorological investigations, the potential temperature is often the parameter.

emanating power [NUCLEO] The fraction of radon atoms, formed in a solid or solution, which escape.

emanation *See* radioactive emanation.

emanometer [ENG] An instrument for the measurement of the radon content of the atmosphere: radon is removed from a sample of air by condensation or adsorption on a surface, and is then placed in an ionization chamber and its activity determined.

emanometry [NUCLEO] The collective techniques for ionization-chamber determination of the amounts of radioactive gases escaping into the lower atmosphere from the earth's surface; in emanometric measurements, the objective is to count, by typical ionization-chamber methods, all of the ions produced by the alpha particles emitted by the one or more radioactive gases contained in the chamber.

embedding [MATH] An injective homomorphism between two algebraic systems of the same type.

embrittlement [MECH] Reduction or loss of ductility or toughness in a metal or plastic with little change in other mechanical properties.

emf *See* electromotive force.

emission [ELECTROMAG] Any radiation of energy by means of electromagnetic waves, as from a radio transmitter.

emission characteristics [ELECTR] Relation, usually shown by a graph, between the emission and a factor controlling the emission, such as temperature, voltage, or current of the filament or heater.

emission electron microscope [ELECTR] An electron microscope in which thermionic, photo, secondary, or field electrons emitted from a metal surface are projected on a fluorescent screen, with or without focusing.

emission lines [SPECT] Spectral lines resulting from emission of electromagnetic radiation by atoms, ions, or molecules during changes from excited states to states of lower energy.

emission spectrometer [SPECT] A spectrometer that measures percent concentrations of preselected elements in samples of metals and other materials; when the sample is vaporized by an electric spark or arc, the characteristic wavelengths of light emitted by each element are measured with a diffraction grating and an array of photodetectors.

emission spectrum [SPECT] Electromagnetic spectrum produced when radiations from any emitting source, excited by any of various forms of energy, are dispersed.

emissive power *See* emittance.

emissivity [THERMO] The ratio of the radiation emitted by a surface to the radiation emitted by a perfect blackbody radiator at the same temperature. Also known as thermal emissivity.

emittance [THERMO] The power radiated per unit area of a radiating surface. Also known as emissive power; radiating power.

emitter [ELECTR] A transistor region from which charge carriers that are minority carriers in the base are injected into the base, thus controlling the current flowing through the collector; corresponds to the cathode of an electron tube. Symbolized E. Also known as emitter region.

emitter barrier [ELECTR] One of the regions in which rectification takes place in a transistor, lying between the emitter region and the base region.

emitter bias [ELECTR] A bias voltage applied to the emitter electrode of a transistor.

emitter-coupled logic [ELECTR] A form of current-mode logic in which the emitters of two transistors are connected to a single current-carrying resistor in such a way that only one transistor conducts at a time. Abbreviated ECL.

emitter follower [ELECTR] A grounded-collector transistor amplifier which provides less than unity voltage gain but high input resistance and low output resistance, and which is similar to a cathode follower in its operations.

emitter junction [ELECTR] A transistor junction normally biased in the low-resistance direction to inject minority carriers into a base.

emitter region *See* emitter.

emitter resistance [ELECTR] The resistance in series with the emitter lead in an equivalent circuit representing a transistor.

E mode *See* transverse magnetic mode.

empirical curve [MATH] A smooth curve drawn through or close to points representing measured values of two variables on a graph.

empirical probability [STAT] The ratio of the number of times an event has occurred to the total number of trials performed. Also known as a posteriori probability.

empty set [MATH] The set with no elements.

emu *See* electromagnetic system of units.

emulsion jet [NUCLEO] A jetlike formation in a nuclear emulsion caused by an incident particle with very high energy, greater than 100 GeV.

enantiomer *See* enantiomorph.

enantiomorph [CHEM] One of an isomeric pair of crystalline forms or compounds whose molecules are nonsuperimpos-

EMITTER FOLLOWER

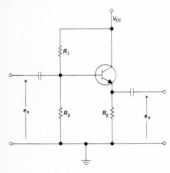

Circuit diagram of a typical emitter follower.

able mirror images. Also known as enantiomer; optical antipode; optical isomer.

enantiomorphism [CHEM] A phenomenon of mirror-image relationship exhibited by right-handed and left-handed crystals or by the molecular structures of two stereoisomers.

enantiotropy [CHEM] The relation of crystal forms of the same substance in which one form is stable above the transition-point temperature and the other stable below it, so that the forms can change reversibly one into the other.

Encke roots [MATH] For any two numbers a_1 and a_2, the numbers $-x_1$ and $-x_2$, where x_1 and x_2 are the roots of the equation $x^2 + a_1 x + a_2 = 0$, with $|x_1| < |x_2|$.

encoder [ELECTR] In an electronic computer, a network or system in which only one input is excited at a time and each input produces a combination of outputs.

end correction [ACOUS] A correction that must be made to the assumption that an antinode exists at an open end of a pipe in which air is vibrating, in order to take into account the radiation of sound waves from the pipe.

end effect [ELECTROMAG] The effect of capacitance at the ends of an antenna; it requires that the actual length of a half-wave antenna be about 5% less than a half wavelength.

end-fire antenna *See* end-fire array.

end-fire array [ELECTROMAG] A linear array whose direction of maximum radiation is along the axis of the array; it may be either unidirectional or bidirectional; the elements of the array are parallel and in the same plane, as in a fishbone antenna. Also known as end-fire antenna.

end loss [ELECTROMAG] The difference between the actual and effective lengths of a radiating antenna element.   [MET] That portion remaining after designated lengths of bar have been cut into multiples.

endo- [SCI TECH] Prefix denoting within or inside.

endoergic *See* endothermic.

endoergic collision *See* collision of the first kind.

endogenous variables [MATH] In a mathematical model, the dependent variables; their values are to be determined by the solution of the model equations.

endomorphism [MATH] A function from a set with some structure (such as a group, ring, vector space, or topological space) to itself which preserves this structure.

end-on position [ELECTROMAG] The position of a point which lies on the magnetic axis of a magnet. Also known as Gauss A position.

ENDOR *See* electron nuclear double resonance.

endotherm [PHYS CHEM] In differential thermal analysis, a graph of the temperature difference between a sample compound and a thermally inert reference compound (commonly aluminum oxide) as the substances are simultaneously heated to elevated temperatures at a predetermined rate, and the sample compound undergoes endothermal or exothermal processes.

endothermic [PHYS CHEM] Pertaining to a reaction which absorbs heat. Also known as endoergic.

end product [PHYS] The final product of a chemical or nuclear reaction or process.

end radiation *See* quantum limit.

endurance limit *See* fatigue limit.

endurance ratio *See* fatigue ratio.

endurance strength *See* fatigue strength.

energetics [PHYS] The study of energy and of its transformation from one form to another.

energetic solar particles [ASTROPHYS] Electrons and atomic nuclei produced in association with solar flares, with energies mostly in the range 1–100 MeV, but occasionally as high as 15 BeV. Also known as solar cosmic rays.

energized [ELEC] Electrically connected to a voltage source. Also known as alive; hot; live.

energy [PHYS] The capacity for doing work.

energy absorption [PHYS] Conversion of mechanical or radiant energy into the internal potential energy or heat energy of a system.

energy balance [PHYS] The arithmetic balancing of energy inputs versus outputs for an object, reactor, or other processing system; it is positive if energy is released, and negative if it is absorbed.

energy conservation *See* conservation of energy.

energy conversion [PHYS] The process of changing energy from one form to another.

energy density [PHYS] The energy per unit volume of a medium; in the case of an electric or magnetic field, the energy needed to set up the field is thought of as residing in the field.

energy diagram *See* energy level diagram.

energy dissipation *See* dissipation.

energy eigenstate *See* energy state.

energy ellipsoid *See* momental elllipsoid.

energy flux [PHYS] A vector quantity whose component perpendicular to any surface equals the energy transported across that surface by some medium per unit area per unit time.

energy gap [SOLID STATE] A range of forbidden energies in the band theory of solids.

energy gradient [PHYS] Any change in energy over time or space.

energy head [FL MECH] The elevation of the hydraulic grade line at any section of a waterway plus the velocity head of the mean velocity of the water in that section.

energy integral [MECH] A constant of integration resulting from integration of Newton's second law of motion in the case of a conservative force; equal to the sum of the kinetic energy of the particle and the potential energy of the force acting on it.

energy level [QUANT MECH] An allowed energy of a physical system; there may be several allowed states at one level.

energy-level diagram [QUANT MECH] A diagram in which the energy levels of a quantized system are indicated by distances of horizontal lines from a zero energy level. Also known as energy diagram; level scheme.

energy momentum tensor [PHYS] A tensor whose 16 elements give the energy density, momentum density, and stresses in a distribution of matter or radiation.

energy of a charge [ELEC] Charge energy measured in ergs according to the equation $E = QV$, where Q is the charge and V is the potential in electrostatic units.

energy of rotation [PHYS] Kinetic energy of a mass with moment of inertia I rotating with angular velocity ω about the axis, expressed as $E = \frac{1}{2}I\omega^2$.

energy operator [QUANT MECH] The operator corresponding to the energy or Hamiltonian of a classical system. Also known as Hamiltonian operator.

energy product curve [ELECTROMAG] Curve obtained by plotting the product of the values of magnetic induction B and demagnetizing force H for each point on the demagnetization curve of a permanent magnet material; usually shown with the demagnetization curve.

energy spectrum [PHYS] Any plot, display, or photographic record of the intensity of some type of radiation as a function of its energy.

ENERGY-LEVEL DIAGRAM

Energy levels of the hydrogen atom, classified by the orbital angular momentum of the electron, expressed in units of ħ.

energy spread [QUANT MECH] The width in energy of a wave packet or metastable state.

energy state [QUANT MECH] An eigenstate of the energy (Hamiltonian) operator, so that the energy has a definite stationary value. Also known as eigenstate; energy eigenstate; quantum state; stationary state.

engine cycle [THERMO] Any series of thermodynamic phases constituting a cycle for the conversion of heat into work; examples are the Otto cycle, Stirling cycle, and Diesel cycle.

engineering [SCI TECH] The science by which the properties of matter and the sources of power in nature are made useful to humans in structures, machines, and products.

engineer's system of units *See* British gravitational system of units.

Engler viscometer [ENG] An instrument used in the measurement of the degree Engler, a measure of viscosity; the kinematic viscosity v in stokes for this instrument is obtained from the equation $v = 0.00147t - 3.74/t$, where t is the efflux time in seconds.

enhanced line *See* enhanced spectral line.

enhanced spectral line [SPECT] A spectral line of a very hot source, such as a spark, whose intensity is much greater than that of a line in a flame or arc spectrum. Also known as enhanced line.

enhancement [ELECTR] An increase in the density of charged carriers in a particular region of a semiconductor.

enlargement loss [FL MECH] Energy loss by friction in a flowing fluid when it moves into a cross-sectional area of sudden enlargement.

enlarger [OPTICS] An optical projector used to project an enlarged image of a photograph's negative onto photosensitized film or paper. Also known as photoenlarger.

enriched material [NUC ENG] Material in which the amount of one or more isotopes has been increased above that occurring in nature, such as uranium in which the abundance of U^{235} is increased.

enriched reactor [NUCLEO] A nuclear reactor in which the fuel is an enriched material.

enrichment [NUCLEO] A process that changes the isotopic ratio in a material; for uranium, for example, the ratio of U^{235} to U^{238} may be increased by gaseous diffusion of uranium hexafluoride.

enrichment factor [NUCLEO] The ratio of the abundance of a particular isotope in an enriched material to its abundance in the original material.

ensemble [STAT MECH] A collection of systems of particles used to describe an individual system; time averages of quantities describing the individual system are found by averaging over the systems in the ensemble at a fixed time.

Enskog theory *See* Chapman-Enskog theory.

enthalpy [THERMO] The sum of the internal energy of a system plus the product of the system's volume multiplied by the pressure exerted on the system by its surroundings. Also known as heat content; sensible heat; total heat.

enthalpy-entropy chart [THERMO] A graph of the enthalpy of a substance versus its entropy at various values of temperature, pressure, or specific volume; useful in making calculations about a machine or process in which this substance is the working medium.

enthalpy of reaction [PHYS CHEM] The change in enthalpy accompanying a chemical reaction.

enthalpy of transition [PHYS CHEM] The change of enthalpy accompanying a phase transition.

enthalpy of vaporization *See* heat of vaporization.

enthalpy-pressure chart *See* pressure-enthalpy chart.

entire function [MATH] A function of a complex variable which is analytic throughout the entire complex plane. Also known as integral function.

entire ring *See* integral domain.

entire series [MATH] A power series which converges for all values of its variable; a power series with an infinite radius of convergence.

entrained fluid [FL MECH] Fluid in the form of mist, fog, or droplets that is carried out of a column or vessel by a rising gas or vapor stream.

entrance loss [FL MECH] Energy loss by friction in a flowing fluid when it moves into a cross-sectional area of sudden contraction, as at the entrance of a pipe or a suddenly reduced area of a duct.

entrance pupil [OPTICS] The image of the aperture stop of an optical system formed in the object space by rays emanating from a point on the optical axis in the image space.

entrance slit [SPECT] Narrow slit through which passes the light entering a spectrometer.

entropy [MATH] In a mathematical context, this concept is attached to dynamical systems, transformations between measure spaces, or systems of events with probabilities; it expresses the amount of disorder inherent or produced. [STAT MECH] Measure of the disorder of a system, equal to the Boltzmann constant times the natural logarithm of the number of microscopic states corresponding to the thermodynamic state of the system; this statistical-mechanical definition can be shown to be equivalent to the thermodynamic definition. [THERMO] Function of the state of a thermodynamic system whose change in any differential reversible process is equal to the heat absorbed by the system from its surroundings divided by the absolute temperature of the system. Also known as thermal charge.

entropy of activation [PHYS CHEM] The difference in entropy between the activated complex in a chemical reaction and the reactants.

entropy of a partition [MATH] If ξ is a finite partition of a probability space, the entropy of ξ is the negative of the sum of the products of the probabilities of elements in ξ with the logarithm of the probability of the element.

entropy of a transformation *See* Kolmogorov-Sinai invariant.

entropy of a transformation given a partition [MATH] If T is a measure-preserving transformation on a probability space and ξ is a finite partition of the space, the entropy of T given ξ is the limit $n \to \infty$ of $1/n$ times the entropy of the partition which is the common refinement of ξ, $T^{-1}\xi$, . . . , $T^{-n+1}\xi$.

entropy of mixing [PHYS CHEM] After mixing substances, the difference between the entropy of the mixture and the sum of the entropies of the components of the mixture.

entropy of transition [PHYS CHEM] The heat absorbed or liberated in a phase change divided by the absolute temperature at which the change occurs.

entry ballistics [MECH] That branch of ballistics which pertains to the entry of a missile, spacecraft, or other object from outer space into and through an atmosphere.

enumerable *See* countable.

envelope [MATH] **1.** The envelope of a one-parameter family of curves is a curve which has a common tangent with each member of the family. **2.** The envelope of a one-parameter family of surfaces is the surface swept out by the characteristic curves of the family.

envelope detector *See* detector.

envelope of a family of curves [MATH] A curve which is tangent to every member of the given family.

environment [ENG] The aggregate of all natural, operational, or other conditions that affect the operation of equipment or components. [PHYS] The aggregate of all the conditions and the influences that determine the behavior of a physical system.

eolian sounds [ACOUS] Sounds produced by eddying motions of air in the lee of obstacles, such as wires, twigs, and even the ear itself, when wind blows over those obstacles.

eon [MECH] A unit of time, equal to 10^9 years.

eötvös [GEOPHYS] A unit of horizontal gradient of gravitational acceleration, equal to a change in gravitational acceleration of 10^{-9} galileo over a horizontal distance of 1 centimeter.

Eötvös constant [PHYS] A constant that appears in an expression for the behavior of the surface tension γ of a liquid as the temperature T drops to a critical temperature T_c at which the surface tension disappears, equal to $\gamma(M/\rho)^{2/3}/(T_c - T)$, where M is the molecular weight and ρ the density of the liquid.

Eötvös experiment [RELAT] An experiment which tests the equality of inertial mass and gravitational mass by balancing on a given body the earth's gravitational attraction against the kinetic reaction arising from the rotation of the earth.

Eötvös number *See* Bond number.

Eötvös rule [THERMO] The rule that the rate of change of molar surface energy with temperature is a constant for all liquids; deviations are encountered in practice.

Eötvös torsion balance [ENG] An instrument which records the change in the acceleration of gravity over the horizontal distance between the ends of a beam; used to measure density variations of subsurface rocks.

ephemeris [ASTRON] A periodical publication tabulating the predicted positions of celestial bodies at regular intervals, such as daily, and containing other data of interest to astronomers. Also known as astronomical ephemeris.

ephemeris day [ASTRON] A unit of time equal to 86,400 ephemeris seconds (International System of Units).

ephemeris second [ASTRON] The fundamental unit of time of the International System of Units of 1960, equal to 1/31556925.9747 of the tropical year defined by the mean motion of the sun in longitude at the epoch 1900 January 0 day 12 hours.

ephemeris time [ASTRON] The uniform measure of time defined by the laws of dynamics and determined in principle from the orbital motions of the planets, specifically the orbital motion of the earth as represented by Newcomb's Tables of the Sun. Abbreviated E.T.

epicadmium [NUCLEO] Energy above the greatest level at which cadmium shows a large neutron cross section, about 0.3 electron volt.

epicycle [MATH] The circle which generates an epicycloid or hypocycloid.

epicycloid [MATH] The curve traced by a point on a circle as it rolls along the outside of a fixed circle.

epidiascope [OPTICS] 1. An optical projection system for forming an enlarged real image of a flat opaque object, in which light is reflected from the object and then from a mirror before being focused by a projection lens. Also known as episcope. 2. An optical projection system which can easily be altered to project either transparent or opaque objects.

episcope *See* epidiascope.

episcotister [OPTICS] A device for reducing the intensity of light by a known fraction, consisting of a rapidly rotating disk with transparent and opaque sectors.

epitaxial diffused-junction transistor [ELECTR] A junction

EPIDIASCOPE

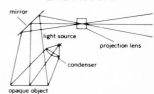

Diagram of an epidiascope (def. 1).

transistor produced by growing a thin, high-purity layer of semiconductor material on a heavily doped region of the same type.

epitaxial diffused-mesa transistor [ELECTR] A diffused-mesa transistor in which a thin, high-resistivity epitaxial layer is deposited on the substrate to serve as the collector.

epitaxial layer [SOLID STATE] A semiconductor layer having the same crystalline orientation as the substrate on which it is grown.

epitaxy [CRYSTAL] Growth of one crystal on the surface of another crystal in which the growth of the deposited crystal is oriented by the lattice structure of the substrate.

epithermal neutron [NUCLEO] A neutron having an energy in the range immediately above the thermal range, roughly between 0.02 and 100 electron volts.

epithermal reactor [NUCLEO] A nuclear reactor in which a substantial fraction of fissions is induced by neutrons having more than thermal energy.

epithermal thorium reactor [NUCLEO] A sodium-cooled reactor based on operation with neutrons in the high epithermal energy range; a uranium-thorium fuel mixture is used, with graphite or beryllium as moderator.

epitrochoid [MATH] A curve traced by a point rigidly attached to a circle at a point other than the center when the circle rolls without slipping on the outside of a fixed circle.

E-plane antenna [ELECTROMAG] An antenna which lies in a plane parallel to the electric field vector of the radiation that it emits.

E-plane bend *See* E bend.

E-plane T junction [ELECTROMAG] Waveguide T junction in which the change in structure occurs in the plane of the electric field. Also known as series T junction.

Eppley pyrheliometer [ENG] A pyrheliometer of the thermoelectric type; radiation is allowed to fall on two concentric silver rings, the outer covered with magnesium oxide and the inner covered with lampblack; a system of thermocouples (thermopile) is used to measure the temperature difference between the rings; attachments are provided so that measurements of direct and diffuse solar radiation may be obtained.

EPR *See* electron paramagnetic resonance.

epsilon meson [PARTIC PHYS] Neutral, scalar, meson resonance having positive charge conjugation parity and G parity, a mass of about 1200 MeV, and a width of about 600 MeV; decays to two pions.

epsilon neighborhood [MATH] The set of all points in a metric space whose distance from a given point is less than some number; this number is designated ε.

epsilon structure [SOLID STATE] The hexagonal close-packed structure of the ε-phase of an electron compound.

equal [MATH] Being the same in some sense determined by context.

equal-arm balance [MECH] A simple balance in which the distances from the point of support of the balance-arm beam to the two pans at the end of the beam are equal.

equal-energy source [PHYS] Electromagnetic or sound source of energy which emits the same amount of energy for each frequency of the spectrum.

equality [MATH] The state of being equal.

equal listener response scale [ACOUS] An arbitrary scale of noisiness which measures the average response of a listener to a noise when allowance is made for the apparent increase of intensity of a noise as its frequency increases. Abbreviated ELR scale.

equal loudness contour [ACOUS] A curve on a graph of sound intensity in decibels versus frequency at each point

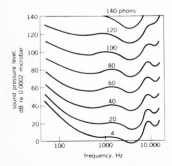

EQUAL LOUDNESS CONTOUR

Equal loudness level contours.

along which sound appears to be equally loud to a listener. Also known as Fletcher-Munson contour.

equally likely cases [STAT] All simple events in a trial have the same probability.

equally tempered scale [ACOUS] A musical scale formed by dividing the octave into 12 equal intervals and selecting from the resulting notes; thus, the frequency ratio between any two successive notes is exactly $2^{1/12}$ or $2^{1/6}$. Also known as equitempered scale.

equal ripple property [MATH] For any continuous function $f(x)$ on the interval $[-1,1]$, and for any positive integer n, a property of the polynomial $p_n(x)$, which is the best possible approximation to $f(x)$ in the sense that the maximum absolute value of $e_n(x) = f(x) - p_n(x)$ is as small as possible; namely, that $e_n(x)$ assumes its extreme values at least $n + 2$ times, with the consecutive extrema having opposite signs.

equal sets [MATH] Sets with precisely the same elements.

equal tails test [STAT] A technique for choosing two critical values for use in a two-sided test; it consists of selecting critical values c and d so that the probability of acceptance of the null hypothesis if the test statistic does not exceed c is the same as the probability of acceptance of the null hypothesis if the test statistic is not smaller than d.

equate [MATH] To state algebraically that two expressions are equal to one another.

equation [MATH] A statement that each of two expressions is equal to the other.

equation of mixed type [MATH] A partial differential equation which is of hyperbolic, parabolic, or elliptic type in different parts of a region. Also known as mixed-type boundary-value problem.

equation of motion [FL MECH] One of a set of hydrodynamical equations representing the application of Newton's second law of motion to a fluid system; the total acceleration on an individual fluid particle is equated to the sum of the forces acting on the particle within the fluid. [MECH] **1.** Equation which specifies the coordinates of particles as functions of time. **2.** A differential equation, or one of several such equations, from which the coordinates of particles as functions of time can be obtained if the initial positions and velocities of the particles are known. [QUANT MECH] A differential equation which enables one to predict the statistical distribution of the results of any measurement upon a system at any time if the initial dynamical state of the system is known.

equation of piezotropy [THERMO] An equation obeyed by certain fluids which states that the time rate of change of the fluid's density equals the product of a function of the thermodynamic variables and the time rate of change of the pressure.

equation of state [PHYS CHEM] A mathematical expression which defines the physical state of a homogeneous substance (gas, liquid, or solid) by relating volume to pressure and absolute temperature for a given mass of the material.

equation of time [ASTRON] The addition of a quantity to mean solar time to obtain apparent solar time; formerly, when apparent solar time was in common use, the opposite convention was used; apparent solar time has annual variation as a result of the sun's inclination in the ecliptic and the eccentricity of the earth's elliptical orbit.

equatorial acceleration [ASTROPHYS] A state in which the equatorial atmosphere of a celestial body has a larger absolute angular velocity than the more poleward portions of the atmosphere; exhibited by the sun, Jupiter, and Saturn.

equatorial electrojet [GEOPHYS] A concentration of electric current in the atmosphere found in the magnetic equator.

EQUATION OF TIME

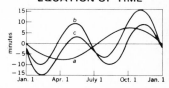

The equation of time; c is the sum of the elliptical motion effect a and the inclination effect b.

equatorial horizontal parallax [ASTRON] The parallax of a member of the solar system measured from positional observations made at the same time at two stations on earth, whose distance apart is the earth's equatorial radius.

equatorial mounting [ENG] The mounting of an equatorial telescope; it has two perpendicular axes, the polar axis (parallel to the earth's axis) that turns on fixed bearings, and the declination axis, supported by the polar axis.

equatorial orbit [ASTRON] An orbit in the plane of the earth's equator.

equatorial plane [ASTRON] The plane passing through the equator of the earth, or of an other celestial body, perpendicular to its axis of rotation and equidistant from its poles. [MECH] A plane perpendicular to the axis of rotation of a rotating body and equidistant from the intersections of this axis with the body's surface, provided that the body is symmetric about the axis of rotation and is symmetric under reflection through this plane. [OPTICS] *See* sagittal plane.

equatorial system [ASTRON] A set of celestial coordinates based on the celestial equator as the primary great circle; usually declination and hour angle or sidereal hour angle. Also known as celestial equator system of coordinates; equinoctial system of coordinates.

equatorial telescope [ENG] An astronomical telescope that revolves about an axis parallel to the earth's axis and automatically keeps a star on which it has been fixed in its field of view.

equiangular polygon [MATH] A polygon all of whose interior angles are equal.

equiangular spiral *See* logarithmic spiral.

equiangular spiral antenna [ELECTROMAG] A frequency-independent broad-band antenna, cut from sheet metal, that radiates a very broad, circularly polarized beam on both sides of its surface; this bidirectional radiation pattern is its chief limitation.

equicontinuous family of functions [MATH] A family of functions with the property that given any $\varepsilon > 0$ there is a $\delta > 0$ such that whenever $|x - y| < \delta$, $|f(x) - f(y)| < \varepsilon$ for every function $f(x)$ in the family.

equicontinuous flow [MATH] A transformation group (G, X, π) on a compact metric space X with metric $d(x,y)$, such that for each $\varepsilon > 0$ there exists $\delta > 0$ such that $d[\pi(g,x), \pi(g,y)] < \varepsilon$ for all g in G, whenever $d(x,y) < \delta$.

equidistant [MATH] Being the same distance from some given object.

equilateral polygon [MATH] A polygon all of whose sides are the same length.

equilateral polyhedron [MATH] A polyhedron all of whose faces are identical.

equilibrant [MECH] A single force which cancels the vector sum of a given system of forces acting on a rigid body and whose torque cancels the sum of the torques of the system.

equilibrium [MECH] Condition in which a particle, or all the constituent particles of a body, are at rest or in unaccelerated motion in an inertial reference frame. Also known as static equilibrium. [PHYS] Condition in which no change occurs in the state of a system as long as its surroundings are unaltered. [STAT MECH] Condition in which the distribution function of a system is time-independent.

equilibrium constant [CHEM] A constant at a given temperature such that when a reversible chemical reaction $cC + bB = gG + hH$ has reached equilibrium, the value of this constant K^0 is equal to

$$\frac{a_G^g a_H^h}{a_C^c a_B^b}$$

where a_G, a_H, a_C, and a_B represent chemical activities of the species G, H, C, and B at equilibrium.

equilibrium diagram [PHYS CHEM] A phase diagram of the equilibrium relationship between temperature, pressure, and composition in any system.

equilibrium orbit [NUCLEO] A path, such as a circular path in a synchrotron, or a point moving through space, such as in a linear accelerator, about which the particles in a particle accelerator oscillate, experiencing an effective restoring force toward the path or point. Also known as stable orbit.

equilibrium point [MATH] For the vector differential equation $d\mathbf{x}/dt = \mathbf{f}(\mathbf{x})$, where $\mathbf{f}$ is continuously differentiable, describing the motion of points in euclidean n-space, a point $\bar{\mathbf{x}}$ such that $\mathbf{f}(\bar{\mathbf{x}}) = 0$.

equilibrium prism [PHYS CHEM] Three-dimensional (solid) diagram for multicomponent mixtures to show the effects of composition changes on some key property, such as freezing point.

equilibrium ratio [PHYS CHEM] **1.** In any system, relation of the proportions of the various components (gas, liquid) at equilibrium conditions. **2.** *See* equilibrium vaporization ratio.

equilibrium vaporization ratio [PHYS CHEM] In a liquid-vapor equilibrium mixture, the ratio of the mole fraction of a component in the vapor phase (y) to the mole fraction of the same component in the liquid phase (x), or $y/x = K$ (the K factor). Also known as equilibrium ratio.

equilibrium vapor pressure [PHYS] The vapor pressure of a system in which two or more phases of water coexist in equilibrium.

equinoctial *See* celestial equator.

equinoctial colure [ASTRON] The great circle of the celestial sphere through the celestial poles and the equinoxes; the hour circle of the vernal equinox.

equinoctial point *See* equinox.

equinoctial system of coordinates *See* equatorial system.

equinox [ASTRON] **1.** Either of the two points of intersection of the ecliptic and the celestial equator, occupied by the sun when its declination is 0°. Also known as equinoctial point. **2.** That instant when the sun occupies one of the equinoctial points.

equipartition law [STAT MECH] In a classical ideal gas, the average kinetic energy per molecule associated with any degree of freedom which occurs as a quadratic term in the expression for the mechanical energy, is equal to half of Boltzmann's constant times the absolute temperature.

equiphase wave surface [PHYS] Any surface in a wave over which the field vectors at the same instant are in the same phase or 180° out of phase.

equipollent [MECH] Of two systems of forces, having the same vector sum and the same total torque about an arbitrary point.

equipotential cathode *See* indirectly heated cathode.

equipotential surface [ELEC] A surface on which the electric potential is the same at every point. [GEOPHYS] A surface characterized by the potential being constant everywhere on it for the attractive forces concerned. [MECH] A surface which is always normal to the lines of force of a field and on which the potential is everywhere the same.

equisignal surface [ELECTROMAG] Surface around an antenna formed by all points at which, for transmission, the field strength (usually measured in volts per meter) is constant.

equitempered scale *See* equally tempered scale.

equivalence [MATH] A logic operator having the property that if P, Q, R, etc., are statements, then the equivalence of P, Q, R, etc., is true if and only if all statements are true or all statements are false.

equivalence classes [MATH] The collection of pairwise disjoint subsets determined by an equivalence relation on a set; two elements are in the same equivalence class if and only if they are equivalent under the given relation.

equivalence law of ordered sampling [STAT] If a random ordered sample of size s is drawn from a population of size N, then on any particular one of the s draws each of the N items has the same probability, $1/N$, of appearing.

equivalence principle [RELAT] In general relativity, the principle that the observable local effects of a gravitational field are indistinguishable from those arising from acceleration of the frame of reference. Also known as Einstein's equivalency principle; principle of equivalence.

equivalence relation [MATH] A relation which is reflexive, symmetric, and transitive.

equivalence transformation [MATH] A mapping which associates with each square matrix A the matrix $B = SAT$, where S and T are nonsingular matrices. Also known as equivalent transformation.

equivalent absorption area [ACOUS] Area of perfectly absorbing surface that will absorb sound energy at the same rate as the given object under the same conditions; the acoustic unit of equivalent absorption is the sabin.

equivalent bending moment [MECH] A bending moment which, acting alone, would produce in a circular shaft a normal stress of the same magnitude as the maximum normal stress produced by a given bending moment and a given twisting moment acting simultaneously.

equivalent circuit [ELEC] A circuit whose behavior is identical to that of a more complex circuit or device over a stated range of operating conditions.

equivalent continued fractions [MATH] Continued fractions whose values to n terms are the same for $n = 1, 2, 3, \ldots$.

equivalent electrons [ATOM PHYS] Electrons in an atom which have the same principal and orbital quantum numbers, but not necessarily the same magnetic orbital and magnetic spin quantum numbers.

equivalent elements *See* associates.

equivalent focal length [OPTICS] The focal length of a thin lens which forms images that most nearly duplicate those of a given compound lens, thick lens, or system of lenses.

equivalent footcandle *See* foot-lambert.

equivalent magnetic length *See* effective magnetic length.

equivalent matrices [MATH] Two square matrices A and B for which there exist nonsingular square matrices S and T such that $B = SAT$.

equivalent noise conductance [ELECTR] Spectral density of a noise current generator measured in conductance units at a specified frequency.

equivalent noise resistance [ELECTR] Spectral density of a noise voltage generator measured in ohms at a specified frequency.

equivalent noise temperature [ELECTR] Absolute temperature at which a perfect resistor, of equal resistance to the component, would generate the same noise as does the component at room temperature.

equivalent normal towers [MATH] Two normal towers of subgroups of a group G, $G = G_1, G_2, \ldots, G_n = \{e\}$, and $G = H_1, H_2, \ldots, H_m = \{e\}$ (where e is the unit element), for which $n = m$ and there exists a permutation of the indices $i = 1, \ldots,$

$n - 1$, written $i \rightarrow i'$, such that the quotient group G_i/G_{i+1} is isomorphic to $H_{i'}/H_{i'+1}$.

equivalent nuclei [PHYS CHEM] A set of nuclei in a molecule which are transformed into each other by rotations, reflections, or combinations of these operations, leaving the molecule invariant.

equivalent piston [ACOUS] For a diaphragm vibrating at a frequency below its fundamental frequency, a piston that vibrates with the same velocity as a specified point on the diaphragm (usually the center) and has an area such that its strength as a source of sound is the same as that of the diaphragm.

equivalent positions [CRYSTAL] The collection of points that are generated by applying the operations of a given space group to a single point in a crystal.

equivalent resistance [ELEC] Concentrated or lumped resistance that would cause the same power loss as the actual small resistance values distributed throughout a circuit.

equivalent sine wave [PHYS] A sine wave whose root-mean-square value and period are the same as that of a given periodic wave.

equivalent twisting moment [MECH] A twisting moment which, if acting alone, would produce in a circular shaft a shear stress of the same magnitude as the shear stress produced by a given twisting moment and a given bending moment acting simultaneously.

equivalent width [PHYS] A measure of the total absorption of radiant energy as indicated by an absorption line or absorption band.

equivoluminal wave *See* S wave.

Er *See* erbium.

ERA *See* electron ring accelerator.

eradiation *See* terrestrial radiation.

erbium [CHEM] A trivalent metallic rare-earth element, symbol Er, of the yttrium subgroup, found in euxenite, gadolinite, fergusonite, and xenotine; atomic number 68, atomic weight 167.26, specific gravity 9.051; insoluble in water, soluble in acids; melts at 1400–1500°C.

Erdélyi-Kober operations [MATH] A pair of integral operators which carry out fractional integration of a general kind and are used in the study of dual integral equations; they may be written

$$I_{p,q}f(x) = \frac{2x^{-2p-2q}}{\Gamma(q)} \int_0^x (x^2 - y^2)^{q-1} y^{2p+1} f(y) \, dy$$

$$K_{p,q}f(x) = \frac{2x^{2p}}{\Gamma(q)} \int_x^\infty (y^2 - x^2)^{q-1} y^{-2p-2q+1} f(y) \, dy$$

for Re $q > 0$, Re $p > -\frac{1}{2}$.

Erdélyi's function [MATH] The confluent hypergeometric function

$$_2F_0(a,b; -1/z) = z^a \Psi(a, a - b + 1; z)$$

where Ψ is Tricomi's function.

erect image [OPTICS] An image in which directions are the same as those in the object, in contrast to an inverted image.

erecting lens [OPTICS] An eyepiece sometimes used in Kepler telescopes that consists of four lenses and provides an erect image, which is more convenient for viewing terrestrial objects than the inverted image provided by simpler eyepieces.

erecting prism [OPTICS] A system of prisms that converts the inverted image formed by most types of astronomical telescopes into an erect image.

erg [PHYS] A unit of energy or work in the centimeter-gram-

ERBIUM

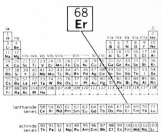

Periodic table of the chemical elements showing the position of erbium.

second system of units, equal to the work done by a force of magnitude of 1 dyne when the point at which the force is applied is displaced 1 centimeter in the direction of the force. Also known as dyne centimeter (dyne-cm).

ergodic [STAT] **1.** Property of a system or process in which averages computed from a data sample over time converge, in a probabilistic sense, to ensemble or special averages. **2.** Pertaining to such a system or process.

ergodic theory [MATH] The study of measure-preserving transformations. [STAT MECH] Mathematical theory which attempts to show that the various possible microscopic states of a system are equally probable, and that the system is therefore ergodic.

ergodic transformation [MATH] A measure-preserving transformation on X with the property that whenever X is written as a union of two disjoint invariant subsets, one of these must have measure zero.

ergon [QUANT MECH] A quantum of energy; for any oscillator it is equal to the product of the oscillator's frequency and Planck's constant.

ergosphere [RELAT] A region of space-time in which a particle can never be at rest as seen from infinity; associated with rotating bodies in general relativity theory.

Ericsson cycle [THERMO] An ideal thermodynamic cycle consisting of two isobaric processes interspersed with processes which are, in effect, isothermal, but each of which consists of an infinite number of alternating isentropic and isobaric processes.

eriometer [OPTICS] A device used to measure diameters of small particles or fibers by observing the diameter of the diffraction pattern produced by them in light coming from a small hole in a metal plate.

error coefficient [CONT SYS] The steady-state value of the output of a control system, or of some derivative of the output, divided by the steady-state actuating signal. Also known as error constant.

error constant *See* error coefficient.

error equation [STAT] The equation of a normal distribution.

error function [MATH] The real function defined as the integral from 0 to x of $e^{-t^2} \, dt$ or $e^{t^2} \, dt$, or the integral from x to ∞ of $e^{-t^2} \, dt$.

error mean-square [STAT] The residual or error sum of squares divided by the number of degrees of freedom of the sum; gives an estimate of the error or residual variance.

error of the first kind *See* type I error.

error of the second kind *See* type II error.

error range [STAT] The difference between the highest and lowest error values; a measure of the uncertainty associated with a number.

error signal [CONT SYS] In an automatic control device, a signal whose magnitude and sign are used to correct the alignment between the controlling and the controlled elements. [ELEC] *See* error voltage. [ELECTR] A voltage that depends on the signal received from the target in a tracking system, having a polarity and magnitude dependent on the angle between the target and the center of the scanning beam.

error sum of squares [STAT] In analysis of variance, the sum of squares of the estimates of the contribution from the stochastic component. Also known as residual sum of squares.

error voltage [ELEC] A voltage, usually obtained from a selsyn, that is proportional to the difference between the angular positions of the input and output shafts of a servosystem; this voltage acts on the system to produce a motion that tends to reduce the error in position. Also known as error signal.

eruptive prominence [ASTRON] A prominence on the sun that is formed from active material above the chromosphere and reaches high altitudes on the sun at great speed.

eruptive star [ASTRON] A star that has a rapid change in its intensity because of the physical change it undergoes; examples are flare stars, recurrent novae, novae, supernovae, and nebular variables.

Es *See* einsteinium.

ESCA *See* x-ray photoelectron spectroscopy.

escape probability [NUCLEO] The proportion of neutrons produced in a reactor which eventually leaves the reactor without being absorbed.

Esclangon effect [OPTICS] Bending of a reflected light ray caused by movement of the mirror in a direction making an acute angle with its surface.

ESR *See* electron paramagnetic resonance.

Essen coefficient [ELEC] The torque exerted on the moving part of an electric rotating machine divided by the volume enclosed by the air gap.

essential singularity [MATH] An isolated singularity of a complex function which is neither removable nor a pole.

estimator [STAT] A random variable or a function of it used to estimate population parameters.

esu *See* electrostatic units.

E.T. *See* ephemeris time.

etalon [OPTICS] **1.** Two adjustable parallel mirrors mounted so that either one may serve as one of the mirrors in a Michelson interferometer; used to measure distances in terms of wavelengths of spectral lines. **2.** An instrument similar to the Fabry-Perot interferometer, except that the distance between the plates is fixed. Also known as Fabry-Perot etalon.

eta meson [PARTIC PHYS] Neutral pseudoscalar meson having zero isotopic spin and hypercharge, positive charge parity and G parity, and a mass of about 549 MeV; decays via electromagnetic interactions.

eta-prime meson *See* chi meson.

ether [ELECTROMAG] The medium postulated to carry electromagnetic waves, similar to the way a gas carries sound waves.

ether drag [ELECTROMAG] The hypothesis, advanced unsuccessfully to account for results of the Michelson-Morley experiment, that ether is dragged along with matter.

ether drift [ELECTROMAG] Hypothetical motion of the ether relative to the earth.

ether thermoscope [PHYS] A device for detecting radiant heat; consists of an evacuated U-shaped tube, with ether at the bottom of the tube and a bulb at each end, one bulb being blackened.

E transformer [ELECTROMAG] A transformer consisting of two coils wound around a laminated iron core in the shape of an E, with the primary and secondaries occupying the center and outside legs respectively.

Ettingshausen coefficient [PHYS] A measure of the strength of the Ettingshausen effect, equal to the ratio of the temperature gradient to the product of the current density and magnetic field strength which produce this gradient.

Ettingshausen effect [PHYS] The phenomenon that, when a metal strip is placed with its plane perpendicular to a magnetic field and an electric current is sent longitudinally through the strip, corresponding points on opposite edges of the strip have different temperatures.

Ettingshausen-Nernst coefficient [PHYS] A measure of the

ETALON

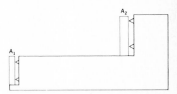

An etalon with two adjustable mirrors A_1 and A_2 which are to be used with the Michelson interferometer in distance measurement.

strength of the Ettingshausen-Nernst effect, equal to the ratio of the electric field to the product of the temperature gradient and magnetic field strength which produce this field.

Ettingshausen-Nernst effect [PHYS] The phenomenon that, when a conductor or semiconductor is subjected to a temperature gradient and to a magnetic field perpendicular to the temperature gradient, an electric field arises perpendicular to both the temperature gradient and the magnetic field. Also known as Nernst effect.

Eu *See* europium.

euclidean algorithm [MATH] A method of finding the greatest common divisor of a pair of integers.

euclidean geometry [MATH] The study of the properties preserved by isometries of two- and three-dimensional euclidean space.

euclidean space [MATH] A space consisting of all ordered sets $(x_1, \ldots, x_n)$ of n numbers with the distance between $(x_1, \ldots, x_n)$ and $(y_1, \ldots, y_n)$ being given by

$$\left[\sum_{i=1}^{n} (x_i - y_i)^2\right]^{\frac{1}{2}};$$

the number n is called the dimension of the space.

Euler angles [MECH] Three angular parameters that specify the orientation of a body with respect to reference axes.

Euler characteristic of a topological space X [MATH] The number $\chi(X) = \sum (-1)^q \beta_q$, where β_q is the qth-Betti number of X.

Euler diagram [MATH] A diagram consisting of closed curves, used to represent relations between logical propositions or sets; similar to a Venn diagram.

Euler equation [MECH] Expression for the energy removed from a gas stream by a rotating blade system (as a gas turbine), independent of the blade system (as a radial- or axial-flow system).

Euler equations of motion [MECH] A set of three differential equations expressing relations between the force moments, angular velocities, and angular accelerations of a rotating rigid body.

Euler force [MECH] The greatest load that a long, slender column can carry without buckling, according to the Euler formula for long columns.

Euler formula for long columns [MECH] A formula which gives the greatest axial load that a long, slender column can carry without buckling, in terms of its length, Young's modulus, and the moment of inertia about an axis along the center of the column.

Euler graph [MATH] A graph for which an Euler path exists.

Eulerian coordinates [FL MECH] Any system of coordinates in which properties of a fluid are assigned to points in space at each given time, without attempt to identify individual fluid parcels from one time to the next; a sequence of synoptic charts is a Eulerian representation of the data.

Eulerian correlation [FL MECH] The correlation between the properties of a flow at various points in space at a single instant of time. Also known as synoptic correlation.

Eulerian equation [FL MECH] A mathematical representation of the motions of a fluid in which the behavior and the properties of the fluid are described at fixed points in a coordinate system.

Euler-Lagrange equation [MATH] A partial differential equation arising in the calculus of variations, which provides a necessary condition that $y(x)$ minimize the integral over some finite interval of $f(x,y,y')\, dx$, where $y' = dy/dx$; the equation is $(f(x,y,y')/\partial y) - (d/dx)(\partial f(x,y,y')/\partial y') = 0$. Also known as Euler's equation.

Euler-Maclaurin formula [MATH] A formula used in the numerical evaluation of integrals, which states that the value of an integral is equal to the sum of the value given by the trapezoidal rule and a series of terms involving the odd-numbered derivatives of the function at the end points of the interval over which the integral is evaluated.

Euler method [MECH] A method of studying fluid motion and the mechanics of deformable bodies in which one considers volume elements at fixed locations in space, across which material flows; the Euler method is in contrast to the Lagrangian method.

Euler number 1 [FL MECH] A dimensionless number used in the study of fluid friction in conduits, equal to the pressure drop due to friction divided by the product of the fluid density and the square of the fluid velocity.

Euler number 2 [FL MECH] A dimensionless number equal to two times the Fanning friction factor.

Euler path [MATH] A path along the edges of a graph that traverses every edge exactly once and terminates at its starting point.

Euler-Rodrigues parameter [MECH] One of four numbers which may be used to specify the orientation of a rigid body; they are components of a quaternion.

Euler's constant [MATH] The limit, as n approaches infinity, of $1 + 1/2 + 1/3 + \cdots + 1/n - \ln n$, equal to approximately 0.5772. Denoted γ. Also known as Mascheroni's constant.

Euler's equation *See* Euler-Lagrange equation.

Euler's expansion [FL MECH] The transformation of a derivative (d/dt) describing the behavior of a moving particle with respect to time, into a local derivative $(\delta/\delta t)$ and three additional terms that describe the changing motion of a fluid as it passes through a fixed point.

Euler's formula [MATH] The formula $e^{ix} = \cos x + i \sin x$, where $i = \sqrt{-1}$.

Euler's functional equation [MATH] The equation $f(xz,yz) = z^a f(x,y)$ for $z \neq 0$.

Euler's numbers [MATH] The numbers E_{2n} defined by the equation

$$\frac{1}{\cos z} = \sum_{n=0}^{\infty} (-1)^n \frac{E_{2n}}{(2n)!} z^{2n}.$$

Euler's relation [MATH] The relation $e^{i\pi} + 1 = 0$, connecting the numbers e, π, and $i = \sqrt{-1}$; it is a special case of Euler's formula.

Euler's theorem [MATH] For any polyhedron, $V - E + F = 2$, where V, E, F represent the number of vertices, edges, and faces respectively.

Euler's theorem for homogeneous functions [MATH] The value of a homogeneous function of degree r in the variables $x_1, x_2, \ldots, x_n$, multiplied by r, is equal to the sum over $i = 1, 2, \ldots, n$ of x_i times the partial derivative of the function with respect to x_i.

Euler transformation [MATH] A method of obtaining from a given convergent series a new series which converges faster to the same limit, and for defining sums of certain divergent series; the transformation carries the series $a_0 - a_1 + a_2 - a_3 + \cdots$ into a series whose nth term is

$$\sum_{r=0}^{n-1} (-1)^r \binom{n-1}{r} a_r/2^n.$$

Euler transform of the first kind *See* Riemann-Liouville fractional integral.

Euler transform of the second kind *See* Weyl fractional integral.

EUROPIUM

Periodic table of the chemical elements showing the position of europium.

europium [CHEM] A member of the rare-earth elements in the cerium subgroup, symbol Eu, atomic number 63, atomic weight 151.96, steel gray and malleable, melting at 1100–1200°C.

eV *See* electron volt.

evaporation [PHYS] Conversion of a liquid to the vapor state by the addition of latent heat.

evaporation gage *See* atmometer.

evaporimeter *See* atmometer.

evection [ASTROPHYS] A perturbation of the moon in its orbit due to the attraction of the sun.

E vector [ELECTROMAG] Vector representing the electric field of an electromagnetic wave.

even-even nucleus [NUC PHYS] A nucleus which has an even number of neutrons and an even number of protons.

even function [MATH] A function with the property that $f(x) = f(-x)$ for each number x.

even harmonic [PHYS] A harmonic that is an even multiple of the fundamental frequency.

even number [MATH] An integer which is a multiple of 2.

even-odd nucleus [NUC PHYS] A nucleus which has an even number of protons and an odd number of neutrons.

even permutation [MATH] A permutation which may be represented as a product of an even number of transpositions.

event [PHYS] A point in space-time. [STAT] A mathematical model of the result of a conceptual experiment; this model is a measurable subset of a probability space.

event horizon [RELAT] The boundary of a region of space-time from which it is not possible to escape to infinity. Symbolized $\mathscr{E}^+$.

Everett's interpolation formula [MATH] A formula for estimating the value of a function at an intermediate value of the independent variable, when its value is known at a series of equally spaced points (such as those that appear in a table), in terms of the central differences of the function of even order only and coefficients which are polynomial functions of the independent variable.

Evershed product [ELECTROMAG] A criterion for the value of a material as a permanent magnet, based on the maximum value of the energy product $\vec{B} \cdot \vec{H}$ on the demagnetization curve.

Eve's constant [NUCLEO] A measure of a substance's intensity of radioactivity, equal to the number of ions produced per cubic centimeter per second in air by 1 gram of the substance at a distance of 1 centimeter.

Evjen method [SOLID STATE] Method of calculating lattice sums in which groups of charges whose total charge is zero are taken together, so that the contribution of each group is small and the series rapidly converges.

evolute [MATH] The locus of the centers of curvature of a curve.

evolution equation [MATH] An equation of the form $(du/dt) + Au(t) = f(t)$, where t ranges over the positive real numbers, u is a function from the positive real numbers into a Banach space B, A is an operator on B, and f is a prescribed function from the positive real numbers into B.

Ewald-Kornfeld method [SOLID STATE] An extension of the Ewald method to calculate Coulomb energies of dipole arrays.

Ewald method [SOLID STATE] Method of calculating lattice sums in which certain mathematical techniques are employed to make series converge rapidly.

Ewald sphere [SOLID STATE] A sphere superimposed on the reciprocal lattice of a crystal, used to determine the directions in which an x-ray or other beam will be reflected by a crystal lattice.

E wave *See* transverse magnetic wave.

Ewing's hysteresis tester [ENG] An instrument for determining the hysteresis loss of a specimen of magnetic material by measuring the deflection of a horseshoe magnet when the specimen is rapidly rotated between the poles of the magnet and the magnet is allowed to rotate about an axis that is aligned with the axis of rotation of the specimen.

Ewing theory of ferromagnetism [SOLID STATE] Theory of ferromagnetic phenomena which assumes each atom is a permanent magnet which can turn freely about its center under the influence of applied fields and other magnets.

exact differential equation [MATH] A differential equation obtained by setting the exact differential of some function equal to zero.

exact differential form [MATH] A differential form which is the differential of some other form.

exact sequence [MATH] A sequence of homomorphisms with the property that the kernel of each homomorphism is precisely the image of the previous homomorphism.

except [MATH] A logical operator which has the property that if P and Q are two statements, then the statement "P except Q" is true only when P alone is true; it is false for the other three combinations (P false Q false, P false Q true, and P true Q true).

except gate [ELECTR] A gate that produces an output pulse only for a pulse on one or more input lines and the absence of a pulse on one or more other lines.

exceptional group [MATH] One of five Lie groups which leave invariant certain forms constructed out of the Cayley numbers; they are Lie groups with maximum symmetry in the sense that, compared with other simple groups with the same rank (number of independent invariant operators), they have maximum dimension (number of generators).

exceptional Jordan algebra [MATH] A Jordan algebra that cannot be written as a symmetrized product over a matrix algebra; used in formulating a generalization of quantum mechanics.

exceptional space [QUANT MECH] A space used to describe a system with a finite number of degrees of freedom in a generalization of quantum mechanics; this generalization is achieved by reformulating quantum mechanics in terms of a Jordan algebra of observables and states, and then generalizing this to the exceptional Jordan algebra realized by the algebra of 3×3 Hermitian matrices over the Cayley numbers.

excess conduction [SOLID STATE] Electrical conduction by excess electrons in a semiconductor.

excess electron [SOLID STATE] Electron introduced into a semiconductor by a donor impurity and available for conduction.

excess reactivity [NUCLEO] The amount of surplus reactivity over that needed to achieve criticality; it is built into a reactor (by using extra fuel) in order to compensate for fuel burnup and the accumulation of fission-product poisons during operation.

exchange [QUANT MECH] **1.** Operation of exchanging the space and spin coordinates in a Schrödinger-Pauli wave function representing two identical particles; this operation must leave the wave function unchanged, except possibly for sign. **2.** Process of exchanging a real or virtual particle between two other particles.

exchange anisotropy [ELECTROMAG] Phenomenon observed in certain mixtures of magnetic materials under certain conditions, in which magnetization is favored in some direction (rather than merely along some axis); thought to be caused by exchange coupling across the interface between compounds when one is ferromagnetic and one is antiferromagnetic.

exchange broadening [SPECT] The broadening of a spectral line by some type of chemical or spin exchange process which limits the lifetime of the absorbing or emitting species and produces the broadening via the Heisenberg uncertainty principle.

exchange coefficient [FL MECH] A coefficient of eddy flux in turbulent flow, defined in analogy to those coefficients of the kinetic theory of gases. Also known as austausch coefficient; eddy coefficient; interchange coefficient.

exchange current [ELEC] The magnitude of the current which flows through a galvanic cell when it is operating in a reversible manner.

exchange degeneracy [PARTIC PHYS] Coincidence of two Regge trajectories for particles having the same quantum numbers (except for parity, charge parity, and G parity) where one would have expected separate trajectories for alternate Regge recurrences. [QUANT MECH] An exchange process that leads back to the original configuration. Also known as exchange symmetry.

exchange force [QUANT MECH] The force arising in an exchange interaction.

exchange integral [QUANT MECH] Integral over the coordinates of two identical particles which can be thought of as the interaction between a given state and a second state in which the coordinates of the particles are exchanged.

exchange interaction [QUANT MECH] **1.** An interaction represented by a potential involving exchange of space or spin coordinates, or both, of the particles involved; can be visualized physically in terms of exchange of particles. **2.** Any interaction which can be looked upon as due to exchange of particles.

exchange narrowing [SPECT] The phenomenon in which, when a spectral line is split and thereby broadened by some variable perturbation, the broadening may be narrowed by a dynamic process that exchanges different values of the perturbation.

exchange operator [QUANT MECH] An operator which exchanges the spatial coordinates of the particles in a wave function, or their spins, or both positions and spins.

exchange reaction [CHEM] Reaction in which two atoms or ions exchange places either in two different molecules or in the same molecule.

exchange symmetry *See* exchange degeneracy.

excimer laser [OPTICS] A laser containing a noble gas, such as helium or neon, which is based on a transition between an excited state in which a metastable bond exists between two gas atoms and a rapidly dissociating ground state.

excitation [ATOM PHYS] A process in which an atom or molecule gains energy from electromagnetic radiation or by collision, raising it to an excited state. [CONT SYS] The application of energy to one portion of a system or apparatus in a manner that enables another portion to carry out a specialized function; a generalization of the electricity and electronics definitions. [ELEC] The application of voltage to field coils to produce a magnetic field, as required for the operation of an excited-field loudspeaker or a generator. [ELECTR] **1.** The signal voltage that is applied to the control electrode of an electron tube. Also known as drive. **2.** Application of signal power to a transmitting antenna. [QUANT

EXCITATION

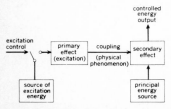

Diagram showing function of excitation in a control system.

MECH] The addition of energy to a particle or system of particles at ground state to produce an excited state.

excitation curve [NUC PHYS] A curve showing the relative yield of a specified nuclear reaction as a function of the energy of the incident particles or photons. Also known as excitation function.

excitation energy [QUANT MECH] The minimum energy required to change a system from its ground state to a particular excited state.

excitation function [ATOM PHYS] 1. The cross section for an incident electron to excite an atom to a particular excited state expressed as a function of the electron energy. 2. *See* excitation curve.

excitation index [SPECT] In emission spectroscopy, the ratio of intensities of a pair of extremely nonhomologous spectra lines; used to provide a sensitive indication of variation in excitation conditions.

excitation loss *See* core loss.

excitation potential [QUANT MECH] Electric potential which gives the excitation energy when multiplied by the magnitude of the electron charge.

excitation spectrum [SPECT] The graph of luminous efficiency per unit energy of the exciting light absorbed by a photoluminescent body versus the frequency of the exciting light.

excitation voltage [ELEC] Nominal voltage required for excitation of a circuit.

excitation volume [PHYS] In electron-probe microanalysis, the volume of the x-ray source used to penetrate and diffuse into the target sample.

excited state [QUANT MECH] A stationary state of higher energy than the lowest stationary state or ground state of a particle or system of particles.

exciting line [SPECT] The frequency of electromagnetic radiation, that is, the spectral line from a noncontinuous source, which is absorbed by a system in connection with some particular process.

exciton [SOLID STATE] An excited state of an insulator or semiconductor which allows energy to be transported without transport of electric charge; may be thought of as an electron and a hole in a bound state.

exclusion area [NUCLEO] The area around a nuclear operation (reactor, bomb test, and so on) where human habitation is restricted.

exclusion principle [QUANT MECH] The principle that no two fermions of the same kind may simultaneously occupy the same quantum state. Also known as Pauli exclusion principle.

exclusive or [MATH] A logic operator which has the property that if P is a statement and Q is a statement, then P exclusive or Q is true if either but not both statements are true, false if both are true or both are false.

excursion [NUCLEO] A sudden, very rapid rise in the power level of a nuclear reactor caused by supercriticality.

exhaust velocity [FL MECH] The velocity of gaseous or other particles in the exhaust stream of the nozzle of a reaction engine, relative to the nozzle.

existential quantifier [MATH] A logical relation, often symbolized $\exists$, that may be expressed by the phrase "there is a" or "there exists"; if P is a predicate, the statement $(\exists x)P(x)$ is true if there exists at least one value of x in the domain of P for which $P(x)$ is true, and is false otherwise.

exit pupil [OPTICS] The image of the aperture stop of an optical system formed in the image space by rays emanating from a point on the optical axis in the object space.

exline correction [FL MECH] Calculation of fluid-flow friction loss through annular sections with a correction for the flow eccentricity in the laminar-flow range.

exmeridian altitude [ASTRON] An altitude of a celestial body near the celestial meridian of the observer to which a correction is to be applied to determine the meridian altitude. Also known as circummeridian altitude.

exmeridian observation [ASTRON] **1.** Measurement of the altitude of a celestial body near the celestial meridian of the observer, for conversion to a meridian altitude. **2.** The altitude so measured.

exo- [SCI TECH] A prefix denoting outside or outer.

exoelectrons [PHYS] Electrons emitted from the surfaces of metals and certain ceramics after these surfaces have been freshly formed by a process such as abrasion or fracture; electrons obtain energy required for emission from processes such as establishment of surface films and rearrangement of disturbed atoms.

exoergic *See* exothermic.

exoergic collision *See* collision of the second kind.

exogenous electrification [ELEC] The separation of electric charge in a conductor placed in a preexisting electric field, especially applied to the charge separation observed on metal-covered aircraft, resulting from induction effects, and by itself does not create any net total charge on the conductor.

exogenous variables [MATH] In a mathematical model, the independent variables, which are predetermined and given outside the model.

exothermic [PHYS] Indicating liberation of heat. Also known as exoergic.

expansion [ELECTR] A process in which the effective gain of an amplifier is varied as a function of signal magnitude, the effective gain being greater for large signals than for small signals; the result is greater volume range in an audio amplifier and greater contrast range in facsimile. [PHYS] Process in which the volume of a constant mass of a substance increases.

expansion chamber *See* cloud chamber.

expansion coefficient *See* coefficient of cubical expansion.

expansion ellipsoid [SOLID STATE] An ellipsoid whose axes have lengths which are proportional to the coefficient of linear expansion in the corresponding direction in a crystal.

expansion ratio [FL MECH] For the calculation of the mass flow of a gas out of a nozzle or other expanding duct, the ratio of the nozzle exit section area to the nozzle throat area, or the ratio of final to initial volume.

expansion wave [FL MECH] A pressure wave or shock wave that decreases the density of air as the air passes through it.

expansivity *See* coefficient of cubical expansion.

expectation *See* expected value.

expectation value [QUANT MECH] The average of the results of a large number of measurements of a quantity made on a system in a given state; in case the measurement disturbs the state, the state is reprepared before each measurement.

expected value [MATH] **1.** For a random variable x with probability density function $f(x)$, this is the integral from $-\infty$ to ∞ of $xf(x)\, dx$. Also known as expectation. **2.** For a random variable x on a probability space (Ω, P), the integral of x with respect to the probability measure P.

experiment [SCI TECH] The test of a hypothesis under controlled conditions.

experimental breeder reactor [NUCLEO] A fast, heterogeneous nuclear reactor used for research and breeding; its core consists of enriched U^{235} surrounded by a blanket of natural uranium.

experimental design [STAT] A pattern for setting up experiments and making observations about the relationship between several variables in which one attempts to obtain as much information as possible for a fixed expenditure level.

experimental reactor [NUCLEO] A reactor to test the design of a new reactor concept.

exploring coil [ELECTROMAG] A small coil used to measure a magnetic field or to detect changes produced in a magnetic field by a hidden object; the coil is connected to an indicating instrument either directly or through an amplifier. Also known as magnetic test coil; search coil.

explosion method [THERMO] Method of measuring the specific heat of a gas at constant volume by enclosing the gas with an explosive mixture, whose heat of reaction is known, in a chamber closed with a corrugated steel membrane which acts as a manometer, and by deducing the maximum temperature reached on ignition of the mixture from the pressure change.

explosive variable *See* cataclysmic variable.

exponent [MATH] A number or symbol placed to the right and above some given mathematical expression.

exponential curve [MATH] A graph of the function $y = a^x$, where a is a positive constant.

exponential decay [PHYS] The decrease of some physical quantity according to the exponential law $N = N_0 e^{-t/\tau}$.

exponential density function [MATH] A probability density function obtained by integrating a function of the form $\exp(-|x - m|/\sigma)$, where m is the mean and σ the standard deviation.

exponential distribution [STAT] A continuous probability distribution whose density function is given by $f(x) = ae^{-ax}$, where $a > 0$ for $x > 0$, and $f(x) = 0$ for $x \leq 0$; the mean and standard deviation are both $1/a$.

exponential equation [MATH] An equation containing e^x (the Naperian base raised to a power) as a term.

exponential experiment [NUCLEO] A nuclear experiment involving a subcritical assembly of fissionable and moderator material.

exponential function [MATH] The function $f(x) = e^x$, written $f(x) = \exp(x)$.

exponential integral [MATH] The function defined to be the integral from x to ∞ of $(e^{-t}/t)\, dt$ for x positive.

exponential law [MATH] *See* law of exponents. [PHYS] The principle that growth or decay of some physical quantity is at a rate such that its value at a certain time or place is the initial value times e raised to a power equal to a constant times some convenient coordinate, such as the elapsed time or the distance traveled by a wave; there is growth if the constant is positive, decay if it is negative.

exponential of an operator [MATH] For a bounded linear operator A on a Banach space, the sum of a series which is formally the exponential series in A.

exponential pulse [PHYS] Variation of some quantity with time similar to the displacement of a critically damped harmonic oscillator which is initially given an impulse in its equilibrium position.

exponential series [MATH] The Maclaurin series expansion of e^x, namely,

$$e^x = 1 + \sum_{n=1}^{\infty} \frac{x^n}{n!}.$$

exponential transmission line [ELEC] A two-conductor transmission line whose characteristic impedance varies exponentially with electrical length along the line.

exposure [NUCLEO] **1.** The total quantity of radiation at a given point, measured in air. **2.** The cumulative amount of radiation exposure to which nuclear fuel has been subjected in a nuclear reactor; usually expressed in terms of the thermal energy produced by the reactor per ton of fuel initially present, as megawatt days per ton. [OPTICS] *See* light exposure; radiant exposure.

exposure factor [NUCLEO] A quantity f used to specify radiographic exposure equal to st/d^2, where s is the intensity of radioactive source, t is the time, and d is the source to film distance.

exposure meter [OPTICS] An instrument used to measure the intensity of light reflected from an object, for the purpose of determining proper camera exposure.

exposure time [PHYS] The amount of time a material is illuminated or irradiated.

exsecant [MATH] The trigonometric function defined by subtracting unity from the secant, that is exsec $\theta = \sec \theta - 1$.

extended dislocation [CRYSTAL] A dislocation in a close-packed structure consisting of a strip of stacking fault edged by two lines across which slip through a fraction of a lattice constant, into one of the alternative stacking positions, has occurred.

extended-interaction tube [ELECTR] Microwave tube in which a moving electron stream interacts with a traveling electric field in a long resonator; bandwidth is between that of klystrons and traveling-wave tubes.

extended real numbers [MATH] The real numbers together with the quantities $+\infty$ and $-\infty$, with the topology defined by the metric

$$d(x,y) = |f(x) - f(y)|$$

with $f(x) = x/(1 + |x|)$, $f(+\infty) = 1$, and $f(-\infty) = -1$.

extended state [QUANT MECH] A state of motion in which an electron may be found anywhere within a region of a material of linear extent equal to that of the material itself.

extended x-ray absorption fine structure [PHYS] A variation in the x-ray absorption of a substance as a function of energy, at energies just above that required for photons to liberate core electrons into the continuum; it is due to interference between the outgoing photoelectron waves and electron waves backscattered from atoms adjacent to the absorbing atoms. Abbreviated EXAFS.

extensibility [MECH] The amount to which a material can be stretched or distorted without breaking.

extension field [MATH] An extension field of a given field E is a field F such that E is a subfield of F.

extension map [MATH] An extension map of a map f from a set A to a set L is a map g from a set B to L such that A is a subset of B and the restriction of g to A equals f.

extension of a field [MATH] Any field containing the original field.

extensive air shower *See* Auger shower.

extensive property [PHYS CHEM] A noninherent property of a system, such as volume or internal energy, that changes with the quantity of material in the system; the quantitative value equals the sum of the values of the property for the individual constituents.

exterior algebra [MATH] An algebra whose structure is analogous to that of the collection of differential forms on a Riemannian manifold. Also known as Grassmann algebra.

exterior angle [MATH] **1.** An angle between one side of a polygon and the prolongation of an adjacent side. **2.** An angle made by a line (the transversal) that intersects two other lines, and either of the latter on the outside.

EXPOSURE METER

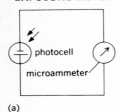

(a)

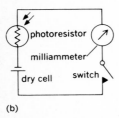

(b)

Basic exposure meter circuits.
(a) Photovoltaic-type circuit.
(b) Photoconductive-type circuit.

exterior ballistics [MECH] The science concerned with behavior of a projectile after leaving the muzzle of the firing weapon.

exterior of a set [MATH] The largest open set contained in the complement of a given set.

external angle [MATH] The angle defined by an arc around the boundaries of an internal angle or included angle.

external beam [NUCLEO] A beam of particles which originate in a particle accelerator and are directed outside the accelerator so that they can be used for experiments with external apparatus.

external force [MECH] A force exerted on a system or on some of its components by an agency outside the system.

external line [QUANT MECH] A component of a Feynman graph (in the diagrammatic presentation of perturbative quantum field theory) describing an incoming or outgoing particle in a scattering.

external photoelectric effect *See* photoemission.

external Q [ELECTR] The inverse of the difference between the loaded and unloaded Q values of a microwave tube.

external wave [FL MECH] **1.** A wave in fluid motion having its maximum amplitude at an external boundary such as a free surface. **2.** Any surface wave on the free surface of a homogeneous incompressible fluid is an external wave.

external work [THERMO] The work done by a system in expanding against forces exerted from outside.

extinction [OPTICS] Phenomenon in which plane polarized light is almost completely absorbed by a polarizer whose axis is perpendicular to the plane of polarization. [PHYS CHEM] *See* absorbance.

extinction coefficient *See* absorptivity.

extinction meter [OPTICS] An exposure meter in which light intensity is measured by gradually attenuating the light by a known fraction until a selected design is just visible or disappears.

extinction voltage [ELECTR] The lowest anode voltage at which a discharge is sustained in a gas tube.

extract a root of a number [MATH] To determine a root of the number, usually a positive real root, or a negative real odd root of a negative number.

extraordinary component *See* extraordinary wave.

extraordinary index [OPTICS] The index of refraction of the extraordinary wave propagating in a direction perpendicular to the optical axis of a uniaxial crystal.

extraordinary ray [OPTICS] One of two rays into which a ray incident on an anisotropic uniaxial crystal is split; its deviation at the crystal's surface depends on the orientation of the crystal, and it is deviated even in the case of normal incidence.

extraordinary singular point [MATH] A singular point on a curve at which at least two tangents to branches at the point coincide.

extraordinary wave [GEOPHYS] Magnetoionic wave component which, when viewed below the ionosphere in the direction of propagation, has clockwise or counterclockwise elliptical polarization respectively, accordingly as the earth's magnetic field has a positive or negative component in the same direction. Also known as X wave. [OPTICS] Component of electromagnetic radiation propagating in an anisotropic uniaxial crystal whose electric displacement vector lies in the plane containing the optical axis and the direction normal to the wavefront; it gives rise to the extraordinary ray. Also known as extraordinary component.

extrapolated boundary [NUCLEO] A hypothetical surface outside a medium sustaining a neutron chain reaction on which the neutron flux would be 0 if it could be extrapolated

EXTRAORDINARY RAY

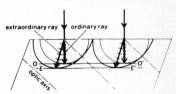

Huygens construction for a plane wave incident normally on transparent calcite showing the ordinary and extraordinary ray. *(From F. A. Jenkins and H. E. White, Fundamentals of Optics, 3d ed., McGraw-Hill, 1957)*

from the flux a few mean free paths inside the medium.

extrapolation [MATH] Estimating a function at a point which is larger than (or smaller than) all the points at which the value of the function is known.

extrapolation distance [NUCLEO] The distance from the boundary of a medium sustaining a neutron chain reaction to the extrapolated boundary.

extrapolation ionization chamber [NUCLEO] An ionization chamber so designed that volume, electrode separation, or some other factor can be varied in suitable steps for measurement purposes; the resulting measured values are plotted in appropriate form and the desired result is obtained by extrapolation of the curve.

extraterrestrial noise [ELECTROMAG] Cosmic and solar noise; radio disturbances from other sources other than those related to the earth.

extremally disconnected space [MATH] A topological space in which the closure of any open set is open.

extremals [MATH] For a variational problem in the calculus of variations entailing use of the Euler-Lagrange equation, the extremals are the solutions of this equation.

extreme *See* extremum.

extreme and mean ratio *See* golden section.

extreme narrowing approximation [SPECT] A mathematical approximation in the theory of spectral-line shapes to the effect that the exchange narrowing of a perturbation is complete.

extreme point [MATH] **1.** A maximum or minimum value of a function. **2.** A point in a convex subset K of a vector space is called extreme if it does not lie on the interior of any line segment contained in K.

extreme relativistic limit [PHYS] Limit which a formula describing a particle's behavior approaches when the speed of the particle approaches the speed of light.

extreme terms [MATH] The first and last terms in a proportion.

extreme value problem [MATH] A set of mathematical conditions which may be met by values that are less than or greater than an upper or a lower bound, that is, an extreme value.

extremum [MATH] A maximum or minimum value of a function. Also known as extreme.

extrinsic photoemission [ELECTR] Photoemission by an alkali halide crystal in which electrons are ejected directly from negative ion vacancies, forming color centers. Also known as direct ionization.

extrinsic properties [ELECTR] The properties of a semiconductor as modified by impurities or imperfections within the crystal.

extrinsic semiconductor [ELECTR] A semiconductor whose electrical properties are dependent on impurities added to the semiconductor crystal, in contrast to an intrinsic semiconductor, whose properties are characteristic of an ideal pure crystal.

extrinsic sol [PHYS CHEM] A colloid whose stability is attributed to electric charge on the surface of the colloidal particles.

eye lens [OPTICS] The lens in a two-lens eyepiece which is nearer to the eye.

eyepiece [OPTICS] A lens or optical system which offers to the eye the image originating from another system (the objective) at a suitable viewing distance. Also known as ocular.

Eykman formula [OPTICS] An empirical formula which relates the molal refraction of a liquid at a given optical frequency to its index of refraction, density, and molecular weight.

Eyring equation [PHYS CHEM] An equation, based on statistical mechanics, which gives the specific reaction rate for a chemical reaction in terms of the heat of activation, entropy of activation, the temperature, and various constants.

Eyring formula [FL MECH] A formula, based on the Eyring theory of rate processes, which relates shear stress acting on a liquid and the resulting rate of shear.

Eyring molecular system [FL MECH] Theory to account for liquid properties; assumes that each liquid molecule can move freely within a certain free volume. Also known as Eyring theory.

Eyring theory *See* Eyring molecular system.

F

f [PHYS] Notation representing the Coriolis parameter.

F *See* farad; fluorine.

fA *See* femtoampere.

Faber flaw [SOLID STATE] A deformation in a superconducting material that acts as a nucleation center for the growth of a superconducting region.

Faber polynomials [MATH] Let D be a bounded, closed set of points in the complex plane whose complement is a simply connected region; the Faber polynomials of D are the set of polynomials $P_n(z)$, $n = 0, 1, 2, \ldots$, such that the equation

$$\frac{tf'(t)}{f(t) - z} = \sum_{n=0}^{\infty} P_n(z)\frac{1}{t^n}$$

is satisfied for all z and t, where $f(t)$ is the function that maps the region $|t| > 1$ one-to-one conformally onto the complement of D.

Fabry-Barot method [OPTICS] Method of determining the index of refraction of a prism in which the prism is set up so that the incident beam is perpendicular to the emergent face, and the index of refraction is calculated from the angle of the prism and the angle of deviation.

Fabry lens [OPTICS] A lens placed between the image of a star formed by a telescope and a light receptor to provide an image that is more uniform and more suited to photometric measurement.

Fabry-Perot etalon *See* etalon.

Fabry-Perot filter [OPTICS] An optical interference filter, similar to the Fabry-Perot interferometer except that the space between the partially reflecting surfaces is only a few thousand angstroms.

Fabry-Perot fringes [OPTICS] Series of rings observed when a monochromatic light source is viewed through a Fabry-Perot interferometer.

Fabry-Perot interferometer [OPTICS] An interferometer having two parallel glass plates (whose separation of a few centimeters may be varied), silvered on their inner surfaces so that the incoming wave is multiply reflected between them and ultimately transmitted.

face [CRYSTAL] *See* crystal face. [ELECTR] *See* faceplate. [MATH] A face of a simplex is the subset obtained by setting one or more of the coordinates a_i, defining the simplex, equal to 0; for example, the faces of a triangle are its sides and vertices.

face angle [MATH] An angle between two successive edges of a polyhedral angle.

face-centered cubic lattice [CRYSTAL] A lattice whose unit cells are cubes, with lattice points at the center of each face of the cube, as well as at the vertices. Abbreviated fcc lattice.

face-centered orthorhombic lattice [CRYSTAL] An ortho-

FABRY-PEROT FILTER

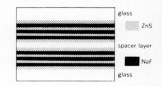

Schematic diagram of seven-layer solid Fabry-Perot filter. *(From D. E. Gray, ed., American Institute of Physics Handbook, McGraw-Hill, 1957)*

rhombic lattice which has lattice points at the center of each face of a unit cell, as well as at the vertices.

face of a geometric simplex [MATH] Any simplex whose vertices are also vertices of the original simplex.

faceplate [ELECTR] The transparent or semitransparent glass front of a cathode-ray tube, through which the image is viewed or projected; the inner surface of the face is coated with fluorescent chemicals that emit light when hit by an electron beam. Also known as face.

factor [STAT] **1.** A quantity or a variable being studied in an experiment as a possible cause of variation. **2.** In factor analysis, observables are considered functions of latent factors, usually linear functions.

factorable polynomial [MATH] A polynomial which has polynomial factors other than itself.

factor analysis [MATH] Given sets of variables which are related linearly, factor analysis studies techniques of approximating each set relative to the others; usually the variables denote numbers.

factor group *See* quotient group.

factorial design [STAT] A design for an experiment that allows the experimenter to find out the effect levels of each factor on levels of all the other factors.

factorial of a positive integer n [MATH] The product of all positive integers less than or equal to n; written $n!$; by convention $0! = 1$.

factorial ring *See* unique factorization domain.

factoring [MATH] Finding the factors of an integer or polynomial.

factoring of the secular equation [MATH] Factoring the polynomial that results from expanding the secular determinant of a matrix, in order to find the roots of this polynomial, which are the eigenvalues of the matrix.

factorization into irreducible elements [MATH] An element $a \neq 0$ in a commutative ring with identity R is said to have a factorization into irreducible elements if it can be written $a = up_1 p_2 \cdots p_n$, where u is a unit and $p_1, \ldots, p_n$ are irreducible elements in R.

factor model [STAT] Any one of the probability models which goes into the construction of a product model.

factor module [MATH] The factor module of a module M over a ring R by a submodule N is the quotient group M/N, where the product of a coset $x + N$ by an element a in R is defined to be the coset $ax + N$.

factor of an integer [MATH] Any integer which when multiplied by another integer gives the original integer.

factor of a polynomial [MATH] Any polynomial which when multiplied by another polynomial gives the original polynomial.

factor of proportionality [MATH] Two quantities A and B are related by a factor of proportionality μ if either $A = \mu B$ or $B = \mu A$.

factor of safety [MECH] **1.** The ratio between the breaking load on a member, appliance, or hoisting rope and the safe permissible load on it. Also known as safety factor. **2.** *See* factor of stress intensity.

factor of stress concentration [MECH] Any irregularity producing localized stress in a structural member subject to load. Also known as fatigue-strength reduction factor.

factor of stress intensity [MECH] The ratio of the maximum stress to which a structural member can be subjected, to the maximum stress to which it is likely to be subjected. Also known as factor of safety.

factor-reversal test [STAT] A test for index numbers in which an index number of quantity, obtained if symbols for price and quantity are interchanged in an index number of price, is multiplied by the original price index to give an index of changes in total value.

factor ring   *See* quotient ring.

factor theorem of algebra   [MATH] A polynomial $f(x)$ has $(x - a)$ as a factor if and only if $f(a) = 0$.

facula   [ASTRON] Any of the large patches of bright material forming a veined network in the vicinity of sunspots; faculae appear to be more permanent than sunspots and are probably due to elevated clouds of luminous gas.

Fahrenheit scale   [THERMO] A temperature scale; the temperature in degrees Fahrenheit (°F) is the sum of 32 plus 9/5 the temperature in degrees Celsius; water at a pressure of 1 atmosphere (101,325 newtons per square meter) freezes very near 32°F and boils very near 212°F.

Fahrenheit's hydrometer   [ENG] A type of hydrometer which carries a pan at its upper end in which weights are placed; the relative density of a liquid is measured by determining the weights necessary to sink the instrument to a fixed mark, first in water and then in the liquid being studied.

failure   [MECH] Condition caused by collapse, break, or bending, so that a structure or structural element can no longer fulfill its purpose.

fair game   [MATH] A game in which all of the participants have equal expectation of gain.

faithful module   [MATH] A module M over a commutative ring R such that if a is an element in R for which $am = 0$ for all m in M, then $a = 0$.

faithful representation   [MATH] A homomorphism h of a group onto some group of matrices or linear operators such that h is an injection.

Falker-Skan equation   [MATH] The ordinary differential equation $d^3y/dx^3 + y \, d^2y/dx^2 + a[1 - (dy/dx)^2] = 0$; used in the theory of viscous flow.

fall   [ASTRON] **1.** Of a spacecraft or spatial body, to drop toward a spatial body under the influence of its gravity. **2.** *See* autumn.

fallback   [NUCLEO] The system, electronic or manual, which is substituted for the computer system in case of breakdown. [NUCLEO] That part of the material carried into the air by an atomic explosion which ultimately drops back to the earth or water at the site of the explosion.

falling-ball viscometer   *See* falling-sphere viscometer.

falling body   [MECH] A body whose motion is accelerated toward the center of the earth by the force of gravity, other forces acting on it being negligible by comparison.

falling-drop method   [PHYS] Technique for measurement of liquid densities in which the time of fall of a drop of the sample liquid through a reference liquid is measured.

falling film   [FL MECH] A theoretical liquid film that moves downward in even flow on a vertical surface in laminar flow; the concept is used for heat- and mass-transfer calculations.

falling-sphere viscometer   [ENG] A viscometer which measures the speed of a spherical body falling with constant velocity in the fluid whose viscosity is to be determined. Also known as falling-ball viscometer.

fallout   [NUCLEO] The material that descends to the earth or water well beyond the site of a surface or subsurface nuclear explosion. Also known as atomic fallout; radioactive fallout.

fallout area   [NUCLEO] The area on which radioactive materials have settled out, or the area on which it is predicted from weather conditions that radioactive materials may settle out.

false acceptance   [STAT] Accepting on the basis of a statistical test a hypothesis which is wrong.

false body   [PHYS CHEM] The property of certain colloidal substances, such as paints and printing inks, of solidifying when left standing.

false pyroelectricity   *See* tertiary pyroelectricity.

false rejection   [STAT] Rejecting on the basis of a statistical test a hypothesis which is correct.

false white rainbow   *See* fogbow.

faltung   [MATH] A family of functions where the convolution of any two members of the family is also a member of the family. Also known as convolution family.

fan antenna   [ELECTROMAG] An array of folded dipoles of different length forming a wide-band ultra-high-frequency or very-high-frequency antenna.

fanned-beam antenna   [ELECTROMAG] Unidirectional antenna so designed that transverse cross sections of the major lobe are approximately elliptical.

Fanning friction factor   [FL MECH] A dimensionless number used in studying fluid friction in pipes, equal to the pipe diameter times the drop in pressure in the fluid due to friction as it passes through the pipe, divided by the product of the pipe length and the kinetic energy of the fluid per unit volume. Symbolized f.

Fanning's equation   [FL MECH] The equation expressing that frictional pressure drop of fluid flowing in a pipe is a function of the Reynolds number, rate of flow, acceleration due to gravity, and length and diameter of the pipe.

Fanno flow   [FL MECH] An ideal flow used to study the flow of fluids in long pipes; the flow obeys the same simplifying assumptions as Rayleigh flow except that the assumption there is no friction is replaced by the requirement the flow be adiabatic.

Fano plane   [MATH] A projective plane in which the points of intersection of the three possible pairs of opposite sides of a quadrilateral are collinear.

Fano's axiom   [MATH] The points of intersection of the three possible pairs of opposite sides of any quadrilateral in a given projective plane are noncollinear; thus a projective plane satisfying Fano's axiom is not a Fano plane, and a Fano plane does not satisfy Fano's axiom.

farad   [ELEC] The unit of capacitance in the meter-kilogram-second system, equal to the capacitance of a capacitor which has a potential difference of 1 volt between its plates when the charge on one of its plates is 1 coulomb, there being an equal and opposite charge on the other plate. Symbolized F.

faraday   [PHYS] The electric charge required to liberate 1 gram-equivalent of a substance by electrolysis; experimentally equal to $96,487.0 \pm 1.6$ coulombs. Also known as Faraday constant.

Faraday birefringence   [OPTICS] Difference in the indices of refraction of left and right circularly polarized light passing through matter parallel to an applied magnetic field; it is responsible for the Faraday effect.

Faraday cage   *See* Faraday shield.

Faraday constant   *See* faraday.

Faraday cylinder   [ELEC] **1.** A closed, or nearly closed, hollow conductor, usually grounded, within which apparatus is placed to shield it from electrical fields. **2.** A nearly closed, insulated, hollow conductor, usually shielded by a second grounded cylinder, used to collect and detect a beam of charged particles.

Faraday dark space   [ELECTR] The relatively nonluminous region that separates the negative glow from the positive column in a cold-cathode glow-discharge tube.

Faraday disk machine   [ELECTROMAG] A device for demonstrating electromagnetic induction, consisting of a copper disk in which a radial electromotive force is induced when the disk is rotated between the poles of a magnet. Also known as Faraday generator.

Faraday effect [OPTICS] Rotation of polarization of a beam of linearly polarized light when it passes through matter in the direction of an applied magnetic field; it is the result of Faraday birefringence. Also known as Faraday rotation; Kundt effect; magnetic rotation.

Faraday generator *See* Faraday disk machine.

Faraday rotation *See* Faraday effect.

Faraday rotation experiment [ELECTROMAG] An experiment in which a wire dipping in a pool of mercury surrounding a magnet rotates around the magnet when a current passes through it, demonstrating the effect of a magnetic field on a current-carrying conductor.

Faraday rotation isolator *See* ferrite isolator.

Faraday screen *See* Faraday shield.

Faraday shield [ELEC] Electrostatic shield composed of wire mesh or a series of parallel wires, usually connected at one end to another conductor which is grounded. Also known as Faraday cage; Faraday screen.

Faraday's law of electromagnetic induction [ELECTROMAG] The law that the electromotive force induced in a circuit by a changing magnetic field is equal to the negative of the rate of change of the magnetic flux linking the circuit. Also known as law of electromagnetic induction.

Faraday's laws of electrolysis [PHYS CHEM] **1.** The amount of any substance dissolved or deposited in electrolysis is proportional to the total electric charge passed. **2.** The amounts of different substances dissolved or desposited by the passage of the same electric charge are proportional to their equivalent weights.

Faraday tube [ELEC] A tube of force for electric displacement which is of such size that the integral over any surface across the tube of the component of electric displacement perpendicular to that surface is unity.

faradic current [ELEC] An intermittent and nonsymmetrical alternating current like that obtained from the secondary winding of an induction coil; used in electrobiology.

far field *See* Fraunhofer region.

far-infrared maser [ENG] A gas maser that generates a beam having a wavelength well above 100 micrometers, and ranging up to the present lower wavelength limit of about 500 micrometers for microwave oscillators.

far-infrared radiation [ELECTROMAG] Infrared radiation the wavelengths of which are the longest of those in the infrared region, about 50–1000 micrometers; requires diffraction gratings for spectroscopic analysis.

Farmer dosimeter [NUCLEO] A small ionization chamber with an air wall, used for routine measurements of radiation.

far point [OPTICS] The farthest point from an eye at which an object is distinctly seen; for a normal eye it is theoretically at infinity. Also known as punctum remotum.

far region *See* Fraunhofer region.

far-ultraviolet radiation [ELECTROMAG] Ultraviolet radiation in the wavelength range of 200–300 nanometers; germicidal effects are greatest in this range.

far zone *See* Fraunhofer region.

fast axis [OPTICS] The direction of the electrical displacement vector of light propagating in an anisotropic crystal with the greatest possible phase velocity corresponding to a specified direction of propagation.

fast breeder reactor [NUCLEO] A type of fast reactor using highly enriched fuel in the core, fertile material in the blanket, and a liquid-metal coolant, such as sodium; high-speed neutrons fission the fuel in the compact core, and the excess neutrons convert fertile material to fissionable isotopes; the breeding ratio is 1.0 or larger. Abbreviated FBR.

fast-burst reactor [NUCLEO] A nuclear reactor that supplies microsecond pulses of fast neutrons for use in biomedical research.

fast chemical reaction [PHYS CHEM] A reaction with a half-life of milliseconds or less; such reactions occur so rapidly that special experimental techniques are required to observe their rate.

fast effect [NUCLEO] The reactivity change (increase in neutrons) due to fissions caused by fast neutrons in a thermal reactor.

fast fission [NUC PHYS] Fission caused by fast neutrons.

fast-fission factor [NUCLEO] The ratio of the total number of fast neutrons produced by fission in a nuclear reactor to the number produced by fission resulting from thermal neutrons only.

fast Fourier transform [MATH] A Fourier transform employing the Cooley-Tukey algorithm to reduce the number of operations.

fast neutron [NUCLEO] A neutron having energy much greater than some arbitrary lower limit (that may be only a few thousand electron volts).

fast-neutron spectrometry [NUC PHYS] Neutron spectrometry in which nuclear reactions are produced by or yield fast neutrons; such reactions are more varied than in the slow-neutron case.

fast reactor [NUCLEO] A nuclear reactor in which most of the fissions are produced by fast neutrons, with little or no moderator to slow down the neutrons.

fast time constant [ELEC] An electric circuit which combines resistance and capacitance to give a short time constant for capacitor discharge through the resistor. [ELECTR] Circuit with short time constant used to emphasize signals of short duration to produce discrimination against low-frequency components of clutter in radar.

fast-vibration direction [OPTICS] The direction of the electric field vector of the ray of light that travels with the greatest velocity in an anisotropic crystal and therefore corresponds to the minimum refractive index.

Fata Morgana [OPTICS] A complex mirage characterized by multiple distortions of images, generally in the vertical, so that such objects as cliffs or cottages are distorted and magnified into fantastic castles.

fatigue [ELECTR] The decrease of efficiency of a luminescent or light-sensitive material as a result of excitation. [MECH] Failure of a material by cracking resulting from repeated or cyclic stress.

fatigue life [MECH] The number of applied repeated stress cycles a material can endure before failure.

fatigue limit [MECH] The maximum stress that a material can endure for an infinite number of stress cycles without breaking. Also known as endurance limit.

fatigue ratio [MECH] The ratio of the fatigue limit or fatigue strength to the static tensile strength. Also known as endurance ratio.

fatigue strength [MECH] The maximum stress a material can endure for a given number of stress cycles without breaking. Also known as endurance strength.

fatigue-strength reduction factor *See* factor of stress concentration.

Fatou-Lebesgue lemma [MATH] Given a sequence f_n of positive measurable functions on a measure space (X, μ), then

$$\int_X (\lim_n \inf f_n)\, d\mu \le \lim_n \inf \int_X f_n\, d\mu.$$

fault [ELEC] A defect, such as an open circuit, short circuit, or ground, in a circuit, component, or line. Also known as electrical fault; faulting. [ELECTR] Any physical condition that causes a component of a data-processing system to fail in performance.

Faure storage battery [ELEC] A storage battery in which the plates consist of lead-antimony supporting grids covered with a lead oxide paste, immersed in weak sulfuric acid. Also known as pasted-plate storage battery.

F band [SOLID STATE] The optical absorption band arising from *F* centers.

FBR *See* fast breeder reactor.

fcc lattice *See* face-centered cubic lattice.

F center [SOLID STATE] A color center consisting of an electron trapped by a negative ion vacancy in an ionic crystal, such as an alkali halide or an alkaline-earth fluoride or oxide.

F′ center [SOLID STATE] A color center that gives rise to a broad absorption band at longer wavelengths than the band of the F center; probably an F center that has trapped an additional electron.

F corona [ASTRON] The outer layer of the sun's corona. Also known as Fraunhofer corona.

F curve *See* atomic F curve.

F distribution [STAT] The ratio of two independent chi-square variables each divided by its degree of freedom; used to test hypotheses in the analysis of variance and hypotheses about whether or not two normal populations have the same variance.

Fe *See* iron.

feasible solution [ADP] In linear programming, any set of values for the variables $x_j, j = 1, 2, \ldots, n$, that (1) satisfy the set of restrictions

$$\sum_{j=1}^{n} a_{ij}x_j \le b_i, \; i = 1, 2, \ldots, m$$

$$\left(\text{alternatively,} \; \sum_{j=1}^{n} a_{ij}x_j \ge b_i, \; \text{or} \; \sum_{j=1}^{n} a_{ij}x_j = b_i \right)$$

where the b_i are numerical constants known collectively as the right-hand side and the a_{ij} are coefficients of the variables x_j, and (2) satisfy the restrictions $x_j \ge 0$.

Feather analysis [NUCLEO] A technique for determining the range in aluminum of the beta rays of a species by comparing the absorption curve of that species with the absorption curve of a reference species.

Fechner color [OPTICS] A sensation of color caused by achromatic stimuli at intervals in time.

Fechner fraction [PHYSIO] The smallest difference in the brightness of two sources that can be detected by the human eye divided by the brightness of one of them.

Fedorov stage *See* universal stage.

feedback [ELECTR] The return of a portion of the output of a circuit or device to its input. [SCI TECH] The control of input as a function of output by returning a portion of the output to the input.

feedback admittance [ELECTR] Short-circuit transadmittance from the output electrode to the input electrode of an electron tube.

feedback branch [CONT SYS] A branch in a signal-flow graph that belongs to a feedback loop.

feedback circuit [ELECTR] A circuit that returns a portion of the output signal of an electronic circuit or control system to the input of the circuit or system.

feedback compensation [CONT SYS] Improvement of the re-

F BAND

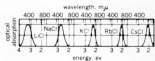

F bands in different alkali halide crystals.

FEEDBACK CIRCUIT

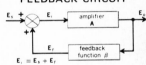

Block diagram of a feedback circuit: E_s is sinusoidal input signal; E_i is actuating signal; E_o is output signal; E_f is feedback signal; A is amplifier gain; and β is feedback function.

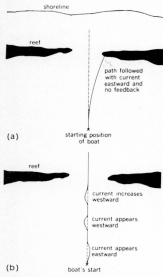

**FEEDBACK
CONTROL SYSTEM**

shoreline

reef

path followed
with current
eastward and
no feedback

(a)   starting position
of boat

reef

current increases
westward

current appears
westward

current appears
eastward

(b)   boat's start

Use of a feedback control
system in a navigation problem.
(a) Course with no feedback.
(b) Course with feedback.

sponse of a feedback control system by placing a compensator in the feedback path, in contrast to cascade compensation. Also known as parallel compensation.

feedback control loop   *See* feedback loop.

feedback control signal   [CONT SYS] The portion of an output signal which is retransmitted as an input signal.

feedback control system   [CONT SYS] A system in which the value of some output quantity is controlled by feeding back the value of the controlled quantity and using it to manipulate an input quantity so as to bring the value of the controlled quantity closer to a desired value. Also known as closed-loop control system.

feedback loop   [CONT SYS] A closed transmission path or loop that includes an active transducer and consists of a forward path, a feedback path, and one or more mixing points arranged to maintain a prescribed relationship between the loop input signal and the loop output signal. Also known as feedback control loop.

feedback regulator   [CONT SYS] A feedback control system that tends to maintain a prescribed relationship between certain system signals and other predetermined quantities.

feedback transfer function   [CONT SYS] In a feedback control loop, the transfer function of the feedback path.

feedforward control   [CONT SYS] Process control in which changes are detected at the process input and an anticipating correction signal is applied before process output is affected.

feed materials   [NUCLEO] Refined uranium or thorium metal or their pure compounds in a form suitable for use in nuclear reactor fuel elements or as feed for uranium-enrichment processes.

Fejér kernel   [MATH] **1.** The function

$$k_n(t) = n^{-1}(\sin nt/\sin t)^2$$

which appears in Fejér's integral. **2.** The function

$$K(t) = \pi^{-1}(\sin t/t)^2.$$

Fejér's integral   [MATH] One of the partial sums which appears in the Cesaro summation of the Fourier series; it may be written

$$\sigma_n(x) = \frac{1}{\pi} \int_0^{\pi/2} [f(x + 2t) + f(x - 2t)]k_n(t)\, dt$$

where $k_n(t)$ is the Fejér kernel.

Fejér's theorem   [MATH] The Fourier series of a function $f(x)$, defined and integrable on the interval $-\pi \leq x \leq \pi$, is Cesaro-summable $(C,1)$ to the limit $\frac{1}{2}[f(x + 0) + f(x - 0)]$ at all points x at which limits from the right and left, $f(x + 0)$ and $f(x - 0)$, exist; furthermore, if $f(x)$ is continuous in a subinterval (a,b), then the first Cesaro sums converge uniformly to $f(x)$ in any interval (c,d) where $a < c < d < b$.

female connector   [ELEC] A connector having one or more contacts set into recessed openings; jacks, sockets, and wall outlets are examples.

femitrons   [ELECTR] Class of field-emission microwave devices.

femto-   [SCI TECH] A prefix representing 10^{-15}, which is 0.000 000 000 000 001, or one-thousandth of a millionth of a millionth.

femtoampere   [ELEC] A unit of current equal to 10^{-15} ampere. Abbreviated fA.

femtometer   [MECH] A unit of length, equal to 10^{-15} meter; used particularly in measuring nuclear distances. Abbreviated fm. Also known as fermi.

femtovolt [ELEC] A unit of voltage equal to 10^{-15} volt. Abbreviated fV.

Fermat's last theorem [MATH] The conjecture that there are no positive integer solutions of the equation $x^n + y^n = z^n$ for $n \geq 3$.

Fermat's principle [OPTICS] The principle that an electromagnetic wave will take a path that involves the least travel time when propagating between two points. Also known as least-time principle; stationary time principle.

Fermat's theorem [MATH] If p is a prime number and a is a positive integer which is not divisible by p, then $a^{p-1} - 1$ is divisible by p.

fermi *See* femtometer.

Fermi age [NUCLEO] The value calculated for the slowing-down area in the Fermi age model; it has the dimensions of area, not time. Also known as age; neutron age; symbolic age of neutrons.

Fermi age equation [NUCLEO] An equation in the Fermi age model which states that the Laplacian of the slowing-down density equals the partial derivative of the slowing-down density with respect to the Fermi age.

Fermi age model [NUCLEO] A model used in studying the slowing down of neutrons by elastic collisions; it is assumed that the slowing down takes place by a very large number of very small energy changes.

Fermi beta-decay theory [NUC PHYS] Theory in which a nucleon source current interacts with an electron-neutrino field to produce beta decay, in a manner analogous to the interaction of an electric current with an electromagnetic field during the emission of a photon of electromagnetic radiation.

Fermi constant [NUC PHYS] A universal constant, introduced in beta-disintegration theory, that expresses the strength of the interaction between the transforming nucleon and the electron-neutrino field.

Fermi derivative [RELAT] A generalization of covariant differentiation along a curve that reduces to covariant differentiation when the curve is geodesic; an orthonormal tetrad constructed at each point along a timelike curve such that the Fermi derivative of the tetrad along the curve is zero has (1) its timelike basis vector equal to the curve's unit tangent vector and (2) its spatial basis vectors nonrotating along the curve.

Fermi-Dirac distribution function [STAT MECH] A function specifying the probability that a member of an assembly of independent fermions, such as electrons in a semiconductor or metal, will occupy a certain energy state when thermal equilibrium exists.

Fermi-Dirac gas *See* Fermi gas.

Fermi-Dirac statistics [STAT MECH] The statistics of an assembly of identical half-integer spin particles; such particles have wave functions antisymmetrical with respect to particle interchange and satisfy the Pauli exclusion principle.

Fermi distribution [SOLID STATE] Distribution of energies of electrons in a semiconductor or metal as given by the Fermi-Dirac distribution function; nearly all energy levels below the Fermi level are filled, and nearly all above this level are empty.

Fermi energy [STAT MECH] **1.** The average energy of electrons in a metal, equal to $\frac{3}{5}$ of the Fermi level. **2.** *See* Fermi level.

Fermi gas [STAT MECH] An assembly of independent particles that obey Fermi-Dirac statistics, and therefore obey the Pauli exclusion principle; this concept is used in the free-electron theory of metals and in one model of the behavior of the nucleons in a nucleus. Also known as Fermi-Dirac gas.

Fermi hole [SOLID STATE] A region surrounding an electron in a solid in which the energy band theory predicts that the probability of finding other electrons is less than the average over the volume of the solid.

Fermi interaction [PARTIC PHYS] The direct interaction between four Dirac fields at a single point in space-time, postulated in conventional theories of the weak interactions.

Fermi level [STAT MECH] The energy level at which the Fermi-Dirac distribution function of an assembly of fermions is equal to one-half. Also known as Fermi energy.

Fermi liquid [CRYO] A liquid of particles which have Fermi-Dirac statistics; an example is the liquid phase of helium-3, in which the atoms belong to the isotope with mass number 3.

fermion [QUANT MECH] A particle, such as the electron, proton, or neutron, which obeys the rule that the wave function of several identical particles changes sign when the coordinates of any pair are interchanged; it therefore obeys the Pauli exclusion principle.

fermion anticommutation relations [QUANT MECH] The symmetric operator relations, $a_r a_s + a_s a_r = 0$ and $a_r a_s\dagger + a_s\dagger a_r = \delta_{rs}$, between variables and their Hermitian conjugates that is central to the canonical formalism for fermions in quantum theory.

fermion field [QUANT MECH] An operator defined at each point in space-time that creates or annihilates a particular type of fermion and its antiparticle.

Fermi plot *See* Kurie plot.

Fermi-propagated [RELAT] A vector field is said to be Fermi-propagated along a curve γ when it is constructed so that its Fermi derivative along γ is 0.

Fermi resonance [PHYS CHEM] In a polyatomic molecule, the relationship of two vibrational levels that have in zero approximation nearly the same energy; they repel each other, and the eigenfunctions of the two states mix.

Fermi selection rules [NUC PHYS] Selection rules for beta decay in a Fermi transition; that is, there is no change in total angular momentum or parity of the nucleus in an allowed transition.

Fermi's golden rules [QUANT MECH] The equations giving the first-order (rule number 2) and second-order (rule number 1) contributions to the transition probability per unit time induced by a perturbation Hamiltonian, in terms of matrix elements of the perturbation Hamiltonian.

Fermi sphere [STAT MECH] The Fermi surface of an assembly of fermions in the approximation that the fermions are free particles.

Fermi surface [SOLID STATE] A constant-energy surface in the space containing the wave vectors of states of members of an assembly of independent fermions, such as electrons in a semiconductor or metal, whose energy is that of the Fermi level.

Fermi temperature [STAT MECH] The energy of the Fermi level of an assembly of fermions divided by Boltzmann's constant, which appears as a parameter in the Fermi-Dirac distribution function.

Fermi transition [NUC PHYS] Beta decay subject to Fermi selection rules.

fermium [CHEM] A synthetic radioactive element, symbol Fm, with atomic number 100; discovered in debris of the 1952 hydrogen bomb explosion, and now made in nuclear reactors.

Ferrers diagram [MATH] An array of dots associated with an integer partition $n = a_1 + \cdots + a_k$, whose ith row contains a_i dots.

ferrimagnetism [SOLID STATE] A type of magnetism in which the magnetic moments of neighboring ions tend to align

FERMIUM

Periodic table of the chemical elements showing the position of fermium.

FERRIMAGNETISM

Schematic representation of arrangement of magnetic moments of neighboring ions in collinear ferrimagnetism.

nonparallel, usually antiparallel, to each other, but the moments are of different magnitudes, so there is an appreciable resultant magnetization.

ferrite [SOLID STATE] Any ferrimagnetic material having high electrical resistivity which has a spinel crystal structure and the chemical formula XFe_2O_4, where X represents any divalent metal ion whose size is such that it will fit into the crystal structure.

ferrite isolator [ELECTROMAG] A device consisting of a ferrite rod, centered on the axis of a short length of circular waveguide, located between rectangular-waveguide sections displaced 45° with respect to each other, which passes energy traveling through the waveguide in one direction while absorbing energy from the opposite direction. Also known as Faraday rotation isolator.

ferrite rotator [ELECTROMAG] A gyrator consisting of a ferrite cylinder surrounded by a ring-type permanent magnet, inserted in a waveguide to rotate the plane of polarization of the electromagnetic wave passing through the waveguide.

ferrite switch [ELECTROMAG] A ferrite device that blocks the flow of energy through a waveguide by rotating the electric field vector 90°; the switch is energized by sending direct current through its magnetizing coil; the rotated electromagnetic wave is then reflected from a reactive mismatch or absorbed in a resistive card.

ferroelectric [SOLID STATE] A crystalline substance displaying ferroelectricity, such as barium titanate, potassium dihydrogen phosphate, and Rochelle salt; used in ceramic capacitors, acoustic transducers, and dielectric amplifiers. Also known as seignette-electric.

ferroelectric converter [ELEC] A converter that transforms thermal energy into electric energy by utilizing the change in the dielectric constant of a ferroelectric material when heated beyond its Curie temperature.

ferroelectric crystal [SOLID STATE] A crystal of a ferroelectric material.

ferroelectric domain [SOLID STATE] A region of a ferroelectric material within which the spontaneous polarization is constant.

ferroelectric hysteresis [ELEC] The dependence of the polarization of ferroelectric materials not only on the applied electric field but also on their previous history; analogous to magnetic hysteresis in ferromagnetic materials. Also known as dielectric hysteresis; electric hysteresis.

ferroelectric hysteresis loop [ELEC] Graph of polarization or electric displacement versus applied electric field of a material displaying ferroelectric hysteresis.

ferroelectricity [SOLID STATE] Spontaneous electric polarization in a crystal; analogous to ferromagnetism.

ferroelectric shutter [OPTICS] A shutter consisting of a slab of ferroelectric crystal located between polarizers whose planes are at right angles; opens to pass light when activated by a pulse of up to 100 volts.

ferrohydrodynamics [PHYS] The study of the motion of strongly magnetizable fluids subjected to magnetic fields.

ferromagnetic ceramic See ceramic magnet.

ferromagnetic crystal [SOLID STATE] A crystal of a ferromagnetic material. Also known as polar crystal.

ferromagnetic domain [SOLID STATE] A region of a ferromagnetic material within which atomic or molecular magnetic moments are aligned parallel. Also known as magnetic domain.

ferromagnetic film See magnetic thin film.

ferromagnetism [SOLID STATE] A property, exhibited by certain metals, alloys, and compounds of the transition (iron

FERROELECTRIC HYSTERESIS LOOP

Hysteresis loop showing the relation between the resulting polarization P of the ferroelectric crystal and the externally applied electric field E; P_s is permanent or spontaneous magnetization.

FERROMAGNETIC DOMAIN

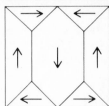

Drawing of a uniaxial crystal showing the ferromagnetic domains. The arrows show the direction of the magnetic field in each domain.

group), rare-earth, and actinide elements, in which the internal magnetic moments spontaneously organize in a common direction; gives rise to a permeability considerably greater than that of vacuum, and to magnetic hysteresis.

ferrum [CHEM] Latin term for iron; derivation of the symbol Fe.

fertile material [NUCLEO] A material, such as thorium-232 or uranium-238, which is capable of being transformed into a fissionable material by capture of a neutron.

Féry spectrograph [SPECT] A spectrograph whose only optical element consists of a back-reflecting prism with cylindrically curved faces.

FET *See* field-effect transistor.

Feynman diagram [QUANT MECH] A diagram which gives an intuitive picture of a term in a perturbation expansion of a scattering matrix element or other physical quantity associated with interactions of particles; each line represents a particle, each vertex an interaction.

Feynman integral [QUANT MECH] A term in a perturbation expansion of a scattering matrix element; it is an integral over the Minkowski space of various particles (or over the corresponding momentum space) of the product of propagators of these particles and quantities representing interactions between the particles.

Feynman propagator [QUANT MECH] A factor $(p+m)/(p^2-m^2+i\varepsilon)$ in a transition amplitude corresponding to a line that connects two vertices in a Feynman diagram, and that represents a virtual particle.

Feynman's rules [QUANT MECH] Rules for carrying out perturbation expansions in quantum field theory codified by Feynman diagrams.

Feynman's superfluidity theory [CRYO] Microscopic theory of superfluid helium which accounts for the spectrum of elementary excitations assumed by Landau's superfluidity theory.

FG achromatism *See* actinic achromatism.

fiber bundle [MATH] A bundle whose total space is a G-space X and whose base is the homomorphic image of the orbit space of X and whose fibers are isomorphic to the orbits of points in the base space under the action of G. [OPTICS] A flexible bundle of glass or other transparent fibers, parallel to each other, used in fiber optics to transmit a complete image from one end of the bundle to the other.

fiber diagram [SOLID STATE] The x-ray diffraction pattern of a collection of crystallites that have one crystallographic axis approximately parallel to a common direction but are otherwise randomly oriented

fiber of a bundle [MATH] The set of points in the total space of a bundle which are sent into the same element of the base of the bundle by the projection map.

fiber optics [OPTICS] The technique of transmitting light through long, thin, flexible fibers of glass, plastic, or other transparent materials; bundles of parallel fibers can be used to transmit complete images.

fiberscope [OPTICS] An arrangement of parallel glass fibers with an objective lens on one end and an eyepiece at the other; the assembly can be bent as required to view objects that are inaccessible for direct viewing.

fiber stress [MECH] **1.** The tensile or compressive stress on the fibers of a fiber metal or other fibrous material, especially when fiber orientation is parallel with the neutral axis. **2.** Local stress through a small area (a point or line) on a section where the stress is not uniform, as in a beam under bending load.

Fibonacci sequence [MATH] The sequence 1, 1, 2, 3, 5, 8, 13,

FEYNMAN DIAGRAM

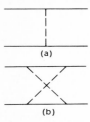

Typical Feynman diagrams for two-particle scattering. *(a)* Second-order diagram, *(b)* fourth-order diagram. Solid lines represent scattered particles; broken lines, particles which transmit force between them.

21, . . . , or any sequence where each entry is the sum of the two previous entries.

fibrous fracture [MECH] Failure of a material resulting from a ductile crack; broken surfaces are dull and silky. Also known as ductile fracture.

Fick's law [PHYS] The law that the rate of diffusion of matter across a plane is proportional to the negative of the rate of change of the concentration of the diffusing substance in the direction perpendicular to the plane.

fictitious year [ASTRON] The period between successive returns of the fictitious mean sun to a sidereal hour angle of 80° (right ascension 18 hours 40 minutes; about January 1); the length of the fictitious year is the same as that of the tropical year, since both are based upon the position of the sun with respect to the vernal equinox. Also known as Besselian year.

fiducial inference [STAT] A type of inference whose purpose is to make probabilistic statements about values of unknown parameters; based on the distribution of population values about which the inference is to be made.

fiducial limits [STAT] The boundaries within which a parameter is considered to be located; a concept in fiducial inference.

fiducial point [OPTICS] A mark, or one of several marks, visible in the field of view of an optical instrument, used as a reference or for measurement. Also known as fiduciary point.

fiduciary point *See* fiducial point.

field [ELEC] That part of an electric motor or generator which produces the magnetic flux which reacts with the armature, producing the desired machine action. [ELECTR] One of the equal parts into which a frame is divided in interlaced scanning for television; includes one complete scanning operation from top to bottom of the picture and back again. [MATH] An algebraic system possessing two operations which have all the properties that addition and multiplication of real numbers have. [OPTICS] *See* field of view. [PHYS] **1.** An entity which acts as an intermediary in interactions between particles, which is distributed over part or all of space, and whose properties are functions of space coordinates and, except for static fields, of time; examples include gravitational field, sound field, and the strain tensor of an elastic medium. **2.** The quantum-mechanical analog of this entity, in which the function of space and time is replaced by an operator at each point in space-time.

field brightness *See* adaptation luminance.

field coil [ELECTROMAG] A coil used to produce a constant-strength magnetic field in an electric motor, generator, or excited-field loudspeaker; depending on the type of motor or generator, the field core may be on the stator or the rotor. Also known as field winding.

field desorption [SOLID STATE] A technique which tears atoms from a surface by an electric field applied at a sharp dip to produce very well-ordered, clean, plane surfaces of many crystallographic orientations.

field-desorption microscope [ELECTR] A type of field-ion microscope in which the tip specimen is imaged by ions that are field-desorbed or field-evaporated directly from the surface rather than by ions obtained from an externally supplied gas.

field distortion [ELECTROMAG] Any alteration in the direction of an electric or magnetic field; in particular, distortion of the magnetic fields between the north and south poles of a generator due to the counter electromotive force in the armature winding.

field-effect capacitor [ELECTR] A capacitor in which the

FIELD COIL

A rotor of a synchronous motor showing the field coils. *(Allis-Chalmers)*

FIELD-EFFECT TRANSISTOR

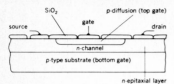

One type of field-effect transistor: the junction-gate field-effect transistor.

FIELD EMISSION

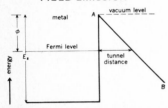

Diagram of energy level scheme for field emission from a metal at absolute zero. Dashed vacuum level represents energy of an electron at rest in free space. Energy of conduction electron at rest in metal lies below vacuum level by amount E_s. Energy levels below Fermi level are occupied by conduction electrons whereas those above are empty. AB is the potential of an electron outside the metal in presence of a strong constant field produced by a positive electrode. ϕ is the work function.

FIELD-EMISSION MICROSCOPE

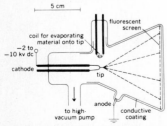

Electron-operated field-emission microscope.

effective dielectric is a region of semiconductor material that has been depleted or inverted by the field effect.

field-effect device [ELECTR] A semiconductor device whose properties are determined largely by the effect of an electric field on a region within the semiconductor.

field-effect diode [ELECTR] A semiconductor diode in which the charge carriers are of only one polarity.

field-effect tetrode [ELECTR] Four-terminal device consisting of two independently terminated semiconducting channels so displaced that the conductance of each is modulated along its length by the voltage conditions in the other.

field-effect transistor [ELECTR] A transistor in which the resistance of the current path from source to drain is modulated by applying a transverse electric field between grid or gate electrodes; the electric field varies the thickness of the depletion layer between the gates, thereby reducing the conductance. Abbreviated FET.

field-effect-transistor resistor [ELECTR] A field-effect transistor in which the gate is generally tied to the drain; the resultant structure is used as a resistance load for another transistor.

field-effect varistor [ELECTR] A passive, two-terminal, nonlinear semiconductor device that maintains constant current over a wide voltage range.

field emission [ELECTR] The emission of electrons from the surface of a metallic conductor into a vacuum (or into an insulator) under influence of a strong electric field; electrons penetrate through the surface potential barrier by virtue of the quantum-mechanical tunnel effect. Also known as cold emission.

field-emission microscope [ELECTR] A device that uses field emission of electrons or of positive ions (field-ion microscope) to produce a magnified image of the emitter surface on a fluorescent screen.

field-emission tube [ELECTR] A vacuum tube within which field emission is obtained from a sharp metal point; must be more highly evacuated than an ordinary vacuum tube to prevent contamination of the point.

field-enhanced emission [ELECTR] An increase in electron emission resulting from an electric field near the surface of the emitter.

field flattener [OPTICS] A thin planoconvex lens placed in front of the photographic plate in some telescopes that have a curved focal plane so as to focus light on the flat plate.

field-free emission current [ELECTR] Electron emitted by a cathode when the electric field at the surface of the cathode is zero. Also known as zero-field emission.

field gradient [PHYS] 1. A vector obtained by applying the del operator to a scalar field. 2. A tensor obtained by dyadic multiplication of the del operator with a vector field.

field index [NUCLEO] The constant n for a betatron in which the magnetic field strength at radius r is equal to $B_0 (r/R)^{-n}$, where R is the radius of equilibrium orbit of an electron, and B_0 is the corresponding magnetic field. Also known as n value.

field intensity See field strength.

field ionization [ELECTR] The ionization of gaseous atoms and molecules by an intense electric field, often at the surface of a solid.

field-ion microscope [ELECTR] A microscope in which atoms are ionized by an electric field near a sharp tip; the field then forces the ions to a fluorescent screen, which shows an enlarged image of the tip, and individual atoms are made visible; this is the most powerful microscope yet produced. Also known as ion microscope.

field lens [OPTICS] The lens in a two-lens eyepiece which is farther from the eye.

field line reconnection [ASTRON] A topological rearrangement of the magnetic field lines surrounding an astronomical body, for example, the transfer of lines between open and closed configurations in the terrestrial magnetotail; a possible source of the energy released explosively in solar flares and magnetospheric substorms. Also known as field line annihilation; magnetic merging.

field luminance *See* adaptation luminance.

field magnet [ELECTROMAG] The magnet which creates a magnetic field in an electric machine or device.

field of planes on a manifold [MATH] A continuous assignment of a vector subspace of tangent vectors to each point in the manifold. Also known as plane field.

field of vectors on a manifold [MATH] A continuous assignment of a tangent vector to each point in the manifold. Also known as vector field.

field of view [OPTICS] The area or solid angle which can be viewed through an optical instrument. Also known as field.

field operator [QUANT MECH] An operator function of space and time for the annihilation or creation of a particle.

field pattern *See* radiation pattern.

field pole [ELECTROMAG] A structure of magnetic material on which a field coil of a loudspeaker, motor, generator, or other electromagnetic device may be mounted.

field quenching [SOLID STATE] Decrease in the emission of light of a phosphor excited by ultraviolet radiation, x-rays, alpha particles, or cathode rays when an electric field is simultaneously applied.

field stop [OPTICS] An opening, usually circular, in an opaque screen, whose edges determine the limits of the field of view of an optical instrument.

field strength [PHYS] A vector characterizing a field. Also known as field intensity.

field theory [MATH] The study of fields and their extensions. [PHYS] A theory in which the basic quantities are fields; classically the equations governing the fields may be given; in quantum field theory the commutation rules satisfied by the field operators also are specified.

field waveguide [ELECTROMAG] A single wire, threaded or coated with dielectric, which guides an electromagnetic field. Also known as G string.

field winding *See* field coil.

Fierz interference [NUC PHYS] Interference between the axial vector and tensor parts of the weak interaction of nucleon and lepton (electron-neutrino) fields in beta decay; measurements of the beta-particle energy spectrum indicate that it vanishes.

fifteen-degrees calorie *See* calorie.

figure of merit [ELECTR] A performance rating that governs the choice of a device for a particular application; for example, the figure of merit of a magnetic amplifier is the ratio of usable power gain to the control time constant.

figuring [OPTICS] Grinding or polishing of surfaces of optical components to remove aberrations.

filament [ELEC] Metallic wire or ribbon which is heated in an incandescent lamp to produce light, by passing an electric current through the filament. [ELECTR] A cathode made of resistance wire or ribbon, through which an electric current is sent to produce the high temperature required for emission of electrons in a thermionic tube. Also known as directly heated cathode; filamentary cathode; filament-type cathode. [SCI TECH] A long, flexible object with a small cross section.

filamentary cathode *See* filament.

FILAMENT

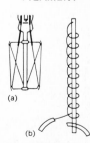

(a)

(b)

Two types of filaments: (a) in incandescent lamp; (b) in thermionic tube. (*General Electric Co.*)

filament current [ELECTR] The current supplied to the filament of an electron tube for heating purposes.

filament emission [ELECTR] Liberation of electrons from a heated filament wire in an electron tube.

filament lamp *See* incandescent lamp.

filament saturation *See* temperature saturation.

filament-type cathode *See* filament.

filled band [SOLID STATE] An energy band, each of whose energy levels is occupied by an electron.

filled-system thermometer [ENG] A thermometer which has a bourdon tube connected by a capillary tube to a hollow bulb; the deformation of the bourdon tube depends on the pressure of a gas (usually nitrogen or helium) or on the volume of a liquid filling the system. Also known as filled thermometer.

filled thermometer *See* filled-system thermometer.

fillet lightning *See* ribbon lightning.

film badge [NUCLEO] A device worn for the purpose of indicating the absorbed dose of radiation received by the wearer; usually made of metal, plastic, or paper and loaded with one or more pieces of x-ray film. Also known as badge meter.

film balance [ENG] An instrument that employs a torsion balance to measure the two-dimensional pressure of a film spread over the surface of a liquid.

film boiling [THERMO] Boiling in which a continuous film of vapor forms at the hot surface of the container holding the boiling liquid, reducing heat transfer across the surface.

film coefficient [THERMO] For a fluid confined in a vessel, the rate of flow of heat out of the fluid, per unit area of vessel wall divided by the difference between the temperature in the interior of the fluid and the temperature at the surface of the wall. Also known as convection coefficient.

film cooling [THERMO] The cooling of a body or surface, such as the inner surface of a rocket combustion chamber, by maintaining a thin fluid layer over the affected area.

film dosimetry [NUCLEO] The determination of radiation dose by measurement of the darkening of a photographic film which is exposed to radiation and developed under controlled conditions.

film pressure [PHYS] The difference between the surface tension of a pure liquid and the surface tension of the liquid with a unimolecular layer of a given substance adsorbed on it. Also known as surface pressure.

film resistor [ELEC] A fixed resistor in which the resistance element is a thin layer of conductive material on an insulated form; the conductive material does not contain binders or insulating material.

film theory [PHYS] A theory of the transfer of material or heat across a phase boundary, where one or both of the phases are flowing fluids, the main controlling factor being resistance to heat conduction or mass diffusion through a relatively stagnant film of the fluid next to the surface. Also known as boundary-layer theory.

filmwise condensation [THERMO] Condensation of a vapor on a surface when the surface is insulated by a film of condensate.

filter [CONT SYS] *See* compensator. [ELECTR] Any transmission network used in electrical systems for the selective enhancement of a given class of input signals. Also known as electric filter; electric-wave filter. [ENG ACOUS] A device employed to reject sound in a particular range of frequencies while passing sound in another range of frequencies. Also known as acoustic filter. [MATH] A family of subsets of a set S: it does not include the empty set, the intersection of any

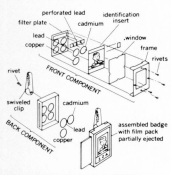
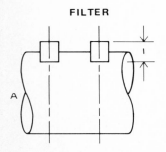

two members of the family is also a member, and any subset of S containing a member is also a member. [OPTICS] An optical element that partially absorbs incident electromagnetic radiation in the visible, ultraviolet, or infrared spectra, consisting of a pane of glass or other partially transparent material, or of films separated by narrow layers; the absorption may be either selective or nonselective with respect to wavelength. Also known as optical filter. [SCI TECH] In general, a selective device that transmits a desired range of matter or energy while substantially attenuating all other ranges.

filter base [MATH] A family of subsets of a given set with the property that it does not include the empty set, and the intersection of any finite number of members of the famiy includes another member.

filter capacitor [ELEC] A capacitor used in a power-supply filter system to provide a low-reactance path for alternating currents and thereby suppress ripple currents, without affecting direct currents.

filter choke [ELEC] An iron-core coil used in a power-supply filter system to pass direct current while offering high impedance to pulsating or alternating current.

filter crystal [ELECTR] Quartz crystal which is used in an electrical circuit designed to pass energy of certain frequencies.

filter discrimination [ELECTR] Difference between the minimum insertion loss at any frequency in a filter attenuation band and the maximum insertion loss at any frequency in the operating range of a filter transmission band.

filter factor [OPTICS] The number of times the exposure must be increased when a filter is used on a camera, because the filter absorbs some of the light.

filter impedance compensator [ELECTR] Impedance compensator which is connected across the common terminals of electric wave filters when the latter are used in parallel to compensate for the effects of the filters on each other.

filter pass band *See* filter transmission band.

filter reactor [ELEC] A reactor used for reducing the harmonic components of voltage in an alternating-current or direct-current circuit.

filter section [ELEC] A simple RC, RL, or LC network used as a broad-band filter in a power supply, grid-bias feed, or similar device.

filter slot [ELECTROMAG] Choke in the form of a slot designed to suppress unwanted modes in a waveguide.

filter spectrophotometer [SPECT] Spectrophotographic analyzer of spectral radiations in which a filter is used to isolate narrow portions of the spectrum.

filter transmission band [ELECTR] Frequency band of free transmission; that is, frequency band in which, if dissipation is neglected, the attenuation constant is zero. Also known as filter pass band.

final-value theorem [MATH] The theorem that if $f(t)$ is a function which has a Laplace transform $F(s)$, and if the derivative of $f(t)$ with respect to t is also Laplace transformable, and if the limit of $f(t)$ as t approaches infinity exists, then this limit is equal to the limit of $sF(s)$ as s approaches zero.

finder [OPTICS] A small telescope having a wide-angle lens and low power, which is attached to a larger telescope and points in the same direction; used to locate objects that are to be viewed in the larger telescope.

fine structure [ATOM PHYS] The splitting of spectral lines in atomic and molecular spectra caused by the spin angular momentum of the electrons and the coupling of the spin to the orbital angular momentum.

fine-structure constant [PHYS] A fundamental dimensionless constant, equal to 2π times the square of the electron charge in electrostatic units, divided by the product of the speed of light and Planck's constant; mathematically, equal to $(7.297351 \pm 0.000011) \times 10^{-3}$, approximately $1/137$. Also known as Sommerfeld fine-structure constant.

finite cosine transform [MATH] For a function $f(x)$, defined on an interval $[0,a]$, this is the function $F(n)$, $n = 0, 1, 2, \ldots$, equal to the integral from 0 to a of $f(x) \cos (n\pi x/a) \, dx$.

finite difference [MATH] The difference between the values of a function at two discrete points, used to approximate the derivative of the function.

finite difference equations [MATH] Equations arising from differential equations by substituting difference quotients for derivatives, and then using these equations to approximate a solution.

finite elasticity theory *See* finite strain theory.

finite extension [MATH] An extension field F of a given field E such that F, viewed as a vector space over E, has finite dimension.

finite Fourier transform [MATH] A Fourier series written as a finite transform by a change in notation, such as the finite sine transform of the finite cosine transform.

finite group [MATH] A group which contains a finite number of distinct elements.

finite Hankel transform [MATH] For a function $f(x)$, defined on an interval $[0,a]$, this is the function $F(n)$, $n = 0, 1, 2, \ldots$, equal to the integral 0 to a of $xf(x)J_\nu(\lambda_n x) \, dx$, where J_ν is the Bessel function of the first kind of order ν, and $\lambda_1 < \lambda_2 < \lambda_3 < \cdots$ are the roots of the equation $J_\nu(\lambda_n a) = 0$.

finite intersection property of a family of sets [MATH] If the intersection of any finite number of them is nonempty, then the intersection of all the members of the family is nonempty.

finite Legendre transform [MATH] For a function $f(x)$, $-1 \leq x \leq 1$, this is the function $F(n)$, $n = 0, 1, 2, \ldots$, equal to the integral from -1 to 1 of $f(x)P_n(x) \, dx$, where $P_n(x)$ is the Legendre polynomial of degree n.

finitely generated extension [MATH] A finitely generated extension of a field k is the smallest field which contains k and some finite set of elements.

finite mathematics [MATH] **1.** Those parts of mathematics which deal with finite sets. **2.** Those fields of mathematics which make no use of the concept of limit.

finite matrix [MATH] A matrix with a finite number of rows and columns.

finite measure space [MATH] A measure space in which the measure of the entire space is a finite number.

finite moment theorem [MATH] The theorem that if $f(x)$ is a continuous function, and if the integral of $f(x) \, x^n$ over a finite interval is zero for all positive integers n, then $f(x)$ is identically zero in that interval.

finite plane [MATH] In projective geometry, a plane with a finite number of points and lines.

finite population [STAT] A population of finite individuals or elements.

finite quantity [MATH] Any bounded quantity.

finite set [MATH] A set whose elements can be indexed by integers $1, 2, 3, \ldots, n$ inclusive.

finite sine transform [MATH] For a function $f(x)$, defined on an interval $[0,a]$, this is the function $F(n) = 0, 1, 2, \ldots$, equal to the integral from 0 to a of $f(x) \sin (n\pi x/a) \, dx$.

finite strain theory [MECH] A theory of elasticity, appropriate for high compressions, in which it is not assumed that strains are infinitesimally small. Also known as finite elasticity theory.

finite transform [MATH] A mapping which associates with each function $f(x)$, defined on a finite interval (a,b), the function

$$F(n) = \int_a^b f(x)g_n(x)\ dx,$$

defined on the positive integers, where the $g_n(x)$, $n = 1, 2,$ . . . , are a complete, orthonormal set of functions on (a,b).

finsen unit [ELECTROMAG] A unit of intensity of ultraviolet radiation, equal to the intensity of ultraviolet radiation at a specified wavelength whose energy flux is 100,000 watts per square meter; the wavelength usually specified is 296.7 nanometers. Abbreviated FU.

Finsler geometry [MATH] The study of the geometry of a manifold in terms of the various possible metrics on it by means of Finsler structures.

Finsler structure on a manifold [MATH] A family of metrics varying continuously from point to point.

fin waveguide [ELECTROMAG] Waveguide containing a thin longitudinal metal fin that serves to increase the wavelength range over which the waveguide will transmit signals efficiently; usually used with circular waveguides.

fireball [ASTRON] A bright meteor with luminosity equal to or exceeding that of the brightest planets. [NUCLEO] The luminous sphere of hot gases that forms a few millionths of a second after a nuclear explosion.

fired state [ELECTR] The "on" state of a silicon controlled rectifier or other semiconductor switching device, occurring when a suitable triggering pulse is applied to the gate.

firing [ELECTR] **1.** The gas ionization that initiates current flow in a gas-discharge tube. **2.** Excitation of a magnetron or transmit-receive tube by a pulse. **3.** The transition from the unsaturated to the saturated state of a saturable reactor.

firing point *See* critical grid voltage.

firing potential [ELECTR] Controlled potential at which conduction through a gas-filled tube begins.

firmoviscosity [MECH] Property of a substance in which the stress is equal to the sum of a term proportional to the substance's deformation, and a term proportional to its rate of deformation.

first-class current [PARTIC PHYS] A weak-interaction current whose charge symmetry (or G parity) properties are the same as those of currents which arise in the Fermi theory of beta decay.

first countable topological space [MATH] A topological space in which every point has a countable number of open neighborhoods so that any neighborhood of this point contains one of these.

first derived curve *See* derived curve.

first detector *See* mixer.

first Fresnel zone [ELECTROMAG] Circular portion of a wavefront transverse to the line between an emitter and a more distant point, where the resultant disturbance is being observed, whose center is the intersection of the front with the direct ray, and whose radius is such that the shortest path from the emitter through the periphery to the receiving point is one-half wavelength longer than the direct ray.

first fundamental form [MATH] For a surface with curvilinear coordinates u^i, $i = 1, 2$, this is the quadratic differential form

$$\sum_{ij} a_{ij}\ du^i\ du^j$$

where the a_{ij} are fundamental magnitudes of first order; equal to the square of a line element from u^i to $u^i + du^i$. Also known as first ground form.

first ground form *See* first fundamental form.

first harmonic *See* fundamental.

first integral [MATH] For an nth-order differential equation, a function of the independent variable, the dependent variable, and its first $n-1$ derivatives which is a constant for any solution of the equation.

first law of motion *See* Newton's first law.

first law of the mean *See* mean value theorem.

first law of the mean for integrals [MATH] The definite integral of a continuous function over an interval equals the length of the interval multiplied by the value of the function at some point in the interval.

first law of thermodynamics [THERMO] The law that heat is a form of energy, and the total amount of energy of all kinds in an isolated system is constant; it is an application of the principle of conservation of energy.

first-order reaction [PHYS CHEM] A chemical reaction in which the rate of decrease of concentration of component A with time is proportional to the concentration of A.

first-order spectrum [SPECT] A spectrum, produced by a diffraction grating, in which the difference in path length of light from adjacent slits is one wavelength.

first-order theory [MATH] A logical theory in which predicates are not allowed to have other functions or predicates as arguments and in which predicate quantifiers and function quantifiers are not permitted. [OPTICS] *See* Gaussian optics. [PHYS] A theory which takes into account only the most important terms, such as the term proportional to the independent variable in the series expansion of a function appearing in the theory.

first-order transition [THERMO] A change in state of aggregation of a system accompanied by a discontinuous change in enthalpy, entropy, and volume at a single temperature and pressure.

first point of Aries *See* vernal equinox.

first point of Cancer *See* summer solstice.

first quarter [ASTRON] The phase of the moon when it is near east quadrature, when the western half of it is visible to an observer on the earth.

first radiation constant [STAT MECH] A constant appearing in the Planck radiation formula; its value depends on the form of the formula used; in the formula for power emitted by a blackbody per unit area per unit wavelength interval, it is 2π times Planck's constant, times the square of the speed of light, or approximately 3.7415×10^{-16} watt (meter)^2. Symbolized c_1; C_1.

first sound [CRYO] Ordinary sound in helium II, in which pressure and density variations are propagated; in contrast to second sound.

first variation [MATH] The first variation of the integral

$$I = \int_a^b f(x, y, dy/dx)\, dx,$$

with respect to the variation $\phi(x)$ of $y(x)$, is the quantity

$$\delta I = \frac{d}{d\varepsilon}\int_a^b f(x, y + \varepsilon\phi, dy/dx + \varepsilon\, d\phi/dx)\, dx,$$

evaluated at $\varepsilon = 0$.

Fischer-Hinnen method [ELEC] Method of analysis of a complex waveform which has like loops above and below the time axis, in which the amplitude and phase of the nth harmonic is determined from the ordinates of the resultant wave at a series of times which divide the half wave into $2n$ equal time intervals.

Fischer's distribution [STAT] Given data from a normal

population with S_1^2 and S_2^2 two independent estimates of variance, the distribution $\frac{1}{2} \log (S_1^2 / S_2^2)$.

Fisher's ideal index [STAT] The geometric mean of Laspeyres and Paasche index numbers. Also known as ideal index number.

Fisher's inequality [MATH] The number b of blocks in a balanced incomplete block design is equal to or greater than the number v of elements arranged among the blocks.

Fischer-Yates test [STAT] A test of independence of data arranged in a 2×2 contingency table.

fish-bone antenna [ELECTROMAG] **1.** Antenna consisting of a series of coplanar elements arranged in collinear pairs, loosely coupled to a balanced transmission line. **2.** Directional antenna in the form of a plane array of doublets arranged transversely along both sides of a transmission line.

Fisher-Irwin test [STAT] A method for testing the null hypothesis in an experiment with quantal response.

fissile *See* fissionable.

fission [NUC PHYS] The division of an atomic nucleus into parts of comparable mass; usually restricted to heavier nuclei such as isotopes of uranium, plutonium, and thorium. Also known as atomic fission; nuclear fission.

fissionable [NUCLEO] **1.** A property of material whose nuclei are capable of undergoing fission. Also known as fissile. **2.** A material capable of fission.

fission bomb *See* atomic bomb.

fission chamber [NUCLEO] An ionization chamber used to detect slow neutrons; the inside wall has a thin coating of uranium, in which a slow neutron produces a fission; the resulting highly ionizing fission fragments produce a count in the chamber. Also known as fission counter.

fission counter *See* fission chamber.

fission cross section [NUC PHYS] The cross section for a bombarding neutron, gamma ray, or other particle to induce fission of a nucleus.

fission detector [NUCLEO] Device for detecting spontaneous fission, consisting of a mica or special glass which is placed near the sample and which is subsequently chemically etched, making fission tracks visible.

fission fraction [NUCLEO] The fraction of the total yield of a nuclear weapon that is due to fission; for thermonuclear weapons the average value is about 50%.

fission fragments [NUCLEO] The nuclear species first produced when an atom such as uranium-238 or plutonium-239 undergoes fission. Also known as primary fission products.

fission fuel *See* nuclear fuel.

fission-fusion bomb [NUCLEO] An explosive device which derives its energy in comparable amounts from nuclear fission and nuclear fusion.

fission neutron [NUC PHYS] A neutron emitted as a result of nuclear fission.

fission product [NUC PHYS] Any radioactive or stable nuclide resulting from fission, including both primary fission fragments and their radioactive decay products.

fission-product poisoning [NUCLEO] Inhibition of a nuclear chain reaction by fission products which have large cross sections for slow neutrons, and thus capture these neutrons before they can cause fission.

fission reactor *See* nuclear reactor.

fission spectrum [NUC PHYS] The energy distribution of neutrons arising from fission.

fission spike [NUCLEO] A displacement spike produced by fission fragments.

fission threshold [NUC PHYS] The minimum kinetic energy

of a bombarding neutron required to induce fission of a nucleus.

fission yield [NUCLEO] The amount of energy released by fission in a nuclear explosion, as distinct from that released by fusion. [NUC PHYS] The percent of fissions that gives a particular nuclide or group of isobars.

fissium [NUCLEO] An equilibrium mixture of fission products in reactor fuel that can improve the stability of uranium and uranium-plutonium fuel alloys under fast-neutron irradiation.

FitzGerald-Lorentz contraction [RELAT] The contraction of a moving body in the direction of its motion when its speed is comparable to the speed of light. Also known as Lorentz contraction; Lorentz-FitzGerald contraction.

five-dimensional space [MATH] A vector space whose basis has five vectors.

five-fourths power law [THERMO] The rate of heat loss from a body by free convection is proportional to the five-fourths power of the difference between the temperature of the body and that of its surroundings.

fixed-base index [STAT] In a time series, an index number whose base period for computing the index number is constant throughout the lifetime of the index.

fixed bias [ELECTR] A constant value of bias voltage, independent of signal strength.

fixed-bias transistor circuit [ELECTR] A transistor circuit in which a current flowing through a resistor is independent of the quiescent collector current.

fixed capacitor [ELEC] A capacitor having a definite capacitance value that cannot be adjusted.

fixed end [MECH] An end of a structure, such as a beam, that is clamped in place so that both its position and orientation are fixed.

fixed end moment *See* fixing moment.

fixed-field accelerator [NUCLEO] A circular particle accelerator whose magnetic fields do not vary with time, such as an ordinary cyclotron or a fixed-field, alternating-gradient synchrotron.

fixed-focus lens [OPTICS] A lens whose focus is invariable, as on inexpensive cameras with no mechanism for adjusting focus but so designed that all objects from a few feet away to infinity are tolerably in focus.

fixed inductor [ELEC] An inductor whose coils are wound in such a manner that the turns remain fixed in position with respect to each other, and which either has no magnetic core or has a core whose air gap and position within the coil are fixed.

fixed point [MATH] For a function f mapping a set S to itself, any element of S which f sends to itself.

fixed-point theorem [MATH] Any theorem, such as the Brouwer theorem or Schauder's fixed-point theorem, which states that a certain type of mapping of a set into itself has at least one fixed point.

fixed radix notation [MATH] A form of positional notation in which successive digits are interpreted as coefficients of successive powers of an integer called the base or radix.

fixed resistor [ELEC] A resistor that has no provision for varying its resistance value.

fixing moment [MECH] The bending moment at the end support of a beam necessary to fix it and prevent rotation. Also known as fixed end moment.

Fizeau fringes [OPTICS] 1. Interference fringes of monochromatic light from interference in a geometrical situation other than plane parallel plates. Also known as fringes of equal

FIZEAU FRINGES

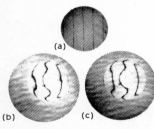

Fizeau multiple-beam fringe patterns (def. 2). *(a)* Narrow-gap plane mirrors. *(b)* Wide-gap test lens without end mirrors imaged on each other; *(c)* with mirrors imaged on each other.

thickness. **2.** Interference fringes in light from a Fizeau interferometer.

Fizeau interferometer [OPTICS] Interferometer in which light from a point source is collimated and multiply reflected between a plane mirror and the partially silvered inner surface of a parallel plane plate, and is viewed in reflection.

Fizeau toothed wheel [OPTICS] Rapidly rotating toothed wheel which was used to measure the speed of light by adjusting the rotation speed until light passing through one tooth opening and reflected from a distant mirror would pass through the next tooth opening on return.

flame emission spectroscopy [SPECT] A flame photometry technique in which the solution containing the sample to be analyzed is optically excited in an oxyhydrogen or oxyacetylene flame.

flame excitation [SPECT] Use of a high-temperature flame (such as oxyacetylene) to excite spectra emission lines from alkali and alkaline-earth elements and metals.

flame laser [OPTICS] A molecular gas laser in which gases such as carbon disulfide and oxygen are mixed at low pressures and ignited; the flame is then self-sustaining and produces carbon monoxide laser emission.

flame photometer [SPECT] One of several types of instruments used in flame photometry, such as the emission flame photometer and the atomic absorption spectrophotometer, in each of which a solution of the chemical being analyzed is vaporized; the spectral lines resulting from the light source going through the vapors enters a monochromator that selects the band or bands of interest.

flame photometry [SPECT] A branch of spectrochemical analysis in which samples in solution are excited to produce line emission spectra by introduction into a flame.

flame spectrometry [SPECT] A procedure used to measure the spectra or to determine wavelengths emitted by flame-excited substances.

flame spectrophotometry [SPECT] A method used to determine the intensity of radiations of various wavelengths in a spectrum emitted by a chemical inserted into a flame.

flame spectrum [SPECT] An emission spectrum obtained by evaporating substances in a nonluminous flame.

flap attenuator [ELECTROMAG] A waveguide attenuator in which a contoured sheet of dissipative material is moved into the guide through a nonradiating slot to provide a desired amount of power absorption. Also known as vane attenuator.

flare [ASTRON] A bright eruption from the sun's chromosphere; flares may appear within minutes and fade within an hour, cover a wide range of intensity and size, and tend to occur between sunspots or over their penumbrae. [ELECTR] A radar screen target indication having an enlarged and distorted shape due to excessive brightness. [ELECTROMAG] *See* horn antenna.

flare spot [OPTICS] A small, diffuse, brightly illuminated region produced by multiple reflections of light from the various surfaces of an optical system.

flare stars *See* UV Ceti stars.

flash arc [ELECTR] A sudden increase in the emission of large thermionic vacuum tubes, probably due to irregularities in the cathode surface.

flash factor [OPTICS] In photography using a photoflash lamp, a number dependent on the lamp and the film speed, equal to the product of the distance of the lamp from the subject and the correct *f* number for that distance.

flashing [OPTICS] The apparent filling of a curved mirror or lens with light when viewed from a distance, as a result of the

**FLAME EMISSION
SPECTROSCOPY**

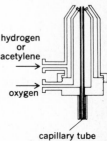

hydrogen
or
acetylene →

oxygen →

capillary tube
for introduction of
sample solution

Beckman aspirator burner used to aspirate the solution directly into oxyhydrogen or oxyacetylene flame.

FLAME PHOTOMETER

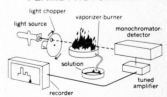

Schematic diagram of atomic absorption spectrophotometer for determining metal concentrations.

FLASH LAMP

Elementary circuit of electronic flash lamp *L*.

production of a parallel beam by a light source at the focus.

flash lamp [ELECTR] A gaseous-discharge lamp used in a photoflash unit to produce flashes of light of short duration and high intensity for stroboscopic photography. Also known as stroboscopic lamp.

flash magnetization [ELECTROMAG] Magnetization of a ferromagnetic object by a current impulse of short duration.

flashover [ELEC] An electric discharge around or over the surface of an insulator.

flash photography *See* stroboscopic photography.

flash photolysis [PHYS CHEM] A method of studying fast photochemical reactions in gas molecules; a powerful lamp is discharged in microsecond flashes near a reaction vessel holding the gas, and the products formed by the flash are observed spectroscopically.

flash spectroscopy [SPECT] The study of the electronic states of molecules after they absorb energy from an intense, brief light flash.

flash spectrum [ASTRON] The emission spectrum of the sun's chromosphere, observed for a few seconds just before and just after a total solar eclipse.

flat line [ELECTROMAG] A radio-frequency transmission line, or part thereof, having essentially 1-to-1 standing wave ratio.

flat space [MATH] A Riemannian space for which a coordinate system exists such that the components of the metric tensor are constants throughout the space; equivalently, a space in which the Riemann-Christoffel tensor vanishes throughout the space.

flat space-time [RELAT] Space-time in which the Riemann-Christoffel tensor vanishes; geometry is then equivalent to that of the Minkowski universe used in special relativity.

flat spin [MECH] Motion of a projectile with a slow spin and a very large angle of yaw, happening most frequently in fin-stabilized projectiles with some spin-producing moment, when the period of revolution of the projectile coincides with the period of its oscillation; sometimes observed in bombs and in unstable spinning projectiles.

flat-top antenna [ELECTROMAG] An antenna having two or more lengths of wire parallel to each other and in a plane parallel to the ground, each fed at or near its midpoint.

flat top response *See* band-pass response.

flat trajectory [MECH] A trajectory which is relatively flat, that is, described by a projectile of relatively high velocity.

flavor [PARTIC PHYS] A label used to distinguish different types of leptons (the electron, electron neutrino, muon, muon neutrino, and possibly others) and different color triplets of quarks (the up, down, strange, and charmed quarks, and possibly others).

F layer [GEOPHYS] An ionized layer in the F region of the ionosphere which consists of the F_1 and F_2 layers in the day hemisphere, and the F_2 layer alone in the night hemisphere; it is capable of reflecting radio waves to earth at frequencies up to about 50 megahertz.

F_1 layer [GEOPHYS] The ionosphere layer beneath the F_2 layer during the day, at a virtual height of 200–300 kilometers, being closest to earth around noon; characterized by a distinct maximum of free-electron density, except at high latitudes during winter, when the layer is not detectable.

F_2 layer [GEOPHYS] The highest constantly observable ionosphere layer, characterized by a distinct maximum of free-electron density at a virtual height from about 225 kilometers in the polar winter to more than 400 kilometers in daytime near the magnetic equator. Also known as Appleton layer.

fl dr *See* fluid dram.

Fleming-Kennelly law [ELECTROMAG] The reluctivity of a

ferromagnetic substance varies linearly with magnetic field strength at points near magnetic saturation.

Fleming's rule *See* left-hand rule; right-hand rule.

Fleming tube [ELECTR] The original diode, consisting of a heated filament and a cold metallic electrode in an evacuated glass envelope; negative current flows from the filament to the cold electrode, but not in the reverse direction.

Fletcher-Munson contour *See* equal loudness contour.

flexibility [MECH] The quality or state of being able to be flexed or bent repeatedly.

flexible waveguide [ELECTROMAG] A waveguide that can be bent or twisted without appreciably changing its electrical properties.

flexural modulus [MECH] A measure of the resistance of a beam of specified material and cross section to bending, equal to the product of Young's modulus for the material and the square of the radius of gyration of the beam about its neutral axis.

flexural rigidity [MECH] The ratio of the sideward force applied to one end of a beam to the resulting displacement of this end, when the other end is clamped.

flexural strength [MECH] Strength of a material in blending, that is, resistance to fracture.

flexure [MECH] **1.** The deformation of any beam subjected to a load. **2.** Any deformation of an elastic body in which the points originally lying on any straight line are displaced to form a plane curve.

flexure theory [MECH] Theory of the deformation of a prismatic beam having a length at least 10 times its depth and consisting of a material obeying Hooke's law, in response to stresses within the elastic limit.

flicker [OPTICS] A visual sensation produced by periodic fluctuations in light at rates ranging from a few cycles per second to a few tens of cycles per second.

flicker effect [ELECTR] Random variations in the output current of an electron tube having an oxide-coated cathode, due to random changes in cathode emission.

flicker photometer [OPTICS] A photometer in which a single field of view is alternately illuminated by the light sources to be compared, and the rate of alternation is such that color flicker is absent but brightness flicker is not; disappearance of flicker signifies equality of luminance.

F line [SPECT] A green-blue line in the spectrum of hydrogen, at a wavelength of 486.133 nanometers.

flip coil [ELECTROMAG] A small coil used to measure the strength of a magnetic field; it is placed in the field, connected to a ballistic galvanometer or other instrument, and suddenly flipped over 180°; alternatively, the coil may be held stationary and the magnetic field reversed.

flip-flop circuit *See* bistable multivibrator.

flip-over process *See* Umklapp process.

float barograph [ENG] A type of siphon barograph in which the mechanically magnified motion of a float resting on the lower mercury surface is used to record atmospheric pressure on a rotating drum.

floating [ELECTR] The condition wherein a device or circuit is not grounded and not tied to an established voltage supply.

floating arithmetic *See* floating-point arithmetic.

floating-decimal arithmetic *See* floating-point arithmetic.

floating grid [ELECTR] Vacuum-tube grid that is not connected to any circuit; it assumes a negative potential with respect to the cathode. Also known as free grid.

floating input [ELEC] Isolated input circuit not connected to ground at any point.

floating neutral [ELEC] Neutral conductor whose voltage to ground is free to vary when circuit conditions change.

floating-point arithmetic [MATH] A method of performing arithmetical operations, used especially by automatic computers, in which numbers are expressed as integers multiplied by the radix raised to an integral power, as 87×10^{-4} instead of 0.0087. Also known as floating arithmetic; floating-decimal arithmetic.

floating reticle [OPTICS] A reticle the image of which is movable within the field of view.

flocculus [ASTRON] A patch in the sun's surface seen in the light of calcium or hydrogen; the patch may be bright or dark and is usually in the vicinity of sunspots.

Floquet theorem [MATH] A second-order linear differential equation whose coefficients are periodic single-valued functions of an independent variable x, has a solution of the form $e^{\mu x}P(x)$ where μ is a constant and $P(x)$ a periodic function.

florentium See promethium-147.

flotation analysis [PHYS] Technique to measure liquid density in which a float of known density is adjusted with weights to match that of the liquid.

flow [FL MECH] The forward continuous movement of a fluid, such as gases, vapors, or liquids, through closed or open channels or conduits. [MATH] A topological transformation group (G,X,π) in which the phase group G consists of the real numbers. [PHYS] The movement of electric charges, gases, liquids, or other materials or quantities.

flowability [FL MECH] Capability of a liquid or loose particulate solid to move by flow.

flow birefringence [OPTICS] Birefringence of a flowing liquid caused by orientation of anisotropic molecules as a result of the flow, or of small particles in the liquid.

flow coefficient [FL MECH] An experimentally determined proportionality constant, relating the actual velocity of fluid flow in a pipe, duct, or open channel to the theoretical velocity expected under certain assumptions.

flow counter See gas-flow counter tube.

flow curve [FL MECH] A graph of the total shear of a fluid as a function of time. [MECH] The stress-strain curve of a plastic material.

flow distribution See flow field.

flow equation [FL MECH] Equation for the calculation of fluid (gas, vapor, liquid) flow through conduits or channels; consists of an interrelation of fluid properties (such as density or viscosity), environmental conditions (such as temperature or pressure), and conduit or channel geometry and conditions (such as diameter, cross-sectional shape, or surface roughness).

flow field [FL MECH] The velocity and the density of a fluid as functions of position and time. Also known as flow distribution.

flow figure See strain figure.

flow graph See signal-flow graph.

flowing-temperature factor [THERMO] Calculation correction factor for gases flowing at temperatures other than that for which a flow equation is valid, that is, other than 60°F (15.5°C).

flowmeter [ENG] An instrument used to measure pressure, flow rate, and discharge rate of a liquid, vapor, or gas flowing in a pipe. Also known as fluid meter.

flow net [FL MECH] A diagram used in studying the flow of a fluid through a permeable substance (such as water through a soil structure) having two nests of curves, one representing the flow lines, which follow the path of the fluid, and the other the equipotential lines, which connect points of equal head.

flow pattern [FL MECH] Pattern of two-phase flow in a con-

FLOW NET

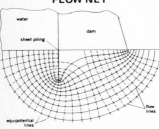

Flow net indicated under the cutoff wall of a dam. *(From D. P. Krynine, Soil Mechanics, 2d ed., McGraw-Hill, 1947)*

duit or channel pipe, taking into consideration the ratio of gas to liquid and conditions of flow resistance and liquid holdup.

flow rate [FL MECH] Also known as rate of flow. **1.** Time required for a given quantity of flowable material to flow a measured distance. **2.** Weight or volume of flowable material flowing per unit time.

flow resistance [FL MECH] **1.** Any factor within a conduit or channel that impedes the flow of fluid, such as surface roughness or sudden bends, contractions, or expansions. **2.** *See* viscosity.

flow separation *See* boundary-layer separation.

flow stress [MECH] The stress along one axis at a given value of strain that is required to produce plastic deformation.

fl oz *See* fluid ounce.

fluctuating current [ELEC] Direct current that changes in value but not at a steady rate.

fluctuation [SCI TECH] **1.** Variation, especially back and forth between successive values in a series of observations. **2.** Variation of data points about a smooth curve passing among them.

fluctuation noise *See* random noise.

fluctuation theory [OPTICS] The theory proposed by M. von Smoluchowski and A. Einstein which states that the scattering of light occurs in pure water because random molecular motion causes density variations which effect changes in the refraction of light.

fluctuation velocity *See* eddy velocity.

fluence [NUCLEO] The time integral of the flux density of atomic or nuclear particles. [PHYS] A measure of time-integrated particle flux, expressed in particles per square centimeter.

fluid [PHYS] An aggregate of matter in which the molecules are able to flow past each other without limit and without fracture planes forming.

fluid density [FL MECH] The mass of a fluid per unit volume.

fluid dram [MECH] Abbreviated fl dr. **1.** A unit of volume used in the United States for measurement of liquid substances, equal to 1/8 fluid ounce, or $3.6966911953125 \times 10^{-6}$ cubic meter. **2.** A unit of volume used in the United Kingdom for measurement of liquid substances and occasionally of solid substances, equal to 1/8 fluid ounce or approximately 3.55163×10^{-6} cubic meter.

fluid dynamics [FL MECH] The science of fluids in motion.

fluid friction [FL MECH] Conversion of mechanical energy in fluid flow into heat energy.

fluid fuel reactor [NUCLEO] A type of reactor (for example, a fused-salt reactor) whose fuel is in fluid form.

fluidics [ENG] A control technology that employs fluid dynamic phenomena to perform sensing, control, information processing, and actuation functions without the use of moving mechanical parts.

fluidity [FL MECH] The reciprocal of viscosity; expresses the ability of a substance to flow.

fluidized-bed reactor *See* fluidized reactor.

fluidized reactor [NUCLEO] A nuclear reactor in which the fuel has been given the properties of a quasi-fluid, such as by suspension of fine fuel particles in a carrying gas or liquid. Also known as fluidized-bed reactor.

fluid mechanics [MECH] The science concerned with fluids, either at rest or in motion, and dealing with pressures, velocities, and accelerations in the fluid, including fluid deformation and compression or expansion.

fluid meter *See* flowmeter.

fluid ounce [MECH] Abbreviated fl oz. **1.** A unit of volume used in the United States for measurement of liquid substances, equal to 1/16 liquid pint, or 231/128 cubic inches, or $2.95735295625 \times 10^{-5}$ cubic meter. **2.** A unit of volume used in the United Kingdom for measurement of liquid substances, and occasionally of solid substances, equal to 1/20 pint or approximately 2.84130×10^{-5} cubic meter.

fluid resistance [FL MECH] The force exerted by a gas or liquid opposing the motion of a body through it. Also known as resistance.

fluid statics [FL MECH] The determination of pressure intensities and forces exerted by fluids at rest.

fluid stress [MECH] Stress associated with plastic deformation in a solid material.

fluid ton [MECH] A unit of volume equal to 32 cubic feet or approximately 9.0614×10^{-2} cubic meter; used for many hydrometallurgical, hydraulic, and other industrial purposes.

fluophor *See* luminophor.

fluor *See* luminophor.

fluorescence [ATOM PHYS] **1.** Emission of electromagnetic radiation that is caused by the flow of some form of energy into the emitting body and which ceases abruptly when the excitation ceases. **2.** Emission of electromagnetic radiation that is caused by the flow of some form of energy into the emitting body and whose decay, when the excitation ceases, is temperature-independent. [NUC PHYS] Gamma radiation scattered by nuclei which are excited to and radiate from an excited state.

fluorescence microscope [OPTICS] A variation of the compound laboratory light microscope which is arranged to admit ultraviolet, violet, and sometimes blue radiations to a specimen, which then fluoresces.

fluorescence spectra [SPECT] Emission spectra of fluorescence in which an atom or molecule is excited by absorbing light and then emits light of characteristic frequencies.

fluorescence x-rays [ATOM PHYS] Characteristic x-rays emitted as the result of the absorption of x-rays of higher frequency.

fluorescence yield [ATOM PHYS] The probability that an atom in an excited state will emit an x-ray photon in its first transition rather than an Auger electron.

fluorescent lamp [ELECTR] A tubular discharge lamp in which ionization of mercury vapor produces radiation that activates the fluorescent coating on the inner surface of the glass.

fluorine [CHEM] A gaseous or liquid chemical element, symbol F, atomic number 9, atomic weight 18.998; a member of the halide family, it is the most electronegative element and the most chemically energetic of the nonmetallic elements; highly toxic, corrosive, and flammable; used in rocket fuels and as a chemical intermediate.

fluorod [NUCLEO] A rod made from silver-activated phosphate glass and used in solid-state dosimeters; under irradiation the rod absorbs ultraviolet light and emits orange fluorescent light; measurement of the intensity of the emitted light with a photomultiplier gives a measure of the absorbed dose of radiation.

fluorometer [ENG] An instrument that measures the fluorescent radiation emitted by a sample which is exposed to monochromatic radiation, usually radiation from a mercury-arc lamp or a tungsten or molybdenum x-ray source that has passed through a filter; used in chemical analysis, or to determine the intensity of the radiation producing fluorescence. Also spelled fluorimeter.

fluoroscope [ENG] A fluorescent screen designed for use with an x-ray tube to permit direct visual observation of x-ray

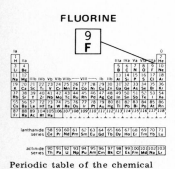

FLUORINE

Periodic table of the chemical elements showing the position of fluorine.

shadow images of objects interposed between the x-ray tube and the screen.

fluoroscopic image intensifier [ELECTR] An electron-beam tube that converts a relatively feeble fluoroscopic image on the fluorescent input phosphor into a much brighter image on the output phosphor.

flutter [FL MECH] An aeroelastic self-excited vibration in which the external source of energy is the airstream and which depends on the elastic, inertial, and dissipative forces of the system in addition to the aerodynamic forces.

flutter echo [ACOUS] A multiple echo in which the reflections rapidly follow each other.

flux [ELECTROMAG] The electric or magnetic lines of force in a region. [NUCLEO] The product of the number of particles per unit volume and their average velocity; a special case of the physics definition. Also known as flux density. [PHYS] **1.** The integral over a given surface of the component of a vector field (for example, the magnetic flux density, electric displacement, or gravitational field) perpendicular to the surface; by definition, it is proportional to the number of lines of force crossing the surface. **2.** The amount of some quantity flowing across a given area (often a unit area perpendicular to the flow) per unit time; the quantity may be, for example, mass or volume of fluid, electromagnetic energy, or number of particles.

fluxball [ELECTROMAG] A type of magnetic test coil in which the wire is wound into the form of a solid spherical winding by combining a series of coaxial cylindrical windings of different lengths; it gives accurate values of the magnetic flux density (or its variation) at its center, even in a nonuniform magnetic field.

flux-closure domain *See* closure domain.

flux density [NUCLEO] *See* flux. [PHYS] Any vector field whose flux is a significant physical quantity; examples are magnetic flux density, electric displacement, gravitational field, and the Poynting vector.

flux density threshold *See* threshold illuminance.

flux gate [ENG] A detector that gives an electric signal whose magnitude and phase are proportional to the magnitude and direction of the external magnetic field acting along its axis; used to indicate the direction of the terrestrial magnetic field.

flux-gate magnetometer [ELECTROMAG] A magnetometer in which the degree of saturation of the core by an external magnetic field is used as a measure of the strength of the earth's magnetic field; the essential element is the flux gate.

flux jumping *See* Meissner effect.

flux leakage [ELECTROMAG] Magnetic flux that does not pass through an air gap or other part of a magnetic circuit where it is required.

flux line *See* line of force.

flux linkage [ELECTROMAG] The product of the number of turns in a coil and the magnetic flux passing through the coil. Also known as linkage.

flux mapping [NUCLEO] The process of measuring the radiation flux at representative points within a nuclear reactor or around some other radiation source.

flux of energy [PHYS] The energy which passes through a surface per unit area per unit time.

flux path [ELECTROMAG] A path which is followed by magnetic lines of force and in which the magnetic flux density is significant.

flux pump [CRYO] A cryogenic direct-current generator that converts a small alternating-current input to a large direct-current output when cooled to about 4 K; the output current builds up in a series of steps, much like the action of a pump.

FLUX PUMP

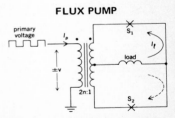

A flux-pump circuit (simplified).

flux refraction [ELECTROMAG] The abrupt change in direction of magnetic flux lines at the boundary between two media having different permeabilities, or of the electric flux lines at the boundary between two media having different dielectric constants, when these lines are oblique to the boundary.

flux unit [ASTROPHYS] A unit of energy flux density of radio-astronomical sources, equal to 10^{-26} watt per square meter per hertz. Abbreviated fu.

flying-spot microscope [OPTICS] A microscope in which a minute spot of light, produced in the lens system, passes through a specimen while sweeping over it systematically, and falls on a photocell; the image is produced on a cathode-ray tube that is scanned in synchronization with the spot.

Fm *See* fermium.

FM *See* frequency modulation.

F martingale [MATH] A stochastic process $\{X_t, t>0\}$ such that the conditional expectation of X_t given F_s equals X_s whenever $s<t$, where $F=\{F_t, t\geq 0\}$ is an increasing family of sigma algebras that represents the amount of information increasing with time.

F meson [PARTIC PHYS] A charged meson that carries both strangeness and charm and has a mass of approximately 2.03 GeV, spin 0, and negative parity.

fnp *See* fusion point.

f number [OPTICS] A lens rating obtained by dividing the lens's focal length by its effective maximum diameter; the larger the f number, the less exposure is given. Also known as focal ratio; stop number.

foam [FL MECH] A froth of bubbles on the surface of a liquid, often stabilized by organic contaminants, as found at sea or along shore.

foaminess [PHYS] The volume of foam produced in a liquid, in cubic centimeters, produced by passing air through it divided by the rate of flow of air, in cubic centimeters per second.

focal distance *See* focal length.

focal length [OPTICS] The distance from the focal point of a lens or curved mirror to the principal point; for a thin lens it is approximately the distance from the focal point to the lens. Also known as focal distance.

focal lines *See* astigmatic foci.

focal plane [OPTICS] A plane perpendicular to the axis of an optical system and passing through the focal point of the system.

focal-plane shutter [OPTICS] A camera shutter consisting of a blind containing a slot; the blind is pulled rapidly across the film, exposing it through the slot.

focal point [OPTICS] The point to which rays that are initially parallel to the axis of a lens, mirror, or other optical system are converged or from which they appear to diverge. Also known as principal focus.

focal ratio *See* f number.

Foch space [QUANT MECH] An infinite-dimensional vector space in which the state of a quantum-mechanical system with a variable number of particles is represented by an infinite number of wave functions, each of which corresponds to a fixed number of particles.

focometer [ENG] An instrument for measuring focal lengths of optical systems.

focus [ELECTR] To control convergence or divergence of the electron paths within one or more beams, usually by adjusting a voltage or current in a circuit that controls the electric or magnetic fields through which the beams pass, in order to obtain a desired image or a desired current density within the beam. [MATH] A point in the plane which together with a line (directrix) defines a conic section. [NUCLEO] To guide particles along a desired path in a particle accelerator by means of electric or magnetic fields. [OPTICS] **1.** The point or small region at which rays converge or from which they appear to diverge. **2.** To move an optical lens toward or away from a screen or film to obtain the sharpest possible image of a desired object.

focused collision sequence [PHYS] A cascade of interatomic collisions, initiated by the bombardment of a crystal with energetic particles, that propagates in a particular direction along a closely packed row of atoms in the crystal.

focusing anode [ELECTR] An anode used in a cathode-ray tube to change the size of the electron beam at the screen; varying the voltage on this anode alters the paths of electrons in the beam and thus changes the position at which they cross or focus.

focusing coil [ELECTR] A coil that produces a magnetic field parallel to an electron beam for the purpose of focusing the beam.

focusing electrode [ELECTR] An electrode to which a potential is applied to control the cross-sectional area of the electron beam in a cathode-ray tube.

focusing glass [OPTICS] A magnifying glass designed to enlarge the image thrown on the ground glass of the viewfinder of a camera, to help achieve exact focusing.

focusing magnet [ELECTR] A permanent magnet used to produce a magnetic field for focusing an electron beam.

focusing scale [OPTICS] A graduated scale to indicate appropriate lens-to-image plane positions for given lens-to-object plane distances.

focus lamp [ELEC] **1.** A lamp whose filament has a spiral or zigzag form in order to reduce its size, so that it can be brought into the focus of a lens or mirror. **2.** An arc lamp whose feeding mechanism is designed to hold the arc in a constant position with respect to an optical system that is used to focus its rays.

fogbow [OPTICS] A faintly colored circular arc similar to a rainbow but formed on fog layers containing drops whose diameters are of the order of 100 micrometers or less. Also known as false white rainbow; mistbow; white rainbow.

fog chamber *See* cloud chamber.

fog track [NUCLEO] A line of condensation, produced in supersaturated water vapor by the passage of charged particles; used in studying the courses and collisions of particles in cloud chambers.

foil dosimeter [NUCLEO] A device for measuring the amount of radiation exposure by means of the degree of activation created in a metal foil inserted in the radiation field.

Fokker-Planck equation [STAT MECH] An equation for the distribution function of a gas, analogous to the Boltzmann equation but applying where the forces are long-range and the collisions are not binary.

folded cavity [ELECTR] Arrangement used in a klystron repeater to make the incoming wave act on the electron stream from the cathode at several places and produce a cumulative effect.

folded dipole *See* folded-dipole antenna.

folded-dipole antenna [ELECTROMAG] A dipole antenna whose outer ends are folded back and joined together at the center; the impedance is about 300 ohms, as compared to 70 ohms for a single-wire dipole; widely used with television and frequency-modulation receivers. Also known as folded dipole.

folded horn [ENG ACOUS] An acoustic horn in which the path from throat to mouth is folded or curled to give the longest possible path in a given volume.

folium of Descartes [MATH] A plane cubic curve whose equation in cartesian coordinates x and y is $x^3 + y^3 = 3axy$, where a is a constant. Also known as leaf of Descartes.

foot [MECH] The unit of length in the British systems of units, equal to exactly 0.3048 meter. Abbreviated ft.

footcandle [OPTICS] A unit of illumination, equal to the illumination of a surface, 1 square foot in area, on which there is a luminous flux of 1 lumen uniformly distributed, or equal to the illumination of a surface all points of which are at a distance of 1 foot from a uniform point source of 1 candela; equal to approximately 10.7639 lux. Abbreviated ftc.

footlambert [OPTICS] A unit of luminance (photometric brightness), equal to $1/\pi$ candela per square foot, or to the uniform luminance of a perfectly diffusing surface emitting or reflecting light at the rate of 1 lumen per square foot; equal to approximately 3.42625 nits. Abbreviated ft-L. Also known as equivalent footcandle.

foot-pound [MECH] **1.** Unit of energy or work in the English gravitational system, equal to the work done by 1 pound of force when the point at which the force is applied is displaced 1 foot in the direction of the force; equal to approximately 1.355818 joules. Abbreviated ft-lb; ft-lbf. **2.** Unit of torque in the English gravitational system, equal to the torque produced by 1 pound of force acting at a perpendicular distance of 1 foot from an axis of rotation. Also known as pound-foot. Abbreviated lbf-ft.

foot-poundal [MECH] **1.** A unit of energy or work in the English absolute system, equal to the work done by a force of magnitude 1 poundal when the point at which the force is applied is displaced 1 foot in the direction of the force; equal to approximately 0.04214011 joule. Abbreviated ft-pdl. **2.** A unit of torque in the English absolute system, equal to the torque produced by a force of magnitude 1 poundal acting at a perpendicular distance of 1 foot from the axis of rotation. Also known as poundal-foot. Abbreviated pdl-ft.

foot-pound-second system of units *See* British absolute system of units.

Forbes bar [THERMO] A metal bar which has one end immersed in a crucible of molten metal and thermometers placed in holes at intervals along the bar; measurement of temperatures along the bar together with measurement of cooling of a short piece of the bar enables calculation of the thermal conductivity of the metal.

forbidden band [SOLID STATE] A range of unallowed energy levels for an electron in a solid.

forbidden line [ATOM PHYS] A spectral line associated with a transition forbidden by selection rules; optically this might be a magnetic dipole or electric quadrupole transition.

forbidden transition [QUANT MECH] A transition between two states of a quantum-mechanical system which is considerably less probable than a competing allowed transition.

Forbush decrease [ASTROPHYS] A sudden decrease in cosmic-ray intensity which occurs a day or two after a solar flare, and at the same time as the commencement of magnetic storms and auroral activity.

force [MECH] That influence on a body which causes it to accelerate; quantitatively it is a vector, equal to the body's time rate of change of momentum.

force constant [MECH] The ratio of the force to the deformation of a system whose deformation is proportional to the applied force. [PHYS CHEM] An expression for the force acting to restrain the relative displacement of the nuclei in a molecule.

forced convection [THERMO] Heat convection in which fluid motion is maintained by some external agency.

forced oscillation [MECH] An oscillation produced in a simple oscillator or equivalent mechanical system by an external periodic driving force. Also known as forced vibration.

forced vibration *See* forced oscillation.

forced wave [FL MECH] Any wave which is required to fit irregularities at the boundary of a system or satisfy some impressed force within the system; the forced wave will not in general be a characteristic mode of oscillation of the system.

force feedback [CONT SYS] A method of error detection in which the force exerted on the effector is sensed and fed back to the control, usually by mechanical, hydraulic, or electric transducers.

force polygon [MECH] A closed polygon whose sides are vectors representing the forces acting on a body in equilibrium.

forecast [STAT] To assess the magnitude that a quantity will have at a specified time in the future. Also known as predict.

forensic physics [PHYS] The application of physics for discussion, debate, argumentative, or legal purposes.

formal logic [MATH] The study of the permissible relationships between propositions, a study that concerns the form rather than the content.

formant [ACOUS] A set of resonances of a musical instrument or voice mechanism that form partials of sounds produced by the instrument, independent of the fundamental frequency, and give these sounds their quality.

form birefringence [OPTICS] Birefringence of a liquid caused by the orientation of rod-shaped particles in the liquid whose thickness and separation are much smaller than a wavelength of light.

form drag [FL MECH] **1.** The drag from all causes resulting from the particular shape of a body relative to its direction of motion, as of fuselage, wing, or nacelle. **2.** At supersonic speed, the drag caused by losses due to shock waves, exclusive of losses due to skin friction.

form factor [ELEC] **1.** The ratio of the effective value of a periodic function, such as an alternating current, to its average absolute value. **2.** A factor that takes the shape of a coil into account when computing its inductance. Also known as shape factor. [MECH] The theoretical stress concentration factor for a given shape, for a perfectly elastic material. [PHYS] A function which describes the internal structure of a particle, allowing calculations to be made even though the structure is unknown. [QUANT MECH] An expression used in studying the scattering of electrons or radiation from atoms, nuclei, or elementary particles, which gives the deviation from point particle scattering due to the distribution of charge and current in the target.

formula [CHEM] **1.** A combination of chemical symbols that expresses a molecule's composition. **2.** A reaction formula showing the interrelationship between reactants and products. [MATH] An equation or rule relating mathematical objects or quantities.

formula weight [CHEM] **1.** The gram-molecular weight of a substance. **2.** In the case of a substance of uncertain molecular weight such as certain proteins, the molecular weight calculated from the composition, assuming that the element present in the smallest proportion is represented by only one atom.

fors *See* G; gram-force.

Fortrat parabola [SPECT] Graph of wave numbers of lines in a molecular spectral band versus the serial number of the successive lines.

forward bias [ELECTR] A bias voltage that is applied to a *pn* junction in the direction that causes a large current flow; used in some semiconductor diode circuits.

forward current [ELECTR] Current which flows upon application of forward voltage.

forward difference [MATH] One of a series of quantities obtained from a function whose values are known at a series of equally spaced points by repeatedly applying the forward difference operator to these values; used in interpolation or numerical calculation and integration of functions.

forward difference operator [MATH] A difference operator, denoted Δ, defined by the equation $\Delta f(x) = f(x + h) - f(x)$, where h is a constant indicating the difference between successive points of interpolation or calculation.

forward direction [ELECTR] Of a semiconductor diode, the direction of lower resistance to the flow of steady direct current.

forward drop [ELECTR] The voltage drop in the forward direction across a rectifier.

forward path [CONT SYS] The transmission path from the loop actuating signal to the loop output signal in a feedback control loop.

forward recovery time [ELECTR] Of a semiconductor diode, the time required for the forward current or voltage to reach a specified value after instantaneous application of a forward bias in a given circuit.

forward scattering [PHYS] **1.** Scattering in which there is no change in the direction of motion of the scattered particles. **2.** Scattering in which the angle between the initial and final directions of motion of the scattered particles is less than 90°.

forward-scatter propagation *See* scatter propagation.

forward shift operator *See* displacement operator.

forward transfer function [CONT SYS] In a feedback control loop, the transfer function of the forward path.

forward wave [ELECTR] Wave whose group velocity is the same direction as the electron stream motion.

Foster's reactance theorem [CONT SYS] The theorem that the most general driving point impedance or admittance of a network, in which every mesh contains independent inductance and capacitance, is a meromorphic function whose poles and zeros are all simple and occur in conjugate pairs on the imaginary axis, and in which these poles and zeros alternate.

Foucault current *See* eddy current.

Foucault knife-edge test [OPTICS] Test of a lens or a concave mirror in which a pinhole source is placed at twice the focal length behind the lens or at the mirror's center of curvature, the eye is placed at the image of the pinhole, and defects in the lens or mirror result in irregular darkening of the image when a knife edge is moved across the image immediately in front of the eye.

Foucault mirror [OPTICS] Experiment for measuring the speed of light in which light is reflected from a rapidly rotating mirror to a distant mirror and back, and the speed of light is deduced from the displacement of the beam after its second reflection from the rotating mirror, the angular speed of the rotating mirror, and the distance the light travels.

Foucault pendulum [MECH] A swinging weight supported by a long wire, so that the wire's upper support restrains the wire only in the vertical direction, and the weight is set swinging with no lateral or circular motion; the plane of the pendulum gradually changes, demonstrating the rotation of the earth on its axis.

fountain effect [FL MECH] The effect occurring when two containers of superfluid helium are connected by a capillary tube and one of them is heated, so that helium flows through the tube in the direction of higher temperature.

four-color problem [MATH] The problem of proving the statment that, given any map in the plane, it is possible to color the regions with four colors so that any two regions with a common boundary have different colors.

four-current density [RELAT] A four-vector whose three space components are those of the ordinary current density and whose time component is the charge density.

four-degree calorie [CHEM] The heat needed to change the temperature of 1 gram of water from 3.5 to 4.5°C.

four-factor formula [NUCLEO] The principle that the multiplication factor of a thermal reactor with no leakage is the product of the average number of fast neutrons emitted when a nucleus in the fuel material captures a thermal neutron, the fast fission factor, the fraction of neutrons which are not captured while being slowed down, and the number of thermal neutrons absorbed in the fuel divided by the total number of neutrons absorbed in the fuel and the moderator.

four-force [RELAT] A four-vector equal to the product of the rest mass of a particle and the rate of change of its four-momentum with respect to its proper time.

fourier *See* thermal ohm.

Fourier analysis [MATH] The study of convergence of Fourier series and when and how a function is approximated by its Fourier series or transform.

Fourier analyzer [ENG] A digital spectrum analyzer that provides push-button or other switch selection of averaging, coherence function, correlation, power spectrum, and other mathematical operations involved in calculating Fourier transforms of time-varying signal voltages for such applications as identification of underwater sounds, vibration analysis, oil prospecting, and brain-wave analysis.

Fourier-Bessel integrals [MATH] Given a function $F(v,\theta)$ independent of θ where v,θ are the polar coordinates in the plane, these integrals have the form

$$\int_0^\infty u \, du \int_0^\infty F(r) \, J_m(ur) r \, dr$$

where J_m is a Bessel function order m.

Fourier-Bessel series [MATH] For a function $f(x)$, the series whose mth term is $a_m J_0(j_m x)$, where $j_1, j_2, \ldots$ are positive zeros of the Bessel function J_0 arranged in ascending order, and a_m is the product of $2/J_1^2 (j_m)$ and the integral over t from 0 to 1 of $tf(t)J_0(j_m t)$; J_1 is a Bessel function.

Fourier-Bessel transform *See* Hankel transform.

Fourier cosine transform [MATH] For a function $f(t)$, the function $F(x)$ equal to $\sqrt{2/\pi}$ times the integral from 0 to ∞ of $f(t) \cos(tx) \, dt$.

Fourier expansion *See* Fourier series.

Fourier heat equation *See* Fourier law of heat conduction; heat equation.

Fourier integrals [MATH] For a function $f(x)$ the Fourier integrals are

$$\frac{1}{\pi}\int_0^\infty du \int_{-\infty}^\infty f(t) \cos u(x-t) \, dt$$

$$\frac{1}{\pi}\int_0^\infty du \int_{-\infty}^\infty f(t) \sin u(x-t) \, dt.$$

Fourier kernel [MATH] Any kernel $K(x,y)$ of an integral transform which may be written in the form $K(x,y) = k(xy)$ and which is identical with the kernel of the inverse transform.

Fourier law of heat conduction [THERMO] The law that the rate of heat flow through a substance is proportional to the area normal to the direction of flow and to the negative of the rate of change of temperature with distance along the direction of flow. Also known as Fourier heat equation.

Fourier-Legendre series [MATH] Given a function $f(x)$, the series from $n = 0$ to infinity of $a_n P_n(x)$, where $P_n(x)$, $n = 0, 1, 2, \ldots$ are the Legendre polynomials, and a_n is the product of $(2n + 1)/2$ and the integral over x from -1 to 1 of $f(x) P_n(x)$.

Fourier number [FL MECH] A dimensionless number used in unsteady-state flow problems, equal to the product of the dynamic viscosity and a characteristic time divided by the product of the fluid density and the square of a characteristic length. Symbolized Fo_f. [PHYS] A dimensionless number used in the study of unsteady-state mass transfer, equal to the product of the diffusion coefficient and a characteristic time divided by the square of a characteristic length. Symbolized N_{Fo_m}. [THERMO] A dimensionless number used in the study of unsteady-state heat transfer, equal to the product of the thermal conductivity and a characteristic time, divided by the product of the density, the specific heat at constant pressure, and the distance from the midpoint of the body through which heat is passing to the surface. Symbolized N_{Fo_h}.

Fourier series [MATH] The Fourier series of a function $f(x)$ is

$$\frac{1}{2}a_0 + \sum_{n=1}^{\infty} (a_n \cos nx + b_n \sin nx)$$

with

$$a_n = \frac{1}{\pi} \int_{-\pi}^{\pi} f(x) \cos nx \, dx$$

$$b_n = \frac{1}{\pi} \int_{-\pi}^{\pi} f(x) \sin nx \, dx.$$

Also known as Fourier expansion.

Fourier sine transform [MATH] For a function $f(t)$, the function $F(x)$ equal to $\sqrt{2/\pi}$ times the integral from 0 to ∞ of $f(t) \sin(tx) \, dt$.

Fourier space [MATH] The space in which the Fourier transform of a function is defined.

Fourier spectrum [PHYS] A plot of the magnitude and phase of the Fourier transform of a function.

Fourier's theorem [MATH] If $f(x)$ satisfies the Dirichlet conditions on the interval $-\pi < x < \pi$, then its Fourier series converges to $f(x)$ for all values of x in this interval at which $f(x)$ is continuous, and approaches $\frac{1}{2}[f(x + 0) + f(x - 0)]$ at points at which $f(x)$ is discontinuous, where $f(x - 0)$ is the limit on the left of f at x and $f(x + 0)$ is the limit on the right of f at x.

Fourier-Stieltjes series [MATH] For a function $f(x)$ of bounded variation on the interval $[0, 2\pi]$, the series from $n = 0$ to infinity of $c_n \exp(inx)$, where c_n is $1/2\pi$ times the integral from $x = 0$ to $x = 2\pi$ of $\exp(-inx) \, df(x)$.

Fourier-Stieltjes transform [MATH] For a function $f(y)$ of bounded variation on the interval $(-\infty, \infty)$, the function $F(x)$ equal to $1/\sqrt{2\pi}$ times the integral from $y = -\infty$ to $y = \infty$ of $\exp(-ixy) \, df(y)$.

Fourier synthesis [MATH] The determination of a periodic function from its Fourier components.

Fourier transform [MATH] For a function $f(t)$, the function $F(x)$ equal to $1/\sqrt{2\pi}$ times the integral over t from $-\infty$ to ∞ of $f(t) \exp(itx)$.

Fourier transform spectroscopy [SPECT] A method of finding an electromagnetic spectrum by passing the entire frequency range of interest through an interferometer, recording the output signal as a function of path difference in the interferometer, and then performing a Fourier transform on this function in a digital computer. Abbreviated FTS.

four laws of black hole mechanics [RELAT] Four laws of general relativity theory describing black holes, which are closely analogous to the four laws of classical thermodynamics.

four-layer device [ELECTR] A *pnpn* semiconductor device, such as a silicon controlled rectifier, that has four layers of alternating *p*- and *n*-type material to give three *pn* junctions.

four-layer diode [ELECTR] A semiconductor diode having three junctions, terminal connections being made to the two outer layers that form the junctions; a Shockley diode is an example.

four-layer transistor [ELECTR] A junction transistor having four conductivity regions but only three terminals; a thyristor is an example.

four-level laser [PHYS] A laser in which the lowest level for a laser transition is an excited state rather than the ground level.

four-pi counter [ENG] An instrument which measures the radiation that a radioactive material emits in all directions.

four-point [MATH] A set of four points in a plane, no three of which are collinear. Also known as complete four-point.

fourth dimension [RELAT] Time in the theory of relativity, in which space and time are conceived as particular aspects of a four-dimensional world.

fourth-power law *See* Stefan-Boltzmann law.

four-vector [RELAT] A set of four quantities which transform under a Lorentz transformation in the same way as the three space coordinates and the time coordinate of an event. Also known as Lorentz four-vector.

Fowler-DuBridge theory [SOLID STATE] Theory of photoelectric emission from a metal based on the Sommerfeld model, which takes into account the thermal agitation of electrons in the metal and predicts the photoelectric yield and the energy spectrum of photoelectrons as functions of temperature and the frequency of incident radiation.

Fowler function [SOLID STATE] A mathematical function used in the Fowler-DuBridge theory to calculate the photoelectric yield.

fp *See* freezing point.

F process [MATH] A stochastic process $\{X_t, t > 0\}$ whose value at time t is determined by the information up to time t; more precisely, the events $\{X_t \leq a\}$ belong to F_t for every t and a, where $F = \{F_t, t \geq 0\}$ is an increasing family of sigma algebras that represents the amount of information increasing with time.

fps system of units *See* British absolute system of units.

Fr *See* francium; statcoulomb.

fraction [MATH] An expression which is the product of a real number or complex number with the multiplicative inverse of a real or complex number. [SCI TECH] A portion of a mixture which represents a discrete unit and can be isolated from the whole system.

fractional equation [MATH] **1.** Any equation that contains fractions. **2.** An equation in which the unknown variable appears in the denominator of one or more terms.

fractional factorial experiment [STAT] An experiment in which certain properly chosen levels of factors are left out. Also known as fractional replicate.

fractional ideal [MATH] A submodule of the quotient field of an integral domain.

fractional integration [MATH] A generalization of the procedure of repeatedly integrating a function f over some interval, which results in an integral over some finite or infinite interval of the form $[\Gamma(\alpha)]^{-1}|x - t|^{\alpha-1}f(t)\,dt$, where the real part of α is strictly positive and x is one of the limits of integration.

fractional replicate *See* fractional factorial experiment.

fractional sine wave [PHYS] A pulse train whose waveform is a truncated sine wave.

fractionation [NUCLEO] Alterations in the isotopic composition of substances found in nature or in radioactive weapon debris, which result from small differences in the physical and chemical properties of isotopes of an element.

fracture strength *See* fracture stress.

fracture stress [MECH] The minimum tensile stress that will cause fracture. Also known as fracture strength.

fracture wear [MECH] The wear on individual abrasive grains on the surface of a grinding wheel caused by fracture.

frame of reference [PHYS] A coordinate system for the purpose of assigning positions and times to events. Also known as reference frame.

framework structure [SOLID STATE] A crystalline structure in which there are strong interatomic bonds which are not confined to a single plane, in contrast to a layer structure.

framing [ELECTR] 1. Adjusting a television picture to a desired position on the screen of the picture tube. 2. Adjusting a facsimile picture to a desired position in the direction of line progression. Also known as phasing.

Francis formula [FL MECH] An equation for the calculation of water flow rate over a rectangular weir in terms of length and head.

FRANCIUM

Periodic table of the chemical elements showing the position of francium.

francium [CHEM] A radioactive alkali-metal element, symbol Fr, atomic number 87, atomic weight distinguished by nuclear instability; exists in short-lived radioactive forms, the chief isotope being francium-223.

Franck-Condon principle [PHYS CHEM] The principle that in any molecular system the transition from one energy state to another is so rapid that the nuclei of the atoms involved can be considered to be stationary during the transition.

Franck-Hertz experiment [ELECTR] Experiment for measuring the kinetic energy lost by electrons in inelastic collisions with atoms; it established the existence of discrete energy levels in atoms, and can be used to determine excitation and ionization potentials.

Franck-Rabinowitch hypothesis [PHYS CHEM] The hypothesis that the decreased quantum efficiencies of certain photochemical reactions observed in the dissolved or liquid state are due to the formation of a cage of solvent molecules around the molecule which has been excited by absorption of a photon.

franklin *See* statcoulomb.

franklin centimeter [ELEC] A unit of electric dipole moment, equal to the dipole moment of a charge distribution consisting of positive and negative charges of 1 statcoulomb separated by a distance of 1 centimeter.

Franklin equation [ENG ACOUS] An equation for intensity of sound in a room as a function of time after shutting off the source, involving the volume and exposed surface area of the room, the speed of sound, and the mean sound-absorption coefficient.

Frank partial dislocation [CRYSTAL] A partial dislocation whose Burger's vector is not parallel to the fault plane, so that it can only diffuse and not glide, in contrast to a Schockley partial dislocation.

Franz-Keldysh effect [OPTICS] A shift to longer wavelength in the spectrum transmitted by a semiconductor when a strong electric field is applied.

fraunhofer [SPECT] A unit for measurement of the reduced width of a spectrum line such that a spectrum line's reduced width in fraunhofers equals 10^6 times its equivalent width divided by its wavelength.

Fraunhofer corona *See* F corona.

Fraunhofer diffraction [OPTICS] Diffraction of a beam of parallel light observed at an effectively infinite distance from the diffracting object, usually with the aid of lenses which collimate the light before diffraction and focus it at the point of observation.

Fraunhofer lines [SPECT] The dark lines constituting the Fraunhofer spectrum.

Fraunhofer region [ELECTROMAG] The region far from an antenna compared to the dimensions of the antenna and the wavelength of the radiation. Also known as far field; far region; far zone; radiation zone.

Fraunhofer spectrum [SPECT] The absorption lines in sunlight, due to the cooler outer layers of the sun's atmosphere.

Fréchet derivative [MATH] The Fréchet derivative of a functional or transformation F at a point u of a vector space X is a linear functional or transformation L_u such that $F(u + h) - F(u) = L_u(h) + R_u(h)$ for all vectors h in X, where $\|R_u(h)\|/\|h\|$ approaches 0 as $\|h\|$ approaches 0 (provided such a linear functional or transformation exists).

Fréchet differential [MATH] The Fréchet differential of a functional F at a point u of a vector space X is the linear form $L_u(h)$, where L_u is the Fréchet derivative of F at u and h is an arbitrary element of X.

Fréchet space [MATH] **1.** A quasi-normed linear space in which every Cauchy sequence converges to a point in the space. **2.** A quasi-normed linear space which is locally convex under the topology generated by the norm, and in which every Cauchy sequence converges to a point in the space. **3.** A complete metrizable locally convex topological space.

Fredholm determinant [MATH] A power series obtained from the function $K(x,y)$ of the Fredholm equation which provides solutions to the equation under certain conditions.

Fredholm integral equations [MATH] Given functions $f(x)$ and $K(x,y)$, the Fredholm integral equations with unknown function y are

$$\text{type 1: } f(x) = \int_a^b K(x,t)y(t)\, dt$$

and $\quad$ type 2: $y(x) = f(x) + \lambda \int_a^b K(x,t)y(t)\, dt.$

Fredholm operator [MATH] A linear operator between Banach spaces which has closed range, and both the Fredholm operator and its adjoint have finite dimensional null space.

Fredholm theorem [MATH] A Fredholm equation of type 2 with continuous $f(x)$ has a unique continuous solution, or else the corresponding equation of type 1 has a positive number of linearly independent solutions.

Fredholm theory [MATH] The study of the solutions of the Fredholm equations.

free Abelian group [MATH] An Abelian group whose elements are arbitrary sums of elements in a given set S; more precisely, the set of all mappings f from S into the integers such that $f(x) = 0$ for all but a finite number of elements x of S, with addition defined by the formula $(f + g)(x) = f(x) + g(x)$; the mapping whose value on x_i is

FRAUNHOFER DIFFRACTION

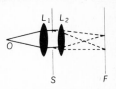

Diagram of Fraunhofer diffraction, with circular aperture. Light source O lies at the principal focus of lens L_1, which renders light parallel as it falls on aperture S. The second lens L_2 focuses parallel diffracted beams on observing screen F.

FRAUNHOFER SPECTRUM

Two sections of the Fraunhofer spectrum, showing bright continuum and dark absorption lines. The wavelength range covered by each strip is approximately 85 angstroms. *(Sacramento Peak Observatory, operated by the Association of Universities for Research in Astronomy, Inc.)*

$f(x_i) = f_i$, $i = 1, \ldots, n$, and 0 elsewhere is written $f_1 x_1 + \cdots + f_n x_n$.

free-air ionization chamber [NUCLEO] An ionization chamber in which the ionizing radiation is limited by a diaphragm, so that ionization is detected in a region of accurately known volume, away from the electrodes and other internal parts of the equipment.

free algebra [MATH] The free algebra of a monoid G over a commutative ring R is the algebra consisting of all sums of the form

$$\sum_{x \in G} a_x x$$

where the sum ranges over elements x in G, and a_x are arbitrary elements of R, with $a_x = 0$ for all but a finite number of elements, and addition and multiplication are defined by the formulas

$$\sum_x a_x x + \sum_x b_x x = \sum_x (a_x + b_x) x$$

$$\left(\sum_x a_x x \right) \left(\sum_y b_y y \right) = \sum_{x,y} \left(a_x b_y \right) (xy).$$

free atom [ATOM PHYS] An atom, as in a gas, whose properties, such as spectrum and magnetic moment, are not significantly affected by other atoms, ions, or molecules nearby.

free charge [ELEC] Electric charge which is not bound to a definite site in a solid, in contrast to the polarization charge.

free convection *See* natural convection.

free convection number *See* Grashof number.

free electromagnetic field [ELECTROMAG] An electromagnetic field in empty space that does not interact with matter.

free electron [PHYS] An electron that is not constrained to remain in a particular atom, and is therefore able to move in matter or in a vacuum when acted on by external electric or magnetic fields.

free-electron laser [OPTICS] A high-power tunable laser in which a beam of high-energy electrons passes through a spatially periodic magnetic field, causing the kinetic energy of the electrons to be converted directly into light.

free-electron paramagnetism [ELECTROMAG] Paramagnetism of certain metals that results from the magnetic moments of nearly free electrons in their conduction bands. Also known as Pauli paramagnetism.

free-electron theory of metals [SOLID STATE] A model of a metal in which the free electrons, that is, those giving rise to the conductivity, are regarded as moving in a potential (due to the metal ions in the lattice and to all the remaining free electrons) which is approximated as constant everywhere inside the metal. Also known as Sommerfeld model; Sommerfeld theory.

free energy [THERMO] **1.** The internal energy of a system minus the product of its temperature and its entropy. Also known as Helmholtz free energy; Helmholtz function; Helmholtz potential; thermodynamic potential at constant volume; work function. **2.** *See* Gibbs free energy.

free enthalpy *See* Gibbs free energy.

free fall [MECH] The ideal falling motion of a body acted upon only by the pull of the earth's gravitational field.

free field [ACOUS] An isotropic, homogeneous sound field that is free from all bounding surfaces. [PHYS] A field in empty space not interacting with other fields or sources.

free-field room *See* anechoic chamber.

free flight [MECH] Unconstrained or unassisted flight.

free-flight angle [MECH] The angle between the horizontal and a line in the direction of motion of a flying body, especially a rocket, at the beginning of free flight.

free-flight trajectory [MECH] The path of a body in free fall.

free gas [PHYS] Any gas at any pressure not in solution, or mechanically held in the liquid hydrocarbon phase.

free grid *See* floating grid.

free group [MATH] A group whose generators satisfy the equation $x \circ y = e$ (e the identity element in the group) only when $x = y^{-1}$ or $y = x^{-1}$.

free gyroscope [ENG] A gyroscope that uses the property of gyroscopic rigidity to sense changes in altitude of a machine, such as an airplane; the spinning wheel or rotor is isolated from the airplane by gimbals; when the plane changes from level flight, the gyro remains vertical and gives the pilot an artificial horizon reference.

free hole [SOLID STATE] Any hole which is not bound to an impurity or to an exciton.

free impedance [ELECTR] Impedance at the input of the transducer when the impedance of its load is made 0. Also known as normal impedance.

free ion [PHYS CHEM] An ion, such as found in an ionized gas, whose properties, such as spectrum and magnetic moment, are not significantly affected by other atoms, ions, or molecules nearby.

free module [MATH] A module which is a free group with respect to its additive group.

free molecule [PHYS CHEM] A molecule, as in a gas, whose properties, such as spectrum and magnetic moment, are not affected by other atoms, ions, and molecules nearby.

free molecule flow [PHYS] Flow of a gas in which the mean free path of the molecules is long compared to a characteristic dimension of the flow field, such as the diameter of a tube through which gas is flowing. Also known as Knudsen flow.

free motional impedance [ELECTR] Of a transducer, the complex remainder after the blocked impedance has been subtracted from the free impedance.

free oscillation [PHYS] The oscillation of a physical system with no externally applied stimuli. Also known as free vibration.

free-pendulum clock *See* Shortt clock.

free-piston gage [ENG] An instrument for measuring high fluid pressures in which the pressure is applied to the face of a small piston that can move in a cylinder and the force needed to keep the piston stationary is determined. Also known as piston gage.

free product [MATH] A group G is a free product of nonempty set S of subgroups of G if each nonunit element of G can be written in exactly one way in the form $g = x_1 \ldots x_k$, where x_i is a nonunit element of subgroup H_i in S and $H_i \neq H_{i+1}, i = 1, \ldots, k-1$.

free progressive wave [PHYS] A wave in a medium or in vacuum, free from boundary effects. Also known as free wave.

free radical [CHEM] An atom or a diatomic or polyatomic molecule which possesses at least one unpaired electron.

free-running multivibrator *See* astable multivibrator.

free space [PHYS] A region of space in which there are no particles of matter and no electromagnetic or gravitational fields other than those whose behavior is under consideration.

free-space field intensity [ELECTROMAG] Radio field intensity that would exist at a point in a uniform medium in the absence of waves reflected from the earth or other objects.

free-space loss [ELECTROMAG] The theoretical radiation loss, depending only on frequency and distance, that would occur if all variable factors were disregarded when transmitting energy between two antennas.

free-space propagation [ELECTROMAG] Propagation of electromagnetic radiation over a straight-line path in a vacuum or ideal atmosphere, sufficiently removed from all objects that affect the wave in any way.

free-space radiation pattern [ELECTROMAG] Radiation pattern that an antenna would have if it were in free space where there is nothing to reflect, refract, or absorb the radiated waves.

free-space wave [ELECTROMAG] An electromagnetic wave propagating in a vacuum, free from boundary effects.

free streamline [FL MECH] A streamline separating fluid in motion from fluid at rest.

free surface [FL MECH] A boundary between two homogeneous fluids.

free-traveling wave *See* progressive wave.

free variable [MATH] In logic, a variable that has an occurrence which is not within the scope of a quantifier and thus can be replaced by a constant.

free vector [MECH] A vector whose direction in space is prescribed but whose point of application and line of application are not prescribed.

free vibration *See* free oscillation.

free volume [STAT MECH] In a lattice theory of a dense gas or liquid, the volume of the cage in which a given molecule is free to wander when its nearest neighbors are fixed at their lattice positions.

free vortex [FL MECH] Two-dimensional fluid flow in which the fluid moves in concentric circles at speeds inversely proportional to the radii of the circles.

free wave *See* free progressive wave.

freeze [PHYS CHEM] To solidify a liquid by removal of heat.

freezing mixture [PHYS CHEM] A mixture of substances whose freezing point is lower than that of its constituents.

freezing point [PHYS CHEM] The temperature at which a liquid and a solid may be in equilibrium. Abbreviated fp.

freezing-point depression [PHYS CHEM] The lowering of the freezing point of a solution compared to the pure solvent; the depression is proportional to the active mass of the solute in a given amount of solvent.

Frégier's theorem [MATH] A theorem of projective geometry which states that the chords of a conic which subtend a right angle at a fixed point P on the conic pass through a fixed point on the normal to the conic at P.

F region [GEOPHYS] The general region of the ionosphere in which the F_1 and F_2 layers tend to form.

french [MECH] A unit of length used to measure small diameters, especially those of fiber optic bundles, equal to $\frac{1}{3}$ millimeter.

Frenet-Serret formulas [MATH] Formulas in the theory of space curves, which give the directional derivatives of the unit vectors along the tangent, principal normal and binormal of a space curve in the direction tangent to the curve. Also known as Serret-Frenet formulas.

Frenkel defect [SOLID STATE] A crystal defect consisting of a vacancy and an interstitial which arise when an atom is plucked out of a normal lattice site and forced into an interstitial position. Also known as Frenkel pair.

Frenkel exciton [SOLID STATE] A tightly bound exciton in which the electron and the hole are usually on the same atom, although the pair can travel anywhere in the crystal.

Frenkel-Halsey-Hill isotherm equation [PHYS] An equation for the volume v of a gas adsorbed on a surface at a given temperature, $\ln (p/p_o) = k/v^s$, where p is the pressure of the gas, p_o is the vapor pressure, and k and s are constants.

Frenkel pair *See* Frenkel defect.

frequency [PHYS] The number of cycles completed by a periodic quantity in a unit time. [STAT] The number of times an event or item falls into or is expected to fall into a certain class or category.

frequency band [PHYS] A continuous range of frequencies extending between two limiting frequencies.

frequency bridge [ELECTR] A bridge in which the balance varies with frequency in a known manner, such as the Wien bridge; used to measure frequency.

frequency characteristic *See* frequency-response curve.

frequency compensation *See* compensation.

frequency curve [STAT] A graphical representation of a continuous frequency distribution; the value of the variable is the abscissa and the frequency is the ordinate.

frequency cutoff [ELECTR] The frequency at which the current gain of a transistor drops 3 decibels below the low-frequency gain value.

frequency distribution [MATH] A function which measures the relative frequency or probability that a variable can take on a set of values.

frequency domain [CONT SYS] Pertaining to a method of analysis, particularly useful for fixed linear systems, in which one does not deal with functions of time explicitly, but with their Laplace or Fourier transforms, which are functions of frequency.

frequency factor [PHYS CHEM] The constant A (or ν) in the Arrhenius equation, which is the relation between reaction rate and absolute temperature T; the equation is $k = Ae - \Delta H_{act}/RT$, where k is the specific rate constant, ΔH_{act} is the heat of activation, and R is the gas constant.

frequency function *See* probability density function.

frequency locus [CONT SYS] The path followed by the frequency transfer function or its inverse, either in the complex plane or on a graph of amplitude against phase angle; used in determining zeros of the describing function.

frequency-modulated cyclotron *See* synchrocyclotron.

frequency-modulated laser [OPTICS] A helium-neon or other laser in which an ultrasonic modulation cell is used to impress a frequency-modulated video signal on the output beam of the laser.

frequency modulation [COMMUN] Modulation in which the instantaneous frequency of the modulated wave differs from the carrier frequency by an amount proportional to the instantaneous value of the modulating wave. Abbreviated FM.

frequency-modulation laser [OPTICS] Conventional laser containing a phase modulator inside its Fabry-Perot cavity; characterized by the lack of noise resulting from the random fluctuation in the phase in the various modes.

frequency polygon [STAT] A graph obtained from a frequency distribution by joining with straight lines points whose abscissae are the midpoints of successive class intervals and whose ordinates are the corresponding class frequencies.

frequency probabilities *See* objective probabilities.

frequency response [ENG] A measure of the effectiveness with which a circuit, device, or system transmits the different frequencies applied to it; it is a phasor whose magnitude is the ratio of the magnitude of the output signal to that of a sine-wave input, and whose phase is that of the output with respect to the input. Also known as amplitude-frequency response; sine-wave response.

frequency-response curve [ENG] A graph showing the mag-

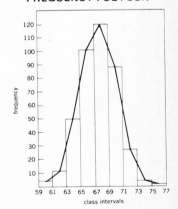

FREQUENCY POLYGON

The conversion of a histogram into a frequency polygon by connecting the midpoint value (at the top of each rectangle) with the adjacent midpoint value by straight lines.

nitude or the phase of the frequency response of a device or system as a function of frequency. Also known as frequency characteristic.

frequency-response trajectory [CONT SYS] The path followed by the frequency-response phasor in the complex plane as the frequency is varied.

frequency-selective device *See* electric filter.

frequency spectrum [PHYS] A plot of the distribution of the intensity of some type of electromagnetic or acoustic radiation as a function of frequency. [SYS ENG] In the analysis of a random function of time, such as the amplitude of noise in a system, the limit as T approaches infinity of $1/2\pi T$ times the ensemble average of the squared magnitude of the amplitude of the Fourier transform of the function from $-T$ to T. Also known as power-density spectrum; power spectrum; spectral density.

frequency table [STAT] A tabular arrangement of the distribution of an event or item according to some specified category or class intervals.

frequency transformation [CONT SYS] A transformation used in synthesizing a band-pass network from a low-pass prototype, in which the frequency variable of the transfer function is replaced by a function of the frequency. Also known as low-pass band-pass transformation.

fresnel [PHYS] A unit of frequency, equal to 10^{12} hertz.

Fresnel-Arago laws [OPTICS] The three laws stating that two rays of polarized light interfere in the same way as ordinary light if they are polarized in the same plane, but do not interfere if they are polarized at right angles; two rays polarized from ordinary light at right angles do not interfere in the ordinary sense when they are brought into the same plane of polarization; and two rays polarized at right angles from plane polarized light, and then brought into the same polarization plane, interfere.

Fresnel biprism [OPTICS] A very flat triangular prism which has two very acute angles and one very obtuse angle; used to observe the interference of light from a slit passing through the two halves of the prism.

Fresnel diffraction [OPTICS] Diffraction in which the source of light or the observing screen are at a finite distance from the aperture or obstacle.

Fresnel drag coefficient [OPTICS] The quantity $1 - (1/n^2)$, where n is the index of diffraction of a transparent medium, believed by Fresnel to be the ratio of the velocity with which ether was dragged along in the medium to the velocity of the medium itself.

Fresnel ellipsoid [OPTICS] An ellipsoid whose three perpendicular axes are proportional to the principal values of the wave velocity of light in an anisotropic medium. Also known as ray ellipsoid.

Fresnel equations [OPTICS] Equations which give the intensity of each of the two polarization components of light which is reflected or transmitted at the boundary between two media with different indices of refraction.

Fresnel fringe [OPTICS] One of a series of light and dark bands that appear near the edge of a shadow in Fresnel diffraction.

Fresnel-Huygens principle *See* Huygens-Fresnel principle.

Fresnel integrals [MATH] Given a parameter x, the integrals over t from 0 to x of $\sin t^2$ and of $\cos t^2$ or from x to ∞ of $(\cos t)/t^{1/2}$ and of $(\sin t)/t^{1/2}$.

Fresnel lens [OPTICS] A thin lens constructed with stepped setbacks so as to have the optical properties of a much thicker lens.

Fresnel mirrors [OPTICS] Two plane mirrors which are in-

FRESNEL BIPRISM

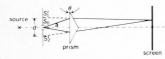

Light from slit S_0 is transmitted through two halves of the Fresnel biprism to the screen. The beam from each half strikes the screen at a different angle and will appear to come from a source (S_1', S_2') slightly displaced from the slit. Separation of S_1' and S_2' equals $2a(\mu - 1)\theta$, where a = distance of slit from biprism, θ = acute angle of prism, μ = index of refraction of prism material.

clined to each other on the order of a degree and used to observe the interference of light which originates from a slit and is reflected from both mirrors.

Fresnel ovaloid [OPTICS] For an anisotropic crystal, an ovaloid whose central section normal to the propagation direction of an electromagnetic wave gives the axes of polarization of the displacement vector and the associated wave velocities.

Fresnel reflection formula [OPTICS] The Fresnel equations for light reflected from a boundary.

Fresnel region [ELECTROMAG] The region between the near field of an antenna (close to the antenna compared to a wavelength) and the Fraunhofer region.

Fresnel rhomb [OPTICS] A glass rhomb which has an acute angle of about 52°; light which is incident normal to the end of the rhomb undergoes two internal reflections, and if it is initially linearly polarized at an angle of 45° to the plane of incidence, it emerges circularly polarized.

Fresnel theory of double refraction [OPTICS] The theory which explains double refraction of a crystal in terms of nonspherical wave surfaces.

Fresnel zones [ELECTROMAG] Circular portions of a wavefront transverse to a line between an emitter and a point where the disturbance is being observed; the nth zone includes all paths whose lengths are between $n-1$ and n half-wavelengths longer than the line-of-sight path. Also known as half-period zones.

Freundlich isotherm equation [PHYS] Equation which states that the volume of gas adsorbed on a surface at a given temperature is proportional to the pressure of the gas raised to a constant power.

Fricke dosimeter [NUCLEO] A radiation dosimeter in which the energy of ionizing radiation is determined from the amount of ferrous ions converted to ferric ions in an aerated acidic ferrous sulfate solution.

friction [MECH] A force which opposes the relative motion of two bodies whenever such motion exists or whenever there exist other forces which tend to produce such motion.

frictional electricity [ELEC] The electric charges produced on two different objects, such as silk and glass or catskin and ebonite, by rubbing them together. Also known as triboelectricity.

frictional grip [MECH] The adhesion between the wheels of a locomotive and the rails of the railroad track.

frictional secondary flow *See* secondary flow.

friction coefficient *See* coefficient of friction.

friction damping [MECH] The conversion of the mechanical vibrational energy of solids into heat energy by causing one dry member to slide on another.

friction factor [FL MECH] Any of several dimensionless numbers used in studying fluid friction in pipes, equal to the Fanning friction factor times some dimensionless constant.

friction flow [FL MECH] Fluid flow in which a significant amount of mechanical energy is dissipated into heat by action of viscosity.

friction head [FL MECH] The head lost by the flow in a stream or conduit due to frictional disturbances set up by the moving fluid and its containing conduit and by intermolecular friction.

frictionless flow *See* inviscid flow.

friction loss [MECH] Mechanical energy lost because of mechanical friction between moving parts of a machine.

friction torque [MECH] The torque which is produced by frictional forces and opposes rotational motion, such as that associated with journal or sleeve bearings in machines.

friction-tube viscometer [ENG] Device to determine liquid

FRESNEL RHOMB

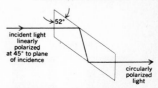

Graphic demonstration of the Fresnel rhomb.

FRICTION DAMPING

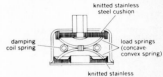

A frictional damper, effective in vertical and horizontal directions.

viscosity by measurement of pressure drop through a friction tube with the liquid in viscous flow; gives direct solution to Poiseuille's equation.

Friedel's law [CRYSTAL] The law that x-ray or electron diffraction measurements cannot determine whether or not a crystal has a center of symmetry.

Friedman solution [RELAT] A solution of Einstein's equations of general relativity with flat spatial sections describing a cosmological model.

friendship theorem [MATH] Among a finite set of people, if every pair of people has exactly one common friend, then there is someone who knows everyone else.

frigorie [THERMO] A unit of rate of extraction of heat used in refrigeration, equal to 1000 fifteen-degree calories per hour, or 1.16264 ± 0.00014 watts.

frigorimeter [ENG] A thermometer which measures low temperatures.

fringe [OPTICS] One of the light or dark bands produced by interference or diffraction of light.

fringe magnetic field [ELECTROMAG] The part of the magnetic field of a horseshoe magnet that extends outside the space between its poles.

fringes of equal thickness *See* Fizeau fringes.

fringe value [OPTICS] A quantity used in photoelastic work, equal to the stress which must be applied to a material, in pounds per square inch (1 pound per square inch equals approximately 6.89476 kilopascals), to produce a relative retardation between the components of a linearly polarized light beam of 1 wavelength when the light passes through a thickness of 1 inch (2.54 centimeters) in a direction perpendicular to the stress.

fringing fields [ELECTR] The electric fields produced by scattered electrons in an electron microscope.

Frobenius endomorphism [MATH] An endomorphism of a field which takes each element x into x^{p^r}, where p is the characteristic of the field and r is some integer.

Frobenius method [MATH] A method of finding a series solution near a point for a linear homogeneous ordinary differential equation.

frontier of a set in a topological space [MATH] All points in the closure of the set but not in its interior. Also known as boundary of a set.

front pinacoid [CRYSTAL] The {100} pinacoid in an orthorhombic, monoclinic, or triclinic crystal. Also known as macropinacoid; orthopinacoid.

front-to-back ratio [ELECTROMAG] Ratio of the effectiveness of a directional antenna, loudspeaker, or microphone toward the front and toward the rear. [SOLID STATE] Ratio of resistance of a crystal to current flowing in the normal direction to current flowing in the opposite direction.

Froude number 1 [FL MECH] A dimensionless number used in studying the motion of a body floating on a fluid with production of surface waves and eddies; equal to the ratio of the square of the relative speed to the product of the acceleration of gravity and a characteristic length of the body. Symbolized N_{Fr_1}.

Froude number 2 [FL MECH] A dimensionless number, equal to the ratio of the speed of flow of a fluid in an open channel to the speed of very small gravity waves, the latter being equal to the square root of the product of the acceleration of gravity and a characteristic length. Symbolized N_{Fr_2}.

frozen flux [PL PHYS] The lines of force of a frozen-in field.

frozen-in field [PL PHYS] A magnetic field in a plasma which has negligible electrical resistance; it can be shown that the

FRINGE

A Fraunhofer diffraction pattern, for a slit, or fringe, photographed with visible light. (*F. S. Harris*)

lines of force of this field are constrained to move with the material.

Frullani's integral [MATH] The integral

$$\int_0^\infty \frac{f(bx) - f(ax)}{x} \, dx = (A - B) \log (b/a)$$

where $f(x)$ is a differentiable function, $f(x)$ approaches A as x approaches ∞, and $f(x)$ approaches B and x approaches 0.

frustrated internal reflectance *See* attenuated total reflectance.

frustum [MATH] The part of a solid between two cutting parallel planes.

F star [ASTRON] A star whose spectral type is F; surface temperature is 7000 K, and color is yellowish.

f stop [OPTICS] An aperture setting for a camera lens; indicated by the f number.

f-sum rule [ATOM PHYS] The rule that the sum of the f values (or oscillator strengths) of absorption transitions of an atom in a given state, minus the sum of the f values of the emission transitions in that state, equals the number of electrons which take part in these transitions. Also known as Thomas-Reiche-Kuhn sum rule.

ft *See* foot.

ftc *See* footcandle.

F test *See* variance ratio test.

ft-L *See* foot lambert.

ft-lb *See* foot-pound.

ft-lbf *See* foot-pound.

ft-pdl *See* foot-poundal.

fu *See* flux unit.

FU *See* finsen unit.

Fubini's theorem [MATH] The theorem stating conditions under which

$$\int \int f(u,v) \, du \, dv = \int du \int f(u,v) \, dv = \int dv \int f(u,v) \, du.$$

Fuchsian differential equation [MATH] A homogeneous, linear differential equation whose coefficients are analytic functions whose only singularities, if any, are poles of order one.

Fuchsian group [MATH] A Kleinian group G for which there is a region D in the complex plane, consisting of either the interior of a circle or the portion of the plane on one side of a straight line, such that D is mapped onto itself by every element of G.

Fuchs's theorem [MATH] A singular point z_0 of the homogeneous differential equation

$$d^n w/dz^n + a_1(z) \, d^{n-1}w/dz^{n-1} + \cdots a_n(z)w = 0$$

is a regular singular point if and only if the functions $(z - z_0)^k a_k(z)$ are analytic in a neighborhood of z_0 for $k = 1, 2, \ldots, n$.

fuel assembly [NUCLEO] A combination of fuel and structural materials, used in some nuclear reactors to facilitate assembly of the core.

fuel cycle *See* reactor fuel cycle.

fuel decanner [NUCLEO] A machine used for removing the stainless steel or other metal cans that enclose the enriched uranium fuel rods of a nuclear reactor; the cans are removed by a chipless machining operation in which the tubing is sheared into a spiral strip.

fuel element [NUCLEO] A rod, tube, plate, or other geometrical form into which nuclear fuel is fabricated for use in a reactor.

fuel pellet [NUCLEO] A small pellet of frozen deuterium and

FUEL ELEMENT

A plate-type fuel element.
(From Robert Laws, Salon of Photography)

tritium that would be used as fuel in a laser-induced fusion power plant.

fuel plate [NUCLEO] A form of nuclear fuel element consisting of a flat or slightly curved sheet of fuel, which is usually a sandwich of uranium fuel protected by metallic cladding.

fuel reprocessing [NUCLEO] The processing of nuclear reactor fuel to recover the unused fissionable material.

fuel rod [NUCLEO] A long, rod-shaped fuel assembly.

fuel seed *See* fuel spike.

fuel spike [NUCLEO] Nuclear fuel that is more highly enriched than the majority of the other fuel in a nuclear reactor. Also known as fuel seed.

fugacity [THERMO] A function used as an analog of the partial pressure in applying thermodynamics to real systems; at a constant temperature it is proportional to the exponential of the ratio of the chemical potential of a constituent of a system divided by the product of the gas constant and the temperature, and it approaches the partial pressure as the total pressure of the gas approaches zero.

Fulcher bands [SPECT] A group of bands in the spectrum of molecular hydrogen that are preferentially excited by a low-voltage discharge.

fulcrum [MECH] The rigid point of support about which a lever pivots.

full linear group [MATH] The group of all nonsingular linear transformations of a vector space whose group operation is composition.

full moon [ASTRON] The moon at opposition, with a phase angle of $0°$, when it appears as a round disk to an observer on the earth because the illuminated side is toward the observer.

full section filter [ELECTR] A filter network whose graphical representation has the shape of the Greek letter pi, connoting capacitance in the upright legs and inductance or reactance in the horizontal member.

full-wave rectification [ELECTR] Rectification in which output current flows in the same direction during both half cycles of the alternating input voltage.

full-wave rectifier [ELECTR] A double-element rectifier that provides full-wave rectification; one element functions during positive half cycles and the other during negative half cycles.

funal *See* sthène.

function [MATH] A mathematical rule between two sets which assigns to each member of the first set exactly one member of the second.

functional [MATH] Any function from a vector space into its scalar field.

functional analysis [MATH] A branch of analysis which studies the properties of mappings of classes of functions from one topological vector space to another.

functional constraint [MATH] A mathematical equation which must be satisfied by the independent parameters in an optimization problem, representing some physical principle which governs the relationship among these parameters.

functional equation [MATH] An equation whose terms are formed from a finite number of unknown functions and finite number of independent variables; for example, the equations $f(x + y) = f(x) + f(y)$ and $f(xz,yz) = z^k f(x,y)$, where f is the unknown function and x, y, and z are the independent variables.

function algebra [MATH] An algebra whose elements are functions and in which the multiplication of two elements f and g is specified by the rule $(fg)(x) = f(x)g(x)$.

functional switching circuit [ELECTR] One of a relatively small number of types of circuits which implements a Boolean function and constitutes a basic building block of a switching system; examples are the AND, OR, NOT, NAND, and NOR circuits.

function space [MATH] A metric space whose elements are functions.

function table [MATH] A table that lists the values of a function for various values of the variable.

functor [MATH] A function between categories which associates objects with objects and morphisms with morphisms.

fundamental [PHYS] The lowest frequency component of a complex wave. Also known as first harmonic; fundamental component.

fundamental affine connection [MATH] An affine connection whose coefficients arise from the covariant and contravariant metric tensors of a space.

fundamental component *See* fundamental.

fundamental forms of a surface [MATH] Differential forms which express the area and curvature of the surface.

fundamental frequency [PHYS] **1.** The lowest frequency at which a system vibrates freely. **2.** The lowest frequency in a complex wave.

fundamental group of a topological space [MATH] The group of homotopy classes of all closed paths about a point; this group yields information about the number and type of "holes" in a surface.

fundamental interaction [PARTIC PHYS] One of the fundamental forces that act between the elementary particles of matter.

fundamental interval [THERMO] **1.** The value arbitrarily assigned to the difference in temperature between two fixed points (such as the ice point and steam point) on a temperature scale, in order to define the scale. **2.** The difference between the values recorded by a thermometer at two fixed points; for example, the difference between the resistances recorded by a resistance thermometer at the ice point and steam point.

fundamental magnitudes of first order [MATH] The components of the metric tensor on a surface, given by $a_{ij} = (\partial \mathbf{r}/\partial u^i) \cdot (\partial \mathbf{r}/\partial u^j)$, where the u^i, $i = 1, 2$, are curvilinear coordinates on the surface, and $\mathbf{r}$ is the vector from the origin to the point u^i on the surface.

fundamental magnitudes of second order [MATH] For a surface with curvilinear coordinates u^i, $i = 1, 2$, these are the quantities $b_{ij} = -\frac{1}{2}[(\partial \mathbf{n}/\partial u^i) \cdot (\partial \mathbf{r}/\partial u^j) + (\partial \mathbf{n}/\partial u^j) \cdot (\partial \mathbf{r}/u^i)]$, where $\mathbf{r}$ is the vector from the origin to the point u^i on the surface, and $\mathbf{n}$ is the unit normal to the surface.

fundamental magnitudes of third order [MATH] For a surface parametrized by curvilinear coordinates, the quantities

$$c_{ij} = \sum_{k,l} a^{kl} b_{ik} b_{jl},$$

where the b_{ik} are fundamental magnitudes of second order and the a^{kl} are components of the contravariant metric tensor, defined by the equation

$$\sum_j a^{ij} a_{jk} = \delta_k^i,$$

where the a_{ik} are fundamental magnitudes of first order and δ_k^i is the Kronecker delta.

fundamental mode [ELECTROMAG] The waveguide mode having the lowest critical frequency. Also known as dominant mode; principal mode. [PHYS] The normal mode of vibration having the lowest frequency.

fundamental particle *See* elementary particle.

fundamental quantity *See* base quantity.

fundamental region [MATH] Any region in the complex

plane that can be mapped conformally onto all of the complex plane.

fundamental sequence *See* Cauchy sequence.

fundamental series [SPECT] A series occurring in the line spectra of many atoms and ions having one, two, or three electrons in the outer shell, in which the total orbital angular momentum quantum number changes from 3 to 2.

fundamental tensor *See* metric tensor.

fundamental theorem of algebra [MATH] Every polynomial of degree n with complex coefficients has exactly n roots counted according to multiplicity.

fundamental theorem of arithmetic [MATH] Every positive integer greater than 1 can be factored uniquely into the form $P_1{}^{n_1} \ldots P_k{}^{n_1} \ldots P_k{}^{n_k}$, where the P_i are primes, the n_i positive integers.

fundamental theorem of calculus [MATH] Given a continuous function $f(x)$ on the closed interval $[a,b]$ the functional

$$F(x) = \int_a^x f(t)\, dt$$

is differentiable on $[a,b]$ and $F'(x) = f(x)$ for every x in $[a,b]$, and if G is any function on $[a,b]$ such that $G'(x) = f(x)$ for all x in $[a,b]$, then

$$\int_a^b f(t)\, dt = G(b) - G(a).$$

fundamental tone [ACOUS] The component tone of lowest pitch in a complex tone.

fundamental unit *See* base unit.

fundamental wavelength [PHYS] Of an oscillatory device, that wavelength corresponding to its fundamental frequency.

funicular polygon [MECH] **1.** The figure formed by a light string hung between two points from which weights are suspended at various points. **2.** A force diagram for such a string, in which the forces (weights and tensions) acting on points of the string from which weights are suspended are represented by a series of adjacent triangles.

furlong [MECH] A unit of length, equal to 1/8 mile, 660 feet, or 201.168 meters.

Furry theorem [QUANT MECH] In quantum electrodynamics, the theorem that the contribution of a Feynman diagram, consisting of a closed polygon of fermion lines connected to an odd number of photon lines, vanishes.

fused junction *See* alloy junction.

fused-junction diode *See* alloy-junction diode.

fused-junction transistor *See* alloy-junction transistor.

fused-salt electrolysis [PHYS CHEM] Electrolysis with use of purified fused salts as raw material and as an electrolyte.

fused-salt reactor *See* molten-salt reactor.

fused semiconductor [ELECTR] Junction formed by recrystallization on a base crystal from a liquid phase of one or more components and the semiconductor.

fusibility [THERMO] The quality or degree of being capable of being liquefied by heat.

fusible resistor [ELEC] A resistor designed to protect a circuit against overload; its resistance limits current flow and thereby protects against surges when power is first applied to a circuit; its fuse characteristic opens the circuit when current drain exceeds design limits.

fusion [NUC PHYS] Combination of two light nuclei to form a heavier nucleus (and perhaps other reaction products) with release of some binding energy. Also known as atomic fusion; nuclear fusion. [PHYS CHEM] A change of the state of

FUSION

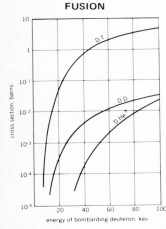

Cross sections versus bombarding energy for three simple fusion reactions. D-T = deuterium-tritium reaction; D-D = deuterium deuterium reaction; D-He³ = deuterium-helium 3 reaction. *(From R. F. Post, Fusion power, Sci. Amer., 197(6):73–84, copyright © 1957 by Scientific American, Inc.; all rights reserved)*

a substance from the solid phase to the liquid phase. Also known as melting.

fusion fuel [NUCLEO] A substance which may generate energy in a fusion reaction, such as deuterium, deuterium and tritium, or deuterium and helium-3.

fusion point [NUCLEO] The temperature of a plasma above which the rate of energy generation by nuclear fusion reactions exceeds the rate of energy loss from the plasma, so that the fusion reaction can be self-sustaining. Abbreviated fnp.

fusion reactor [NUCLEO] Proposed device in which controlled, self-sustaining nuclear fusion reactions would be carried out in order to produce useful power.

future asymptotically predictable [RELAT] A mathematical restriction on the global nature of an asymptotically flat space-time such that Cauchy data set on a spacelike surface (partial Cauchy surface) S will determine the evolution of the space-time to the future of S; naked singularities to the future of S are thereby ruled out.

future Cauchy development [RELAT] The set of points p relative to a surface S in a space-time such that every past-directed inextendible timelike or null curve through p intersects S. Symbolized $D^+(s)$. Also known as domain of dependence.

future horismos [RELAT] The set of points p relative to a surface S that can be causally affected by events in S; that is, the set of points in the future of S which can be reached from S by future-directed timelike curves.

future of an event [RELAT] Those events which can be reached by a signal emitted at the event and which move at a speed less than or equal to the speed of light in a vacuum.

future trapped set [RELAT] A set of points in a space-time such that no two points of the set have timelike separation and the future horismos is compact.

fV *See* femtovolt.

f value *See* oscillator strength.

G

g *See* gram.

G [ELEC] *See* conductance. [MECH] A unit of acceleration equal to the standard acceleration of gravity, 9.80665 meters per second per second, or approximately 32.1740 feet per second per second. Also known as fors; grav. [SCI TECH] *See* giga-.

Ga *See* gallium.

Gabor trolley [ENG] A small three-wheel trolley with knife-edge wheels, used in constructing trajectories of charged particles in an electric field.

gadolinium [CHEM] A rare-earth element, symbol Gd, atomic number 64, atomic weight 157.25; highly magnetic, especially at low temperatures.

gage [ELECTROMAG] One of the family of possible choices for the electric scalar potential and magnetic vector potential, given the electric and magnetic fields. Also spelled gauge.

gage-fixing term [QUANT MECH] A term added to a Lagrangian in quantum field theory that breaks gage invariance.

gage invariance [ELECTROMAG] The invariance of electric and magnetic fields and electrodynamic interactions under gage transformations. [QUANT MECH] An invariance of a Lagrangian based on an internal gage group, such as U(1) for electromagnetism or U(1) × SU(2) for the Weinberg-Salam unified model of weak and electromagnetic interactions.

gage theory [PHYS] Any field theory in which, as the result of the conservation of some quantity, it is possible to perform a transformation in which the phase of the fields is altered by a function of space and time without altering any measurable physical quantity, so that the fields obtained by any such transformation give a valid description of a given physical situation.

gage transformation [ELECTROMAG] The addition of the gradient of some function of space and time to the magnetic vector potential, and the addition of the negative of the partial derivative of the same function with respect to time, divided by the speed of light, to the electric scalar potential; this procedure gives different potentials but leaves the electric and magnetic fields unchanged.

gaging [NUCLEO] The measurement of the thickness, density, or quantity of material by the amount of radiation it absorbs; this is the most common use of radioactive isotopes in industry. Also spelled gauging.

gain [ELECTR] **1.** The increase in signal power that is produced by an amplifier; usually given as the ratio of output to input voltage, current, or power, expressed in decibels. Also known as transmission gain. **2.** *See* antenna gain.

gain asymptotes [CONT SYS] Asymptotes to a logarithmic graph of gain as a function of frequency.

gain-bandwidth product [ELECTR] The midband gain of an amplifier stage multiplied by the bandwidth in megacycles.

GADOLINIUM

64
Gd

Periodic table of the chemical elements showing the position of gadolinium.

gain control [ELECTR] A device for adjusting the gain of a system or component.

gain-crossover frequency [CONT SYS] The frequency at which the magnitude of the loop ratio is unity.

gain margin [CONT SYS] The reciprocal of the magnitude of the loop ratio at the phase crossover frequency, frequently expressed in decibels.

gain reduction [ELECTR] Diminution of the output of an amplifier, usually achieved by reducing the drive from feed lines by use of equalizer pads or reducing amplification by a volume control.

gal [MECH] **1.** The unit of acceleration in the centimeter-gram-second system, equal to 1 centimeter per second squared; commonly used in geodetic measurement. With numerical quantities, symbolized Gal. Also known as galileo. **2.** *See* gallon.

Gal *See* gal.

galactic center [ASTRON] The gravitational center of the Milky Way Galaxy; the sun and other stars of the Galaxy revolve about this center.

galactic circle *See* galactic equator.

galactic cluster *See* open cluster.

galactic coordinates *See* galactic system.

galactic disk [ASTRON] The flat distribution of stars and interstellar matter in the spiral arms and plane of the Milky Way Galaxy.

galactic equator [ASTRON] A great circle of the celestial sphere, inclined 62° to the celestial equator, coinciding approximately with the center line of the Milky Way, and constituting the primary great circle for the galactic system of coordinates; it is everywhere 90° from the galactic poles. Also known as galactic circle.

galactic halo [ASTRON] The spherical distribution of oldest stars that are centered about the galactic center of the Milky Way Galaxy.

galactic latitude [ASTRON] Angular distance north or south of the galactic equator; the arc of a great circle through the galactic poles, between the galactic equator and a point on the celestial sphere, measured northward or southward from the galactic equator through 90° and labeled N or S to indicate the direction of measurement.

galactic longitude [ASTRON] Angular distance east of sidereal hour angle 94°.4 along the galactic equator; the arc of the galactic equator or the angle at the galactic pole between the great circle through the intersection of the galactic equator and the celestial equator in Sagittarius (SHA 94°.4) and a great circle through the galactic poles, measured eastward from the great circle through SHA 94°.4 through 360°.

galactic nebula [ASTRON] A nebula that is in or near the galactic system known as the Milky Way.

galactic noise [ASTRON] Radio-frequency noise that originates outside the solar system; it is similar to thermal noise and is strongest in the direction of the Milky Way.

galactic nova [ASTRON] One of the novae that are concentrated largely in a band 10° on each side of the plane of the galaxy and are most frequent toward the center of the galaxy.

galactic nucleus [ASTRON] The center area in the galaxy about which there is a large spherical distribution of stars and from which the spiral arms emanate.

galactic plane [ASTRON] The plane that may be drawn through the galactic equator; the plane of the Milky Way Galaxy.

galactic pole [ASTRON] On the celestial sphere, either of the two points 90° from the galactic equator.

galactic rotation [ASTRON] The rotation of the Milky Way about an axis through the center and perpendicular to the plane of the Galaxy; the rotation is apparent from the highly flattened shape and from relative stellar motion.

galactic system [ASTRON] An astronomical coordinate system using latitude measured north and south from the galactic equator, and longitude measured in the sense of increasing right ascension from 0 to 360°. Also known as galactic coordinates.

galactic windows [ASTROPHYS] The regions near the equator of the Milky Way where there is low absorption of light by interstellar clouds so that some distant external galaxies may be seen through them.

galaxy [ASTRON] A large-scale aggregate of stars, gas, and dust; the aggregate is a separate system of stars covering a mass range from 10^7 to 10^{12} solar masses and ranging in diameter from 1500 to 300,000 light-years (1.419×10^{19} to 2.838×10^{21} m).

Galaxy *See* Milky Way Galaxy.

Galerkin solution [MECH] A solution of the equations of elastic equilibrium in the absence of body forces, in which the displacement **u** is given by $\mu u = (1 - \sigma)\nabla^2 G - \text{grad div } G$, where μ is the rigidity modulus, σ Poisson's ratio, and **G** a vector whose components are biharmonic functions.

Galerkin vector [MECH] The vector **G** that appears in a Galerkin solution.

Galilean glass *See* Galilean telescope.

Galilean telescope [OPTICS] A refracting telescope whose objective is a converging (convex) lens and whose eyepiece is a diverging (concave) lens; it forms erect images. Also known as Galilean glass.

Galilean transformation [MECH] A mathematical transformation used to relate the space and time variables of two uniformly moving (inertial) reference systems in nonrelativistic kinematics.

galileo *See* gal.

Galileo number [FL MECH] A dimensionless number used in studying the circulation of viscous liquids, equal to the cube of a characteristic dimension, times the acceleration of gravity, times the square of the liquid's density, divided by the square of its viscosity. Symbolized N_{Ga}.

Galileo's law of inertia *See* Newton's first law.

Galitzin pendulum [MECH] A massive horizontal pendulum that is used to measure variations in the direction of the force of gravity with time, and thus serves as the basis of a seismograph.

gallium [CHEM] A chemical element, symbol Ga, atomic number 31, atomic weight 69.72.

gallium arsenide semiconductor [SOLID STATE] A semiconductor having a forbidden-band gap of 1.4 electron volts and a maximum operating temperature of 400°C when used in a transistor.

gallium arsenide laser [OPTICS] A laser that emits light at right angles to a junction region in gallium arsenide, at a wavelength of 9000 angstroms (900 nanometers); can be modulated directly at microwave frequencies; cryogenic cooling is required.

gallium phosphide semiconductor [SOLID STATE] A semiconductor having a forbidden-band gap of 2.4 electron volts and a maximum operating temperature of 870°C when used in a transistor.

gallon [MECH] Abbreviated gal. **1.** A unit of volume used in the United States for measurement of liquid substances, equal to 231 cubic inches, or to $3.785\,411\,784 \times 10^{-3}$ cubic meter, or to 3.785 411 784 liters; equal to 128 fluid ounces. **2.** A unit of volume used in the United Kingdom for measurement of

GALAXY

Great Spiral NGC 224 in Andromeda, which resembles the Milky Way Galaxy. Above it is elliptical galaxy NGC 205. Photograph is by the 48-inch (1.2-meter) Schmidt telescope. *(Hale Observatories)*

GALILEAN TELESCOPE

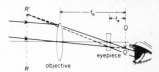

Diagram of Galilean telescope. QQ' is real inverted image which would be formed by objective alone. Eyepiece forms enlarged, erect virtual image, RR', which is viewed by observer; f_o and f_e are focal lengths of objective and eyepiece respectively. For distant object magnification is f_o/f_e.

GALLIUM

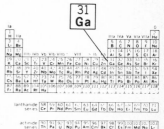

Periodic table of the chemical elements showing the position of gallium.

liquid and solid substances, usually the former; equal to the volume occupied by 10 pounds of weight of water of density 0.998859 gram per milliliter in air of density 0.001217 gram per milliliter against weights of density 8.136 grams per milliliter (the milliliter here has its old definition of 1.000028×10^{-6} cubic meter); approximately equal to 277.420 cubic inches, or to 4.54609×10^{-3} cubic meter, or to 4.54609 liters; equal to 160 fluid ounces.

Galois extension [MATH] Any algebraic extension field which is both normal and separable; it is usually obtained from considering the coefficients and roots of a given polynomial.

Galois field [MATH] A field which consists of a finite number of elements.

Galois group [MATH] The group of automorphisms of a Galois extension E of a field F whose restrictions to F are equal to the identity map on F.

Galois theory [MATH] The study of the Galois field and Galois group corresponding to a polynomial.

Galtonian curve [STAT] A graph showing the variation of any quantity from its normal value.

Galton whistle [ENG ACOUS] A short cylindrical pipe with an annular nozzle, which is set into resonant vibration in order to generate ultrasonic sound waves.

galvanic [ELEC] Pertaining to electricity flowing as a result of chemical action.

galvanoluminescence [PHYS] Light emission which may occur when electrodes of certain metals, such as aluminum or tantalum, are immersed in suitable electrolytes and current is passed between them.

galvanomagnetic effect [ELECTROMAG] One of the electrical or thermal phenomena occurring when a current-carrying conductor or semiconductor is placed in a magnetic field; examples are the Hall effect, Ettingshausen effect, transverse magnetoresistance, and Nernst effect. Also known as magnetogalvanic effect.

galvanometer [ENG] An instrument for indicating or measuring a small electric current by means of a mechanical motion derived from electromagnetic or electrodynamic forces produced by the current.

galvanometer constant [ELEC] Number by which a certain function of the reading of a galvanometer must be multiplied to obtain the current value in ordinary units.

galvanometer shunt [ELEC] Resistor connected in parallel with a galvanometer to increase its range under certain conditions; it allows only a known fraction of the current to pass through the galvanometer.

gambler's ruin [STAT] A game of chance which can be considered to be a series of Bernoulli trials at which each player wins a specified sum of money for every success and loses another sum for every failure; play goes on until the initial capital is lost and the player is ruined.

game [MATH] A mathematical model expressing a contest between two players under specified rules.

game theory [MATH] The mathematical study of games or abstract models of conflict situations from the viewpoint of determining an optimal policy or strategy. Also known as theory of games.

gamma [ELECTROMAG] A unit of magnetic field strength, equal to 10 microoersteds, or 0.00001 oersted. [MECH] A unit of mass equal to 10^{-6} gram or 10^{-9} kilogram.

gamma-absorption gage *See* gamma gage.

gamma counter [ENG] A device for detecting gamma radiation, primarily through the detection of fast electrons produced by the gamma rays; it either yields information about integrated intensity within a time interval or detects each photon separately.

gamma cross section [NUC PHYS] The cross section for absorption or scattering of gamma rays by a nucleus or atom.

gamma decay *See* gamma emission.

gamma distribution [STAT] A normal distribution whose frequency function involves a gamma function.

gamma emission [NUC PHYS] A quantum transition between two energy levels of a nucleus in which a gamma ray is emitted. Also known as gamma decay.

gamma flux density [NUC PHYS] The number of gamma rays passing through a unit area in a unit time.

gamma function [MATH] The complex function given by the integral with respect to t from 0 to ∞ of $e^{-t}t^{z-1}$; this function helps determine the general solution of Gauss' hypergeometric equation.

gamma gage [NUCLEO] A penetration-type thickness gage that measures the thickness or density of a sample by measuring its absorption of gamma rays. Also known as gamma-absorption gage.

gamma heating [NUCLEO] Heating resulting from absorption of gamma-ray energy by a material.

gamma irradiation [NUCLEO] Exposure of a material to gamma rays.

gamma matrix *See* Dirac matrix.

gamma radiography [NUCLEO] Radiography by means of gamma rays.

gamma ray [NUC PHYS] A high-energy photon, especially as emitted by a nucleus in a transition between two energy levels.

gamma-ray astronomy [ASTRON] The study of gamma rays from extraterrestrial sources, especially gamma-ray bursts.

gamma-ray bursts [ASTRON] Intense blasts of soft gamma rays of unknown origin, which range in duration from a tenth of a second to tens of seconds and occur several times a year from sources widely distributed over the sky.

gamma-ray laser [PHYS] A hypothetical device which would generate coherent radiation in the range 0.005–0.5 nanometer by inducing isomeric radiative transitions between isomeric nuclear states. Also known as graser.

gamma-ray scattering *See* Compton scattering.

gamma-ray spectrometry [NUCLEO] **1.** Determination of the energy distribution of gamma rays emitted by nuclei. Also known as gamma-ray spectroscopy. **2.** In particular, a variation of neutron activation analysis in which the induced radiation from the sample is gamma rays instead of neutrons.

gamma-ray spectroscopy *See* gamma-ray spectrometry.

gamma-ray spectrum [SPECT] The set of wavelengths or energies of gamma rays emitted by a given source.

gamma-ray transformation [NUC PHYS] A radioactive decay in which gamma rays are emitted.

gamma scanning [NUCLEO] The scanning of a fuel rod in a nuclear reactor for gamma activity by moving the rod past a slit in a lead block; photons emerging from the slit are detected by a scintillation spectrometer and recorded as a function of rod position.

gamma structure [SOLID STATE] A Hume-Rothery designation for structurally analogous phases or intermetallic phases having 21 valence electrons to 13 atoms, analogous to the γ-brass structure.

gamma transition *See* glass transition.

gammil [CHEM] A unit of concentration, equal to a concentration of 1 milligram of solute in 1 liter of solvent. Also known as micril; microgammil.

Gamow barrier [NUC PHYS] The potential barrier which

retards the escape of alpha particles from the nucleus according to the Gamow-Condon-Gurney theory.

Gamow-Condon-Gurney theory [NUC PHYS] An early quantum-mechanical theory of alpha-particle decay according to which the alpha particle penetrates a potential barrier near the surface of the nucleus by a tunneling process.

Gamow-Teller interaction [NUC PHYS] Interaction between a nucleon source current and a lepton field which has an axial vector or tensor form.

Gamow-Teller selection rules [NUC PHYS] Selection rules for beta decay caused by the Gamow-Teller interaction; that is, in an allowed transition there is no parity change of the nuclear state, and the spin of the nucleus can either remain unchanged or change by ± 1; transitions from spin 0 to spin 0 are excluded, however.

gap [ELEC] The spacing between two electric contacts. [ELECTROMAG] A break in a closed magnetic circuit, containing only air or filled with a nonmagnetic material.

gap factor [ELECTR] Ratio of the maximum energy gained in volts to the maximum gap voltage in a tube employing electron accelerating gaps, that is, a traveling-wave tube.

gas [PHYS] A phase of matter in which the substance expands readily to fill any containing vessel; characterized by relatively low density.

gas adsorption [PHYS CHEM] The concentration of a gas upon the surface of a solid substance by attractive forces between the surface and the gas molecules.

gas amplification [NUCLEO] The ratio of the charge collected to the charge liberated by the initial ionizing event in a radiation-counter tube.

gas capacitor [ELEC] A capacitor consisting of two or more electrodes separated by a gas, other than air, that serves as a dielectric.

gas-cell frequency standard [ATOM PHYS] An atomic frequency standard in which the frequency-determining element is a gas cell containing rubidium, cesium, or sodium vapor.

gas centrifuge process [NUCLEO] A method of isotope separation in which a mixture of isotopes in the gaseous state is spun at high speeds in a centrifuge, and centrifugal forces cause a concentration of heavy isotopes near the walls and light isotopes near the center.

gas constant [THERMO] The constant of proportionality appearing in the equation of state of an ideal gas, equal to the pressure of the gas times its molar volume divided by its temperature. Also known as gas-law constant.

gas-cooled reactor [NUCLEO] A nuclear reactor in which a gas, such as air, carbon dioxide, or helium, is used as a coolant.

gas counter [NUCLEO] A counter in which the radioactive sample is prepared in the form of a gas and introduced into the counter tube.

gas current [ELECTR] A positive-ion current produced by collisions between electrons and residual gas molecules in an electron tube. Also known as ionization current.

gas cycle [THERMO] A sequence in which a gaseous fluid undergoes a series of thermodynamic phases, ultimately returning to its original state.

gas-deviation factor *See* compressibility factor.

gas discharge [ELECTR] Conduction of electricity in a gas, due to movements of ions produced by collisions between electrons and gas molecules.

gas-discharge lamp *See* discharge lamp.

gas-discharge laser [OPTICS] A gas laser in which optical pumping is caused by nonequilibrium processes in a gas discharge.

gas dynamic laser [OPTICS] A gas laser that converts thermal energy directly into coherent radiation at an efficiency high enough to offer promise of wireless power transmission.

gas dynamics [PHYS] The study of the motion of gases, and of its causes, which takes into account thermal effects generated by the motion.

gaseous diffusion plant [NUCLEO] A facility where the gaseous diffusion process is used in the separation of uranium isotopes in order to provide fissionable fuel for nuclear power plants and weapons.

gaseous diffusion process [NUCLEO] A method of separating isotopes in which an isotopic mixture of gases is allowed to diffuse through a porous wall; the lighter molecules pass through the porous wall more readily than the heavier molecules.

gaseous nebulae [ASTRON] Clouds of gas, such as the Network Nebula in Cygnus, that are members of the Milky Way galactic system and are small compared with its overall dimensions.

gas-filled diode [ELECTR] A gas tube which is a diode, such as a cold-cathode rectifier or phanotron.

gas-filled radiation counter [NUCLEO] A gas tube used to detect radiation by means of gas ionization.

gas-filled rectifier *See* cold-cathode rectifier.

gas-filled triode [ELECTR] A gas tube which has a grid or other control element, such as a thyratron or ignitron.

gas-flow counter tube [NUCLEO] A radiation-counter tube in which an appropriate atmosphere is maintained by a flow of gas through the tube. Also known as flow counter; gas-flow radiation counter.

gas-flow radiation counter *See* gas-flow counter tube.

gas focusing [ELECTR] A method of concentrating an electron beam by utilizing the residual gas in a tube; beam electrons ionize the gas molecules, forming a core of positive ions along the path of the beam which attracts beam electrons and thereby makes the beam more compact. Also known as ionic focusing.

gas ionization [ELECTR] Removal of the planetary electrons from the atoms of gas filling an electron tube, so that the resulting ions participate in current flow through the tube.

gas kinematics [FL MECH] The motion of a gas considered by itself, without regard for the causes of motion.

Gaskin's theorem [MATH] A theorem in projective geometry which states that if a circle circumscribes a triangle which is identical with its conjugate triangle with respect to a given conic, then the tangent to the circle at either of its intersections with the director circle of the conic is perpendicular to the tangent to the director circle at the same intersection.

gas laser [OPTICS] A laser in which the active medium is a discharge in a gas contained in a glass or quartz tube with a Brewster-angle window at each end; the gas can be excited by a high-frequency oscillator or direct-current flow between electrodes inside the tube; the function of the discharge is to pump the medium, to obtain population inversion.

gas law [THERMO] Any law relating the pressure, volume, and temperature of a gas.

gas-law constant *See* gas constant.

gas lens [OPTICS] An optical lens formed by a flow of gas which gives rise to gradients of refractive index that bring about the focusing of light.

gas magnification [ELECTR] Increase in current through a phototube due to ionization of the gas in the tube.

gas manometer [ENG] A gage for determining the difference in pressure of two gases, usually by measuring the difference in height of liquid columns in the two sides of a U-tube.

gas maser [PHYS] A maser in which the microwave electromagnetic radiation interacts with the molecules of a gas such as ammonia; used chiefly in highly stable oscillator applications, as in atomic clocks.

gas mechanics [FL MECH] The action of forces on gases.

gas phototube [ELECTR] A phototube into which a quantity of gas has been introduced after evacuation, usually to increase its sensitivity.

gas scattering [ELECTR] The scattering of electrons or other particles in a beam by residual gas in the vacuum system.

gassiness [ELECTR] Presence of unwanted gas in a vacuum tube, usually in relatively small amounts, caused by the leakage from outside or evolution from the inside walls or elements of the tube.

gassing [ELEC] The evolution of gas in the form of small bubbles in a storage battery when charging continues after the battery has been completely charged.

gas slippage [FL MECH] Phenomenon of gas bypassing liquids that occurs when the diameter of capillary openings approaches the mean free path of the gas; occurs not only in capillary tubing, but in porous oil-reservoir formations.

gas solubility [PHYS CHEM] The extent that a gas dissolves in a liquid to produce a homogeneous system.

gassy tube [ELECTR] A vacuum tube that has not been fully evacuated or has lost part of its vacuum due to release of gas by the electrode structure during use, so that enough gas is present to impair operating characteristics appreciably. Also known as soft tube.

gas thermometer [ENG] A device to measure temperature by measuring the pressure exerted by a definite amount of gas enclosed in a constant volume; the gas (preferably hydrogen or helium) is enclosed in a glass or fused-quartz bulb connected to a mercury manometer. Also known as constant-volume gas thermometer.

gas tube [ELECTR] An electron tube into which a small amount of gas or vapor is admitted after the tube has been evacuated; ionization of gas molecules during operation greatly increases current flow.

gas vacuum breakdown [ELECTR] Ionization of residual gas in a vacuum, causing reverse conduction in an electron tube.

gas viscosity [FL MECH] The internal fluid function of a gas.

gate [ELECTR] **1.** A circuit having an output and a multiplicity of inputs and so designed that the output is energized only when a certain combination of pulses is present at the inputs. **2.** A circuit in which one signal, generally a square wave, serves to switch another signal on and off. **3.** One of the electrodes in a field-effect transistor. **4.** An output element of a cryotron. **5.** To control the passage of a pulse or signal. **6.** In radar, an electric waveform which is applied to the control point of a circuit to alter the mode of operation of the circuit at the time when the waveform is applied. Also known as gating waveform. [NUCLEO] A movable barrier of shielding material used for closing a hole in a nuclear reactor.

Gâteaux derivative [MATH] The Gâteaux derivative of a functional F at a point u of a vector space X is the functional which maps each element h of X into the limit, as t approaches 0, of $[F(u + th) - F(u)]/t$ (provided such a limit exists for all h).

Gâteaux differential [MATH] The Gâteaux differential of a functional F at a point u of a vector space X is the form $G_u(h)$, where G_u is the Gâteaux derivative of F at u and h is an arbitrary element of X.

gate-controlled rectifier [ELECTR] A three-terminal semiconductor device, such as a silicon controlled rectifier, in which the unidirectional current flow between the rectifier terminals is controlled by a signal applied to a third terminal called the gate.

gate-controlled switch [ELECTR] A semiconductor device that can be switched from its nonconducting or "off" state to its conducting or "on" state by applying a negative pulse to its gate terminal and that can be turned off at any time by applying reverse drive to the gate. Abbreviated GCS.

gate equivalent circuit [ELECTR] A unit of measure for specifying relative complexity of digital circuits, equal to the number of individual logic gates that would have to be interconnected to perform the same function as the digital circuit under evaluation.

gate pulse [ELECTR] A pulse that triggers a gate circuit so it will pass a signal.

gating [ELECTR] The process of selecting those portions of a wave that exist during one or more selected time intervals or that have magnitudes between selected limits.

gating waveform *See* gate.

gauge *See* gage.

gauging *See* gaging.

gauss [ELECTROMAG] Unit of magnetic induction in the electromagnetic and Gaussian systems of units, equal to 1 maxwell per square centimeter, or 10^{-4} weber per square meter. Also known as abtesla (abT).

Gauss A position *See* end-on position.

Gauss B position *See* broadside-on position.

Gauss-Codazzi equations [MATH] Equations dealing with the components of the fundamental tensor and Riemann-Christoffel tensor of a surface.

Gauss' error curve *See* normal distribution.

Gauss eyepiece [OPTICS] A Ramsden eyepiece which has a thin glass plate between the two lenses, making an angle of $45°$ with the optical axis; used to set a telescope perpendicular to a plane reflecting surface.

Gauss formulas [MATH] Formulas dealing with the sine and cosine of angles in a spherical triangle.

Gauss' hypergeometric equation [MATH] The differential equation, commonly arising in many physical contexts, $x(1 - x)y'' + [c - (a + b + 1)x]y' - aby = 0$.

Gaussian coefficients [MATH] The numbers

$$\begin{bmatrix} n \\ k \end{bmatrix}_q$$

giving the number of subspaces of dimension k in a vector space of dimension n over a finite field with q elements.

Gaussian complex integers [MATH] Complex numbers whose real and imaginary parts are both integers.

Gaussian constant [ASTRON] The acceleration caused by the attraction of the sun at the mean distance of the earth from the sun.

Gaussian curvature [MATH] The invariant of a surface specified by Gauss' theorem. Also known as total curvature.

Gaussian curve [STAT] The bell-shaped curve corresponding to a population which has a normal distribution. Also known as normal curve.

Gaussian distribution *See* normal distribution.

Gaussian formulas [MATH] Formulas for calculating approximate values of definite integrals, having the form

$$\int_{-1}^{+1} f(x)\, dx \cong \sum_{i=1}^{n} a_i f(x_i)$$

where the x_i are zeros of the Legendre polynomial $P_n(x)$ and

$$a_i = \frac{1}{P'_n(x_i)} \int \frac{P_n(x)\, dx}{x - x_i}$$

Gaussian noise [COMMUN] Noise that has a frequency distribution which follows the Gaussian curve. [MATH] *See* Wiener process.

Gaussian optics [OPTICS] An approximation which describes rays which are very close to the axis of an optical system and are nearly parallel to this axis, so that only the linear terms of Taylor series for the distance of a point from the axis or the angle which a ray makes with the axis need be considered. Also known as first-order theory.

Gaussian pulse [PHYS] A pulse for which the graph of intensity as a function of time is a Gaussian curve.

Gaussian reduction [MATH] A procedure of simplification of the rows of a matrix which is based upon the notion of solving a system of simultaneous equations. Also known as Gauss-Jordan elimination.

Gaussian system [ELECTROMAG] A combination of the electrostatic and electromagnetic systems of units (esu and emu), in which electrostatic quantities are expressed in esu and magnetic and electromagnetic quantities in emu, with appropriate use of the conversion constant c (the speed of light) between the two systems. Also known as Gaussian units.

Gaussian units *See* Gaussian system.

Gauss image point [OPTICS] A point through which pass all paraxial rays from a specified point source in an optical system.

Gauss-Jordan elimination *See* Gaussian reduction.

Gauss' law of flux [ELEC] The law that the total electric flux which passes out from a closed surface equals (in rationalized units) the total charge within the surface.

Gauss' law of the arithmetic mean [MATH] The law that a harmonic function can attain its maximum value only on the boundary of its domain of definition, unless it is a constant.

Gauss-Legendre rule [MATH] An approximation technique of definite integrals by a finite series which uses the zeros and derivatives of the Legendre polynomials.

Gauss lemma [MATH] The product of the contents of two polynomials f and g with coefficients in the quotient field of a unique factorization domain is equal to the content of the polynomial fg.

Gauss lens system *See* Celor lens system.

Gauss-Markov theorem [STAT] An unbiased linear estimator of a parameter has the minimum variance, that is, is the best estimator when it is determined by the least-squares method.

Gauss' mean value theorem [MATH] The value of a harmonic function at a point in a planar region is equal to its integral about a circle centered at the point.

gaussmeter [ENG] A magnetometer whose scale is graduated in gauss or kilogauss, and usually measures only the intensity, and not the direction, of the magnetic field.

Gauss method of weighing *See* double weighing.

Gauss objective lens *See* Celor lens system.

Gauss point *See* cardinal point.

Gauss' principle of least constraint [MECH] The principle that the motion of a system of interconnected material points subjected to any influence is such as to minimize the constraint on the system; here the constraint, during an infinitesimal period of time, is the sum over the points of the product of the mass of the point times the square of its deviation from the position it would have occupied at the end of the time period if it had not been connected to other points.

Gauss-Seidel method *See* Seidel method.

Gauss test [MATH] In an infinite series with general term a_n, if $a_{n+1}/a_n = 1 - (x/n) - [f(n)/n^\lambda]$ where x and λ are greater than 1, and $f(n)$ is a particular integer function, then the series converges.

Gauss' theorem [MATH] **1.** The assertion, under certain light restrictions, that the volume integral through a volume V of the divergence of a vector function is equal to the surface integral of the exterior normal component of the vector function over the boundary surface of V. Also known as divergence theorem. **2.** At a point on a surface the product of the principal curvatures is an invariant of the surface, called the Gaussian curvature.

gauze tones *See* howling tones.

Gay-Lussac's law *See* Charles' law; combining-volumes principle.

gc *See* gigahertz.

g-cal *See* calorie.

gcd *See* greatest common divisor.

G center *See* N center.

g-cm *See* gram-centimeter.

g-completeness *See* geodesic completeness.

g constant [SOLID STATE] The ratio of the induced electric field in a piezoelectric material to the applied force that produces this field.

GCS *See* gate-controlled switch.

Gd *See* gadolinium.

Ge *See* germanium.

Gedanken experiment [PHYS] German for "thought" experiment; a hypothetical experiment which is possible in principle and is analyzed (but not performed) to test some hypothesis.

geepound *See* slug.

Gegenbauer functions [MATH] Solutions of a special case of the Gauss hypergeometric equations, $(x^2 - 1)y'' + (2\nu + 1)xy' - a(a + 2\nu)y = 0$; for integral values of a they become Gegenbauer polynomials.

Gegenbauer polynomials [MATH] A family of polynomials solving a special case of the Gauss hypergeometric equation. Also known as ultraspherical polynomials.

gegenions *See* counterions.

gegenschein [ASTRON] A round or elongated, faint, ill-defined spot of light in the sky at a point 180° from the sun. Also known as counterglow; zodiacal counterglow.

Geiger-Briggs rule [NUCLEO] The rule that the range of an alpha ray in dry air, at 15°C and 1 atmosphere (101,325 newtons per square meter), above 5 centimeters, is proportional to its initial velocity raised to the 3.26 power; an improvement on the Geiger formula.

Geiger counter *See* Geiger-Müller counter.

Geiger counter tube *See* Geiger-Müller tube.

Geiger formula [NUCLEO] A formula which states that the range of an alpha particle in dry air, at 15°C and 1 atmosphere (101,325 newtons per square meter), is proportional to the cube of its initial velocity.

Geiger-Müller counter [NUCLEO] **1.** A radiation counter that uses a Geiger-Müller tube in appropriate circuits to detect and count ionizing particles; each particle crossing the tube produces ionization of gas in the tube which is roughly independent of the particle's nature and energy, resulting in a uniform discharge across the tube. Abbreviated GM counter. Also known as Geiger counter. **2.** *See* Geiger-Müller tube.

Geiger-Müller counter tube *See* Geiger-Müller tube.

Geiger-Müller region *See* Geiger region.

Geiger-Müller tube [NUCLEO] A radiation-counter tube operated in the Geiger region; it usually consists of a gas-filled cylindrical metal chamber containing a fine-wire anode at its axis. Also known as Geiger counter tube; Geiger-Müller counter; Geiger-Müller counter tube.

GEIGER-MÜLLER TUBE

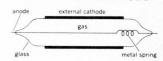

Cylindrical external-cathode Geiger-Müller tube, with thin soda glass and central wire of 0.003-inch-diameter (0.08-millimeter) tungsten. Metal spring keeps central wire taut.

Geiger-Nutall rule [NUC PHYS] The rule that the logarithm of the decay constant of an alpha emitter is linearly related to the logarithm of the range of the alpha particles emitted by it.

Geiger region [NUCLEO] The range of operating voltages of a radiation counter tube within which the output charge per count does not depend on the nature of the initial ionizing event. Also known as Geiger-Müller region.

Geiger threshold [NUCLEO] The lower limit of the Geiger region.

Geissler tube [ELECTR] An experimental discharge tube with two electrodes at opposite ends, used to demonstrate and study the luminous effects of electric discharges through various gases at low pressures.

Gellerstedt's equation [MATH] The partial differential equation $y^{2n+1} \, \partial^2 u/\partial x^2 + \partial^2 u/\partial y^2 = 0$.

Gell-Mann–Nishijima scheme [PARTIC PHYS] A classification of elementary particles according to hypercharge, total isotopic spin, and its third component (which distinguishes between members of an isospin multiplet).

Gell-Mann–Okubo mass formula [PARTIC PHYS] A formula, based on SU(3) symmetry, giving the masses of the members of a unitary multiplet of mesons or baryons in terms of their total isotopic spin, hypercharge, and coefficients characteristic of the multiplet considered.

Gell-Mann relation [PARTIC PHYS] A relation, derived from the Gell-Mann–Okubo mass formula, between the masses of the pi, eta, and K mesons of a meson octet.

gel point [PHYS CHEM] Stage at which a liquid begins to exhibit elastic properties and increased viscosity.

gemmho [ELEC] A unit of conductance, equal to the conductance of a substance which has a resistance of 10^6 ohms, or to 10^{-6} mho.

general integral *See* general solution of an ordinary differential equation.

generalized analytic function [MATH] A function $w(z)$ of the complex variable $z = x + iy$ whose real and imaginary parts, $u(x,y)$ and $v(x,y)$, satisfy the equations $\partial u/\partial x - \partial v/\partial y + au + bv = e$, and $\partial u/\partial y + \partial v/\partial x + cu + dv = f$, where a, b, c, d, e, and f are constants. Also known as pseudoanalytic function.

generalized binomial trials model [STAT] A product model in which the nth factor model has two simple events with probabilities p_n and $q_n = 1 - p_n$. Also known as Poisson binomial trials model.

generalized coordinates [MECH] A set of variables used to specify the position and orientation of a system, in principle defined in terms of cartesian coordinates of the system's particles and of the time in some convenient manner; the number of such coordinates equals the number of degrees of freedom of the system Also known as Lagrangian coordinates.

generalized force [MECH] The generalized force corresponding to a generalized coordinate is the ratio of the virtual work done in an infinitesimal virtual displacement, which alters that coordinate and no other, to the change in the coordinate.

generalized function *See* distribution.

generalized Hooke's law [MECH] The law that each of the six components of stress at a point in a solid is a linear function of the six components of strain at that point.

generalized hydrostatic equation [GEOPHYS] The vertical component of the vector equation of motion in natural coordinates when the acceleration of gravity is replaced by the virtual gravity; for most purposes it is identical to the hydrostatic equation.

generalized hypergeometric function [MATH] A generalization of the hypergeometric series having the form

$$_pF_q(a_1, a_2, \ldots, a_p; b_1, b_2, \ldots, b_q; z)$$

$$= \sum_{n=0}^{\infty} \frac{(a_1;n)(a_2;n) \ldots (a_p;n)}{(b_1;n)(b_2;n) \ldots (b_q;n)} \frac{z^n}{n!}$$

where $(a;0) = 1$ and $(a;n) = a(a + 1) \cdots (a + n - 1)$.

generalized momentum *See* conjugate momentum.

generalized Poincaré conjecture [MATH] The question as to whether every closed n-manifold which has the homotopy type of the n-sphere is homeomorphic to the n-sphere.

generalized power [MATH] For a positive number a and an irrational number x, the number a^x defined by the equation $a^x = e^{x \log a}$, where e is the base of the natural logarithms and $\log a$ is taken to that base.

generalized transmission function [GEOPHYS] In atmospheric-radiation theory, a set of values, variable with wavelength, each one of which represents an average transmission coefficient for a small wavelength interval and for a specified optical path through the absorbing gas in question.

generalized velocity [MECH] The derivative with respect to time of one of the generalized coordinates of a particle. Also known as Lagrangian generalized velocity.

general precession [ASTRON] The resultant motion of the components causing precession of the equinoxes westward along the ecliptic at the rate of about 50".3 per year.

general relativistic collapse [RELAT] Process in which a star undergoing gravitational collapse cannot release its kinetic energy to the outside universe, but continues to collapse into a general relativistic singularity of infinite density.

general relativity [RELAT] The theory of Einstein which generalizes special relativity to noninertial frames of reference and incorporates gravitation, and in which events take place in a curved space.

general solution of an ordinary differential equation [MATH] For an nth-order differential equation, a function of the independent variables of the equation and of n parameters such that assignment of any numerical values to the parameters yields a solution to the equation. Also known as general integral.

general term [MATH] The general term of a sequence or series is an expression subscripted by an integer which determines any desired entry.

generating flow [FL MECH] For a liquid allowed to flow smoothly into a duct, the flow while the boundary layer, which starts at the entrance and grows until it fills the duct, is growing.

generating function [MATH] A function $g(x,y)$ corresponding to a family of orthogonal polynomials $f_0(x), f_1(x), \ldots$, where a Taylor series expansion of $g(x,y)$ in powers of y will have the polynomial $f_n(x)$ as the coefficient for the term y^n.

generating magnetometer [ENG] A magnetometer in which a coil is rotated in the magnetic field to be measured with the resulting generated voltage being proportional to the strength of the magnetic field.

generation rate [ELECTR] In a semiconductor, the time rate of creation of electron-hole pairs.

generation time [NUCLEO] The mean time required for a neutron arising from a fission to produce a new fission.

generator [ELEC] A machine that converts mechanical energy into electrical energy; in its commonest form, a large number of conductors are mounted on an armature that is rotated in a magnetic field produced by field coils. Also known as dynamo; electric generator. [ELECTR] **1.** A vacuum-tube oscillator or any other nonrotating device that generates an alternating voltage at a desired frequency when

GENERAL RELATIVISTIC COLLAPSE

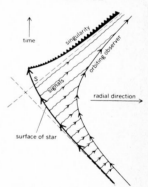

Space-time diagram for the collapse of a star past its Schwarzschild radius (S) and into a general relativistic singularity of infinite density.

energized with direct-current power or low-frequency alternating-current power. **2.** A circuit that generates a desired repetitive or nonrepetitive waveform, such as a pulse generator. [MATH] **1.** One of the set of elements of an algebraic system such as a group, ring, or module which determine all other elements when all admissible operations are performed upon them. **2.** *See* generatrix.

generator of a flow [MATH] A vector field F on a differentiable manifold X generates the flow $(\mathbf{R}, X, \pi)$ if $(d/dt)\pi(t,x) = F[\pi(t,x)]$; that is, for every continuously differentiable function f on X, $(d/dt)f[\pi(t,x)]$, evaluated at $t = 0$, is equal to $F(x)f(x)$.

generator resistance [ELEC] The resistance of the current source in a network; usually much smaller than the load but taken into account in some network calculations.

generatrix [MATH] The straight line generating a ruled surface.

genus [MATH] An integer associated to a surface which measures the number of holes in the surface.

geoacoustics [ACOUS] Study of the acoustic properties of rock, mainly to study possible use of the rock system as a carrier of seismic signals in a communications system.

geocentric [ASTRON] Relative to the earth as a center; that is, measured from the center of the earth.

geocentric coordinates [ASTRON] Coordinates that define the position of a point with respect to the center of the earth; can be either cartesian (x, y, and z) or spherical (latitude, longitude, and radial distance). Also known as geocentric coordinate system; geocentric position.

geocentric coordinate system *See* geocentric coordinates.

geocentric latitude [ASTRON] The latitude of a celestial body from the center of the earth.

geocentric longitude [ASTRON] The celestial longitude of the position of a body projected on the celestial sphere when the body is viewed from the center of the earth.

geocentric parallax [ASTRON] The difference in the apparent direction or position of a celestial body, measured in seconds of arc, as determined from the center of the earth and from a point on its surface; this varies with the body's altitude and distance from the earth. Also known as diurnal parallax.

geocentric position *See* geocentric coordinates.

geocentric zenith [ASTRON] The point where a line from the center of the earth through a point on its surface meets the celestial sphere.

geodesic [MATH] A curve joining two points in a Riemannian manifold which has minimum length.

geodesic completeness [RELAT] A space-time is geodesically complete if all timelike and null geodesics can be extended to arbitrary values of their affine parameter. Also known as g-completeness.

geodesic coordinates [RELAT] Coordinates in the neighborhood of a point P such that the gradient of the metric tensor is zero at P.

geodesic curvature [MATH] For a point on a curve lying on a surface, the curvature of the orthogonal projection of the curve onto the tangent plane to the surface at the point; it measures the departure of the curve from a geodesic. Also known as tangential curvature.

geodesic flow [MATH] The flow $(\mathbf{R}, T_1M, \pi)$, where T_1M is the bundle of tangent vectors of unit length on a differentiable manifold M and, for any tangent vector v_p at a point p, $\pi(t, v_p)$ is equal to dx/ds evaluated at $s = t$, where $x(t)$ is the unique geodesic, parametrized by arc length, with $x(0) = p$, and dx/ds, evaluated at $s = 0$ equal to v_p.

geodesic incompleteness [RELAT] A space-time is geodesically incomplete if there exists at least one timelike or null geodesic that cannot be extended to arbitrarily large values of its affine parameter; such a space-time contains a singularity. Also known as g-incompleteness.

geodesic line [MATH] The shortest line between two points on a mathematically derived surface.

geodesic motion [RELAT] Motion of a particle along a geodesic path in the four dimensional space-time continuum; according to general relativity, this is the motion which occurs in the absence of nongravitational forces.

geodetic triangle *See* spheroidal triangle.

geodynamic height *See* dynamic height.

geoid [GEOD] The figure of the earth considered as a sea-level surface extended continuously over the entire earth's surface.

geoisotherm [GEOPHYS] The locus of points of equal temperature in the interior of the earth; a line in two dimensions or a surface in three dimensions. Also known as geotherm; isogeotherm.

geomagnetic coordinates [GEOPHYS] A system of spherical coordinates based on the best fit of a centered dipole to the actual magnetic field of the earth.

geomagnetic cutoff [GEOPHYS] The minimum energy of a cosmic-ray particle able to reach the top of the atmosphere at a particular geomagnetic latitude.

geomagnetic dipole [GEOPHYS] The magnetic dipole caused by the earth's magnetic field.

geomagnetic equator [GEOPHYS] That terrestrial great circle which is 90° from the geomagnetic poles.

geomagnetic field [GEOPHYS] The earth's magnetic field.

geomagnetic latitude [GEOPHYS] The magnetic latitude that a location would have if the field of the earth were to be replaced by a dipole field closely approximating it.

geomagnetic longitude [GEOPHYS] Longitude that is determined around the geomagnetic axis instead of around the rotation axis of the earth.

geomagnetic meridian [GEOPHYS] A circle passing around the earth and through the geomagnetic poles.

geomagnetic pole [GEOPHYS] Either of two antipodal points marking the intersection of the earth's surface with the extended axis of a powerful bar magnet assumed to be located at the center of the earth and having a field approximating the actual magnetic field of the earth.

geomagnetic reversal [GEOPHYS] Reversed magnetization of the earth's geomagnetic dipole.

geomagnetic secular variation *See* secular variation.

geomagnetic storm *See* magnetic storm.

geomagnetic variation [GEOPHYS] Temporal changes in the geomagnetic field, both long-term (secular) and short-term (transient).

geomagnetism [GEOPHYS] **1.** The magnetism of the earth. Also known as terrestrial magnetism. **2.** The branch of science that deals with the earth's magnetism.

geometrical acoustics *See* ray acoustics.

geometrical isomerism [PHYS CHEM] The phenomenon in which isomers contain atoms attached to each other in the same order and with the same bonds but with different spatial, or geometrical, relationships; the explicit geometry imposed upon a molecule by, say, a double bond between carbon atoms makes possible the existence of these isomers.

geometrical optics [OPTICS] The geometry of paths of light rays and their imagery through optical systems.

geometrical similarity [FL MECH] Property of two fluid flows for which a simple alteration of scales of length and velocity transforms one into the other.

geometric attenuation [NUCLEO] That part of the reduction in the intensity of ionizing radiation with distance from a source which is associated with spreading out of the radiation and is independent of the interaction of the radiation with matter.

geometric average *See* geometric mean.

geometric distribution [STAT] A discrete probability distribution whose probability function is given by $p(x) = p (1-p)^{x-1}$ for x any positive integer, $p(x) = 0$ otherwise, when $0 \le p \le 1$; the mean is $1/p$.

geometric mean [MATH] The geometric mean of n given quantities is the nth root of their product. Also known as geometric average.

geometric moment of inertia [MATH] The geometric moment of inertia of a plane figure about an axis in or perpendicular to the plane is the integral over the area of the figure of the square of the distance from the axis. Also known as second moment of area.

geometric number theory [MATH] The branch of number theory studying relationships among numbers by examining the geometric properties of ordered pair sets of such numbers.

geometric programming [SYS ENG] A nonlinear programming technique in which the relative contribution of each of the component costs is first determined; only then are the variables in the component costs determined.

geometric progression [MATH] A sequence which has the form $a, ar, ar^2, ar^3, \ldots$.

geometric series [MATH] An infinite series of the form $a + ar + ar^2 + ar^3 + \cdots$.

geometrodynamics [RELAT] A theory involving only geometry which attempts to combine gravitational and electromagnetic theory; characterized by a multiply connected spacetime manifold containing structures, descriptively called wormholes, associated with electric charge.

geometry [MATH] The qualitative study of shape and size.

geon [PHYS] A hypothetical electromagnetic field that is held together by its own gravitational attraction.

geophysics [GEOL] The physics of the earth and its environment, that is, earth, air, and (by extension) space.

geopotential [PHYS] The potential energy of a unit mass relative to sea level, numerically equal to the work that would be done in lifting the unit mass from sea level to the height at which the mass is located, against the force of gravity.

geopotential unit [GEOPHYS] A unit of gravitational potential used in describing the earth's gravitational field; it is equal to the difference in gravitational potential of two points separated by a distance of 1 meter when the gravitational field has a strength of 10 meters per second squared and is directed along the line joining the points. Abbreviated gpu.

geostrophic [GEOPHYS] Pertaining to deflecting force resulting from the earth's rotation.

geostrophic approximation [GEOPHYS] The assumption that the geostrophic current can represent the actual horizontal current. Also known as geostrophic assumption.

geostrophic assumption *See* geostrophic approximation.

geostrophic current [GEOPHYS] A current defined by assuming the existence of an exact balance between the horizontal pressure gradient force and the Coriolis force.

geostrophic equation [GEOPHYS] An equation, used to compute geostrophic current speed, which represents a balance between the horizontal pressure gradient force and the Coriolis force.

geostrophic equilibrium [GEOPHYS] A state of motion of a nonviscous fluid in which the horizontal Coriolis force exactly balances the horizontal pressure force at all points of the field so described.

geostrophic flow [GEOPHYS] A form of gradient flow where the Coriolis force exactly balances the horizontal pressure force.

geotherm *See* geoisotherm.

geothermal [GEOPHYS] Pertaining to heat within the earth.

germanium [CHEM] A brittle, water-insoluble, silvery-gray metallic element in the carbon family, symbol Ge, atomic number 32, atomic weight 72.59, melting at 959°C.

German R unit [NUCLEO] A unit of radiation dose rate due to x-rays, equal to approximately 2.5 Solomon R units, or approximately 1.5 roentgens per second. Also known as R unit.

Gershgorin's method [MATH] A method of obtaining bounds on the eigenvalue of a matrix, based on the fact that the absolute value of any eigenvalue is equal to or less than the maximum over the rows of the matrix of the sum of the absolute values of the entries in a row, and is also equal to or less than the maximum over the columns of the matrix of the sum of the absolute values of the entries in a column.

Gerstner wave [FL MECH] A rotational gravity wave of finite amplitude.

getter [PHYS CHEM] **1.** A substance, such as thallium, that binds gases on its surface and is used to maintain a high vacuum in a vacuum tube. **2.** A special metal alloy that is placed in a vacuum tube during manufacture and vaporized after the tube has been evacuated; when the vaporized metal condenses, it absorbs residual gases. Also known as degasser.

getter-ion pump [ENG] A high-vacuum pump that employs chemically active metal layers which are continuously or intermittently deposited on the wall of the pump, and which chemisorb active gases while inert gases are "cleaned up" by ionizing them in an electric discharge and drawing the positive ions to the wall, where the neutralized ions are buried by fresh deposits of metal. Also known as sputter-ion pump.

GeV *See* gigaelectronvolt.

gf *See* gram-force.

g factor *See* Landé g factor.

G factor *See* G value.

g force [PHYS] A force such that a body subjected to it would have the acceleration of gravity at sea level; used as a unit of measurement for bodies undergoing the stress of acceleration.

ghost crystal *See* phantom crystal.

ghost mode [ELECTROMAG] Waveguide mode having a trapped field associated with an imperfection in the wall of the waveguide; a ghost mode can cause trouble in a waveguide operating close to the cutoff frequency of a propagation mode.

giant planets [ASTRON] The planets Jupiter, Saturn, Uranus, and Neptune.

giant pulse laser *See* Q-switched laser.

giant star [ASTRON] One of a class of stars that is 20 or 30 or more times larger than the sun and over 100 times more luminous.

Giaque-Debye method *See* adiabatic demagnetization.

Giaque's temperature scale [THERMO] The internationally accepted scale of absolute temperature, in which the triple point of water is defined to have a temperature of 273.16 K.

gibbous [MATH] Bounded by convex curves.

gibbous moon [ASTRON] The shape of the moon's visible surface when the sun is illuminating more than half of the side facing the earth.

gibbs [PHYS] A unit of amount of adsorption, equal to a surface concentration of 10^{-6} mole per square meter.

GERMANIUM

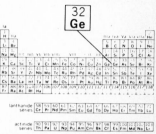

Periodic table of the chemical elements showing the position of germanium.

Gibbs adsorption equation [PHYS CHEM] A formula for a system involving a solvent and a solute, according to which there is an excess surface concentration of solute if the solute decreases the surface tension, and a deficient surface concentration of solute if the solute increases the surface tension.

Gibbs adsorption isotherm [PHYS CHEM] An equation for the surface pressure of surface monolayers,

$$\phi = RT \int_0^p \Gamma \, d(\ln p),$$

where ϕ is surface pressure, T is absolute temperature, R is the gas constant, Γ is the number of molecules adsorbed per gram per unit surface area, and p is the pressure of the gas.

Gibbs-Duhem equation [PHYS CHEM] A relation that imposes a condition on the composition variation of the set of chemical potentials of a system of two or more components,

$$S \, dT - V \, dP + \sum_{i=1}^r n_i \, d\mu_i = 0,$$

where S is entropy, T absolute temperature, P pressure, n_i the number of moles of the ith component, and μ_i is the chemical potential of the ith component. Also known as Duhem's equation.

Gibbs elasticity [PHYS] The elasticity of a film of liquid, equal to twice the product of the surface area and the derivative of the surface tension with respect to surface area.

Gibbs free energy [THERMO] The thermodynamic function $G = H - TS$, where H is enthalpy, T absolute temperature, and S entropy; Also known as free energy; free enthalpy; Gibbs function.

Gibbs function *See* Gibbs free energy.

Gibbs-Helmholtz equation [PHYS CHEM] An expression for the influence of temperature upon the equilibrium constant of a chemical reaction, $(d \ln K^0 / dT)_P = \Delta H^0 / RT^2$, where K^0 is the equilibrium constant, ΔH^0 the standard heat of reaction at the absolute temperature T, and R the gas constant. [THERMO] **1.** Either of two thermodynamic relations that are useful in calculating the internal energy U or enthalpy H of a system; they may be written $U = F - T(\partial F/\partial T)_V$ and $H = G - T(\partial G/\partial T)_P$, where F is the free energy, G is the Gibbs free energy, T is the absolute temperature, V is the volume, and P is the pressure. **2.** Any of the similar equations for changes in thermodynamic potentials during an isothermal process.

Gibbs paradox [STAT MECH] The paradox in which there is an increase in entropy when two separate volumes of gases of the same kind, at the same temperature and pressure, are mixed.

Gibbs phase rule [PHYS CHEM] A relation describing the nature of a heterogeneous chemical system at equilibrium, $F = C + 2 - P$, where F is the degrees of freedom, P the number of phases, and C the number of components. Also known as Gibbs rule.

Gibbs' phenomenon [MATH] A convergence phenomenon occurring when a function with a jump discontinuity is approximated by a finite number of terms from a Fourier series, in which the length of the interval formed by the Gibbs set is greater than the magnitude of the jump.

Gibbs-Poynting equation [PHYS CHEM] An expression relating the effect of the total applied pressure P upon the vapor pressure p of a liquid, $(dp/dP)_T = V_l/V_g$, where V_l and V_g are molar volumes of the liquid and vapor.

Gibbs ratio [MATH] For a sequence $\{f_n(x)\}$ of partial sums of the Fourier series of a function $f(x)$, the Gibbs ratio at a point $x = x_0$, at which $f(x)$ has a jump discontinuity, is the ratio of the length of the interval formed by the Gibbs set of $\{f_n(x)\}$ at $x = x_0$ to the magnitude of the jump of $f(x)$ at $x = x_0$.

Gibbs rule *See* Gibbs phase rule.

Gibbs set [MATH] For a sequence $\{f_n(x)\}$ of continuous functions that converges to a function $f(x)$, the Gibbs set at a point $x = x_0$ is the set of all possible limit points of $f_n(x)$ as n approaches infinity and x approaches x_0 through appropriate values.

Gibbs system [STAT MECH] **1.** A hypothetical replica of a physical system. **2.** A set of such replicas forming an ensemble.

giga- [SCI TECH] A prefix representing 10^9, which is 1,000,-000,000, or a billion. Abbreviated G. Also known as kilomega- (deprecated usage).

gigacycle *See* gigahertz.

gigaelectronvolt [PHYS] A unit of energy, used primarily in high-energy physics, equal to 10^9 electron volts or $(1.60210 \pm 0.00007) \times 10^{-10}$ joule. Abbreviated GeV.

gigahertz [COMMUN] Unit of frequency equal to 10^9 hertz. Abbreviated GHz. Also known as gigacycle (gc); kilomegacycle; kilomegahertz.

gigawatt [ELEC] One billion watts, or 10^9 watts. Abbreviated GW.

gigohm [ELEC] One thousand megohms, or 10^9 ohms.

gilbert [ELECTROMAG] The unit of magnetomotive force in the electromagnetic system, equal to the magnetomotive force of a closed loop of one turn in which there is a current of $1/4\pi$ abamp.

gill [MECH] **1.** A unit of volume used in the United States for the measurement of liquid substances, equal to 1/4 U.S. liquid pint, or to $1.1829411825 \times 10^{-4}$ cubic meter. **2.** A unit of volume used in the United Kingdom for the measurement of liquid substances, and occasionally of solid substances, equal to 1/4 U.K. pint, or to approximately 1.42065×10^{-4} cubic meter.

g-incompleteness *See* geodesic incompleteness.

Ginzburg-Landau equation [MATH] The partial differential equation for the function $u(x,t)$: $i \, \partial u/\partial t + \Delta \mu + f(|u|^2)u = 0$, where u is specified at $t = 0$, $i = \sqrt{-1}$, Δ represents the Laplace operator, and f is a specified function.

Ginzburg-Landau theory [CRYO] A phenomenological theory of superconductivity which accounts for the coherence length; the ordered state of a superconductor is described by a complex order parameter which is similar to a Schrödinger wave function, but describes all the condensed superelectrons, rather than a single charged particle. Also known as Landau-Ginzburg theory.

Ginzburg-London superconductivity theory [SOLID STATE] A modification of the London superconductivity theory to take into account the boundary energy.

Giorgi system *See* meter-kilogram-second-ampere system.

give-and-take lines [MATH] Straight lines which are used to approximate the boundary of an irregular, curvilinear figure for the purpose of approximating its area; they are placed so that small portions excluded from the area under consideration are balanced by other small portions outside the boundary.

Givens's method [MATH] A transformation method for finding the eigenvalues of a matrix, in which each of the orthogonal transformations that reduce the original matrix to a triple-diagonal matrix makes one pair of elements, a_{ij} and a_{ji}, lying off the principal diagonal and the diagonals immediately above and below it, equal to zero, without affecting zeros obtained earlier.

given-year method *See* Paasche's index.

Gladstone-Dale law [OPTICS] A law for the variation of the index of refraction n of a substance, according to which $n + 1$ is proportional to its density.

glancing angle [PHYS] The angle between a surface and a beam of particles or radiation incident upon it; it is the complement of the angle of incidence.

glare [OPTICS] **1.** Discomfort produced in an observer by one or more visible sources of light. Also known as discomfort glare. **2.** Visual disability caused by visible sources or areas of luminance which are in an observer's field of view but do not assist in viewing. Also known as disability glare. **3.** Dazzling brightness of the atmosphere, caused by excessive reflection and scattering of light by particles in the line of sight.

glass dosimeter [NUCLEO] A dosimeter using as its radiation-sensing element a fluorod of special glass that fluoresces under ultraviolet light following gamma irradiation.

glass fission detector [NUCLEO] A piece of glass in which fission fragments, flying apart with high energy, can create narrow but continuous, submicroscopic trails of altered material, which can be seen in an ordinary microscope after the altered material has been dissolved by a chemical reagent.

glass laser [OPTICS] A solid laser in which glass serves as the host for laser ions of such materials as erbium, holmium, neodymium, and ytterbium. Also known as amorphous laser.

glass seal [ENG] An airtight seal made by molten glass.

glass switch [ELECTR] An amorphous solid-state device used to control the flow of electric current. Also known as ovonic device.

glass transition [PHYS CHEM] The change in an amorphous region of a partially crystalline polymer from a viscous or rubbery condition to a hard and relatively brittle one; usually brought about by changing the temperature. Also known as gamma transition; glassy transition.

glass-tube manometer [ENG] A manometer for simple indication of difference of pressure, in contrast to the metallic-housed mercury manometer, used to record or control difference of pressure or fluid flow.

glassy transition *See* glass transition.

Glauert number [FL MECH] The quantity $(1 - M^2)^{-1/2}$, where M is the Mach number.

glb *See* greatest lower bound.

glide *See* slip.

glide plane [CRYSTAL] A lattice plane in a crystal on which translation or twin gliding occurs. Also known as slip plane.

G line [ELECTROMAG] A single dielectric-coated, round wire used for transmitting microwave energy.

glisette [MATH] A curve, such as Watt's curve, traced out by a point attached to a curve which moves so that it always touches two fixed curves, or the envelope of any line or curve attached to the moving curve.

glissile dislocation *See* Shockley partial dislocation.

glitch [ASTRON] A sudden change in the period of a pulsar, believed to result from a phenomenon analogous to an earthquake that changes the pulsar's moment of inertia.

glitter [OPTICS] The spots of light reflected from a point source by the surface of the sea or wave facets, that is, specular reflection.

Glivenko-Cantelli lemma [MATH] The empirical distribution functions of a random variable converge uniformly in probability to the distribution function of the random variable.

g load [PHYS] The numerical ratio of any applied force to the gravitational force at the earth's surface.

globally hyperbolic [RELAT] Property of a space-time M that satisfies certain causality conditions ensuring that the solution to the wave equation for a delta function source at a point *p* in *M* is unique and vanishes outside the causal future of *p*.

globe lightning   *See* ball lightning.

globular star cluster   [ASTRON] A group of many thousands of stars that are much closer to each other than the stars around the group and that are traveling through space together; a globular cluster has a slightly flattened spheroidal shape.

globule   [ASTRON] A black volume of cosmic dust viewed against the brighter background of bright nebulae.

glory   [OPTICS] A set of concentric, colored rings of light around the shadow cast by an observer or his head onto a cloud or fog bank.

glory hole   *See* beam hole.

gloss   [OPTICS] The ratio of the light specularly reflected from a surface to the total light reflected.

glossimeter   [ENG] An instrument, often photoelectric, for measuring the ratio of the light reflected from a surface in a definite direction to the total light reflected in all directions. Also known as glossmeter.

glossmeter   *See* glossimeter.

glossy   [OPTICS] Property of a surface from which much more light is specularly reflected than is diffusely reflected.

glow discharge   [ELECTR] A discharge of electricity through gas at relatively low pressure in an electron tube, characterized by several regions of diffuse, luminous glow and a voltage drop in the vicinity of the cathode that is much higher than the ionization voltage of the gas. Also known as cold-cathode discharge.

glow-discharge cold-cathode tube   *See* glow-discharge tube.

glow-discharge tube   [ELECTR] A gas tube that depends for its operation on the properties of a glow discharge. Also known as glow-discharge cold-cathode tube; glow tube.

glow-discharge voltage regulator   [ELECTR] Gas tube that varies in resistance, depending on the value of the applied voltage; used for voltage regulation.

glow lamp   [ELECTR] A two-electrode electron tube containing a small quantity of an inert gas, in which light is produced by a negative glow close to the negative electrode when voltage is applied between the electrodes.

glow potential   [ELECTR] The potential across a glow discharge, which is greater than the ionization potential and less than the sparking potential, and is relatively constant as the current is varied across an appreciable range.

glow tube   *See* glow-discharge tube.

glug   [MECH] A unit of mass, equal to the mass which is accelerated by 1 centimeter per second per second by a force of 1 gram-force, or to 980.665 grams.

gluon   [PARTIC PHYS] One of eight hypothetical massless particles with spin quantum number and negative parity that mediate strong interactions between quarks.

gm   *See* gram.

GM counter   *See* Geiger-Müller counter.

GMT   *See* Greenwich mean time.

gnd   *See* ground.

gnomon   [MATH] A geometric figure formed by removing from a parallelogram a similar parallelogram that contains one of its corners.

gnomonic projection   [CRYSTAL] A projection for displaying the poles of a crystal in which the poles are projected radially from the center of a reference sphere onto a plane tangent to the sphere.

Gödel's universe   [RELAT] An exact solution of the nonvacuum equations of general relativity with matter in the form of dust; there are closed timelike lines in this solution.

Goertler parameter   [FL MECH] A dimensionless number used in studying boundary-layer flow on curved surfaces, equal to the Reynolds number, where the characteristic length is the boundary-layer momentum thickness, times the square root of this thickness, divided by the square root of the surface's radius of curvature.

Golay cell [ENG] A radiometer in which radiation absorbed in a gas chamber heats the gas, causing it to expand and deflect a diaphragm in accordance with the amount of radiation.

gold [CHEM] A chemical element, symbol Au, atomic number 79, atomic weight 196.967; soluble in aqua regia; melts at 1065°C.

gold-198 [NUC PHYS] The radioisotope of gold, with atomic number 198 and a half-life of 2.7 days; used in medical treatment of tumors by injecting it in colloidal form directly into tumor tissue.

Goldbach conjecture [MATH] The unestablished conjecture that every even number save the number 2 is the sum of two primes.

Goldberg-Mohn friction [FL MECH] A force proportional to the velocity of a current and the density of the medium; used as a first approximation in estimating frictional effects in the atmosphere and the ocean.

golden section [MATH] The division of a line so that the ratio of the whole line to the larger interval equals the ratio of the larger interval to the smaller. Also known as extreme and mean ratio.

Goldhaber triangle [PARTIC PHYS] A plot describing a high-energy reaction leading to four or more particles; its coordinates are the invariant masses of two intermediate-state quasi-particle composites, and its kinematical limits form a right-angled isosceles triangle; resonances in the quasi-particle composites appear as horizontal and vertical bands.

gold-leaf electroscope [ELEC] An electroscope in which two narrow strips of gold foil or leaf suspended in a glass jar spread apart when charged; the angle between the strips is related to the charge.

gold point [THERMO] The temperature of the freezing point of gold at a pressure of 1 standard atmosphere (101,325 newtons per square meter); used to define the International Practical Temperature Scale of 1968, on which it is assigned a value of 1337.58 K or 1064.43°C.

Goldschmidt's law [SOLID STATE] The law that crystal structure is determined by the ratios of the numbers of the constituents, the ratios of their sizes, and their polarization properties.

Goldstone bosons [PHYS] Particles with zero mass and zero spin which accompany spontaneous breaking of exact fundamental symmetries.

Gompertz curve [STAT] A curve similar to the exponential curve except that the constant a is raised to the b^x power instead of the x power; used in fitting a trend line to a nonlinear time series.

gon *See* grade.

goniometer [ELECTROMAG] An instrument for determining the direction of maximum response to a received radio signal, or selecting the direction of maximum radiation of a transmitted radio signal; consists of two fixed perpendicular coils, each attached to one of a pair of loop antennas which are also perpendicular, and a rotatable coil which bears the same space relationship to the coils as the direction of the signal to the antennas.

goniophotometer [OPTICS] A photometer designed to measure the intensity of light reflected from a surface at various angles.

GOLD

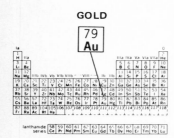

Periodic table of the chemical elements showing the position of gold.

GOLDHABER TRIANGLE

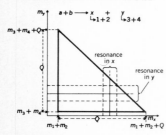

Goldhaber triangle for reaction between two particles, a and b, yielding four particles, 1, 2, 3, and 4, with masses m_1, m_2, m_3, and m_4. Coordinates are invariant masses m_x and m_y of intermediate-state quasi-particle composites x and y which decay into two particles each as indicated at top of figure. Q is total center of mass energy of a and b minus sum of m_1, m_2, m_3, and m_4.

good geometry [NUC PHYS] An arrangement of source and detecting equipment such that little error is produced by the finite sizes of the source and the detector aperture.

goodness of fit [STAT] The degree to which the observed frequencies of occurrence of events in an experiment correspond to the probabilities in a model of the experiment. Also known as best fit.

googol [MATH] A name for 10 to the power 100.

googolplex [MATH] A name for 10 to the power googol.

G orbit of a point [MATH] For a point x in a G-space X, the set of points y in X such that for some s in G, $y = xs$.

Göthert's β^2 rule [FL MECH] A rule for relating subsonic flow (treated according to the linearized theory of fluid flow) to incompressible flow; it states that the perturbation pressures on the surface of a body in subsonic flow are $1/\beta^2$ times the perturbation pressures at corresponding points in an incompressible flow past a body whose cross-stream dimensions are β times those of the original body, where $\beta = \sqrt{1 - M^2}$, M being the Mach number of the undisturbed flow.

gouy [PHYS CHEM] An electrokinetic unit equal to the product of the electrokinetic potential and the electric displacement divided by 4π times the polarization of the electrolyte.

g parameter [ELECTR] One of a set of four transistor equivalent-circuit parameters; they are the inverse of the h parameters.

G parity [PARTIC PHYS] The eigenvalue of a system under the operation of inversion in isotopic spin space; it is conserved by the strong interactions. Also known as isotopic parity.

gpu *See* geopotential unit.

gr *See* grain.

grade [MATH] A unit of plane angle, equal to 0.01 right angle, or $\pi/200$ radians, or 0.9°. Also known as gon.

graded-junction transistor *See* rate-grown transistor.

graded Lie algebra [MATH] A generalization of a Lie algebra in which both commutators and anticommutators occur.

gradient [MATH] A vector obtained from a real function $f(x_1, x_2, \ldots, x_n)$ whose components are the partial derivatives of f; this measures the maximum rate of change of f in a given direction.

gradient coupling [PARTIC PHYS] A hypothetical interaction of particles in which the interaction Hamilton depends explicitly on first derivatives of wave functions associated with the particles with respect to position and time.

gradient method [MATH] A finite iterative procedure for solving a system of n equations in n unknowns.

gradient operator [MATH] For a functional f defined on a Hilbert space H, any operator A satisfying the equation $f(u + h) - f(u) = (Au,h) + r_u(h)$ for all u and h in H, where $|r_u(h)|/\|h\|$ approaches 0 as $\|h\|$ approaches 0.

gradient projection method [MATH] Computational method used in nonlinear programming when constraint functions are linear.

gradiometer [ENG] Any instrument that measures the gradient of some physical quantity, such as certain types of magnetometers which are designed to measure the gradient of magnetic field, or the Eötövs torsion balance and related instruments which measure the gradient of gravitational field.

Graeffe's method [MATH] A method of solving algebraic equations by means of squaring the exponents and making appropriate substitutions.

Graetz number [THERMO] A dimensionless number used in the study of streamline flow, equal to the mass flow rate of a fluid times its specific heat at constant pressure divided by the product of its thermal conductivity and a characteristic length. Also spelled Grätz number. Symbolized N_{Gz}.

Graetz problem [FL MECH] The problem of determining the steady-state temperature field in a fluid flowing in a circular tube when the wall of the tube is held at a uniform temperature and the fluid enters the tube at a different uniform temperature.

Graham's law of diffusion [FL MECH] The law that the rate of diffusion of a gas is inversely proportional to the square root of its density.

Graham's pendulum [DES ENG] A type of compensated pendulum having a hollow bob containing mercury whose thermal expansion balances the thermal expansion of the pendulum rod.

grain [MECH] A unit of mass in the United States and United Kingdom, common to the avoirdupois, apothecaries', and troy systems, equal to 1/7000 of a pound, or to 6.479891×10^{-5} kilogram. Abbreviated gr.

gram [MECH] The unit of mass in the centimeter-gram-second system of units, equal to 0.001 kilogram. Abbreviated g; gm.

gram-atomic weight [CHEM] The atomic weight of an element expressed in grams, that is, the atomic weight on a scale on which the atomic weight of carbon-12 isotope is taken as 12 exactly.

gram-calorie *See* calorie.

gram-centimeter [MECH] A unit of energy in the centimeter-gram-second gravitational system, equal to the work done by a force of magnitude 1 gram force when the point at which the force is applied is displaced 1 centimeter in the direction of the force. Abbreviated g-cm.

Gram determinant [MATH] The Gram determinant of vectors $\mathbf{v}_1, \ldots, \mathbf{v}_n$ from an inner product space is the determinant of the $n \times n$ matrix with the inner product of $\mathbf{v}_i$ and $\mathbf{v}_j$ as entry in the ith column and jth row; its vanishing is a necessary and sufficient condition for linear dependence.

gram-equivalent weight [CHEM] The equivalent weight of an element or compound expressed in grams on a scale in which carbon-12 has an equivalent weight of 3 grams in those compounds in which its formal valence is 4.

gram-force [MECH] A unit of force in the centimeter-gram-second gravitational system, equal to the gravitational force on a 1-gram mass at a specified location. Abbreviated gf. Also known as fors; gram-weight; pond.

gram-molecular volume [CHEM] The volume occupied by a gram-molecular weight of a chemical in the gaseous state at 0°C and pressure of 760 millimeters of mercury (101,325 newtons per square meter).

gram-molecular weight [CHEM] The molecular weight of compound expressed in grams, that is, the molecular weight on a scale on which the atomic weight of carbon-12 isotope is taken as 12 exactly.

gram-rad [NUCLEO] A unit of integral absorbed dose of radiation, equal to 100 ergs.

gram-roentgen [NUCLEO] A unit of energy conversion, equal to a dose of 1 roentgen delivered to 1 gram of air.

Gram-Schmidt orthogonalization process [MATH] A process by which an orthogonal set of vectors is obtained from a linearly independent set of vectors in an inner product space.

Gram's theorem [MATH] A set of vectors are linearly dependent if and only if their Gram determinant vanishes.

gram-weight *See* gram-force.

grand canonical ensemble [STAT MECH] A collection of systems of particles used to describe an individual system which is allowed to exchange both energy and particles with its environment.

granulation [ASTRON] The small "rice grain" markings on the sun's photosphere. Also known as photospheric granulation.

graph [MATH] **1.** The planar object, formed from points and line segments between them, used in the study of circuits and networks. **2.** The graph of a function f is the set of all ordered pairs $[x, f(x)]$, where x is in the domain of f. **3.** *See* graphical representation.

graph component [MATH] A particular type of maximal connected subgraph of a graph.

graphical analysis [MATH] The study of interdependent phenomena by analyzing graphical representations.

graphical representation [MATH] The plot of the points in the plane which constitute the graph of a given real function or a pictorial diagram depicting interdependence of variables. Also known as graph.

graphical statics [MECH] A method of determining forces acting on a rigid body in equilibrium, in which forces are represented on a diagram by straight lines whose lengths are proportional to the magnitudes of the forces.

graph theory [MATH] **1.** The mathematical study of the structure of graphs and networks. **2.** The body of techniques used in graphing functions in the plane.

graser *See* gamma-ray laser.

Grashof formula [FL MECH] A formula, $m = 0.0165 A_2 p_1^{0.97}$, used to express the discharge m of saturated steam, where A_2 is the area of the orifice in square inches, and p_1 is reservoir pressure in pounds per square inch.

Grashof number [FL MECH] A dimensionless number used in the study of free convection of a fluid caused by a hot body, equal to the product of the fluid's coefficient of thermal expansion, the temperature difference between the hot body and the fluid, the cube of a typical dimension of the body and the square of the fluid's density, divided by the square of the fluid's dynamic viscosity. Also known as free convection number.

Grassmann algebra *See* exterior algebra.

Grassmannian *See* Grassmann manifold.

Grassmann manifold [MATH] The differentiable manifold whose points are all k-dimensional planes passing through the origin in n-dimensional euclidean space. Also known as Grassmannian.

Grassot fluxmeter [ENG] A type of fluxmeter in which a light coil of wire is suspended in a magnetic field in such a way that it can rotate; the ends of the suspended coil are connected to a search coil of known area penetrated by the magnetic flux to be measured; the flux is determined from the rotation of the suspended coil when the search coil is moved.

graticule [OPTICS] A scale at the focal plane of an optical instrument to aid in the measurement of objects.

grating [ELECTROMAG] **1.** An arrangement of fine, parallel wires used in waveguides to pass only a certain type of wave. **2.** An arrangement of crossed metal ribs or wires that acts as a reflector for a microwave antenna and offers minimum wind resistance. [SPECT] *See* diffraction grating.

grating constant [OPTICS] The distance between consecutive diffraction centers of an ultrasonic wave which is producing a light diffraction spectrum. [SPECT] The distance between consecutive grooves of a diffraction grating.

grating spectrograph [SPECT] A grating spectroscope provided with a photographic camera or other device for recording the spectrum.

grating spectroscope [SPECT] A spectroscope which em-

GRATING SPECTROSCOPE

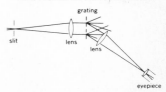

Transmission grating spectroscope.

ploys a transmission or reflection grating to disperse light, and usually also has a slit, a mirror or lenses to collimate the light sent through the slit and to focus the light dispersed by the grating into spectrum lines, and an eyepiece for viewing the spectrum.

Grätz number   *See* Graetz number.

grav   *See* G.

gravimeter   [ENG] A highly sensitive weighing device used for relative measurement of the force of gravity by detecting small weight differences of a constant mass at different points on the earth. Also known as gravity meter.

gravitation   [PHYS] The mutual attraction between all masses in the universe. Also known as gravitational attraction.

gravitational acceleration   [PHYS] The acceleration imparted to a body by the attraction of the earth; approximately equal to 980.7 cm/sec^2, or 32.2 ft/sec^2.

gravitational astronomy   *See* celestial mechanics.

gravitational attraction   *See* gravitation.

gravitational clustering   [ASTRON] A theory that attributes the hierarchy structure of the universe to growth of density fluctuations in a statistically uniform and isotropic universe.

gravitational collapse   [ASTRON] The implosion of a star or other astronomical body from an initial size to a size hundreds or thousands of times smaller.

gravitational constant   [MECH] The constant of proportionality in Newton's law of gravitation, equal to the gravitational force between any two particles times the square of the distance between them, divided by the product of their masses. Also known as constant of gravitation.

gravitational convection   *See* natural convection.

gravitational displacement   [MECH] The gravitational field strength times the gravitational constant. Also known as gravitational flux density.

gravitational energy   *See* gravitational potential energy.

gravitational field   [MECH] The field in a region in space in which a test particle would experience a gravitational force; quantitatively, the gravitational force per unit mass on the particle at a particular point.

gravitational-field theory   [RELAT] A theory in which gravity is treated as a field, as opposed to a theory in which the force acts instantaneously at a distance.

gravitational flux density   *See* gravitational displacement.

gravitational force   [MECH] The force on a particle due to its gravitational attraction to other particles.

gravitational geon   [PHYS] A hypothetical gravitational field that is held together by its own gravitational attraction.

gravitational instability   [MECH] Instability of a dynamic system in which gravity is the restoring force.

gravitational mass   [PHYS] The mass of a particle as it determines the force it experiences in a gravitational field; equal to inertial mass according to the equivalence principle.

gravitational potential   [MECH] The amount of work which must be done against gravitational forces to move a particle of unit mass to a specified position from a reference position, usually a point at infinity.

gravitational potential energy   [MECH] The energy that a system of particles has by virtue of their positions, equal to the work that must be done against gravitational forces to assemble the particles from some reference configuration, such as mutually infinite separation. Also known as gravitational energy.

gravitational pressure   *See* hydrostatic pressure.

gravitational radiation   *See* gravitational wave.

gravitational radius   *See* Schwarzschild radius.

gravitational red shift   [RELAT] A displacement of spectral

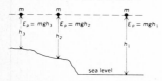

GRAVITATIONAL POTENTIAL ENERGY

The gravitational potential energy E_p of a body of weight mg (m = mass, g = force of gravity) for different reference levels (h_1, h_2, h_3).

lines toward the red when the gravitational potential at the observer of the light is greater than at its source.

gravitational repulsion [PHYS] Hypothetical repulsion of matter and antimatter; however, experimental results indicate that matter and antimatter attract according to the same laws as matter and matter.

gravitational systems of units [MECH] Systems in which length, force, and time are regarded as fundamental, and the unit of force is the gravitational force on a standard body at a specified location on the earth's surface.

gravitational wave [RELAT] A propagating gravitational field predicted by general relativity, which is produced by some change in the distribution of matter; it travels at the speed of light, exerting forces on masses in its path. Also known as gravitational radiation.

gravitometer *See* densimeter.

graviton [PHYS] A theoretically deduced particle postulated as the quantum of the gravitational field, having a rest mass and charge of zero and a spin of 2.

gravity [MECH] The gravitational attraction at the surface of a planet or other celestial body.

gravity anomaly [MECH] A correction which must be added to a theoretical model describing gravity on the surface of the earth, resulting in the main from surface irregularities.

gravity meter [ENG] **1.** U-tube-manometer type of device for direct reading of solution specific gravities in semimicro quantities. **2.** An electrical device for measuring variations in gravitation through different geologic formations; used in mineral exploration. **3.** *See* gravimeter.

gravity pendulum *See* pendulum.

gravity vector [MECH] The force of gravity per unit mass at a given point. Symbolized **g**.

gravity wave [FL MECH] **1.** A wave at a gas-liquid interface which depends primarily upon gravitational forces, surface tension and viscosity being of secondary importance. **2.** A wave in a fluid medium in which restoring forces are provided primarily by buoyancy (that is, gravity) rather than by compression.

gray [NUCLEO] The International System unit of absorbed dose, equal to the energy imparted by ionizing radiation to a mass of matter corresponding to 1 joule per kilogram. Symbolized Gy.

graybody [THERMO] An energy radiator which has a blackbody energy distribution, reduced by a constant factor, throughout the radiation spectrum or within a certain wavelength interval. Also known as nonselective radiator.

gray filter *See* neutral-density filter.

gray scale [OPTICS] A series of achromatic tones having varying proportions of white and black, to give a full range of grays between white and black; a gray scale is usually divided into 10 steps.

grazing angle [PHYS] A very small glancing angle.

grazing incidence [PHYS] Incidence at a small glancing angle.

grease spot photometer [OPTICS] A photometer in which the light sources whose intensities are to be compared illuminate a thin sheet of opaque paper with a translucent spot at the center.

great circle [MATH] The circle on the two-sphere produced by a plane passing through the center of the sphere.

greatest common divisor [MATH] **1.** The greatest common divisor of integers $n_1, n_2, \ldots, n_k$ is the largest of all integers that divide each n_i. Abbreviated gcd. Also known as highest common factor (hcf). **2.** More generally, a nonzero element d of a commutative ring R is the greatest common divisor of

GREAT CIRCLE

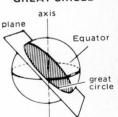

Diagram of a great circle described by a plane through the center of the earth.

elements $a_1, \ldots, a_n$ if d is a divisor of $a_1, \ldots, a_n$, and if any nonzero element b that is a divisor of $a_1, \ldots, a_n$ is also a divisor of d.

greatest elongation [ASTRON] The maximum angular distance of a body of the solar system from the sun, as observed from the earth.

greatest lower bound [MATH] The greatest lower bound of a set of numbers S is the largest number among the lower bounds of S. Abbreviated glb. Also known as infimum (inf).

Great Nebula of Orion *See* Orion Nebula.

great year [ASTRON] The period of one complete cycle of the equinoxes around the ecliptic, about 25,800 years. Also known as platonic year.

Greco-Latin square [STAT] An arrangement of combinations of two sets of letters (one set Greek, the other Roman) in a square array, in such a way that no letter occurs more than once in the array. Also known as orthogonal Latin square.

green [OPTICS] The hue evoked in an average observer by monochromatic radiation having a wavelength in the approximate range from 492 to 577 nanometers; however, the same sensation can be produced in a variety of other ways.

green flash [ASTRON] A brilliant green coloration of the upper limb of the sun occasionally observed just as the sun's apparent disk is about to sink below a distant clear horizon. Also known as blue flash; blue-green flame; green segment; green sun.

green laser [OPTICS] A gas laser using mercury and argon to generate a green line at 5225 angstroms (522.5 nanometers), corresponding to the wavelength that is most readily transmitted through seawater.

Green's dyadic [MATH] A vector operator which plays a role analogous to a Green's function in a partial differential equation expressed in terms of vectors.

green segment *See* green flash.

Green's function [MATH] A function, associated with a given boundary value problem, which appears as an integrand for an integral representation of the solution to the problem.

Green's identities [MATH] Formulas, obtained from Green's theorem, which relate the volume integral of a function and its gradient to a surface integral of the function and its partial derivatives.

Green's theorem [MATH] Under certain general conditions, an integral along a closed curve C involving the sum of functions $P(x,y)$ and $Q(x,y)$ is equal to a surface integral, over the region D enclosed by C, of the partial derivatives of P and Q; namely,

$$\int_C P\,dx + Q\,dy = \int\int_D \left(\frac{\partial Q}{\partial x} - \frac{\partial P}{\partial y} \right)\,dx\,dy.$$

green sun *See* green flash.

Greenwich civil time *See* Greenwich mean time.

Greenwich mean time [ASTRON] Mean solar time at the meridian of Greenwich. Abbreviated GMT. Also known as Greenwich civil time; universal time; Z time; zulu time.

Greenwich meridian [GEOD] The meridian passing through Greenwich, England, and serving as the reference for Greenwich time; it also serves as the origin of measurement of longitude.

Gregorian telescope [OPTICS] A reflecting telescope having a paraboloidal mirror with a hole in the center and a small secondary (concave ellipsoidal) mirror placed beyond the focus of the primary mirror; light is reflected to the secondary mirror and back to an eyepiece at the hole; the telescope produces an erect image but a small field of view.

Gregory formula [MATH] A formula used in the numerical evaluation of integrals derived from the Newton formula.

Gregory's series [MATH] An expression for π: $\pi/4 = 1 - \frac{1}{3} + \frac{1}{5} - \frac{1}{7} + \frac{1}{9} - \cdots$.

Greninger chart [CRYSTAL] A chart that enables angular relations between planes and zones in a crystal to be read directly from an x-ray diffraction photograph.

grenz ray [NUCLEO] An x-ray produced at the long-wavelength end of the x-ray spectrum, involving wavelengths of the order of 1 to 10 angstroms (0.1 to 1 nanometer), by using special x-ray tubes that operate at voltages from only 5000 to 15,000 volts.

grid [ELEC] **1.** A metal plate with holes or ridges, used in a storage cell or battery as a conductor and a support for the active material. **2.** Any systematic network, such as of telephone lines or power lines. [ELECTR] An electrode located between the cathode and anode of an electron tube, which has one or more openings through which electrons or ions can pass, and serves to control the flow of electrons from cathode to anode.

grid-anode transconductance *See* transconductance.

grid battery *See* C battery.

grid bias [ELECTR] The direct-current voltage applied between the control grid and cathode of an electron tube to establish the desired operating point. Also known as bias; C bias; direct grid bias.

grid-bias cell *See* bias cell.

grid blocking [ELECTR] **1.** Method of keying a circuit by applying negative grid bias several times cutoff value to the grid of a tube during key-up conditions; when the key is down, the blocking bias is removed and normal current flows through the keyed circuit. **2.** Blocking of capacitance-coupled stages in an amplifier caused by the accumulation of charge on the coupling capacitors due to grid current passed during the reception of excessive signals.

grid blocking capacitor *See* grid capacitor.

grid capacitor [ELECTR] A small capacitor used in the grid circuit of an electron tube to pass signal current while blocking the direct-current anode voltage of the preceding stage. Also known as grid blocking capacitor; grid condenser.

grid cathode capacitance [ELECTR] Capacitance between the grid and the cathode in a vacuum tube.

grid characteristic [ELECTR] Relationship of grid current to grid voltage of a vacuum tube.

grid circuit [ELECTR] The circuit connected between the grid and cathode of an electron tube.

grid condenser *See* grid capacitor.

grid conductance *See* electrode conductance.

grid control [ELECTR] Control of anode current of an electron tube by variation (control) of the control grid potential with respect to the cathode of the tube.

grid driving power [ELECTR] Average product of the instantaneous value of the grid current and of the alternating component of the grid voltage over a complete cycle; this comprises the power supplied to the biasing device and to the grid.

grid-glow tube [ELECTR] A glow-discharge tube in which one or more control electrodes initiate but do not limit the anode current except under certain operating conditions.

gridistor [ELECTR] Field-effect transistor which uses the principle of centripetal striction and has a multichannel structure, combining advantages of both field effect transistors and minority carrier injection transistors.

grid leak [ELECTR] A resistor used in the grid circuit of an electron tube to provide a discharge path for the grid capacitor and for charges built up on the control grid.

grid locking [ELECTR] Defect of tube operation in which the grid potential becomes continuously positive due to excessive grid emission.

grid-plate transconductance *See* transconductance.

grid-rectification meter [ENG] A type of vacuum-tube voltmeter in which the grid and cathode of a tube act as a diode rectifier, and the rectified grid voltage, amplified by the tube, operates a meter in the plate circuit.

grid resistor [ELECTR] A general term used to denote any resistor in the grid circuit.

grid return [ELECTR] External conducting path for the return grid current to the cathode.

grid spectrometer [SPECT] A grating spectrometer in which a large increase in light flux without loss of resolution is achieved by replacing entrance and exit slits with grids consisting of opaque and transparent areas, patterned to have large transmittance only when the entrance grid image coincides with that of the exit grid.

grid voltage [ELECTR] The voltage between a grid and the cathode of an electron tube.

Griebe-Schiebe method [SOLID STATE] A method of observing the piezoelectric behavior of small crystals, in which the crystals are placed between two electrodes connected to the resonant circuit of an oscillator, and tuning of the resonant circuit results in jumps in the oscillator frequency which produce clicks in headphones or a loudspeaker attached to the plate circuit of the oscillator.

Griffith's criterion [MECH] A criterion for the fracture of a brittle material under biaxial stress, based on the theory that the strength of such a material is limited by small cracks.

Griffiths' method [THERMO] A method of measuring the mechanical equivalent of heat in which the temperature rise of a known mass of water is compared with the electrical energy needed to produce this rise.

Gronwall's lemma [MATH] If

$$f(x) \leq g(x) + \int_a^x h(y)f(y) \, dy,$$

for all x on an interval $[a,b]$, where f, g, and h are real-valued piecewise continuous functions on $[a,b]$ and h is nonnegative, then

$$f(x) \leq g(x) + \int_a^x h(y)g(y) \exp\left[\int_y^x h(u) \, du\right] dy$$

for all x in $[a,b]$.

gross errors [STAT] Errors that occur when a measurement process is subject occasionally to being very far off.

gross ton *See* ton.

Grothendieck group [MATH] Let $F(M)$ be the free Abelian group generated by a commutative monoid M, and let $[x]$ denote the generator of $F(M)$ corresponding to x in M; then the Grothendieck group $K(M)$ is the quotient group of $K(M)$ over the subgroup generated by all elements of the type $[x + y] - [x] - [y]$, where x and y are in M.

ground [ELEC] **1.** A conducting path, intentional or accidental, between an electric circuit or equipment and the earth, or some conducting body serving in place of the earth. Abbreviated gnd. Also known as earth (British usage); earth connection. **2.** To connect electrical equipment to the earth or to some conducting body which serves in place of the earth.

ground current *See* earth current.

ground discharge *See* cloud-to-ground discharge.

grounded-anode amplifier *See* cathode follower.

grounded-base connection [ELECTR] A transistor circuit in which the base electrode is common to both the input and output circuits; the base need not be directly connected to circuit ground. Also known as common-base connection.

grounded-collector connection [ELECTR] A transistor circuit in which the collector electrode is common to both the input and output circuits; the collector need not be directly connected to circuit ground. Also known as common-collector connection.

grounded-emitter connection [ELECTR] A transistor circuit in which the emitter electrode is common to both the input and output circuits; the emitter need not be directly connected to circuit ground. Also known as common-emitter connection.

grounded-plate amplifier *See* cathode follower.

ground fallout plot [NUCLEO] Time lines plotted within a radioactive fallout plot that approximate the distances to which local fallout will have spread at the end of succeeding hours; these lines are determined by using the rate of fall of the radioactive material and the mean wind vector (fallout wind) in the layer of air through which the material is falling.

ground glass [OPTICS] A sheet of matte-surfaced glass on the back of a view camera or process camera so that the image of the subject can be focused on it; it is exactly in the film plane.

grounding [ELEC] Intentional electrical connection to a reference conducting plane, which may be earth, but which more generally consists of a specific array of interconnected electrical conductors referred to as the grounding conductor.

ground junction *See* grown junction.

ground plane [ELEC] A grounding plate, aboveground counterpoise, or arrangement of buried radial wires required with a ground-mounted antenna that depends on the earth as the return path for radiated radio-frequency energy.

ground plane antenna [ELECTROMAG] Vertical antenna combined with a grounded horizontal disk, turnstile element, or similar ground plane simulation; such antennas may be mounted several wavelengths above the ground, and provide a low radiation angle.

ground-reflected wave [ELECTROMAG] Component of the ground wave that is reflected from the ground.

ground state [QUANT MECH] The stationary state of lowest energy of a particle or a system of particles.

ground system [ELECTROMAG] The portion of an antenna that is closely associated with an extensive conducting surface, which may be the earth itself.

ground-to-cloud discharge [GEOPHYS] A lightning discharge in which the original streamer processes start upward from an object located on the ground.

ground wave [GEOPHYS] A radio wave that is propagated along the earth and is ordinarily affected by the presence of the ground and the troposphere; includes all components of a radio wave over the earth except ionospheric and tropospheric waves. Also known as surface wave.

group [MATH] A set G with an associative binary operation where $g_1 \cdot g_2$ always exists and is an element of G; each g has an inverse element g^{-1}, and G contains an identity element.

group algebra [MATH] **1.** The free algebra of a group over a field. **2.** An algebra whose elements are functions and in which the multiplication of two elements f and g is defined by convolution; that is, $f(x)$ equals the integral of $f(x - t)g(t)$ with respect to t.

group diffusion method [NUCLEO] An approximation used in studying the diffusion of neutrons in a nuclear reactor, in which the range of neutron energies, from source energy to thermal energy, is divided into a finite number of intervals, or groups, and the neutrons in each group are assumed to diffuse with no loss of energy, until they have undergone the average

GROUNDED-EMITTER CONNECTION

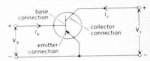

Schematic diagram of grounded-emitter connection of an alloy junction transistor.

GROUNDING

to the electrical load being served

Grounding used in a single-phase three-wire 240/120-volt service; A and B conduct the current; N is the grounding wire.

number of collisions needed to reduce their energy to that of the next lower group.

group frequency [ELECTROMAG] Frequency corresponding to group velocity of propagated waves in a transmission line or waveguide.

group of stars [ASTRON] A number of stars moving in the same direction with the same speed.

groupoid [MATH] A set having a binary relation everywhere defined.

group velocity [PHYS] The velocity of the envelope of a group of interfering waves having slightly different frequencies and phase velocities.

grown-diffused transistor [ELECTR] A junction transistor in which the final junctions are formed by diffusion of impurities near a grown junction.

grown junction [ELECTR] A junction produced by changing the types and amounts of donor and acceptor impurities that are added during the growth of a semiconductor crystal from a melt. Also known as ground junction.

grown-junction photocell [ELECTR] A photodiode consisting of a bar of semiconductor material having a pn junction at right angles to its length and an ohmic contact at each end of the bar.

grown-junction transistor [ELECTR] A junction transistor in which different impurities are placed in the melt in sequence as the silicon or germanium seed crystal is slowly withdrawn, to produce the alternate pn and np junctions.

growth curve [NUCLEO] A curve showing how some quantity associated with a radioactive transformation or induced nuclear reaction increases with time.

growth index [MATH] For a function of bounded growth f, the smallest real number a such that for some positive real constant M the quantity Me^{ax} is greater than the absolute value of $f(x)$ for all positive x; for a function that is not of bounded growth, the quantity $+\infty$.

growth spiral [CRYSTAL] A structure on a crystal surface, observed after growth, consisting of a growth step winding downward and outward in an Archimedean spiral which may be distorted by the crystal structure.

growth step [CRYSTAL] A ledge on a crystal surface, one or more lattice spacings high, where crystal growth can take place.

Grüneisen constant [SOLID STATE] Three times the bulk modulus of a solid times its linear expansion coefficient, divided by its specific heat per unit volume; it is reasonably constant for most cubic crystals. Also known as Grüneisen gamma.

Grüneisen equation of state [SOLID STATE] An equation of state for a solid, $pV + G(V) = \gamma E$, where p is the pressure on the solid, V is its molar volume, $G(V)$ is its potential energy, γ is the Grüneisen constant, and E is the thermal energy, equal to the integral over temperature, from absolute zero to the temperature of the solid, of the molar specific heat at constant volume.

Grüneisen gamma *See* Grüneisen constant.

Grüneisen relation [SOLID STATE] The relation stating that the electrical resistivity of a very pure metal is proportional to a mathematical function which depends on the ratio of the temperature to a characteristic temperature.

Grüneisen's first rule [SOLID STATE] An approximate formula, derived from the Grüneisen equation of state, which may be written $\Delta V = \gamma K_0 E$, where ΔV is the difference between the volume of the solid and its equilibrium volume, γ the Grüneisen constant, K_0 the isothermal compressibility at absolute zero, and E the thermal energy.

GROWTH SPIRAL

Crystal-growth spirals on the surface of a silicon carbide crystal magnified 250 diameters. Each spiral step originates in a screw dislocation defect. From the side the surface looks like an ascending ramp wound around a flat cone. *(General Electric)*

Grüneisen's second rule [SOLID STATE] An approximate formula, derived from the Grüneisen equation of state, which may be written $\beta = \gamma K_0 C_v / V$, where β is the volume coefficient of thermal expansion, γ the Grüneisen constant, K_0 the isothermal compressibility at absolute zero, C_v the molar specific heat at constant volume, and V the volume.

G space [MATH] A topological space X together with a topological group G and a continuous function on the cartesian product of X and G to X such that if the values of this function at (x,g) are denoted by xg, then $x(g_1g_2) = (xg_1)g_2$ and $xe = x$ where e is the identity in G and g_1, g_2 are elements in G.

G star [ASTRON] A star of spectral type G; many metallic lines are seen in the spectra, with hydrogen and potassium being strong; G stars are yellow stars, with surface temperatures of 5500–4200 K for giants, 6000–5000 K for dwarfs.

G string *See* field waveguide.

guarding [ELEC] A method of eliminating surface-leakage effects from measurements of electrical resistance which employs a low-resistance conductor in the vicinity of one of the terminals or a portion of the measuring circuit.

guard ring [THERMO] A device used in heat flow experiments to ensure an even distribution of heat, consisting of a ring that surrounds the specimen and is made of a similar material.

Gudden-Pohl effect [ELECTR] The momentary illumination produced when an electric field is applied to a phosphor previously excited by ultraviolet radiation.

Gudermannian [MATH] The function y of the variable x satisfying $\tan y = \sinh x$ or $\sin y = \tanh x$; written $\text{gd} x$.

guided wave [ELECTROMAG] A wave whose energy is concentrated near a boundary or between substantially parallel boundaries separating materials of different properties and whose direction of propagation is effectively parallel to these boundaries; waveguides transmit guided waves.

guide wavelength [ELECTROMAG] Wavelength of electromagnetic energy conducted in a waveguide; guide wavelength for all air-filled guides is always longer than the corresponding free-space wavelength.

guiding center [ELECTROMAG] A slowly moving point about which a charged particle rapidly revolves; this is used in an approximation for the motion of a charged particle in slowly varying electric and magnetic fields.

guiding telescope [OPTICS] A telescope that is mounted so that it remains parallel to a photographic telescope and is used by a person observing through it to supplement the clock motion in keeping the image of a celestial body motionless on a photographic plate.

Guillemin effect [ELECTROMAG] The tendency of a bent magnetorestrictive rod to straighten in a magnetic field parallel to its length.

Gukhman number [THERMO] A dimensionless number used in studying convective heat transfer in evaporation, equal to $(t_0 - t_m)/T_0$, where t_0 is the temperature of a hot gas stream, t_m is the temperature of a moist surface over which it is flowing, and T_0 is the absolute temperature of the gas stream. Symbolized Gu; N_{Gu}.

Guldberg and Waage law *See* mass action law.

Gunn amplifier [ELECTR] A microwave amplifier in which a Gunn oscillator functions as a negative-resistance amplifier when placed across the terminals of a microwave source.

Gunn diode *See* Gunn oscillator.

Gunn effect [ELECTR] Development of a rapidly fluctuating current in a small block of a semiconductor (perhaps n-type gallium arsenide) when a constant voltage above a critical value is applied to contacts on opposite faces.

Gunn oscillator [ELECTR] A microwave oscillator utilizing the Gunn effect. Also known as Gunn diode.

Gurevich effect [SOLID STATE] An effect observed in electric conductors in which phonon-electron collisions are important, in the presence of a temperature gradient, in which phonons carrying a thermal current tend to drag the electrons with them from hot to cold.

G value [NUCLEO] The number of molecules produced or destroyed for each 100 electron volts absorbed by a substance from ionizing radiation. Also known as G factor.

GW *See* gigawatt.

Gy *See* gray.

gyration tensor [SOLID STATE] A tensor characteristic of an optically active crystal, whose product with a unit vector in the direction of propagation of a light ray gives the gyration vector.

gyration vector [OPTICS] For light propagating in an optically active medium, a vector whose cross product with the time derivative of the electric displacement vector gives a negative contribution to the electric field.

gyrator [ELECTROMAG] A waveguide component that uses a ferrite section to give zero phase shift for one direction of propagation and 180° phase shift for the other direction; in other words, it causes a reversal of signal polarity for one direction of propagation but not for the other direction. Also known as microwave gyrator.

gyrator filter [ELECTR] A highly selective active filter that uses a gyrator which is terminated in a capacitor so as to have an inductive input impedance.

gyro *See* gyroscope.

gyrodynamics [MECH] The study of rotating bodies, especially those subject to precession.

gyrofrequency *See* cyclotron frequency.

gyromagnetic effect [ELECTROMAG] The rotation induced in a body by a change in its magnetization, or the magnetization resulting from a rotation.

gyromagnetic radius *See* Larmor radius.

gyromagnetic ratio [PHYS] **1.** The ratio of the magnetic dipole moment to the angular momentum for a classical, atomic, or nuclear system. **2.** Occasionally, the reciprocal of the quantity in the first definition.

gyromagnetics [ELECTROMAG] The study of the relation between the angular momentum and the magnetization of a substance as exhibited in the gyromagnetic effect.

gyroscope [ENG] An instrument that maintains an angular reference direction by virtue of a rapidly spinning, heavy mass; all applications of the gyroscope depend on a special form of Newton's second law, which states that a massive, rapidly spinning body rigidly resists being disturbed and tends to react to a disturbing torque by precessing (rotating slowly) in a direction at right angles to the direction of torque. Also known as gyro.

gyroscopic precession [MECH] The turning of the axis of spin of a gyroscope as a result of an external torque acting on the gyroscope; the axis always turns toward the direction of the torque.

gyroscopics [MECH] The branch of mechanics concerned with gyroscopes and their use in stabilization and control of ships, aircraft, projectiles, and other objects.

GYRATOR

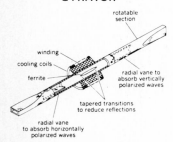

A practical nonreciprocal gyrator.

GYROSCOPE

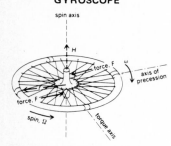

Illustration of gyroscope principle. Bicycle wheel with high spin velocity Ω has angular momentum $H = mr^2 \Omega$, where m and r are mass and radius of wheel. Torque T resulting from force F produces precession with small angular velocity $\omega = T/H$ about axis perpendicular to both spin axis and torque axis.

h   *See* Planck's constant.

H   *See* henry; hydrogen.

ha   *See* hectare.

Haar measure   [MATH] A measure on the Borel subsets of a locally compact topological group whose value on a Borel subset U is unchanged if every member of U is multiplied by a fixed element of the group.

Haar system   [MATH] A sequence of functions $f_1, f_2, \ldots, f_n$ such that $f_1, f_2, \ldots, f_k$ is a Chebyshev system for $k = 1, 2, \ldots, n$. Also known as complete T system.

habit   *See* crystal habit.

habit plane   [CRYSTAL] The crystallographic plane or system of planes along which certain phenomena such as twinning occur.

Hadamard product   [MATH] The Hadamard product of two power series,

$$P = \sum_{n=0}^{\infty} a_n x^n \text{ and } Q = \sum_{n=0}^{\infty} b_n x^n,$$

is the power series

$$P \times Q = \sum_{n=0}^{\infty} a_n b_n x^n.$$

Hadamard's example   [MATH] A solution of Laplace's equation in two variables, $u(x,t) = n^{-k} \sin(nx) \sinh(nt)$, whose properties show that the Cauchy problem for Laplace's equation is without physical significance because small errors in the specification of the initial conditions can distort the solution to an arbitrarily large extent.

Hadamard's inequality   [MATH] An inequality that gives an upper bound for the square of the absolute value of the determinant of a matrix in terms of the squares of the matrix entries; the upper bound is the product, over the rows of the matrix, of the sum of the squares of the absolute values of the entries in a row.

hadron   [PARTIC PHYS] An elementary particle which has strong interactions.

hadronic atom   [ATOM PHYS] An atom consisting of a negatively charged, strongly interacting particle orbiting around an ordinary nucleus.

hafnium   [CHEM] A metallic element, symbol Hf, atomic number 72, atomic weight 178.49; melting point 2000°C, boiling point above 5400°C.

Hagen-Poiseuille law   [FL MECH] In the case of laminar flow of fluid through a circular pipe, the loss of head due to fluid friction is 32 times the product of the fluid's viscosity, the pipe length, and the fluid velocity, divided by the product of the acceleration of gravity, the fluid density, and the square of the pipe diameter.

Hagen-Rubens relation   [OPTICS] An equation for the reflec-

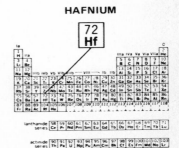

HAFNIUM

Periodic table of the chemical elements showing the position of hafnium.

tivity of a solid surface in terms of the frequency of radiation of the conductivity of the solid; it applies at wavelengths long enough that the product of the frequency and the relaxation time is much less than unity.

Hahn-Banach extension theorem [MATH] The theorem that every continuous linear functional defined on a subspace or linear manifold in a normed linear space X may be extended to a continuous linear functional defined on all of X.

Hahn decomposition [MATH] The Hahn decomposition of a measurable space X with signed measure m consists of two disjoint subsets A and B of X such that the union of A and B equals X, A is positive with respect to m, and B is negative with respect to m.

Hahn polynomials [MATH] Families of polynomials which are orthogonal with respect to probability distributions whose frequencies are

$$(\alpha + 1)(\alpha + 2) \cdots (\alpha + N)(\beta + 1)(\beta + 2) \cdots$$
$$(\beta + N - x)/x!(N - x)!$$

for $x = 0, 1, \ldots, N$.

Hahn technique [SOLID STATE] A method of studying changes in solids under various treatments that involves incorporating small amounts of radium into the solid and measuring the emanating power.

Haidinger brushes [OPTICS] Faint yellow, brushlike patterns that are observed when a bright surface is viewed through a polarizer such as a rotating Nicol prism or sheet of Polaroid film; believed to be caused by birefringence of fibers at the fovea of the eye.

Haidinger fringes [OPTICS] Interference fringes produced by nearly normal incidence of light on thick, flat plates. Also known as constant-angle fringes; constant-deviation fringes.

hair hygrometer [ENG] A hygrometer in which the sensing element is a bundle of human hair, which is held under slight tension by a spring and which expands and contracts with changes in the moisture of the surrounding air or gas.

hairline *See* air line.

halation [ELECTR] An area of glow surrounding a bright spot on a fluorescent screen, due to scattering by the phosphor or to multiple reflections at front and back surfaces of the glass faceplate. [OPTICS] A halo on a photographic image of a bright object caused by light reflected from the back of the film or plate.

half-angle formulas [MATH] In trigonometry, formulas that express the trigonometric functions of half an angle in terms of trigonometric functions of the angle.

half-bridge [ELEC] A bridge having two power supplies, located in two of the bridge arms, to replace the single power supply of a conventional bridge.

half-life [NUCLEO] The average time interval required for one-half of any quantity of identical radioactive atoms to undergo radioactive decay. Also known as half-value period; radioactive half-life.

half line *See* ray.

half-moon [ASTRON] The moon as seen in the first quarter and the last, or third, quarter.

half-period zones *See* Fresnel zones.

half plane [MATH] The portion of a plane lying on one side of some line in the plane; in particular, all points of the complex plane either above or below the real axis.

half-power frequency [ELECTR] One of the two values of frequency, on the sides of an amplifier response curve, at which the voltage is $1/\sqrt{2}$ (70.7%) of a midband or other reference value. Also known as half-power point.

half-power point [ELECTR] **1.** A point on the graph of some quantity in an antenna, network, or control system, versus frequency, distance, or some other variable at which the power is half that of a nearby point at which power is a maximum. **2.** *See* half-power frequency.

half-shade plate [OPTICS] A half-wave plate that is placed near the polarizer of a polariscope, between it and the analyzer.

half-side formulas [MATH] In trigonometry, formulas that express the tangents of one-half of each of the sides of a spherical triangle in terms of its angles.

half-silvered surface [OPTICS] A surface covered with metallic film of a thickness such that approximately half the light falling on it at normal incidence is reflected and half is transmitted.

half space [MATH] A space bounded only by an infinite plane.

half step *See* semitone.

half thickness [PHYS] The thickness of a sheet of material which reduces the intensity of a beam of radiation passing through it to one-half its initial value. Also known as half-value layer; half-value thickness.

half time [NUCLEO] The time during which half the radioactive material resulting from a nuclear explosion remains in the atmosphere.

half-value layer *See* half thickness.

half-value period *See* half-life.

half-value thickness *See* half thickness.

half-wave [ELEC] Pertaining to half of one cycle of a wave. [ELECTROMAG] Having an electrical length of a half wavelength.

half-wave antenna [ELECTROMAG] An antenna whose electrical length is half the wavelength being transmitted or received.

half-wave dipole *See* dipole antenna.

half-wavelength [ELECTROMAG] The distance corresponding to an electrical length of half a wavelength at the operating frequency of a transmission line, antenna element, or other device.

half-wave plate [OPTICS] A thin section of a doubly refracting crystal, of a thickness such that the ordinary and extraordinary components of normally incident light emerge from it with a phase difference corresponding to an odd number of half wavelengths.

half-wave rectification [ELECTR] Rectification in which current flows only during alternate half cycles.

half-wave rectifier [ELECTR] A rectifier that provides half-wave rectification.

half-width [MATH] For a function which has a maximum and falls off rapidly on either side of the maximum, the difference between the two values of the independent variable for which the dependent variable has one-half its maximum value.

Hall accelerator [PL PHYS] A plasma accelerator based on the Hall effect.

Hall angle [ELECTROMAG] The electric field, resulting from the Hall effect, perpendicular to a current, divided by the electric field generating the current.

Hall coefficient [ELECTROMAG] A measure of the Hall effect, equal to the transverse electric field (Hall field) divided by the product of the current density and the magnetic induction. Also known as Hall constant.

Hall constant *See* Hall coefficient.

Hall effect [ELECTROMAG] The development of a transverse electric field in a current-carrying conductor placed in a magnetic field; ordinarily the conductor is positioned so that the magnetic field is perpendicular to the direction of current flow and the electric field is perpendicular to both.

Hall-effect gaussmeter [ENG] A gaussmeter that consists of a thin piece of silicon or other semiconductor material which is inserted between the poles of a magnet to measure the magnetic field strength by means of the Hall effect.

Hall-effect isolator [ELECTROMAG] An isolator that makes use of the Hall effect in a semiconductor plate mounted in a magnetic field, to provide greater loss in one direction of signal travel through a waveguide than in the other direction.

Hall mobility [SOLID STATE] The product of conductivity and the Hall constant for a conductor or semiconductor; a measure of the mobility of the electrons or holes in a semiconductor.

Hall plane [MATH] A finite plane defined by a Hall system.

Hall's theorem *See* marriage theorem.

Hall system [MATH] The cartesian product $F \times F$, where F is a Galois field of order greater than 2, on which an addition and a multiplication operation are defined such that for any a, b, c, and d in F the sum of (a,b) and (c,d) is $(a+c, b+d)$ and the product of (a,b) and (c,d) equals (ac,bd) if $d = 0$, and equals $(ac - bd^{-1}f(c), ad - bc + br)$, if $d \neq 0$, where $f(x) = x^2 - rx - s$ is a specified quadratic polynomial that is irreducible over F.

Hall voltage [ELECTR] The no-load voltage developed across a semiconductor plate due to the Hall effect, when a specified value of control current flows in the presence of a specified magnetic field.

Hallwachs effect [PHYS] The ability of ultraviolet radiation to discharge a negatively charged body in a vacuum.

halo [ELECTR] An undesirable bright or dark ring surrounding an image on the fluorescent screen of a television cathode-ray tube; generally due to overloading or maladjustment of the camera tube. [OPTICS] A ring around the photographic image of a bright source caused by light scattering in any one of a number of possible ways.

halogen counter [NUCLEO] A Geiger counter in which the self-quenching action is provided by a halogen gas, such as chlorine or bromine.

Hamel basis [MATH] For a normed space, a collection of vectors with every finite subset linearly independent, while any vector of the space is a linear combination of at most countably many vectors from this subset.

Hamilton-Cayley theorem *See* Cayley-Hamilton theorem.

Hamiltonian cycle *See* Hamiltonian path.

Hamiltonian function [MECH] A function of the generalized coordinates and momenta of a system, equal in value to the sum over the coordinates of the product of the generalized momentum corresponding to the coordinate, and the coordinate's time derivative, minus the Lagrangian of the system; it is numerically equal to the total energy if the Lagrangian does not depend on time explicitly; the equations of motion of the system are determined by the functional dependence of the Hamiltonian on the generalized coordinates and momenta.

Hamiltonian graph [MATH] A graph which has a Hamiltonian path.

Hamiltonian operator *See* energy operator.

Hamiltonian path [MATH] A path along the edges of a graph that traverses every vertex exactly once and terminates at its starting point. Also known as Hamiltonian cycle.

Hamilton-Jacobi equation [MATH] A particular partial differential equation useful in studying certain systems of ordinary equations arising in the calculus of variations, dynamics, and optics: $H(q_1, \ldots, q_n, \partial\phi/\partial q_1, \ldots, \partial\phi/\partial q_n, t) + \partial\phi/\partial t = 0$, where $q_1, \ldots, q_n$ are generalized coordinates, t is the time coordinate, H is the Hamiltonian function, and ϕ is a function that generates a transformation by means of which the generalized coordinates and momenta may be expressed in terms of new generalized coordinates and momenta which are constants of motion.

Hamilton-Jacobi theory [MATH] The study of the solutions of the Hamilton-Jacobi equation and the information they provide concerning solutions of the related systems of ordinary differential equations. [MECH] A theory that provides a means for discussing the motion of a dynamic system in terms of a single partial differential equation of the first order, the Hamilton-Jacobi equation.

Hamilton's equations of motion [MECH] A set of first-order, highly symmetrical equations describing the motion of a classical dynamical system, namely $\dot{q}_j = \partial H/\partial p_j$, $\dot{p}_j = -\partial H/\partial q_j$; here q_j ($j = 1, 2, \ldots$) are generalized coordinates of the system, p_j is the momentum conjugate to q_j, and H is the Hamiltonian. Also known as canonical equations of motion.

Hamilton's principle [MECH] A variational principle which states that the path of a conservative system in configuration space between two configurations is such that the integral of the Lagrangian function over time is a minimum or maximum relative to nearby paths between the same end points and taking the same time.

Hammerstein equation [MATH] The nonlinear integral equation

$$x(t) = y(t) + \int_a^b K(t,t')f[t',x(t')]\,dt'.$$

hammer track [NUCLEO] A hammer or T-shaped track in a nuclear emulsion that is formed by a nucleus that comes to rest in the emulsion and decays into two fragments that travel in opposite directions.

Hampson process [CRYO] A process for liquefying gases which resembles the Linde process except that the Joule-Thomson expansion reduces the gas pressure to approximately atmospheric pressure.

hand-and-foot counter *See* hand-and-foot monitor.

hand-and-foot monitor [NUCLEO] An instrument routinely used to monitor the hands and feet of atomic energy workers as they leave locations in which radioactive materials are handled. Also known as hand-and-foot counter.

handedness [PHYS] A division of objects, such as coordinate systems, screws, and circularly polarized light beams, into two classes (right and left), which distinguishes an object from a mirror image but not from a rotated object.

hand lens *See* simple microscope.

Hankel functions [MATH] The Bessel functions of the third kind, occurring frequently in physical studies.

Hankel's formula [MATH] A formula stating that $1/\Gamma(z)$ equals the integral of $(1/2\pi i)e^t t^{-z}$ over a contour which starts at $-\infty$, circles the origin in a counterclockwise direction, and returns to $-\infty$.

Hankel's symbol [MATH] The symbol (ν, m), where ν is any number and m is a positive integer, used to represent the quantity $[(\nu - m + \frac{1}{2})(\nu - m + \frac{3}{2}) \cdots (\nu + m - \frac{1}{2})]/m!$.

Hankel transform [MATH] The Hankel transform of order m of a real function $f(t)$ is the function $F(s)$ given by the integral from 0 to ∞ of $f(t)tJ_m(st)dt$, where J_m denotes the mth-order Bessel function. Also known as Bessel transform; Fourier-Bessel transform.

hard cosmic ray [NUCLEO] A cosmic-radiation component that penetrates a moderate thickness of an absorber, such as 4 inches (10 centimeters) of lead.

hard data [SCI TECH] Data in the form of numbers or graphs, as opposed to qualitative information.

hardness [ELECTROMAG] That quality which determines the penetrating ability of x-rays; the shorter the wavelength, the harder and more penetrating the rays.

hard radiation [PHYS] Radiation whose particles or photons have a high energy and, as a result, readily penetrate all kinds of materials, including metals.

hard superconductor [CRYO] A superconductor that requires a strong magnetic field, over 1000 oersteds (79,577 amperes per meter), to destroy superconductivity; niobium and vanadium are examples.

hard tube *See* high-vacuum tube.

hard x-ray [ELECTR] An x-ray having high penetrating power.

Hardy-Littlewood fractional integral [MATH] A generalization of repeated integration of a function f from a fixed lower limit a to a larger number x, given by the integral from a to x of $[\Gamma(\alpha)]^{-1}(x-t)^{\alpha-1}f(t) \, dt$, where the real part of α is positive.

Hardy's formula [MATH] The integral of a real-valued function f on an interval $[a,b]$ is approximated by h $[0.28f(a) + 1.62f(a + h) + 2.2f(a + 3h) + 1.62f(a+5h) + 0.28f(b)]$, where $h = (b - a)/6$.

Hardy space [MATH] One of a collection of normed linear spaces, denoted by H^p, whose points are complex analytic functions on the interior of the unit disk; the norm in the Hardy space H^p is given

by $\lim_{r \to 1} \left\{ \dfrac{1}{2\pi} \int_{-\pi}^{\pi} |f(re^{i\theta})| \, d\theta \right\}^{1/p}$ if $0 < p < \infty$;

by $\lim_{r \to 1} \left\{ \dfrac{1}{2\pi} \int_{-\pi}^{\pi} \log^+ |f(re^{i\theta})| \, d\theta \right\}$ if $p = 0$,

where $\log^+$ is the positive part of the logarithmic function;

and by $\lim_{r \to 1} \sup \{ |f(re^{i\theta})| : -\pi \leq 0 \leq \pi \}$ if $p = \infty$.

Hare's hygrometer [ENG] A type of hydrometer in which the ratio of the densities of two liquids is determined by measuring the heights to which they rise in two vertical glass tubes, connected at their upper ends, when suction is applied.

Haring cell [PHYS CHEM] An electrolytic cell with four electrodes used to measure electrolyte resistance and polarization of electrodes.

Harker-Kasper inequalities [SOLID STATE] Inequalities used in the analysis of crystal structure by x-ray diffraction which relate the structure factors and help to determine their phase factors.

Harkin's rule [PHYS] An empirical rule for the calculation of the nuclear abundances of an element's isotopes stating that isotopes with an odd mass number are less abundant than their even-mass-number neighbors.

harmonic [ACOUS] One of a series of sounds, each of which has a frequency which is an integral multiple of some fundamental frequency. [MATH] A solution of Laplace's equation which is separable in a specified coordinate system. [PHYS] A sinusoidal component of a periodic wave, having a frequency that is an integral multiple of the fundamental frequency. Also known as harmonic component.

harmonic analysis [MATH] A study of functions by attempting to represent them as infinite series or integrals which involve functions from some particular well-understood family; it subsumes studying a function via its Fourier series. [PHYS] Any method of identifying and evaluating the harmonics that make up a complex waveform of sound pressure, voltage, current, or some other varying quantity.

harmonic antenna [ELECTROMAG] An antenna whose electrical length is an integral multiple of a half-wavelength at the operating frequency of the transmitter or receiver.

harmonic component *See* harmonic.

harmonic conjugates [MATH] **1.** Two points, P_3 and P_4, that are collinear with two given points, P_1 and P_2, such that P_3 lies in the line segment $P_1 P_2$ while P_4 lies outside it, and, if x_1, x_2, x_3, and x_4 are the abscissas of the points, $(x_3 - x_1)/(x_3 - x_2) = -(x_4 - x_1)/(x_4 - x_2)$. **2.** A pair of harmonic functions, u and v, such that $u + iv$ is an analytic function, or, equivalently, u and v satisfy the Cauchy-Riemann equations.

harmonic content [PHYS] The components remaining after the fundamental frequency has been removed from a complex wave.

harmonic detector [ELECTR] Voltmeter circuit so arranged as to measure only a particular harmonic of the fundamental frequency.

harmonic distortion [ELECTR] Nonlinear distortion in which undesired harmonics of a sinusoidal input signal are generated because of circuit nonlinearity.

harmonic division [MATH] The division of a line segment externally and internally in the same ratio; that is, the division of a line segment by the harmonic conjugates of its end points.

harmonic echo [ACOUS] An echo that appears to be higher in pitch than the original sound, due to enhancement of harmonics in the original complex tone.

harmonic fields [ELECTROMAG] The sinusoidal Fourier components of a magnetic or other field confined to a finite region of space; their half-wavelengths are integral divisors of the length of the space in which the field is confined.

harmonic frequency [PHYS] An integral multiple of the fundamental frequency of a periodic wave.

harmonic function [MATH] **1.** A function of two real variables which is a solution of Laplace's equation in two variables. **2.** A function of three real variables which is a solution of Laplace's equation in three variables.

harmonic mean [MATH] For n positive numbers $x_1, x_2, \ldots, x_n$ their harmonic mean is the number $n/(1/x_1 + 1/x_2 + \cdots + 1/x_n)$.

harmonic measure [MATH] Let D be a domain in the complex plane bounded by a finite number of Jordan curves Γ, and let Γ be the disjoint union of α and β, where α and β are Jordan arcs; the harmonic measure of α with respect to D is the harmonic function on D which assumes the value 1 on α and the value 0 on β.

harmonic motion [MECH] A periodic motion that is a sinusoidal function of time, that is, motion along a line given by the equation $x = a \cos(kt + \theta)$, where t is the time parameter, and a, k, and θ are constants. Also known as harmonic vibration; simple harmonic motion (SHM).

harmonic oscillator [MECH] Any physical system that is bound to a position of stable equilibrium by a restoring force or torque proportional to the linear or angular displacement from this position. [PHYS] Anything which has equations of motion that are the same as the system in the mechanics definition. Also known as linear oscillator; simple oscillator.

harmonic pencil [MATH] The configuration of four lines, passing through a single point, such that any line that is not parallel to one of the four cuts the four lines at points which are harmonic conjugates.

harmonic progression [MATH] A sequence of numbers whose reciprocals form an arithmetic progression.

harmonic range [MATH] The configuration of four collinear points which are harmonic conjugates.

harmonic series [MATH] A series whose terms form a harmonic progression.

harmonic synthesizer [MECH] A machine which combines elementary harmonic constituents into a single periodic function; a tide-predicting machine is an example.

harmonic vibration *See* harmonic motion.

harmonic vibration-rotation band [SPECT] A vibration-rotation band of a molecule in which the harmonic oscillator approximation holds for the vibrational levels, so that the vibrational levels are equally spaced.

harmonic wave [PHYS] A transverse waveform obtained by mapping onto a time base the periodic up and down excursions of simple harmonic motion.

Harris flow [ELECTR] Electron flow in a cylindrical beam in which a radial electric field is used to overcome space charge divergence.

Harrison's gridiron pendulum [DES ENG] A type of compensated pendulum that has five iron rods and four brass rods arranged so that the effects of their thermal expansion cancel.

Hartley oscillator [ELECTR] A vacuum-tube oscillator in which the parallel-tuned tank circuit is connected between grid and anode; the tank coil has an intermediate tap at cathode potential, so the grid-cathode portion of the coil provides the necessary feedback voltage.

Hartmann dispersion formula [OPTICS] A semiempirical formula relating the index of refraction n and wavelengths λ; $n = n_0 + a/(\lambda - \lambda_0)$, where n_0, a, and λ_0 are empirical constants. Also known as Cornu-Hartmann formula.

Hartmann flow [PL PHYS] The steady flow of an electrically conducting fluid between two parallel plates when there is a uniform applied magnetic field normal to the plates.

Hartmann generator [ENG ACOUS] A device in which shock waves generated at the edges of a nozzle by a supersonic gas jet resonate with the opening of a small cylindrical pipe, placed opposite the nozzle, to produce powerful ultrasonic sound waves.

Hartmann number [PL PHYS] A dimensionless number which gives a measure of the relative importance of drag forces resulting from magnetic induction and viscous forces in Hartmann flow, and determines the velocity profile for such flow.

Hartmann test [OPTICS] A test for telescope mirrors in which the mirror is covered with a screen with regularly spaced holes, and a photographic plate is placed near the focus; for a perfect mirror, this results in regularly spaced dots on the plate. [SPECT] A test for spectrometers in which light is passed through different parts of the entrance slit; any resulting changes of the spectrum indicate a fault in the instrument.

hartree [ATOM PHYS] A unit of energy used in studies of atomic spectra and structure, equal (in centimeter-gram-second units) to $4\pi^2 me^4/h^2$, where e and m are the charge and mass of the electron, and h is Planck's constant; equal to approximately 27.21 electron volts or 4.360×10^{-18} joule.

Hartree equation [ELECTR] An equation which gives the lowest anode voltage at which it is theoretically possible to maintain oscillation in the different modes of a magnetron.

Hartree-Fock approximation [QUANT MECH] A refinement of the Hartree method in which one uses determinants of single-particle wave functions rather than products, thereby introducing exchange terms into the Hamiltonian.

Hartree method [QUANT MECH] An iterative variational method of finding an approximate wave function for a system of many electrons, in which one attempts to find a product of single-particle wave functions, each one of which is a solution of the Schrödinger equation with the field deduced from the charge density distribution due to all the other electrons. Also known as self-consistent field method.

Hartree units [ATOM PHYS] A system of units in which the unit of angular momentum is Planck's constant divided by 2π, the unit of mass is the mass of the electron, and the unit of charge is the charge of the electron. Also known as atomic units.

Hausdorff maximal principle [MATH] The principle that every partially ordered set has a linearly ordered subset S which is maximal in the sense that S is not a proper subset of another linearly ordered subset.

Hausdorff space [MATH] A topological space where each pair of distinct points can be enclosed in disjoint open neighborhoods.

Hausdorff summability [MATH] The property of a series which has a Hausdorff transformation that converges.

Hausdorff transformation [MATH] A transformation of a series s into a series t having the form $t = (\delta\mu\delta)s$, where δ is the transformation that transforms the series $\{s_n\}$ into the series $\{t_m\}$ given by

$$t_m = \sum_{n=0}^{m} (-1)^n [m!/n!(m - n)!]s_n,$$

and μ is a transformation which multiplies each term in a series by a specified constant.

Haüy law [CRYSTAL] The law that for a given crystal there is a set of ratios such that the ratios of the intercepts of any crystal plane on the crystal axes are rational fractions of these ratios.

hav *See* haversine.

Havelock's law [OPTICS] The law that in a substance displaying the Kerr effect, $n_p - n = 2(n_s - n)$. where n is the index of refraction in the absence of an electric field, and n_p and n_s are the indices of refraction of light whose magnetic vector is parallel and perpendicular to the applied electric field.

haversine [MATH] The haversine of an angle A is half of the versine of A, or is $\frac{1}{2}(1 - \cos A)$. Abbreviated hav.

Hay bridge [ELEC] A four-arm alternating-current bridge used to measure inductance in terms of capacitance, resistance, and frequency; bridge balance depends on frequency.

haze [OPTICS] The degree of cloudiness in a solution, cured plastic material, or coating material.

hazemeter *See* transmissometer.

h-bar [QUANT MECH] A fundamental constant equal to $h/2\pi$. where h is Planck's constant. Symbolized $\hbar$. Also known as Dirac h; h-line.

H bomb *See* hydrogen bomb.

hcp structure *See* hexagonal close-packed structure.

He *See* helium.

head *See* coma; pressure head.

head loss [FL MECH] The drop in the sum of pressure head, velocity head, and potential head between two points along the path of a flowing fluid, due to causes such as fluid friction.

health physics [NUCLEO] The protection of personnel from harmful effects of ionizing radiation by such means as routine radiation surveys, area and personnel monitoring, and protective equipment and procedures.

heat [THERMO] Energy in transit due to a temperature difference between the source from which the energy is coming and a sink toward which the energy is going; other types of energy in transit are called work.

heat balance [THERMO] The equilibrium which is known to

exist when all sources of heat gain and loss for a given region or body are accounted for.

heat budget [THERMO] The statement of the total inflow and outflow of heat for a planet, spacecraft, biological organism, or other entity.

heat capacity [THERMO] The quantity of heat required to raise a system one degree in temperature in a specified way, usually at constant pressure or constant volume. Also known as thermal capacity.

heat conduction [THERMO] The flow of thermal energy through a substance from a higher- to a lower-temperature region.

heat conductivity *See* thermal conductivity.

heat content *See* enthalpy.

heat convection [THERMO] The transfer of thermal energy by actual physical movement from one location to another of a substance in which thermal energy is stored. Also known as thermal convection.

heat cycle *See* thermodynamic cycle.

heat death [THERMO] The condition of any isolated system when its entropy reaches a maximum, in which matter is totally disordered and at a uniform temperature, and no energy is available for doing work.

heat energy *See* internal energy.

heat engine [THERMO] A thermodynamic system which undergoes a cyclic process during which a positive amount of work is done by the system; some heat flows into the system and a smaller amount flows out in each cycle.

heat equation [THERMO] A parabolic second-order differential equation for the temperature of a substance in a region where no heat source exists: $\partial t / \partial \tau = (k / \rho c)(\partial^2 t / \partial x^2 + \partial^2 t / \partial y^2 + t^2 / \partial z^2)$, where x, y, and z are space coordinates, τ is the time, $t(x,y,z,\tau)$ is the temperature, k is the thermal conductivity of the body, ρ is its density, and c is its specific heat; this equation is fundamental to the study of heat flow in bodies. Also known as Fourier heat equation; heat flow equation.

heater-type cathode *See* indirectly heated cathode.

heat filter [OPTICS] Special glass in condenser lens systems to keep heat from film.

heat flow [THERMO] Heat thought of as energy flowing from one substance to another; quantitatively, the amount of heat transferred in a unit time. Also known as heat transmission.

heat flow equation *See* heat equation.

heat flux [THERMO] The amount of heat transferred across a surface of unit area in a unit time. Also known as thermal flux.

heating value *See* heat of combustion.

heat of ablation [THERMO] A measure of the effective heat capacity of an ablating material, numerically the heating rate input divided by the mass loss rate which results from ablation.

heat of activation [PHYS CHEM] The increase in enthalpy when a substance is transformed from a less active to a more reactive form at constant pressure.

heat of adsorption [THERMO] The increase in enthalpy when 1 mole of a substance is adsorbed upon another at constant pressure.

heat of aggregation [THERMO] The increase in enthalpy when an aggregate of matter, such as a crystal, is formed at constant pressure.

heat of association , [PHYS CHEM] Increase in enthalpy accompanying the formation of 1 mole of a coordination compound from its constituent molecules or other particles at constant pressure.

heat of combustion [PHYS CHEM] The amount of heat released in the oxidation of 1 mole of a substance at constant pressure, or constant volume. Also known as heat value; heating value.

heat of compression [THERMO] Heat generated when air is compressed.

heat of condensation [THERMO] The increase in enthalpy accompanying the conversion of 1 mole of vapor into liquid at constant pressure and temperature.

heat of cooling [THERMO] Increase in enthalpy during cooling of a system at constant pressure, resulting from an internal change such as an allotropic transformation.

heat of crystallization [THERMO] The increase in enthalpy when 1 mole of a substance is transformed into its crystalline state at constant pressure.

heat of decomposition [PHYS CHEM] The change in enthalpy accompanying the decomposition of 1 mole of a compound into its elements at constant pressure.

heat of dilution [PHYS CHEM] **1.** The increase in enthalpy accompanying the addition of a specified amount of solvent to a solution of constant pressure. Also known as integral heat of dilution; total heat of dilution. **2.** The increase in enthalpy when an infinitesimal amount of solvent is added to a solution at constant pressure. Also known as differential heat of dilution.

heat of dissociation [PHYS CHEM] The increase in enthalpy at constant pressure, when molecules break apart or valence linkages rupture.

heat of emission [ELECTR] Additional heat energy that must be supplied to an electron-emitting surface to maintain it at a constant temperature.

heat of evaporation *See* heat of vaporization.

heat of formation [PHYS CHEM] The increase in enthalpy resulting from the formation of 1 mole of a substance from its elements at constant pressure.

heat of fusion [THERMO] The increase in enthalpy accompanying the conversion of 1 mole, or a unit mass, of a solid to a liquid at its melting point at constant pressure and temperature. Also known as latent heat of fusion.

heat of hydration [PHYS CHEM] The increase in enthalpy accompanying the formation of 1 mole of a hydrate from the anhydrous form of the compound and from water at constant pressure.

heat of ionization [PHYS CHEM] The increase in enthalpy when 1 mole of a substance is completely ionized at constant pressure.

heat of linkage [PHYS CHEM] The bond energy of a particular type of valence linkage between atoms in a molecule, as determined by the energy required to dissociate all bonds of the type in 1 mole of the compound divided by the number of such bonds in a compound.

heat of mixing [THERMO] The difference between the enthalpy of a mixture and the sum of the enthalpies of its components at the same pressure and temperature.

heat of reaction [PHYS CHEM] **1.** The negative of the change in enthalpy accompanying a chemical reaction at constant pressure. **2.** The negative of the change in internal energy accompanying a chemical reaction at constant volume.

heat of solidification [THERMO] The increase in enthalpy when 1 mole of a solid is formed from a liquid or, less commonly, a gas at constant pressure and temperature.

heat of solution [PHYS CHEM] The enthalpy of a solution minus the sum of the enthalpies of its components. Also known as integral heat of solution; total heat of solution.

heat of sublimation [THERMO] The increase in enthalpy accompanying the conversion of 1 mole, or unit mass, of a solid to a vapor at constant pressure and temperature. Also known as latent heat of sublimation.

heat of transformation [THERMO] The increase in enthalpy of a substance when it undergoes some phase change at constant pressure and temperature.

heat of vaporization [THERMO] The quantity of energy required to evaporate 1 mole, or a unit mass, of a liquid, at constant pressure and temperature. Also known as enthalpy of vaporization; heat of evaporation; latent heat of vaporization.

heat of wetting [THERMO] **1.** The heat of adsorption of water on a substance. **2.** The additional heat required, above the heat of vaporization of free water, to evaporate water from a substance in which it has been absorbed.

heat quantity [THERMO] A measured amount of heat; units are the small calorie, normal calorie, mean calorie, and large calorie.

heat radiation [THERMO] The energy radiated by solids, liquids, and gases in the form of electromagnetic waves as a result of their temperature. Also known as thermal radiation.

heat resistance *See* thermal resistance.

heat sink [AERO ENG] **1.** A type of protective device capable of absorbing heat and used as a heat shield. **2.** In nuclear propulsion, any thermodynamic device, such as a radiator or condenser, that is designed to absorb the excess heat energy of the working fluid. [Also known as heat dump. [ELEC] A mass of metal that is added to a device for the purpose of absorbing and dissipating heat; used with power transistors and many types of metallic rectifiers. Also known as dissipator. [THERMO] Any (gas, solid, or liquid) region where heat is absorbed.

heat-transfer coefficient [THERMO] The amount of heat which passes through a unit area of a medium or system in a unit time when the temperature difference between the boundaries of the system is 1 degree.

heat transmission *See* heat flow.

heat transport [THERMO] Process by which heat is carried past a fixed point or across a fixed plane, as in a warm current.

heat value *See* heat of combustion.

heat wave [ELECTROMAG] Infrared radiation, much higher in frequency than radio waves.

heavenly body *See* celestial body.

Heaviside calculus [MATH] A type of operational calculus that is used to completely analyze a linear dynamical system which represents some vibrating physical system.

Heaviside layer *See* E layer.

Heaviside-Lorentz system [ELECTROMAG] A system of electrical units which is the same as the Gaussian system except that the units of charge and current are smaller by a factor of $1/\sqrt{4\pi}$, and those of electric and magnetic field are larger by a factor by $\sqrt{4\pi}$. Also known as Lorentz-Heaviside system.

Heaviside's expansion theorem [MATH] A theorem providing an infinite series representation for the inverse Laplace transforms of functions of a particular type.

Heaviside unit function [MATH] The real function $f(x)$ whose value is 0 if x is negative and whose value is 1 otherwise.

heavy hydrogen [NUC PHYS] Hydrogen consisting of isotopes whose mass number is greater than one, namely deuterium or tritium.

heavy-ion linac *See* heavy-ion linear accelerator.

heavy-ion linear accelerator [NUCLEO] A linear accelerator which produces a beam of heavy particles of high intensity and sharp energy; used to produce transuranic elements and short-lived isotopes, and to study nuclear reactions, nuclear spectroscopy, and the absorption of heavy ions in matter. Also known as heavy-ion linac; hilac.

heavy-liquid bubble chamber   [NUCLEO] A bubble chamber which contains deuterium or an organic liquid such as propane or Freon.

heavy particle   *See* baryon.

heavy water   [INORG CHEM] A compound of hydrogen and oxygen containing a higher proportion of the hydrogen isotope deuterium than does naturally occurring water. Also known as deuterium oxide.

heavy-water reactor   [NUCLEO] A nuclear reactor in which heavy water serves as moderator and sometimes also as coolant.

hectare   [MECH] A unit of area in the metric system equal to 100 ares or 10,000 square meters. Abbreviated ha.

hecto-   [SCI TECH] A prefix representing 10^2 or 100.

hectogram   [MECH] A unit of mass equal to 100 grams. Abbreviated hg.

hectoliter   [MECH] A metric unit of volume equal to 100 liters or to 0.1 cubic meter. Abbreviated hl.

hectometer   [MECH] A unit of length equal to 100 meters. Abbreviated hm.

HEED   *See* high-energy electron diffraction.

Hefner candle   [OPTICS] A luminous intensity standard, formerly used in Germany, equal to 0.9 international candle; produced by a Hefner lamp burning under standard conditions. Abbreviated HK. Also known as Hefnerkerze.

Hefnerkerze   *See* Hefner candle.

height   [MATH] **1.** The perpendicular distance between horizontal lines or planes passing through the top and bottom of an object. **2.** The height of a rational number q is the maximum of $|m|$ and $|n|$, where m and n are relatively prime integers such that $q = m/n$.

heiligenschein   [OPTICS] A diffuse white ring surrounding the shadow cast by the observer's head upon a dew-covered lawn when the solar elevation is low and, therefore, the distance from observer to shadow is great.

Heine-Borel theorem   [MATH] The theorem that the only compact subsets of the real line are those which are closed and bounded.

Heine-Sommerfeld function   [MATH] A function of the form $f(z) = (\pi/2z)^{1/2}B_{n+1/2}(z)$, where n is a nonnegative integer and $B_{n+1/2}$ is a Bessel function of the first kind or a Hankel function.

Heisenberg algebra   [QUANT MECH] The Lie algebra formed by the operators of position and momentum.

Heisenberg equation of motion   [QUANT MECH] An equation which gives the rate of change of an operator corresponding to a physical quantity in the Heisenberg picture.

Heisenberg exchange coupling   [SOLID STATE] The exchange forces between electrons in neighboring atoms which give rise to ferromagnetism in the Heisenberg theory.

Heisenberg force   [NUC PHYS] A force between two nucleons derivable from a potential with an operator which exchanges both the positions and the spins of the particles.

Heisenberg picture   [QUANT MECH] A mode of description of a system in which dynamic states are represented by stationary vectors and physical quantities are represented by operators which evolve in the course of time. Also known as Heisenberg representation.

Heisenberg representation   *See* Heisenberg picture.

Heisenberg theory of ferromagnetism   [SOLID STATE] A theory in which exchange forces between electrons in neighboring atoms are shown to depend on relative orientations of electron spins, and ferromagnetism is explained by the assumption that parallel spins are favored so that all the spins in a lattice have a tendency to point in the same direction.

Heisenberg uncertainty principle *See* uncertainty principle.

Heisenberg uncertainty relation *See* uncertainty relation.

Heitler-London covalence theory [PHYS CHEM] A calculation of the binding energy and the distance between the atoms of a diatomic hydrogen molecule, which assumes that the two electrons are in atomic orbitals about each of the nuclei, and then combines these orbitals into a symmetric or antisymmetric function.

helical [MATH] Pertaining to a cylindrical spiral, for example, a screw thread.

helical angle [MECH] In the study of torsion, the angular displacement of a longitudinal element, originally straight on the surface of an untwisted bar, which becomes helical after twisting.

helical potentiometer [ELEC] A multiturn precision potentiometer in which a number of complete turns of the control knob are required to move the contact arm from one end of the helically wound resistance element to the other end.

helicity [QUANT MECH] The component of the spin of a particle along its momentum.

helicoid [MATH] A surface generated by a curve which is rotated about a straight line and also is translated in the direction of the line at a rate that is a constant multiple of its rate of rotation.

helicon [ELECTROMAG] A low-frequency, circularly polarized electromagnetic wave that is propagated in a metal in the presence of an external magnetic field.

helimagnet [SOLID STATE] A metal, alloy, or salt that possesses helimagnetism.

helimagnetism [SOLID STATE] A property possessed by some metals, alloys, and salts of transition elements or rare earths, in which the atomic magnetic moments, at sufficiently low temperatures, are arranged in ferromagnetic planes, the direction of the magnetism varying in a uniform way from plane to plane.

heliocentric [ASTRON] Relative to the sun as a center.

heliocentric coordinates [ASTRON] A coordinate system relative to the sun as a center.

heliocentric latitude [ASTRON] Sun-centered coordinate of angular distance perpendicular to the ecliptic plane.

heliocentric longitude [ASTRON] The angular distance east or west from a given point on the sun's equator.

heliocentric orbit [ASTRON] An orbit relative to the sun as a center.

heliographic latitude [ASTRON] On the sun, angular distance north or south of its equator.

heliographic longitude [ASTRON] On the sun, angular distance east or west from given point on the equator of the sun.

heliometer [OPTICS] A split-lens telescope used to measure the sun's diameter as well as small distances between stars or other celestial bodies.

helioscope [OPTICS] A telescope for observing the sun that protects the observer's eyes from the sun's glare.

heliostat [ENG] A clock-driven instrument mounting which automatically and continuously points in the direction of the sun; it is used with a pyrheliometer when continuous direct solar radiation measurements are required.

helium [CHEM] A gaseous chemical element, symbol He, atomic number 2, and atomic weight 4.0026; one of the noble gases in group 0 of the periodic table.

helium I [CRYO] The phase of liquid helium-4 which is stable at temperatures above the lambda point (about 2.2 K) and has the properties of a normal liquid, except low density.

HELIUM

Periodic table of the chemical elements showing the position of helium.

helium II [CRYO] The phase of liquid helium-4 which is stable at temperatures between absolute zero and the lambda point (about 2.2 K), and has many remarkable properties such as vanishing viscosity, extremely high heat conductivity, and the fountain effect.

helium-3 [NUC PHYS] The isotope of helium with mass number 3, constituting approximately 1.3 parts per million of naturally occurring helium.

helium-4 [NUC PHYS] The isotope of helium with mass number 4, constituting nearly all naturally occurring helium.

helium-cadmium laser [OPTICS] A metal-vapor ion laser in which cadmium vapor, produced by heat or other means, migrates through a high-voltage glow discharge in helium, generating a continuous laser beam at wavelengths in the ultraviolet and blue parts of the spectrum (about 0.3 to 0.5 micrometer).

helium film [CRYO] A superfluid film that covers any surface in contact with helium II. Also known as Rollin film.

helium liquefier [CRYO] Any one of several machines which liquefy helium by causing it to undergo adiabatic expansion and to do external work.

helium magnetometer [PHYS] A device for measuring magnetic fields by observing the Zeeman effect in the lowest triplet level of helium atoms subjected to the field.

helium-3 maser [PHYS] A gas maser in which the gas used is helium-3.

helium-neon laser [OPTICS] A laser using a combination of helium and neon gases.

helium spectrometer [SPECT] A small mass spectrometer used to detect the presence of helium in a vacuum system; for leak detection, a jet of helium is applied to suspected leaks in the outer surface of the system.

helium stars [ASTRON] The class B stars.

helix [ELEC] A spread-out, single-layer coil of wire, either wound around a supporting cylinder or made of stiff enough wire to be self-supporting. [MATH] A curve traced on a cylindrical or conical surface where all points of the surface are cut at the same angle; this screwlike winding curve is parametrically given by equations $x = a \cos \theta$, $y = a \sin \theta$, $z = b\theta$, where θ is the parameter and a and b are constants.

helix angle [MATH] The constant angle between the tangent to a helix and a generator of the cylinder upon which the helix lies.

Hellman-Feynman theorem [QUANT MECH] A theorem which states that in the Born-Oppenheimer approximation the forces on nuclei in molecules or solids are those which would arise electrostatically if the electron probability density were treated as a static distribution of negative electric charge.

helmholtz [ELEC] A unit of dipole moment per unit area, equal to 1 Debye unit per square angstrom, or approximately 3.335×10^{-10} coulomb per meter.

Helmholtz coils [ELECTROMAG] A pair of flat, circular coils having equal numbers of turns and equal diameters, arranged with a common axis, and connected in series; used to obtain a magnetic field more nearly uniform than that of a single coil.

Helmholtz double layer [PHYS] An electrical double layer of positive and negative charges one molecule thick which occurs at a surface where two bodies of different materials are in contact, or at the surface of a metal or other substance capable of existing in solution as ions and immersed in a dissociating solvent.

Helmholtz equation [MATH] A partial differential equation obtained by setting the Laplacian of a function equal to the function multiplied by a negative constant. [OPTICS] An equation which relates the linear and angular magnifications

HELMHOLTZ COILS

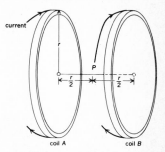

current

P

$\frac{r}{2}$ $\frac{r}{2}$

coil A   coil B

Helmholtz coils showing the optimum arrangement in which the distance between the two coils is equal to the radius r of each of the coils. P = point of nearly uniform field strength. *(From L. B. Loeb, Fundamentals of Electricity and Magnetism, 3d ed., copyright © 1947 by John Wiley & Sons, Inc.; reprinted by permission)*

of a spherical refracting interface. Also known as Lagrange-Helmholtz equation. [PHYS CHEM] The relationship stating that the emf (electromotive force) of a reversible electrolytic cell equals the work equivalent of the chemical reaction when charge passes through the cell plus the product of the temperature and the derivative of the emf with respect to temperature.

Helmholtz flow [FL MECH] Flow with free streamlines or vortex sheets.

Helmholtz free energy *See* free energy.

Helmholtz function *See* free energy.

Helmholtz instability [FL MECH] The hydrodynamic instability arising from a shear, or discontinuity, in current speed at the interface between two fluids in two-dimensional motion; the perturbation gains kinetic energy at the expense of that of the basic currents. Also known as shearing instability.

Helmholtz-Keteller formula [OPTICS] A dispersion formula in which the difference between the square of the index of refraction and unity is set equal to a sum of terms each of which is associated with a resonant wavelength of the medium.

Helmholtz potential *See* free energy.

Helmholtz resonator [ENG ACOUS] An enclosure having a small opening consisting of a straight tube of such dimensions that the enclosure resonates at a single frequency determined by the geometry of the resonator.

Helmholtz's theorem [ELEC] *See* Thévenin's theorem. [FL MECH] The theorem that in the isentropic flow of a nonviscous fluid which is not subject to body forces, individual vortices always consist of the same fluid particles. [MATH] The theorem determining a general class of vector fields as being everywhere expressible as the sum of an irrotational vector with a divergence-free vector.

Helmholtz wave [FL MECH] An unstable wave in a system of two homogeneous fluids with a velocity discontinuity at the interface.

hemicycle [MATH] A curve in the form of a semicircle.

hemihedral symmetry [CRYSTAL] The possession by a crystal of only half of the elements of symmetry which are possible in the crystal system to which it belongs.

hemiholohedral [CRYSTAL] Of hemihedral form but with half the octants having the full number of planes.

hemimorphic crystal [CRYSTAL] A crystal with no transverse plane of symmetry and no center of symmetry; composed of forms belonging to only one end of the axis of symmetry.

hemiprism [CRYSTAL] A pinacoid that cuts two crystallographic axes.

hemisphere [MATH] One of the two pieces of a sphere divided by a great circle.

hemispherical candlepower [OPTICS] Luminous intensity of a hemispherical light source.

hemispherical pyrheliometer [ENG] An instrument for measuring the total solar energy from the sun and sky striking a horizontal surface, in which a thermopile measures the temperature difference between white and black portions of a thermally insulated horizontal target within a partially evacuated transparent sphere or hemisphere.

hemispheroid [MATH] One of the halves into which a spheroid is divided by a plane of symmetry.

hemitropic [CRYSTAL] Pertaining to a twinned structure in which, if one part were rotated 180°, the two parts would be parallel.

Henderson equation for pH [PHYS CHEM] An equation for the pH of an acid during its neutralization: $pH = pK_a + \log [salt]/[acid]$ where pK_a is the logarithm to base 10 of the

HELMHOLTZ RESONATOR

area S

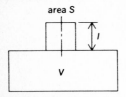

Schematic drawing of a Helmholtz resonator. Here l = length of straight tube, S = cross-sectional area of tube, V = closed volume. Resonant frequency is $f_0 = 1/2\pi \sqrt{c^2 S/l_e V}$ where c = speed of sound and l_e = effective length of tube, approximately $l + 0.8 \sqrt{S}$.

reciprocal of the dissociation constant of the acid; the equation is found to be useful for the pH range 4–10, providing the solutions are not too dilute.

henry [ELECTROMAG] The mks unit of self and mutual inductance, equal to the self-inductance of a circuit or the mutual inductance between two circuits if there is an induced electromotive force of 1 volt when the current is changing at the rate of 1 ampere per second. Symbolized H.

Henry's law [PHYS CHEM] The law that at sufficiently high dilution in a liquid solution, the fugacity of a nondissociating solute becomes proportional to its concentration.

heptagon [MATH] A seven-sided polygon.

heptakaidecagon [MATH] A polygon with 17 sides.

heptode [ELECTR] A seven-electrode electron tube containing an anode, a cathode, a control electrode, and four additional electrodes that are ordinarily grids. Also known as pentagrid.

Herbrand quotient [MATH] For a finite cyclic group of order n, generated by an element s and operating on an Abelian group A, the index of the image of g in the kernel of f divided by the index of the image of f in the kernel of g, where f and g are endomorphisms of A given by $f(x) = sx - x$ and $g(x) = x + sx + \cdots + s^{n-1}x$.

hereditary mechanics [MECH] A field of mechanics in which quantities, such as stress, depend not only on other quantities, such as strain, at the same instant but also on integrals involving the values of such quantities at previous times.

Hermann-Mauguin symbols [CRYSTAL] Symbols representing the 32 symmetry classes, consisting of series of numbers giving the multiplicity of symmetry axes in descending order, with other symbols indicating inversion axes and mirror planes.

Hermite polynomials [MATH] A family of orthogonal polynomials which arise as solutions to Hermite's differential equation, a particular case of the hypergeometric differential equation.

Hermite's differential equation [MATH] A particular case of the hypergeometric equation; it has the form $w'' - 2zw' + 2nw = 0$, where n is an integer.

Hermitian conjugate of a matrix [MATH] The transpose of the complex conjugate of a matrix. Also known as adjoint of a matrix; associate matrix.

Hermitian conjugate operator *See* adjoint operator.

Hermitian form [MATH] **1.** A polynomial in n real or complex variables where the matrix constructed from its coefficients is Hermitian. **2.** More generally, a sesquilinear form g such that $g(x,y) = \overline{g(y,x)}$ for all values of the independent variables x and y, where $\overline{g(x,y)}$ is the image of $g(x,y)$ under the automorphism of the underlying ring.

Hermitian inner product *See* inner product.

Hermitian kernel [MATH] A kernel $K(x,t)$ of an integral transformation or integral equation is Hermitian if $K(x,t)$ equals its adjoint kernel, $K^*(t,x)$.

Hermitian matrix [MATH] A matrix which equals its conjugate transpose matrix, that is, is self-adjoint.

Hermitian operator [MATH] A linear operator A on vectors in a Hilbert space, such that if x and y are in the range of A then the inner products (Ax,y) and (x,Ay) are equal.

Hermitian scalar product *See* inner product.

Hermitian space *See* inner product space.

herpolhode [MECH] The curve traced out on the invariable plane by the point of contact between the plane and the inertia ellipsoid of a rotating rigid body not subject to external torque.

herpolhode cone *See* space cone.

Herschel-Cassegrain telescope [OPTICS] A modification of a Cassegrain telescope in which the primary paraboloidal mirror is slightly inclined to the optical axis, and both the secondary hyperboloidal mirror and the eyepiece are located off the axis, so that it is not necessary to pierce the primary.

Herschel-Quincke tube [ACOUS] A device for demonstrating the interference of sound in which sound waves from a common source travel through two tubes of different lengths and recombine, producing reinforcement or cancellation of sound depending on the difference in path length; used to demonstrate the interference of sound and as a wave filter. Also known as Quincke tube.

Herschel-type venturi tube [ENG] A type of venturi tube in which the converging and diverging sections are cones, the throat section is relatively short, the diverging cone is long, and the pressures preceding the inlet cone and in the throat are transferred through multiple openings into annular openings, called piezometer rings.

hertz [PHYS] Unit of frequency; a periodic oscillation has a frequency of n hertz if in 1 second it goes through n cycles. Also known as cycle per second (cps). Symbolized Hz.

Hertz antenna [ELECTROMAG] An ungrounded half-wave antenna.

Hertz effect [ELECTR] Increase in the length of a spark induced across a spark gap when the gap is irradiated with ultraviolet light.

Hertzian oscillator [ELECTROMAG] **1.** A generator of electric dipole radiation; consists of two capacitors joined by a conducting rod having a small spark gap; an oscillatory discharge occurs when the two halves of the oscillator are raised to a sufficiently high potential difference. **2.** A dumb-bell-shaped conductor in which electrons oscillate from one end to the other, producing electric dipole radiation.

Hertzian wave *See* electric wave.

Hertz's law [MECH] A law which gives the radius of contact between a sphere of elastic material and a surface in terms of the sphere's radius, the normal force exerted on the sphere, and Young's modulus for the material of the sphere.

Hertzsprung-Russell diagram [ASTRON] A plot showing the relation between the luminosity and surface temperature of stars; other related quantities frequently used in plotting this diagram are the absolute magnitude for luminosity, and spectral type or color index for the surface temperatures. Abbreviated H-R diagram.

Hertz vector *See* polarization potential.

Hesse's theorem [MATH] A theorem in projective geometry which states that, from the three pairs of lines containing the two pairs of opposite sides and the diagonals of a quadrilateral, if any two pairs are conjugate lines with respect to a given conic, then so is the third.

Hessian [MATH] For a function $f(x_1, \ldots, x_n)$ of n real variables, the real-valued function of $(x_1, \ldots, x_n)$ given by the determinant of the matrix with entry $\partial^2 f/\partial x_i x_j$ in the ith row and jth column; used for analyzing critical points.

Hess's law [PHYS CHEM] The law that the evolved or absorbed heat in a chemical reaction is the same whether the reaction takes one step or several steps. Also known as law of constant heat summation.

heterochromatic photometry [OPTICS] The branch of photometry concerned with comparing the illuminating powers of light sources with different colors.

heterodesmic [CRYSTAL] Pertaining to those atoms bonded in more than one way in crystals.

heterodyne [ELECTR] To mix two alternating-current signals of different frequencies in a nonlinear device for the purpose

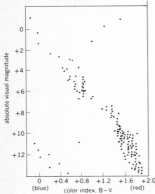

HERTZSPRUNG-RUSSELL DIAGRAM

absolute visual magnitude

0 +0.4 +0.8 +1.2 +1.6 +2.0
(blue)   color index, B−V   (red)

H-R diagram limited to all the stars within 10 parsecs from the sun. *(Adapted from H. L. Johnson and W. W. Morgan, Astrophys. J., 117:313 copyright © 1953 by the University of Chicago Press)*

of producing two new frequencies, the sum of and difference between the two original frequencies.

heterodyne modulator *See* mixer.

heterodyne oscillator [ELECTR] **1.** A separate variable-frequency oscillator used to produce the second frequency required in a heterodyne detector for code reception. **2.** *See* beat-frequency oscillator.

heterodyne reception [ELECTR] Radio reception in which the incoming radio-frequency signal is combined with a locally generated rf signal of different frequency, followed by detection. Also known as beat reception.

heterogeneous [MATH] Pertaining to quantities having different degrees or dimensions. [SCI TECH] Composed of dissimilar or nonuniform constituents.

heterogeneous fluid [FL MECH] A fluid within which the density varies from point to point; for most purposes the atmosphere must be treated as heterogeneous, particularly with regard to the decrease of density with height.

heterogeneous radiation [PHYS] Radiation having a number of different frequencies, different particles, or different particle energies.

heterogeneous reactor [NUCLEO] A nuclear reactor in which fissionable material and moderator are arranged in a regular pattern of discrete bodies with dimensions such that a nonhomogeneous medium is presented to neutrons.

heterogeneous strain [MECH] A strain in which the components of the displacement of a point in the body cannot be expressed as linear functions of the original coordinates.

heterojunction [ELECTR] The boundary between two different semiconductor materials, usually with a negligible discontinuity in the crystal structure.

heteromorphic transformation [THERMO] A change in the values of the thermodynamic variables of a system in which one or more of the component substances also undergo a change of state.

heteropolar bond [PHYS CHEM] A covalent bond whose total dipole moment is not 0.

heterostatic [ELEC] Pertaining to the measurement of one electrostatic potential by means of a different potential.

Heulinger equations [PHYS] Equations which relate the values of various quantities, such as the Hall coefficient, thermoelectric power, electrical resistivity, and thermal conductivity, in isothermal and adiabatic thermoelectric and thermomagnetic effects.

Heun's equation [MATH] The second-order linear differential equation

$$\frac{d^2y}{dz^2} + \left[\frac{\gamma}{z} + \frac{\delta}{z-1} + \frac{\varepsilon}{z-a}\right]\frac{dy}{dz} + \frac{\alpha\beta(z-q)}{z(z-1)(z-a)}\,y = 0$$

where $1 + \alpha + \beta - \gamma - \delta - \varepsilon = 0$ and $|a| > 1$.

heuristic method [MATH] A method of solving a problem in which one tries each of several approaches or methods and evaluates progress toward a solution after each attempt.

hexad axis [CRYSTAL] A rotation axis whose multiplicity is equal to 6.

hexadecimal [MATH] Pertaining to a number system using the base 16. Also known as sexadecimal.

hexadecimal number system [MATH] A digital system based on powers of 16, as compared with the use of powers of 10 in the decimal number system. Also known as sexadecimal number system.

hexagon [MATH] A six-sided polygon.

hexagonal close-packed structure [CRYSTAL] Close-packed crystal structure characterized by the regular alternation of

**HEXAGONAL
CLOSE-PACKED STRUCTURE**

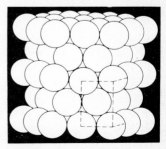

Hexagonal close packing of spheres that simulates the arrangement of the atoms in a crystal that is a hexagonal close-packed structure.

two layers; the atoms in each layer lie at the vertices of a series of equilateral triangles, and the atoms in one layer lie directly above the centers of the triangles in neighboring layers. Abbreviated hcp structure.

hexagonal lattice [CRYSTAL] A Bravais lattice whose unit cells are right prisms with hexagonal bases and whose lattice points are located at the vertices of the unit cell and at the centers of the bases.

hexagonal system [CRYSTAL] A crystal system that has three equal axes intersecting at 120° and lying in one plane; a fourth, unequal axis is perpendicular to the other three.

hexahedron [MATH] A polyhedron with six faces.

hexoctahedron [CRYSTAL] A cubic crystal form that has 48 equal triangular faces, each of which cuts the three crystallographic axes at different distances.

hexode [ELECTR] A six-electrode electron tube containing an anode, a cathode, a control electrode, and three additional electrodes that are ordinarily grids.

hextetrahedron [CRYSTAL] A 24-faced form of crystal in the tetrahedral group of the isometric system.

Hf *See* hafnium.

HFIR *See* high-flux isotope reactor.

hfs *See* hyperfine structure.

HFS *See* type II superconductor.

hg *See* hectogram.

Hg *See* mercury.

Higgs bosons [PARTIC PHYS] Massive scalar mesons whose existence is predicted by certain unified gage theories of the weak and electromagnetic interactions; they are not eliminated by the Higgs mechanism.

Higgs mechanism [PARTIC PHYS] The feature of the spontaneously broken gage symmetries that the Goldstone bosons do not appear as physical particles, but instead constitute the zero helicity states of nonzero gage vector bosons of nonzero mass (such as the intermediate vector boson). [QUANT MECH] A mathematical procedure in which particles in a field theory gain or lose mass due to spontaneous breakdown of symmetry.

high-energy astrophysics [ASTROPHYS] A science concerned with studies of acceleration of charged particles to high energies in space, cosmic rays, radio galaxies, pulsars, and quasi-stellar sources.

high-energy bond [PHYS CHEM] Any chemical bond yielding a decrease in free energy of at least 5 kilocalories per mole.

high-energy electron diffraction [PHYS] Diffraction of electrons with high energies, usually in the range 30,000–70,000 electron volts, mainly to study the structure of atoms and molecules in gases and liquids. Abbreviated HEED.

high-energy particle [PARTIC PHYS] An elementary particle having an energy of hundreds of MeV or more.

high-energy physics *See* particle physics.

high-energy scattering [PARTIC PHYS] Collisions of particles with energies of hundreds of MeV or more, sufficient to produce new particles.

high-epithermal neutron range [NUCLEO] The neutron energy range of 1000 to 100,000 electron volts.

higher plane curve [MATH] Any algebraic curve whose degree exceeds 2.

highest common factor *See* greatest common divisor.

high-field superconductor *See* type II superconductor.

high-flux isotope reactor [NUCLEO] A thermal research reactor at the Oak Ridge National Laboratory, used mainly in the production of elements with atomic numbers greater than that of plutonium. Abbreviated HFIR.

high-frequency resistance [ELEC] The total resistance offered by a device in an alternating-current circuit, including the direct-current resistance and the resistance due to eddy current, hysteresis, dielectric, and corona losses. Also known as alternating-current resistance; effective resistance; radio-frequency resistance.

high-frequency voltmeter [ELECTR] A voltmeter designed to measure currents alternating at high frequencies.

high heat [THERMO] Heat absorbed by the cooling medium in a calorimeter when products of combustion are cooled to the initial atmospheric (ambient) temperature.

high-impedance voltmeter [ELEC] A voltage-measuring device with a high-impedance input to reduce load on the unit under test; a vacuum-tube voltmeter is one type.

high-K capacitor [ELEC] A capacitor whose dielectric material is a ferroelectric having a high dielectric constant, up to about 6000.

high-mu tube [ELECTR] A tube having a very high amplification factor.

high-pass filter [ELECTR] A filter that transmits all frequencies above a given cutoff frequency and substantially attenuates all others.

high-pressure chemistry [PHYS CHEM] The study of chemical reactions and phenomena that occur at pressures exceeding 10,000 bars (a bar is nearly equivalent to a kilogram per square centimeter), mainly concerned with the properties of the solid state.

high-pressure cloud chamber [NUCLEO] A cloud chamber in which the gas is maintained at high pressure to reduce the range of high-energy particles and thereby increase the probability of observing events.

high-pressure mercury-vapor lamp [ELECTR] A discharge tube containing an inert gas and a small quantity of liquid mercury; the initial glow discharge through the gas heats and vaporizes the mercury, after which the discharge through mercury vapor produces an intensely brilliant light.

high-pressure physics [PHYS] The study of the effects of high pressure on the properties of matter.

high Q [ELECTR] A characteristic wherein a component has a high ratio of reactance to effective resistance, so that its Q factor is high.

high-Q cavity [ELECTROMAG] A cavity resonator which has a large Q factor, and thus has a small energy loss. Also known as high-Q resonator.

high-Q resonator *See* high-Q cavity.

high-resistance voltmeter [ELEC] A voltmeter having a resistance considerably higher than 1000 ohms per volt, so that it draws little current from the circuit in which a measurement is made.

high-resolution electron microscope [ELECTR] An electron microscope in which lens aberrations are minimized and lens currents and the accelerating voltage are maintained with a high degree of stability, in order to achieve extremely high resolution.

high-speed oscilloscope [ELECTR] An oscilloscope with a very fast sweep, capable of observing signals with rise times or periods on the order of nanoseconds.

high-temperature chemistry [PHYS CHEM] The study of chemical phenomena occurring above about 500 K.

high-temperature gas-cooled reactor [NUCLEO] A prototype gas-cooled reactor in which the coolant is pressurized helium gas with an inlet temperature of about 325°C and an outlet temperature of about 750°C, and the fuel consists of fully enriched uranium and thorium. Abbreviated HTGR.

high-temperature phenomena [PHYS] Phenomena occurring at temperatures above about 500 K.

high vacuum [PHYS] A vacuum with a pressure between 1×10^{-3} and 1×10^{-6} mmHg (0.1333224 and 0.0001333 newton per square meter).

high-vacuum rectifier [ELECTR] Vacuum-tube rectifier in which conduction is entirely by electrons emitted from the cathode.

high-vacuum tube [ELECTR] Electron tube evacuated to such a degree that its electrical characteristics are essentially unaffected by gaseous ionization. Also known as hard tube.

high-velocity stars [ASTRON] Those stars moving across the galactic track along which the majority of the stars execute their galactic rotation, thus exhibiting high velocity with respect to the sun, low velocity with respect to the galactic center.

high-voltage electron microscope [ELECTR] An electron microscope whose accelerating voltage is on the order of 10^6 volts, as compared with 40–100 kilovolts for an ordinary electron microscope.

hilac *See* heavy-ion linear accelerator.

Hilbert cube [MATH] The topological space which is the cartesian product of a countable number of copies of I, the unit interval.

Hilbert problem [MATH] The problem of finding a sectionally holomorphic function $F(z)$ defined on the complex plane cut along a set of arcs C, of the order of z^n at infinity for some finite n, which satisfies the boundary condition $F^+(z) - aF^-(z) = f(z)$ at all points z of C, where $f(z)$ is a specified function, a is a constant, and $F^+(z)$ and $F^-(z)$ are the limiting values of $F(z)$ on opposite sides of the arc.

Hilbert-Schmidt theory [MATH] A body of theorems which investigates the kernel of an integral equation via its eigenfunctions, and then applies these functions to help determine solutions of the equation.

Hilbert's nullstellensatz [MATH] A theorem which states that if I is an ideal in the ring of polynomials in n variables over a field F, and if f is a polynomial in n variables such that $f(c_1, \ldots, c_n) = 0$ whenever $(c_1, \ldots, c_n)$ is a 0 of I in the algebraic closure of F, then there exists a nonnegative integer m such that f^m is in I.

Hilbert space [MATH] A Banach space which also is an inner-product space with the inner product of a vector with itself being the same as the square of the norm of the vector.

Hilbert's theorem [MATH] The ring of polynomials with coefficients in a commutative Noetherian ring is itself a Noetherian ring.

Hilbert's theorem 90 [MATH] An element b of a cyclic extension E of a field F has norm from E to F equal to 1 if and only if there exists a nonzero element a in E such that $b = a/f(a)$, where f is a generator of the Galois group of E over F.

Hilbert transform [MATH] The transform of a function $f(x)$ realized by taking the integral of $f(x)[1 + \cot (y-x)/2] \, dx$.

Hildebrand function [THERMO] The heat of vaporization of a compound as a function of the molal concentration of the vapor; it is nearly the same for many compounds.

hill bandwidth [ELECTR] The difference between the upper and lower frequencies at which the gain of an amplifier is 3 decibels less than its maximum value.

hill-climbing [MATH] Any numerical procedure for finding the maximum or maxima of a function.

Hille-Yosida-Phillips theorem [MATH] The fundamental theorem in the theory of strongly continuous semigroups; it states that a necessary and sufficient condition for a closed linear operator T on a Banach space with dense domain to be the infinitesimal generator of a strongly continuous semigroup is that there exist numbers M and a such that for all

$\lambda > a$ the operator $\lambda I - A$ has an inverse (where I is the identity operator) and $\|(\lambda I - A)^{-n}\| \leq M(\lambda - a)^{-n}$ for $n = 1, 2, \ldots$.

Hill's equation [MATH] A differential equation of the form $d^2y/dx^2 + [\lambda + g(x)]y = 0$, where λ is a parameter and $g(x)$ is a specified function of period π.

Hiltner-Hall effect [ASTRON] The polarization of the light received from distant stars; this effect is thought to take place in interstellar space.

Hittorf dark space *See* cathode dark space.

Hittorf method [PHYS CHEM] A procedure for determining transference numbers in which one measures changes in the composition of the solution near the cathode and near the anode of an electrolytic cell, due to passage of a known amount of electricity.

Hittorf principle [ELECTR] The principle that a discharge between electrodes in a gas at a given pressure does not necessarily occur between the closest points of the electrodes if the distance between these points lies to the left of the minimum on a graph of spark potential versus distance. Also known as short-path principle.

Hjelmslev plane *See* affine Hjelmslev plane.

HK *See* Hefner candle.

hl *See* hectoliter.

h-line *See* h-bar.

hm *See* hectometer.

H mode *See* transverse electric mode.

Ho *See* holmium.

hodograph [PHYS] **1.** The curve traced out in the course of time by the tip of a vector representing some physical quantity. **2.** In particular, the path traced out by the velocity vector of a given particle.

hodograph method [FL MECH] A method for studying two-dimensional steady fluid flow in which the independent variables are taken as the components of the velocity with respect to cartesian or polar coordinates, rather than the coordinates themselves.

hodoscope [NUCLEO] An array of small Geiger counters, scintillation counters, or other radiation counters used in tracing paths of high-energy particles in experiments with particle accelerators or in cosmic rays. [OPTICS] *See* cono-scope.

Hoffmann electrometer [ENG] A variant of the quadrant electrometer that has two sections instead of four.

hoghorn antenna *See* horn antenna.

hohlraum *See* blackbody.

hold circuit [ELECTR] A circuit in a sampled-data control system that converts the series of impulses, generated by the sampler, into a rectangular function, in order to smooth the signal to the motor or plant.

Hölder condition [MATH] **1.** A function $f(x)$ satisfies the Hölder condition in a neighborhood of a point x_0 if $|f(x) - f(x_0)| \leq c|(x - x_0)|^n$, where c and n are constants. **2.** A function $f(x)$ satisfies a Hölder condition in an interval or in a region of the plane if $|f(x) - f(y)| \leq c|x - y|^n$ for all x and y in the interval or region, where c and n are constants.

Hölder condition at infinity [MATH] A function $f(x)$, defined on the real numbers or on an arc that extends to infinity, satisfies a Hölder condition at infinity if $f(x)$ tends to a finite value, denoted $f(\infty)$, as x approaches infinity, and if there is a number M such that $|f(x) - f(\infty)| \leq c|x|^{-a}$ when $|x| \geq M$, where c and a are constants, with $a > 0$.

Hölder's inequality [MATH] A generalization of the Schwarz inequality; for real functions it says that $|\int f(x)g(x)\,dx| \leq (\int |f(x)|^p\,dx)^{1/p}(\int |g(x)|^q\,dx)^{1/q}$ where $1/p + 1/q = 1$.

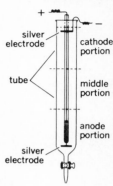

HITTORF METHOD

The cell used in the Hittorf method.

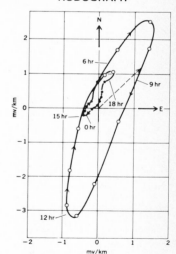

HODOGRAPH

Hodograph representing the potential-gradient vector of the average diurnal variation of earth currents at Tucson, Arizona, for 10 quiet days each month for 1932 to 1942 inclusive.

Hölder summation [MATH] A method of attributing a sum to certain divergent series in which a new series is formed, each of whose partial sums is the average of the first n partial sums of the original series, and this process is repeated until a stage is reached where the limit of this average exists.

holding magnet *See* lifting magnet.

hole [SOLID STATE] A vacant electron energy state near the top of an energy band in a solid; behaves as though it were a positively charged particle. Also known as electron hole.

hole conduction [ELECTR] Conduction occurring in a semiconductor when electrons move into holes under the influence of an applied voltage and thereby create new holes.

hole injection [ELECTR] The production of holes in an n-type semiconductor when voltage is applied to a sharp metal point in contact with the surface of the material.

hole mobility [ELECTR] A measure of the ability of a hole to travel readily through a semiconductor, equal to the average drift velocity of holes divided by the electric field.

hole theory [QUANT MECH] A theory about the significance of negative energy states in the Dirac theory which leads to the prediction of the existence of the positron and, by extension, to that of other antiparticles.

hole trap [ELECTR] A semiconductor impurity capable of releasing electrons to the conduction or valence bands, equivalent to trapping a hole.

hollow cathode [ELECTR] A cathode which is hollow and closed at one end in a discharge tube filled with inert gas, designed so that radiation is emitted from the cathode glow inside the cathode.

holmium [CHEM] A rare-earth element belonging to the yttrium subgroup, symbol Ho, atomic number 67, atomic weight 164.93, melting point 1400–1525°C.

holoaxial [CRYSTAL] Having all possible axes of symmetry.

hologram [OPTICS] The special photographic plate used in holography; when this negative is developed and illuminated from behind by a coherent gas-laser beam, it produces a three-dimensional image in space. Also known as hologram interferometer.

hologram interferometer *See* hologram.

holography [PHYS] A technique for recording, and later reconstructing, the amplitude and phase distributions of a wave disturbance; widely used as a method of three-dimensional optical image formation, and also with acoustical and radio waves; in optical image formation, the technique is accomplished by recording on a photographic plate the pattern of interference between coherent light reflected from the object of interest, and light that comes directly from the same source or is reflected from a mirror.

holohedral [CRYSTAL] Pertaining to a crystal structure having the highest symmetry in each crystal class. Also known as holosymmetric; holosystemic.

holohedron [CRYSTAL] A crystal form of the holohedral class, having all the faces needed for complete symmetry.

holomicrography [OPTICS] The use of holography to produce three-dimensional images with various types of microscopes.

holomorphic function *See* analytic function.

holonomic system [MECH] A system in which the constraints are such that the original coordinates can be expressed in terms of independent coordinates and possibly also the time.

holophotal [OPTICS] **1.** Pertaining to a holophote. **2.** Reflecting all the light from a source in one direction.

holophote [OPTICS] An optical system consisting of lenses or reflectors that collect a large amount of the light from a source (such as the lamp of a lighthouse) and send it in a desired direction.

HOLMIUM

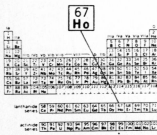

Periodic table of the chemical elements showing the position of holmium.

HOLOGRAPHY

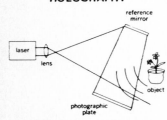

Interference pattern between light from laser source reflected from object of interest and light from same source reflected from mirror is recorded on photographic plate which, when developed, forms a hologram.

holosymmetric *See* holohedral.

holosystemic *See* holohedral.

Holtz machine *See* Toepler-Holtz machine.

Holzer's method [MECH] A method of determining the shapes and frequencies of the torsional modes of vibration of a system, in which one imagines the system to consist of a number of flywheels on a massless flexible shaft and, starting with a trial frequency and motion for one flywheel, determines the torques and motions of successive flywheels.

homenergic flow [THERMO] Fluid flow in which the sum of kinetic energy, potential energy, and enthalpy per unit mass is the same at all locations in the fluid and at all times.

homentropic flow [FL MECH] Fluid flow in which the entropy per unit mass is the same at all locations in the fluid and at all times.

homeomorphic spaces [MATH] Two topological spaces with a homeomorphism existing between them; intuitively one can be obtained from the other by stretching, twisting, or shrinking.

homeomorphism [MATH] A continuous map between topological spaces which is one-to-one, onto, and its inverse function is continuous. Also known as bicontinuous function; topological mapping.

hometaxial-base transistor [ELECTR] Transistor manufactured by a single-diffusion process to form both emitter and collector junctions in a uniformly doped silicon slice; the resulting homogeneously doped base region is free from accelerating fields in the axial (collector-to-emitter) direction, which could cause undesirable high current flow and destroy the transistor.

homing antenna [ELECTROMAG] A directional antenna array used in flying directly to a target that is emitting or reflecting radio or radar waves.

homocentric [OPTICS] Pertaining to rays which have the same focal point, or which are parallel. Also known as stigmatic.

homodesmic [CRYSTAL] Of a crystal, having atoms bonded in a single way.

homogeneity [PHYS] Quality of a substance whose properties are independent of position. [STAT] Equality of the distribution functions of several populations.

homogeneous [MATH] Pertaining to a group of mathematical symbols of uniform dimensions or degree. [SCI TECH] Uniform in structure or composition.

homogeneous coordinates [MATH] To a point in the plane with cartesian coordinates (x,y) there corresponds the homogeneous coordinates (x_1,x_2,x_3), where $x_1/x_3 = x$, $x_2/x_3 = y$; any polynomial equation in cartesian coordinates becomes homogeneous if a change into these coordinates is made.

homogeneous differential equation [MATH] A differential equation where every scalar multiple of a solution is also a solution.

homogeneous equation [MATH] An equation that can be rewritten into the form having zero on one side of the equal sign and a homogeneous function of all the variables on the other side.

homogeneous function [MATH] A real function $f(x_1,x_2,\ldots,x_n)$ is homogeneous of degree r if $f(ax_1,ax_2,\ldots,ax_n) = a^r f(x_1,x_2,\ldots,x_n)$ for every real number a.

homogeneous Hilbert problem [MATH] The problem of finding a sectionally holomorphic function $F(z)$, defined on the complex plane cut along a set of arcs C, which approaches 0 as $|z|$ approaches infinity and which satisfies the boundary condition $F^+(z) = aF^-(z)$ at all points z of C, where a is a constant and $F^+(z)$ and $F^-(z)$ are the limiting values of $F(z)$ on the opposite sides of the arc.

homogeneous integral equation [MATH] An integral equation where every scalar multiple of a solution is also a solution.

homogeneous polynomial [MATH] A polynomial all of whose terms have the same total degree; equivalently it is a homogenous function of the variables involved.

homogeneous radiation [PHYS] Radiation having an extremely narrow band of frequencies, or a beam of monoenergetic particles of a single type, so that all components of the radiation are alike.

homogeneous reactor [NUCLEO] A nuclear reactor in which fissionable material and moderator (if used) are intimately mixed to form an effectively homogeneous medium for neutrons.

homogeneous space [MATH] A topological space having a group of transformations acting upon it, that is, a transformation group, where for any two points x and y some transformation from the group will send x to y.

homogeneous strain [MECH] A strain in which the components of the displacement of any point in the body are linear functions of the original coordinates.

homogeneous transformation *See* linear transformation.

homographic transformations *See* Möbius transformations.

homological algebra [MATH] The study of the structure of modules, particularly by means of exact sequences; it has application to the study of a topological space via its homology groups.

homologous transformation [ASTRON] A mathematical transformation in the study of stellar models.

homology group [MATH] Associated to a topological space X, one of a sequence of Abelian groups $H_n(X)$ that reflect how n-dimensional simplicial complexes can be used to fill up X and also help determine the presence of n-dimensional holes appearing in X. Also known as Betti group.

homology theory [MATH] Theory attempting to compare topological spaces and investigate their structures by determining the algebraic nature and interrelationships appearing in the various homology groups.

homometric pair [CRYSTAL] A pair of crystal structures whose x-ray diffraction patterns are identical.

homomorphism [MATH] A function between two algebraic systems of the same type which preserves the algebraic operations.

homomorphous transformation [THERMO] A change in the values of the thermodynamic variables of a system in which none of the component substances undergoes a change of state.

homopolar [ELEC] **1.** Electrically symmetrical. **2.** Having equal distribution of charge.

homopolar bond [PHYS CHEM] A covalent bond whose total dipole moment is zero.

homopolar crystal [SOLID STATE] A crystal in which the bonds are all covalent.

homoscedastic [STAT] **1.** Pertaining to two or more distributions whose variances are equal. **2.** Pertaining to a variate in a bivariate distribution whose variance is the same for all values of the other variate.

homothetic curves [MATH] For a given point, a set of curves such that any straight line through the point intersects all the curves in the set at the same angle.

homothetic figures [MATH] Similar figures which are placed so that lines joining corresponding points pass through a common point and are divided in a constant ratio by this point.

homotopy [MATH] Between two mappings of the same topological spaces, a continuous function representing how, in a step-by-step fashion, the image of one mapping can be continuously deformed onto the image of the other.

homotopy groups [MATH] Associated to a topological space X, the groups appearing for each positive integer n, which reflect the number of different ways (up to homotopy) than an n-dimensional sphere may be mapped to X.

homotopy theory [MATH] The study of the topological structure of a space by examining the algebraic properties of its various homotopy groups.

honeycomb coil [ELECTROMAG] A coil wound in a crisscross manner to reduce distributed capacitance. Also known as duolateral coil; lattice-wound coil.

hook [ELECTR] A circuit phenomenon occurring in four-zone transistors, wherein hole or electron conduction can occur in opposite directions to produce voltage drops that encourage other types of conduction.

hook collector transistor [ELECTR] A transistor in which there are four layers of alternating n- and p-type semiconductor material and the two interior layers are thin compared to the diffusion length. Also known as hook transistor; pn hook transistor.

Hookean deformation [MECH] Deformation of a substance which is proportional to the force applied to it.

Hookean solid [MECH] An ideal solid which obeys Hooke's law exactly for all values of stress, however large.

Hooke number *See* Cauchy number.

Hooke's law [MECH] The law that the stress of a solid is directly proportional to the strain applied to it.

hook transistor *See* hook collector transistor.

Hope's apparatus [THERMO] An apparatus consisting of a vessel containing water, a freezing mixture in a tray surrounding the vessel, and thermometers inserted in the water at points above and below the freezing mixture; used to show that the maximum density of water lies at about 4°C.

Hopkinson's coefficient [ELECTROMAG] The average magnetic flux per turn of an induction coil divided by the average flux per turn of another coil linked with it.

horizon system of coordinates [ASTRON] A set of celestial coordinates based on the celestial horizon as the primary great circle.

horizontal deflection electrode [ELECTR] One of a pair of electrodes that move the electron beam horizontally from side to side on the fluorescent screen of a cathode-ray tube employing electrostatic deflection.

horizontal field-strength diagram [ELECTROMAG] Representation of the field strength at a constant distance from an antenna and in a horizontal plane; unless otherwise specified, this plane is that passing through the antenna.

horizontal force instrument [ENG] An instrument used to make a comparison between the intensity of the horizontal component of the earth's magnetic field and the magnetic field at the compass location on board a craft; basically, it consists of a magnetized needle pivoted in a horizontal plane, as a dry-card compass; it settles in some position which indicates the direction of the resultant magnetic field; if the needle is started swinging, it damps down with a certain period of oscillation dependent upon the strength of the magnetic field. Also known as horizontal vibrating needle.

horizontal intensity variometer [ENG] Essentially a declination variometer with a larger, stiffer fiber than in the standard model; there is enough torsion in the fiber to cause the magnet to turn 90° out of the magnetic meridian; the magnet is aligned with the magnetic prime vertical to within 0.5° so it does not respond appreciably to changes in declination. Also known as H variometer.

horizontal magnetometer [ENG] A measuring instrument for ascertaining changes in the horizontal component of the magnetic field intensity.

horizontal pendulum [MECH] A pendulum that moves in a horizontal plane, such as a compass needle turning on its pivot.

horizontal pressure force [GEOPHYS] The horizontal pressure gradient per unit mass, $-\alpha\nabla_H p$, where α is the specific volume, p the pressure, and ∇_H the horizontal component of the del operator; this force acts normal to the horizontal isobars toward lower pressure; it is one of the three important forces appearing in the horizontal equations of motion, the others being the Coriolis force and friction.

horizontal sweep [ELECTR] The sweep of the electron beam from left to right across the screen of a cathode-ray tube.

horizontal vee [ELECTROMAG] An antenna consisting of two linear radiators in the form of the letter V, lying in a horizontal plane.

horizontal vibrating needle See horizontal force instrument.

horn [ELECTROMAG] See horn antenna. [ENG ACOUS] A tube whose cross-sectional area increases from one end to the other, used to radiate or receive sound waves and to intensify and direct them. Also known as acoustic horn.

horn angle [MATH] A geometric figure formed by two tangent plane curves that lie on the same side of their mutual tangent line in the neighborhood of the point of tangency.

horn antenna [ELECTROMAG] A microwave antenna produced by flaring out the end of a circular or rectangular waveguide into the shape of a horn, for radiating radio waves directly into space. Also known as electromagnetic horn; flare (British usage); hoghorn antenna (British usage); horn; horn radiator.

horn equation [ACOUS] A second-order partial differential equation for the velocity potential as a function of time and of distance along an acoustic horn.

Horner's method [MATH] A technique for approximating the real roots of an algebraic equation; a root is located between consecutive integers, then a successive search is performed.

horn radiator See horn antenna.

horology [SCI TECH] The science of measuring time and the technology of constructing instruments for this measurement.

horsepower [MECH] The unit of power in the British engineering system, equal to 550 foot-pounds per second, approximately 745.7 watts. Abbreviated hp.

horseshoe magnet [ELECTROMAG] A permanent magnet or electromagnet in which the core is horseshoe-shaped or U-shaped, to bring the two poles near each other.

hot [ELEC] See energized. [NUCLEO] Being highly radioactive.

hot atom [NUCLEO] An atom that has high internal or kinetic energy as a result of a nuclear process such as beta decay or neutron capture.

hot carrier [ELECTR] A carrier, which may be either an electron or a hole, that has relatively high energy with respect to the carriers normally found in majority-carrier devices such as thin-film transistors.

hot-carrier diode See Schottky barrier diode.

hot cathode [ELECTR] A cathode in which electron or ion emission is produced by heat. Also known as thermionic cathode.

hot-cathode gas-filled tube See thyratron.

HORN ANTENNA

feed

Horn antenna, a type of direct-aperture antenna.

hot cell *See* cave.

hot electron [ELECTR] An electron that is in excess of the thermal equilibrium number and, for metals, has an energy greater than the Fermi level; for semiconductors, the energy must be a definite amount above that of the edge of the conduction band.

hot-electron triode [ELECTR] Solid-state, evaporated thin-film structure directly equivalent to a vacuum triode.

hot-filament ionization gage [ELECTR] An ionization gage in which electrons emitted by an incandescent filament, and attracted toward a positively charged grid electrode, collide with gas molecules to produce ions which are then attracted to a negatively charged electrode; the ion current is a measure of the number of gas molecules.

hot hole [ELECTR] A hole that can move at much greater velocity than normal holes in a semiconductor.

hot junction [ELECTR] The heated junction of a thermocouple.

hot laboratory [NUCLEO] A laboratory designed for research with radioactive materials that have such high strengths that special handling precautions are required.

hot spot [NUCLEO] 1. A surface area of higher than average radioactivity. 2. A part of a reactor fuel surface element that has become overheated. [PHYS] A localized region with temperature higher than the surroundings.

hot strength *See* tensile strength.

hot-wire instrument [ENG] An instrument that depends for its operation on the expansion by heat of a wire carrying a current.

hour [MECH] A unit of time equal to 3600 seconds. Abbreviated hr.

hour angle [ASTRON] Angular distance west of a celestial meridian or hour circle; the arc of the celestial equator, or the angle at the celestial pole, between the upper branch of a celestial meridian or hour circle and the hour circle of a celestial body or the vernal equinox, measured westward through 360°.

hour angle difference *See* meridian angle difference.

hour circle [ASTRON] An imaginary great circle passing through the celestial poles on the celestial sphere above which declination is measured. Also known as circle of declination; circle of right ascension.

Householder's method [MATH] A transformation method for finding the eigenvalues of a symmetric matrix, in which each of the orthogonal transformations that reduce the original matrix to a triple-diagonal matrix reduces one complete row to the required form.

howling tones [ACOUS] The sounds produced by a howling tube. Also known as gauze tones.

howling tube [ACOUS] A vertical open tube with a piece of gauze in the lower half which is placed over a flame to make the tube produce powerful sound waves with many overtones.

hp *See* horsepower.

h parameter [ELECTR] One of a set of four transistor equivalent-circuit parameters that conveniently specify transistor performance for small voltages and currents in a particular circuit. Also known as hybrid parameter.

H plane [ELECTROMAG] The plane of an antenna in which lies the magnetic field vector of linearly polarized radiation.

H-plane T junction [ELECTROMAG] Waveguide T junction in which the change in structure occurs in the plane of the magnetic field. Also known as shunt T junction.

hr *See* hour.

H-R diagram *See* Hertzsprung-Russell diagram.

HOT-FILAMENT IONIZATION GAGE

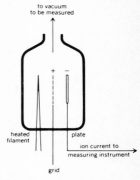

Diagram of hot-filament ionization gage.

HTGR *See* high-temperature gas-cooled reactor.

H theorem of Boltzmann *See* Boltzmann H theorem.

hubble [ASTRON] A unit of astronomical distance equal to 10^9 light-years or 9.4605×10^{24} meters.

Hubble constant [ASTROPHYS] The rate at which the velocity of recession of the galaxies increases with distance; the value is about 30 kilometers per second per million light-years (or 3.2×10^{-18} sec^{-1}) with an uncertainty of about ± 15 km/sec.

Hubble effect *See* red shift.

Hubble law [ASTRON] The linear relation between red shift and distance for not too distant galaxies predicted by general relativity and verified by observation.

Hübner rhomb [OPTICS] A glass rhombohedron used in photometry to compare two illuminated surfaces.

hue [OPTICS] The name of a color, such as red, yellow, green, blue, or purple, as perceived subjectively.

Hughes plane [MATH] A finite projective plane with nine points on each line that can be represented by a nonlinear ternary ring generated by a four-point in the plane.

Hugoniot function [PHYS] A function specifying the locus of states which are possible immediately after the passage of a shock front; gives the state's pressure as a function of its specific volume.

Humbert's symbol [MATH] The symbol $\Phi(\alpha, \rho; z)$, used to denote Kummer's function $_1F_1(\alpha; \rho; z)$.

Humphreys series [SPECT] A series of lines in the infrared spectrum of atomic hydrogen whose wave numbers are given by $R_H[(1/36) - (1/n^2)]$, where R_H is the Rydberg constant for hydrogen, and n is any number greater than 6.

Humphries equation [THERMO] An equation which gives the ratio of specific heats at constant pressure and constant volume in moist air as a function of water vapor pressure.

Hund coupling cases [ATOM PHYS] Five ways of combining the electron-spin angular momentum, electron-orbital angular momentum, and nuclear-rotation angular momentum to form the total angular momentum of a molecule.

hundredweight [SCI TECH] Abbreviated cwt. **1.** A unit of weight, in common use in the United States, equal to 100 pounds or the weight of 45.359237 kilograms. Also known as cental; centner; kintal; quintal; short hundredweight. **2.** A unit of weight in common use in the United Kingdom equal to 112 pounds or to the weight of 50.80234544 kilograms. Also known as long hundredweight. **3.** A unit of weight in troy measure equal to 100 troy pounds or the weight of 37.32417216 kilograms.

Hund rules [ATOM PHYS] Two rules giving the order in energy of atomic states formed by equivalent electrons: of the terms given by equivalent electrons, the ones with greatest multiplicity have the least energy, and of these the one with greatest orbital angular momentum is lowest; the state of a multiplet with lowest energy is that in which the total angular momentum is the least possible, if the shell is less than half-filled, and the greatest possible, if more than half filled.

hunting [CONT SYS] Undesirable oscillation of an automatic control system, wherein the controlled variable swings on both sides of the desired value. [ELECTR] Operation of a selector in moving from terminal to terminal until one is found which is idle.

Hurwitz polynomial [MATH] A polynomial whose zeros all have negative real parts.

Hurwitz's criterion [MATH] A criterion that determines whether a polynomial is a Hurwitz polynomial, based on the signs of a set of determinants formed from the polynomial's coefficients.

Huttig equation [THERMO] An equation which states that the ratio of the volume of gas adsorbed on the surface of a nonporous solid at a given pressure and temperature to the volume of gas required to cover the surface completely with a unimolecular layer equals $(1 + r) c^r/(1 + c^r)$, where r is the ratio of the equilibrium gas pressure to the saturated vapor pressure of the adsorbate at the temperature of adsorption, and c is the product of a constant and the exponential of $(q - q_l)/RT$, where q is the heat of adsorption into a first layer molecule, q_l is the heat of liquefaction of the adsorbate, T is the temperature, and R is the gas constant.

Huygens' approximation [MATH] The length of a small circular arc is approximately $\frac{1}{3}(8c' - c)$, where c is the chord of the arc and c' is the chord of half the arc.

Huygens eyepiece [OPTICS] An eyepiece in which there are two plano-convex lenses, and the plane sides of both lenses face the eye.

Huygens-Fresnel principle [OPTICS] A modification of Huygens' principle according to which the amplitude of secondary waves falls off in proportion to the cosine of the angle between the normals to the original and secondary waves, and the secondary waves interfere with each other according to the principle of superposition. Also known as Fresnel-Huygens principle.

Huygens' principle [OPTICS] The principle that each point on a light wavefront may be regarded as a source of secondary waves, the envelope of these secondary waves determining the position of the wavefront at a later time.

Huygens wavelet [OPTICS] A secondary wave as used in Huygens' principle.

H variometer *See* horizontal intensity variometer.

H vector [ELECTROMAG] A vector that is the magnetic field. For a plane wave in free space, it is perpendicular to the E vector and to the direction of propagation.

H wave *See* transverse electric wave.

hybrid computer [ADP] A computer designed to handle both analog and digital data. Also known as analog-digital computer.

hybrid electromagnetic wave [ELECTROMAG] Wave which has both transverse and longitudinal components of displacement.

hybridized orbital [PHYS CHEM] A molecular orbital which is a linear combination of two or more orbitals of comparable energy (such as $2s$ and $2p$ orbitals), is concentrated along a certain direction in space, and participates in formation of a directed valence bond.

hybrid junction [ELECTR] A transformer, resistor, or waveguide circuit or device that has four pairs of terminals so arranged that a signal entering at one terminal pair divides and emerges from the two adjacent terminal pairs, but is unable to reach the opposite terminal pair. Also known as bridge hybrid.

hybrid parameter *See* h parameter.

hybrid tee [ELECTROMAG] A microwave hybrid junction composed of an E-H tee with internal matching elements; it is reflectionless for a wave propagating into the junction from any arm when the other three arms are match-terminated. Also known as magic tee.

hybrid wave function [QUANT MECH] A linear combination of wave functions of one problem used as an approximation to the wave function in another problem; for example, a linear combination of atomic orbitals used to represent a molecular bond.

hydrated electron [PHYS CHEM] An electron released during ionization of a water molecule by water and surrounded by

HUYGENS EYEPIECE

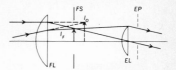

Arrangement of lenses in Huygens eyepiece. FL = field lens; EL = eye lens; FS = field stop; EP = exit pupil or eye point; I_O = image formed by preceding system; I_f = image formed by preceding system and the field lens.

HUYGENS' PRINCIPLE

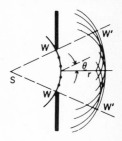

The construction for a spherical wave. WW = wave originating at S; $W'W'$ = envelope of secondary waves; θ = angle between the normal to the original wavefront and any point on the secondary wave; r = radius representing the distance wave would travel in time t.

HYBRID COMPUTER

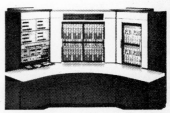

A hybrid computer console with the special feature of double patch bays: one bay for the analog circuits and one bay for the digital logic circuits. (*Applied Dynamics Computer Systems, Division of Reliance Electric Co.*)

HYDRAULIC ANALOG TABLE

Shadowgraph of airfoil in hydraulic analog table.

HYDROGEN

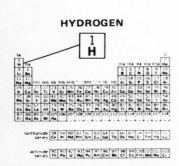

Periodic table of the chemical elements showing the position of hydrogen.

water molecules oriented so that the electron cannot escape. Also known as aqueous electron.

hydraulic analog table [FL MECH] An experimental facility based on the hydraulic analogy; the water flows over a smooth horizontal surface and is bounded by vertical walls geometrically similar to the boundaries of the corresponding compressible gas flow; flow patterns are easily observed, and boundary changes may be made rapidly and inexpensively during exploratory studies.

hydraulic analogy [FL MECH] The analogy between the flow of a shallow liquid and the flow of a compressible gas; various phenomena such as shock waves occur in both systems; the analogy requires neglect of vertical accelerations in the liquid, and restrictions on the ratio of specific heats for the gas.

hydraulic conductivity *See* permeability coefficient.

hydraulic friction [FL MECH] Resistance to flow which is exerted on the surface of contact between a stream and its conduit and which induces a loss of energy.

hydraulic grade line [FL MECH] **1.** In a closed channel, a line joining the elevations that water would reach under atmospheric pressure. **2.** The free water surface in an open channel.

hydraulic gradient [FL MECH] With regard to an aquifer, the rate of change of pressure head per unit of distance of flow at a given point and in a given direction.

hydraulic jump [FL MECH] A steady-state, finite-amplitude disturbance in a channel, in which water passes turbulently from a region of (uniform) low depth and high velocity to a region of (uniform) high depth and low velocity; when applied to hydraulic jumps, the usual hydraulic formulas governing the relations of velocity and depth do not conserve energy.

hydraulic loss [FL MECH] The loss in fluid power due to flow friction within the system.

hydraulic radius [FL MECH] The ratio of the cross-sectional area of a conduit in which a fluid is flowing to the inner perimeter of the conduit.

hydraulics [FL MECH] The branch of science and technology concerned with the mechanics of fluids, especially liquids.

hydroacoustics *See* underwater acoustics.

hydrodynamic equations [FL MECH] Three equations which express the net acceleration of a unit water particle as the sum of the partial accelerations due to pressure gradient force, frictional force, earth's deflecting force, gravitational force, and other factors.

hydrodynamic pressure [FL MECH] The difference between the pressure of a fluid and the hydrostatic pressure; this concept is useful chiefly in problems of the steady flow of an incompressible fluid in which the hydrostatic pressure is constant for a given elevation (as when the fluid is bounded above by a rigid plate), so that the external force field (gravity) may be eliminated from the problem.

hydrodynamics [FL MECH] The study of the motion of a fluid and of the interactions of the fluid with its boundaries, especially in the incompressible inviscid case.

hydroelasticity [FL MECH] **1.** Theory of elasticity of a fluid. **2.** The interaction between the flow of water or other liquid and the elastic behavior of a body immersed in it.

hydrogen [CHEM] The first chemical element, symbol H, in the periodic table, atomic number 1, atomic weight 1.00797; under ordinary conditions it is a colorless, odorless, tasteless gas composed of diatomic molecules, H_2; used in manufacture of ammonia and methanol, for hydrofining, for desulfurization of petroleum products, and to reduce metallic oxide ores.

hydrogen bomb [ORD] A device in which heavy hydrogen nuclei, under intense heat and pressure, undergo an uncontrolled, self-sustaining fusion reaction to produce an explosion. Also known as H bomb.

hydrogen burning [ASTROPHYS] Thermonuclear reactions occurring in the cores of main-sequence stars, in which nuclei of hydrogen fuse to form helium nuclei.

hydrogen-discharge lamp [ELECTR] A discharge lamp containing hydrogen and used as a source of ultraviolet radiation.

hydrogen electrode [PHYS CHEM] A noble metal (such as platinum) of large surface area covered with hydrogen gas in a solution of hydrogen ion saturated with hydrogen gas; metal is used in a foil form and is welded to a wire sealed in the bottom of a hollow glass tube, which is partially filled with mercury; used as a standard electrode with a potential of zero to measure hydrogen ion activity.

hydrogen laser [OPTICS] A molecular gas laser in which hydrogen is used to generate coherent wavelengths near 0.6 micrometer in the vacuum ultraviolet region.

hydrogen line [SPECT] A spectral line emitted by neutral hydrogen having a frequency of 1420 megahertz and a wavelength of 21 centimeters; radiation from this line is used in radio astronomy to study the amount and velocity of hydrogen in the Galaxy.

hydrogen maser [PHYS] A maser in which hydrogen gas is the basis for providing an output signal with a high degree of stability and spectral purity.

hydrokinematics [FL MECH] The study of the motion of a liquid apart from the cause of motion.

hydrokinetics [FL MECH] The study of the forces produced by a liquid as a consequence of its motion.

hydromagnetic instability See magnetohydrodynamic instability.

hydromagnetics See magnetohydrodynamics.

hydromagnetic stability See magnetohydrodynamic stability.

hydromagnetic wave See magnetohydrodynamic wave.

hydromechanics [FL MECH] The study of liquids, traditionally water, as a medium for the transmission of forces.

hydrometer [ENG] A direct-reading instrument for indicating the density, specific gravity, or some similar characteristic of liquids.

hydrometry [FL MECH] The science and technology of measuring specific gravities, particularly of liquids.

hydrophone [ENG ACOUS] A device which receives underwater sound waves and converts them to electric waves.

hydrophotometer [OPTICS] An instrument for measuring the attenuation coefficient of collimated light in sea water, in which light from a collimated source is directed through a column of sea water and is measured by a photocell or other electronic device at the other end of the column.

hydroscope [OPTICS] An instrument designed to observe objects an appreciable distance below the surface of water, consisting of a series of mirrors enclosed in a steel tube.

hydrostatic analogy [PHYS] An analogy between the relations among current, potential difference, and resistance in an electric circuit and the relations among corresponding quantities describing water flowing under a hydrostatic head.

hydrostatic balance [MECH] An equal-arm balance in which an object is weighed first in air and then in a beaker of water to determine its specific gravity.

hydrostatic equation [PHYS] The form assumed by the vertical component of the vector equation of motion when all Coriolis force, earth curvature, frictional, and vertical acceleration terms are considered negligible compared with those

HYDROGEN MASER

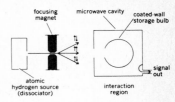

focusing magnet · microwave cavity · coated-wall storage bulb · signal out · atomic hydrogen source (dissociator) · interaction region

Basic elements of the hydrogen maser. The hydrogen atoms in the upper hyperfine level are focused into the coated bulb, and the energy coupled out through a loop and coaxial cable.

HYDROMETER

A plain hydrometer. *(Taylor Instrument Co.)*

involving the vertical pressure force and the force of gravity.

hydrostatic equilibrium [PHYS] The state of a fluid whose surfaces of constant mass (or density) coincide and are horizontal throughout; complete balance exists between the force of gravity and the pressure force; the relation between the pressure and the geometic height is given by the hydrostatic equation.

hydrostatic modulus *See* bulk modulus of elasticity.

hydrostatic pressure [FL MECH] 1. The pressure at a point in a fluid at rest due to the weight of the fluid above it. Also known as gravitational pressure. 2. The negative of the stress normal to a surface in a fluid.

hydrostatics [FL MECH] The study of liquids at rest and the forces exerted on them or by them.

hydrostatic strength [MECH] The ability of a body to withstand hydrostatic stress.

hydrostatic stress [MECH] The condition in which there are equal compressive stresses or equal tensile stresses in all directions, and no shear stresses on any plane.

hydrostatic weighing [FL MECH] A method of determining the density of a sample in which the sample is weighed in air, and then weighed in a liquid of known density; the volume of the sample is equal to the loss of weight in the liquid divided by the density of the liquid.

hygrodeik [ENG] A form of psychrometer with wet-bulb and dry-bulb thermometers mounted on opposite edges of a specially designed graph of the psychrometric tables, arranged so that the intersections of two curves determined by the wet-bulb and dry-bulb readings yield the relative humidity, dew-point, and absolute humidity.

hygrogram [ENG] The record made by a hygrograph.

hygrograph [ENG] A recording hygrometer.

hygrometer [ENG] An instrument for giving a direct indication of the amount of moisture in the air or other gas, the indication usually being in terms of relative humidity as a percentage which the moisture present bears to the maximum amount of moisture that could be present at the location temperature without condensation taking place.

hygrometry [ENG] The study which treats of the measurement of the humidity of the atmosphere and other gases.

hygroscopic [CHEM] 1. Possessing a marked ability to accelerate the condensation of water vapor; applied to condensation nuclei composed of salts which yield aqueous solutions of a very low equilibrium vapor pressure compared with that of pure water at the same temperature. 2. Pertaining to a substance whose physical characteristics are appreciably altered by effects of water vapor. 3. Pertaining to water absorbed by dry soil minerals from the atmosphere; the amounts depend on the physicochemical character of the surfaces, and increase with rising relative humidity.

hyl *See* metric-technical unit of mass.

Hylleraas coordinates [ATOM PHYS] Coordinates for two particles used in studying the helium atom; they comprise the distance between the two particles, the sum of the distances of the particles from the origin, and the difference of the distances of the particles from the origin.

hyperbola [MATH] The plane curve obtained by intersecting a circular cone of two nappes with a plane parallel to the axis of the cone.

hyperbolic antenna [ELECTROMAG] A radiator whose reflector in cross section describes a half hyperbola.

hyperbolic cosecant [MATH] A function whose value is equal to the reciprocal of the value of the hyperbolic sine. Abbreviated csch.

hyperbolic cosine [MATH] A function whose value at the

HYGROMETER

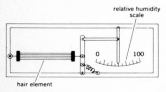

relative humidity scale

0   100

hair element

Hygrometer which uses hair as the sensing element. *(From D. M. Considine, ed., Process Instruments and Controls Handbook, McGraw-Hill, 1957)*

complex number z is one-half the sum of the exponential of z and the exponential of $-z$. Abbreviated cosh.

hyperbolic differential equation [MATH] A general type of second-order partial differential equation which includes the wave equation and has the form

$$\sum_{i,j=1}^{n} A_{ij}(\partial^2 u/\partial x_i \partial x_j) + \sum_{i=1}^{n} B_i(\partial u/\partial x_i) + Cu + F = 0$$

where the A_{ij}, B_i, C, and F are suitably differentiable real functions of $x_1, x_2, \ldots, x_n$, and there exists at each point $(x_1, x_2, \ldots, x_n)$ a real linear transformation on the x_i which reduces the quadratic form

$$\sum_{i,j=1}^{n} A_{ij} x_i x_j$$

to a sum of n squares not all of the same sign.

hyperbolic distance [ELECTROMAG] A function of pairs of points within a unit circle, where the interior of this circle is a conformal or projective representation of a hyperbolic space used in transmission line theory and waveguide analysis.

hyperbolic form [MATH] A nondegenerate, symmetric or alternating form on a vector space E such that E is a hyperbolic space under this form.

hyperbolic functions [MATH] The real or complex functions sinh (x), cosh (x), tanh (x), coth (x), sech (x), csch (x); they are related to the hyperbola in somewhat the same fashion as the trigonometric functions are related to the circle, and have properties analogous to those of the trigonometric functions.

hyperbolic geometry *See* Lobachevski geometry.

hyperbolic logarithm *See* logarithm.

hyperbolic paraboloid [MATH] A surface which can be so situated that sections parallel to one coordinate plane are parabolas while those parallel to the other plane are hyperbolas.

hyperbolic plane [MATH] A two-dimensional vector space E on which there is a nondegenerate, symmetric or alternating form $f(x,y)$ such that there exists a nonzero element w in E for which $f(w,w) = 0$.

hyperbolic point [FL MECH] A singular point in a streamline field which constitutes the intersection of a convergence line and a divergence line; it is analogous to a col in the field of a single-valued scalar quantity. Also known as neutral point. [MATH] A point on a surface where the Gaussian curvature is strictly negative.

hyperbolic secant [MATH] A function whose value is equal to the reciprocal of the value of the hyperbolic cosine. Abbreviated sech.

hyperbolic sine [MATH] A function whose value at the complex number z is one-half the difference between the exponential of z and the exponential of $-z$. Abbreviated sinh.

hyperbolic space [MATH] **1.** A space described by hyperbolic rather than cartesian coordinates. **2.** An orthogonal sum of hyperbolic planes.

hyperbolic spiral [MATH] A plane curve for which the radius vector is inversely proportional to the polar angle. Also known as reciprocal spiral.

hyperbolic structure [MATH] A hyperbolic structure for a flow (R, X, π), where X is a compact Riemannian manifold on which the length of a tangent vector is denoted $\|v\|$, is a decomposition of the tangent bundle TX into a Whitney sum of three bundles, E, E^u, and E^s, each of which is flow-invariant (for the derivative of the flow) and such that (1) E is tangent to the flow (that is, to the orbits) and (2) there exist constants $c < \infty$ and $\lambda > 0$ such that $\|d\pi(t,x)v_x\| \leq ce^{-\lambda t}\|v_x\|$ for v_x in $E_x{}^s$, $t \geq 0$, and (3) $\|d\pi(t,x)v_x\| \leq ce^{\lambda t}$ for v_x in $E_x{}^u$ and $t \leq 0$.

hyperbolic tangent [MATH] A function whose value is equal to the value of the hyperbolic sine divided by the value of the hyperbolic cosine. Abbreviated tanh.

hyperboloid of revolution [MATH] A surface generated by rotating a hyperboloid about one of its axes.

hyperboloids [MATH] Quadric surfaces given by equations of the form $(x^2/a^2) \pm (y^2/b^2) - (z^2/c^2) = 1$; in certain cases they can be realized by spinning the pieces of a hyperbola about an appropriate axis.

hypercharge [PARTIC PHYS] A quantum number conserved by strong interactions, equal to twice the average of the charges of the members of an isospin multiplet.

hypercircle method [MATH] A geometric method of obtaining approximate solutions of linear boundary value problems of mathematical physics that cannot be solved exactly, in which a correspondence is made between physical variables and vectors in a function space.

hypercomplex number *See* quaternion.

hypercomplex system *See* algebra.

hyperconjugation [PHYS CHEM] An arrangement of bonds in a molecule that is similar to conjugation in its formulation and manifestations, but the effects are weaker; it occurs when a CH_2 or CH_3 group (or in general, an AR_2 or AR_3 group where A may be any polyvalent atom and R any atom or radical) is adjacent to a multiple bond or to a group containing an atom with a lone π-electron, π-electron pair or quartet, or π-electron vacancy; it can be sacrificial (relatively weak) or isovalent (stronger).

hyperelliptic Riemann surface [MATH] A Riemann surface of a function f having the form $f(z) = (z - a_1)^{1/2}(z - a_2)^{1/2} \cdots (z - a_n)^{1/2}$, where $a_1, \ldots, a_n$ are distinct complex numbers.

hyperfine structure [SPECT] A splitting of spectral lines due to the spin of the atomic nucleus or to the occurrence of a mixture of isotopes in the element. Abbreviated hfs.

hyperfocal distance [OPTICS] The distance from the camera lens to the nearest object in acceptable focus when the lens is focused on infinity.

hyperfragment [NUC PHYS] An unstable nucleus in which one or more nucleons are replaced by hyperons. Also known as hypernucleus.

hyperfrequency waves [ELECTROMAG] Microwaves having wavelengths in the range from 1 centimeter to 1 meter.

hypergeometric differential equation *See* Gauss' hypergeometric equation.

hypergeometric distribution [STAT] The distribution of the number D of special items in a random sample of size s drawn from a population of size N that contains r of the special items:

$$P(D = d) = \binom{r}{d}\binom{N-r}{s-d} \bigg/ \binom{N}{s}.$$

hypergeometric function [MATH] A function which is a solution to the hypergeometric equation and obtained as an infinite series expansion.

hypergeometric series [MATH] A particular infinite series which in certain cases is a solution to the hypergeometric equation, and having the form

$$1 + \frac{ab}{c}z + \frac{1}{2!}\frac{a(a+1)b(b+1)}{c(c+1)}z^2 + \cdots.$$

hypernucleus *See* hyperfragment.

hyperon [PARTIC PHYS] **1.** An elementary particle which has baryon number $B = +1$, that is, which can be transformed into a nucleon and some number of mesons or lighter particles, and which has nonzero strangeness number. **2.** A hyperon (as in the first definition) which is semistable (the lifetime is much longer than 10^{-22} second).

hyperplane [MATH] A hyperplane is an $(n-1)$-dimensional subspace of an n-dimensional vector space.

hypersonic [ACOUS] Pertaining to frequencies above 500 megahertz. [FL MECH] Pertaining to hypersonic speeds, or air currents moving at hypersonic speeds.

hypersonics [ACOUS] Production and utilization of sound waves of frequencies above 500 megahertz.

hypersonic speed [FL MECH] A speed of an object greater than about five times the speed of sound in the fluid through which the object is moving.

hypervelocity [MECH] **1.** Muzzle velocity of an artillery projectile of 3500 feet per second (1067 meters per second) or more. **2.** Muzzle velocity of a small-arms projectile of 5000 feet per second (1524 meters per second) or more. **3.** Muzzle velocity of a tank-cannon projectile in excess of 3350 feet per second (1021 meters per second).

hypobaric [PHYS] Having less weight or pressure.

hypocycloid [MATH] The curve which is traced in the plane as a given point fixed on a circle moves while this circle rolls along the inside of another circle.

hypotenuse [MATH] On a right triangle, the side opposite the right angle.

hypothesis [SCI TECH] **1.** A proposition which is assumed to be true in proving another proposition. **2.** A proposition which is thought to be true because its consequences are found to be true. [STAT] A statement which specifies a population or distribution, and whose truth can be tested by sample evidence.

hypothesis testing [STAT] The branch of statistics which considers the problem of choosing between two actions on the basis of the observed value of a random variable whose distribution depends on a parameter, the value of which would indicate the correct action.

hypothetical parallax *See* dynamic parallax.

hypotrochoid [MATH] A curve traced by a point rigidly attached to a circle at a point other than the center when the circle rolls without slipping on the inside of a fixed circle.

hysteresis [ELECTR] An oscillator effect wherein a given value of an operating parameter may result in multiple values of output power or frequency. [ELECTROMAG] *See* magnetic hysteresis. [NUCLEO] A temporary change in the counting-rate-voltage characteristic of a radiation counter tube, caused by its previous operation. [PHYS] The dependence of the state of a system on its previous history, generally in the form of a lagging of a physical effect behind its cause.

hysteresis coefficient [PHYS] A constant, characteristic of a particular material, in a formula for hysteresis loss.

hysteresis damping [MECH] Damping of a vibration due to energy lost through mechanical hysteresis.

hysteresis error [PHYS] The maximum separation due to hysteresis between upscale-going and downscale-going indications of a measured variable.

hysteresis heating [PHYS] **1.** Supply of heat to a material through hysteresis loss. **2.** In particular, supply of a controlled amount of heat to a thermally isolated paramagnetic sample at temperatures below 1 kelvin by taking it through a magnetic hysteresis loop.

hysteresis loop [PHYS] The closed curve followed by a material displaying hysteresis (such as a ferromagnet or ferroelectric) on a graph of a driven variable (such as magnetic flux density or electric polarization) versus the driving variable (such as magnetic field or electric field).

HYSTERESIS LOOP

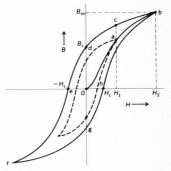

Hysteresis loop. H = magnetizing force; B = flux density; Oab = normal magnetization curve; H_c = coercive force; B_r = retentivity; B_m = maximum value of flux density; $bdefghb$ = the hysteresis loop.

hysteresis loss [PHYS] The energy converted to heat in a material because of magnetic or other hysteresis, accompanying cyclic variation of the magnetic field or other driving variable.

hysteretic damping [MECH] Damping of a vibrating system in which the retarding force is proportional to the velocity and inversely proportional to the frequency of the vibration.

Hz *See* hertz.

I   *See* iodine.

IA   *See* international angstrom.

IAT   *See* international atomic time.

IC   *See* integrated circuit.

ice line   [THERMO] A graph of the freezing point of water as a function of pressure.

ice point   [PHYS CHEM] The true freezing point of water; the temperature at which a mixture of air-saturated pure water and pure ice may exist in equilibrium at a pressure of 1 standard atmosphere (101,325 newtons per square meter).

iconocenter   [ELECTROMAG] The image of the reflection coefficient of a matched load as plotted on an Argand diagram.

iconometer   [OPTICS] **1.** An instrument used to find the size of an object of known distance or the distance of an object of known size by measurement of the image of it produced by a lens whose focal length is known. **2.** A direct viewfinder with a metal frame.

icosahedron   [MATH] A 20-sided polyhedron.

icositetrahedron   *See* trapezohedron.

ICT   *See* International Critical Tables.

ideal   [MATH] A subset I of a ring R where $x - y$ is in I for every x,y in I and either rx is in I for every r in R and x in I or xr is in I for every r in R and x in I; in the first case I is called a left ideal, and in the second a right ideal; an ideal is two-sided if it is both a left and a right ideal.

ideal aerodynamics   [FL MECH] A branch of aerodynamics that deals with simplifying assumptions that help explain some airflow problems and provide approximate answers. Also known as ideal fluid dynamics.

ideal bunching   [ELECTR] Theoretical condition in which the bunching of electrons in a velocity-modulated tube would give a single infinitely large current peak during each cycle.

ideal crystal   *See* perfect crystal.

ideal dielectric   [ELEC] Dielectric in which all the energy required to establish an electric field in the dielectric is returned to the source when the field is removed. Also known as perfect dielectric.

ideal exhaust velocity   [FL MECH] The theoretical maximum velocity, relative to the nozzle, of the gas flow as it passes from a given nozzle inlet temperature and pressure to a given ambient pressure, when the combustion gas has a given mean molecular weight.

ideal flow   [FL MECH] **1.** Fluid flow which is incompressible, two-dimensional, irrotational, steady, and nonviscous. **2.** *See* inviscid flow.

ideal fluid   [FL MECH] **1.** A fluid which has ideal flow. **2.** *See* inviscid fluid.

ideal fluid dynamics   *See* ideal aerodynamics.

ideal gas   [THERMO] Also known as perfect gas. **1.** A gas whose molecules are infinitely small and exert no force on each other. **2.** A gas that obeys Boyle's law (the product of the pressure and volume is constant at constant temperature) and Joule's law (the internal energy is a function of the temperature alone).

ideal gas law [THERMO] The equation of state of an ideal gas which is a good approximation to real gases at sufficiently high temperatures and low pressures; that is, $PV = RT$, where P is the pressure, V is the volume per mole of gas, T is the temperature, and R is the gas constant.

ideal index number *See* Fisher's ideal index.

ideal line [MATH] The collection of all ideal points, each corresponding to a given family of parallel lines. Also known as line at infinity.

ideal point [MATH] In projective geometry, all lines parallel to a given line are hypothesized to meet at a point at infinity, called an ideal point. Also known as point at infinity.

ideal propeller [FL MECH] A propeller which is considered as acting alone on an inviscid, incompressible fluid stream.

ideal radiator *See* blackbody.

idem factor [MATH] The dyadic $I = ii + jj + kk$ such that scalar multiplication of I by any vector yields that vector.

idempotent [MATH] **1.** An element x of an algebraic system satisfying the equation $x^2 = x$. **2.** An algebraic system in which every element x satisfies $x^2 = x$.

idempotent law [MATH] A law which states that an element x of an algebraic system satisfies $x^2 = x$.

idempotent matrix [MATH] A matrix E satisfying the equation $E^2 = E$.

identification [CONT SYS] The procedures for deducing a system's transfer function from its response to a step-function input or to an impulse.

identity [MATH] An equation satisfied for all possible choices of values for the variables involved.

identity element [MATH] The unique element e of a group where $g \cdot e = e \cdot g = g$ for every element g of the group. Also known as identity of a group.

identity function [MATH] The function of a set to itself which assigns to each element the same element. Also known as identity operator.

identity matrix [MATH] The square matrix all of whose entries are zero save along the principal diagonal where they all are 1.

identity of a group *See* identity element.

identity operator *See* identity function.

idle current *See* reactive current.

if and only if operation *See* biconditional operation.

if-then operation *See* implication.

ignition point *See* ignition temperature.

ignition temperature [CHEM] The lowest temperature at which combustion begins and continues in a substance when it is heated in air. Also known as autogenous ignition temperature; ignition point. [PL PHYS] The lowest temperature at which the fusion energy generated in a plasma exceeds the energy lost through bremsstrahlung radiation.

ignorable coordinate *See* cyclic coordinate.

IGY *See* International Geophysical Year.

I²L *See* integrated injection logic.

illinium *See* promethium-147.

illuminance [OPTICS] The density of the luminous flux on a surface. Also known as illumination; luminous flux density.

illumination [ELECTROMAG] **1.** The geometric distribution of power reaching various parts of a dish reflector in an antenna system. **2.** The power distribution to elements of an antenna array. [OPTICS] **1.** The science of the application of lighting. **2.** *See* illuminance.

illumination distribution [OPTICS] The manner in which light is dispersed on a surface.

illuminometer [OPTICS] A portable photometer which is used in the field or outside the laboratory and yields results of lower accuracy than a laboratory photometer.

image [ACOUS] *See* acoustic image. [ELEC] *See* electric image. [ELECTROMAG] The input reflection coefficient corresponding to the reflection coefficient of a specified load when the load is placed on one side of a waveguide junction and a slotted line is placed on the other. [OPTICS] An optical counterpart of a self-luminous or illuminated object formed by the light rays that traverse an optical system; each point of the object has a corresponding point in the image from which rays diverge or appear to diverge. [PHYS] Any reproduction of an object produced by means of focusing light, sound, electron radiation, or other emanations coming from the object or reflected by the object.

image admittance [ELECTR] The reciprocal of image impedance.

image antenna [ELECTROMAG] A fictitious electrical counterpart of an actual antenna, acting mathematically as if it existed in the ground directly under the real antenna and served as the direct source of the wave that is reflected from the ground by the actual antenna.

image attenuation constant [ELECTR] The real part of the image transfer constant. [ENG] The real part of the transfer constant.

image converter [ELECTR] *See* image tube. [OPTICS] A converter that uses a fiber optic bundle to change the form of an image, for more convenient recording and display or for the coding of secret messages.

image converter camera [ELECTR] A camera consisting of an image tube and an optical system which focuses the image produced on the phosphorescent screen of the tube onto photographic film.

image dissection photography [ELECTR] A method of high-speed photography in which an image is split in any one of various ways into interlaced space and time elements which can be unscrambled or played back through the system either to be viewed or to give a master negative.

image effect [ELECTROMAG] Effect produced on the field of an antenna due to the presence of the earth; electromagnetic waves are reflected from the earth's surface, and these reflections often are accounted for by an image antenna at an equal distance below the earth's surface.

image force [ELEC] The electrostatic force on a charge in the neighborhood of a conductor, which may be thought of as the attraction to the charge's electric image.

image impedance [ELECTR] One of the impedances that, when connected to the input and output of a transducer, will make the impedances in both directions equal at the input terminals and at the output terminals.

image intensifier *See* light amplifier.

image parameter design [ELECTR] A method of filter design using image impedance and image transfer functions as the fundamental network functions.

image parameter filter [ELECTR] A filter constructed by image parameter design.

image phase constant [ELECTR] The imaginary part of the image transfer constant. [ENG] The imaginary part of the transfer constant.

image plane [OPTICS] The plane in which an image produced by an optical system is formed; if the object plane is perpendicular to the optical axis, the image plane will ordinarily also be perpendicular to the axis.

image potential [ELEC] The potential set up by an electric image.

image space [OPTICS] The region of space where real or virtual images are formed by an optical system.

image surface [OPTICS] A surface on which lie images of points on a given plane perpendicular to the axis of an optical system.

image transfer constant [ELECTR] One-half the natural logarithm of the complex ratio of the steady-state apparent power entering and leaving a network terminated in its image impedance.

image tube [ELECTR] An electron tube that reproduces on its fluorescent screen an image of the optical image or other irradiation pattern incident on its photosensitive surface. Also known as electron image tube; image converter.

imaginary axis [MATH] All complex numbers $x + iy$ where $x = 0$; the vertical coordinate axis for the complex plane.

imaginary number [MATH] A complex number of the form $a + bi$, with b not equal to zero, where a and b are real numbers, and $i = \sqrt{-1}$; some writers require also that $a = 0$. Also known as imaginary quantity.

imaginary part [MATH] For a complex number $x + iy$ the imaginary part is the real number y.

imaginary quantity *See* imaginary number.

imaging [PHYS] The formation of images of objects.

imbedding [MATH] A homeomorphism of one topological space to a subspace of another topological space.

immersion [ASTRON] The disappearance of a celestial body either by passing behind another or passing into another's shadow. [MATH] A mapping f of a topological space X into a topological space Y such that for every $x \, \varepsilon \, X$ there exists a neighborhood N of x, such that f is a homeomorphism of N onto $f(N)$. [SCI TECH] Placement into or within a fluid, usually water.

immersion electron lens [ELECTR] An electron lens in which the object, usually the cathode, lies deep within the electric field so that the index of refraction varies rapidly in its vicinity.

immersion electron microscope [ELECTR] An emission electron microscope in which the specimen is a flat conducting surface which may be heated, illuminated, or bombarded by high-velocity electrons or ions so as to emit low-velocity thermionic, photo-, or secondary electrons; these are accelerated to a high velocity in an immersion objective or cathode lens and imaged as in a transmission electron microscope.

immersion electrostatic lens *See* bipotential electrostatic lens.

immersion lens *See* immersion objective.

immersion objective [OPTICS] A high-power microscope objective designed to work with the space between the objective and the cover glass over the object filled with an oil whose index of refraction is nearly the same as that of the objective and the cover glass, in order to reduce reflection losses and increase the index of refraction of the object space. Also known as immersion lens.

immersion refractometer [OPTICS] Device to measure refractive indices by immersing the prism portion in the sample being checked. Also known as dipping refractometer.

immittance [ELEC] A term used to denote both impedance and admittance, as commonly applied to transmission lines, networks, and certain types of measuring instruments.

immittance bridge [ELECTROMAG] A modification of an admittance bridge which compares the output current of a four-terminal device with admittance standards in a T configuration in order to measure transfer admittance by a null method.

impact [MECH] A forceful collision between two bodies which is sufficient to cause an appreciable change in the momentum of the system on which it acts. Also known as impulsive force.

impact avalanche and transit time diode *See* IMPATT diode.

impact energy [MECH] The energy necessary to fracture a material. Also known as impact strength.

impact ionization [ELECTR] Ionization produced by the impact of a high-energy charge carrier on an atom of semiconductor material; the effect is an increase in the number of charge carriers.

impact law [PHYS] The relationship of fluid density, particle density, and fluid viscosity in the settling velocity of large particles in a given liquid: settling velocity is directly proportional to the square root of the particle diameter.

impact loss [FL MECH] Loss of head in a flowing stream due to the impact of water particles upon themselves or some bounding surface.

impact parameter [NUC PHYS] In a nuclear collision, the perpendicular distance from the target nucleus to the initial line of motion of the incident particle.

impact pressure *See* dynamic pressure.

impact strength [MECH] **1.** Ability of a material to resist shock loading. **2.** *See* impact energy.

impact stress [MECH] Force per unit area imposed on a material by a suddenly applied force.

impact tube *See* pitot tube.

impact velocity [MECH] The velocity of a projectile or missile at the instant of impact. Also known as striking velocity.

IMPATT diode [ELECTR] A *pn* junction diode that has a depletion region adjacent to the junction, through which electrons and holes can drift, and is biased beyond the avalanche breakdown voltage. Derived from impact avalanche and transit time diode.

impedance [ELEC] *See* electrical impedance. [PHYS] **1.** The ratio of a sinusoidally varying quantity to a second quantity which measures the response of a physical system to the first, both being considered in complex notation; examples are electrical impedance, acoustic impedance, and mechanical impedance. Also known as complex impedance. **2.** The ratio of the greatest magnitude of a sinusoidally varying quantity to the greatest magnitude of a second quantity which measures the response of a physical system to the first; equal to the magnitude of the quantity in the first definition.

impedance-admittance matrix [ELECTR] A four-element matrix used to describe analytically a transistor in terms of impedances or admittances.

impedance bridge [ELEC] A device similar to a Wheatstone bridge, used to compare impedances which may contain inductance, capacitance, and resistance.

impedance coil [ELEC] A coil of wire designed to provide impedance in an electric circuit.

impedance compensator [ELEC] Electric network designed to be associated with another network or a line with the purpose of giving the impedance of the combination a desired characteristic with frequency over a desired frequency range.

impedance component [ELEC] **1.** Resistance or reactance. **2.** A device such as a resistor, inductor, or capacitor designed to provide impedance in an electric circuit.

impedance coupling [ELEC] Coupling of two signal circuits with an impedance.

impedance drop [ELEC] The total voltage drop across a component or conductor of an alternating-current circuit, equal to the phasor sum of the resistance drop and the reactance drop.

impedance magnetometer [ENG] An instrument for determining local variations in magnetic field by measuring the change in impedance of a high-permeability nickel-iron wire.

impedance match [ELEC] The condition in which the external impedance of a connected load is equal to the internal impedance of the source or to the surge impedance of a transmission line, thereby giving maximum transfer of energy from source to load, minimum reflection, and minimum distortion. [ENG ACOUS] The condition in which the impedance of the air in the chamber of an acoustic horn is adjusted to that of the diaphragm, so that sound is transmitted from the horn to the atmosphere with minimum reflection back into the horn.

impedance matrix [ELEC] A matrix Z whose elements are the mutual impedances between the various meshes of an electrical network; satisfies the matrix equation $V = ZI$, where V and I are column vectors whose elements are the voltages and currents in the meshes.

impedance meter *See* electrical impedance meter.

impedometer [ELECTROMAG] An instrument used to measure impedances in waveguides.

imperfect crystal [CRYSTAL] A crystal in which the regular, periodic structure is interrupted by various defects.

imperfect gas *See* real gas.

imperial pint *See* pint.

implication [MATH] **1.** The logical relation between two statements p and q, usually expressed as "if p then q." **2.** A logic operator having the characteristic that if p and q are statements, the implication of p and q is false if p is true and q is false, and is true otherwise. Also known as conditional implication; if-then operation; material implication.

implicit function [MATH] A function defined by an equation $f(x,y) = 0$, when x is considered as an independent variable and y, called an implicit function of x, as a dependent variable.

implicit function theorem [MATH] A theorem that gives conditions under which an equation in variables x and y may be solved so as to express y directly as a function of x; it states that if $F(x,y)$ and $\partial F(x,y)/\partial y$ are continuous in a neighborhood of the point (x_0,y_0) and if $F(x,y) = 0$ and $\partial F(x,y)/\partial y \neq 0$, then there is a number $\varepsilon > 0$ such that there is one and only one function $f(x)$ that is continuous and satisfies $F[x, f(x)] = 0$ for $|x - x_0| < \varepsilon$, and satisfies $f(x_0) = y_0$.

implosion [PHYS] A bursting inward, as in the inward collapse of an evacuated container (such as the glass envelope of a cathode-ray tube) or the compression of fissionable material by ordinary explosives in a nuclear weapon.

impressed voltage [ELEC] Voltage applied to a circuit or device.

imprisoned incompleteness [RELAT] The property of incomplete geodesics in a space-time being confined to a compact neighborhood.

improper divisor [MATH] An improper divisor of an element x in a commutative ring with identity is any unit of the ring or any associate of x.

improper fraction [MATH] **1.** In arithmetic, the quotient of two integers in which the numerator is greater than or equal to the denominator. **2.** In algebra, the quotient of two polynomials in which the degree of the numerator is greater than or equal to that of the denominator.

improper integral [MATH] Any integral in which either the integrand becomes unbounded on the domain of integration, or the domain of integration is itself unbounded.

improper orthogonal transformation [MATH] An orthogonal transformation such that the determinant of its matrix is −1.

impulse [MECH] The integral of a force over an interval of time. [PHYS] A pulse which lasts for so short a time that its duration can be thought of as infinitesimal.

impulse approximation [PHYS] An approximation for studying the collision of an incident particle with a bound target particle, in which the binding forces on the target particle during the collision are ignored.

impulse function [MATH] An idealized or generalized function defined not by its values but by its behavior under integration, such as the (Dirac) delta function.

impulse generator [ELEC] An apparatus which produces very short surges of high-voltage or high-current power by discharging capacitors in parallel or in series. Also known as pulse generator.

impulse modulation [CONT SYS] Modulation of a signal in which it is replaced by a series of impulses, equally spaced in time, whose strengths (integrals over time) are proportional to the amplitude of the signal at the time of the impulse.

impulse response [CONT SYS] The response of a system to an impulse which differs from zero for an infinitesimal time, but whose integral over time is unity; this impulse may be represented mathematically by a Dirac delta function.

impulse strength [ELEC] Voltage breakdown of insulation under voltage surges on the order of microseconds in duration.

impulse train [CONT SYS] An input consisting of an infinite series of unit impulses, equally separated in time.

impulsive force *See* impact.

impulsive sound [ACOUS] A sound that lasts for a short period of time and includes frequencies over a large portion of the acoustic spectrum, such as a hammer blow or hand clap.

impulsive sound equation [ACOUS] An equation which states that the total sound energy produced by a short burst of sound in a room is an exponentially decreasing function of time, whose decay constant depends on the speed of sound, the sound absorption coefficient, and the volume and surface area of the room.

impurity [SOLID STATE] A substance that, when diffused into semiconductor metal in small amounts, either provides free electrons to the metal or accepts electrons from it.

impurity scattering [SOLID STATE] Scattering of electrons by holes or phonons in the crystal.

impurity semiconductor [SOLID STATE] A semiconductor whose properties are due to impurity levels produced by foreign atoms.

In *See* indium.

in. *See* inch.

incandescence [OPTICS] The emission of visible radiation by a hot body.

incandescent lamp [ELEC] An electric lamp that produces light when a metallic filament is heated white-hot in a vacuum by passing an electric current through the filament. Also known as filament lamp; light bulb.

inch [MECH] A unit of length in common use in the United States and Great Britain, equal to $\frac{1}{12}$ foot or 2.54 centimeters. Abbreviated in.

inch of mercury [MECH] The pressure exerted by a 1-inch-high (2.54-centimeter-high) column of mercury that has a density of 13.5951 grams per cubic centimeter when the acceleration of gravity has the standard value of 9.80665 m/sec² or approximately 32.17398 ft/sec²; equal to

INCANDESCENT LAMP

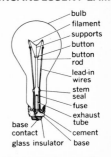

Parts of an incandescent lamp.
(General Electric Co.)

3386.388640341 newtons per square meter; used as a unit in the measurement of atmospheric pressure.

incidence angle *See* angle of incidence.

incidence matrix [MATH] In a graph the p by q matrix (b_{ij}) for which $b_{ij} = 1$ if the ith vertex is an end point of the jth edge, and $b_{ij} = 0$ otherwise.

incidence plane *See* plane of incidence.

incident light [OPTICS] The direct light that falls on a surface.

incident wave [ELECTR] A current or voltage wave that is traveling through a transmission line in the direction from source to load. [PHYS] A wave that impinges on a discontinuity, particle, or body, or on a medium having different propagation characteristics.

inclination [GEOPHYS] In magnetic inclination, the dip angle of the earth's magnetic field. Also known as magnetic dip. [MATH] **1.** The inclination of a line in a plane is the angle made with the positive x axis. **2.** The inclination of a line in space with respect to a plane is the smaller angle the line makes with its orthogonal projection in the plane. **3.** The inclination of a plane with respect to a given plane is the smaller of the dihedral angles which it makes with the given plane. [SCI TECH] **1.** Angular deviation of a direction or surface from the true vertical or horizontal. **2.** The angle which a direction or surface makes with the vertical or horizontal. **3.** A surface which deviates from the vertical or horizontal.

inclination of planetary orbits [ASTRON] The angle between the plane of the orbit and the plane of the ecliptic, which is the plane of the earth's orbit.

inclined extinction [OPTICS] Extinction in which the vibration directions are inclined to a crystal axis or direction of cleavage. Also known as oblique extinction.

inclined plane [MECH] A plane surface at an angle to some force or reference line.

inclusion relation [MATH] **1.** A set theoretic relation, usually denoted by the symbol $\subset$, such that, if A and B are two sets, $A \subset B$ if and only if every element of A is an element of B. **2.** Any relation on a Boolean algebra which is reflexive, antisymmetric, and transitive.

inclusive or *See* or.

incoherent light [OPTICS] Electromagnetic radiant energy not all of the same phase, and possibly also consisting of various wavelengths.

incoherent scattering [PHYS] Scattering of particles or photons in which the scattering elements act independently of one another, so that there are no definite phase relationships among the different parts of the scattered beam.

incoherent waves [PHYS] Waves having no fixed phase relationship.

incommensurable numbers [MATH] Two numbers whose ratio is irrational.

incompatible equations [MATH] Two or more equations that are not satisfied by any set of values for the variables appearing. Also known as inconsistent equations.

incomplete Anger function [MATH] A generalization $u_p{}^r(\sigma)$ of the Anger function $J_r(x)$; it is equal to the integral from 0 to σ of $\cos{(r \sin{\theta} - p\theta)}\ d\theta$, so that $u_r{}^x(\pi) = \pi J_r(x)$.

incomplete beta function [MATH] The function $\beta_x(p,q)$ defined by

$$\beta_x(p,q) = \int_0^x t^{p-1}(1 - t)^{q-1}\ dt$$

where $0 \le x \le 1$, $p > 0$, and $q > 0$.

incomplete fusion *See* quasifission.

INCLINED PLANE

(a)

(b)

Weight resting on an inclined plane *(a)* with principal forces applied, and *(b)* their resolution into normal force. θ is angle of inclination of plane, W is weight of body, F_p is force parallel to the surface, F_n is force normal to the surface.

incomplete gamma function [MATH] Either of the functions $\gamma(a,x)$ and $\Gamma(a,x)$ defined by

$$\gamma(a,x) = \int_0^x t^{a-1} e^{-t} \, dt$$

$$\Gamma(a,x) = \int_x^\infty t^{a-1} e^{-t} \, dt$$

where $0 \leq x \leq \infty$ and $a > 0$.

incomplete Latin square *See* Yonden square.

incompressibility [MECH] Quality of a substance which maintains its original volume under increased pressure.

incompressibility condition [FL MECH] The condition prevailing when dp/dt, the time rate of change of the density of a fluid, is zero; this is a valid assumption for most problems in dynamic oceanography.

incompressible flow [FL MECH] Fluid motion without any change in density.

incompressible fluid [FL MECH] A fluid which is not reduced in volume by an increase in pressure.

inconsistent equations *See* incompatible equations.

increment [MATH] A change in the argument or values of a function, usually restricted to being a small positive or negative quantity. [SCI TECH] A small change in the value of a variable.

incremental hysteresis loss [ELECTROMAG] Hysteresis loss when a magnetic material is subjected to a pulsating magnetizing force.

incremental induction [ELECTROMAG] The quantity lying between the highest and lowest value of a magnetic induction at a point in a polarized material, when subjected to a small cycle of magnetization.

incremental permeability [ELECTROMAG] The ratio of a small cyclic change in magnetic induction to the corresponding cyclic change in magnetizing force when the average magnetic induction is greater than zero.

indefinite integral [MATH] An indefinite integral of a function $f(x)$ is a function $F(x)$ whose derivative equals $f(x)$. Also known as antiderivative; integral.

independent axioms [MATH] A list of axioms such that no axiom can be deduced as a theorem from the others.

independent equations [MATH] A system of equations such that no one of them is necessarily satisfied by a solution to the rest.

independent events [STAT] Two events in probability such that the occurrence of one of them does not affect the probability of the occurrence of the other.

independent functions [MATH] A set of functions such that knowledge of the values obtained by all but one of them at a point is insufficient to determine the value of the remaining function.

independent random variables [STAT] The discrete random variables $X_1, X_2, \ldots, X_n$ are independent if for arbitrary values $x_1, x_2, \ldots, x_n$ of the variables the probability that $X_1 = x_1$ and $X_2 = x_2$, etc., is equal to the product of the probabilities that $X_i = x_i$ for $i = 1, 2, \ldots, n$; random variables which are unrelated.

independent variable [MATH] In an equation $y = f(x)$, the input variable x. Also known as argument.

indeterminacy principle *See* uncertainty principle.

indeterminate equations [MATH] A set of equations possessing an infinite number of solutions.

indeterminate forms [MATH] Products, quotients, differences, or powers of functions which are undefined when the argument of the function has a certain value, because one or both of the functions are zero or infinite; however, the limit of the product, quotient, and so on as the argument approaches this value is well defined.

index   [MATH] **1.** Unity of a logarithmic scale, as the C scale of a slide rule. **2.** A subscript or superscript used to indicate a specific element of a set or sequence. **3.** For a subgroup of a finite group, the order of the group divided by the order of the subgroup. **4.** For a continuous complex-valued function defined on a closed plane curve, the change in the amplitude of the function when traversing the curve in a counterclockwise direction, divided by 2π.   [PHYS] A numerical quantity, usually dimensionless, denoting the magnitude of some physical effect, such as the refractive index.

index ellipsoid   [OPTICS] An ellipsoid whose three perpendicular axes are proportional in length to the principal values of the index of refraction of light in an anisotropic medium and point in the direction of the corresponding electric vector. Also known as ellipsoidal of wave normals; ellipsoid of wave normals; indicatrix; optical indicatrix; polarizability ellipsoid; reciprocal ellipsoid.

index line   *See* isopleth.

index liquid   [OPTICS] A liquid whose index of refraction is known, used to find the index of refraction of powdered substances with a microscope.

index number   [STAT] A number indicating change in magnitude, as of cost or of volume of production, as compared with the magnitude at a specified time, usually taken as 100; for example, if production volume in 1970 was two times as much as the volume in 1950 (taken as 100), its index number is 200.

index of absorption   *See* absorption index.

index of a radical   [MATH] The number above and to the left of the radical sign indicating the root to be extracted.

index of a subgroup   [MATH] The quotient of the order of the group by the order of a subgroup.

index of precision   [STAT] The constant h in the normal curve $y = K \exp [-h^2(x-u)^2]$; a large value of h indicates a high precision, or small standard deviation.

index of refraction   [OPTICS] The ratio of the phase velocity of light in a vacuum to that in a specified medium. Also known as refractive index; refracture index.

indicating instrument   [ENG] An instrument in which the present value of the quantity being measured is visually indicated.

indicator function   *See* characteristic function.

indicatrix   *See* index ellipsoid.

indifferent equilibrium   *See* neutral equilibrium.

indirect cycle   [NUCLEO] A nuclear reactor cycle in which a heat exchanger transfers heat from the reactor coolant to a second fluid, which then drives a prime mover.

indirectly heated cathode   [ELECTR] A cathode to which heat is supplied by an independent heater element in a thermionic tube; this cathode has the same potential on its entire surface, whereas the potential along a directly heated filament varies from one end to the other. Also known as equipotential cathode; heater-type cathode; unipotential cathode.

indirect wave   [PHYS] Any radio wave which arrives by an indirect path, having undergone an abrupt change of direction by refraction or reflection.

indium   [CHEM] A metallic element, symbol In, atomic number 49, atomic weight 114.82; soluble in acids; melts at 156°C, boils at 1450°C.

induced anisotropy   [SOLID STATE] A type of uniaxial anisotropy in a magnetic material produced by annealing the magnetic material in a magnetic field.

induced capacity   *See* absolute permeability.

induced current   [ELECTROMAG] A current produced in a

INDIUM

Periodic table of the chemical elements showing the position of indium.

conductor by a time-varying magnetic field, as in induction heating.

induced dipole [ELEC] An electric dipole produced by application of an electric field.

induced electromotive force [ELECTROMAG] An electromotive force resulting from the motion of a conductor through a magnetic field, or from a change in the magnetic flux that threads a conductor.

induced emission *See* stimulated emission.

induced fission [NUC PHYS] Fission which takes place only when a nucleus is bombarded with neutrons, gamma rays, or other carriers of energy.

induced magnetism [ELECTROMAG] The magnetism acquired by magnetic material while it is in a magnetic field.

induced moment [ELEC] The average electric dipole moment per molecule which is produced by the action of an electric field on a dielectric substance.

induced orientation [MATH] An orientation of a face of a simplex S opposite a vertex p_i obtained by deleting p_i from the ordering defining the orientation of S.

induced potential *See* induced voltage.

induced radioactivity [NUCLEO] Radioactivity created by bombarding a substance with radiation. Also known as artificial radioactivity.

induced voltage [ELECTROMAG] A voltage produced by electromagnetic or electrostatic induction. Also known as induced potential.

inductance [ELECTROMAG] **1.** That property of an electric circuit or of two neighboring circuits whereby an electromotive force is generated (by the process of electromagnetic induction) in one circuit by a change of current in itself or in the other. **2.** Quantitatively, the ratio of the emf (electromotive force) to the rate of change of the current. **3.** *See* coil.

inductance bridge [ELECTROMAG] **1.** A device, similar to a Wheatstone bridge, for comparing inductances. **2.** A four-coil alternating-current bridge circuit used for transmitting a mechanical movement to a remote location over a three-wire circuit; half of the bridge is at each location.

inductance coil *See* coil.

inductance measurement [ELECTROMAG] The determination of the self-inductance of a circuit or the mutual inductance of two circuits.

inductance meter [ELECTROMAG] A device which measures the self-inductance of a circuit or the mutual inductance of two circuits.

inductance standards [ELECTROMAG] Two equal, multilayer coils, wound on toroidal cores of nonmagnetic materials, connected in series and located so that their interactions with external fields tend to cancel one another.

induction *See* electromagnetic induction; electrostatic induction.

induction accelerator *See* betatron.

induction coil [ELECTROMAG] A device for producing high-voltage alternating current or high-voltage pulses from low-voltage direct current, in which interruption of direct current in a primary coil, containing relatively few turns of wire, induces a high voltage in a secondary coil, containing many turns of wire wound over the primary.

induction field [ELECTROMAG] A component of an electromagnetic field associated with an alternating current in a loop, coil, or antenna which carries energy alternately away from and back into the source, with no net loss, and which is responsible for self-inductance in a coil or mutual inductance with neighboring coils.

induction force [PHYS CHEM] A type of van der Waals force

INDUCTANCE BRIDGE

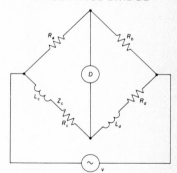

Circuit diagram for a general inductance bridge (def. 1); R_{a-d} are resistors; L_{c-d} are inductances; Z_c is impedance; V is source of ac voltage; and D is a null detector.

resulting from the interaction of the dipole moment of a polar molecule and the induced dipole moment of a nonpolar molecule. Also known as Debye force.

induction inclinometer *See* earth inductor.

induction motor [ELEC] An alternating-current motor in which a primary winding on one member (usually the stator) is connected to the power source, and a secondary winding on the other member (usually the rotor) carries only current induced by the magnetic field of the primary.

induction period [PHYS CHEM] A time of acceleration of a chemical reaction from zero to a maximum rate.

inductive capacities [ELECTROMAG] The permeability and permittivity of a substance.

inductive charge [ELEC] The charge that exists on an object as a result of its being near another charged object.

inductive circuit [ELEC] A circuit containing a higher value of inductive reactance than capacitive reactance.

inductive coupler [ELEC] A mutual inductance that provides electrical coupling between two circuits; used in radio equipment.

inductive coupling [ELEC] Coupling of two circuits by means of the mutual inductance provided by a transformer. Also known as transformer coupling.

inductive filter [ELECTR] A low-pass filter used for smoothing the direct-current output voltage of a rectifier; consists of one or more sections in series, each section consisting of an inductor on one of the pair of conductors in series with a capacitor between the conductors. Also known as LC filter.

inductive load [ELEC] A load that is predominantly inductive, so that the alternating load current lags behind the alternating voltage of the load. Also known as lagging load.

inductive-output tube [ELECTR] A tube in which output energy is obtained from the electron stream by electric induction between a cylindrical output electrode and the electron stream that flows through but does not touch the electrode.

inductive post [ELECTROMAG] Metal post or screw extending across a waveguide parallel to the E field, to add inductive susceptance in parallel with the waveguide for tuning or matching purposes.

inductive reactance [ELEC] Reactance due to the inductance of a coil or circuit.

inductive susceptance [ELEC] In a circuit containing almost no resistance, the part of the susceptance due to inductance.

inductive waveform [ELEC] A graph or trace of the effect of current buildup across an inductive network; proportional to the exponential of the product of a negative constant and the time.

inductive window [ELECTROMAG] Conducting diaphragm extending into a waveguide from one or both sidewalls of the waveguide, to give the effect of an inductive susceptance in parallel with the waveguide.

inductometer [ELECTROMAG] A coil of wire of known inductance; the inductance may be fixed as in the case of primary standards, adjustable by means of switches, or continuously variable by means of a movable-coil construction.

inductor *See* coil.

inelastic [MECH] Not capable of sustaining a deformation without permanent change in size or shape.

inelastic buckling [MECH] Sudden increase of deflection or twist in a column when compressive stress reaches the elastic limit but before elastic buckling develops.

inelastic collision [MECH] A collision in which the total kinetic energy of the colliding particles is not the same after the collision as before it.

INDUCTOMETER

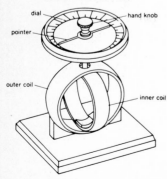

A variable inductometer showing the movable inner coil.

inelastic cross section [PHYS] The cross section for an inelastic collision.

inelastic scattering [PHYS] Scattering that results from inelastic collisions.

inelastic stress [MECH] A force acting on a solid which produces a deformation such that the original shape and size of the solid are not restored after removal of the force.

inequality [MATH] A statement that one quantity is less than, less than or equal to, greater than, or greater than or equal to another quantity.

inequality of Clausius *See* Clausius inequality.

inert [SCI TECH] Lacking an activity, reactivity, or effect.

inert gas *See* noble gas.

inertia [MECH] That property of matter which manifests itself as a resistance to any change in the momentum of a body.

inertia ellipsoid [MECH] An ellipsoid used in describing the motion of a rigid body; it is fixed in the body, and the distance from its center to its surface in any direction is inversely proportional to the square root of the moment of inertia about the corresponding axis. Also known as Poinsot ellipsoid.

inertial-confinement fusion *See* pellet fusion.

inertial coordinate system *See* inertial reference frame.

inertial flow [FL MECH] Flow in which no external forces are exerted on a fluid. [GEOPHYS] Frictionless flow in a geopotential surface in which there is no pressure gradient; the centrifugal and Coriolis accelerations must therefore be equal and opposite, and the constant inertial wind speed V_i is given by $V_i = fR$, where f is the Coriolis parameter and R the radius of curvature of the path.

inertial force [MECH] The fictitious force acting on a body as a result of using a noninertial frame of reference; examples are the centrifugal and Coriolis forces that appear in rotating coordinate systems. Also known as effective force.

inertial instability [FL MECH] **1.** Generally, instability in which the only form of energy transferred between the steady state and the disturbance in the fluid is kinetic energy. **2.** The hydrodynamic instability arising in a rotating fluid mass when the velocity distribution is such that the kinetic energy of a disturbance grows at the expense of kinetic energy of the rotation. Also known as dynamic instability.

inertial mass [MECH] The mass of an object as determined by Newton's second law, in contrast to the mass as determined by the proportionality to the gravitational force.

inertial reference frame [MECH] A coordinate system in which a body moves with constant velocity as long as no force is acting on it. Also known as inertial coordinate system.

inertial size *See* aerodynamic size.

inertia matrix [MECH] A matrix **M** used to express the kinetic energy T of a mechanical system during small displacements from an equilibrium position, by means of the equation $T = \frac{1}{2}\dot{q}^T\mathbf{M}\dot{q}$, where $\dot{q}$ is the vector whose components are the derivatives of the generalized coordinates of the system with respect to time, and $\dot{q}^T$ is the transpose of $\dot{q}$.

inertia of energy [RELAT] The principle that the inertial properties of matter both determine and are determined by its total energy content.

inertia tensor [MECH] A tensor associated with a rigid body whose product with the body's rotation vector yields the body's angular momentum.

inertia wave [FL MECH] **1.** Any wave motion in which no form of energy other than kinetic energy is present; in this general sense, Helmholtz waves, barotropic disturbances, Rossby waves, and so forth, are inertia waves. **2.** More restrictedly, a wave motion in which the source of kinetic energy of the disturbance is the rotation of the fluid about some given axis; in the atmosphere a westerly wind system is such a source, the inertia waves here being, in general, stable.

inextensional deformation [MECH] A bending of a surface that leaves unchanged the length of any line drawn on the surface and the curvature of the surface at each point.

infall process [ASTROPHYS] A process in which gas falls upon a very compact object such as a neutron star or black hole, reaching a high velocity and forming a hot plasma; postulated as a model for x-ray sources such as Centaurus X-1 and Hercules X-1.

inferior conjunction [ASTRON] A type of configuration in which two celestial bodies have their least apparent separation; the smaller body is nearer the observer than the larger body, about which it orbits; for example, Venus is closest to the earth at its inferior conjunction.

inferior mirage [OPTICS] A spurious image of an object formed below the true position of that object by abnormal refraction conditions along the line of sight; one of the most common types of mirage, and the opposite of a superior mirage.

inferior planet [ASTRON] A planet that circles the sun in an orbit that is smaller than the earth's.

infimum *See* greatest lower bound.

infinite [MATH] Larger than any fixed number.

infinite extension [MATH] An extension field F of a given field E such that F, viewed as a vector space over E, has infinite dimension.

infinite integral [MATH] An integral at least one of whose limits of integration is infinite.

infinite multiplication factor [NUCLEO] The multiplication factor of a theoretical system from which there is no leakage of neutrons, that is, a reactor of infinite size.

infinite population [STAT] A universe which contains an infinite number of elements; it can be continuous or discrete.

infinite product [MATH] The product of an infinite number of quantities, written $a_1 a_2 a_3 \ldots$ or

$$\prod_{k=1}^{\infty} a_k \, ;$$

its value is taken to be the limit of the sequence of partial products

$$\prod_{k=1}^{n} a_k$$

as n goes to infinity, provided this limit exists.

infinite sequence *See* sequence.

infinite series [MATH] An indicated sum of an infinite sequence of quantities, written $a_1 + a_2 + a_3 + \cdots$, or

$$\sum_{k=1}^{\infty} a_k .$$

infinite set [MATH] A set with more elements than any fixed integer; such a set can be put into a one to one correspondence with a proper subset of itself.

infinitesimal [MATH] A function whose value approaches 0 as its argument approaches some specified limit.

infinitesimal generator [MATH] A closed linear operator defined relative to some semigroup of operators and which uniquely determines that semigroup.

infinity [MATH] The concept of a value larger than any finite value.

infinity method [OPTICS] Method of adjusting two lines of sight to make them parallel; lines are adjusted on an object at great distance, for example, a star.

infix notation [MATH] A method of forming mathematical or logical expressions in which operators are written between the operands on which they act.

inflectional tangent [MATH] A tangent to a curve at a point of inflection.

inflection point *See* point of inflection.

influence line [MECH] A graph of the shear, stress, bending moment, or other effect of a movable load on a structural member versus the position of the load.

information function of a partition [MATH] If ξ is a finite partition of a probability space, the information function of ξ is a step function whose sets of constancy are the elements of ξ and whose value on an element of ξ is the negative of the logarithm of the probability of this element.

information processing [ADP] **1.** The manipulation of data so that new data (implicit in the original) appear in a useful form. **2.** *See* data processing.

information theory [MATH] The branch of probability theory concerned with the likelihood of the transmission of messages, accurate to within specified limits, when the bits of information composing the message are subject to possible distortion.

infrared astronomy [ASTROPHYS] The study of electromagnetic radiation in the spectrum between 0.75 and 1000 micrometers emanating from astronomical sources.

infrared binoculars [OPTICS] An instrument for viewing an enlarged infrared image with both eyes; it has two infrared telescopes whose lens systems are similar to those of ordinary binoculars.

infrared bolometer [ELECTR] A bolometer adapted to detecting infrared radiation, as opposed to microwave radiation.

infrared catastrophe [QUANT MECH] The logarithmic divergence in the cross section (which one would expect to be finite) for the emission of low-energy photons in bremsstrahlung and in the double Compton effect, according to quantum electrodynamics; the difficulty is resolved by taking radiative corrections to elastic scattering into account. Also known as infrared problem.

infrared dome *See* irdome.

infrared filter [OPTICS] A substance or device which is highly transparent to infrared radiation at certain wavelengths while absorbing other types of electromagnetic radiation.

infrared galaxy [ASTRON] A galaxy or quasar whose nucleus emits enormous amounts of infrared radiation, in some cases more than 1000 times the output of the entire Milky Way Galaxy at all wavelengths.

infrared image converter [ELECTR] A device for converting an invisible infrared image into a visible image, consisting of an infrared-sensitive, semitransparent photocathode on one end of an evacuated envelope and a phosphor screen on the other, with an electrostatic lens system between the two. Also known as infrared image tube.

infrared image tube *See* infrared image converter.

infrared lamp [ELEC] An incandescent lamp which operates at reduced voltage with a filament temperature of 4000°F (2200°C) so that it radiates electromagnetic energy primarily in the infrared region.

infrared laser [PHYS] A laser which emits infrared radiation, especially in the near- and intermediate-infrared regions.

infrared maser [PHYS] A laser which emits infrared radiation, especially in the far-infrared region, or which is pumped with radiation at infrared frequencies and emits radiation at millimeter wavelengths.

infrared microscope [OPTICS] A type of reflecting microscope which uses radiation of wavelengths greater than 700 nanometers and is used to reveal detail in materials that are

INFRARED IMAGE CONVERTER

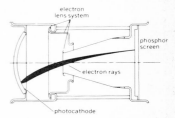

Components of the infrared image converter.

INFRARED LAMP

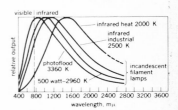

Special distribution of energy from various types of infrared lamps. *(From Illuminating Engineering Society, IES Lighting Handbook, 5th ed., 1968)*

opaque to light, such as molybdenum, wood, corals, and many red-dyed materials.

infrared optical material [ELECTROMAG] A material which is transparent to infrared radiation.

infrared phosphor [SOLID STATE] A phosphor which, when exposed to infrared radiation during or even after decay of luminescence resulting from its usual or dominant activator, emits light having the same spectrum as that of the dominant activator; sulfide and selenide phosphors are the most important examples.

infrared photoconductor [ELECTR] A conductor whose conductivity increases when it is exposed to infrared radiation.

infrared problem *See* infrared catastrophe.

infrared radiation [ELECTROMAG] Electromagnetic radiation whose wavelengths lie in the range from 0.75 or 0.8 micrometer (the long-wavelength limit of visible red light) to 1000 micrometers (the shortest microwaves).

infrared spectrometer [SPECT] Device used to identify and measure the concentrations of heteroatomic compounds in gases, in many nonaqueous liquids, and in some solids by arc or spark excitation and subsequent measurement of the electromagnetic emissions in the wavelength range of 0.78 to 300 micrometers.

infrared spectrophotometry [SPECT] Spectrophotometry in the infrared region, usually for the purpose of chemical analysis through measurement of absorption spectra associated with rotational and vibrational energy levels of molecules.

infrared spectroscopy [SPECT] The study of the properties of material systems by means of their interaction with infrared radiation; ordinarily the radiation is dispersed into a spectrum after passing through the material.

infrared spectrum [ELECTROMAG] **1.** The range of wavelengths of infrared radiation. **2.** A display or graph of the intensity of infrared radiation emitted or absorbed by a material as a function of wavelength or some related parameter.

infrared telescope [OPTICS] An instrument that converts an invisible infrared image into a visible image and enlarges this image, consisting of an infrared image converter tube, an objective lens for imaging the scene to be viewed onto the photocathode of the tube, and an ocular for viewing the phosphor screen of the tube.

infrared thermistor [ELECTR] A thermistor used to measure the power of infrared radiation.

infrared window [GEOPHYS] A frequency region in the infrared where there is good transmission of electromagnetic radiation through the atmosphere.

infrasonic [ACOUS] Pertaining to signals, equipment, or phenomena involving frequencies below the range of human hearing, hence below about 15 hertz. Also known as subsonic (deprecated usage).

infrasound [ACOUS] Vibrations of the air at frequencies too low to be perceived as sound by the human ear, below about 15 hertz.

Ingen-Hausz apparatus [THERMO] An apparatus for comparing the thermal conductivities of different conductors; specimens consisting of long wax-coated rods of equal length are placed with one end in a tank of boiling water covered with a radiation shield, and the lengths along the rods from which the wax melts are compared.

ingress [ASTRON] The entrance of the moon into the shadow of the earth in an eclipse, of a planet into the disk of the sun, or of a satellite (or its shadow) onto the disk of the parent planet. [SCI TECH] The act of entering, as of air into the lungs or a liquid into an orifice.

inhibiting signal [ELECTR] A signal, which when entered into a specific circuit will prevent the circuit from exercising its normal function; for example, an inhibit signal fed into an AND gate will prevent the gate from yielding an output when all normal input signals are present.

inhibition [SCI TECH] The act of repressing or restraining a physical or chemical action.

inhour [NUCLEO] A unit of reactivity of a reactor; 1 inhour is the reactivity that will give the reactor a period of 1 hour. Derived from inverse hour.

inhour equation [NUCLEO] An equation relating the reactivity of a nuclear reactor to the parameters of the delayed-neutron emitters and the neutron lifetime of the reactor.

initial nuclear radiation [NUCLEO] Radiation emitted from the fireball of a nuclear explosive during the first minute (an arbitrary time interval) after detonation.

initial permeability [ELECTROMAG] The limit of the normal permeability as the magnetic induction and magnetic field strength approach 0.

initial-value problem [FL MECH] A dynamical problem whose solution determines the state of a system at all times subsequent to a given time at which the state of the system is specified by given initial conditions; the initial-value problem is contrasted with the steady-state problem, in which the state of the system remains unchanged in time. Also known as transient problem. [MATH] An nth order ordinary or partial differential equation in which the solution and its first $(n - 1)$ derivatives are required to take on specified values at a particular value of a given independent variable.

initial-value theorem [MATH] The theorem that, if a function $f(t)$ and its first derivative have Laplace transforms, and if $g(s)$ is the Laplace transform of $f(t)$, and if the limit of $sg(s)$ as s approaches infinity exists, then this limit equals the limit of $f(t)$ as t approaches 0.

initial velocity [PHYS] The velocity of anything at the beginning of a specific phase of its motion.

injection [MATH] A mapping f from a set A into a set B which has the property that for any element b of B there is at most one element a of A for which $f(a) = b$. Also known as injective mapping; one-to-one mapping.

injection electroluminescence [ELECTR] Radiation resulting from recombination of minority charge carriers injected in a pn or pin junction that is biased in the forward direction. Also known as Lossev effect; recombination electroluminescence.

injection grid [ELECTR] Grid introduced into a vacuum tube in such a way that it exercises control over the electron stream without causing interaction between the screen grid and control grid.

injection laser [OPTICS] A laser in which a forward-biased gallium arsenide diode converts direct-current input power directly into coherent light, without optical pumping.

injection locking [ELECTR] The capture or synchronization of a free-running oscillator by a weak injected signal at a frequency close to the natural oscillator frequency or to one of its subharmonics; used for frequency stabilization in IMPATT or magnetron microwave oscillators, gas-laser oscillators, and many other types of oscillators.

injection luminescent diode [ELECTR] Gallium arsenide diode, operating in either the laser or the noncoherent mode, that can be used as a visible or near-infrared light source for triggering such devices as light-activated switches.

injective mapping *See* injection.

injector [ELECTR] An electrode through which charge carriers (holes or electrons) are forced to enter the high-field region in a spacistor.

inner automorphism [MATH] An automorphism h of a group where $h(g) = g_0^{-1} \cdot g \cdot g_0$, for every g in the group with g_0 some fixed group element.

inner bremsstrahlung [NUC PHYS] The emission of a photon during beta decay or electron capture by a nucleus. Also known as internal bremsstrahlung.

inner effect [PHYS] Effects on x-ray diffraction that occur within a particular atom or molecule.

inner function [MATH] **1.** A continuous open mapping of a topological space X into a topological space Y where the inverse image of each point in Y is zero dimensional. **2.** A complex function which is a member of the Hardy space H^∞ and whose radial limits have absolute value 1 on the unit circle.

inner planet [ASTRON] Any of the four planets (Mercury, Venus, Earth, and Mars) in the solar system whose orbits are closest to the sun.

inner potential [SOLID STATE] The average value of the electrostatic potential, taken over the volume of a crystal.

inner product [MATH] **1.** A scalar valued function of pairs of vectors from a vector space, denoted by (x,y) where x and y are vectors, and with the properties that (x,x) is always positive and is zero only if $x = 0$, that $(ax + by, z) = a(x,z) + b(y,z)$ for any scalars a and b, and that $(x,y) = (y,x)$ if the scalars are real numbers, $(x,y) = \overline{(y,x)}$ if the scalars are complex numbers. Also known as Hermitian inner product; Hermitian scalar product. **2.** The inner product of vectors $(x_1,\ldots,x_n)$ and $(y_1,\ldots,y_n)$ from n-dimensional euclidean space is the sum of $x_i y_i$ as i ranges from 1 to n. Also known as dot product; scalar product.

inner product of functions [MATH] For two functions f and g of a real or complex variable, this is defined to be the integral of $f(x)\overline{g(x)}\ dx$, where $\overline{g(x)}$ denotes the conjugate of $g(x)$.

inner product of tensors [MATH] The inner product of two tensors is the contracted tensor obtained from their product by means of pairing contravariant indices of one with covariant indices of the other.

inner product space [MATH] A vector space that has an inner product defined on it. Also known as Hermitian space; unitary space.

inner quantum number [ATOM PHYS] A quantum number J which gives an atom's total angular momentum, excluding the nuclear spin.

inorganic liquid laser [OPTICS] A liquid laser in which an inorganic liquid such as neodymium-selenium oxychloride or neodymium-doped phosphorus chloride is used as the active material. Also known as neodymium liquid laser.

in phase [PHYS] Having waveforms that are of the same frequency and that pass through corresponding values at the same instant.

in-pile [NUCLEO] Designating experiments or equipment inside a reactor.

in-pile loop [NUCLEO] An experiment inserted directly in a nuclear reactor (pile) incorporating a closed circuit (loop) of fluid usually for cooling purposes.

input [ADP] The information that is delivered to a data-processing device from the external world, the process of delivering these data, or the equipment that performs this process. [ELECTR] **1.** The power or signal fed into an electrical or electronic device. **2.** The terminals to which the power or signal is applied. [SCI TECH] Those resources and other environmental factors converted by a system.

input capacitance [ELECTR] The short-circuited transfer capacitance that exists between the input terminals and all other terminals of an electron tube (except the output terminal) connected together.

input gap [ELECTR] An interaction gap used to initiate a variation in an electron stream; in a velocity-modulated tube it is in the buncher resonator.

input impedance [ELEC] The impedance across the input terminals of a four-terminal network when the output terminals are short-circuited.

input/output relation [SYS ENG] The relation between two vectors whose components are the inputs (excitations, stimuli) of a system and the outputs (responses) respectively.

input resistance See transistor input resistance.

input resonator See buncher resonator.

inradius [MATH] The radius of a circle or sphere inscribed in a given geometric figure.

inscribed [MATH] A polygon is inscribed in a circle or some curve if every vertex of the polygon lies on the circle or curve.

inseparable degree [MATH] Let E be a finite extension of a field F; the inseparable degree of E over F is the dimension of E viewed as a vector space over F divided by the separable degree of E over F.

insolation [ASTRON] **1.** Exposure of an object to the sun. **2.** Solar energy received, often expressed as a rate of energy per unit horizontal surface.

inspection by variables [IND ENG] A quality-control inspection method in which the sampled articles are evaluated on the basis of quantitative criteria.

instability [CONT SYS] A condition of a control system in which excessive positive feedback causes persistent, unwanted oscillations in the output of the system. [PHYS] A property of the steady state of a system such that certain disturbances or perturbations introduced into the steady state will increase in magnitude, the maximum perturbation amplitude always remaining larger than the initial amplitude.

instantaneous axis [MECH] The axis about which a rigid body is carrying out a pure rotation at a given instant in time.

instantaneous center [MECH] A point about which a rigid body is rotating at a given instant in time. Also known as instant center.

instantaneous condition [PHYS] The condition of a system at a particular instant in time.

instantaneous recovery [MECH] The immediate reduction in the strain of a solid when a stress is removed or reduced, in contrast to creep recovery.

instantaneous strain [MECH] The immediate deformation of a solid upon initial application of a stress, in contrast to creep strain.

instantaneous value [PHYS] The value of a sinusoidal or otherwise varying quantity at a particular instant.

instant center See instantaneous center.

instaton [PARTIC PHYS] A solution of equations of classical gage field theories in four-dimensional space in which energy in the fields is concentrated at a particular point in space and at a particular time.

instruction [ADP] A pattern of digits which signifies to a computer that a particular operation is to be performed and which may also indicate the operands (or the locations of operands) to be operated on.

instrument [ENG] A device for measuring and sometimes also recording and controlling the value of a quantity under observation.

instrument multiplier [ELEC] A highly accurate resistor used in series with a voltmeter to extend its voltage range. Also known as voltage multiplier; voltage-range multiplier.

INSTRUMENT TRANSFORMER

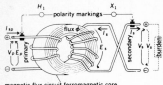

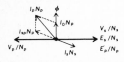

A simple instrument transformer and its phasor diagram. Core-loss components and impedance drops have been omitted from the diagram. Subscripts p and s = primary and secondary windings respectively; E = induced voltage; V = applied voltage; N = number of turns; I = current. *(General Electric Co.)*

instrument resistor [ELEC] A high-accuracy, four-terminal resistor used to bypass the major portion of currents around the low-current elements of an instrument, such as a direct-current ammeter.

instrument shunt [ELEC] A resistor designed to be connected in parallel with an ammeter to extend its current range.

instrument transformer [ELEC] A transformer that transfers primary current, voltage, or phase values to the secondary circuit with sufficient accuracy to permit connecting an instrument to the secondary rather than the primary; used so only low currents or low voltages are brought to the instrument.

insulated [ELEC] Separated from other conducting surfaces by a nonconducting material.

insulating strength [ELEC] Measure of the ability of an insulating material to withstand electric stress without breakdown; it is defined as the voltage per unit thickness necessary to initiate a disruptive discharge; usually measured in volts per centimeter.

insulation [ELEC] A material having high electrical resistivity and therefore suitable for separating adjacent conductors in an electric circuit or preventing possible future contact between conductors. Also known as electrical insulation.

insulation resistance [ELEC] The electrical resistance between two conductors separated by an insulating material.

insulator [ELEC] A device having high electrical resistance and used for supporting or separating conductors to prevent undesired flow of current from them to other objects. Also known as electrical insulator. [SOLID STATE] A substance in which the normal energy band is full and is separated from the first excitation band by a forbidden band that can be penetrated only by an electron having an energy of several electron volts, sufficient to disrupt the substance.

integer [MATH] Any positive or negative counting number or zero.

integer partition [MATH] For a positive integer n, a nonincreasing sequence of positive integers whose sum equals n.

integer programming [SYS ENG] A series of procedures used in operations research to find maxima or minima of a function subject to one or more constraints, including one which requires that the values of some or all of the variables be whole numbers.

integer spin [QUANT MECH] Property of a particle whose spin angular momentum is a whole number times Planck's constant divided by 2π; bosons have this property; in contrast, fermions have half-integer spin.

integral [MATH] **1.** A solution of a differential equation is sometimes called an integral of the equation. **2.** An element a of a ring B is said to be integral over a ring A contained in B if it is the root of a polynomial with coefficients in A and with leading coefficient 1. **3.** *See* definite Riemann integral; indefinite integral.

integral absorbed dose *See* integral dose.

integral action [CONT SYS] A control action in which the rate of change of the correcting force is proportional to the deviation.

integral calculus [MATH] The study of integration and its applications to finding areas, volumes, or solutions of differential equations.

integral closure [MATH] The integral closure of a subring A of a ring B is the set of all elements in B that are integral over A.

integral compensation [CONT SYS] Use of a compensator whose output changes at a rate proportional to its input.

integral control [CONT SYS] Use of a control system in which the control signal changes at a rate proportional to the error signal.

integral discriminator [ELECTR] A circuit which accepts only pulses greater than a certain minimum height.

integral domain [MATH] A commutative ring with identity where the product of nonzero elements is never zero. Also known as entire ring.

integral dose [NUCLEO] The total energy imparted to an irradiated body by an ionizing radiation; usually expressed in gram-rads or gram-roentgens. Also known as integral absorbed dose; volume dose.

integral equation [MATH] **1.** An equation where the unknown function occurs under an integral sign. **2.** An integral equation for an element x over a commutative ring R is an equation of the form $x^n + a_{n-1} x^{n-1} + \cdots + a_0 = 0$, where $a_0, \ldots, a_{n-1}$ are elements of R.

integral extension [MATH] An integral extension of a commutative ring A is a commutative ring B containing A such that every element of B is integral over A.

integral function [MATH] **1.** A function taking on integer values. **2.** *See* entire function.

integral heat of dilution *See* heat of dilution.

integral heat of solution *See* heat of solution.

integrally closed ring [MATH] An integral domain which is equal to its integral closure in its quotient field.

integral map [MATH] A homomorphism from a commutative ring A into a commutative ring B such that B is an integral extension of $f(A)$.

integral-mode controller [CONT SYS] A controller which produces a control signal proportional to the integral of the error signal.

integral network [CONT SYS] A compensating network which produces high gain at low input frequencies and low gain at high frequencies, and is therefore useful in achieving low steady-state errors. Also known as lagging network; lag network.

integral operator [MATH] A rule for transforming one function into another function by means of an integral; this often is in context a linear transformation on some vector space of functions.

integral square error [CONT SYS] A measure of system performance formed by integrating the square of the system error over a fixed interval of time; this performance measure and its generalizations are frequently used in linear optimal control and estimation theory.

integral test [MATH] If $f(x)$ is a function that is positive and decreasing for positive x, then the infinite series with nth term $f(n)$ and the integral of $f(x)$ from 1 to ∞ are either both convergent (finite) or both infinite.

integral transformation [MATH] A transform of a function $F(x)$ given by the function

$$f(y) = \int_a^b K(x,y)F(x)\,dx$$

where $K(x,y)$ is some function. Also known as integral transform.

integrand [MATH] The function which is being integrated in a given integral.

integraph [ENG] An instrument for drawing a graph of the integral of a function from a graph of the function itself.

integrated circuit [ELECTR] An interconnected array of active and passive elements integrated with a single semiconductor substrate or deposited on the substrate by a continuous series of compatible processes, and capable of performing at

INTEGRATED CIRCUIT

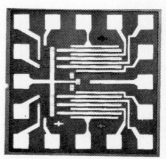

Photomicrograph of a simple MOS (metal oxide semiconductor) integrated circuit, a three-input logic gate circuit.

least one complete electronic circuit function. Abbreviated IC. Also known as integrated semiconductor.

integrated-circuit capacitor [ELECTR] A capacitor that can be produced in a silicon substrate by conventional semiconductor production processes.

integrated-circuit resistor [ELECTR] A resistor that can be produced in or on an integrated-circuit substrate as part of the manufacturing process.

integrated injection logic [ELECTR] Integrated-circuit logic that uses a simple and compact bipolar transistor gate structure which makes possible large-scale integration on silicon for logic arrays, memories, watch circuits, and many other analog and digital applications. Abbreviated I²L. Also known as merged-transistor logic.

integrated neutron flux [NUCLEO] A measure of radiation exposure, equal to the product of the number of free neutrons per unit volume, the average speed of neutrons, and the exposure time.

integrated optics [OPTICS] A thin-film device containing tiny lenses, prisms, and switches to transmit very thin laser beams, and serving the same purposes as the manipulation of electrons in thin-film devices of integrated electronics.

integrated reflection [PHYS] The intensity of a beam of x-rays or neutrons reflected from a given atomic plane of a crystal, integrated over a small range of angles about the general direction of the beam.

integrated semiconductor *See* integrated circuit.

integrating amplifier [ELECTR] An operational amplifier with a shunt capacitor such that mathematically the waveform at the output is the integral (usually over time) of the input.

integrating factor [MATH] A factor which when multiplied into a differential equation makes the portion involving derivatives an exact differential.

integrating galvanometer [ENG] A modification of the d'Arsonval galvanometer which measures the integral of current over time; it is designed to be able to measure changes of flux in an exploring coil which last over periods of several minutes.

integrating gyroscope [ENG] A gyroscope that senses the rate of angular displacement and measures and transmits the time integral of this rate.

integrating ionization chamber [NUCLEO] An ionization chamber in which the collected charge is stored on a capacitor for subsequent measurement.

integrating network [ELECTR] A circuit or network whose output waveform is the time integral of its input waveform. Also known as integrator.

integrating-sphere photometer [OPTICS] An instrument for measuring the total luminous flux of a lamp or luminaire; the source is placed inside a sphere whose inside surface has a diffusely reflecting white finish, and the light reflected from this surface onto a window is measured by an ordinary photometer. Also known as sphere photometer.

integration [MATH] The process of finding a definite or indefinite integral.

integration by parts [MATH] A technique used to find the integral of the product of two functions by means of an identity involving another simpler integral; for functions of one variable the identity is

$$\int_a^b fg'\,dx + \int_a^b gf'\,dx = f(b)g(b) - f(a)g(a);$$

for functions of several variables the technique is tantamount to using Stokes' theorem or the divergence theorem.

INTEGRATING-SPHERE
PHOTOMETER

Uhlbricht sphere for measuring luminous flux and efficiency. *(Westinghouse Electric Corp.)*

integrator [ELECTR] **1.** A computer device that approximates the mathematical process of integration. **2.** *See* integrating network.

integrodifferential equation [MATH] An equation relating a function, its derivatives, and its integrals.

intensifier electrode [ELECTR] An electrode used to increase the velocity of electrons in a beam near the end of their trajectory, after deflection of the beam. Also known as post-accelerating electrode; post-deflection accelerating electrode.

intensifying screen [GRAPHICS] A layer of material, such as a salt screen or a metal screen, that is placed next to an x-ray film to increase the effect of x-rays on the film.

intensity [PHYS] **1.** The strength or amount of a quantity, as of electric field, current, magnetization, radiation, or radioactivity. **2.** The power transmitted by a light or sound wave across a unit area perpendicular to the wave.

intensity control *See* brightness control.

intensity level [PHYS] The logarithm of the ratio of two intensities, powers or energies, usually expressed in decibels.

intensity of magnetization *See* intrinsic induction.

interaction [FL MECH] With respect to wave components, the nonlinear action by which properties of fluid flow (such as momentum, energy, vorticity) are transferred from one portion of the wave spectrum to another, or viewed in another manner, between eddies of different size-scales. [PHYS] A process in which two or more bodies exert mutual forces on each other. [STAT] The phenomenon which causes the response to applying two treatments not to be the simple sum of the responses to each treatment.

interaction picture [QUANT MECH] A mode of description of a system in which the time dependence is carried partly by the operators and partly by the state vectors, the time dependence of the state vectors being due entirely to that part of the Hamiltonian arising from interactions between particles. Also known as interaction representation.

interaction representation *See* interaction picture.

interaction space [ELECTR] A region of an electron tube in which electrons interact with an alternating electromagnetic field.

interbase current [ELECTR] The current that flows from one base connection of a junction tetrode transistor to the other, through the base region.

intercept [CRYSTAL] One of the distances cut off a crystal's reference axis by planes. [MATH] The point where a straight line crosses one of the axes of a cartesian coordinate system.

interchange coefficient *See* exchange coefficient.

intercloud discharge *See* cloud-to-cloud discharge.

intercombination lines [ATOM PHYS] Lines in atomic spectra resulting from transitions between energy levels with different multiplicities, that is, with different total spin quantum numbers.

interdiffusion [PHYS CHEM] The self-mixing of two fluids, initially separated by a diaphragm.

interdigital structure [ELECTR] A structure in which the length of the region between two electrodes is increased by an interlocking-finger design for metallization of the electrodes. Also known as interdigitated structure.

interdigitated structure *See* interdigital structure.

interelectrode capacitance [ELECTR] The capacitance between one electrode of an electron tube and the next electrode on the anode side. Also known as direct interelectrode capacitance.

interelectrode transit time [ELECTR] Time required for an electron to traverse the distance between the two electrodes.

interface [SCI TECH] A shared boundary; it may be a piece of

INTERFACE RESISTANCE

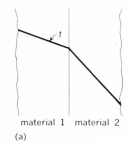

material 1   material 2

(a)

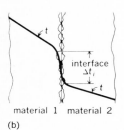

interface
Δt_i

material 1   material 2

(b)

Distribution of temperature t through composite wall. *(a)* With perfect interface contact. *(b)* For typical actual surfaces. Δt_i is temperature difference across interface. Interface resistance equals Δt_i divided by heat flux.

INTERFERENCE FRINGES

Interference fringes formed with Fresnel biprism and mercury-arc light source.

hardware used between two pieces of equipment, a portion of computer storage accessed by two or more programs, or a surface that forms the boundary between two types of materials.

interface resistance [THERMO] **1.** Impairment of heat flow caused by the imperfect contact between two materials at an interface. **2.** Quantitatively, the temperature difference across the interface divided by the heat flux through it.

interfacial angle [CRYSTAL] The angle between two crystal faces.

interfacial energy [PHYS] The free energy of the surfaces at an interface, resulting from differences in the tendencies of each phase to attract its own molecules; equal to the surface tension. Also known as surface energy.

interfacial force *See* interfacial tension.

interfacial polarization [ELEC] *See* space-charge polarization. [OPTICS] Polarization of light by reflection from the surface of a dielectric at Brewster's angle.

interfacial tension [PHYS] A kind of surface tension, occurring at the interface between two liquids. Also known as interfacial force.

interference [COMMUN] Any undesired energy that tends to interfere with the reception of desired signals. Also known as electrical interference; radio interference. [PHYS] The variation with distance or time of the amplitude of a wave which results from the superposition (algebraic or vector addition) of two or more waves having the same, or nearly the same, frequency. Also known as wave interference.

interference colors [OPTICS] Colors formed by interference of a beam of light passed through a thin section of a mineral placed in a polarizing microscope.

interference figure [OPTICS] A pattern of light and dark areas observed with a conoscope when a birefringent crystal is placed in a convergent beam of linearly polarized light.

interference filter [ELECTR] **1.** A filter used to attenuate artificial interference signals entering a receiver through its power line. **2.** A filter used to attenuate unwanted carrier-frequency signals in the tuned circuits of a receiver. [OPTICS] An optical filter in which the wavelengths that are not transmitted are removed by interference phenomena rather then by absorbtion or scattering.

interference fringes [OPTICS] A series of light and dark bands produced by interference of light waves.

interference microscope [OPTICS] A microscope used for visualizing and measuring differences in phase or optical paths in transparent or reflecting specimens; it differs from the phase contrast microscope in that the incident and diffracted waves are not separated, but interference is produced between the transmitted wave and another wave which originates from the same source.

interference pattern [ELECTR] Pattern produced on a radarscope by interference signals. [PHYS] Resulting space distribution of pressure, particle density, particle velocity, energy density, or energy flux when progressive waves of the same frequency and kind are superimposed.

interference spectrum [ELECTR] Frequency distribution of the jamming interference in the propagation medium external to the receiver. [SPECT] A spectrum that results from interference of light, as in a very thin film.

interferometer [OPTICS] An instrument in which light from a source is split into two or more beams, which are subsequently reunited after traveling over different paths and display interference.

interferometry [OPTICS] The design and use of optical inferometers; uses include precise measurement of wavelength, measurement of very small distances and thicknesses, study of hyperfine structure of spectral lines, precise measurement of indices of refraction, and determination of separations of binary stars and diameters of very large stars.

intergalactic matter [ASTRON] The material between the galaxies.

interior angle [MATH] **1.** An angle between two adjacent sides of a polygon that lies within the polygon. **2.** For a line (called the transversal) that intersects two other lines, an angle between the transversal and one of the two lines that lies within the space between the two lines.

interior of a set [MATH] The set of all interior points of a set in a topological space.

interior point [MATH] A point p in a topological space is an interior point of a set S if there is some open neighborhood of p which is contained in S.

interleaved windings [ELEC] An arrangement of winding coils around a transformer core in which the coils are wound in the form of a disk, with a group of disks for the low-voltage windings stacked alternately with a group of disks for the high-voltage windings.

intermediate-infrared radiation [ELECTROMAG] Infrared radiation having a wavelength between about 2.5 micrometers and about 50 micrometers; this range includes most molecular vibrations.

intermediate neutron [NUCLEO] A neutron having energy in a range from about 100 to 100,000 electron volts.

intermediate reactor [NUCLEO] A reactor in which the chain reaction is sustained mainly by intermediate neutrons.

intermediate state [CRYO] A state of partial superconductivity that occurs when a magnetic field of approximate strength is applied to a superconducting material below its critical temperature. [QUANT MECH] A state through which a system may pass during transition from an initial state to a final state.

intermediate value theorem [MATH] If $f(x)$ is a continuous real-valued function on the closed interval from a to b, then, for any y between the least upper bound and the greatest lower bound of the values of f, there is an x between a and b with $f(x) = y$.

intermediate vector boson [PARTIC PHYS] One of the hypothetical particles with spin quantum number 1 and negative parity, which would interact with weak currents and mediate the weak interactions in the same way that photons interact with electromagnetic currents and mediate the electromagnetic interactions.

intermodulation [ELECTR] Modulation of the components of a complex wave by each other, producing new waves whose frequencies are equal to the sums and differences of integral multiples of the component frequencies of the original complex wave.

internal absorptance [ELECTROMAG] The value of absorptance, corrected to eliminate the effects of scattering and of reflection from the surfaces of the substance; that is, the ratio of the radiant power absorbed between the entry and exit surfaces of the substance to the radiant power leaving the entry surface.

internal bremmsstrahlung *See* inner bremsstrahlung.

internal conversion [NUC PHYS] A nuclear deexcitation process in which energy is transmitted directly from an excited nucleus to an orbital electron, causing ejection of that electron from the atom.

internal dielectric field *See* dielectric field.

internal energy [THERMO] A characteristic property of the state of a thermodynamic system, introduced in the first law

INTERLEAVED WINDINGS

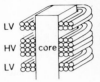

Interleaved windings arrangement. LV = low voltage, HV = high voltage.

of thermodynamics; it includes intrinsic energies of individual molecules, kinetic energies of internal motions, and contributions from interactions between molecules, but excludes the potential or kinetic energy of the system as a whole; it is sometimes erroneously referred to as heat energy.

internal force [MECH] A force exerted by one part of a system on another.

internal friction [FL MECH] *See* viscosity. [MECH] Conversion of mechanical strain energy to heat within a material subjected to fluctuating stress.

internal line [QUANT MECH] A component of a Feynman graph (in the diagrammatic presentation of perturbative quantum field theory) describing the propagation of a virtual particle whose momentum is integrated over all possible values.

internal photoelectric effect [SOLID STATE] A process in which the absorption of a photon in a semiconductor results in the excitation of an electron from the valence band to the conduction band.

internal pressure *See* intrinsic pressure.

internal reflectance spectroscopy *See* attenuated total reflectance.

internal resistance [ELEC] The resistance within a voltage source, such as an electric cell or generator.

internal standard [SPECT] The principal line in spectrum analysis by the logarithmic sector method, a quantitative spectroscopy procedure.

internal stress [MECH] A stress system within a solid that is not dependent on external forces. Also known as residual stress.

internal transmittance [ELECTROMAG] The value of transmittance, corrected to eliminate the effects of scattering and of reflection from the surfaces of the substance; that is, the ratio of the radiant power reaching the exit surface of the substance to the radiant power leaving the entry surface.

internal wave [FL MECH] A wave motion of a stably stratified fluid in which the maximum vertical motion takes place below the surface of the fluid.

internal work [THERMO] The work done in separating the particles composing a system against their forces of mutual attraction.

international ampere [ELEC] The current that, when flowing through a solution of silver nitrate in water, deposits silver at a rate of 0.001118 gram per second; it has been superseded by the ampere as a unit of current, and is equal to approximately 0.999850 ampere.

international angstrom [PHYS] A unit of length, equal to 1/6438.4696 of the wavelength of the red cadmium line in dry air at standard atmospheric pressure, at a temperature of 15°C containing 0.03% by volume of carbon dioxide; equal to 1.0000002 angstroms. Abbreviated IA.

international atomic time [HOROL] Time based on atomic clocks operating in conformity with the definition of the second as the International System unit of time. Abbreviated IAT.

international candle [OPTICS] A unit of luminous intensity, now replaced by the candela; as defined in the United States, it was a specified fraction of the average luminous intensity radiated in a horizontal direction by a group of 45 carbon-filament lamps preserved at the National Bureau of Standards when the lamps were operated at a specified voltage. Also known as standard candle.

International Critical Tables [PHYS] A seven-volume series of tables of numerical data in physics, chemistry, and technology, published in 1926–1930, prepared by experts who gave the "best" value which could be derived from all the data available at the time. Abbreviated ICT.

International Geophysical Year [GEOPHYS] An internationally accepted period, extending from July 1957 through December 1958, for concentrated and coordinated geophysical exploration, primarily of the solar and terrestrial atmospheres. Abbreviated IGY.

international gravity formula [GEOD] A formula for the acceleration of gravity at the earth's surface, stating that the acceleration of gravity is equal to $9.780318[1 + 5.3024 \times 10^{-3} \sin^2 \phi - 5.8 \times 10^{-6} \sin^2 2\phi]$ m/s^2, where ϕ is the latitude.

international henry [ELECTROMAG] A unit of electrical inductance which has been superseded by the henry, and is equal to 1.00049 henries. Also known as quadrant; secohm.

international ohm [ELEC] A unit of resistance, equal to that of a column of mercury of uniform cross section that has a length of 160.3 centimeters and a mass of 14.4521 grams at the temperature of melting ice; it has been superseded by the ohm, and is equal to 1.00049 ohms.

international practical temperature scale [THERMO] Temperature scale based on six points; the water triple point, the boiling points of oxygen, water, sulfur, and the solidification points of silver and gold; designated as °C, degrees Celsius, or t_{int}.

International Quiet Sun Year [GEOPHYS] An international cooperative effort, similar to the International Geophysical Year and extending through 1964 and 1965, to study the sun and its terrestrial and planetary effects during the minimum of the 11-year cycle of solar activity. Abbreviated IQSY. Also known as the International Year of the Quiet Sun.

international system of electrical units [ELEC] System of electrical units based on agreed fundamental units for the ohm, ampere, centimeter, and second, in use between 1893 and 1947, inclusive; in 1948, the Giorgi, or meter-kilogram-second-absolute system, was adopted for international use.

International System of Units [PHYS] A system of physical units in which the fundamental quantities are length, time, mass, electric current, temperature, luminous intensity, and amount of substance, and the corresponding units are the meter, second, kilogram, ampere, kelvin, candela, and mole; it has been given official status and recommended for universal use by the General Conference on Weights and Measures. Also known (in French) as Système International d'Unités. Abbreviated SI (in all languages).

international table British thermal unit *See* British thermal unit.

international table calorie *See* calorie.

international volt [ELEC] A unit of potential difference or electromotive force, equal to 1/1.01858 of the electromotive force of a Weston cell at 20°C; it has been superseded by the volt, and is equal to 1.00034 volts.

internuclear distance [PHYS CHEM] The distance between two nuclei in a molecule.

interpenetration twin [CRYSTAL] Two or more individual crystals so twinned that they appear to have grown through one another. Also known as penetration twin.

interplanar spacing [CRYSTAL] The perpendicular distance between successive parallel planes of atoms in a crystal.

interpolation [MATH] A process used to estimate an intermediate value of one (dependent) variable which is a function of a second (independent) variable when values of the dependent variable corresponding to several discrete values of the independent variable are known.

interquartile range [STAT] The distance between the top of the lower quartile and the bottom of the upper quartile of a distribution.

Intersecting Storage Rings [NUCLEO] Proton storage rings, located at Geneva, Switzerland, in which counterrotating protons with energies of up to 31 GeV injected from a proton synchrotron are made to undergo nearly head-on collisions. Abbreviated ISR.

intersection [MATH] The intersection of two sets is the set consisting of all elements common to both of the two sets.

intersection of Boolean matrices [MATH] The intersection of two Boolean matrices A and B, with the same number of rows and columns, is the Boolean matrix whose element c_{ij} in row i and column j is the intersection of corresponding elements a_{ij} in A and b_{ij} in B.

interstellar lines [ASTRON] Dark, narrow lines in the spectra of stars, caused by absorption of radiation by a gaseous medium in space.

interstice [SOLID STATE] A space or volume between atoms of a lattice, or between groups of atoms or grains of a solid structure.

interstitial [CRYSTAL] A crystal defect in which an atom occupies a position between the regular lattice positions of a crystal. [SCI TECH] Of, pertaining to, or situated in a space between two things.

interstitial atom [CRYSTAL] A displaced atom which is forced into a nonequilibrium site within a crystal lattice.

interstitial compound [SOLID STATE] A binary compound in which atoms of one element (usually a light, nonmetallic element) occupy spaces between atoms of the crystal lattice formed by the other element (usually a heavy, metallic element).

interval [ACOUS] The spacing in pitch or frequency between two sounds; the frequency interval is the ratio of the frequencies or the logarithm of this ratio. [MATH] A set of numbers which consists of those numbers that are greater than one fixed number and less than another, and that may also include one or both of the end numbers. [PHYS] The time separating two events, or the distance between two objects. [RELAT] **1.** In special relativity, the Lorentz invariant quantity $c^2(\Delta t)^2 - (\Delta x)^2 - (\Delta y)^2 - (\Delta z)^2$, where c is the speed of light, Δt is the difference in the time coordinates of two specified events, and Δx, Δy, and Δz are the differences in their x, y, and z coordinates, respectively. **2.** In general relativity, a generalization of this concept, namely the sum over the indices μ and ν of $g_{\mu\nu} \, dx^\mu \, dx^\nu$, where dx^μ and dx^ν are the differences in the x^μ and x^ν coordinates of two specified neighboring events, and $g_{\mu\nu}$ is an element of the metric tensor.

interval estimate [STAT] An estimate which specifies a range of values for a population parameter.

interval estimation [STAT] A technique that expresses uncertainty about an estimate by defining an interval, or range of values, and indicates the certain degree of confidence with which the population parameter will fall within the interval.

interval measurement [STAT] A method of measuring quantifiable data that assumes an exact knowledge of the quantitative difference between the objects being scaled. Also known as cardinal measurement.

interval of Sturm *See* astigmatic interval.

interval scale [STAT] A rule or system for assigning numbers to objects in such a way that the difference between any two objects is reflected in the difference in the numbers assigned to them; used in interval measurement.

intracloud discharge *See* cloud discharge.

intrinsic-barrier diode [ELECTR] A *pin* diode, in which a thin region of intrinsic material separates the *p*-type region and the *n*-type region.

intrinsic-barrier transistor [ELECTR] A *pnip* or *npin* transistor, in which a thin region of intrinsic material separates the base and collector.

intrinsic conductivity [SOLID STATE] The conductivity of a semiconductor or metal in which impurities and structural defects are absent or have a very low concentration.

intrinsic electric strength [ELEC] The extremely high dielectric strength displayed by a substance at low temperatures.

intrinsic equations of a curve [MATH] The equations describing the radius of curvature and torsion of a curve as a function of arc length; these equations determine the curve up to its position in space.

intrinsic flux density *See* intrinsic induction.

intrinsic geometry of a surface [MATH] The description of the intrinsic properties of a surface.

intrinsic impedance *See* characteristic acoustic impedance.

intrinsic induction [ELECTROMAG] The vector difference between the magnetic flux density at a given point and the magnetic flux density which would exist there, for the same magnetic field strength, if the point were in a vacuum. Symbolized B_i. Also known as intensity of magnetization; intrinsic flux density; magnetic polarization.

intrinsic layer [ELECTR] A layer of semiconductor material whose properties are essentially those of the pure undoped material.

intrinsic luminosity [ASTROPHYS] The total amount of radiation emitted by a star over a specified range of wavelengths.

intrinsic mobility [SOLID STATE] The mobility of the electrons in an intrinsic semiconductor.

intrinsic parity [PARTIC PHYS] A quantum number, equal to $+1$ or -1, which is assigned to particles so that the product of the intrinsic parities of the particles composing a system times the parity of the system's wave function yields the total parity.

intrinsic photoemission [SOLID STATE] Photoemission which can occur in an ideally pure and perfect crystal, in contrast to other types of photoemission which are associated with crystal defects.

intrinsic pressure [PHYS] Pressure in a fluid resulting from inward forces on molecules near the fluid surface, caused by attraction between molecules. Also known as internal pressure.

intrinsic properties of a surface [MATH] All of the properties of a surface which can be described without reference to the surrounding space.

intrinsic property [SOLID STATE] A property of a substance that is not seriously affected by impurities or imperfections in the crystal structure.

intrinsic semiconductor [SOLID STATE] A semiconductor in which the concentration of charge carriers is characteristic of the material itself rather than of the content of impurities and structural defects of the crystal. Also known as *i*-type semiconductor.

intrinsic temperature range [SOLID STATE] In a semiconductor, the temperature range in which its electrical properties are essentially not modified by impurities or imperfections within the crystal.

intrinsic tracer [NUC PHYS] An isotope that is present naturally in a form suitable for tracing a given element through chemical and physical processes.

intrinsic variable star [ASTRON] A star that is variable for a reason other than an eclipse.

intrinsic viscosity [PHYS CHEM] The ratio of a solution's specific viscosity to the concentration of the solute, extrapolated to zero concentration. Also known as limiting viscosity number.

in vacuo [PHYS] In a vacuum.

invariable line [MECH] A line which is parallel to the angular momentum vector of a body executing Poinsot motion, and which passes through the fixed point in the body about which there is no torque.

invariable plane [MECH] A plane which is perpendicular to the angular momentum vector of a rotating rigid body not subject to external torque, and which is always tangent to its inertia ellipsoid.

invariance [MATH] *See* invariant property. [OPTICS] Any property of a light beam that remains constant when the light is reflected or refracted at one or more surfaces. [PHYS] The property of a physical quantity or physical law of being unchanged by certain transformations or operations, such as reflection of spatial coordinates, time reversal, charge conjugation, rotations, or Lorentz transformations. Also known as symmetry.

invariance principle [PHYS] Any principle which states that a physical quantity or physical law possesses invariance under certain transformations. Also known as symmetry law; symmetry principle. [RELAT] In general relativity, the principle that the laws of motion are the same in all frames of reference, whether accelerated or not.

invariant [MATH] **1.** An element x of a set E is said to be invariant with respect to a group G of mappings acting on E if $g(x) = x$ for all g in G. **2.** A subset F of a set E is said to be invariant with respect to a group G of mappings acting on E if $g(x)$ is in F for all x in F and all g in G.

invariant function [MATH] A function f on a set S is said to be invariant under a transformation T of S into itself if $f(Tx) = f(x)$ for all x in S.

invariant measure [MATH] A Borel measure m on a topological space X is invariant for a transformation group (G,X,π) if for all Borel sets A in X and all elements g in G, $m(A_g) = m(A)$, where A_g is the set of elements equal to $\pi(g,x)$ for some x in A.

invariant property [MATH] A mathematical property of some space unchanged after the application of any member from some given family of transformations. Also known as invariance.

invariant subgroup *See* normal subgroup.

inverse beta decay [NUC PHYS] A reaction providing evidence for the existence of the neutrino, in which an antineutrino (or neutrino) collides with a proton (or neutron) to produce a neutron (or proton) and a positron (or electron).

inverse Compton effect [QUANT MECH] A process in which relativistic particles give up some of their energy to long-wavelength radiation, converting it to shorter-wavelength radiation.

inverse electrode current [ELECTR] Current flowing through an electrode in the direction opposite to that for which the tube is designed.

inverse element [MATH] In a group G the inverse of an element g is the unique element g^{-1} such that $g \cdot g^{-1} = g^{-1} \cdot g = e$, where $\cdot$ denotes the group operation and e is the identity element.

inverse feedback *See* negative feedback.

inverse function [MATH] An inverse function for a function f is a function g whose domain is the range of f and whose range is the domain of f with the property that both f composed with g and g composed with f give the identity function.

inverse function theorem [MATH] If f is a continuously differentiable function of euclidean n-space to itself and at a point x_0 the matrix with the entry $\partial f_i / \partial x_j^{(x_0)}$ in the ith row and jth column is nonsingular, then there is a continuously differentiable function $g(y)$ defined in a neighborhood of $f(x_0)$ which is an inverse function for $f(x)$ at all points near x_0.

inverse hour   *See* inhour.

inverse logarithm of a number   *See* antilogarithm of a number.

inverse matrix   [MATH] The inverse of a nonsingular matrix A is the matrix A^{-1} where $A \cdot A^{-1} = A^{-1} \cdot A = I$, the identity matrix.

inverse network   [ELEC] Two two-terminal networks are said to be inverse when the product of their impedances is independent of frequency within the range of interest.

inverse of a fractional ideal   [MATH] The inverse of a fractional ideal I of an integral domain R is the set of all elements x in the quotient field K of R such that xy is in I for all y in I.

inverse of a number   [MATH] The additive inverse of a real or complex number a is the number which when added to a gives 0; the multiplicative inverse of a is the number which when multiplied with a gives 1.

inverse operator   [MATH] The inverse of an operator L is the operator which is the inverse function of L.

inverse peak voltage   [ELECTR] **1.** The peak value of the voltage that exists across a rectifier tube or x-ray tube during the half cycle in which current does not flow. **2.** The maximum instantaneous voltage value that a rectifier tube or x-ray tube can withstand in the inverse direction (with anode negative) without breaking down and becoming conductive.

inverse piezoelectric effect   [SOLID STATE] The contraction or expansion of a piezoelectric crystal under the influence of an electric field, as in crystal headphones; also occurs at *pn* junctions in some semiconductor materials.

inverse points   [MATH] A pair of points lying on a diameter of a circle or sphere such that the product of the distances of the points from the center equals the square of the radius.

inverse probability principle   *See* Bayes' theorem.

inverse problem   [CONT SYS] The problem of determining, for a given feedback control law, the performance criteria for which it is optimal.

inverse ranks   [STAT] Ranking responses to treatments from largest response to smallest response.

inverse-square law   [PHYS] Any law in which a physical quantity varies with distance from a source inversely as the square of that distance.

inverse Stark effect   [SPECT] The Stark effect as observed with absorption lines, in contrast to emission lines.

inverse voltage   [ELECTR] The voltage that exists across a rectifier tube or x-ray tube during the half cycle in which the anode is negative and current does not normally flow.

inverse Zeeman effect   [SPECT] A splitting of the absorption lines of atoms or molecules in a static magnetic field; it is the Zeeman effect observed with absorption lines.

inversion   [CRYSTAL] A change from one crystal polymorph to another. Also known as transformation.   [MATH] Given a point O lying in a plane or in space, a mapping of the plane or of space, excluding the point O, into itself in which every point P is mapped into the point Q lying on OP such that the product of OP and OQ is a constant k^2; that is, the mapping of every point into its inverse with respect to a circle or sphere centered at O with radius k.   [OPTICS] The formation of an inverted image by an optical system.   [PHYS] The simultaneous reflection of all three directions in space, so that each coordinate is replaced by the negative of itself. Also known as space inversion.   [SOLID STATE] The production of a layer at the surface of a semiconductor which is of opposite

INVERSE-SQUARE LAW

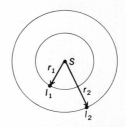

Point source S emitting energy of intensity I. The inverse-square law states that $I_1 / I_2 = r_2^2 / r_1^2$.

type from that of the bulk of the semiconductor, usually as the result of an applied electric field. [THERMO] A reversal of the usual direction of a variation or process, such as the change in sign of the expansion coefficient of water at 4°C, or a change in sign in the Joule-Thomson coefficient at a certain temperature.

inversion axis   *See* rotation-inversion axis.

inversion spectrum   [SPECT] Lines in the microwave spectra of certain molecules (such as ammonia) which result from the quantum-mechanical analog of an oscillation of the molecule between two configurations which are mirror images of each other.

inversion symmetry   [PHYS] The principle that the laws of physics are unchanged by the operation of inversion; it is violated by the weak interactions.

inversion temperature   [ENG] The temperature to which one junction of a thermocouple must be raised in order to make the thermoelectric electromotive force in the circuit equal to zero, when the other junction of the thermocouple is held at a constant low temperature. [THERMO] The temperature at which the Joule-Thomson effect of a gas changes sign.

inverted amplifier   [ELECTR] A two-tube amplifier in which the control grids are grounded and the input signal is applied between the cathodes; the grid then serves as a shield between the input and output circuits.

inverted image   [OPTICS] An image in which up and down, as well as left and right, are interchanged; that is, an image that results from rotating a line from the object 180° about a line from the object to the observer; such images are formed by most astronomical telescopes. Also known as reversed image.

inverted L antenna   [ELECTROMAG] An antenna consisting of one or more horizontal wires to which a connection is made by means of a vertical wire at one end.

inverted microscope   [OPTICS] A microscope in which the body of the microscope, including the objective and the ocular, are below the stage, the illumination for transmitted light is above the stage, and with opaque materials, the vertical illuminator is used under the stage near the objective.

inverted vee   [ELECTROMAG] **1.** A directional antenna consisting of a conductor which has the form of an inverted V, and which is fed at one end and connected to ground through an appropriate termination at the other. **2.** A center-fed horizontal dipole antenna whose arms have ends bent downward 45°.

inverter   *See* phase inverter.

inverter circuit   *See* NOT circuit.

invertible ideal   [MATH] A fractional ideal I of an integral domain R such that R is equal to the set of elements of the form xy, where x is in I and y is in the inverse of I.

inverting amplifier   [ELECTR] Amplifier whose output polarity is reversed as compared to its input; such an amplifier obtains its negative feedback by a connection from output to input, and with high gain is widely used as an operational amplifier.

inverting telescope   [OPTICS] A telescope that inverts the usual telescopic image, allowing the object to be seen right side up.

inviscid flow   [FL MECH] Flow of an inviscid fluid. Also known as frictionless flow; ideal flow; nonviscous flow.

inviscid fluid   [FL MECH] A fluid which has no viscosity; it therefore can support no shearing stress, and flows without energy dissipation. Also known as ideal fluid; nonviscous fluid; perfect fluid.

involute   [MATH] A curve produced by any point of a perfectly flexible inextensible thread that is kept taut as it is wound upon or unwound from another curve.

involution [MATH] A mapping $x \to x^*$ of a real or complex algebra onto itself, such that $(x^*)^* = x$, $(x+y)^* = x^* + y^*$, and $(xy)^* = y^*x^*$ for all x and y in the algebra, and $(ax)^* = ax^*$ in the case of a real algebra, while $(ax)^* = \bar{a}x^*$ in the case of a complex algebra, where a is an element of the underlying field and $\bar{a}$ is the complex conjugate of a.

iodine [CHEM] A nonmetallic halogen element, symbol I, atomic number 53, atomic weight 126.9044; melts at 114°C, boils at 184°C; the poisonous, corrosive, dark plates or granules are readily sublimed; insoluble in water, soluble in common solvents; used as germicide and antiseptic, in dyes, tinctures, and pharmaceuticals, in engraving lithography, and as a catalyst and analytical reagent.

iodine -131 [NUC PHYS] A radioactive, artificial isotope of iodine, mass number 131; its half-life is 8 days with beta and gamma radiation; used in medical and industrial radioactive tracer work; moderately radiotoxic.

Ioffe bars [PL PHYS] Heavy current-carrying bars that are used to increase plasma stability in some types of controlled fusion reactor.

ion [CHEM] An isolated electron or positron or an atom or molecule which by loss or gain of one or more electrons has acquired a net electric charge.

ion accelerator [NUCLEO] A linear accelerator in which ions are accelerated by an electric field in a standing-wave pattern that is set up in a resonant cavity by external oscillators or amplifiers.

ion-acoustic wave [PL PHYS] A longitudinal compression wave in the ion density of a plasma which can occur at high electron temperatures and low frequencies, caused by a combination of ion inertia and electron pressure.

ion atmosphere *See* ion cloud.

ion backscattering [SOLID STATE] Large-angle elastic scattering of monoenergetic ions in a beam directed at a metallized film on silicon or some other thin multilayer system.

ion-beam scanning [ELECTR] The process of analyzing the mass spectrum of an ion beam in a mass spectrometer either by changing the electric or magnetic fields of the mass spectrometer or by moving a probe.

ion burn *See* ion spot.

ion chamber *See* ionization chamber.

ion cloud [GEOPHYS] An inhomogeneity or patch of unusually great ion density in one of the regular regions of the ionosphere; such patches occur quite often in the E region. [PHYS CHEM] A slight preponderance of negative ions around a positive ion in an electrolyte, and vice versa, according to the Debye-Hückel theory. Also known as ion atmosphere.

ion column [GEOPHYS] The trail of ionized gases in the trajectory of a meteoroid entering the upper atmosphere; a part of the composite phenomenon known as a meteor. Also known as meteor trail.

ion concentration *See* ion density.

ion counter *See* ionization counter.

ion current [PHYS] The electric current resulting from motion of ions.

ion density [PHYS] The number of ions per unit volume. Also known as ion concentration.

ion emission [PHYS] The ejection of ions from the surface of a substance into the surrounding space.

ion exchange [PHYS CHEM] A chemical reaction in which mobile hydrated ions of a solid are exchanged, equivalent for equivalent, for ions of like charge in solution; the solid has an open, fishnetlike structure, and the mobile ions neutralize the

IODINE

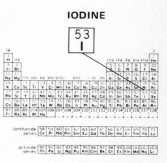

Periodic table of the chemical elements showing the position of iodine.

charged, or potentially charged, groups attached to the solid matrix; the solid matrix is termed the ion exchanger.

ion gage   *See* ionization gage.

ion gun   *See* ion source.

ionic bond   [PHYS CHEM] A type of chemical bonding in which one or more electrons are transferred completely from one atom to another, thus converting the neutral atoms into electrically charged ions; these ions are approximately spherical and attract one another because of their opposite charge. Also known as electrovalent bond.

ionic charge   [PHYS] **1.** The total charge of an ion. **2.** The charge of an electron; the charge of any ion is equal to this electron charge in magnitude, or is an integral multiple of it.

ionic conductance   [PHYS CHEM] The contribution of a given type of ion to the total equivalent conductance in the limit of infinite dilution.

ionic conduction   [SOLID STATE] Electrical conduction of a solid due to the displacement of ions within the crystal lattice.

ionic conductivity   [SOLID STATE] The portion of the electrical conductivity of a solid that results from ionic conduction.

ionic crystal   [CRYSTAL] A crystal in which the lattice-site occupants are charged ions held together primarily by their electrostatic interaction.

ionic dissociation   [PHYS CHEM] Dissociation that results in the production of ions.

ionic equilibrium   [PHYS CHEM] The condition in which the rate of dissociation of nonionized molecules is equal to the rate of combination of the ions.

ionic focusing   *See* gas focusing.

ionic-heated cathode   [ELECTR] Hot cathode heated primarily by ionic bombardment of the emitting surface.

ionic lattice   [CRYSTAL] The lattice of an ionic crystal.

ionic mobility   [PHYS] The ratio of the average drift velocity of an ion in a liquid or gas to the electric field.

ionic radii   [PHYS CHEM] Radii which can be assigned to ions because the rapid variation of their repulsive interaction with distance makes them repel like hard spheres; these radii determine the dimensions of ionic crystals.

ionic semiconductor   [SOLID STATE] A solid whose electrical conductivity is due primarily to the movement of ions rather than that of electrons and holes.

ionic solid   [SOLID STATE] A solid made up of ions held together primarily by their electrostatic interaction.

ionic strength   [PHYS CHEM] A measure of the average electrostatic interactions among ions in an electrolyte; it is equal to one-half the sum of the terms obtained by multiplying the molality of each ion by its valence squared.

ion implantation   [ENG] A process of introducing impurities into the near-surface regions of solids by directing a beam of ions at the solid.

ionium   [NUC PHYS] A naturally occurring radioisotope, symbol Io, of thorium, atomic weight 230.

ionization   [CHEM] A process by which a neutral atom or molecule loses or gains electrons, thereby acquiring a net charge and becoming an ion; occurs as the result of the dissociation of the atoms of a molecule in solution ($NaCl \rightarrow Na^+ + Cl^-$) or of a gas in an electric field ($H_2 \rightarrow 2H^+$).

ionization chamber   [NUCLEO] A particle detector which measures the ionization produced in the gas filling the chamber by the fast-moving charged particles as they pass through. Also known as ion chamber.

ionization constant   [PHYS CHEM] Analog of the dissociation constant, where $k = [H^+][A^-]/[HA]$; used for the application of the law of mass action to ionization; in the equation HA represents the acid, such as acetic acid.

ION IMPLANTATION

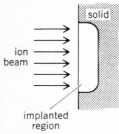

ion beam

solid

implanted region

Ion implantation into a solid.

ionization counter [NUCLEO] An ionization chamber in which there is no internal amplification by gas multiplication; used for counting ionizing particles. Also known as ion counter.

ionization cross section [PHYS] The cross section for a particle or photon to undergo a collision with an atom, thus removing or adding one or more electrons to the atom.

ionization current *See* gas current.

ionization degree [PHYS CHEM] The proportion of potential ionization that has taken place for an ionizable material in a solution or reaction mixture.

ionization density [ELECTR] The density of ions in a gas.

ionization energy [ATOM PHYS] The amount of energy needed to remove an electron from a given kind of atom or molecule to an infinite distance; usually expressed in electron volts, and numerically equal to the ionization potential in volts.

ionization gage [ELECTR] An instrument for measuring low gas densities by ionizing the gas and measuring the ion current. Also known as ion gage; ionization vacuum gage.

ionization potential [ATOM PHYS] The energy per unit charge needed to remove an electron from a given kind of atom or molecule to an infinite distance; usually expressed in volts. Also known as ion potential.

ionization radiation *See* ionizing radiation.

ionization source *See* ion source.

ionization spectrometer *See* Bragg spectrometer.

ionization temperature [STAT MECH] The temperature at which the average kinetic energy of gas molecules having a Maxwell distribution equals the ionization energy.

ionization time [ELECTR] Of a gas tube, the time interval between the initiation of conditions for and the establishment of conduction at some stated value of tube voltage drop.

ionization vacuum gage *See* ionization gage.

ionized atom [CHEM] An atom with an excess or deficiency of electrons, so that it has a net charge.

ionized layers [GEOPHYS] Layers of increased ionization within the ionosphere produced by cosmic radiation; responsible for absorption and reflection of radio waves and important in connection with communications and tracking of satellites and other space vehicles.

ionizing event [PHYS] Any occurrence in which an ion or group of ions is produced; for example, by passage of charged particles through matter.

ionizing radiation [NUCLEO] **1.** Particles or photons that have sufficient energy to produce ionization directly in their passage through a substance. Also known as ionization radiation. **2.** Particles that are capable of nuclear interactions in which sufficient energy is released to produce ionization.

ion kinetic energy spectrometry [SPECT] A spectrometric technique that uses a beam of ions of high kinetic energy passing through a field-free reaction chamber from which ionic products are collected and energy analyzed; it is a generalization of metastable ion studies in which both unimolecular and bimolecular reactions are considered.

ion laser [OPTICS] A gas laser in which stimulated emission takes place between two energy levels of an ion; gases used include argon, krypton, neon, and xenon; examples include helium-cadmium lasers and metal vapor lasers.

ion mean life [PHYS CHEM] The average time between the ionization of an atom or molecule and its recombination with one or more electrons, or its loss of excess electrons.

ion microprobe *See* secondary ion mass spectrometer.

ion microprobe mass spectrometer [ENG] A type of secondary ion mass spectrometer in which primary ions are focused on a spot 1–2 micrometers in diameter, mass-charge separation of secondary ions is carried out by a double focusing mass spectrometer or spectrograph, and a magnified image of elemental or isotopic distributions on the sample surface is produced using synchronous scanning of the primary ion beam and an oscilloscope.

ion microscope *See* field-ion microscope.

ionosphere [GEOPHYS] That part of the earth's upper atmosphere which is sufficiently ionized by solar ultraviolet radiation so that the concentration of free electrons affects the propagation of radio waves; its base is at about 70 or 80 kilometers and it extends to an indefinite height.

ion pair [NUCLEO] A positive ion and an equal-charge negative ion, usually an electron, that are produced by the action of radiation on a neutral atom or molecule.

ion potential *See* ionization potential.

ion probe *See* secondary ion mass spectrometer.

ion pump [ELECTR] A vacuum pump in which gas molecules are first ionized by electrons that have been generated by a high voltage and are spiraling in a high-intensity magnetic field, and the molecules are then attracted to a cathode, or propelled by electrodes into an auxiliary pump or an ion trap.

ion source [ELECTR] A device in which gas ions are produced, focused, accelerated, and emitted as a narrow beam. Also known as ion gun; ionization source.

ion spot [ELECTR] Of a cathode-ray tube screen, an area of localized deterioration of luminescence caused by bombardment with negative ions. Also known as ion burn.

ion trap [ELECTR] **1.** An arrangement whereby ions in the electron beam of a cathode-ray tube are prevented from bombarding the screen and producing an ion spot, usually employing a magnet to bend the electron beam so that it passes through the tiny aperture of the electron gun, while the heavier ions are less affected by the magnetic field and are trapped inside the gun. **2.** A metal electrode, usually of titanium, into which ions in an ion pump are absorbed.

IQSY *See* International Quiet Sun Year.

Ir *See* iridium.

irdome [OPTICS] A dome used to protect an infrared detector and its optical elements, generally made from quartz, silicon, germanium, sapphire, calcium aluminate, or other material having high transparency to infrared radiation. Derived from infrared dome.

IR drop *See* resistance drop.

iridescence [OPTICS] A rainbow color effect exhibited in various bodies as a result of interference in a thin film (as of soap bubbles or mother of pearl) or of diffraction of light reflected from a ribbed surface (as of the plumage of some birds).

iridium [CHEM] A metallic element, symbol Ir, atomic number 77, atomic weight 192.2, in the platinum group; insoluble in acids, melting at 2454°C.

iridium-192 [NUC PHYS] Radioactive isotope of iridium with a 75-day half-life; β and γ radiation; used in cancer treatment and for radiography of light metal castings.

iris [ELECTROMAG] A conducting plate mounted across a waveguide to introduce impedance; when only a single mode can be supported, an iris acts substantially as a shunt admittance and may be used for matching the waveguide impedance to that of a load. Also known as diaphragm; waveguide window. [OPTICS] A circular mechanical device, whose diameter can be varied continuously, which controls the amount of light reaching the film of a camera. Also known as iris diaphragm.

iris diaphragm *See* iris.

IRIDIUM

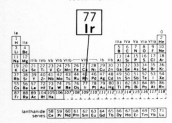

Periodic table of the chemical elements showing the position of iridium.

IRM *See* isothermal remanent magnetization.

iron [CHEM] A silvery-white metallic element, symbol Fe, atomic number 26, atomic weight 55.847, melting at 1530°C.

iron-55 [NUC PHYS] Radioactive isotope of iron, symbol Fe^{55}, with a 2.91-year half-life; highly toxic.

iron-59 [NUC PHYS] Radioactive isotope of iron, symbol Fe^{59}, 46.3-day half-life; β and γ radiation; highly toxic; used to study metallic welds, corrosion mechanisms, engine wear, and bodily functions.

iron core [ELECTROMAG] A core made of solid or laminated iron, or some other magnetic material which may contain very little iron.

iron-core choke *See* iron-core coil.

iron-core coil [ELECTROMAG] A coil in which solid or laminated iron or other magnetic material forms part or all of the magnetic circuit linking its winding. Also known as iron-core choke; magnet coil.

iron-core transformer [ELECTROMAG] A transformer in which laminations of iron or other magnetic material make up part or all of the path for magnetic lines of force that link the transformer windings.

iron loss *See* core loss.

irradiance *See* radiant flux density.

irradiation [ENG] The exposure of a material, object, or patient to x-rays, gamma rays, ultraviolet rays, or other ionizing radiation. [OPTICS] An optical illusion which makes bright objects appear larger than they really are.

irrational equation [MATH] An equation having an unknown raised to some fractional power. Also known as radical equation.

irrationality of dispersion [OPTICS] The effect whereby spectra produced by prisms of different types of glass are not geometrically similar.

irrational number [MATH] A number which is not the quotient of two integers.

irreducible element [MATH] An element x of a ring which is not a unit and such that every divisor of x is improper.

irreducible equation [MATH] An equation that is equivalent to one formed by setting an irreducible polynomial equal to zero.

irreducible function *See* irreducible polynomial.

irreducible polynomial [MATH] A polynomial is irreducible over a field K if it cannot be written as the product of two polynomials of lesser degree whose coefficients come from K. Also known as irreducible function.

irreducible representation of a group [MATH] A representation of a group as a family of linear operators of a vector space V where there is no proper closed subspace of V invariant under these operators.

irreducible tensor [MATH] A tensor that cannot be written as the inner product of two tensors of lower degree.

irregular galaxy [ASTRON] A galaxy which shows no definite order or shape, except that of a general flattened appearance.

irregular variable star [ASTRON] A star with no fixed period.

irreversible energy loss [THERMO] Energy transformation process in which the resultant condition lacks the driving potential needed to reverse the process; the measure of this loss is expressed by the entropy increase of the system.

irreversible process [THERMO] A process which cannot be reversed by an infinitesimal change in external conditions.

irreversible thermodynamics *See* nonequilibrium thermodynamics.

irrotational flow [FL MECH] Fluid flow in which the curl of the velocity function is zero everywhere, so that the circula-

IRON

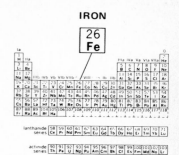

Periodic table of the chemical elements showing the position of iron.

IRRADIATION

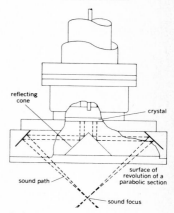

Reflector-type focusing irradiator, a device for producing irradiation by sound waves in the megahertz range. The sound waves emanating from the circular crystal plate impinge on a 90° included angle cone, from which surface they are directed to a surface of revolution generated by a section of a parabola. Focusing is obtained by reflection of the waves from this surface.

tion of the velocity about any closed curve vanishes. Also known as acyclic motion; irrotational motion.

irrotational motion   *See* irrotational flow.

irrotational vector field   [MATH] A vector field whose curl is identically zero; every such field is the gradient of a scalar function. Also known as lamellar vector field.

irrotational wave   *See* compressional wave.

isenergic flow   [THERMO] Fluid flow in which the sum of the kinetic energy, potential energy, and enthalpy of any part of the fluid does not change as that part is carried along with the fluid.

isenthalpic expansion   [THERMO] Expansion which takes place without any change in enthalpy.

isenthalpic process   [THERMO] A process that is carried out at constant enthalpy.

isentrope   [THERMO] A line of equal or constant entropy.

isentropic   [THERMO] Having constant entropy; at constant entropy.

isentropic compression   [THERMO] Compression which occurs without any change in entropy.

isentropic expansion   [THERMO] Expansion which occurs without any change in entropy.

isentropic flow   [THERMO] Fluid flow in which the entropy of any part of the fluid does not change as that part is carried along with the fluid.

isentropic process   [THERMO] A change that takes place without any increase or decrease in entropy, such as a process which is both reversible and adiabatic.

Ising coupling   [SOLID STATE] A model of coupling between two atoms in a lattice, used to study ferromagnetism, in which the spin component of each atom along some axis is taken to be $+1$ or -1, and the energy of interaction is proportional to the negative of the product of the spin components along this axis.

Ising model   [SOLID STATE] A crude model of a ferromagnetic material or an analogous system, used to study phase transitions, in which atoms in a one-, two-, or three-dimensional lattice interact via Ising coupling between nearest neighbors, and the spin components of the atoms are coupled to a uniform magnetic field.

isobar   [NUC PHYS] One of two or more nuclides having the same number of nucleons in their nuclei but differing in their atomic numbers and chemical properties.   [PHYS] **1.** A line connecting points of equal pressure along a given surface in a physical system. **2.** A line connecting points of equal pressure on a graph plotting thermodynamic variables.

isobaric   [THERMO] Of equal or constant pressure, with respect to either space or time.

isobaric process   [THERMO] A thermodynamic process of a gas in which the heat transfer to or from the gaseous system causes a volume change at constant pressure.

isobaric spin   *See* isotopic spin.

isocandle diagram   [OPTICS] A diagram showing the distribution of light from a lighting system in various directions by means of contours connecting directions of equal luminous intensity, projected in a suitable manner.

isochor   *See* isochore.

isochore   [PHYS] A graph that shows the variation of one quantity with another; for example, the variation of pressure with temperature, when the volume of the substance is held constant. Also known as isochor; isometric.

isochoric   [PHYS] Taking place without change in volume. Also known as isovolumic.

isochromatic   [OPTICS] **1.** Pertaining to a variation of certain quantities related to light (such as density of the medium through which the light is passing, index of refraction), in which the color or wavelength of the light is held constant. **2.** Pertaining to lines connecting points of the same color.

isochromatic fringe pattern [OPTICS] A pattern of bands, each of uniform color, observed when a plate is placed in a polariscope and subjected to stress, making it birefringent.

isochrone [PHYS] A line on a chart connecting all points having the same time of occurrence of particular phenomena or of a particular value of a quantity.

isochronism [MECH] The property of having a uniform rate of operation or periodicity, for example, of a pendulum or watch balance.

isochronous [PHYS] Having a fixed frequency or period.

isochronous circuits [ELEC] Circuits having the same resonant frequency.

isoclinal *See* isoclinic line.

isoclinal chart [GEOPHYS] A chart showing isoclinic lines. Also known as isoclinic chart.

isoclinic chart *See* isoclinal chart.

isoclinic line [GEOPHYS] A line connecting points on the earth's surface which have the same magnetic dip. Also known as isoclinal. [SOLID STATE] A line joining points in a plate at which the principal stresses have parallel directions.

isodesmic structure [SOLID STATE] An ionic crystal structure in which all bonds are of the same strength, so that no distinct groups of atoms are formed.

isodiaspheres [NUC PHYS] Nuclides which have the same difference in the number of neutrons and protons.

isodisperse [CHEM] **1.** Having dispersed particles, of colloidal dimensions, that are all of the same size. **2.** Dispersible in solutions with the same pH value.

isodose curve [NUCLEO] A curve, drawn on a chart of an object, connecting points receiving equal doses of radiation.

isodynamic [MECH] Pertaining to equality of two or more forces or to constancy of a force.

isodynamic line [GEOPHYS] One of the lines on a map of a magnetic field that connect points having equal strengths of the earth's field.

isoelectric point [PHYS CHEM] The pH value of the dispersion medium of a colloidal suspension at which the colloidal particles do not move in an electric field.

isoelectronic [ATOM PHYS] Pertaining to atoms having the same number of electrons outside the nucleus of the atom.

isoelectronic principle [PHYS CHEM] The principle that two molecules having the same number of electrons distributed over similar arrangements of atoms will have similar molecular orbitals.

isoelectronic sequence [SPECT] A set of spectra produced by different chemical elements ionized so that their atoms or ions contain the same number of electrons.

isofootcandle *See* isolux.

isogam [GEOPHYS] A line joining points on the earth's surface having the same value of the acceleration of gravity.

isogeotherm *See* geoisotherm.

isogonal transformation [MATH] A mapping of the plane into itself which leaves the magnitudes of angles between intersecting lines unchanged but may reverse their sense.

isogyre [OPTICS] A dark band in an interference figure located at those points that correspond to directions of transmission through the crystal plate in which the polarization of the incident light is not affected by passing through the plate.

isolated point [MATH] A point p in a topological space is an isolated point of a set if p is in the set and there is a neighborhood of p which contains no other points of the set. Also known as acnode.

Isochromatic fringe pattern for plate with hole. *(From M. M. Frocht, Photoelasticity, vol. 2, Wiley, 1948)*

ISOELECTRIC POINT

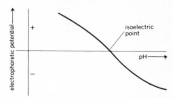

Graph showing the isoelectric point where particles are electrophoretically inert.

isolated set [MATH] A set consisting entirely of isolated points.

isolated subgroup [MATH] An isolated subgroup of a totally ordered Abelian group G is a subgroup of G which is also a segment of G.

isolator [ELECTR] A passive attenuator in which the loss in one direction is much greater than that in the opposite direction; a ferrite isolator for waveguides is an example.

isolith [ELECTR] Integrated circuit of components formed on a single silicon slice, but with the various components interconnected by beam leads and with circuit parts isolated by removal of the silicon between them.

isolux [OPTICS] A curve or surface connecting points at which light intensity is the same. Also known as isofootcandle; isophot.

isomagnetic [GEOPHYS] Of or pertaining to lines connecting points of equality in some magnetic element.

isomer [CHEM] One of two or more chemical substances having the same elementary percentage composition and molecular weight but differing in structure, and therefore in properties; there are many ways in which such structural differences occur; one example is provided by the compounds n-butane, $CH_3(CH_2)_2CH_3$, and isobutane, $CH_3CH(CH_3)_2$. [NUC PHYS] One of two or more nuclides having the same mass number and atomic number, but existing for measurable times in different quantum states with different energies and radioactive properties.

isomeric shift [PHYS CHEM] Shift in the Mössbauer resonance caused by the effect of the valence of the atom on the interaction of the electron density at the nucleus with the nuclear charge. Also known as chemical shift.

isomeric transition [NUC PHYS] A radioactive transition from one nuclear isomer to another of lower energy.

isomerism [CHEM] The phenomenon whereby certain chemical compounds have structures that are different although the compounds possess the same elemental composition. [NUC PHYS] The occurrence of nuclear isomers.

isometric *See* isochore.

isometric forms [MATH] Two bilinear forms f and g on vector spaces E and F for which there exists a linear isomorphism of E onto F such that $f(x,y) = g(\sigma x, \sigma y)$ for all x and y in E.

isometric process [THERMO] A constant-volume, frictionless thermodynamic process in which the system is confined by mechanically rigid boundaries.

isometric space [MATH] Two spaces between which an isometry exists.

isometric system [CRYSTAL] The crystal system in which the forms are referred to three equal, mutually perpendicular axes. Also known as cubic system.

isometry [MATH] **1.** A mapping f from a metric space X to a metric space Y where the distance between any two points of X equals the distance between their images under f in Y. **2.** A linear isomorphism σ of a vector space E onto itself such that, for a given bilinear form g, $g(\sigma x, \sigma y) = g(x,y)$ for all x and y in E.

isometry class [MATH] A set consisting of all bilinear forms (on vector spaces over a given field) which are isometric to a given form.

isomorphic systems [MATH] Two algebraic structures between which an isomorphism exists.

isomorphism [MATH] A one to one function of an algebraic structure (for example, group, ring, module, vector space) onto another of the same type, preserving all algebraic relations; its inverse function behaves likewise. [PHYS CHEM] A condition present when an ion at high dilution is incorporated by mixed crystal formation into a precipitate, even though such formation would not be predicted on the basis of crystallographic and ionic radii; an example is coprecipitation of lead with potassium chloride. [SCI TECH] The quality or state of being identical or similar in form, shape, or structure, such as between organisms resulting from evolutionary convergence, or crystalline forms of similar composition.

isoperimetric figures [MATH] Figures whose perimeters are equal.

isoperimetric inequality [MATH] The area enclosed by a plane curve is equal to or less than the square of its perimeter divided by 4π.

isoperimetric problem [MATH] In the calculus of variations, this problem deals with finding a closed curve in the plane which encloses the greatest area, given its length as fixed.

isophot *See* isolux.

isophotometer [OPTICS] A direct-recording photometer that automatically scans and measures optical density of all points in a film transparency or plate, and plots the measured density values in a quantitative two-dimensional isodensity tracing of the scanned areas.

isopiestic [PHYS] Denoting equal or constant pressure.

isopleth [MATH] The straight line which cuts the three scales of a nomograph at values satisfying some equation. Also known as index line.

isopor [GEOPHYS] An imaginary line connecting points on the earth's surface having the same annual change in a magnetic element.

isopycnic [PHYS] Of equal or constant density, with respect to either space or time.

isosceles triangle [MATH] A triangle with two sides of equal length.

isoseismal [GEOPHYS] Pertaining to points having equal intensity of earthquake shock, or to a line on a map of the earth's surface connecting such points.

isospin *See* isotopic spin.

isospin multiplet [PARTIC PHYS] A collection of elementary particles which have approximately the same mass and the same quantum numbers except for charge, but have a sequence of charge values, $Y/2 - I$, $Y/2 - I + 1$, . . . , $Y/2 + I$ times the proton charge, where Y is an integer known as the hypercharge, and I is an integer or half-integer known as the isospin; examples are the pions ($Y = 0$, $I = 1$) and the nucleons ($Y = 1$, $I = \frac{1}{2}$). Also known as charge multiplet; particle multiplet.

isostasy [GEOPHYS] A theory of the condition of approximate equilibrium in the outer part of the earth, such that the gravitational effect of masses extending above the surface of the geoid in continental areas is approximately counterbalanced by a deficiency of density in the material beneath those masses, while deficiency of density in ocean waters is counterbalanced by an excess in density of the material under the oceans.

isostatics [MECH] In photoelasticity studies of stress analyses, those curves to which the tangents represent the progressive change in principal-plane directions. Also known as stress lines; stress trajectories.

isostatic surface [MECH] A surface in a three-dimensional elastic body such that at each point of the surface one of the principal planes of stress at that point is tangent to the surface.

isoteniscope [ENG] An instrument for measuring the vapor pressure of a liquid, consisting of a U tube containing the liquid, one arm of which connects with a closed vessel containing the same liquid, while the other connects with a pressure gage where the pressure is adjusted until the levels in the arms of the U tube are equal.

isosteric [CHEM] Referring to similar electronic arrangements in chemical compounds. [PHYS] Of equal or constant specific volume with respect to either time or space.

isosterism [PHYS CHEM] A similarity in the physical properties of ions, compounds, or elements, as a result of electron arrangements that are identical or similar.

isostructural [CRYSTAL] Pertaining to crystalline materials that have corresponding atomic positions, and have a considerable tendency for ionic substitution.

isotherm [GEOPHYS] A line on a chart connecting all points of equal or constant temperature. [THERMO] A curve or formula showing the relationship between two variables, such as pressure and volume, when the temperature is held constant. Also known as isothermal.

isothermal [THERMO] **1.** Having constant temperature; at constant temperature. **2.** *See* isotherm.

isothermal calorimeter [THERMO] A calorimeter in which the heat received by a reservoir, containing a liquid in equilibrium with its solid at the melting point or with its vapor at the boiling point, is determined by the change in volume of the liquid.

isothermal chart [GEOPHYS] A map showing the distribution of air temperature (or sometimes sea-surface or soil temperature) over a portion of the earth or at some level in the atmosphere; places of equal temperature are connected by lines called isotherms.

isothermal compression [THERMO] Compression at constant temperature.

isothermal equilibrium [THERMO] The condition in which two or more systems are at the same temperature, so that no heat flows between them.

isothermal expansion [THERMO] Expansion of a substance while its temperature is held constant.

isothermal flow [THERMO] Flow of a gas in which its temperature does not change.

isothermal layer [THERMO] A layer of fluid, all points of which have the same temperature.

isothermal magnetization [THERMO] Magnetization of a substance held at constant temperature; used in combination with adiabatic demagnetization to produce temperatures close to absolute zero.

isothermal process [THERMO] Any constant-temperature process, such as expansion or compression of a gas, accompanied by heat addition or removal from the system at a rate just adequate to maintain the constant temperature.

isothermal remanent magnetization [GEOPHYS] Remanent magnetization that has been acquired under normal temperature conditions in a short period of time as a result of the application of a magnetic field. Abbreviated IRM.

isothermal transformation [THERMO] Any transformation of a substance which takes place at a constant temperature.

isotone [NUC PHYS] One of several nuclides having the same number of neutrons in their nuclei but differing in the number of protons.

isotone theorem [MATH] If $f(x_1, \ldots, x_n)$ is an expression involving the operations of union, intersection, and parenthesis on the elements $x_1, \ldots, x_n$ of a Boolean algebra and if x_i includes y_i for $i = 1, \ldots, n$, then $f(x_1, \ldots, x_n)$ includes $f(y_1, \ldots, y_n)$.

isotope [NUC PHYS] One of two or more atoms having the same atomic number but different mass number.

isotope abundance [NUC PHYS] The ratio of the number of

atoms of a particular isotope in a sample of an element to the number of atoms of a specified isotope, or to the total number of atoms of the element.

isotope dilution [NUCLEO] The introduction of a radioisotope into stable isotopes of an element in order to make volume, mass, and age measurements of the element.

isotope effect [PHYS CHEM] The effect of difference of mass between isotopes of the same element on nonnuclear physical and chemical properties, such as the rate of reaction or position of equilibrium, of chemical reactions involving the isotopes. [SOLID STATE] Variation of the transition temperatures of the isotopes of a superconducting element in inverse proportion to the square root of the atomic mass.

isotope exchange [NUCLEO] **1.** Exchange of places by two atoms, but different isotopes, of the same element in two different molecules, or in different locations of the same molecule. **2.** The transfer of isotopically tagged atoms from one chemical form or valence state to another, without net chemical reaction.

isotope fractionation [NUCLEO] Natural or artificial alteration of the isotopic composition of an element via processes of diffusion, evaporation, and chemical exchange, utilizing small differences in physical and chemical properties of isotopes.

isotope lamp [ELECTR] A discharge lamp containing gas of a single isotope and thus producing highly monochromatic light.

isotope separation [NUCLEO] The physical separation of different stable isotopes of an element from one another.

isotope shift [SPECT] A displacement in the spectral lines due to the different isotopes of an element.

isotopic age determination *See* radiometric dating.

isotopic carrier *See* carrier.

isotopic chronometer [NUCLEO] A method of determining the age of geological, archeological, or other samples by measuring the amount of a particular radioisotope and of its daughter isotope in a sample.

isotopic element [NUC PHYS] An element which has more than one naturally occurring isotope.

isotopic enrichment [NUCLEO] The process by which the relative abundances of the isotopes of a given element are altered in a batch, thus producing a form of the element enriched in a particular isotope.

isotopic exchange [PHYS CHEM] A process in which two atoms belonging to different isotopes of the same element exchange valency states or locations in the same molecule or different molecules.

isotopic incoherence [PHYS] Incoherence in the scattering of neutrons from a crystal lattice due to differences in scattering lengths of different isotopes of the same element.

isotopic indicator *See* isotopic tracer.

isotopic irradiation [NUCLEO] The subjection of a material to radiation from radioactive isotopes for therapeutic or other purposes.

isotopic label *See* isotopic tracer.

isotopic molecule [NUCLEO] A molecule in which the nucleus of one of the atoms is a special isotope.

isotopic number *See* neutron excess.

isotopic parity *See* G parity.

isotopic spin [NUC PHYS] A quantum-mechanical variable, resembling the angular momentum vector in algebraic structure whose third component distinguished between members of groups of elementary particles, such as the nucleons, which apparently behave in the same way with respect to strong nuclear forces, but have different charges. Also known as isobaric spin; isospin; i-spin.

isotopic tracer [CHEM] An isotope of an element, either radioactive or stable, a small amount of which may be incorporated into a sample material (the carrier) in order to follow the course of that element through a chemical, biological, or physical process, and also follow the larger sample. Also known as isotopic indicator; isotopic label; label; tag.

isotron [NUCLEO] A device for sorting isotopes of an element in which ions are accelerated to a fixed energy in a strong electric field, and a radio-frequency field then selects ions according to their velocity, which is inversely proportional to the square root of their mass.

isotropic [PHYS] Having identical properties in all directions.

isotropic antenna *See* unipole.

isotropic dielectric [ELEC] A dielectric whose polarization always has a direction that is parallel to the applied electric field, and a magnitude which does not depend on the direction of the electric field.

isotropic fluid [FL MECH] A fluid whose properties are not dependent on the direction along which they are measured.

isotropic flux [PHYS] Radiation, or a flow of particles or matter, which reaches a location from all directions with equal intensity.

isotropic gain of an antenna *See* absolute gain of an antenna.

isotropic material [PHYS] A material whose properties are not dependent on the direction along which they are measured.

isotropic noise [ELECTROMAG] Random noise radiation which reaches a location from all directions with equal intensity.

isotropic plasma [PL PHYS] A plasma whose properties, such as pressure, are not dependent on the direction along which they are measured.

isotropic radiation [ELECTROMAG] Radiation which is emitted by a source in all directions with equal intensity, or which reaches a location from all directions with equal intensity.

isotropic radiator [PHYS] An energy source that radiates uniformly in all directions.

isotropic turbulence [FL MECH] Turbulence whose properties, especially statistical correlations, do not depend on direction.

isotropic universe [ASTRON] A universe postulated to have the same properties when viewed from all directions.

isotropy [PHYS] The quality of a property which does not depend on the direction along which it is measured, or of a medium or entity whose properties do not depend on the direction along which they are measured.

isotropy group [MATH] For an operation of a group G on a set S, the isotropy group of an element s of S is the set of elements g in G such that $gs = s$.

isotypic [CRYSTAL] Pertaining to a crystalline substance whose chemical formula is analogous to, and whose structure is like, that of another specified compound.

isovalent hyperconjugation [PHYS CHEM] An arrangement of bonds in a hyperconjugated molecule such that the number of bonds is the same in the two resonance structures but the second structure is energetically less favorable than the first structure; examples are $H_3{\equiv}C{-}C^+H_2$ and $H_3{\equiv}C{-}CH_2$.

isovolumic *See* isochoric.

i-spin *See* isotopic spin.

ISR *See* Intersecting Storage Rings.

Israel's theorem [RELAT] A theorem of general relativity essentially proving that the Schwarzschild solution is the unique solution of Einstein's equations describing nonrotating black holes in empty space and that the Reissner-Nordstrom solution is the unique solution describing nonrotating charged black holes.

iterated integral [MATH] An integral over an area or volume designated to be performed by successive integrals over line segments.

iterated kernel [MATH] One of a series of kernels $K_n(x,y)$, $n = 0, 1, 2, \ldots$, constructed from a given kernel of an integral equation, $K(x,y)$, by setting $K_1(x,y) = K(x,y)$ and, for each value of n, setting $K_{n+1}(x,y)$ equal to the integral of $K(x,t)K_n(t,y)\,dt$.

iteration *See* iterative method.

iterative filter [ELECTR] Four-terminal filter that provides iterative impedance.

iterative impedance [ELECTR] Impedance that, when connected to one pair of terminals of a four-terminal transducer, will cause the same impedance to appear between the other two terminals.

iterative method [MATH] Any process of successive approximation used in such problems as numerical solution of algebraic equations, differential equations, or the interpolation of the values of a function. Also known as iteration.

iterative process [MATH] A process for calculating a desired result by means of a repeated cycle of operations, which comes closer and closer to the desired result; for example, the arithmetical square root of N may be approximated by an iterative process using additions, subtractions, and divisions only.

Itô's formula *See* stochastic chain rule.

Itô's integral *See* stochastic integral.

i-type semiconductor *See* intrinsic semiconductor.

Ixion [PHYS] An experimental magnetic-mirror device used for research on controlled fusion; involves study of plasma rotation in a magnetic-mirror confinement system using crossed electric and magnetic fields.

J *See* joule.

jacket [NUCLEO] A thin container for one or more fuel slugs, used to prevent the fuel from escaping into the coolant of a reactor. Also known as can; cartridge.

Jacobian [MATH] The Jacobian of functions f_i $(x_1, x_2, \ldots, x_n)$, $i = 1, 2, \ldots, n$, of real variables x_i is the determinant of the matrix whose ith row lists all the first-order partial derivatives of the function $f_i(x_1, x_2, \ldots, x_n)$. Also known as Jacobian determinant.

Jacobian determinant *See* Jacobian.

Jacobian elliptic function [MATH] For m a real number between 0 and 1, and u a real number, let ϕ be that number such that

$$\int_0^\phi d\theta / (1 - m \sin^2 \theta)^{1/2} = u;$$

the 12 Jacobian elliptic functions of u with parameter m are sn $(u|m) = \sin \phi$, cn $(u|m) = \cos \phi$, dn $(u|m) = (1 - m \sin^2 \phi)^{1/2}$, the reciprocals of these three functions, and the quotients of any two of them.

Jacobian matrix [MATH] The matrix used to form the Jacobian.

Jacobi condition [MATH] In the calculus of variations, a differential equation used to study the extremals in a variational problem.

Jacobi ellipsoid [MECH] The ellipsoid formed by a rapidly rotating liquid body, composed of shells of uniform density and subject to no forces other than its own gravitation.

Jacobi polynomials [MATH] Polynomials that are constructed from the hypergeometric function and satisfy the differential equation $(1-x^2)y'' + [\beta-\alpha-(\alpha+\beta+2)x]y' + n(\alpha + \beta + n + 1)y = 0$, where n is an integer and α and β are constants greater than -1; in certain cases these generate the Legendre and Chebyshev polynomials.

Jacobi's formula [MATH] The equation

$$\int_0^{\pi/2} \left[F(-n, p+q+n+1; q+1; \sin^2 x) \right]^2$$
$$\times \cos^{2p+1} x \, \sin^{2q+1} x \, dx$$
$$= \frac{1}{2} \frac{n! \Gamma(p+n+1) [\Gamma(q+1)]^2}{(p+q+2n+1) \, \Gamma(p+q+n+1) \, \Gamma(q+n+1)}$$

where $p > -1$, $q > -1$, and $F(a,b;c;z)$ is the hypergeometric series.

Jacobi's lemma [MATH] The equation

$$\frac{d^{n-1}}{dx^{n-1}} \sin^{2n-1} \theta = \frac{(-1)^{n-1}}{n} \frac{(2n)!}{2^n n!} \sin n\theta$$

where $x = \cos \theta$.

Jacobi's method [MATH] **1.** A method of determining the eigenvalues of a Hermitian matrix. **2.** A method for finding a complete integral of the general first-order partial differential equation in two independent variables; it involves solving a set of six ordinary differential equations.

Jacobi's theorem [MATH] A periodic, analytic function of a complex variable is simply periodic or doubly periodic.

Jacobi's transformations [MATH] Transformations of Jacobian elliptic functions to other Jacobian elliptic functions given by change of parameter and variable.

Jacobi variety [MATH] If M is a compact Riemann surface of genus g, and $w_1, \ldots, w_g$ is a basis of Abelian differential forms on M, then the Jacobi variety of M is a complex torus, of complex dimension g, given by the quotient C^g/Γ, where Γ is the g-dimensional lattice consisting of points in C^g of the form

$$\left(\int_\gamma w_1, \ldots, \int_\gamma w_g \right)$$

where γ is any closed curve in M.

Jacobi zeta function [MATH] A function $Z(u_1, k)$ which is used in evaluating elliptic integrals and is related to the theta functions; it is given by the integral from 0 to u_1 of $[dn^2(u|k^2) - E(k)/K(k)] \, du$, where $dn(u|k^2)$ is a Jacobian elliptic function, and $K(k)$ and $E(k)$ are complete elliptic integrals.

Jacobson radical *See* radical.

Jaeger method [FL MECH] A method of determining surface tension of a liquid in which one measures the pressure required to cause air to flow from a capillary tube immersed in the liquid.

Jaeger-Steinwehr method [THERMO] A refinement of the Griffiths method for determining the mechanical equivalent of heat, in which a large mass of water, efficiently stirred, is used, the temperature rise of the water is small, and the temperature of the surroundings is carefully controlled.

Jamin effect [FL MECH] Resistance to flow of a column of liquid divided by air bubbles in a capillary tube, even when subjected to a substantial pressure difference between the ends of the tube.

Jamin refractometer [OPTICS] An instrument for measuring the index of refraction of a gas in which two light beams from a common source are each passed through an evacuated tube and recombined, and the displacement of interference fringes is noted as gas is slowly admitted into one of the tubes.

jar [ELEC] A unit of capacitance equal to 1000 statfarads, or approximately 1.11265×10^{-9} farad; it is approximately equal to the capacitance of a Leyden jar; this unit is now obsolete.

Jeans viscosity equation [THERMO] An equation which states that the viscosity of a gas is proportional to the temperature raised to a constant power, which is different for different gases.

Jensen's equation [MATH] The functional equation

$$\tfrac{1}{2}\,[f(x) + f(y)] = f\!\left(\frac{x+y}{2}\right).$$

Jensen's inequality [MATH] **1.** A general inequality satisfied by a convex function

$$f\!\left(\sum_{i=1}^{n} a_i x_i \right) \leq \sum_{i=1}^{n} a_i f(x_i)$$

where the x_i are any numbers in the region where f is convex and the a_i are nonnegative numbers whose sum is equal to 1. **2.** If $a_1, a_2, \ldots, a_n$ are positive numbers and $s > t > 0$, then $(a_1{}^s + a_2{}^s + \cdots + a_n{}^s)^{1/s}$ is less than or equal to $(a_1{}^t + a_2{}^t + \cdots + a_n{}^t)^{1/t}$.

jerk [MECH] **1.** The rate of change of acceleration; it is the third derivative of position with respect to time. **2.** A unit of rate of change of acceleration, equal to 1 foot (30.48 centimeters) per second squared per second.

jet [FL MECH] A strong, well-defined stream of compressible fluid, either gas or liquid, issuing from an orifice or nozzle or moving in a contracted duct. [PARTIC PHYS] A group of particles issuing in approximately the same direction from a high-energy collision of elementary particles, believed to consist of decay products of a member of a quark-antiquark pair created in the collision.

jet-membrane method [NUCLEO] A method of uranium isotope separation in which a rarefied vapor jet interacts with a background gas of uranium hexafluoride, causing preferential diffusion of uranium-235 into the jet; the enriched gas is extracted from the interaction region between the jet and the uranium hexafluoride by using small channels.

jet tones [ACOUS] Unsteady tones produced when a stream of air issues into still air from an orifice.

JFET *See* junction field-effect transistor.

jitter [ELECTR] Small, rapid variations in a waveform due to mechanical vibrations, fluctuations in supply voltages, control-system instability, and other causes.

j-j coupling [ATOM PHYS] A process for building up many-electron wave functions; the spin and orbital functions of each particle are combined to form eigenfunctions of the particle's total angular momentum, and then the wave functions of all the particles are combined to form eigenfunctions of the total angular momentum of the system; this coupling is used when the spin-orbit interaction is strong compared to the electrostatic interaction.

jog [CRYSTAL] A shift in a dislocation from one crystal plane to another.

Johann crystal geometry [CRYSTAL] The focusing shape of a diffracting crystal for x-ray dispersion used in electron-probe microanalysis; less stringent than Johannson crystal geometry.

Johannson crystal geometry [CRYSTAL] The full-focusing shape of a diffracting crystal for x-ray dispersion used in electron-probe microanalyzers; more stringent than Johann crystal geometry.

Johnson and Lark-Horowitz formula [SOLID STATE] A formula according to which the resistivity of a metal or degenerate semiconductor resulting from impurities which scatter the electrons is proportional to the cube root of the density of impurities.

Johnson noise *See* thermal noise.

Johnson-Rahbeck effect [PHYS] An increase in frictional force between two electrodes in contact with a semiconductor that arises when a potential difference is applied between the electrodes.

joint [ELEC] A juncture of two wires or other conductive paths for current.

joint distribution of two random variables [STAT] The distribution which gives the probability that $Z = z$ and $W = w$ for all values z and w of the random variables Z and W respectively.

joint marginal distribution [STAT] The distribution obtained by summing the joint distribution of three random variables over all possible values of one of these variables.

Jolly balance [ENG] A spring balance used to measure specific gravity of mineral specimens by weighing a specimen when in the air and when immersed in a liquid of known density.

Joly photometer [OPTICS] A photometer consisting of two equal paraffin wax or opal glass blocks separated by a thin opaque sheet; the positions of two light sources under comparison are adjusted until the two blocks appear equally bright. Also known as wax-block photometer.

Joly steam calorimeter [ENG] **1.** A calorimeter in which the mass of steam that condenses on a specimen and a pan holding it is measured, as well as the mass of steam that condenses on an empty pan. **2.** *See* differential steam calorimeter.

Jordan algebra [MATH] A nonassociative algebra over a field in which products satisfy the Jordan identity $(xy)x^2 = x(yx^2)$.

Jordan arc *See* simple arc.

Jordan curve *See* simple closed curve.

Jordan curve theorem [MATH] The theorem that in the plane every simple closed curve separates the plane into two parts.

Jordan-Hölder theorem [MATH] The theorem that for a group any two composition series have the same number of subgroups listed, and both series produce the same quotient groups.

Jordan lag [ELECTROMAG] A type of magnetic viscosity in which the angular lag of the magnetic induction behind a sinusoidally varying magnetic field strength, and also the energy loss per cycle, is independent of frequency.

Jordan matrix [MATH] A matrix whose elements are equal and nonzero on the principal diagonal, equal to 1 on the diagonal immediately above, and equal to 0 everywhere else.

Jordan's lemma [MATH] If $f(z)$ is continuous on the upper half of the complex plane (Im $z \geq 0$) and if $|f(z)|$ tends to 0 as $z \rightarrow \infty$, uniformly as regards arg z (the argument of z), $0 \leq \arg z \leq \pi$, then the integral $J(R) = \int e^{si\theta} f(\theta) \, d\theta$, for $s > 0$, taken along that half of the circle $|z| = R$ which is above the real axis, tends to 0 as $R \rightarrow \infty$.

Jordan-Wigner commutation rules [QUANT MECH] Rules obtained by replacing commutators of creation and destruction operators by anticommutators; applicable to fermion fields in a quantized field theory.

Josephson current [CRYO] The current across a Josephson junction in the absence of voltage across the junction, resulting from the Josephson effect.

Josephson effect [CRYO] The tunneling of electron pairs through a thin insulating barrier between two superconducting materials. Also known as Josephson tunneling.

Josephson equation [CRYO] An equation according to which the Josephson current is a sinusoidally varying function of the applied magnetic field.

Josephson junction [CRYO] A thin insulator separating two superconducting materials; it displays the Josephson effect.

Josephson tunneling *See* Josephson effect.

Joshi effect [ELECTR] The change in the current passing through a gas or vapor when the gas or vapor is irradiated with visible light.

Joukowski profile [FL MECH] An airfoil profile with a cusp-shaped trailing edge, resulting from the Joukowski transformation of a circle which passes through the point $z = a$ and which is located so that the point $z = -a$ does not lie outside the circle.

Joukowski transformation [FL MECH] A conformal mapping used to transform circles into airfoil profiles for the purpose of studying fluid flow past the airfoil profiles; it assigns to each complex number z the number $w = z + (a^2/z)$.

joule [MECH] The unit of energy or work in the meter-kilogram-second system of units, equal to the work done by a force of magnitude of 1 newton when the point at which the force is applied is displaced 1 meter in the direction of the force. Symbolized J. Also known as newton-meter of energy.

Joule and Playfairs' experiment [THERMO] An experiment in which the temperature of the maximum density of water is

JOSEPHSON JUNCTION

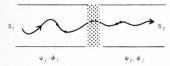

$\psi_1, \phi_1 \qquad \psi_2, \phi_2$

Josephson junction showing two superconductors S_1 and S_2 separated by a thin insulator (shaded area). The phases ϕ_1 and ϕ_2 of the order parameters ψ_1 and ψ_2 of the two superconductors are coherent.

measured by taking the mean of the temperatures of water in two columns whose densities are determined to be equal from the absence of correction currents in a connecting trough.

Joule calorimeter [ENG] Any electrically heated calorimeter, such as that used in the Griffiths method.

Joule-Clausius velocity [STAT MECH] A quantity used in the description of the kinetic behavior of a gas, equal to the square root of the ratio of the pressure of the gas to one-third of its density.

Joule cycle *See* Brayton cycle.

Joule effect [PHYS] **1.** The heating effect produced by the flow of current through a resistance. **2.** A change in the length of a ferromagnetic substance which occurs parallel to an applied magnetic field. Also known as Joule magnetorestriction; longitudinal magnetorestriction.

Joule equivalent [THERMO] The numerical relation between quantities of mechanical energy and heat; the present accepted value is 1 fifteen-degrees calorie equals 4.1855 ± 0.0005 joules. Also known as mechanical equivalent of heat.

Joule experiment [THERMO] **1.** An experiment to detect intermolecular forces in a gas, in which one measures the heat absorbed when gas in a small vessel is allowed to expand into a second vessel which has been evacuated. **2.** An experiment to measure the mechanical equivalent of heat, in which falling weights cause paddles to rotate in a closed container of water whose temperature rise is measured by a thermometer.

Joule heat [ELEC] The heat which is evolved when current flows through a medium having electrical resistance, as given by Joule's law.

Joule-Kelvin effect *See* Joule-Thomson effect.

Joule magnetorestriction *See* Joule effect.

Joule's law [ELEC] The law that when electricity flows through a substance, the rate of evolution of heat in watts equals the resistance of the substance in ohms times the square of the current in amperes. [THERMO] The law that at constant temperature the internal energy of a gas tends to a finite limit, independent of volume, as the pressure tends to zero.

Joule-Thomson coefficient [THERMO] The ratio of the temperature change to the pressure change of a gas undergoing isenthalpic expansion.

Joule-Thomson effect [THERMO] A change of temperature in a gas undergoing Joule-Thomson expansion. Also known as Joule-Kelvin effect.

Joule-Thomson expansion [THERMO] The adiabatic, irreversible expansion of a fluid flowing through a porous plug or partially opened valve. Also known as Joule-Thomson process.

Joule-Thomson inversion temperature [THERMO] A temperature at which the Joule-Thomson coefficient of a given gas changes sign.

Joule-Thomson process *See* Joule-Thomson expansion.

Joule-Thomson valve [CRYO] A valve through which a gas is allowed to expand adiabatically, resulting in lowering of its temperature; used in production of liquid hydrogen and helium.

Jovian planet [ASTRON] Any of the four major planets (Jupiter, Saturn, Uranus, and Neptune) that are at a greater distance from the sun than the terrestrial planets (Mercury, Venus, Earth, and Mars).

Jovian Van Allen belts [ASTROPHYS] The extended belts of high-energy charged particles that are trapped in Jupiter's magnetic field and cause the microwave nonthermal emission of radio waves observed in the band from about 3 to 70 centimeters.

JOVIAN VAN ALLEN BELTS

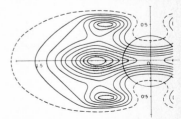

Contours of radio emission from Jupiter originating from one-half of the Van Allen belt, and extending far beyond the planet's disk.

J particle [PARTIC PHYS] A neutral meson which has a mass of 3095 megaelectronvolts, spin quantum number 1, and negative parity and charge parity; it has an anomalously long lifetime of approximately 10^{-20} second (corresponding to a width of approximately 70 kiloelectronvolts). Also known as psi particle (symbolized ψ).

J-shaped distribution [STAT] A frequency distribution that is extremely asymmetrical in that the initial (or final) frequency group contains the highest frequency, with succeeding frequencies becoming smaller (or larger) elsewhere; the shape of the curve roughly approximates the letter "J" lying on its side.

judgment sample [STAT] Sample selection in which personal views or opinions of the individual doing the sampling enter into the selection.

Julian day [ASTRON] The number of each day, as reckoned consecutively since the beginning of the present Julian period on January 1, 4713 B.C.; it is used primarily by astronomers to avoid confusion due to the use of different calendars at different times and places; the Julian day begins at noon, 12 hours later than the corresponding civil day.

Julian ephemeris century [ASTRON] The unit of ephemeris time (ET) in Simon Newcomb's formulas which relate the orbital position of the earth to ephemeris time; the Julian ephemeris century is subdivided into 36,525 days, and 1 ephemeris day = 86,400 ephemeris seconds.

jump discontinuity [MATH] A point a where for a real-valued function $f(x)$ the limit on the left of $f(x)$ as x approaches a and the limit on the right both exist but are distinct.

jump function [MATH] A function used to represent a sampled data sequence arising in the numerical study of linear difference equations.

jump phenomenon [CONT SYS] A phenomenon occurring in a nonlinear system subjected to a sinusoidal input at constant frequency, in which the value of the amplitude of the forced oscillation can jump upward or downward as the input amplitude is varied through either of two fixed values, and the graph of the forced amplitude versus the input amplitude follows a hysteresis loop.

jump resonance [CONT SYS] A jump discontinuity occurring in the frequency response of a nonlinear closed-loop control system with saturation in the loop.

junction [ELEC] *See* major node. [ELECTR] A region of transition between two different semiconducting regions in a semiconductor device, such as a *pn* junction, or between a metal and a semiconductor. [ELECTROMAG] A fitting used to join a branch waveguide at an angle to a main waveguide, as in a tee junction. Also known as waveguide junction.

junction battery [NUCLEO] A nuclear-type battery in which a radioactive material, such as strontium-90, irradiates a *pn* silicon junction.

junction capacitor [ELECTR] An integrated-circuit capacitor that uses the capacitance of a reverse-biased *pn* junction.

junction detector [NUCLEO] A reverse-biased semiconductor junction functioning as a solid ionization chamber to produce an electric output pulse whose amplitude is linearly proportional to the energy deposited in the junction depletion layer by the incident ionizing radiation.

junction diode [ELECTR] A semiconductor diode in which the rectifying characteristics occur at an alloy, diffused, electrochemical, or grown junction between *n*-type and *p*-type semiconductor materials. Also known as junction rectifier.

junction field-effect transistor [ELECTR] A field-effect transistor in which there is normally a channel of relatively low-conductivity semiconductor joining the source and drain, and

JUNCTION DETECTOR

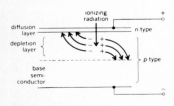

Silicon junction detector.

this channel is reduced and eventually cut off by junction depletion regions, reducing the conductivity, when a voltage is applied between the gate electrodes. Abbreviated JFET. Also known as depletion-mode field-effect transistor.

junction isolation [ELECTR] Electrical isolation of a component on an integrated circuit by surrounding it with a region of a conductivity type that forms a junction, and reverse-biasing the junction so it has extremely high resistance.

junction laser [OPTICS] A laser in which a junction in a semiconductor serves as the source of the coherent laser beam.

junction phenomena [ELECTR] Phenomena which occur at the boundary between two semiconductor materials, or a semiconductor and a metal, such as the existence of an electrostatic potential in the absence of current flow, and large injection currents which may arise when external voltages are applied across the junction in one direction.

junction point *See* branch point.

junction rectifier *See* junction diode.

junction transistor [ELECTR] A transistor in which emitter and collector barriers are formed between semiconductor regions of opposite conductivity type.

Jupiter [ASTRON] The largest planet in the solar system, and the fifth in order of distance from the sun; semimajor axis = 485×10^6 miles (780×10^6 kilometers); sidereal revolution period = 11.86 years; mean orbital velocity = 8.2 miles per second (13.2 kilometers per second); inclination of orbital plane to ecliptic = 1.03; equatorial diameter = 88,700 miles (142,700 kilometers); polar diameter = 82,800 miles (133,300 kilometers); mass = about 318.4 (earth = 1).

Jurin rule [FL MECH] The rule that a height to which a liquid rises in a capillary tube is twice the liquid's surface tension times the cosine of its contact angle with the capillary, divided by the product of the liquid's weight density and the internal radius of the tube.

just scale [ACOUS] A diatonic scale rendered in the just tuning system.

just ton *See* ton.

just tuning [ACOUS] A tuning system generated by octave rearrangements of the notes of three consecutive triads, each having the frequency ratio 4:5:6, with the highest note of one triad serving as the lowest note of the next.

JUPITER

Telescopic appearance of Jupiter showing bands and Red Spot. *(Hale Observatories)*

k *See* kilo-.

K *See* cathode; potassium.

kA *See* kiloampere.

K-A age [GEOL] The radioactive age of a rock determined from the ratio of potassium-40 (K^{40}) to argon-40 (A^{40}) present in the rock.

K-A decay [NUC PHYS] Radioactive decay of potassium-40 (K^{40}) to argon-40 (A^{40}), as the nucleus of potassium captures an orbital electron and then decays to argon-40; the ratio of K^{40} to A^{40} is used to determine the age of rock (K-A age).

kalium *See* potassium.

Kalman filter [CONT SYS] A linear system in which the mean squared error between the desired output and the actual output is minimized when the input is a random signal generated by white noise.

Kaluza theory [RELAT] An attempted unified field theory in which the four-dimensional world that one observes is taken to be a projection of a five-dimensional continuum.

kaon *See* K meson.

kaonic atom [ATOM PHYS] An atom consisting of a negatively charged kaon orbiting around an ordinary nucleus.

Kapetyn selected areas [ASTRON] Certain areas in the Milky Way Galaxy that the astronomer J. C. Kapetyn suggested be studied intensively in order to determine the structure of the galaxy. Also known as selected areas.

Kapteyn series [MATH] A series whose nth term is of the form $a_n J_{\nu + n}[(\nu + n)z]$, where $J_{\nu + n}$ is a Bessel function and ν and the a_n are constants.

Kapitza balance [ENG] A magnetic balance for measuring susceptibilities of materials in large magnetic fields that are applied for brief periods.

Kármán constant [FL MECH] A dimensionless number formed from the velocity of turbulent flow parallel to a plane wall, the distance from the wall, the shear stress, and the density of the fluid; for a wide range of flow patterns it has a constant value.

Kármán-Trefftz transformation [FL MECH] A generalization of the Joukowski transformation that maps circles into airfoil profiles that have a nonzero angle between the upper and lower edges at the trailing edge; it maps complex numbers z into complex numbers w so as to satisfy the equation $(w - 2a)/(w + 2a) = [(z - a)/(z + a)]^b$.

Kármán-Tsien method [FL MECH] A method of approximating equations for two-dimensional compressible flow which yields a simple rule for estimating compressibility effects of subsonic flow.

Kármán vortex street [FL MECH] A double row of line vortices in a fluid which, under certain conditions, is shed in the wake of cylindrical bodies when the relative fluid velocity is perpendicular to the axis of the cylinder.

KÁRMÁN VORTEX STREET

Kármán vortex street showing double row of line vortices. Here U = stream speed, h = perpendicular distance between the two lines of vortices, and a = distance between successive vortices on the same line.

Karnaugh map [ELECTR] A truth table that has been rearranged to show a geometrical pattern of functional relationships for gating configurations; with this map, essential gating requirements can be recognized in their simplest form.

Kata thermometer [ENG] An alcohol thermometer used to measure low velocities in air circulation, by heating the large bulb of the thermometer above 100°F (38°C) and noting the time it takes to cool from 100 to 95°F (38 to 35°C) or some other interval above ambient temperature, the time interval being a measure of the air current at that location.

Kater's reversible pendulum [MECH] A gravity pendulum designed to measure the acceleration of gravity and consisting of a body with two knife-edge supports on opposite sides of the center of mass.

katharometer [ENG] An instrument for detecting the presence of small quantities of gases in air by measuring the resulting change in thermal conductivity of the air. Also known as thermal conductivity cell.

katoptric system [OPTICS] An optical system such that, when the object is displaced in a direction parallel to the axis, the image is displaced in the opposite direction (in contrast to a dioptric system). Also known as contracurrent system.

kayser [SPECT] A unit of reciprocal length, especially wave number, equal to the reciprocal of 1 centimeter. Also known as rydberg.

kb *See* kilobar.

K band [COMMUN] A band of radio frequencies extending from 10,900 to 36,000 megahertz, corresponding to wavelengths of 2.75 to 0.834 centimeters. [SOLID STATE] An optical absorption band which appears together with an F band and has a lower intensity and shorter wavelength than the latter.

K-beta filter *See* beta filter.

kc *See* kilohertz.

kcal *See* kilocalorie.

K capture [NUC PHYS] A type of beta interaction in which a nucleus captures an electron from the K shell of atomic electrons (the shell nearest the nucleus) and emits a neutrino.

K corona [ASTRON] The inner portion of the sun's corona, having a continuous spectrum caused by electron scattering.

keeper [ELECTROMAG] A bar of iron or steel placed across the poles of a permanent magnet to complete the magnetic circuit when the magnet is not in use, to avoid the self-demagnetizing effect of leakage lines. Also known as magnet keeper.

Keesom force *See* orientation force.

kei function [MATH] One of the functions that is defined by $\ker_n (z) \pm i \, \text{kei}_n (z) = i^{\pm n} K_n (z e^{\pm \pi i/4})$, where K_n is the nth modified Bessel function of the second kind.

Keldysh theory [ATOM PHYS] A theory of multiphoton ionization, in which an atom is ionized by rapid absorption of a sufficient number of photons; it predicts that the ionization rate depends primarily upon the ratio of the mean binding electric field to the peak strength of the incident electromagnetic field, and upon the ratio of the binding energy to the energy of photons in the field.

K electron [ATOM PHYS] An electron in the K shell.

Kellner eyepiece [OPTICS] A Ramsden eyepiece with an achromatic eye lens.

Kellogg equation [THERMO] An equation of state for a gas, of the form

$$p = RT\rho + \sum_{n=2}^{\infty} (b_n T - a_n - C_n/T^2)\rho^n$$

where p is the pressure, T the absolute temperature, ρ the density, R the gas constant, and a_n, b_n, and c_n are constants.

kelvin [ELEC] A name formerly given to the kilowatt-hour. Also known as thermal volt. [THERMO] A unit of absolute temperature equal to 1/273.16 of the absolute temperature of the triple point of water. Symbolized K. Formerly known as degree Kelvin.

Kelvin absolute temperature scale [THERMO] A temperature scale in which the ratio of the temperatures of two reservoirs is equal to the ratio of the amount of heat absorbed from one of them by a heat engine operating in a Carnot cycle to the amount of heat rejected by this engine to the other reservoir; the temperature of the triple point of water is defined as 273.16 K. Also known as Kelvin temperature scale.

Kelvin balance [ELECTROMAG] An ammeter in which the force between two coils in series that carry the current to be measured, one coil being attached to one arm of a balance, is balanced against a known weight at the other end of the balance arm.

Kelvin body [MECH] An ideal body whose shearing (tangential) stress is the sum of a term proportional to its deformation and a term proportional to the rate of change of its deformation with time. Also known as Voigt body.

Kelvin bridge [ELEC] A specialized version of the Wheatstone bridge network designed to eliminate, or greatly reduce, the effect of lead and contact resistance, and thus permit accurate measurement of low resistance. Also known as double bridge; Kelvin network; Thomson bridge.

Kelvin equation [THERMO] An equation giving the increase in vapor pressure of a substance which accompanies an increase in curvature of its surface; the equation describes the greater rate of evaporation of a small liquid droplet as compared to that of a larger one, and the greater solubility of small solid particles as compared to that of larger particles.

Kelvin functions [MATH] A collective name for the ber function, bei function, ker function, and kei function. Also known as Thomson functions.

Kelvin guard-ring capacitor [ELEC] A capacitor with parallel circular plates, one of which has a guard ring separated from the plate by a narrow gap; it is used as a standard, whose capacitance can be accurately calculated from its dimensions.

Kelvin-Helmholtz contraction [ASTROPHYS] A contraction of a star once it is formed and before it is hot enough to ignite its hydrogen; the contraction converts gravitational potential energy into heat, some of which is radiated, with the remainder used to raise the internal temperature of the star.

Kelvin network *See* Kelvin bridge.

Kelvin relations *See* Thomson relations.

Kelvin replenisher [ELEC] A simple electrostatic generator in which curved metal plates attached to an insulating arm rotate between larger curved plates, and the contacts of the smaller plates with wipers connecting them to the larger plates and to each other result in the accumulation of charge on the smaller plates, energy being supplied by the rotation of the arm.

Kelvin scale [THERMO] The basic scale used for temperature definition; the triple point of water (comprising ice, liquid, and vapor) is defined as 273.16 K; given two reservoirs, a reversible heat engine is built operating in a cycle between them, and the ratio of their temperatures is defined to be equal to the ratio of the heats transferred.

Kelvin's circulation theorem [FL MECH] The theorem that, if the external forces acting on an inviscid fluid are conservative and if the fluid density is a function of the pressure only, then the circulation along a closed curve which moves with the fluid does not change with time.

KELVIN BRIDGE

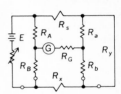

Circuit of the Kelvin bridge. E = battery; R_A, R_B = main ratio resistors; R_a, R_b = auxiliary ratio; R_x = unknown; R_s = standard; R_y = a heavy copper yoke of low resistance connected between the unknown and standard resistors; R_G = resistance in series with galvanometer G.

Kelvin's formula *See* Thomson formula.

Kelvin's inversion theorem [MATH] If $f(r,\theta,\phi)$ is a harmonic function in a domain D, where r, θ, and ϕ are spherical coordinates and a is a constant, then $(a/r)f(a^2/r, \theta, \phi)$ is also a harmonic function in the domain into which D is mapped by inversion with respect to a sphere of radius a centered at the origin.

Kelvin skin effect *See* skin effect.

Kelvin's minimum-energy theorem [FL MECH] The theorem that the irrotational motion of an incompressible, inviscid fluid occupying a simply connected region has less kinetic energy than any other fluid motion consistent with the boundary condition of zero relative velocity normal to the boundaries of the region.

Kelvin's statement of the second law of thermodynamics [THERMO] The statement that it is *not* possible that, at the end of a cycle of changes, heat has been extracted from a reservoir and an equal amount of work has been produced without producing some other effect.

Kelvin's theorem on harmonic functions [MATH] If $V(x,y,z)$ is a homogeneous function of x, y, and z of degree n which satisfies Laplace's equation, then $r^{-2n-1}V(x,y,z)$, where $r^2 = x^2 + y^2 + z^2$, also satisfies Laplace's equation.

Kelvin temperature scale [THERMO] **1.** An International Practical Temperature Scale which agrees with the Kelvin absolute temperature scale within the limits of experimental determination. **2.** *See* Kelvin absolute temperature scale.

Kendall's rank correlation coefficient [STAT] A statistic used as a measure of correlation in nonparametric statistics when the data are in ordinal form. Also known as Kendall's tau.

Kendall's tau *See* Kendall's rank correlation coefficient.

Kennard packet [QUANT MECH] A wave packet for which the product of the root-mean-square deviations of position and momentum from their respective mean values is as small as possible, being equal to Planck's constant divided by 4π.

Kennedy and Pancu circle [MECH] For a harmonic oscillator subject to hysteretic damping and subjected to a sinusoidally varying force, a plot of the in-phase and quadrature components of the displacement of the oscillator as the frequency of the applied vibration is varied.

Kennelly-Heaviside layer *See* E layer.

Keplerian ellipse *See* Keplerian orbit.

Keplerian motion [ASTRON] Orbital movement of a body about another that is not disturbed by the presence of a third celestial body.

Keplerian orbit [ASTRON] An elliptical orbit of a celestial body about another, the latter at a focus of the ellipse. Also known as Keplerian ellipse.

Keplerian telescope [OPTICS] A telescope that forms a real intermediate image in the focal plane and can be used for introducing a reticle or a scale into the focal plane.

Kepler's equations [ASTRON] The mathematical relationship between two different systems of angular measurements of the position of a body in an ellipse.

Kepler's laws [ASTRON] Three laws, determined by Johannes Kepler, that describe the motions of planets in their orbits: the orbits of the planets are ellipses with the sun at a common focus; the line joining a planet and the sun sweeps over equal areas during equal intervals of time; the squares of the periods of revolution of any two planets are proportional to the cubes of their mean distances from the sun.

keratoid [MATH] A plane curve whose equation in cartesian coordinates x and y is $y^2 = x^2y + x^5$.

keratoid cusp [MATH] A cusp of a curve which has one

branch of the curve on each side of the common tangent. Also known as single cusp of the first kind.

ker function [MATH] One of the functions that is defined by $\ker_n (z) \pm i \, \text{kei}_n(z) = i^{\pm n} K_n(ze^{\pm \pi i/4})$, where K_n is the nth modified Bessel function of the second kind.

kerma [NUCLEO] The kinetic energy imparted to charged particles in a unit mass of material by uncharged particles such as neutrons; it may be expressed as joules per kilogram or ergs per gram.

kernel [ATOM PHYS] An atom that has been stripped of its valence electrons, or a positively charged nucleus lacking the outermost orbital electrons. [MATH] *See* null space.

kernel of a homomorphism [MATH] For a homomorphism h from a group G to a group H, this consists of all elements of G which h sends to the identity element of H.

kernel of a linear transformation [MATH] All those vectors which a linear transformation maps to the zero vector.

kernel of a mapping [MATH] For any mapping f from a group A to a group B, the kernel of f, denoted ker f, is the set of all elements a of A such that $f(a)$ equals the identity element of B.

kernel of an integral equation [MATH] For Fredholm and Volterra type of equations, this is the function $K(x,t)$.

kernel of an integral transform [MATH] The function $K(x,t)$ in the transformation which sends the function $f(x)$ to the function $\int K(x,t)f(t)\,dt = F(x)$.

Kerr cell [OPTICS] A glass cell containing a dielectric liquid that exhibits the Kerr effect, such as nitrobenzene, in which is inserted the two plates of a capacitor, used to observe the Kerr effect on light passing through the cell.

Kerr constant [OPTICS] A measure of the strength of the Kerr effect in a substance, equal to the difference between the extraordinary and ordinary indices of refraction divided by the product of the light's wavelength and the square of the electric field.

Kerr effect *See* electrooptical Kerr effect.

Kerr magnetooptical effect *See* magnetooptic Kerr effect.

Kerr-Newman solution [RELAT] The unique solution to the equations of general relativity describing a charged, rotating black hole.

Ketteler formula [ELECTROMAG] The case of Sellmeier's equation where only two characteristic frequencies are involved.

ket vector [QUANT MECH] A vector in Hilbert space specifying the state of a system (opposed to bra vector); represented by the symbol $|\rangle$, with a letter or one or more indices inserted to distinguish it from other vectors.

keV *See* kiloelectronvolt.

Keyes equation [THERMO] An equation of state of a gas which is designed to correct the van der Waals equation for the effect of surrounding molecules on the term representing the volume of a molecule.

K factor [NUCLEO] A measure of the energy of the gamma rays produced by a particular type of emitter; it is the gamma-ray dose rate in roentgens per hour at a distance of 1 centimeter from a source having a radioactive disintegration rate of 1 millicurie (3.7×10^7 disintegrations per second).

k-fold transitive group *See* k-ply transitive group.

kg *See* kilogram; kilogram force.

kg-cal *See* kilocalorie.

kgf *See* kilogram force.

kgf-m *See* meter-kilogram.

kg-wt *See* kilogram force.

kHz *See* kilohertz.

kick-sorter *See* pulse-height analyzer.

Kikuchi lines [CRYSTAL] A pattern consisting of pairs of white and dark parallel lines, obtained when an electron beam is scattered (diffracted) by a crystalline solid; the pattern gives information on the structure of the crystal.

killer [SOLID STATE] An impurity that inhibits luminescence in a solid.

Killing's equation [MATH] The equations for an isometry-generating vector field in a geometry.

Killing vector [MATH] An element of a vector field in a geometry that generates an isometry.

kilo- [SCI TECH] A prefix representing 10^3 or 1000. Abbreviated k.

kiloampere [ELEC] A metric unit of current flow equal to 1000 amperes. Abbreviated kA.

kilobar [MECH] A unit of pressure equal to 1000 bars (100 megapascals). Abbreviated kb.

kilocalorie [THERMO] A unit of heat energy equal to 1000 calories. Abbreviated kcal. Also known as kilogram-calorie (kg-cal); large calorie (Cal).

kilocycle *See* kilohertz.

kiloelectronvolt [PHYS] A unit of energy, equal to 1000 electron volts. Abbreviated keV.

kilogram [MECH] **1.** The unit of mass in the meter-kilogram-second system, equal to the mass of the international prototype kilogram stored at Sèvres, France. Abbreviated kg. **2.** *See* kilogram force.

kilogram-calorie *See* kilocalorie.

kilogram-equivalent weight [CHEM] A unit of mass 1000 times the gram-equivalent weight.

kilogram force [MECH] A unit of force equal to the weight of a 1-kilogram mass at a point on the earth's surface where the acceleration of gravity is 9.80665 meters/sec^2. Abbreviated kgf. Also known as kilogram (kg); kilogram weight (kg-wt).

kilogram-meter *See* meter-kilogram.

kilogram weight *See* kilogram force.

kilohertz [PHYS] A unit of frequency equal to 1000 hertz. Abbreviated kHz. Also known as kilocycle (kc).

kilohm [ELEC] A unit of electrical resistance equal to 1000 ohms. Abbreviated K; kohm.

kilojoule [PHYS] A unit of energy or work equal to 1000 joules. Abbreviated kJ.

kiloliter [MECH] A unit of volume equal to 1000 liters or to 1 cubic meter. Abbreviated kl.

kilomega- *See* giga-.

kilomegacycle *See* gigahertz.

kilomegahertz *See* gigahertz.

kilometer [MECH] A unit of length equal to 1000 meters. Abbreviated km.

kiloparsec [ASTRON] A distance of 1000 parsecs (3260 light-years).

kiloton [PHYS] A unit used in specifying the yield of a fission or fusion bomb, equal to the explosive power of 1000 metric tons of trinitrotoluene (TNT). Abbreviated kt.

kilovar [ELEC] A unit of 1000 volt-amperes reactive. Abbreviated kvar.

kilovolt [ELEC] A unit of potential difference equal to 1000 volts. Abbreviated kV.

kilovolt-ampere [ELEC] A unit of apparent power in an alternating-current circuit, equal to 1000 volt-amperes. Abbreviated kVA.

kilovoltmeter [ELEC] A voltmeter which measures potential differences on the order of several kilovolts.

kilovolts peak [ELECTR] The peak voltage applied to an x-ray tube, expressed in kilovolts. Abbreviated kVp.

kilowatt [PHYS] A unit of power equal to 1000 watts. Abbreviated kW.

kilowatt-hour [ELEC] A unit of energy or work equal to 1000 watt-hours. Abbreviated kWh; kWhr. Also known as Board of Trade Unit.

kinematically admissible motion [MECH] Any motion of a mechanical system which is geometrically compatible with the constraints.

kinematic boundary condition [FL MECH] The condition that the component of fluid velocity perpendicular to a solid boundary must vanish on the boundary itself; when the boundary is a fluid surface, the condition applies to the vector difference of velocities across the interface.

kinematic fluidity [FL MECH] The reciprocal of the kinematic viscosity.

kinematics [MECH] The study of the motion of a system of material particles without reference to the forces which act on the system.

kinematic similarity [FL MECH] A relationship between fluid-flow systems in which corresponding fluid velocities and velocity gradients are in the same ratios at corresponding locations.

kinematic viscosity [FL MECH] The absolute viscosity of a fluid divided by its density. Also known as coefficient of kinematic viscosity.

kinetic energy [MECH] The energy which a body possesses because of its motion; in classical mechanics, equal to one-half of the body's mass times the square of its speed.

kinetic equilibrium *See* dynamic equilibrium.

kinetic friction [MECH] The friction between two surfaces which are sliding over each other.

kinetic momentum [MECH] The momentum which a particle possesses because of its motion; in classical mechanics, equal to the particle's mass times its velocity.

kinetic potential *See* Lagrangian.

kinetic pressure [FL MECH] The kinetic energy per unit volume of a fluid, equal to one-half the product of its density and the square of its velocity.

kinetic reaction [MECH] The negative of the mass of a body multiplied by its acceleration.

kinetics [MECH] The dynamics of material bodies.

kinetic stress [STAT MECH] A stress which arises, in a theory taking the motions of individual molecules into account, from the existence of a velocity distribution of molecules, an example is the pressure of an ideal gas.

kinetic theory [STAT MECH] A theory which attempts to explain the behavior of physical systems on the assumption that they are composed of large numbers of atoms or molecules in vigorous motion; it is further assumed that energy and momentum are conserved in collisions of these particles, and that statistical methods can be applied to deduce the particles' average behavior. Also known as molecular theory.

kink instability [PL PHYS] A type of hydromagnetic instability in which the ionized gas and its magnetic confining field tend to form a loop or kink, which then grows steadily larger. Also known as sausage instability.

kintal *See* hundredweight.

kip [MECH] A 1000-pound (453.6-kilogram) load.

Kirchhoff formula [THERMO] A formula for the dependence of vapor pressure p on temperature T, valid over limited temperature ranges; it may be written $\log p = A - (B/T) - C \log T$, where A, B, and C are constants.

Kirchhoff's current law [ELEC] The law that at any given instant the sum of the instantaneous values of all the currents

KINK INSTABILITY

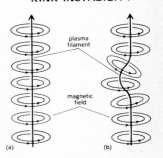

Plasma-magnetic-field configuration for cylindrical pinch. *(a)* Equilibrium configuration of plasma filament and magnetic field generated by axial current flow through plasma. *(b)* Onset of kink instability.

flowing toward a point is equal to the sum of instantaneous values of all the currents flowing away from the point. Also known as Kirchhoff's first law.

Kirchhoff's equations [THERMO] Equations which state that the partial derivative of the change of enthalpy (or of internal energy) during a reaction, with respect to temperature, at constant pressure (or volume) equals the change in heat capacity at constant pressure (or volume).

Kirchhoff's first law *See* Kirchhoff's current law.

Kirchhoff's law [ELEC] Either of the two fundamental laws dealing with the relation of currents at a junction and voltages around closed loops in an electric network; they comprise Kirchhoff's current law and Kirchhoff's voltage law. [THERMO] The law that the ratio of the emissivity of a heat radiator to the absorptivity of the same radiator is the same for all bodies, depending on frequency and temperature alone, and is equal to the emissivity of a blackbody. Also known as Kirchhoff's principle.

Kirchhoff's principle *See* Kirchhoff's law.

Kirchhoff's second law *See* Kirchhoff's voltage law.

Kirchhoff's voltage law [ELEC] The law that at each instant of time the algebraic sum of the voltage rises around a closed loop in a network is equal to the algebraic sum of the voltage drops, both being taken in the same direction around the loop. Also known as Kirchhoff's second law.

Kirchhoff theory [OPTICS] A theory of diffraction of light which gives a mathematical formulation of Huygens' principle, based on the wave equation and Green's theorem, and enables quantitative determination of the amplitude and phase at any point to a very close approximation.

Kirchhoff vapor pressure formula [THERMO] An approximate formula for the variation of vapor pressure p with temperature T, valid over a limited temperature range; it is $\ln p = A - B/T - C \ln T$, where A, B, and C are constants.

Kirkman triple system [MATH] A resolvable balanced incomplete block design with block size k equal to 3.

Kirkwood gaps [ASTRON] Regions in the main zone of asteroids where almost no asteroids are found.

Kistiakowsky-Fishtine equation [PHYS CHEM] An equation to calculate latent heats of vaporization of pure compounds; useful when vapor pressure and critical data are not available.

Kiwi nuclear reactor [NUCLEO] One of several test reactors for the nuclear engine for a rocket vehicle.

kJ *See* kilojoule.

kl *See* kiloliter.

Klein bottle [MATH] The nonorientable surface having only one side with no inside or outside; it resembles a bottle pulled into itself.

Klein-Gordon equation [QUANT MECH] A wave equation describing a spinless particle which is consistent with the special theory of relativity. Also known as Schrödinger-Klein-Gordon equation.

Kleinian group [MATH] A group of conformal mappings of a Riemann surface onto itself which is discontinuous at one or more points and is not discontinuous at more than two points.

Klein-Nishina formula [QUANT MECH] A formula, based on the Dirac electron theory without radiative correction, for the differential cross section for scattering of a photon by an unbound electron.

Klein paradox [QUANT MECH] The paradox whereby, according to the Dirac electron theory, an electron can penetrate into a potential barrier which is greater than twice the rest energy of the electron (about 1 MeV) by making a transition from a positive energy state to a negative energy state, provided the potential change occurs over a distance on the order of a Compton wavelength or less.

Klein-Rydberg method [PHYS CHEM] A method for determining the potential energy function of the distance between the nuclei of a diatomic molecule from the molecule's vibrational and rotational levels.

Klein's four-group [MATH] The noncyclic group of order four.

Klein's hypothesis [ASTRON] A theory of the overall structure of the universe that regards the visible universe as part of a large but finite astronomical system called a metagalaxy, which may itself belong to a much larger bounded system.

K/L ratio [NUC PHYS] The ratio of the number of internal conversion electrons emitted from the K shell of an atom during de-excitation of a nucleus to the number of such electrons emitted from the L shell.

klystron [ELECTR] An evacuated electron-beam tube in which an initial velocity modulation imparted to electrons in the beam results subsequently in density modulation of the beam; used as an amplifier in the microwave region or as an oscillator.

km *See* kilometer.

K meson [PARTIC PHYS] **1.** Collective name for four pseudoscalar mesons, having masses of about 495 MeV and decaying via weak interactions: K^+, K^-, K_S^0, and K_L^0; they consist of two isotopic spin doublets, the (K^+, K^0) doublet and its antiparticle doublet $(K^-, \bar{K}^0)$, having hypercharge or strangeness of $+1$ and -1 respectively, where K^0 and $\bar{K}^0$ are certain combinations of K_L^0 and K_S^0 states. Also known as kaon. **2.** Collective name for any meson resonance belonging to an isotopic doublet with hypercharge $+1$ or -1, denoted $K_{JP}(m)$ or $\bar{K}_{JP}(m)$ respectively, where m is the mass, and J and P are the spin and parity.

knee frequency *See* break frequency.

Knight shift [PHYS] The fractional increase in the frequency for nuclear magnetic resonance of a given nuclide in a metal relative to that for the same nuclide in a nonmetallic compound in the same external magnetic field, caused by orientation of the conduction electrons in the metal.

knocked-on atom [NUCLEO] An atom in a solid which recoils from a collision with an energetic particle moving through the solid.

knot [MATH] In the general case, a knot consists of an embedding of an n-dimensional sphere in an $(n+2)$-dimensional sphere; classically, it is an interlaced closed curve, homeomorphic to a circle. [PHYS] A speed unit of 1 nautical mile (1.852 kilometers) per hour, equal to approximately 0.51444 meter per second.

knot theory [MATH] The topological and algebraic study of knots emphasizing their classification and how one may be continuously deformed into another.

Knudsen cell [PHYS CHEM] A vessel used to measure very low vapor pressures by measuring the mass of vapor which escapes when the vessel contains a liquid in equilibrium with its vapor.

Knusden cosine law [PHYS] A law which states that the probability of a gas molecule leaving a solid surface in a given direction within a solid angle $d\omega$ is proportional to $\cos\theta\, d\omega$, where θ is the angle between the direction and the normal to the surface. [STAT MECH] The number of gas molecules striking or leaving a wall, per unit area per unit solid angle, is proportional to the cosine of the angle between the direction of the molecule's motion and the normal to the surface.

Knudsen flow *See* free molecule flow.

Knudsen gage [ENG] An instrument for measuring very low pressures, which measures the force of a gas on a cold plate beside which there is an electrically heated plate.

Knudsen number [FL MECH] The ratio of the mean free path length of the molecules of a fluid to a characteristic length; used to describe the flow of low-density gases.

Knudsen's equation [PHYS] An equation for the amount of gas that flows through a tube in free molecule flow, $q = \sqrt{2\pi}\Delta p d^3/6l\sqrt{\rho}$, where q is the volume of gas measured at unit pressure that flows through the tube per second, Δp is the difference between the pressures at the ends of the tube, d is the inside diameter of the tube, l is the length of the tube, and ρ is the density of the gas at unit pressure.

Knudsen vacuum gage [ENG] Device to measure negative gas pressures; a rotatable vane is moved by the pressure of heated molecules, proportionately to the concentration of molecules in the system.

Kobayashi potential [MATH] A solution of Laplace's equation in three dimensions constructed by superposition of the solutions obtained by separation of variables in cylindrical coordinates.

Kohler illumination [OPTICS] A method of illumination for the optical microscope used with coiled filaments or other sources of irregular form or brightness; an image of the filament large enough to fill the iris opening is focused on the condenser which is focused so that the image of the iris diaphragm on the lamp is in focus with the specimen, and the lamp iris is opened only enough to fill the field of view; the iris of the microscope is opened only enough to illuminate the back aperture of the objective; no ground glass is used.

Kohlrausch law [PHYS CHEM] **1.** Every ion contributes a definite amount to the equivalent conductance of an electrolyte in the limit of infinite dilution, regardless of the presence of other ions. **2.** The equivalent conductance of a very dilute solution of a strong electrolyte is a linear function of the concentration.

kohm *See* kilohm.

Kolmogorov consistency conditions [MATH] For each finite subset F of the real numbers or integers, let P_F denote a probability measure defined on the Borel subsets of the cartesian product of $k(F)$ copies of the real line indexed by elements in F, where $k(F)$ denotes the number of elements in F; the family $\{P_F\}$ of measures satisfy the Kolmogorov consistency conditions if given any two finite sets F_1 and F_2 with F_1 contained in F_2, the restriction of P_{F_2} to those sets which are independent of the coordinates in F_2 which are not in F_1 coincides with P_{F_1}.

Kolmogorov inequalities [MATH] For each integer K let X_k be a random variable with finite variance σ_k and suppose $\{X_k\}$ is an independent sequence which is uniformly bounded by some constant c; then for every $\varepsilon > 0$, and integer n,

$$1 - (\varepsilon + 2c)^2 \Big/ \sum_{k=1}^{n} \sigma_k^2 \leq \text{Prob} \{ \max_{k \leq n} |S_k + ES_k| \geq \varepsilon \}$$

$$\text{and } \frac{1}{\varepsilon^2} \sum_{k=1}^{n} \sigma_k^2 \geq \text{Prob} \{ \max_{k \leq n} |S_k + ES_k| \geq \varepsilon \};$$

here $S_k = \sum_{i=1}^{k} X_i$ and ES_k denotes the expected value of S_k.

Kolmogorov-Seliverstov-Plessner theorem [MATH] If the series whose nth term is $(a_n^2 + b_n^2) \log n$ converges, then the series whose nth term is $a_n \cos nx + b_n \sin nx$ converges for almost all values of x.

Kolmogorov-Sinai invariant [MATH] An isomorphism invariant of measure-preserving transformations; if T is a measure-preserving transformation on a probability space, the Kolmogorov-Sinai invariant is the least upper bound of the set of entropies of T given each finite partition of the probability space. Also known as entropy of a transformation.

Kolmogorov-Smirnov test [STAT] A procedure used to measure goodness of fit of sample data to a specified population; critical values exist to test goodness of fit.

Kolosov-Muskhelishvili formulas [MECH] Formulas which express plane strain and plane stress in terms of two holomorphic functions of the complex variable $z = x + iy$, where x and y are plane coordinates.

konig [OPTICS] The X tristimulus value.

Königsberg bridge problem [MATH] The problem of walking across seven bridges connecting four land masses in a specified manner exactly once and returning to the starting point; this is the original problem which gave rise to graph theory.

Konig's theorem [MATH] In combinatorial theory, the theorem that the minimum number m of lines is the same as the maximum number m of independent points.

Konowaloff rule [PHYS CHEM] An empirical rule which states that in the vapor over a liquid mixture there is a higher proportion of that component which, when added to the liquid, raises its vapor pressure, than of other components.

Kontorovich-Lebedev transform [MATH] For a function $f(x)$, this is the function $F(y)$ equal to the integral from 0 to ∞ of $x^{-1}f(x)K_{iy}(x)\,dx$, where $K_{iy}(x)$ is a modified Hankel function.

Kopp's law [PHYS CHEM] The law that for solids the molal heat capacity of a compound at room temperature and pressure approximately equals the sum of heat capacities of the elements in the compound.

Korteweg-de Vries equation [MATH] A nonlinear, partial differential equation, $\partial u/\partial t + u\ \partial u/\partial x + \partial^3 u/\partial x^3 = 0$, which describes shallow-water waves in a one-dimensional channel.

Kossel effect [PHYS] The production of a series of cones of reflected x-rays by characteristic x-rays generated by atoms in a single crystal.

Kossel lines [PHYS] Conic sections recorded on a flat film from the cones generated in the Kossel effect.

Kossel-Sommerfeld law [SPECT] The law that the arc spectra of the atom and ions belonging to an isoelectronic sequence resemble each other, especially in their multiplet structure.

Kozeny-Carmen equation [FL MECH] Equation for streamline flow of fluids through a powdered bed.

k-ply transitive group [MATH] For a positive integer k, a group of permutations of a set S of n elements, $n \geq k$, such that for any two ordered k-tuples $a_1, \ldots, a_k$ and $b_1, \ldots, b_k$ of elements in S (with $a_i \neq a_j$, $b_i \neq b_j$ for $i \neq j$), there exists an element of the group which takes a_i into b_i. Also known as k-fold transitive group.

Kr *See* krypton.

Kramers-Kronig relation [OPTICS] A relation between the real and imaginary parts of the index of refraction of a substance, based on the causality principle and Cauchy's theorem.

Kramer's theorem [SOLID STATE] The theorem that the states of a system consisting of an odd number of electrons in an external electrostatic field are at least twofold degenerate.

Krawtchouk polynomials [MATH] Families of polynomials which are orthogonal with respect to binomial distributions.

Krein-Milman theorem [MATH] The theorem that in a locally convex topological vector space, any compact convex set K is identical with the intersection of all convex sets containing the extreme points of K.

Krigar-Menzel law [MECH] A generalization of the second Young-Helmholtz law which states that when a string is

KRONIG-PENNEY MODEL

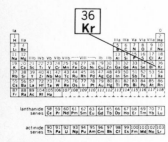

Potential energy $V(x)$ which is assumed for the one-dimensional Kronig-Penney model. Here x is directed distance from origin O. Square wells of depth V_O and width a are separated by distance b.

KRYPTON

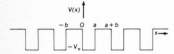

Periodic table of the chemical elements showing the position of krypton.

bowed at a point which is at a distance of p/q times the string's length from one of the ends, where p and q are relative primes, then the string moves back and forth with two constant velocities, one of which is $q-1$ times as large as the other.

Kronecker delta [MATH] The function or symbol δ_{ij} dependent upon the subscripts i and j which are usually integers; its value is 1 if $i = j$ and 0 if $i \neq j$.

Kronecker product [MATH] Given two different representations of the same group, their Kronecker product is a representation of the group constructed by taking direct products of matrices from the respective representations.

Kronig-Penney model [SOLID STATE] An idealized one-dimensional model of a crystal in which the potential energy of an electron is an infinite sequence of periodically spaced square wells.

Krull dimension [MATH] The Krull dimension of a commutative ring R is the largest number n for which there exists a sequence $P_0, P_1, \ldots, P_n$ of prime ideals in R such that P_i is a proper subset of P_{i-1}, $i = 1, 2, \ldots, n$.

Krull domain [MATH] An integral domain R with quotient field K for which there exists a family I of discrete valuations on K with rank one such that (1) R is equal to the intersection of the valuation rings of the valuations in I and (2) for any nonzero element x in R, there are at most a finite number of valuations v in I such that x is not a unit in the valuation ring of v.

Krull's theorem [MATH] If I is an ideal that is contained in every maximal ideal of a commutative Noetherian ring R, and E is a finitely generated module over R, then 0 is the only element of E that belongs to all the sets $I^n E$, where $I^n E$ denotes the set of all elements in E that can be written in the form $a_1 \ldots a_n x$, where $a_1, \ldots, a_n$ are in I and X is in E.

Kruskal coordinates [RELAT] Coordinate system used in general relativity to describe in a nonsingular manner the geometry of a nonrotating black hole in empty space.

Kruskal extension [RELAT] An extended coordinate system, used to describe a nonrotating black hole in empty space, that is valid from infinity, across the event horizon, and to the singularity at the black hole's center.

Krylov-Bogolinbov method [MATH] A method of obtaining an approximate solution to the ordinary differential equation $\partial^2 y / \partial x^2 + af(y, \partial y / \partial x) + b^2 y = 0$, where a and b are constants.

krypton [CHEM] A colorless, inert gaseous element, symbol Kr, atomic number 36, atomic weight 83.80; it is odorless and tasteless; used to fill luminescent electric tubes.

krypton-86 [NUC PHYS] An isotope of krypton, atomic mass 86; used in measurement of the standard meter.

krypton lamp [ELEC] An arc lamp filled with krypton; one type pierces fog for 1000 feet (300 meters) or more and is used to light airplane runways at night.

K shell [ATOM PHYS] The innermost shell of electrons surrounding the atomic nucleus, having electrons characterized by the principal quantum number 1.

kt *See* kiloton.

Kubelka-Munk model [OPTICS] A widely used theoretical model of reflectance; the model supposes that some light passing through a homogeneous sample is scattered and absorbed so that the light is attenuated in both directions.

Kummer relation [MATH] The equation

$$F(a,b; 1+a-b; -1)$$

$$= \Gamma(1+a-b)\Gamma(1+\tfrac{1}{2}a)/\Gamma(1+a)\Gamma(1+\tfrac{1}{2}a-b)$$

where $F(a,b;c;z)$ is the hypergeometric series and Γ denotes the gamma function.

Kummer's equation [MATH] The differential equation $x(d^2y/dx^2) + (c-y)(dy/dx) - ay = 0$, satisfied by confluent hypergeometric functions.

Kummer's function [MATH] The confluent hypergeometric function

$$_1F_1(\alpha;\rho;z) = \sum_{n=0}^{\infty} \frac{\alpha(\alpha+1)(\alpha+2)\cdots(\alpha+n-1)}{\rho(\rho+1)(\rho+2)\cdots(\rho+n-1)n!}.$$

Kummer's nonlinear differential equation [MATH] The equation $-(y''/2y')' + (y''/2y')^2 + Q(y)(y')^2 = q(t)$, where $Q(y)$ and $q(t)$ are specified continuous functions, and primes denote differentiation with respect to t.

Kummer's test [MATH] A general test for the convergence of a positive series Σa_n: the series converges if the limit inferior of $c_n \equiv b_n(a_n/a_{n+1}) - b_{n+1}$ is positive for some positive sequence b_n for which the series $\Sigma(1/b_n)$ diverges; the series diverges if the limit superior of c_n is negative for some positive sequence b_n for which the series $\Sigma(1/b_n)$ diverges.

Kundt effect *See* Faraday effect.

Kundt rule [SPECT] The rule that the optical absorption bands of a solution are displaced toward the red when its refractive index increases because of changes in composition or other causes.

Kundt's constant [OPTICS] A measure of the strength of the Faraday effect in a material, equal to the ratio of the Faraday rotation to the product of the path length and the magnetization of the material; it depends only on the temperature for any magnetic material.

Kundt tube [ACOUS] A tube used to measure the speed of sound; it is filled with air or other gas and contains a light powder which becomes lumped at nodes, giving the length of standing waves generated in the tube.

Kuratowski graphs [MATH] Two graphs which appear in Kuratowski's theorem, the complete graph K_5 with five vertices and the bipartite graph $K_{3,3}$.

Kuratowski's lemma [MATH] Each linearly ordered subset of a partially ordered set is contained in a maximal linearly ordered subset.

Kuratowski's theorem [MATH] A graph is nonplanar if and only if it has a subgraph which is either a Kuratowski graph or a subdivision of a Kuratowski graph.

Kurie plot [NUC PHYS] Graph used in studying beta decay, in which the square root of the number of beta particles whose momenta (or energy) lie within a certain narrow range, divided by a function worked out by Fermi, is plotted against beta-particle energy; it is a straight line for allowed transitions and some forbidden transitions, in accord with the Fermi beta-decay theory. Also known as Fermi plot.

kurtosis [STAT] The extent to which a frequency distribution is concentrated about the mean or peaked; it is sometimes defined as the ratio of the fourth moment of the distribution to the square of the second moment.

Kutta-Joukowski condition [FL MECH] A boundary condition or fluid flow about an airfoil which requires that the circulation of the flow be such that a streamline leaves the trailing edge of the airfoil smoothly, or, equivalently, that the fluid velocity at the trailing edge be finite.

Kutta-Joukowski equation [FL MECH] An equation which states that the lift force exerted on a body by an ideal fluid, per unit length of body perpendicular to the flow, is equal to the product of the mass density of the fluid, the linear velocity of the fluid relative to the body, and the fluid circulation. Also known as Kutta-Joukowski theorem.

Kutta-Joukowski theorem   *See* Kutta-Joukowski equation.

kV   *See* kilovolt.

kVA   *See* kilovolt-ampere.

kvar   *See* kilovar.

kVp   *See* kilovolts peak.

kW   *See* kilowatt.

kWh   *See* kilowatt-hour.

kWhr   *See* kilowatt-hour.

L

l *See* liter.

L *See* lambert.

La *See* lanthanum.

label *See* isotopic tracer.

laboratory coordinate system [MECH] A reference frame attached to the laboratory of the observer, in contrast to the center-of-mass system.

ladar [OPTICS] A missile-tracking system that uses a visible light beam in place of a microwave radar beam to obtain measurements of speed, altitude, direction, and range of missiles. Derived from laser detecting and ranging. Also known as colidar; laser radar.

ladder network [ELECTR] A network composed of a sequence of H, L, T, or pi networks connected in tandem; chiefly used as an electric filter. Also known as series-shunt network.

Ladenburg f value *See* oscillator strength.

lag [ELECTR] A persistence of the electric charge image in a camera tube for a small number of frames. [PHYS] **1.** The difference in time between two events or values considered together. **2.** *See* lag angle.

lag angle [PHYS] The negative of phase difference between a sinusoidally varying quantity and a reference quantity which varies sinusoidally at the same frequency, when this phase difference is negative. Also known as angle of lag; lag.

lag coefficient *See* time constant.

lag correlation [STAT] The strength of the relationship between two elements in an ordered series, usually a time series, where one element lags a specific number of places behind the other elements.

lagging load *See* inductive load.

lagging network *See* integral network.

lag-lead network *See* lead-lag network.

lag network *See* integral network.

Lagrange bracket [MECH] Given two functions of coordinates and momenta in a system, their Lagrange bracket is an expression measuring how coordinates and momenta change jointly with respect to the two functions.

Lagrange function *See* Lagrangian.

Lagrange-Hamilton theory [MECH] The formalized study of continuous systems in terms of field variables where a Lagrangian density function and Hamiltonian density function are introduced to produce equations of motion.

Lagrange-Helmholtz equation *See* Helmholtz equation.

Lagrange method [MATH] A method for solving a linear partial differential equation with two independent variables x and y, having the form $P(\partial z/\partial x) + Q(\partial z/\partial y) = R$, where P, Q, and R are functions of x, y, and z.

Lagrange's equations [MECH] Equations of motion of a mechanical system for which a classical (non-quantum-mechanical) description is suitable, and which relate the kinetic energy of the system to the generalized coordinates, the generalized forces, and the time. Also known as Lagrangian equations of motion.

Lagrange's form of remainder [MATH] An expression for the difference R_n between the value of a function $f(x)$ and the sum of the first $n + 1$ terms of its Taylor series about a point a; it may be written $R_n = f^{(n+1)}(X)(x-a)^{n+1}/(n+1)!$, where X is some number between a and x.

Lagrange's formula *See* mean value theorem.

Lagrange's interpolation formula [MATH] A formula by use of which a polynomial $f(x)$ of degree less than or equal to n can be completely determined provided that one knows $n + 1$ distinct values of it; if $x_0, x_1, x_2, \ldots, x_n$ are the values of x for which the values of $f(x)$ are known, the formula is

$$f(x) = \frac{(x-x_1)(x-x_2) \cdots (x-x_n)}{(x_0-x_1)(x_0-x_2) \cdots (x_0-x_n)} f(x_0)$$

$$+ \frac{(x-x_0)(x-x_2) \cdots (x-x_n)}{(x_1-x_0)(x_1-x_2) \cdots (x_1-x_n)} f(x_1) + \cdots$$

$$+ \frac{(x-x_0)(x-x_1) \cdots (x-x_{n-1})}{(x_n-x_0)(x_n-x_1) \cdots (x_n-x_{n-1})} f(x_n).$$

Lagrange's theorem [MATH] In a group of finite order, the order of any subgroup must divide the order of the entire group.

Lagrange stream function [FL MECH] A scalar function of position used to describe steady, incompressible two-dimensional flow; constant values of this function give the streamlines, and the rate of flow between a pair of streamlines is equal to the difference between the values of this function on the streamlines. Also known as current function; stream function.

Lagrangian [MECH] **1.** The difference between the kinetic energy and the potential energy of a system of particles, expressed as a function of generalized coordinates and velocities from which Lagrange's equations can be derived. Also known as kinetic potential; Lagrange function. **2.** For a dynamical system of fields, a function which plays the same role as the Lagrangian of a system of particles; its integral over a time interval is a maximum or a minimum with respect to infinitesimal variations of the fields, provided the initial and final fields are held fixed.

Lagrangian coordinates *See* generalized coordinates.

Lagrangian density [MECH] For a dynamical system of fields or continuous media, a function of the fields, of their time and space derivatives, and the coordinates and time, whose integral over space is the Lagrangian.

Lagrangian equations of motion *See* Lagrange's equations.

Lagrangian function [MECH] The function which measures the difference between the kinetic and potential energy of a dynamical system.

Lagrangian generalized velocity *See* generalized velocity.

Lagrangian method [FL MECH] A method of studying fluid motion and the mechanics of deformable bodies in which one considers volume elements which are carried along with the fluid or body, and across whose boundaries material does not flow; in contrast to Euler method.

Lagrangian multipliers [MATH] A technique whereby potential extrema of functions of several variables are obtained. Also known as undetermined multipliers.

Lagrangian points [ASTRON] In planetary orbits, two positions in which the motion of a body of negligible mass (such as an asteroid) is stable under the gravitational influence of two other bodies (such as the sun and Jupiter), one of which is moving about the other in an approximately circular orbit; the two positions are located on the orbit, 60° ahead of or behind the orbiting body, so that the three bodies form an equilateral triangle.

Laguerre polynomials [MATH] A sequence of orthogonal polynomials which are solutions of Laguerre's differential equation with the parameter α equal to a positive integer.

Laguerre's differential equation [MATH] The equation $xy'' + (1 - x)y' + \alpha y = 0$, where α is a constant.

Laguerre transform [MATH] An integral transform which takes a function f into the function $F(n)$ defined as the integral from 0 to infinity of $e^{-t}L_n(x)f(x)\,dx$, where L_n is the Laguerre function of order n.

Lalande cell [ELEC] A type of wet cell that uses a zinc anode and cupric oxide cathode cast as flat plates or hollow cylinders, and an electrolyte of sodium hydroxide in aqueous solution (caustic soda).

Lalescu-Picard equation [MATH] The integral equation

$$f(x) = g(x) + \lambda \int_{-\infty}^{\infty} e^{-|x-y|}f(y)\,dy$$

where $g(x)$ is a known function and $f(x)$ is to be determined.

lambda [MECH] A unit of volume equal to 10^{-6} liter or 10^{-9} cubic meter.

lambda hyperon [PARTIC PHYS] **1.** A quasi-stable baryon, forming an isotopic singlet, having zero charge and hypercharge, a spin of ½, positive parity and mass of 1115.5 MeV. Designated Λ. Also known as lambda particle. **2.** Any baryon resonance having zero hypercharge and total isotopic spin; designated $\Lambda_J P(m)$, where m is the mass of the baryon in MeV, and J and P are its spin and parity (if known).

lambda leak [CRYO] A leak of liquid helium II through small holes where normal liquids cannot pass. Also known as superleak.

lambda particle *See* lambda hyperon.

lambda point [CRYO] The temperature (2.1780 K), at atmospheric pressure, at which the transformation between the liquids helium I and helium II takes place; a special case of the thermodynamics definition. [THERMO] A temperature at which the specific heat of a substance has a sharply peaked maximum, observed in many second-order transitions.

lambert [OPTICS] A unit of luminance (photometric brightness) that is equal to $1/\pi$ candela per square centimeter, or to the uniform luminance of a perfectly diffusing surface emitting or reflecting light at the rate of 1 lumen per square centimeter. Abbreviated L.

Lambert-Beer law *See* Bouguer-Lambert-Beer law.

Lambert's law [OPTICS] **1.** The law that the illumination of a surface by a light ray varies as the cosine of the angle of incidence between the normal to the surface and the incident ray. **2.** The law that the luminous intensity in a given direction radiated or reflected by a perfectly diffusing plane surface varies as the cosine of the angle between that direction and the normal to the surface. **3.** *See* Bouguer-Lambert law.

Lambert surface [THERMO] An ideal, perfectly diffusing surface for which the intensity of reflected radiation is independent of direction.

Lamb shift [ATOM PHYS] A small shift in the energy levels of a hydrogen atom, and of hydrogenlike ions, from those predicted by the Dirac electron theory, in accord with principles of quantum electrodynamics.

Lamb wave [ACOUS] *See* plate wave. [ELECTROMAG] Electromagnetic wave propagated over the surface of a solid whose thickness is comparable to the wavelength of the wave.

LAMBDA POINT

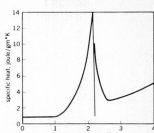

Specific heat of liquid helium computed as a function of temperature. Helium I is the high-temperature liquid; helium II occurs below the lambda point.

Lamé constants [MECH] Two constants which relate stress to strain in an isotropic, elastic material.

Lamé functions [MATH] Functions that arise when Laplace's equation is separated in ellipsoidal coordinates.

lamellar crystal [CRYSTAL] A polycrystalline substance whose grains are in the form of thin sheets.

lamellar vector field *See* irrotational vector field.

Lamé polynomials [MATH] Polynomials which result when certain parameters of Lamé functions assume integral values, and which are used to express physical solutions of Laplace's equation in ellipsoidal coordinates.

Lamé's equations [MATH] A general collection of second-order differential equations which have five regular singularities.

Lamé's relations [MATH] Six independent relations which when satisfied by the covariant metric tensor of a three-dimensional space provide necessary and sufficient conditions for the space to be euclidean.

Lamé wave functions [MATH] Functions which arise when the wave equation is separated in ellipsoidal coordinates. Also known as ellipsoidal wave functions.

laminar boundary layer [FL MECH] A thin layer over the surface of a body immersed in a fluid, in which the fluid velocity relative to the surface increases rapidly with distance from the surface and the flow is laminar.

laminar flow [FL MECH] Streamline flow of an incompressible, viscous Newtonian fluid; all particles of the fluid move in distinct and separate lines.

laminar sublayer [FL MECH] The laminar boundary layer underlying a turbulent boundary layer.

laminated core [ELECTROMAG] An iron core for a coil transformer, armature, or other electromagnetic device, built up from laminations stamped from sheet iron or steel and more or less insulated from each other by surface oxides and sometimes also by application of varnish.

laminography *See* sectional radiography.

Lami's theorem [MECH] When three forces act on a particle in equilibrium, the magnitude of each is proportional to the sine of the angle between the other two.

Lanchester's rule [MECH] A torque applied to a rotating body along an axis perpendicular to the rotation axis will produce precession in a direction such that, if the body is viewed along a line of sight coincident with the torque axis, then a point on the body's circumference, which initially crosses the line of sight, will appear to describe an ellipse whose sense is that of the torque.

Lanczos's method [MATH] A transformation method for diagonalizing a matrix in which the matrix used to transform the original matrix to triple-diagonal form is formed from a set of column vectors that are determined by a recursive process.

Landau fluctuations [NUCLEO] Variations in the losses of energy of different particles in a thin detector, resulting from random variations in the number of collisions and in the energies lost in each collision of the particle.

Landau-Ginzburg theory *See* Ginzburg-Landau theory.

Landau levels [SOLID STATE] Energy levels of conduction electrons which occur in a metal subjected to a magnetic field at very low temperatures and which are quantized because of the quantization of the electron motion perpendicular to the field.

Landé g factor [ATOM PHYS] Also known as g factor. **1.** The negative ratio of the magnetic moment of an electron or atom, in units of the Bohr magneton, to its angular momentum, in units of Planck's constant divided by 2π. **2.** The ratio of the

LAMINAR FLOW

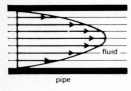

pipe

Laminar flow in a circular pipe. In this case the velocity adjacent to the wall is zero and increases to a maximum in the center of the pipe.

difference in energy between two energy levels which differ only in magnetic quantum number to the product of the Bohr magneton, the applied magnetic field, and the difference between the magnetic quantum numbers of the levels; identical to the first definition for free atoms. Also known as Landé splitting factor; spectroscopic splitting factor. [NUC PHYS] The ratio of the magnetic moment of a nucleon, in units of the nuclear magneton, to its angular momentum in units of Planck's constant divided by 2π.

Landé interval rule [ATOM PHYS] The rule that when the spin-orbit interaction is weak enough to be treated as a perturbation, an energy level having definite spin angular momentum and orbital angular momentum is split into levels of differing total angular momentum, so that the interval between successive levels is proportional to the larger of their total angular momentum values.

Landen transformation [MATH] **1.** The formula $F(\phi_1,k_1) = (1+k')F(\phi,k)$, where F is a Legendre normal elliptic integral, $k' = (1-k^2)^{1/2}$, $k_1 = (1-k')/(1+k')$, and $\sin \phi_1 = (1+k') \sin \phi \cos \phi (1-k^2 \sin^2 \phi)^{-1/2}$. **2.** Any of several related formulas among Legendre normal elliptic intergrals and Jacobi elliptic functions.

Landé Γ-permanence rule [ATOM PHYS] The rule that the sum of the shifts of energy levels produced by the spin-orbit interaction, over a series of states having the same spin and orbital angular momentum quantum numbers (or the same total angular momentum quantum numbers for individual electrons) but different total angular momenta, and having the same total magnetic quantum number, is independent of the strength of an applied magnetic field.

Landé splitting factor *See* Landé g factor.

Landholt fringe [OPTICS] A black fringe that crosses the darkened field which is produced when a brilliant source of light is viewed through two Nicol prisms oriented with their principal axes at right angles to one another.

land mile *See* mile.

Lane's law [ASTROPHYS] For the contraction of a star that is assumed to be a sphere of perfect gas, the law that the temperature of the perfect-gas sphere is inversely proportional to its radius.

Langevin-Debye formula [STAT MECH] A formula for the polarizability of a dielectric material or the paramagnetic susceptibility of a magnetic material, in which these quantities are the sum of a temperature-independent contribution and a contribution arising from the partial orientation of permanent electric or magnetic dipole moments which varies inversely with the temperature. Also known as Langevin-Debye law.

Langevin-Debye law *See* Langevin-Debye formula.

Langevin function [ELECTROMAG] A mathematical function, $L(x)$, which occurs in the expressions for the paramagnetic susceptibility of a classical (non-quantum-mechanical) collection of magnetic dipoles, and for the polarizability of molecules having a permanent electric dipole moment; given by $L(x) = \coth x - 1/x$.

Langevin ion-mobility theories [ELECTR] Two theories developed to calculate the mobility of ions in gases; the first assumes that atoms and ions interact through a hard-sphere collision and have a constant mean free path, while the second assumes that there is an attraction between atoms and ions arising from the polarization of the atom in the ion's field, in addition to hard-sphere repulsion for close distances of approach.

Langevin ion-recombination theory [ELECTR] A theory predicting the rate of recombination of negative with positive

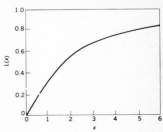

LANGEVIN FUNCTION

Plot of Langevin function.

ions in an ionized gas on the assumption that ions of opposite sign approach one another under the influence of mutual attraction, and that their relative velocities are determined by ion mobilities; applicable at high pressures, above 1 or 2 atmospheres (101,325 or 202,650 newtons per square meter).

Langevin radiation pressure [ACOUS] A measure of acoustic radiation pressure, equal to the difference between the mean pressure on an absorbing or reflecting wall and that in the same acoustic medium, at rest, behind the wall.

Langevin theory of diamagnetism [ELECTROMAG] A theory based on the idea that diamagnetism results from electronic currents caused by Larmor precession of electrons inside atoms.

Langevin theory of paramagnetism [ELECTROMAG] A theory which treats a substance as a classical (non-quantum-mechanical) collection of permanent magnetic dipoles with no interactions between them, having a Boltzmann distribution with respect to energy of interaction with an applied field.

langley [PHYS] A unit of energy per unit area commonly employed in radiation theory; equal to 1 gram-calorie per square centimeter.

Langmuir-Child equation *See* Child's law.

Langmuir dark space [ELECTR] A nonluminous region surrounding a negatively charged probe inserted in the positive column of a glow discharge.

Langmuir diffusion pump [ENG] A type of diffusion pump in which the mercury vapor emerges from a nozzle, giving it motion in a direction away from the high-vacuum side of the pump.

Langmuir effect [SOLID STATE] The ionization of atoms of low ionization potential that come into contact with a hot metal with a high work function.

Langmuir isotherm equation [PHYS CHEM] An equation, useful chiefly for gaseous systems, for the amount of material adsorbed on a surface as a function of pressure, while the temperature is held constant, assuming that a single layer of molecules is adsorbed; it is $f = ap/(1 + ap)$, where f is the fraction of surface covered, p is the pressure, and a is a constant.

Langmuir plasma frequency [PL PHYS] The frequency of nonpropagating oscillations in a plasma; in rationalized mks units, it is $(ne^2/\varepsilon_0 m)^{1/2}$, where e and m are the charge and mass of the oscillating electrons or ions, n is their number density, and ε_0 is the permittivity of empty space. Also known as plasma frequency.

Langmuir probe [PL PHYS] A device for measuring the temperature and electron density of a plasma, consisting of an electrode in contact with the plasma whose potential is varied while the resulting collection currents are measured.

language theory [MATH] A branch of automata theory which attempts to formulate the grammar of a language in mathematical terms; it has been applied to automatic language translation and to the construction of higher-level programming languages and systems such as the propositional calculus, nerve networks, sequential machines, and programming schemes.

lanthanide contraction [ATOM PHYS] A phenomenon encountered in the rare-earth elements; the radii of the atoms of the members of the series decrease slightly as the atomic numbers increase; starting with element 58 in the periodic table, the balancing electron fills in an inner incomplete 4*f* shell as the charge on the nucleus increases.

lanthanide series [CHEM] Rare-earth elements of atomic numbers 57 through 71; their chemical properties are similar to those of lanthanum, atomic number 57.

lanthanum [CHEM] A chemical element, symbol La, atomic number 57, atomic weight 138.91; it is the second most abundant element in the rare-earth group.

lanthanum-doped lead zirconate–lead titanate *See* lead lanthanum zirconate titanate.

Laplace-Carson transform [MATH] The Laplace-Carson transform of a function $f(x)$ is the function $G(y) = yF(y)$, where $F(y)$ is the Laplace transform of $f(x)$. Also known as Carson transform.

Laplace irrotational motion [FL MECH] Irrotational flow of an inviscid, incompressible fluid.

Laplace law *See* Ampère law.

Laplace operator [MATH] The linear operator defined on differentiable functions which gives for each function the sum of all its nonmixed second partial derivatives. Also known as Laplacian.

Laplace's equation [ACOUS] An equation for the speed c of sound in a gas; it may be written $c = \sqrt{\gamma p/\rho}$, where p is the pressure, ρ is the density, and γ is the ratio of specific heats. [MATH] The partial differential equation which states that the sum of all the nonmixed second partial derivatives equals 0; the potential functions of many physical systems satisfy this equation.

Laplace's expansion [MATH] An expansion by means of which the determinant of a matrix may be computed in terms of the determinants of all possible smaller square matrices contained in the original.

Laplace's measure of dispersion [STAT] The expected value of the absolute value of the difference between a random variable and its mean.

Laplace-Stieltjes transform [MATH] An integral transform which takes a function of bounded variation g into the function $G(s)$ defined as the Stieltjes integral from 0 to infinity of $e^{-sx} dg(x)$, where s is a complex number.

Laplace transform [MATH] For a function $f(x)$ its Laplace transform is the function $F(y)$ defined as the integral over x from 0 to ∞ of the function $e^{-yx}f(x)$.

Laplacian *See* Laplace operator.

Laplacian speed of sound [FL MECH] The phase speed of a sound wave in a compressible fluid under the assumption that the expansions and compressions are adiabatic.

Laporte selection rule [ATOM PHYS] The rule that an electric dipole transition can occur only between states of opposite parity.

large calorie *See* kilocalorie.

large dyne *See* newton.

Large Magellanic Cloud [ASTRON] An irregular cloud of stars in the constellation Doradus; it is 150,000 light-years (1.42×10^{21} meters) away and nearly 30,000 light-years (2.84×10^{20} meters) in diameter. Abbreviated LMC.

large-scale integrated circuit [ELECTR] A very complex integrated circuit, which contains well over 100 interconnected individual devices, such as basic logic gates and transistors, placed on a single semiconductor chip. Abbreviated LSI circuit. Also known as chip circuit; multiple-function chip.

Larmor formula [ELECTROMAG] The rate at which energy is radiated by a nonrelativistic, accelerated charge is $2q^2a^2/3c^3$, where q is the particle's charge in esu (electrostatic units), a is its acceleration, and c is the speed of light.

Larmor frequency [ELECTROMAG] The angular frequency of the Larmor precession, equal in esu (electrostatic units) to the negative of a particle's charge times the magnetic induction divided by the product of twice the particle's mass and the speed of light.

Larmor orbit [ELECTROMAG] The motion of a charged parti-

LANTHANUM

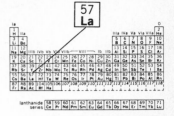

Periodic table of the chemical elements showing the position of lanthanum.

LARGE-SCALE INTEGRATED CIRCUIT

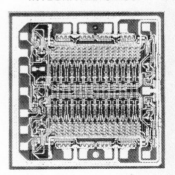

Monolithic silicon 128-bit read-only memory unit. *(Courtesy of Motorola Semiconductor Products Inc.)*

cle in a uniform magnetic field, which is a superposition of uniform circular motion in a plane perpendicular to the field, and uniform motion parallel to the field.

Larmor precession [ELECTROMAG] A common rotation superposed upon the motion of a system of charged particles, all having the same ration of charge to mass, by a magnetic field.

Larmor radius [ELECTROMAG] For a charged particle moving transversely in a uniform magnetic field, the radius of curvature of the projection of its path on a plane perpendicular to the field. Also known as gyromagnetic radius.

Larmor's theorem [ELECTROMAG] The theorem that for a system of charged particles, all having the same ratio of charge to mass, moving in a central field of force, the motion in a uniform magnetic induction B is, to first order in B, the same as a possible motion in the absence of B except for the superposition of a common precession of angular frequency equal to the Larmor frequency.

Larson-Miller parameter [MECH] The effects of time and temperature on creep which is defined empirically as $P = T(C + \log t) \times 10^{-3}$, where T = test temperature in degrees Rankine (degrees Fahrenheit + 460) and t = test time in hours; the constant C depends upon the material but is frequently taken to be 20.

laser [OPTICS] An active electron device that converts input power into a very narrow, intense beam of coherent visible or infrared light; the input power excites the atoms of an optical resonator to a higher energy level, and the resonator forces the excited atoms to radiate in phase. Derived from light amplification by stimulated emission of radiation.

laser camera [OPTICS] An airborne camera system for night photography in which a laser beam is split into two beams; one beam, which is almost invisible, scans the ground, while the second beam is modulated by a detector of light reflected from the ground area being scanned, and is in turn swept back and forth over a moving film by the same scanner.

laser detecting and ranging *See* ladar.

laser diode *See* semiconductor laser.

laser drill [OPTICS] A drill in which concentrated light from a ruby laser generates intense heat for drilling holes as small as 0.0001 inch (0.00025 centimeter) in diameter in tungsten, gemstones, and other hard materials.

laser extensometer [OPTICS] A device which uses interference of laser beams to measure small changes in distance; it can operate between points as much as 1 kilometer apart, and has been used to measure effects produced by earth tides.

laser flash tube [ELECTR] A high-power, air-cooled or water-cooled xenon flash tube designed to produce high-intensity flashes for pumping applications.

laser fusion [NUCLEO] The use of an intense beam of laser light to heat a small pellet of deuterium and tritium to a temperature of about 100,000,000°C, as required for initiating a fusion reaction. Also known as laser-induced fusion.

laser gyro [ENG] A gyro in which two laser beams travel in opposite directions over a ring-shaped path formed by three or more mirrors; rotation is thus measured without the use of a spinning mass. Also known as ring laser.

laser-induced fusion *See* laser fusion.

laser infrared radar *See* lidar.

laser interferometer [OPTICS] An interferometer which uses a laser as a light source; because of the monochromaticity and high intrinsic brilliance of laser light, it can operate with path differences in the interfering beams of hundreds of meters, in contrast to a maximum of about 20 centimeters for classical interferometers.

laser radar *See* ladar.

LASER

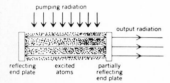

pumping radiation

output radiation

reflecting end plate · excited atoms · partially reflecting end plate

Structure of a parallel-plate laser.

laser rangefinder [OPTICS] A portable rangefinder using a battery-powered ruby laser in combination with an optical telescope to aim a laser beam and a photomultiplier for picking up the laser beam reflected from the target.

laser spectroscopy [SPECT] A branch of spectroscopy in which a laser is used as an intense, monochromatic light source; in particular, it includes saturation spectroscopy, as well as the application of laser sources to Raman spectroscopy and other techniques.

laser spectrum [PHYS] The spectrum that includes all optical wavelengths, ranging from infrared through visible light to ultraviolet, in which coherent radiation can be produced by various types of lasers.

laser threshold [ELECTR] The minimum pumping energy required to initiate lasing action in a laser.

laser velocimeter *See* diffraction velocimeter.

lasing [OPTICS] Generation of visible or infrared light waves having very nearly a single frequency by pumping or exciting electrons into high-energy states in a laser.

Laspeyre's index [STAT] A weighted aggregate price index with base-year quantity weights. Also known as base-year method.

last quarter [ASTRON] The phase of the moon at western quadrature, half of the illuminated hemisphere being visible from the earth; has the characteristic half-moon shape.

latent heat [THERMO] The amount of heat absorbed or evolved by 1 mole, or a unit mass, of a substance during a change of state (such as fusion, sublimation or vaporization) at constant temperature and pressure.

latent heat of fusion *See* heat of fusion.

latent heat of sublimation *See* heat of sublimation.

latent heat of vaporization *See* heat of vaporization.

latent root *See* eigenvalue.

lateral aberration [OPTICS] **1.** The distance from the axis of an optical system at which a ray intersects a plane perpendicular to the axis through the focus of paraxial rays. **2.** The difference between the reciprocals of the image distances for paraxial and rim rays. **3.** For chromatic aberration, the difference in sizes of the images of an object for two different colors.

lateral area [MATH] The area of a surface with the bases (if any) excluded.

lateral chromatic aberration *See* chromatic difference of magnification.

lateral face [MATH] The lateral face for a prism or pyramid is any edge or face which is not part of a base.

lateral magnification [OPTICS] The ratio of some linear dimension, perpendicular to the optical axis, of an image formed by an optical system, to the corresponding linear dimension of the object. Also known as magnification.

lateral mirage [OPTICS] A very rare type of mirage in which the apparent position of an object appears displaced to one side of its true position.

lateral quadrupole [ACOUS] A sound source resulting from a variation of a component of the velocity of matter in a direction that is perpendicular to the velocity component. [ELECTROMAG] An electric or magnetic quadrupole which produces a field equivalent to that of two equal and opposite electric or magnetic dipoles separated by a small distance perpendicular to the direction of the dipoles.

Latin square [MATH] An $n \times n$ square array of n different symbols, each symbol appearing once in each row and once in each column; these symbols prove useful in ordering the observations of an experiment.

lattice [CRYSTAL] A regular periodic arrangement of points in three-dimensional space; it consists of all those points P for which the vector from a given fixed point to P has the form $n_1 \mathbf{a} + n_2 \mathbf{b} + n_3 \mathbf{c}$, where $n_1, n_2,$ and n_3 are integers, and $\mathbf{a}, \mathbf{b},$ and $\mathbf{c}$ are fixed, linearly independent vectors. Also known as periodic lattice; space lattice. [MATH] A partially ordered set in which each pair of elements has both a greatest lower bound and least upper bound. [NUCLEO] An orderly array or pattern of nuclear fuel elements and moderator in a reactor or critical assembly.

lattice constant [CRYSTAL] A parameter defining the unit cell of a crystal lattice, that is, the length of one of the edges of the cell or an angle between edges. Also known as lattice parameter.

lattice defect *See* crystal defect.

lattice dynamics [SOLID STATE] The study of the thermal vibrations of a crystal lattice. Also known as crystal dynamics.

lattice energy [SOLID STATE] The energy required to separate ions in an ionic crystal an infinite distance from each other.

lattice filter [ELECTR] An electric filter consisting of a lattice network whose branches have LC parallel-resonant circuits shunted by quartz crystals.

lattice network [ELEC] A network that is composed of four branches connected in series to form a mesh; two nonadjacent junction points serve as input terminals, and the remaining two junction points serve as output terminals.

lattice parameter *See* lattice constant.

lattice polarization [SOLID STATE] Electric polarization of a solid due to displacement of ions from equilibrium positions in the lattice.

lattice reactor [NUCLEO] A heterogeneous nuclear reactor in which both fuel and moderator are in the form of long rods.

lattice scattering [SOLID STATE] Scattering of electrons by collisions with vibrating atoms in a crystal lattice, reducing the mobility of charge carriers in the crystal and thereby affecting its conductivity.

lattice vibration [SOLID STATE] A periodic oscillation of the atoms in a crystal lattice about their equilibrium positions.

lattice wave [SOLID STATE] A disturbance propagated through a crystal lattice in which atoms oscillate about their equilibrium positions.

lattice-wound coil *See* honeycomb coil.

latus rectum [MATH] The length of a chord through the focus and perpendicular to the axis of symmetry in a conic section.

Laue camera [CRYSTAL] The apparatus used in the Laue method; the x-ray beam usually enters through a hole in the x-ray film, which records beams bent through an angle of nearly 180° by the crystal; less commonly, the film is placed beyond the crystal.

Laue condition [CRYSTAL] **1.** The condition for a vector to lie in a Laue plane: its scalar product with a specified vector in the reciprocal lattice must be one-half of the scalar product of the latter vector with itself. **2.** *See* Laue equations.

Laue equations [CRYSTAL] Three equations which must be satisfied for an x-ray beam of specified wavelength to be diffracted through a specified angle by a crystal; they state that the scaler products of each of the crystallographic axial vectors with the difference between unit vectors in the directions of the incident and scattered beams, are integral multiples of the wavelength. Also known as Laue condition.

Laue method [CRYSTAL] A method of studying crystalline structures by x-ray diffraction, in which a finely collimated beam of polychromatic x-rays falls on a single crystal whose orientation can be set as desired, and diffracted beams are recorded on a photographic film.

LAUE CAMERA

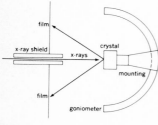

Schematic diagram of back-reflection Laue camera. Orientation of single crystal can be set as desired by system of goniometer circles. Diffracted beams, bent through angles of nearly 180°, are recorded on film.

Laue pattern [CRYSTAL] The characteristic photographic record obtained in the Laue method.

Laue plane [CRYSTAL] A plane which is the perpendicular bisector of a vector in the reciprocal lattice; such planes form the boundaries of Brillouin zones.

laurence [OPTICS] A shimmering seen over a hot surface on a calm, cloudless day, caused by the unequal refraction of light by innumerable convective air columns of different temperatures and densities.

Laurent expansion [MATH] An infinite series in which an analytic function $f(z)$ defined on an annulus about the point z_0 may be expanded, with nth term $a_n(z - z_0)^n$, n ranging from $-\infty$ to ∞, and $a_n = 1/2\pi i$ times the integral of $f(t)/(t - z_0)^{n+1}$ along a simple closed curve interior to the annulus. Also known as Laurent series.

Laurent half-shade plate [OPTICS] A device used to determine the direction of polarization of plane polarized light; it consists of a quartz plate of special thickness that covers half of the plane polarized beam, followed by a plane polarization analyzer.

Laurent series *See* Laurent expansion.

Lauritsen electroscope [ELEC] A rugged and sensitive electroscope in which a metallized quartz fiber is the sensitive element.

Lavrent'ev's equation [MATH] The partial differential equation $(\partial^2 z/\partial x^2) + \text{sgn } y \, (\partial^2 z/\partial y^2) = 0$.

law of action and reaction *See* Newton's third law.

law of constant angles [CRYSTAL] The law that the angles between the faces of a crystal remain constant as the crystal grows.

law of constant heat summation *See* Hess's law.

law of corresponding states [CHEM] The law that when, for two substances, any two ratios of pressure, temperature, or volume to their respective critical properties are equal, the third ratio must equal the other two.

law of corresponding times [MECH] The principle that the times for corresponding motions of dynamically similar systems are proportional to L/V and also to $\sqrt{L/G}$, where L is a typical dimension of the system, V a typical velocity, and G a typical force per unit mass.

law of cosines [MATH] Given a triangle with angles A, B, and C and sides a, b, c opposite these angles respectively: $a^2 = b^2 + c^2 - 2bc \cos A$.

law of definite composition *See* law of definite proportion.

law of definite proportion [CHEM] The law that a given chemical compound always contains the same elements in the same fixed proportion by weight. Also known as law of definite composition.

law of electromagnetic induction *See* Faraday's law of electromagnetic induction.

law of electrostatic attraction *See* Coulomb's law.

law of exponents [MATH] Any one of the laws $a^m a^n = a^{m+n}$, $a^m/a^n = a^{m-n}$, $(a^m)^n = a^{mn}$, $(ab)^n = a^n b^n$, $(a/b)^n = a^n/b^n$; these laws are valid when m and n are any integers, or when a and b are positive and m and n are any real numbers. Also known as exponential law.

law of gravitation *See* Newton's law of gravitation.

law of large numbers [STAT] The law that if, in a collection of independent identical experiments, $N(B)$ represents the number of occurrences of an event B in n trials, and p is the probability that B occurs at any given trial, then for large enough n it is unlikely that $N(B)/n$ differs from p by very much. Also known as Bernoulli theorem.

law of parallel solenoids [PHYS] The law that under stationary conditions isopycnals and isobars must be parallel at all levels, and isobars and isopycnals at one level must be parallel to those at all other levels.

law of partial pressures   *See* Dalton's law.

law of rational intercepts   *See* Miller law.

law of reflection   *See* reflection law.

law of signs   [MATH] The product or quotient of two numbers is positive if the numbers have the same sign, negative if they have opposite signs.

law of sines   [MATH] Given a triangle with angles A, B, and C and sides a, b, c opposite these angles respectively: $\sin A/a = \sin B/b = \sin C/c$.

law of tangents   [MATH] Given a triangle with angles A, B, and C and sides a, b, c opposite these angles respectively: $(a - b)/(a + b) = [\tan \frac{1}{2}(A - B)]/[\tan \frac{1}{2}(A + B)]$.

law of the mean   *See* mean value theorem.

lawrencium   [CHEM] A chemical element, symbol Lr, atomic number 103; the isotope of longest known half-life has mass number 260.

laws of refraction   *See* Snell laws of refraction.

Lawson criterion   [PL PHYS] The requirement for the energy produced by fusion in a plasma to exceed that required to produce the confined plasma; it states that for a mixture of deuterium and tritium in the temperature range from 1×10^8 to 5×10^8 degrees Celsius, the product of the ionic density and the confinement time must be about 10^{14} seconds per cubic centimeter.

layer capacitance   *See* cathode interface capacitance.

layer impedance   *See* cathode interface impedance.

layer lattice   *See* layer structure.

layer structure   [CRYSTAL] A crystalline structure found in substances such as graphites and clays, in which the atoms are largely concentrated in a set of parallel planes, with the regions between the planes comparatively vacant. Also known as layer lattice.

lb   *See* pound.

lbf   *See* pound.

lbf-ft   *See* foot-pound.

LCAO   *See* linear combination of atomic orbitals.

L capture   [NUC PHYS] A type of generalized beta interaction in which a nucleus captures an electron from the L shell of atomic electrons (the shell second closest to the nucleus).

LC filter   *See* inductive filter.

lcm   *See* least common multiple.

LC ratio   [ELEC] The inductance of a circuit in henrys divided by capacitance in farads.

lead   [CHEM] A chemical element, symbol Pb, atomic number 82, atomic weight 207.19.   [PHYS] *See* lead angle.

lead-208   [NUC PHYS] Lead isotope, atomic mass number of 208, formed by the radioactive decay of thorium.

lead-acid battery   [ELEC] A storage battery in which the electrodes are grids of lead containing lead oxides that change in composition during charging and discharging, and the electrolyte is dilute sulfuric acid.

lead angle   [PHYS] The phase difference between a sinusoidally varying quantity and a reference quantity which varies sinusoidally at the same frequency, when this phase difference is positive. Also known as angle of lead; lead; phase lead.

lead compensation   [CONT SYS] A type of feedback compensation primarily employed for stabilization or for improving a system's transient response; it is generally characterized by a series compensation transfer function of the type

$$G_c(s) = K\frac{(s - z)}{(s - p)}$$

where $z < p$ and K is a constant.

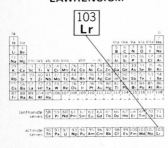

LAWRENCIUM

$$\boxed{\begin{array}{c} 103 \\ \textbf{Lr} \end{array}}$$

Periodic table of the chemical elements showing the position of lawrencium.

LEAD

$$\boxed{\begin{array}{c} 82 \\ \textbf{Pb} \end{array}}$$

Periodic table of the chemical elements showing the position of lead.

lead equivalent [NUCLEO] The thickness of lead that gives the same reduction in radiation dose rate as the material in question.

lead-I-lead junction [SOLID STATE] A Josephson junction consisting of two pieces of lead separated by a thin insulating barrier of lead oxide. Abbreviated Pb-I-Pb junction.

leading edge [PHYS] The major portion of the rise of a pulse.

leading load [ELEC] Load that is predominately capacitive, so that its current leads the voltage applied to the load.

leading zeros [MATH] Zeros preceding the first nonzero integer of a number.

lead-lag network [CONT SYS] Compensating network which combines the characteristics of the lag and lead networks, and in which the phase of a sinusoidal response lags a sinusoidal input at low frequencies and leads it at high frequencies. Also known as lag-lead network.

lead lanthanum zirconate titanate [MATER] A ferroelectric, ceramic, electrooptical material whose optical properties can be changed by an electric field or by being placed in tension or compression; used in optoelectronic storage and display devices. Abbreviated PLZT. Also known as lanthanum-doped lead zirconate–lead titanate.

lead network *See* derivative network.

lead zirconate titanate [MATER] A ferroelectric, ceramic, electrooptic material that has lower optical transparency than lead lanthanum zirconate titanate but similar other properties. Abbreviated PZT.

leaf of Descartes *See* folium of Descartes.

league [MECH] A unit of length equal to 3 miles or 4828.032 meters.

leakage coefficient *See* leakage factor.

leakage current [ELEC] **1.** Undesirable flow of current through or over the surface of an insulating material or insulator. **2.** The flow of direct current through a poor dielectric in a capacitor. [ELECTR] The alternating current that passes through a rectifier without being rectified.

leakage factor [ELECTROMAG] The total magnetic flux in an electric rotating machine or transformer divided by the useful flux that passes through the armature or secondary winding. Also known as leakage coefficient.

leakage flux [ELECTROMAG] Magnetic lines of force that go beyond their intended path and do not serve their intended purpose. [NUCLEO] The number of neutrons which pass outward through a unit area at the surface of a reactor core in a unit time, and are not reflected back into the core.

leakage resistance [ELEC] The resistance of the path over which leakage current flows; it is normally high.

leaky [ELEC] Pertaining to a condition in which the leakage resistance has dropped so much below its normal value that excessive leakage current flows; usually applied to a capacitor.

least-action principle *See* principle of least action.

least common denominator [MATH] The least common multiple of the denominators of a collection of fractions.

least common multiple [MATH] Abbreviated lcm. **1.** The least common multiple of a set of integers is the smallest integer which is divisible by each of them. **2.** More generally, a least common multiple of elements $a_1, \ldots, a_n$ of a unique factorization domain R is an element c in R which is divisible by $a_1, \ldots, a_n$, and such that if d is divisible by $a_1, \ldots, a_n$, then d is also divisible by c.

least-energy principle [MECH] The principle that the potential energy of a system in stable equilibrium is a minimum relative to that of nearby configurations.

least squares estimate   [STAT] An estimate obtained by the least squares method.

least squares method   [STAT] A technique of fitting a curve close to some given points which minimizes the sum of the squares of the deviations of the given points from the curve.

least-time principle   *See* Fermat's principle.

least upper bound   [MATH] The least upper bound of a subset A of a set S with ordering $<$ is the smallest element of S which is greater than or equal to every element of A. Abbreviated lub. Also known as supremum (sup).

least-work theory   [MECH] A theory of statically indeterminate structures based on the fact that when a stress is applied to such a structure the individual parts of it are deflected so that the energy stored in the elastic members is minimized.

Lebesgue density theorem   [MATH] If S is a measurable subset of the real numbers, the metric density of S exists and is equal to 1 at every point of S except for a set of measure zero.

Lebesgue integral   [MATH] The generalization of Riemann integration of real valued functions, which allows for integration over more complicated sets, existence of the integral even though the function has many points of discontinuity, and convergence properties which are not valid for Riemann integrals.

Lebesgue measure   [MATH] A measure defined on subsets of euclidean space which expresses how one may approximate a set by coverings consisting of intervals.

Lebesgue number   [MATH] The Lebesgue number of an open cover of a compact metric space X is a positive real number so that any subset of X whose diameter is less than this number must be completely contained in a member of the cover.

Lebesgue-Stieltjes integral   [MATH] A Lebesgue integral of the form

$$\int_b^a f(x)\, d\phi(x)$$

where ϕ is of bounded variation; if $\phi(x) = x$, it reduces to the Lebesgue integral of $f(x)$; if $\phi(x)$ is differentiable, it reduces to the Lebesgue integral of $f(x)\phi'(x)$.

Le Chatelier's principle   [PHYS] The principle that when an external force is applied to a system at equilibrium, the system adjusts so as to minimize the effect of the applied force.

Lecher line   *See* Lecher wires.

Lecher wires   [ELECTROMAG] Two parallel wires that are several wavelengths long and a small fraction of a wavelength apart, used to measure the wavelength of a microwave source that is connected to one end of the wires; a shorting bar which slides along the wires is used to determine the position of standing-wave nodes. Also known as Lecher line; Lecher wire wavemeter.

Lecher wire wavemeter   *See* Lecher wires.

Leclanché cell   [ELEC] The common dry cell, which is a primary cell having a carbon positive electrode and a zinc negative electrode in an electrolyte of sal ammoniac and a depolarizer.

LED   *See* light-emitting diode.

Ledoux bell meter   [ENG] A type of manometer used to measure the difference in pressure between two points generated by any one of several types of flow measurement devices such as a pitot tube; it is equipped with a shaped plug which makes the reading of the meter directly proportional to the flow rate.

Leduc effect   *See* Righi-Leduc effect.

Leduc law   *See* Amagat-Leduc rule.

LECLANCHÉ CELL

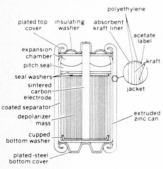

plated top cover
insulating washer
absorbent kraft liner
polyethylene
acetate label
expansion chamber
pitch seal
kraft
seal washers
sintered carbon electrode
jacket
coated separator
depolarizer mass
extruded zinc can
cupped bottom washer
plated-steel bottom cover

Modern Leclanché dry cell.
(Bright Star Industries Inc.)

LEDOUX BELL METER

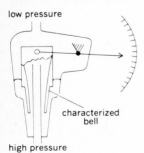

low pressure

characterized bell

high pressure

Ledoux bell meter with uniform flow rate scales.

LEED *See* low-energy electron diffraction.

lee eddies [FL MECH] The small, irregular motions or eddies produced immediately in the rear of an obstacle in a turbulent fluid.

Lee's disk [THERMO] A device for determining the thermal conductivity of poor conductors in which a thin, cylindrical slice of the substance under study is sandwiched between two copper disks, a heating coil is placed between one of these disks and a third copper disk, and the temperatures of the three copper disks are measured.

Leeson disk [OPTICS] The screen sometimes used in a grease-spot photometer, in which the translucent spot at the center is star-shaped to provide a fine line of demarcation.

lee wave [FL MECH] Any wave disturbance which is caused by, and is therefore stationary with respect to, some barrier in the fluid flow.

left-continuous function [MATH] A function $f(x)$ of a real variable is left-continuous at a point c if $f(x)$ approaches $f(c)$ as x approaches c from the left, that is, $x < c$ only.

left coset [MATH] A left coset of a subgroup H of a group G is a subset of G consisting of all elements of the form ah, where a is a fixed element of G and h is any element of H.

left-hand derivative [MATH] The limit of the difference quotient $[f(x) - f(c)]/[x - c]$ as x approaches c from the left, that is, $x < c$ only.

left-handed [CRYSTAL] Having a crystal structure with a mirror-image relationship to a right-handed structure.

left-handed coordinate system [MATH] **1.** A three-dimensional rectangular coordinate system such that when the thumb of the left hand extends in the positive direction of the first (or x) axis, the fingers fold in the direction in which the second (or y) axis could be rotated about the first axis to coincide with the third (or z) axis. **2.** A coordinate system of a Riemannian space which has negative scalar density function.

left-hand limit *See* limit on the left.

left-hand polarization [ELECTROMAG] In elementary-particle discussions, circular or elliptical polarization of an electromagnetic wave in which the electric field vector at a fixed point in space rotates in the left-hand sense about the direction of propagation; in optics, the opposite convention is used; in facing the source of the beam, the electric vector is observed to rotate counterclockwise.

left-hand rule [ELECTROMAG] **1.** For a current-carrying wire, the rule that if the fingers of the left hand are placed around the wire so that the thumb points in the direction of electron flow, the fingers will be pointing in the direction of the magnetic field produced by the wire. **2.** For a current-carrying wire in a magnetic field, such as a wire on the armature of a motor, the rule that if the thumb, first, and second fingers of the left hand are extended at right angles to one another, with the first finger representing the direction of magnetic lines of force and the second finger representing the direction of current flow, the thumb will be pointing in the direction of force on the wire. Also known as Fleming's rule.

left-hand taper [ELEC] A taper in which there is greater resistance in the counterclockwise half of the operating range of a rheostat or potentiometer (looking from the shaft end) than in the clockwise half.

left identity [MATH] In a set on which a binary operation ∘ is defined, an element e with the property that $e ∘ a = a$ for every element a in the set.

Legendre condition [MATH] A necessary condition for an extremal $y(x)$ to minimize the integral over some finite interval of $f(x,y,y')\,dx$, namely that $\partial^2 f/\partial y'^2 \geq 0$ everywhere on the extremal.

Legendre contact transformation *See* Legendre transformation.

Legendre equation [MATH] The second-order linear homogeneous differential equation $(1-x^2)y'' - 2xy' + \nu(\nu+1)y = 0$, where ν is real and nonnegative.

Legendre function [MATH] Any solution of the Legendre equation.

Legendre normal elliptic integrals [MATH] Three standard elliptic integrals having the property that any elliptic integral is a linear combination of these integrals and elementary functions; they are the integrals from 0 to sin ϕ of dx/y, $(1-k^2x^2)\,dx/y$ and $dx/(1+nx^2)y$, designated $F(\phi,k)$, $E(\phi,k)$, and $\Pi(\phi,n,k)$, respectively, where y and k are the radical and modulus in Legendre normal form, and ϕ and n are real or complex numbers.

Legendre normal form [MATH] **1.** The standard form, $y = \sqrt{(1-x^2)(1-k^2x^2)}$, to which the radical in an elliptic integral can be reduced by a transformation on the integration variable x, where k is a real or complex number with $|k| \leq 1$, $k^2 \neq 1$, and the substitution $x = \sin\phi$ is often used. **2.** The form of the Legendre normal elliptic integrals.

Legendre polynomials [MATH] A collection of orthogonal polynomials which provide solutions to the Legendre equation for nonnegative integral values of the parameter.

Legendre's associated equation [MATH] A second-order homogeneous differential equation that generalizes the Legendre equation: $(1-x^2)y'' - 2xy' + [\nu(\nu+1) - \mu^2/(1-x^2)]y = 0$, where μ and ν are parameters.

Legendre series [MATH] For a function f, the infinite series whose nth term is $(n+\tfrac{1}{2})a_n P_n(x)$, where P_n is the Legendre polynomial of degree n, and a_n is the integral from -1 to $+1$ of $f(t)P_n(t)\,dt$; it converges to $f(x)$ if f satisfies certain criteria.

Legendre transformation [FL MECH] A version of the hodograph method for compressible flow in which a replacement is made not only of the independent variables but also of the dependent variables, that is, of the velocity potential and the stream function. [MATH] A mathematical procedure in which one replaces a function of several variables with a new function which depends on partial derivatives of the original function with respect to some of the original independent variables. Also known as Legendre contact transformation.

Leibnitz's rule [MATH] A formula to compute the nth derivative of the product of two functions f and g:

$$d^n(f{\cdot}g)/dx^n = \sum_{k=0}^{n} \binom{n}{k}\, d^{n-k}f/dx^{n-k}{\cdot}d^k g/dx^k$$

where $\binom{n}{k} = n!/(n-k)!\,k!$

Leibnitz's test [MATH] If the sequence of positive numbers a_n approaches zero monotonically, then the series

$$\sum_{n=1}^{\infty} (-1)^n a_n \text{ is convergent.}$$

Leibniz's series [MATH] A series for π: $\pi/4 = 1 - 1/3 + 1/5 - 1/7 + \cdots$.

Leidenfrost point [THERMO] The lowest temperature at which a hot body submerged in a pool of boiling water is completely blanketed by a vapor film; there is a minimum in the heat flux from the body to the water at this temperature.

Leidenfrost's phenomenon [THERMO] A phenomenon in which a liquid dropped on a surface that is above a critical temperature becomes insulated from the surface by a layer of vapor, and does not wet the surface as a result.

LEIT *See* light emission via inelastic tunneling.

LEIT device [ELECTR] A light source consisting of two crossed, thin metal-film strips separated by a very thin insulating layer and attached to a battery to produce light emission via inelastic tunneling (LEIT).

L electron [ATOM PHYS] An electron in the L shell.

lemma [MATH] A mathematical fact germane to the proof of some theorem.

lemma of duBois-Reymond [MATH] A continuous function $f(x)$ is constant in the interval (a,b) if for certain functions g whose integral over (a,b) is zero, the integral over (a,b) of f times g is zero.

lemniscate of Bernoulli [MATH] The locus of points (x,y) in the plane satisfying the equation $(x^2 + y^2)^2 = a^2(x^2 - y^2)$ or, in polar coordinates (r,θ), the equation $r^2 = a^2 \cos 2\theta$.

Lenard rays [ELECTR] Cathode rays produced in air by a Lenard tube.

Lenard's mass absorption law *See* mass absorption law.

Lenard spiral [ENG] A type of magnetometer consisting of a spiral of bismuth wire and a Wheatstone bridge to measure changes in the resistance of the wire produced by magnetic fields and as a result of the transverse magnetoresistance of bismuth.

Lenard tube [ELECTR] An early experimental electron-beam tube that had a thin glass or metallic foil window at the end opposite the cathode, through which the electron beam could pass into the atmosphere.

length [MECH] Extension in space.

lengthened dipole [ELECTROMAG] An antenna element with lumped inductance to compensate an end loss.

length of a curve [MATH] A curve represented by $x = x(t)$, $y = y(t)$ for $t_1 \leq t \leq t_2$, with $x(t_1) = x_1$, $x(t_2) = x_2$, $y(t_1) = y_1$, $y(t_2) = y_2$, has length from (x_1,y_1) to (x_2,y_2) given by the integral from t_1 to t_2 of the function $\sqrt{(dx/dt)^2 + (dy/dt)^2}$.

Lennard-Jones potential [PHYS CHEM] A semiempirical approximation to the potential of the force between two molecules, given by $V = (A/R^{12}) - (B/R^6)$, where R is the distance between the centers of the molecules, and A and B are constants.

lens [ELECTR] *See* electron lens. [ELECTROMAG] *See* magnetic lens. [MATER] *See* acoustic lens. [OPTICS] A curved piece of ground and polished or molded material, usually glass, used for the refraction of light, its two surfaces having the same axis; or two or more such surfaces cemented together. Also known as optical lens.

lens antenna [ELECTROMAG] A microwave antenna in which a dielectric lens is placed in front of the dipole or horn radiator to concentrate the radiated energy into a narrow beam or to focus received energy on the receiving dipole or horn.

lens element [OPTICS] A separate component lens of a multielement lens.

lens equation [OPTICS] Any equation which relates the distance of a point object from some well-defined reference point in an optical system to the distance of its image from a similar point.

lens shim [OPTICS] Thin piece of material used to position and focus a lens.

lens stop *See* diaphragm.

lenticular [OPTICS] Of or pertaining to a lens. [SCI TECH] Having the shape of a lentil or double convex lens.

lentor *See* stoke.

Lenz's law [ELECTROMAG] The law that whenever there is an induced electromotive force (emf) in a conductor, it is always in such a direction that the current it would produce would oppose the change which causes the induced emf.

leo [MECH] A unit of acceleration, equal to 10 meters per second per second; it has rarely been employed.

leptokurtic distribution [STAT] A distribution in which the ratio of the fourth moment to the square of the second moment is greater than 3, which is the value for a normal distribution; it appears to be more heavily concentrated about the mean, or more peaked, than a normal distribution.

lepton [PARTIC PHYS] A fermion having a mass smaller than the proton mass; leptons interact with electromagnetic and gravitational fields, but beyond this they interact only through weak interactions.

lepton conservation [PARTIC PHYS] The principle that the number of electrons and e-neutrinos minus the number of positrons and e-antineutrinos is unchanged in any interaction; similarly, the number of negatively charged muons and μ-neutrinos minus the number of positively charged muons and μ-antineutrinos is unchanged.

leptonic decay [PARTIC PHYS] Decay of an elementary particle in which at least some of the products are leptons.

lepton number [PARTIC PHYS] A conserved quantum number, equal to the number of leptons minus the number of antileptons in a system.

Lerch's theorem [MATH] **1.** If $f(x)$ is continuous for $0 \le x \le a$ and if the integral from 0 to a of $x^n f(x)\ dx$ vanishes for all positive integral values of n, then $f(x)$ is identically 0 for $0 \le x \le a$. **2.** Two continuous functions with identical Laplace transforms are identical for all positive values of the argument.

Leslie cube [THERMO] A metal box, with faces having different surface finishes, in which water is heated and next to which a thermopile is placed in order to compare the heat emission properties of different surfaces.

lethargy [NUCLEO] A measure of the energy which has been lost by a neutron, equal to the natural logarithm of the ratio of the initial energy of a neutron to its energy at any given point in the slowing-down process.

leucitohedron *See* trapezohedron.

level [ELEC] A single bank of contacts, as on a stepping relay. [ELECTR] **1.** The difference between a quantity and an arbitrarily specified reference quantity, usually expressed as the logarithm of the ratio of the quantities. **2.** A charge value that can be stored in a given storage element of a charge storage tube and distinguished in the output from other charge values.

level converter [ELECTR] An amplifier that converts nonstandard positive or negative logic input voltages to standard DTL or other logic levels.

level of a factor [STAT] In factorial experiments, the quantitative or qualitative intensity at which a particular value of a factor is held fixed during an experiment.

level of significance of a test [STAT] The probability of false rejection of the null hypothesis. Also known as significance level.

level point *See* point of fall.

level scheme *See* energy-level diagram.

level width [QUANT MECH] A measure of the spread in energy of an unstable state, equal to the difference between the energies at which intensity of emission or absorption of photons or particles, or the cross section for a reaction, is one-half its maximum value.

leverage [MECH] The multiplication of force or motion achieved by a lever.

Leverett function [FL MECH] A dimensionless number used in studying two-phase flow in porous mediums, equal to $(\xi/e)^{1/2}(p/\sigma)$, where ξ is the permeability of a medium (as defined by Darcy's law), e is the medium's porosity, σ is the surface tension between two liquids flowing through it, and p is the capillary pressure.

Levi-Civita symbol [MATH] A symbol $\varepsilon_{i,j,\ldots,s}$ where $i,j,\ldots,s$ are n indices, each running from 1 to n; the symbol equals zero if any two indices are identical, and 1 or -1 otherwise, depending on whether $i,j,\ldots,s$ form an even or an odd permutation of $1, 2, \ldots, n$.

levo form [PHYS CHEM] An optical isomer which induces levorotation in a beam of plane polarized light.

levorotation [OPTICS] Rotation of the plane of polarization of plane polarized light in a counterclockwise direction, as seen by an observer facing in the direction of light propagation. Also known as levulorotation.

levulorotation *See* levorotation.

Lewis number [PHYS] **1.** A dimensionless number used in studies of combined heat and mass transfer, equal to the thermal diffusivity divided by the diffusion coefficient. Symbolized Le; N_{Le}. **2.** Sometimes, the reciprocal of this quantity.

Lewis-Rayleigh afterglow [ELECTR] A golden yellow light emitted by nitrogen gas following the passage of an electric discharge, associated with recombination of nitrogen atoms.

lexicographic order [MATH] Given sets A and B with a common ordering $<$, one defines an ordering between all sequences (finite or infinite) of elements of A and of elements of B by $(a_1, a_2, \ldots) < (b_1, b_2, \ldots)$ if either $a_i = b_i$ for every i, or $a_n < b_n$, where n is the first place in which they differ; this is the way words are ordered in a dictionary.

Leyden jar [ELEC] An early type of capacitor, consisting simply of metal foil sheets on the inner and outer surfaces of a glass jar.

l'Hôpital's cubic *See* Tschirnhausen's cubic.

l'Hôpital's rule [MATH] A rule useful in evaluating indeterminate forms: if both the functions $f(x)$ and $g(x)$ and all their derivatives up to order $(n-1)$ vanish at $x = a$, but the nth derivatives both do not vanish or both become infinite at $x = a$, then $\lim_{x \to a} f(x)/g(x) = f^{(n)}(a)/g^{(n)}(a)$, $f^{(n)}$ denoting the nth derivative.

l'Huilier's equation [MATH] An equation used in the solution of a spherical triangle, involving tangents of various functions of its angles and sides.

Li *See* lithium.

Liapunov function *See* Lyapunov function.

Libby counter *See* black carbon counter.

libration [PHYS] Any oscillatory rotational motion, such as that of the moon, or of a molecule in a solid which does not have enough energy to make full rotations.

libration in latitude *See* lunar libration.

Lichenberger figures *See* Lichtenberg figures.

Lichtenberg figures [ELEC] Patterns produced on a photographic emulsion, or in fine powder spread over the surface of a solid dielectric, by an electric discharge produced by a high transient voltage. Also known as Lichenberger figures.

lidar [OPTICS] An instrument in which a ruby laser generates intense infrared pulses in beam widths as small as 30 seconds of arc; beam reflections and scattering effects of clouds, smog layers, and some atmospheric discontinuities are measured by radar techniques; it can also be used for tracking weather balloons, smoke puffs, and rocket trails. Derived from laser infrared radar.

Lie algebra [MATH] The algebra of vector fields on a manifold with additive operation given by pointwise sum and multiplication by the Lie bracket.

Liebmann effect [OPTICS] The effect whereby it is more

difficult to visually distinguish contrasting forms when they have the same luminance and different chromaticities than when they have different luminances and the same chromaticity.

Lie bracket [MATH] Given vector fields X, Y on a manifold M, their Lie bracket is the vector field whose value is the difference between the values of XY and YX.

Lie group [MATH] A topological group which is also a differentiable manifold in such a way that the group operations are themselves analytic functions.

Liénard's equation [MATH] The second-order, nonlinear differential equation $d^2y/dx^2 + f(y) \, dy/dx + y = 0$.

Liénard-Wiechert potentials [ELECTROMAG] The retarded and advanced electromagnetic scalar and vector potentials produced by a moving point charge, expressed in terms of the (retarded or advanced) position and velocity of the charge.

lifetime *See* mean life.

lift *See* aerodynamic lift.

lifting [MATH] **1.** Given a fiber bundle $(\bar{X}, B, p)$ and a continuous map of a topological space $\bar{Y}$ to B, $g: \bar{Y} \rightarrow B$, lifting entails finding a continuous map $\bar{g}: \bar{Y} \rightarrow \bar{X}$ such that the function g is the composition $p \circ \bar{g}$. **2.** *See* translation.

lifting magnet [ELECTROMAG] A type of electromagnet in which a material to be held or moved is initially placed in contact with the magnet, in contrast to a traction magnet. Also known as holding magnet.

light [OPTICS] Electromagnetic radiation with wavelengths capable of causing the sensation of vision, ranging approximately from 4000 (extreme violet) to 7700 angstroms (extreme red), or 400 to 770 nanometers. Also known as light radiation; visible radiation. **2.** More generally, electromagnetic radiation of any wavelength; thus, the term is sometimes applied to infrared and ultraviolet radiation.

light absorption [OPTICS] The process in which energy of light radiation is transferred to a medium through which it is passing.

light amplification by stimulated emission of radiation *See* laser.

light amplifier [ELECTR] Any electronic device which, when actuated by a light image, reproduces a similar image of enhanced brightness, and which is capable of operating at very low light levels without introducing spurious brightness variations (noise) into the reproduced image. Also known as image intensifier.

light-beam galvanometer *See* d'Arsonval galvanometer.

light-beam oscillograph [ELECTROMAG] An oscillograph in which a beam of light, focused to a point by a lens, is reflected from a tiny mirror attached to the moving coil of a galvanometer onto a photographic film moving at constant speed.

light bulb *See* incandescent lamp.

light chopper [ELECTR] A rotating fan or other mechanical device used to interrupt a light beam that is aimed at a phototube, to permit alternating-current amplification of the phototube output and to make its output independent of strong, steady ambient illumination.

light cone [RELAT] The surface generated by the set of all null vectors emanating from a point in space-time. Also known as null cone.

light curve [ASTROPHYS] A graph showing the variations in brightness of a celestial object; the stellar magnitude is usually shown on the vertical axis, and time is the horizontal coordinate.

light-distribution photometer [OPTICS] A device which measures the luminous intensity of a light source in various directions; the light source is fixed, and a mirror system is

LIGHT-BEAM OSCILLOGRAPH

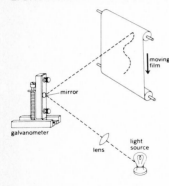

Essential components of a light-beam oscillograph. *(General Electric Co.)*

LIGHT-DISTRIBUTION PHOTOMETER

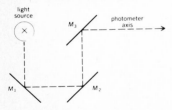

Mirror system in a light-distribution photometer. M_1, M_2, M_3 are mirrors.

rotated about an axis passing through the centers of the light source and a photocell so that the light emitted by the source in any direction perpendicular to this axis is reflected to the photocell. Also known as distribution photometer.

light emission via inelastic tunneling [ELECTR] A process in which electrons tunneling through a thin insulating layer separating two metals excite surface plasmons which then scatter from surface and structural discontinuities, radiating visible light. Abbreviated LEIT.

light-emitting diode [ELECTR] A semiconductor diode that converts electric energy efficiently into spontaneous and noncoherent electromagnetic radiation at visible and near-infrared wavelengths by electroluminescence at a forward-biased *pn* junction. Abbreviated LED. Also known as solid-state lamp.

light exposure [OPTICS] A measure of the total amount of light falling on a surface; equal to the integral over time of the luminance of the surface. Also known as exposure.

light filter *See* color filter.

light guide *See* optical fiber.

light hydrogen *See* protium.

light intensity *See* luminous intensity.

light microscope *See* optical microscope.

light microsecond [ELECTROMAG] Distance a light wave travels in free space in one-millionth of a second.

light-negative [ELECTR] Having negative photoconductivity, hence decreasing in conductivity (increasing in resistance) under the action of light.

lightning [GEOPHYS] The large spark produced by an abrupt discontinuous discharge of electricity through the air, resulting most often from the creation and separation of electric charge in cumulonimbus clouds.

lightning channel [GEOPHYS] The irregular path through the air along which a lightning discharge occurs.

lightning flash [GEOPHYS] In atmospheric electricity, the total observed luminous phenomenon accompanying a lightning discharge.

lightning stroke [GEOPHYS] Any one of a series of repeated discharges comprising a single lightning discharge (or lightning flash); specifically, in the case of the cloud-to-ground discharge, a leader plus its subsequent return streamer.

light-of-the-night-sky *See* airglow.

light pipe [OPTICS] A solid, transparent plastic rod that transmits light from one end to the other even when bent.

light-positive [ELECTR] Having positive photoconductivity; selenium ordinarily has this property.

light quantum *See* photon.

light radiation *See* light.

light ratio [ASTROPHYS] A number (2.512) that expresses the ratio of a star's light to that of another star that is one magnitude fainter or brighter.

light ray [OPTICS] A beam of light having a small cross section.

light scattering [OPTICS] The process in which energy is removed from a beam of light radiation and reemitted without appreciable change in wavelength.

light-sensitive [ELECTR] Having photoconductive, photoemissive, or photovoltaic characteristics. Also known as photosensitive.

light-sensitive tube *See* phototube.

light transmission [OPTICS] The process in which light travels through a medium without being absorbed or scattered.

light valve [ELECTR] A device whose light transmission can be made to vary in accordance with an externally applied electrical quantity, such as voltage, current, electric field, or magnetic field, or an electron beam.

light water [MATER] A water solution of perfluorocarbon compounds mixed with a polyoxyethylene thickener; used as a fire-fighting agent. [NUCLEO] Water in which both the hydrogen atoms in each molecule are of the isotope protium.

light-water reactor [NUCLEO] A nuclear reactor that uses ordinary water as moderator, in contrast to heavy water.

light watt [OPTICS] A unit of luminous power equal to the luminous power of light of a single wavelength λ whose radiant power is $1/V_\lambda$ watts, where V_λ is the value of the luminosity function at λ.

light-year [ASTROPHYS] A unit of measurement of astronomical distance; it is the distance light travels in one sidereal year and is equivalent to 9.461×10^{12} kilometers or 5.879×10^{12} miles.

likelihood [MATH] The likelihood of a sample of independent values of $x_1, x_2, \ldots, x_n$, with $f(x)$ the probability function, is the product $f(x_1) \circ f(x_2) \circ \cdots \circ f(x_n)$.

likelihood ratio [STAT] The probability of a random drawing of a specified sample from a population, assuming a given hypothesis about the parameters of the population, divided by the probability of a random drawing of the same sample, assuming that the parameters of the population are such that this probability is maximized.

likelihood ratio test [STAT] A procedure used in hypothesis testing based on the ratio of the values of two likelihood functions, one derived from the hypothesis being tested and one without the constraints of the hypothesis under test.

limacon [MATH] The locus of points of the plane which in polar coordinates (r,θ) satisfy the equation $r = a \cos \theta + b$. Also known as Pascal's limacon.

limb [ASTRON] The circular outer edge of a celestial body; the half with the greater altitude is called the upper limb, and the half with the lesser altitude, the lower limb.

limb darkening [ASTROPHYS] An observed darkening near the surface of the sun's limb as compared to its brighter center.

liminal contrast *See* threshold contrast.

limit cycle of a differential equation [MATH] A closed trajectory C in the plane (corresponding to a periodic solution of the equation) where every point of C has a neighborhood so that every trajectory through it spirals toward C.

limited proportionality region [NUCLEO] The range of operating voltages of a radiation counter within which the charge collected is proportional to the charge liberated by the initial events, but saturates for larger initial events.

limiter [ELECTR] An electronic circuit used to prevent the amplitude of an electronic waveform from exceeding a specified level while preserving the shape of the waveform at amplitudes less than the specified level. Also known as amplitude limiter; amplitude-limiting circuit; automatic peak limiter; clipper; clipping circuit; limiter circuit; peak limiter.

limiter circuit *See* limiter.

limit inferior [MATH] **1.** The limit inferior of a sequence whose nth term is a_n is the limit as N approaches infinity of the greatest lower bound of the terms a_n for which n is greater than N; denoted by $\liminf_{n \to \infty} a_n$ or $\varliminf_{n \to \infty} a_n$. **2.** The limit inferior of a function f at a point c is the limit as ε approaches zero of the greatest lower bound of $f(x)$ for $|x - c| < \varepsilon$ and $x \neq c$; denoted by $\liminf_{x \to c} f(x)$ or $\varliminf_{x \to c} f(x)$.

limiting [ELECTR] A desired or undesired amplitude-limiting action performed on a signal by a limiter. Also known as clipping.

limiting current density [PHYS CHEM] The maximum current density to achieve a desired electrode reaction before hydrogen or other extraneous ions are discharged simultaneously.

limiting viscosity number *See* intrinsic viscosity.

limit of a function [MATH] A function $f(x)$ has limit L as x tends to c if given any positive number ε (no matter how small) there is a positive number δ such that if x is in the domain of f, x is not c, and $|x - c| < \delta$, then $|f(x) - L| < \varepsilon$; written $\lim_{x \to c} f(x) = L$.

limit of a sequence [MATH] A sequence $\{a_n : n = 1, 2, \dots \}$ has limit L if given a positive number ε (no matter how small), there is a positive integer N such that for all integers n greater than N, $|a_n - L| < \varepsilon$.

limit of resolution [OPTICS] The minimum distance or angular separation between two point objects which allows them to be resolved according to the Rayleigh criterion.

limit on the left [MATH] The limit on the left of the function f at a point c is the limit of f at c which would be obtained if only values of x less than c were taken into account; more precisely, it is the number L which has the property that for any positive number ε, there is a positive number δ so that if x is the domain of f and $0 < c - x < \delta$ then $|f(x) - L| < \varepsilon$; denoted by $\lim_{x \to c^-} f(x) = L$, or $f(c^-) = L$. Also known as left-hand limit.

limit on the right [MATH] The limit on the right of the function $f(x)$ at a point c is the limit of f at c which would be obtained if only values of x greater than c were taken into account; more precisely, it is the number L which has the property that for any positive number ε there is a positive number δ so that if x is in the domain of f and $0 < x - c < \delta$, then $|f(x) - L| < \varepsilon$; denoted by $\lim_{x \to c^+} f(x) = L$ or $f(c^+) = L$. Also known as right-hand limit.

limit point *See* alpha-limit point; cluster point; omega-limit point.

limit set [MATH] For a vector differential equation, the set of all alpha-limit points and omega-limit points of a given point.

limits of integration [MATH] The end points of the interval over which a function is being integrated.

limit superior [MATH] **1.** The limit superior of a sequence whose nth term is a_n is the limit as N approaches infinity of the least upper bound of the terms a_n for which n is greater than N; denoted by $\lim \sup_{n \to \infty} a_n$ or $\overline{\lim}_{n \to \infty} a_n$. **2.** The limit superior of a function f at a point c is the limit as ε approaches zero of the least upper bound of $f(x)$ for $|x - c| < \varepsilon$ and $x \neq c$; denoted by $\lim \sup_{x \to c} f(x)$ or $\overline{\lim}_{x \to c} f(x)$.

linac *See* linear accelerator.

Lindeck potentiometer [ELEC] A potentiometer in which an unknown potential difference is balanced against a known potential difference derived from a fixed resistance carrying a variable current; the converse of most potentiometers.

Lindelöf space [MATH] A topological space where if a family of open sets covers the space, a countable number of these sets also covers the space.

Lindelöf theorem [MATH] There is a countable subcover of each open cover of a subset of a space whose topology has a countable base.

Lindemann electrometer [ELEC] A variant of the quadrant electrometer, designed for portability and insensitivity to changes in position, in which the quadrants are two sets of plates about 6 millimeters apart, mounted on insulating quartz pillars; a needle rotates about a taut silvered quartz suspension toward the oppositely charged plates when volt-

LINDEMANN ELECTROMETER

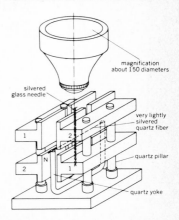

magnification about 150 diameters

silvered glass needle

very lightly silvered quartz fiber

quartz pillar

quartz yoke

Lindemann electrometer. Here 1 and 2 indicate the quadrants. *(From F. A. Laws, Electrical Measurements, 2d ed., McGraw-Hill, 1938)*

age is applied to it, and its movement is observed through a microscope.

Lindemann's theorem [MATH] If $x_1, x_2, \ldots, x_n$ are distinct, real, or complex algebraic numbers, and $y_1, y_2, \ldots, y_n$ are algebraic numbers which are not equal to 0, then the sum $y_1e^{x_1} + y_2e^{x_2} + \cdots + y_ne^{x_n}$ is not equal to 0; it follows from this theorem that π is transcendental.

Lindemann theory [SOLID STAT] A theory of the melting point of solids according to which solids melt when the amplitude of oscillation of the atoms becomes so great that neighboring atoms collide.

Linde process [CRYO] A cyclic process for liquefying gases in which compressed gas is cooled by Joule-Thomson expansion through a valve to a pressure of about 40 atmospheres (4 megapascals), further cools the incoming gas in a heat exchanger, and is compressed for the next cycle.

Linde's rule [SOLID STATE] The rule that the increase in electrical resistivity of a monovalent metal produced by a substitutional impurity per atomic percent impurity is equal to $a + b(v - 1)^2$, where a and b are constants for a given solvent metal and a given row of the periodic table for the impurity, and v is the valence of the impurity.

line [ELECTR] **1.** The path covered by the electron beam of a television picture tube in one sweep from left to right across the screen. **2.** One horizontal scanning element in a facsimile system. **3.** *See* trace. [MATH] The set of points $(x_1, \ldots, x_n)$ in euclidean space, each of whose coordinates is a linear function of a single parameter t; $x_i = f_i(t)$.

lineage structure [CRYSTAL] An imperfection structure characterizing a crystal, parts of which have slight differences in orientation.

linear [CONT SYS] Having an output that varies in direct proportion to the input. [SCI TECH] **1.** Of or relating to a line. **2.** Having a single dimension.

linear accelerator [NUCLEO] A particle accelerator which accelerates electrons, protons, or heavy ions in a straight line by the action of alternating voltages. Also known as linac.

linear algebra [MATH] The study of vector spaces and linear transformations.

linear algebraic equation [MATH] An equation in some algebraic system where the unknowns occur linearly, that is, to the first power.

linear amplifier [ELECTR] An amplifier in which changes in output current are directly proportional to changes in applied input voltage.

linear array [ELECTROMAG] An antenna array in which the dipole or other half-wave elements are arranged end to end on the same straight line. Also known as collinear array.

linear birefringence [OPTICS] Birefringence effects which are proportional to applied stresses.

linear circuit *See* linear network.

linear combination [MATH] A linear combination of vectors $\mathbf{v}_1, \ldots, \mathbf{v}_n$ in a vector space is any expression of the form $a_1\mathbf{v}_1 + a_2\mathbf{v}_2 + \cdots + a_n\mathbf{v}_n$, where the a_i are scalars.

linear combination of atomic orbitals [PHYS] A method of constructing approximate wave functions for molecular orbitals or for electrons in solids, by taking sums of atomic orbitals of the component atoms, each centered on an atom in the structure and multiplied by a coefficient, and then varying these coefficients to minimize the energy of the wave function. Abbreviated LCAO.

linear conductor antenna [ELECTROMAG] An antenna consisting of one or more wires which all lie along a straight line.

linear control [ELEC] Rheostat or potentiometer having uniform distribution of graduated resistance along the entire length of its resistance element.

linear control system [CONT SYS] A linear system whose inputs are forced to change in a desired manner as time progresses.

linear dependence [MATH] The property enjoyed by a set of vectors $\mathbf{v}_1, \ldots, \mathbf{v}_n$ for which there exist scalars $a_1, \ldots, a_n$ not all equal to zero such that $a_1\mathbf{v}_1 + a_2\mathbf{v}_2 + \cdots + a_n\mathbf{v}_n$ equals the zero vector.

linear differential equation [MATH] A differential equation in which all derivatives occur linearly, and all coefficients are functions of the independent variable.

linear discriminant function [STAT] A function, used in conjunction with a set of threshold values in a classification procedure, whose values are linear combinations of the values of selected variables.

linear energy transfer *See* stopping power.

linear equation [MATH] A linear equation in the variables $x_1, \ldots, x_n$, and y is any equation of the form $a_1x_1 + a_2x_2 + \cdots + a_nx_n = y$.

linear expansion [PHYS] Expansion of a body in one direction.

linear extrapolation distance [NUCLEO] The extrapolation distance of a medium sustaining a neutron chain reaction, based on extrapolation of the neutron flux density just inside the medium by a linear function.

linear feedback control [CONT SYS] Feedback control in a linear system.

linear fractional transformations *See* Möbius transformations.

linear function *See* linear transformation.

linear functional [MATH] A linear transformation from a vector space to its scalar field.

linear hypothesis *See* linear model.

linear independence [MATH] The property of a set of vectors $\mathbf{v}_1, \ldots, \mathbf{v}_n$ in a vector space where if $a_1\mathbf{v}_1 + a_2\mathbf{v}_2 + \cdots + a_n\mathbf{v}_n = 0$, then all the scalars $a_i = 0$.

linear inequalities [MATH] A collection of relations among variables x_i, where at least one relation has the form $\Sigma_i a_i x_i \geq 0$.

linear interpolation [MATH] A process to find a value of a function between two known values under the assumption that the three plotted points lie on a straight line.

linearity [MATH] The property whereby a mathematical system is well behaved (in the context of the given system) with regard to addition and scalar multiplication. [PHYS] The relationship that exists between two quantities when a change in one of them produces a directly proportional change in the other.

linearity control [ELECTR] A cathode-ray-tube control which varies the distribution of scanning speed throughout the trace interval. Also known as distribution control.

linearizable [MATH] A nonlinear functional or transformation is linearizable at a given point of a vector space if it has a Fréchet derivative there.

linearization [CONT SYS] **1.** The modification of a system so that its outputs are approximately linear functions of its inputs, in order to facilitate analysis of the system. **2.** The mathematical approximation of a nonlinear system, whose departures from linearity are small, by a linear system corresponding to small changes in the variables about their average values.

linearized theory of fluid flow [FL MECH] An approximate method for solving aerodynamic problems; it treats the flow of an inviscid gas past a body whose geometry and motion are such that the disturbance velocities caused by its introduction into some previously known flow are small compared with the speed of sound; as a result, the equations of motion can be approximated by retaining only those terms which are linear in disturbance or perturbation velocities, pressures, densities, and so forth.

linearly disjoint extensions [MATH] Two extension fields E and F of a field k contained in a common field L, such that any finite set of elements in E that is linearly independent when E is regarded as a vector space over k remains linearly independent when E is regarded as a vector space over F.

linearly independent quantities [MATH] Quantities which do not jointly satisfy a homogeneous linear equation unless all coefficients are zero.

linearly ordered set [MATH] A set with an ordering $\leq$ such that for any two elements a and b either $a \leq b$ or $b \leq a$. Also known as chain; serially ordered set; simply ordered set.

linear manifold [MATH] A subset of a vector space which is itself a vector space with the induced operations of addition and scalar multiplication.

linear model [STAT] A mathematical model in which linear equations connect the random variables and the parameters. Also known as linear hypothesis.

linear molecule [PHYS CHEM] A molecule whose atoms are arranged so that the bond angle between each is 180°; an example is carbon dioxide, CO_2.

linear momentum *See* momentum.

linear motion *See* rectilinear motion.

linear network [ELEC] A network in which the parameters of resistance, inductance, and capacitance are constant with respect to current or voltage, and in which the voltage or current of sources is independent of or directly proportional to other voltages and currents, or their derivatives, in the network. Also known as linear circuit.

linear operator *See* linear transformation.

linear order [MATH] Any order $<$ on a set S with the property that for any two elements a and b in S, either $a < b$ or $b < a$. Also known as complete order; simple order; total order.

linear oscillator *See* harmonic oscillator.

linear polarization [OPTICS] Polarization of an electromagnetic wave in which the electric vector at a fixed point in space remains pointing in a fixed direction, although varying in magnitude. Also known as plane polarization.

linear programming [MATH] The study of maximizing or minimizing a linear function $f(x_1, \ldots, x_n)$ subject to given constraints which are linear inequalities involving the variables x_i.

linear-quadratic-Gaussian problem [CONT SYS] An optimal-state regulator problem, containing Gaussian noise in both the state and measurement equations, in which the expected value of the quadratic performance index is to be minimized. Abbreviated LQG problem.

linear regression [STAT] The straight line running among the points of a scatter diagram about which the amount of scatter is smallest, as defined, for example, by the least squares method.

linear space *See* vector space.

linear Stark effect [ATOM PHYS] A splitting of spectral lines of hydrogenlike atoms placed in an electric field; each energy level of principal quantum number n is split into $2n - 1$ equidistant levels of separation proportional to the field strength.

linear stopping power *See* stopping power.

linear strain [MECH] The ratio of the change in the length of a body to its initial length. Also known as longitudinal strain.

linear sweep [ELECTR] A cathode-ray sweep in which the beam moves at constant velocity from one side of the screen to the other, then suddenly snaps back to the starting side.

linear-sweep delay circuit [ELECTR] A widely used form of linear time-delay circuit in which the input signal initiates action by a linear sawtooth generator, such as the bootstrap or Miller integrator, whose output is then compared with a calibrated direct-current reference voltage level.

linear-sweep generator [ELECTR] An electronic circuit that provides a voltage or current that is a linear function of time; the waveform is usually recurrent at uniform periods of time.

linear system [CONT SYS] A system in which the outputs are components of a vector which is equal to the value of a linear operator applied to a vector whose components are the inputs. [MATH] A system where all the interrelationships among the quantities involved are expressed by linear equations which may be algebraic, differential, or integral.

linear taper [ELEC] A taper that gives the same change in resistance per degree of rotation over the entire range of a potentiometer.

linear ternary ring [MATH] A ternary ring (R,t) such that $t(a,b,c) = t[1,t(a,b,0),c]$ for all a, b, and c in R.

linear time base [ELECTR] A time base that makes the electron beam of a cathode-ray tube move at a constant speed along the horizontal time scale.

linear topological space *See* topological vector space.

linear transducer [ELECTR] A transducer for which the pertinent measures of all the waves concerned are linearly related.

linear transformation [MATH] A function T defined in a vector space E and having its values in another vector space over the same field, such that if f and g are vectors in E, and c is a scalar, then $T(f+g) = Tf + Tg$ and $T(cf) = c(Tf)$. Also known as homogeneous transformation; linear function; linear operator.

linear trend [STAT] A first step in analyzing a time series, to determine whether a linear relationship provides a good approximation to the long-term movement of the series; computed by the method of semiaverages or by the method of least squares.

linear velocity *See* velocity.

line at infinity *See* ideal line.

line-balance converter *See* balun.

line defect *See* dislocation.

line displacement [ASTROPHYS] Widening or shifting of spectral lines of celestial objects arising from several causes, such as gas under high pressure.

line graph [MATH] A graph in which successive points representing the value of a variable at selected values of the independent variable are connected by straight lines.

line integral [MATH] **1.** For a curve in a vector space defined by $\mathbf{x} = \mathbf{x}(t)$, and a vector function $\mathbf{V}$ defined on this curve, the line integral of $\mathbf{V}$ along the curve is the integral over t of the scalar product of $\mathbf{V}[\mathbf{x}(t)]$ and $d\mathbf{x}/dt$; this is written $\int \mathbf{V} \cdot d\mathbf{x}$. **2.** For a curve defined by $x = x(t)$, $y = y(t)$, and a scalar function f depending on x and y, the line integral of f along the curve is the integral over t of $f[x(t),y(t)]\sqrt{(dx/dt)^2 + (dy/dt)^2}$; this is written $\int f\, ds$, where $ds = \sqrt{(dx)^2 + (dy)^2}$ is an infinitesimal element of length along the curve. **3.** For a curve in the complex plane defined by $z = z(t)$, and a function f depending on z, the line integral of f along the curve is the integral over t of $f[z(t)]\,(dz/dt)$; this is written $\int f\, dz$.

line lengthener [ELECTROMAG] Device for altering the electrical length of a waveguide or transmission line without

LINEAR-SWEEP DELAY CIRCUIT

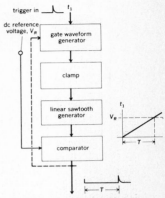

Elements of linear-sweep delay circuit. T = delay time; V = reference voltage; t_1 = time.

LINEAR-SWEEP GENERATOR

Sawtooth waveform of a linear-sweep generator. Current i or voltage v is plotted against time t.

altering other electrical characteristics, or the physical length.

line of apsides [ASTRON] **1.** The line connecting the two points of an orbit that are nearest and farthest from the center of attraction, as the perigee and apogee of the moon or the perihelion and aphelion of a planet. **2.** The length of this line.

line of collimation [OPTICS] In a surveying telescope, the imaginary line through the optical center of the object glass and the cross-hair intersection in the diaphragm.

line of curvature [MATH] A curve on a surface whose tangent lies along a principal direction at each point.

line of electrostatic induction [ELEC] A unit of electric flux equal to the electric flux associated with a charge of 1 statcoulomb.

line of fall [MECH] The line tangent to the ballistic trajectory at the level point.

line of flight [MECH] The line of movement, or the intended line of movement, of an aircraft, guided missile, or projectile in the air.

line of flux *See* line of force.

line of force [PHYS] An imaginary line in a field of force (such as an electric, magnetic, or gravitational field) whose tangent at any point gives the direction of the field at that point; the lines are spaced so that the number through a unit area perpendicular to the field represents the intensity of the field. Also known as flux line; line of flux.

line of impact [MECH] A line tangent to the trajectory of a missile at the point of impact.

line of magnetic induction *See* maxwell.

line of sight [ELECTROMAG] The straight line for a transmitting radar antenna in the direction of the beam. [SCI TECH] A straight, unobstructed path or line between two points, as between an observer's eye and a target.

line-of-sight velocity *See* radial velocity.

line of thrust [MECH] Locus of the points through which the resultant forces pass in an arch or retaining wall.

line pair [SPECT] In spectrographic analysis, a particular spectral line and the internal standard line with which it is compared to determine the concentration of a substance.

line profile [ASTROPHYS] A curve that indicates the internal variation in intensity of a spectral line of a celestial body.

line segment [MATH] A connected piece of a line.

line source [OPTICS] An idealized source of light consisting of an infinitely long line from which light is emitted with uniform intensity.

line spectrum [SPECT] **1.** A spectrum of radiation in which the quantity being studied, such as frequency or energy, takes on discrete values. **2.** Conventionally, the spectra of atoms, ions, and certain molecules in the gaseous phase at low pressures; distinguished from band spectra of molecules, which consist of a pattern of closely spaced spectral lines which could not be resolved by early spectroscopes.

line-turn *See* maxwell-turn.

line vortex [FL MECH] A type of fluid motion in which fluid flows approximately in circles about a line, at speeds inversely proportional to the distance from the line, so that there is an infinite concentration of vorticity on the line, and vorticity vanishes elsewhere.

linkage *See* flux linkage.

link relatives method [STAT] A method for computing indexes by dividing the value of a magnitude in one period by the value in the previous period.

Linnik interference microscope [OPTICS] A type of interference microscope used for studying the surface structure of reflecting specimens; light from a source is divided by a semireflecting mirror into two beams, one of which is focused

LINNIK INTERFERENCE MICROSCOPE

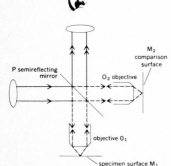

The Linnik interference microscope for reflecting specimens.

through an objective onto the specimen surface, the other onto a comparison surface; after reflection from the respective surfaces, the beams are reunited by the mirror.

Liouville equation [STAT MECH] An equation which states that the density of points representing an ensemble of systems in phase space which are in the neighborhood of some given system does not change with time.

Liouville-Neumann series [MATH] An infinite series of functions constructed from the given functions in the Fredholm equation which under certain conditions provides a solution. Also known as Neumann series.

Liouville's theorem [MATH] Every function of a complex variable which is bounded and analytic in the entire complex plane must be constant.

Lippich prism [OPTICS] A Nicol prism which is placed in the eyepiece of a polarimeter, covering half the field of view, to identify the character of polarized light emerging from the instrument.

Lippmann effect [PHYS] A change in surface tension that results from a potential difference across the interface between two immiscible liquid conductors.

Lippmann electrometer *See* capillary electrometer.

Lippmann fringes [OPTICS] Interference fringes in standing electromagnetic waves generated when light is reflected by a mercury coating at the back of a special fine-grained photographic emulsion; originally used in color photography.

Lipschitz condition [MATH] **1.** A function f satisfies such a condition at a point b if $|f(x) - f(b)| \leq K|x - b|$, with K a constant, for all x in some neighborhood of b. **2.** A function $f(x)$ satisfies the Lipschitz condition in an interval or in a region of the plane if $|f(x) - f(y)| \leq K|x - y|$, with K a constant, for all x and y in the interval or region.

liquefaction [PHYS] A change in the phase of a substance to the liquid state; usually, a change from the gaseous to the liquid state, especially of a substance which is a gas at normal pressure and temperature.

liquid [PHYS] A state of matter intermediate between that of crystalline substances and gases in which a substance has the capacity to flow under extremely small shear stresses and conforms to the shape of a confining vessel, but is relatively incompressible, lacks the capacity to expand without limit, and can possess a free surface.

liquid-bubble tracer [FL MECH] A method of observing the motion of a liquid by following tiny particles of an immiscible liquid of the same density as the moving liquid.

liquid-column gage *See* U-tube manometer.

liquid crystal [PHYS CHEM] A liquid which is not isotropic; it is birefringent and exhibits interference patterns in polarized light; this behavior results from the orientation of molecules parallel to each other in large clusters.

liquid degeneracy [STAT MECH] A process in which a liquid cooled below a certain temperature loses the entropy associated with disordered motion of its molecules, without becoming a solid.

liquid-drop model of nucleus [NUC PHYS] A model of the nucleus in which it is compared to a drop of incompressible liquid, and the nucleons are analogous to molecules in the liquid; used to study binding energies, fission, collective motion, decay, and reactions. Also known as drop model of nucleus.

liquid extraction *See* solvent extraction.

liquid flow [FL MECH] The flow or movement of materials in the liquid phase.

liquid helium [CRYO] The state of helium which exists at atmospheric pressure at temperatures below −268.95°C (4.2

LIQUID HELIUM

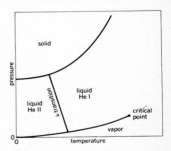

Phase diagram for He⁴ (not to scale). There are two liquid phases, helium I and helium II; the transition between them is called the λ-transition. Solid phase is never in equilibrium with vapor phase, and occurs only at elevated pressures. Critical point has temperature of 5.2 K and pressure of 2.26 atmospheres.

K), and for temperatures near absolute zero at pressures up to about 25 atmospheres (2.53 megapascals); has two phases, helium I and helium II.

liquid holdup [FL MECH] A condition in two-phase flow through a vertical pipe; when gas flows at a greater linear velocity than the liquid, slippage takes place and liquid holdup occurs.

liquid hydrogen [CRYO] Hydrogen that exists as a liquid at atmospheric pressure, at −252.7°C (20.46 K); used for high-impulse rocket fuels.

liquid-hydrogen bubble chamber [NUCLEO] A bubble chamber in which the active liquid is hydrogen; particularly useful in research on elementary particles produced in high-energy interactions, because the hydrogen provides a dense target of protons.

liquid ionization chamber [NUCLEO] A particle detector in which the gas filling a conventional ionization chamber is replaced by an exceedingly pure liquid, usually a liquefied noble gas.

liquid junction emf [PHYS CHEM] The emf (electromotive force) generated at the area of contact between the salt bridge and the test solution in a pH cell electrode.

liquid junction potential *See* diffusion potential.

liquid laser [OPTICS] A laser whose active material is dissolved in a liquid contained in a transparent cylindrical shell; rare-earth ions in suitable dissolved molecules and organic dye solutions are used.

liquid measure [MECH] A system of units used to measure the volumes of liquid substances in the United States; the units are the fluid dram, fluid ounce, gill, pint, quart, and gallon.

liquid-metal MHD generator [ELEC] A system for generating electric power in which the kinetic energy of a flowing, molten metal is converted to electric energy by magnetohydrodynamic (MHD) interaction.

liquid-metal nuclear fuel [NUCLEO] A nuclear fuel consisting of a solution of uranium or plutonium in a molten metal such as bismuth.

liquid nitrogen [CRYO] Nitrogen that exists as a liquid at atmospheric pressure, at −195°C (77.4 K); used in research work, cryogenics, and cryosurgery.

liquid oxygen [CRYO] Oxygen that exists as a liquid at 59 atmospheres (5.978 megapascals), at −113°C (160.2 K); a pale-blue, transparent, mobile liquid.

liquid pint *See* pint.

liquid poison [NUCLEO] A neutron-absorbing liquid that can be injected quickly into the cooling system of a nuclear reactor by explosive-actuated valves; used for automatic or manual scramming to shut down a reactor.

liquid scintillation detector [NUCLEO] Scintillation counter in which the sensitive material is a liquid, such as *p*-terphenyl dissolved in toluene, placed in a glass or metal container.

liquid semiconductor [ELECTR] An amorphous material in solid or liquid state that possesses the properties of varying resistance induced by charge carrier injection.

liquidus line [THERMO] For a two-component system, a curve on a graph of temperature versus concentration which connects temperatures at which fusion is completed as the temperature is raised.

Lissajous figure [PHYS] The path of a particle moving in a plane when the components of its position along two perpendicular axes each undergo simple harmonic motions and the ratio of their frequencies is a rational number.

liter [MECH] A unit of volume or capacity, equal to 1 decimeter cubed, or 0.001 cubic meter, or 1000 cubic centimeters. Abbreviated l.

literal constant [MATH] A letter denoting a constant.

literal expression [MATH] An expression or equation in which the constants are represented by letters.

literal notation [MATH] The use of letters to denote numbers, known or unknown.

liter-atmosphere [PHYS] A unit of energy equal to the work done on a piston by a fluid at a pressure of 1 standard atmosphere (101,325 newtons per square meter) when the piston sweeps out a volume of 1 liter; equal to 101.325 joules.

lithium [CHEM] A chemical element, symbol Li, atomic number 3, atomic weight 6.939; an alkali metal.

lithium cell [CHEM] An electrolytic cell for the production of metallic lithium. [ELEC] A primary cell for producing electrical energy using lithium metal for one electrode immersed in usually an organic electrolyte.

lithium fluoride dosimetry [NUCLEO] A method of dosimetry in which the radiation dose received by a sample of the phosphor lithium fluoride is determined by measuring the thermoluminescent output of the phosphor upon heating, following the irradiation.

Littrow grating spectrograph [SPECT] A spectrograph having a plane grating at an angle to the axis of the instrument, and a lens in front of the grating which both collimates and focuses the light.

Littrow mounting [SPECT] The arrangement of the grating and other components of a Littrow grating spectrograph, which is analogous to that of a Littrow quartz spectrograph.

Littrow prism [OPTICS] A prism having angles of 30, 60, and 90°, silvered on the side opposite the 60° angle; a lens used with it can serve both as a telescope and as a collimator.

Littrow quartz spectrograph [SPECT] A spectrograph in which dispersion is accomplished by a Littrow quartz prism with a rear reflecting surface that reverses the light; a lens in front of the prism acts as both collimator and focusing lens.

lituus [MATH] The trumpet-shaped plane curve whose points in polar coordinates (r,θ) satisfy the equation $r^2 = a/\theta$.

Litzendraht wire *See* Litz wire.

Litz wire [ELEC] Wire consisting of a number of separately insulated strands woven together so each strand successively takes up all possible positions in the cross section of the entire conductor, to reduce skin effect and thereby reduce radio-frequency resistance. Derived from Litzendraht wire.

live *See* energized.

live load [MECH] A moving load or a load of variable force acting upon a structure, in addition to its own weight.

Livens's theorem [MECH] For a general path of a classical dynamical system, the condition that the variation of the integral, over time, from t_0 to t_1, of $\Sigma p_i\dot{q}_i - H(p_i,q_i t)$ depend only on terms at the limits $t = t_0$ and $t = t_1$ when the path and the termini are varied simultaneously, with the q_i and p_i allowed to vary independently, is given by Hamilton's equations of motion.

livre [MECH] A unit of mass, used in France, equal to 0.5 kilogram.

LLL circuit *See* low-level logic circuit.

Lloyd's mirror interference [OPTICS] The interference pattern produced when part of the light from a slit falls directly on a screen, and part is reflected from a mirror whose surface makes a small angle with the incident beam.

lm *See* lumen.

L/M [NUC PHYS] The ratio of the number of internal conversion electrons emitted from the L shell in the deexcitation of

LITHIUM

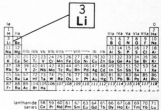

Periodic table of the chemical elements showing the position of lithium.

LITTROW QUARTZ SPECTROGRAPH

Littrow quartz spectrograph, typical arrangement for double-beam recording instrument. S, slit; P_1, totally reflecting quartz prism; L, autocollimating quartz lens; P_2, Littrow quartz prism; C, camera; RV, red to violet spectrum.

LLOYD'S MIRROR INTERFERENCE

Splitting of a light source with Lloyd's mirror. The slit S_1 and its virtual image S_2' constitute the double source. Part of the light falls directly on the screen at P, and part is reflected at grazing incidence from a plane mirror.

a nucleus to the number of such electrons emitted from the *M* shell.

L meson [PARTIC PHYS] A *K*-meson resonance having a mass of 1781 ± 14 MeV and a width of 70 MeV; its decay modes are dominantly $K\pi\pi$, with some $K\omega$ (5%).

lm-hr *See* lumen-hour.

lm-sec *See* lumen-second.

lm/w *See* lumen per watt.

L network [ELECTR] A network composed of two branches in series, with the free ends connected to one pair of terminals; the junction point and one free end are connected to another pair of terminals.

load [ELEC] **1.** A device that consumes electric power. **2.** The amount of electric power that is drawn from a power line, generator, or other power source. **3.** The material to be heated by an induction heater or dielectric heater. Also known as work. [ELECTR] The device that receives the useful signal output of an amplifier, oscillator, or other signal source. [MECH] **1.** The weight that is supported by a structure. **2.** Mechanical force that is applied to a body. **3.** The burden placed on any machine, measured by units such as horsepower, kilowatts, or tons.

load cell [ELEC] A device which measures large pressures by applying the pressure to a piezoelectric crystal and measuring the voltage across the crystal; the cell plus a recording mechanism constitutes a strain gage.

load characteristic [ELECTR] Relation between the instantaneous values of a pair of variables such as an electrode voltage and an electrode current, when all direct electrode supply voltages are maintained constant. Also known as dynamic characteristic.

load compensation [CONT SYS] Compensation in which the compensator acts on the output signal after it has generated feedback signals. Also known as load stabilization.

loaded concrete [MATER] Concrete to which elements of high atomic number or capture cross section have been added to increase its effectiveness as a radiation shield in nuclear reactors.

loaded motional impedance *See* motional impedance.

loaded Q [ELEC] The Q factor of an impedance which is connected or coupled under working conditions. Also known as working Q. [ELECTROMAG] The Q factor of a specific mode of resonance of a microwave tube or resonant cavity when there is external coupling to that mode.

load impedance [ELECTR] The complex impedance presented to a transducer by its load.

loading [ELEC] The addition of inductance to a transmission line to improve its transmission characteristics throughout a given frequency band. Also known as electrical loading. [NUCLEO] Placing fuel in a nuclear reactor.

loading disk [ELECTROMAG] Circular metal piece mounted at the top of a vertical antenna to increase its natural wavelength.

load line [ELECTR] A straight line drawn across a series of tube or transistor characteristic curves to show how output signal current will change with input signal voltage when a specified load resistance is used.

load loss [ELEC] The sum of the copper loss of a transformer, due to resistance in the windings, plus the eddy current loss in the winding, plus the stray loss.

load stabilization *See* load compensation.

Lobachevski geometry [MATH] A system of planar geometry in which the euclidean parallel postulate fails; any point *p* not on a line *L* has at least two lines through it parallel to *L*. Also known as Bolyai geometry; hyperbolic geometry.

LOAD LINE

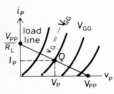

Load line drawn across characteristic curves giving plate current, i_P, as function of plate voltage, v_P, for various values of grid supply voltage, V_{GG}. V_{PP} = plate supply voltage, R_L = load resistance, v_G = grid voltage. Quiescent point, Q, determines quiescent plate current, I_P, and quiescent plate voltage, V_P.

lobe [ELECTROMAG] A part of the radiation pattern of a directional antenna representing an area of stronger radio-signal transmission. Also known as radiation lobe.

lobe-half-power width [ELECTROMAG] In a plane containing the direction of the maximum energy of a lobe, the angle between the two directions in that plane about the maximum in which the radiation intensity is one-half the maximum value of the lobe.

local algebra [MATH] An algebra A over a field F which is the sum of the radical of A and the subalgebra consisting of products of elements of F with the multiplicative identity of A.

local apparent noon [ASTRON] Twelve o'clock local apparent time, or the instant the apparent sun is over the upper branch of the local meridian.

local apparent time [ASTRON] The arc of the celestial equator, or the angle at the celestial pole, between the lower branch of the local celestial meridian and the hour circle of the apparent or true sun, measured westward from the lower branch of the local celestial meridian through 24 hours.

local base [MATH] For a point x in a topological space, a family of neighborhoods of x such that every neighborhood of x contains a member of the family. Also known as base for the neighborhood system.

local buckling [MECH] Buckling of thin elements of a column section in a series of waves or wrinkles.

local cell [ELEC] A galvanic cell resulting from differences in potential between adjacent areas on the surface of a metal immersed in an electrolyte.

local cluster of stars *See* local star system.

local coefficient [MATH] By using fiber bundles where the fiber is a group, one may generalize cohomology theory for spaces; one uses such bundles as the algebraic base for such a theory and calls the bundle a system of local coefficients.

local coefficient of heat transfer [THERMO] The heat transfer coefficient at a particular point on a surface, equal to the amount of heat transferred to an infinitesimal area of the surface at the point by a fluid passing over it, divided by the product of this area and the difference between the temperatures of the surface and the fluid.

local coordinate system [MATH] The coordinate system about a point which is induced when the global space is locally euclidean.

local derivative [FL MECH] The rate of change of a quantity with respect to time at a fixed point of a fluid, $\partial f/\partial t$; it is related to the individual derivative df/dt through the expression $\partial f/\partial t = df/dt - V \cdot \nabla f$, where f is a thermodynamic property $f(x,y,z,t)$ of the fluid, V the vector velocity of the fluid, and ∇ the del operator.

local group [ASTRON] A group of at least 20 known galaxies in the vicinity of the sun; the Andromeda Spiral is the largest of the group, and the Milky Way Galaxy is the second largest.

local hour angle [ASTRON] Angular distance west of the local celestial meridian.

localized state [QUANT MECH] A state of motion in which an electron may be found anywhere within a region of a material of linear extent smaller than that of the material.

localized vector [MECH] A vector whose line of application or point of application is prescribed, in addition to its direction.

local lunar time [ASTRON] The arc of the celestial equator, or the angle at the celestial pole, between the lower branch of the local celestial meridian and the hour circle of the moon, measured westward from the lower branch of the local celestial meridian through 24 hours; local hour angle of the moon, expressed in time units, plus 12 hours; local lunar time at the Greenwich meridian is called Greenwich lunar time.

locally arcwise connected topological space [MATH] A topological space in which every point has an arcwise connected neighborhood, that is, an open set any two points of which can be joined by an arc.

locally compact topological space [MATH] A topological space in which every point lies in a compact neighborhood.

locally connected topological space [MATH] A topological space in which every point has a connected neighborhood.

locally convex space [MATH] A Hausdorff topological vector space E such that every neighborhood of any point x belonging to E contains a convex neighborhood of x.

locally euclidean topological space [MATH] A topological space in which every point has a neighborhood which is homeomorphic to a euclidean space.

locally integrable function [MATH] A function is said to be locally integrable on an open set S in n-dimensional euclidean space if it is defined almost everywhere in S and has a finite integral on compact subsets S.

locally one to one [MATH] A function is locally one to one if it is one to one in some neighborhood of each point.

locally trivial bundle [MATH] A bundle for which each point in the base has a neighborhood U whose inverse image under the projection map is isomorphic to a cartesian product of U with a space isomorphic to the fibers of the bundle.

local maximum [MATH] A local maximum of a function f is a value $f(c)$ of f where $f(x) \leq f(c)$ for all x in some neighborhood of c; if $f(c)$ is a local maximum, f is said to have a local maximum at c.

local mean noon [ASTRON] Twelve o'clock local mean time, or the instant the mean sun is over the upper branch of the local meridian; local mean noon at the Greenwich meridian is called Greenwich mean noon.

local mean time [ASTRON] The arc of the celestial equator, or the angle at the celestial pole, between the lower branch of the local celestial meridian and the hour circle of the mean sun, measured westward from the lower branch of the local celestial meridian through 24 hours.

local meridian [ASTRON] The meridian through any particular position which serves as the reference for local time.

local minimum [MATH] A local minimum of a function f is a value $f(c)$ of f where $f(x) \geq f(c)$ for all x in some neighborhood of c; if $f(c)$ is a local minimum, f is said to have a local minimum at c.

local noon [ASTRON] Noon at the local meridian.

local quasi-F martingale [MATH] A stochastic process $\{X_t\}$ such that the process obtained from $\{X_t\}$ by stopping it when it reaches n or $-n$ is a quasi-F martingale for each integer n.

local ring [MATH] A ring with only one maximal ideal.

local sidereal noon [ASTRON] Zero hour local sidereal time, or the instant the vernal equinox is over the upper branch of the local meridian; local sidereal noon at the Greenwich meridian is called Greenwich sidereal noon.

local sidereal time [ASTRON] The arc of the celestial equator, or the angle at the celestial pole which is between the upper branch of the local celestial meridian and the hour circle of the vernal equinox.

local solution [MATH] A function which solves a system of equations only in a neighborhood of some point.

local star cloud *See* local star system.

local star system [ASTRON] The group of stars of which the sun is a member. Also known as local cluster of stars; local star cloud.

local time [ASTRON] **1.** Time based upon the local meridian as reference, as contrasted with that based upon a zone meridian, or the meridian of Greenwich. **2.** Any time kept locally.

located vector [MATH] An ordered pair of points in n-dimensional euclidean space.

locked oscillator [ELECTR] A sine-wave oscillator whose frequency can be locked by an external signal to the control frequency divided by an integer.

lock-in [ELECTR] Shifting and automatic holding of one or both of the frequencies of two oscillating systems which are coupled together, so that the two frequencies have the ratio of two integral numbers.

lock-in amplifier [ELECTR] An amplifier that uses some form of automatic synchronization with an external reference signal to detect and measure very weak electromagnetic radiation at radio or optical wavelengths in the presence of very high noise levels.

locking [ELECTR] Controlling the frequency of an oscillator by means of an applied signal of constant frequency.

locus [MATH] A collection of points in a euclidean space whose coordinates satisfy one or more algebraic conditions.

locus of an equation [MATH] A collection of points, all of which satisfy a single equation.

logarithm [MATH] **1.** The real-valued function log u defined by log $u = v$ if $e^v = u$, e^v denoting the exponential function. Also known as hyperbolic logarithm; Naperian logarithm; natural logarithm. **2.** An analog in complex variables relative to the function e^z.

logarithmic amplifier [ELECTR] An amplifier whose output signal is a logarithmic function of the input signal.

logarithmic coordinate paper [MATH] Paper ruled with two sets of mutually perpendicular, parallel lines spaced according to the logarithms of consecutive numbers, rather than the numbers themselves.

logarithmic coordinates [MATH] In the plane, logarithmic coordinates are defined by two coordinate axes, each marked with a scale where the distance between two points is the difference of the logarithms of the two numbers.

logarithmic curve [MATH] A curve whose equation in cartesian coordinates is $y = \log ax$, where a is greater than 1.

logarithmic decrement [PHYS] The natural logarithm of the ratio of the amplitude of one oscillation to that of the next which has the same polarity, when no external forces are applied to maintain the oscillation.

logarithmic derivative [MATH] The logarithmic derivative of a function $f(z)$ of a real (complex) variable is the ratio $f'(z)/f(z)$, that is, the derivative of log $f(z)$.

logarithmic differentiation [MATH] A technique often helpful in computing the derivatives of a differentiable function $f(x)$; set $g(x) = \log f(x)$ where $f(x) \neq 0$, then $g'(x) = f'(x)/f(x)$, and if there is some other way to find $g'(x)$, then one also finds $f'(x)$.

logarithmic distribution [STAT] A frequency distribution whose value at any integer $n = 1, 2, \ldots$ is $\lambda^n/(-n) \log (1-\lambda)$, where λ is fixed.

logarithmic equation [MATH] An equation which involves a logarithmic function of some variable.

logarithmic integral [MATH] A function whose value at x is equal to the integral from 0 to x of $(1/\ln t) \, dt$. Denoted li; Li.

logarithmic multiplier [ELECTR] A multiplier in which each variable is applied to a logarithmic function generator, and the outputs are added together and applied to an exponential function generator, to obtain an output proportional to the product of two inputs.

logarithmic potential [PHYS] A potential function that is proportional to the logarithm of some coordinate; for example, a straight, electrically charged cylinder of circular cross section and effectively infinite length gives rise to an electrostatic potential that is the sum of a constant and a term proportional to the logarithm of the distance from the cylinder's axis.

logarithmic profile of velocity [FL MECH] The mean velocity parallel to a boundary of a fluid in turbulent motion as a function of distance from the boundary, on the assumption that the shearing stress is independent of distance from the boundary, and the mixing length is proportional either to the distance from the boundary or to the ratio of the first derivative of the profile of velocity itself to the second derivative.

logarithmic scale [MATH] A scale in which the distances that numbers are at from a reference point are proportional to their logarithms.

logarithmic series [MATH] The Maclaurin series expansion of the natural logarithm of $1 + x$, valid for $-1 < x \leq 1$, namely $\log(1 + x) = x - (x^2/2) + (x^3/3) - (x^4/4) + \cdots$.

logarithmic spiral [MATH] The spiral plane curve whose points in polar coordinates (r, θ) satisfy the equation $\log r = a\theta$. Also known as equiangular spiral.

logarithmic transformation [STAT] The replacement of a variate y with a new variate $z = \log y$ or $z = \log(y + c)$, where c is a constant; this operation is often performed when the resulting distribution is normal, or if the resulting relationship with another variable is linear.

logic [ELECTR] **1.** The basic principles and applications of truth tables, interconnections of on/off circuit elements, and other factors involved in mathematical computation in a computer. **2.** General term for the various types of gates, flip-flops, and other on/off circuits used to perform problem-solving functions in a digital computer. [MATH] The subject that investigates, formulates, and establishes principles of valid reasoning.

logical addition [MATH] The additive binary operation of a Boolean algebra.

logical connectives [MATH] Symbols which link mathematical statements; these symbols represent the terms "and," "or," "implication," and "negation."

logical function *See* predicate.

logical gate *See* switching gate.

logical multiplication [MATH] The multiplicative binary operation of a Boolean algebra.

logistic curve [STAT] **1.** A type of growth curve, representing the size of a population y as a function of time t: $y = k/(1 + e^{-kbt})$, where k and b are positive constants. Also known as Pearl-Reed curve. **2.** More generally, a curve representing a function of the form $y = k/(1 + e^{cf(t)})$, where c is a constant and $f(t)$ is some function of time.

log-mean temperature difference [THERMO] The log-mean temperature difference

$$T_{LM} = \frac{(T_2 - T_1)}{\ln T_2/T_1}$$

where T_2 and T_1 are the absolute (K or °R) temperatures of the two extremes being averaged; used in heat transfer calculations in which one fluid is cooled or heated by a second held separate by pipes or process vessel walls.

log-periodic antenna [ELECTROMAG] A broad-band antenna which consists of a sheet of metal with two wedge-shaped cutouts, each with teeth cut into its radii along circular arcs; characteristics are repeated at a number of frequencies that are equally spaced on a logarithmic scale.

Lommel integral [MATH] The integral of $[(a^2 - b^2)x + (\mu^2 - \nu^2)/x]B_\nu(ax)B_\mu(bx)\ dx$, where $B_\nu(x)$ is a solution of Bessel's equation with parameter ν.

Lommel's functions [MATH] The functions

$$U_\nu(x,y) = \sum_{n=0}^{\infty} (-1)^n (x/y)^{\nu+2n} J_{\nu+2n}(y)$$

$$V_\nu(x,y) = \cos(x/2 + y/2x + \nu\pi/2) + U_{2-\nu}(x,y)$$

where J_ν is the Bessel function of the first kind of order ν.

London dispersion force *See* van der Waals force.

London equations [SOLID STATE] Equations for the time derivative and the curl of the current in a superconductor in terms of the electric and magnetic field vectors respectively, derived in the London superconductivity theory.

London penetration depth [SOLID STATE] A measure of the depth which electric and magnetic fields can penetrate beneath the surface of a superconductor from which they are otherwise excluded, according to the London superconductivity theory.

London superconductivity theory [SOLID STATE] An extension of the two-fluid model of superconductivity, in which it is assumed that superfluid electrons behave as if the only force acting on them arises from applied electric fields, and that the curl of the superfluid current vanishes in the absence of a magnetic field.

London superfluidity theory [CRYO] A theory, based on the fact that helium-4 obeys Bose-Einstein statistics, in which helium-4 is treated as an ideal Bose-Einstein gas, and its superfluid component is equated with the finite fraction of the atoms of such a gas which are in the ground state at very low temperatures.

long-conductor antenna *See* long-wire antenna.

long discharge [ELEC] **1.** A capacitor or other electrical charge accumulator which takes a long time to leak off. **2.** A gaseous electrical discharge in which the length of the discharge channel is very long compared with its diameter; lightning discharges are natural examples of long discharges. Also known as long spark.

long hundredweight *See* hundredweight.

longitudinal aberration [OPTICS] **1.** The distance along the optical axis from the focus of paraxial rays to the point where rays coming from the outer edges of its lens or reflecting surface intersect this axis. **2.** In chromatic aberration, the distance along the optical axis between the foci of two standard colors.

longitudinal acceleration [MECH] The component of the linear acceleration of an aircraft, missile, or particle parallel to its longitudinal, or X, axis.

longitudinal magnetorestriction *See* Joule effect.

longitudinal mass [RELAT] The ratio of a force acting on a relativistic particle in the direction of its velocity to the resulting acceleration; equal to $m_0(1 - v^2/c^2)^{-3/2}$, where m_0 is the particle's rest mass, v is its speed, and c is the speed of light.

longitudinal quadrupole [ACOUS] A sound source resulting from a variation of a component of the velocity of matter in a direction that is parallel to the velocity component. [ELECTROMAG] An electric or magnetic quadrupole which produces a field equivalent to that of two equal and opposite electric or magnetic dipoles separated by a small distance parallel to the direction of the dipoles. Also known as axial quadrupole.

longitudinal strain *See* linear strain.

longitudinal vibration [MECH] A continuing periodic change in the displacement of elements of a rod-shaped object in the direction of the long axis of the rod.

longitudinal wave [PHYS] A wave in which the direction of some vector characteristic of the wave, for example, the displacement of particles of the transmitting medium, is along the direction of propagation.

long period variable [ASTRON] A variable star with a period from about 100 to more than 600 days.

long-range order [SOLID STATE] A tendency for some property of atoms in a lattice (such as spin orientation or type of atom) to follow a pattern which is repeated every few unit cells.

long run frequency [STAT] The ratio of the number of occurrences of an event in a large number of trials to the number of trials.

long spark *See* long discharge.

long-tail pair [ELECTR] A two-tube or transistor circuit that has a common resistor (tail resistor) which gives strong negative feedback.

long-time trend *See* secular trend.

long ton *See* ton.

long wave [COMMUN] An electromagnetic wave having a wavelength longer than the longest broadcast-band wavelength of about 545 meters, corresponding to frequencies below about 550 kilohertz.

long-wire antenna [ELECTROMAG] An antenna whose length is a number of times greater than its operating wavelength, so as to give a directional radiation pattern. Also known as long-conductor antenna.

Loomis-Wood diagram [SPECT] A graph used to assign lines in a molecular spectrum to the various branches of rotational bands when these branches overlap, in which the difference between observed wave numbers and wave numbers extrapolated from a few lines that apparently belong to one branch are plotted against arbitrary running numbers for that branch.

loop [ELEC] **1.** A closed path or circuit over which a signal can circulate, as in a feedback control system. **2.** Commercially, the portion of a connection from central office to subscriber in a telephone system. **3.** *See* mesh. [ELECTROMAG] *See* coupling loop; loop antenna. [PHYS] **1.** A closed curve on a graph, such as a hysteresis loop. **2.** *See* antinode.

loop antenna [ELECTROMAG] A directional-type antenna consisting of one or more complete turns of a conductor, usually tuned to resonance by a variable capacitor connected to the terminals of the loop. Also known as loop.

loop coupling [ELECTROMAG] A method of transferring energy between a waveguide and an external circuit, by inserting a conducting loop into the waveguide, oriented so that electric lines of flux pass through it.

loop gain [CONT SYS] The ratio of the magnitude of the primary feedback signal in a feedback control system to the magnitude of the actuating signal. [ELECTR] Total usable power gain of a carrier terminal or two-wire repeater; maximum usable gain is determined by, and may not exceed, the losses in the closed path.

loop ratio *See* loop transfer function.

loop transfer function [CONT SYS] For a feedback control system, the ratio of the Laplace transform of the primary feedback signal to the Laplace transform of the actuating signal. Also known as loop ratio.

loop transmittance [CONT SYS] **1.** The transmittance between the source and sink created by the splitting of a specified node in a signal flow graph. **2.** The transmittance between the source and sink created by the splitting of a node which has been inserted in a specified branch of a signal flow graph in

LOOP ANTENNA

Shape of loop antenna.

such a way that the transmittance of the branch is unchanged.

Lorentz-Boltzmann equation [STAT MECH] An approximation to the Boltzmann transport equation for states that are near equilibrium, which shows that the Maxwell-Boltzmann distribution applies at equilibrium.

Lorentz contraction *See* FitzGerald-Lorentz contraction.

Lorentz electron [ELECTROMAG] A model of the electron as a damped harmonic oscillator; used to explain the variation of the real and imaginary parts of the index of refraction of a substance with frequency.

Lorentz equation [ELECTROMAG] The equation of motion for a charged particle, which sets the rate of change of its momentum equal to the Lorentz force.

Lorentz-FitzGerald contraction *See* FitzGerald-Lorentz contraction.

Lorentz force [ELECTROMAG] The force on a charged particle moving in electric and magnetic fields, equal to the particle's charge times the sum of the electric field and the cross product of particle's velocity with the magnetic flux density.

Lorentz force density [ELECTROMAG] The force per unit volume on a charge density and current density, assuming that these densities arise from large numbers of charged particles experiencing a Lorentz force.

Lorentz four-vector *See* four-vector.

Lorentz frame [RELAT] Any of the family of inertial coordinate systems, with three space coordinates and one time coordinate, used in the special theory of relativity; each frame is in uniform motion with respect to all the other Lorentz frames, and the interval between any two events is the same in all frames.

Lorentz gage [ELECTROMAG] Any gage in which the sum of the divergence of the vector potential and the partial derivative of the scalar potential divided by the speed of light (in Gaussian units) vanishes identically; it is always possible to find a gage satisfying this condition.

Lorentz gas [ELECTR] A model of completely ionized gas in which ions are assumed to be stationary and interactions between electrons are neglected.

Lorentz group [MATH] The group of all Lorentz transformations of euclidean four-space with composition as the operation.

Lorentz-Heaviside system *See* Heaviside-Lorentz system.

Lorentz invariance [RELAT] The property, possessed by the laws of physics and of certain physical quantities, of being the same in any Lorentz frame, and thus unchanged by a Lorentz transformation.

Lorentz line-splitting theory [ATOM PHYS] A theory predicting that when a light source is placed in a strong magnetic field, its spectral lines are each split into three components, one of them retaining the zero-field frequency, and the other two shifted upward and downward in frequency by the Larmor frequency (the normal Zeeman effect).

Lorentz local field [ELEC] In a theory of electric polarization, the average electric field due to the polarization at a molecular site that is calculated under the assumption that the field due to polarization by molecules inside a small sphere centered at the site may be neglected. Also known as Mosotti field.

Lorentz-Lorenz equation [OPTICS] The equation that results from replacing the relative dielectric constant with the square of the index of refraction in the Clausius-Mosotti equation.

Lorentz-Lorenz molar refraction *See* molar refraction.

Lorentz matrix [RELAT] A matrix whose product with a vector whose components are the space and time coordinates of an event yields a vector whose components are new coordinates derived from the original ones by a Lorentz transformation.

Lorentz number [PL PHYS] The ratio of the velocity of a fluid to the velocity of light. Symbolized N_{Lo}. [SOLID STATE] The thermal conductivity of a metal divided by the product of its temperature and its electrical conductivity, according to the Wiedemann-Franz law.

Lorentz polarization factor [OPTICS] A geometric factor in the equation for the intensity of x-rays or other radiation diffracted through a given angle by a crystalline substance.

Lorentz relation *See* Wiedemann-Franz law.

Lorentz theory of light sources [ATOM PHYS] A theory according to which light is emitted by vibrations of electrons, which are damped harmonic oscillators attached to atoms.

Lorentz transformation [MATH] Any linear transformation of euclidean four space which preserves the quadratic form $q(x,y,z,t) = t^2 - x^2 - y^2 - z^2$. [RELAT] Any of the family of mathematical transformations used in the special theory of relativity to relate the space and time variables of different Lorentz frames.

Lorentz unit [SPECT] A unit of reciprocal length used to measure the difference, in wave numbers, between a (zero field) spectrum line and its Zeeman components; equal to $eH/4\pi mc^2$, where H is the magnetic field strength, c is the speed of light, and e and m are the charge and mass of the electron respectively (Gaussian units).

Lorenz curve [STAT] A graph for showing the concentration of ownership of economic quantities such as wealth and income; it is formed by plotting the cumulative distribution of the amount of the variable concerned against the cumulative frequency distribution of the individuals possessing the amount.

Loschmidt number [PHYS] The number of molecules in 1 cubic centimeter of an ideal gas at 1 atmosphere pressure (101,325 newtons per square meter) and 0°C, equal to approximately 2.687×10^{19}.

loss angle [ELECTROMAG] A measure of the power loss in an inductor or a capacitor, equal to the amount by which the angle between the phasors denoting voltage and current across the inductor or capacitor differs from 90°.

loss cone [PL PHYS] A cone in the velocity space of particles in a plasma confined by magnetic mirrors; particles with velocities in the cone are not trapped by the mirrors and are lost out of the system.

loss-cone instability [PL PHYS] An instability in a plasma confined between magnetic mirrors.

loss current [ELEC] The current which passes through a capacitor as a result of the conductivity of the dielectric and results in power loss in the capacitor. [ELECTROMAG] The component of the current across an inductor which is in phase with the voltage (in phasor notation) and is associated with power losses in the inductor.

Lossev effect *See* injection electroluminescence.

loss factor [ELEC] The power factor of a material multiplied by its dielectric constant; determines the amount of heat generated in a material.

loss function [MATH] In decision theory, the function, dependent upon the decision and the true underlying distributions, which expresses the loss produced in taking the decision.

lossless junction [ELECTROMAG] A waveguide junction in which all the power incident on the junction is reflected from it.

lossless material [PHYS] An ideal material that dissipates none of the energy of electromagnetic or acoustic waves passing through it.

loss of head [FL MECH] Energy decrease between two points in a hydraulic system due to such causes as friction, bends, obstructions, or expansions.

lossy attenuator [ELECTROMAG] In waveguide technique, a length of waveguide deliberately introducing a transmission loss by the use of some dissipative material.

lossy material [PHYS] A material that dissipates energy of electromagnetic or acoustic energy passing through it.

loudness level [ACOUS] The level of a sound, in phons, equal to the sound pressure level in decibels, relative to 0.0002 microbar, of a pure 1000-hertz tone that is judged to be equally loud by listeners.

loudness unit [ACOUS] A unit of loudness equal to the loudness of a sound having a loudness level of 0 phon; the loudness unit has been replaced by the sone.

Love wave [GEOPHYS] A horizontal dispersive surface wave, multireflected between internal boundaries of an elastic body, applied chiefly in the study of seismic waves in the earth's crust.

Love-Young fractional integral [MATH] A generalization of repeated integration of a function f from a number x to a fixed upper limit b that is larger than x, given by the integral from x to b of $[\Gamma(\alpha)]^{-1}(t - x)^{\alpha-1}f(t)\,dt$, where the real part of α is positive.

Lovibond tintometer [OPTICS] A colorimeter which compares a solution or object under examination with a series of slides of each of three colors.

low-angle scattering _See_ small-angle scattering.

low-energy electron diffraction [SOLID STATE] A technique for studying the atomic structure of single crystal surfaces, in which electrons of uniform energy in the approximate range 5–500 electron volts are scattered from a surface, and those scattered electrons that have lost no energy are selected and accelerated to a fluorescent screen where the diffraction pattern from the surface can be observed. Abbreviated LEED.

low-energy np scattering [NUC PHYS] An elastic collision of a neutron, having an energy from less than 1 electron volt to 10,000,000 electron volts, with a proton (usually the nucleus of a hydrogen atom).

low-energy physics [PHYS] That part of physics which studies microscopic phenomena involving energies of several million electron volts or less, such as the arrangement of electrons in an atom or a solid, and the arrangement of protons and neutrons within the atomic nucleus, and the nature of forces between these particles.

low-energy pp scattering [NUC PHYS] An elastic collision of a proton, having an energy of less than 10,000,000 electron volts, with another proton (usually the nucleus of a hydrogen atom).

lower bound [MATH] A lower bound of a subset A of a set S is a point of S which is smaller than every element of A.

lower branch [ASTRON] That half of a meridian or celestial meridian from pole to pole which passes through the antipode or nadir of a place.

lower control limit [IND ENG] The horizontal line drawn on a control chart at a specified distance below the central line; points plotted below the lower control limit indicate that the process may be out of control.

lower culmination _See_ lower transit.

lower Darboux integral [MATH] For a bounded function $f(x)$ defined on a closed interval $[a,b]$, the limit specified by

LOW-ENERGY ELECTRON DIFFRACTION

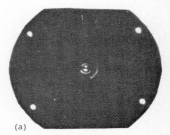

(a)

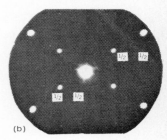

(b)

Low-energy electron diffraction patterns from a (110) tungsten surface. _(a)_ Clean, 54 volts. _(b)_ Half monolayer of oxygen atoms, 53 volts. Fractions refer to Miller indices of superstructure produced by oxygen absorption. _(From L. H. Germer and J. W. May, Diffraction study of oxygen absorption on a (110) tungsten face, Surface Sci., 4:452, 1966)_

Darboux's theorem for the sum s derived from greatest lower bounds $f(x)$ on subintervals of $[a,b]$; written

$$\int_a^b f(x)\ dx.$$

lower derivative [MATH] The lower derivative of a real-valued function of a real variable $f(x)$ at a point x_0 is the limit inferior of $[f(x) - f(x_0)]/(x - x_0)$ as x approaches x_0.

lower half-power frequency [ELECTR] The frequency on an amplifier response curve which is smaller than the frequency for peak response and at which the output voltage is $1/\sqrt{2}$ of its midband or other reference value.

lower heating value *See* low heat value.

lower metric density [MATH] The lower metric density of a measurable subset S of the real numbers at a point x is the limit as n approaches infinity of $\phi_n(x)$, where $\phi_n(x)$ is the greatest lower bound, over all intervals I containing x and having length $l(I)$ less than $1/n$, of $m(S \cap I)/l(I)$, where $m(S \cap I)$ is the measure of the intersection of S with I.

lower pitch limit [ACOUS] Minimum frequency, for a sinusoidal sound wave, that will produce a pitch sensation.

lower transit [ASTRON] Transit across the lower branch of the celestial meridian. Also known as lower culmination.

lower variation [MATH] The lower variation of a signed measure m is a set function m^- defined for every measurable set E by $m^-(E) = -m(E \cap B)$, where B is the member of the Hahn decomposition which is negative with respect to m.

low-frequency cutoff [ELECTR] A frequency below which the gain of a system or device decreases rapidly.

low-frequency spectrum [SPECT] Spectrum of atoms and molecules in the microwave region, arising from such causes as the coupling of electronic and nuclear angular momenta, and the Lamb shift.

low-frequency transconductance [ELECTR] The change in the plate current of a vacuum tube divided by the change in the control-grid voltage that produces it, at frequencies small enough for these two quantities to be considered in phase.

low-frequency tube [ELECTR] An electron tube operated at frequencies small enough so that the transit time of an electron between electrodes is much smaller than the period of oscillation of the voltage.

low heat value [THERMO] The heat value of a combustion process assuming that none of the water vapor resulting from the process is condensed out, so that its latent heat is not available. Also known as lower heating value; net heating value.

low-level counting [NUCLEO] The measurement of very small amounts of radioactivity, such as that generated by long-lived natural radioactive isotopes, and isotopes produced by cosmic rays and nuclear explosions.

low-level logic circuit [ELECTR] A modification of a diode-transistor logic circuit in which a resistor and capacitor in parallel are replaced by a diode, with the result that a relatively small voltage swing is required at the base of the transistor to switch it on or off. Abbreviated LLL circuit.

low-pass band-pass transformation *See* frequency transformation.

low-pass filter [ELEC] A filter that transmits alternating currents below a given cutoff frequency and substantially attenuates all other currents.

low-pressure fluid flow [FL MECH] Flow of fluids below atmospheric pressures, particularly gases and vapors following ideal gas laws, in pipes, fittings, and other common configurations.

low-Q filter [ELECTR] A filter in which the energy dissipated in each cycle is a fairly large fraction of the energy stored in the filter.

low-reflection film [OPTICS] A transparent film covering a glass surface, designed so that a small proportion of the light incident will be reflected and a correspondingly large proportion transmitted into the glass.

low-temperature hygrometry [ENG] The study that deals with the measurement of water vapor at low temperatures; the techniques used differ from those of conventional hygrometry because of the extremely small amounts of moisture present at low temperatures and the difficulties imposed by the increase of the time constants of the standard instruments when operated at these temperatures.

low-temperature physics [CRYO] A study of the properties of gross matter at low temperature's especially at temperatures so low that the quantum character of the substance becomes observable in effects such as superconductivity, superfluid liquid helium, magnetic cooling, and nuclear orientation.

low-temperature production [CRYO] Production of temperatures from about 80 K down to about 10^{-6} K by techniques such as isentropic expansion of gases, refrigeration cycles, and adiabatic demagnetization.

low-temperature thermometry [CRYO] The assignment of numbers on the Kelvin absolute temperature scale to achievable and reproducible low-temperature states, and the choice and calibration of suitable instruments for the practical measurement of low temperatures, such as thermocouples, and resistance, vapor-pressure, gas, and magnetic thermometers.

low-velocity layer [GEOPHYS] A layer in the solid earth in which seismic wave velocity is lower than the layers immediately below or above.

loxodromic spiral [MATH] A curve on a surface of revolution which cuts the meridians at a constant angle other than 90°.

Lr *See* lawrencium.

LSA diode [ELECTR] A microwave diode in which a space charge is developed in the semiconductor by the applied electric field and is dissipated during each cycle before it builds up appreciably, thereby limiting transit time and increasing the maximum frequency of oscillation. Derived from limited space-charge accumulation diode.

LS coupling *See* Russell-Saunders coupling.

L shell [ATOM PHYS] The second shell of electrons surrounding the nucleus of an atom, having electrons whose principal quantum number is 2.

LSI circuit *See* large-scale integrated circuit.

L^p-space [MATH] For a given positive number p and a given measure space (S,μ) the space $L^p(S)$ consists of all functions f defined on S such that the integral over S of $|f(x)|^p \, d\mu$ converges.

Lu *See* lutetium.

lub *See* least upper bound.

Luckiesh-Moss visibility meter [ENG] A type of photometer that consists of two variable-density filters (one for each eye) that are adjusted so that an object seen through them is just barely discernible; the reduction in visibility produced by the filters is read on a scale of relative visibility related to a standard task.

lumberg [OPTICS] A unit of luminous energy equal to the luminous energy corresponding to a radiant energy of $1/K$ ergs, where K is the luminous efficiency in lumens per watt. Formerly known as lumerg.

lumen [OPTICS] The unit of luminous flux, equal to the

**LUCKIESH-MOSS
VISIBILITY METER**

Photograph of Luckiesh-Moss visibility meter. *(General Electric Co.)*

luminous flux emitted within a unit solid angle (1 steradian) from a point source having a uniform intensity of 1 candela, or to the luminous flux received on a unit surface, all points of which are at a unit distance from such a source. Symbolized lm. [SCI TECH] The space within a tube.

lumen-hour [OPTICS] A unit of quantity of light (luminous energy), equal to the quantity of light radiated or received for a period of 1 hour by a flux of 1 lumen. Abbreviated lm-hr.

lumen per watt [OPTICS] The unit of luminosity factor and of luminous efficacy. Abbreviated lm/w.

lumen-second [OPTICS] A unit of quantity of light (luminous energy), equal to the quantity of light radiated or received for a period of 1 second by a flux of 1 lumen. Abbreviated lm-sec.

lumerg *See* lumberg.

luminance [OPTICS] The ratio of the luminous intensity in a given direction of an infinitesimal element of a surface containing the point under consideration, to the orthogonally projected area of the element on a plane perpendicular to the given direction. Formerly known as brightness.

luminance factor [OPTICS] The ratio of the luminance of a body when illuminated and observed under certain conditions to that of a perfect diffuser under the same conditions.

luminescence [PHYS] Light emission that cannot be attributed merely to the temperature of the emitting body, but results from such causes as chemical reactions at ordinary temperatures, electron bombardment, electromagnetic radiation, and electric fields.

luminescent [PHYS] Capable of exhibiting luminescence.

luminescent center [SOLID STATE] A point-lattice defect in a transparent crystal that exhibits luminescence.

luminescent screen [ELECTR] The screen in a cathode-ray tube, which becomes luminous when bombarded by an electron beam and maintains its luminosity for an appreciable time.

luminophor [PHYS] A luminescent material that converts part of the absorbed primary energy into emitted luminescent radiation. Also known as fluophor; fluor; phosphor.

luminosity *See* luminosity factor.

luminosity classes [ASTRON] A classification of stars in an orderly sequence according to their absolute brightness.

luminosity curve *See* luminosity function.

luminosity factor [OPTICS] The ratio of luminous flux in lumens emitted by a source at a particular wavelength to the corresponding radiant flux in watts at the same wavelength; thus this is a measure of the visual sensitivity of the eye. Also known as luminosity.

luminosity function [ASTRON] The functional relationship between stellar magnitude and the number and distribution of stars of each magnitude interval. Also known as relative luminosity factor. [OPTICS] A standard measure of the response of an eye to monochromatic light at various wavelengths; the function is normalized to unity at its maximum value. Also known as luminosity curve; spectral luminous efficiency; visibility function.

luminous coefficient [OPTICS] A measure of the fraction of the radiant power of a light source which contributes to its luminous properties, equal to the average of the luminosity function at various wavelengths, weighted according to the spectral intensity of the source. Also known as luminous efficiency.

luminous efficacy [OPTICS] **1.** The ratio of the total luminous flux in lumens emitted by a light source over all wavelengths to the total radiant flux in watts. Formerly known as luminous efficiency. **2.** The ratio of the total luminous flux

LUMINOSITY FUNCTION

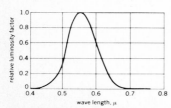

Human-eye luminosity function. *(From W. B. Boast, Illumination Engineering, 2d ed., McGraw-Hill, 1953)*

emitted by a light source to the power input of the source; expressed in lumens per watt.

luminous efficiency   *See* luminous coefficient; luminous efficacy.

luminous emittance   [OPTICS] The emittance of visible radiation weighted to take into account the different response of the human eye to different wavelengths of light; in photometry, luminous emittance is always used as a property of a self-luminous source, and therefore should be distinguished from luminance. Also known as luminous exitance.

luminous energy   [OPTICS] The total radiant energy emitted by a source, evaluated according to its capacity to produce visual sensation; measured in lumen-hours or lumen-seconds.

luminous exitance   *See* luminous emittance.

luminous flux   [OPTICS] The time rate of flow of radiant energy, evaluated according to its capacity to produce visual sensations; measured in lumens.

luminous flux density   *See* illuminance.

luminous intensity   [OPTICS] The luminous flux incident on a small surface which lies in a specified direction from a light source and is normal to this direction, divided by the solid angle (in steradians) which the surface subtends at the source of light. Also known as light intensity.

luminous nebula   [ASTRON] A nebula made bright by radiation from stars in the vicinity.

luminous quantities   [OPTICS] Physical quantities used in photometry, such as luminous intensity and luminance, which are based on the response of the human eye, and are thus weighted to take into account the difference in response at different wavelengths of light.

Lummer-Brodhun sight box   [OPTICS] A device, having a series of prisms, for viewing simultaneously the two sides of a white diffuse plaster screen illuminated by light sources whose luminous intensities are being compared.

Lummer-Gehrcke plate   [OPTICS] An interferometer consisting of a glass or quartz plate with parallel surfaces and sizable thickness in which multiple reflections take place.

lumped constant   [ELEC] A single constant that is electrically equivalent to the total of that type of distributed constant existing in a coil or circuit. Also known as lumped parameter.

lumped-constant network   [ELEC] An analytical tool in which distributed constants (inductance, capacitance, and resistance) are represented as hypothetical components.

lumped discontinuity   [ELECTROMAG] An analytical tool in the study of microwave circuits in which the effective values of inductance, capacitance, and resistance representing a discontinuity in a waveguide are shown as discrete components of equivalent value.

lumped parameter   *See* lumped constant.

lunar appulse   [ASTRON] An eclipse of the moon in which the penumbral shadow of the earth falls on the moon. Also known as penumbral eclipse.

lunar atmosphere   [ASTROPHYS] The volatile elements postulated to have been present on the moon's surface at one time.

lunar day   [ASTRON] The time interval between two successive crossings of the meridian by the moon.

lunar ephemeris   [ASTRON] A computed list of positions the moon will occupy in the sky on certain dates.

lunar geology   *See* selenology.

lunar inequality   [ASTRON] Variation in the moon's motion in its orbit, due to attraction by other bodies of the solar system. [GEOPHYS] A minute fluctuation of a magnetic needle from its mean position, caused by the moon.

lunar interval   [ASTRON] The difference in time between the

LUMMER-BRODHUN SIGHT BOX

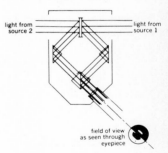

Lummer-Brodhun contrast sight box.

transit of the moon over the Greenwich meridian and a local meridian; the lunar interval equals the difference between the Greenwich and local intervals of a tide or current phase.

lunar libration [ASTRON] **1.** The effect wherein the face of the moon appears to swing east and west about 8° from its central position each month. Also known as apparent libration in longitude. **2.** The state wherein the inclination of the moon's polar axis allows an observer on earth to see about 59% of the moon's surface. Also known as libration in latitude. **3.** The small oscillation with which the moon rocks back and forth about its mean rotation rate. Also known as physical libration of the moon.

lunar month [ASTRON] The period of revolution of the moon about the earth, especially a synodical month.

lunar rainbow *See* moonbow.

lunar time [ASTRON] **1.** Time based upon the rotation of the earth relative to the moon; it may be designated as local or Greenwich, as the local or Greenwich meridian is used as the reference. **2.** Time on the moon.

lune [MATH] A section of a plane bounded by two circular arcs, or of a sphere bounded by two great circles.

Luneberg lens [ELECTROMAG] A type of antenna consisting of a dielectric sphere whose index of refraction varies with distance from the center of the sphere so that a beam of parallel rays falling on the lens is focused at a point on the lens surface diametrically opposite from the direction of incidence, and, conversely, energy emanating from a point on the surface is focused into a plane wave.

lunisolar precession [ASTROPHYS] Precession of the earth's equinox caused by the gravitational attraction of the sun and moon.

Lunt's theorem [MATH] If X is an $n \times n$ Boolean matrix each of whose diagonal elements is the universal element of the underlying Boolean algebra, then the matrix X^{p+1} includes X^p for $p = 1, 2, \ldots$, and X^{p+1} equals X^p for $p \geq n - 1$.

lusec [PHYS] A unit used for the measurement of power of evacuation of a vacuum pump, equal to the power associated with a leak rate of 1 liter per second at a pressure of 1 millitorr, or to approximately 1.33322×10^{-4} watt.

Lusin's theorem [MATH] Given a measurable function f which is finite almost everywhere in a euclidean space, then for every number $\varepsilon > 0$ there is a continuous function g which agrees with f, except on a set of measure less than ε.

luster [OPTICS] The appearance of a surface dependent on reflected light; types include metallic, vitreous, resinous, adamantine, silky, pearly, greasy, dull, and earthy; applied to minerals, textiles, and many other materials.

lutetium [CHEM] A chemical element, symbol Lu, atomic number 71, atomic weight 174.97; a very rare metal and the heaviest member of the rare-earth group.

lux [OPTICS] A unit of illumination, equal to the illumination on a surface 1 square meter in area on which there is a luminous flux of 1 lumen uniformly distributed, or the illumination on a surface all points of which are at a distance of 1 meter from a uniform point source of 1 candela. Symbolized lx. Also known as meter-candle.

luxon *See* troland.

L wave [GEOPHYS] A phase designation for an earthquake wave that is a surface wave, without respect to type.

lx *See* lux.

Lyapunov function [MATH] A function of a vector and of time which is positive-definite and has a negative-definite derivative with respect to time for nonzero vectors, is identically zero for the zero vector, and approaches infinity as the norm of the vector approaches infinity; used in determining the

LUNEBERG LENS

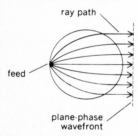

Luneberg lens with dielectric sphere between feed and plane-phase wavefront.

LUTETIUM

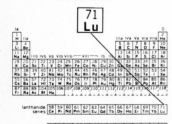

Periodic table of the chemical elements showing the position of lutetium

stability of control systems. Also spelled Liapunov function.

Lyapunov stability criterion [CONT SYS] A method of determining the stability of systems (usually nonlinear) by examining the sign-definitive properties of an associated Lyapunov function.

Lyddane-Sachs-Teller relation [SOLID STATE] For an infinite ionic crystal, the relation $\varepsilon(0)/\varepsilon(\infty) = \omega_L^2/\omega_T^2$, where $\varepsilon(0)$ is the crystal's static dielectric constant, $\varepsilon(\infty)$ is the dielectric constant at a frequency at which electronic polarizability is effective but ionic polarizability is not, ω_L is the frequency of longitudinal optical phonons with zero wave vectors, and ω_T is the frequency of transverse optical phonons with large wave vector.

Lyman-alpha radiation [SPECT] Radiation emitted by hydrogen associated with the spectral line in the Lyman series whose wavelength is 1215 angstrom units (121.5 nanometers).

Lyman band [SPECT] A band in the ultraviolet spectrum of molecular hydrogen, extending from 125 to 161 nanometers.

Lyman continuum [SPECT] A continuous range of wavelengths (or wave numbers or frequencies) in the spectrum of hydrogen at wavelengths less than the Lyman limit, resulting from transitions between the ground state of hydrogen and states in which the single electron is freed from the atom.

Lyman ghost [SPECT] A false line observed in a spectroscope as a result of a combination of periodicities in the ruling.

Lyman limit [SPECT] The lower limit of wavelengths of spectral lines in the Lyman series (912 angstrom units or 91.2 nanometers), or the corresponding upper limit in frequency, energy of quanta, or wave number (equal to the Rydberg constant for hydrogen).

Lyman series [SPECT] A group of lines in the ultraviolet spectrum of hydrogen covering the wavelengths of 1215–912 angstrom units (121.5–91.2 nanometers).

Lyot filter *See* birefringent filter.

m *See* meter; milli-.

M *See* mega-; molarity.

mA *See* milliampere.

Macbeth illuminometer [OPTICS] A type of portable visual photometer in which the light to be measured is balanced by a Lummer-Brodhun sight box against a comparison lamp, whose apparent brightness can be varied by moving it along a tube; a control box supplies a calibrated current to the comparison lamp, and calibrated optical filters can be placed in the light paths to correct for color differences in the comparison and measured sources and to extend the range of the instrument.

MacCullagh's formula [PHYS] A formula for the potential due to a distribution of mass or charge at an external point: the potential V at a point P resulting from a distribution of mass or positive charge centered about a point O is $V = kM/r + (k/2r^3)(A + B + C - 3I) + O(1/r^4)$, where r is the distance from O to P, k is the gravitational or electrostatic constant, M is the total mass or charge, A, B, and C are the principal moments about O, I is the moment about OP, and $O(1/r^4)$ is a quantity that falls off at least as rapidly as $1/r^4$.

MacDonald functions *See* modified Hankel functions.

MacDonald's formula [MATH] An approximate equation, $P_\nu(\cos\theta) = J_0[(2\nu + 1)\sin\frac{1}{2}\theta] + O[(\sin\frac{1}{2}\theta)^2]$, where P_ν is the Legendre function of order ν, J_0 is the Bessel function of the first kind of order 0, and $O[(\sin\frac{1}{2}\theta)^2]$ is a function of order $(\sin\frac{1}{2}\theta)^2$.

Mach angle [FL MECH] The vertex half angle of the Mach cone generated by a body in supersonic flight.

Mach cone [FL MECH] **1.** The cone-shaped shock wave theoretically emanating from an infinitesimally small particle moving at supersonic speed through a fluid medium; it is the locus of the Mach lines. **2.** The cone-shaped shock wave generated by a sharp-pointed body, as at the nose of a high-speed aircraft.

Mach front *See* Mach stem.

Mach indicator *See* Machmeter.

machine language [ADP] The set of instructions available to a particular digital computer, and by extension the format of a computer program in its final form, capable of being executed by a computer.

Machin's formula [MATH] The equation $\pi/4 = 4\tan^{-1}(1/5) - \tan^{-1}(1/239)$, useful for calculating π to large numbers of decimal places.

Mach line [FL MECH] ·**1.** A line representing a Mach wave. **2.** *See* Mach wave.

Machmeter [ENG] An instrument that measures and indicates the speed of sound, that is, indicates the Mach number. Also known as Mach indicator.

Mach number [FL MECH] The ratio of the speed of a body or

MACH CONE

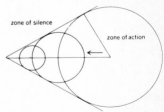

zone of silence

zone of action

Generation of Mach cone (def. 1) by particle moving through fluid at supersonic speed. Arrow indicates direction of particle's motion. Cone is tangent to spherical surfaces of disturbances generated by particle at successive times, and separates zones of action and silence.

of a point on a body with respect to the surrounding air or other fluid, or the ratio of the speed of a fluid, to the speed of sound in the medium. Symbolized Ma; N_{Ma} Also known as relative Mach number.

Mach principle [RELAT] The principle that the motion of a particle is only meaningful when referred to the rest of the matter in the universe; this motion is thus determined by the distribution of this matter and is not an intrinsic property of an absolute space.

Mach reflection [FL MECH] The reflection of a shock wave from a rigid wall in which the shock strength of the reflected wave and the angle of reflection both have the smaller of the two values which are theoretically possible.

Mach refractometer *See* Mach-Zehnder interferometer.

Mach stem [FL MECH] A shock wave or front formed above the surface of the earth by the fusion of direct and reflected shock waves resulting from an airburst bomb. Also known as Mach front.

Mach wave [FL MECH] Also known as Mach line. **1.** A shock wave theoretically occurring along a common line of intersection of all the pressure disturbances emanating from an infinitesimally small particle moving at supersonic speed through a fluid medium, with such a wave considered to exert no changes in the condition of the fluid passing through it. **2.** A very weak shock wave appearing, for example, at the nose of a very sharp body, where the fluid undergoes no substantial change in direction.

Mach-Zehnder interferometer [OPTICS] A variation of the Michelson interferometer used mainly in measuring the spatial variation of the index of refraction of a gas; the device has two semitransparent mirrors and two wholly reflecting mirrors at alternate corners of a rectangle, and half the beam travels along each side of the rectangle. Also known as Mach refractometer.

Maclaurin expansion [MATH] The power series representation of a function arising from Maclaurin's theorem.

Maclaurin series [MATH] The power series in the Maclaurin expansion.

Maclaurin's theorem [MATH] The theorem giving conditions when a function, which is infinitely differentiable, may be represented in a neighborhood of the origin as an infinite series with nth term $(1/n!) \cdot f^{(n)}(0) \cdot x^n$, where $f^{(n)}$ denotes the nth derivative.

Macleod equation [FL MECH] An equation which states that the fourth root of the surface tension of a liquid is proportional to the difference between the densities of the liquid and of its vapor.

MacMichael degree [FL MECH] An arbitrary unit used in measuring viscosity with a type of Couette viscometer; its size depends on the stiffness of the suspension of the inner cylinder of the viscometer.

macro- [SCI TECH] Prefix meaning large.

macroanalytical balance [ENG] A relatively large type of analytical balance that can weigh loads of up to 200 grams to the nearest 0.1 milligram.

MacRobert's E function [MATH] A function formed from the generalized hypergeometric function $_pF_q$, gamma functions of the parameters of $_pF_q$, and powers, which is defined for all finite values of the argument z except possibly $z = 0$; used to give meaning to $_pF_q$ when $p > q + 1$.

macrodome [CRYSTAL] Dome of a crystal in which planes are parallel to the longer lateral axis.

macrometer [OPTICS] Instrument that has two mirrors and a focusing telescope with which the ranges of distant objects can be found.

MACH-ZEHNDER INTERFEROMETER

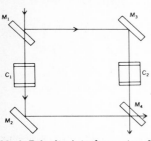

Mach-Zehnder interferometer. M_1, M_4 are semitransparent mirrors; M_2, M_3 are reflecting mirrors; C_1, C_2 are cells containing substances whose index of refraction is to be measured.

macropinacoid *See* front pinacoid.

macrorheology [MECH] A branch of rheology in which materials are treated as homogeneous or quasi-homogeneous, and processes are treated as isothermal.

macroscopic cross section [PHYS] The sum of the cross sections of an atom in a substance.

macroscopic property [NUCLEO] A nuclear reactor property that can be treated independently of other factors. [THERMO] *See* thermodynamic property.

macroscopic state [STAT MECH] Any state of a system as described by actual or hypothetical observations of its macroscopic statistical properties. Also known as macrostate.

macroscopic theory [PHYS] A theory concerning only phenomena observable with the naked eye or with an ordinary light microscope, and not with the behavior of atoms, molecules, or their constituents which may underlie these phenomena.

macrosonics [ACOUS] The technology of sound at signal amplitudes so large that linear approximations are not valid, as in the use of ultrasonics for cleaning or drilling.

macrostate *See* macroscopic state.

Madelung constant [SOLID STATE] A dimensionless constant which determines the electrostatic energy of a three-dimensional periodic crystal lattice consisting of a large number of positive and negative point charges when the number and magnitude of the charges and the nearest-neighbor distance between them is specified.

madistor [ELECTR] A cryogenic semiconductor device in which injection plasma can be steered or controlled by transverse magnetic fields, to give the action of a switch.

MADT *See* microalloy diffused transistor.

MAG *See* maximum available gain.

magamp *See* magnetic amplifier.

Magellanic Clouds [ASTRON] Two irregular clouds of stars that are the nearest galaxies to the galactic system; both the Large and Small Magellanic Clouds are identified as Irregular in the classification of E. P. Hubble. Also known as Nubeculae.

Maggi-Righi-Leduc effect [PHYS] A phenomenon in which the thermal conductivity of a conductor changes when it is placed in a magnetic field.

magic numbers [NUC PHYS] The integers 8, 20, 28, 50, 82, 126; nuclei in which the number of protons, neutrons, or both is magic have a stability and binding energy which is greater than average, and have other special properties.

magic square [MATH] A square array of integers where the sum of the entries of each row, each column, and each diagonal is the same.

magic tee *See* hybrid tee.

magn [ELECTROMAG] A unit of absolute permeability equal to 1 henry per meter; it has been proposed by the Soviet Union but has not won general acceptance.

magnesium [CHEM] A metallic element, symbol Mg, atomic number 12, atomic weight 24.312.

magnesium cell [ELEC] A primary cell in which the negative electrode is made of magnesium or one of its alloys.

magnesium–silver chloride cell [ELEC] A reserve primary cell that is activated by adding water; active elements are magnesium and silver chloride.

magnet [ELECTROMAG] A piece of ferromagnetic or ferrimagnetic material whose domains are sufficiently aligned so that it produces a net magnetic field outside itself and can experience a net torque when placed in an external magnetic field.

magnet coil *See* iron-core coil.

MAGNESIUM

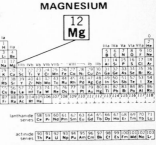

Periodic table of the chemical elements showing the position of the element magnesium.

MAGNET

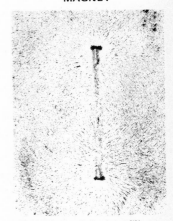

Photograph of the iron-filing map of the magnetic field of a permanent bar magnet. Note that the magnetic flux lines can be traced by the lines of iron filings, which act like tiny compass needles.

magnetic amplifier [ELECTR] A device that employs saturable reactors to modulate the flow of alternating-current electric power to a load in response to a lower-energy-level direct-current input signal. Abbreviated magamp. Also known as transductor.

magnetic anisotropy [ELECTROMAG] The dependence of the magnetic properties of some materials on direction.

magnetic annual change [GEOPHYS] The amount of secular change in the earth's magnetic field which occurs in 1 year. Also known as annual magnetic change.

magnetic annual variation [GEOPHYS] The small, systematic temporal variation in the earth's magnetic field which occurs after the trend for secular change has been removed from the average monthly values. Also known as annual magnetic variation.

magnetic axis [ELECTROMAG] A line through the center of a magnet such that the torque exerted on the magnet by a magnetic field in the direction of this line equals 0. [PL PHYS] The single line of force that closes on itself after one revolution in a magnetic field with a rotational transform.

magnetic balance [ENG] **1.** A device for determining the repulsion or attraction between magnetic poles, in which one magnet is suspended and the forces needed to cancel the effects of bringing a pole of another magnet close to one end are measured. **2.** Any device for measuring the small forces involved in determining paramagnetic or diamagnetic susceptibility.

magnetic bay [GEOPHYS] A small magnetic disturbance whose magnetograph resembles an indentation of a coastline; on earth, magnetic bays occur mainly in the polar regions and have a duration of a few hours.

magnetic bias [ELECTROMAG] A steady magnetic field applied to the magnetic circuit of a relay or other magnetic device.

magnetic blowout [ELECTROMAG] **1.** A permanent magnet or electromagnet used to produce a magnetic field that lengthens the arc between opening contacts of a switch or circuit breaker, thereby helping to extinguish the arc. **2.** See blowout.

magnetic bottle [PL PHYS] A magnetic field used to confine or contain a plasma in controlled fusion experiments.

magnetic bubble [SOLID STATE] A cylindrical stable (nonvolatile) region of magnetization produced in a thin-film magnetic material by an external magnetic field; direction of magnetization is perpendicular to the plane of the material. Also known as bubble.

magnetic character figure See C index.

magnetic circuit [ELECTROMAG] A group of magnetic flux lines each forming a closed path, especially when this circuit is regarded as analogous to an electric circuit because of the similarity of its magnetic field equations to direct-current circuit equations.

magnetic coercive force See coercive force.

magnetic confinement [PL PHYS] The containment of a plasma within a region of space by the forces of magnetic fields on the charged particles in the gas.

magnetic constant [ELECTROMAG] The absolute permeability of empty space, equal to 1 electromagnetic unit in the centimeter-gram-second system, and to $4\pi \times 10^{-7}$ henry per meter or, numerically, to 1.25664×10^{-6} henry per meter in Intenational System units. Symbolized μ_0.

magnetic cooling See adiabatic demagnetization.

magnetic core Also known as core. [ELECTR] A configuration of magnetic material, usually a mixture of iron oxide or ferrite particles mixed with a binding agent and formed into

MAGNETIC CONFINEMENT

Stable magnetic confinement by a convex magnetic field. The lines of force are convex to the plasma; arrow indicates decrease of magnetic field strength.

a tiny doughnutlike shape, that is placed in a spatial relationship to current-carrying conductors, and is used to maintain a magnetic polarization for the purpose of storing data, or for its nonlinear properties as a logic element. Also known as memory core. [ELECTROMAG] A quantity of ferrous material placed in a coil or transformer to provide a better path than air for magnetic flux, thereby increasing the inductance of the coil and increasing the coupling between the windings of a transformer.

magnetic coupling [ELECTROMAG] For a pair of particles or systems, the effect of the magnetic field created by one system on the magnetic moment or angular momentum of the other.

magnetic Curie temperature [SOLID STATE] The temperature below which a magnetic material exhibits ferromagnetism, and above which ferromagnetism is destroyed and the material is paramagnetic.

magnetic daily variation See magnetic diurnal variation.

magnetic damping [ELECTROMAG] Damping of a mechanical motion by means of the reaction between a magnetic field and the current generated by the motion of a coil through the magnetic field.

magnetic declination See declination.

magnetic diffusivity [ELECTROMAG] A measure of the tendency of a magnetic field to diffuse through a conducting medium at rest; it is equal to the partial derivative of the magnetic field strength wth respect to time divided by the Laplacian of the magnetic field, or to the reciprocal of $4\pi\mu\sigma$, where μ is the magnetic permeability and σ is the conductivity in electromagnetic units.

magnetic dip See inclination.

magnetic dipole [ELECTROMAG] An object, such as a permanent magnet, current loop, or particle with angular momentum, which experiences a torque in a magnetic field, and itself gives rise to a magnetic field, as if it consisted of two magnetic poles of oposite sign separated by a small distance.

magnetic dipole antenna [ELECTROMAG] Simple loop antenna capable of radiating an electromagnetic wave in response to a circulation of electric current in the loop.

magnetic dipole density See magnetization.

magnetic dipole moment [ELECTROMAG] A vector associated with a magnet, current loop, particle, or such, whose cross product with the magnetic induction (or alternatively, the magnetic field strength) of a magnetic field is equal to the torque exerted on the system by the field. Also known as dipole moment; magnetic moment.

magnetic displacement See magnetic induction.

magnetic diurnal variation [GEOPHYS] Oscillations of the earth's magnetic field which have a periodicity of about a day and which depend to a close approximation only on local time and geographic latitude. Also known as magnetic daily variation.

magnetic domain See ferromagnetic domain.

magnetic double refraction [OPTICS] The double refraction of light passing through certain substances when the substance is placed in a transverse magnetic field.

magnetic energy [ELECTROMAG] The energy required to set up a magnetic field.

magnetic equator [GEOPHYS] That line on the surface of the earth connecting all points at which the magnetic dip is zero. Also known as aclinic line.

magnetic field [ELECTROMAG] **1.** One of the elementary fields in nature; it is found in the vicinity of a magnetic body or current-carrying medium and, along with electric field, in a light wave; charges moving through a magnetic field experience the Lorentz force. **2.** See magnetic field strength.

magnetic field intensity *See* magnetic field strength.

magnetic field strength [ELECTROMAG] An auxiliary vector field, used in describing magnetic phenomena, whose curl, in the case of static charges and currents, equals (in meter-kilogram-second units) the free current density vector, independent of the magnetic permeability of the material. Also known as magnetic field; magnetic field intensity; magnetic force; magnetic intensity; magnetizing force.

magnetic film *See* magnetic thin film.

magnetic flux [ELECTROMAG] **1.** The integral over a specified surface of the component of magnetic induction perpendicular to the surface. **2.** *See* magnetic lines of force.

magnetic flux density *See* magnetic induction.

magnetic focusing [ELECTROMAG] Focusing a beam of electrons or other charged particles by using the action of a magnetic field.

magnetic force *See* magnetic field strength.

magnetic force parameter [PL PHYS] A dimensionless number used in magnetofluid dynamics, equal to the product of the square of the magnetic permeability, the square of the magnetic field strength, the electrical conductivity, and a characteristic length, divided by the product of the mass density and the fluid velocity. Symbolized N.

magnetic groups *See* Shubnikov groups.

magnetic hysteresis [ELECTROMAG] Lagging of changes in the magnetization of a substance behind changes in the magnetic field as the magnetic field is varied. Also known as hysteresis.

magnetic induction [ELECTROMAG] A vector quantity that is used as a quantitative measure of magnetic field; the force on a charged particle moving in the field is equal to the particle's charge times the cross product of the particle's velocity with the magnetic induction (mks units). Also known as magnetic displacement; magnetic flux density; magnetic vector.

magnetic induction gyroscope [ENG] A gyroscope without moving parts, in which alternating- and direct-current magnetic fields act on water doped with salts which exhibit nuclear paramagnetism.

magnetic intensity *See* magnetic field strength.

magnetic leakage [ELECTROMAG] Passage of magnetic flux outside the path along which it can do useful work.

magnetic lens [ELECTROMAG] A magnetic field with axial symmetry, capable of converging beams of charged particles of uniform velocity and of forming images of objects placed in the path of such beams; the field may be produced by solenoids, electromagnets, or permanent magnets. Also known as lens.

magnetic lines of flux *See* magnetic lines of force.

magnetic lines of force [ELECTROMAG] Lines used to represent the magnetic induction in a magnetic field, selected so that they are parallel to the magnetic induction at each point, and so that the number of lines per unit area of a surface perpendicular to the induction is equal to the induction. Also known as magnetic flux; magnetic lines of flux.

magnetic Mach number [PL PHYS] A dimensionless number equal to the ratio of the velocity of a fluid to the velocity of Alfvén waves in the fluid. Symbolized M_{Ma}.

magnetic meridian [GEOPHYS] A line which is at any point in the direction of horizontal magnetic force of the earth; a compass needle without deviation lies in the magnetic meridian.

magnetic mirror [PL PHYS] A magnetic field used in controlled-fusion experiments to reflect charged particles into the central region of a magnetic bottle; reflection occurs in the

MAGNETIC LENS

Two types of magnetic lens. *(a)* Uniform magnetic field. *(b)* Short magnetic lens formed at gap in soft-iron casing about coil. Arrows show direction of movement of a beam of charged particles. *(From E. G. Ramberg and G. A. Morton, J. Appl. Phys., vol. 10, 1939, and J. Hillier and E. G. Ramberg, J. Appl. Phys., vol. 18, 1947)*

region where the magnetic field increases abruptly in strength.

magnetic moment   *See* magnetic dipole moment.

magnetic monopole   [ELECTROMAG] A hypothetical particle carrying magnetic charge; it would be a source for magnetic field in the same way that a charged particle is a source for electric field. Also known as monopole.

magnetic multipole   [ELECTROMAG] One of a series of types of static or oscillating distributions of magnetization, which is a magnetic multipole of order 2; the electric and magnetic fields produced by a magnetic multipole of order 2^n are equivalent to those of two magnetic multipoles of order 2^{n-1} of equal strength but opposite sign, separated from each other by a short distance.

magnetic multipole field   [ELECTROMAG] The electric and magnetic fields generated by a static or oscillating magnetic multipole.

magnetic needle   [ELECTROMAG] **1.** A bar magnet or collection of bar magnets which is hung so as to show the direction of the magnetic field. **2.** In particular, a slender bar magnet, pointed at both ends, that is pivoted or freely suspended in a magnetic compass.

magnetic north   [GEOPHYS] At any point on the earth's surface, the horizontal direction of the earth's magnetic lines of force (direction of a magnetic meridian) toward the north magnetic pole; a particular direction indicated by the needle of a magnetic compass.

magnetic number   [PL PHYS] A dimensionless number used in magnetofluid dynamics, equal to the square root of the magnetic force parameter. Symbolized R_M.

magnetic octupole moment   [ELECTROMAG] A quantity characterizing a distribution of magnetization; obtained by integrating the product of the divergence of the magnetization, the third power of the distance from the origin, and a spherical harmonic Y^*_{3m} over the magnetization distribution.

magnetic oscillograph   [ELECTROMAG] An instrument that records a trace measuring one component of the earth's magnetic field.

magnetic Oseen number   [PL PHYS] A dimensionless number used in magnetofluid dynamics, equal to $\frac{1}{2}(1 - N_{AL}^2)R_M$, where N_{AL} is the Alfvén number, and R_M is the magnetic number. Symbolized k.

magnetic pendulum   [ELECTROMAG] A bar magnet which is hung by a thread or balanced on a pivot so that it oscillates in a horizontal plane when disturbed and released in a magnetic field having a horizontal component.

magnetic pinch   *See* pinch effect.

magnetic polarization   *See* intrinsic induction.

magnetic pole   [ELECTROMAG] **1.** One of two regions located at the ends of a magnet that generate and respond to magnetic fields in much the same way that electric charges generate and respond to electric fields. **2.** A particle which generates and responds to magnetic fields in exactly the same way that electric charges generate and respond to electric fields; the particle probably does not have physical reality, but it is often convenient to imagine that a magnetic dipole consists of two magnetic poles of opposite sign, separated by a small distance. [GEOPHYS] In geomagnetism, either of the two points on the earth's surface where the magnetic meridians converge, that is, where the magnetic field is vertical. Also known as dip pole.

magnetic pole strength   [ELECTROMAG] The magnitude of a (fictional) magnetic pole, equal to the force exerted on the pole divided by the magnetic induction (or, alternatively, by the magnetic field strength). Also known as pole strength.

magnetic potential *See* magnetic scalar potential.

magnetic potentiometer [ENG] Instrument that measures magnetic potential differences.

magnetic pressure [PL PHYS] A function, proportional to the square of the magnetic induction, such that the force exerted by a magnetic field on an electrically conducting fluid (excluding the force associated with curvature of magnetic flux lines) is the same as the force that would be exerted by a hydrostatic pressure equal to this function.

magnetic pressure transducer [ENG] A type of pressure transducer in which a change of pressure is converted into a change of magnetic reluctance or inductance when one part of a magnetic circuit is moved by a pressure-sensitive element, such as a bourdon tube, bellows, or diaphragm.

magnetic probe [ELECTROMAG] A small coil inserted in a magnetic field to measure changes in field strength.

magnetic pumping [ELECTROMAG] A method of moving a conducting liquid by applying a magnetic field which varies with time. [PL PHYS] A method of heating a plasma to a high ion temperature by applying an oscillating electromagnetic field.

magnetic quadrupole lens [ELECTROMAG] A magnetic field generated by four magnetic poles of alternating sign arranged in a circle; used to focus beams of charged particles in devices such as electron microscopes and particle accelerators.

magnetic quantum number [ATOM PHYS] The eigenvalue of the component of an angular momentum operator in a specified direction, such as that of an applied magnetic field, in units of Planck's constant divided by 2π.

magnetic refrigerator [CRYO] A device for keeping substances cooled to about 0.2 K, in which a working substance consisting of a paramagnetic salt undergoes a cycle of processes which approximates a Carnot cycle between a high-temperature reservoir consisting of a liquid-helium bath at 1.2 K and a low-temperature reservoir consisting of the substance to be cooled, and isentropic cooling of the working substance is accomplished by demagnetization.

magnetic relaxation [PHYS] The approach of a magnetic system to an equilibrium or steady-state condition, over a period of time.

magnetic reluctance *See* reluctance.

magnetic reluctivity *See* reluctivity.

magnetic resonance [PHYS] A phenomenon exhibited by the magnetic spin systems of certain atoms whereby the spin systems absorb energy at specific (resonant) frequencies when subjected to magnetic fields alternating at frequencies which are in synchronism with natural frequencies of the system. Also known as spin resonance.

magnetic Reynolds number [PL PHYS] A dimensionless number used to compare the transport of magnetic lines of force in a conducting fluid to the leakage of such lines from the fluid, equal to a characteristic length of the fluid times the fluid velocity, divided by the magnetic diffusivity. Symbolized R_M.

magnetic rigidity [ELECTROMAG] A measure of the momentum of a particle moving perpendicular to a magnetic field, equal to the magnetic induction times the particle's radius of curvature. [PL PHYS] The existence of restoring forces which resist displacements of a conducting fluid when a magnetic field is present.

magnetic rotation *See* Faraday effect.

magnetics [ELECTROMAG] The study of magnetic phenomena, comprising magnetostatics and electromagnetism.

magnetic saturation [ELECTROMAG] The condition in which,

MAGNETIC QUADRUPOLE LENS

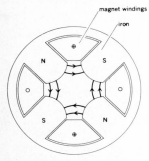

magnet windings

iron

Section normal to the lens axis of magnetic quadrupole lens.

after a magnetic field strength becomes sufficiently large, further increase in the magnetic field strength produces no additional magnetization in a magnetic material. Also known as saturation.

magnetic scalar potential [ELECTROMAG] The work which must be done against a magnetic field to bring a magnetic pole of unit strength from a reference point (usually at infinity) to the point in question. Also known as magnetic potential.

magnetic scattering [PHYS] Scattering of neutrons as a result of the interaction of the magnetic moment of the neutron with the magnetic moments of atoms or other particles.

magnetic secular change [GEOPHYS] The gradual variation in the value of a magnetic element which occurs over a period of years.

magnetic shell [ELECTROMAG] Two layers of magnetic charge of opposite sign, separated by an infinitesimal distance.

magnetic shielding *See* magnetostatic shielding.

magnetic shunt [ELECTROMAG] Piece of iron, usually adjustable as to position, used to divert a portion of the magnetic lines of force passing through an air gap in an instrument or other device.

magnetic spark chamber [NUCLEO] A spark chamber in a magnetic field up to 20,000 gauss, in which the sign of the charge and the momentum of charged particles can be measured by measuring the curvature of their trajectories.

magnetic spectrograph [NUCLEO] A magnetic spectrometer that provides a permanent record of the distribution of intensity versus momentum of a beam of charged particles.

magnetic spectrometer [NUCLEO] A device for measuring the momentum of charged particles, or their distribution of intensity versus momentum, by passing the particles through a magnetic field which bends their paths in proportion to their momentum.

magnetic storm [GEOPHYS] A worldwide disturbance of the earth's magnetic field; frequently characterized by a sudden onset, in which the magnetic field undergoes marked changes in the course of an hour or less, followed by a very gradual return to normalcy, which may take several days. Also known as geomagnetic storm.

magnetic strain energy [SOLID STATE] The potential energy of a magnetic domain, subject to both a tensile stress and a magnetic field, associated with the domain's magnetostriction expansion.

magnetic stress [PL PHYS] The force which acts across a surface in a conducting fluid because of curving or stretching of magnetic flux lines.

magnetic stress tensor [PL PHYS] A second-rank tensor, proportional to the dyad product of the magnetic induction with itself, whose divergence gives that part of the force of a magnetic field on a unit volume of conducting fluid which is due to curvature or stretching of magnetic flux lines.

magnetic susceptibility [ELECTROMAG] The ratio of the magnetization of a material to the magnetic field strength; it is a tensor when these two quantities are not parallel; otherwise it is a simple number. Also known as susceptibility.

magnetic test coil *See* exploring coil.

magnetic thermometer [SOLID STATE] A sample of a paramagnetic salt whose magnetic susceptibility is measured and whose temperature is then calculated from the inverse relationship between the two quantities; useful at temperatures below about 1 K.

magnetic thin film [SOLID STATE] A sheet or cylinder of magnetic material less than 5 micrometers thick, usually possessing uniaxial magnetic anisotropy; used mainly in

MAGNETIC SPARK CHAMBER

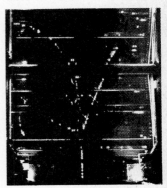

A photograph taken through a pair of cylindrical field lenses of an event in a narrow-gap spark-chamber system in a magnetic field. *(From A. Roberts, Spark chambers, Encyclopaedic Dictionary of Physics, Pergamon Press, 1962–1964)*

MAGNETIC THIN FILM

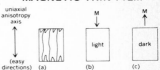

Typical domain patterns in a thin film viewed by use of the Kerr magnetooptic technique. *(a)* Magnetic film demagnetized. *(b)* Magnetization pointing downward. *(c)* Magnetization pointing upward.

MAGNETIC TRANSDUCER

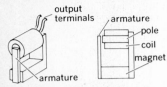

Perspective and sectional view
of a single-pole and armature
magnetic transducer.

computer storage and logic elements. Also known as ferromagnetic film; magnetic film.

magnetic transducer [ELECTROMAG] A device for transforming mechanical into electrical energy, which consists of a magnetic field including a variable-reluctance path and a coil surrounding all or a part of this path, so that variation in reluctance leads to a variation in the magnetic flux through the coil and a corresponding induced emf (electromotive force).

magnetic vector *See* magnetic induction.

magnetic vector potential *See* vector potential.

magnetic viscosity [ELECTROMAG] The existence of a time delay between a change in the magnetic field applied to a ferromagnetic material and the resulting change in magnetic induction which is too great to be explained by the existence of eddy currents. [PL PHYS] The effect, possessed by a magnetic field in the absence of sizable mechanical forces or electric fields, of damping motions of a conducting fluid perpendicular to the field similar to ordinary viscosity.

magnetic wave [SOLID STATE] The spread of magnetization from a small portion of a substance where an abrupt change in the magnetic field has taken place.

magnetic well [PL PHYS] A configuration of magnetic fields used to contain a plasma in controlled fusion experiments, in which the plasma is confined in a central region surrounded by fields which keep it from escaping in any direction.

magnetism [PHYS] Phenomena involving magnetic fields and their effects upon materials.

magnetization [ELECTROMAG] **1.** The property and in particular, the extent of being magnetized; quantitatively, the magnetic moment per unit volume of a substance. Also known as magnetic dipole density; magnetization intensity. **2.** The process of magnetizing a magnetic material.

magnetization curve *See* B-H curve; normal magnetization curve.

magnetization intensity *See* magnetization.

magnetizing force *See* magnetic field strength.

magnet keeper *See* keeper.

magnetoacoustics [PHYS] The study of the effects of magnetic fields on acoustical phenomena, such as various oscillations in the attenuation of ultrasonic sound waves by a crystal placed in a magnetic field at a very low temperature, as the magnetic field strength or sound frequency is varied.

magnetoaerodynamics [PL PHYS] Study of the properties and characteristics of, and the forces exerted by, highly ionized air and other gases; applied principally to study of reentering ballistic missiles and spacecraft.

magneto anemometer [ENG] A cup anemometer with its shaft mechanically coupled to a magnet; both the frequency and amplitude of the voltage generated are proportional to the wind speed, and may be indicated or recorded by suitable electrical instruments.

magnetocaloric effect [THERMO] The reversible change of temperature accompanying the change of magnetization of a ferromagnetic material.

magnetochemistry [PHYS CHEM] A branch of chemistry which studies the interrelationship between the bulk magnetic properties of a substance and its atomic and molecular structure.

magnetodamping [SOLID STATE] An increase in the internal friction experienced by vibrations of a ferromagnetic substance when the substance is placed in a strong magnetic field.

magnetoelastic coupling [SOLID STATE] The interaction between the magnetization and the strain of a magnetic material.

magnetoelasticity [SOLID STATE] Phenomenon in which an elastic strain alters the magnetization of a ferromagnetic material.

magnetoelectricity [ELECTROMAG] Magnetic techniques for generating voltages, such as in an ordinary generator. [SOLID STATE] The appearance of an electric field in certain substances, such as chromic oxide (Cr_2O_3), when they are subjected to a static magnetic field.

magnetofluid [PHYS CHEM] A Newtonian or shear-thinning fluid whose flow properties become viscoplastic when it is modulated by a magnetic field.

magnetofluid dynamics [PHYS] **1.** The study of the motion of an electrically conducting metal, such as mercury, in the presence of electric and magnetic fields. **2.** *See* magnetohydrodynamics.

magnetogalvanic effect *See* galvanomagnetic effect.

magnetogas dynamics [PL PHYS] The science of motion in a plasma under the influence of mechanical, electric, and magnetic forces.

magnetograph [ELECTROMAG] A set of three variometers attached to a suitable recording unit, which records the components of the magnetic field vector in each of three perpendicular directions.

magnetohydrodynamic instability [PL PHYS] An instability of a plasma in which the plasma expands while moving into a region of weaker magnetic field, until it is expelled from the field. Also known as hydromagnetic instability.

magnetohydrodynamics [PHYS] The study of the dynamics or motion of an electrically conducting fluid, such as an ionized gas or liquid metal, interacting with a magnetic field. Abbreviated MHD. Also known as hydromagnetics; magnetofluid dynamics.

magnetohydrodynamic stability [PL PHYS] The condition of a plasma in which fluctuations in density, pressure, velocity, or the distribution of particles in phase space, die out rather than increase. Also known as hydromagnetic stability.

magnetohydrodynamic turbulence [PL PHYS] Motion of a plasma in which velocities and pressures fluctuate irregularly.

magnetohydrodynamic wave [PHYS] Wave motion in an electrically conducting fluid, such as plasma or liquid metal, in a strong magnetic field at a frequency much less than that of the ion cyclotron frequency. Also known as hydromagnetic wave.

magnetoionic duct [GEOPHYS] Duct along the geomagnetic lines of force which exhibits waveguide characteristics for radio-wave propagation between conjugate points on the earth's surface.

magnetoionic wave component [GEOPHYS] Either of the two elliptically polarized wave components into which a linearly polarized wave incident on the ionosphere is separated because of the earth's magnetic field.

magnetomechanical factor [PHYS] The gyromagnetic ratio of an atom or substance (magnetic dipole moment divided by angular momentum) divided by the quantity $e/2mc$, where e and m are the charge (in esu, or electrostatic units) and mass of the electron respectively, and c is the speed of light. Also known as g factor.

magnetomechanics [PHYS] The study of the effects which the magnetization of a material and its strain have on each other.

magnetometer [ENG] An instrument for measuring the magnitude and sometimes also the direction of a magnetic field, such as the earth's magnetic field.

magnetomotive force [ELECTROMAG] The work that would

MAGNETOMETER

Observatory-type magnetometer, capable of accuracy of 3–4 γ.
(*U.S. Coast and Geodetic Survey*)

be required to carry a magnetic pole of unit strength once around a magnetic circuit. Abbreviated mmf.

magneton [PHYS] A unit of magnetic moment used for atomic, molecular, or nuclear magnets, such as the Bohr magneton, Weiss magneton, or nuclear magneton.

magneton number [PHYS] The ratio of the magnetic moment per atom, ion, or molecule of a paramagnetic or ferromagnetic material to the Bohr magneton.

magnetooptical shutter [OPTICS] A device in which light passes through crossed Nicol prisms and a glass cell containing a liquid displaying the Faraday effect between the prisms; light can pass through the system only when a magnetic field is applied to the cell at an angle of 45° to the polarization planes of both prisms.

magnetooptic Kerr effect [OPTICS] Changes produced in the optical properties of a reflecting surface of a ferromagnetic substance when the substance is magnetized; this applies especially to the elliptical polarization of reflected light, when the ordinary rules of metallic reflection would give only plane polarized light. Also known as Kerr magnetooptical effect.

magnetooptic material [OPTICS] A material whose optical properties are changed by an applied magnetic field.

magnetooptics [OPTICS] The study of the effect of a magnetic field on light passing through a substance in the field.

magnetopause [GEOPHYS] A boundary that marks the transition from the earth's magnetosphere to the interplanetary medium.

magnetoplasmadynamics [ELECTROMAG] The generation of electric current by shooting a beam of ionized gas through a magnetic field, to give the same effect as moving copper bars near a magnet.

magnetoresistance [ELECTROMAG] The change in electrical resistance produced in a current-carrying conductor or semiconductor on application of a magnetic field.

magnetoresistivity [ELECTROMAG] The change in resistivity produced in a current-carrying conductor or semiconductor on application of a magnetic field.

magnetosheath [GEOPHYS] The relatively thin region between the earth's magnetopause and the shock front in the solar wind.

magnetosphere [GEOPHYS] The region of the earth in which the geomagnetic field plays a dominant part in controlling the physical processes that take place; it is usually considered to begin at an altitude of about 100 kilometers and to extend outward to a distant boundary that marks the beginning of interplanetary space.

magnetospheric plasma [GEOPHYS] A low-energy plasma with particle energies less than a few electron volts that permeates the entire region of the earth's magnetosphere.

magnetospheric ring current [GEOPHYS] A belt of charged particles around the earth whose perturbations give rise to ionospheric storms.

magnetospheric substorm [GEOPHYS] A disturbance of particles and magnetic fields in the magnetosphere; occurs intermittently, lasts 1 to 3 hours, and is accompanied by various phenomena sensible from the earth's surface, such as intense auroral displays and magnetic disturbances, particularly in the nightside polar regions.

magnetostatic [ELECTROMAG] Pertaining to magnetic properties that do not depend upon the motion of magnetic fields.

magnetostatic mode [SOLID STATE] A spin wave in a magnetic material whose wavelength is greater than about one-tenth the size of the sample.

magnetostatics [ELECTROMAG] The study of magnetic fields

MAGNETOSHEATH

Configuration of the magnetosphere in the plane containing the sun-earth line and the geomagnetic axis; position of magnetosheath is shown.

that remain constant with time. Also known as static magnetism.

magnetostatic shielding [ELECTROMAG] The use of an enclosure made of a high-permeability magnetic material to prevent a static magnetic field outside the enclosure from reaching objects inside it, or to confine a magnetic field within the enclosure. Also known as magnetic shielding.

magnetostriction [ELECTROMAG] The dependence of the state of strain (dimensions) of a ferromagnetic sample on the direction and extent of its magnetization.

magnetotail [GEOPHYS] The portion of the magnetosphere extending from earth in the direction away from the sun for a variable distance of the order of 1000 earth radii.

magnetron [ELECTR] One of a family of crossed-field microwave tubes, wherein electrons, generated from a heated cathode, move under the combined force of a radial electric field and an axial magnetic field in such a way as to produce microwave radiation in the frequency range 1–40 gigahertz; a pulsed microwave radiation source for radar, and continuous source for microwave cooking.

magnetron vacuum gage [ELECTR] A vacuum gage that is essentially a magnetron operated beyond cutoff in the vacuum being measured.

magnification [OPTICS] **1.** A measure of the effectiveness of an optical system in enlarging or reducing an image; the magnification may be lateral, longitudinal, or angular. **2.** See lateral magnification.

magnifier See simple microscope.

magnifying glass [OPTICS] **1.** Any device that uses a simple lens which enlarges the object being viewed. **2.** See simple microscope.

magnifying power [OPTICS] The ratio of the tangent of the angle subtended at the eye by an image formed by an optical system, to the tangent of the angle subtended at the eye by the corresponding object at a distance for convenient viewing.

magnistor [ELECTR] A device that utilizes the effects of magnetic fields on injection plasmas in semiconductors such as indium antimonide.

magnitude [ASTRON] The relative luminance of a celestial body; the smaller (algebraically) the number indicating magnitude, the more luminous the body. Also known as stellar magnitude. [GEOPHYS] A measure of the amount of energy released by an earthquake.

magnitude of a complex number See absolute value of a complex number.

magnitude of a real number See absolute value of a real number.

magnitude of a vector See absolute value of a vector.

magnitude system [ASTRON] A system for designating the relative brightness of stars when photography is used; emulsions of different color sensitivities, used with color filters, permit measurements of starlight of different wavelengths with corresponding determination of magnitude at these wavelengths.

magnon [SOLID STATE] A quasi-particle which is introduced to describe small departures from complete ordering of electronic spins in ferro-, ferri-, antiferro-, and helimagnetic arrays. Also known as quantized spin wave.

Magnus effect [FL MECH] A force on a rotating cylinder in a fluid flowing perpendicular to the axis of the cylinder; the force is perpendicular to both flow direction and cylinder axis. Also known as Magnus force.

Magnus force See Magnus effect.

Magnus moment [FL MECH] A torque associated with the Magnus effect, such as moments about the pitch and yaw axes

MAGNETRON

tuning mechanism

permanent magnet

air cooling fins

heater-cathode voltage terminals

output waveguide

Coaxial cavity magnetron with horseshoe-shaped magnets. *(From J. W. Gewartowski and H. A. Watson, Principles of Electron Tubes, Van Nostrand, 1965)*

of a missile or aircraft due to rotation about the roll axis.

main diagonal   *See* principal diagonal.

main effect   [STAT] The effect of the change in level of one factor in a factorial experiment measured independently of other variables.

main lobe   *See* major lobe.

main sequence   [ASTRON] The band in the spectrum luminosity diagram which has the great majority of stars; their energy derives from core burning of hydrogen into helium.

main sequence star   [ASTRON] **1.** Any of those stars in the smooth curve termed the main sequence in a Hertzsprung-Russell diagram. **2.** *See* dwarf star.

main stroke   *See* return streamer.

Majorana force   [NUC PHYS] A force between two nucleons postulated to explain various phenomena, which can be derived from a potential containing an operator which exchanges the nucleons' positions but not their spins.

Majorana neutrino   [PARTIC PHYS] A particle described by a wave function that satisfies the Dirac equation with mass equal to zero, and that is self–charge-conjugate.

major arc   [MATH] The longer of the two arcs produced by a secant of a circle.

major axis   [MATH] The longer of the two axes with respect to which an ellipse is symmetric.

major diatonic scale   [ACOUS] A diatonic scale in which the relative sizes of the sequence of intervals are approximately 2, 2, 1, 2, 2, 2, 1.

majority   [MATH] A logic operator having the property that if P,Q,R are statements, then the function (P, Q, R, . . .) is true if more than half the statements are true, or false if half or less are true.

majority carrier   [ELECTR] The type of carrier, that is, electron or hole, that constitutes more than half the carriers in a semiconductor.

majority emitter   [ELECTR] Of a transistor, an electrode from which a flow of minority carriers enters the interelectrode region.

major lobe   [ELECTROMAG] Antenna lobe indicating the direction of maximum radiation or reception. Also known as main lobe.

major node   [ELEC] A point in an electrical network at which three or more elements are connected together. Also known as junction.

major planet   [ASTRON] Any of the four planets that are larger than earth: Jupiter, Saturn, Neptune, and Uranus.

Maksutov system   [OPTICS] A catadioptric telescope optical system capable of covering a large field (60° and more); used to survey large areas of the sky.

Malter effect   [SOLID STATE] A phenomenon in which a metal with a nonconducting surface film has a large coefficient of secondary electron emission; this is particularly notable in aluminum whose surface has been oxidized and then coated with cesium oxide.

maltese cross   [MATH] A plane curve whose equation in cartesian coordinates x and y is $xy(x^2 - y^2) = x^2 + y^2$.

Malus cosine-squared law   [OPTICS] The law that if a beam of plane polarized light passes through a Nicol prism, the intensity of light emerging from the prism is proportional to the square of the cosine between the plane of polarization of the incident light and the plane of polarization of the prism.

Malus' law of rays   [OPTICS] The law that an orthotomic system of rays is still orthotomic after the rays have been reflected and refracted any number of times.

Mandelstam plane   [PARTIC PHYS] A method of plotting energy versus scattering angle of three reactions, each having

MAKSUTOV SYSTEM

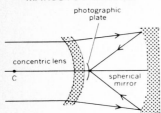

photographic plate

concentric lens

C

spherical mirror

Optics of Maksutov system; *C* is the center of curvature of the mirror.

two particles both before and after scattering, which can be derived from each other by the crossing principle; the three reactions are on an equal footing, and poles in the scattering amplitude representing exchanged particles lie along straight lines.

Mandelstam representation [PARTIC PHYS] For a reaction in which there are two particles both before and after scattering: an expression, containing several integrals, for a function related to the scattering amplitude; the arguments of the function are the center-of-mass energy and scattering angle, extended to complex values; the function is conjectured to be analytic in these variables except for certain cuts and to have values along these cuts which give the scattering amplitude of the reaction, and of the two reactions derivable from it by the crossing principle.

manganese [CHEM] A metallic element, symbol Mn, atomic weight 54.938, atomic number 25; a transition element whose properties fall between those of chromium and iron.

Mangin mirror [OPTICS] A negative meniscus lens whose shallower surface is silvered to act as a spherical mirror while the other surface corrects for spherical aberration of the reflecting surface; used in searchlights and aircraft gunsights.

manifest covariance [RELAT] Property of an expression composed of Lorentz invariant numbers and operators, four-vectors, and tensors in such a way that its Lorentz covariance is immediately obvious.

manifold of states [ATOM PHYS] A set of states sufficient to form a representation of an operator or a Lie group of operators.

Mann-Whitney test [STAT] A procedure used in nonparametric statistics to determine whether the means of two populations are equal.

manocryometer [THERMO] An instrument for measuring the change of a substance's melting point with change in pressure; the height of a mercury column in a U-shaped capillary supported by an equilibrium between liquid and solid in an adjoining bulb is measured, and the whole apparatus is in a thermostat.

manometer [ENG] A double-leg liquid-column gage used to measure the difference between two fluid pressures.

manometric capsule [ACOUS] A device for studying air vibrations in a pipe or resonator, consisting of a rubber membrane which is stretched over a hole in the pipe, or over the end of a flange attached to such a hole, and apparatus for measuring vibrations of the membrane.

manometry [ENG] The use of manometers to measure gas and vapor pressures.

mantissa [ADP] A fixed point number composed of the most significant digits of a given floating point number. Also known as fixed-point part; floating-point coefficient. [MATH] The positive decimal part of a common logarithm.

many-body force [PHYS] A force exerted on a particle, in the presence of two or more other particles, which differs from the vector sum of the forces which would be exerted on it if each of the other particles were present alone.

many-body problem [MECH] The problem of predicting the motions of three or more objects obeying Newton's laws of motion and attracting each other according to Newton's law of gravitation. Also known as n-body problem.

many-body theory [PHYS] A scheme for calculating physical quantities for systems with large numbers of particles, without finding details of each particle's motion, often at temperatures close to absolute zero.

map *See* mapping.

mapping [MATH] **1.** Any function or multiple-valued rela-

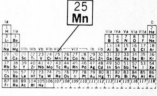

MANGANESE

Periodic table of the chemical elements showing the position of manganese.

MARS

Photograph of Mars through yellow filter at Lick Observatory.

MASKING

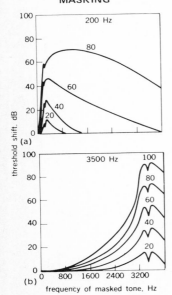

Masking of pure tones by pure tones at masking frequencies of (a) 200 hertz, (b) 3500 hertz. The number above each curve is the level in decibels above the threshold of audibility of each masking tone. (From H. Fletcher, Speech and Hearing in Communication, 2d ed., Van Nostrand, 1953)

tion. Also known as map. **2.** In topology, a continuous function.

March equinox *See* vernal equinox.

Marconi antenna [ELECTROMAG] Antenna system of which the ground is an essential part, as distinguished from a Hertz antenna.

mare [ASTRON] **1.** One of the large, dark, flat areas on the lunar surface. **2.** One of the less well-defined areas on Mars.

marginal dimensionality [STAT MECH] The largest number of spatial dimensions for which nonlinear effects are important in calculating the behavior of a substance near a critical point.

marginally outer trapped surface [RELAT] A spacelike, two-dimensional surface in a space-time such that outgoing null rays perpendicular to the surface are neither diverging nor converging.

marginal probability [STAT] Probability expressed by the two conditional probability distributions which arise from the joint distribution of two random variables.

Margoulis number *See* Stanton number.

Mariotte's law *See* Boyle's law.

Markov chain [MATH] A Markov process whose state space is finite or countably infinite.

Markov inequality [STAT] If x is a random variable with probability P and expectation E, then $P(|x| \geq a) \leq E(|x|^n / a^n)$.

Markov process [MATH] A stochastic process which assumes that in a series of random events the probability of an occurrence of each event depends only on the immediately preceding outcome.

marriage theorem [MATH] A family of n subsets of a set S with n elements is a system of distinct representatives for S if any k of the subsets, $k = 1, 2, \ldots, n$, together contain at least k distinct elements. Also known as Hall's theorem.

Mars [ASTRON] The planet fourth in distance from the sun; it is visible to the naked eye as a bright red star, except for short periods when it is near its conjunction with the sun; its diameter is about 6700 kilometers.

Martens wedge [OPTICS] A wedge-shaped piece of quartz used to rotate the plane of polarization of linearly polarized light.

martingale [STAT] A sequence of random variables $x_1, x_2, \ldots,$ where the conditional expected value of x_{n+1}, given $x_1, x_2, \ldots, x_n$, equals x_n.

Marx effect [SOLID STATE] The effect wherein the energy of photoelectrons emitted from an illuminated surface is decreased when the surface is simultaneously illuminated by light of lower frequency than that causing the emission.

mA s *See* milliampere-second.

Mascheroni's constant *See* Euler's constant.

mascon [GEOL] A large, high-density mass concentration below a ringed mare on the surface of the moon.

maser [PHYS] A device for coherent amplification or generation of electromagnetic waves in which an ensemble of atoms or molecules, raised to an unstable energy state, is stimulated by an electromagnetic wave to radiate excess energy at the same frequency and phase as the stimulating wave. Derived from microwave amplification by stimulated emission of radiation. Also known as paramagnetic amplifier.

maser amplifier [ELECTR] A maser which is used to increase the power produced by another maser.

mask [ELECTR] A thin sheet of metal or other material containing an open pattern, used to shield selected portions of a semiconductor or other surface during a deposition process.

masking [ACOUS] The amount by which the threshold of audibility of a sound is raised by the presence of another sound; the unit customarily used is the decibel. Also known as audio masking; aural masking. [ELECTR] **1.** Using a covering or coating on a semiconductor surface to provide a masked area for selective deposition or etching. **2.** A programmed procedure for eliminating radar coverage in areas where such transmissions may be of use to the enemy for navigation purposes, by weakening the beam in appropriate directions or by use of additional transmitters on the same frequency at suitable sites to interfere with homing; also used to suppress the beam in areas where it would interfere with television reception.

mass [MECH] A quantitative measure of a body's resistance to being accelerated; equal to the inverse of the ratio of the body's acceleration to the acceleration of a standard mass under otherwise identical conditions.

mass absorption coefficient [PHYS] The linear absorption coefficient divided by the density of the medium.

mass absorption law [NUCLEO] The absorption of electrons with speeds greater than one-fifth that of light depends only on the mass of absorbing matter in the electron's path and not on its chemical composition. Also known as Lenard's mass absorption law.

mass action law [PHYS CHEM] The law that the rate of a chemical reaction for a uniform system at constant temperature is proportional to the concentrations of the substances reacting. Also known as Guldberg and Waage law.

mass-analyzed ion kinetic energy spectrometry [SPECT] A type of ion kinetic energy spectrometry in which the ionic products undergo mass analysis followed by energy analysis. Abbreviated MIKES.

mass attraction vertical [GEOPHYS] The vertical which is a function only of the distribution of mass and is unaffected by forces resulting from the motions of the earth.

mass defect [NUC PHYS] The difference between the mass of an atom and the sum of the masses of its individual components in the free (unbound) state.

mass divergence [FL MECH] The divergence of the momentum field, a measure of the rate of net flux of mass out of a unit volume of a system; in symbols, $\nabla \cdot \rho \mathbf{V}$, where ρ is the fluid density, $\mathbf{V}$ the velocity vector, and ∇ the del operator.

mass-energy conservation [RELAT] The principle that energy cannot be created or destroyed; however, one form of energy is that which a particle has because of its rest mass, equal to this mass times the square of the speed of light.

Massenfilter *See* quadrupole spectrometer.

Massey formula [ATOM PHYS] A formula for the probability that an excited atom approaching the surface of a metal will emit secondary electrons.

mass flow [FL MECH] The mass of a fluid in motion which crosses a given area in a unit time.

mass formula [NUC PHYS] An equation giving the atomic mass of a nuclide as a function of its atomic number and mass number.

Massieu function [THERMO] The negative of the Helmholtz free energy divided by the temperature.

mass-luminosity relation [ASTROPHYS] A relation between stellar magnitudes and mass of the stars; when the absolute magnitudes of stars are plotted versus the logarithms of their masses, the points fall closely along a smooth curve.

mass number [NUC PHYS] The sum of the numbers of protons and neutrons in the nucleus of an atom or nuclide. Also known as nuclear number; nucleon number.

mass operator [QUANT MECH] An operator which is added to the Lagrangian in a quantized field theory in order to eliminate certain infinite quantities, and whose sum with the mechanical mass gives the observed mass.

mass reactance *See* acoustic mass reactance.

mass renormalization [QUANT MECH] The mathematical operation of adding the mass which a particle possesses because of its self interaction, to its mechanical mass in order to obtain its measured mass.

mass resistivity [ELEC] The product of the electrical resistance of a conductor and its mass, divided by the square of its length; the product of the electrical resistivity and the density.

mass spectrograph [ENG] A mass spectroscope in which the ions fall on a photographic plate which after development shows the distribution of particle masses.

mass spectrometer [ENG] A mass spectroscope in which a slit moves across the paths of particles with various masses, and an electrical detector behind it records the intensity distribution of masses.

mass spectrometry [ANALY CHEM] An analytical technique for identification of chemical structures, determination of mixtures, and quantitative elemental analysis, based on application of the mass spectrometer.

mass spectroscope [ENG] An instrument used for determining the masses of atoms or molecules, in which a beam of ions is sent through a combination of electric and magnetic fields so arranged that the ions are deflected according to their masses.

mass spectrum [PARTIC PHYS] A plot of masses of elementary particles, including unstable states. Also known as particle spectrum. [PHYS] A display, record, or plot of the distribution in mass, or in mass-to-charge ratio, of ionized atoms, molecules, or molecular fragments.

mass stopping power [NUCLEO] The decrease, per unit surface density traversed, in kinetic energy of an ionizing particle passing through matter; equal to the linear energy transfer (energy loss per unit path length) divided by the density of the material.

mass susceptibility [PHYS CHEM] Magnetic susceptibility of a compound per gram. Also known as specific susceptibility.

mass-to-charge ratio [ANALY CHEM] In analysis by mass spectroscopy, the measurement of the sample mass as a ratio to its ionic charge.

mass-transfer rate [PHYS] The measurement of the movement of matter as a function of time.

mass transport [FL MECH] **1.** Carrying of loose materials in a moving medium such as water or air. **2.** The movement of fluid, especially water, from one place to another.

mass units [MECH] Units of measurement having to do with masses of materials, such as pounds or grams.

mass velocity [FL MECH] The weight flow rate of a fluid divided by the cross-sectional area of the enclosing chamber or conduit; for example, lb/hr ft².

master equation [ATOM PHYS] An equation which determines the rate of change of the population of an energy level in terms of the populations of other levels and transition probabilities.

master/slave manipulator [NUCLEO] A pair of mechanical hands used to handle radioactive materials by remote control by an operator behind a protective shield.

Matano-Boltzmann solution [PHYS] A solution of Fick's law and the continuity equation, in which the initial concentration at time $t = 0$ is $\rho(x) = a$ for $x < 0$ and $\rho(x) = b$ for $x > 0$, and the solution is given in terms of the variable $\lambda = x'/\sqrt{t}$, where $x' = x - a$ and $x = a$ is the location of the Matano interface.

Matano interface [PHYS] For a solution of Fick's law and the continuity equation in which the initial concentration is $\rho(x) = a$ for $x < 0$, and $\rho(x) = b$ for $x > 0$, this is a plane normal to

MASS SPECTROMETER

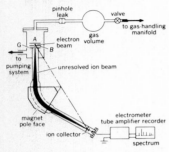

Schematic diagram of mass spectrometer tube. Electric field caused by potential difference of several volts between plates *A* and *B* draws ions through slit in *B*. Ions are further accelerated by potential difference of hundreds or thousands of volts between *B* and *G*.

the x axis such that the area between the curve $\rho(x)$ and the line $\rho = a$ on one side of the interface is equal to the area between the curve $\rho(x)$ and the line $\rho = b$ on the other side.

match *See* biconditional operation.

matched pairs [STAT] The design of an experiment for paired comparison in which the assignment of subjects to treatment or control is not completely at random, but the randomization is restricted to occur separately within each pair.

matching [ELEC] Connecting two circuits or parts together with a coupling device in such a way that the maximum transfer of energy occurs between the two circuits, and the impedance of either circuit will be terminated in its image.

matching diaphragm [ELECTROMAG] Diaphragm consisting of a slit in a thin sheet of metal, placed transversely across a waveguide for matching purposes; the orientation of the slit with respect to the long dimension of the waveguide determines whether the diaphragm acts as a capacitive or inductive reactance.

matching distribution [STAT] The distribution of number of matches obtained if N tickets labeled 1 to N are drawn at random one at a time and laid in a row, and a match is counted when a ticket's label matches its position.

material implication *See* implication.

materialization [PHYS] The direct conversion of energy into mass, as in pair production.

material particle [MECH] An object which has rest mass and an observable position in space, but has no geometrical extension, being confined to a single point. Also known as particle.

materials testing reactor [NUCLEO] A nuclear reactor designed mainly for studying the behavior of materials and equipment subjected to large fluxes of neutrons and other radiation.

mathematical analysis *See* analysis.

mathematical induction [MATH] A general method of proof of statements concerning a positive integral variable: if a statement is proven true for $x = 1$, and if it is proven that, if the statement is true for $x = 1, \ldots, n$, then it is true for $x = n + 1$, then it follows that the statement is true for any integer. Also known as complete induction.

mathematical logic [MATH] The study of mathematical theories from the viewpoint of model theory, recursive function theory, proof theory, and set theory.

mathematical model [MATH] **1.** A mathematical representation of a process, device, or concept by means of a number of variables which are defined to represent the inputs, outputs, and internal states of the device or process, and a set of equations and inequalities describing the interaction of these variables. **2.** A mathematical theory or system together with its axioms.

mathematical physics [PHYS] The study of the mathematical systems which represent physical phenomena; particular areas are, for example, quantum and statistical mechanics and field theory.

mathematical probability [MATH] The ratio of the number of mutually exclusive, equally likely outcomes of interest to the total number of such outcomes when the total is exhaustive. Also known as a priori probability.

mathematical programming *See* optimization theory.

mathematical table [MATH] A listing of the values of a function of one or several variables at a series of values of the arguments, usually equally spaced.

mathematics [SCI TECH] The deductive study of shape, quantity, and dependence; the two main areas are applied mathematics and pure mathematics, the former arising from the study of physical phenomena, the latter the intrinsic study of mathematical structures.

Mathieu equation [MATH] A differential equation of the form $y'' + (a + b \cos 2x)y = 0$, whose solution depends on periodic functions.

Mathieu functions [MATH] Any solution of the Mathieu equation which is periodic and an even or odd function.

m-atm *See* meter-atmosphere.

matrix [MATH] A rectangular array of numbers or scalars from a vector space.

matrix algebra [MATH] An algebra whose elements are matrices and whose operations are addition and multiplication of matrices.

matrix calculus [MATH] The treatment of matrices whose entries are functions as functions in their own right with a corresponding theory of differentiation; this has application to the study of multidimensional derivatives of functions of several variables.

matrix element [MATH] One of the set of numbers which form a matrix. [QUANT MECH] The scalar product of a member of a complete, orthogonal set of vectors, representing states, with a vector which results from applying a specified operator to another member of this set.

matrix game [MATH] A game involving two persons, which gives rise to a matrix representing the amount received by the two players. Also known as rectangular game.

matrix mechanics [QUANT MECH] The theory of quantum mechanics developed by using the Heisenberg picture and representing operators by their matrix elements between eigenfunctions of the Hamiltonian operator; Heisenberg's original formulation of quantum mechanics.

matrix of a linear transformation [MATH] A unique matrix A, such that for a specified linear transformation L from one vector space to another, and for specified finite bases in each space, L applied to a vector is equal to A times that vector.

matrix spectrophotometry [SPECT] Spectrophotometric analysis in which the specimen is irradiated in sequence at more than one wavelength, with the visible spectrum evaluated for the energy leaving for each wavelength of irradiation.

matrix theory [MATH] The algebraic study of matrices and their use in evaluating linear processes.

matter wave *See* de Broglie wave.

Matteuci effect [PHYS] A phenomenon in which an electric potential difference appears between the ends of a ferromagnet that is twisted in a magnetic field.

Matthias' rules [SOLID STATE] Several empirical rules giving the dependence of the transition temperatures of superconducting metals and alloys on the position of the metals in the periodic table and in the composition of the alloys.

Matthiessen sinker method [THERMO] A method of determining the thermal expansion coefficient of a liquid, in which the apparent weight of a sinker when immersed in the liquid is measured for two different temperatures of the liquid.

Matthiessen's rule [SOLID STATE] An empirical rule which states that the total resistivity of a crystalline metallic specimen is the sum of the resistivity due to thermal agitation of the metal ions of the lattice and the resistivity due to imperfections in the crystal.

mattress array *See* billboard array.

maunder minimum [ASTRON] A period of time from about 1650 to 1710 when the sun did not appear to have sunspots.

Maupertius' principle [MECH] The principle of least action is sufficient to determine the motion of a mechanical system.

mavar *See* parametric amplifier.

maximal analytic extension [RELAT] An extension, in a real analytic manner, past all coordinate singularities of a solution to Einstein's equations of general relativity.

maximal element *See* maximal member.

maximal ideal [MATH] An ideal I in a ring R which is not equal to R, and such that there is no ideal containing I and not equal to I or R.

maximal member [MATH] In a partially ordered set a maximal member is one for which no other element follows it in the ordering. Also known as maximal element.

maximax criterion [MATH] In decision theory, one of several possible prescriptions for making a decision under conditions of uncertainty; it prescribes the strategy which will maximize the maximum possible profit.

maxim criterion [MATH] One of several prescriptions for making a decision under conditions of uncertainty; it prescribes the strategy which will maximize the minimum profit. Also known as maximin criterion.

maximin criterion *See* maxim criterion.

maximizing a function [MATH] Finding the largest value assumed by a function.

maximum [MATH] The maximum of a real-valued function is the greatest value it assumes. Abbreviated max.

maximum available gain [ELECTR] The theoretical maximum power gain available in a transistor stage; it is seldom achieved in practical circuits because it can be approached only when feedback is negligible. Abbreviated MAG.

maximum credible accident [NUCLEO] The most serious nuclear reactor accident that can be hypothesized from an adverse combination of equipment malfunction, operating errors, and other reasonable foreseen causes.

maximum likelihood method [STAT] A technique in statistics where the likelihood distribution is so maximized as to produce an estimate to the random variables involved.

maximum-minimum principle *See* min-max theorem.

maximum-modulus theorem [MATH] For a complex analytic function in a closed bounded simply connected region its modulus assumes its maximum value on the boundary of the region.

maximum ordinate [MECH] Difference in altitude between the origin and highest point of the trajectory of a projectile.

maxwell [ELECTROMAG] A centimeter-gram-second electromagnetic unit of magnetic flux, equal to the magnetic flux which produces an electromotive force of 1 abvolt in a circuit of one turn linking the flux, as the flux is reduced to zero in 1 second at a uniform rate. Abbreviated Mx. Also known as abweber (abWb); line of magnetic induction.

Maxwell body *See* Maxwell liquid.

Maxwell-Boltzmann density function *See* Maxwell-Boltzmann distribution.

Maxwell-Boltzmann distribution [STAT MECH] Any function giving the probability (or some function proportional to it) that a molecule of a gas in thermal equilibrium will have values of certain variables within given infinitesimal ranges, assuming that the gas molecules obey classical mechanics, and possibly making other assumptions; examples are the Maxwell distribution and the Boltzmann distribution. Also known as Maxwell-Boltzmann density function.

Maxwell-Boltzmann equation *See* Boltzmann transport equation.

Maxwell-Boltzmann statistics [STAT MECH] The classical statistics of identical particles, as opposed to the Bose-Einstein or Fermi-Dirac statistics. Also known as Boltzmann statistics.

Maxwell bridge [ELEC] A four-arm alternating-current bridge used to measure inductance (or capacitance) in terms of resistance and capacitance (or inductance); bridge balance is independent of frequency. Also known as Maxwell-Wien bridge; Wien-Maxwell bridge.

Maxwell distribution [STAT MECH] A function giving the number of molecules of a gas in thermal equilibrium whose velocities lie within a given, infinitesimal range of values, assuming that the molecules obey classical mechanics, and do not interact. Also known as Maxwellian distribution.

Maxwell effect [OPTICS] Double refraction of a viscous liquid having anisotropic molecules, which results from components of the velocity gradient perpendicular to the fluid velocity itself.

Maxwell equal area rule [THERMO] At temperatures for which the theoretical isothermal of a substance, on a graph of pressure against volume, has a portion with positive slope (as occurs in a substance with liquid and gas phases obeying the van der Waals equation), a horizontal line drawn at the equilibrium vapor pressure and connecting two parts of the isothermal with negative slope has the property that the area between the horizontal and the part of the isothermal above it is equal to the area between the horizontal and the part of the isothermal below it.

Maxwell equations *See* Maxwell field equations.

Maxwell field equations [ELECTROMAG] Four differential equations which relate the electric and magnetic fields to electric charges and currents, and form the basis of the theory of electromagnetic waves. Also known as electromagnetic field equations; Maxwell equations.

Maxwellian distribution *See* Maxwell distribution.

Maxwellian distribution law [STAT MECH] Equation relating the statistical distribution of speeds and energies of molecules of a pure gas at a uniform temperature where there are no convection currents.

Maxwellian equilibrium [STAT MECH] Thermal equilibrium of a gas, or of some group of particles, in which the velocity distribution of the particles is the Maxwell distribution corresponding to the temperature of the object with which they are in equilibrium.

Maxwellian gas [STAT MECH] A gas whose molecules have the Maxwell distribution of velocities.

Maxwellian view [OPTICS] A method of using an optical instrument in which a real image of a light source is focused on the pupil of the eye, instead of using an eyepiece.

Maxwell liquid [FL MECH] A liquid whose rate of deformation is the sum of a term proportional to the shearing stress acting on it and a term proportional to the rate of change of this stress. Also known as Maxwell body.

Maxwell primaries [OPTICS] The primary colors in a system of colorimetry devised by J. C. Maxwell; they are cyan, green, and magenta.

Maxwell relation [ELECTROMAG] According to Maxwell's electromagnetic theory, that relation wherein the dielectric constant of a substance equals the square of its index of refraction. [THERMO] One of four equations for a system in thermal equilibrium, each of which equates two partial derivatives, involving the pressure, volume, temperature, and entropy of the system.

Maxwell's coefficient of diffusion [FL MECH] A number in an equation for the difference between mean velocities of two gases which are allowed to mix, which determines the contribution to this quantity of the concentration gradient.

Maxwell's cyclic currents *See* mesh currents.

Maxwell's demon *See* demon of Maxwell.

Maxwell's displacement current *See* displacement current.

Maxwell's electromagnetic theory [ELECTROMAG] A mathematical theory of electric and magnetic fields which predicts the propagation of electromagnetic radiation, and is valid for electromagnetic phenomena where effects on an atomic scale can be neglected.

Maxwell's law [ELECTROMAG] A movable portion of a circuit will always move in such a direction as to give maximum magnetic flux linkages through the circuit.

Maxwell's stress functions [MECH] Three functions of position, ϕ_1, ϕ_2, and ϕ_3, in terms of which the elements of the stress tensor σ of a body may be expressed, if the body is in equilibrium and is not subjected to body forces; the elements of the stress tensor are given by $\sigma_{11} = \partial^2\phi_2/\partial x_3{}^2 + \partial^2\phi_3/\partial x_2{}^2$, $\sigma_{23} = -\partial^2\phi_1/\partial x_2\,\partial x_3$, and cyclic permutations of these equations.

Maxwell's stress tensor [ELECTROMAG] A second-rank tensor whose product with a unit vector normal to a surface gives the force per unit area transmitted across the surface by an electromagnetic field.

Maxwell's theorem [MECH] If a load applied at one point A of an elastic structure results in a given deflection at another point B, then the same load applied at B will result in the same deflection at A.

Maxwell's theory of light [OPTICS] An application of Maxwell's electromagnetic theory in which light is treated as a propagating electromagnetic wave.

Maxwell triangle [OPTICS] Color-matching chromaticity values plotted on an x,y diagram. Also known as x,y chromaticity diagram.

maxwell-turn [ELECTROMAG] A centimeter-gram-second electromagnetic unit of flux linkage, equal to the flux linkage of a coil consisting of one complete loop of wire through which passes a magnetic flux of one maxwell. Also known as line-turn.

Maxwell-Wagner mechanism [ELEC] A capacitor consisting of two parallel metal plates with two layers of material between them, one with vanishing conductivity, the other with finite conductivity and vanishing electric susceptibility.

Maxwell-Wien bridge *See* Maxwell bridge.

mayer [THERMO] A unit of heat capacity equal to the heat capacity of a substance whose temperature is raised 1° centigrade by 1 joule.

Mayer condensation theory [STAT MECH] A theory of the condensation and critical state of a system of chemically saturated molecules, in which the system is assumed to consist of independent clusters of molecules.

Mayer problem [MATH] In the calculus of variations, the problem of finding an admissible arc that minimizes the functional $I = G(t_f, x_1(t_f), \ldots, x_m(t_f)) - G(t_i, x_1(t_i), \ldots, x_m(t_i))$, where $G(t, x_1, \ldots, x_m)$ is a specified function of the state variables $x_1, \ldots, x_m$ and the independent variable t which ranges from initial value t_i to final value t_f.

Mayer's formula [THERMO] The difference between the specific heat of a gas at constant pressure and its specific heat at constant volume is equal to the gas constant divided by the molecular weight of the gas.

mb *See* millibar; millibarn.

MBE *See* molecular beam epitaxy.

mc *See* millihertz.

Mc *See* megahertz.

mcd *See* millicurie-destroyed.

M center [SOLID STATE] A color center consisting of an F center combined with two ion vacancies.

McLeod gage [FL MECH] A type of instrument used to measure vacuum by measuring the height of a column of mercury supported by the gas whose pressure is to be measured, when

McLEOD GAGE

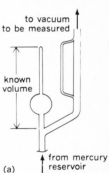

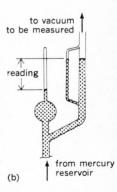

McLeod gage in *(a)* filling (charging) position, and *(b)* measuring position.

McMATH TELESCOPE

McMath 60-inch (1.524-meter) solar telescope, Kitt Peak, Arizona.

this gas is trapped and compressed into a capillary tube.

McMath telescope [OPTICS] A unique 60-inch (1.524-meter) solar telescope at Kitt Peak, Arizona, that has an unconventional configuration; the sun's light is reflected from an 80-inch (2.032-meter) mirror into a long, fixed tube.

M contour [CONT SYS] A line on a Nyquist diagram connecting points having the same magnitude of the primary feedback ratio.

md *See* millidarcy.

Md *See* mendelevium.

M-derived filter [ELECTR] A filter consisting of a series of T or pi sections whose impedances are matched at all frequencies, even though the sections may have different resonant frequencies.

mean [MATH] A single number that typifies a set of numbers, such as the arithmetic mean, the geometric mean, or the expected value. Also known as mean value.

mean British thermal unit *See* British thermal unit.

mean calorie [THERMO] One-hundredth of the heat needed to raise 1 gram of water from 0 to 100°C.

mean curvature [MATH] Half the sum of the principal curvatures at a point on a surface.

mean deviation *See* average deviation.

mean difference [STAT] The average of the absolute values of the $n(n - 1)/2$ differences between pairs of elements in a statistical distribution that has n elements.

mean ergodic theorem [MATH] If T is a measure-preserving transformation of a finite measure space S and f is a function that is Lebesgue-integrable over S, then the sequence whose nth term is $(f(x) + f(Tx) + \cdots + f(T^n x)) / (n + 1)$ converges in mean of order 2 to a function f^* which is Lesbesgue-integrable over S; furthermore f^* is an invariant function under T, and the integral of f^* over S equals that of f.

mean free path [ACOUS] For sound waves in an enclosure, the average distance sound travels between successive reflections in the enclosure. [PHYS] The average distance traveled between two similar events, such as elastic collisions of molecules in a gas, of electrons or phonons in a crystal, or of neutrons in a moderator.

mean life [PHYS] The average time during which a system, such as an atom, nucleus, or elementary particle, exists in a specified form; for a radionuclide or an excited state of an atom or nucleus, it is the reciprocal of the decay constant. Also known as average life; lifetime.

mean noon [ASTRON] Twelve o'clock mean time, or the instant the mean sun is over the upper branch of the meridian; it may be either local or Greenwich, depending upon the reference meridian.

mean normal stress [MECH] In a system stressed multiaxially, the algebraic mean of the three principal stresses.

mean proportional [MATH] For two numbers a and b, a number x, such that $x/a = b/x$.

mean range [SCI TECH] The average difference in the extreme values of a variable quantity.

mean rank method [STAT] A method of handling data which has the same observed frequency occurring at two or more consecutive ranks; it consists of assigning the average of the ranks as the rank for the common frequency.

mean sidereal time [ASTRON] Sidereal time adjusted for nutation, to eliminate slight irregularities in the rate.

mean solar day [ASTRON] The duration of one rotation of the earth on its axis, with respect to the mean sun; the length of the mean solar day is 24 hours of mean solar time or $24^h 03^m 56.555^s$ of mean sidereal time.

mean solar second [ASTRON] A unit equal to 1/86,400 of a mean solar day.

mean solar time [ASTRON] Time that has the mean solar second as its unit, and is based on the mean sun's motion.

mean specific heat [THERMO] The average over a specified range of temperature of the specific heat of a substance.

mean spherical intensity [OPTICS] The luminous intensity of a light source averaged over all directions.

mean square [STAT] The arithmetic mean of the squares of the differences of a set of values from some given value.

mean-square deviation [STAT] A measure of the extent to which a collection $v_1, v_2, \ldots, v_n$ of numbers is unequal; it is given by the expression $(1/n)[(v_1 - \bar{v})^2 + \cdots + (v_n - \bar{v})^2]$, where $\bar{v}$ is the mean of the numbers.

mean-square-error criterion [CONT SYS] Evaluation of the performance of a control system by calculating the square root of the average over time of the square of the difference between the actual output and the output that is desired.

mean-square velocity [PHYS] The average value of the square of the velocities of a group of particles, such as the molecules of a gas.

mean stress [MECH] **1.** The algebraic mean of the maximum and minimum values of a periodically varying stress. **2.** *See* octahedral normal stress.

mean sun [ASTRON] A fictitious sun conceived to move eastward along the celestial equator at a rate that provides a uniform measure of time equal to the average apparent time; used as a reference for reckoning time, such as mean time or zone time.

mean time [ASTRON] Time based on the rotation of the earth relative to the mean sun.

mean-tone scale [ACOUS] A musical scale formed by giving the major third a ratio of exactly 5:4 and adapting other intervals to equalize them.

mean value [MATH] **1.** For a function $f(x)$ defined on an interval (a,b), the integral from a to b of $f(x)\,dx$ divided by $b - a$. **2.** *See* mean.

mean value theorem [MATH] If a function $f(x)$ is continuous on the closed interval $[a,b]$ and differentiable on the open interval (a,b), then there exists x_0, $a < x_0 < b$, such that $f(b) - f(a) = (b - a)f'(x_0)$. Also known as first law of the mean; Lagrange's formula; law of the mean.

mean velocity [PHYS] The average value of the velocities of a group of particles, such as the molecules of a gas.

measurable function [MATH] **1.** A real valued function f defined on a measurable space X, where for every real number a all those points x in X for which $f(x) \geq a$ form a measurable set. **2.** A function on a measurable space to a measurable space such that the inverse image of a measurable set is a measurable set.

measurable set [MATH] A member of the sigma algebra of subsets of a measurable space.

measurable space [MATH] A set together with a sigma algebra of subsets of this set.

measure [MATH] A nonnegative real valued function m defined on a sigma-algebra of subsets of a set S whose value is zero on the empty set, and whose value on a countable union of disjoint sets is the sum of its values on each set.

measurement ton *See* ton.

measure of location [STAT] A statistic, such as the mean, median, quartile, or mode; it has the property for the mean that if a constant is added to each value the same constant must also be added to the location measure.

measure-preserving transformation [MATH] A transformation T of a measure space S into itself such that if E is a measurable subset of S then so is $T^{-1}E$ (the set of points mapped into E by T) and the measure of $T^{-1}E$ is then equal to that of E.

measure space [MATH] A set together with a sigma algebra of subsets of the set and a measure defined on this sigma algebra.

measure zero [MATH] **1.** A set has measure zero if it is measurable and the measure of it is zero. **2.** A subset of euclidean n-dimensional space which has the property that for any positive number ε there is a covering of the set by n-dimensional rectangles such that the sum of the volumes of the rectangles is less than ε.

mechanical birefringence [OPTICS] A change in the double refraction of a solid material when it is subjected to stress. Also known as stress birefringence.

mechanical equivalent of heat [THERMO] The amount of mechanical energy equivalent to a unit of heat.

mechanical equivalent of light [OPTICS] The ratio of the radiant power emitted by a monochromatic light source whose wavelength is that at which the sensitivity of phototopic vision is greatest (about 555 nanometers), to its luminous flux measured in lumens.

mechanical hygrometer [ENG] A hygrometer in which an organic material, most commonly a bundle of human hair, which expands and contracts with changes in the moisture in the surrounding air or gas is held under slight tension by a spring, and a mechanical linkage actuates a pointer.

mechanical hysteresis [MECH] The dependence of the strain of a material not only on the instantaneous value of the stress but also on the previous history of the stress; for example, the elongation is less at a given value of tension when the tension is increasing than when it is decreasing.

mechanical impedance [MECH] The complex ratio of a phasor representing a sinusoidally varying force applied to a system to a phasor representing the velocity of a point in the system.

mechanical mass [QUANT MECH] The part of a particle's mass which is supposed to exist in the absence of any interaction of the particle with itself through a field.

mechanical ohm [MECH] A unit of mechanical resistance, reactance, and impedance, equal to a force of 1 dyne divided by a velocity of 1 centimeter per second.

mechanical oscillograph *See* direct-writing recorder.

mechanical property [MECH] A property that involves a relationship between stress and strain or a reaction to an applied force.

mechanical reactance [MECH] The imaginary part of mechanical impedance.

mechanical resistance *See* resistance.

mechanical rotational impedance *See* rotational impedance.

mechanical rotational reactance *See* rotational reactance.

mechanical rotational resistance *See* rotational resistance.

mechanical units [MECH] Units of length, time, and mass, and of physical quantities derivable from them.

mechanics [PHYS] **1.** In the original sense, the study of the behavior of physical systems under the action of forces. **2.** More broadly, the branch of physics which seeks to formulate general rules for predicting the behavior of a physical system under the influence of any type of interaction with its environment.

mechanocaloric effect [CRYO] An effect resulting from the fact that a temperature gradient in helium II is invariably accompanied by a pressure gradient, and conversely; examples are the fountain effect, and the heating of liquid helium left behind in a container when part of it leaks out through a small orifice.

mechanomotive force [MECH] The root-mean-square value of a periodically varying force.

mechanooptical vibrometer [ENG] A vibrometer in which the motion given to a probe by a surface whose vibration amplitude is to be measured is used to rock a mirror; a light beam reflected from the mirror and focused onto a scale provides an indication of the vibration amplitude.

median [MATH] Any line in a triangle which joins a vertex to the midpoint of the opposite side. [SCI TECH] Located in the middle. [STAT] An average of a series of quantities or values; specifically, the quantity or value of that item which is so positioned in the series, when arranged in order of numerical quantity or value, that there are an equal number of items of greater magnitude and lesser magnitude.

medium [ADP] The material, or configuration thereof, on which data are recorded; usually not applied to disk, drum, or core, but to storable, removable media, such as paper tape, cards, and magnetic tape. [PHYS] That entity in which objects exist and phenomena take place; examples are free space and various fluids and solids.

medium-scale integration [ELECTR] Solid-state integrated circuits having more than about 12 gate-equivalent circuits. Abbreviated MSI.

mega- [SCI TECH] A prefix representing 10^6, or one million. Abbreviated M.

megacycle *See* megahertz.

megaelectronvolt *See* million electron volts.

megaelectronvolt–curie [NUCLEO] A unit of radioactive power, equal to the power generated by 1 curie emitting an average energy of 10^6 electron volts per disintegration; equal to approximately 5.92777×10^{-3} watt. Abbreviated MeV Ci.

megagauss physics [PHYS] The production, measurement, and application of megagauss fields, as produced by discharge of capacitor banks or explosive flux-compression techniques.

megahertz [PHYS] Unit of frequency, equal to 1,000,000 hertz. Abbreviated MHz. Also known as megacycle (Mc).

megaparsec [ASTRON] A unit equal to 1,000,000 parsecs.

megaphone [ACOUS] A conical or rectangular horn used to amplify or direct the sound of a speaker's voice.

megasecond [MECH] A unit of time, equal to 1,000,000 seconds. Abbreviated Msec.

megaton [PHYS] The energy released by 1,000,000 metric tons of chemical high explosive calculated at a rate of 1000 calories per gram, or a total of 4.18×10^{15} joules; used principally in expressing the energy released by a nuclear bomb. Abbreviated MT.

megavolt [ELEC] A unit of potential difference or emf (electromotive force), equal to 1,000,000 volts. Abbreviated MV.

megawatt [MECH] A unit of power, equal to 1,000,000 watts. Abbreviated MW.

megawatt-day per ton [NUCLEO] A unit used for expressing the burnup of fuel in a reactor; specifically, the number of megawatt-days of heat output per metric ton of fuel in the reactor.

megawatt electric [NUCLEO] Unit of the electric power of a nuclear reactor, as opposed to thermal power. Abbreviated MW(E).

megawatt thermal [NUCLEO] Unit of the thermal power of a nuclear reactor, as opposed to electric power. Abbreviated MW(Th).

megawatt year of electricity [ELEC] A unit of electric energy, equal to the energy delivered by a power of 1,000,000 watts

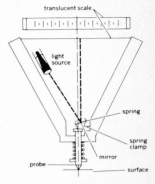

MECHANOOPTICAL VIBROMETER

Diagram of mechanooptical vibrometer. *(General Electric Co.)*

over a period of 1 tropical year, or to 3.1557×10^{13} joules. Abbreviated MWYE.

megohmmeter [ELEC] A high-range, permanent-magnet, moving-coil, direct-reading ohmmeter, commonly used as a portable instrument for measuring the high resistance of electrical materials of the order of 20,000 megohms at 1000 volts.

Mehler-Fock transform [MATH] An integral transform which associates to a function f the function whose value $F(y)$ is equal to the integral from 1 to infinity of $f(x)P_{-\frac{1}{2}+iy}(x)\,dx$, where $P_{-\frac{1}{2}+iy}$ is the Legendre function of the first kind of order $-\frac{1}{2}+iy$.

Mehler's integral [MATH] **1.** The integral from 0 to t of $(\sqrt{2}/\pi)\cos[(\nu+\frac{1}{2})x](\cos x - \cos t)^{-1/2}\,dx$; it is equal to $P_\nu(\cos t)$, where P_ν is a Legendre function of order ν. **2.** The integral from 0 to t of $(\sqrt{2}/\pi)\cos(yx)(\cosh t - \cosh x)^{-1/2}\,dx$; it is equal to $P_{-\frac{1}{2}+iy}(\cosh t)$.

Meijer's G function [MATH] A function formed from Mellin-Barnes integrals which provides an interpretation of the generalized hypergeometric function ${}_pF_q$ for $p>q+1$ in agreement with that of MacRobert's E function, and provides solutions for the generalized hypergeometric differential equation $[xP(\delta) - Q(\delta)]y = 0$, where P and Q are polynomials and δ is the differential operator xd/dx.

Meijer transform [MATH] The Meijer transform of a function $f(x)$ is the function $F(y)$ defined as the integral from 0 to ∞ of $\sqrt{xy}\,K_\nu(xy)f(x)\,dx$, where K_ν is a modified Bessel function.

Meinzer unit *See* permeability coefficient.

Meissner effect [SOLID STATE] The expulsion of magnetic flux from the interior of a piece of superconducting material as the material undergoes the transition to the superconducting phase. Also known as flux jumping; Meissner-Ochsenfeld effect.

Meissner-Ochsenfeld effect *See* Meissner effect.

Meixner polynomials [MATH] Families of polynomials which are orthogonal with respect to probability distributions whose frequencies are $\beta(\beta+1)\ \cdots\ (\beta+x-1)c^x/x!$, where $x=0, 1, 2, \ldots$.

Meixner's function [MATH] The confluent hypergeometric function

$$F_1(a,c,z) = e^{i\pi a}\frac{\Gamma(c)}{\Gamma(c-a)}\,\Psi\,(a,c;z)$$

where Ψ is Tricomi's function.

mel [ACOUS] A unit of pitch, equal to one-thousandth of the pitch of a simple tone whose frequency is 1000 hertz and whose loudness is 40 decibels above a listener's threshold.

Melde's experiment [MECH] An experiment to study transverse vibrations in a long, horizontal thread when one end of the thread is attached to a prong of a vibrating tuning fork, while the other passes over a pulley and has weights suspended from it to control the tension in the thread.

M electron [ATOM PHYS] An electron whose principal quantum number is 3.

Mellin-Barnes integral [MATH] Any contour integral of the form

$$\frac{1}{2\pi i}\int_c \frac{\Gamma(a_1 + A_1 s)\cdots\Gamma(a_m + A_m s)}{\Gamma(c_1 + C_1 s)\cdots\Gamma(c_p + C_p s)}$$

$$\times\ \frac{\Gamma(b_1 - B_1 s)\cdots\Gamma(b_n - B_n s)}{\Gamma(d_1 - D_1 s)\cdots\Gamma(d_q - D_q s)}\,z^s\,ds$$

where the integration contour is a straight line, parallel to the imaginary axis, with identations if necessary to avoid the poles of the integrand.

Mellin transform [MATH] The transform $F(s)$ of a function $f(t)$ defined as the integral over t from 0 to ∞ of $f(t)t^{s-1}$.

meltback transistor [ELECTR] A junction transistor in which the junction is made by melting a properly doped semiconductor and allowing it to solidify again.

melting *See* fusion.

melting point [THERMO] **1.** The temperature at which a solid of a pure substance changes to a liquid. Abbreviated mp. **2.** For a solution of two or more components, the temperature at which the first trace of liquid appears as the solution is heated.

member [MATH] An element of a set.

membrane analogy [MECH] **1.** A formal identity between the differential equation and boundary conditions for a stress function for torsion of an elastic prismatic bar, and those for the deflection of a uniformly stretched membrane with the same boundary as the cross section of the bar, subjected to a uniform pressure. **2.** A formal identity between the differential equations and boundary conditions for a stress function for bending of an elastic prismatic bar, and those for the deflection of a uniformly stretched membrane with the same boundary as the cross section of the bar, subjected to a pressure distribution determined from the bar problem.

memory core *See* magnetic core.

memory switch *See* ovonic memory switch.

mendelevium [CHEM] Synthetic radioactive element, symbol Md, with atomic number 101; made by bombarding lighter elements with light nuclei accelerated in cyclotrons.

Menelaus' theorem [MATH] If ABC is a triangle and PQR is a straight line that cuts AB, CA, and the extension of BC at P, Q, and R, respectively, then $(AP/PB)(CQ/QA)(BR/CR) = 1$.

Menger's theorem [MATH] A theorem in graph theory which states that if G is a connected graph and A and B are disjoint sets of points of G, then the minimum number of points whose deletion separates A and B is equal to the maximum number of disjoint paths between A and B.

menisc-, menisco- [SCI TECH] A combining form denoting crescentic, sickle-shaped, semilunar.

meniscus [FL MECH] The free surface of a liquid which is near the walls of a vessel and which is curved because of surface tension.

meniscus lens [OPTICS] A lens with one convex surface and one concave surface.

mensuration [MATH] The measurement of geometric quantities; for example, length, area, and volume. [SCI TECH] The act or process of measuring.

mer-, mero- [SCI TECH] A combining form meaning part or partial.

mercury [CHEM] A metallic element, symbol Hg, atomic number 80, atomic weight 200.59, existing at room temperature as a silvery, heavy liquid. Also known as quicksilver.

Mercury [ASTRON] The planet nearest to the sun; it is visible to the naked eye shortly after sunset or before sunrise when it is nearest to its greatest angular distance from the sun.

mercury arc [ELECTR] An electric discharge through ionized mercury vapor, giving off a brilliant bluish-green light containing strong ultraviolet radiation.

mercury barometer [ENG] An instrument which determines atmospheric pressure by measuring the height of a column of mercury which the atmosphere will support; the mercury is in a glass tube closed at one end and placed, open end down, in a well of mercury. Also known as Torricellian barometer.

mercury cell [ELEC] A primary dry cell that delivers an essentially constant output voltage throughout its useful life by means of a chemical reaction between zinc and mercury

MENDELEVIUM

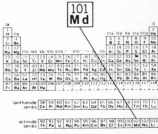

Periodic table of the chemical elements showing the position of mendelevium.

MENISCUS LENS

(a)   (b)

Two types of meniscus lens. (a) Positive. (b) Negative. (*From F. A. Jenkins and H. E. White, Fundamentals of Optics, 3d ed., McGraw-Hill, 1957*)

MERCURY

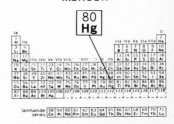

Periodic table of the chemical elements showing the position of mercury.

oxide; widely used in hearing aids. Also known as mercury oxide cell.

mercury jet magnetometer [ENG] A type of magnetometer in which the magnetic field strength is determined by measuring the electromotive force between electrodes at opposite ends of a narrow pipe made of insulating material, through which mercury is forced to flow.

mercury lamp *See* mercury-vapor lamp.

mercury manometer [ENG] A manometer in which the instrument fluid is mercury; used to record or control difference of pressure or fluid flow.

mercury thermometer [ENG] A liquid-in-glass thermometer or a liquid-in-metal thermometer using mercury as the liquid.

mercury tube *See* mercury-vapor tube.

mercury-vapor lamp [ELECTR] A lamp in which light is produced by an electric arc between two electrodes in an ionized mercury-vapor atmosphere; it gives off a bluish-green light rich in ultraviolet radiation. Also known as mercury lamp.

mercury-vapor tube [ELECTR] A gas tube in which the active gas is mercury vapor. Also known as mercury tube.

merged-transistor logic *See* integrated injection logic.

meridian [ASTRON] **1.** A great circle passing through the poles of the axis of rotation of a planet or satellite. **2.** *See* celestial meridian. [GEOD] A north-south reference line, particularly a great circle through the geographical poles of the earth.

meridian angle [ASTRON] Angular distance east or west of the local celestial meridian; the arc of the celestial equator, or the angle at the celestial pole, between the upper branch of the local celestial meridian and the hour circle of a celestial body, measured eastward or westward from the local celestial meridian through 180°, and labeled E or W to indicate the direction of measurement.

meridian angle difference [ASTRON] The difference between two meridian angles, particularly between the meridian angle of a celestial body and the value used as an argument for entering into a table. Also called hour angle difference.

meridian circle *See* transit circle.

meridian observation [ASTRON] Measurement of the altitude of a celestial body on the celestial meridian of the observer, or the altitude so measured.

meridian passage [ASTRON] The passage of a celestial body across an observer's meridian.

meridian telescope [OPTICS] Any telescope used to make observations in the plane of the meridian, such as a transit telescope or zenith telescope.

meridian transit *See* transit; transit circle.

meridional focus *See* primary focus.

meridional plane [OPTICS] A plane containing the axis of an optical system. Also known as tangential plane.

meridional ray [OPTICS] A ray that lies within a plane which also contains the axis of an optical system.

merit [ELECTR] A performance rating that governs the choice of a device for a particular application; it must be qualified to indicate type of rating, as in gain-bandwidth merit or signal-to-noise merit.

merohedral [CRYSTAL] Of a crystal class in a system, having a general form with only one-half, one-fourth, or one-eighth the number of equivalent faces of the corresponding form in the holohedral class of the same system. Also known as merosymmetric.

meromorphic function [MATH] A function of complex variables which is analytic in its domain of definition save at a finite number of points which are poles.

MERCURY-VAPOR LAMP

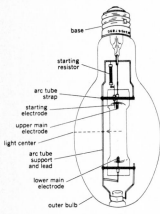

base

starting resistor

arc tube strap

starting electrode

upper main electrode

light center

arc tube support and lead

lower main electrode

outer bulb

Parts of the mercury-vapor lamp.

Mersenne's law [MECH] The fundamental frequency of a vibrating string is proportional to the square root of the tension and inversely proportional both to the length and the square root of the mass per unit length.

Merton grating [OPTICS] A type of diffraction grating which is produced by a process in which a helical thread is cut on a cylinder, and errors are smoothed by cutting a second thread further along the same cylinder with a Merton nut.

mes-, meso- [SCI TECH] A combining form denoting mid-, middle, medial; medium, moderate, intermediate.

mesa diode [ELECTR] A diode produced by diffusing the entire surface of a large germanium or silicon wafer and then delineating the individual diode areas by a photoresist-controlled etch that removes the entire diffused area except the island or mesa at each junction site.

mesa transistor [ELECTR] A transistor in which a germanium or silicon wafer is etched down in steps so the base and emitter regions appear as physical plateaus above the collector region.

MESFET *See* metal semiconductor field-effect transistor.

mesh [ELEC] A set of branches forming a closed path in a network so that if any one branch is omitted from the set, the remaining branches of the set do not form a closed path. Also known as loop.

mesh analysis [ELEC] A method of electrical circuit analysis in which the mesh currents are taken as independent variables and the potential differences around a mesh are equated to 0.

mesh currents [ELEC] The currents which are considered to circulate around the meshes of an electric network, so that the current in any branch of the network is the algebraic sum of the mesh currents of the meshes to which that branch belongs. Also known as cyclic currents; Maxwell's cyclic currents.

mesh impedance [ELEC] The ratio of the voltage to the current in a mesh when all other meshes are open. Also known as self-impedance.

mesic atom [PARTIC PHYS] An atom in which one of the electrons is replaced by a negative muon or meson orbiting close to or within the nucleus. Also known as mesonic atom.

mesic molecule [PARTIC PHYS] A molecule in which one of the electrons is replaced by a negative muon or meson orbiting close to or within one of the nuclei. Also known as mesonic molecule.

mesomorphism [PHYS CHEM] A state of matter intermediate between a crystalline solid and a normal isotropic liquid, in which long rod-shaped organic molecules contain dipolar and polarizable groups.

meson [PARTIC PHYS] Any elementary (noncomposite) particle with strong nuclear interactions and baryon number equal to zero.

meson capture [PARTIC PHYS] Process in which an atomic nucleus acquires a negative muon or meson which circles it in a tightly bound orbit until it decays.

mesonic atom *See* mesic atom.

mesonic molecule *See* mesic molecule.

mesonic x-ray [PARTIC PHYS] An x-ray emitted by a mesic atom when the muon or meson makes a transition from one bound state to another.

meson resonance [PARTIC PHYS] Any elementary particle with a baryon number of zero which decays through strong interactions, and therefore has an extremely short lifetime on the order of 10^{-23} second.

Messier number [ASTRON] A number by which star clusters and nebulae are listed in Messier's catalog; for example, the Andromeda Galaxy is M 31.

MESA DIODE

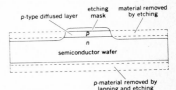

Mesa structure of a high-speed diffused silicon diode.

Messier's catalog [ASTRON] A listing of 103 star clusters and nebulae compiled in 1784.

metacenter [FL MECH] The intersection of a vertical line through the center of buoyancy of a floating body, slightly displaced from its equilibrium position, with a line connecting the center of gravity and the equilibrium center of buoyancy; the floating body is stable if the metacenter lies above the center of gravity.

metagalaxy [ASTRON] The total assemblage of recognized galaxies; essentially this represents the entire material universe.

metal-air battery *See* air depolarized battery.

metal antenna [ELECTROMAG] An antenna which has a relatively small metal surface, in contrast to a slot antenna.

metal-film resistor [ELEC] A resistor in which the resistive element is a thin film of metal or alloy, deposited on an insulating substrate of an integrated circuit.

metalimnion *See* thermocline.

metal-insulator semiconductor [SOLID STATE] Semiconductor construction in which an insulating layer, generally a fraction of a micrometer thick, is deposited on the semiconducting substrate before the pattern of metal contacts is applied. Abbreviated MIS.

metallic [OPTICS] Having a brilliant mineral luster characteristic of metals. [SCI TECH] Pertaining to metals.

metallic bond [PHYS CHEM] The type of chemical bond that is present in all metals, and may be thought of as resulting from a sea of valence electrons which are free to move throughout the metal lattice.

metallic-disk rectifier *See* metallic rectifier.

metallic electrode arc lamp [ELEC] A type of arc lamp in which light is produced by luminescent vapor introduced into the arc by evaporation from the cathode; the anode is solid copper, and the cathode is formed of magnetic iron oxide with titanium as the light-producing element and other chemicals to control steadiness and vaporization.

metallic nuclear fuel [NUCLEO] A fissionable isotope of a metallic element, or an alloy containing such an isotope, used as the energy source for a nuclear reactor.

metallic rectifier [ELECTR] A rectifier consisting of one or more disks of metal under pressure-contact with semiconductor coatings or layers, such as a copper oxide, selenium, or silicon rectifier. Also known as contact rectifier; dry-disk rectifier; dry-plate rectifier; metallic-disk rectifier; semiconductor rectifier.

metallized capacitor [ELEC] A capacitor in which a film of metal is deposited directly on the dielectric to serve in place of a separate foil strip; has self-healing characteristics.

metallized-paper capacitor [ELEC] A modification of a paper capacitor in which metal foils are replaced by extremely thin films of metal deposited on the paper; if a breakdown occurs, these films burn away in the area of the breakdown.

metallized resistor [ELEC] A resistor made by depositing a thin film of high-resistance metal on the surface of a glass or ceramic rod or tube.

metallograph [OPTICS] An optical microscope equipped with a camera for both visual observation and photography of the structure and constitution of a metal or alloy.

metal-nitride-oxide semiconductor [SOLID STATE] A semiconductor structure that has a double insulating layer; typically, a layer of silicon dioxide (SiO_2) is nearest the silicon substrate, with a layer of silicon nitride (Si_3N_4) over it. Abbreviated MNOS.

metal oxide resistor [ELEC] A metal-film resistor in which an oxide of a metal such as tin is deposited as a film onto an insulating substrate.

metal oxide semiconductor [SOLID STATE] A metal insulator semiconductor structure in which the insulating layer is an oxide of the substrate material; for a silicon substrate, the insulating layer is silicon dioxide (SiO_2). Abbreviated MOS.

metal oxide semiconductor field-effect transistor [ELECTR] A field-effect transistor having a gate that is insulated from the semiconductor substrate by a thin layer of silicon dioxide. Abbreviated MOSFET; MOST; MOS transistor. Formerly known as insulated-gate field-effect transistor.

metal screen [GRAPHICS] An intensifying screen consisting of a metal which emits secondary electrons and x-rays when bombarded by x-rays.

metal semiconductor field-effect transistor [ELECTR] A field-effect transistor that uses a thin film of gallium arsenide, with a Schottky barrier gate formed by depositing a layer of metal directly onto the surface of the film. Abbreviated MESFET.

metal vapor laser [OPTICS] An ion laser based on vaporization of a solid or liquid metal, such as cadmium, calcium, copper, lead, manganese, selenium, strontium, and tin, vaporized with a buffer gas such as helium.

metamathematics [MATH] The study of the principles of deductive logic as they are used in mathematical logic.

metarheology [MECH] A branch of rheology whose approach is intermediate between those of macrorheology and microrheology; certain processes that are not isothermal are taken into consideration, such as kinetic elasticity, surface tension, and rate processes.

metastable equilibrium [PHYS] A condition in which a system returns to equilibrium after small (but not large) displacements; it may be represented by a ball resting in a small depression on top of a hill. [PHYS CHEM] A state of pseudo-equilibrium having higher free energy than the true equilibrium state.

metastable ion [ANALY CHEM] In mass spectroscopy, an ion formed by a secondary dissociation process in the analyzer tube (formed after the parent or initial ion has passed through the accelerating field).

metastable phase [PHYS CHEM] Existence of a substance as either a liquid, solid, or vapor under conditions in which it is normally unstable in that state.

metastable state [QUANT MECH] An excited stationary energy state whose lifetime is unusually long.

metastasis [PHYS] A transition of an electron or nucleon from one bound state to another in an atom or molecule, or the capture of an electron by a nucleus.

meteor [ASTRON] The phenomena which accompany a body from space (a meteoroid) in its passage through the atmosphere, including the flash and streak of light and the ionized trail.

meteoric ionization [ASTROPHYS] Ionization resulting from collisional interactions of a meteoroid and its vaporization products with the air.

meteorite [GEOL] Any meteoroid that has fallen to the earth's surface.

meteoroid [ASTRON] Any solid object moving in interplanetary space that is smaller than a planet or asteroid but larger than a molecule.

meteorological optics [OPTICS] A branch of atmospheric physics or physical meteorology in which optical phenomena occurring in the atmosphere are described and explained. Also known as atmospheric optics.

meteor shower [ASTRON] A number of meteors with approximately parallel trajectories.

meteor stream [ASTRON] A group of meteoric bodies with nearly identical orbits.

meteor trail *See* ion column.

meter [MECH] The international standard unit of length, equal to 1,650,763.73 times the wavelength of the orange light emitted when a gas consisting of the pure krypton isotope of mass number 86 is excited in an electrical discharge. Abbreviated m.

meter-atmosphere [PHYS] The depth of an equivalent atmosphere of a given gas, in meter-atmospheres, is equal to the depth in meters that the atmosphere would have if it were composed entirely of the gas in question and in the same amount as exists in the actual atmosphere, and had a uniform temperature and pressure of 0°C and 1 standard atmosphere (101,325 newtons per square meter). Abbreviated m-atm. Also known as atmo-meter.

meter-candle *See* lux.

meter factor [ENG] A factor used with a meter to correct for ambient conditions, for example, the factor for a fluid-flow meter to compensate for such conditions as liquid temperature change and pressure shrinkage.

meter-kilogram [MECH] **1.** A unit of energy or work in a meter-kilogram-second gravitational system, equal to the work done by a kilogram-force when the point at which the force is applied is displaced 1 meter in the direction of the force; equal to 9.80665 joules. Abbreviated m-kgf. Also known as meter kilogram-force. **2.** A unit of torque, equal to the torque produced by a kilogram-force acting at a perpendicular distance of 1 meter from the axis of rotation. Also known as kilogram-meter (kgf-m).

meter kilogram-force *See* meter-kilogram.

meter-kilogram-second-ampere system [PHYS] A system of electrical and mechanical units in which length, mass, time, and electric current are the fundamental quantities, and the units of these quantities are the meter, the kilogram, the second, and the ampere respectively. Abbreviated mksa system. Also known as Giorgi system; practical system.

meter-kilogram-second system [MECH] A metric system of units in which length, mass, and time are fundamental quantities, and the units of these quantities are the meter, the kilogram, and the second respectively. Abbreviated mks system.

meter sizing factor [FL MECH] A dimensionless number used in calculating the rate of flow of fluid through a pipe from the readings of a flowmeter that measures the drop in pressure when the fluid is forced to flow through a circular orifice; it is equal to $K(d/D)^2$, where K is the flow coefficient, d is the orifice bore diameter, and D is the internal diameter of the pipe.

meter-ton-second system [MECH] A modification of the meter-kilogram-second system in which the metric ton (1000 kilograms) replaces the kilogram as the unit of mass.

method of images [ELEC] In electrostatics, a method of determining the electric fields and potentials set up by charges in the vicinity of a conductor, in which the conductor and its induced surface charges are replaced by one or more fictitious charges. [PHYS] Any method of solving magnetostatic, hydrodynamic, and other problems involving boundary conditions at the interface between two media, in which fictitious objects, such as magnetic dipoles and sources and sinks of fluid, are introduced to satisfy the boundary conditions; these methods are generalizations of the method in electrostatics.

method of mixtures [THERMO] A method of determining the heat of fusion of a substance whose specific heat is known, in which a known amount of the solid is combined with a known amount of the liquid in a calorimeter, and the decrease in the liquid temperature during melting of the solid is measured.

method of moments [STAT] A procedure for estimating the parameters of a distribution; at one time it was heavily used for fitting Pearson-type frequency distributions.

method of moving averages [STAT] A series of averages where each average is the mean value of the time series over a fixed interval of time, and where all possible averages of the length are included in the analysis; used to smooth data in a time series.

method of semiaverages [STAT] Data are divided into two equal sets and the means of the two sets or two other points representative of each set are determined and a straight line drawn through them; used to provide a quick estimate of a linear regression line.

metonic cycle [ASTRON] A time period of 235 lunar months, or 19 years 11 days; after this period the phases of the moon occur on the same days of the month.

metric [MATH] A real valued "distance" function on a topological space X satisfying four rules: for x, y, and z in X, the distance from x to itself is zero; the distance from x to y is positive if x and y are different; the distance from x to y is the same as the distance from y to x; and the distance from x to y is less than or equal to the distance from x to z plus the distance from z to y (triangle inequality).

metricate [SCI TECH] To use the metric system in expressing all physical quantities.

metric centner [MECH] **1.** A unit of mass equal to 50 kilograms. **2.** A unit of mass equal to 100 kilograms. Also known as quintal.

metric density [MATH] Let S be a measurable subset of the real numbers, and x a real number at which the upper metric density $\overline{\phi}(x)$ and the lower metric density $\underline{\phi}(x)$ of S are equal; then the metric density of S at x is the number $\phi(x) = \overline{\phi}(x)$.

metric grain [MECH] A unit of mass, equal to 50 milligrams; used in commercial transactions in precious stones.

metric horsepower [PHYS] A unit of power, equal to 75 meter kilogram-force per second; equal to 753.49875 watts.

metric line *See* millimeter.

metric ounce *See* mounce.

metric slug *See* metric-technical unit of mass.

metric space [MATH] Any topological space which has a metric defined on it.

metric system [MECH] A system of units used in scientific work throughout the world and employed in general commercial transactions and engineering applications; its units of length, time, and mass are the meter, second, and kilogram respectively, or decimal multiples and submultiples thereof.

metric-technical unit of mass [MECH] A unit of mass, equal to the mass which is accelerated by 1 meter per second per second by a force of 1 kilogram-force; it is equal to 9.80665 kilograms. Abbreviated TME. Also known as hyl; metric slug.

metric tensor [MATH] A second rank tensor of a Riemannian space whose components are functions which help define magnitude and direction of vectors about a point. Also known as fundamental tensor.

metric ton *See* tonne.

metrizable space [MATH] A topological space on which can be defined a metric whose topological structure is equivalent to the original one.

metrology [PHYS] The science of measurement.

Meusnier's theorem [MATH] A theorem stating that the curvature of a surface curve equals the curvature of the normal section through the tangent to the curve divided by the cosine of the angle between the plane of this normal section and the osculating plane of the curve.

MeV *See* million electron volts.

MeV Ci *See* megaelectronvolt curie.

Meyer atomic volume curve [ATOM PHYS] A graph of the atomic volumes of the elements versus their atomic numbers; it reveals a periodicity, with peaks at the alkali elements and valleys at the transition elements.

mF *See* millifarad.

mg *See* milligram.

mG *See* milligauss.

Mg *See* magnesium.

mGal *See* milligal.

mg h *See* milligram-hour.

mH *See* millihenry.

MHD *See* magnetohydrodynamics.

mho *See* siemens.

mHz *See* millihertz.

MHz *See* megahertz.

mi *See* mile.

Michaelson actinograph [ENG] A pyrheliometer of the bimetallic type used to measure the intensity of direct solar radiation; the radiation is measured in terms of the angular deflection of a blackened bimetallic strip which is exposed to the direct solar beams.

Michel parameter [PARTIC PHYS] A number appearing in an equation for the momentum spectrum of muon decay, which depends on the nature of the weak interactions; the number is equal to 3/4 in any two-component neutrino theory before radiative corrections are taken into account.

Michelson interferometer [OPTICS] An interferometer in which light strikes a partially reflecting plate at an angle of 45°, the light beams reflected and transmitted by the plate are both reflected back to the plate by mirrors, and the beams are recombined at the plate, interfering constructively or destructively depending on the distances from the plate to the two mirrors.

Michelson-Morley experiment [OPTICS] An experiment which uses a Michelson interferometer to determine the difference between the speeds of light in two perpendicular directions.

Michelson stellar interferometer [OPTICS] An instrument for measuring angular diameters of astronomical objects, in which a system of mirrors directs two parallel beams of light into a telescope, and angular diameter is determined from the maximum distance between the beams at which interference fringes are observable.

micril *See* gammil.

micro- [MATH] A prefix representing 10^{-6}, or one-millionth. [SCI TECH] **1.** A prefix indicating smallness, as in microwave. **2.** A prefix indicating extreme sensitivity, as in microradiometer and microphone.

microalloy diffused transistor [ELECTR] A microalloy transistor in which the semiconductor wafer is first subjected to gaseous diffusion to produce a nonuniform base region. Abbreviated MADT.

microalloy transistor [ELECTR] A transistor in which the emitter and collector electrodes are formed by etching depressions, then electroplating and alloying a thin film of the impurity metal to the semiconductor wafer, somewhat as in a surface-barrier transistor.

MICHELSON INTERFEROMETER

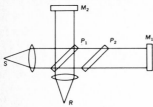

Michelson interferometer, S = narrow angle source; R = receiver; M_1, M_2 = mirrors; P_1 = 50% partially reflecting plate; P_2 = reflector plate which compensates for thickness of P_1. (*From A. C. Hardy and F. H. Perrin, The Principles of Optics, McGraw-Hill, 1932*)

MICHELSON STELLAR INTERFEROMETER

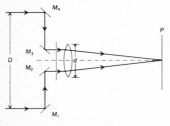

Michelson stellar interferometer, used for measuring the diameters of stars. M_2, M_3 = two fixed mirrors; M_1, M_4 = two mirrors that are movable so as to vary their separation D; d = diameter of lens; P = focal plane of lens.

microammeter [ELEC] An ammeter whose scale is calibrated to indicate current values in microamperes.

microampere [ELEC] A unit of current equal to one-millionth of an ampere. Abbreviated μA.

microangstrom [MECH] A unit of length equal to one-millionth of an angstrom, or 10^{-16} meter. Abbreviated μA.

microbalance [ENG] A small, light type of analytical balance that can weigh loads of up to 0.1 gram to the nearest microgram.

microbar *See* barye.

microbarm [GEOPHYS] That portion of the record of a microbarograph between any two or a specified small number of the successive crossings of the average pressure level in the same direction; analogous to microseism.

microbarogram [ENG] The record or trace made by a microbarograph.

microbarograph [ENG] A type of aneroid barograph designed to record atmospheric pressure variations of very small magnitude.

microcalorimeter [ENG] A calorimeter for measuring very small amounts of heat, in which the heat source and a small heating coil are placed in identical vessels and the amount of current through the coil is varied until the temperatures of the vessels are identical, as indicated by thermocouples.

microcanonical ensemble [STAT MECH] A collection of systems describing a single isolated system of specified energy; its members are uniformly distributed over a part of phase space whose energies lie within an infinitesimal range.

microcapacitor [ELECTR] Any very small capacitor used in microelectronics, usually consisting of a thin film of dielectric material sandwiched between electrodes.

microchannel plate [ELECTR] A plate that consists of extremely small cylinder-shaped electron multipliers mounted side by side, to provide image intensification factors as high as 100,000.

microcircuitry [ELECTR] Electronic circuit structures that are orders of magnitude smaller and lighter than circuit structures produced by the most compact combinations of discrete components. Also known as microelectronic circuitry; microminiature circuitry.

microcoulomb [ELEC] A unit of electric charge equal to one-millionth of a coulomb. Abbreviated μC.

microcrystalline [CRYSTAL] Composed of or containing crystals that are visible only under the microscope.

microdensitometer [SPECT] A high-sensitivity densitometer used in spectroscopy to detect spectrum lines too faint on a negative to be seen by the human eye.

microdiffusiometer [ENG] A type of diffusiometer in which diffusion is measured over microscopic distances, greatly reducing the time required for the measurement and the effects of vibration and temperature changes.

microelectronic circuitry *See* microcircuitry.

microelectronics [ELECTR] The technology of constructing circuits and devices in extremely small packages by various techniques. Also known as microminiaturization; microsystem electronics.

microelement [ELECTR] Resistor, capacitor, transistor, diode, inductor, transformer, or other electronic element or combination of elements mounted on a ceramic wafer 0.025 centimeter thick and about 0.75 centimeter square; individual microelements are stacked, interconnected, and potted to form micromodules.

microfarad [ELEC] A unit of capacitance equal to one-millionth of a farad. Abbreviated μF.

microfluid [FL MECH] A fluid in which the effects of local

MICROCIRCUITRY

Size comparison of solid-state circuit, at right, and discrete-component printed circuit. Each is a binary full-serial adder. Solid-state circuit occupies 0.02 cubic inch (0.33 cubic centimeter) and weighs 1.5 grams compared with 4-square-inch (26-square-centimeter) size and 42-gram weight for the discrete-component printed circuit. *(Texas Instruments, Inc.)*

motion of contained material particles on properties and behavior of the fluid are not disregarded.

microgammil *See* gammil.

microgram [MECH] A unit of mass equal to one-millionth of a gram. Abbreviated μg.

microhm [ELEC] A unit of resistance, reactance, and impedance, equal to 10^{-6} ohm.

microhysteresis effect [SOLID STATE] Hysteresis that results from the motion of domain walls lagging behind an applied magnetic or elastic stress when these walls are held up by dislocations and other imperfections in the material.

microinterferometer [OPTICS] Functional combination of a microscope with an interferometer; used to study thin films, platings, or transparent coatings.

microlite [CRYSTAL] A microscopic crystal which polarizes light. Also known as microlith.

microlith *See* microlite.

micromanometer [ENG] Any manometer that is designed to measure very small pressure differences.

micrometeorite [ASTRON] A very small meteorite or meteoritic particle with a diameter generally less than a millimeter.

micrometeoroid [ASTRON] A very small meteoroid with diameter generally less than a millimeter.

micrometer [ENG] 1. An instrument attached to a telescope or microscope for measuring small distances or angles. 2. A caliper for making precise measurements; a spindle is moved by a screw thread so that it touches the object to be measured; the dimension can then be read on a scale. Also known as micrometer caliper. [MECH] A unit of length equal to one-millionth of a meter. Abbreviated μm. Also known as micron (μ).

micrometer caliper *See* micrometer.

micrometer of mercury *See* micron.

micromicro- *See* pico-.

micromicrofarad *See* picofarad.

micromicrosecond *See* picosecond.

micromicrowatt *See* picowatt.

microminiature circuitry *See* microcircuitry.

microminiaturization *See* microelectronics.

micromodule [ELECTR] Cube-shaped, plug-in, miniature circuit composed of potted microelements; each microelement can consist of a resistor, capacitor, transistor, or other element, or a combination of elements.

micron [MECH] 1. A unit of pressure equal to the pressure exerted by a column of mercury 1 micrometer high, having a density of 13.5951 grams per cubic centimeter, under the standard acceleration of gravity; equal to 0.133322387415 pascal; it differs from the millitorr by less than one part in seven million. Also known as micrometer of mercury. 2. *See* micrometer.

microphonics [ELECTR] Noise caused by mechanical vibration of the elements of an electron tube, component, or system. Also known as microphonism.

microphonism *See* microphonics.

microphotometer [ENG] A photometer that provides highly accurate illumination measurements; in one form, the changes in illumination are picked up by a phototube and converted into current variations that are amplified by vacuum tubes.

microprobe [SPECT] An instrument for chemical microanalysis of a sample, in which a beam of electrons is focused on an area less than a micrometer in diameter, and the characteristic x-rays emitted as a result are dispersed and analyzed in a crystal spectrometer to provide a qualitative and quantitative evaluation of chemical composition.

MICROMETER

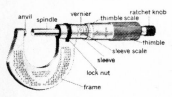

Machinist's outside caliper with micrometer reading 0.250 inch (6.35 millimeters), showing component parts. (*L. S. Starrett Co.*)

micropulsation [GEOPHYS] A short-period geomagnetic variation in the range of about 0.2–600 seconds, typically exhibiting an oscillatory waveform.

micropycnometer [ENG] A small-volume pycnometer with a capacity from 0.25 to 1.6 milliliters; weighing precision is 1 part in 10,000, or better.

microradiogram [PHYS] A two-dimensional x-ray image of a sample, produced by one type of x-ray microscope used in microradiography; all levels of the sample object are imaged into essentially a single focal plane for subsequent microphotographic enlargement.

microradiograph [GRAPHICS] An enlarged radiographic image on photographic film produced either by increasing the distance from specimen to photographic plate to secure inherent enlargement of divergent x-ray beams, or by optical enlargement of a developed image.

microradiography [ANALY CHEM] Technique for the study of surfaces of solids by monochromatic-radiation (such as x-ray) contrast effects shown via projection or enlargement of a contact radiograph. [GRAPHICS] The radiography of small objects having details too fine to be seen by the unaided eye, with optical enlargement of the resulting negative.

microradiometer [ELECTR] A radiometer used for measuring weak radiant power, in which a thermopile is supported on and connected directly to the moving coil of a galvanometer. Also known as radiomicrometer.

micro-reciprocal-degree *See* mired.

microrefractometry [OPTICS] The measurement of refractive indices of microscopic objects; this is often done by immersing an object in a series of mediums of graded refractive index until one is found that makes the object invisible in a phase-contrast microscope.

microrheology [MECH] A branch of rheology in which the heterogeneous nature of dispersed systems is taken into account.

microscope [OPTICS] An instrument through which minute objects are enlarged by means of a lens or lens system; principal types include optical, electron, and x-ray.

microscopic [OPTICS] *See* microscopical. [SCI TECH] Of extremely small size.

microscopical [OPTICS] Also known as microscopic. **1.** Of or pertaining to the microscope. **2.** Visible only under a microscope.

microscopic reversibility [STAT MECH] A principle which requires that in a system at equilibrium any molecular process and its reverse take place at the same average rate. Also known as reversibility principle.

microscopic state [STAT MECH] The state of a system as specified by the actual properties of each individual, elemental component, in the ultimate detail permitted by the uncertainty principle. Also known as microstate.

microscopic theory [PHYS] A theory concerned with the interactions of atoms, molecules, or their constituents, involving distances on the order of 10^{-10} meter or less, which underlie observable phenomena.

microscopy [OPTICS] The interpretive application of microscope magnification to the study of materials that cannot be properly seen by the unaided eye.

microsecond [MECH] A unit of time equal to one-millionth of a second. Abbreviated μs.

microseism [GEOPHYS] A weak, continuous, oscillatory motion in the earth having a period of 1–9 seconds and caused by a variety of agents, especially atmospheric agents; not related to an earthquake.

microspectrograph [SPECT] A microspectroscope provided

MICRORADIOGRAPHY

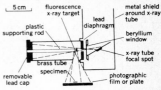

Basic elements of microradiograph used for producing radiographic images on photographic film.

with a photographic camera or other device for recording the spectrum.

microspectrophotometer [SPECT] A split-beam or double-beam spectrophotometer including a microscope for the localization of the object under study, and capable of carrying out spectral analyses within the dimensions of a single cell.

microspectroscope [SPECT] An instrument for analyzing the spectra of microscopic objects, such as living cells, in which light passing through the sample is focused by a compound microscope system, and both this light and the light which has passed through a reference sample are dispersed by a prism spectroscope, so that the spectra of both can be viewed simultaneously.

microstate *See* microscopic state.

microstrip [ELECTROMAG] A strip transmission line that consists basically of a thin-film strip in intimate contact with one side of a flat dielectric substrate, with a similar thin-film ground-plane conductor on the other side of the substrate.

microsystem electronics *See* microelectronics.

microtron [NUCLEO] A type of circular particle accelerator for accelerating electrons to energies of several million electron volts, in which the time of successive revolutions of the particles increases by exactly one cycle of the accelerating radio-frequency voltage, so that synchronism is maintained.

microvolt [ELEC] A unit of potential difference equal to one-millionth of a volt. Abbreviated μV.

microvoltmeter [ELECTR] A voltmeter whose scale is calibrated to indicate voltage values in microvolts.

microwatt [MECH] A unit of power equal to one-millionth of a watt. Abbreviated μW.

microwave [ELECTROMAG] An electromagnetic wave which has a wavelength between about 0.3 and 30 centimeters, corresponding to frequencies of 1-100 gigahertz; however, there are no sharp boundaries distinguishing microwaves from infrared and radio waves.

microwave acoustics [ACOUS] The production and study of elastic vibrations in materials at microwave frequencies, on the order of 10^9 to 10^{11} hertz, such as in single-crystal delay lines used in radar systems.

microwave amplification by stimulated emission of radiation *See* maser.

microwave attenuator [ELECTROMAG] A device that causes the field intensity of microwaves in a waveguide to decrease by absorbing part of the incident power; usually consists of a piece of lossy material in the waveguide along the direction of the electric field vector.

microwave background [ASTRON] The homogeneous, isotropic thermal radiation filling the universe that is thought to be a relic of the big bang.

microwave bridge [ELECTROMAG] A microwave circuit equivalent to an ordinary electrical bridge and used to measure impedance; consists of six waveguide sections arranged to form a multiple junction.

microwave cavity *See* cavity resonator.

microwave circuit [ELECTROMAG] Any particular grouping of physical elements, including waveguides, attenuators, phase changers, detectors, wavemeters, and various types of junctions, which are arranged or connected together to produce certain desired effects on the behavior of microwaves.

microwave circulator *See* circulator.

microwave detector [ELECTR] A device that can demonstrate the presence of a microwave by a specific effect that the wave produces, such as a bolometer, or a semiconductor crystal making a pinpoint contact with a tungsten wire.

MICROWAVE

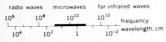

A portion of the electromagnetic spectrum showing the part occupied by microwaves.

microwave filter [ELECTROMAG] A device which passes microwaves of certain frequencies in a transmission line or waveguide while rejecting or absorbing other frequencies; consists of resonant cavity sections or other elements.

microwave frequency [PHYS] A frequency on the order of 10^9–10^{11} hertz.

microwave generator *See* microwave oscillator.

microwave gyrator *See* gyrator.

microwave optics [ELECTROMAG] The study of those properties of microwaves which are analogous to the properties of light waves in optics.

microwave oscillator [ELECTR] A type of electron tube or semiconductor device used for generating microwave radiation or voltage waveforms with microwave frequencies. Also known as microwave generator.

microwave radiometer *See* radiometer.

microwave reflectometer [ELECTROMAG] A pair of single detector couplers on opposite sides of a waveguide, one of which is positioned to monitor transmitted power, and the other to measure power reflected from a single discontinuity in the line.

microwave refractometer [ELECTROMAG] An instrument that measures the index of refraction of the atmosphere by measuring the travel time of microwave signals through each of two precision microwave transmission cavities, one of which is hermetically sealed to serve as a reference.

microwave resonance cavity *See* cavity resonator.

microwave spectrometer [SPECT] An instrument which makes a graphical record of the intensity of microwave radiation emitted or absorbed by a substance as a function of frequency, wavelength, or some related variable.

microwave spectroscope [SPECT] An instrument used to observe the intensity of microwave radiation emitted or absorbed by a substance as a function of frequency, wavelength, or some related variable.

microwave spectroscopy [SPECT] The methods and techniques of observing and the theory for interpreting the selective absorption and emission of microwaves at various frequencies by solids, liquids, and gases.

microwave spectrum [ELECTROMAG] The range of wavelengths or frequencies of electromagnetic radiation that are designated microwaves. [SPECT] A display, photograph, or plot of the intensity of microwave radiation emitted or absorbed by a substance as a function of frequency, wavelength, or some related variable.

microwave tube [ELECTR] A high-vacuum tube designed for operation in the frequency region from approximately 3000 to 300,000 megahertz.

microwave waveguide *See* waveguide.

microwave wavemeter [ELECTROMAG] Any device for measuring the free-space wavelengths (or frequencies) of microwaves; usually made of a cavity resonator whose dimensions can be varied until resonance with the microwaves is achieved.

middle-ultraviolet lamp [ELECTR] A mercury-vapor lamp designed to produce radiation in the wavelength band from 2800 to 3200 angstrom units (280 to 320 nanometers) such as sunlamps and photochemical lamps.

midpoint [MATH] The midpoint of a line segment is the point which separates the segment into two equal parts.

midrange [STAT] The mean of the highest and lowest observed value.

Mie-Grüneisen equation [THERMO] An equation of state particularly useful at high pressure, which states that the volume of a system times the difference between the pressure and the pressure at absolute zero equals the product of a number which depends only on the volume times the difference between the internal energy and the internal energy at absolute zero.

Mie scattering [OPTICS] The scattering of light by a sphere of dielectric material.

Mie's double plate [ELEC] A device consisting of two small metal disks with insulating handles; they are held in contact in an electric field and then separated, and the charge on one of the disks is then measured to determine the electric displacement.

migration [SOLID STATE] **1.** The movement of charges through a semiconductor material by diffusion or drift of charge carriers or ionized atoms. **2.** The movement of crystal defects through a semiconductor crystal under the influence of high temperature, strain, or a continuously applied electric field.

migration area [NUCLEO] One-sixth the mean square distance that a neutron travels in a medium from its birth in fission until its absorption.

migration length [NUCLEO] The square root of the migration area.

MIKES *See* mass-analyzed ion kinetic energy spectrometry.

mil [MATH] A unit of angular measure which, due to nonuniformity of usage, may have any one of three values: 0.001 radian or approximately 0.0572958°; 1/6400 of a full revolution or 0.5625°; 1/1000 of a right angle or 0.09°. [MECH] **1.** A unit of length, equal to 0.001 inch, or to 2.54×10^{-5} meter. Also known as milli-inch; thou. **2.** *See* milliliter.

mile [MECH] A unit of length in common use in the United States, equal to 5280 feet, or 1609.344 meters. Abbreviated mi. Also known as land mile; statute mile.

Milky Way [ASTRON] The faint band of light which encircles the sky and results from the combined light of the many stars near the plane of our galaxy.

Milky Way Galaxy [ASTRON] The large aggregation of stars and interstellar gas and dust of which the sun is a member. Also known as Galaxy.

Miller bridge [ELECTR] Type of bridge circuit for measuring amplification factors of vacuum tubes.

Miller effect [ELECTR] The increase in the effective grid-cathode capacitance of a vacuum tube due to the charge induced electrostatically on the grid by the anode through the grid-anode capacitance.

Miller indices [CRYSTAL] Three integers identifying a type of crystal plane; the intercepts of a plane on the three crystallographic axes are expressed as fractions of the crystal parameters; the reciprocals of these fractions, reduced to integral proportions, are the Miller indices. Also known as crystal indices.

Miller law [CRYSTAL] If the edges formed by the intersections of three faces of a crystal are taken as the three reference axes, then the three quantities formed by dividing the intercept of a fourth face with one of these axes by the intercept of a fifth face with the same axis are proportional to small whole numbers, rarely exceeding 6. Also known as law of rational intercepts.

milli- [MATH] A prefix representing 10^{-3}, or one-thousandth. Abbreviated m.

milliammeter [ELEC] An ammeter whose scale is calibrated to indicate current values in milliamperes.

milliampere [ELEC] A unit of current equal to one-thousandth of an ampere. Abbreviated mA.

milliampere-second [NUCLEO] A unit of radiation dose resulting from exposure to x-rays, equal to the dose produced

MILKY WAY GALAXY

A part of the Milky Way Galaxy. Dark line is caused by obscuring matter. *(Hale Observatories)*

MILLER INDICES

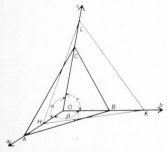

The reference system used to define Miller indices. a, b and c are chosen as crystallographic axes, and ABC as unit plane, defined by angles α, β and γ, and ratio $OA:OB:OC$. Miller indices of another plane, HKL, are integers proportional to OA/OH, OB/OK and OC/OL.

by an electron beam, carrying a current of 1 milliampere, bombarding the target of an x-ray tube for 1 second. Abbreviated mA s.

millibar [MECH] A unit of pressure equal to one-thousandth of a bar. Abbreviated mb. Also known as vac.

millibarn [NUC PHYS] A unit of cross section equal to one-thousandth of a barn. Abbreviated mb.

millicurie-destroyed [NUCLEO] A unit of radiation dose, equal to the radiation emitted by a sample over a period during which its activity decreases by 1 millicurie; for radon-222 (for which this unit is most often used), it is approximately 133 milligram-hours. Abbreviated mcd.

millidarcy [PHYS] A unit of fluid permeability equal to one-thousandth of a darcy. Abbreviated md.

milliequivalent [CHEM] One-thousandth of a compound's or an element's equivalent weight.

millier *See* tonne.

millifarad [ELEC] A unit of capacitance equal to one-thousandth of a farad. Abbreviated mF.

milligal [MECH] A unit of acceleration commonly used in geodetic measurements, equal to 10^{-3} gal, or 10^{-5} meter per second per second. Abbreviated mGal.

milligauss [ELECTROMAG] A unit of magnetic flux density equal to one-thousandth of a gauss. Abbreviated mG.

milligram [MECH] A unit of mass equal to one-thousandth of a gram. Abbreviated mg.

milligram-hour [NUCLEO] A unit of radiation dose, equal to the radiation emitted by a source with an equivalent radium content of 1 milligram for a period of 1 hour. Abbreviated mg h.

millihenry [ELECTROMAG] A unit of inductance equal to one-thousandth of a henry. Abbreviated mH.

millihertz [PHYS] A unit of frequency equal to one-thousandth of a hertz. Abbreviated mHz. Also known as millicycle (mc).

millihg *See* millimeter of mercury.

milli-inch *See* mil.

milli-k [NUCLEO] A unit of reactivity; the reactivity of a reactor in milli-k is equal to $1000(k - 1)$, where k is the effective multiplication factor.

Millikan meter [ELECTR] An integrating ionization chamber in which a gold-leaf electroscope is charged a known amount and ionizing events reduce this charge, so that the resulting angle through which the gold leaf is repelled at any given time indicates the number of ionizing events that have occurred.

Millikan oil-drop experiment [ATOM PHYS] A method of determining the charge on an electron, in which one measures the terminal velocities of rise and fall of oil droplets in an electric field after the droplets have picked up charge from ionization in the surrounding gas produced by an x-ray beam.

milliliter [MECH] A unit of volume equal to 10^{-3} liter or 10^{-6} cubic meter. Abbreviated ml. Also known as mil.

milli-mass-unit [PHYS] 1/1000 of an atomic mass unit. Abbreviated mmu.

millimeter [MECH] A unit of length equal to one-thousandth of a meter. Abbreviated mm. Also known as metric line; strich.

millimeter of mercury [MECH] A unit of pressure, equal to the pressure exerted by a column of mercury 1 millimeter high with a density of 13.5951 grams per cubic centimeter under the standard acceleration of gravity; equal to 133.322387415 pascals; it differs from the torr by less than 1 part in 7,000,000. Abbreviated mmHg. Also known as millihg.

millimeter of water [MECH] A unit of pressure, equal to the pressure exerted by a column of water 1 millimeter high with a density of 1 gram per cubic centimeter under the standard acceleration of gravity; equal to 9.80665 pascals. Abbreviated mmH$_2$O.

millimeter wave [ELECTROMAG] An electromagnetic wave having a wavelength between 1 millimeter and 1 centimeter, corresponding to frequencies between 30 and 300 gigahertz. Also known as millimetric wave.

millimetric wave *See* millimeter wave.

milli-micro- *See* nano-.

millimicron *See* nanometer.

Millington reverberation formula [ACOUS] A formula that states that the reverberation time of a chamber in seconds is 0.05 times its volume in cubic feet, divided by the sum over the surfaces of the chamber of the product of the surface's area in square feet by the natural logarithm of 1 minus its absorption coefficient.

million [MATH] The number 10^6.

million electron volts [PHYS] A unit of energy commonly used in nuclear and particle physics, equal to the energy acquired by an electron in falling through a potential of 10^6 volts. Abbreviated MeV. Also known as megaelectronvolt.

millirad [NUCLEO] A unit of absorbed ionizing radiation dose equal to one-thousandth of a rad. Abbreviated mrad.

milliroentgen [NUCLEO] A unit of radioactive dose of electromagnetic radiation equal to one-thousandth of a roentgen. Abbreviated mr.

millisecond [MECH] A unit of time equal to one-thousandth of a second. Abbreviated msec.

millivolt [ELEC] A unit of potential difference or emf equal to one-thousandth of a volt. Abbreviated mV.

millivoltmeter [ELEC] A voltmeter whose scale is calibrated to indicate voltage values in millivolts.

milliwatt [MECH] A unit of power equal to one-thousandth of a watt. Abbreviated mW.

Milne method [MATH] A technique which provides numerical solutions to ordinary differential equations.

mimetic [CRYSTAL] Pertaining to a crystal that is twinned or malformed but whose crystal symmetry appears to be of a higher grade than it actually is.

minimal equation [MATH] An algebraic equation whose zeros define a minimal surface.

minimal polynomial [MATH] The polynomial of least degree which both divides the characteristic polynomial of a matrix and has the same roots.

minimal realization [CONT SYS] In linear system theory, a set of differential equations, of the smallest possible dimension, which have an input/output transfer function matrix equal to a given matrix function $G(s)$.

minimal surface [MATH] A surface whose mean curvature is identically zero.

minimal surface equation [MATH] A nonlinear, partial differential equation,

$$[1 + (\partial u/\partial y)^2]\partial^2 u/\partial x^2 - 2(\partial u/\partial x)(\partial u/\partial y)(\partial^2 u/\partial x\partial y)$$

$$+ [1 + (\partial u/\partial x)^2]\partial^2 u/\partial y^2 = 0,$$

whose solution represents surfaces of minimal surface area, for example, soap films spanning wire frames.

minimal transformation group [MATH] A transformation group such that every orbit is dense in the phase space.

minimax criterion [STAT] A concept in game theory and decision theory which requires that losses or expected losses associated with a variable that can be controlled be minimized, and thus maximizes the losses or expected losses associated with the variable that cannot be controlled.

minimax estimator [STAT] A random variable obtained by applying the minimax criterion to a risk function associated with a loss function.

mini-maxi regret [CONT SYS] In decision theory, a criterion which selects that strategy which has the smallest maximum difference between its payoff and that of the best hindsight choice.

minimax technique *See* min-max technique.

minimum [MATH] The least value that a real valued function assumes.

minimum deviation [OPTICS] For a prism, the smallest possible angle between the incident and refracted rays; this angle is realized when refraction is symmetrical.

minimum ionizing speed [ATOM PHYS] The smallest speed at which a charged particle passing through a gas can ionize an atom or molecule.

minimum-phase system [CONT SYS] A linear system for which the poles and zeros of the transfer function all have negative or zero real parts.

minimum-variance estimator [STAT] An estimator that possesses the least variance among the members of a defined class of estimators.

Minkowski electrodynamics [ELECTROMAG] An electromagnetic theory, compatible with the special theory of relativity, which takes into account the presence of matter with electric and magnetic polarization.

Minkowski's inequality [MATH] **1.** An inequality involving powers of sums of sequences of real or complex numbers, a_k and b_k:

$$\left[\sum_{k=1} \left| a_k + b_k \right|^s \right]^{1/s} \le \left[\sum_{k=1} \left| a_k \right|^s \right]^{1/s} + \left[\sum_{k=1}^{\infty} \left| b_k \right|^s \right]^{1/s}$$

provided $s \ge 1$. **2.** An inequality involving powers of integrals of real or complex functions f and g over an interval or region R:

$$\left[\int_R \left| f(x) + g(x) \right|^s dx\right]^{1/s} \le \left[\int_R \left| f(x) \right|^s dx\right]^{1/s}$$
$$+ \left[\int_R \left| g(x) \right|^s dx\right]^{1/s}$$

provided $s \ge 1$ and the integrals involved exist.

Minkowski space-time [RELAT] The space-time of special relativity; it is completely flat and contains no gravitating matter. Also known as Minkowski universe.

Minkowski universe *See* Minkowski space-time.

min-max technique [MATH] A method of approximation of a function f by a function g from some class where the maximum of the modulus of $f-g$ is minimized over this class. Also known as Chebyshev approximation; minimax technique.

min-max theorem [MATH] The theorem that provides information concerning the nth eigenvalue of a symmetric operator on an inner product space without necessitating knowledge of the other eigenvalues. Also known as maximum-minimum principle.

minor [MATH] The minor of an entry of a matrix is the determinant of the matrix obtained by removing the row and column containing the entry. Also known as cofactor; complementary minor.

minor arc [MATH] The smaller of the two arcs on a circle produced by a secant.

minor axis [MATH] The smaller of the two axes of an ellipse.

minor bend [ELECTROMAG] Rectangular waveguide bent so

that throughout the length of a bend a longitudinal axis of the guide lies in one plane which is parallel to the narrow side of the waveguide.

minor diatonic scale [ACOUS] A diatonic scale in which the relative sizes of the sequence of intervals are approximately 2, 1, 2, 2, 2, 2, 1.

minority carrier [SOLID STATE] The type of carrier, electron, or hole that constitutes less than half the total number of carriers in a semiconductor.

minority emitter [ELECTR] Of a transistor, an electrode from which a flow of minority carriers enters the interelectrode region.

minor lobe [ELECTROMAG] Any lobe except the major lobe of an antenna radiation pattern. Also known as secondary lobe; side lobe.

minor loop [CONT SYS] A portion of a feedback control system that consists of a continuous network containing both forward elements and feedback elements.

minor planet [ASTRON] **1.** Those planets smaller than the earth, specifically Mercury, Venus, Mars, and Pluto. **2.** *See* asteroid.

minuend [MATH] The quantity from which another quantity is to be subtracted.

minus [MATH] A minus B means that the quantity B is to be subtracted from the quantity A.

minute [MATH] A unit of measurement of angle, equal to $\frac{1}{60}$ of a degree. Symbolized ′. Also known as arcmin. [MECH] A unit of time, equal to 60 seconds.

Mira [ASTRON] The first star recognized to be a periodic variable; has a period of 332 ± 9 days and its spectrum changes from M5e at maximum to M9e at minimum; it is the prototype of long-period variable stars.

mirage [OPTICS] Any one of a variety of unusual images of distant objects seen as a result of the bending of light rays in the atmosphere during abnormal vertical distribution of air density.

Mira variables [ASTRON] A group of over 1300 stars having the same type of variability as the star Mira.

mired [THERMO] A unit used to measure the reciprocal of color temperature, equal to the reciprocal of a color temperature of 10^6 kelvins. Derived from micro-reciprocal-degree.

mirror [OPTICS] A surface which specularly reflects a large fraction of incident light.

mirror coating [OPTICS] A thin film of highly reflective material spread over a correctly shaped glass surface to produce a mirror; aluminum is usually used in the visible region. Also known as reflective coating.

mirror galvanometer [ELEC] A galvanometer having a small mirror attached to the moving element, to permit use of a beam of light as an indicating pointer. Also known as reflecting galvanometer.

mirror interferometer [ENG] An interferometer used in radio astronomy, in which the sea surface acts as a mirror to reflect radio waves up to a single antenna, where the reflected waves interfere with the waves arriving directly from the source. [OPTICS] Any interferometer which makes use of mirror interference.

mirror machine [PL PHYS] A device which confines plasma in a tube with magnetic mirrors at each end to prevent it from escaping.

mirror nuclei [NUC PHYS] A pair of atomic nuclei, each of which would be transformed into the other by changing all its neutrons into protons, and vice versa.

mirror optics [OPTICS] The science and technology of mir-

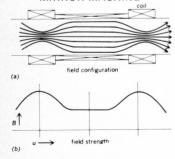

MIRROR MACHINE

Mirror machine. *(a)* Longitudinal section. Lines represent magnetic lines of force. *(b)* Curve represents magnetic field strength B along path u in the tube. Regions where B has a large value are magnetic mirrors produced by coils and located at constrictions of field lines at each end; they confine plasma to central part.

rors which, by means of reflecting rays of light, either revert optical bundles or focus them to form images.

mirror plane of symmetry   *See* plane of mirror symmetry.

mirror reflection   *See* specular reflection.

MIS   *See* metal-insulator semiconductor.

miscibility   [CHEM] The tendency or capacity of two or more liquids to form a uniform blend, that is, to dissolve in each other; degrees are total miscibility, partial miscibility, and immiscibility.

missing mass spectrometer   [PARTIC PHYS] An apparatus which measures the momentum of the recoil protons in a reaction such as $\pi^- + p \rightarrow p + (MM)^-$, in order to determine the distribution of masses of the MM system, without any detailed observations on this system.

mist   [FL MECH] Fine liquid droplets suspended in or falling through a moving or stationary gas atmosphere.

mistbow   *See* fogbow.

mistuning   [MECH] The difference between the square of the natural frequency of vibration of a vibrating system, without the effect of damping, and the square of the frequency of an external, oscillating force.

MIT bag model   [PARTIC PHYS] A model describing quark confinement in hadrons in which a hadron is viewed as a bubble of gas in a uniform, isotropic, perfect fluid, with the thermodynamic pressure of the gas replaced by the quantum pressure of quarks. Derived from Massachusetts Institute of Technology bag model.

Mittag-Leffler's theorem   [MATH] A theorem that enables one to explicitly write down a formula for a meromorphic complex function with given poles; for a function $f(z)$ with poles at $z = z_i$, having order m_i and principal parts

$$\sum_{j=1}^{m_i} a_{ij}(z - z_i)^{-j},$$

the formula is

$$f(z) = \sum_{k} \left[\sum_{j=1}^{m_i} a_{ij}(z - z_i)^{-j} + p_i(z) \right] + g(z),$$

where the $p_i(z)$ are polynomials, $g(z)$ is an entire function, and the series converges uniformly in every bounded region where $f(z)$ is analytic.

mix crystal   *See* mixed crystal.

mixed-base notation   [MATH] A computer number system in which a single base, such as 10 in the decimal system, is replaced by two number bases used alternately, such as 2 and 5.

mixed-base number   [MATH] A number in mixed-base notation. Also known as mixed-radix number.

mixed crystal   [CRYSTAL] A crystal whose lattice sites are occupied at random by different ions or molecules of two different compounds. Also known as mix crystal.

mixed decimal   [MATH] Any decimal plus an integer.

mixed model   [STAT] **1.** A model having both determinate and stochastic elements in its equations. **2.** A model having both difference and differential equations. **3.** A model containing both endogenous and exogenous elements. **4.** In analysis of variance for a two-way layout, the combined rows and columns.

mixed number   [MATH] The sum of an integer and a fraction.

mixed radix   [MATH] Pertaining to a numeration system using more than one radix, such as the biquinary system.

mixed-radix number   *See* mixed-base number.

mixed sampling   [STAT] The use of two or more methods of sampling; for example, in multistage sampling, if samples are drawn at random at one stage and drawn by a systematic method at another.

mixed strategy [MATH] A method of playing a matrix game in which the player attaches a probability weight to each of the possible options, the probability weights being nonnegative numbers whose sum is unity, and then operates a chance device that chooses among the options with probabilities equal to the corresponding weights. [STAT] A concept in game theory which allows a player more than one choice of action which is determined by a chance mechanism.

mixed-type boundary-value problem *See* equation of mixed type.

mixer [ELECTR] **1.** A device having two or more inputs, usually adjustable, and a common output; used to combine separate audio or video signals linearly in desired proportions to produce an output signal. **2.** The stage in a superheterodyne receiver in which the incoming modulated radio-frequency signal is combined with the signal of a local rf oscillator to produce a modulated intermediate-frequency signal. Also known as first detector; heterodyne modulator; mixer-first detector. [OPTICS] A nonlinear device in which two light beams are combined to form new beams having frequencies equal to the sum or the difference of the input wavelengths.

mixer-first detector *See* mixer.

mixing length [PHYS] A mean length of travel, characteristic of a particular motion, over which an eddy maintains its identity; it is analogous to the mean free path of a molecule; physically, the idea implies that mixing occurs by discontinuous steps, that fluctuations which arise as eddies with different characteristics wander about, and that the mixing is done almost entirely by the small eddies.

mixing transformation [MATH] A function of a measure space which moves the measurable sets in such a manner that, asymptotically as regards measure, any measurable set is distributed uniformly throughout the space.

m-kgf *See* meter-kilogram.

mksa system *See* meter-kilogram-second-ampere system.

mks system *See* meter-kilogram-second system.

ml *See* milliliter.

mm *See* millimeter.

mmf *See* magnetomotive force.

mmHg *See* millimeter of mercury.

mmH₂O *See* millimeter of water.

mmu *See* milli-mass-unit.

Mn *See* manganese.

MNOS *See* metal-nitride-oxide semiconductor.

Mo *See* molybdenum.

mobility [FL MECH] The reciprocal of the plastic viscosity of a Bingham plastic. [PHYS] Freedom of particles to move, either in random motion or under the influence of fields or forces. [SOLID STATE] *See* drift mobility.

mobility coefficient [PHYS CHEM] The average speed of motion of molecules in a solution in the direction of the concentration gradient, at unit concentration and unit osmotic pressure gradient.

mobility tensor [PL PHYS] A second-rank tensor whose product with the electric field vector for a plane wave in a plasma gives a vector equal to the average velocity of electrons or ions; components of both vectors are in phasor notation.

Möbius band [MATH] The nonorientable surface obtained from a rectangular strip by twisting it once and then gluing the two ends. Also known as Möbius strip.

Möbius function [MATH] **1.** The function μ of the positive integers where $\mu(1) = 1$, $\mu(n) = (-1)^r$ if n factors into r

distinct primes, and $\mu(n) = 0$ otherwise; also, $\mu(n)$ is the sum of the primitive nth roots of unity. **2.** For any lattice, the function $\mu(s,t)$ specified in Möbius inversion.

Möbius inversion [MATH] The principle that for any lattice L there is a function $\mu(s,t)$ defined for pairs of elements (s,t) in L such that, for any two functions f and g defined on the lattice, the equation

$$g(s) = \sum_{s \leq t} f(t)$$

implies that
$$f(s) = \sum_{s \leq t} \mu(s,t)g(t).$$

Möbius resistor [ELEC] A nonreactive resistor made by placing strips of aluminum or other metallic tape on opposite sides of a length of dielectric ribbon, twisting the strip assembly half a turn, joining the ends of the metallic tape, then soldering leads to opposite surfaces of the resulting loop.

Möbius strip *See* Möbius band.

Möbius transformations [MATH] These are the most commonly used conformal mappings of the complex plane; their form is $f(z) = (az + b)/(cz + d)$ where the real numbers a, b, c, and d satisfy $ad - bc \neq 0$. Also known as linear fractional transformations. Also known as bilinear transformations; homographic transformations.

mock moon *See* paraselene.

mode [ELECTROMAG] A form of propagation of guided waves that is characterized by a particular field pattern in a plane transverse to the direction of propagation. Also known as transmission mode. [PHYS] A state of an oscillating system that corresponds to a particular field pattern and one of the possible resonant frequencies of the system. [STAT] The most frequently occurring member of a set of numbers.

mode filter [ELECTROMAG] A waveguide filter designed to separate waves of the same frequency but of different transmission modes.

mode-locked laser [OPTICS] A laser designed so that several modes of oscillation with closely spaced wavelengths, in which the laser would normally oscillate, are synchronized so that a pulse of light, lasting for as little as a picosecond, is generated.

model reference system [CONT SYS] An ideal system whose response is agreed to be optimum; computer simulation in which both the model system and the actual system are subjected to the same stimulus is carried out, and parameters of the actual system are adjusted to minimize the difference in the outputs of the model and the actual system.

model theory [MATH] The general qualitative study of the structure of a mathematical theory.

mode number [ELECTR] **1.** The number of complete cycles during which an electron of average speed is in the drift space of a reflex klystron. **2.** The number of radians of phase in the microwave field of a magnetron divided by 2π as one goes once around the anode.

mode of oscillation *See* mode of vibration.

mode of vibration [MECH] A characteristic manner in which a system which does not dissipate energy and whose motions are restricted by boundary conditions can oscillate, having a characteristic pattern of motion and one of a discrete set of frequencies. Also known as mode of oscillation.

moderator [NUCLEO] The material used in a nuclear reactor to moderate or slow down neutrons from the high velocities at which they are created in the fission process.

modern algebra [MATH] The study of algebraic systems such as groups, rings, modules, and fields.

modified Bessel functions [MATH] The functions defined by $I_\nu(x) = \exp(-i\nu\pi/2) J_\nu(ix)$, where J_ν is the Bessel function of order ν, and x is real and positive.

modified exponential curve [STAT] The equation resulting when a constant is added to the exponential curve equation; used to estimate trend in a nonlinear time series.

modified Hankel functions [MATH] The functions defined by $K_\nu(x) = (i\pi/2) \exp(i\nu\pi/2) H_\nu^{(1)}(ix)$, where $H_\nu^{(1)}$ is the first Hankel function of order ν, and x is real and positive. Also known as MacDonald functions.

modified mean [STAT] A mean computed after elimination of observations judged to be atypical.

modified operator of Hankel transforms [MATH] An integral operator used in the study of Erdélyi-Kober operators, fractional integration, and dual integral equations; it may be written

$$S_{p,q}f(x) = 2^q x^{-q} \int_0^\infty y^{1-q} J_{2p+q}(xy) f(y)\, dy$$

where $J_\nu(y)$ is the Bessel function of the first kind of order ν.

modular function [MATH] A single-valued, meromorphic function in the upper half of the complex plane which is automorphic with respect to the modular group or a specified nontrivial subgroup of the modular group.

modular group [MATH] The group of transformations of the form $z' = (az + b)/(cz + d)$ on the complex number z, where $a, b, c,$ and d are integers, and $ad - bc \neq 0$.

modulate [ELECTR] To vary the amplitude, frequency, or phase of a wave, or vary the velocity of the electrons in an electron beam in some characteristic manner.

modulated Raman scattering [SPECT] Application of modulation spectroscopy to the study of Raman scattering; in particular, use of external perturbations to lower the symmetry of certain crystals and permit symmetry-forbidden modes, and the use of wavelength modulation to analyze second-order Raman spectra.

modulation [COMMUN] The process or the result of the process by which some parameter of one wave is varied in accordance with some parameter of another wave.

modulator [ELECTR] **1.** The transmitter stage that supplies the modulating signal to the modulated amplifier stage or that triggers the modulated amplifier stage to produce pulses at desired instants as in radar. **2.** A device that produces modulation by any means, such as by virtue of a nonlinear characteristic or by controlling some circuit quantity in accordance with the waveform of a modulating signal. **3.** One of the electrodes of a spacistor.

modulator crystal [OPTICS] Crystal which is used to modulate a polarized light beam by the use of the Pockel's effect; useful as a modulator in laser systems.

module [ELECTR] A packaged assembly of wired components, built in a standardized size and having standardized plug-in or solderable terminations. [MATH] A vector space in which the scalars are a ring rather than a field.

modulo [MATH] **1.** A group G modulo a subgroup H is the quotient group G/H of cosets of H in G. **2.** A technique of identifying elements in an algebraic structure in such a manner that the resulting collection of identified objects is the same type of structure.

modulo N [MATH] Two integers are said to be congruent modulo N (where N is some integer) if they have the same remainder when divided by N.

modulo N arithmetic [MATH] Calculations in which all integers are replaced by their remainders after division by N (where N is some fixed integer.)

modulus of a complex number *See* absolute value of a complex number.

modulus of a congruence [MATH] A number a, such that two specified numbers b and c give the same remainder when divided by a; b and c are then said to be congruent, modulus a (or congruent, modulo a).

modulus of a logarithm [MATH] The modulus of a logarithm with a given base is the factor by which a logarithm with a second base must be multiplied to give the first logarithm.

modulus of an elliptic function [MATH] The square root of the parameter m specified in the definition of a Jacobian elliptic function.

modulus of an elliptic integral [MATH] The real or complex number k specified in the definition of Legendre normal form.

modulus of compression *See* bulk modulus of elasticity.

modulus of continuity [MATH] For a real valued continuous function f, this is the function whose value at a real number r is the maximum of the modulus of $f(x) - f(y)$ where the modulus of $x - y$ is less than r; this function is useful in approximation theory.

modulus of decay [MECH] The time required for the amplitude of oscillation of an underdamped harmonic oscillator to drop to $1/e$ of its initial value; the reciprocal of the damping factor.

modulus of distance [ASTRON] The quantity $m - M$, where M is the absolute magnitude of a given star and m is its apparent magnitude. Also known as distance modulus.

modulus of elasticity [MECH] The ratio of the increment of some specified form of stress to the increment of some specified form of strain, such as Young's modulus, the bulk modulus, or the shear modulus. Also known as coefficient of elasticity; elasticity modulus; elastic modulus.

modulus of elasticity in shear [MECH] A measure of a material's resistance to shearing stress, equal to the shearing stress divided by the resultant angle of deformation expressed in radians. Also known as coefficient of rigidity; modulus of rigidity; shear modulus.

modulus of resilience [MECH] The maximum mechanical energy stored per unit volume of material when it is stressed to its elastic limit.

modulus of rigidity *See* modulus of elasticity in shear.

modulus of rupture in bending [MECH] The maximum stress per unit area that a specimen can withstand without breaking when it is bent, as calculated from the breaking load under the assumption that the specimen is elastic until rupture takes place.

modulus of rupture in torsion [MECH] The maximum stress per unit area that a specimen can withstand without breaking when its ends are twisted, as calculated from the breaking load under the assumption that the specimen is elastic until rupture takes place.

modulus of simple longitudinal extension *See* axial modulus.

modulus of torsion *See* torsional modulus.

modulus of volume elasticity *See* bulk modulus of elasticity.

mohm [MECH] A unit of mechanical mobility, equal to the reciprocal of 1 mechanical ohm.

Moho *See* Mohorovičić discontinuity.

Mohorovičić discontinuity [GEOPHYS] A seismic discontinuity that separates the earth's crust from the subjacent mantle, inferred from travel time curves indicating that seismic waves undergo a sudden increase in velocity. Also known as Moho.

MOHOROVIČIĆ DISCONTINUITY

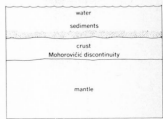

(a)

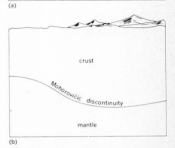

(b)

Mohorovičić discontinuity. *(a)* At average depths of about 10–12 kilometers beneath ocean basins. *(b)* At average subcontinental depths of 33–35 kilometers.

MOIRÉ EFFECT

Two simple gratings crossed at a small angle.

Mohr liter [MECH] A unit of volume, equal to 1000 Mohr cubic centimeters.

Mohr's circle [MECH] A graphical construction making it possible to determine the stresses in a cross section if the principal stresses are known.

moiré effect [OPTICS] The effect whereby, when one family of curves is superposed on another family of curves so that the curves cross at angles of less than about 45°, a new family of curves appears which pass through intersections of the original curves.

moiré fringes [OPTICS] The bands which appear in the moiré effect.

moisture flux *See* eddy flux.

moisture-vapor transmission [FL MECH] The rate at which water vapor permeates a porous film (such as plastic or paper) or a wall.

mol *See* mole.

molal elevation of the boiling point *See* ebullioscopic constant.

molal heat capacity *See* molar heat capacity.

molality [CHEM] Concentration given as moles per 1000 grams of solvent.

molal quantity [CHEM] The number of moles (gram-molecular weights) present, expressed with weight in pounds, grams, or such units, numerically equal to the molecular weight; for example, pound-mole, gram-mole.

molal solution [CHEM] Concentration of a solution expressed in moles of solute divided by 1000 grams of solvent.

molal specific heat *See* molar specific heat.

molal volume *See* molar volume.

molar dispersion [OPTICS] In refractometry, the difference in molar refraction (refractive index) of a compound at two different light-beam wavelengths.

molar heat capacity [PHYS CHEM] The amount of heat required to raise 1 mole of a substance 1° in temperature. Also known as molal heat capacity; molecular heat capacity.

molarity [CHEM] Measure of the number of gram-molecular weights of a compound present (dissolved) in 1 liter of solution; it is indicated by M, preceded by a number to show solute concentration.

molar magnetic rotation [OPTICS] A measure of the strength of the Faraday effect in a substance, equal to $M\alpha\rho'/M'\alpha'\rho$, where α is the angle of rotation, M is the molecular weight of the substance, ρ is its density, and α', M', and ρ' are corresponding quantities for water.

molar refraction [OPTICS] Equation for the refractive index of a compound modified by the compound's molecular weight and density. Also known as the Lorentz-Lorenz molar refraction.

molar solution [CHEM] Aqueous solution that contains 1 mole (gram-molecular weight) of solute in 1 liter of the solution.

molar specific heat [PHYS CHEM] The ratio of the amount of heat required to raise the temperature of 1 mole of a compound 1°, to the amount of heat required to raise the temperature of 1 mole of a reference substance, such as water, 1° at a specified temperature. Also known as molal specific heat; molecular specific heat.

molar susceptibility [PHYS CHEM] Magnetic susceptibility of a compound per gram-mole of that compound.

molar volume [PHYS CHEM] The volume occupied by one mole of a substance in the form of a solid, liquid, or gas. Also known as molal volume; mole volume.

mole [CHEM] An amount of substance of a system which contains as many elementary units as there are atoms of carbon in 0.012 kilogram of the pure nuclide carbon-12; the elementary unit must be specified and may be an atom, molecule, ion, electron, photon, or even a specified group of such units. Symbolized mol.

molectronics *See* molecular electronics.

molecular adhesion [PHYS CHEM] A particular manifestation of intermolecular forces which causes solids or liquids to adhere to each other; usually used with reference to adhesion of two different materials, in contrast to cohesion.

molecular association [PHYS CHEM] The formation of double molecules or polymolecules from a single species as a result of specific and moderately strong intermolecular forces.

molecular asymmetry *See* asymmetry.

molecular beam [PHYS] A beam of neutral molecules whose directions of motion lie within a very small solid angle.

molecular-beam apparatus [PHYS] A device in which a molecular beam in a vacuum is subjected to magnetic fields, oscillating fields, or other influences, and a detector measures the resulting intensity of the beam at some location; used primarily in radio-frequency spectroscopy.

molecular beam epitaxy [SOLID STATE] A technique of growing single crystals in which beams of atoms or molecules are made to strike a single-crystalline substrate in a vacuum, giving rise to crystals whose crystallographic orientation is related to that of the substrate. Abbreviated MBE.

molecular binding [SOLID STATE] The force which holds a molecule at some site on the surface of a crystal.

molecular circuit [ELECTR] A circuit in which the individual components are physically indistinguishable from each other.

molecular conductivity [PHYS CHEM] The conductivity of a volume of electrolyte containing 1 mole of dissolved substance.

molecular crystal [CRYSTAL] A solid consisting of a lattice array of molecules such as hydrogen, methane, or more complex organic compounds, bound by weak van der Waals forces, and therefore retaining much of their individuality.

molecular diamagnetism [PHYS CHEM] Diamagnetism of compounds, especially organic compounds whose susceptibilities can often be calculated from the atoms and chemical bonds of which they are composed.

molecular diameter [PHYS CHEM] The diameter of a molecule, assuming it to be spherical; has a numerical value of 10^{-8} centimeter multiplied by a factor dependent on the compound or element.

molecular diffusion [FL MECH] The transfer of mass between adjacent layers of fluid in laminar flow.

molecular dipole [PHYS CHEM] A molecule having an electric dipole moment, whether it is permanent or produced by an external field.

molecular drag pump [ENG] A vacuum pump in which pumping is accomplished by imparting a high momentum to the gas molecules by impingement of a body rotating at very high speeds, as much as 16,000 revolutions per minute; such pumps achieve a vacuum as high as 10^{-6} torr.

molecular effusion [FL MECH] Mass-transfer flow mechanism of free-molecule transfer through pores or orifices.

molecular electronics [ELECTR] The branch of electronics that deals with the production of complex electronic circuits in microminiature form by producing semiconductor devices and circuit elements integrally while growing multizoned crystals in a furnace. Also known as molectronics.

molecular energy level [PHYS CHEM] One of the states of motion of nuclei and electrons in a molecule, having a definite energy, which is allowed by quantum mechanics.

molecular field theory *See* Weiss theory.

MOLECULAR BINDING

A model of a crystal surface showing different types of molecular binding sites. Molecule *A* resting on completed plane of molecules is more weakly bound than molecule *B* at ledge formed by incomplete plane. *(General Electric Co.)*

molecular flow [FL MECH] Gas-flow phenomenon at low pressures or in small channels when the mean free path is of the same order of magnitude as the channel diameter; a gas molecule thus migrates along the channel independent of other gas molecules present.

molecular gage [ENG] Any instrument, such as a rotating viscometer gage or a decrement gage, that uses the dependence of the viscosity of a gas on its pressure to measure pressures on the order of 1 pascal or less. Also known as viscosity gage; viscosity manometer.

molecular gas laser [OPTICS] Any gas laser in which the gas consists of molecules rather than atoms; such a laser can be operated on a large number of rotational-vibrational lines, and, at a sufficiently high pressure, these lines overlap and a wide gain region is obtained. Also known as molecular laser.

molecular heat [THERMO] The heat capacity per mole of a substance.

molecular heat capacity *See* molar heat capacity.

molecular heat diffusion [THERMO] Transfer of heat through the motion of molecules.

molecular laser *See* molecular gas laser.

molecular magnet [PHYS CHEM] A molecule having a non-vanishing magnetic dipole moment, whether it is permanent or produced by an external field.

molecular optics [OPTICS] The study of the propagation of light and associated phenomena, such as refraction, absorption, and scattering, through collections of molecules in gases, liquids, and solids.

molecular orbital [PHYS CHEM] A wave function describing an electron in a molecule.

molecular paramagnetism [PHYS CHEM] Paramagnetism of molecules, such as oxygen, some other molecules, and a large number of organic compounds.

molecular physics [PHYS] The study of the behavior and structure of molecules, including the quantum-mechanical explanation of several kinds of chemical binding between atoms in a molecule, directed valence, the polarizability of molecules, the quantization of vibrational, rotational, and electronic motions of molecules, and the phenomena arising from intermolecular forces.

molecular polarizability [PHYS CHEM] The electric dipole moment induced in a molecule by an external electric field, divided by the magnitude of the field.

molecular relaxation [PHYS CHEM] Transition of a molecule from an excited energy level to another excited level of lower energy or to the ground state.

molecular rotation [OPTICS] In a solution of an optically active compound, the specific rotation (angular rotation of polarized light) multiplied by the compound's molecular weight.

molecular specific heat *See* molar specific heat.

molecular spectroscopy [SPECT] The production, measurement, and interpretation of molecular spectra.

molecular spectrum [SPECT] The intensity of electromagnetic radiation emitted or absorbed by a collection of molecules as a function of frequency, wave number, or some related quantity.

molecular stopping power [NUCLEO] For an ionizing particle passing through a compound, the particle's energy loss per molecule within a unit area normal to the particle's path; equal to the linear energy transfer (energy loss per unit path length) divided by the number of molecules per unit volume.

molecular structure [PHYS CHEM] The manner in which electrons and nuclei interact to form a molecule, as elucidated by quantum mechanics and a study of molecular spectra.

molecular theory   *See* kinetic theory.

molecular vibration   [PHYS CHEM] The theory that all atoms within a molecule are in continuous motion, vibrating at definite frequencies specific to the molecular structure as a whole as well as to groups of atoms within the molecule; the basis of spectroscopic analysis.

molecular volume   [CHEM] The volume that is occupied by 1 mole (gram-molecular weight) of an element or compound; equals the molecular weight divided by the density.

molecular weight   [CHEM] The sum of the atomic weights of all the atoms in a molecule.

molecule   [CHEM] A group of atoms held together by chemical forces; the atoms in the molecule may be identical as in H_2, S_2, and S_8, or different as in H_2O and CO_2; a molecule is the smallest unit of matter which can exist by itself and retain all its chemical properties.

mole fraction   [CHEM] The ratio of the number of moles of a substance in a mixture or solution to the total number of moles of all the components in the mixture or solution.

Molenbroeck-Chaplygin transformation   [FL MECH] A version of the hodograph method for compressible flow in which only the independent variables are replaced and no change is made in the dependent variables, that is, the velocity potential and stream function.

mole percent   [CHEM] Percentage calculation expressed in terms of moles rather than weight.

mole volume   *See* molar volume.

Møller scattering   [QUANT MECH] Scattering of electrons by electrons.

Mollier diagram   [THERMO] Graph of enthalpy versus entropy of a vapor on which isobars, isothermals, and lines of equal dryness are plotted.

Moll thermopile   [ENG] A thermopile used in some types of radiation instruments; alternate junctions of series-connected manganan-constantan thermocouples are embedded in a shielded nonconducting plate having a large heat capacity; the remaining junctions, which are blackened, are exposed directly to the radiation; the voltage developed by the thermocouple is proportional to the intensity of radiation.

molten-salt reactor   [NUCLEO] A nuclear reactor in which the fissile and fertile material, in the form of fluoride salts, is dissolved in the coolant, which is a molten mixture of salts such as lithium fluoride and beryllium fluoride. Abbreviated MSR. Also known as fused-salt reactor.

molybdenum   [CHEM] A chemical element, symbol Mo, atomic number 42, and atomic weight 95.95.

moment   [MECH] Static moment of some quantity, except in the term "moment of inertia." [STAT] The nth moment of a distribution $f(x)$ about a point x_0 is the expected value of $(x - x_0)^n$, that is, the integral of $(x - x_0)^n \, df(x)$, where $df(x)$ is the probability of some quantity's occurrence; the first moment is the mean of the distribution, while the variance may be found in terms of the first and second moments.

momental ellipsoid   [MECH] An inertia ellipsoid whose size is specified to be such that the tip of the angular velocity vector of a freely rotating object, with origin at the center of the ellipsoid, always lies on the ellipsoid's surface. Also known as energy ellipsoid.

moment diagram   [MECH] A graph of the bending moment at a section of a beam versus the distance of the section along the beam.

moment of force   *See* torque.

moment of inertia   [MECH] The sum of the products formed by multiplying the mass (or sometimes, the area) of each

MOLYBDENUM

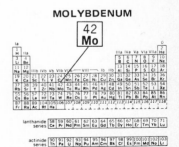

Periodic table of the chemical elements showing the position of molybdenum.

element of a figure by the square of its distance from a specified line. Also known as rotational inertia.

moment of momentum   *See* angular momentum.

momentum   [MECH] Also known as linear momentum; vector momentum. **1.** For a single nonrelativistic particle, the product of the mass and the velocity of a particle. **2.** For a single relativistic particle, $m\mathbf{v}(1 - v^2/c^2)^{1/2}$, where m is the rest mass, $\mathbf{v}$ the velocity, and c the speed of light. **3.** For a system of particles, the vector sum of the momenta (as in the first or second definiton) of the particles.

momentum conservation   *See* conservation of momentum.

momentum density   [PHYS] The momentum per unit volume of any given field.

momentum thickness   [FL MECH] A measure of the thickness of a boundary-layer flow, given by the expression

$$\int_0^\infty (u/V)(1 - u/V)\, dy$$

where u is the fluid velocity at a distance y from the surface of the body, and V is the velocity outside the body. Symbolized θ.

momentum thrust   *See* thrust.

momentum-transport hypothesis   [FL MECH] The hypothesis that the principle of conservation of momentum is valid in turbulent eddy transfer.

momentum wave function   [QUANT MECH] A function of the momenta of a system of particles and of time which results from taking Fourier transforms, over the coordinates of all the particles, of the Schrödinger wave function; the absolute value squared is proportional to the probability that the particles will have given momenta at a given time.

Monge form   [MATH] An equation of a surface of the form $z = f(x,y)$, where x, y, and z are cartesian coordinates.

Monge's theorem   [MATH] For three coplanar circles, and for radii of these circles which are parallel to each other, the three outer centers of similitude of the circles taken in pairs lie on a single straight line, and any two inner centers of similitude lie on a straight line with one of the outer centers.

monochord   *See* sonotone.

monochromatic   [OPTICS] Pertaining to the color of a surface which radiates light having an extremely small range of wavelengths. [PHYS] Consisting of electromagnetic radiation having an extremely small range of wavelengths, or particles having an extremely small range of energies.

monochromatic emissivity   [THERMO] The ratio of the energy radiated by a body in a very narrow band of wavelengths to the energy radiated by a blackbody in the same band at the same temperature. Also known as color emissivity.

monochromatic filter   *See* birefringent filter.

monochromatic interference   [OPTICS] Interference between beams coming from a source of monochromatic light.

monochromatic light   [OPTICS] Light of one color, having wavelengths confined to an extremely narrow range.

monochromatic neutron beam   [NUCLEO] A beam of neutrons whose energies are confined to an extremely narrow range of values.

monochromatic radiation   [ELECTROMAG] Electromagnetic radiation having wavelengths confined to an extremely narrow range.

monochromatic temperature scale   [THERMO] A temperature scale based upon the amount of power radiated from a blackbody at a single wavelength.

monochromator   [SPECT] A spectrograph in which a detector is replaced by a second slit, placed in the focal plane, to isolate a particular narrow band of wavelengths for refocusing on a detector or experimental object.

monochrome [OPTICS] Having only one chromaticity.

monoclinic system [CRYSTAL] One of the six crystal systems characterized by a single, two-fold symmetry axis or a single symmetry plane.

monodisperse system [PHYS CHEM] A colloidal system whose particles are approximately the same size.

monodromy theorem [MATH] If a complex function is analytic at a point of a bounded simply connected domain and can be continued analytically along every curve from the point, then it represents a single-valued analytic function in the domain.

monoenergetic gamma rays [PHYS] A beam of gamma rays whose energies are confined to an extremely narrow range.

monoid [MATH] A semigroup which has an identity element.

monolithic [SCI TECH] Constructed from a single crystal or other single piece of material.

monolithic ceramic capacitor [ELECTR] A capacitor that consists of thin dielectric layers interleaved with staggered metal-film electrodes; after leads are connected to alternate projecting ends of the electrodes, the assembly is compressed and sintered to form a solid monolithic block.

monomial [MATH] A polynomial of degree one.

monopole *See* magnetic monopole.

monopole antenna [ELECTROMAG] An antenna, usually in the form of a vertical tube or helical whip, on which the current distribution forms a standing wave, and which acts as one part of a dipole whose other part is formed by its electrical image in the ground or in an effective ground plane. Also known as spike antenna.

monostable circuit [ELECTR] A circuit having only one stable condition, to which it returns in a predetermined time interval after being triggered.

monostable multivibrator [ELECTR] A multivibrator with one stable state and one unstable state; a trigger signal is required to drive the unit into the unstable state, where it remains for a predetermined time before returning to the stable state. Also known as one-shot multivibrator; single-shot multivibrator; start-stop multivibrator; univibrator.

monotone convergence theorem [MATH] The integral of the limit of a monotone increasing sequence of nonnegative measurable functions is equal to the limit of the integrals of the functions in the sequence.

monotone function [MATH] A function which is either monotone nondecreasing or monotone nonincreasing. Also known as monotonic function.

monotone nondecreasing function [MATH] A function which never decreases, that is, if $x \leq y$ then $f(x) \leq f(y)$. Also known as increasing function; monotonically nondecreasing function.

monotone nonincreasing function [MATH] A function which never increases, that is, if $x \leq y$ then $f(x) \geq f(y)$. Also known as decreasing function; monotonically nonincreasing function.

monotone set [MATH] A subset X of the product of a Banach space B with its dual space B^* such that if x_1 and x_2 are elements of B, f_1 and f_2 are elements of B^*, and (x_1, f_1) and (x_2, f_2) are in X, then the functional $f_1 - f_2$ evaluated at the point $x_1 - x_2$ is equal to or greater than 0.

monotonic function *See* monotone function.

monotrophic [CRYSTAL] Of crystal pairs, having one of the pair always metastable with respect to the other.

monotropic [PHYS] Pertaining to an element which may exist

MONOPOLE ANTENNA

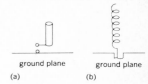

ground plane   ground plane
(a)   (b)

Types of monopole antenna with horizontal and vertical patterns. *(a)* Vertical tube. *(b)* Helical.

in two or more forms, but in which one form is the stable modification at all temperatures and pressures.

monotropy coefficient [FL MECH] A coefficient ν related to the ratio of velocity coefficients, A_y/A_x, in an equation developed by P. Raethjen for the velocity profile in a fluid.

Monte Carlo method [STAT] A technique which obtains a probabilistic approximation to the solution of a problem by using statistical sampling techniques.

month [ASTRON] **1.** The period of the revolution of the moon around the earth (sidereal month). **2.** The period of the phases of the moon (synodic month). **3.** The month of the calendar (calendar month).

Moody friction factor [FL MECH] Modification of the friction factor–Reynolds number–fluid flow relationship into which a roughness factor has been incorporated.

moon [ASTRON] **1.** The natural satellite of the earth. **2.** A natural satellite of any planet.

moonbow [OPTICS] A rainbow formed by light from the moon; the colors in a moonbow are usually very difficult to detect. Also known as lunar rainbow.

moon illusion [OPTICS] An optical illusion whereby the moon appears larger when it is close to the horizon than when it is higher up.

Moore-Smith convergence [MATH] Convergence of a net.

Morera's stress functions [MECH] Three functions of position, ψ_1, ψ_2, and ψ_3, in terms of which the elements of the stress tensor σ of a body may be expressed, if the body is in equilibrium and is not subjected to body forces; the elements of the stress tensor are given by $\sigma_{11} = -2\partial^2\psi_1/\partial x_2\,\partial x_3$, $\sigma_{23} = \partial^2\psi_2/\partial x_1\,\partial x_2 + \partial^2\psi_3/\partial x_1\,\partial x_3$, and cyclic permutations of these equations.

Morera's theorem [MATH] If a function of a complex variable is continuous in a simply connected domain D, and if the integral of the function about every simply connected curve in D vanishes, then the function is analytic in D.

Morgan equation [THERMO] A modification of the Ramsey-Shields equation, in which the expression for the molar surface energy is set equal to a quadratic function of the temperature rather than to a linear one.

morphism [MATH] The class of elements which together with objects form a category; in most cases, morphisms are functions which preserve some structure on a set.

morphotropism [CRYSTAL] Similarity of structure, axial ratios, and angles between faces of one or more zones in crystalline substances whose formulas can be derived one from another by substitution.

Morse equation [PHYS CHEM] An equation according to which the potential energy of a diatomic molecule in a given electronic state is given by a Morse potential.

Morse potential [PHYS CHEM] An approximate potential associated with the distance r between the nuclei of a diatomic molecule in a given electronic state; it is $V(r) = D\{1 - \exp[-a(r-r_e)]\}^2$, where r_e is the equilibrium distance, D is the dissociation energy, and a is a constant.

Morse theory [MATH] The study of differentiable mappings of differentiable manifolds, which by examining critical points shows how manifolds can be constructed from one another.

MOS *See* metal oxide semiconductor.

mosaic [ELECTR] A light-sensitive surface used in television camera tubes, consisting of a thin mica sheet coated on one side with a large number of tiny photosensitive silver-cesium globules, insulated from each other. [SCI TECH] A surface pattern made by the assembly and arrangement of many small pieces.

mosaic structure [CRYSTAL] In crystals, a substructure in which neighboring regions are oriented slightly differently.

Moseley's law [SPECT] The law that the square-root of the frequency of an x-ray spectral line belonging to a particular series is proportional to the difference between the atomic number and a constant which depends only on the series.

MOSFET *See* metal oxide semiconductor field-effect transistor.

Mosotti field *See* Lorentz local field.

Mössbauer effect [NUC PHYS] The emission and absorption of gamma rays by certain nuclei, bound in crystals, without loss of energy through nuclear recoil, with the result that radiation emitted by one such nucleus can be absorbed by another.

Mössbauer spectroscopy [SPECT] The study of Mössbauer spectra, for example, for nuclear hyperfine structure, chemical shifts, and chemical analysis.

Mössbauer spectrum [SPECT] A plot of the absorption, by nuclei bound in a crystal lattice, of gamma rays emitted by similar nuclei in a second crystal, as a function of the relative velocity of the two crystals.

MOST *See* metal oxide semiconductor field-effect transistor.

most powerful test [STAT] If two tests have the same level of significance, then the test with a smaller-size type II error is the most powerful test of the two at that significance level.

MOS transistor *See* metal oxide semiconductor field-effect transistor.

motion [MECH] A continuous change of position of a body.

motional electromotive force [ELECTROMAG] An electromotive force in a circuit that results from the motion of all or part of the circuit through a magnetic field.

motional impedance [ELECTR] Of a transducer, the complex remainder after the blocked impedance has been subtracted from the loaded impedance. Also known as loaded motional impedance.

motional induction [ELECTROMAG] The production of an electromotive force in a circuit by motion of all or part of the circuit through a magnetic field in such a way that the circuit cuts across the magnetic flux.

motor [ELEC] A machine that converts electric energy into mechanical energy by utilizing forces produced by magnetic fields on current-carrying conductors. Also known as electric motor.

motor effect [ELECTROMAG] The mutually repulsive force exerted by neighboring conductors that carry current in opposite directions.

Mott scattering [QUANT MECH] 1. The scattering of identical particles due to a Coulomb force. 2. The scattering of a relativistic electron by a Coulomb field.

mounce [MECH] A unit of mass, equal to 25 grams. Also known as metric ounce.

moving cluster [ASTRON] 1. A star cluster with common motions. 2. An open star cluster near the sun such that measurements may be made of the individual proper motions of the stars.

moving-coil galvanometer [ENG] Any galvanometer, such as the d'Arsonval galvanometer, in which the current to be measured is sent through a coil suspended or pivoted in a fixed magnetic field, and the current is determined by measuring the resulting motion of the coil.

moving-coil meter [ELEC] A meter in which a pivoted coil is the moving element.

moving-coil voltmeter [ENG] A voltmeter in which the current, produced when the voltage to be measured is applied across a known resistance, is sent through coils pivoted in the

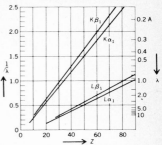

MOSELEY'S LAW

Plot of Moseley's law showing dependence of characteristic x-ray line wavelengths on atomic number. *(Philips Tech. Rev., vol. 17, no. 10, 1956)*

MÖSSBAUER EFFECT

Experiment demonstrating Mössbauer effect. Source is moved toward stationary absorber at various known velocities to Doppler-shift the gamma-ray energy received by absorber and study absorption by absorber of gamma rays as a function of energy.

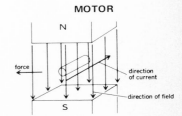

MOTOR

Relative directions of field flux, current, and force in an electric motor.

magnetic field of permanent magnets, and the resulting torque on the coils is balanced by control springs so that the deflection of a pointer attached to the coils is proportional to the current.

moving-coil wattmeter *See* electrodynamic wattmeter.

moving constraint [MECH] A constraint that changes with time, as in the case of a system on a moving platform.

moving-iron voltmeter [ENG] A voltmeter in which a field coil is connected to the voltage to be measured through a series resistor; current in the coil causes two vanes, one fixed and one attached to the shaft carrying the pointer, to be similarly magnetized; the resulting torque on the shaft is balanced by control springs.

moving load [MECH] A load that can move, such as vehicles or pedestrians.

moving-magnet voltmeter [ENG] A voltmeter in which a permanent magnet aligns itself with the resultant magnetic field produced by the current in a field coil and another permanent control magnet.

moving totals [STAT] The sum of the year's figures and those of some years before and after it.

mp *See* melting point.

mr *See* milliroentgen.

mrad *See* millirad.

M region [ASTROPHYS] Any of the areas on the surface of the sun that are theoretically responsible for magnetic disturbances on the earth.

msec *See* millisecond.

Msec *See* megasecond.

M shell [ATOM PHYS] The third layer of electrons about the nucleus of an atom, having electrons characterized by the principal quantum number 3.

MSI *See* medium-scale integration.

MSR *See* molten salt reactor.

M star [ASTRON] A spectral classification for a star whose spectrum is characterized by the presence of titanium oxide bands; M stars have surface temperatures of 3000 K for giants and 3400 K for dwarfs.

MT *See* megaton.

Mueller matrices [OPTICS] Matrix operators in a calculus used to treat polarized light; in this calculus, the light vector is split into four components one of which is the intensity of the light, and unpolarized light can be treated directly.

mu factor [ELECTR] Ratio of the change in one electrode voltage to the change in another electrode voltage under the conditions that a specified current remains unchanged and that all other electrode voltages are maintained constant; a measure of the relative effect of the voltages on two electrodes upon the current in the circuit of any specified electrode.

Muller method [MATH] A method for finding zeros of a function $f(x)$, in which one repeatedly evaluates $f(x)$ at three points, $x_1, x_2,$ and x_3, fits a quadratic polynomial to $f(x_1), f(x_2),$ and $f(x_3)$, and uses $x_2, x_3,$ and the root of this quadratic polynomial nearest to x_3 as three new points to repeat the process.

mull technique [SPECT] Method for obtaining infrared spectra of materials in the solid state; material to be scanned is first pulverized, then mulled with mineral oil.

multi- [SCI TECH] A prefix meaning many.

multicellular horn [ELECTROMAG] A cluster of horn antennas having mouths that lie in a common surface and that are fed from openings spaced one wavelength apart in one face of a common waveguide. [ENG ACOUS] A combination of individual horn loudspeakers having individual driver units or

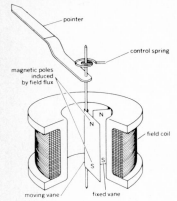

MOVING-IRON VOLTMETER

A representative form of the moving-iron voltmeter. *(General Electric Co.)*

joined in groups to a common driver unit. Also known as cellular horn.

multichannel analyzer *See* pulse-height analyzer.

multichannel field-effect transistor [ELECTR] A field-effect transistor in which appropriate voltages are applied to the gate to control the space within the current flow channels.

multicollinearity [STAT] A concept in regression analysis describing the situation where, because of the high degree of correlation between two or more independent variables, it is not possible to separate accurately the effect of each individual independent variable upon the dependent variable.

multidimensional derivative [MATH] The generalized derivative of a function of several variables which is usually represented as a matrix involving the various partial derivatives of the function.

multielement parasitic array [ELECTROMAG] Antennas consisting of an array of driven dipoles and parasitic elements, arranged to produce a beam of high directivity.

multigroup diffusion [NUCLEO] Diffusion of neutrons in a material as it is regarded in the multigroup model.

multigroup model [NUCLEO] A model for the behavior of neutrons in a material in which they are grouped into several energy ranges, taking into account differences in spatial behavior of neutrons in the various groups, and transfer of neutrons between groups.

multilinear algebra [MATH] The study of functions of several variables which are linear relative to each variable.

multilinear form [MATH] A multilinear form of degree n is a polynomial expression which is linear in each of n variables.

multimeter *See* volt-ohm-milliammeter.

multimodal distribution [STAT] A frequency distribution that has several relative maxima.

multinomial [MATH] An algebraic expression which involves the sum of at least two terms.

multinomial distribution [MATH] The joint distribution of the set of random variables which are the number of occurrences of the possible outcomes in a sequence of multinomial trials.

multinomial trials [STAT] Unrelated trials with more than two possible outcomes the probabilities of which do not change from trial to trial.

multipactor [ELECTR] A high-power, high-speed microwave switching device in which a thin electron cloud is driven back and forth between two parallel plane surfaces in a vacuum by a radio-frequency electric field.

multiphase sampling [STAT] A sampling method in which certain items of information are drawn from the whole units of a sample and certain other items of information are taken from the subsample.

multiple [ELEC] **1.** Group of terminals arranged to make a circuit or group of circuits accessible at a number of points at any one of which connection can be made. **2.** To connect in parallel. **3.** *See* parallel.

multiple-beam antenna [ELECTROMAG] An antenna or antenna array which radiates several beams in different directions.

multiple-beam interference [OPTICS] Interference which arises when part of a beam is reflected several times back and forth between a pair of strongly reflecting surfaces before being reflected or transmitted from the pair.

multiple-beam interferometer [OPTICS] An interferometer in which a beam is reflected several times back and forth between a pair of parallel plane surfaces; examples are the Fizeau interferometer and the Fabry-Perot interferometer.

multiple coefficient of determination [STAT] A statistic that measures the proportion of total variation which is explained by the regression line; computed by taking the square root of the coefficient of multiple correlation.

multiple decay *See* branching.

multiple discharge *See* composite flash.

multiple disintegration *See* branching.

multiple-function chip *See* large-scale integrated circuit.

multiple integral [MATH] An integral over a subset of n-dimensional space.

multiple linear correlation [STAT] An index for estimating the strength of the linear relationship between one dependent variable and two or more independent variables.

multiple linear regression [STAT] A technique for determining the linear relationship between one dependent variable and two or more independent variables.

multiple-loop system [CONT SYS] A system whose block diagram has at least two closed paths, along each of which all arrows point in the same direction.

multiple-purpose tester *See* volt-ohm-milliammeter.

multiple reflection [OPTICS] Reflection of light back and forth several times between a pair of strongly reflecting surfaces.

multiple root [MATH] A polynomial $f(x)$ has c as a multiple root if $(x - c)^n$ is a factor for some $n > 1$. Also known as repeated root.

multiple scattering [PHYS] Process in which a particle undergoes a large number of collisions, and the total change in its momentum is the sum of the many small changes occurring during individual collisions.

multiple star [ASTRON] A system of three or more stars which appear to the naked eye as a single star.

multiple stratification [STAT] Division of a population into two or more parts with respect to two or more variables.

multiplet [QUANT MECH] A collection of relatively closely spaced energy levels which result from the splitting of a single energy level by an interaction which is relatively weak; examples are spin-orbit multiplets and isospin multiplets. [SPECT] A collection of relatively closely spaced spectral lines resulting from transitions to or from the members of a multiplet (as in the quantum-mechanics definition).

multiplet intensity rules [SPECT] Rules for the relative intensities of spectral lines in a spin-orbit multiplet, stating that the sum of the intensities of all lines which start from a common initial level, or end on a common final level, is proportional to $2J + 1$, where J is the total angular momentum of the initial level or final level respectively.

multiple tropopause [GEOPHYS] A frequent condition in which the tropopause appears not as a continuous single "surface" of discontinuity between the troposphere and the stratosphere, but as a series of quasi-horizontal "leaves," which are partly overlapping in steplike arrangement.

multiple-tuned antenna [ELECTROMAG] Low-frequency antenna having a horizontal section with a multiplicity of tuned vertical sections.

multiple-valued [MATH] A relation between sets is multiple-valued if it associates to an element of one more than one element from the other; sometimes functions are allowed to be multiple-valued.

multiplicand [MATH] If a number x is to be multiplied by a number y, then x is called the multiplicand.

multiplication [ELECTR] An increase in current flow through a semiconductor due to increased carrier activity. [MATH] Any algebraic operation analogous to multiplication of real numbers. [NUCLEO] The ratio of neutron flux in a subcritical reactor to that supplied by a neutron source; it is the factor by which, in effect, the reactor multiplies the source strength.

multiplication constant *See* multiplication factor.

multiplication factor [NUCLEO] The ratio of the number of neutrons present in a reactor in any one neutron generation to that in the immediately preceding generation. Also known as multiplication constant; neutron multiplication factor.

multiplication formula [MATH] The formula for the gamma function:

$$\prod_{m=0}^{n-1} \Gamma(z + m/n) = (2\pi)^{(n-1)/2} n^{-nz+\frac{1}{2}} \Gamma(nz).$$

multiplicative acoustic array [ACOUS] An acoustic array of receiving elements which is divided into two parts, the signal voltages obtained from them being multiplied together. Also known as correlation array.

multiplicative subset [MATH] A subset S of a commutative ring such that if x and y are in S then so is xy.

multiplicity [MATH] **1.** A root of a polynomial $f(x)$ has multiplicity n if $(x - a)^n$ is a factor of $f(x)$ and n is the largest possible integer for which this is true. **2.** The geometric multiplicity of an eigenvalue λ of a linear transformation T is the dimension of the null space of the transformation $T - \lambda I$, where I denotes the identity transformation. **3.** The algebraic multiplicity of an eigenvalue λ of a linear transformation T on a finite-dimensional vector space is the multiplicity of λ as a root of the characteristic polynomial of T. [PHYS] In a system having Russell-Saunders coupling, the quantity $2S + 1$, where S is the total spin quantum number.

multiplier [ELEC] A resistor used in series with a voltmeter to increase the voltage range. Also known as multiplier resistor. [ELECTR] **1.** A device that has two or more inputs and an output that is a representation of the product of the quantities represented by the input signals; voltages are the quantities commonly multiplied. **2.** *See* electron multiplier. [MATH] If a number x is to be multiplied by a number y, then y is called the multiplier.

multiplier phototube [ELECTR] A phototube with one or more dynodes between its photocathode and the output electrode; the electron stream from the photocathode is reflected off each dynode in turn, with secondary emission adding electrons to the stream at each reflection. Also known as electron-multiplier phototube; photoelectric electron-multiplier tube; photomultiplier; photomultiplier tube.

multiplier resistor *See* multiplier.

multiplier tube [ELECTR] Vacuum tube using secondary emission from a number of electrodes in sequence to obtain increased output current; the electron stream is reflected, in turn, from one electrode of the multiplier to the next.

multiply connected region [MATH] An open set in the plane which has holes in it.

multiply transitive group [MATH] A group of permutations on a set of n elements which is k-ply transitive for some integer k, $2 \leq k \leq n$.

multipolar machine [ELECTROMAG] An electric machine that has a field magnet with more than one pair of poles.

multipole [ELECTROMAG] One of a series of types of static or oscillating distributions of charge or magnetization; namely, an electric multipole or a magnetic multipole.

multipole fields [ELECTROMAG] The electric and magnetic fields generated by static or oscillating electric or magnetic multipoles.

multipole radiation [PHYS] **1.** Electromagnetic radiation which has characteristics equivalent to those of radiation generated by an oscillating electric or magnetic multipole, and is made up of photons of well-defined angular momentum and parity. **2.** Internal conversion electrons, or positron-

MULTIPLIER PHOTOTUBE

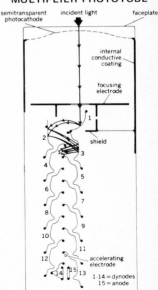

Typical multiplier phototube construction.

electron pairs having similar characteristics, emitted from an atom when the nucleus makes a transition between two energy states.

multipole transition [PHYS] A transition between two energy states of an atom or nucleus in which a quantum of multipole radiation is emitted or absorbed.

multistage amplifier *See* cascade amplifier.

multistage sampling [STAT] A sampling method in which the population is divided into a number of groups or primary stages from which samples are drawn; these are then divided into groups or secondary stages from which samples are drawn, and so on.

multistator watt-hour meter [ELEC] An induction type of watt-hour meter in which several stators exert a torque on the rotor.

multivariable system [CONT SYS] A dynamical system in which the number of either inputs or outputs is greater than 1.

multivariate analysis [STAT] The study of random variables which are multidimensional.

multivariate distribution *See* joint distribution.

multivibrator [ELECTR] A relaxation oscillator using two tubes, transistors, or other electron devices, with the output of each coupled to the input of the other through resistance-capacitance elements or other elements to obtain in-phase feedback voltage.

mu meson *See* muon.

Munsell chroma *See* chroma.

Munsell color system [OPTICS] A system for designating colors which employs three perceptually uniform scales (Munsell hue, Munsell value, Munsell chroma) defined in terms of daylight reflectance.

Munsell hue [OPTICS] The dimension of the Munsell system of color that determines whether a color is blue, green, yellow, red, purple, or the like, without regard to its lightness or saturation.

Munsell value [OPTICS] The dimension, in the Munsell system of object-color specificiation, that indicates the apparent luminous transmittance or reflectance of the object on a scale having approximately equal perceptual steps under the usual conditions of observation.

muon [PARTIC PHYS] Collective name for two semistable elementary particles with positive and negative charge, designated μ^+ and μ^- respectively, which are leptons and have a spin of $\frac{1}{2}$ and a mass of approximately 105.7 MeV. Also known as mu meson.

muonic atom [PARTIC PHYS] An atom in which an electron is replaced by a negatively charged muon orbiting close to or within the nucleus.

muonium [PARTIC PHYS] An atom consisting of an electron bound to a positively charged muon by their mutual Coulomb attraction, just as an electron is bound to a proton in the hydrogen atom.

Murphy's formula [MATH] **1.** The equation

$$x'_i = \sum_{j=1}^{3} x_j \cos (x'_i, x_j), \ i = 1, 2, 3,$$

giving the cartesian coordinates x'_i with respect to one set of perpendicular axes in terms of the coordinates x_j with respect to another set of axes, where (x'_i, x_j) is the angle between the x'_i and x_j axes. **2.** The equation $P\nu(z) = F(-\nu, \nu + 1; 1; \frac{1}{2} - \frac{1}{2}z)$, expressing the Legendre function of order ν in terms of the hypergeometric series.

musical echo [ACOUS] A musical tone produced by the reflection of an impulsive sound from a stepped structure such as a picket fence, when reflections from successive steps reach the observer with suitable frequency.

musical quality *See* timbre.

Muskhelishvili's method [MECH] A method of solving problems concerning the elastic deformation of a planar body that involves using methods from the theory of functions of a complex variable to calculate analytic functions which determine the plane strain of the body.

mutual admittance [ELEC] For two meshes of a network carrying alternating current, the ratio of the complex current in one mesh to the complex voltage in the other, when the voltage in all meshes besides these two is 0.

mutual branch *See* common branch.

mutual capacitance [ELEC] The accumulation of charge on the surfaces of conductors of each of two circuits per unit of potential difference between the circuits.

mutual conductance *See* transconductance.

mutual exclusion rule [PHYS CHEM] The rule that if a molecule has a center of symmetry, then no transition is allowed in both its Raman scattering and infrared emission (and absorption), but only in one or the other.

mutual impedance [ELEC] For two meshes of a network carrying alternating current, the ratio of the complex voltage in one mesh to the complex current in the other, when all meshes besides the latter one carry no current.

mutual inductance [ELECTROMAG] Property of two neighboring circuits, equal to the ratio of the electromotive force induced in one circuit to the rate of change of current in the other circuit.

mutual induction [ELECTROMAG] The generation of a voltage in one circuit by a varying current in another.

mutuality of phases [CHEM] The rule that if two phases, with respect to a reaction, are in equilibrium with a third phase at a certain temperature, then they are in equilibrium with respect to each other at that temperature.

mV *See* millivolt.

MV *See* megavolt.

mW *See* milliwatt.

MW *See* megawatt.

MW(E) *See* megawatt electric.

MW(Th) *See* megawatt thermal.

MWYE *See* megawatt year of electricity.

Mx *See* maxwell.

myria- [SCI TECH] A prefix representing 10^4 or 10,000.

myriotic field [QUANT MECH] A quantized field that has creation and annihilation operators satisfying specified commutation rules, but no vacuum state.

N *See* newton; nitrogen; normality.

Na *See* sodium.

N.A. *See* numerical aperture.

nabla *See* del operator.

Nakayama's lemma [MATH] If R is a commutative ring, I is an ideal contained in all maximal ideals of R, and M is a finitely generated module over R, and if $IM = M$, where IM denotes the set of all elements of the form am with a in I and m in M, then $M = 0$.

naked singularity [RELAT] A singularity in space-time without an associated event horizon; it is visible from infinity.

NAND [MATH] A logic operator having the characteristic that if P, Q, R, . . . are statements, then the NAND of P, Q, R, . . . is true if at least one statement is false, false if all statements are true. Derived from NOT-AND. Also known as sheffer stroke.

NAND circuit [ELECTR] A logic circuit whose output signal is a logical 1 if any of its inputs is a logical 0, and whose output signal is a logical 0 if all of its inputs are logical 1.

nano- [MATH] A prefix representing 10^{-9}, which is 0.000000001 or one-billionth of the unit adjoined. Also known as milli-micro- (deprecated usage).

nanogram [MECH] One-billionth (10^{-9}) of a gram. Abbreviated ng.

nanometer [MECH] A unit of length equal to one-billionth of a meter, or 10^{-9} meter. Abbreviated nm. Also known as millimicron ($m\mu$); nanon.

nanon *See* nanometer.

nanosecond [MECH] A unit of time equal to one-billionth of a second, or 10^{-9} second.

Naperian logarithm *See* logarithm.

napier *See* neper.

Napierian logarithm *See* logarithm.

Napier's analogies [MATH] Formulas which enable one to study the relationships between the sides and the angles of a spherical triangle.

Napier's rules [MATH] Two rules which give the formulas necessary in the solution of right spherical triangles.

nappe [MATH] One of the two parts of a conical surface defined by the vertex.

narrow-band pyrometer [ENG] A pyrometer in which light from a source passes through a color filter, which passes only a limited band of wavelengths, before falling on a photoelectric detector. Also known as spectral pyrometer.

narrow beam [PHYS] In measurements of the attenuation of a beam of ionizing radiation, a beam in which the scattered radiation does not reach the detector.

narrow-beam antenna [ELECTROMAG] An antenna which radiates most of its power in a cone having a radius of only a few degrees.

NARROW-BAND PYROMETER

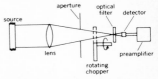

Components of a narrow-band pyrometer.

narrow cut filter [OPTICS] An optical filter which displays an abrupt change from high transmission to complete absorption over a narrow wavelength region.

narrow-gap spark chamber [NUCLEO] A type of spark chamber in which the plates are only 6 to 10 millimeters apart, so that sparks usually follow the electric field perpendicular to the plates, and coincide with the particle track at one point; the great majority of spark chambers are of this type.

n-ary composition [MATH] A function that associates an element of a set with every sequence of n elements of the set.

native uranium [GEOCHEM] Uranium as found in nature; a mixture of the fertile uranium-238 isotope (99.3%), the fissionable uranium-235 isotope (0.7%), and a minute percentage of other uranium isotopes. Also known as natural uranium; normal uranium.

natrium [CHEM] Latin name for sodium; source of the symbol Na.

natural abundance [NUCLEO] The abundance ratio of an isotope in a naturally occurring terrestrial sample of an element.

natural antenna frequency [ELECTROMAG] Lowest resonant frequency of an antenna without added inductance or capacitance.

natural boundary [MATH] Those points of the boundary of a region where an analytic function is defined through which the function cannot be continued analytically.

natural circulation reactor [NUCLEO] A reactor in which the coolant (usually water) circulates without pumping, owing to the different densities of its cold and reactor-heated portions.

natural convection [THERMO] Convection in which fluid motion results entirely from the presence of a hot body in the fluid, causing temperature and hence density gradients to develop, so that the fluid moves under the influence of gravity. Also known as free convection; gravitational convection.

natural coordinates [FL MECH] An orthogonal, or mutually perpendicular, system of curvilinear coordinates for the description of fluid motion, consisting of an axis t tangent to the instantaneous velocity vector and an axis n normal to this velocity vector to the left in the horizontal plane, to which a vertically directed axis z may be added for the description of three-dimensional flow; such a coordinate system often permits a concise formulation of atmospheric dynamical problems, especially in the Lagrangian system of hydrodynamics.

natural draft [FL MECH] Unforced gas flow through a chimney or vertical duct, directly related to chimney height and the temperature difference between the ascending gases and the atmosphere, and not dependent upon the use of fans or other mechanical devices.

natural frequency [ELECTR] The lowest resonant frequency of an antenna, circuit, or component. [PHYS] The frequency with which a system oscillates in the absence of external forces; or, for a system with more than one degree of freedom, the frequency of one of the normal modes of vibration.

natural fuel reactor *See* natural uranium reactor.

natural function [MATH] A trigonometric function, as opposed to its logarithm.

natural numbers [MATH] The integers 1, 2, 3, ...

natural period [PHYS] Period of the free oscillation of a body or system; when the period varies with amplitude, the natural period is the period when the amplitude approaches zero.

natural radiation *See* background radiation.

natural radioactivity [NUCLEO] Radioactivity exhibited by naturally occurring radionuclides.

natural remanent magnetization [GEOPHYS] The magnetization of rock which exists in the absence of a magnetic field and has been acquired from the influence of the earth's magnetic field at the time of their formation or, in certain cases, at later times. Abbreviated NRM.

natural resonance [PHYS] Resonance in which the period or frequency of the applied agency maintaining oscillation is the same as the natural period of oscillation of a system.

natural uranium *See* native uranium.

natural uranium reactor [NUCLEO] A nuclear reactor in which natural unenriched uranium is the principal fissionable material. Also known as natural fuel reactor.

natural wavelength [ELECTROMAG] Wavelength corresponding to the natural frequency of an antenna or circuit.

natural width of energy level [PHYS] A measure of the spread in energy of an excited state of a quantized system due to spontaneous transitions to other states; quantitatively, it is the difference between the energies for which the intensity of emission from or absorption by the state, or of the scattering cross section associated with it, is one-half its maximum value, in the absence of any external influence on the system.

nautical chain [MECH] A unit of length equal to 15 feet or 4.572 meters.

nautical twilight [ASTRON] The interval of incomplete darkness between sunrise or sunset and the time at which the center of the sun's disk is 12° below the celestial horizon.

navel point *See* umbilical point.

Navier's equation [MECH] A vector partial differential equation for the displacement vector of an elastic solid in equilibrium and subjected to a body force.

Navier-Stokes equations [FL MECH] The equations of motion for a viscous fluid which may be written $d\mathbf{V}/dt = -(1/\rho)\nabla p + \mathbf{F} + \nu\nabla^2\mathbf{V} + (1/3)\nu\nabla(\nabla\cdot\mathbf{V})$, where p is the pressure, ρ the density, F the total external force per unit mass, $\mathbf{V}$ the fluid velocity, and ν the kinematic viscosity; for an incompressible fluid, the term in $\nabla\cdot\mathbf{V}$ (divergence) vanishes, and the effects of viscosity then play a role analogous to that of temperature in thermal conduction and to that of density in simple diffusion.

Nb *See* niobium.

n-body problem *See* many-body problem.

NC *See* numerical control.

N/C *See* numerical control.

N center [SOLID STATE] A color center which arises from continued exposure to light in the F band or to x-rays and which produces a faint absorption band on the long-wavelength side of the M band. Also known as G center.

n-channel [ELECTR] A conduction channel formed by electrons in an n-type semiconductor, as in an n-type field-effect transistor.

n-component [PARTIC PHYS] Cosmic-ray particles that can take part in nuclear interactions, that is, nucleons, pions, and other baryons and mesons.

N curve [ELECTR] A plot of voltage against current for a negative-resistance device; its slope is negative for some values of current or voltage.

Nd *See* neodymium.

n-dimensional space [MATH] A vector space whose basis has n vectors.

Ne *See* neon.

nearest neighbors [CRYSTAL] Any pair of atoms in a crystal lattice which are as close to each other, or closer to each other, than any other pair.

near field [ACOUS] The acoustic radiation field that is close to an acoustic source such as a loudspeaker. [ELECTROMAG] The electromagnetic field that exists within one wavelength of a source of electromagnetic radiation, such as a transmitting antenna.

near-infrared radiation [ELECTROMAG] Infrared radiation having a relatively short wavelength, between 0.75 and about 2.5 micrometers (some scientists place the upper limit from 1.5 to 3 micrometers), at which radiation can be detected by photoelectric cells, and which corresponds in frequency range to the lower electronic energy levels of molecules and semiconductors. Also known as photoelectric infrared radiation.

near-infrared spectrophotometry [ANALY CHEM] Spectrophotometry at wavelengths in the near-infrared region, generally using instruments with quartz prisms in the monochromators and lead sulfide photoconductor cells as detectors to observe absorption bands which are harmonics of bands at longer wavelengths.

nearly free electron method [SOLID STATE] A method of approximating the energy levels of electrons in a crystal lattice by considering the potential energy resulting from atomic nuclei and from other electrons in the lattice as a perturbation on free electron states. Abbreviated NFE method.

near ring [MATH] An algebraic system with two binary operations called multiplication and addition; the system is a group (not necessarily commutative) relative to addition, and multiplication is associative, and is left-distributive with respect to addition, that is, $x(y + z) = xy + xz$ for any x, y, and z in the near ring.

near stars [ASTRON] Those stars in the celestial neighborhood of the sun, sometimes taken as those 22 stars within 13 light-years (123×10^{15} meters) of the sun.

near-ultraviolet radiation [ELECTROMAG] Ultraviolet radiation having relatively long wavelength, in the approximate range from 300 to 400 nanometers.

nebula [ASTRON] Interstellar clouds of gas or small particles; an example is the Horsehead Nebula in Orion.

nebular hypothesis [ASTROPHYS] A theory, proposed in 1796 by P. S. Laplace, supposing that the planets originated from the solar nebula surrounding the proto-sun; as the sun cooled, it contracted, rotated faster, and thus caused a ringlike bulging at the equator; this bulge eventually broke off and formed the planets; Laplace further theorized that the sun and other stars formed from clouds of nebulous matter; the theory in this form is not accepted.

nebular lines [ASTROPHYS] The spectral lines formed in the glow of bright nebulae; they arise from forbidden atomic transition which can take place because of the very low pressure in the nebula itself.

nebular red shift [ASTROPHYS] A systematic shift observed in the spectra of all distant galaxies; the wavelength shift toward the red increases with the distance of the galaxies from the earth.

nebular transitions [ASTROPHYS] Those electronic transitions for doubly ionized argon and chlorine that yield the nebular lines seen in the spectra of gaseous nebulae.

nebular variable *See* T Tauri star.

Néel point *See* Néel temperature.

Néel's theory [SOLID STATE] A theory of the behavior of antiferromagnetic and other ferrimagnetic materials in which the crystal lattice is divided into two or more sublattices; each atom in one sublattice responds to the magnetic field generated by nearest neighbors in other sublattices, with the result that magnetic moments of all the atoms in any sublattice are parallel, but magnetic moments of two different sublattices can be different.

NEBULA

Horsehead Nebula in Orion (IC 434, Bernard 33), which was photographed in red light with a 200-inch (508-centimeter) telescope. *(Hale Observatories)*

Néel temperature [SOLID STATE] A temperature, characteristic of certain metals, alloys, and salts, below which spontaneous nonparalleled magnetic ordering takes place so that they become antiferromagnetic, and above which they are paramagnetic. Also known as Néel point.

Néel wall [SOLID STATE] The boundary between two magnetic domains in a thin film in which the magnetization vector remains parallel to the faces of the film in passing through the wall.

negation [MATH] The negation of a proposition P is a proposition which is true if and only if P is false; this is often written $\sim$ P.

negative acceleration [MECH] Acceleration in a direction opposite to the velocity, or in the direction of the negative axis of a coordinate system.

negative angle [MATH] The angle subtended by moving a ray in the clockwise direction.

negative charge [ELEC] The type of charge which is possessed by electrons in ordinary matter, and which may be produced in a resin object by rubbing with wool. Also known as negative electricity.

negative crystal [OPTICS] A uniaxial crystal in which the extraordinary wave travels faster than the ordinary wave, such as calcite.

negative electricity See negative charge.

negative electrode See cathode; negative plate.

negative electron See electron.

negative energy states [QUANT MECH] Electron states with negative energy (considering the Hamiltonian to be the energy operator) that occur in the Dirac electron theory; for a vacuum they are all occupied, and a vacancy in one of them represents the existence of a positron.

negative feedback [CONT SYS] Feedback in which a portion of the output of a circuit, device, or machine is fed back 180° out of phase with the input signal, resulting in a decrease of amplification so as to stabilize the amplification with respect to time or frequency, and a reduction in distortion and noise. Also known as inverse feedback; reverse feedback; stabilized feedback. [SCI TECH] Feedback which tends to reduce the output in a system.

negative g [MECH] In designating the direction of acceleration on a body, the opposite of positive g; for example, the effect of flying an outside loop in the upright seated position.

negative glow [ELECTR] The luminous flow in a glow-discharge cold-cathode tube occurring between the cathode dark space and the Faraday dark space.

negative impedance [ELECTR] An impedance such that when the current through it increases, the voltage drop across the impedance decreases.

negative integer [MATH] The additive inverse of a positive integer relative to the additive group structure of the integers.

negative ion [CHEM] An atom or group of atoms which by gain of one or more electrons has acquired a negative electric charge. [PHYS] An electron or negatively charged subatomic particle.

negative-ion vacancy [CRYSTAL] A point defect in an ionic crystal in which a negative ion is missing from its lattice site.

negative lens See diverging lens.

negative meniscus lens [OPTICS] A lens having one convex and one concave surface, with the radius of curvature of the convex surface greater than that of the concave surface. Also known as diverging meniscus lens.

negative part [MATH] For a real-valued function f, this is the function, denoted f^-, for which $f^-(x) = f(x)$ if $f(x) \leq 0$ and $f^-(x) = 0$ if $f(x) > 0$.

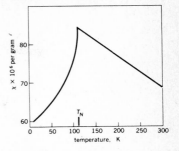

Magnetic susceptibility of powdered manganese oxide. T_N is the Néel temperature. (After H. Bizette, C. F. Squire, and B. Tsai, 1938)

Shape of a negative meniscus lens. (From F. A. Jenkins and H. E. White, Fundamentals of Optics, 3d ed., McGraw-Hill, 1957)

negative pedal [MATH] The negative pedal of a curve with respect to a point O is the envelope of the line drawn through a point P of the curve perpendicular to OP.

negative pion [PARTIC PHYS] A pion having a negative electric charge.

negative plate [ELEC] The internal plate structure that is connected to the negative terminal of a storage battery. Also known as negative electrode.

negative potential [ELEC] An electrostatic potential which is lower than that of the ground, or of some conductor or point in space that is arbitrarily assigned to have zero potential.

negative pressure [PHYS] A way of expressing vacuum; a pressure less than atmospheric or the standard 760 mmHg (101,325 newtons per square meter).

negative principal planes [OPTICS] Two planes perpendicular to the optical axis such that objects in one plane form images in the other with a lateral magnification of −1. Also known as antiprincipal planes.

negative principal point [OPTICS] The intersection of a negative principal plane with the optical axis. Also known as antiprincipal point.

negative resistance [ELECTR] The resistance of a negative-resistance device.

negative-resistance device [ELECTR] A device having a range of applied voltages within which an increase in this voltage produces a decrease in the current.

negative-resistance oscillator [ELECTR] An oscillator in which a parallel-tuned resonant circuit is connected to a vacuum tube so that the combination acts as the negative resistance needed for continuous oscillation.

negative skewness [MATH] Skewness in which the mean is smaller than the mode.

negative temperature coefficient [PHYS] Condition wherein the resistance, length, or some other characteristic of a material decreases when temperature increases.

negative terminal [ELEC] The terminal of a battery or other voltage source that has more electrons than normal; electrons flow from the negative terminal through the external circuit to the positive terminal.

negative with respect to a measure [MATH] A set A is negative with respect to a signed measure m if, for every measurable set B, the intersection of A and B, $A \cap B$, is measurable and $m(A \cap B) \leq 0$.

negatron *See* dynatron; electron.

neighbor [CRYSTAL] One of a pair of atoms or ions in a crystal which are close enough to each other for their interaction to be of significance in the physical problem being studied.

neighborhood of a point [MATH] A set in a topological space which contains an open set which contains the point; in euclidean space, an example of a neighborhood of a point is an open (without boundary) ball centered at that point.

Neil's parabola [MATH] The graph of the equation $y = ax^{3/2}$, where a is a constant.

N electron [ATOM PHYS] An electron in the fourth (N) shell of electrons surrounding the atomic nucleus, having the principal quantum number 4.

nematic phase [PHYS CHEM] A phase of a liquid crystal in the mesomorphic state, in which the liquid has a single optical axis in the direction of the applied magnetic field, appears to be turbid and to have mobile threadlike structures, can flow readily, has low viscosity, and lacks a diffraction pattern.

neodymium [CHEM] A metallic element, symbol Nd, with atomic weight 144.24, atomic number 60; a member of the rare-earth group of elements.

NEODYMIUM

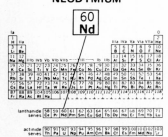

Periodic table of the chemical elements showing the position of neodymium.

neodymium glass laser [OPTICS] An amorphous solid laser in which glass is doped with neodymium; characteristics are comparable with those of a pulsed ruby laser, but the wavelength of radiation is outside the visible range.

neodymium liquid laser *See* inorganic liquid laser.

neon [CHEM] A gaseous element, symbol Ne, atomic number 10, atomic weight 20.183; a member of the family of noble gases in the zero group of the periodic table.

neon-helium laser [OPTICS] A continuous-wave gas laser using a combination of neon and helium gases to obtain a 6328-angstrom (632.8 nanometer) visible red beam.

neper [PHYS] Abbreviated Np. Also known as napier. **1.** A unit used for expressing the ratio of two currents, voltages, or analogous quantities; the number of nepers is the natural logarithm of this ratio. **2.** A unit used for expressing the ratio of two powers (even when this ratio is not the square of the corresponding current or voltage ratio); the number of nepers is the natural logarithm of the square root of this ratio; to avoid confusion, this usage should be accompanied by a specific statement.

nephelometer [OPTICS] A type of instrument that measures, at more than one angle, the scattering function of particles suspended in a medium; information obtained may be used to determine the size of the suspended particles and the visual range through the medium.

nephroid [MATH] A plane curve with parametric equation $x = a(3 \cos t - \cos 3t)$, $y = a(3 \sin t - \sin 3t)$, where a is a constant.

nepit *See* nit.

Neptune [ASTRON] The outermost of the four giant planets, and the next to last planet, from the sun; it is 30 astronomical units from the sun, and the sidereal revolution period is 164.8 years.

neptunium [CHEM] A chemical element, symbol Np, atomic number 93, atomic weight 237.0482; a member of the actinide series of elements.

neptunium decay series [CHEM] Little-known radioactive elements with short lives; produced as successive series of decreasing atomic weight when uranium-237 and plutonium-241 decay radioactively through neptunium-237 to bismuth-209.

Nernst approximation formula [THERMO] An equation for the equilibrium constant of a gas reaction based on the Nernst heat theorem and certain simplifying assumptions.

Nernst bridge [ELEC] A four-arm bridge containing capacitors instead of resistors, used for measuring capacitance values at high frequencies.

Nernst effect *See* Ettingshausen-Nernst effect.

Nernst equation [PHYS CHEM] The relationship showing that the electromotive force developed by a dry cell is determined by the activities of the reacting species, the temperature of the reaction, and the standard free-energy change of the overall reaction.

Nernst glower *See* Nernst lamp.

Nernst heat theorem [THERMO] The theorem expressing that the rate of change of free energy of a homogeneous system with temperature, and also the rate of change of enthalpy with temperature, approaches zero as the temperature approaches absolute zero.

Nernst lamp [ELEC] An electric lamp consisting of a short, slender rod of zirconium oxide in open air, heated to brilliant white incandescence by current. Also known as Nernst glower.

Nernst-Lindemann calorimeter [ENG] A calorimeter for measuring specific heats at low temperatures, in which the

NEON

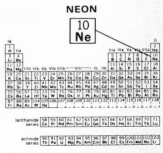

Periodic table of the chemical elements showing the position of neon.

NEPTUNE

Telescopic appearance of Neptune.

NEPTUNIUM

Periodic table of the chemical elements showing the position of neptunium.

heat reservoir consists of a metal of high thermal conductivity such as copper, to promote rapid temperature equalization; none of the material under study is more than a few millimeters from a metal surface, and the whole apparatus is placed in an evacuated vessel and heated by current through a platinum heating coil.

Nernst-Lindemann theory of specific heat [SOLID STATE] A theory of the specific heat of solids based on the assumption that atoms can vibrate at one of two frequencies, one of which is twice the other.

Nernst-Simon statement of the third law of thermodynamics [THERMO] The statement that the change in entropy which occurs when a homogeneous system undergoes an isothermal reversible process approaches zero as the temperature approaches absolute zero.

Nernst-Thomson rule [PHYS CHEM] The rule that in a solvent having a high dielectric constant the attraction between anions and cations is small so that dissociation is favored, while the reverse is true in solvents with a low dielectric constant.

Nernst zero of potential [PHYS CHEM] An electrode potential corresponding to the reversible equilibrium between hydrogen gas at a pressure of 1 standard atmosphere (101,325 newtons per square meter) and hydrogen ions at unit activity.

nesistor [ELECTR] A negative-resistance semiconductor device that is basically a bipolar field-effect transistor.

nested sets [MATH] A family of sets where, given any two of its sets, one is contained in the other.

net [MATH] **1.** A set whose members are indexed by elements from a directed set; this is a generalization of a sequence. **2.** A nondegenerate partial plane satisfying the parallel axiom.

net head [FL MECH] The difference in elevation between the last free water surface in a power conduit above the waterwheel and the first free water surface in the conduit below the waterwheel, less the friction losses in the conduit.

net heating value *See* low heat value.

net power flow [ELECTROMAG] The difference between the power carried by electromagnetic waves traveling in a given direction along a waveguide and the power carried by waves traveling in the opposite direction.

net radiometer [ENG] A Moll thermopile modified so that both sides are sensitive to radiation and the resulting electromotive force is proportional to the difference in intensities of radiation incident on the two sides; used to measure the difference in intensity between radiation entering and leaving the earth's surface.

net ton *See* ton.

network [ELEC] A collection of electric elements, such as resistors, coils, capacitors, and sources of energy, connected together to form several interrelated circuits. Also known as electric network. [MATH] The name given to a graph in applications in management and the engineering sciences; to each segment linking points in the graph, there is usually associated a direction and a capacity on the flow of some quantity.

network admittance [ELEC] The admittance between two terminals of a network under specified conditions.

network analysis [ELEC] Derivation of the electrical properties of a network, from its configuration, element values, and driving forces.

network constant [ELEC] One of the resistance, inductance, mutual inductance, or capacitance values involved in a circuit or network; if these values are constant, the network is said to be linear.

network filter [ELEC] A combination of electrical elements (for example, interconnected resistors, coils, and capacitors) that represents relatively small attenuation to signals of a certain frequency, and great attenuation to all other frequencies.

network input impedance [ELEC] The impedance between the input terminals of a network under specified conditions.

network structure [MET] A crystal structure in a metal in which one constituent occurs primarily at grain boundaries er.veloping the grains made up of other constituents.

network synthesis [ELEC] Derivation of the configuration and element values of a network with given electrical properties.

network theory [ELEC] The systematizing and generalizing of the relations between the currents, voltages, and impedances associated with the elements of an electrical network.

network transfer admittance [ELEC] The current that would flow through a short circuit between one pair of terminals in a network if a unit voltage were applied across the other pair.

Neuber-Papkovich solution *See* Papkovich-Neuber solution.

Neumann boundary condition [MATH] The boundary condition imposed on the Neumann problem in potential theory.

Neumann function [MATH] 1. One of a class of Bessel functions arising in the study of the solutions to Bessel's differential equation. 2. A harmonic potential function in potential theory occurring in the study of Neumann's problem.

Neumann-Kopp rule [THERMO] The rule that the heat capacity of 1 mole of a solid substance is approximately equal to the sum over the elements forming the substance of the heat capacity of a gram atom of the element times the number of atoms of the element in a molecule of the substance.

Neumann line [MATH] The generalization of the concept of a line occurring in Neumann's study of continuous geometry.

Neumann polynomials [MATH] The polynomials

$$O_0(t) = 1/t$$

$$O_n(t) = (\tfrac{1}{4}) \sum_{m=0}^{[n/2]} \frac{n(n - m - 1)!}{m! \, (\tfrac{1}{2}t)^{n-2m+1}}, \, n = 1, 2, 3, \ldots$$

(where $[n/2]$ denotes the greatest integer equal to or less than $n/2$), used in expanding the function $1/(t - z)$ in Bessel functions $J_n(z)$.

Neumann problem [MATH] The determination of a harmonic function within a finite region of three-dimensional space enclosed by a closed surface when the normal derivatives of the function on the surface are specified.

Neumann series [MATH] 1. An infinite series which takes the form of a sum over $n = 0, 1, 2, \ldots$ of $a_n J_{v+n}(z)$, where the a_n are constants and J_{v+n} are Bessel functions of the first kind of order $v + n$. 2. *See* Liouville-Neumann series.

Neumann's formula [ELECTROMAG] A formula for the mutual inductance M_{12} between two closed circuits C_1 and C_2; it is

$$M_{12} = \frac{\mu_0}{4\pi} \int_{C_1} \int_{C_2} \frac{ds_1 \, ds_2}{r}$$

where r is the distance between line elements ds_1 and ds_2, and μ_0 is the permeability of the empty space.

Neumann's principle [CRYSTAL] The principle that the symmetry elements of the point group of a crystal are included among the symmetry elements of any property of the crystal.

Neumann's triangle [FL MECH] A triangle whose sides have lengths proportional to the surface tensions of two immiscible liquids and their interfacial tension, and directions parallel to the free surfaces of the liquids and the interface between them at a line where these three surfaces meet, when one liquid is placed on the surface of the other.

neuristor [ELECTR] A device that behaves like a nerve fiber in having attenuationless propagation of signals; one goal of research is development of a complete artificial nerve cell, containing many neuristors, that could duplicate the function of the human eye and brain in recognizing characters and other visual images.

neutral [CHEM] Property of a solution which is neither acidic nor basic, having the same concentration of hydrogen ions as water. [ELEC] Having no net electric charge.

neutral atom [ATOM PHYS] An atom in which the number of electrons that surround the nucleus is equal to the number of protons in the nucleus, so that there is no net electric charge.

neutral axis [MECH] In a beam bent downward, the line of zero stress below which all fibers are in tension and above which they are in compression.

neutral beam [PHYS] A stream of uncharged particles.

neutral-density filter [OPTICS] An optical filter that reduces the intensity of light without appreciably changing its color; used on a camera when the lens cannot be stopped down sufficiently for use with a given film. Also known as gray filter; neutral filter.

neutral equilibrium [FL MECH] A property of the steady state of a system which exhibits neither instability nor stability according to the particular criterion under consideration; a disturbance introduced into such an equilibrium will thus be neither amplified nor damped. Also known as indifferent equilibrium.

neutral fiber [MECH] A line of zero stress in cross section of a bent beam, separating the region of compressive stress from that of tensile stress.

neutral filter *See* neutral-density filter.

neutralize [CHEM] To make a solution neutral (neither acidic nor basic, pH of 7) by adding a base to an acidic solution, or an acid to a basic solution. [ELECTR] To nullify oscillation-producing voltage feedback from the output to the input of an amplifier through tube interelectrode capacitances; an external feedback path is used to produce at the input a voltage that is equal in magnitude but opposite in phase to that fed back through the interelectrode capacitance. [OPTICS] To place a lens in contact with other lenses of equal and opposite power so that the combination has zero power.

neutralizing power of a lens [OPTICS] The power of a lens, measured by neutralizing it with trial lenses of equal and opposite power; for a spectacle lens, it may differ significantly from back vertex power.

neutral molecule [PHYS CHEM] A molecule in which the number of electrons surrounding the nuclei is the same as the total number of protons in the nuclei, so that there is no net electric charge.

neutral particle [PARTIC PHYS] A particle that carries no electric charge.

neutral point [ELEC] Point which has the same potential as the point of junction of a group of equal nonreactive resistances connected at their free ends to the appropriate main terminals or lines of the system. [FL MECH] *See* hyperbolic point. [OPTICS] In atmospheric optics, one of several points in the sky for which the degree of polarization of diffuse sky radiation is zero.

neutral stability [CONT SYS] Condition in which the natural motion of a system neither grows nor decays, but remains at its initial amplitude.

neutral surface [MECH] A surface in a bent beam along which material is neither compressed nor extended.

neutral temperature [ELECTR] The temperature of the hot junction of a thermocouple at which the electromotive force of the thermocouple attains its maximum value, when the cold junction is maintained at a constant temperature of 0°C.

neutral wave [PHYS] Any wave whose amplitude does not change with time; in most contexts the wave is referred to as a stable wave, the term "neutral wave" being used when it is important to emphasize that the wave is neither damped nor amplified.

neutrino [PHYS] A neutral particle having zero rest mass and spin $1/2 (h/2\pi)$, where h is Planck's constant; experimentally, there are two such particles known as the e-neutrino (ν_e) and the μ-neutrino (ν_μ).

neutron [PHYS] An elementary particle which has approximately the same mass as the proton but lacks electric charge, and is a constituent of all nuclei having mass number greater than 1.

neutron absorber [NUCLEO] A material in which a significant number of neutrons passing through combine with nuclei and are not reemitted.

neutron absorption *See* neutron capture.

neutron activation analysis [NUCLEO] Activation analysis in which the specimen is bombarded with neutrons; identification is made by measuring the resulting radio isotopes.

neutron age *See* Fermi age.

neutron albedo [NUCLEO] The probability, under specified conditions, that a neutron entering into a region through a surface will return through that surface.

neutron binding energy [NUC PHYS] The energy required to remove a single neutron from a nucleus.

neutron capture [NUC PHYS] A process in which the collision of a neutron with a nucleus results in the absorption of the neutron into the nucleus with the emission of one or more prompt gamma rays; in certain cases, beta decay or fission of the nucleus results. Also known as neutron absorption; neutron radiative capture.

neutron-capture cross section [NUC PHYS] The cross section for neutron capture by nuclei in a material; it is a measure of the probability that this reaction will occur.

neutron chopper [NUCLEO] A device that interrupts the output beam of neutrons from a nuclear reactor mechanically, to provide bursts or pulses of neutrons for research purposes.

neutron counter [NUCLEO] A neutron detector which counts the number of neutrons passing through the detecting medium.

neutron cross section [NUC PHYS] A measure of the probability that an interaction of a given kind will take place between a nucleus and an incident neutron; it is an area such that the number of interactions which occur in a sample exposed to a beam of neutrons is equal to the product of the number of nuclei in the sample and the number of neutrons in the beam that would pass through this area if their velocities were perpendicular to it.

neutron cycle [NUCLEO] The life history of the neutrons in a nuclear reactor, extending from the initial fission process until all the neutrons have been absorbed or have leaked out.

neutron detector [NUCLEO] Any device which detects passing neutrons, for example, by observing the charged particles or gamma rays released in nuclear reactions induced by the neutrons, or observing the recoil of charged particles caused by collisions with neutrons.

neutron diffraction [PHYS] The phenomenon associated with the interference processes which occur when neutrons are scattered by the atoms within solids, liquids, and gases.

neutron diffraction analysis [PHYS] The study of the atomic

NEUTRON CHOPPER

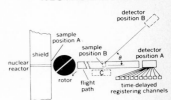

Schematic diagram of a neutron chopper and time-of-flight apparatus. Registering channels record at successive times following passage of pulse of neutrons through rotor slit to measure neutron time-of-flight. Sample and detector position A used to measure total cross section. Sample and position B used to measure differential cross section for scattering through angle θ. Detector position C used to measure absorption cross section.

structure of solids, liquids, and gases by passing high-flux beams of thermal neutrons through them and measuring the intensity of scattered neutrons in various directions.

neutron diffractometer [PHYS] A diffractometer in which a beam of neutrons is used for diffraction analysis, and the intensities of the diffracted beams at different angles are measured with an ionization chamber or radiation counter.

neutron excess [NUC PHYS] The number of neutrons in a nucleus in excess of the number of protons. Also known as difference number; isotopic number.

neutron flux [NUCLEO] The intensity of neutron radiation, expressed as the number of neutrons passing through a unit area per unit time. Also known as neutron flux density.

neutron flux density See neutron flux.

neutron hardening [NUCLEO] The increase which occurs in the average energy of thermal neutrons diffusing in a medium whose absorption cross section decreases as energy increases.

neutron howitzer [NUCLEO] A device which produces a neutron beam; consists of a neutron source contained in a block of moderating material with a small hole from the source to the surface of the block, from which the beam emerges.

neutron inelastic scattering reactions [NUCLEO] Emissions of low-energy neutrons when materials have been bombarded by fast neutrons during neutron activation analysis. Also known as n-n' reactions.

neutron irradiation [NUCLEO] The exposure of a material or object to neutrons.

neutron magnetic moment [NUC PHYS] A vector whose scalar product with the magnetic flux density gives the negative of the energy of interaction of a neutron with a magnetic field.

neutron multiplication factor See multiplication factor.

neutron number [NUC PHYS] The number of neutrons in the nucleus of an atom.

neutron optics [PHYS] The study of certain phenomena, for example, crystal diffraction, in which the wave character of neutrons dominates and leads to behavior similar to that of light.

neutron radiative capture See neutron capture.

neutron radiography [NUCLEO] Radiography that uses a neutron beam generated by a nuclear reactor; the neutrons are detected by placing a conventional x-ray film next to a converter screen composed of potentially radioactive materials or prompt emission materials which convert the neutron radiation to other types of radiation more easily detected by the film.

neutron reflection [PHYS] Specular reflection of neutrons, either from lattice planes of crystalline substances according to the Bragg law for their de Broglie wavelength, or from highly polished surfaces of certain substances at an angle smaller than their critical angle.

neutron scattering [NUCLEO] The change in direction of neutrons caused by collision with nuclei in a material.

neutron spectrometer [NUCLEO] An instrument used to determine the energies of neutrons and the relative intensities of neutrons of different energies in a neutron beam.

neutron spectrometry [NUC PHYS] A method of observing excited states of nuclei in which neutrons are used to bombard a target, causing nuclei to be transmuted into excited states by various nuclear reactions; the resultant excited states are determined by observing resonances in the reaction cross sections or by observing spectra of emitted particles or gamma rays. Also known as neutron spectroscopy.

neutron spectroscopy See neutron spectrometry.

neutron spectrum [NUC PHYS] A plot or display of the

number of neutrons at various energies, such as the neutrons emitted in a nuclear reaction, or the neutrons in a nuclear reactor.

neutron star [ASTRON] A star that is supposed to occur in the final stage of stellar evolution; it consists of a superdense mass mainly of neutrons, and has a strong gravitational attraction from which only neutrinos and high-energy photons could escape so that the star is invisible.

neutron transport theory [NUCLEO] Theory of diffusion of neutrons in a material based on the Boltzmann transport equation.

neutron velocity selector [NUCLEO] An instrument that isolates and detects neutrons having a particular range of velocities.

Nevanlinna class of functions [MATH] The algebra of functions f that are analytic on the open unit disk and such that the function

$$T(r,f) = (2\pi)^{-1} \int_0^{2\pi} \log^+ |f(re^{it})| \, dt$$

is bounded for $0 \leq r < 1$, where $\log^+$ is the positive part of the logarithmic function.

new achromat [OPTICS] An achromatic lens in which the component lenses are made of glasses chosen from a relatively broad selection, among which refractive index and dispersive power may vary inversely, permitting better correction of optical errors.

new candle *See* candela.

Newman-Penrose formalism [RELAT] A formalism in general relativity, based on spinor analysis, convenient for dealing with gravitational perturbations of space-time.

new moon [ASTRON] The moon at conjunction, when little or none of it is visible to an observer on the earth because the illuminated side is turned away.

newton [MECH] The unit of force in the meter-kilogram-second system, equal to the force which will impart an acceleration of 1 meter/second^2 to the International Prototype Kilogram mass. Symbolized N. Formerly known as large dyne.

Newton-Cotes formulas [MATH] Approximation formulas for the integral of a function along a small interval in terms of the values of the function and its derivatives.

Newton formula for the stress *See* Newtonian friction law.

Newtonian attraction [MECH] The mutual attraction of any two particles in the universe, as given by Newton's law of gravitation.

Newtonian-Cassegrain telescope [OPTICS] A modification of a Cassegrain telescope in which the light reflected from the hyperboloidal secondary mirror is again reflected from a diagonal plane mirror and focused at a point on the side of the telescope, avoiding the need to pierce the primary mirror and making the eyepiece more accessible. Also known as Cassegrain-Newtonian telescope.

Newtonian flow [FL MECH] Flow system in which the fluid performs as a Newtonian fluid, that is, shear stress is proportional to shear rate.

Newtonian fluid [FL MECH] A simple fluid in which the state of stress at any point is proportional to the time rate of strain at that point; the proportionality factor is the viscosity coefficient.

Newtonian focus [OPTICS] The position in a Newtonian telescope at which the image is formed, located at the side of the tube near its open end.

Newtonian friction law [FL MECH] The law that shear stress in a fluid is proportional to the shear rate; it holds only for some fluids, which are then called Newtonian. Also known as Newton formula for the stress.

Newtonian mechanics [MECH] The system of mechanics based upon Newton's laws of motion in which mass and energy are considered as separate, conservative, mechanical properties, in contrast to their treatment in relativistic mechanics.

Newtonian potential [PHYS] A potential which is associated with an inverse square law of force (such as an electrostatic force), and therefore varies with distance in the same manner as a gravitational potential.

Newtonian reference frame [MECH] One of a set of reference frames with constant relative velocity and within which Newton's laws hold; the frames have a common time, and coordinates are related by the Galilean transformation rule.

Newtonian speed of sound [ACOUS] An approximation to the speed of sound in a perfect gas given by the relation $c^2 = p/\rho$, where p is pressure and ρ the density.

Newtonian telescope [OPTICS] A reflecting telescope in which the light reflected from a concave mirror is reflected again by a plane mirror making an angle of 45° with the telescope axis, so that it passes through a hole in the side of the telescope containing the eyepiece.

Newtonian velocity [MECH] The velocity of an object in a Newtonian reference frame, S, which can be determined from the velocity of the object in any other such frame, S', by taking the vector sum of the velocity of the object in S' and the velocity of the frame S' relative to S.

Newtonian viscosity [FL MECH] The viscosity of a Newtonian fluid.

newton-meter of energy *See* joule.

newton-meter of torque [MECH] The unit of torque in the meter-kilogram-second system, equal to the torque produced by 1 newton of force acting at a perpendicular distance of 1 meter from an axis of rotation. Abbreviated N-m.

Newton-Raphson formula [MATH] If c is an approximate value of a root of the equation $f(x) = 0$, then a better approximation is the number $c - [f(c)/f'(c)]$.

Newton's equations of motion [MECH] Newton's laws of motion expressed in the form of mathematical equations.

Newton's first law [MECH] The law that a particle not subjected to external forces remains at rest or moves with constant speed in a straight line. Also known as first law of motion; Galileo's law of inertia.

Newton's identities [MATH] Relations between the coefficients of a polynomial and quantities s_k defined by $s_k = r_1{}^k + r_2{}^k + \cdots + r_n{}^k$, where $r_1, \ldots, r_n$ are the roots of the polynomial; for the polynomial $a_0 x^n + a_1 x^{n-1} + \cdots + a_n$, the relations are $a_0 s_k + a_1 s_{k-1} + \cdots + a_{k-1} s_1 + k a_k = 0$, for $k < n$, and $a_0 s_k + a_1 s_{k-1} + \cdots + a_n s_{k-n} = 0$, for $k > n$.

Newton's interpolation formula [MATH] One of two formulas used to estimate the value of a function at an intermediate value of the independent variable when its value is known at a series of equally spaced points (such as those that appear in a table), in terms of the forward differences of the function or in terms of the backward differences of the function; if $y_i = f(x_0 + ih)$, with i running over the integers, are the known values of the function, the formulas state that $f(x_0 + uh)$ is approximated by a series whose kth term is $u(u - 1) \cdots (u - k) \Delta^k y_0/k!$, or by a series whose kth term is $u(u + 1) \cdots (u + k) \nabla^k y_0/k!$, where Δ and ∇ are the forward and backward difference operators.

Newton's law of cooling [THERMO] The law that the rate of heat flow out of an object by both natural convection and radiation is proportional to the temperature difference be-

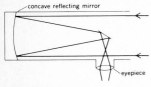

NEWTONIAN TELESCOPE

concave reflecting mirror

eyepiece

Simplified diagram of a Newtonian telescope.

tween the object and its environment, and to the surface area of the object.

Newton's law of gravitation [MECH] The law that every two particles of matter in the universe attract each other with a force that acts along the line joining them, and has a magnitude proportional to the product of their masses and inversely proportional to the square of the distance between them. Also known as law of gravitation.

Newton's law of resistance [FL MECH] The law that the force opposing the motion of an object through a fluid at moderate velocities is proportional to the square of the velocity.

Newton's laws of motion [MECH] Three fundamental principles (called Newton's first, second, and third laws) which form the basis of classical, or Newtonian, mechanics, and have proved valid for all mechanical problems not involving speeds comparable with the speed of light and not involving atomic or subatomic particles.

Newton's lens formula [OPTICS] The product of the distances of two conjugate points from the respective principal foci of a lens or mirror is equal to the square of the focal length.

Newton's method [MATH] A technique to approximate the roots of an equation by the methods of the calculus.

Newton's rings [OPTICS] A series of circular bright and dark bands which appear about the point of contact between a glass plate and a convex lens which is pressed against it and illuminated with monochromatic light.

Newton's second law [MECH] The law that the acceleration of a particle is directly proportional to the resultant external force acting on the particle and is inversely proportional to the mass of the particle. Also known as second law of motion.

Newton's square-root method [MATH] A technique for the estimation of the roots of an equation exhibiting faster convergence than Newton's method; this involves calculus methods and the square-root function.

Newton's theory of lift [FL MECH] A theory of the forces acting on an airfoil in a fluid current in which these forces are assumed to result from the impact of particles of the fluid on the body.

Newton's theory of light *See* corpuscular theory of light.

Newton's third law [MECH] The law that, if two particles interact, the force exerted by the first particle on the second particle (called the action force) is equal in magnitude and opposite in direction to the force exerted by the second particle on the first particle (called the reaction force). Also known as law of action and reaction; third law of motion.

Neyman-Pearson theory [STAT] A theory that determines what is the best test to use to examine a statistical hypothesis.

NFE method *See* nearly free electron method.

ng *See* nanogram.

N galaxy [ASTRON] A galaxy that has the optical appearance of a strongly concentrated object with a semistellar nucleus surrounded by a faint halo or extension.

Ni *See* nickel.

nicad battery *See* nickel-cadmium battery.

Nichol's chart [CONT SYS] A plot of curves along which the magnitude M or argument α of the frequency control ratio is constant on a graph whose ordinate is the logarithm of the magnitude of the open-loop transfer function, and whose abscissa is the open-loop phase angle.

Nicholson's hydrometer [ENG] A modification of Fahrenheit's hydrometer in which the lower end of the instrument carries a scale pan to permit the determination of the relative density of a solid.

Nichols radiometer [ENG] An instrument, used to measure the pressure exerted by a beam of light, in which there are two

NICKEL

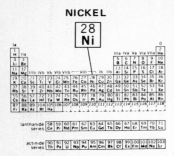

Periodic table of the chemical elements showing the position of nickel.

NICOL PRISM

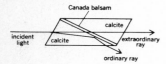

Drawing of path of incident light through a Nicol prism. The extraordinary ray is plane-polarized.

NIOBIUM

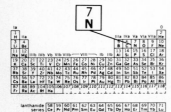

Periodic table of the chemical elements showing the position of niobium.

NITROGEN

Periodic table of the chemical elements showing the position of nitrogen.

small, silvered glass mirrors at the ends of a light rod that is suspended at the center from a fine quartz fiber within an evacuated enclosure.

nickel [CHEM] A chemical element, symbol Ni, atomic number 28, atomic weight 58.71.

nickel-63 [NUC PHYS] Radioactive nickel with beta radiation and 92-year half-life; derived by pile-irradiation of nickel; used in radioactive composition studies and tracer studies.

nickel-cadmium battery [ELEC] A sealed storage battery having a nickel anode, a cadmium cathode, and an alkaline electrolyte; widely used in cordless appliances; without recharging, it can serve as a primary battery. Also known as cadmium-nickel storage cell; nicad battery.

nickel-iron battery See Edison battery.

Nicol prism [OPTICS] A device for producing plane-polarized light, consisting of two pieces of transparent calcite (a birefringent crystal) which together form a parallelogram and are cemented together with Canada balsam.

night [ASTRON] The period of darkness between sunset and sunrise.

nightglow [GEOPHYS] A subdivision of airglow in which energy comes from reactions of atomic oxygen between 70 and 100 kilometers, and from ionic recombination around 300 kilometers.

night-sky light See airglow.

night-sky luminescence See airglow.

nighttime visual range See night visual range.

night visual range [OPTICS] The greatest distance at which a point source of light of a given candlepower can be perceived at night by an observer under given atmospheric conditions. Also known as nighttime visual range; penetration range; transmission range.

nile [NUCLEO] A unit of reactivity; the reactivity of a reactor in niles is equal to $100(1-1/k)$, where k is the effective multiplication factor.

nilmanifold [MATH] The factor space of a connected nilpotent Lie group by a closed subgroup.

nilpotent [MATH] An element of some algebraic system which vanishes when raised to a certain power.

nine-j-symbol [QUANT MECH] A coefficient used in the general recoupling of four angular momenta, as in a transformation from L-S to j-j coupling in a two-electron system. Also known as X coefficient.

nine-point circle [MATH] The circle that passes through the midpoints of the sides of a triangle, the feet of the perpendiculars from the vertices to the opposite sides, and the midpoints of the line segments between the vertices and the orthocenter.

niobium [CHEM] A chemical element, symbol Nb, atomic number 41, atomic weight 92.906.

nit [OPTICS] A unit of luminance, equal to 1 candela per square meter. Abbreviated nt.

nitrogen [CHEM] A chemical element, symbol N, atomic number 7, atomic weight 14.0067; it is a gas, diatomic (N_2) under normal conditions; about 78% of the atmosphere is N_2; in the combined form the element is a constituent of all proteins.

nitrogen cycle See carbon-nitrogen cycle.

N-level logic [ELECTR] An arrangement of gates in a digital computer in which not more than N gates are connected in series.

N line [SPECT] One of the characteristic lines in an atom's x-ray spectrum, produced by excitation of an N electron.

N-m See newton-meter of torque.

NMR See nuclear magnetic resonance.

n-net [MATH] A finite net in which *n* lines pass through each point.

nn junction [ELECTR] In a semiconductor, a region of transition between two regions having different properties in *n*-type semiconducting material.

n-n′ reactions *See* neutron inelastic scattering reactions.

No *See* nobelium.

nobelium [CHEM] A chemical element, symbol No, atomic number 102, atomic weight 254 when the element is produced in the laboratory; a synthetic element, in the actinium series.

noble gas [CHEM] A gas in group 0 of the periodic table of the elements; it is monatomic and, with limited exceptions, chemically inert. Also known as inert gas.

noble potential [PHYS CHEM] A potential equaling or approaching that of the noble elements, such as gold, silver, or copper, of the electromotive series.

nocturnal radiation *See* effective terrestrial radiation.

nodal line [ASTRON] The line passing through the ascending and descending nodes of the orbit of a celestial body.   [PHYS] **1.** A line or curve in a two-dimensional standing-wave system, such as a vibrating diaphragm, where some specified characteristic of the wave, such as velocity of pressure, does not oscillate. **2.** A line which remains fixed during some deformation or rotation of a body or coordinate system.

nodal points [ELEC] Junction points in a transmission system; the automatic switches and switching centers are the nodal points in automated systems. [OPTICS] A pair of points on the axis of an optical system such that an incident ray passing through one of them results in a parallel emergent ray passing through the other.

node [ASTRON] **1.** One of two points at which the orbit of a planet, planetoid, or comet crosses the plane of the ecliptic. **2.** One of two points at which a satellite crosses the equatorial plane of its primary. [ELEC] *See* branch point. [ELECTR] A junction point in a network. [MATH] *See* crunode. [PHYS] A point, line, or surface in a standing-wave system where some characteristic of the wave has essentially zero amplitude.

node voltage [ELEC] The voltage at a given point in an electric network with respect to that at a node.

nodical month [ASTRON] The average period of revolution of the moon about the earth with respect to the moon's ascending node, a period of 27 days 5 hours 5 minutes 35.8 seconds, or approximately 27¼ days. Also known as draconitic month.

Noetherian module [MATH] A module in which every nested family of submodules possesses only a finite number of members.

Noetherian ring [MATH] A ring in which every nested family of ideals possesses only a finite number of members.

no-hair theorems [RELAT] Popular name for general relativistic theorems proving that black holes are uniquely described by their mass, charge, and angular momentum.

noise [ACOUS] Sound which is unwanted, either because of its effect on humans, its effect on fatigue or malfunction of physical equipment, or its interference with the perception or detection of other sounds.   [ELEC] Interfering and unwanted currents or voltages in an electrical device or system.

noise analysis [PHYS] Determination of the frequency components that make up a particular noise being studied.

noise analyzer [ELECTR] A device used for noise analysis.

noise factor [ELECTR] The ratio of the total noise power per unit bandwidth at the output of a system to the portion of the noise power that is due to the input termination, at the

NOBELIUM

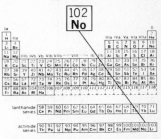

Periodic table of the chemical elements showing the position of nobelium.

NOISE-POWER MEASUREMENT

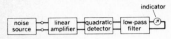

Diagram of setup for
noise-power measurement.

NOMARSKI MICROSCOPE

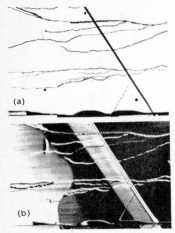

(a)

(b)

Application of the Nomarski
interference microscope to
a metallurgical specimen.
(a) Cleavage surface of a zinc
crystal, showing mechanical
twins and irregular cleavage
steps; vertical illumination.
(b) As *a* with slight surface
tilts ($\sim 1°$), much fine detail
is revealed between the large
cleavage steps. Magnification
×60. *(Courtesy of A. R. Entwisle)*

NOMOGRAPH

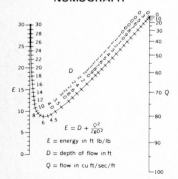

$$E = D + \frac{Q^2}{2gD^2}$$

E = energy in ft lb/lb

D = depth of flow in ft

Q = flow in cu ft/sec/ft

Nomograph for energy content
of a rectangular channel with
uniform flow.

standard noise temperature of 290 K. Also known as noise
figure.

noise figure   *See* noise factor.

noise generator   [ELECTR] A device which produces (usually
random) electrical noise, for use in tests of the response of
electrical systems to noise, and in measurements of noise
intensity. Also known as noise source.

noise level   [PHYS] The intensity of unwanted sound, or the
magnitude of unwanted currents or voltages, averaged over a
specified frequency range and time interval, and weighted
with frequency in a specified manner; usually expressed in
decibels relative to a specified reference.

noise-power measurement   [ELECTR] Measurement of the
power carried by electrical noise averaged over some brief
interval of time, usually by amplifying noise from the source
in a linear amplifier and then using a quadratic detector
followed by a low-pass filter and an indicating device.

noise rating number   [ACOUS] The perceived noise level of
the noise that can be tolerated under specified conditions; for
example, the noise rating number of a bedroom is 25, that of
a workshop is 65.

noise reduction coefficient   [ACOUS] The average over the
logarithm of frequency, in the frequency range from 256 to
2048 hertz inclusive, of the sound absorption coefficient of a
material.

noise source   *See* noise generator.

noise temperature   [ELEC] The temperature at which the
thermal noise power of a passive system per unit bandwidth
would be equal to the actual noise at the actual terminals; the
standard reference temperature for noise measurements is
290 K.

Nomarski microscope   [OPTICS] A type of interference microscope that is used to study reflecting specimens, such as
metallic surfaces or metallized replicas of surfaces, and that
gives a true relief image uncomplicated by variations in
refractive index.

nomenclature   [SCI TECH] A systematic arrangement of the
distinctive names employed in any science.

nominal scale measurement   [STAT] A method for sorting
objects into categories according to some distinguishing
characteristic and attaching a name or label to each category;
considered the weakest type of measurement.

nominal value   [ELEC] The value of some property (such as
resistance, capacitance, or impedance) of a device at which it
is supposed to operate, under normal conditions, as opposed
to actual value.

nomogram   *See* nomograph.

nomograph   [MATH] A chart which represents an equation
containing three variables by means of three scales so that a
straight line cuts the three scales in values of the three
variables satisfying the equation. Also known as abac; alignment chart; nomogram.

nonagon   [MATH] A nine-sided polygon.

nonanticipatory system   *See* causal system.

non-Archimedean valuation   [MATH] A real-valued function
v on a field K such that (1) $v(x) > 0$ for all x in K and $v(x) =$
0 if and only if $x = 0$, (2) for all x and y in K, $v(xy) = v(x)v(y)$,
and (3) for all x and y in K, $v(x + y) \leq$ max $[v(x),v(y)]$.

nonassociative algebra   [MATH] A generalization of the concept of an algebra; it is a nonassociative ring R which is a
vector space over a field F satisfying $a(xy) = (ax)y = x(ay)$ for
all a in F and x and y in R.

nonassociative ring   [MATH] A generalization of the concept
of a ring; it is an algebraic system with two binary operations
called addition and multiplication such that the system is a

commutative group relative to addition, and multiplication is distributive with respect to addition, but multiplication is not assumed to be associative.

nonatomic Boolean algebra [MATH] A Boolean algebra in which there is no element x with the property that if $y \cdot x = y$ for some y, then $y = 0$.

nonatomic measure space [MATH] A measure space in which no point has positive measure.

nonblackbody [THERMO] A body that reflects some fraction of the radiation incident upon it; all real bodies are of this nature.

noncentral chi-square distribution [STAT] The distribution of the sum of squares of independent normal random variables, each with unit variance and nonzero mean; used to determine the power function of the chi-square test.

noncentral distribution [STAT] A distribution of random variables which is not normal.

noncentral F distribution [STAT] The distribution of the ratio of two independent random variables, one with a noncentral chi-square distribution and one with a central chi-square distribution; used to determine the power of the F test in the analysis of variance.

noncentral force [PHYS] A force between two particles, other than an attraction or repulsion, directed along the line connecting them; for example, a tensor force between two nucleons.

noncentral quadric [MATH] A quadric surface that does not have a point about which the surface is symmetrical; namely, an elliptic or hyperbolic paraboloid, or a quadric cylinder.

noncentral t distribution [STAT] A particular case of a noncentral F distribution; used to test the power of the t test.

noncontacting piston *See* choke piston.

noncontacting plunger *See* choke piston.

noncritical region [STAT] The set of sample values in testing hypotheses that lead to acceptance of the hypotheses tested.

noncrossing rule [PHYS CHEM] The rule that when the potential energies of two electronic states of a diatomic molecule are plotted as a function of distance between the nuclei, the resulting curves do not cross, unless the states have different symmetry.

nondegenerate bilinear form [MATH] A bilinear form, the sum of terms of the form $a_{ij}x_i y_j$, $i = 1, 2, \ldots, m, j = 1, 2, \ldots, n$, is nondegenerate if, for any values of the variables $x_1, \ldots, x_m$ not all equal to 0, values of the variables $y_1, \ldots, y_n$ can be found such that the sum is not 0 and, for any values of $y_1, \ldots, y_n$ not all 0 values of $x_1, \ldots, x_m$ can be found such that the sum is not 0.

nondegenerate form [MATH] A bilinear form $f(x,y)$ such that if, for a given element $a, f(a,y) = 0$ for all y, then $a = 0$, and if, for a given element $b, f(x,b) = 0$ for all x, then $b = 0$.

nondegenerate plane [MATH] In projective geometry, a plane in which to every line L there are at least two distinct points that do not lie on L, and to every point p there are at least two distinct lines which do not pass through p.

nondegenerative basic feasible solution [ADP] In linear programming, a basic feasible solution with exactly m positive variables x_i, where m is the number of constraint equations.

nondenumerable set [MATH] A set that cannot be put into one-to-one correspondence with the positive integers or any subset of the positive integers.

nondestructive breakdown [ELECTR] Breakdown of the barrier between the gate and channel of a field-effect transistor without causing failure of the device; in a junction field-effect transistor, avalanche breakdown occurs at the pn junction.

nondestructive testing [ENG] Any testing method which does not involve damaging or destroying the test sample; includes use of x-rays, ultrasonics, radiography, and magnetic flux.

nondeterministic [SCI TECH] Unpredictable in terms of observable antecedents and known laws; this is a relative term pertaining to a given state of knowledge but not necessarily implying ultimate unpredictability.

nondeviated absorption [PHYS] Absorption that occurs without any appreciable slowing up of waves.

nondimensional parameter *See* dimensionless number.

nonequilibrium thermodynamics [THERMO] A quantitative treatment of irreversible processes and of rates at which they occur. Also known as irreversible thermodynamics.

noneuclidean geometry [MATH] A geometry in which one or more of the axioms of euclidean geometry are modified or discarded.

nonholonomic constraints [MATH] A nonintegrable set of differential equations which describe the restrictions on the motion of a system or in optimization.

nonholonomic system [MECH] A system of particles which is subjected to restraints of such a nature that the system cannot be described by independent coordinates; examples are a rolling hoop, or an ice skate which must point along its path.

nonhoming [CONT SYS] Not returning to the starting or home position, as when the wipers of a stepping relay remain at the last-used set of contacts instead of returning to their home position.

noninductive capacitor [ELEC] A capacitor constructed so it has practically no inductance; foil layers are staggered during winding, so an entire layer of foil projects at either end for contact-making purposes; all currents then flow laterally rather than spirally around the capacitor.

noninductive resistor [ELEC] A wire-wound resistor constructed to have practically no inductance, either by using a hairpin winding or be reversing connections to adjacent sections of the winding.

noninductive winding [ELEC] A winding constructed so that the magnetic field of one turn or section cancels the field of the next adjacent turn or section.

noninteracting control [CONT SYS] A feedback control in a system with more than one input and more than one output, in which feedback transfer functions are selected so that each input influences only one output.

noninverting amplifier [ELECTR] An operational amplifier in which the input signal is applied to the ungrounded positive input terminal to give a gain greater than unity and make the output voltage change in phase with the input voltage.

nonlinear [PHYS] Pertaining to a response which is other than directly or inversely proportional to a given variable.

nonlinear amplifier [ELECTR] An amplifier in which a change in input does not produce a proportional change in output.

nonlinear capacitor [ELEC] Capacitor having a mean charge characteristic or a peak charge characteristic that is not linear, or a reversible capacitance that varies with bias voltage.

nonlinear circuit [ELEC] A circuit in which the current and voltage in any element that results from two sources of energy acting together is not equal to the sum of the currents or voltages that result from each of the sources acting alone.

nonlinear circuit component [ELECTR] An electrical device for which a change in applied voltage does not produce a proportional charge in current. Also known as nonlinear device; nonlinear element.

nonlinear coil [ELECTROMAG] Coil having an easily saturable core, possessing high impedance at low or zero current and low impedance when current flows and saturates the core.

nonlinear control [CONT SYS] An on-line process control which can stabilize operations by means of nonlinear compensatory functions.

nonlinear crystal [SOLID STATE] A crystal in which some influence (such as stress, electric field, or magnetic field) produces a response (such as strain, electric polarization, or magnetization) which is not proportional to the influence.

nonlinear device *See* nonlinear circuit component.

nonlinear dielectric [ELEC] A dielectric whose polarization is not proportional to the applied electric field.

nonlinear distortion [ELECTR] Distortion in which the output of a system or component does not have the desired linear relation to the input.

nonlinear element *See* nonlinear circuit component.

nonlinear equation [MATH] An equation in variables $x_1, \ldots, x_n, y$ which cannot be put into the form $a_1 x_1 + \cdots + a_n x_n = y$.

nonlinear feedback control system [CONT SYS] Feedback control system in which the relationships between the pertinent measures of the system input and output signals cannot be adequately described by linear means.

nonlinear inductance [ELEC] The behavior of an inductor for which the voltage drop across the inductor is not proportional to the rate of change of current, such as when the inductor has a core of magnetic material in which magnetic induction is not proportional to magnetic field strength.

nonlinear material [PHYS] A material in which some specified influence (such as stress, electric field, or magnetic field) produces a response (such as strain, electric polarization, or magnetization) which is not proportional to the influence.

nonlinear network [ELEC] A network in which the current or voltage in any element that results from two sources of energy acting together is not equal to the sum of the currents or voltages that result from each of the sources acting alone.

nonlinear optics [OPTICS] The study of the interaction of radiation with matter in which certain variables describing the response of the matter (such as electric polarization or power absorption) are not proportional to variables describing the radiation (such as electric field strength or energy flux).

nonlinear programming [MATH] A branch of applied mathematics concerned with finding the maximum or minimum of a function of several variables, when the variables are constrained to yield values of other functions lying in a certain range, and either the function to be maximized or minimized, or at least one of the functions whose value is constrained, is nonlinear.

nonlinear reactance [ELECTR] The behavior of a coil or capacitor whose voltage drop is not proportional to the rate of change of current through the coil, or the charge on the capacitor.

nonlinear resistance [ELECTR] The behavior of a substance (usually a semiconductor) which does not obey Ohm's law but has a voltage drop across it that is proportional to some power of the current.

nonlinear system [MATH] A system in which the interrelationships among the quantities involved are expressed by equations, some of which are not linear. [SCI TECH] A system in which outputs are not linear functions of vectors whose components represent the inputs.

nonlinear taper [ELEC] Nonuniform distribution of resistance throughout the element of a potentiometer or rheostat.

nonlinear vibration [MECH] A vibration whose amplitude is large enough so that the elastic restoring force on the vibrating object is not proportional to its displacement.

nonlinear viscoelasticity [FL MECH] The behavior of a fluid which does not obey a first-order differential equation in stress and strain.

nonloaded Q [ELEC] Of an electric impedance, the Q value of the impedance without external coupling or connection. Also known as basic Q.

nonlocalized electron [PHYS] An electron whose wave function is not confined to the vicinity of one or two nuclei, but is spread out over a molecule or a crystal lattice.

nonmagnetic [ELECTROMAG] Not magnetizable, and therefore not affected by magnetic fields.

non-minimum-phase system [CONT SYS] A linear system whose transfer function has one or more poles or zeros with positive, nonzero real parts.

nonnegative semidefinite *See* positive semidefinite.

non-Newtonian fluid [FL MECH] A fluid whose flow behavior departs from that of a Newtonian fluid, so that the rate of shear is not proportional to the corresponding stress. Also known as non-Newtonian system.

non-Newtonian fluid flow [FL MECH] The flow behavior of non-Newtonian fluids, whose study has applications in many important problems of practical significance such as flow in tubes, extrusion, flow through dies, coating operations, rolling operations, and mixing of fluids.

non-Newtonian system *See* non-Newtonian fluid.

non-Newtonian viscosity [FL MECH] The behavior of a fluid which, when subjected to a constant rate of shear, develops a stress which is not proportional to the shear. Also known as anomalous viscosity.

nonparametric statistics [STAT] A class of statistical methods applicable to a large set of probability distributions used to test for correlation, location, independence, and so on.

nonpolar [CHEM] Pertaining to an element or compound which has no permanent electric dipole moment.

nonprobabilistic sampling [STAT] A process in which some criterion other than the laws of probability determines the elements of the population to be included in the sample.

nonquantum mechanics [MECH] The classical mechanics of Newton and Einstein as opposed to the quantum mechanics of Heisenberg, Schrödinger, and Dirac; particles have definite position and velocity, and they move according to Newton's laws.

nonreactive [ELEC] Pertaining to a circuit, component, or load that has no capacitance or impedance, so that an alternating current is in phase with the corresponding voltage.

nonrelativistic approximation [PHYS] The approximation in which it is assumed that speeds of objects are small compared to the speed of light.

nonrelativistic kinematics [MECH] The study of motions of systems of objects at speeds which are small compared to the speed of light, without reference to the forces which act on the system.

nonrelativistic mechanics [MECH] The study of the dynamics of systems in which all speeds are small compared to the speed of light.

nonrelativistic particle [RELAT] A particle whose velocity is small with respect to that of light.

nonrelativistic quantum mechanics [QUANT MECH] The modern theory of matter and its interaction with radiation, applicable to systems of material particles which move slowly compared to the speed of light, which are neither created or destroyed, and whose internal structure (except for spin)

NON-NEWTONIAN FLUID

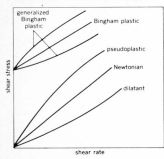

Typical flow curves, giving rate of shear as a function of shear stress, for a Newtonian fluid and for the three types of time-independent non-Newtonian fluid: Bingham plastic, pseudoplastic fluid, and dilatant fluid.

either does not change or is irrelevant to the description of the system.

nonremovable discontinuity [MATH] A point at which a function is not continuous or is undefined, and cannot be made continuous by being given a new value at the point.

nonresonant antenna [ELECTROMAG] A long-wire or traveling-wave antenna which does not have natural frequencies of oscillation and responds equally well to radiation over a broad range of frequencies.

nonresonant line [ELECTROMAG] Transmission line having no reflected waves and neither current nor voltage standing waves.

nonselective radiator *See* graybody.

nonsingular matrix [MATH] A matrix which has an inverse; equivalently, its determinant is not 0.

nonsingular transformation [MATH] A linear transformation which has an inverse; equivalently, it has null space kernel consisting only of the zero vector.

nonsinusoidal waveform [ELEC] The representation of a wave which does not vary in a sinusoidal manner and therefore contains harmonics.

nonthermal decimetric emission [ELECTROMAG] A radio-wave emission above the 4-centimeter wavelength from the planet Jupiter that has a nearly constant flux between 5-centimeter and 1-meter wavelength. Also known as DIM.

nonthermal radiation [PHYS] Electromagnetic radiation emitted by accelerated charged particles that are not in thermal equilibrium; aurora light and fluorescent-lamp light are examples.

nontrivial solution [MATH] A solution to a homogeneous equation which is not the zero solution.

nonuniform flow [FL MECH] Fluid flow which does not have the same velocity at all points in a medium, at a given instant.

nonviscous flow *See* inviscid flow.

nonviscous fluid *See* inviscid fluid.

noon [ASTRON] The instant at which a time reference is over the upper branch of the reference meridian.

noon interval [ASTRON] The predicted time interval between a given instant, usually the time of a morning observation, and local apparent noon; it is used to predict the time for observing the sun on the celestial meridian.

NOR [MATH] A logic operator having the property that if P, Q, R, . . . are statements, then the NOR of P, Q, R, . . . is true if all statements are false, false if at least one statement is true. Derived from NOT-OR. Also known as Peirce stroke relationship.

Nordheim's rule [SOLID STATE] The rule that the residual resistivity of a binary alloy that contains mole fraction x of one element and $1 - x$ of the other is proportional to $x(1 - x)$.

norm [MATH] **1.** A scalar valued function on a vector space with properties analogous to those of the modulus of a complex number; namely: the norm of the zero vector is zero, all other vectors have positive norm, the norm of a scalar times a vector equals the absolute value of the scalar times the norm of the vector, and the norm of a sum is less than or equal to the sum of the norms. **2.** Let E be a finite extension of a field F, and let a be an element of E; the norm of a from E to F is the element

$$\left[\prod_{j=1}^{p} f_j(a) \right]^q$$

where p is the separable degree of E over F, q is the inseparable degree of E over F, and f_j, $j = 1, \ldots, p$, are the distinct embeddings of E in $\hat{F}$ over F, where $\hat{F}$ is the algebraic closure of F. [QUANT MECH] **1.** The square of the modulus of a Schrödinger-Pauli wave function, integrated over the space coordinates and summed over the spin coordinates of the particles it describes. **2.** The square root of this quantity.

normal acceleration [MECH] **1.** The component of the linear acceleration of an aircraft or missile along its normal, or Z, axis. **2.** The usual or typical acceleration.

normal barometer [ENG] A barometer of such accuracy that it can be used for the determination of pressure standards; an instrument such as a large-bore mercury barometer is usually used.

normal bundle [MATH] If A is a manifold and B is a submanifold of A, then the normal bundle of B in A is the set of pairs (x,y), where x is in B, y is a tangent vector to A, and y is orthogonal to B.

normal coordinates [MECH] A set of coordinates for a coupled system such that the equations of motion each involve only one of these coordinates.

normal curvature [MATH] The normal curvature at a point on a surface is the curvature of the normal section to the point.

normal curve *See* Gaussian curve.

normal density function [STAT] A normally distributed frequency distribution of a random variable x with mean e and variance σ is given by $(1/\sigma\sqrt{2\pi})\exp[-(x-e)^2/2\sigma^2]$.

normal derivative [MATH] The directional derivative of a function at a point on a given curve or surface in the direction of the normal to the curve or surface.

normal distribution [STAT] The most commonly occurring probability distributions have the form

$$(1/\sigma\sqrt{2\pi})\int_{-\infty}^{u}\exp(-u^2/2)\,du,\ u = (x-e)/\sigma,$$

where e is the mean and σ is the variance. Also known as Gauss' error curve; Gaussian distribution.

normal divisor *See* normal subgroup.

normal equations [STAT] The set of equations arising in the least squares method whose solutions give the constants that determine the shape of the estimated function.

normal extension [MATH] An algebraic extension of K of a field k, contained in the algebraic closure $\bar{k}$ of k, such that every injective homomorphism of K into $\bar{k}$, inducing the identity on k, is an automorphism of K; equivalently, K is the splitting field of a family of polynomials in k; equivalently, every irreducible polynomial in k which has a root in K splits into linear factors in K.

normal family [MATH] A family of complex functions analytic in a common domain where every sequence of these functions has a subsequence converging uniformly on compact subsets of the domain to an analytic function on the domain or to $+\infty$.

normal fluid [CRYO] The component of liquid helium II, postulated in the two-fluid theory, that has viscosity and behaves like an ordinary fluid.

normal frequencies [MECH] The frequencies of the normal modes of vibration of a system.

normal function *See* normalized function.

normal impact [MECH] **1.** Impact on a plane perpendicular to the trajectory. **2.** Striking of a projectile against a surface that is perpendicular to the line of flight of the projectile.

normal impedance *See* free impedance.

normal-incidence pyrheliometer [ENG] An instrument that measures the energy in the solar beam; it usually measures the radiation that strikes a target at the end of a tube equipped with a shutter and baffles to collimate the beam.

normal incidence reflectivity [ELECTROMAG] The ratio of the energy of electromagnetic radiation reflected from the interface between two media to the energy of the incident radiation when the incident radiation travels in a direction perpendicular to the surface.

normal induction [ELECTROMAG] Limiting induction, either positive or negative, in a magnetic material that is under the influence of a magnetizing force which varies between two specific limits.

normal inspection [IND ENG] The number of items inspected as specified by the sampling inspection plan at the outset; if the quality of the product improves, the number of units to be inspected is reduced; if quality deteriorates, the number of units inspected is increased.

normality [CHEM] Measure of the number of gram-equivalent weights of a compound per liter of solution. Abbreviated N.

normalize [MATH] To multiply a quantity by a suitable constant or scalar so that it then has norm one; that is, its norm is then equal to one. [QUANT MECH] To multiply a wave function by a constant so that its norm is equal to unity. [STAT] To carry out a normal transformation on a variate.

normalized admittance [ELECTROMAG] The reciprocal of the normalized impedance.

normalized coupling coefficient [ELECTROMAG] Mutual inductance, expressed on a scale running from zero to one.

normalized current [ELECTROMAG] The current divided by the square root of the characteristic admittance of a waveguide or transmission line.

normalized function [MATH] A function with norm one; the norm is usually given by an integral $(\int |f|^p \, d\mu)^{1/p}$, $1 \leq p < \infty$. Also known as normal function.

normalized impedance [ELECTROMAG] An impedance divided by the characteristic impedance of a transmission line or waveguide.

normalized Q [ELEC] The ratio of the reactive component of the impedance of a filter section to the resistive component.

normalized standard scores [STAT] A procedure in which each set of original scores is converted to some standard scale under the assumption that the distribution of scores approximates that of a normal.

normalized susceptance [ELECTROMAG] The susceptance of an element of a waveguide or transmission line divided by the characteristic admittance.

normalized voltage [ELECTROMAG] The voltage divided by the square root of the characteristic impedance of a waveguide or transmission line.

normalizer [MATH] The normalizer of a subset S of a group G is the subgroup of G consisting of all elements x such that xsx^{-1} is in S whenever s is in S.

normally distributed observations [STAT] Any set of observations whose histogram looks like the normal curve.

normal magnetization curve [ELECTROMAG] Curve traced on a graph of magnetic induction versus magnetic field strength in an originally unmagnetized specimen, as the magnetic field strength is increased from zero. Also known as magnetization curve.

normal map [MATH] A planar map in which no more than three regions meet any one point and no region completely encloses another. Also known as regular map.

normal matrix [MATH] A matrix is normal if multiplying it on the right by its adjoint is the same as multiplying it on the left.

normal-mode helix [ELECTROMAG] A type of helical antenna whose diameter and electrical length are considerably less

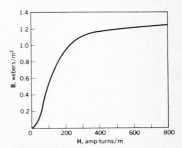

**NORMAL
MAGNETIZATION CURVE**

Normal magnetization curve; the magnetic induction is the ordinate and the magnetic field strength is the abscissa.

than a wavelength, and which has a radiation pattern with greatest intensity normal to the helix axis.

normal mode of vibration [MECH] Vibration of a coupled system in which the value of one of the normal coordinates oscillates and the values of all the other coordinates remain stationary.

normal operator [MATH] A linear operator where composing it with its adjoint operator in either order gives the same result. Also known as normal transformation.

normal permeability [ELECTROMAG] The permeability of a specimen whose magnetic induction and magnetic field strength lie on the normal magnetization curve. Also known as cyclic permeability.

normal plane [MATH] For a point P on a curve in space, the plane passing through P which is perpendicular to the tangent to the curve at P.

normal-plate anemometer [ENG] A type of pressure-plate anemometer in which the plate, restrained by a stiff spring, is held perpendicular to the wind; the wind-activated motion of the plate is measured electrically; the natural frequency of this system can be made high enough so that resonance magnification does not occur.

normal pressure *See* standard pressure.

normal probability paper [STAT] Graph paper with the abscissa ruled in uniform increments and the ordinate ruled in such a way that the plot of a cumulative normal distribution is a straight line.

normal reaction [MECH] The force exerted by a surface on an object in contact with it which prevents the object from passing through the surface; the force is perpendicular to the surface, and is the only force that the surface exerts on the object in the absence of frictional forces.

normal section [MATH] Relative to a surface, this is a planar section produced by a plane containing the normal to a point.

normal series [MATH] A normal series of a group G is a normal tower of subgroups of G, G_0, G_1, . . . , G_n in which $G_0 = G$ and G_n is the trivial group containing only the identity element.

normal space [MATH] A topological space in which any two disjoint closed sets may be covered respectively by two disjoint open sets.

normal state [NUC PHYS] A term sometimes used for ground state.

normal subgroup [MATH] A subgroup N of a group G where every expression $g^{-1}ng$ is in N for every g in G and every n in N. Also known as invariant subgroup; normal divisor.

normal surface [OPTICS] The surface that is generated by taking, at each point of the ray surface, the intersection of the tangent plane to the ray surface at that point with the perpendicular from the origin to this plane.

normal temperature and pressure *See* standard conditions.

normal to a curve [MATH] The normal to a curve at a point is the line perpendicular to the tangent line at the point.

normal to a surface [MATH] The normal to a surface at a point is the line perpendicular to the tangent plane at that point.

normal tower [MATH] A tower of subgroups, G_0, G_1, . . . , G_n, such that each G_{i+1} is normal in G_i, $i = 1, 2, . . . , n - 1$.

normal transformation [MATH] *See* normal operator. [STAT] A transformation on a variate that converts it into a variate which has a normal distribution.

normal uranium *See* native uranium.

normal volume *See* standard volume.

normed linear space [MATH] A vector space which has a norm defined on it. Also known as normed vector space.

normed vector space   *See* normed linear space.

norm of a matrix   [MATH] The square root of the sum of the squares of the moduli of the matrix entries.

Norris-Eyring reverberation formula   [ACOUS] The reverberation time of a chamber, in seconds, is equal to 0.05 times its volume, in cubic feet, divided by the product of its surface area, in square feet, and the negative of the natural logarithm of 1 minus the absorption coefficient averaged over the surface.

North American Nebula   [ASTRON] A cloud of dust and gas in the constellation Cygnus; the density of this gas and dust is possibly a thousand times greater than the average density of interstellar gas; a much denser cloud of dust between the nebula and earth obscures portions of the emission nebula to create the appearance of the "Gulf of Mexico" and the "Atlantic Ocean."

northbound node   *See* ascending node.

northern lights   *See* aurora borealis.

north pole   [ASTRON] The north celestial pole that indicates the zenith of the heavens when viewed from the north geographic pole.   [ELECTROMAG] The pole of a magnet at which magnetic lines of force are considered as leaving the magnet; the lines enter the south pole; if the magnet is freely suspended, its north pole points toward the north geomagnetic pole.   [GEOPHYS] The geomagnetic pole in the Northern Hemisphere, at approximately latitude 78.5°N, longitude 69°W. Also known as north magnetic pole; north geomagnetic pole.

Norton's theorem   [ELEC] The theorem that the voltage across an element that is connected to two terminals of a linear network is equal to the short-circuit current between these terminals in the absence of the element, divided by the sum of the admittances between the terminals associated with the element and the network respectively.

nose   [FL MECH] The dense, forward part of a turbidity current.

notation   [MATH] **1.** The use of symbols to denote quantities or operations. **2.** *See* positional notation.

notch antenna   [ELECTROMAG] Microwave antenna in which the radiation pattern is determined by the size and shape of a notch or slot in a radiating surface.

notch filter   [ELECTR] A band-rejection filter that produces a sharp notch in the frequency response curve of a system; used in television transmitters to provide attenuation at the low-frequency end of the channel, to prevent possible interference with the sound carrier of the next lower channel.

NOT circuit   [ELECTR] A logic circuit with one input and one output that inverts the input signal at the output; that is, the output signal is a logical 1 if the input signal is a logical 0, and vice versa. Also known as inverter circuit.

NOT function   [MATH] A logical operator having the property that if P is a statement, then the NOT of P is true if P is false, and false if P is true.

NOT-OR   *See* NOR.

nova   [ASTRON] A star that suddenly becomes explosively bright, the term is a misnomer because it does not denote a new star but the brightening of an existing faint star.

nowhere dense set   [MATH] A set in a topological space whose closure has empty interior.

nox   [OPTICS] A unit of illumination, used in measuring low-level illumination, equal to 10^{-3} lux.

noy   [ACOUS] A unit of perceived noisiness equal to the perceived noisiness of random noise occupying the frequency band 910–1090 hertz at a sound pressure level of 40 decibels above 0.0002 microbar; a sound that is *n* times as noisy as this

NOT CIRCUIT

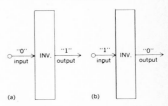

NOT circuit, also known as inverter circuit. *(a)* A "0" input produces a "1" output. *(b)* A "1" input produces a "0" output.

sound has a perceived noisiness of n noys, under the assumption that the perceived noisiness of a sound increases with physical intensity at the same rate as the loudness.

nozzle-divergence loss factor [FL MECH] The ratio between the momentum of the gases in a nozzle and the momentum of an ideal nozzle.

nozzle process [NUCLEO] A method of separating isotopes by allowing a gaseous compound to exhaust through a properly shaped nozzle.

Np *See* neptunium.

npin transistor [ELECTR] An *npn* transistor which has a layer of high-purity germanium between the base and collector to extend the frequency range.

npnp diode *See* pnpn diode.

npnp transistor [ELECTR] An *npn*-junction transistor having a transition or floating layer between p and n regions, to which no ohmic connection is made. Also known as *pnpn* transistor.

npn semiconductor [ELECTR] Double junction formed by sandwiching a thin layer of p-type material between two layers of n-type material of a semiconductor.

npn transistor [ELECTR] A junction transistor having a p-type base between an n-type emitter and an n-type collector; the emitter should then be negative with respect to the base, and the collector should be positive with respect to the base.

np semiconductor [ELECTR] Region of transition between n- and p-type material.

NRM *See* natural remanent magnetization.

N shell [ATOM PHYS] The fourth layer of electrons about the nucleus of an atom, having electrons characterized by the principal quantum number 4.

n space [MATH] A vector space over the real numbers whose basis has n vectors.

N star [ASTRON] An obsolete classification for a star in the carbon sequence; has about the same temperature as an M star in the Draper catalog.

nt *See* nit.

NTP *See* standard conditions.

n-type conduction [ELECTR] The electrical conduction associated with electrons, as opposed to holes, in a semiconductor.

N-type crystal rectifier [ELECTR] Crystal rectifier in which forward current flows when the semiconductor is negative with respect to the metal.

n-type semiconductor [ELECTR] An extrinsic semiconductor in which the conduction electron density exceeds the hole density.

Nubeculae *See* Magellanic Clouds.

Nubecula Major *See* Large Magellanic Cloud.

Nubecula Minor *See* Small Magellanic Cloud.

nuclear [CHEM] Pertaining to a group of atoms joined directly to the central group of atoms or central ring of a molecule. [NUCLEO] Pertaining to nuclear energy. [NUC PHYS] Pertaining to the atomic nucleus.

nuclear age determination *See* radiometric dating.

nuclear angular momentum *See* nuclear spin.

nuclear battery [NUCLEO] A primary battery in which the energy of radioactive material is converted into electric energy by solar cells or other energy converters. Also known as atomic battery; radioisotope battery; radioisotopic generator.

nuclear binding energy [NUC PHYS] The energy required to separate an atom into its constituent protons, neutrons, and electrons.

nuclear boiler [NUCLEO] A nuclear reactor in which water is the primary coolant and is converted to steam, such as a pressurized-water reactor or a boiling-water reactor.

nuclear bomb   *See* atomic bomb.

nuclear breeder   [NUCLEO] A nuclear reactor in which more fissionable material is formed in each generation than is used up in fission.

nuclear capture   [NUC PHYS] Any process in which a particle, such as a neutron, proton, electron, muon, or alpha particle, combines with a nucleus.

nuclear chain reaction   [NUCLEO] A succession of generation after generation of acts of nuclear division such that the neutrons set free in the nuclear disruptions of the nth generation split the fissile nuclei (U^{233}, U^{235}, Pu^{239}) of the $(n + 1)$st generation. Also known as chain reaction.

nuclear chemistry   [ATOM PHYS] Study of the atomic nucleus, including fission and fusion reactions and their products.

nuclear collision   [NUC PHYS] A collision between an atomic nucleus and another nucleus or particle.

nuclear converter   *See* converter.

nuclear cross section   [NUC PHYS] A measure of the probability for a reaction to occur between a nucleus and a particle; it is an area such that the number of reactions which occur in a sample exposed to a beam of particles equals the product of the number of nuclei in the sample and the number of incident particles which would pass through this area if their velocities were perpendicular to it.

nuclear decay mode   [NUC PHYS] One of the ways in which a nucleus can undergo radioactive decay, distinguished from other decay modes by the resulting isotope and the particles emitted.

nuclear density   [NUC PHYS] The mass per unit volume of a nucleus as a function of distance from the center of the nucleus, as determined by a number of different types of experiments which are in reasonably good agreement.

nuclear emulsion   [NUCLEO] A photographic emulsion specially designed to register individual tracks of ionizing particles.

nuclear emulsion counter   [NUCLEO] A device used to measure the intensity of ionizing radiation by counting the number of tracks in an emulsion which has been exposed to the radiation.

nuclear energy   [NUCLEO] Energy released by nuclear fission or nuclear fusion. Also known as atomic energy.

nuclear engine   [NUCLEO] A type of thermal engine utilizing nuclear fission or fusion reactions to heat a working fluid for propulsive purposes.

nuclear engineering   [NUCLEO] The branch of technology that deals with the utilization of the nuclear fission process, and is concerned with the design and construction of nuclear reactors and auxiliary facilities, the development and fabrication of special materials, and the handling and processing of reactor products.

nuclear explosion   [NUCLEO] An explosion for which the energy is produced by a nuclear transformation, either fission or fusion.

nuclear fission   *See* fission.

nuclear force   [NUC PHYS] That part of the force between nucleons which is not electromagnetic; it is much stronger than electromagnetic forces, but drops off very rapidly at distances greater than about 10^{-13} centimeter; it is responsible for holding the nucleus together.

nuclear fuel   [NUCLEO] A fissionable or fertile isotope with a reasonably long half-life, used as a source of energy in a nuclear reactor. Also known as fission fuel; reactor fuel.

nuclear fuel cycle   *See* reactor fuel cycle.

NUCLEAR DENSITY

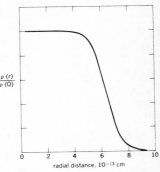

Distribution of charge in a nucleus of gold; this is believed to be approximately proportional to the nuclear mass density, on the assumption that the distributions of protons and neutrons are approximately the same; $\rho(r) =$ density at distance r from the center; $\rho(O) =$ density at the center.

NUCLEAR FORCE

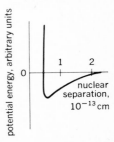

The potential operating between two nucleons due to the nuclear forces.

nuclear fuel element [NUCLEO] A piece of nuclear fuel which has been formed and coated, and is ready to be placed in a reactor fuel assembly. Also known as reactor fuel element.

nuclear fuel pebble *See* nuclear fuel pellet.

nuclear fuel pellet [NUCLEO] A piece of nuclear fuel usually in the shape of a sphere or cylinder, used in pebble-bed reactors, inserted in graphite blocks, or used in metallic tubular fuel elements. Also known as fuel ball; nuclear fuel pebble; reactor fuel pellet.

nuclear fuel plate [NUCLEO] A flat or curved sandwich of metallic cladding, with nuclear fuel inside.

nuclear fuel reprocessing [NUCLEO] The periodic chemical, physical, and metallurgical treatment of materials used as fuel elements in nuclear reactors, to recover and purify the residual fissionable and fertile materials.

nuclear fusion *See* fusion.

nuclear ground state [NUC PHYS] The stationary state of lowest energy of an isotope.

nuclear gyroscope [ENG] A gyroscope in which the conventional spinning mass is replaced by the spin of atomic nuclei and electrons; one version uses optically pumped mercury isotopes, and another uses nuclear magnetic resonance techniques.

nuclear heat [NUCLEO] The heat released in a nuclear reactor due to the fission process.

nuclear induction [PHYS] Magnetic induction originating in the magnetic moments of nuclei; the effect depends on the unequal population of energy states available when the material is placed in a magnetic field.

nuclear laser [OPTICS] A gas laser in which the gas molecules are excited by high-energy fission particles produced by a pulsed nuclear reactor.

nuclear magnetic moment [NUC PHYS] The magnetic dipole moment of an atomic nucleus; a vector whose scalar product with the magnetic flux density gives the negative of the energy of interaction of a nucleus with a magnetic field.

nuclear magnetic resonance [PHYS] A phenomenon exhibited by a large number of atomic nuclei, in which nuclei in a static magnetic field absorb energy from a radio-frequency field at certain characteristic frequencies. Abbreviated NMR. Also known as magnetic nuclear resonance.

nuclear magnetic resonance spectrometer [SPECT] A spectrometer in which nuclear magnetic resonance is used for the analysis of protons and nuclei and for the study of changes in chemical and physical quantities over wide frequency ranges.

nuclear magnetism [PHYS] The phenomena associated with the magnetic dipole, octupole, and higher moments of a nucleus, including the magnetic field generated by the nucleus, the force on the nucleus in an inhomogeneous magnetic field, and the splitting of nuclear energy levels in a magnetic field.

nuclear magnetometer [ENG] Any magnetometer which is based on the interaction of a magnetic field with nuclear magnetic moments, such as the proton magnetometer. Also known as nuclear resonance magnetometer.

nuclear magneton [NUC PHYS] A unit of magnetic dipole moment used to express magnetic moments of nuclei and baryons; equal to the electron charge times Planck's constant divided by the product of 4π, the proton mass, and the speed of light.

nuclear mass [NUC PHYS] The mass of an atomic nucleus, which is usually measured in atomic mass units; it is less than the sum of the masses of its constituent protons and neutrons by the binding energy of the nucleus divided by the square of the speed of light.

nuclear moment [NUC PHYS] One of the various static electric or magnetic multipole moments of a nucleus.

nuclear number *See* mass number.

nuclear paramagnetism [PHYS] Paramagnetism in which a substance develops a net magnetic moment because the magnetic moments of nuclei tend to point in the direction of the field.

nuclear physics [PHYS] The study of the characteristics, behavior, and internal structures of the atomic nucleus.

nuclear pile *See* nuclear reactor.

nuclear polarization [NUC PHYS] For a nucleus in a mixed state, with spin I and probability $p(I_z)$ that the I_z substate is populated, the polarization is the sum over allowed values of I_z of $I_z p(I_z)/I$.

nuclear potential [NUC PHYS] The potential energy of a nuclear particle as a function of its position in the field of a nucleus or of another nuclear particle.

nuclear potential energy [NUC PHYS] The average total potential energy of all the protons and neutrons in a nucleus due to the nuclear forces between them, excluding the electrostatic potential energy.

nuclear potential scattering [NUC PHYS] That part of elastic scattering of particles by a nucleus which may be treated by studying the scattering of a wave which obeys the Schrödinger equation with a potential determined by the properties of the nucleus.

nuclear power [NUCLEO] Power whose source is nuclear fission or fusion

nuclear quadrupole moment [NUC PHYS] The electric quadrupole moment of an atomic nucleus.

nuclear quadrupole resonance [PHYS] The phenomenon in which certain nuclei in a static, inhomogeneous electric field absorb energy from a radio-frequency field.

nuclear radiation [NUC PHYS] A term used to denote alpha particles, neutrons, electrons, photons, and other particles which emanate from the atomic nucleus as a result of radioactive decay and nuclear reactions.

nuclear radiation spectroscopy [NUC PHYS] Study of the distribution of energies or momenta of particles emitted by nuclei.

nuclear radius [NUC PHYS] The radius of a sphere within which the nuclear density is large, and at the surface of which it falls off sharply.

nuclear reaction [NUC PHYS] A reaction involving a change in an atomic nucleus, such as fission, fusion, neutron capture, or radioactive decay, as distinct from a chemical reaction, which is limited to changes in the electron structure surrounding the nucleus. Also known as reaction.

nuclear reactor [NUCLEO] A device containing fissionable material in sufficient quantity and so arranged as to be capable of maintaining a controlled, self-sustaining nuclear fission chain reaction. Also known as atomic pile (deprecated usage); atomic reactor; fission reactor; nuclear pile (deprecated); pile (deprecated); reactor.

nuclear recoil [NUC PHYS] The imparting of motion to an atomic nucleus during its emission of particles in radioactive decay, or during its collision with another particle, according to the principle of conservation of momentum.

nuclear relaxation [PHYS] The approach of a system of nuclear spins to a steady-state or equilibrium condition over a period of time, following a change in the applied magnetic field.

nuclear resonance [NUC PHYS] **1.** An unstable excited state formed in the collision of a nucleus and a bombarding particle, and associated with a peak in a plot of cross section versus energy. **2.** The absorption of energy by nuclei from radio-frequency fields at certain frequencies when these nuclei are also subjected to certain types of static fields, as in magnetic resonance and nuclear quadrupole resonance.

nuclear resonance magnetometer *See* nuclear magnetometer.

nuclear scattering [NUC PHYS] The change in directions of particles as a result of collisions with nuclei.

nuclear spallation *See* spallation.

nuclear species *See* nuclide.

nuclear spectrum [NUC PHYS] **1.** The relative number of particles emitted by atomic nuclei as a function of energy or momenta of these particles. **2.** The graphical display of data from devices used to measure these quantities.

nuclear spin [NUC PHYS] The total angular momentum of an atomic nucleus, resulting from the coupled spin and orbital angular momenta of its constituent nuclei. Also known as nuclear angular momentum. Symbolized I.

nuclear spontaneous reaction *See* radioactive decay.

nuclear stability [NUC PHYS] The ability of an isotope to resist decay or fission.

nuclear star *See* star.

nuclear superheating [NUCLEO] Superheating the steam produced in a reactor by using additional heat from a reactor; two methods may be used: recirculating the steam through the same core in which it is first produced (integral superheating) or passing the steam through a second and separate reactor.

nuclear thermionic converter [NUCLEO] A thermionic converter whose heat source is a nuclear reactor or radioisotope.

nuclear transformation *See* transmutation.

nuclear triode detector [ELECTR] A type of junction detector that has two outputs which together determine the precise location on the detector where the ionizing radiation was incident, as well as the energy of the ionizing particle.

nuclear Zeeman effect [SPECT] A splitting of atomic spectral lines resulting from the interaction of the magnetic moment of the nucleus with an applied magnetic field.

nucleate boiling [THERMO] Boiling in which bubbles of vapor form at the hot surface of the container holding the boiling liquid.

nucleon [PHYS] A collective name for a proton or a neutron; these particles are the main constituents of atomic nuclei, have approximately the same mass, have a spin of $\frac{1}{2}$, and can transform into each other through the process of beta decay.

nucleonics [ENG] The technology based on phenomena of the atomic nucleus such as radioactivity, fission, and fusion; includes nuclear reactors, various applications of radioisotopes and radiation, particle accelerators, and radiation-detection devices.

nucleonium [ATOM PHYS] A bound state of a nucleus and an antinucleus.

nucleon number *See* mass number.

nucleor [PARTIC PHYS] A hypothetical core of a nucleon, surrounded by a hypothetical cloud of pions.

nucleus [ASTRON] The small permanent body of a comet, believed to have a diameter between one and a few tens of kilometers, and to be composed of water and volatile hydrocarbons. [NUC PHYS] The central, positively charged, dense portion of an atom. Also known as atomic nucleus. [SCI TECH] A central mass about which accretion takes place.

nuclide [NUC PHYS] A species of atom characterized by the number of protons, number of neutrons, and energy content in the nucleus, or alternatively by the atomic number, mass number, and atomic mass; to be regarded as a distinct nuclide, the atom must be capable of existing for a measurable lifetime, generally greater than 10^{-10} second. Also known as nuclear species; species.

nuisance parameter [STAT] A parameter to be estimated by a statistic which arises in the distribution of the statistic under some hypothesis to be tested about the parameter.

null [MATH] A term meaning that an object is nonexistent or a quantity is zero.

nullary composition [MATH] The selection of a particular element of a set.

null-balance recorder [ENG] An instrument in which a motor-driven slide wire in a measuring circuit is continuously adjusted so that the voltage or current to be measured will be balanced against the voltage or current from this circuit; a pen linked to the slide wire makes a graphical record of its position as a function of time.

null cone *See* light cone.

null-current circuit [ELECTR] A circuit used to measure current, in which the unknown current is opposed by a current resulting from applying a voltage controlled by a slide wire across a series resistor, and the slide wire is continuously adjusted so that the resulting current, as measured by a direct-current detector amplifier, is equal to zero.

null detection [ELEC] Altering of adjustable bridge circuit components, to obtain zero current.

null detector *See* null indicator.

null direction [ELECTROMAG] A direction in which the power radiated or received per unit solid angle by an antenna array is zero.

null geodesic [MATH] In a Riemannian space, a minimal geodesic curve. [RELAT] A curve in space-time which has the property that the infinitesimal interval between any two neighboring points on the curve equals zero; it represents a possible path of a light ray. Also known as zero geodesic.

null hypothesis [STAT] The hypothesis that there is no validity to the specific claim that two variations (treatments) of the same thing can be distinguished by a specific procedure.

null indicator [ENG] A galvanometer or other device that indicates when voltage or current is zero; used chiefly to determine when a bridge circuit is in balance. Also known as null detector.

nullity [MATH] The dimension of the null space of a linear transformation.

null matrix [MATH] The matrix all of whose entries are zero.

null method [ENG] A method of measurement in which the measuring circuit is balanced to bring the pointer of the indicating instrument to zero, as in a Wheatstone bridge, and the settings of the balancing controls are then read. Also known as balance method; zero method.

null sequence [MATH] A sequence of numbers or functions which converges to the number zero or the zero function.

null set [MATH] The empty set; the set which contains no elements.

null space [MATH] For a linear transformation, the vector subspace of all vectors which the transformation sends to the zero vector. Also known as kernel.

null surface [RELAT] A surface in space-time whose normal vector is everywhere null.

null vector [MATH] A vector whose invariant length, that is, the sum over the coordinates of the vector space of the product of its covariant component and contravariant component, is equal to zero. [RELAT] In special relativity, a four vector whose spatial part in any Lorentz frame has a magnitude equal to the speed of light multiplied by its time part in that frame; a special case of the mathematics definition.

number [MATH] **1.** Any real or complex number. **2.** The number of elements in a set is the cardinality of the set.

number class modulo N [MATH] The class of all numbers which differ from a given number by a multiple of N.

number scale [MATH] Representation of points on a line with numbers arranged in some order.

number theory [MATH] The study of integers and relations between them.

numeral [MATH] A symbol used to denote a number.

numeral system *See* numeration system.

numeration [MATH] The listing of numbers in their natural order.

numeration system [MATH] An orderly method of representing numbers by numerals in which each numeral is associated with a unique number. Also known as numeral system.

numerator [MATH] In a fraction a/b, the numerator is the quantity a.

numerical [MATH] Pertaining to numbers.

numerical analysis [MATH] The study of approximation techniques using arithmetic for solutions of mathematical problems.

numerical aperture [OPTICS] A measure of the resolving power of a microscope objective, equal to the product of the refractive index of the medium in front of the objective and the sine of the angle between the outermost ray entering the objective and the optical axis. Abbreviated N.A.

numerical control [CONT SYS] A control system for machine tools and some industrial processes, in which numerical values corresponding to desired positions of tools or controls are recorded on punched paper tapes, punched cards, or magnetic tapes so that they can be used to control the operation automatically. Abbreviated NC; N/C.

numerical decrement *See* decrement.

numerical equation [MATH] An equation all of whose constants and coefficients are numbers.

numerical integration [MATH] The process of using a set of approximate values of a function to calculate its integral to comparable accuracy.

numerical range [MATH] For a linear operator T of a Hilbert space into itself, the set of values assumed by the inner product of Tx with x as x ranges over the set of vectors with norm equal to 1.

numerical value of a real number *See* absolute value of a real number.

numeric character *See* digit.

N unit [OPTICS] A unit of index of refraction; a mathematical simplification designed to replace rather awkward numbers involved in the values of the index of refraction n for the atmosphere; it is defined by the relation $N = (n-1)10^6$.

nuplex [NUCLEO] Proposed nuclear reactor which would both produce electrical power and heat water for desalination and other industrial benefits.

Nusselt equation [THERMO] Dimensionless equation used to calculate convection heat transfer for heating or cooling of fluids outside a bank of 10 or more rows of tubes to which the fluid flow is normal.

Nusselt number [PHYS] A dimensionless number used in the study of mass transfer, equal to the mass-transfer coefficient times the thickness of a layer through which mass transfer is taking place divided by the molecular diffusivity. Symbolized Nu_m; N_{Nu_m}. Also known as Sherwood number (N_{Sh}). [THERMO] A dimensionless number used in the study of forced convection which gives a measure of the ratio of the total heat transfer to conductive heat transfer, and is equal to the heat-

NUMERICAL CONTROL

A typical numerically controlled machine tool, showing the numerical control system or "director" at the right. (*Cintimatic Div., Cincinnati Milacron Co.*)

transfer coefficient times a characteristic length divided by the thermal conductivity. Symbolized N_{Nu}.

nutation [ASTRON] A slight, slow, nodding motion of the earth's axis of rotation which is superimposed on the precession of the equinoxes; it is the combination of a number of perturbations (lunar, solar, and fortnightly nutation). [MECH] A bobbing or nodding up-and-down motion of a spinning rigid body, such as a top, as it precesses about its vertical axis.

nu value [OPTICS] The reciprocal of the dispersive power of a medium. Also known as constringence.

n value *See* field index.

Nyquist contour [CONT SYS] A directed closed path in the complex frequency plane used in constructing a Nyquist diagram, which runs upward, parallel to the whole length of the imaginary axis at an infinitesimal distance to the right of it, and returns from $+j\infty$ to $-j\infty$ along a semicircle of infinite radius in the right half-plane.

Nyquist diagram [CONT SYS] A plot in the complex plane of the open-loop transfer function as the complex frequency is varied along the Nyquist contour; used to determine stability of a control system.

Nyquist stability criterion *See* Nyquist stability theorem.

Nyquist stability theorem [CONT SYS] The theorem that the net number of counterclockwise rotations about the origin of the complex plane carried out by the value of an analytic function of a complex variable, as its argument is varied around the Nyquist contour, is equal to the number of poles of the variable in the right half-plane minus the number of zeros in the right half-plane. Also known as Nyquist stability criterion.

Nyquist's theorem [ELECTR] The mean square noise voltage across a resistance in thermal equilibrium is four times the product of the resistance, Boltzmann's constant, the absolute temperature, and the frequency range within which the voltage is measured.

O *See* oxygen.

OASM system [PHYS] A system of electrical and mechanical units in which the fundamental quantities are electric resistance, electric current, time, and length, and the base units of these quantities are the ohm, ampere, second, and meter, respectively.

obelisk [MATH] A frustrum of a regular, rectangular pyramid.

object [OPTICS] A collection of points which may be regarded as a source of light rays in an optical system, whether it actually has this function (as in a real object) or does not (as in a virtual object).

object contrast [OPTICS] The ratio of the difference between the brightness of an object and of the background to the brightness of the background in an image or reproduction.

object glass *See* objective.

objective [OPTICS] The first lens, lens system, or mirror through which light passes or from which it is reflected in an optical system; many scientists exclude mirrors from the definition. Also known as object glass.

objective function [MATH] In nonlinear programming, the function, expressing given conditions for a system, which one seeks to minimize subject to given constraints.

objective grating [OPTICS] A series of equally spaced parallel wires placed over the objective lens of a telescope; photographic magnitudes of stars are calculated from the relative brightnesses of images in the resulting diffraction pattern.

objective prism [OPTICS] A large prism, usually having a small angle, which is placed in front of the objective of a photographic telescope to make spectroscopic observations.

objective probabilities [STAT] Probabilities determined by the long-run relative frequency of an event. Also known as frequency probabilities.

object space [OPTICS] The region of space where objects are located so that a given optical system can form images of them.

oblate ellipsoid *See* oblate spheroid.

oblate spheroid [MATH] The surface or ellipsoid generated by rotating an ellipse about one of its axes so that the diameter of its equatorial circle exceeds the length of the axis of revolution. Also known as oblate ellipsoid.

oblate spheroidal coordinate system [MATH] A three-dimensional coordinate system whose coordinate surfaces are the surfaces generated by rotating a plane containing a system of confocal ellipses and hyperbolas about the minor axis of the ellipses, together with the planes passing through the axis of rotation.

oblique angle [MATH] An angle that is neither a right angle nor a multiple of a right angle.

oblique astigmatism *See* radial astigmatism.

OBLATE SPHEROID

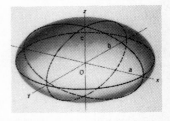

Drawing of an oblate spheroid generated by rotating an ellipse about its minor axis lying along z axis of coordinate system, with center of ellipse at origin of coordinates, O. Diameters $2a$ and $2b$ along x and y axes are equal to each other, and greater than axis of revolution $2c$.

oblique coordinates [MATH] Magnitudes defining a point relative to two intersecting nonperpendicular lines, called axes; the magnitudes indicate the distance from each axis, measured along a parallel to the other axis; oblique coordinates are a form of cartesian coordinates.

oblique extinction *See* inclined extinction.

oblique-incidence reflectivity [OPTICS] The reflectivity of an interface between two media when the direction of propagation of the incident electromagnetic radiation is not perpendicular to the interface; it differs for the component whose electric vector lies in the plane containing the perpendicular to the surface and the propagation direction, and the component for which this vector is perpendicular to this plane.

oblique lines [MATH] Lines that are neither perpendicular nor parallel.

oblique shock *See* oblique shock wave.

oblique shock wave [FL MECH] A shock wave inclined at an oblique angle to the direction of flow in a supersonic flow field. Also known as oblique shock.

oblique triangle [MATH] A triangle that does not contain a right angle.

oblique visibility *See* oblique visual range.

oblique visual range [OPTICS] The greatest distance at which a specified target can be perceived when viewed along a line of sight inclined to the horizontal. Also known as oblique visibility; slant visibility.

obliquity factor [OPTICS] A function which is proportional to the amplitudes of secondary waves propagating in various directions according to Huygens' principle; it is $1 + \cos \theta$, where θ is the angle between the normal to the original wavefront and the normal to the secondary wavefront.

obliquity of the ecliptic [ASTRON] The acute angle between the plane of the ecliptic and the plane of the celestial equator, about $23°27'$.

observability [CONT SYS] Property of a system for which observation of the output variables at all times is sufficient to determine the initial values of all the state variables.

observable operator [QUANT MECH] A Hermitian operator with a complete, orthonormal set of eigenfunctions on the Hilbert space representing the states of a physical system; such operators are postulated to represent the observable quantities of the system.

observable quantity [PHYS] A measurable physical quantity.

observer [CONT SYS] A linear system B driven by the inputs and outputs of another linear system A which produces an output that converges to some linear function of the state of system A. Also known as state estimator; state observer.

obtuse angle [MATH] An angle of more than 90° and less than 180°.

obtuse triangle [MATH] A triangle having one obtuse angle.

occlusion [PHYS] Adhesion of gas or liquid on a solid mass, or the trapping of a gas or liquid within a mass.

occultation [ASTRON] The disappearance of the light of a celestial body by intervention of another body of larger apparent size; especially, a lunar eclipse of a star or planet.

OC curve *See* operating characteristic curve.

octagon [MATH] A polygon with eight sides.

octahedral coordination [CRYSTAL] An atomic structure where six cations surround every anion, and vice versa.

octahedral normal stress [MECH] The normal component of stress across the faces of a regular octahedron whose vertices lie on the principal axes of stress; it is equal in magnitude to the spherical stress across any surface. Also known as mean stress.

octahedral plane [CRYSTAL] The plane in a cubic lattice having three numerically equal Miller indices.

octahedral shear stress [MECH] The tangential component of stress across the faces of a regular octahedron whose vertices lie on the principal axes of stress; it is a measure of the strength of the deviatoric stress.

octahedron [MATH] A polyhedron having eight faces, each of which is an equilateral triangle.

octal [MATH] Pertaining to the octal number system.

octal digit [MATH] The symbol 0, 1, 2, 3, 4, 5, 6, or 7 used as a digit in the octal number system.

octal number system [MATH] A number system in which a number r is written as $n_k n_{k-1} \cdots n_1$ where $r = n_1 8^0 + n_2 8^1 + \cdots + n_k 8^{k-1}$.

octave [ACOUS] The interval in pitch between two tones such that one tone may be regarded as duplicating at the next higher pitch the basic musical import of the other tone; the sounds producing these tones then have a frequency ratio of 2 to 1. [PHYS] The interval between any two frequencies having a ratio of 2 to 1.

octave frequency band [PHYS] A band of frequencies whose highest frequency is twice its lowest frequency.

octet [ATOM PHYS] A collection of eight valence electrons in an atom or ion, which form the most stable configuration of the outermost, or valence, electron shell. [PARTIC PHYS] A multiplet of eight elementary particles, corresponding to a representation of the approximate unitary symmetry (SU$_3$) of the strong interactions.

octillion [MATH] **1.** The number 10^{27}. **2.** In British and German usage, the number 10^{48}.

octonions *See* Cayley numbers.

octupole [PHYS] **1.** Two electric or magnetic quadrupoles having charge distributions of opposite signs and separated from each other by a small distance. **2.** Any device for controlling beams of electrons or other charged particles, consisting of eight electrodes or magnetic poles arranged in a circular pattern, with alternating polarities; commonly used to correct aberrations of quadrupole systems.

ocular *See* eyepiece.

ocular prism [OPTICS] The prism employed in a range finder to bend the line of sight through the instrument into the eyepiece.

OD *See* optical density.

odd-even nucleus [NUC PHYS] A nucleus which has an odd number of protons and an even number of neutrons.

odd function [MATH] A function $f(x)$ is odd if, for every x, $f(-x) = -f(x)$.

odd number [MATH] A natural number not divisible by 2.

odd-odd nucleus [NUC PHYS] A nucleus that has an odd number of protons and an odd number of neutrons.

odd parity [QUANT MECH] Property of a system whose state vector is multiplied by -1 under the operation of space inversion, that is, the simultaneous reflection of all spatial coordinates through the origin.

odds ratio [STAT] The ratio of the probability of occurrence of an event to the probability of the event not occurring.

odd term [ATOM PHYS] A term of an atom or molecule for which the sum of the angular-momentum quantum numbers of all the electrons is odd, so that the states have odd parity; designated by a superscript o or u.

O electron [ATOM PHYS] An electron in the fifth (O) shell of electrons surrounding the atomic nucleus, having the principal quantum number 5.

oersted [ELECTROMAG] The unit of magnetic field strength in the centimeter-gram-second electromagnetic system of units, equal to the field strength at the center of a plane circular coil of one turn and 1-centimeter radius, when there is a current of $1/2\pi$ abamp in the coil.

Oersted experiment [ELECTROMAG] The deflection of a magnetic needle when placed near a wire carrying an electric current.

offense against the sine condition [OPTICS] A numerical measure of coma, equal to the sagittal coma divided by the perpendicular distance from the image point to the optical axis.

offset [CONT SYS] The steady-state difference between the desired control point and that actually obtained in a process control system. [MECH] The value of strain between the initial linear portion of the stress-strain curve and a parallel line that intersects the stress-strain curve of an arbitrary value of strain; used as an index of yield stress; a value of 0.2% is common.

offset yield strength [MECH] That stress at which the strain surpasses by a specific amount (called the offset) an extension of the initial proportional portion of the stress-strain curve; usually expressed in pounds per square inch.

ohm [ELEC] The unit of electrical resistance in the rationalized meter-kilogram-second system of units, equal to the resistance through which a current of 1 ampere will flow when there is a potential difference of 1 volt across it. Symbolized Ω.

ohmic contact [ELEC] A region where two materials are in contact, which has the property that the current flowing through it is proportional to the potential difference across it.

ohmic dissipation [ELECTR] Loss of electric energy when a current flows through a resistance due to conversion into heat. Also known as ohmic loss.

ohmic loss *See* ohmic dissipation.

ohmic resistance [ELEC] Property of a substance, circuit, or device for which the current flowing through it is proportional to the potential difference across it.

ohmmeter [ENG] An instrument for measuring electric resistance; scale may be graduated in ohms or megohms.

Ohm's law [ELEC] The law that the direct current flowing in an electric circuit is directly proportional to the voltage applied to the circuit; it is valid for metallic circuits and many circuits containing an electrolytic resistance.

ohms per volt [ENG] Sensitivity rating for measuring instruments, obtained by dividing the resistance of the instrument in ohms at a particular range by the full-scale voltage value at that range.

Olbers' paradox [ASTRON] If the universe were static, of infinite age, and the galaxies distributed isotropically, the distance attenuation of their light would be exactly balanced by the increase in number in successive spherical shells centered at the earth; hence the night sky would be of daylight brightness instead of dark.

old achromat [OPTICS] An achromatic lens in which the component lenses are made of glasses chosen from a limited selection, in which refractive index and dispersive power vary roughly together.

omega hyperon [PARTIC PHYS] A semistable baryon with a mass of approximately 1672 MeV, negative charge, spin of 3/2, and positive parity; constitutes an isotopic spin singlet. Also known as omega particle. Symbolized Ω^-.

omega-limit point [MATH] For the vector differential equation $dx/dt = \mathbf{f}(x)$, where $\mathbf{f}$ is continuously differentiable, describing the motion of points in euclidean n space, an ω-limit point of $\mathbf{x}$ is a point $\mathbf{y}$ for which there exists a sequence

$t_n \to \infty$ such that $\mathbf{x}(t_n) \to \mathbf{y}$ as $n \to \infty$. Also known as limit point.

omega meson [PARTIC PHYS] An unstable, neutral vector meson having a mass of about 783 MeV, a width of about 10 MeV, and negative charge parity and G parity. Symbolized $\omega(783)$.

omega particle *See* omega hyperon.

omegatron [ELECTR] A miniature mass spectrograph, about the size of a receiving tube, that can be sealed to another tube and used to identify the residual gases left after evacuation.

omission solid solution [CRYSTAL] A crystal with certain atomic sites incompletely filled.

OMS *See* ovonic memory switch.

onde de choc [ACOUS] The first sound heard as the result of the passage of a high-speed projectile.

ondograph [ELECTR] An instrument that draws the waveform of an alternating-current voltage step by step; a capacitor is charged momentarily to the amplitude of a point on the voltage wave, then discharged into a recording galvanometer, with the action being repeated a little further along on the waveform at intervals of about 0.01 second.

one-dimensional flow [FL MECH] Fluid flow in which all flow is parallel to some straight line, and characteristics of flow do not change in moving perpendicular to this line.

one-dimensional lattice [CRYSTAL] A simplified model of a crystal lattice consisting of particles lying along a straight line at either equal or periodically repeating distances.

one-group model [NUCLEO] A neutron-behavior model in which neutrons of all energies are treated as having the same characteristics.

one-parameter semigroup [MATH] A semigroup with which there is associated a bijective mapping from the positive real numbers onto the semigroup.

one-particle exchange [PARTIC PHYS] A model for the interaction of two particles in which the interaction results entirely from a single virtual particle being emitted by one interacting particle and absorbed by the other.

one-point compactification [MATH] The one-point compactification $\hat{X}$ of a topological space X is the union of X with a set consisting of a single element, with the topology of $\hat{X}$ consisting of the open subsets of X and all subsets of $\hat{X}$ whose complements in $\hat{X}$ are closed compact subsets in X. Also known as Alexandroff compactification.

one-sample problem [STAT] The problem of testing the hypothesis that the average of a sequence of observations or measurement of the same kind has a specified value.

one-shot multivibrator *See* monostable multivibrator.

one-sided abrupt junction [ELECTR] An abrupt junction that is realized by giving one side of the junction a high doping level compared with the other; that is, a n^+p or p^+n junction.

one-sided limit [MATH] Either a limit on the left or a limit on the right.

one-sided test [STAT] A test statistic T which rejects a hypothesis only for $T \geq d$ or $T \leq c$ but not for both (here d and c are critical values).

one-to-one mapping *See* injection.

O network [ELEC] Network composed of four impedance branches connected in series to form a closed circuit, two adjacent junction points serving as input terminals, the remaining two junction points serving as output terminals.

one-way classification [STAT] The basis for the simplest case of the analysis of variance; a set of observations are categorized according to values of one variable or one characteristic.

on-line operation [ADP] Computer operation in which input data are fed into the computer directly from observing instruments or other input equipment, and computer results are obtained during the progress of the event.

on-off control [CONT SYS] A simple control system in which the device being controlled is either full on or full off, with no intermediate operating positions. Also known as on-off system.

on-off system *See* on-off control.

Onsager equation [PHYS CHEM] An equation which relates the measured equivalent conductance of a solution at a certain concentration to that of the pure solvent.

Onsager reciprocal relations [THERMO] A set of conditions which state that the matrix, whose elements express various fluxes of a system (such as diffusion and heat conduction) as linear functions of the various conjugate affinities (such as mass and temperature gradients) for systems close to equilibrium, is symmetric when certain definitions are chosen for these fluxes and affinities.

Onsager theory of dielectrics [ELEC] A theory for calculating the dielectric constant of a material with polar molecules in which the local field at a molecule is calculated for an actual spherical cavity of molecular size in the dielectric using Laplace's equation, and the polarization catastrophe of the Lorentz field theory is thereby avoided.

opacity [OPTICS] The light flux incident upon a medium divided by the light flux transmitted by it.

opalescence [OPTICS] The milky, iridescent appearance of a dense, transparent medium or colloidal system when it is illuminated by polychromatic radiation in the visible range, such as sunlight.

opaque medium [OPTICS] A medium which is impervious to rays of light, that is, not transparent to the human eye. [PHYS] **1.** A medium which does not transmit electromagnetic radiation of a specified type, such as that in the infrared, x-ray, ultraviolet, and microwave regions. **2.** A medium which prevents the passage of particles of a specified type.

open ball [MATH] In a metric space, an open set about a point x which consists of all points within a fixed distance from x.

open circuit [ELEC] An electric circuit that has been broken, so that there is no complete path for current flow.

open-circuited line [ELECTROMAG] A microwave discontinuity which reflects an infinite impedance.

open-circuit impedance [ELEC] Of a line or four-terminal network, the driving-point impedance when the far end is open.

open cluster [ASTRON] One of the groupings of stars that are concentrated along the central plane of the Milky Way; most have an asymmetrical appearance and are loosely assembled, and the stars are concentrated in their central region; they may contain from a dozen to many hundreds of stars. Also known as galactic cluster.

open cycle [THERMO] A thermodynamic cycle in which new mass enters the boundaries of the system and spent exhaust leaves it; the automotive engine and the gas turbine illustrate this process.

open-cycle reactor system [NUCLEO] A reactor system in which the coolant passes through the reactor core only once and is then discarded.

open-ended class [STAT] The first or last class interval in a frequency distribution having no upper or lower limit.

open-flame arc [ELECTR] An electric arc which causes the anode to evaporate and be ejected as a flame.

open form [CRYSTAL] A crystal form in which the crystal faces do not entirely enclose a space.

open interval [MATH] An open interval of real numbers,

denoted by (a,b), consists of all numbers strictly greater than a and strictly less than b.

open-loop control system [CONT SYS] A control system in which the system outputs are controlled by system inputs only, and no account is taken of actual system output.

open map [MATH] A function between two topological spaces which sends each open set of one to an open set of the other.

open mapping theorem [MATH] A continuous linear function between Banach spaces which has closed range must be an open map.

open set [MATH] A set included in a topology; equivalently, a set which is a neighborhood of each of its points; a topology on a space is determined by a collection of subsets which are called open.

open system [THERMO] A system across whose boundaries both matter and energy may pass.

open-window unit *See* sabin.

opera glasses [OPTICS] Small binocular telescopes, usually of the Galilean type, adapted for use where magnification and field of view are secondary to compactness and cost.

operating characteristic curve [STAT] In hypothesis testing, a plot of the probability of accepting the hypothesis against the true state of nature. Abbreviated OC curve.

operating stress [MECH] The stress to which a structural unit is subjected in service.

operational amplifier [ELECTR] An amplifier having high direct-current stability and high immunity to oscillation, generally achieved by using a large amount of negative feedback; used to perform analog-computer functions such as summing and integrating.

operational calculus [MATH] A technique by which problems in analysis, in particular differential equations, are transformed into algebraic problems, usually the problem of solving a polynomial equation. Also known as operational analysis.

operation of a group [MATH] An operation of a group G on a set S is a mapping which associates to each ordered pair (g,s), where g is in G and s is in S, another element in S, denoted gs, such that, for any g,h in G and s in S, $(gh)s = g(hs)$, and $es = s$, where e is the identity element of G; the operation may also be regarded as a mapping which associates with each element g in G the bijection which maps each s in S into gs and, if S has a specified algebraic structure, these bijections are taken to be automorphisms of this structure.

operations research [MATH] The mathematical study of systems with input and output from the viewpoint of optimization subject to given constraints. [SCI TECH] The application of objective and quantitative criteria to decision making previously undertaken by empirical methods.

operator [MATH] A function between vector spaces.

operator algebra [MATH] An algebra whose elements are functions and in which the multiplication of two elements f and g is defined by composition; that is, $(fg)(x) = (f \circ g)(x) = f[g(x)]$.

operator theory [MATH] The general qualitative study of operators in terms of such concepts as eigenvalues, range, domain, and continuity.

ophthalmometer [OPTICS] 1. An instrument for measuring refractive errors, especially astigmatism. 2. An instrument for measuring the capacity of the chamber of the eye. 3. An instrument for measuring the eye as a whole.

ophthalmoscope [OPTICS] An instrument, consisting essentially of a concave mirror with a hole in it and fitted with lenses of different powers, for examining the interior of the eye through the pupil.

Oppenheimer-Phillips reaction [NUC PHYS] A type of stripping reaction which can occur when a deuteron passes near a nucleus, in which the proton in the deuteron experiences Coulomb repulsion from the nucleus while the neutron is attracted to the nucleus by nuclear forces, with the result that the neutron-proton bond in the deuteron is broken, the neutron is absorbed into the nucleus, and the proton is repelled.

opponent-colors theory [OPTICS] A theory of color vision according to which various processes in the visual system are capable of responding in two opposite ways; the Hering theory is an example.

opposition [ASTRON] The situation of two celestial bodies having either celestial longitudes or sidereal hour angles differing by 180°; the term is usually used only in relation to the position of a superior planet or the moon with reference to the sun. [PHYS] The condition in which the phase difference between two periodic quantities having the same frequency is 180°, corresponding to one half-cycle.

optic [OPTICS] Pertaining to the lenses, prisms, and mirrors of a camera, microscope, or other conventional optical instrument.

optical [OPTICS] Pertaining to or utilizing visible or near-visible light; the extreme limits of the optical spectrum are about 100 nanometers (0.1 micrometer or 3×10^{15} hertz) in the far ultraviolet and 30,000 nanometers (30 micrometers or 10^{13} hertz) in the far infrared.

optical aberration [OPTICS] Deviation from perfect image formation by an optical system; examples are spherical aberration, coma, astigmatism, curvature of field, distortion, and chromatic aberration. Also known as aberration.

optical achromatism *See* visual achromatism.

optical activity [OPTICS] The behavior of substances which rotate the plane of polarization of plane-polarized light, as it passes through them. Also known as rotary polarization.

optical analysis [OPTICS] Study of properties of a substance or medium, such as its chemical composition or the size of particles suspended in it, through observation of effects on transmitted light, such as scattering, absorption, refraction, and polarization.

optical anisotropy [OPTICS] The behavior of a medium, or of a single molecule, whose effect on electromagnetic radiation depends on the direction of propagation of the radiation.

optical antipode *See* enantiomorph.

optical aspherical surface [OPTICS] An optical surface that does not form part of a sphere, such as a paraboloidal or ellipsoidal surface.

optical axis [OPTICS] **1.** A line passing through a radially symmetrical optical system such that rotation of the system about this line does not alter it in any detectable way. **2.** *See* optic axis.

optical bench [ENG] A rigid horizontal bar or track for holding optical devices in experiments; it allows device positions to be changed and adjusted easily.

optical branch [SOLID STATE] The vibrations of an optical mode plotted on a graph of frequency versus wave number; it is separated from, and has higher frequencies than, the acoustic branch.

optical center [OPTICS] A point on the axis of a lens so that, for any ray passing through this point, the incident part and the emergent part are parallel.

optical coating [OPTICS] Either a mirror coating, or a film of the proper thickness and refractive index applied to the air-glass surface of a lens to reduce reflection.

optical contact [OPTICS] Contact between two surfaces in which the surfaces are separated by a distance much less than a wavelength of light, so that interference fringes are not formed.

optical crystal [CRYSTAL] Any natural or synthetic crystal, such as sodium chloride, calcium fluoride, silver chloride, potassium iodide, or stilbene, that is used in infrared and ultraviolet optics and for its piezoelectric effects.

optical density [OPTICS] The degree of opacity of a translucent medium expressed by $\log I_0/I$, where I_0 is the intensity of the incident ray, and I is the intensity of the transmitted ray. Abbreviated OD.

optical diffraction velocimeter *See* diffraction velocimeter.

optical dispersion [OPTICS] Separation of different colors of light such as occurs when it passes from one medium to another or is reflected from a diffraction grating.

optical distance *See* optical path.

optical Doppler effect [ELECTROMAG] A change in the observed frequency of light or other electromagnetic radiation caused by relative motion of the source and observer.

optical double star [ASTRON] Two stars not formally a physical system but that appear to be a typical double star; a false binary star whose components happen to lie nearby in the same line of sight.

optical element [OPTICS] A part of an optical instrument which acts upon the light passing through the instrument, such as a lens, prism, or mirror.

optical fiber [OPTICS] A long, thin thread of fused silica, or other transparent substance, used to transmit light. Also known as light guide.

optical filter *See* filter.

optical flat [OPTICS] **1.** A disk of high-grade quartz glass approximately 2 centimeters thick, with a deviation in flatness usually not exceeding 0.05 micrometer all over, and a surface quality of 5 microfinish or less; used in determinations of surface contour and in comparison of lineal measurement. **2.** A plane surface, with deviations from a plane surface generally not exceeding $\frac{1}{10}$ of a wavelength of light, used to redirect light in a telescope or other optical instrument.

optical fluid-flow measurement [ENG] Any method of measuring the varying densities of a fluid in motion, such as schlieren, interferometer, or shadowgraph, which depends on the fact that light passing through a flow field of varying density is retarded differently through the field, resulting in refraction of the rays, and in a relative phase shift among different rays.

optical frequency [PHYS] A frequency comparable to that of electromagnetic waves in the optical region, above about 3×10^{11} hertz.

optical galaxy [ASTRON] One of the galaxies that appear as nearly starlike, generally having compact nuclei.

optical glass [MATER] A type of glass which is free from imperfections, such as unmelted particles, bubbles, and chemical inhomogeneities, which would affect its transmission of light.

optical harmonic [SOLID STATE] Light, generated by passing a laser beam with a power density on the order of 10^{10} watts/cm^2 or more through certain transparent materials, which has a frequency which is an integral multiple of that of the incident laser light.

optical haze *See* terrestrial scintillation.

optical indicatrix *See* index ellipsoid.

optical instrument [OPTICS] An optical system which acts on light in some desired way, such as to form a real or virtual image, to form an optical spectrum, or to produce light with a specified polarization or wavelength.

optical isomer *See* enantiomorph.

optical isomerism [PHYS CHEM] Existence of two forms of a molecule such that one is a mirror image of the other; the two molecules differ in that rotation of light is equal but in opposite directions.

optical length *See* optical path.

optical lens *See* lens.

optical maser *See* laser.

optical material [MATER] A material which is transparent to light or to infrared, ultraviolet, or x-ray radiation, such as glass and certain single crystals, polycrystalline materials (chiefly for the infrared), and plastics.

optical measurement [OPTICS] Measurement of the intensity, spectral distribution, polarization, or other characteristics of light or of infrared or ultraviolet radiation, which is emitted by or reflected from an object or passes through some medium.

optical microscope [OPTICS] An instrument used to obtain an enlarged image of a small object, utilizing visible light; in general it consists of a light source, a condenser, an objective lens, and an ocular or eyepiece, which can be replaced by a recording device. Also known as light microscope; photon microscope.

optical mode [SOLID STATE] A type of vibration of a crystal lattice whose frequency varies with wave number only over a limited range, and in which neighboring atoms or molecules in different sublattices move in opposition to each other.

optical model *See* cloudy crystal-ball model.

optical moment [OPTICS] For a ray of light passing through an optical system, the triple product of a vector from an arbitrary origin on the optical axis to a point on the ray, a vector tangent to the ray at that point whose length equals the refractive index, and a unit vector along the optical axis; it does not depend on the point on the ray.

optical monochromator [SPECT] A monochromator used to observe the intensity of radiation at wavelengths in the visible, infrared, or ultraviolet regions.

optical null method [SPECT] In infrared spectrometry, the adjustment of a reference beam's energy transmission to match that of a beam that has been passed through a sample being analyzed.

optical parametric oscillator [OPTICS] A device, employing a nonlinear dielectric, which when pumped by a laser can generate coherent light whose wavelength can be varied continuously over a wide range.

optical path [OPTICS] For a ray of light traveling along a path between two points, the optical path is the integral, over elements of length along the path, of the refractive index. Also known as optical distance; optical length.

optical-path difference *See* retardation.

optical phenomena [ELECTROMAG] Phenomena associated with the generation, transmission, and detection of electromagnetic radiation in the visible, infrared, or ultraviolet regions.

optical phonon [SOLID STATE] A quantum of an optical mode of vibration of a crystal lattice.

optical prism *See* prism.

optical projection system [OPTICS] An optical system which forms a real image of a suitably illuminated object so that it can be viewed, photographed, or otherwise observed. Also known as optical projector; projector.

optical projector   *See* optical projection system.

optical properties   [ELECTROMAG] The effects of a substance or medium on light or other electromagnetic radiation passing through it, such as absorption, scattering, refraction, and polarization.

optical pumping   [OPTICS] The process of causing strong deviations from thermal equilibrium populations of selected quantized states of different energy in atomic or molecular systems by the use of electromagnetic radiation in or near the visible region.

optical pyrometer   [ENG] An instrument which determines the temperature of a very hot surface from its incandescent brightness; the image of the surface is focused in the plane of an electrically heated wire, and current through the wire is adjusted until the wire blends into the image of the surface. Also known as disappearing filament pyrometer.

optical quenching   [OPTICS] Reduction in the intensity of luminescent radiation by long-wavelength, visible or infrared radiation.

optical reflectometer   [ENG] An instrument which measures on surfaces the reflectivity of electromagnetic radiation at wavelengths in or near the visible region.

optical rotary dispersion   [OPTICS] Specific rotation, considered as a function of wavelength. Abbreviated ORD.

optical rotation   [OPTICS] Rotation of the plane of polarization of plane-polarized light, or of the major axis of the polarization ellipse of elliptically polarized light by transmission through a substance or medium.

optical sight   [OPTICS] A sight with lenses, prisms, or mirrors that is used in laying weapons, for aerial bombing, or for surveying.

optical spectra   [SPECT] Electromagnetic spectra for wavelengths in the ultraviolet, visible and infrared regions, ranging from about 10 nanometers to 1 millimeter, associated with excitations of valence electrons of atoms and molecules, and vibrations and rotations of molecules.

optical spectrograph   [SPECT] An optical spectroscope provided with a photographic camera or other device for recording the spectrum made by the spectroscope.

optical spectrometer   [SPECT] An optical spectroscope that is provided with a calibrated scale either for measurement of wavelength or for measurement of refractive indices of transparent prism materials.

optical spectroscope   [SPECT] An optical instrument, consisting of a slit, collimator lens, prism or grating, and a telescope or objective lens, which produces an optical spectrum arising from emission or absorption of radiant energy by a substance, for visual observation.

optical spectroscopy   [SPECT] The production, measurement, and interpretation of optical spectra arising from either emission or absorption of radiant energy by various substances.

optical spherical surface   [OPTICS] An optical surface which forms part of a sphere.

optical staining   *See* Rheinberg illumination.

optical superposition principle   [OPTICS] The principle that the optical rotation produced by a compound which is made up of two radicals of opposite optical activity is the algebraic sum of the rotations of each radical alone; not always valid.

optical surface   [OPTICS] An interface between two media, such as between air and glass, which is used to reflect or refract light.

optical system   [OPTICS] A collection of mirrors, lens, prisms, and other devices, placed in some specified configuration, which reflect, refract, disperse, absorb, polarize, or otherwise act on light.

optical thickness [OPTICS] The thickness of an optical material times its index of refraction.

optical transition [PHYS] A process in which an atom or molecule changes from one energy state to another and emits or absorbs electromagnetic radiation in the visible, infrared, or ultraviolet region.

optical twinning [CRYSTAL] Growing together of two crystals which are the same except that the structure of one is the mirror image of the structure of the other. Also known as chiral twinning.

optical waveguide [ELECTROMAG] A waveguide in which a light-transmitting material such as a glass or plastic fiber is used for transmitting information from point to point at wavelengths somewhere in the ultraviolet, visible-light, or infrared portions of the spectrum. Also known as fiber waveguide; optical-fiber cable.

optic axis [OPTICS] The axis in a doubly refracting medium in which the ordinary and extraordinary waves propagate with the same velocity, and double refraction vanishes. Also known as optical axis; principal axis.

optics [PHYS] **1.** Narrowly, the science of light and vision. **2.** Broadly, the study of the phenomena associated with the generation, transmission, and detection of electromagnetic radiation in the spectral range extending from the long-wave edge of the x-ray region to the short-wave edge of the radio region, or in wavelength from about 1 nanometer to about 1 millimeter.

optimal control theory [MATH] A generalized calculus of variations dealing with the analysis of dynamical systems with the viewpoint of finding optimizing conditions.

optimal policy [MATH] In optimization problems of systems, a sequence of decisions changing the states of a system in such a manner that a given criterion function is minimized.

optimal smoother [CONT SYS] An optimal filer algorithm which generates the best estimate of a dynamical variable at a certain time based on all available data, both past and future.

optimal strategy [MATH] One of the pair of mixed strategies carried out by the two players of a matrix game when each player adjusts strategy so as to minimize the maximum loss that an opponent can inflict.

optimal systems [MATH] Systems where the variables representing the various states are so determined that a given criterion function is minimized subject to given constraints.

optimization [MATH] The maximizing or minimizing of a given function possibly subject to some type of constraints. [SYS ENG] **1.** Broadly, the efforts and processes of making a decision, a design, or a system as perfect, effective, or functional as possible. **2.** Narrowly, the specific methodology, techniques, and procedures used to decide on the one specific solution in a defined set of possible alternatives that will best satisfy a selected criterion. Also known as system optimization.

optimization theory [MATH] The specific methodology, techniques, and procedures used to decide on the one specific solution in a defined set of possible alternatives that will best satisfy a selected criterion; includes linear and nonlinear programming, stochastic programming, and control theory. Also known as mathematical programming.

optimum allocation [STAT] A procedure used in stratified sampling to allocate numbers of sample units to different strata to either maximize precision at a fixed cost or minimize cost for a selected level of precision.

optimum bunching [ELECTR] Bunching condition required for maximum output in a velocity modulation tube.

optimum coupling *See* critical coupling.

optimum filter [ELECTR] An electric filter in which the mean square value of the error between a desired output and the actual output is at a minimum.

optimum reverberation time [ACOUS] The reverberation time which is most desirable for a given room size and a given use, such as speech, chamber music, or symphony orchestra.

optoacoustic effect [PHYS] A phenomenon in which a periodically interrupted beam of light generates sound in a gas through which it is passing; this results from energy in the light beam being transformed first into internal motions of the gas molecules, then into random translational motions of these molecules, or heat, and finally into periodic pressure fluctuations or sound. Also known as thermoacoustic effect.

optoacoustic modulator *See* acoustooptic modulator.

optoelectronics [ELECTR] The branch of electronics that deals with solid-state and other electronic devices for generating, modulating, transmitting, and sensing electromagnetic radiation in the ultraviolet, visible-light, and infrared portions of the spectrum.

optoelectronic shutter [ENG] A shutter that uses a Kerr cell to modulate a beam of light.

OPW method *See* orthogonalized plane-wave method.

or [MATH] A logical operation whose result is false (or zero) only if every one of its operands is false, and true (or one) otherwise. Also known as inclusive or.

orange [OPTICS] The hue evoked in an average observer by monochromatic radiation having a wavelength in the approximate range from 597 to 622 nanometers; however, the same sensation can be produced in a variety of other ways.

orange spectrometer [SPECT] A type of beta-ray spectrometer that consists of a number of modified double-focusing spectrometers employing a common source and a common detector, and has exceptionally high transmission.

O ray *See* ordinary ray.

orbit [MATH] Let G be a group which operates on a set S; the orbit of an element s of S under G is the subset of S consisting of all elements gs where g is in G. [PHYS] **1.** Any closed path followed by a particle or body, such as the orbit of a celestial body under the influence of gravity, the elliptical path followed by electrons in the Bohr theory, or the paths followed by particles in a circular particle accelerator. **2.** More generally, any path followed by a particle, such as helical paths of particles in a magnetic field, or the parabolic path of a comet.

orbital [ATOM PHYS] The space-dependent part of the Schrödinger wave function of an electron in an atom or molecule in an approximation such that each electron has a definite wave function, independent of the other electrons.

orbital angular momentum [MECH] The angular momentum associated with the motion of a particle about an origin, equal to the cross product of the position vector with the linear momentum. Also known as orbital momentum. [QUANT MECH] The angular momentum operator associated with the motion of a particle about an origin, equal to the cross product of the position vector with the linear momentum, as opposed to the intrinsic spin angular momentum. Also known as orbital moment.

orbital electron [ATOM PHYS] An electron which has a high probability of being in the vicinity (at distances on the order of 10^{-10} meter or less) of a particular nucleus, but has only a very small probability of being within the nucleus itself. Also known as planetary electron.

orbital elements [PHYS] A set of seven parameters defining the orbit of a body attracted by a central, inverse-square force.

orbital magnetic moment [QUANT MECH] The magnetic dipole moment associated with the motion of a charged particle about an origin, rather than with its intrinsic spin.

orbital moment *See* orbital angular momentum.

orbital momentum *See* orbital angular momentum.

orbital motion [PHYS] Continuous motion of a body in a closed path, such as a circle or an ellipse, about some point.

orbital node [ASTRON] One of the two points at which the orbit of a planet or satellite crosses the plane of the ecliptic or equator.

orbital parity [QUANT MECH] The parity associated with the wave function of a particle, or system of particles, as a function of spatial coordinates; it is opposed to intrinsic parity; if the orbital angular momentum quantum number is l, the orbital parity is $(-1)^l$.

orbital period [ASTRON] The interval between successive passages of a satellite through the same specified point in its orbit.

orbital plane [MECH] The plane which contains the orbit of a body or particle in a central force field; it passes through the center of force.

orbital symmetry [PHYS CHEM] The property of certain molecular orbitals of being carried into themselves or into the negative of themselves by certain geometrical operations, such as a rotation of 180° about an axis in the plane of the molecule, or reflection through this plane.

orbital velocity [ASTRON] The instantaneous velocity at which an earth satellite or other orbiting body travels around the origin of its central force field.

orbit space of a G space [MATH] The orbit space of a G space X is the topological space whose points are equivalence classes obtained by identifying points in X which have the same G orbit and whose topology is the largest topology that makes the function which sends x to its orbit continuous.

OR circuit *See* OR gate.

ORD *See* optical rotary dispersion.

order [PHYS] A range of magnitudes of a quantity (and of all other quantities having the same physical dimensions) extending from some value of the quantity to some small multiple of the quantity (usually 10). Also known as order of magnitude.

order at a prime [MATH] **1.** The order of an element a of a unique factorization domain at a prime p is the exponent of p in the unique factorization of a. Denoted $\text{ord}_p a$. **2.** The order of an element a of the quotient field of a unique factorization domain at a prime p is the integer r in the equation $a = p^r b$, where b is an element of the quotient field such that p does not divide the numerator or denominator of b; if $a = 0$ its order at p is defined to be $-\infty$. Denoted $\text{ord}_p a$. **3.** The order of a polynomial f, with coefficients in the quotient field or a unique factorization domain, at a prime p is the minimum of the orders of the nonzero coefficients of f at p; if $f = 0$, its order at p is defined to be $-\infty$. Denoted $\text{ord}_p f$.

order-disorder transition [SOLID STATE] The transition of an alloy or other solid solution between a state in which atoms of one element occupy certain regular positions in the lattice of another element, and a state in which this regularity is not present.

ordered field [MATH] A field with an ordering as a set analogous to the properties of less than or equal for real numbers relative to addition and multiplication.

ordered pair [MATH] A pair of elements x and y from a set, written (x,y), where x is distinguished as first and y as second.

ordered rings [MATH] Rings which have an ordering on them as sets in a manner analogous to the behavior of the usual ordering of the real numbers relative to addition and multiplication.

ordering [MATH] A binary relation $\leq$ among the elements of a set such that $a \leq b$ and $b \leq c$ implies $a \leq c$, and $a \leq b$, $b \leq a$ implies $a = b$; it need not be the case that either $a \leq b$ or $b \leq a$. Also known as order relation; partial ordering. [SOLID STATE] A solid-state transformation in certain solid solutions, in which a random arrangement in the lattice is transformed into a regular ordered arrangement of the atoms with respect to one another; a so-called superlattice is formed.

order of a differential equation [MATH] An equation has order n if the derivatives of a function appear up to the nth derivative.

order of a function [MATH] 1. A function $f(x)$ is of the order of $g(x)$ as x approaches x_0 if there exists a neighborhood of $x = x_0$ in which $|f(x)/g(x)|$ is bounded. Symbolized $f(x) = O[g(x)]$. 2. The order of an analytic function $f(z)$ at a point z_0 is the lowest exponent in its Taylor series about z_0.

order of a group [MATH] The number of elements contained within the group.

order of a matrix [MATH] A square matrix with n rows and n columns has order n.

order of an elliptic function [MATH] The number of poles an elliptic function has in a parallelogram region where it repeats its values.

order of an infinitesimal [MATH] A characteristic of infinitesimals used in their comparison.

order of a pole [MATH] For a pole z_0 of a meromorphic function f, the negative of the exponent of the lowest power in the expansion of $f(z)$ in powers of $z - z_0$.

order of a polynomial [MATH] The largest exponent appearing in the polynomial.

order of degeneracy *See* degree of degeneracy.

order of interference [OPTICS] The difference in the number of wavelengths along the paths of two constructively interfering rays of light.

order of magnitude *See* order.

order of phase transition [THERMO] A phase transition in which there is a latent heat and an abrupt change in properties, such as in density, is a first-order transition; if there is not such a change, the order of the transition is one greater than the lowest derivative of such properties with respect to temperature which has a discontinuity.

order relation *See* ordering.

order statistics [STAT] Variate values arranged in ascending order of magnitude; for example, first-order statistic is the smallest value of sample observations.

ordinal number [MATH] A generalized number which expresses the size of a set, in the sense of "how many" elements.

ordinal scale measurement [STAT] A method of measuring quantifiable data in nonparametric statistics that is considered to be stronger than nominal scale; it expresses the relationship of order by characterizing objects by relative rank.

ordinary differential equation [MATH] An equation involving functions of one variable and their derivatives.

ordinary index [OPTICS] The index of refraction of the ordinary ray in a crystal.

ordinary point of a curve [MATH] A point where a curve does not cross itself and where there is a smoothly turning tangent.

ordinary ray [CRYSTAL] *See* O ray. [OPTICS] One of two rays into which a ray incident on an anisotropic uniaxial crystal is split; it obeys the ordinary laws of refraction, in contrast to the extraordinary ray.

ordinary singular point [MATH] A singular point at which the tangents to all branches at the point are distinct.

ordinary-wave component [GEOPHYS] One of the two components into which an electromagnetic wave entering the ionosphere is divided under the influence of the earth's magnetic field; it has characteristics more nearly like those expected in the absence of a magnetic field. Also known as O-wave component. [OPTICS] The component of electromagnetic radiation propagating in an anisotropic uniaxial crystal whose electric displacement vector is perpendicular to the optical axis and the direction normal to the wavefront; gives rise to the ordinary ray.

ordinate [MATH] The perpendicular distance of a point (x,y) of the plane from the x axis.

organic-cooled reactor [NUCLEO] A reactor that uses organic chemicals, such as mixtures of polyphenyls (diphenyls and terphenyls), as coolant.

organic electrolyte cell [ELEC] A type of wet cell that is based on the use of particularly reactive metals such as lithium, calcium, or magnesium in conjunction with organic electrolytes; the best-known type is the lithium-cupric fluoride.

organic-moderated reactor [NUCLEO] A nuclear reactor in which organic compounds are used as moderator and coolant.

organic semiconductor [MATER] An organic material having unusually high conductivity, often enhanced by the presence of certain gases, and other properties commonly associated with semiconductors; an example is anthracene.

OR gate [ELECTR] A multiple-input gate circuit whose output is energized when any one or more of the inputs is in a prescribed state; performs the function of the logical inclusive-or; used in digital computers. Also known as OR circuit.

orientability of a sound signal [ACOUS] The property of a sound signal by virtue of which a listener can estimate the direction of the location of the apparatus producing the signal.

orientation [CRYSTAL] The directions of the axes of a crystal lattice relative to the surfaces of the crystal, to applied fields, or to some other planes or directions of interest. [ELECTROMAG] The physical positioning of a directional antenna or other device having directional characteristics. [MATH] **1.** A choice of sense or direction in a topological space. **2.** An ordering $p_0, p_1, \ldots, p_n$ of the vertices of a simplex, two such orderings being regarded as equivalent if they differ by an even permutation. [PHYS] **1.** The direction of some vector or set of vectors, such as the direction of the electric vector and the propagation direction of plane polarized light, or the direction of a preponderance of nuclear spins in a crystal near absolute zero, relative to some other directions of interest. **2.** Any process in which vectors associated with atoms or molecules in the substance are organized relative to some direction, rather than pointed at random; examples include dipole moments of polar molecules in an electric field, and nuclear spins in a crystal in a magnetic field at temperatures near absolute zero.

orientation effect [ELEC] Those bulk properties of a material which result from orientation polarization. [PHYS CHEM] A method of determining attractive forces among molecules, or components of these forces, from the interaction energy associated with the relative orientation of molecular dipoles.

orientation force [PHYS CHEM] A type of van der Waals force, resulting from interaction of the dipole moments of two polar molecules. Also known as dipole-dipole force; Keesom force.

orientation polarization [ELEC] Polarization arising from the orientation of molecules which have permanent dipole moments arising from an asymmetric charge distribution. Also known as dipole polarization.

orifice [ELECTROMAG] Opening or window in a side or end wall of a waveguide or cavity resonator through which energy is transmitted. [SCI TECH] An aperture or hole.

origin [MATH] The center of a coordinate system, where all coordinate axes meet, usually denoted by $(0,0, \ldots ,0)$.

Orion Nebula [ASTRON] A luminous cloud surrounding Ori, the northern star in Orion's dagger; visible to the naked eye as a hazy object. Also known as Great Nebula of Orion.

Ornstein-Uhlenbeck process [STAT] A stochastic process used as a theoretical model for Brownian motion.

orthoaxis [CRYSTAL] The diagonal or lateral axis perpendicular to the vertical axis in the monoclinic system.

orthobaric density [PHYS] The density of a liquid and of a saturated vapor with which it is at equilibrium at a given temperature.

orthocenter [MATH] The point at which the altitudes of a triangle intersect.

orthogonal [MATH] Perpendicular, or some concept analogous to it.

orthogonal basis [MATH] A basis for an inner product space consisting of mutually orthogonal vectors.

orthogonal complement [MATH] In an inner product space, the orthogonal complement of a vector **v** consists of all vectors orthogonal to **v**; the orthogonal complement of a subset S consists of all vectors orthogonal to each vector in S.

orthogonal crystal [CRYSTAL] A crystal whose axes are mutually perpendicular.

orthogonal family *See* orthogonal system.

orthogonal functions [MATH] Two real-valued functions are orthogonal if their inner product vanishes.

orthogonal group [MATH] The group of matrices arising from the orthogonal transformations of a euclidean space.

orthogonality [MATH] Two geometric objects have this property if they are perpendicular.

orthogonalization [MATH] A procedure in which, given a set of linearly independent vectors in an inner product space, a set of orthogonal vectors is recursively obtained so that each set spans the same subspace.

orthogonalized plane-wave method [SOLID STATE] A method of approximating the energy states of electrons in a crystal lattice: trial wave functions (the orthogonalized plane waves) are constructed which are linear combinations of plane waves and Bloch functions based on core states, and which are orthogonal to the Bloch functions, and linear combinations of these trial functions are then determined by the variational method. Abbreviated OPW method.

orthogonal Latin squares [MATH] Two $n \times n$ Latin squares which, when superposed, have the property that the n^2 cells contain each of the n^2 possible pairs of symbols exactly once. [STAT] *See* Greco-Latin square.

orthogonal lines [MATH] Lines which are perpendicular.

orthogonal matrix [MATH] A matrix whose inverse and transpose are identical.

orthogonal polynomial [MATH] Orthogonal polynomials are various families of polynomials, which arise as solutions to differential equations related to the hypergeometric equation, and which are mutually orthogonal as functions.

orthogonal projection [MATH] A continuous linear map P of a Hilbert space H onto a subspace M such that if **h** is any vector in H, $\mathbf{h} = P\mathbf{h} + \mathbf{w}$, where **w** is in the orthogonal complement of M. Also known as orthographic projection.

orthogonal series [MATH] An infinite series each term of which is the product of a member of an orthogonal family of functions and a coefficient; the coefficients are usually chosen so that the series converges to a desired function.

orthogonal spaces [MATH] Two subspaces F and F' of a vector space E with a scalar product g such that $g(x,x') = 0$ for any x in F and x' in F'.

orthogonal sum [MATH] **1.** A vector space E with a scalar product is said to be the orthogonal sum of subspaces F and F' if E is the direct sum of F and F' and if F and F' are orthogonal spaces. **2.** A scalar product g on a vector space E is said to be the orthogonal sum of scalar products f and f' on subspaces F and F' if E is the orthogonal sum of F and F' (in the sense of the first definition) and if $g(x + x', y + y') = f(x,y) + f'(x', y')$ for all x,y in F and x',y' in F'.

orthogonal system [MATH] **1.** A system made up of n families of curves on an n-dimensional manifold in an $n + 1$ dimensional euclidean space, such that exactly one curve from each family passes through every point in the manifold, and, at each point, the tangents to the n curves that pass through that point are mutually perpendicular. **2.** A set of real-valued functions, the inner products of any two of which vanish. Also known as orthogonal family.

orthogonal trajectory [MATH] A curve that intersects all the curves of a given family at right angles.

orthogonal transformation [MATH] A linear transformation between real inner product spaces which preserves the length of vectors.

orthogonal vectors [MATH] In an inner product space, two vectors are orthogonal if their inner product vanishes.

orthographic projection [CRYSTAL] A method of displaying the positions of the poles of a crystal by dropping perpendiculars from the poles onto a plane passing through the center of the reference sphere. [MATH] *See* orthogonal projection.

orthohelium [ATOM PHYS] Those states of helium atoms in which the spins of the two electrons are parallel.

orthohydrogen [ATOM PHYS] Those states of hydrogen molecules in which the spins of the two nuclei are parallel.

orthonormal coordinates [MATH] In an inner product space, the coordinates for a vector expressed relative to an orthonormal basis.

orthonormal functions [MATH] Orthogonal functions f_1, $f_2, \ldots$ with the additional property that the inner product of $f_n(x)$ with itself is 1.

orthonormal vectors [MATH] A collection of mutually orthogonal vectors, each having length 1.

orthopinacoid *See* front pinacoid.

orthopositronium [PARTIC PHYS] The state of positronium in which the positron and electron have parallel spins.

orthorhombic lattice [CRYSTAL] A crystal lattice in which the three axes of a unit cell are mutually perpendicular, and no two have the same length. Also known as rhombic lattice.

orthorhombic system [CRYSTAL] A crystal system characterized by three axes of symmetry that are mutually perpendicular and of unequal length. Also known as rhombic system.

orthoscopic eyepiece [OPTICS] An eyepiece that consists of a single lens, made up of three cemented elements, to which a planoconvex lens is added; designed to minimize distortion and spherical aberration.

orthoscopic system [OPTICS] An optical system that has been corrected so that distortion and spherical aberration are eliminated. Also known as rectilinear system.

orthotomic [MATH] The orthotomic of a curve with respect to a point is the envelope of the circles which pass through the point and whose centers lie on the curve.

orthotomic system [OPTICS] An optical system in which all the rays may be interesected at right angles by a suitably chosen surface.

orthotropic [MECH] Having elastic properties such as those of timber, that is, with considerable variations of strength in two or more directions perpendicular to one another.

Os *See* osmium.

oscillation [CONT SYS] *See* cycling. [PHYS] Any effect that varies periodically back and forth between two values.

oscillation of a function [MATH] **1.** The oscillation of a real-valued function on an interval is the difference between its least upper bound and greatest lower bound there. **2.** The oscillation of a real-valued function at a point x is the limit of the oscillation of the function on the interval $[x - e, x + e]$ as e approaches 0. Also known as saltus.

oscillation photography [SOLID STATE] A method of x-ray diffraction analysis in which a single crystal is made to oscillate through a small angle about an axis perpendicular to a beam of monochromatic x-rays or particles.

oscillator [ELECTR] **1.** An electronic circuit that converts energy from a direct-current source to a periodically varying electric output. **2.** The stage of a superheterodyne receiver that generates a radio-frequency signal of the correct frequency to mix with the incoming signal and produce the intermediate-frequency value of the receiver. **3.** The stage of a transmitter that generates the carrier frequency of the station or some fraction of the carrier frequency. [PHYS] Any device (mechanical or electrical) which, in the absence of external forces, can have a periodic back-and-forth motion, the frequency determined by the properties of the oscillator.

oscillator strength [ATOM PHYS] A quantum-mechanical analog of the number of dispersion electrons having a given natural frequency in an atom, used in an equation for the absorption coefficient of a spectral line; it need not be a whole number. Also known as f value; Ladenburg f value.

oscillator circuit [ELEC] Circuit containing inductance or capacitance, or both, and resistance, connected so that a voltage impulse will produce an output current which periodically reverses or oscillates.

oscillatory extinction *See* undulatory extinction.

oscillatory shear [FL MECH] Application of small-amplitude oscillations to produce shear in viscoelastic fluids for the study of dynamic viscosity.

oscillatory twinning [CRYSTAL] Repeated, parallel twinning.

oscillistor [ELECTR] A bar of semiconductor material, such as germanium, that will oscillate much like a quartz crystal when it is placed in a magnetic field and is carrying direct current that flows parallel to the magnetic field.

oscillogram [ENG] The permanent record produced by an oscillograph, or a photograph of the trace produced by an oscilloscope.

oscillograph [ENG] A measurement device for determining waveform by recording the instantaneous values of a quantity such as voltage as a function of time.

oscilloscope *See* cathode-ray oscilloscope.

osculating circle [MATH] For a plane curve C at a point p, the limiting circle obtained by taking the circle that is tangent to C at p and passes through a variable point q on C, and then letting q approach p.

osculating orbit [ASTRON] The orbit which would be followed by a body such as an asteroid or comet if, at a given time, all the planets suddenly disappeared, and it then moved under the gravitational force of the sun alone.

OSCILLATOR

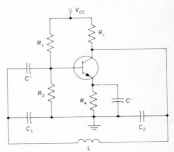

Circuit diagram of a transistor Colpitts oscillator, a commonly used type of oscillator.

osculating plane [MATH] For a curve C at some point p, this is the limiting plane obtained from taking planes through the tangent to C at p and containing some variable point p' and then letting p' approach p along C.

osculating sphere [MATH] For a curve C at a point p, the limiting sphere obtained by taking the sphere that passes through p and three other points on C and then letting these three points approach p independently along C.

O shell [ATOM PHYS] The fifth layer of electrons about the nucleus of an atom, having electrons characterized by the principal quantum number 5.

osmium [CHEM] A chemical element, symbol Os, atomic number 76, atomic weight 190.2.

Ostwald's adsorption isotherm [THERMO] An equation stating that at a constant temperature the weight of material adsorbed on an adsorbent dispersed through a gas or solution, per unit weight of adsorbent, is proportional to the concentration of the adsorbent raised to some constant power.

Ostwald viscometer [ENG] A viscometer in which liquid is drawn into the higher of two glass bulbs joined by a length of capillary tubing, and the time for its meniscus to fall between calibration marks above and below the upper bulb is compared with that for a liquid of known viscosity.

OTS *See* ovonic threshold switch.

Otto cycle [THERMO] A thermodynamic cycle for the conversion of heat into work, consisting of two isentropic phases interspersed between two constant-volume phases. Also known as spark-ignition combustion cycle.

Otto-Lardillon method [MECH] A method of computing trajectories of missiles with low velocities (so that drag is proportional to the velocity squared) and quadrant angles of departure that may be high, in which exact solutions of the equations of motion are arrived at by numerical integration and are then tabulated.

O-type star [ASTRON] A spectral-type classification in the Draper catalog of stars; a star having spectral type O; a very hot, blue star in which the spectral lines of ionized helium are prominent.

ounce [MECH] **1.** A unit of mass in avoirdupois measure equal to 1/16 pound or to approximately 0.0283495 kilogram. Abbreviated oz. **2.** A unit of mass in either troy or apothecaries' measure equal to 480 grains or exactly 0.0311034768 kilogram. Also known as apothecaries' ounce or troy ounce (abbreviations are oz ap and oz t in the United States, and oz apoth and oz tr in the United Kingdom).

ouncedal [MECH] A unit of force equal to the force which will impart an acceleration of 1 foot per second per second to a mass of 1 ounce; equal to 0.0086409346485 newton.

outer automorphism [MATH] Any element of the quotient group formed from the group of automorphisms of a group and the subgroup of inner automorphisms.

outer bremsstrahlung [PHYS] Bremsstrahlung involving the acceleration of a charged particle coming from outside the atom whose nucleus produces the acceleration, and in which the energy loss by radiation is much greater than that by ionization, usually seen in electrons with energies greater than about 50 MeV.

outer effects [PHYS] Effects on x-ray diffraction that involve neighboring atoms or molecules.

outer measure [MATH] A function with the same properties as a measure except that it is only countably subadditive rather than countably additive; usually defined on the collection of all subsets of a given set.

outer orbital complex [PHYS CHEM] A metal coordination compound in which the d orbital used in forming the

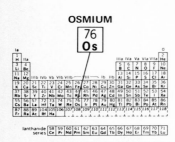

OSMIUM

Periodic table of the chemical elements showing the position of osmium.

coordinate bond is at the same energy level as the *s* and *p* orbitals.

outer planets [ASTRON] The planets with orbits larger than that of Mars: Jupiter, Saturn, Uranus, Neptune, and Pluto.

outer product [MATH] For any two tensors *R* and *S*, a tensor *T* each of whose indices corresponds to an index of *R* or an index of *S*, and each of whose components is the product of the component of *R* and the component of *S* with identical values of the corresponding indices.

outer-shell electron *See* conduction electron.

outer space [ASTRON] A general term for any region that is beyond the earth's atmosphere.

outer trapped surface [RELAT] A compact, spacelike, two-dimensional surface in a space-time, such that outgoing light rays perpendicular to the surface are not diverging; whether ingoing light rays are converging or not is immaterial.

outlier [STAT] In a set of data, a value so far removed from other values in the distribution that its presence cannot be attributed to the random combination of chance causes.

out of phase [PHYS] Having waveforms that are of the same frequency but do not pass through corresponding values at the same instant.

output [ELECTR] **1.** The current, voltage, power, driving force, or information which a circuit or device delivers. **2.** Terminals or other places where a circuit or device can deliver current, voltage, power, driving force, or information.

output capacitance [ELECTR] Of an *n*-terminal electron tube, the short-circuit transfer capacitance between the output terminal and all other terminals, except the input terminal, connected together.

output impedance [ELECTR] The impedance presented by a source to a load.

output power [ELEC] Power delivered by a system or transducer to its load.

output resistance [ELECTR] The resistance across the output terminals of a circuit or device.

output stage [ELECTR] The final stage in any electronic equipment.

oval [MATH] A curve shaped like a section of an egg.

oval of Cassini [MATH] An ovallike curve similar to a lemniscate obtained as the locus corresponding to a general type of quadratic equation in two variables *x* and *y*; namely, $[(x + a)^2 + y^2] [(x - a)^2 + y^2] = k^4$, where *a* and *k* are constants.

overall response [ELECTR] The ratio between system input and output.

over a map [MATH] A map *f* from a set *A* to a set *L* is said to be over a map *g* from a set *B* to *L* if *B* is a subset of *A* and the restriction of *f* to *B* equals *g*.

over a set [MATH] A map *f* from a set *A* to a set *L* is said to be over a set *B* if *B* is a subset of both *A* and *L* and if the restriction of *F* to *B* is the identity map on *B*.

overbunching [ELECTR] In velocity-modulated streams of electrons, the bunching condition produced by the continuation of the bunching process beyond the optimum condition.

overcoupled circuits [ELECTR] Two resonant circuits which are tuned to the same frequency but coupled so closely that two response peaks are obtained; used to attain broad-band response with substantially uniform impedance.

overdamping [PHYS] Damping greater than that required for critical damping.

overgrowth [CRYSTAL] A crystal growth in optical and crystallographic continuity around another crystal of different composition.

Overhauser effect [ATOM PHYS] The effect whereby, if a radio frequency field is applied to a substance in an external magnetic field, whose nuclei have spin $\frac{1}{2}$ and which has unpaired electrons, at the electron spin resonance frequency, the resulting polarization of the nuclei is as great as if the nuclei had the much larger electron magnetic moment.

overlap integral [QUANT MECH] The integral over space of the product of the wave function of a particle and the complex conjugate of the wave function of another particle.

overlapping orbitals [ATOM PHYS] Two orbitals (usually of electrons associated with different atoms in a molecule) for which there is a region of space where both are of appreciable magnitude.

overload [ELECTR] A load greater than that which a device is designed to handle; may cause overheating of power-handling components and distortion in signal circuits.

overpotential *See* overvoltage.

overpressure [FL MECH] The transient pressure, usually expressed in pounds per square inch, exceeding existing atmospheric pressure and manifested in the blast wave from an explosion.

override [CONT SYS] To cancel the influence of an automatic control by means of a manual control.

overriding process control [CONT SYS] Process control in which any one of several controllers associated with one control valve can be made to override another in accordance with a priority requirement of the process.

overstability [PL PHYS] Condition in which the restoring forces acting on an oscillation of a plasma or other conducting fluid drive the fluid back to its equilibrium state at a speed greater than its original outward speed, resulting in continually greater oscillation.

over-the-horizon propagation *See* scatter propagation.

overtone [MECH] One of the normal modes of vibration of a vibrating system whose frequency is greater than that of the fundamental mode. [PHYS] A harmonic other than the fundamental component.

overtone band [SPECT] The spectral band associated with transitions of a molecule in which the vibrational quantum number changes by 2 or more.

overvoltage [ELEC] A voltage greater than that at which a device or circuit is designed to operate. Also known as overpotential. [ELECTR] The amount by which the applied voltage exceeds the Geiger threshold in a radiation counter tube.

ovoid [MATH] A plane curve whose equation in cartesian coordinates r and θ is $r = a \cos^3 \theta$, where a is a constant.

ovonic device *See* glass switch.

ovonic memory switch [ELECTR] A glass switch which, after being brought from the highly resistive state to the conducting state, remains in the conducting state until a current pulse returns it to its highly resistive state. Abbreviated OMS. Also known as memory switch.

ovonic threshold switch [ELECTR] A glass switch which, after being brought from the highly resistive state to the conducting state, returns to the highly resistive state when the current falls below a holding current value. Abbreviated OTS.

Ovshinsky effect [ELECTR] The characteristic of a special thin-film solid-state switch that responds identically to both positive and negative polarities so that current can be made to flow in both directions equally.

O-wave component *See* ordinary-wave component.

Owen bridge [ELECTR] A four-arm alternating-current bridge used to measure self-inductance in terms of capacitance and resistance; bridge balance is independent of frequency.

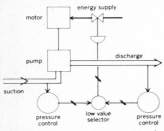

OVERRIDING PROCESS CONTROL

Diagram of overriding process control in an oil pipeline. Motor driving the pipeline is controlled by suction pressure when this pressure approaches the atmospheric value, and by the discharge pressure when this is close to the maximum allowable value. The low-value selector passes the lower of its inputs, in order to prevent the noncritical controlled value from increasing the speed of the pump.

oxidation potential [PHYS CHEM] The difference in potential between an atom or ion and the state in which an electron has been removed to an infinite distance from this atom or ion.

oxide-coated cathode [ELECTR] A cathode that has been coated with oxides of alkaline-earth metals to improve electron emission at moderate temperatures. Also known as Wehnelt cathode.

oxide fuel reactor [NUCLEO] A nuclear fission reactor with fuel in the form UO_2 or PuO_2.

oxide nuclear fuel [NUCLEO] The fissionable nuclear fuel UO_2 or PuO_2.

oxygen [CHEM] A gaseous chemical element, symbol O, atomic number 8, and atomic weight 15.9994; an essential element in cellular respiration and in combustion processes; the most abundant element in the earth's crust, and about 20% of the air by volume.

oxygen-18 [NUC PHYS] Oxygen isotope with atomic weight 18; found 8 parts to 10,000 of oxygen-16 in water, air, and rocks; used in tracer experiments. Also known as heavy oxygen.

oxygen bomb calorimeter [ENG] Device to measure heat of combustion; the sample is burned with oxygen in a closed vessel, and the temperature rise is noted.

oxygen cell *See* aeration cell.

oxygen point [THERMO] The temperature at which liquid oxygen and its vapor are in equilibrium, that is, the boiling point of oxygen, at standard atmospheric pressure; it is taken as a fixed point on the International Practical Temperature Scale of 1968, at $-182.962°C$.

oz *See* ounce.

oz ap *See* ounce.

oz apoth *See* ounce.

oz t *See* ounce.

oz tr *See* ounce.

OXYGEN

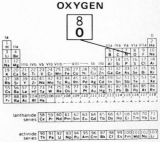

Periodic table of the chemical elements showing the position of oxygen.

P *See* phosphorus; poise.

pA *See* picoampere.

Pa *See* pascal; protactinium.

Paasche's index [STAT] A weighted aggregate price index with given-year quantity weights. Also known as given-year method.

pachimeter [ENG] An instrument for measuring the limit beyond which shear of a solid ceases to be elastic.

package power reactor [NUC PHYS] A nuclear power plant designed to be crated in packages small enough for transportation to remote locations.

packing [CRYSTAL] The arrangement of atoms or ions in a crystal lattice.

packing fraction [NUC PHYS] The quantity $(M - A)/A$, where M is the mass of an atom in atomic mass units and A is its atomic number.

packing index [CRYSTAL] The volume of ion divided by the volume of the unit cell in a crystal.

packing radius [CRYSTAL] One-half the smallest approach distance of atoms or ions.

Pade table [MATH] A table associated to a power series having in its pth row and qth column the ratio of a polynomial of degree q by one of degree p so that this fraction expanded into a power series agrees with the original up to the $p + q$ term.

p-adic field [MATH] The set of all p-adic numbers, with addition and multiplication defined by the equations $a/p^k + b/p^m = (ap^m + bp^k)/p^{k+m}$ and $(a/p^k)(b/p^m) = ab/p^{k+m}$.

p-adic integer [MATH] An object determined by a sequence of integers $\{x_n\} = \{x_0, x_1, \ldots, x_n, \ldots\}$, satisfying

$$x_n \equiv x_{n-1} \pmod{p^n}$$

for all $n \geq 1$, where p is a fixed prime number; two such sequences, $\{x_n\}$ and $\{y_n\}$, are said to determine the same p-adic integer if and only if $x_n \equiv y_n \pmod{p^{n+1}}$ for all $n \geq 0$.

p-adic number [MATH] A fraction of the form a/p^k, where p is a fixed prime number, a is a p-adic integer, and k is a nonnegative integer; two fractions, a/p^k and b/p^m, are said to determine the same p-adic number if and only if ap^m and bp^k are the same p-adic integer.

pair [ELEC] Two like conductors employed to form an electric circuit. [SCI TECH] A set of two things that are identical or nearly so, or are designed to function as a unit.

paired comparison [STAT] A method used where order relations are more easily determined than measurements, such as studying taste preferences; in the comparison of a group of objects, each pair of objects is tested with either one or the other or neither preferred.

paired electron [PHYS CHEM] One of two electrons that form a valence bond between two atoms.

PALLADIUM

46
Pd

Ia																	0
1 H	IIa											IIIa	IVa	Va	VIa	VIIa	2 He
3 Li	4 Be											5 B	6 C	7 N	8 O	9 F	10 Ne
11 Na	12 Mg	IIIb	IVb	Vb	VIb	VIIb	— VIII —			Ib	IIb	13 Al	14 Si	15 P	16 S	17 Cl	18 Ar
19 K	20 Ca	21 Sc	22 Ti	23 V	24 Cr	25 Mn	26 Fe	27 Co	28 Ni	29 Cu	30 Zn	31 Ga	32 Ge	33 As	34 Se	35 Br	36 Kr
37 Rb	38 Sr	39 Y	40 Zr	41 Nb	42 Mo	43 Tc	44 Ru	45 Rh	46 Pd	47 Ag	48 Cd	49 In	50 Sn	51 Sb	52 Te	53 I	54 Xe
55 Cs	56 Ba	57 La	72 Hf	73 Ta	74 W	75 Re	76 Os	77 Ir	78 Pt	79 Au	80 Hg	81 Tl	82 Pb	83 Bi	84 Po	85 At	86 Rn
87 Fr	88 Ra	89 Ac	104 Rf	105 Ha	106	107	108	109	110	111	112	113	114	115	116	117	118

lanthanide series	58 Ce	59 Pr	60 Nd	61 Pm	62 Sm	63 Eu	64 Gd	65 Tb	66 Dy	67 Ho	68 Er	69 Tm	70 Yb	71 Lu

actinide series	90 Th	91 Pa	92 U	93 Np	94 Pu	95 Am	96 Cm	97 Bk	98 Cf	99 Es	100 Fm	101 Md	102 No	103 Lr

Periodic table of the chemical elements showing the position of palladium.

pair production [PHYS] The conversion of a photon into an electron and a positron when the photon traverses a strong electric field, such as that surrounding a nucleus or an electron.

palladium [CHEM] A chemical element, symbol Pd, atomic number 46, atomic weight 106.4.

palpable coordinate [MECH] A generalized coordinate that appears explicitly in the Lagrangian of a system.

panchratic eyepiece [OPTICS] A telescope eyepiece whose magnifying power can be varied by moving the erecting lens while keeping the focus at infinity.

panoramic [OPTICS] Property of a lens or optical instrument that has a wide field of view.

paper capacitor [ELEC] A capacitor whose dielectric material consists of oiled paper sandwiched between two layers of metallic foil.

Papkovich-Neuber solution [MECH] A solution to the equations of equilibrium for an isotropic, elastic solid obeying a generalized Hooke's law, having the form $2\mu u = 4(1 - \sigma)A$ − grad (ϕ + $r \cdot A$), where u is the displacement vector at the point with position vector r, μ is the rigidity modulus, σ is Poisson's ratio, and the scalar ϕ and vector A are functions that satisfy Laplace's equation in the absence of body forces. Also known as Boussinesq-Papkovich solution; Neuber-Papkovich solution.

Papkovich-Neuber stress functions [MECH] The scalar (ϕ) and vector (A) functions used in the Papkovich-Neuber solution.

Pappian plane [MATH] Any projective plane in which points and lines satisfy Pappus' theorem (third definition).

Pappus' theorem [MATH] 1. The area of a surface of revolution generated by rotating a plane curve about an axis in its own plane which does not intersect it is equal to the length of the curve multiplied by the length of the path of its centroid. 2. The volume of a solid of revolution generated by rotating a plane area about an axis in its own plane which does not intersect it is equal to the area multiplied by the length of the path of its centroid. 3. A theorem of projective geometry which states that if A, B, and C are collinear points and A', B', and C' are also collinear points, then the intersection of AB' with $A'B$, the intersection of AC' with $A'C$, and the intersection of BC' with $B'C$ are collinear. 4. A theorem of projective geometry which states that if A, B, C, and D are fixed points on a conic and P is a variable point on the same conic, then the product of the perpendiculars from P to AB and CD divided by the product of the perpendiculars from P to AD and BC is constant.

parabola [MATH] The U-shaped curve in the plane given by an equation of the form $y = ax^2 + bx + c$.

parabolic antenna [ELECTROMAG] Antenna with a radiating element and a parabolic reflector that concentrates the radiated power into a beam.

parabolic coordinate system [MATH] 1. A two-dimensional coordinate system determined by a system of confocal parabolas. 2. A three-dimensional coordinate system whose coordinate surfaces are the surfaces generated by rotating a plane containing a system of confocal parabolas about the axis of symmetry of the parabolas, together with the planes passing through the axis of rotation.

parabolic cylinder functions [MATH] Solutions to the Weber differential equation, which results from separation of variables of the Laplace equation in parabolic cylindrical coordinates.

parabolic cylindrical coordinate system [MATH] A three-dimensional coordinate system in which two of the coordinates depend on the x and y coordinates in the same manner as parabolic coordinates and are independent of the z coordinate, while the third coordinate is directly proportional to the z coordinate.

parabolic differential equation [MATH] A general type of second-order partial differential equation which includes the heat equation and has the form

$$\sum_{i,j=1}^{n} A_{ij}(\partial^2 u / \partial x_i \, \partial x_j) + \sum_{i=1}^{n} B_i(\partial u / \partial x_i) + Cu + F = 0$$

where the A_{ij}, B_i, C, and F are suitably differentiable real functions of $x_1, x_2, \ldots, x_n$, and there exists at each point $(x_1, \ldots, x_n)$ a real linear transformation on the x_i which reduces the quadratic form

$$\sum_{i,j=1}^{n} A_{ij} x_i x_j$$

to a sum of $n - 1$ squares all of which have the same sign, while the same transformation does not reduce the B_i to 0. Also known as parabolic partial differential equation.

parabolic orbit [ASTRON] An orbit whose overall shape is like a parabola; the orbit represents the least eccentricity for escape from an attracting body.

parabolic partial differential equation *See* parabolic differential equation.

parabolic point of a surface [MATH] A point on a surface where the total curvature vanishes.

parabolic reflector [ELECTROMAG] An antenna having a concave surface which is generated either by translating a parabola perpendicular to the plane in which it lies (in a cylindrical parabolic reflector), or rotating it about its axis of symmetry (in a paraboloidal reflector). Also known as dish. [OPTICS] *See* paraboloidal reflector.

parabolic rule *See* Simpson's rule.

parabolic segment [MATH] The line segment given by a chord perpendicular to the axis of a parabola.

parabolic spiral [MATH] The curve whose equation in polar coordinates is $r^2 = a\theta$.

parabolic velocity [ASTRON] The velocity attained by a celestial body in a parabolic orbit.

paraboloid [MATH] A surface where sections through one of its axes are ellipses, and sections through the other are parabolas.

paraboloidal antenna *See* paraboloidal reflector.

paraboloidal coordinate system [MATH] A three-dimensional coordinate system in which the coordinate surfaces form families of confocal elliptic and hyperbolic paraboloids.

paraboloidal reflector [ELECTROMAG] An antenna having a concave surface which is a paraboloid of revolution; it concentrates radiation from a source at its focal point into a beam. Also known as paraboloidal antenna. [OPTICS] A concave mirror which is a paraboloid of revolution and produces parallel rays of light from a source located at the focus of the parabola. Also known as parabolic reflector.

paraboloid of revolution [MATH] The surface obtained by spinning a parabola about its axis.

parachor [PHYS] The molecular weight of a liquid times the fourth root of its surface tension, divided by the difference between the density of the liquid and the density of the vapor in equilibrium with it; essentially constant over wide ranges of temperature.

paracompact space [MATH] A Hausdorff space with the property that for every open covering F there is a locally finite open covering G such that every element of G is a subset of an element F.

PARABOLOIDAL REFLECTOR

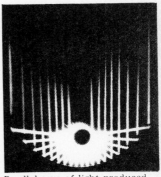

Parallel rays of light produced from paraboloidal reflector; dark circle is the light source.

paradox [SCI TECH] An argument which gives a contradictory conclusion.

parahelium [ATOM PHYS] Those states of helium in which the spins of the two electrons are antiparallel, in contrast to orthohelium. Also spelled parhelium.

parahydrogen [ATOM PHYS] Those states of hydrogen molecules in which the spins of the two nuclei are antiparallel.

parallax [OPTICS] The change in the apparent relative orientations of objects when viewed from different positions.

parallax error [OPTICS] Error in reading an instrument employing a scale and pointer because the observer's eye and pointer are not in a line perpendicular to the plane of the scale.

parallax-second *See* parsec.

parallel [MATH] **1.** Lines are parallel in a euclidean space if they lie in a common plane and do not intersect. **2.** Planes are parallel in a euclidean three-dimensional space if they do not intersect. **3.** A circle parallel to the primary great circle of a sphere or spheroid. [PHYS] Of two or more displacements or other vectors, having the same direction.

parallel axiom [MATH] The axiom of an affine plane which states that if p and L are a point and line in the plane such that p is not on L, then there exists exactly one line that passes through p and does not intersect L.

parallel axis theorem [MECH] A theorem which states that the moment of inertia of a body about any given axis is the moment of inertia about a parallel axis through the center of mass, plus the moment of inertia that the body would have about the given axis if all the mass of the body were located at the center of mass. Also known as Steiner's theorem.

parallel circuit [ELEC] An electric circuit in which the elements, branches (having elements in series), or components are connected between two points, with one of the two ends of each component connected to each point.

parallel compensation *See* feedback compensation.

parallel displacement [MATH] A vector A at a point P of an affine space is said to be obtained from a vector B at a point Q of the space by a parallel displacement with respect to a curve connecting A and B if a vector $V(X)$ can be associated with each point X on the curve in such a manner that $A=V(P)$, $B = V(Q)$, and the values of V at neighboring points of the curve are parallel as specified by the affine connection.

parallelepiped [MATH] A polyhedron all of whose faces are parallelograms.

parallel extinction [OPTICS] Nearly total absorption of light that is propagating in an anisotropic crystal in a direction parallel to crystal outlines or traces of cleavage planes.

parallel feed [ELECTR] Application of a direct-current voltage to the plate or grid of a tube in parallel with an alternating-current circuit, so that the direct-current and the alternating-current components flow in separate paths. Also known as shunt feed.

parallel growth *See* parallel intergrowth.

parallel impedance [ELEC] One of two or more impedances that are connected to the same pair of terminals.

parallel intergrowth [CRYSTAL] Intergrowth of two or more crystals in such a way that one or more axes in each crystal are approximately parallel. Also known as parallel growth.

parallel of altitude [ASTRON] A circle on the celestial sphere parallel to the horizon connecting all points of equal altitude. Also known as almucantar; altitude circle.

parallel of declination [ASTRON] A small circle of the celestial sphere parallel to the celestial equator. Also known as celestial parallel; circle of equal declination.

parallel of latitude *See* circle of longitude.

PARALLEL CIRCUIT

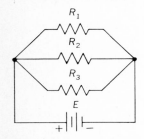

Schematic diagram of a simple parallel circuit in which the resistors, R_1, R_2, and R_3, are connected in parallel between terminals of battery which supplies voltage E.

parallelogram [MATH] A four-sided polygon with each pair of opposite sides parallel.

parallelogram of vectors [MATH] A parallelogram whose sides form two vectors to be added and whose diagonal is the sum of the two vectors.

parallel operation [ELECTR] The connecting together of the outputs of two or more batteries or other power supplies so that the sum of their output currents flows to a common load.

parallelotope [MATH] A parallelepiped with sides in proportion of 1, ½, and ¼.

parallel-plate capacitor [ELEC] A capacitor consisting of two parallel metal plates, with a dielectric filling the space between them.

parallel-plate laser [OPTICS] A laser which has two small parallel plates facing each other at a distance which is large compared with their diameters; one of them reflects light and the other is partially reflecting, so that light can bounce back and forth between the plates enough to build up a strong pulse.

parallel-plate waveguide [ELECTROMAG] Pair of parallel conducting planes used for propagating uniform circularly cylindrical waves having their axes normal to the plane.

parallel resonance Also known as antiresonance. [ELEC] **1.** The frequency at which the inductive and capacitive reactances of a parallel resonant circuit are equal. **2.** The frequency at which the parallel impedance of a parallel resonant circuit is a maximum. **3.** The frequency at which the parallel impedance of a parallel resonant circuit has a power factor of unity.

parallel resonant circuit [ELEC] A circuit in which an alternating-current voltage is applied across a capacitor and a coil in parallel. Also known as antiresonant circuit.

parallel series [ELEC] Circuit in which two or more parts are connected together in parallel to form parallel circuits, and in which these circuits are then connected together in series so that both methods of connection appear.

parallel-slit interferometer [OPTICS] A type of stellar interferometer consisting of a screen with two narrow, parallel slits whose separation is adjustable, placed over the objective of a refracting telescope.

parallel-T network [ELEC] A network used in capacitance measurements at radio frequencies, having two sets of three impedances, each in the form of the letter T, with the arms of the two Ts joined to common terminals, and the source and detector each connected between two of these terminals. Also known as twin-T network.

parallel-tuned circuit [ELEC] A circuit with two parallel branches, one having an inductance and a resistance in series, the other a capacitance and a resistance in series.

paramagnetic [ELECTROMAG] Exhibiting paramagnetism.

paramagnetic amplifier *See* maser.

paramagnetic cooling *See* adiabatic demagnetization.

paramagnetic crystal [ELECTROMAG] A crystal whose permeability is slightly greater than that of vacuum and is independent of the magnetic field strength.

paramagnetic Faraday effect [OPTICS] The Faraday effect observed in paramagnetic salts at frequencies near an absorption line of the salt which is split due to splitting of the lower energy level.

paramagnetic material [ELECTROMAG] A material within which an applied magnetic field is increased by the alignment of electron orbits.

paramagnetic relaxation [ELECTROMAG] The approach of a system, which displays paramagnetism because of electronic magnetic moments of atoms or ions, to an equilibrium or

PARALLEL-PLATE CAPACITOR

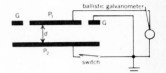

Cross section of parallel-plate capacitor with all parts of capacitor at ground potential. P_1 and P_2 are the parallel plates; G is the guard ring, used to reduce edge effects; d is the distance between plates. Distance between guard ring and plate P_1 is exaggerated.

PARALLEL-T NETWORK

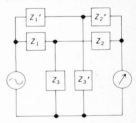

Schematic drawing of the network showing the two sets of three impedances: Z_1, Z_2, Z_3, and Z_1', Z_2', Z_3'.

PARAMAGNETIC FARADAY EFFECT

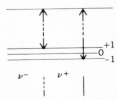

Schematic energy level diagram of ions of a paramagnetic salt. The lengths of the arrows are proportional to the frequencies of absorption lines for right-handed (ν^+) and left-handed (ν^-) circularly polarized light.

steady-state condition over a period of time, following a change in the magnetic field.

paramagnetic resonance *See* electron paramagnetic resonance.

paramagnetic salt [ELECTROMAG] A salt whose permeability is slightly greater than that of vacuum and is independent of magnetic field strength; used in adiabatic demagnetization.

paramagnetic spectra [SPECT] Spectra associated with the coupling of the electronic magnetic moments of atoms or ions in paramagnetic substances, or in paramagnetic centers of diamagnetic substances, to the surrounding liquid or crystal environment, generally at microwave frequencies.

paramagnetic susceptibility [ELECTROMAG] The susceptibility of a paramagnetic substance, which is a positive number and is, in general, much smaller than unity.

paramagnetism [ELECTROMAG] A property exhibited by substances which, when placed in a magnetic field, are magnetized parallel to the field to an extent proportional to the field (except at very low temperatures or in extremely large magnetic fields).

parameter [CRYSTAL] Any of the axial lengths or interaxial angles that define a unit cell. [ELEC] **1.** The resistance, capacitance, inductance, or impedance of a circuit element. **2.** The value of a transistor or tube characteristic. [MATH] An arbitrary constant or variable so appearing in a mathematical expression that changing it gives various cases of the phenomenon represented. [PHYS] A quantity which is constant under a given set of conditions, but may be different under other conditions.

parameterization [SCI TECH] The representation, in a dynamic model, of physical effects in terms of admittedly oversimplified parameters, rather than realistically requiring such effects to be consequences of the dynamics of the system.

parametric acoustic array [ACOUS] A device for generating very sharp beams of sound devoid of side lobes, consisting of a source of well-collimated high-frequency sound modulated at the frequency of the sound which is to be generated.

parametric amplifier [ELECTR] A highly sensitive ultra-high-frequency or microwave amplifier having as its basic element an electron tube or solid-state device whose reactance can be varied periodically by an alternating-current voltage at a pumping frequency. Also known as mavar; paramp; reactance amplifier. [OPTICS] A device consisting of an optically nonlinear crystal in which an optical or infrared beam draws power from a laser beam at a higher frequency and is amplified.

parametric converter [ELECTR] Inverting or noninverting parametric device used to convert an input signal at one frequency into an output signal at a different frequency.

parametric curves [MATH] On a surface determined by equations $x = f(u,v)$, $y = g(u,v)$, and $z = h(u,v)$, these are families of curves obtained by setting the parameters u and v equal to various constants.

parametric device [ELECTR] Electronic device whose operation depends essentially upon the time variation of a characteristic parameter usually understood to be a reactance.

parametric equation [MATH] An equation where coordinates of points appear dependent on parameters, such as the parametric equation of a curve or a surface.

parametric excitation [ENG] The method of exciting and maintaining oscillation in either an electrical or mechanical dynamic system, in which excitation results from a periodic variation in an energy storage element in a system such as a capacitor, inductor, or spring constant.

parametric mixing [OPTICS] In a medium possessing optical

PARAMETRIC AMPLIFIER

Schematic of negative-resistance-type parametric amplifier that uses a varactor diode as the variable reactor. f_p = pumping frequency. Input signal, with frequency f_s, is reflected back down transmission line with increased power. Wave at idler frequency $f_p - f_s$ is not utilized outside the amplifier.

nonlinearities, the mixing of electromagnetic waves to form waves with frequencies linearly related to the frequency of incident radiation.

parametric oscillator [ELECTR] An oscillator in which the reactance parameter of an energy-storage device is varied to obtain oscillation. [OPTICS] A device consisting of an optically nonlinear crystal surrounded by a pair of mirrors to which is applied a relatively high-frequency laser beam and a relatively low-frequency signal, resulting in a low-frequency output whose frequency can be varied, usually by varying the indices of refraction.

paramp *See* parametric amplifier.

paraphase amplifier [ELECTR] An amplifier that provides two equal output signals 180° out of phase.

parapositronium [PARTIC PHYS] The state of positronium in which the positron and electron have antiparallel spins.

parasitic absorption *See* parasitic capture.

parasitic antenna *See* parasitic element.

parasitic capture [NUCLEO] Any absorption of a neutron that does not result in a fission or the production of a desired element. Also known as parasitic absorption.

parasitic element [ELECTROMAG] An antenna element that serves as part of a directional antenna array but has no direct connection to the receiver or transmitter and reflects or reradiates the energy that reaches it, in a phase relationship such as to give the desired radiation pattern. Also known as parasitic antenna; parasitic reflector; passive element.

parasitic reflector *See* parasitic element.

parastate [ATOM PHYS] A state of a diatomic molecule in which the spins of the nuclei are antiparallel.

paraxial [SCI TECH] Lying near the axis.

paraxial rays [OPTICS] Rays which are close enough to the opical axis of a system, and thus whose directions are sufficiently close to being parallel to it, so that sines of angles between the rays and the optical axis may be replaced by the angles themselves in calculations.

paraxial trajectory [ELEC] A trajectory of a charged particle in an axially symmetric electric or magnetic field in which both the distance of the particle from the axis of symmetry and the angle between this axis and the tangent to the trajectory are small for all points on the trajectory.

parent [NUC PHYS] A radionuclide that upon disintegration yields a specified nuclide, the daughter, either directly, or indirectly as a later member of a radioactive series.

parent of a state [QUANT MECH] If an n-electron state is written as a sum of products of 1-electron states and $(n - 1)$-electron states, the $(n - 1)$-electron states are called the parents of the n-electron states.

parfocal eyepieces [OPTICS] Eyepieces whose lower focal points lie in the same plane, so that they can be interchanged without changing the focus of the instrument with which they are used.

parhelium *See* parahelium.

parity [MATH] Two integers have the same parity if they are both even or both odd. [QUANT MECH] A physical property of a wave function which specifies its behavior under an inversion, that is, under simultaneous reflection of all three spatial coordinates through the origin; if the wave function is unchanged by inversion, its parity is 1 (or even); if the function is changed only in sign, its parity is minus 1 (or odd). Also known as space reflection symmetry.

parity conservation *See* conservation of parity.

parity selection rules [QUANT MECH] Rules which specify whether or not a change in parity occurs during a given type of transition of an atom, molecule, or nucleus; for example, the Laporte selection rule, or the rule that there is no parity change in an allowed β-decay transition of a nucleus.

parker *See* rep.

Parker-Washburn boundary [SOLID STATE] A surface which separates two regions in a solid in which the crystal axes point in different directions, and which is made up of a single array of dislocations.

Parry arcs [OPTICS] A class of halos appearing as faintly colored arcs above and below the sun; these refraction phenomena are produced by ice crystals which exhibit a preferred orientation, and are correspondingly more unusual than those associated with randomly oriented crystals.

parsec [ASTRON] The distance at which a star would have a parallax equal to 1 second of arc; 1 parsec equals 3.258 light-years or 3.08572×10^{13} kilometers. Derived from parallax-second.

Parseval's equation [MATH] The equation which states that the square of the length of a vector in an inner product space is equal to the sum of the squares of the inner products of the vector with each member of a complete orthonormal base for the space. Also known as Parseval's identity; Parseval's relation.

Parseval's identity *See* Parseval's equation.

Parseval's relation *See* Parseval's equation.

Parseval's theorem [MATH] A theorem that gives the integral of a product of two functions, $f(x)$ and $F(x)$, in terms of their respective Fourier coefficients; if the coefficients are defined by

$$a_n = (1/\pi) \int_0^{2\pi} f(x) \cos nx \, dx$$

$$b_n = (1/\pi) \int_0^{2\pi} f(x) \sin nx \, dx$$

and similarly for $F(x)$, the relationship is

$$\int_0^{2\pi} f(x)F(x) \, dx = \pi [\tfrac{1}{2}a_0 A_0 + \sum_{n=1}^{\infty} (a_n A_n + b_n B_n)].$$

partial [ACOUS] Also known as partial tone. **1.** A simple sinusoidal physical component of a complex tone. **2.** A sound sensation component that is distinguishable as a simple tone, cannot be further analyzed by the ear, and contributes to the character of the complex sound; the frequency of a partial may be higher or lower than the basic frequency and may be an integral multiple or submultiple of the basic frequency.

partial Cauchy surface [RELAT] A spacelike surface S which is intersected only once by each timelike or null curve; "partial" means that only a portion of the future history of the space-time can be predicted from S, that is, there exists a Cauchy horizon.

partial correlation [STAT] The strength of the linear relationship between two random variables where the effect of other variables is held constant.

partial correlation analysis [STAT] A technique used to measure the strength of the relationship between the dependent variable and one independent variable in such a way that variations in other independent variables are taken into account.

partial correlation coefficient [STAT] A measure of the strength of association between a dependent variable and one independent variable when the effect of all other independent variables is removed; equal to the square root of the partial coefficient of determination.

partial derivative [MATH] A derivative of a function of several variables taken with respect to one variable while holding the others fixed.

partial differential equation [MATH] An equation that involves more than one independent variable and partial derivatives with respect to those variables.

partial dislocation [CRYSTAL] The line at the edge of an extended dislocation where a slip through a fraction of a lattice constant has occurred.

partial fractions [MATH] A collection of fractions which when added are a given fraction whose numerator and denominator are usually polynomials; the partial fractions are usually constants or linear polynomials divided by factors of the denominator of the given fraction.

partially balanced incomplete block design [STAT] An experimental design in which, while all treatments are not represented in each block, each treatment is tested the same number of times and certain aspects of the design satisfy conditions which simplify the least squares analysis.

partially ionic bond [PHYS CHEM] A chemical bond that is neither wholly ionic nor wholly covalent in character.

partially ordered set [MATH] A set on which a partial order is defined.

partial ordering *See* ordering.

partial plane [MATH] In projective geometry, a plane in which at most one line passes through any two points.

partial pressure [PHYS] The pressure that would be exerted by one component of a mixture of gases if it were present alone in a container.

partial regression coefficient [STAT] Statistics in the population multiple linear regression equation that indicate the effect of each independent variable on the dependent variable with the influence of all the remaining variables held constant; each coefficient is the slope between the dependent variable and each of the independent variables.

partial sum [MATH] A partial sum of an infinite series is the sum of its first n-terms for some n.

partial tone *See* partial.

particle [MECH] *See* material particle. [PARTIC PHYS] *See* elementary particle. [PHYS] **1.** Any very small part of matter, such as a molecule, atom, or electron. **2.** Any relatively small subdivision of matter, ranging in diameter from a few angstroms (as with gas molecules) to a few millimeters (as with large raindrops).

particle accelerator [NUCLEO] A device which accelerates electrically charged atomic or subatomic particles, such as electrons, protons, or ions, to high energies. Also known as accelerator; atom smasher.

particle beam [PHYS] A concentrated, nearly unidirectional flow of particles.

particle detector [NUCLEO] A device used to indicate the presence of fast-moving charged atomic or nuclear particles by observation of the electrical disturbance created by a particle as it passes through the device. Also known as radiation detector.

particle dynamics [MECH] The study of the dependence of the motion of a single material particle on the external forces acting upon it, particularly electromagnetic and gravitational forces.

particle emission [NUC PHYS] The ejection of a particle other than a photon from a nucleus, in contrast to gamma emission.

particle energy [MECH] For a particle in a potential, the sum of the particle's kinetic energy and potential energy. [RELAT] For a relativistic particle the sum of the particle's potential energy, kinetic energy, and rest energy; the last is equal to the product of the particle's rest mass and the square of the speed of light.

particle horizon [RELAT] The spatial boundary beyond

which, in certain universe models, it is impossible for an observer at a given time to receive a signal.

particle lens [PHYS] An electric or magnetic field, or a combination thereof, which acts upon an electron beam in a manner analogous to that in which an optical lens acts upon a light beam.

particle mechanics [MECH] The study of the motion of a single material particle.

particle multiplet *See* isospin multiplet.

particle physics [PHYS] The branch of physics concerned with understanding the properties and behavior of elementary particles, especially through study of collisions or decays involving energies of hundreds of MeV or more. Also known as high-energy physics.

particle properties [PARTIC PHYS] The various quantities which characterize the behavior of an elementary particle, such as mass, charge, baryon number, spin, parity, hypercharge, and isospin.

particle spectrum *See* mass spectrum.

particle track [PHYS] Any visible phenomenon along the path of an ionizing particle, such as a trail of bubbles, water droplets, or sparks in a bubble chamber, cloud chamber, or spark chamber respectively, or of altered material in an emulsion or in glass.

particle velocity [ACOUS] The instantaneous velocity of a given infinitesimal part of a medium, with reference to the medium as a whole, due to the passage of a sound wave.

particular integral *See* particular solution of an ordinary differential equation.

particular solution of an ordinary differential equation [MATH] A solution to a differential equation obtained by assigning numerical values to the parameters in the general solution. Also known as particular integral.

partition function [STAT MECH] **1.** The integral, over the phase space of a system, of the exponential of $(-E/kT)$, where E is the energy of the system, k is Boltzmann's constant, and T is the temperature; from this function all the thermodynamic properties of the system can be derived. **2.** In quantum statistical mechanics, the sum over allowed states of the exponential of $(-E/kT)$. Also known as sum of states; sum over states.

partition of an integer [MATH] Any collection of positive integers whose sum is the given integer.

partition of a positive integer [MATH] Any sum of positive integers equaling the integer.

partition of a set [MATH] A finite collection of disjoint sets whose union is the given set.

partition of unity [MATH] On a topological space X, this is a covering by open sets U_α with continuous functions f_α from X to $[0,1]$, where each f_α is zero on all but a finite number of the U_α, and the sum of all these f_α at any point equals 1.

parton [PARTIC PHYS] One of the very singular (or hard), small charged particles of which hadrons are proposed to be constructed, according to a theory developed to account for the scattering of very-high-energy electrons from protons at large angles and with large momentum transfers.

pascal [MECH] A unit of pressure equal to the pressure resulting from a force of 1 newton acting uniformly over an area of 1 square meter. Symbolized Pa. Also known as torr.

Pascal's law [FL MECH] The law that a confined fluid transmits externally applied pressure uniformly in all directions, without change in magnitude.

Pascal's limacon *See* limacon.

Pascal's theorem [MATH] The theorem that when one in-

PARTICLE TRACK

Photograph of particle tracks consisting of trails of bubbles in a liquid-hydrogen bubble chamber, produced by negative pions with momentum 16 GeV/c, where c is the speed of light. (CERN)

scribes a simple hexagon in a conic, the three pairs of opposite sides meet in collinear points.

Pascal's triangle [MATH] A triangular array of the binomial coefficients, bordered by ones, where the sum of two adjacent entries from a row equals the entry in the next row directly below. Also known as binomial array.

Paschen-Back effect [SPECT] An effect on spectral lines obtained when the light source is placed in a very strong magnetic field; the anomalous Zeeman effect obtained with weaker fields changes over to what is, in a first approximation, the normal Zeeman effect.

Paschen-Runge mounting [SPECT] A diffraction grating mounting in which the slit and grating are fixed, and photographic plates are clamped to a fixed track running along the corresponding Rowland circle.

Paschen series [SPECT] A series of lines in the infrared spectrum of atomic hydrogen whose wave numbers are given by $R_H [(1/9) - (1/n^2)]$, where R_H is the Rydberg constant for hydrogen, and n is any integer greater than 3.

Paschen's law [ELECTR] The law that the sparking potential between two parallel plate electrodes in a gas is a function of the product of the gas density and the distance between the electrodes. Also known as Paschen's rule.

Paschen's rule *See* Paschen's law.

passband [ELECTR] A frequency band in which the attenuation of a filter is essentially zero.

passivation [ELECTR] Growth of an oxide layer on the surface of a semiconductor to provide electrical stability by isolating the transistor surface from electrical and chemical conditions in the environment; this reduces reverse-current leakage, increases breakdown voltage, and raises power dissipation rating.

passive component *See* passive element.

passive element [ELEC] An element of an electric circuit that is not a source of energy, such as a resistor, inductor, or capacitor. Also known as passive component. [ELECTRO-MAG] *See* parasitic element.

passive filter [ELEC] An electric filter composed of passive elements, such as resistors, inductors, or capacitors, without any active elements, such as vacuum tubes or transistors.

passive network [ELEC] A network that has no source of energy.

passive reflector [ELECTROMAG] A flat reflector used to change the direction of a microwave or radar beam; often used on microwave relay towers to permit placement of the transmitter, repeater, and receiver equipment on the ground, rather than at the tops of towers. Also known as plane reflector.

past Cauchy development [RELAT] The set of points p relative to a surface S in space-time such that every future-directed timelike or null curve through p intersects S. Symbolized $D^-(S)$.

paste [ELEC] In batteries, the medium in the form of a paste or jelly, containing an electrolyte; it is positioned adjacent to the negative electrode of a dry cell; in an electrolytic cell, the paste serves as one of the conducting plates.

pasted-plate storage battery *See* Faure storage battery.

past of an event [RELAT] All events from which a signal could be emitted that could reach the event in question by traveling at speeds less than or equal to the speed of light.

path [MATH] In a topological space, a path is a continuous curve joining two points.

path integral [QUANT MECH] An integral of a functional over function space; central to a formulation of quantum mechanics developed by R. Feynman.

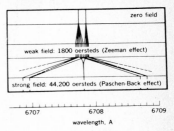

PASCHEN-BACK EFFECT

zero field

weak field: 1800 oersteds (Zeeman effect)

strong field: 44,200 oersteds (Paschen-Back effect)

6707 6708 6709
wavelength, A

Zeeman and Paschen-Back effects of red lithium doublet, whose natural separation is 0.175 angstrom.

Patterson function [SOLID STATE] A function of three spatial coordinates, constructed in the Patterson-Harker method, which has peaks at all vectors between two atoms in a crystal, the heights of the peaks being approximately proportional to the product of the atomic numbers of the corresponding atoms.

Patterson-Harker method [SOLID STATE] A method of analyzing the structure of a crystal from x-ray diffraction results; a Fourier series involving squares of the absolute values of the structure factors, which are directly observable, is used to construct a vectorial representation of interatomic distances in the crystal (Patterson map).

Pauli anomalous moment term [QUANT MECH] An additional term inserted in the Dirac equation to provide for a g-value of the particle different from 2.

Pauli electron correlation [QUANT MECH] Correlation in space of electrons as a result of the Pauli exclusion principle.

Pauli exclusion principle *See* exclusion principle.

Pauli-Fermi principle [QUANT MECH] The principle that each level of a quantized system can include one, two, or no electrons; if there are two electrons, they must have spins in opposite directions.

Pauli g-permanence rule [ATOM PHYS] For given L, S, and M_J in LS coupling, the sum, over J, of the weak-field g factors is equal to the sum of the strong-field factors.

Pauli g-sum rule [ATOM PHYS] For all the states arising from a given electron configuration, the sum of the g factors for levels with the same J value is a constant, independent of the coupling scheme.

Pauling rule [SOLID STATE] A rule governing the number of ions of opposite charge in the neighborhood of a given ion in an ionic crystal, in accordance with the requirement of local electrical neutrality of the structure.

Pauli paramagnetism *See* free-electron paramagnetism.

Pauli spin matrices [QUANT MECH] Three anticommuting matrices, each having two rows and two columns, which represent the components of the electron spin operator:

$$\sigma_x = \begin{pmatrix} 0 & 1 \\ 1 & 0 \end{pmatrix}, \ \sigma_y = \begin{pmatrix} 0 & -i \\ i & 0 \end{pmatrix}, \ \sigma_z = \begin{pmatrix} 1 & 0 \\ 0 & -1 \end{pmatrix}.$$

Pauli spin space [QUANT MECH] A two-dimensional vector space over the complex numbers, whose vectors describe orientations of the electron spin.

Pauli spin susceptibility [SOLID STATE] The susceptibility of free electrons in a metal due to the tendency of their spins to align with a magnetic field.

Pauli-Weisskopf equation [QUANT MECH] The equation resulting from second quantization of the Klein-Gordon equation.

payoff matrix [MATH] A matrix arising from certain two-person games which gives the amount gained by a player.

Pb *See* lead.

Pb-I-Pb junction *See* lead-I-lead junction.

Pd *See* palladium.

PD *See* potential difference.

PDA *See* postacceleration.

pdl-ft *See* foot-poundal.

peak amplitude [PHYS] The maximum amplitude of an alternating quantity, measured from its zero value.

peak factor *See* crest factor.

peak inverse anode voltage [ELECTR] Maximum instantaneous anode voltage in the direction opposite to that in which the tube or other device is designed to pass current.

peak inverse voltage [ELECTR] Maximum instantaneous anode-to-cathode voltage in the reverse direction which is actually applied to the diode in an operating circuit.

peak limiter   *See* limiter.

peak-to-peak amplitude   [PHYS] Amplitude of an alternating quantity measured from positive peak to negative peak.

peak value   [ELEC] The maximum instantaneous value of a varying current, voltage, or power during the time interval under consideration. Also known as crest value.

Peano continuum   [MATH] A compact, connected, and locally connected metric space.

Peano curve   [MATH] A surjective continuous mapping from the closed interval $I = [0,1]$ onto the cartesian product $I \times I$. Also known as space-filling curve.

Peano's postulates   [MATH] The five axioms by which the natural numbers may be formally defined; they state that (1) there is a natural number 1; (2) every natural number n has a successor n^+; (3) no natural number has 1 as its successor; (4) every set of natural numbers which contains 1 and the successor of every member of the set contains all the natural numbers; (5) if $n^+ = m^+$, then $n = m$.

Pearl-Reed curve   *See* logistic curve.

Pearson Type I distribution   *See* beta distribution.

Peaucellier linkage   [MECH ENG] A mechanical linkage to convert circular motion exactly into straight-line motion.

pebble-bed reactor   [NUCLEO] A nuclear reactor in which the fuel consists of small spheres or pellets stacked in the core; the reaction rate is controlled by coolant flow and by loading and unloading pellets.

peck   [MECH] Abbreviated pk. **1.** A unit of volume used in the United States for measurement of solid substances, equal to 8 dry quarts, or ¼ bushel, or 537.605 cubic inches, or 0.00880976754172 cubic meter. **2.** A unit of volume used in the United Kingdom for measurement of solid and liquid substances, although usually the former, equal to 2 gallons, or approximately 0.00909218 cubic meter.

peculiar velocity   [ASTRON] Superposed on the systematic rotation of the galaxy are individual motions of the stars; each star moves in a somewhat elliptical orbit and therefore shows a velocity of its own (peculiar velocity) to the local standard of rest, the standard moving in a circular orbit around the galactic center.

pedal curve   [MATH] The pedal curve of a given curve C with respect to a fixed point P is the locus of the foot of the perpendicular from P to a variable tangent to C.

pedal point   [MATH] The fixed point with respect to which a pedal curve is defined.

pedal triangle   [MATH] **1.** The triangle whose vertices are located at the feet of the perpendiculars from some given point to the sides of a specified triangle. **2.** In particular, the triangle whose vertices are located at the feet of the altitudes of a given triangle.

pedion   [CRYSTAL] A crystal form with only one face; member of the asymmetric class of the triclinic system.

Peierls-Nabarro force   [SOLID STATE] The force required to displace a dislocation along its slip plane.

Peirce stroke relationship   *See* NOR.

p electron   [ATOM PHYS] In the approximation that each electron has a definite central-field wave function, an atomic electron that has an orbital angular momentum quantum number of unity.

Pell equation   [MATH] The diophantine equation $x^2 - Dy^2 = 1$, with D a positive integer that is not a perfect square.

pellet fusion   [NUCLEO] A method of controlled fusion in which the rapid implosion of a fuel pellet, produced by laser, electron, or ion beams, raises the temperature and density of the pellet core to levels at which nuclear fusion can take place before the pellet flies apart. Also known as inertial-confinement fusion.

pelletron [NUCLEO] A type of electrostatic accelerator that utilizes a charging system consisting of steel cylinders joined by links of solid insulating material such as nylon to form a chain; the metal cylinders are charged as they leave a pulley at ground potential, and the charge is removed as they pass over a pulley in the high-potential terminal.

Peltier coefficient [PHYS] The ratio of the rate at which heat is evolved or absorbed at a junction of two metals in the Peltier effect to the current passing through the junction.

Peltier effect [PHYS] Heat is evolved or absorbed at the junction of two dissimilar metals carrying a small current, depending upon the direction of the current.

penalty function [MATH] A function used in treating maxima and minima problems subject to constraints.

pencil [MATH] A family of geometric objects which share a common property. [OPTICS] A bundle of rays that emanate from or converge to a common point.

pencil beam [ELECTROMAG] A beam of radiant energy concentrated in an approximately conical or cylindrical portion of space of relatively small diameter; this type of beam is used for many revolving navigational lights and radar beams.

pendant-drop method [PHYS] Method for the measurement of liquid surface tension by the elongation of a hanging drop of the liquid.

pendulous gyroscope [MECH] A gyroscope whose axis of rotation is constrained by a suitable weight to remain horizontal; it is the basis of one type of gyrocompass.

pendulum [PHYS] A rigid body mounted on a fixed horizontal axis, about which it is free to rotate under the influence of gravity. Also known as gravity pendulum.

pendulum day [PHYS] The time required for the plane of a freely suspended (Foucault) pendulum to complete an apparent rotation about the local vertical.

penetrating shower [NUC PHYS] A cosmic-ray shower, consisting mainly of muons, that can penetrate 15 to 20 centimeters of lead.

penetration depth [CRYO] The depth beneath the surface of superconductor in a magnetic field at which the magnetic field strength has fallen to $1/e$ of its value at the surface. [ELEC] In induction heating, the thickness of a layer, extending inward from a conductor's surface, whose resistance to direct current equals the resistance of the whole conductor to alternating current of a given frequency.

penetration frequency *See* critical frequency.

penetration probability [QUANT MECH] The probability that a particle will pass through a potential barrier, that is, through a finite region in which the particle's potential energy is greater than its total energy. Also known as transmission coefficient.

penetration range *See* night visual range.

penetration twin *See* interpenetration twin.

Penning gage *See* Philips ionization gage.

Penning ionization [ATOM PHYS] The ionization of gas atoms or molecules in collisions with metastable atoms.

pennyweight [MECH] A unit of mass equal to $\frac{1}{20}$ troy ounce or to 1.55517384 grams; the term is employed in the United States and in England for the valuation of silver, gold, and jewels. Abbreviated dwt; pwt.

Penrose diagram [RELAT] A diagram of a space-time where the causal and infinity structure is displayed through the use of conformal transformations. Also known as conformal diagram.

PENDULUM

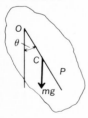

Schematic diagram of a pendulum. O = axis about which pendulum rotates, C = center of mass, and P = center of oscillation. Line OC makes instantaneous angle θ with the vertical. Gravitational force on pendulum is equivalent to downward force at C with magnitude mg, where m is pendulum mass, g is acceleration of gravity.

Penrose-Hawking theorems [RELAT] The general relativistic theorems proving that singularities must occur in space-times, such as the universe, based on reasonable assumptions such as causality and dependent on the existence of a trapped surface.

pentagon [MATH] A polygon with five sides.

pentagonal dodecahedron *See* pyritohedron.

pentagonal prism [MATH] A prism with two pentagonal sides, parallel and congruent.

pentagrid *See* heptode.

pentane candle [OPTICS] A unit of luminous intensity equal to one-tenth of the luminous intensity of a standard pentane lamp, and approximately equal to 1 candela.

pentode [ELECTR] A five-electrode electron tube containing an anode, a cathode, a control electrode, and two additional electrodes that are ordinarily grids.

pentode transistor [ELECTR] Point-contact transistor with four-point-contact electrodes; the body serves as a base with three emitters and one collector.

penumbra [OPTICS] That portion of a shadow illuminated by only part of a radiating source.

penumbral eclipse *See* lunar appulse.

PeP reaction *See* proton-electron-proton reaction.

perceived noise decibel [ACOUS] A unit of perceived noise level. Abbreviated PNdB.

perceived noise level [ACOUS] In perceived noise decibels, the noise level numerically equal to the sound pressure level, in decibels, of a band of random noise of width one-third to one octave centered on a frequency of 1000 hertz which is judged by listeners to be equally noisy.

percent [MATH] A quantitative term whereby n-percent of a number is n one-hundredths of the number. Symbolized %.

percentage [MATH] The result obtained by taking a given percent of a given quantity.

percentage depth dose [NUCLEO] The percentage of the absorbed dose from ionizing radiation at a given depth within a body to the absorbed dose at a reference point, usually the position of the peak absorbed dose.

percentage distribution [STAT] A frequency distribution in which the individual class frequencies are expressed as a percentage of the total frequency equated to 100. Also known as relative frequency distribution; relative frequency table.

percentile [STAT] A value in the range of a set of data which separates the range into two groups so that a given percentage of the measures lies below this value.

perch [MECH] Also known as pole; rod. **1.** A unit of length, equal to 5.5 yards, or 16.5 feet, or 5.0292 meters. **2.** A unit of area, equal to 30.25 square yards, or 272.25 square feet, or 25.29285264 square meters.

percussion figure [CRYSTAL] Radiating lines on a crystal section produced by a sharp blow.

perfect crystal [CRYSTAL] A single crystal considered as if it were constructed by the infinite periodic repetition in space of units that are identical in structure and orientation. Also known as ideal crystal.

perfect cube [MATH] A number or polynomial which is the exact cube of another number or polynomial.

perfect dielectric *See* ideal dielectric.

perfect fluid *See* inviscid fluid.

perfect gas *See* ideal gas.

perfectly diffuse radiator [OPTICS] A body that emits radiant energy in accordance with Lambert's law.

perfectly diffuse reflector [OPTICS] A body that reflects radiant energy in such a manner that the reflected energy may be

PENTODE

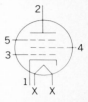

Circuit symbol for a pentode; the five electrodes are indicated by number. X's indicate leads to heater for indirectly heated cathode.

treated as if it were being emitted (radiated) in accordance with Lambert's law.

perfectly inelastic collision [PHYS] A collision in which as much translational kinetic energy is converted into internal energy of the colliding systems as is consistent with the conservation of momentum. Also known as completely inelastic collision.

perfect number [MATH] An integer which equals the sum of all its factors other than itself.

perfect set [MATH] A set in a topological space which equals its set of accumulation points.

perfect square [MATH] A number or polynomial which is the exact square of another number or polynomial.

perfect vacuum *See* absolute vacuum.

periapsis [ASTRON] The orbital point nearest the center of attraction of an orbiting body.

periastron [ASTRON] The coordinates and time when the two stars of a binary star system are nearest to each other in their orbits.

pericline twin law [CRYSTAL] A parallel twin law in triclinic feldspars, in which the b axis is the twinning axis and the composition surface is a rhombic section.

perigee [ASTRON] The point in the orbit of the moon or other satellite when it is nearest the earth.

perigee-to-perigee period *See* anomalistic period.

perihelion [ASTRON] That orbital point nearest the sun when the sun is the center of attraction.

perimeter [MATH] The total length of a closed curve; for example, the perimeter of a polygon is the total length of its sides.

period [MATH] **1.** A number T such that $f(x + T) = f(x)$ for all x, where $f(x)$ is a specified function of a real or complex variable. **2.** The period of an element a of a group G is the smallest positive integer n such that a^n is the identity element; if there is no such integer, a is said to be of infinite period. [NUCLEO] The time required for exponentially rising or falling neutron flux in a nuclear reactor to change by a factor of e (2.71828). [PHYS] The duration of a single repetition of a cyclic phenomenon.

periodic [SCI TECH] Repeating itself identically at regular intervals.

periodic damping [PHYS] Damping which is less than critical damping.

periodic field focusing [ELECTR] Focusing of an electron beam where the electrons follow a trochoidal path and the focusing field interacts with them at selected points.

periodic function [MATH] A function $f(x)$ of a real or complex variable is periodic with period T if $f(x + T) = f(x)$ for every value of x.

periodicity [MATH] The property of periodic functions.

periodic lattice *See* lattice.

periodic law [CHEM] The law that the properties of the chemical elements and their compounds are a periodic function of their atomic weights.

periodic motion [MECH] Any motion that repeats itself identically at regular intervals.

periodic perturbation [ASTRON] Small deviations from the computed orbit of a planet or satellite; the deviations extend through cycles that generally do not exceed a century. [MATH] A perturbation which is periodic as a function.

periodic quantity [PHYS] Oscillating quantity, the values of which recur for equal increments of the independent variable.

periodic table [CHEM] A table of the elements, written in sequence in the order of atomic number or atomic weight and arranged in horizontal rows (periods) and vertical columns (groups) to illustrate the occurrence of similarities in the properties of the elements as a periodic function of the sequence.

periodic wave [PHYS] A wave whose displacement has a periodic variation with time or distance, or both.

period-luminosity relation [ASTRON] Relation between the periods of Cepheid variable stars and their absolute magnitude; the absolutely brighter the star, the longer the period.

period of a variable star [ASTRON] The average time interval for a variable star to complete a cycle of its variations.

period of vibration [PHYS] The time for one complete cycle of a vibration.

periodogram [STAT] A graph used in harmonic analysis of a series that oscillates, such as a time series consisting potentially of several cycles differing in length; the square of the amplitude or intensity for each curve covering a length of time is plotted against the lengths of the various curves.

period parallelogram [MATH] For a doubly periodic function $f(z)$ of a complex variable, a parallelogram with vertices at z_0, $z_0 + a$, $z_0 + a + b$, and $z_0 + b$, where z_0 is any complex number, and a and b are periods of $f(z)$ but are not necessarily primitive periods.

periphery [MATH] The bounding curve of a surface or the surface of a solid.

periscope [OPTICS] **1.** An optical instrument used to provide a raised line of vision where it may not be practical or possible, as in entrenchments, tanks, or submarines; the raised line of vision is obtained by the use of mirrors or prisms within the structure of the item; it may have single or dual optical systems. **2.** A thin astigmatic lens which approximates a meniscus shape and has a base curve of ± 1.25 diopters.

peristaltic charge-coupled device [ELECTR] A high-speed charge-transfer integrated circuit in which the movement of the charges is similar to the peristaltic contractions and dilations of the digestive system.

permanent aurora *See* airglow.

permanent gas [THERMO] A gas at a pressure and temperature far from its liquid state.

permanent magnet [ELECTROMAG] A piece of hardened steel or other magnetic material that has been strongly magnetized and retains its magnetism indefinitely. Abbreviated PM.

permanent-magnet moving-coil instrument [ENG] An ammeter or other electrical instrument in which a small coil of wire, supported on jeweled bearings between the poles of a permanent magnet, rotates when current is carried to it through spiral springs which also exert a restoring torque on the coil; the position of the coil is indicated by an attached pointer.

permanent-magnet moving-iron instrument [ENG] A meter that depends for its operation on a movable iron vane that aligns itself in the resultant magnetic field of a permanent magnet and adjacent current-carrying coil.

permanent wave [FL MECH] A wave (in a fluid) which moves with no change in streamline pattern, and which, therefore, is a stationary wave relative to a coordinate system moving with the wave.

permeability [ELECTROMAG] A factor, characteristic of a material, that is proportional to the magnetic induction produced in a material divided by the magnetic field strength; it is a tensor when these quantities are not parallel.   [FL MECH] **1.** The ability of a membrane or other material to permit a substance to pass through it. **2.** Quantitatively, the amount of substance which passes through the material under given conditions.

**PERMANENT-MAGNET
MOVING-COIL INSTRUMENT**

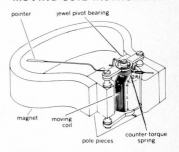

Mechanism of a permanent-magnet moving-coil instrument. L = length of active conductors, R = radius of action of these conductors. *(Weston Instruments, Division of Sangamo Weston, Inc.)*

PERMEAMETER

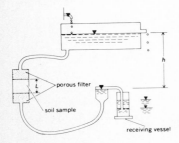

Constant-head permeameter (def. 2). Head water level, h, is kept constant. Water percolates through soil sample of thickness L. Porous filters hold soil in place. The tail water level is kept constant by overflow. Volume of discharge is measured in receiving vessel. *(From G. P. Tschebotarioff, Soil Mechanics, Foundations and Earth Structures, McGraw-Hill, 1951)*

PERPETUAL MOTION MACHINE OF THE THIRD KIND

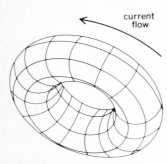

If a direct current is caused to flow in a superconducting ring, this current will continue to flow undiminished in time without application of any external force.

permeability coefficient [FL MECH] The rate of water flow in gallons per day through a cross section of 1 square foot under a unit hydraulic gradient, at the prevailing temperature or at 60°F (16°C). Also known as coefficient of permeability; hydraulic conductivity; Meinzer unit.

permeability tuning [ELEC] Process of tuning a resonant circuit by varying the permeability of an inductor; it is usually accomplished by varying the amount of magnetic core material of the inductor by slug movement.

permeameter [ENG] **1.** Device for measurement of the average size or surface area of small particles; consists of a powder bed of known dimension and degree of packing through which the particles are forced; pressure drop and rate of flow are related to particle size, and pressure drop is related to surface area. **2.** A device for measuring the coefficient of permeability by measuring the flow of fluid through a sample across which there is a pressure drop produced by gravity. **3.** An instrument for measuring the magnetic flux or flux density produced in a test specimen of ferromagnetic material by a given magnetic intensity, to permit computation of the magnetic permeability of the material.

permeance [ELECTROMAG] A characteristic of a portion of a magnetic circuit, equal to magnetic flux divided by magnetomotive force; the reciprocal of reluctance. Symbolized P.

permissible dose [NUCLEO] The amount of radiation that may be safely received by an individual within a specified period. Formerly known as tolerance dose.

permittivity [ELEC] The dielectric constant multiplied by the permittivity of empty space, where the permittivity of empty space (ε_0) is a constant appearing in Coulomb's law, having the value of 1 in centimeter-gram-second electrostatic units, and of 8.854×10^{-12} farad/meter in rationalized meter-kilogram-second units. Symbolized ε.

permutation [MATH] A function which rearranges a finite number of symbols; more precisely, a one-to-one function of a finite set onto itself.

permutation group [MATH] The group whose elements are permutations of some set of symbols where the product of two permutations is the permutation arising from successive application of the two.

permutation tensor *See* determinant tensor.

perpendicular [MATH] Geometric objects are perpendicular if they intersect in an angle of 90°.

perpendicular axis theorem [MECH] A theorem which states that the sum of the moments of inertia of a plane lamina about any two perpendicular axes in the plane of the lamina is equal to the moment of inertia about an axis through their intersection perpendicular to the lamina.

perpetual motion machine of the first kind [PHYS] A mechanism which, once set in motion, continues to do useful work without an input of energy, or which produces more energy than is absorbed in its operation; it violates the principle of conservation of energy.

perpetual motion machine of the second kind [PHYS] A device that extracts heat from a source and then converts this heat completely into other forms of energy; it violates the second law of thermodynamics.

perpetual motion machine of the third kind [PHYS] A device which has a component that can continue moving forever; an example is a superconductor.

Perron-Frobenius theorem [MATH] If M is a matrix with positive entries, then its largest eigenvalue λ is positive and simple; moreover, there exist vectors v and w with positive components such that $vM = \lambda v$ and $Mw = \lambda w$, and if the inner product of v with w is 1, then the limit of $\lambda - n$ times the i,jth entry of M^n as n goes to infinity is the product of the ith component of w and the jth component of v.

Perron-Frobenius theory [MATH] The study of positive matrices and their eigenvalues; in particular, application of the Perron-Frobenius theorem.

persistence [ELECTR] **1.** A measure of the length of time that the screen of a cathode-ray tube remains luminescent after excitation is removed; ranges from 1 for short persistence to 7 for long persistence. **2.** A faint luminosity displayed by certain gases for some time after the passage of an electric discharge.

persistent current [CRYO] A magnetically induced current that flows undiminished in a superconducting material or circuit.

personal probability [STAT] A number between 0 and 1 assigned to an event based upon personal views concerning whether the event will occur or not; it is obtained by deciding whether one would accept a bet on the event at odds given by this number. Also known as subjective probability.

personnel monitoring [NUCLEO] Determination of the degree of radioactive contamination on individuals, using standard survey meters, and determination of the dose received by means of dosimeters.

perturbation [ASTRON] A deviation of an astronomical body from its computed orbit because of the attraction of another body or bodies. [MATH] A function which produces a small change in the values of some given function. [PHYS] Any effect which makes a small modification in a physical system, especially in case the equations of motion could be solved exactly in the absence of this effect.

perturbation equation [PHYS] Any equation governing the behavior of a perturbation; often this will be a linear differential equation.

perturbation motion [PHYS] The motion of a disturbance (usually but not necessarily assumed infinitesimal), as opposed to the motion of the system on which the perturbation is superimposed.

perturbation quantity [PHYS] Any characteristic of a system which may be assumed to be a perturbation from an established value.

perturbation theory [MATH] The study of the solutions of differential and partial differential equations from the viewpoint of perturbation of solutions. [PHYS] The theory of obtaining approximate solutions to the equations of motion of a physical system when these equations differ by a small amount from equations which can be solved exactly.

perveance [ELECTR] The space-charge-limited cathode current of a diode divided by the 3/2 power of the anode voltage.

peta- [SCI TECH] A prefix that represents 10^{15}. Abbreviated P.

Peters' formula [STAT] An approximate formula for the probable error in the value of a quantity determined from several equally careful, independent measurements of the value of the quantity.

petrographic microscope [OPTICS] A polarizing microscope used for analysis of petrographic thin sections.

Petrov classification [RELAT] An algebraic classification of space-times based on eigenvalues of the curvature tensor.

Petzval condition [OPTICS] The condition whereby an optical system will eliminate the aberration of curvature of field only if the Petzval curvature vanishes.

Petzval curvature [OPTICS] The axial curvature of the image of a plane object produced by an optical system, equal to the sum over all the optical surfaces in the system of $R(1/n' - 1/n)$, where R is the curvature of the surface, and n and n' are the refractive indices before and after the surface. Also known as Petzval sum.

Petzval lens [OPTICS] A photographic objective which consists of four lenses ordered in two pairs widely separated from each other, with the first pair cemented together and the second usually having a small air space.

Petzval sum *See* Petzval curvature.

Petzval surface [OPTICS] A paraboloidal surface on which point images of point objects are formed by a doublet lens whose separation is such that astigmatism is eliminated.

Pexider's equations [MATH] The functional equations $f(x + y) = g(x) + h(y), f(x + y) = g(x)h(y), f(xy) = g(x) + h(y)$, and $f(xy) = g(x)h(y)$, which generalize Cauchy's functional equations.

pf *See* power factor.

pF *See* picofarad.

Pfaffian differential equation [MATH] The first-order linear total differential equation $P(x,y,z) dx + Q(x,y,z) dy + R(x,y,z) dz = 0$, where the functions P, Q, and R are continuously differentiable.

PFE *See* photoferroelectric effect.

p-form [MATH] A totally antisymmetric covariant tensor of rank p.

Pfund series [SPECT] A series of lines in the infrared spectrum of atomic hydrogen whose wave numbers are given by $R_H[(1/25) - (1/n^2)]$, where R_H is the Rydberg constant for hydrogen, and n is any integer greater than 5.

pH [CHEM] A term used to describe the hydrogen-ion activity of a system; it is equal to $-\log a_H{}^+$; here $a_H{}^+$ is the activity of the hydrogen ion; in dilute solution, activity is essentially equal to concentration and pH is defined as $-\log_{10} [H^+]$, where $[H^+]$ is hydrogen-ion concentration in moles per liter; a solution of pH 0 to 7 is acid, pH of 7 is neutral, pH over 7 to 14 is alkaline.

phantom [NUCLEO] A volume of material approximating as closely as possible the density and effective atomic number of living tissue, used in biological experiments involving radiation.

phantom crystal [CRYSTAL] A crystal containing an earlier stage of crystallization outlined by dust, minute inclusions, or bubbles. Also known as ghost crystal.

phase [ASTRON] One of the cyclically repeating appearances of the moon or other orbiting body as seen from earth. [CHEM] Portion of a physical system (liquid, gas, solid) that is homogeneous throughout, has definable boundaries, and can be separated physically from other phases. [MATH] An additive constant in the argument of a trigonometric function. [PHYS] **1.** The fractional part of a period through which the time variable of a periodic quantity (alternating electric current, vibration) has moved, as measured at any point in time from an arbitrary time origin; usually expressed in terms of angular measure, with one period being equal to 360° or 2π radians. **2.** For a sinusoidally varying quantity, the phase (first definition) with the time origin located at the last point at which the quantity passed through a zero position from a negative to a positive direction. **3.** The argument of the trigonometric function describing the space and time variation of a sinusoidal disturbance, $y = A \cos [(2\pi/\lambda) (x - vt)]$, where x and t are the space and time coordinates, v is the velocity of propagation, and λ is the wavelength. [THERMO] The type of state of a system, such as solid, liquid, or gas.

phase angle [PHYS] The difference between the phase of a sinusoidally varying quantity and the phase of a second quantity which varies sinusoidally at the same frequency. Also known as phase difference.

PHASE ANGLE

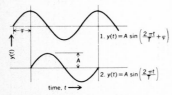

An illustration of the meaning of phase for a sinusoidal wave, $y(t)$. The difference in phase between waves 1 and 2 is φ and is called the phase angle. For each wave, A is the amplitude and T is the period.

phase-angle meter *See* phase meter.

phase boundary [PHYS] The interface between two or more separate phases, such as liquid-gas, liquid-solid, gas-solid, or, for immiscible materials, liquid-liquid or solid-solid.

phase change [PHYS] **1.** The metamorphosis of a material or mixture from one phase to another, such as gas to liquid, solid to gas. **2.** *See* phase shift.

phase-change coefficient *See* phase constant.

phase coherence [PHYS] The existence of a statistical or time coherence between the phases of two or more waves.

phase constant [ELECTROMAG] A rating for a line or medium through which a plane wave of a given frequency is being transmitted; it is the imaginary part of the propagation constant, and is the space rate of decrease of phase of a field component (or of the voltage or current) in the direction of propagation, in radians per unit length. Also known as phase-change coefficient; wavelength constant.

phase-contrast microscope [OPTICS] A compound microscope that has an annular diaphragm in the front focal plane of the substage condenser and a phase plate at the rear focal plane of the objective, to make visible differences in phase or optical path in transparent or reflecting media.

phase crossover [CONT SYS] A point on the plot of the loop ratio at which it has a phase angle of 180°.

phased array [ELECTROMAG] An array of dipoles on a radar antenna in which the signal feeding each dipole is varied so that antenna beams can be formed in space and scanned very rapidly in azimuth and elevation.

phase detector [ELECTR] A circuit that provides a direct-current output voltage which is related to the phase difference between an oscillator signal and a reference signal, for use in controlling the oscillator to keep it in synchronism with the reference signal. Also known as phase discriminator.

phase diagram [PHYS CHEM] A graphical representation of the equilibrium relationships between phases (such as vapor-liquid, liquid-solid) of a chemical compound, mixture of compounds, or solution. [THERMO] **1.** A graph showing the pressures at which phase transitions between different states of a pure compound occur, as a function of temperature. **2.** A graph showing the temperatures at which transitions between different phases of a binary system occur, as a function of the relative concentrations of its components.

phase difference *See* phase angle.

phase discriminator *See* phase detector.

phase equilibria [PHYS CHEM] The equilibrium relationships between phases (such as vapor, liquid, solid) of a chemical compound or mixture under various conditions of temperature, pressure, and composition.

phase factor [ELEC] *See* power factor. [SOLID STATE] The argument (phase) of a structure factor; it cannot be directly observed.

phase front [PHYS] A surface of constant phase (or phase angle) of a propagating wave disturbance.

phase group [MATH] The topological group G in the topological transformation group (G,X,π).

phase integral *See* action.

phase integral method *See* Wentzel-Kramers-Brillouin method.

phase inversion [ELECTR] Production of a phase difference of 180° between two similar wave shapes of the same frequency.

phase inverter [ELECTR] A circuit or device that changes the phase of a signal by 180°, as required for feeding a push-pull amplifier stage without using a coupling transformer, or for changing the polarity of a pulse; a triode is commonly used as a phase inverter. Also known as inverter.

phase lead   *See* lead angle.

phase lock   [ELECTR] Technique of making the phase of an oscillator signal follow exactly the phase of a reference signal by comparing the phases between the two signals and using the resultant difference signal to adjust the frequency of the reference oscillator.

phase margin   [CONT SYS] The difference between 180° and the phase of the loop ratio of a stable system at the gain-crossover frequency.

phase meter   [ENG] An instrument for the measurement of electrical phase angles. Also known as phase-angle meter.

phase modulation   [COMMUN] A special kind of modulation in which the linearly increasing angle of a sine wave has added to it a phase angle that is proportional to the instantaneous value of the modulating wave (message to be communicated). Abbreviated PM.

phase plane analysis   [CONT SYS] A method of analyzing systems in which one plots the time derivative of the system's position (or some other quantity characterizing the system) as a function of position for various values of initial conditions.

phase portrait   [CONT SYS] A graph showing the time derivative of a system's position (or some other quantity characterizing the system) as a function of position for various values of initial conditions.

phase quadrature   *See* quadrature.

phaser   [ELECTROMAG] Microwave ferrite phase shifter employing a longitudinal magnetic field along one or more rods of ferrite in a waveguide.

phase resonance   [PHYS] The frequency at which the angular phase difference between the fundamental components of an oscillation and of the applied agency is 90° ($\pi/2$ radians). Also known as velocity resonance.

phase response   [ELECTR] A graph of the phase shift of a network as a function of frequency.

phase reversal   [PHYS] A change of 180°, or one half-cycle, in phase.

phase shift   [ELECTR] The phase angle between the input and output signals of a network or system.   [PHYS] **1.** A change in the phase of a periodic quantity. Also known as phase change. **2.** A change in the phase angle between two periodic quantities.   [QUANT MECH] For a partial wave of a particle scattered by a spherically symmetric potential, the phase shift is the difference between the phase of the wave function far from the scatterer and the corresponding phase of a free particle.

phase-shift oscillator   [ELECTR] An oscillator in which a network having a phase shift of 180° per stage is connected between the output and the input of an amplifier.

phase solubility   [PHYS CHEM] The different solubilities of a sample's solid constituents (phases) in a selected solvent.

phase space   [MATH] In a dynamical system or transformation group, this is the topological space whose points are being moved about by the given transformations.   [STAT MECH] For a system with n degrees of freedom, a euclidean space with $2n$ dimensions, one dimension for each of the generalized coordinates and one for each of the corresponding momenta.

phase speed   *See* phase velocity.

phase splitter   [ELECTR] A circuit that takes a single input signal voltage and produces two output signal voltages 180° apart in phase.

phase stability   [NUCLEO] A principle governing the stability of motion of particles in a synchrotron; the charged particle must be accelerated in each cycle at a time slightly earlier than the peak value of the accelerating potential.

phase transformation [ELEC] A change of polyphase power from three-phase to six-phase, from three-phase to twelve-phase, and so forth, by means of transformers. [PHYS] *See* phase transition.

phase transition [PHYS] A change of a substance from one phase to another. Also known as phase transformation.

phase velocity [PHYS] The velocity of a point that moves with a wave at constant phase. Also known as celerity; phase speed; wave celerity; wave speed; wave velocity.

phasing *See* framing.

phasor [PHYS] **1.** A rotating line used to represent a sinusoidally varying quantity; the length of the line represents the magnitude of the quantity, and its angle with the x axis at any instant represents the phase. **2.** Any quantity (such as impedance or admittance) which is a complex number.

phasotron *See* cyclotron.

Philips ionization gage [ELECTR] An ionization gage in which a high voltage is applied between two electrodes, and a strong magnetic field deflects the resulting electron stream, increasing the length of the electron path and thus increasing the chance for ionizing collisions of electrons with gas molecules. Abbreviated pig. Also known as cold-cathode ionization gage; Penning gage.

phi meson [PARTIC PHYS] A neutral vector meson resonance, having a mass of about 1019 MeV, a width of about 4.2 MeV, and negative charge parity and G parity.

phon [ACOUS] A unit of loudness level; the loudness level, in phons, of a sound is numerically equal to the sound pressure level, in decibels, of a 1000-hertz reference tone which is judged by listeners to be equally loud to the sound under evaluation.

phonon [SOLID STATE] A quantum of an acoustic mode of thermal vibration in a crystal lattice.

phonon-electron interaction [SOLID STATE] An interaction between an electron and a vibration of a lattice, resulting in a change in both the momentum of the particle and the wave vector of the vibration.

phonon emission [SOLID STATE] The production of a phonon in a crystal lattice, which may result from the interaction of other phonons via anharmonic lattice forces, from scattering of electrons in the lattice, or from scattering of x-rays or particles which bombard the crystal.

phosphor *See* luminophor.

phosphorescence [ATOM PHYS] **1.** Luminescence that persists after removal of the exciting source. Also known as afterglow. **2.** Luminescence whose decay, upon removal of the exciting source, is temperature-dependent.

phosphorogen [PHYS] A substance that promotes phosphorescence in another substance, as manganese does in zinc sulfide.

phosphorus [CHEM] A nonmetallic element, symbol P, atomic number 15, atomic weight 30.98; used to manufacture phosphoric acid, in phosphor bronzes, incendiaries, pyrotechnics, matches, and rat poisons; the white (or yellow) allotrope is a soft waxy solid melting at 44.5°C, is soluble in carbon disulfide, insoluble in water and alcohol, and is poisonous and self-igniting in air; the red allotrope is an amorphous powder subliming at 416°C, igniting at 260°C, is insoluble in all solvents, and is nonpoisonous; the black allotrope comprises lustrous crystals similar to graphite, and is insoluble in most solvents.

phot [OPTICS] A unit of illumination equal to the illumination of a surface, 1 square centimeter in area, on which there

PHOSPHORUS

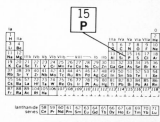

Periodic table of the chemical elements showing the position of phosphorus.

is a luminous flux of 1 lumen, or the illumination on a surface all points of which are at a distance of 1 centimeter from a uniform point source of 1 candela. Also known as centimeter-candle (deprecated usage).

photoacoustic spectroscopy [SPECT] A spectroscopic technique for investigating solid and semisolid materials, in which the sample is placed in a closed chamber filled with a gas such as air and illuminated with monochromatic radiation of any desired wavelength, with intensity modulated at some suitable acoustic frequency; absorption of radiation results in a periodic heat flow from the sample, which generates sound that is detected by a sensitive microphone attached to the chamber. Abbreviated PAS.

photocathode [ELECTR] A photosensitive surface that emits electrons when exposed to light or other suitable radiation; used in phototubes, television camera tubes, and other light-sensitive devices.

photocell [ELECTR] A solid-state photosensitive electron device whose current-voltage characteristic is a function of incident radiation. Also known as electric eye; photoelectric cell.

photochemical reaction [PHYS CHEM] A chemical reaction influenced or initiated by light, particularly ultraviolet light, as in the chlorination of benzene to produce benzene hexachloride.

photochemistry [PHYS CHEM] The study of the effects of light on chemical reactions.

photoconduction [SOLID STATE] An increase in conduction of electricity resulting from absorption of electromagnetic radiation.

photoconductive cell [ELECTR] A device for detecting or measuring electromagnetic radiation by variation of the conductivity of a substance (called a photoconductor) upon absorption of the radiation by this substance. Also known as photoresistive cell; photoresistor.

photoconductive film [ELECTR] A film of material whose current-carrying ability is enhanced when illuminated.

photoconductive gain factor [ELECTR] The ratio of the number of electrons per second flowing through a circuit containing a cube of semiconducting material, whose sides are of unit length, to the number of photons per second absorbed in this volume.

photoconductivity [SOLID STATE] The increase in electrical conductivity displayed by many nonmetallic solids when they absorb electromagnetic radiation.

photoconductor [SOLID STATE] A nonmetallic solid whose conductivity increases when it is exposed to electromagnetic radiation.

photoconductor diode *See* photodiode.

photodetachment [PHYS CHEM] The removal of an electron from a negative ion by absorption of a photon, resulting in a neutral atom or molecule.

photodichroic material [OPTICS] A material which exhibits photoinduced dichroism and birefringence.

photodiffusion effect *See* Dember effect.

photodiode [ELECTR] A semiconductor diode in which the reverse current varies with illumination; examples include the alloy-junction photocell and the grown-junction photocell. Also known as photoconductor diode.

photodisintegration [NUC PHYS] The breakup of an atomic nucleus into two or more fragments as a result of bombardment by gamma radiation. Also known as Chadwick-Goldhaber effect.

photodissociation [PHYS CHEM] The removal of one or more atoms from a molecule by the absorption of a quantum of electromagnetic energy.

photodosimetry [NUCLEO] Determination of the cumulative dose of ionizing radiation by use of photographic film.

photoelastic effect [OPTICS] Changes in optical properties of a transparent dielectric when it is subjected to mechanical stress, such as mechanical birefringence. Also known as photoelasticity.

photoelasticity [OPTICS] **1.** An experimental technique for the measurement of stresses and strains in material objects by means of the phenomenon of mechanical birefringence. **2.** *See* photoelastic effect.

photoelectret [SOLID STATE] An electret produced by the removal of light from an illuminated photoconductor in an electric field.

photoelectric [ELECTR] Pertaining to the electrical effects of light, such as the emission of electrons, generation of voltage, or a change in resistance when exposed to light.

photoelectric absorption [ELECTR] Absorption of photons in one of the several photoelectric effects.

photoelectric cell *See* photocell.

photoelectric color comparator *See* color comparator.

photoelectric constant [ELECTR] The ratio of the frequency of radiation causing emission of photoelectrons to the voltage corresponding to the energy absorbed by a photoelectron; equal to Planck's constant divided by the electron charge.

photoelectric effect *See* photoelectricity.

photoelectric electron-multiplier tube *See* multiplier phototube.

photoelectric infrared radiation *See* near-infrared radiation.

photoelectricity [ELECTR] The liberation of electrons by electromagnetic radiation incident on a substance; includes photoemission, photoionization, photoconduction, the photovoltaic effect, and the Auger effect (an internal photoelectric process). Also known as photoelectric effect; photoelectric process.

photoelectric photometer [ENG] A photometer that uses a photocell, phototransistor, or phototube to measure the intensity of light. Also known as electronic photometer.

photoelectric photometry [OPTICS] In contrast to the methods of visual photometry, an objective approach to the problems of photometry, wherein any of several types of photoelectric devices are used to replace the human eye as the sensing element.

photoelectric process *See* photoelectricity.

photoelectric tube *See* phototube.

photoelectromagnetic effect [ELECTR] The effect whereby, when light falls on a flat surface of an intermetallic semiconductor located in a magnetic field that is parallel to the surface, excess hole-electron pairs are created, and these carriers diffuse in the direction of the light but are deflected by the magnetic field to give a current flow through the semiconductor that is at right angles to both the light rays and the magnetic field.

photoelectromotive force [ELECTR] Electromotive force caused by photovoltaic action.

photoelectron [ELECTR] An electron emitted by the photoelectric effect.

photoelectron spectroscopy [SPECT] The branch of electron spectroscopy concerned with the energy analysis of photoelectrons ejected from a substance as the direct result of bombardment by ultraviolet radiation or x-radiation.

photoemission [ELECTR] The ejection of electrons from a solid (or less commonly, a liquid) by incident electromagnetic radiation. Also known as external photoelectric effect.

PHOTOEMISSION

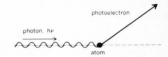

Albert Einstein's approach to photoemission. Light beam behaves like a stream of photons, each of energy $h\nu$, where h is Planck's constant, and ν the frequency of the photon. When a photon interacts with an electron, the electron absorbs entire photon energy and is ejected from emitter if this energy exceeds a well-defined minimum value.

Photographic zenith tube.
*(Official U.S. Naval Observatory
photograph)*

PHOTOLYSIS

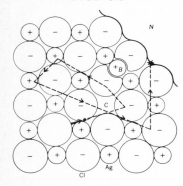

Schematic representation of
Gurney-Mott theory of
photolysis of silver halides.
Broken line shows path of
photoelectron which is
eventually trapped in vicinity of
a silver speck *N*. The interstitial
silver ion *B* has come up to
neutralize the charge of the
electron and thus to add to the
silver speck. *(After J. R. Haynes
and W. Schockley)*

photoemission threshold [ELECTR] The energy of a photon which is just sufficient to eject an electron from a solid or liquid in photoemission.

photoemissive cell [ELECTR] A device which detects or measures radiant energy by measurement of the resulting emission of electrons from the surface of a photocathode.

photoemissive tube photometer [ENG] A photometer which uses a tube made of a photoemissive material; it is highly accurate, but requires electronic amplification, and is used mainly in laboratories.

photoemissivity [ELECTR] The property of a substance that emits electrons when struck by light.

photoemitter [SOLID STATE] A material that emits electrons when sufficiently illuminated.

photoenlarger *See* enlarger.

photoferroelectric effect [SOLID STATE] An effect observed in ferroelectric ceramics such as PLZT materials, in which light at or near the band-gap energy of the material has an effect on the electric field in the material created by an applied voltage, and, at a certain value of the voltage, also influences the degree of ferroelectric remanent polarization. Abbreviated PFE.

photofission [NUC PHYS] Fission of an atomic nucleus that results from absorption by the nucleus of a high-energy photon.

photogoniometer [ENG] A goniometer that uses a phototube or photocell as a sensing device for studying x-ray spectra and x-ray diffraction effects in crystals.

photographic objective [OPTICS] A camera lens designed to form sharp real images of objects on a photographic film.

photographic photometry [SPECT] The use of a comparator-densitometer to analyze a photographed spectrograph spectrum by emulsion density measurements.

photographic zenith tube [OPTICS] A type of zenith telescope in which light is reflected from a pool of mercury, and the photographic plate is held in a carriage just below the objective that is alternately rotated through 180° and moved slowly across the field of view to follow a star image; used for the accurate determination of time.

photoheliograph [OPTICS] A refracting telescope specially designed to photograph the sun's disk.

photoionization [PHYS CHEM] The removal of one or more electrons from an atom or molecule by absorption of a photon of visible or ultraviolet light. Also known as atomic photoelectric effect.

photojunction battery [NUCLEO] A nuclear-type battery in which a radioactive material such as promethium-147 irradiates a phosphor which converts nuclear energy into light; the light is then converted to electrical energy by a small silicon junction.

photology [OPTICS] The scientific study of light.

photoluminescence [ATOM PHYS] Luminescence stimulated by visible, infrared, or ultraviolet radiation.

photolysis [PHYS CHEM] The use of radiant energy to produce chemical changes.

photomagnetic effect [PHYS] **1.** The direct effect of light on the magnetic susceptibility of certain substances. **2.** Paramagnetism displayed by certain substances when they are in a phosphorescent state. [NUC PHYS] Photodisintegration that results from the action of the magnetic field component of electromagnetic radiation.

photomagnetoelectric effect [ELECTROMAG] The generation of a voltage when a semiconductor material is positioned in a magnetic field and one face is illuminated.

photometer [ENG] An instrument used for making measurements of light or electromagnetic radiation, in the visible range.

photometry [OPTICS] The calculation and measurement of quantities describing light, such as luminous intensity, luminous flux, luminous flux density, light distribution, color, absorption factor, spectral distribution, and the reflectance and transmittance of light; sometimes taken to include measurement of near-infrared and near-ultraviolet radiation as well as visible light.

photomultiplier *See* multiplier phototube.

photomultiplier cell [ELECTR] A transistor whose *pn*-junction is exposed so that it conducts more readily when illuminated.

photomultiplier counter [ELECTR] A scintillation counter that has a built-in multiplier phototube.

photomultiplier tube *See* multiplier phototube.

photon [OPTICS] *See* troland. [QUANT MECH] A massless particle, the quantum of the electromagnetic field, carrying energy, momentum, and angular momentum. Also known as light quantum.

photon antibunching [OPTICS] A quantum phenomenon that occurs in certain types of light emission such as resonance fluorescence, in which the emission of one photon reduces the probability that another photon will be emitted immediately afterward.

photon bunching [OPTICS] The tendency of photoelectric pulses from an illuminated photodetector to occur in bunches rather than at random.

photon coupled isolator [ELECTR] Circuit coupling device, consisting of an infrared emitter diode coupled to a photon detector over a short shielded light path, which provides extremely high circuit isolation.

photon coupling [ELECTR] Coupling of two circuits by means of photons passing through a light pipe.

photonegative [ELECTR] Having negative photoconductivity, hence decreasing in conductivity (increasing in resistance) under the action of light; selenium sometimes exhibits photonegativity.

photon emission spectrum [PHYS] The relative numbers of optical photons emitted by a scintillator material per unit wavelength as a function of wavelength; the emission spectrum may also be given in alternative units such as wave number, photon energy, or frequency.

photoneutron [NUC PHYS] A neutron released from a nucleus in a photonuclear reaction.

photon flux [OPTICS] The number of photons in a light beam reaching a surface, such as the surface of the photocathode of a photomultiplier tube, in a unit of time.

photon gas [STAT MECH] An electromagnetic field treated as a collection of photons; it behaves as any other collection of bosons, except that the particles are emitted or absorbed without restriction on their number.

photon microscope *See* optical microscope.

photon theory [QUANT MECH] A theory of photoemission developed by Einstein, according to which a light beam behaves like a stream of particles (called photons) when it delivers energy to a substance displaying photoemission, the particles each having an energy equal to Planck's constant times the frequency of the light.

photonuclear reaction [NUC PHYS] A nuclear reaction resulting from the collision of a photon with a nucleus.

photophoresis [PHYS] Production of unidirectional motion in a collection of very fine particles, suspended in a gas or falling in a vacuum, by a powerful beam of light.

photopositive [ELECTR] Having positive photoconductivity, hence increasing in conductivity (decreasing in resistance) under the action of light; selenium ordinarily has photopositivity.

photoproton [NUC PHYS] A proton released from a nucleus in a photonuclear reaction.

photoresistive cell *See* photoconductive cell.

photoresistor *See* photoconductive cell.

photosensitive *See* light-sensitive.

photosphere [ASTRON] The intensely bright portion of the sun visible to the unaided eye; it is a shell a few hundred miles in thickness marking the boundary between the dense interior gases of the sun and the more diffuse cooler gases in the outer portions of the sun.

photospheric granulation *See* granulation.

photothermoelasticity [OPTICS] Changes in optical properties of a transparent dielectric when it is subjected to mechanical stress, which is, in turn, induced by temperature gradients.

phototransistor [ELECTR] A junction transistor that may have only collector and emitter leads or also a base lead, with the base exposed to light through a tiny lens in the housing; collector current increases with light intensity, as a result of amplification of base current by the transistor structure.

phototube [ELECTR] An electron tube containing a photocathode from which electrons are emitted when it is exposed to light or other electromagnetic radiation. Also known as electric eye; light-sensitive tube; photoelectric tube.

phototube cathode [ELECTR] The photoemissive surface which is the most negative element of a phototube.

photovaristor [ELECTR] Varistor in which the current-voltage relation may be modified by illumination, for example, one in which the semiconductor is cadmium sulfide or lead telluride.

photovoltaic [ELECTR] Capable of generating a voltage as a result of exposure to visible or other radiation.

photovoltaic cell [ELECTR] A device that detects or measures electromagnetic radiation by generating a potential at a junction (barrier layer) between two types of material, upon absorption of radiant energy. Also known as barrier-layer cell; barrier-layer photocell; boundary-layer photocell; photronic photocell.

photovoltaic effect [ELECTR] The production of a voltage in a nonhomogeneous semiconductor, such as silicon, or at a junction between two types of material, by the absorption of light or other electromagnetic radiation.

photovoltaic meter [ELECTR] An exposure cell in which a photovoltaic cell produces a current proportional to the light falling on the cell, and this current is measured by a sensitive microammeter.

photronic photocell *See* photovoltaic cell.

phthalocyanine Q switching [OPTICS] Laser Q switching in which a solution of metal-organic compounds known as phthalocyanines is placed in a cell between an uncoated ruby laser crystal and a high-reflectivity mirror; when the incident ruby light reaches a certain level, the solution suddenly becomes almost perfectly transparent to this light, permitting the release of all the energy stored in the ruby as a giant pulse.

physical adsorption [PHYS CHEM] Reversible adsorption in which the adsorbate is held by weak physical forces.

physical constant [PHYS] A physical quantity which has a fixed and unchanging numerical value.

physical electronics [ELECTR] The study of physical phenomena basic to electronics, such as discharges, thermionic and field emission, and conduction in semiconductors and metals.

PHOTOVOLTAIC CELL

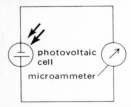

light

metal film
selenium
metal plate

Schematic of a photovoltaic cell; metal plate usually is iron, and thin metal film is either gold or platinum.

PHOTOVOLTAIC METER

photovoltaic cell

microammeter

Circuit diagram of photovoltaic type of exposure meter. Heavy arrows represent light falling on cell.

physical law [PHYS] A property of a physical phenomenon, or a relationship between the various quantities or qualities which may be used to describe the phenomenon, that applies to all members of a broad class of such phenomena, without exception.

physical libration of the moon *See* lunar libration.

physical measurement [PHYS] Quantitative information on a physical condition, property, or relation, generally in the form of the ratio of the measured quantity to a standard quantity, or to some fixed multiple or fraction thereof.

physical optics [OPTICS] The study of the interaction of electromagnetic waves in the optical frequency range with material systems.

physical property [CHEM] Property of a compound that can change without involving a change in chemical composition; examples are the melting point and boiling point.

physical realizability [CONT SYS] For a transfer function, the possibility of constructing a network with this transfer function.

physical system *See* causal system.

physical theory [PHYS] An attempt to explain a certain class of physical phenomena by deducing them as necessary consequences of some primitive assumptions.

physicist [PHYS] A person who does research in physics.

physics [SCI TECH] The study of those aspects of nature which can be understood in a fundamental way in terms of elementary principles and laws.

pi [MATH] The irrational number which is the ratio of the circumference of any circle to its radius; an approximation is 3.14159. Symbolized π.

pi bonding [PHYS CHEM] Covalent bonding in which the greatest overlap between atomic orbitals is along a plane perpendicular to the line joining the nuclei of the two atoms.

Picard method [MATH] A method of successive substitution for solving differential equations.

Picard's big theorem [MATH] The image of every neighborhood of an essential singularity of a complex function is dense in the complex plane. Also known as Picard's second theorem.

Picard's first theorem *See* Picard's little theorem.

Picard's little theorem [MATH] A nonconstant entire function of the complex plane assumes every value save at most one. Also known as Picard's first theorem.

Picard's second theorem *See* Picard's big theorem.

pickup [NUC PHYS] A type of nuclear reaction in which the incident particle takes a nucleon from the target nucleus and proceeds with this nucleon bound to itself.

pico- [MATH] A prefix meaning 10^{-12}; used with metric units. Also known as micromicro-.

picoammeter [ENG] An ammeter whose scale is calibrated to indicate current values in picoamperes.

picoampere [ELEC] A unit of current equal to 10^{-12} ampere, or one-millionth of a microampere. Abbreviated pA.

picofarad [ELEC] A unit of capacitance equal to 10^{-12} farad, or one-millionth of a microfarad. Also known as micromincrofarad (deprecated usage); puff (British usage). Abbreviated pF.

picosecond [MECH] A unit of time equal to 10^{-12} second, or one-millionth of a microsecond. Abbreviated ps; psec. Formerly known as micromicrosecond.

picowatt [MECH] A unit of power equal to 10^{-12} watt, or one-millionth of a microwatt. Abbreviated pW. Formerly known as micromicrowatt.

piecewise-linear [MATH] A continuous curve or function obtained by joining a finite number of linear pieces.

piecewise-linear system [CONT SYS] A system for which one can divide the range of values of input quantities into a finite number of intervals such that the output quantity is a linear function of the input quantity within each of these intervals.

piecewise linear topology *See* combinatorial topology.

pi electron [PHYS CHEM] An electron which participates in pi bonding.

Pierce oscillator [ELECTR] Oscillator in which a piezoelectric crystal unit is connected between the grid and the plate of an electron tube, in what is basically a Colpitts oscillator, with voltage division provided by the grid-cathode and plate-cathode capacitances of the circuit.

pièze [MECH] A unit of pressure equal to 1 sthène per square meter, or to 1000 pascals. Abbreviated pz.

piezoelectric [SOLID STATE] Having the ability to generate a voltage when mechanical force is applied, or to produce a mechanical force when a voltage is applied, as in a piezoelectric crystal.

piezoelectric crystal [SOLID STATE] A crystal which exhibits the piezoelectric effect; used in crystal loudspeakers, crystal microphones, and crystal cartridges for phono pickups.

piezoelectric effect [SOLID STATE] **1.** The generation of electric polarization in certain dielectric crystals as a result of the application of mechanical stress. **2.** The reverse effect, in which application of a voltage between certain faces of the crystal produces a mechanical distortion of the material.

piezoelectric hysteresis [SOLID STATE] Behavior of a piezoelectric crystal whose electric polarization depends not only on the mechanical stress to which the crystal is subjected, but also on the previous history of this stress.

piezoelectricity [SOLID STATE] Electricity or electric polarization resulting from the piezoelectric effect.

piezoelectric oscillator *See* crystal oscillator.

piezoelectric resonator *See* crystal resonator.

piezoelectric semiconductor [SOLID STATE] A semiconductor exhibiting the piezoelectric effect, such as quartz, Rochelle salt, and barium titanate.

piezoelectric vibrator [SOLID STATE] An element cut from piezoelectric material, usually in the form of a plate, bar, or ring, with electrodes attached to or supported near the element to excite one of its resonant frequencies.

piezometer [ENG] Any instrument for measuring pressures, especially high pressures.

piezotropic [FL MECH] Characterized by piezotropy.

piezotropy [FL MECH] The property of a fluid in which processes are characterized by a functional dependence of the thermodynamic functions of state: $d\rho/dt = b(dp/dt)$, where ρ is the density, p the pressure, and b a function of the thermodynamic variables, called the coefficient of piezotropy.

pi filter [ELECTR] A filter that has a series element and two parallel elements connected in the shape of the Greek letter pi (π).

pig [ELECTR] **1.** An ion source based on the same principle as the Philips ionization gage. **2.** *See* Philips ionization gage. [NUCLEO] A heavily shielded container, usually lead, used to ship or store radioisotopes and other radioactive materials.

pigeonhole principle [MATH] If a very large set of elements is partitioned into a small number of blocks, then at least one block contains a rather large number of elements. Also known as Dirichlet drawer principle.

pile *See* nuclear reactor.

pillbox antenna [ELECTROMAG] Cylindrical parabolic reflector enclosed by two plates perpendicular to the cylinder, spaced to permit the propagation of only one mode in the desired direction of polarization.

pilot streamer [GEOPHYS] A relatively slow-moving, nonluminous lightning streamer, the existence of which has been postulated to help account for the observed mode of advance of a stepped leader as it initiates a lightning discharge.

pilot wire regulator [CONT SYS] Automatic device for controlling adjustable gains or losses associated with transmission circuits to compensate for transmission changes caused by temperature variations, the control usually depending upon the resistance of a conductor or pilot wire having substantially the same temperature conditions as the conductors of the circuits being regulated.

pi meson [PARTIC PHYS] **1.** Collective name for three semistable mesons which have charges of $+1, 0$, and -1 times the proton charge, and form a charge multiplet, with an approximate mass of 138 MeV, spin 0, negative parity, negative G parity, and positive charge parity (for the neutral meson). Also known as pion. Symbolized π. **2.** Any meson belonging to an isospin triplet with hypercharge 0, negative G parity, and positive charge parity (for the neutral meson).

pin [ELECTR] A terminal on an electron tube, semiconductor, integrated circuit, plug, or connector. Also known as base pin; prong.

pinacoid [CRYSTAL] An open crystal form that comprises two parallel faces.

pinch effect [ELEC] Manifestation of the magnetic self-attraction of parallel electric currents, such as constriction of ionized gas in a discharge tube, or constriction of molten metal through which a large current is flowing. Also known as cylindrical pinch; magnetic pinch; rheostriction.

pinch-off voltage [ELECTR] Of a field-effect transistor, the voltage at which the current flow between source and drain is blocked because the channel between these electrodes is completely depleted.

pinch resistor [ELECTR] A silicon integrated-circuit resistor produced by diffusing an n-type layer over a p-type resistor; this narrows or pinches the resistive channel, thereby increasing the resistance value.

pincushion distortion [ELECTR] Distortion in which all four sides of a received television picture are concave (curving inward). [OPTICS] Aberration in which the magnification produced by an optical system increases with the distance of the object point from the optical axis, so that the image of a square has concave sides.

pin diode [ELECTR] A diode consisting of a silicon wafer containing nearly equal p-type and n-type impurities, with additional p-type impurities diffused from one side and additional n-type impurities from the other side; this leaves a lightly doped intrinsic layer in the middle, to act as a dielectric barrier between the n-type and p-type regions. Also known as power diode.

pi network [ELEC] An electrical network which has three impedance branches connected in series to form a closed circuit, with the three junction points forming an output terminal, an input terminal, and a common output and input terminal.

pin junction [ELECTR] A semiconductor device having three regions: p-type impurity, intrinsic (electrically pure), and n-type impurity.

pint [MECH] Abbreviated pt. **1.** A unit of volume, used in the United States for measurement of liquid substances, equal to $\frac{1}{8}$ U.S. gallon, or $29\frac{7}{8}$ cubic inches, or $4.73176473 \times 10^{-4}$ cubic meters. Also known as liquid pint (liq pt). **2.** A unit of volume used in the United States for measurement of solid substances, equal to $\frac{1}{64}$ U.S. bushel, or $107{,}521/3200$ cubic inches, or approximately 5.50610×10^{-4} cubic meters. Also

PINCH EFFECT

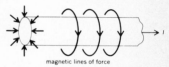

magnetic lines of force

Example of pinch effect. Current I, uniformly distributed in cylindrical wire, gives rise to compression force indicated by arrows at left.

known as dry pint (dry pt). **3.** A unit of volume, used in the United Kingdom for measurement of liquid and solid substances, although usually the former, equal to $\frac{1}{8}$ imperial gallon, or approximately 5.68261×10^{-4} cubic meters. Also known as imperial pint.

pion See pi meson.

pion condenstate [NUC PHYS] A state of nuclear matter compressed to abnormally high densities, in which great numbers of pairs of particles, each consisting of a positive pion and a negative pion, are generated, and interact strongly with the nucleons, causing them to form a coherent spin-isospin structure.

pion double-charge-exchange reaction [NUC PHYS] A nuclear reaction in which a positive pion strikes a target nucleus and is converted into a negative pion while two of the neutrons in the target nucleus are converted into protons.

pip See blip.

pi point [ELEC] Frequency at which the insertion phase shift of an electric structure is 180° or an integral multiple of 180°.

pipper [OPTICS] A small hole in the reticle of an optical sight or computing sight.

pipper image [OPTICS] A spot of light projected through the pipper in an optical or computing sight, used in aiming.

Pirani gage [PHYS] A thermal conductivity gage (where the thermal conductivity of a gas heated by a hot wire varies with pressure) connected to a Wheatstone bridge to measure the resistance of the hot wire, thus the gas pressure; used to measure pressure from 1 to 10^{-3} mmHg (133.32 to 0.13332 newtons per square meter).

pi section filter [ELEC] An electric filter made of several pi networks connected in series.

piston [ELECTROMAG] A sliding metal cylinder used in waveguides and cavities for tuning purposes or for reflecting essentially all of the incident energy. Also known as plunger; waveguide plunger.

piston attenuator [ELECTROMAG] A microwave attenuator inserted in a waveguide to introduce an amount of attenuation that can be varied by moving an output coupling device along its longitudinal axis.

piston gage See free-piston gage.

pistonphone [ENG ACOUS] A small chamber equipped with a reciprocating piston having a measurable displacement and used to establish a known sound pressure in the chamber, as for testing microphones.

piston viscometer [ENG] A device for the measurement of viscosity by the timed fall of a piston through the liquid being tested.

pitch [ACOUS] That psychological property of sound characterized by highness or lowness, depending primarily upon frequency of the sound stimulus, but also upon its sound pressure and waveform. [MECH] **1.** Of an aerospace vehicle, an angular displacement about an axis parallel to the lateral axis of the vehicle. **2.** The rising and falling motion of the bow of a ship or the tail of an airplane as the craft oscillates about a transverse axis. [SCI TECH] The inclination or degree of slope of an object or structure.

pitch acceleration [MECH] The angular acceleration of an aircraft or missile about its lateral, or Y, axis.

pitch attitude [MECH] The attitude of an aircraft, rocket, or other flying vehicle, referred to the relationship between the longitudinal body axis and a chosen reference line or plane as seen from the side.

pitch axis [MECH] A lateral axis through an aircraft, missile, or similar body, about which the body pitches. Also known as pitching axis.

PI SECTION FILTER

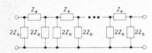

Schematic diagram of pi section filter. Z_a and Z_b represent impedances. These are chosen so that $|Z_a| \gg |Z_b|$ in the frequency range which is to be a stop band and $|Z_a| \ll |Z_b|$ in the desired pass band.

PITCH

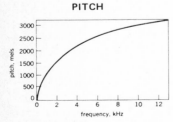

Pitch level of sounds of different frequency, in mels. A tone of 1000 hertz at an intensity level of 40 decibels above absolute threshold has a pitch level of 1000 mels. (After S. S. Stevens and J. Volkman, The relation of pitch to frequency, Amer. J. Psychol., 53(3):329–353, 1940)

pitching axis *See* pitch axis.

pi theorem *See* Buckingham's π theorem.

pitot pressure [FL MECH] Pressure at the open end of a pitot tube.

pitot tube [ENG] An instrument that measures the stagnation pressure of a flowing fluid, consisting of an open tube pointing into the fluid and connected to a pressure-indicating device. Also known as impact tube.

pi-T transformation *See* Y-delta transformation.

pivot [MECH] A short, pointed shaft forming the center and fulcrum on which something turns, balances, or oscillates.

pivotal condensation [MATH] A method of evaluating a determinant that is convenient for determinants of large order, especially when digital computers are used, involving a repeated process in which a determinant of order n is reduced to the product of one of its elements raised to a power and a determinant of order $n - 1$.

pivotal method [STAT] A technique for passing from one set of double inequalities to another in order to find a confidence interval for a parameter.

pk *See* peck.

PKA *See* primary knocked-on atom.

Pl *See* poiseuille.

place [MATH] **1.** A position corresponding to a given power of the base in positional notation. Also known as column. **2.** A homomorphic mapping p of a subring K_p of a field K into a field, such that (1) if x is in K but not in K_p, then $1/x$ is in K_p and $p(1/x) = 0$, and (2) $p(x) \neq 0$ for some x in K_p.

plage [ASTRON] One of the luminous areas that appear in the vicinity of sunspots or disturbed areas on the sun; they may be seen distinctively in spectroheliograms taken in the calcium K line.

plagiohedral [CRYSTAL] Pertaining to obliquely arranged spiral faces; in particular, to a member of a group in the isometric system with 13 axes but no center or planes.

planar [MATH] Lying in or pertaining to a euclidean plane.

planar device [ELECTR] A semiconductor device having planar electrodes in parallel planes, made by alternate diffusion of p- and n-type impurities into a substrate.

planar graph [MATH] A graph which is isomorphic to a graph whose vertex set is a point set in a plane π while the edges are Jordan curves in π such that two different edges have, at most, end points in common.

planar map [MATH] A plane or sphere divided into connected regions by a topological graph.

planar photodiode [ELECTR] A vacuum photodiode consisting simply of a photocathode and an anode; light enters through a window sealed into the base, behind the photocathode.

planar process [ENG] A silicon-transistor manufacturing process in which a fractional-micrometer-thick oxide layer is grown on a silicon substrate; a series of etching and diffusion steps is then used to produce the transistor inside the silicon substrate.

planar transistor [ELECTR] A transistor constructed by an etching and diffusion technique in which the junction is never exposed during processing, and the junctions reach the surface in one plane; characterized by very low leakage current and relatively high gain.

planck [PHYS] A unit of action equal to the product of an energy of 1 joule and a time of 1 second.

Planck distribution law *See* Planck radiation formula.

Planck function [THERMO] The negative of the Gibbs free energy divided by the absolute temperature.

Planck length [PHYS] The length $\sqrt{Gh/2\pi c^3}$ (where G is the

PITOT TUBE

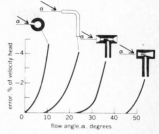

error, % of velocity head

flow angle, α, degrees

Effect of the flow alignment of a pitot tube with streamline on accuracy of measurement. Here α = angle between direction of flow and direction of tube opening.

PLAGE

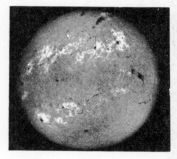

The disk of the sun photographed in H-α light, showing bright plages in centers of activity and dark filaments (prominences). *(Sacramento Peak Observatory, operated by the Association of Universities for Research in Astronomy, Inc.)*

gravitational constant, h is Planck's constant, and c is the speed of light) at which quantum fluctuations are believed to dominate the geometry of space-time; it is equal to 1.616×10^{-35} meter.

Planck mass [PHYS] The mass $\sqrt{hc/2\pi G}$, where h is Planck's constant, c is the speed of light, and G is the gravitational constant; equivalently, the mass of a particle whose reduced Compton wavelength equals the Planck length; it is equal to 21.77 micrograms or 1.221×10^{19} GeV/c^2.

Planck oscillator [QUANT MECH] An oscillator which can absorb or emit energy only in amounts which are integral multiples of Planck's constant times the frequency of the oscillator. Also known as radiation oscillator.

Planck radiation formula [STAT MECH] A formula for the intensity of radiation emitted by a blackbody within a narrow band of frequencies (or wavelengths), as a function of frequency, and of the body's temperature. Also known as Planck distribution law; Planck's law.

Planck's constant [QUANT MECH] A fundamental physical constant, the elementary quantum of action; the ratio of the energy of a photon to its frequency, it is equal to $6.62620 \pm 0.00005 \times 10^{-34}$ joule-second. Symbolized h. Also known as quantum of action.

Planck's law [QUANT MECH] A fundamental law of quantum theory stating that energy associated with electromagnetic radiation is emitted or absorbed in discrete amounts which are proportional to the frequency of radiation. [STAT MECH] *See* Planck radiation formula.

Planck time [PHYS] The constant $(Gh/c^5)^{1/2}$ with dimensions of "time" formed from Planck's constant h, the gravitational constant G, and the speed of light c; approximately 10^{-43} second.

plane [MATH] **1.** A surface containing any straight line through any two of its points. **2.** In projective geometry, a triple of sets (P, L, I) where P denotes the set of points, L the set of lines, and I the incidence relation on points and lines, such that (1) P and L are disjoint sets, (2) the union of P and L is nonnull, and (3) I is a subset of $P \times L$, the cartesian product of P and L.

plane angle [MATH] An angle between lines in the euclidean plane.

plane curve [MATH] Any curve lying entirely within a plane.

plane cyclic curve *See* cyclic curve.

plane dendrite *See* plane-dendritic crystal.

plane-dendritic crystal [CRYSTAL] An ice crystal exhibiting an elaborately branched (dendritic) structure of hexagonal symmetry, with its much larger dimension lying perpendicular to the principal (c axis) of the crystal. Also known as plane dendrite; stellar crystal.

plane field *See* field of planes on a manifold.

plane geometry [MATH] The geometric study of the figures in the euclidean plane such as lines, triangles, and polygons.

plane group [CRYSTAL] The group of operations (rotations, reflections, translations, and combinations of these) which leave a regular, periodic structure in a plane unchanged.

plane lamina [MECH] A body whose mass is concentrated in a single plane.

plane lattice [CRYSTAL] A regular, periodic array of points in a plane.

plane mirror [OPTICS] A mirror whose surface lies in a plane; it forms an image of an object such that the mirror surface is perpendicular to and bisects the line joining all corresponding object-image points.

plane of flotation [FL MECH] The plane in which the surface of a liquid intersects a stationary floating body.

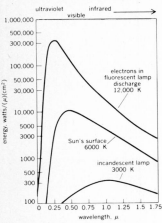

PLANCK RADIATION FORMULA

Graphs of the Planck radiation formula for various temperatures.

plane of incidence [PHYS] A plane containing the direction of propagation of a wave striking a surface and a line perpendicular to the surface. Also known as incidence plane.

plane of mirror symmetry Also known as mirror plane of symmetry; plane of symmetry; reflection plane; symmetry plane. [CRYSTAL] In certain crystals, a symmetry element whereby reflection of the crystal through a certain plane leaves the crystal unchanged. [MATH] An imaginary plane which divides an object into two halves, each of which is the mirror image of the other in this plane.

plane of polarization [ELECTROMAG] Plane containing the electric vector and the direction of propagation of electromagnetic wave.

plane of reflection [CRYSTAL] *See* plane of mirror symmetry. [MATH] *See* plane of mirror symmetry. [OPTICS] A plane containing the direction of propagation of radiation reflected from a surface, and the normal to the surface. Also known as reflection plane.

plane of symmetry *See* plane of mirror symmetry.

plane Poiseuille flow [FL MECH] Rheological (viscosity) measurement in which the fluid of interest is propelled through a narrow slot, and the volumetric flow rate and the pressure gradient are measured simultaneously to determine viscosity.

plane polarization *See* linear polarization.

plane-polarized wave [ELECTROMAG] An electromagnetic wave whose electric field vector at all times lies in a fixed plane that contains the direction of propagation through a homogeneous isotropic medium.

plane polygon [MATH] A polygon lying in the euclidean plane.

plane quadrilateral [MATH] A four-sided polygon lying in the euclidean plane.

plane reflector *See* passive reflector.

plane strain [MECH] A deformation of a body in which the displacements of all points in the body are parallel to a given plane, and the values of these displacements do not depend on the distance perpendicular to the plane.

planet [ASTRON] A relatively small, solid celestial body circulating around a star, in particular the star known as the sun (which has nine planets).

planetary electron *See* orbital electron.

planetary nebula [ASTRON] An oval or round nebula of expanding concentric rings of gas associated with a hot central star.

planetary physics [ASTROPHYS] The study of the structure, composition, and physical and chemical properties of the planets of the solar system, including their atmospheres and immediate cosmic environment.

planetoid *See* asteroid.

plane trigonometry [MATH] The study of triangles in the euclidean plane with the use of functions defined by the ratios of sides of right triangles.

plane wave [PHYS] Wave in which the wavefront is a plane surface; a wave whose equiphase surfaces form a family of parallel planes.

planigraphy *See* sectional radiography.

planoconcave lens [OPTICS] A lens for which one surface is plane and the other is concave.

planoconvex lens [OPTICS] A lens for which one surface is plane and the other is convex.

planocylindrical lens [OPTICS] A lens, one of whose surfaces is a portion of a plane, while the other is a portion of a cylinder.

Plante cell [ELEC] A type of lead-acid cell in which the active

PLANOCONCAVE LENS

Shape of a planoconcave lens. *(From F. A. Jenkins and H. E. White, Fundamentals of Optics, 3d ed., McGraw-Hill, 1957)*

PLANOCONVEX LENS

Shape of a planoconvex lens. *(From F. A. Jenkins and H. E. White, Fundamentals of Optics, 3d ed., McGraw-Hill, 1957)*

material is formed on the plates by electrochemical means during repeated charging and discharging, instead of being applied as a prepared paste.

plasma [PL PHYS] **1.** A highly ionized gas which contains equal numbers of ions and electrons in sufficient density so that the Debye shielding length is much smaller than the dimensions of the gas. **2.** A completely ionized gas, composed entirely of a nearly equal number of positive and negative free charges (positive ions and electrons).

plasma accelerator [PL PHYS] An accelerator that forms a high-velocity jet of plasma by using a magnetic field, an electric arc, a traveling wave, or other similar means.

plasma cathode [ELECTR] A cathode in which the source of electrons is a gas plasma rather than a solid.

plasma frequency *See* Langmuir plasma frequency.

plasma gun [ELECTR] A machine, such as an electric-arc chamber, that will generate very high heat fluxes to convert neutral gases into plasma. [ELECTROMAG] An electromagnetic device which creates and accelerates bursts of plasma.

plasma instability [PL PHYS] A sudden change in the quasistatic distribution of positions or velocities of particles constituting a plasma, and a sudden change in the accompanying electromagnetic field.

plasma-jet excitation [SPECT] The use of a high-temperature plasma jet to excite an element to provide measurable spectra with many ion lines similar to those from spark-excited spectra.

plasma mantle [GEOPHYS] A thick layer of plasma just inside the magnetopause characterized by a tailward bulk flow with a speed of 100 to 200 kilometers per second and by a gradual decrease of density, temperature, and speed as the depth inside the magnetosphere increases.

plasma oscillations [PL PHYS] Various vibrations and wave motions of the electrons and ions in a plasma.

plasma physics [PHYS] The study of highly ionized gases.

plasma pinch [PL PHYS] Application of the pinch effect to plasma in attempts to produce controlled nuclear fusion.

plasma radiation [PL PHYS] Electromagnetic radiation emitted from a plasma, primarily by free electrons undergoing transitions to other free states or to bound states of atoms and ions, but also by bound electrons as they undergo transitions to other bound states.

plasma wave [PL PHYS] A disturbance of a plasma involving oscillation of its constituent particles and of an electromagnetic field, which propagates from one point in the plasma to another without net motion of the plasma.

plasmoid [PHYS] An isolated collection of electrons, ions, and neutral particles which holds together for a duration many times as long as the collision times between particles.

plasmon [SOLID STATE] A quantum of a collective longitudinal wave in the electron gas of a solid.

plastic [MECH] Displaying or associated with plasticity.

plastic collision [MECH] A collision in which one or both of the colliding bodies suffers plastic deformation and mechanical energy is dissipated.

plastic deformation [MECH] Permanent change in shape or size of a solid body without fracture resulting from the application of sustained stress beyond the elastic limit.

plastic film capacitor [ELEC] A capacitor constructed by stacking, or forming into a roll, alternate layers of foil and a dielectric which consists of a plastic, such as polystyrene or Mylar, either alone or as a laminate with paper.

plastic flow [PHYS] Rheological phenomenon in which flowing behavior of the material occurs after the applied stress reaches a critical (yield) value, such as with putty.

plasticity [MECH] The property of a solid body whereby it undergoes a permanent change in shape or size when subjected to a stress exceeding a particular value, called the yield value.

plasticoviscosity [MECH] Plasticity in which the rate of deformation of a body subjected to stresses greater than the yield stress is a linear function of the stress.

plastic viscosity [FL MECH] The difference between the shear stress and the yield stress of a Bingham plastic, divided by the rate of shear.

plate [ELEC] **1.** One of the conducting surfaces in a capacitor. **2.** One of the electrodes in a storage battery. [ELECTR] *See* anode.

plateau [ELECTR] The portion of the plateau characteristic of a counter tube in which the counting rate is substantially independent of the applied voltage.

plateau characteristic [ELECTR] The relation between counting rate and voltage for a counter tube when radiation is constant, showing a plateau after the rise from the starting voltage to the Geiger threshold. Also known as counting rate–voltage characteristic.

plateau problem [MATH] The problem of finding a minimal surface having as boundary a given curve.

Plateau's sphere [FL MECH] A small drop of liquid which follows a larger drop that breaks away and falls.

plate circuit *See* anode circuit.

plate current *See* anode current.

plate efficiency *See* anode efficiency.

plate impedance *See* anode impedance.

plate input power *See* anode input power.

plate-load impedance *See* anode impedance.

plate modulus [MECH] The ratio of the stress component T_{xx} in an isotropic, elastic body obeying a generalized Hooke's law to the corresponding strain component S_{xx}, when the strain components S_{yy} and S_{zz} are 0; the sum of the Poisson ratio and twice the rigidity modulus.

plate resistance *See* anode resistance.

plate saturation *See* anode saturation.

plate wave [ACOUS] A type of ultrasonic vibration generated in a thin solid, such as a sheet of metal having a thickness of less than one wavelength, and usually consisting of a variety of simultaneous modes having different velocities; it is used in metal inspection. Also known as Lamb wave.

platinum [CHEM] A chemical element,' symbol Pt, atomic number 78, atomic weight 195.09.

platinum resistance thermometer [ENG] The basis of the International Practical Temperature Scale of 1968 from 259.35 to 630.74°C; used in industrial thermometers in the range 0 to 650°C; capable of high accuracy because platinum is noncorrosive, ductile, and nonvolatile, and can be obtained in a very pure state. Also known as Callendar's thermometer.

platonic year *See* great year.

playkurtic distribution [STAT] A distribution of a data set which is relatively flat.

play of color [OPTICS] An optical phenomenon consisting of a rapid succession of flashes of a variety of prismatic colors as certain minerals or cabochon-cut gems are moved about; caused by diffraction of light from spherical particles of amorphous silica stacked in an orderly three-dimensional pattern. Also known as schiller.

Plemelj formulas [MATH] Formulas for the limits of the Cauchy integrals of an arc with respect to a point z as z approaches the arc from either side.

Plemelj function [MATH] The solution of the homogeneous

PLATINUM

Periodic table of the chemical elements showing the position of platinum.

Hilbert problem when the set of arcs C is reduced to a single simple arc.

pleochroic halos [OPTICS] Halos of color or color differences that are sometimes observed around inclusions in minerals, resulting from irradiation by alpha particles.

pleochroism [OPTICS] Phenomenon exhibited by certain transparent crystals in which light viewed through the crystal has different colors when it passes through the crystal in different directions. Also known as polychroism.

pleomorphism *See* polymorphism.

pli [MECH] A unit of line density (mass per unit length) equal to 1 pound per inch, or approximately 17.8580 kilograms per meter.

Plowshare [NUCLEO] The U.S. Atomic Energy Commission program of research and development on peaceful uses of nuclear explosives; possible uses include large-scale excavation, such as for canals and harbors, crushing ore bodies, and producing heavy transuranic isotopes.

Plucker equations [MATH] Equations which relate the degree n and class m of an algebraic plane curve, and the number of nodes d, cusps c, inflectional tangents i, and bitangents b which it possesses; the most important equations are $n(n - 1) = m + 2d + 3c$; $m(m - 1) = n + 2b + 3i$; $3n(n - 2) = i + 6d + 8c$; $3m(m - 2) = c + 6b + 8i$; $3(m - n) = i - c$; and $\frac{1}{2}(n - 1)(n - 2) - d - c = \frac{1}{2}m(m - 1)(m - 2) - b - i$.

plumbum [CHEM] Latin name for lead; source of the element symbol Pb.

plume *See* column.

plunger *See* piston.

plural scattering [PHYS] A change in direction of a particle or photon because of a small number of collisions.

plus [MATH] A mathematical symbol; A plus B, where A and B are mathematical quantities, denotes the quantity obtained by taking their sum in an appropriate context.

Pluto [ASTRON] The most distant planet in the solar system; mean distance to the sun is about 5.6×10^9 kilometers; it has no known satellite, and its sidereal revolution period is 248.4 years.

plutonium [CHEM] A reactive metallic element, symbol Pu, atomic number 94, in the transuranium series of elements; the first isotope to be identified was plutonium-239; used as a nuclear fuel, to produce radioactive isotopes for research, and as the fissile agent in nuclear weapons.

plutonium-238 [NUC PHYS] The first synthetic isomer made of plutonium; similar chemically to uranium and neptunium; atomic number 94; formed by bombardment of uranium with deuterons.

plutonium-239 [NUC PHYS] A synthetic element chemically similar to uranium and neptunium; atomic number 94; made by bombardment of uranium-238 with slow electrons in a nuclear reactor; used as nuclear reactor fuel and an ingredient for nuclear weapons.

plutonium reactor [NUCLEO] A nuclear reactor in which plutonium is the principal fissionable material.

PLZT *See* lead lanthanum zirconate titanate.

Pm *See* promethium.

PM *See* permanent magnet; phase modulation.

p.m.f. *See* probability mass function.

PMR *See* projection microradiography.

PNdB *See* perceived noise decibel.

pneumatic [ENG] Pertaining to or operated by air or other gas.

pneumatics [FL MECH] Fluid statics and behavior in closed systems when the fluid is a gas.

PLUTO

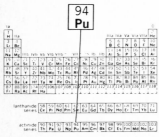

Orbit of Pluto. A perspective view to show the inclination i and eccentricity of the orbit. A, aphelion; P, perihelion; NN', line of nodes. (*From L. Rudaux and G. de Vaucouleurs, Astronomie, Larousse, 1948*)

PLUTONIUM

94
Pu

Periodic table of the chemical elements showing the position of plutonium.

pneumatic servomechanism [CONT SYS] A servomechanism in which power is supplied and transmission of signals is carried out through the medium of compressed air.

pn hook transistor *See* hook collector transistor.

pnip transistor [ELECTR] An intrinsic junction transistor in which the intrinsic region is sandwiched between the *n*-type base and the *p*-type collector.

pn junction [ELECTR] The interface between two regions in a semiconductor crystal which have been treated so that one is a *p*-type semiconductor and the other is an *n*-type semiconductor; it contains a permanent dipole charge layer.

pnpn diode [ELECTR] A semiconductor device consisting of four alternate layers of *p*-type and *n*-type semiconductor material, with terminal connections to the two outer layers. Also known as *npnp* diode.

pnpn transistor *See* npnp transistor.

pnp transistor [ELECTR] A junction transistor having an *n*-type base between a *p*-type emitter and a *p*-type collector.

Po *See* polonium.

Pochhammer's symbol [MATH] The symbol $(a)_n$, where a is any number and n is a positive integer, used to represent the quantity $a(a + 1) \cdots (a + n - 1)$, with $(a)_0 = 1$.

Pockels cell [OPTICS] A crystal that exhibits the Pockels effect, such as potassium dihydrogen phosphate, which is placed between crossed polarizers and has ring electrodes bonded to two faces to allow application of an electric field; used to modulate light beams, especially laser beams.

Pockels effect [OPTICS] Changes in the refractive properties of certain crystals in an applied electric field, which are proportional to the first power of the electric field strength.

Pockels equation [MATH] A partial differential equation which states that the Laplacian of an unknown function, plus the product of the value of the function with a constant, is equal to 0; it arises in finding solutions of the wave equation that are products of time-independent and space-independent functions.

Poggendorff's first method *See* constant-current dc potentiometer.

Poggendorff's second method *See* constant-resistance dc potentiometer.

Poincaré-Bendixson theorem [MATH] For the system of differential equations $dx_1/dt = f_1(x_1, x_2)$, $dx_2/dt = f_2(x_1, x_2)$, describing the motion of points in the (x_1, x_2) plane, where f_1 and f_2 are continuously differentiable, any limit set which does not contain an equilibrium point is the image of a nontrivial, periodic solution of the equations.

Poincaré conjecture [MATH] The question as to whether a compact, simply connected three-dimensional manifold without boundary must be homeomorphic to the three-dimensional sphere.

Poincaré duality theorem [MATH] If X is a topological space that is homeomorphic to an n-dimensional simplicial complex, then the p-dimensional connectivity number of X, R_p, is equal to R_{n-p}, and the torsion group $G_p(X)$ is isomorphic to $G_{n-p-1}(X)$.

Poincaré electron [ELECTROMAG] A classical model of the electron in which nonelectromagnetic forces hold the electron together so that it has zero self-stress; it is unstable and has infinite self-energy in the case of a point electron.

Poincaré recurrence theorem [MATH] **1.** A volume-preserving homeomorphism T of a finite dimensional euclidean space will have, for almost all points x, infinitely many points of the form $T^i(x)$, $i = 1, 2, \ldots$ within any open set containing x. **2.** A measure preserving transformation on a space with finite measure is recurrent.

Poinsot ellipsoid   *See* inertia ellipsoid.

Poinsot motion   [MECH] The motion of a rigid body with a point fixed in space and with zero torque or moment acting on the body about the fixed point.

Poinsot's central axis   [MECH] A line through a rigid body which is parallel to the vector sum **F** of a system of forces acting on the body, and which is located so that the system of forces is equivalent to the force **F** applied anywhere along the line, plus a couple whose torque is equal to the component of the total torque **T** exerted by the system in the direction **F**.

Poinsot's method   [MECH] A method of describing Poinsot motion, by means of a geometrical construction in which the inertia ellipsoid rolls on the invariable plane without slipping.

point   [MATH] **1.** An element in a topological space. **2.** One of the basic undefined elements of geometry possessing position but no nonzero dimension. **3.** In positional notation, the character or the location of an implied symbol that separates the integral part of a numerical expression from its fractional part; for example, it is called the binary point in binary notation and the decimal point in decimal notation.

point at infinity   *See* ideal point.

point biserial correlation coefficient   [STAT] A modification of the biserial correlation coefficient in which one variable is dichotomous and the other is continuous; a product moment correlation coefficient.

point characteristic function   [OPTICS] The integral between two points of $n\, ds$ along some path, where ds is the arc length of an infinitesimal piece of the path and n is the refractive index; according to Fermat's principle, it is a maximum or minimum with respect to nearby paths for the actual path of a light ray.

point-contact diode   [ELECTR] A semiconductor rectifier that uses the barrier formed between a specially prepared semiconductor surface and a metal point to produce the rectifying action.

point-contact transistor   [ELECTR] A transistor having a base electrode and two or more point contacts located near each other on the surface of an n-type semiconductor.

point defect   [CRYSTAL] A departure from crystal symmetry which affects only one, or, in some cases, two lattice sites.

point distal flow   [MATH] A transformation group on a compact metric space for which there exists a distal point with a dense orbit.

point estimates   [STAT] Estimates which produce a single value of the population.

point function   [MATH] A function whose values are points. [PHYS] A quantity whose value depends on the location of a point in space, such as an electric field, pressure, temperature, or density.

point group   [CRYSTAL] A group consisting of the symmetry elements of an object having a single fixed point; 32 such groups are possible.

point-junction transistor   [ELECTR] Transistor having a base electrode and both point-contact and junction electrodes.

point of contraflexure   [MECH] A point at which the direction of bending changes. Also known as point of inflection.

point of fall   [MECH] The point in the curved path of a falling projectile that is level with the muzzle of the gun. Also known as level point.

point of inflection   [MATH] A point where a plane curve changes from the concave to the convex relative to some fixed line; equivalently, if the function determining the curve has a second derivative, this derivative changes sign at this point.

POINSOT'S METHOD

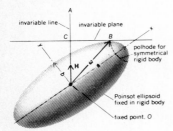

Inertia ellipsoid has axes with lengths a, b, and c (not shown) along principal axes of body, x, y, and z (not shown). OA = invariable line drawn in direction of fixed vector angular momentum **H**. Inertia ellipsoid rolls on invariable plane which is perpendicular to OA and intersects it at C. Ellipsoid is tangent to plane at B, where OB = vector angular momentum $\boldsymbol{\omega}$.

POINT DEFECT

Key:
a = substitutional impurity
b = interstitial impurity
c = vacancy
d = interstitial
e = Frenkel pair
f = interstitial impurity moving from one site through a saddle point to a neighboring site
g = divacancy

Point defects in a lattice. The squares indicate positions that were formerly occupied by atoms displaced as shown by the arrows.

Also known as inflection point. [MECH] *See* point of contraflexure.

point of osculation *See* double cusp.

point projection electron microscope [ELECTR] An electron microscope in which a real or virtual point source of electrons produces a highly magnified shadow.

point set topology [MATH] The general study of topological spaces.

point source [PHYS] A source of radiation having definite position but no extension in space; this is an ideal which is a good approximation for distances from the source sufficiently large compared to the dimensions of the source.

point spectrum [MATH] Those eigenvalues in the spectrum of a linear operator between Banach spaces whose corresponding eigenvectors are nonzero and of finite norm.

pointwise convergence [MATH] A sequence of functions f_1, $f_2, \ldots$ defined on a set S converges pointwise to a function f if the sequence $f_1(x), f_2(x), \ldots$ converges to $f(x)$ for each x in S.

poise [FL MECH] A unit of dynamic viscosity equal to the dynamic viscosity of a fluid in which there is a tangential force 1 dyne per square centimeter resisting the flow of two parallel fluid layers past each other when their differential velocity is 1 centimeter per second per centimeter of separation. Abbreviated P.

poiseuille [FL MECH] A unit of dynamic viscosity of a fluid in which there is a tangential force of 1 newton per square meter resisting the flow of two parallel layers past each other when their differential velocity is 1 meter per second per meter of separation; equal to 10 poise; used chiefly in France. Abbreviated Pl.

Poiseuille flow [FL MECH] The steady flow of an incompressible fluid parallel to the axis of a circular pipe of infinite length, produced by a pressure gradient along the pipe.

Poiseuille's law [FL MECH] The law that the volume flow of an incompressible fluid through a circular tube is equal to $\pi/8$ times the pressure differences between the ends of the tube, times the fourth power of the tube's radius divided by the product of the tube's length and the dynamic viscosity of the fluid.

poison [ATOM PHYS] A substance which reduces the phosphorescence of a luminescent material. [ELECTR] A material which reduces the emission of electrons from the surface of a cathode. [NUCLEO] A substance that absorbs neutrons without any fission resulting, and thereby lowers the reactivity of a nuclear reactor.

Poisson binomial trials model *See* generalized binomial trials model.

Poisson bracket [MECH] For any two dynamical variables, X and Y, the sum, over all degrees of freedom of the system, of $(\partial X/\partial q)(\partial Y/\partial p) - (\partial X/\partial p)(\partial Y/\partial q)$, where q is a generalized coordinate and p is the corresponding generalized momentum.

Poisson constant [PHYS] The ratio k of the gas constant R to the specific heat at constant pressure C_p.

Poisson density functions [STAT] Density functions corresponding to Poisson distributions.

Poisson distribution [STAT] A probability distribution whose mean and variance have a common value k, and whose frequency is $f(x) = k^x e^{-k}/x!$, for $x = 0, 1, 2, \ldots$.

Poisson formula [MATH] If the infinite series of functions $f(2\pi k + t), k$ ranging from $-\infty$ to ∞, converges uniformly to a function of bounded variation, then the infinite series with term $f(2\pi k), k$ ranging from $-\infty$ to ∞, is identical to the series with term the integral of $f(x)e^{-ikx}\,dx, k$ ranging from $-\infty$ to ∞.

Poisson index of dispersion [STAT] An index used for events which follow a Poisson distribution and should have a chi-square distribution.

Poisson integral formula [MATH] This formula gives a solution function for the Dirichlet problem in terms of integrals; an integral representation for the Bessel functions.

Poisson process [STAT] A process given by a discrete random variable which has a Poisson distribution.

Poisson ratio [MECH] The ratio of the transverse contracting strain to the elongation strain when a rod is stretched by forces which are applied at its ends and which are parallel to the rod's axis.

Poisson's equation [MATH] The partial differential equation which states that the Laplacian of an unknown function is equal to a given function.

Poisson transform [MATH] An integral transform which transforms the function $f(t)$ to the function

$$F(x) = (2/\pi) \int_0^\infty [t/(x^2 + t^2)] f(t)\,dt$$

Also known as potential transform.

polar [MATH] **1.** For a conic section, the polar of a point is the line that passes through the points of contact of the two tangents drawn to the conic from the point. **2.** For a quadric surface, the polar of a point is the plane that passes through the curve which is the locus of the points of contact of the tangents drawn to the surface from the point. **3.** For a quardric surface, the polar of a line is the line of intersection of the planes which are tangent to the surface at its points of intersection with the original line.

polar axis [CRYSTAL] An axis of crystal symmetry which does not have a plane of symmetry perpendicular to it. [MATH] The directed straight line relative to which the angle is measured for a representation of a point in the plane by polar coordinates.

polar compound [CHEM] Molecules which contain polar covalent bonds; they can ionize when dissolved or fused; polar compounds include inorganic acids, bases, and salts.

polar coordinates [MATH] A point in the plane may be represented by coordinates (r,θ), where θ is the angle between the positive x axis and the ray from the origin to the point, and r the length of that ray.

polar crystal *See* ferroelectric crystal.

polar diagram [PHYS] A diagram employing polar coordinates to show the magnitude of a quantity in some or all directions from a point; examples include directivity patterns and radiation patterns.

polar distance [ASTRON] Angular distance from a celestial pole; the arc of an hour circle between a celestial pole, usually the elevated pole, and a point on the celestial sphere, measured from the celestial pole through 180°.

polar electrojet [GEOPHYS] An intense current that flows in a relatively narrow band of the auroral zone ionosphere during disturbances of the magnetosphere.

polar form [MATH] A complex number $x + iy$ has as polar form $re^{i\theta}$, where (r,θ) are the polar coordinates corresponding to the point of the plane with rectangular coordinates (x,y), that is, $r = \sqrt{x^2 + y^2}$ and $\theta = \arctan y/x$.

polarimeter [OPTICS] An instrument used to determine the rotation of the plane of polarization of plane polarized light when it passes through a substance; the light is linearly polarized by a polarizer (such as a Nicol prism), passes through the material being analyzed, and then passes through an analyzer (such as another Nicol prism).

polarimetry [OPTICS] Determination of the rotation of the plane of polarization of plane polarized light when it passes through a substance, using a polarimeter.

polariscope [OPTICS] Any of several instruments used to determine the effects of substances on polarized light, in which linearly or elliptically polarized light passes through the substance being studied, and then through an analyzer.

polarity [MATH] Property of a line segment whose two ends are distinguishable. [PHYS] Property of a physical system which has two points with different (usually opposite) characteristics, such as one which has opposite charges or electric potentials, or opposite magnetic poles.

polarizability [ELEC] The electric dipole moment induced in a system, such as an atom or molecule, by an electric field of unit strength.

polarizability catastrophe [ELEC] According to a theory using the Lorentz field concept, the phenomenon where, at a certain temperature, the dielectric constant of a material becomes infinite.

polarizability ellipsoid *See* index ellipsoid.

polarization [ELEC] **1.** The process of producing a relative displacement of positive and negative bound charges in a body by applying an electric field. **2.** A vector quantity equal to the electric dipole moment per unit volume of a material. Also known as dielectric polarization; electric polarization. **3.** A chemical change occurring in dry cells during use, increasing the internal resistance of the cell and shortening its useful life. [PHYS] **1.** Phenomenon exhibited by certain electromagnetic waves and other transverse waves in which the direction of the electric field or the displacement direction of the vibrations is constant or varies in some definite way. **2.** The direction of the electric field or the displacement vector of a wave exhibiting polarization (first definition). **3.** The process of bringing about polarization (first definition) in a transverse wave. **4.** Property of a collection of particles with spin, in which the majority have spin components pointing in one direction, rather than at random.

polarization charge *See* bound charge.

polarization ellipse [PHYS] The ellipse traced out by the tip of the electric field vector or the displacement vector of a polarized wave at a fixed point in space in the course of time.

polarization potential [ELECTROMAG] One of two vectors from which can be derived, by differentiation, an electric scalar potential and magnetic vector potential satisfying the Lorentz condition. Also known as Hertz vector. [PHYS CHEM] The reverse potential of an electrolytic cell which opposes the direct electrolytic potential of the cell.

polarized electromagnetic radiation [ELECTROMAG] Electromagnetic radiation in which the direction of the electric field vector is not random.

polarized light [OPTICS] Polarized electromagnetic radiation whose frequency is in the optical region.

polarized meter [ENG] A meter having a zero-center scale, with the direction of deflection of the pointer depending on the polarity of the voltage or the direction of the current being measured.

polarized neutrons [PHYS] A collection of neutrons in which the majority have spin pointing in one direction rather than at random.

polarizer [OPTICS] A device which produces polarized light, such as a Nicol prism or Polaroid sheet.

polarizing angle *See* Brewster's angle.

polarizing filter [OPTICS] A device which selectively absorbs

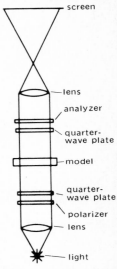

POLARISCOPE

screen
lens
analyzer
quarter-wave plate
model
quarter-wave plate
polarizer
lens
light

A polariscope used in measuring photoelastic stress.

components of electromagnetic radiation passing through it, so that light emerging from it is plane-polarized.

polarizing microscope [OPTICS] A microscope in which an object is viewed in polarized light.

polarizing pyrometer [ENG] A type of pyrometer, such as the Wanner optical pyrometer, in which monochromatic light from the source under investigation and light from a lamp with filament maintained at a constant but unknown temperature are both polarized and their intensities compared.

polar molecule [PHYS CHEM] A molecule having a permanent electric dipole moment.

Polaroid [OPTICS] Trademark of Polaroid Corporation for a sheet material that produces plane-polarized light.

polaron [SOLID STATE] An electron in a crystal lattice together with a cloud of phonons that result from the deformation of the lattice produced by the interaction of the electron with ions or atoms in the lattice.

polar radiation pattern [ELECTROMAG] Diagram showing the relative strength of the radiation from an antenna in all directions in a given plane. [ENG ACOUS] Diagram showing the strength of sound waves radiated from a loudspeaker in various directions in a given plane, or a similar response pattern for a microphone.

polar telescope [OPTICS] A telescope which uses rotating mirrors to enable celestial objects to be observed through a fixed eyepiece.

polar triangle [MATH] A triangle associated to a given spherical triangle obtained from three directed lines perpendicular to the planes associated with the sides of the original triangle.

polar vector *See* vector.

pole [CRYSTAL] **1.** A direction perpendicular to one of the faces of a crystal. **2.** One of the points at which normals to crystal faces or planes intersect a reference sphere at whose center the crystal is located. [ELEC] **1.** One of the electrodes in an electric cell. **2.** An output terminal on a switch; a double-pole switch has two output terminals. [MATH] **1.** An isolated singular point z_0 of a complex function whose Laurent series expansion about z_0 will include finitely many terms of the form $a_n(z - z_0)^{-n}$. **2.** For a great circle on a sphere, the pole of the circle is a point of intersection of the sphere and a line that passes through the center of the sphere and is perpendicular to the plane of the circle. **3.** For a conic section, the pole of a line is the intersection of the tangents to the conic at the points of intersection of the conic with the line. **4.** For a quadric surface, the pole of a plane is the vertex of the cone which is tangent to the surface along the curve where the plane intersects the surface. [MECH] **1.** A point at which an axis of rotation or of symmetry passes through the surface of a body. **2.** *See* perch.

pole dominance [PARTIC PHYS] Property of a scattering amplitude, analytically continued to complex values of energy and scattering angle, whose behavior is dominated by the term or terms of negative power in the Laurent series of a nearby pole.

pole-positioning [CONT SYS] A design technique used in linear control theory in which many or all of a system's closed-loop poles are positioned as required, by proper choice of a linear state feedback law; if the system is controllable, all of the closed-loop poles can be arbitrarily positioned by this technique.

pole strength *See* magnetic pole strength.

pole-zero configuration [CONT SYS] A plot of the poles and zeros of a transfer function in the complex plane; used to study the stability of a system, its natural motion, its frequency response, and its transient response.

polhode [MECH] For a rotating rigid body not subject to external torque, the closed curve traced out on the inertia ellipsoid by the intersection with this ellipsoid of an axis parallel to the angular velocity vector and through the center.

Polhode cone *See* body cone.

Polish space [MATH] A separable metric space which is homeomorphic to a complete metric space.

polonium [CHEM] A chemical element, symbol Po, atomic number 84; all polonium isotopes are radioactive; polonium-210 is the naturally occurring isotope found in pitchblende.

polonium-210 [NUC PHYS] Radioactive isotope of polonium; mass 210, half-life 140 days, α-radiation; used to calibrate radiation counters, and in oil well logging and atomic batteries. Also known as radium F.

Polya counting formula [MATH] A formula which counts the number of functions from a finite set D to another finite set, with two functions f and g assumed to be the same if some element of a fixed group of complete permutations of D takes f into g.

Polya-Eggenberger distribution [STAT] A discrete frequency distribution that was originally considered in connection with contagious distributions.

Polya process [STAT] A type of birth process in which the probability of an increase in population by 1 in a small period Δ at time t is designated by $(1 + an) / (1 + at)$, where a is a constant and n is the population total at t.

polyatomic molecule [CHEM] A chemical molecule with three or more atoms.

polychroism *See* pleochroism.

polychromatic radiation [ELECTROMAG] Electromagnetic radiation that is spread over a range of frequencies.

polydisperse system [PHYS CHEM] A colloidal system whose particles are of various sizes.

polygon [MATH] A figure in the plane given by points p_1, p_2, . . . , p_n and line segments p_1p_2, p_2p_3, . . . , $p_{n-1}p_n$, p_np_1.

polygonization [SOLID STATE] A phenomenon observed during the annealing of plastically bent crystals in which the edge dislocations created by cold working organize themselves vertically above each other so that polygonal domains are formed.

polygon of vectors [MATH] A polygon all but one of whose sides represent vectors to be added, directed in the same sense along the perimeter, and whose remaining side represents the sum of these vectors, directed in the opposite sense.

polygon wall *See* tilt boundary.

polyhedral angle [MATH] The shape formed by the lateral faces of a polyhedron which have a common vertex.

polyhedron [MATH] A solid bounded by planar polygons.

polymorph [CRYSTAL] A crystal form of a polymorphic material. Also known as polymorphic modification.

polymorphic modification *See* polymorph.

polymorphism [CRYSTAL] The property of a chemical substance crystallizing into two or more forms having different structures, such as diamond and graphite. Also known as pleomorphism.

polynomial [MATH] A polynomial in the quantities $x_1, x_2, \ldots, x_n$ is an expression involving a finite sum of terms of the form $bx_1^{p_1}x_2^{p_2} \cdots x_n^{p_n}$, where b is some number, and $p_1, \ldots, p_n$ are integers.

polynomial trend [STAT] A trend line which is best approximated by a polynomial function; used in time series analysis.

polystyrene capacitor [ELEC] A capacitor that uses film polystyrene as a dielectric between rolled strips of metal foil.

polytropic compression curve [PHYS] Graphical relationship between pressure p and volume V for various values of

POLONIUM

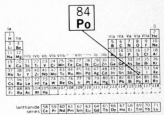

Periodic table of the chemical elements showing the position of polonium.

specific-heat ratios n in the compression formula $pV^n = K$.

polytropic process [THERMO] An expansion or compression of a gas in which the quantity pV^n is held constant, where p and V are the pressure and volume of the gas, and n is some constant.

Pomeranchuk pole *See* Pomeron.

Pomeranchuk theorem [PARTIC PHYS] The theorem that if the total cross section both for scattering of a particle by a given target particle and for scattering of its antiparticle by the same target particle approach a limit at high energies, and do so sufficiently rapidly, then these limits must be the same.

Pomeron [PARTIC PHYS] A Regge pole which is located at $+1$ in the angular momentum plane when the momentum transfer in the crossed channel equals zero, corresponding to the fact that total cross sections of reactions are observed to approach constants at high energies. Also known as Pomeranchuk pole.

poncelet [PHYS] A unit of power equal to the power delivered by a force of 100 kilograms-force when the point at which the force is applied is moved at a rate of 1 meter per second in the direction of the force; equal to 980.665 watts.

pond *See* gram-force.

Pontryagin duality theorem [MATH] Every locally compact Abelian group is the dual group (second definition) of its dual group.

Pontryagin index [MATH] A topological invariant of a geometry that can be represented in terms of an integral of a local curvature invariant.

Pontryagin's maximum principle [MATH] A theorem giving a necessary condition for the solution of optimal control problems: let $\theta(\tau)$, $\tau_0 \leq \tau \leq T$, be a piecewise continuous vector function satisfying certain constraints; in order that the scalar function $S = \Sigma c_i x_i(T)$ be minimum for a process described by the equation $\partial x_i / \partial \tau = (\partial H / \partial z_i)[z(\tau), x(\tau), \theta(\tau)]$ with given initial conditions $x(\tau_0) = x^0$ it is necessary that there exist a nonzero continuous vector function $z(\tau)$ satisfying $dz_i / d\tau = - (\partial H / \partial x_i) [z(\tau), x(\tau), \theta(\tau)]$, $z_i(T) = - c_i$, and that the vector $\theta(\tau)$ be so chosen that $H[z(\tau), x(\tau), \theta(\tau)]$ is maximum for all τ, $\tau_0 \leq \tau \leq T$.

pooling of error [STAT] A method used in the analysis of variance to secure more degrees of freedom for estimating the error variance; the sums of squares of several sets of data considered to be generated under the same model are added together and divided by the sum of the degrees of freedom in the several sets of data.

pool reactor [NUCLEO] A research nuclear reactor in which the core is suspended in a large pool of water that serves as moderator, reflector, coolant, and radiation shield. Also known as swimming-pool reactor.

Popov's stability criterion [CONT SYS] A frequency domain stability test for systems consisting of a linear component described by a transfer function $G(s)$ preceded by a nonlinear component characterized by an input-output function $N(.)$, with a unity gain feedback loop surrounding the series connection.

population [STAT] Any finite or infinite collection of individuals or elements that can be specified or labeled.

population correlation coefficient [STAT] The ratio of the covariance of two random variables to their standard deviations.

population inversion [ATOM PHYS] The condition in which a higher energy state in an atomic system is more heavily populated with electrons than a lower energy state of the same system.

population mean [STAT] The average of the numbers obtained for all members in a population by measuring some quantity associated with each member.

population multiple linear regression equation [STAT] An equation relating the conditional mean of the dependent variable to each one of the independent variables under the assumption that this relationship is linear; for the multivariate, normal distribution linearity always exists.

population of levels [STAT MECH] The number of members of an ensemble which are in each of the allowed energy states of a system.

population variance [STAT] The arithmetic average of the numbers $(v_1 - \bar{v})^2, \ldots, (\bar{v}_N - \bar{v})^2$, where v_i are numbers obtained from a population with N members, one for each member, and $\bar{v}$ is the population mean.

p orbital [ATOM PHYS] The orbital of an atomic electron with an orbital angular momentum quantum number of unity.

porcelain capacitor [ELEC] A fixed capacitor in which the dielectric is a high grade of porcelain, molecularly fused to alternate layers of fine silver electrodes to form a monolithic unit that requires no case or hermetic seal.

pore diffusion [FL MECH] The movement of fluids (gas or liquid) into the interstices of porous solids or membranes; occurs in membrane separation, zeolite adsorption, dialysis, and reverse osmosis.

porosity [PHYS] **1.** Property of a solid which contains many minute channels or open spaces. **2.** The fraction as a percent of the total volume occupied by these channels or spaces; for example, in petroleum engineering the ratio (expressed in percent) of the void space in a rock to the bulk volume of that rock.

Porro prism [OPTICS] One of two identical prisms used in the Porro prism erecting system; it is a right-angle prism with the corners rounded to minimize breakage and simplify assembly.

Porro prism erecting system [OPTICS] A compound erecting system, designed by M. Porro, in which there are four reflections to completely erect the image; two Porro prisms are employed; the line of sight is bent through 360°, is displaced, but is not deviated; used in prism binoculars and some telescope systems.

port [ELEC] An entrance or exit for a network. [ELECTROMAG] An opening in a waveguide component, through which energy may be fed or withdrawn, or measurements made. [NUCLEO] An opening in a research reactor through which objects are inserted for irradiation or from which beams of radiation emerge for experimental use.

positional-error constant [CONT SYS] For a stable unit feedback system, the limit of the transfer function as its argument approaches 0.

positional notation [MATH] Any of several numeration systems in which a number is represented by a sequence of digits in such a way that the significance of each digit depends on its position in the sequence as well as its numeric value. Also known as notation.

positional servomechanism [CONT SYS] A feedback control system in which the mechanical position (as opposed to velocity) of some object is automatically maintained.

positioning action [CONT SYS] Automatic control action in which there is a predetermined relation between the value of a controlled variable and the position of a final control element.

position operator [QUANT MECH] The quantum-mechanical operator corresponding to the classical position variable of a particle.

position representation [QUANT MECH] A representation in which the state functions are eigenfunctions of the position operator. Also known as Schrödinger representation.

position vector [MATH] The position vector of a point in euclidean space is a vector whose length is the distance from the origin to the point and whose direction is the direction from the origin to the point. Also known as radius vector.

positive [ELEC] Having fewer electrons than normal, and hence having the ability to attract electrons. [MATH] Having value greater than zero.

positive acceleration [MECH] **1.** Accelerating force in an upward sense or direction, such as from bottom to top, or from seat to head; **2.** The acceleration in the direction that this force is applied.

positive angle [MATH] The angle swept out by a ray moving in a counterclockwise direction.

positive axis [MATH] The segment of an axis arising from a cartesian coordinate system which is realized by positive values of the coordinate variables.

positive bias [ELECTR] A bias such that the control grid of an electron tube is positive with respect to the cathode.

positive charge [ELEC] The type of charge which is possessed by protons in ordinary matter, and which may be produced in a glass object by rubbing with silk.

positive column [ELECTR] The luminous glow, often striated, that occurs between the Faraday dark space and the anode in a glow-discharge tube. Also known as positive glow.

positive crystal [OPTICS] **1.** Uniaxial anisotropic crystal having the ordinary index of refraction greater than the extraordinary index. **2.** Biaxial anisotropic crystal having the intermediate index of refraction beta closer in value to alpha, and with Z the acute bisectrix.

positive definite [MATH] **1.** A square matrix A of order n is positive definite if

$$\sum_{i,j=1}^{n} A_{ij} x_i \overline{x_j} > 0$$

for every choice of complex numbers $x_1, x_2, \ldots, x_n$, not all equal to 0, where $\overline{x_j}$ is the complex conjugate of x_j. **2.** A linear operator T on an inner product space is positive definite if $<Tu,u>$ is greater than 0 for all nonzero vectors u in the space.

positive electrode *See* anode.

positive electron *See* positron.

positive feedback [CONT SYS] Feedback in which a portion of the output of a circuit or device is fed back in phase with the input so as to increase the total amplification. Also known as reaction (British usage); regeneration; regenerative feedback; retroaction (British usage).

positive glow *See* positive column.

positive-grid oscillator *See* retarding-field oscillator.

positive ion [CHEM] An atom or group of atoms which by loss of one or more electrons has acquired a positive electric charge; occurs on ionization of chemical compounds as H^+ from ionization of hydrochloric acid, HCl.

positive-ion sheath [ELECTR] Collection of positive ions on the control grid of a gas-filled triode tube.

positive linear functional [MATH] A linear functional on some vector space of real-valued functions which takes every nonnegative function into a nonnegative number.

positive logic [ELECTR] Pertaining to a logic circuit in which the more positive voltage (or current level) represents the 1 state; the less positive level represents the 0 state.

positive meniscus lens [OPTICS] A lens having one convex (bulging) and one concave (depressed) surface, with the radius of curvature of the convex surface smaller than that of the concave surface.

positive part [MATH] For a real-valued function f, this is the function, denoted f^+, for which $f^+(x) = f(x)$ if $f(x) \geq 0$ and $f^+(x) = 0$ if $f(x) < 0$.

positive ray [ELECTR] A stream of positively charged atoms or molecules, produced by a suitable combination of ionizing agents, accelerating fields, and limiting apertures.

positive real function [MATH] An analytic function whose value is real when the independent variable is real, and whose real part is positive or 0 when the real part of the independent variable is positive or 0.

positive semidefinite [MATH] Also known as nonnegative semidefinite. **1.** A square matrix A is positive semidefinite if

$$\sum_{i,j=1}^{n} A_{ij} x_i \bar{x}_j \geq 0$$

for every choice of complex numbers $x_1, x_2, \ldots, x_n$, where $\bar{x}_j$ is the complex conjugate of x_j. **2.** A linear operator T on an inner product space is positive semidefinite if $<Tu,u>$ is equal to or greater than 0 for all vectors u in the space.

positive skewness [STAT] A unimodal distribution with a longer tail in the direction of higher values of the random variable.

positive temperature coefficient [THERMO] The condition wherein the resistance, length, or some other characteristic of a substance increases when temperature increases.

positive terminal [ELEC] The terminal of a battery or other voltage source toward which electrons flow through the external circuit.

positive with respect to a signed measure [MATH] A set A is positive with respect to a signed measure m if, for every measurable set B, the intersection of A and B, $A \cap B$, is measurable and $m(A \cap B) \geq 0$.

positron [PARTIC PHYS] An elementary particle having mass equal to that of the electron, and having the same spin and statistics as the electron, but a positive charge equal in magnitude to the electron's negative charge. Also known as positive electron.

positron emission [NUC PHYS] A β-decay process in which a nucleus ejects a positron and a neutrino.

positronium [PARTIC PHYS] The bound state of an electron and a positron.

postaccelerating electrode *See* intensifier electrode.

postacceleration [ELECTR] Acceleration of beam electrons after deflection in an electron-beam tube. Also known as postdeflection acceleration (PDA).

postdeflection accelerating electrode *See* intensifier electrode.

postdeflection acceleration *See* postacceleration.

posterior probabilities [STAT] Probabilities of the outcomes of an experiment after it has been performed and a certain event has occurred.

postmeridian [SCI TECH] After noon, or the period of time between noon (1200) and midnight (2400).

postulate *See* axiom.

pot *See* potentiometer.

potassium [CHEM] A chemical element, symbol K, atomic number 19, atomic weight 39.102; an alkali metal. Also known as kalium.

potassium-42 [NUC PHYS] Radioactive isotope with mass number of 42; half-life is 12.4 hours, with β- and γ-radiation; radiotoxic; used as radiotracer in medicine.

potential [ELEC] *See* electric potential. [PHYS] A function or set of functions of position in space, from whose first

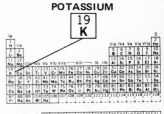

Periodic table of the chemical elements showing the position of potassium.

derivatives a vector can be formed, such as that of a static field intensity.

potential barrier [PHYS] The potential in a region in a field of force where the force exerted on a particle is such as to oppose the passage of the particle through the region. Also known as barrier; potential hill.

potential density [PHYS] The density that would be reached by a compressible fluid if it were adiabatically compressed or expanded to a standard pressure of the bar.

potential difference [ELEC] Between any two points, the work which must be done against electric forces to move a unit charge from one point to the other. Abbreviated PD.

potential divider *See* voltage divider.

potential drop [ELEC] The potential difference between two points in an electric circuit. [FL MECH] The difference in pressure head between one equipotential line and another.

potential energy [MECH] The capacity to do work that a body or system has by virtue of its position or configuration.

potential flow [FL MECH] Flow in which the velocity of flow is the gradient of a scalar function, known as the velocity potential.

potential gradient [ELEC] Difference in the values of the voltage per unit length along a conductor or through a dielectric.

potential head *See* elevation head.

potential hill *See* potential barrier.

potential refractive index *See* potential index of refraction.

potential scattering [QUANT MECH] Scattering of a particle which can be treated as the effect of a potential, representing the particle's potential energy, on the particle's Schrödinger wave function.

potential temperature [THERMO] The temperature that would be reached by a compressible fluid if it were adiabatically compressed or expanded to a standard pressure, usually 1 bar.

potential theory [MATH] The study of the functions arising from Laplace's equation, especially harmonic functions.

potential transform *See* Poisson transform.

potential transformer *See* voltage transformer.

potential transformer phase angle [ELEC] Angle between the primary voltage vector and the secondary voltage vector reversed; this angle is conveniently considered as positive when the reversed, secondary voltage vector leads the primary voltage vector.

potential vorticity [FL MECH] The product of the absolute vorticity and the static stability, conservative in adiabatic flow, given by the expression $(\eta/\theta)(\partial\theta/\partial p)$, where η is the absolute vorticity of a fluid parcel, θ the potential temperature, and p the pressure. Also known as absolute potential vorticity.

potential well [PHYS] For an object in a conservative field of force, a region in which the object has a lower potential energy than in all the surrounding regions.

potentiometer [ELEC] A resistor having a continuously adjustable sliding contact that is generally mounted on a rotating shaft; used chiefly as a voltage divider. Also known as pot (slang). [ENG] A device for the measurement of an electromotive force by comparison with a known potential difference.

potentiometric controller [CONT SYS] A controller that operates on the null balance principle, in which an error signal is produced by balancing the sensor signal against a set-point voltage in the input circuit; the error signal is amplified for use in keeping the load at a desired temperature or other parameter.

POTENTIOMETER

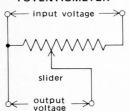

Schematic of potentiometer.

potentiometry [ELEC] Use of a potentiometer to measure electromotive forces, and the applications of such measurements.

pound [MECH] **1.** A unit of mass in the English absolute system of units, equal to 0.45359237 kilogram. Abbreviated lb. Also known as avoirdupois pound; pound mass. **2.** A unit of force in the English gravitational system of units, equal to the gravitational force experienced by a pound mass when the acceleration of gravity has its standard value of 9.80665 meters per second per second (approximately 32.1740 ft/sec²) equal to 4.4482216152605 newtons. Abbreviated lb. Also spelled Pound (Lb). Also known as pound force (lbf). **3.** A unit of mass in the troy and apothecaries' systems, equal to 12 troy or apothecaries' ounces, or 5760 grains, or 5760/7000 avoirdupois pound, or 0.3732417216 kilogram. Also known as apothecaries' pound (abbreviated lb ap in the United States or lb apoth in the United Kingdom), troy pound (abbreviated lb t in the United States or lb tr in the United Kingdom).

poundal [MECH] A unit of force in the British absolute system of units, equal to the force which will impart an acceleration of 1 ft/sec² to a pound mass, or to 0.138254954376 newton.

poundal-foot *See* foot-poundal.

pound-foot *See* foot-pound.

pound force *See* pound.

pound mass *See* pound.

pound per square foot [MECH] A unit of pressure equal to the pressure resulting from a force of 1 pound applied uniformly over an area of 1 square foot. Abbreviated psf.

pound per square inch [MECH] A unit of pressure equal to the pressure resulting from a force of 1 pound applied uniformly over an area of 1 square inch. Abbreviated psi.

Pound-Rebka experiment [RELAT] A terrestrial experiment demonstrating the gravitational red shift of light.

pounds per square inch absolute [MECH] The absolute, thermodynamic pressure, measured by the number of pounds force exerted on an area of 1 square inch. Abbreviated lbf in.⁻² abs; psia.

pounds per square inch gage [MECH] The gage pressure, measured by the number of pounds force exerted on an area of 1 square inch. Abbreviated psig.

pour point [FL MECH] Lowest test temperature at which a liquid will flow.

powder diffraction camera [CRYSTAL] A metal cylinder having a window through which an x-ray beam of known wavelength is sent by an x-ray tube to strike a finely ground powder sample mounted in the center of the cylinder; crystal planes in this powder sample diffract the x-ray beam at different angles to expose a photographic film that lines the inside of the cylinder; used to study crystal structure. Also known as x-ray powder diffractometer.

powder method [SOLID STATE] A method of x-ray diffraction analysis in which a collimated, monochromatic beam of x-rays is directed at a sample consisting of an enormous number of tiny crystals having random orientation, producing a diffraction pattern that is recorded on film or with a counter tube. Also known as x-ray powder method.

powder pattern [ELECTROMAG] The pattern created by very fine powders or colloidal particles, spread over the surface of a magnetic material; reveals the magnetic domains in a single crystal of such material.

power [MATH] **1.** The value that is assigned to a mathematical expression and its exponent. **2.** The power of a set is its cardinality. [OPTICS] The reciprocal of the focal length of a lens or mirror. [PHYS] The time rate of doing work.

power amplification *See* power gain.

power attenuation *See* power loss.

power breeder [NUCLEO] A nuclear reactor designed to produce both useful power and fuel.

power component *See* active component.

power control rod [NUCLEO] A control rod that produces only a small change in reactivity of a nuclear reactor, as required for controlling power level.

power curve [STAT] The graph of the power of a test for various alternatives.

power density [ELECTROMAG] The amount of power per unit area in a radiated microwave or other electromagnetic field, usually expressed in watts per square centimeter. [NUCLEO] The power generation per unit volume of a nuclear-reactor core.

power-density spectrum *See* frequency spectrum.

power diode *See* pin-diode.

power divider [ELECTROMAG] A device used to produce a desired distribution of power at a branch point in a waveguide system.

power excursion [NUCLEO] A sudden increase in the power level of a nuclear reactor, caused by a sudden increase in reactivity.

power factor [ELEC] The ratio of the average (or active) power to the apparent power (root-mean-square voltage times rms current) of an alternating-current circuit. Abbreviated pf. Also known as phase factor.

power flow [ELECTROMAG] The rate at which energy is transported across a surface by an electromagnetic field.

power gain [ELECTR] The ratio of the power delivered by a transducer to the power absorbed by the input circuit of the transducer. Also known as power amplification. [ELECTROMAG] An antenna ratio equal to 4π (12.57) times the ratio of the radiation intensity in a given direction to the total power delivered to the antenna.

power-law fluid [FL MECH] A fluid in which the shear stress at any point is proportional to the rate of shear at that point raised to some power.

power level [ELEC] The ratio of the amount of power being transmitted past any point in an electric system to a reference power value; usually expressed in decibels. [NUCLEO] The power production of a nuclear reactor in watts.

power loss [ELECTR] The ratio of the power absorbed by the input circuit of a transducer to the power delivered to a specified load; usually expressed in decibels. Also known as power attenuation.

power of a test [STAT] One minus the probability that a given test causes the acceptance of the null hypothesis when it is false due to the validity of an alternative hypothesis; this is the same as the probability of rejecting the null hypothesis by the test when the alternative is true.

power of the continuum [MATH] The cardinality of the set of real numbers.

power pack [ELECTR] Unit for converting power from an alternating- or direct-current supply into an alternating- or direct-current power at voltages suitable for supplying an electronic device.

power ratio [ELECTROMAG] The ratio of the maximum power to the minimum power in a waveguide that is improperly terminated.

power reactor [NUCLEO] A nuclear reactor designed to provide useful power, as for submarines, aircraft, ships, vehicles, and power plants.

power series [MATH] An infinite series composed of functions having nth term of the form $a_n(x - x_0)^n$, where x_0 is some point and a_n some constant.

power set [MATH] The set consisting of all subsets of a given set.

power spectrum *See* frequency spectrum.

power station *See* generating station.

power supply [ELECTR] A source of electrical energy, such as a battery or power line, employed to furnish the tubes and semiconductor devices of an electronic circuit with the proper electric voltages and currents for their operation. Also known as electronic power supply.

power transfer theorem [ELEC] The theorem that, in an electrical network which carries direct or sinusoidal alternating current, the greatest possible power is transferred from one section to another when the impedance of the section that acts as a load is the complex conjugate of the impedance of the section that acts as a source, where both impedances are measured across the pair of terminals at which the power is transferred, with the other part of the network disconnected.

Poynting-Robertson effect [ASTRON] The gradual decrease in orbital velocity of a small particle such as a micrometeorite in orbit about the sun due to the absorption and reemission of radiant energy by the particle.

Poynting theorem [ELECTROMAG] A theorem, derived from Maxwell's equations, according to which the rate of loss of energy stored in electric and magnetic fields within a region of space is equal to the sum of the rate of dissipation of electrical energy as heat and the rate of flow of electromagnetic energy outward through the surface of the region.

pp junction [ELECTR] A region of transition between two regions having different properties in p-type semiconducting material.

Pr *See* praseodymium.

practical entropy *See* virtual entropy.

practical system *See* meter-kilogram–second–ampere system.

practical units [ELECTROMAG] The units of the meter-kilogram-second-ampere system.

Prandtl-Glauert rule [FL MECH] The pressure coefficient at any point in the subsonic flow of a fluid about a slender body is equal to the pressure coefficient at that point in the corresponding incompressible fluid flow, divided by $\sqrt{1 - M^2}$, where M is the Mach number far from the body.

Prandtl-Meyer flow [FL MECH] A two-dimensional, supersonic fluid flow in which an initially uniform flow passes a sharp, convex corner in a boundary, resulting in expansion of the fluid.

Prandtl number [FL MECH] A dimensionless number used in the study of diffusion in flowing systems, equal to the kinematic viscosity divided by the molecular diffusivity. Symbolized Pr_m. Also known as Schmidt number 1 (N_{Sc}). [THERMO] A dimensionless number used in the study of forced and free convection, equal to the dynamic viscosity times the specific heat at constant pressure divided by the thermal conductivity. Symbolized N_{Pr}.

Prandtl stress function [MECH] The function $f(x,y) = \psi(x,y) - \frac{1}{2}(x^2 + y^2)$, where $\psi(x,y)$ is the conjugate torsion function of a cylinder undergoing torsion; its value is constant at the surface of the cylinder.

praseodymium [CHEM] A chemical element, symbol Pr, atomic number 59, atomic weight 140.91; a metallic element of the rare-earth group.

precession [MECH] The angular velocity of the axis of spin of a spinning rigid body, which arises as a result of external torques acting on the body.

POWER SUPPLY

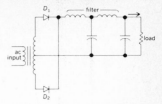

Schematic for a typical direct-current power supply. Alternating-current power line energizes primary of transformer whose secondary is connected to anodes of rectifier diodes D_1 and D_2.

PRASEODYMIUM

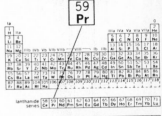

Periodic table of the chemical elements showing the position of praseodymium.

PRECESSION

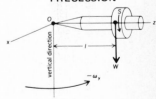

A fast-spinning top supported at point O and released in a horizontal plane precesses about the vertical y-axis. Here $z =$ top's axis of symmetry; $x =$ axis perpendicular to y and z; $S =$ spin velocity of top about z-axis; $W =$ weight of top; $l =$ distance from O to top's center of mass; $\omega_y =$ angular velocity of precession of top about y-axis, given by $\omega_y = -Wl/I_zS$, where $I_z =$ top's moment of inertia about z-axis.

precessional torque [MECH] A torque which causes a rotating body to precess.

precession of the equinoxes [ASTRON] A slow conical motion of the earth's axis about the vertical to the plane of the ecliptic, having a period of 26,000 years, caused by the attractive force of the sun, moon, and other planets on the equatorial protuberance of the earth; it results in a gradual westward motion of the equinoxes.

precipitation electricity [GEOPHYS] **1.** That branch of the study of atmospheric electricity concerned with the electric charges carried by precipitation particles and with the manner in which these charges are acquired. **2.** The electric charge borne by precipitation particles.

precompact set [MATH] A set in a metric space which can always be covered by open balls of any diameter about some finite number of its points. Also known as totally bounded set.

predator-prey equations [MATH] The system of differential equations

$$dx/dt = (A - By)x - ef_1(x,y)$$

$$dy/dt = (Cx - D)y + ef_2(x,y)$$

where A, B, C, D, and e are constants; f_1 and f_2 are specified functions; x and y may be thought of as representing the populations of prey and predator species; and t is time. **2.** In particular, equations with $f_1(x,y) = f_2(x,y) = 0$. Also known as Volterra-Lotka equations.

predicate [MATH] **1.** To affirm or deny, in mathematical logic, one or more subjects. **2.** A function of one or more variables which takes the values "true" or "false." Also known as logical function; propositional function.

predicate calculus [MATH] The mathematical study of logical statements relating to arbitrary sets of objects and involving predicates and quantifiers as well as propositional connectives.

predict *See* forecast.

predictor-corrector methods [MATH] Methods of calculating numerical solutions of differential equations that employ two formulas, the first of which predicts the value of the solution function at a point x in terms of the values and derivatives of the function at previous points where these have already been calculated, enabling approximations to the derivatives at x to be obtained, while the second corrects the value of the function at x by using the newly calculated values.

predissociation [PHYS CHEM] The dissociation of a molecule that has absorbed energy before it can lose energy by radiation.

preece [ELEC] A unit of electrical resistivity equal to 10^{13} times the product of 1 ohm and 1 meter.

pre-Hilbert space [MATH] A linear space which has an inner product defined on it.

prescaler [ELECTR] A scaler that extends the upper frequency limit of a counter by dividing the input frequency by a precise amount, generally 10 or 100.

pressure [MECH] A type of stress which is exerted uniformly in all directions; its measure is the force exerted per unit area.

pressure broadening [SPECT] A spreading of spectral lines when pressure is increased, due to an increase in collision broadening.

pressure coefficient [FL MECH] The quantity $(p - p')/(\frac{1}{2} \rho'v'^2)$, where p is the pressure at a specified point in the flow of fluid around a body, and p', ρ', and v' are the pressure, density, and fluid velocity far from the body. [THERMO] The ratio of the fractional change in pressure to the change in temperature under specified conditions, usually constant volume.

pressure deflection [ENG] In a Bourdon or bellows-type pressure gage, the deflection or movement of the primary sensing element when pressured by the fluid being measured.

pressure drop [FL MECH] The difference in pressure between two points in a flow system, usually caused by frictional resistance to a fluid flowing through a conduit, filter media, or other flow-conducting system.

pressure-drop manometer [ENG] Manometer device (liquid-filled U tube) open at both ends, each end connected by tubing to a different location in a flow system (such as fluid- or gas-carrying pipe) to measure the drop in system pressure between the two points.

pressure effect [SPECT] The effect of changes in pressure on spectral lines in the radiation emitted or absorbed by a substance; namely, pressure broadening and pressure shift.

pressure-enthalpy chart [PHYS] A graph of the pressure versus the enthalpy of a substance at various values of temperature, specific volume, and entropy; especially useful in refrigeration calculations. Also known as enthalpy-pressure chart.

pressure force [FL MECH] The force due to differences of pressure within a fluid mass; the (vector) force per unit volume is equal to the pressure gradient $-\nabla p$, and the force per unit mass (specific force) is equal to the product of the volume force and the specific volume $-\alpha\nabla p$.

pressure gage [ENG] An instrument having metallic sensing element (as in a Bourdon pressure gage or aneroid barometer) or a piezoelectric crystal (as in a quartz pressure gage) to measure pressure.

pressure gradient [FL MECH] The rate of decrease (that is, the gradient) of pressure in space at a fixed time; sometimes loosely used to denote simply the magnitude of the gradient of the pressure field. Also known as barometric gradient.

pressure head [FL MECH] Also known as head. **1.** The height of a column of fluid necessary to develop a specific pressure. **2.** The pressure of water at a given point in a pipe arising from the pressure in it.

pressure loss coefficient [FL MECH] For fluid flowing in a pipe, the dimensionless quantity $d\,\Delta p/(\frac{1}{2}\rho\bar{u}^2)l$, where Δp is the pressure loss over the length l of the pipe, d is the pipe diameter, ρ is the fluid density, and $\bar{u}$ is the mean velocity.

pressure melting [PHYS] The melting of ice due to applied pressure.

pressure pickup [ELECTR] A device that converts changes in the pressure of a gas or liquid into corresponding changes in some more readily measurable quantity such as inductance or resistance.

pressure shift [SPECT] An increase in the wavelength at which a spectral line has maximum intensity, which takes place when pressure is increased.

pressure tensor [PL PHYS] A tensor which plays a role in magnetohydrodynamics analogous to that of the pressure in ordinary fluid mechanics.

pressure transducer [ENG] An instrument component that detects a fluid pressure and produces an electrical signal related to the pressure. Also known as electrical pressure transducer.

pressure-tube reactor [NUCLEO] A nuclear reactor in which the fuel elements are located inside numerous tubes containing coolant circulating at high pressure; the tube assembly is surrounded by a tank containing the moderator at low pressure.

pressure viscosity [FL MECH] Property of petroleum lubricating oils to increase in viscosity when subjected to pressure.

pressure wave *See* compressional wave.

pressurized water reactor [NUCLEO] A nuclear reactor in which water is circulated under enough pressure to prevent it from boiling, while serving as moderator and coolant for the uranium fuel; the heated water is then used to produce steam for a power plant. Abbreviated PWR.

pretersonics *See* acoustoelectronics.

pretravel [CONT SYST] The distance or angle through which the actuator of a switch moves from the free position to the operating position.

Prevost's theory [THERMO] A theory according to which a body is constantly exchanging heat with its surroundings, radiating an amount of energy which is independent of its surroundings, and increasing or decreasing its temperature depending on whether it absorbs more radiation than it emits, or vice versa.

pri *See* primary winding.

price index [STAT] A statistic used primarily in economics to indicate an average level of prices in a time series; combines several series of price data into one index.

price relative [STAT] The ratio of the price of certain goods in a specified period to the price of the same goods in the base period.

primary *See* primary winding.

primary battery [ELEC] A battery consisting of one or more primary cells.

primary cell [ELEC] A cell that delivers electric current as a result of an electrochemical reaction that is not efficiently reversible, so that the cell cannot be recharged efficiently.

primary coil [ELEC] The input coil in an induction coil or transformer.

primary colors [OPTICS] Three colors, red, yellow, and blue, which can be combined in various proportions to produce any other color.

primary cosmic rays *See* cosmic rays.

primary creep [MECH] The initial high strain-rate region in a material subjected to sustained stress.

primary decomposition of a submodule [MATH] A primary decomposition of a submodule N of a module M is an expression of N as a finite intersection of primary submodules of M.

primary detector *See* sensor.

primary electron [ELECTR] An electron which bombards a solid surface, causing secondary emission.

primary emission [ELECTR] Emission of electrons due to primary causes, such as heating of a cathode, and not to secondary effects, such as electron bombardment.

primary extinction [SOLID STATE] A weakening of the stronger beams produced in x-ray diffraction by a very perfect crystal, as compared with the weaker.

primary fission products *See* fission fragments.

primary flow [ELECTR] The current flow that is responsible for the major properties of a semiconductor device.

primary focus [OPTICS] In an astigmatic system, a line at which some of the bundle of rays from an off-axis point meet; this line is perpendicular to a plane which contains the point and the optical axis, and has a smaller image distance than the secondary focus. Also known as meridional focus; tangential focus.

primary fuel cell [ELEC] A fuel cell in which the fuel and oxidant are continuously consumed.

primary knocked-on atom [PHYS] An atom in a solid that recoils from a collision with an energetic particle coming from outside the solid, rather than with another knocked-on atom. Abbreviated PKA.

primary lights [OPTICS] Any three lights used in a system of tristimulus colorimetric analysis of solutions.

primary radiation [PHYS] Radiation arriving directly from its source without interaction with matter.

primary rainbow [OPTICS] The most common of the principal rainbow phenomena, which appears as an arc of angular radius of about 42° about the observer's antisolar point; it is the inner of two rainbows, whose light undergoes only one internal reflection, and which is narrower and brighter than the outer, or secondary, rainbow.

primary scattering [PHYS] Any scattering process in which radiation is received at a detector, such as the eye, after having been scattered just once; distinguished from multiple scattering.

primary standard [SCI TECH] A unit directly defined and established by some authority, against which all secondary standards are calibrated; for example, in analytical chemistry, reference substances or solutions of known chemical purity and concentration are used to standardize laboratory solutions prior to volumetric analysis or titration.

primary submodule [MATH] A submodule N of a module M over a commutative ring R such that $M \neq N$ and, for any a in R, the principal homomorphism of the factor module M/N associated with a, $a_{M/N}$, is either injective or nilpotent.

primary voltage [ELEC] The voltage applied to the terminals of the primary winding of a transformer.

primary winding [ELEC] The transformer winding that receives signal energy or alternating-current power from a source. Also known as primary. Abbreviated pri. Symbolized P.

prime *See* prime element.

prime element [MATH] An irreducible element of a unique factorization domain. Also known as prime.

prime field [MATH] For a field K with multiplicative unit element e, the field consisting of elements of the form $(ne)(me)^{-1}$, where $m \neq 0$ and n are integers.

prime focus [OPTICS] The position in a reflecting telescope at which light from celestial objects is focused by the main mirror, located on the axis of the mirror near the open end of the tube.

prime ideal [MATH] A principal ideal of a ring given by a single element that has properties analogous to those of the prime numbers.

prime number [MATH] A positive integer having no divisors except itself and the integer 1.

prime polynomial [MATH] A polynomial whose only factors are itself and constants.

prime ring [MATH] For a field K with multiplicative unit element e, the ring consisting of elements of the form ne, where n is an integer.

primitive cell [CRYSTAL] A parallelepiped whose edges are defined by the primitive translations of a crystal lattice; it is a unit cell of minimum volume.

primitive element [MATH] A primitive element of an extension E of a field F is an element a in E such that E is generated by a and F.

primitive equations [FL MECH] The Eulerian equations of motion of a fluid in which the primary dependent variables are the fluid's velocity components; these equations govern a wide variety of fluid motions and form the basis of most hydrodynamical analysis; in meteorology, these equations are frequently specialized to apply directly to the cyclonic-scale motions by the introduction of filtering approximations.

primitive period [MATH] **1.** A period a of a simply periodic function $f(x)$ such that any period of $f(x)$ is an integral multiple of a. **2.** Either of two periods a and b of a doubly periodic function $f(x)$ such that any period of $f(x)$ is of the form $ma + nb$, where m and n are integers.

primitive period parallelogram [MATH] For a doubly periodic function $f(z)$ of a complex variable, a parallelogram with vertices at $z_0, z_0 + a, z_0 + a + b$, and $z_0 + b$, where z_0 is any complex number and a and b are primitive periods of $f(z)$.

primitive plane [MATH] A partial plane in which every line passes through at least two points.

primitive polynomial [MATH] A polynomial with integer coefficients which have 1 as their greatest common divisor.

primitive root [MATH] An nth root of unity that is not an mth root of unity for any m less than n.

primitive root of unity [MATH] A primitive root of unity a in a field F is a root of unity in F such that for any other root of unity b in F there exists a positive integer m such that $b = a^m$.

primitive translation [CRYSTAL] For a space lattice, one of three translations which can be repeatedly applied to generate any translation which leaves the lattice unchanged.

principal axis [CRYSTAL] The longest axis in a crystal. [MATH] **1.** One of a set of perpendicular axes such that a quadratic function can be written as a sum of squares of coordinates referred to these axes. **2.** For a conic, a straight line that passes through the midpoints of all the chords perpendicular to it. **3.** For a quadric surface, the intersection of two principal planes. [MECH] One of three perpendicular axes in a rigid body such that the products of inertia about any two of them vanish. [OPTICS] *See* optic axis.

principal axis of strain [MECH] One of the three axes of a body that were mutually perpendicular before deformation. Also known as strain axis.

principal axis of stress [MECH] One of the three mutually perpendicular axes of a body that are perpendicular to the principal planes of stress. Also known as stress axis.

principal branch [MATH] For complex valued functions such as the logarithm which are multiple-valued, a selection of values so as to obtain a genuine single-valued function.

principal curvatures [MATH] For a point on a surface, the absolute maximum and absolute minimum values attained by the normal curvature.

principal diagonal [MATH] For a square matrix, the diagonal extending from the upper left-hand corner to the lower right-hand corner of the matrix; that is, the diagonal containing the elements a_{ij} for which $i = j$. Also known as main diagonal.

principal directions [MATH] For a point on a surface, the directions in which the normal curvature attains its absolute maximum and absolute minimum values.

principal E plane [ELECTROMAG] Plane containing the direction of radiation of electromagnetic waves and arranged so that the electric vector everywhere lies in the plane.

principal focus *See* focal point.

principal function [MECH] The integral of the Lagrangian of a system over time; it is involved in the statement of Hamilton's principle.

principal homomorphism [MATH] Let a be an element of a ring R and M be a module over R; the principal homomorphism of M associated with a, denoted a_M, is the mapping which takes each element x in M into ax.

principal H plane [ELECTROMAG] The plane that contains the direction of radiation and the magnetic vector, and is everywhere perpendicular to the E plane.

principal ideal [MATH] The smallest ideal of a ring which contains a given element of the ring.

principal ideal ring [MATH] A commutative ring with a unit element in which every ideal is a principal ideal.

principal mode *See* fundamental mode.

principal normal [MATH] The line perpendicular to a space curve at some point which also lies in the osculating plane at that point.

principal normal section [MATH] For a point on a surface, a normal section in a direction in which the curvature of this section has a maximum or minimum value.

principal part [MATH] The principal part of an analytic function $f(z)$ defined in an annulus about a point z_0 is the sum of terms in its Laurent expansion about z_0 with negative powers of $(z - z_0)$.

principal plane [MATH] For a quadric surface, a plane that passes through the midpoints of all the chords perpendicular to it. [OPTICS] 1. Two planes perpendicular to the optical axis such that objects in one plane form images in the other with a lateral magnification of unity. 2. *See* principal section.

principal plane of stress [MECH] For a point in an elastic body, a plane at that point across which the shearing stress vanishes.

principal point [OPTICS] The intersection of a principal plane with the optical axis.

principal quantum number [ATOM PHYS] A quantum number for orbital electrons, which, together with the orbital angular momentum and spin quantum numbers, labels the electron wave function; the energy level and the average distance of an electron from the nucleus depend mainly upon this quantum number.

principal radii [MATH] The radii of curvature of the normal sections with maximum and minimum curvature at a given point on a surface; the reciprocals of the principal curvatures.

principal root [MATH] The positive real root of a positive number, or the negative real root in the case of odd roots of negative numbers.

principal section [MATH] A normal section at a given point on a surface whose curvature has a maximum or minimum value. [OPTICS] A plane in a crystal that contains the crystal's optic axis and the ray of light under consideration. Also known as principal plane.

principal series [SPECT] A series occurring in the line spectra of many atoms and ions with one, two, or three electrons in the outer shell, in which the total orbital angular momentum quantum number changes from 1 to 0.

principal strain [MECH] The elongation or compression of one of the principal axes of strain relative to its original length.

principal stress [MECH] A stress occurring at right angles to a principal plane of stress.

principal value *See* Cauchy principal value.

principle of coincidence [ENG] The principle of operation of a vernier, according to which the fraction of the smallest division of the main scale is determined by the division of the vernier which is exactly in line with a division of the main scale.

principle of covariance [RELAT] 1. In classical physics and in special relativity, the principle that the laws of physics take the same mathematical form in all inertial reference frames. 2. In general relativity, the principle that the laws of physics take the same mathematical form in all conceivable curvilinear coordinate systems.

principle of duality *See* duality principle.

principle of dynamical similarity [MECH] The principle that two physical systems which are geometrically and kinematically similar at a given instant, and physically similar in constitution, will retain this similarity at later corresponding instants if and only if the Froude number 1 for each independent type of force has identical values in the two systems. Also known as similarity principle.

principle of equivalence *See* equivalence principle.

principle of inaccessibility *See* Carathéodory's principle.

principle of insufficient reason [STAT] The principle that cases are equally likely to occur unless reasons to the contrary are known.

principle of least action [MECH] The principle that, for a system whose total mechanical energy is conserved, the trajectory of the system in configuration space is that path which makes the value of the action stationary relative to nearby paths between the same configurations and for which the energy has the same constant value. Also known as least-action principle.

principle of least constraint [MECH] If a system of bodies with masses m_i are subjected to external forces $\mathbf{F}_i$, where $i = 1, 2, \ldots, n$, and in addition exert equal and opposite forces on each other, then for any actual motion of the bodies the sum over the bodies of the quantity $(\mathbf{F}_i - m_i\mathbf{a}_i)^2/2m_i$, where $\mathbf{a}_i$ is the acceleration, is a minimum with respect to variations of the acceleration.

principle of optimality [CONT SYS] A principle which states that for optimal systems any portion of the optimal state trajectory is optimal between the states it joins.

principle of reciprocity *See* reciprocity theorem.

principle of superposition [ELEC] **1.** The principle that the total electric field at a point due to the combined influence of a distribution of point charges is the vector sum of the electric field intensities which the individual point charges would produce at that point if each acted alone. **2.** The principle that, in a linear electrical network, the voltage or current in any element resulting from several sources acting together is the sum of the voltages or currents resulting from each source acting alone. Also known as superposition theorem. [MECH] The principle that when two or more forces act on a particle at the same time, the resultant force is the vector sum of the two. [PHYS] Also known as superposition principle. **1.** A general principle applying to many physical systems which states that if a number of independent influences act on the system, the resultant influence is the sum of the individual influences acting separately. **2.** In all theories characterized by linear homogeneous differential equations, such as optics, acoustics, and quantum theory, the principle that the sum of any number of solutions to the equations is another solution.

principle of the maximum [MATH] The principle that for a nonconstant complex analytic function defined in a domain, the absolute value of the function cannot attain its maximum at any interior point of the domain.

principle of the minimum [MATH] The principle that for a nonvanishing nonconstant complex analytic function defined in a domain, the absolute value of the function cannot attain its minimum at any interior point of the domain.

principle of virtual work [MECH] The principle that the total work done by all forces acting on a system in static equilibrium is zero for any infinitesimal displacement from equilibrium which is consistent with the constraints of the system. Also known as virtual work principle.

printed circuit [ELECTR] A conductive pattern that may or may not include printed components, formed in a predetermined design on the surface of an insulating base in an accurately repeatable manner.

prior probabilities [STAT] Probabilities of the outcomes of an experiment before the experiment has been performed.

prism [CRYSTAL] A crystal which has three, four, six, eight, or twelve faces, with the face intersection edges parallel, and which is open only at the two ends of the axis parallel to the intersection edges. [MATH] A polyhedron with two parallel, congruent faces and all other faces parallelograms. [OPTICS] An optical system consisting of two or more usually plane surfaces of a transparent solid or embedded liquid at an angle with each other. Also known as optical prism.

prismatic binoculars *See* prism binoculars.

prismatic error [OPTICS] That error due to lack of parallelism of the two faces of an optical element, such as a mirror or a shade glass.

prismatic surface [MATH] A surface generated by moving a straight line which always meets a broken line lying in a given plane and which is always parallel to some given line not in that plane.

prismatoid [MATH] A polyhedron whose vertices all are in one or the other of two parallel planes.

prism binoculars [OPTICS] A type of binoculars, each half of which is a Kepler telescope that employs a Porro prism erecting system both to erect the image and to reduce the length of the instrument. Also known as prismatic binoculars.

prism diopter [OPTICS] A unit used in measuring the deviating power of a prism; this power in prism diopters is 100 times the tangent of the angle of deviation of a ray of light.

prismoid [MATH] A prismatoid whose two parallel faces are polygons having the same number of sides while the other faces are trapezoids or parallelograms.

prism spectrograph [SPECT] Analysis device in which a prism is used to give two different but simultaneous light wavelengths derived from a common light source; used for the analysis of materials by flame photometry.

privileged direction [OPTICS] One of two mutually perpendicular directions for the plane of polarization of a beam of plane-polarized light falling on a plate of anisotropic material such that the light which emerges from the plate is also plane-polarized.

probabilistic sampling [STAT] A process in which the laws of probability determine which elements are to be included in a sample.

probability [STAT] The probability of an event is the ratio of the number of times it occurs to the large number of trials that take place; the mathematical model of probability is a positive measure which gives the measure of the space the value 1.

probability amplitude *See* Schrödinger wave function.

probability current density [QUANT MECH] A vector whose component normal to a surface gives the probability that a particle will cross a unit area of the surface during a unit time.

probability density [QUANT MECH] The square of the absolute value of the Schrödinger wave function for a particle at a given point; gives the probability per unit volume of finding the particle at that point.

probability density function [STAT] A real-valued function whose integral over any set gives the probability that a random variable has values in this set. Also known as density function; frequency function.

probability deviation *See* probable error.

probability distribution *See* distribution of a random variable.

probability mass function [STAT] A function which gives the relative frequency of each possible value of the random variable in an experiment involving a discrete set of outcomes. Abbreviated p.m.f.

probability measure [MATH] The measure on a probability space.

probability paper [STAT] Graph paper with one axis specially ruled to transform the distribution function of a specified function to a straight line when it is plotted against the variate as the abscissa.

probability ratio test [STAT] Testing a simple hypothesis against a simple alternative by using the ratio of the probability of each simple event under the alternative to the probability of the event under the hypothesis.

probability sampling [STAT] A method of sampling from a finite population where the probability of each set of units being selected is known.

probability space [MATH] A measure space such that the measure of the entire space equals 1.

probability theory [MATH] The study of the mathematical structures and constructions used to analyze the probability of a given set of events from a family of outcomes.

probable error [STAT] The error that is exceeded by a variable with a probability of $\frac{1}{2}$. Also known as probability deviation.

probe [ELECTROMAG] A metal rod that projects into but is insulated from a waveguide or resonant cavity; used to provide coupling to an external circuit for injection or extraction of energy or to measure the standing-wave ratio. Also known as waveguide probe. [PHYS] A small device which can be brought into contact with or inserted into a system in order to make measurements on the system; ordinarily it is designed so that it does not significantly disturb the system.

probit [STAT] A procedure used in dosage-response studies to avoid obtaining negative response values to certain dosages; five is added to the values of the standardized variate which is assumed to be normal; the term is a contraction of probability unit.

Proca equations [QUANT MECH] A set of equations, analogous to Maxwell's equations, relating a four-vector potential and a second-rank tensor field describing a particle of spin 1 and nonzero mass.

process control system [CONT SYS] The automatic control of a continuous operation.

process heat reactor [NUCLEO] A nuclear reactor that produces heat for use in manufacturing processes.

process lens [OPTICS] A highly corrected, apochromatic lens used for precise color-separation work.

product [MATH] **1.** The product of two algebraic quantities is the result of their multiplication relative to an operation analogous to multiplication of real numbers. **2.** The product of a collection of sets $A_1, A_2, \ldots, A_n$ is the set of all elements of the form $(a_1, a_2, \ldots, a_n)$ where each a_i is an element of A_i for each $i = 1, 2, \ldots, n$.

product bundle [MATH] A bundle whose total space is the cartesian product of the base space B and a topological space F and whose projection map sends (b, f) to b.

production reactor [NUCLEO] A nuclear reactor designed primarily for large-scale production of transmutation products, such as plutonium.

product measure [MATH] A measure on a product of measure spaces constructed from the measures on each of the individual spaces by taking the measure of the product of a finite number of measurable sets, one from each of the measure spaces in the product, to be the product of the measures of these sets.

product model [STAT] A model for independent repetition of an experiment, or independent performance of several experiments, obtained by taking the cartesian product of the probability spaces representing the experiments.

product-moment coefficient *See* sample correlation coefficient.

product of Boolean matrices [MATH] The product AB of two Boolean matrices A and B, where the number n of columns in A equals the number of rows in B, is the matrix whose element c_{ij} in row i and column j is the union over $k = 1, 2, \ldots, n$ of the intersection of the elements a_{ik} in A and b_{kj} in B.

product of inertia [MECH] Relative to two rectangular axes, the sum of the products formed by multiplying the mass (or, sometimes, the area) of each element of a figure by the product of the coordinates corresponding to those axes.

product of matrices [MATH] The product AB of two matrices A and B, where the number n of columns in A equals the number of rows in B, is the matrix whose element c_{ij} in row i and column j is the sum over $k = 1, 2, \ldots, n$ of the product of the elements a_{ik} in A and b_{kj} in B.

product-solution method *See* separation of variables.

product topology [MATH] A topology on a product of topological spaces whose open sets are constructed from cartesian products of open sets from the individual spaces.

programmable read-only memory [ELECTR] A large-scale integrated-circuit chip for storing digital data; it can be erased with ultraviolet light and reprogrammed, or can be programmed only once either at the factory or in the field. Abbreviated PROM.

progression [MATH] A sequence or series of mathematical objects or quantities, each entry determined from its predecessors by some algorithm.

progressive wave [PHYS] A wave which transfers energy from one part of a medium to another, in contrast to a standing wave. Also known as free-traveling wave.

progressive-wave antenna *See* traveling-wave antenna.

projected-scale instrument [ENG] An indicating instrument in which a light beam projects an image of the scale on a screen.

projection [MATH] **1.** The continuous map for a fiber bundle. **2.** Geometrically, the image of a geometric object or vector superimposed on some other. **3.** A linear map P from a linear space to itself such that $P \circ P$ is equal to P.

projection chamber *See* projection spark chamber.

projection microradiography [PHYS] Microradiography in which an electron beam, focused into an extremely fine pencil, generates a point source of x-rays, and enlargement is achieved by placing the sample very near this source, and several centimeters from the recording material. Abbreviated PMR. Also known as shadow microscopy; x-ray projection microscopy.

projection microscope [PHYS] An x-ray microscope which magnifies by image projection, either in contact microradiography or in projection microradiography.

projection of a bundle [MATH] The continuous map p in the bundle (E, p, B).

projection optics *See* Schmidt system.

projection printer [OPTICS] An optical, image-enlarging device, used in enlarging photographs.

projection-slice theorem *See* central-slice theorem.

projection spark chamber [NUCLEO] A spark chamber in which the track of the particle is perpendicular, or nearly so, to the electric field, so that each electron of the track produces a streamer across the gap; the resulting curtain contains

PROJECTION MICROSCOPE

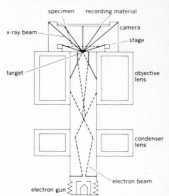

Projection microscope used in projection microradiography, showing components. *(After V. E. Cosslett, A. Engström, and H. H. Pattee, eds., X-Ray Microscopy and Microradiography, Academic Press, 1957)*

information only as to the projection of the track perpendicular to the electric field. Also known as projection chamber.

projective geometry [MATH] The study of those properties of geometric objects which are invariant under projection.

projective group [MATH] A group of transformations arising in the general theory of projective geometry.

projective line [MATH] The line obtained from the stereographic projection of the circle.

projective plane [MATH] **1.** The topological space obtained from the two-dimensional sphere by identifying antipodal points; the space of all lines through the origin in euclidean space. **2.** More generally, a plane (in the sense of projective geometry) such that (1) every two points lie on exactly one line, (2) every two lines pass through exactly one point, and (3) there exists a four-point.

projective point [MATH] The point from which a projection by rays is performed, as in stereographic projection.

projective space [MATH] **1.** The topological space obtained from the n-dimensional sphere under identification of antipodal points. **2.** An n-dimensional projective space over a field K is the set of ordered $(n + 1)$-tuples $(x_0, x_1, \ldots, x_n)$, where x_0, $x_1, \ldots, x_n$ are elements of K, the $(n + 1)$-tuple $(0, 0, \ldots, 0)$ being excluded, and two $(n+1)$-tuples represent the same element if the entries of one are proportional to the entries of the other.

projector *See* optical projection system.

prolate ellipsoid *See* prolate spheroid.

prolate spheroid [MATH] The ellipsoid or surface obtained by revolving an ellipse about one of its axes so that the equatorial circle has a diameter less than the length of the axis of revolution. Also known as prolate ellipsoid.

prolate spheroidal coordinate system [MATH] A three-dimensional coordinate system whose coordinate surfaces are the surfaces generated by rotating a plane containing a system of confocal ellipses and hyperbolas about the major axis of the ellipses, together with the planes passing through the axis of rotation.

promethium [CHEM] A chemical element, symbol Pm, atomic number 61; atomic weight of the most abundant isotope is 147; a member of the rare-earth group of metals.

promethium-147 [NUC PHYS] Artificially produced rare-earth element with atomic number 61 and mass 147; produced during fission of U²³⁵. Also known as florentium; illinium.

promethium cell [NUCLEO] A nuclear energy cell in which beta particles from promethium-147 cause a phosphor to glow; the light output is converted to electric energy by photocells.

prominence [ASTROPHYS] A volume of luminous, predominantly hydrogen gas that appears on the sun above the chromosphere; occurs only in the region of horizontal magnetic fields because these fields support the prominences against solar gravity.

prompt critical [NUCLEO] Capable of sustaining a chain reaction without the aid of delayed neutrons.

prompt neutron [NUC PHYS] A neutron released coincident with the fission process, as opposed to neutrons subsequently released.

prompt radiation [NUC PHYS] Radiation emitted within a time too short for measurement, including γ-rays, characteristic x-rays, conversion and Auger electrons, prompt neutrons, and annihilation radiation.

prong *See* pin.

proof [MATH] A deductive demonstration of a mathematical statement.

proof plane [ELEC] A small metal plate supported by an

PROMETHIUM

Periodic table of the chemical elements showing the position of promethium.

insulating handle, which is used to transfer a small fraction of the electric charge on a body to an electrometer to investigate the charge distribution on the body.

proof resilience [MECH] The tensile strength necessary to stretch an elastomer from zero elongation to the breaking point, expressed in foot-pounds per cubic inch of original dimension.

proof stress [MECH] **1.** The stress that causes a specified amount of permanent deformation in a material. **2.** A specified stress to be applied to a member or structure in order to assess its ability to support service loads.

propagation anomaly [PHYS] Change in propagation characteristics due to a resonance in the medium of propagation.

propagation constant [ELECTROMAG] A rating for a line or medium along or through which a wave of a given frequency is being transmitted; it is a complex quantity; the real part is the attenuation constant in nepers per unit length, and the imaginary part is the phase constant in radians per unit length.

propagation delay [ELECTR] The time required for a signal to pass through a given complete operating circuit; it is generally of the order of nanoseconds, and is of extreme importance in computer circuits.

propagation velocity [ELECTROMAG] Velocity of electromagnetic wave propagation in the medium under consideration.

propeller cavitation [FL MECH] Formation of vapor-filled and air-filled bubbles or cavities in water at or on the surface of a rotating propeller, occurring when the pressure falls below the vapor pressure of water.

proper fraction [MATH] A fraction a/b where the absolute value of a is less than the absolute value of b.

proper function *See* eigenfunction.

proper Lorentz transformation [RELAT] A Lorentz transformation which can be represented by a matrix whose determinant is $+1$.

proper motion [ASTRON] That component of the space motion of a celestial body perpendicular to the line of sight, resulting in the change of a star's apparent position relative to that of other stars; expressed in angular units.

proper orthogonal transformation [MATH] An orthogonal transformation such that the determinant of its matrix is $+1$.

proper rational function [MATH] The quotient of a polynomial P by a polynomial Q whose order is greater than P.

proper subset [MATH] A set X is a proper subset of a set Y if there is an element of Y which is not in X while X is a subset of Y.

proper value *See* eigenvalue.

proportion [MATH] The proportion of two quantities is their ratio.

proportional band [CONT SYS] The range of values of the controlled variable that will cause a controller to operate over its full range.

proportional control [CONT SYS] Control in which the amount of corrective action is proportional to the amount of error; used, for example, in chemical engineering to control pressure, flow rate, or temperature in a process system.

proportional controller [CONT SYS] A controller whose output is proportional to the error signal.

proportional counter [NUCLEO] A radiation counter consisting of a proportional counter tube and its associated circuits; resembles a Geiger-Müller counter, but with a different counting gas (argon methane) and a lower voltage on the tube; used to measure α-, β-, and x-rays; has low sensitivity for γ-radiation.

proportional counter tube [NUCLEO] A radiation-counter

tube operated at voltages high enough to produce ionization by collision and adjusted so the total ionization per count is proportional to the ionization produced by the initial ionizing event.

proportional elastic limit [MECH] The greatest stress intensity for which stress is still proportional to strain.

proportional ionization chamber [ELECTR] An ionization chamber in which the initial ionization current is amplified by electron multiplication in a region of high electric-field strength, as in a proportional counter; used for measuring ionization currents or charges over a period of time, rather than for counting.

proportional limit [MECH] The greatest stress a material can sustain without departure from linear proportionality of stress and strain.

proportional parts [MATH] Numbers in the same proportion as a set of given numbers; such numbers are used in an auxiliary interpolation table based on the assumption that the tabulated quantity and entering arguments differ in the same proportion.

proportional-plus-derivative control [CONT SYS] Control in which the control signal is a linear combination of the error signal and its derivative.

proportional-plus-integral control [CONT SYS] Control in which the control signal is a linear combination of the error signal and its integral.

proportional-plus-integral-plus-derivative control [CONT SYS] Control in which the control signal is a linear combination of the error signal, its integral, and its derivative.

proportional region [NUCLEO] The range of applied voltages in a radiation counter tube in which the gas amplification is greater than 1 and does not depend on the charge liberated in the initial ionizing event.

proposition [MATH] Any problem or theorem.

propositional algebra [MATH] The study of finite configurations of symbols and the interrelationships between them.

propositional calculus [MATH] The mathematical study of logical connectives between propositions and deductive inference. Also known as sentential calculus.

propositional connectives [MATH] The symbols $\sim$, $\wedge$, $\vee$, $\rightarrow$ or $\supset$, and $\longleftrightarrow$ or $\equiv$, denoting logical relations that may be expressed by the phrases "it is not the case that," "and," "or," "if . . . , then," and "if and only if." Also known as sentential connectives.

propositional function *See* predicate.

propulsion [MECH] The process of causing a body to move by exerting a force against it.

protactinium [CHEM] A chemical element, symbol Pa, atomic number 91; the third member of the actinide group of elements; all the isotopes are radioactive; the longest-lived isotope is protactinium-231.

protective colloid [PHYS CHEM] A colloidal substance that protects other colloids from the coagulative effect of electrolytes and other agents.

protium [NUC PHYS] The lightest hydrogen isotope, having a mass number of 1 and consisting of a single proton and electron. Also known as light hydrogen.

protogalaxy [ASTRON] The theoretical precursor of the Galaxy; suggested by James Jeans to be an initial structureless gas cloud, held together by its own gravitation, that broke up into a number of fragments.

proton [PHYS] An elementary particle that is the positively charged constituent of ordinary matter and, together with the neutron, is a building stone of all atomic nuclei; its mass is approximately 938 MeV and spin ½.

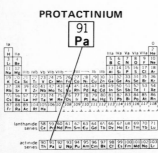

Periodic table of the chemical elements showing the position of protactinium.

proton accelerator [NUCLEO] A particle accelerator which accelerates protons to high energies, as opposed to one which accelerates heavier ions or electrons.

proton capture [NUC PHYS] A nuclear reaction in which a proton combines with a nucleus.

proton-electron-proton reaction [NUC PHYS] A nuclear reaction in which two protons and an electron react to form a deuteron and a neutrino; it is an important source of detectable neutrinos from the sun. Abbreviated PeP reaction.

protonium [ATOM PHYS] A bound state of a proton and an antiproton.

proton magnetometer [ELECTROMAG] A highly sensitive magnetometer which measures the frequency of the proton resonance in ordinary water.

proton microscope [ELECTR] A microscope that is similar to the electron microscope but uses protons instead of electrons as the charged particles.

proton moment [NUC PHYS] The magnetic dipole moment of the proton, a physical constant equal to $(1.41062 \pm 0.00001) \times 10^{-23}$ erg per gauss.

proton number *See* atomic number.

proton-proton chain [NUC PHYS] An energy-releasing nuclear reaction chain which is believed to be of major importance in energy production in hydrogen-rich stars. Also known as deuterium cycle.

proton-recoil counter [NUCLEO] A counter for measuring fast neutrons.

proton resonance [SPECT] A phenomenon in which protons absorb energy from an alternating magnetic field at certain characteristic frequencies when they are also subjected to a static magnetic field; this phenomenon is used in nuclear magnetic resonance quantitative analysis technique.

proton scattering microscope [SOLID STATE] A microscope in which protons produced in a cold-cathode discharge are accelerated and focused on a crystal in a vacuum chamber; protons reflected from the crystal strike a fluorescent screen to give a visual and photographable display that is related to the structure of the target crystal.

proton storage ring [NUCLEO] A machine consisting of magnets and vacuum chambers in which beams of high-energy protons can be stored.

proton synchrotron [NUCLEO] A device for accelerating protons in circular orbits in a time-varying magnetic field, in which the orbit radius is kept constant.

protostar [ASTRON] A flattened mass of gas in space that is hypothesized to form into a star.

protractor [ENG] An instrument used to construct and measure angles formed by lines of a plane; the midpoint of the diameter of the semicircle is marked and serves as the vertex of angles contructed or measured.

proximity effect [ELEC] Redistribution of current in a conductor brought about by the presence of another conductor.

Prüfer domain [MATH] An integral domain in which every nonzero finitely generated ideal is invertible.

ps *See* picosecond.

psec *See* picosecond.

pseudoadiabatic expansion [GEOPHYS] A saturation-adiabatic process in which the condensed water substance is removed from the system, and which therefore is best treated by the thermodynamics of open systems; meteorologically, this process corresponds to rising air from which the moisture is precipitating.

pseudoanalytic function *See* generalized analytic function.

pseudo-Goldstone bosons [PARTIC PHYS] Goldstone bosons which accompany the breakdown of approximate, accidental

PROTON MAGNETOMETER

The water container and biasing coil assembly of the proton vector magnetometer. Instrument measures frequency of voltage induced in coil by the protons in water. (*U.S. Coast and Geodetic Survey*)

PROTON RESONANCE

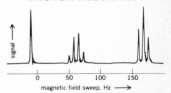

Proton resonance spectra of ethyl alcohol at 40 megahertz. The three main resonance frequencies are due to protons in the OH, CH_2, and CH_3 groups respectively. (*From J. D. Roberts, Nuclear Magnetic Resonance, McGraw-Hill, 1959*)

PROTON SYNCHROTRON

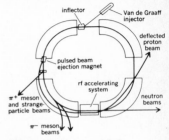

Schematic diagram of the principal components of a proton synchrotron. Note the various possibilities of external neutral beams, charged beams, and extracted primary beams.

symmetries in certain unified gage theories of weak and electromagnetic interactions.

pseudometric *See* semimetric.

pseudoplastic fluid [FL MECH] A fluid whose apparent viscosity or consistency decreases instantaneously with an increase in shear rate.

pseudorandom numbers [ADP] Numbers produced by a definite arithmetic process, but satisfying one or more of the standard tests for randomness.

pseudoscalar [PHYS] A quantity which has magnitude only, and which acts, under Lorentz transformation, like a scalar but with a sign change under space reflection or time reflection, or both.

pseudoscalar coupling [PARTIC PHYS] A type of interaction postulated between a nucleon and a pion in which the interaction energy is a product of the pion's pseudoscalar field and a bilinear pseudoscalar function of the nucleon fields.

pseudoscalar meson [PARTIC PHYS] A meson, such as the pion, which has spin 0 and negative parity, and may be described by a field which is a pseudoscalar quantity. Also known as pseudoscalar particle.

pseudoscalar particle *See* pseudoscalar meson.

pseudoscope [OPTICS] A device that produces reversed stereoscopic effects, for example, by transposing the pictures of a stereoscope.

pseudotensor [PHYS] **1.** A quantity which transforms as a tensor under space rotations, but which transforms as a tensor, together with a change in sign, under space inversion. **2.** A quantity which transforms as a tensor under Lorentz transformations, but with an additional sign change under space reflection or time reflection or both.

pseudovector [PHYS] **1.** A quantity which transforms as a vector under space rotations but which transforms as a vector, together with a change in sign, under a space inversion. Also known as axial vector. **2.** A quantity which transforms as a four-vector under Lorentz transformations, but with an additional sign change under space reflection or time reflection or both.

pseudovector coupling [PARTIC PHYS] A type of interaction postulated between a nucleon and another particle in which the expression for the interaction energy contains a bilinear pseudovector function of nucleon fields.

pseudovector meson [PARTIC PHYS] A meson which has spin quantum number 1 and positive parity and may be described by a field which is a pseudovector quantity.

psf *See* pound per square foot.

P shell [ATOM PHYS] The sixth layer of electrons about the nucleus of an atom, having electrons whose principal quantum number is 6.

psi *See* pound per square inch.

psia *See* pounds per square inch absolute.

psi function [MATH] The special function of a complex variable which is obtained from differentiating the logarithm of the gamma function. [QUANT MECH] *See* Schrödinger wave function.

psig *See* pounds per square inch gage.

psi particle *See* J particle.

psi-prime particle [PARTIC PHYS] A neutral meson which has a mass of 3684 megaelectronvolts, spin quantum number 1, and negative parity and charge parity; it has an anomalously long lifetime. Symbolized ψ'.

psophometer [ENG] An instrument for measuring noise in electric circuits; when connected across a 600-ohm resistance in the circuit under study, the instrument gives a reading that by definition is equal to half of the psophometric electromotive force actually existing in the circuit.

psophometric electromotive force [ELECTR] The true noise voltage that exists in a circuit.

psophometric voltage [ELECTR] The noise voltage as actually measured in a circuit under specified conditions.

psychromatic ratio [THERMO] Ratio of the heat-transfer coefficient to the product of the mass-transfer coefficient and humid heat for a gas-vapor system; used in calculation of humidity or saturation relationships.

psychrometer [ENG] A device comprising two thermometers, one a dry bulb, the other a wet or wick-covered bulb, used in determining the moisture content or relative humidity of air or other gases. Also known as wet and dry bulb thermometer.

psychrometric chart [THERMO] A graph each point of which represents a specific condition of a gas-vapor system (such as air and water vapor) with regard to temperature (horizontal scale) and absolute humidity (vertical scale); other characteristics of the system, such as relative humidity, wet-bulb temperature, and latent heat of vaporization, are indicated by lines on the chart.

psychrometric formula [THERMO] The semiempirical relation giving the vapor pressure in terms of the barometer and psychrometer readings.

psychrometric tables [THERMO] Tables prepared from the psychrometric formula and used to obtain vapor pressure, relative humidity, and dew point from values of wet-bulb and dry-bulb temperature.

psychrometry [ENG] The science and techniques associated with measurements of the water vapor content of the air or other gases.

pt *See* pint.

Pt *See* platinum.

p-type conductivity [ELECTR] The conductivity associated with holes in a semiconductor, which are equivalent to positive charges.

p-type crystal rectifier [ELECTR] Crystal rectifier in which forward current flows when the semiconductor is positive with respect to the metal.

p-type semiconductor [ELECTR] An extrinsic semiconductor in which the hole density exeeds the conduction electron density.

p⁺-type semiconductor [ELECTR] A p-type semiconductor in which the excess mobile hole concentration is very large.

p-type silicon [ELECTR] Silicon to which more impurity atoms of acceptor type (with valence of 3, such as boron) than of donor type (with valence of 5, such as phosphorus) have been added, with the result that the hole density exceeds the conduction electron density.

Pu *See* plutonium.

puff *See* picofarad.

Pulfrich refractometer [OPTICS] A critical angle refractometer in which the material to be tested rests on a prism of material of known higher index of refraction and angle, and the angle of refraction of light which is directed at the interface between the two materials at grazing incidence is observed.

pulling [ELECTR] An effect that forces the frequency of an oscillator to change from a desired value; causes include undesired coupling to another frequency source or the influence of changes in the oscillator load impedance.

pulsar [ASTROPHYS] A celestial radio source, emitting intense short bursts of radio emission; the periods of known pulsars range between 33 milliseconds and 3.75 seconds, and pulse durations range from 2 to about 150 milliseconds with longer-period pulsars generally having a longer pulse duration.

pulsatance [PHYS] Angular velocity in radians, equal to 2π times frequency in hertz.

pulsating current [ELEC] Periodic direct current.

pulsating electromotive force [ELEC] Sum of a direct electromotive force and an alternating electromotive force. Also known as pulsating voltage.

pulsating star [ASTRON] Variable star whose luminosity fluctuates as the star expands and contracts; the variation in brightness is thought to come from the periodic change of radiant energy to gravitational energy and back.

pulsating voltage *See* pulsating electromotive force.

pulse [PHYS] A variation in a quantity which is normally constant; has a finite duration and is usually brief compared to the time scale of interest.

pulse amplifier [ELEC] An amplifier designed specifically to amplify electric pulses without appreciably changing their waveforms.

pulse amplitude [PHYS] The peak, average, effective, instantaneous, or other magnitude of a pulse, usually with respect to the normal constant value; the exact meaning should be specified when giving a numerical value.

pulse circuit [ELECTR] An active electrical network designed to respond to discrete pulses of current or voltage.

pulse-delay network [ELECTR] A network consisting of two or more components such as resistors, coils, and capacitors, used to delay the passage of a pulse.

pulse discriminator [ELECTR] A discriminator circuit that responds only to a pulse having a particular duration or amplitude.

pulsed laser [OPTICS] A laser in which a pulse of coherent light is produced at fixed time intervals, as required for ranging and tracking applications or to permit higher output power than can be obtained with continuous operation.

pulsed light [OPTICS] A beam of light whose intensity is modulated in some prescribed manner; analogous to a radar pulse.

pulsed reactor [NUCLEO] A research nuclear reactor in which continual short, intense surges of power and radiation can be produced; the neutron flux during the surge is much higher than could be tolerated during steady-state operation.

pulsed ruby laser [OPTICS] A laser in which ruby is used as the active material; the extremely high pumping power required is obtained by discharging a bank of capacitors through a special high-intensity flash tube, giving a coherent beam that lasts for about 0.5 millisecond.

pulsed transfer function [CONT SYS] The ratio of the z-transform of the output of a system to the z-transform of the input, when both input and output are trains of pulses. Also known as discrete transfer function; z-transfer function.

pulse form [PHYS] The amplitude of a pulse plotted as a function of time.

pulse-frequency spectrum *See* pulse spectrum.

pulse generator *See* impulse generator.

pulse group *See* pulse train.

pulse height [ELECTR] The strength or amplitude of a pulse, measured in volts.

pulse-height analyzer [NUCLEO] An instrument capable of indicating the number of occurrences of pulses falling within each of one or more specified amplitude ranges; used to obtain the energy spectrum of nuclear radiations. Also known as kick-sorter (British usage); multichannel analyzer.

pulse-height discriminator [ELECTR] A circuit that produces a specified output pulse when and only when it receives an input pulse whose amplitude exceeds an assigned value. Also known as amplitude discriminator.

pulse-height selector [ELECTR] A circuit that produces a specified output pulse only when it receives an input pulse whose amplitude lies between two assigned values. Also known as amplitude selector; diffractional pulse-height discriminator.

pulse-height spectrum [PHYS] Distribution of various pulse wavelengths and strengths (heights) developed during activation analysis.

pulse interval *See* pulse spacing.

pulse ionization chamber [NUCLEO] An ionization chamber that detects individual ionizing events. Also known as counting ionization chamber.

pulse spacing [PHYS] Time between corresponding points of successive pulses. Also known as pulse interval.

pulse spectrum [PHYS] The frequency distribution of the sinusoidal components of a pulse in relative amplitude and in relative phase. Also known as pulse-frequency spectrum.

pulse train [PHYS] A series of regularly recurrent pulses having similar characteristics. Also known as pulse group.

pulse transformer [ELECTR] A transformer capable of operating over a wide range of frequencies, used to transfer nonsinusoidal pulses without materially changing their waveforms.

pumping [PHYS] **1.** The application of optical, infrared, or microwave radiation of appropriate frequency to a laser or maser medium so that absorption of the radiation increases the population of atoms or molecules in higher energy states. Also known as electronic pumping. **2.** The removal of gases and vapors from a vacuum system.

pumping frequency [ELECTR] Frequency at which pumping is provided in a maser, quadrupole amplifier, or other amplifier requiring high-frequency excitation.

pumping radiation [PHYS] Electromagnetic radiation applied to a laser or maser in the process of pumping.

pump oscillator [ELECTR] Alternating-current generator that supplies pumping energy for maser and parametric amplifiers; operates at twice or some higher multiple of the signal frequency.

punch-through [ELECTR] An emitter-to-collector breakdown which can occur in a junction transistor with very narrow base region at sufficiently high collector voltage when the space-charge layer extends completely across the base region.

punctum remotum *See* far point.

puncture [ELEC] Disruptive discharge through insulation involving a sudden and large increase in current through the insulation due to complete failure under electrostatic stress.

puncture voltage [ELEC] The voltage at which a test specimen is electrically punctured.

pure geometry [MATH] Geometry studied from the standpoint of its axioms and postulates rather than its objects.

pure imaginary number [MATH] A complex number $z = x + iy$, where $x = 0$.

purely inseparable [MATH] An element a is said to be purely inseparable over a field F with characteristic p greater than 0 if it is algebraic over F and if there exists a nonnegative integer n such that a^{p^n} lies in F.

purely inseparable extension [MATH] A purely inseparable extension E of a field F is an algebraic extension of F whose separable degree over F equals 1 or, equivalently, an algebraic extension of F in which every element is purely inseparable over F.

pure mathematics [MATH] The intrinsic study of mathematical structures, with no consideration given as to the utility of the results for practical purposes.

pure projective geometry [MATH] The axiomatic study of geometric systems which exhibit invariance relative to a notion of projection.

pure tone *See* simple tone.

purity [OPTICS] The degree to which a primary color is pure and not mixed with the other two primary colors.

purity of state [STAT MECH] Property of a system which is definitely in a certain quantum state, rather than having a certain probability of being in any of several quantum states.

purple boundary [OPTICS] A straight line connecting the ends of the spectrum locus on the chromaticity diagram.

pushing [ELEC] A change in the resonant frequency of a circuit due to changes in the applied voltages.

push-pull amplifier [ELECTR] A balanced amplifier employing two similar electron tubes or equivalent amplifying devices working in phase opposition.

push-pull currents *See* balanced currents.

push-pull oscillator [ELECTR] A balanced oscillator employing two similar electron tubes or equivalent amplifying devices in phase opposition.

push-pull voltages *See* balanced voltages.

push-push amplifier [ELECTR] An amplifier employing two similar electron tubes with grids connected in phase opposition and with anodes connected in parallel to a common load; usually used as a frequency multiplier to emphasize even-order harmonics; transistors may be used in place of tubes.

pW *See* picowatt.

P wave *See* compressional wave.

PWR *See* pressurized water reactor.

pwt *See* pennyweight.

pycnometer [ENG] A container whose volume is precisely known, used to determine the density of a liquid by filling the container with the liquid and then weighing it. Also spelled pyknometer.

pycnometry [PHYS] The determination of liquid density by weighing the liquid in a container (pycnometer) of known volume.

pyknometer *See* pycnometer.

pyramid [CRYSTAL] An open crystal having three, four, six, eight, or twelve nonparallel faces that meet at a point. [MATH] A polyhedron with one face a polygon and all other faces triangles with a common vertex.

pyramidal surface [MATH] A surface generated by a line passing through a fixed point and moving along a broken line in a plane not containing that point.

pyranometer [ENG] An instrument used to measure the combined intensity of incoming direct solar radiation and diffuse sky radiation; compares heating produced by the radiation on blackened metal strips with that produced by an electric current. Also known as solarimeter.

pyritohedron [CRYSTAL] A dodecahedral crystal with 12 irregular pentagonal faces; it is characteristic of pyrite. Also known as pentagonal dodecahedron; pyritoid; regular dodecahedron.

pyritoid *See* pyritohedron.

pyroconductivity [SOLID STATE] Electrical conductivity that develops in a material only at high temperature, chiefly at fusion, in solids that are practically nonconductive at atmospheric temperatures.

pyroelectric crystal [SOLID STATE] A crystal exhibiting pyroelectricity, such as tourmaline, lithium sulfate monohydrate, cane sugar, and ferroelectric barium titanate.

pyroelectricity [SOLID STATE] The property of certain crystals to produce a state of electrical polarity by a change of temperature.

pyrometer [ENG] Any of a broad class of temperature-measuring devices; they were originally designed to measure high temperatures, but some are now used in any temperature range; includes radiation pyrometers, thermocouples, resistance pyrometers, and thermistors.

pyrometry [THERMO] The science and technology of measuring high temperatures.

pyron [PHYS] A unit of area-density of power, equal to the area-density of power resulting from a power of one international table calorie per minute acting uniformly over an area of 1 square centimeter; equal to 697.8 watts per square meter.

Pythagorean numbers [MATH] Positive integers x, y, and z which satisfy the equation $x^2 + y^2 = z^2$.

Pythagorean scale [ACOUS] A musical scale such that the frequency intervals are represented by the ratios of integral powers of the numbers 2 and 3.

Pythagorean theorem [MATH] In a right triangle the square of the length of the hypotenuse equals the sum of the squares of the lengths of the other two sides.

pz *See* pièze.

PZT *See* lead zirconate titanate.

Q [NUC PHYS] *See* disintegration energy. [PHYS] A measure of the ability of a system with periodic behavior to store energy equal to 2π times the average energy stored in the system divided by the energy dissipated per cycle. Also known as Q factor; quality factor; storage factor. [THERMO] A unit of heat energy, equal to 10^{18} British thermal units, or approximately 1.055×10^{21} joules.

Q factor *See* Q.

Q machine [PL PHYS] A device in which a highly ionized, magnetically confined plasma is created by contact ionization of atoms and thermionic emission of electrons.

Q-machine plasma [PLASMA PHYS] A plasma column in a magnetic field created by surface ionization of a cesium beam on a hot tungsten plate.

Q meter [ENG] A direct-reading instrument which measures the Q of an electric circuit at radio frequencies by determining the ratio of inductance to resistance, and which has also been developed to measure many other quantities. Also known as quality-factor meter.

Q multiplier [ELECTR] A filter that gives a sharp response peak or a deep rejection notch at a particular frequency, equivalent to boosting the Q of a tuned circuit at that frequency.

qr *See* quarter.

qr tr *See* quarter.

QSO *See* quasar.

Q-switched laser [OPTICS] A laser whose Q factor is kept at a low value while an ion population inversion is built up, and then is suddenly switched to a high value just before instability occurs, resulting in a very high rate of stimulated emission. Also known as giant pulse laser.

qt *See* quart.

quad [THERMO] A unit of heat energy, equal to 10^{15} British thermal units, or approximately 1.055×10^{18} joules.

quadrangle [MATH] A geometric figure bounded by four straight-line segments called sides, each of which intersects each of two adjacent sides in points called vertices, but fails to intersect the opposite sides. Also known as quadrilateral.

quadrant [ELECTROMAG] *See* international henry. [MATH] **1.** A quarter of a circle; either an arc of 90° or the area bounded by such an arc and the two radii. **2.** Any of the four regions into which the plane is divided by a pair of coordinate axes. [OPTICS] A double-reflecting instrument for measuring angles, used primarily for measuring altitudes of celestial bodies; the instrument was replaced by the sextant.

quadrant angle of fall [MECH] The vertical acute angle at the level point, between the horizontal and the line of fall of a projectile.

quadrant electrometer [ENG] An instrument for measuring electric charge by the movement of a vane suspended on a

Q METER

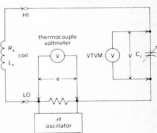

Simplified measurement circuit of a Q meter. Coil being measured, with inductance L_X and resistance R_X, is connected into the circuit by external terminals HI and LO. Calibrated capacitor C_c is tuned to bring the coil into resonance. An input voltage e is supplied by a radio-frequency oscillator and measured by a thermocouple voltmeter. A vacuum-tube voltmeter (VTVM) measures voltage V across calibrated capacitor. Q of coil is determined from the equation $V/e = (1 + Q^2)^{1/2}$.

wire between metal quadrants; the charge is introduced on the vane and quadrants in such a way that there is a proportional twist to the wire.

quadratic [MATH] Any second-degree expression.

quadratic equation [MATH] Any second-degree polynomial equation.

quadratic form [MATH] Any second-degree, homogeneous polynomial.

quadratic formula [MATH] A formula giving the roots of a quadratic equation in terms of the coefficients; for the equation $ax^2 + bx + c = 0$, the roots are $x = (-b \pm \sqrt{b^2 - 4ac})/2a$.

quadratic performance index [CONT SYS] A measure of system performance which is, in general, the sum of a quadratic function of the system state at fixed times, and the integral of a quadratic function of the system state and control inputs.

quadratic polynomial [MATH] A polynomial where the highest degree of any of its terms is 2.

quadratic programming [MATH] A body of techniques developed to find extremals for systems of quadratic inequalities.

quadratic Stark effect [ATOM PHYS] A splitting of spectral lines of atoms in an electric field in which the energy levels shift by an amount proportional to the square of the electric field, and all levels shift to lower energies; observed in lines resulting from the lower energy states of many-electron atoms.

quadratic Zeeman effect [ATOM PHYS] A splitting of spectral lines of atoms in a magnetic field in which the energy levels shift by an amount proportional to the square of the magnetic field.

quadrature [ASTRON] The right-angle physical alignment of the sun, moon, and earth. [MATH] **1.** The construction of a square whose area is equal to that of a given surface. **2.** The process of calculating a definite integral. [PHYS] State of being separated in phase by 90°, or one quarter-cycle. Also known as phase quadrature.

quadrature component [ELEC] **1.** A vector representing an alternating quantity which is in quadrature (at 90°) with some reference vector. **2.** *See* reactive component.

quadrature current *See* reactive current.

quadric cone [MATH] A conical surface whose directrices are conic curves.

quadrics [MATH] Homogeneous, second-degree expressions.

quadric surface [MATH] A surface whose equation is a second-degree algebraic equation.

quadrilateral *See* quadrangle.

quadrillion [MATH] **1.** The number 10^{15}. **2.** In British and German usage, the number 10^{24}.

quadrinomial distribution [STAT] A multinomial distribution with four possible outcomes.

quadruple point [PHYS CHEM] Temperature at which four phases are in equilibrium, such as a saturated solution containing an excess of solute.

quadruple vector product [MATH] **1.** For any four vectors, the dot product of two derived vectors, one of which is the cross product of two of the original vectors, and the other of which is the cross product of the other two. **2.** For any four vectors, the cross product of two derived vectors, one of which is the cross product of two of the original vectors, and the other of which is the cross product of the other two.

quadrupole [ELECTROMAG] A distribution of charge or magnetization which produces an electric or magnetic field equivalent to that produced by two electric or magnetic dipoles whose dipole moments have the same magnitude but point in opposite directions, and which are separated from each other by a small distance.

quadrupole field [ELECTROMAG] **1.** An electric or magnetic field equivalent to that produced by two electric or magnetic dipoles whose dipole moments have the same magnitude but point in opposite directions, and which are separated from each other by a small distance. **2.** The field produced by a quadrupole lens.

quadrupole lens [ELECTROMAG] A device for focusing beams of charged particles which has four electrodes or magnetic poles of alternating sign arranged in a circle about the beam; used in instruments such as electron microscopes and particle accelerators.

quadrupole moment [ELECTROMAG] A quantity characterizing a distribution of charge or magnetization; it is given by integrating the product of the charge density or divergence of magnetization density, the second power of the distance from the origin, and a spherical harmonic Y^*_{2m} over the charge or magnetization distribution.

quadrupole spectrometer [ANALY CHEM] A type of mass spectroscope in which ions pass along a line of symmetry between four parallel cylindrical rods; an alternating potential superimposed on a steady potential between pairs of rods filters out all ions except those of a predetermined mass. Also known as Massenfilter.

quality factor [NUCLEO] The factor by which absorbed dose is to be multiplied to obtain a quantity that expresses on a common scale, for all ionizing radiations, the irradiation incurred by exposed persons. [PHYS] *See* Q.

quality-factor meter *See* Q meter.

quantal response [STAT] Response to treatment which has only two outcomes, all or none.

quantic [MATH] A homogeneous algebraic polynomial with more than one variable.

quantifier [MATH] A logical relation used to form true or false statements from predicates, such as the existential quantifier or the universal quantifier.

quantile [STAT] A value which divides a set of data into equal proportions; examples are quartile and decile.

quantity [MATH] Any expression which is concerned with value rather than relations.

quantization [QUANT MECH] **1.** The restriction of an observable quantity, such as energy or angular momentum, associated with a physical system, such as an atom, molecule, or elementary particle, to a discrete set of values. **2.** The transition from a description of a system of particles or fields in the classical approximation where canonically conjugate variables commute, to a description where these variables are treated as noncommuting operators; quantization (first definition) is a result of this procedure.

quantized spin wave *See* magnon.

quantized vortex [CRYO] A circular flow pattern observed in superfluid helium and type II superconductors, in which a superfluid flows about a normal (nonsuperfluid) cylindrical region or core which has the form of a thin line, and either the circulation or the magnetic flux is quantized.

quantum [QUANT MECH] **1.** For certain physical quantities, a unit such that the values of the quantity are restricted to integral multiples of this unit; for example, the quantum of angular momentum is Planck's constant divided by 2π. **2.** An entity resulting from quantization of a field or wave, having particlelike properties such as energy, mass, momentum, and angular momentum; for example, the photon is the quantum of an electromagnetic field, and the phonon is the quantum of a lattice vibration.

quantum chemistry [PHYS CHEM] A branch of physical chemistry concerned with the explanation of chemical phenomena by means of the laws of quantum mechanics.

quantum detector [PHYS] A detector of electromagnetic radiation which converts a quantum of the radiation into a proportionate signal by some process which is insensitive to quanta of less than a certain energy; examples include photographic emulsions, photoelectric cells, and Geiger counters.

quantum discontinuity [QUANT MECH] The emission or absorption of a definite amount of energy that accompanies a quantum jump.

quantum efficiency [ELECTR] The average number of electrons photoelectrically emitted from a photocathode per incident photon of a given wavelength in a phototube.

quantum electrodynamics [QUANT MECH] The quantum theory of electromagnetic radiation, synthesizing the wave and corpuscular pictures, and of the interaction of radiation with electrically charged matter, in particular with atoms and their constituent electrons. Also known as quantum theory of light; quantum theory of radiation.

quantum electronics [ELECTR] The branch of electronics associated with the various energy states of matter, motions within atoms or groups of atoms, and various phenomena in crystals; examples of practical applications include the atomic hydrogen maser and the cesium atomic-beam resonator.

quantum field theory [QUANT MECH] Quantum theory of physical systems possessing an infinite number of degrees of freedom, such as the electromagnetic field, gravitation field, or wave fields in a medium.

quantum gravity [PHYS] The quantized theory of Einstein's theory of general relativity; not yet definitively formulated.

quantum hydrodynamics [CRYO] The mechanics of a superfluid, such as helium II, investigating phenomena such as the fountain effect and second sound.

quantum hypothesis [QUANT MECH] A hypothesis that some physical quantity can assume only a certain discrete set of values; examples are Planck's law, and the condition in the Bohr-Sommerfeld theory that the action integral of a system must be an integral multiple of Planck's constant.

quantum jump [QUANT MECH] The transition of a quantum system from one stationary state to another, accompanied by emission or absorption of energy.

quantum limit [SPECT] The shortest wavelength present in a continuous x-ray spectrum. Also known as boundary wavelength; end radiation.

quantum-mechanical operator [QUANT MECH] A linear, Hermitian operator associated with some physical quantity; for a physical system in any state, the expectation value of the physical quantity equals the integral over configuration space of $\psi^*(A\psi)$, where $A\psi$ is the result of the operator acting on the wave function of the system, and ψ^* is the complex conjugate of the wave function.

quantum mechanics [PHYS] The modern theory of matter, of electromagnetic radiation, and of the interaction between matter and radiation; it differs from classical physics, which it generalizes and supersedes, mainly in the realm of atomic and subatomic phenomena. Also known as quantum theory.

quantum number [QUANT MECH] One of the quantities, usually discrete with integer or half-integer values, needed to characterize a quantum state of a physical system; they are usually eigenvalues of quantum-mechanical operators or integers sequentially assigned to these eigenvalues.

quantum of action *See* Planck's constant.

quantum state [QUANT MECH] **1.** The condition of a physical system as described by a wave function; the function may be simultaneously an eigenfunction of one or more quantum-mechanical operators; the eigenvalues are then the quantum numbers that label the state. **2.** *See* energy state.

quantum statistics [STAT MECH] The statistical description of particles or systems of particles whose behavior must be described by quantum mechanics rather than classical mechanics.

quantum theory *See* quantum mechanics.

quantum theory of heat capacity [STAT MECH] Application of quantum statistics to calculate heat capacities of various substances; an important result of the theory is the decrease of specific heats at low temperatures to values smaller than their classical values as a result of energy quantization.

quantum theory of light *See* quantum electrodynamics.

quantum theory of radiation [QUANT MECH] **1.** The theory of heat radiation based on Planck's law; its principal result is the Planck radiation formula. **2.** *See* quantum electrodynamics.

quantum theory of spectra [QUANT MECH] The contemporary theory of spectra, based on the idea that an atom, molecule, or nucleus can exist only in certain allowed energy states, that it emits or absorbs energy as it changes from one state to another, and that the frequency of the associated electromagnetic radiation equals the difference in energies of two states divided by Planck's constant.

quantum theory of valence [PHYS CHEM] The theory of valence based on quantum mechanics; it accounts for many experimental facts, explains the stability of a chemical bond, and allows the correlation and prediction of many different properties of molecules not possible in earlier theories.

quantum-wave equation [QUANT MECH] A partial differential equation which relates the spatial and time dependences of the wave function of a system of one or more atomic or subatomic particles; examples are the Schrödinger equation in nonrelativistic quantum mechanics, and the Klein-Gordon, Dirac, Rarita-Schwinger and Proca equations in relativistic quantum mechanics.

quark [PARTIC PHYS] One of the hypothetical basic particles, having charges whose magnitudes are $\frac{1}{3}$ or $\frac{2}{3}$ of the electron charge, from which many of the elementary particles may, in theory, be built up; for example, nucleons may be formed from three quarks and mesons from quark-antiquark combinations; no experimental evidence for the actual existence of free quarks has been found.

quark confinement [PARTIC PHYS] The phenomenon wherein quarks can never be removed from the hadrons they compose, even though the interactions between them are relatively weak.

quart [MECH] Abbreviated qt. **1.** A unit of volume used for measurement of liquid substances in the United States, equal to 2 pints, or $\frac{1}{4}$ gallon, or $57\frac{3}{4}$ cubic inches, or $9.46352946 \times 10^{-4}$ cubic meter. **2.** A unit of volume used for measurement of solid substances in the United States, equal to 2 dry pints, or $\frac{1}{32}$ bushel, or $107,521/1,600$ cubic inches, or approximately 1.10122×10^{-3} cubic meter. **3.** A unit of volume used for measurement of both liquid and solid substances, although mainly the former, in the United Kingdom, equal to 2 U.K. pints, or $\frac{1}{4}$ U.K. gallon, or approximately 1.13652×10^{-3} cubic meter.

quarter [MECH] **1.** A unit of mass in use in the United States, equal to $\frac{1}{4}$ short ton, or 500 pounds, or 226.796185 kilograms. **2.** A unit of mass in troy measure, equal to $\frac{1}{4}$ troy hundredweight, or 25 troy pounds, or 9.33104304 kilograms. Abbreviated qr tr. **3.** A unit of mass used in the United Kingdom,

QUARK

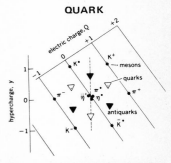

Charge-hypercharge plot showing the Gell-Mann-Nishijima loci of the elementary particles together with the positions of the pseudoscalar meson nonet and the quarks and the antiquarks.

equal to ¼ hundredweight, or 28 pounds, or 12.70058636 kilograms. Abbreviated qr. **4.** A unit of volume used in the United Kingdom for measurement of liquid and solid substances, equal to 8 bushels, or 64 gallons, or approximately 0.290950 cubic meter.

quarter-phase *See* two-phase.

quarter-wave [ELECTROMAG] Having an electrical length of one quarter-wavelength.

quarter-wave plate [OPTICS] A thin sheet of mica or other doubly refracting crystal material of such thickness as to introduce a phase difference of one quarter-cycle between the ordinary and the extraordinary components of light passing through; such a plate converts circularly polarized light into plane-polarized light.

quartic *See* biquadratic.

quartic equation [MATH] Any fourth-degree polynomial equation. Also known as biquadratic equation.

quartile [STAT] The value of any of the three random variables which separate the frequency of a distribution into four equal parts.

quartile deviation [STAT] One-half of the difference between the upper and lower, that is, the third and first, quartiles. Also known as semi-interquartile range.

quartz clock [HOROL] A clock using the piezoelectric property of a quartz crystal, in which the crystal is introduced into an oscillating electric circuit having a frequency nearly equal to the natural frequency of vibration of the crystal.

quartz crystal [ELECTR] A natural or artificially grown piezoelectric crystal composed of silicon dioxide, from which thin slabs or plates are carefully cut and ground to serve as a crystal plate.

quartz-crystal filter [ELECTR] A filter which utilizes a quartz crystal; it has a small bandwidth, a high rate of cutoff, and a higher unloaded Q than can be obtained in an ordinary resonator.

quartz-crystal resonator [ELECTR] A quartz plate whose natural frequency of vibration is used to control the frequency of an oscillator. Also known as quartz resonator.

quartz fiber [ENG] An extremely fine and uniform quartz filament that may be used as a torsion thread or as an indicator in an electroscope or dosimeter.

quartz-fiber dosimeter [ENG] A dosimeter in which radiation dose is determined from the deflection of a quartz fiber that is initially charged, repelling it from its metal support, and has its charge reduced by ionizing radiation, causing a proportional reduction in its deflection.

quartz-fiber electroscope [ELECTR] Electroscope in which a gold-plated quartz fiber serves the same function as the gold leaf of a conventional electroscope.

quartz-fiber manometer *See* decrement gage.

quartz lamp [ELECTR] A mercury-vapor lamp having a transparent envelope made from quartz instead of glass; quartz resists heat, permitting higher currents, and passes ultraviolet rays that are absorbed by ordinary glass.

quartz oscillator [ELECTR] An oscillator in which the frequency of the output is determined by the natural frequency of vibration of a quartz crystal.

quartz plate *See* crystal plate.

quartz pressure gage [ENG] A pressure gage that uses a highly stable quartz crystal resonator whose frequency changes directly with applied pressure.

quartz resonator *See* quartz-crystal resonator.

quartz thermometer [ENG] A thermometer based on the sensitivity of the resonant frequency of a quartz crystal to changes in temperature.

quartz wedge [OPTICS] A very thin wedge of quartz cut parallel to an optic axis; used to determine the sign of double refraction of biaxial crystals, and in other applications involving polarized light and its interaction with matter.

quasar [ASTRON] Quasi-stellar astronomical object, often a radio source; all quasars have large red shifts; they have small optical diameter, but may have large radio diameter. Also known as quasi-stellar object (QSO).

quasi-atom [ATOM PHYS] A system formed by two colliding atoms whose nuclei approach each other so closely that, for a very short time, the atomic electrons arrange themselves as if they belonged to a single atom whose atomic number equals the sum of the atomic numbers of the colliding atoms.

quasi-fission [NUC PHYS] A nuclear reaction induced by heavy ions in which the two product nuclei have kinetic energies typical of fission products, but have masses close to those of the target and projectile, individually. Also known as deep inelastic transfer; incomplete fusion; relaxed peak process; strongly damped collision.

quasi-free-electron theory [SOLID STATE] A modification of the free-electron theory of metals to take into account the periodic variation of the potential acting on a conduction electron, in which these electrons are assigned an effective scalar mass which differs from their real mass.

quasi-linear feedback control system [CONT SYS] Feedback control system in which the relationships between the pertinent measures of the system input and output signals are substantially linear despite the existence of nonlinear elements.

quasi-linear system [CONT SYS] A control system in which the relationships between the input and output signals are substantially linear despite the existence of nonlinear elements.

quasi-molecule [ATOM PHYS] The structure formed by two colliding atoms when their nuclei are close enough for the atoms to interact, but not so close as to form a quasi-atom.

quasi-norm [MATH] A scalar-valued function on a real or complex vector space associating to each element x a scalar denoted $\| x \|$, such that $\| x \| = \| -x \| \geq 0$ for all x in the space, $\| x \| = 0$ if and only if $x = 0$, $\| x + y \| \leq \| x \| + \| y \|$ for all x and y in the space, the sequence $\| a_n x \|$ converges to 0 for any sequence of scalars a_n converging to 0 and any element x in the space, and the sequence $\| a x_n \|$ converges to 0 for any sequence of vectors such that $\| x_n \|$ converges to 0 and any scalar a.

quasi-normed linear space [MATH] A real or complex vector space which has a quasi-norm defined on it. Also known as quasi-normed vector space.

quasi-normed vector space *See* quasi-normed linear space.

quasi-particle [PHYS] An entity used in the description of a system of many interacting particles which has particlelike properties such as mass, energy, and momentum, but which does not exist as a free particle; examples are phonons and other elementary excitations in solids, and "dressed" helium-3 atoms in Landau's theory of liquid helium-3.

quasi-periodic function [MATH] A function $f(t)$ which is of the form $f(t) = F(e^{i\omega_1 t}, e^{i\omega_2 t}, \ldots, e^{i\omega_n t})$ for some continuous function F and real numbers $\omega_1, \omega_2, \ldots, \omega_n$. Also known as conditionally periodic function.

quasi-reflection [OPTICS] A term applied to the very strong return of light produced by dust particles and other suspensoids whose diameters are large compared to the wavelength of the incident radiation.

quasi-stable elementary particle [PARTIC PHYS] A term formerly (before the discovery of charmed particles) used for elementary particles that cannot decay into other particles through strong interactions and that have lifetimes longer than 10^{-20} second. Also known as semistable elementary particle.

quasi-static process *See* reversible process.

quasi-stellar object *See* quasar.

quaternion [MATH] The division algebra over the real numbers generated by elements i, j, k subject to the relations $i^2 = j^2 = k^2 = -1$ and $ij = -ji = k$, $jk = -kj = i$, and $ki = -ik = j$. Also known as hypercomplex number.

quaternion algebra [MATH] Any four-dimensional nonassociative algebra over a field F obtained by the Cayley-Dickinson process from a two-dimensional nonassociative algebra over F consisting of either the direct sum of F with itself or a separable quadratic field over F; this generalizes the concept of quaternions.

quenching [ATOM PHYS] Phenomenon in which a very strong electric field, such as a crystal field, causes the orbit of an electron in an atom to precess rapidly so that the average magnetic moment associated with its orbital angular momentum is reduced to zero. [ELECTR] **1.** The process of terminating a discharge in a gas-filled radiation-counter tube by inhibiting reignition. **2.** Reduction of the intensity of resonance radiation resulting from deexcitation of atoms, which would otherwise have emitted this radiation, in collisions with electrons or other atoms in a gas. [SOLID STATE] Reduction in the intensity of sensitized luminescence radiation when energy migrating through a crystal by resonant transfer is dissipated in crystal defects or impurities rather than being reemitted as radiation.

queuing theory [MATH] The area of stochastic processes emphasizing those processes modeled on the situation of individuals lining up for service.

quicksilver *See* mercury.

quiescent [ELECTR] Condition of a circuit element which has no input signal, so that it does not perform its active function.

quiescent point [ELECTR] The point on the characteristic curve of an amplifier representing the conditions that exist when the input signal equals zero.

quiescent value [ELECTR] The voltage or current value for an electron-tube electrode when no signals are present.

quiet sun [ASTROPHYS] The sun when it is free from unusual radio wave or thermal radiation such as that associated with sunspots.

quiet sun noise [ASTROPHYS] Electromagnetic noise originating in the sun at a time when there is little or no sunspot activity.

Quincke tube *See* Herschel-Quincke tube.

quintal *See* hundredweight; metric centner.

quintic [MATH] A fifth-degree expression.

quintic equation [MATH] A fifth-degree polynomial equation.

quintillion [MATH] **1.** The number 10^{18}. **2.** In British and German usage, the number 10^{30}.

Q unit [THERMO] A unit of energy, used in measuring the heat energy of fuel reserves, equal to 10^{18} British thermal units, or approximately 1.055×10^{21} joules.

quotient [MATH] The result of dividing one quantity by another.

quotient field [MATH] The smallest field containing a given integral domain; obtained by formally introducing all quotients of elements of the integral domain.

quotient group [MATH] A group G/H whose elements are the cosets gH of a given normal subgroup H of a given group G, and the group operation is defined as $g_1H \cdot g_2H \equiv (g_1 \cdot g_2)H$. Also known as factor group.

quotient ring [MATH] **1.** A ring R/I whose elements are the cosets rI of a given ideal I in a given ring R, where the additive and multiplicative operations have the form: $r_1I + r_2I \equiv (r_1 + r_2)I$ and $r_1I \cdot r_2I \equiv (r_1 \cdot r_2)I$. Also known as factor ring; residue class ring. **2.** The quotient ring of a commutative ring R by a multiplicative subset S is the set of quotients r/s, where r is in R and s is in S, two such elements, r/s and r'/s', being considered equivalent if there is an s_1 in S such that $s_1(r's - rs') = 0$; multiplication is defined by the equation $(r/s)(r'/s') = rr'/ss'$, and addition by the equation $(r/s) + (r'/s') = (s'r + sr')/ss'$. Also known as ring of fractions.

quotient set [MATH] The set of all the equivalence classes relative to a given equivalence relation on a given set.

quotient space [MATH] The topological space Y which is the set of equivalence classes relative to some given equivalence relation on a given topological space X; the topology of Y is canonically constructed from that of X.

quotient theorem [MATH] If the quantities

$$T^{i_1 i_2 \cdots i_m}_{j_1 j_2 \cdots j_n} V_{i_1}$$

summed over i_1 are the components of a tensor contravariant of order $m - 1$ and covariant of order n, whenever V_{i_1} are the components of a covariant vector, then the quantities

$$T^{i_1 i_2 \cdots i_m}_{j_1 j_2 \cdots j_n}$$

are the components of a tensor contravariant of order m and covariant of order n.

quotient topology [MATH] If X is a topological space, X/R the quotient space by some equivalence relation on X, the quotient topology on X/R is the smallest topology which makes the function which assigns to each element of X its equivalence class in X/R a continuous function.

r *See* roentgen.

r_s *See* Spearman's rank correlation coefficient.

R *See* roentgen.

Ra *See* radium.

R.A. *See* right ascension.

Raabe's convergence test [MATH] An infinite series with positive terms a_n where, for each n, $a_{n+1}/a_n = 1/(1 + b_n)$) will converge if, after a certain term, nb_n always exceeds a fixed number greater than 1 and will diverge if nb_n always is less than a fixed number less than or equal to 1.

rabbit [NUCLEO] A small container that is propelled, usually pneumatically or hydraulically, through a tube into a nuclear reactor; used to expose samples to the radiation, especially neutron flux, then remove them rapidly for measurements of radioactive atoms having short half-lives. Also known as shuttle.

Racah coefficient [QUANT MECH] A coefficient that appears in the transformation between the modes of coupling eigenfunctions of three angular momenta; they differ only by, at most, a sign from the six-*j* coefficients. Also known as *W* coefficient.

race track [NUCLEO] An assembly of several Calutron isotope separators in the shape of a race track, having a common magnetic field. Also known as track.

rad [NUCLEO] The standard unit of absorbed dose, equal to energy absorption of 100 ergs per gram (0.01 joule per kilogram); supersedes the roentgen as the unit of dosage.

radar [ENG] **1.** A system using beamed and reflected radio-frequency energy for detecting and locating objects, measuring distance or altitude, navigating, homing, bombing, and other purposes; in detecting and ranging, the time interval between transmission of the energy and reception of the reflected energy establishes the range of an object in the beam's path. Derived from radio detection and ranging. **2.** *See* radar set.

radar antenna [ELECTROMAG] A device which radiates radio-frequency energy in a radar system, concentrating the transmitted power in the direction of the target, and which provides a large area to collect the echo power of the returning wave.

radar astronomy [ASTRON] The study of astronomical bodies and the earth's atmosphere by means of radar pulse techniques, including tracking of meteors and the reflection of radar pulses from the moon and the planets.

radar equation [ELECTROMAG] An equation that relates the transmitted and received powers and antenna gains of a primary radar system to the echo area and distance of the radar target.

radar range [ELECTROMAG] The maximum distance at which a radar set is ordinarily effective in detecting objects.

radar range equation [ELECTROMAG] An equation which expresses radar range in terms of transmitted power, minimum detectable signal, antenna gain, and the target's radar cross section.

radar reflectivity [ELECTROMAG] The fraction of electromagnetic energy generated by a radar installation which is reflected by an object.

radarscope camera *See* radar camera.

radar set [ENG] A complete assembly of radar equipment for detecting and ranging, consisting essentially of a transmitter, antenna, receiver, and indicator. Also known as radar.

radarsonde [ENG] **1.** An electronic system for automatically measuring and transmitting high-altitude meteorological data from a balloon, kite, or rocket by pulse-modulated radio waves when triggered by a radar signal. **2.** A system in which radar techniques are used to determine the range, elevation, and azimuth of a radar target carried aloft by a radiosonde.

radar telescope [ENG] A large radar antenna and associated equipment used for radar astronomy.

Rademacher functions [MATH] A sequence of orthonormal functions $\{r_n(x)\}$ on the interval [0,1], where $r_0(x) = 1$, and, for $k \geq 1$, $r_k(x) = +1$ if the kth place to the right of the decimal point in the binary notation for x is 1, and $r_k(x) = -1$ if the kth place to the right of the decimal point is 0.

radiac [NUCLEO] Detection, identification, and measurement of the intensity of nuclear radiation in an area. Derived from radioactivity detection, identification, and computation.

radiac instrument *See* radiac set.

radiacmeter *See* radiac set.

radiac set [NUCLEO] A complete system for detecting, identifying, and measuring radioactivity. Also known as radiac instrument; radiacmeter.

radial [SCI TECH] Directed or diverging from a center.

radial astigmatism [OPTICS] Astigmatism which affects the imaging of points that lie off the axis of an optical system, due to oblique incidence of rays from these points. Also known as oblique astigmatism.

radial distribution function [MATH] A function $F(r)$ equal to the average of a given function of the three coordinates over a sphere of radius r centered at the origin of the coordinate system. [PHYS CHEM] A function $\rho(r)$ equal to the average over all directions of the number density of molecules at distance r from a given molecule in a liquid.

radial Doppler effect [ELECTROMAG] The part of the optical Doppler effect which depends on the direction of the relative velocity of source and observer, and is analogous to the acoustical Doppler effect, in contrast to the transverse Doppler effect.

radial grating [ELECTROMAG] Conformal wire grating consisting of wires arranged radially in a circular frame, like the spokes of a wagon wheel, and placed inside a circular waveguide to obstruct E waves of zero order while passing the corresponding H waves.

radial heat flow [THERMO] Flow of heat between two coaxial cylinders maintained at different temperatures; used to measure thermal conductivities of gases.

radial motion [MECH] Motion in which a body moves along a line connecting it with an observer or reference point; for example, the motion of stars which move toward or away from the earth without a change in apparent position.

radial stress [MECH] Tangential stress at the periphery of an opening.

radial velocity [MECH] The component of the velocity of a body that is parallel to a line from an observer or reference point to the body; the radial velocities of stars are valuable in determining the structure and dynamics of the Galaxy. Also known as line-of-sight velocity.

radial wave equation [MECH] Solutions to wave equations with spherical symmetry can be found by separation of variables; the ordinary differential equation for the radial part of the wave function is called the radial wave equation.

radian [MATH] The central angle of a circle determined by two radii and an arc joining them, all of the same length.

radiance [OPTICS] The radiant flux per unit solid angle per unit of projected area of the source; the usual unit is the watt per steradian per square meter. Also known as steradiancy.

radiancy *See* radiant emittance.

radian frequency *See* angular frequency.

radian length [PHYS] Distance, in a sinusoidal wave, between phases differing by an angle of 1 radian; it is equal to the wavelength divided by 2π.

radiant [ASTRON] A point on the celestial sphere through which pass the backward extensions of the trail of a meteor as observed at various locations, or the backward extensions of trails of a number of meteors traveling parallel to each other. [PHYS] **1.** Pertaining to motion of particles or radiation along radii from a common point or a small region. **2.** A point, region, substance, or entity from which particles or radiations are emitted.

radiant density [PHYS] The instantaneous amount of radiant energy contained in a unit volume of propagation medium.

radiant efficiency [OPTICS] The ratio of the radiant flux emitted by a radiation source to the power consumed by the source.

radiant emittance [ELECTROMAG] The radiant flux per unit area that emerges from a surface. Also known as radiancy; radiant exitance.

radiant energy *See* radiation.

radiant-energy thermometer *See* radiation pyrometer.

radiant exitance *See* radiant emittance.

radiant exposure [OPTICS] A measure of the total radiant energy incident on a surface per unit area; equal to the integral over time of the radiant flux density. Also known as exposure.

radiant flux [OPTICS] The time rate of flow of radiant energy.

radiant flux density [ELECTROMAG] The amount of radiant power per unit area that flows across or onto a surface. Also known as irradiance.

radiant intensity [ELECTROMAG] The energy emitted per unit time per unit solid angle about the direction considered; usually expressed in watts per steradian.

radiant power [ELECTROMAG] The energy carried across or onto a surface by electromagnetic radiation per unit time, or the total radiant energy emitted by a source of electromagnetic radiation per unit time.

radiant quantities [OPTICS] Physical quantities used in photometry, such as radiant flux and radiance, which are based on the energy carried by light, and are thus independent of the response of the human eye.

radiant reflectance [ELECTROMAG] Ratio of reflected radiant power to incident radiant power.

radiant transmittance [ELECTROMAG] Ratio of transmitted radiant power to incident radiant power.

radiated power [ELECTROMAG] The total power emitted by a transmitting antenna.

radiating power *See* emittance.

radiating scattering [PHYS] The diversion of radiation (thermal, electromagnetic, or nuclear) from its orginal path as a result of interactions or collisions with atoms, molecules, or larger particles in the atmosphere or other media between the source of radiation (for example, a nuclear explosion) and a point at some distance away.

radiation [PHYS] **1.** The emission and propagation of waves transmitting energy through space or through some medium; for example, the emission and propagation of electromagnetic, sound, or elastic waves. **2.** The energy transmitted by waves through space or some medium; when unqualified, usually refers to electromagnetic radiation. Also known as radiant energy. **3.** A stream of particles, such as electrons, neutrons, protons, α-particles, or high-energy photons, or a mixture of these.

radiation accident [NUCLEO] Any accident resulting in the spread of radioactive materials or in the exposure of individuals to radiation.

radiation area [NUCLEO] Any accessible area in which the level of radiation is such that a major portion of an individual's body could receive in any 1 hour a dose in excess of 5 millirem or in any 5 consecutive days a dose in excess of 150 millirem.

radiation biochemistry [BIOCHEM] The study of the response of the constituents of living matter to radiation.

radiation biology See radiobiology.

radiation biophysics [BIOPHYS] The study of the response of organisms to ionizing radiations and to ultraviolet light.

radiation budget [GEOPHYS] A quantitative statement of the amounts of radiation entering and leaving a given region of the earth.

radiation chemistry [NUCLEO] The branch of chemistry that is concerned with the chemical effects, including decomposition, of energetic radiation or particles on matter.

radiation condition [MATH] The boundary condition $\partial f/\partial n + kf = 0$, to be satisfied by a solution f of a partial differential at a surface, where $\partial f/\partial n$ is the directional derivative in the direction of the normal to the surface and k is a constant.

radiation correction See cooling correction.

radiation counter [NUCLEO] An instrument used for detecting or measuring nuclear radiation by counting the resultant ionizing events; examples include Geiger counters and scintillation counters. Also known as counter.

radiation counter tube See counter tube.

radiation cytology [CYTOL] An aspect of biology that deals with the effects of radiations on living cells.

radiation damage [NUCLEO] Harmful changes in the properties of liquids, gases, and solids caused by any type of radiation.

radiation damping [ELECTROMAG] Damping of a system which loses energy through electromagnetic radiation. [QUANT MECH] Damping which arises in quantum electrodynamics from the virtual interaction of a particle with its zero point field.

radiation decontamination [NUCLEO] The removal of unwanted radioactive material.

radiation detection instrument [NUCLEO] Any device that detects and records the characteristics of ionizing radiation.

radiation detector See particle detector.

radiation dose [NUCLEO] The total amount of ionizing radiation absorbed by material or tissues, in the sense of absorbed dose (expressed in rads), exposure dose (expressed in roentgens), or dose equivalent (expressed in rems).

radiation dose rate [NUCLEO] The radiation dosage absorbed per unit of time; a radiation dose rate can be set at some particular unit of time (for example, H hour plus 1 hour) and would be called H hour plus 1 radiation dose rate.

radiation dosimetry See dosimetry.

radiation efficiency [ELECTROMAG] Of an antenna, the ratio of the power radiated to the total power supplied to the antenna at a given frequency.

radiation field [ELECTROMAG] The electromagnetic field that breaks away from a transmitting antenna and radiates outward into space as electromagnetic waves; the other type of electromagnetic field associated with an energized antenna is the induction field.

radiation filter [ELECTROMAG] Selectively transparent body, which transmits only certain wavelength ranges.

radiation gage [NUCLEO] An instrument for measuring radiation quantity and intensity.

radiation impedance *See* radiation resistance.

radiation intensity [ELECTROMAG] The power radiated from an antenna per unit solid angle in a given direction. [NUCLEO] The quantity of radiant energy passing perpendicularly through a specified location of unit area in unit time; reported as a number of particles or photons per square centimeter per second, or in energy units such as ergs per square centimeter per second.

radiation ionization [PHYS] Ionization of the atoms or molecules of a gas or vapor by the action of electromagnetic radiation.

radiation laws [PHYS] **1.** The four physical laws which, together, fundamentally describe the behavior of blackbody radiation, Kirchhoff's law, Planck's law, Stefan-Boltzmann law, and Wien's law. **2.** All of the more inclusive assemblage of empirical and theoretical laws describing all manifestations of radiative phenomena.

radiation length [NUCLEO] The mean path length required to reduce the energy of relativistic charged particles by the factor $1/e$, or 0.368, as they pass through matter. Also known as cascade unit; radiation unit.

radiationless transition [PHYS] A transition of a system between two energy states in which energy is given to or taken up from another system or particle, rather than being emitted or absorbed in electromagnetic radiation; examples include internal conversion, the Auger effect, and excitation or deexcitation of atoms or molecules in collisions with other atoms or molecules.

radiation lobe *See* lobe.

radiation microbiology [MICROBIO] A field of basic and applied radiobiology concerned chiefly with the damaging effects of radiation on microorganisms.

radiation monitoring [NUCLEO] Continuous or periodic determination of the amount of radiation present in a given area.

radiation oscillator *See* Planck oscillator.

radiation pattern [ELECTROMAG] Directional dependence of the radiation of an antenna. Also known as antenna pattern; directional pattern; field pattern.

radiation physics [PHYS] The study of ionizing radiation and its effects on matter.

radiation pressure [ACOUS] The average pressure exerted on a surface or interface between two media by a sound wave. [ELECTROMAG] The pressure exerted by electromagnetic radiation on objects on which it impinges.

radiation protection guide [NUCLEO] The officially determined radiation doses which should not be exceeded without careful consideration of the reasons for doing so; these standards, established by the Federal Radiation Council, are equivalent to what was formerly called the maximum permissible dose or maximum permissible exposure.

radiation pyrometer [ENG] An instrument which measures the temperature of a hot object by focusing the thermal radiation emitted by the object and making some observation

RADIATION PYROMETER

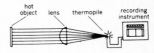

Diagram of an elementary radiation pyrometer. Lens focuses part of radiation emitted by hot object onto thermopile. Resulting heating of thermopile causes it to generate electrical signal which is displayed on recording instrument. *(From D. M. Considine, ed., Process Instruments and Controls Handbook, McGraw-Hill, 1957)*

on it; examples include the total-radiation, optical, and ratio pyrometers. Also known as radiant-energy thermometer; radiation thermometer.

radiation quality [PHYS] The spectrum of radiant energy produced by a given radiation source with respect to its penetration or its suitability for a specific application.

radiation resistance [ACOUS] For a medium, the acoustic impedance of a plane wave in that medium. Also known as radiation impedance. [ELECTROMAG] The total radiated power of an antenna divided by the square of the effective antenna current measured at the point where power is supplied to the antenna.

radiation scattering [PHYS] The diversion of radiation (thermal, electromagnetic, or nuclear) from its original path as a result of interactions or collisions with atoms, molecules, or larger particles in the atmosphere or other media.

radiation shield [ENG] A shield or wall of material interposed between a source of radiation and a radiation-sensitive body, such as a person, radiation-detection instrument, or photographic film, to protect the latter.

radiation sickness [MED] **1.** Illness, usually manifested by nausea and vomiting, resulting from the effects of therapeutic doses of radiation. **2.** Radiation injury following exposure to excessive doses of radiation, such as the explosion of an atomic bomb.

radiation source [NUCLEO] Usually a sealed capsule containing an artificial radioisotope, used in teletherapy, radiography, as a power source for batteries, or in various types of industrial gages; machines such as accelerators, and radioisotopic generators and natural radionuclides may also be considered as sources.

radiation standards [NUCLEO] Exposure standards, permissible concentrations, rules for safe handling, regulations for transportation, regulations for industrial control of radiation, and control of radiation exposure by legislative means.

radiation sterilization [NUCLEO] Exposure of a material, object, or body to ionizing radiation in order to destroy microorganisms.

radiation survey meter [NUCLEO] Portable device to measure the intensity of nuclear radiations in a given region, in such applications as health physics (atomic radiation safety) or supervision of radioactively hot areas.

radiation therapy [MED] The use of ionizing radiation or radioactive substances to treat disease. Also known as actinotherapy; radiotherapy.

radiation thermocouple [ELEC] An infrared detector consisting of several thermocouples connected in series, arranged so that the radiation falls on half of the junctions, causing their temperature to increase so that a voltage is generated.

radiation thermometer *See* radiation pyrometer.

radiation unit *See* radiation length.

radiation vacuum gage [ENG] Vacuum (reduced-pressure) measurement device in which gas ionization from an α-source of radiation varies measurably with changes in the density (molecular concentration) of the gas being measured.

radiation zone *See* Fraunhofer region.

radiative capture [NUC PHYS] A nuclear capture process whose prompt result is the emission of electromagnetic radiation only.

radiative collision [PHYS] A collision between two charged particles in which part of the kinetic energy of the particles is converted into electromagnetic radiation.

radiative correction [QUANT MECH] The change produced in the value of some physical quantity, such as the mass or charge of a particle, as the result of the particle's interactions with various fields.

radiative equilibrium [ASTROPHYS] Energy transfer through a star by radiation, absorption, and reradiation at a rate such that each section of the star is maintained at the appropriate temperature.

radiative recombination [PHYS] Recombination of parts of an atom during which electromagnetic radiation is emitted.

radiative transfer [THERMO] The transmission of heat by electromagnetic radiation.

radiative transition [QUANT MECH] A change of a quantum-mechanical system from one energy state to another in which electromagnetic radiation is emitted.

radiator [ACOUS] A vibrating element of a transducer which radiates sound waves. [ELECTROMAG] **1.** The part of an antenna or transmission line that radiates electromagnetic waves either directly into space or against a reflector for focusing or directing. **2.** A body that emits radiant energy. [PHYS] **1.** In general, a body which emits particles or radiation in any form. **2.** A body placed in a beam of ionizing radiation which, as a result, emits radiation of another kind.

radical [MATH] **1.** In a ring, the intersection of all maximal ideals. Also known as Jacobson radical. **2.** An indicated root of a quantity. Symbolized $\sqrt{\ }$.

radical axis [MATH] The line passing through the two points of intersection of a pair of circles.

radical equation *See* irrational equation.

radio [COMMUN] The transmission of signals through space by means of electromagnetic waves; usually applied to the transmission of sound and code signals, although television and radar also depend on electromagnetic waves.

radio- [ELECTROMAG] A prefix denoting the use of radiant energy, particularly radio waves. [NUCLEO] Chemical prefix designating radiation or radioactivity; used to designate radioactive elements (such as radiocarbon) and substances containing them (such as radiochemicals, radiocolloids, or radio compounds).

radioactinium [NUC PHYS] Conventional name for the isotope of thorium which has mass number 227 and is in the actinium series. Symbolized RdAc.

radioactive [NUC PHYS] Exhibiting radioactivity or pertaining to radioactivity.

radioactive age determination *See* radiometric dating.

radioactive carbon dating *See* carbon-14 dating.

radioactive chain *See* radioactive series.

radioactive clock [NUC PHYS] A radioactive isotope such as potassium-40 which spontaneously decays to a stable end product at a constant rate, allowing absolute geologic age to be determined.

radioactive cloud [NUCLEO] A mass of air and vapor in the atmosphere carrying radioactive debris from a nuclear explosion.

radioactive cobalt [NUC PHYS] Radioactive form of cobalt, such as cobalt-60 with a half-life of 5.3 years.

radioactive collision [NUC PHYS] A nuclear reaction in which a neutron is absorbed by a nucleus and a gamma ray is emitted.

radioactive contaminant [NUCLEO] A radioactive material which has spread to places where it may harm persons, spoil experiments, or make products or equipment unsuitable or unsafe for some specific purpose.

radioactive dating *See* radiometric dating.

radioactive debris [NUCLEO] Radioactive material which is carried through the air from the site of a nuclear explosion.

radioactive decay [NUC PHYS] The spontaneous transforma-

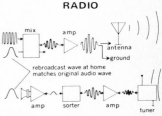

RADIO

mix
amp
antenna
ground

rebroadcast wave at home
matches original audio wave

amp
sorter
amp
tuner

Transmission of audio information by radio.

tion of a nuclide into one or more different nuclides, accompanied by either the emission of particles from the nucleus, nuclear capture or ejection of orbital electrons, or fission. Also known as decay; nuclear spontaneous reaction; radioactive disintegration; radioactive transformation; radioactivity.

radioactive decay constant *See* decay constant.

radioactive decay series *See* radioactive series.

radioactive disintegration *See* radioactive decay.

radioactive displacement law [NUC PHYS] The statement of the changes in mass number A and atomic number Z that take place during various nuclear transformations. Also known as displacement law.

radioactive element [NUC PHYS] An element all of whose isotopes spontaneously transform into one or more different nuclides, giving off various types of radiation; examples include promethium, radium, thorium, and uranium.

radioactive emanation [NUC PHYS] A radioactive gas given off by certain radioactive elements; all of these gases are isotopes of the element radon. Also known as emanation.

radioactive equilibrium [NUCLEO] A condition which may arise in the radioactive decay of a parent having shorter-lived descendants, in which the ratio of the activity of the parent to that of a descendant does not change with time.

radioactive fallout *See* fallout.

radioactive half-life *See* half-life.

radioactive heat [THERMO] Heat produced within a medium as a result of absorption of radiation from decay of radioisotopes in the medium, such as thorium-232, potassium-40, uranium-238, and uranium-235.

radioactive isotope *See* radioisotope.

radioactive metal [NUC PHYS] A luminous metallic element, such as actinium, radium, or uranium, that spontaneously and continuously emits radiation capable in some degree of penetrating matter impervious to ordinary light.

radioactive mineral [MINERAL] Any mineral species that contains uranium or thorium as an essential part of the chemical composition; examples are uraninite, pitchblende, carnotite, coffinite, and autunite.

radioactive salt [NUCLEO] A salt whose molecules have radioactive atoms, such as radium bromide or mesothorium bromide; used in trace amounts to energize self-luminous paints.

radioactive series [NUC PHYS] A succession of nuclides, each of which transforms by radioactive disintegration into the next until a stable nuclide results. Also known as decay chain; decay family; decay series; disintegration chain; disintegration family; disintegration series; radioactive chain; radioactive decay series; series decay; transformation series.

radioactive source [NUCLEO] Any quantity of radioactive material intended for use as a source of ionizing radiation.

radioactive standard [NUCLEO] A sample of radioactive material which contains a known number and type of radioactive atoms at some definite time; used to calibrate radiation measuring instruments.

radioactive tracer [NUCLEO] A radioactive isotope which, when attached to a chemically similar substance or injected into a biological or physical system, can be traced by radiation detection devices, permitting determination of the distribution or location of the substance to which it is attached. Also known as radiotracer.

radioactive transformation *See* radioactive decay.

radioactive waste [NUCLEO] Liquid, solid, or gaseous waste resulting from mining of radioactive ore, production of reactor fuel materials, reactor operation, processing of irradiated reactor fuels, and related operations, and from use of radioactive materials in research, industry, and medicine.

radioactive-waste disposal [NUCLEO] The disposal of waste radioactive materials and of equipment contaminated by radiation; the two basic disposal methods are concentration for burial underground or in the sea, and dilution for controlled dispersion; reprocessing of reactor fuel is a major source of radioactive waste.

radioactivity [NUC PHYS] **1.** A particular type of radiation emitted by a radioactive substance, such as α-radioactivity. **2.** *See* radioactive decay. **3.** *See* activity.

radioactivity analysis *See* activation analysis.

radio antenna *See* antenna.

radio astronomy [ASTRON] The study of celestial objects by measurement and analysis of their emitted electromagnetic radiation in the wavelength range from roughly 1 millimeter to 30 millimeters.

radioautography *See* autoradiography.

radiobiology [BIOL] Study of the scientific principles, mechanisms, and effects of the interaction of ionizing radiation with living matter. Also known as radiation biology.

radiocarbon *See* carbon-14.

radiocarbon dating *See* carbon-14 dating.

radiocesium *See* cesium-137.

radiochemical laboratory [CHEM] A specially equipped and shielded chemical laboratory designed for conducting radiochemical studies without danger to the laboratory personnel.

radiochemistry [CHEM] That area of chemistry concerned with the study of radioactive substances.

radio detection and ranging *See* radar.

radio duct [GEOPHYS] An atmospheric layer, typically shallow and almost horizontal, in which radio waves propagate in an anomalous fashion; ducts occur when, due to sharp inversions of temperature or humidity, the vertical gradient of the radio index of refraction exceeds a critical value.

radio element [NUC PHYS] A radioactive isotope of an element, or a sample consisting of one or more radioactive isotopes of an element.

radio frequency [ELECTROMAG] A frequency at which coherent electromagnetic radiation of energy is useful for communication purposes; roughly the range from 10 kilohertz to 100 gigahertz. Abbreviated rf.

radio-frequency choke [ELEC] A coil designed and used specifically to block the flow of radio-frequency current while passing lower frequencies or direct current.

radio-frequency resistance *See* high-frequency resistance.

radio-frequency spectrometer [SPECT] An instrument which measures the intensity of radiation emitted or absorbed by atoms or molecules as a function of frequency at frequencies from 10^5 to 10^9 hertz; examples include the atomic-beam apparatus, and instruments for detecting magnetic resonance.

radio-frequency spectroscopy [SPECT] The branch of spectroscopy concerned with the measurement of the intervals between atomic or molecular energy levels that are separated by frequencies from about 10^5 to 10^9 hertz, as compared to the frequencies that separate optical energy levels of about 6×10^{14} hertz.

radio galaxy [ASTROPHYS] A galaxy that is emitting much energy in radio frequencies often from regions devoid of visible matter.

radiogenic [NUC PHYS] Pertaining to a material produced by radioactive decay, as the production of lead from uranium decay.

radiogenic age determination *See* radiometric dating.

radiogenic dating *See* radiometric dating.

RADIOGRAPH

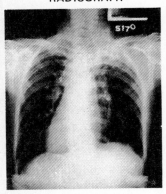

Chest radiograph of foundry worker made with intense beam from rotating-target x-ray tube, showing nodules in lungs which are caused by silicosis, and shadows of skeleton, heart, and stomach.

radiograph [GRAPHICS] The photographic image produced in radiography. Also known as shadowgraph.

radiographic equivalence factor [NUCLEO] The reciprocal of the thickness of a specified material having the same radiographic absorption as a unit thickness of a standard material.

radiographic film [GRAPHICS] The photographic film used in radiography, which must be properly selected for contrast, latitude, and sensitivity.

radiographic sensitivity [NUCLEO] A measure of radiographic quality whereby the minimum discontinuity that may be detected on the film is expressed as a percentage of the base thickness.

radiography [GRAPHICS] The technique of producing a photographic image of an opaque specimen by transmitting a beam of x-rays or γ-rays through it onto an adjacent photographic film; the image results from variations in thickness, density, and chemical composition of the specimen; used in medicine and industry.

radio interference *See* interference.

radioiodine [NUC PHYS] Any radioactive isotope of iodine, especially iodine-131; used as a tracer to determine the activity and size of the thyroid gland, and experimentally, to destroy the thyroid glands of animals.

radioisotope [NUC PHYS] An isotope which exhibits radioactivity. Also known as radioactive isotope; unstable isotope.

radioisotope assay [ANALY CHEM] An analytical technique including procedures for separating and reproducibly measuring a radioactive tracer.

radioisotope battery *See* nuclear battery.

radioisotope thermoelectric generator [NUCLEO] A device for converting nuclear energy to electrical energy in which the heat produced by radioactivity of a radioisotope is used to produce a voltage in a thermocouple circuit; chief use has been in space vehicles and in instruments left on the lunar surface. Abbreviated RTG.

radioisotopic generator *See* nuclear battery.

radiological [NUCLEO] Pertaining to nuclear radiation, radioactivity, and atomic weapons.

radiological dose [NUCLEO] The total amount of ionizing radiation absorbed by an individual exposed to any radiating source.

radiological survey [NUCLEO] Determination of the distribution and dose rates of radiation in an area.

radiology [MED] The medical science concerned with radioactive substances, x-rays, and other ionizing radiations, and the application of the principles of this science to diagnosis and treatment of disease.

radiolucent [ELECTROMAG] Transparent to x-rays and radio waves.

radioluminescence [PHYS] Luminescence produced by x-rays or γ-rays, or by particles emitted in radioactive decay.

radiolysis [PHYS CHEM] The dissociation of molecules by radiation; for example, a small amount of water in a reactor core dissociates into hydrogen and oxygen during operation.

radiometer [ELECTR] A receiver for detecting microwave thermal radiation and similar weak wide-band signals that resemble noise and are obscured by receiver noise; examples include the Dicke radiometer, subtraction-type radiometer, and two-receiver radiometer. Also known as microwave radiometer; radiometer-type receiver. [ENG] An instrument for measuring radiant energy; examples include the bolometer, microradiometer, and thermopile.

radiometer-type receiver *See* radiometer.

radiometric analysis [ANALY CHEM] Quantitative chemical analysis that is based on measurement of the absolute disintegration rate of a radioactive component having a known specific activity.

radiometric dating [NUCLEO] A technique for measuring the age of an object or sample of material by determining the ratio of the concentration of a radioisotope to that of a stable isotope in it; for example, the ratio of carbon-14 to carbon-12 reveals the approximate age of bones, pieces of wood, and other archeological specimens. Also known as isotopic age determination; nuclear age determination; radioactive age determination; radioactive dating; radiogenic age determination; radiogenic dating.

radiometric magnitude [ASTRON] A celestial body's magnitude as calculated from the total amount of radiant energy of all the wavelengths that reach the earth's surface.

radiometric titration [ANALY CHEM] Use of radioactive indicator to track the transfer of material between two liquid phases in equilibrium, such as titration of $Ag^{110}NO_3$ (silver nitrate, with the silver atom having mass number 110) against potassium chloride.

radiometry [PHYS] The detection and measurement of radiant electromagnetic energy, especially that associated with infrared radiation.

radiomicrometer *See* microradiometer.

radio nebula [ASTROPHYS] A nebula that emits nonthermal radio-frequency radiation; derives its luminosity from collisions with the surrounding interstellar medium, or from processes associated with the magnetic fields presumably involved within the nebula; examples are the network nebulae in Cygnus and NGC 443.

radionuclide [NUC PHYS] A nuclide that exhibits radioactivity.

radiopaque [ELECTROMAG] Not appreciably penetrable by x-rays or other forms of radiation.

radiophotoluminescence [PHYS] Luminescence exhibited by minerals such as fluorite and kunzite as a result of irradiation with β- and γ-rays followed by exposure to light.

radio scattering *See* scattering.

radio sextant [ELECTROMAG] An antenna with a high-resolution beam pattern that measures the angle between local direction references and an astronomical radio signal source such as an artificial satellite, the sun, the moon, or a radio star.

radiosonde [ENG] A balloon-borne instrument for the simultaneous measurement and transmission of meteorological data; the instrument consists of transducers for the measurement of pressure, temperature, and humidity, a modulator for the conversion of the output of the transducers to a quantity which controls a property of the radio-frequency signal, a selector switch which determines the sequence in which the parameters are to be transmitted, and a transmitter which generates the radio-frequency carrier.

radio source [ASTROPHYS] A source of extragalactic or interstellar electromagnetic emission in radio wavelengths.

radio star [ASTROPHYS] A discrete celestial radio source.

radio sun [ASTROPHYS] The sun as defined by its electromagnetic radiation in the radio portion of the spectrum.

radio telescope [ENG] An astronomical instrument used to measure the amount of radio energy coming from various directions in the sky, consisting of a highly directional antenna and associated electronic equipment.

radiotherapy *See* radiation therapy.

radiothorium [NUC PHYS] Conventional name of the isotope of thorium which has mass number 228. Symbolized RdTh.

radiotracer *See* radioactive tracer.

radio tube *See* electron tube.

radio wave [ELECTROMAG] An electromagnetic wave pro-

RADIO TELESCOPE

Installation of electronics at the focus of an 85-foot (26-meter) diameter radio telescope.

duced by reversal of current in a conductor at a frequency in the range from about 10 kilohertz to about 300,000 megahertz.

radio wavefront distortion [ELECTROMAG] Change in the direction of advance of a radio wave.

radio-wave propagation [ELECTROMAG] The transfer of energy through space by electromagnetic radiation at radio frequencies.

radio window [GEOPHYS] A band of frequencies extending from about 6 to 30,000 megahertz, in which radiation from the outer universe can enter and travel through the atmosphere of the earth.

radium [CHEM] A radioactive member of group IIA, symbol Ra, atomic number 88; the most abundant naturally occurring isotope has mass number 226 and a half-life of 1620 years. A highly toxic solid that forms water-soluble compounds; decays by emission of α, β, and γ-radiation; melts at 700°C, boils at 1140°C; turns black in air; used in medicine, in industrial radiography, and as a source of neutrons and radon.

radium age [NUCLEO] The age of a mineral as calculated from the numbers of radium atoms present originally, now, and when equilibrium is established with ionium.

radium cell [NUCLEO] A sealed thin-wall tube containing radium.

radium F *See* polonium-210.

radium needle [NUCLEO] A radium cell in the form of a needle, usually of platinum-iridium or gold alloy, designed primarily for insertion in tissue.

radium plaque [NUCLEO] A radium container in which the radium is distributed over a surface; the shielding is usually small in one direction so as to permit transmission of β-rays as well as γ-rays.

radius [MATH] **1.** A line segment joining the center and a point of a circle or sphere. **2.** The length of such a line segment.

radius of convergence [MATH] The positive real number corresponding to a power series expansion about some number a with the property that if $x - a$ has absolute value less than this number the power series converges at x, and if $x - a$ has absolute value greater than this number the power series diverges at x.

radius of curvature [MATH] The radius of the circle of curvature at a point of a curve.

radius of gyration [MATH] The square root of the ratio of the moment of inertia of a plane figure about a given axis to its area. [MECH] The square root of the ratio of the moment of inertia of a body about a given axis to its mass.

radius of spherical curvature [MATH] The radius of the osculating sphere at a specified point on a space curve.

radius ratio [PHYS CHEM] The ratio of the radius of a cation to the radius of an ion; relative ionic radii are pertinent to crystal lattice structure, particularly the determination of coordination number.

radius vector [ASTRON] A line joining the center of an orbiting body with the focus of its orbit located near its primary.

radix *See* base of a number system; root.

radix complement [MATH] A numeral in positional notation that can be derived from another by subtracting the original numeral from the numeral of highest value with the same number of digits, and adding 1 to the difference. Also known as complement; true complement.

radix-minus-one complement [MATH] A numeral in positional notation of base (or radix) B derived from a given numeral by subtracting the latter from the highest numeral

RADIO-WAVE PROPAGATION

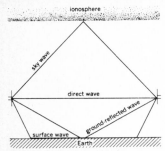

Possible transmission paths of electromagnetic radiation at radio frequencies.

RADIUM

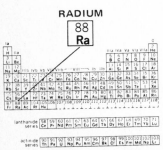

Periodic table of the chemical elements showing the position of radium.

with the same number of digits, that is, from $B - 1$; it is 1 less than the radix complement.

radix notation [MATH] A positional notation in which the successive digits are interpreted as coefficients of successive integral powers of a number called the radix or base; the represented number is equal to the sum of this power series. Also known as base notation.

radix point [MATH] A dot written either on or slightly above the line, used to mark the point at which place values change from positive to negative powers of the radix in a number system; a decimal point is a radix point for radix 10.

radon [CHEM] A chemical element, symbol Rn, atomic number 86; all isotopes are radioactive, the longest half-life being 3.82 days for mass number 222; it is the heaviest element of the noble-gas group, produced as a gaseous emanation from the radioactive decay of radium. [NUC PHYS] The conventional name for radon-222. Symbolized Rn.

radon-220 [NUC PHYS] The isotope of radon having mass number 220, symbol ^{220}Ra, which is a radioactive member of the thorium series with a half-life of 56 seconds.

radon-222 [NUC PHYS] The isotope of radon having mass number 222, symbol ^{222}Ra, which is a radioactive member of the uranium series with a half-life of 3.82 days.

Radon measure *See* regular Borel measure.

rainbow [OPTICS] Colored arc seen in the sky when the sun or moon is illuminating large numbers of falling raindrops.

Raman effect [OPTICS] A phenomenon observed in the scattering of light as it passes through a transparent medium; the light undergoes a change in frequency and a random alteration in phase due to a change in rotational or vibrational energy of the scattering molecules. Also known as Raman scattering.

Raman-Rayleigh ratio [OPTICS] The ratio of Raman scattering (of a light beam passing through a transparent medium) to Rayleigh scattering (of a light beam horizontal to the medium).

Raman scattering *See* Raman effect.

Raman spectrophotometry [SPECT] The study of spectral-line patterns on a photograph taken at right angles through a substance illuminated with a quartz mercury lamp.

Raman spectroscopy [SPECT] Analysis of the intensity of Raman scattering of monochromatic light as a function of frequency of the scattered light.

Raman spectrum [SPECT] A display, record, or graph of the intensity of Raman scattering of monochromatic light as a function of frequency of the scattered light.

Ramanujan's τ-function [MATH] The function $\tau(n)$, $n = 1, 2, 3, \ldots$, that satisfies the equation

$$x[(1 - x)(1 - x^2) \cdots]^{24} = \sum_{n=1}^{\infty} \tau(n)x^n.$$

ramp generator [ELECTR] A circuit that generates a sweep voltage which increases linearly in value during one cycle of sweep, then returns to zero suddenly to start the next cycle.

ramphoid cusp [MATH] A cusp of a curve which has both branches of the curve on the same side of the common tangent. Also known as single cusp of the second kind.

Ramsauer effect [ATOM PHYS] The vanishing of the scattering cross section of electrons from atoms of a noble gas at some value of the electron energy, always below 25 electron volts.

Ramsay-Shields-Eötvös equation [THERMO] An elaboration of the Eötvös rule which states that at temperatures not too near the critical temperature the molar surface energy of a

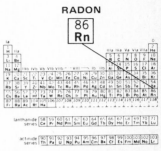

RADON

Periodic table of the chemical elements showing the position of radon.

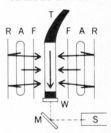

RAMAN EFFECT

Arrangement for excitation of the Raman effect with a noncoherent source. Mercury vapor arcs A are surrounded by appropriate reflectors R to increase intensity of available radiation. Light from arcs passes through filters F which transmit one of several monochromatic frequencies of mercury radiation. Monochromatic radiation enters circular cylindrical tube T in which liquid or vapor is to be studied. Light scattered in direction of vertical arrow passes through window W and to the spectrograph S through optical system M.

liquid is proportional to $t_c - t - 6$ K, where t is the temperature and t_c is the critical temperature.

Ramsay-Young method [THERMO] A method of measuring the vapor pressure of a liquid, in which a thermometer bulb is surrounded by cotton wool soaked in the liquid, and the pressure, measured by a manometer, is reduced until the thermometer reading is steady.

Ramsay-Young rule [THERMO] An empirical relationship which states that the ratio of the absolute temperatures at which two chemically similar liquids have the same vapor pressure is independent of this vapor pressure.

Ramsden circle [OPTICS] A sharp, bright circle of light which appears on a sheet of white paper held near the eyepiece of a telescope focused for infinity and pointed at a bright sky. Also known as Ramsden disk.

Ramsden disk *See* Ramsden circle.

Ramsden eyepiece [OPTICS] An eyepiece consisting of two planoconvex lenses with their plane sides facing outward, having the same power and focal length, and separated by a distance equal to their common focal length.

Ramsey fringes [PHYS] Oscillations in the number of transitions in a molecular beam that passes through two separated radio-frequency fields, as a function of L/v, where L is the separation of the field and v is the speed of the beam.

Ramsey number [MATH] The smallest number N for which Ramsey's theorem is true for a given set of parameters $(r; r_1, \ldots, r_k)$.

Ramsey's theorem [MATH] For any parameters $(r; r_1, \ldots, r_k)$ with each $r_i \geq r$, there exists a number N such that for all $n \geq N$, if S is an n-element set all of whose r-element subsets are partitioned into blocks $A_1, \ldots, A_k$, then there is some r_i element subset S' of S such that all of the r-element subsets of S' are in the block A_i.

random diffusion chamber *See* reverberation chamber.

random digit [STAT] Digit taken from a table of random numbers according to some specified probability rule.

random error [STAT] An error that can be predicted only on a statistical basis.

random experiments [STAT] Experiments which do not always yield the same result when repeated under the same conditions.

random function [MATH] A function whose domain is an interval of the extended real numbers and has range in the set of random variables on some probability space; more precisely, a mapping of the cartesian product of an interval in the extended reals with a probability space to the extended reals so that each section is a random variable.

random interstratification [SOLID STATE] A crystalline structure in which two or more types of layers alternate in a random fashion.

randomization [STAT] Assigning subjects to treatment groups by use of tables of random numbers.

randomized blocks [STAT] An experimental design in which the various treatments are reproduced in each of the blocks and are randomly assigned to the units within the blocks, permitting unbiased estimates of error to be made.

randomized test [STAT] Acceptance or rejection of the null hypothesis by use of a random variable to decide whether an observation causes rejection or acceptance.

random noise [MATH] A form of random stochastic process arising in control theory. [PHYS] Noise characterized by a large number of overlapping transient disturbances occurring at random, such as thermal noise and shot noise. Also known as fluctuation noise.

RAMSDEN EYEPIECE

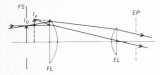

Ramsden eyepiece. FL = field lens; EL = eye lens; FS = field stop; EP = exit pupil or eye point; I_O = image formed by the preceding system; I_F = image formed by the preceding system and the field lens.

random numbers [MATH] A listing of numbers which is nonrepetitive and satisfies no algorithm.

random ordered sample [STAT] An ordered sample of size s drawn from a population of size N such that the probability of any particular ordered sample is the reciprocal of the number of permutations of N things taken s at a time.

random process *See* stochastic process.

random sampling [STAT] A sampling from some population where each entry has an equal chance of being drawn.

random-sampling voltmeter [ENG] A sampling voltmeter which takes samples of an input signal at random times instead of at a constant rate; the synchronizing portions of the instrument can then be simplified or eliminated.

random sequence [MATH] A random sequence $f = (f_1, f_2, \ldots)$ of random variables $f_1, f_2, \ldots$ on a probability space P is a measurable function from P into the space of numerical sequences that maps each element x in P into the sequence $f(x) = (f_1(x), f_2(x), \ldots)$.

random start [STAT] In a systematic sample, the random selection of a starting point in the first sample block followed by taking that value in the same position in every succeeding block.

random structure [CRYSTAL] A crystal structure in which different types of atoms are associated with the various points in a crystal lattice in a random fashion.

random variable [STAT] Any of the values of a variable that occur with some relative frequency; the values occur over a specific range or set of values. Also known as chance variable; stochastic variable; variate.

random vibration [MECH] A varying force acting on a mechanical system which may be considered to be the sum of a large number of irregularly timed small shocks; induced typically by aerodynamic turbulence, airborne noise from rocket jets, and transportation over road surfaces.

random walk [MATH] A succession of movements along line segments where the direction and possibly the length of each move is randomly determined.

range [MATH] The range of a function f from a set X to a set Y consists of those elements y in Y for which there is an x in X with $f(x) = y$. [MECH] The horizontal component of a projectile displacement at the instant it strikes the ground. [NUCLEO] The distance that a given ionizing particle can penetrate a given medium before its energy drops to the point that the particle no longer produces ionization. [PHYS] The greatest distance between two particles at which a given force between them is appreciable. [STAT] The difference between the maximums and minimums of a variable quantity.

rangefinder [ELECTR] A device which determines the distance to an object by measuring the time it takes for a radio wave to travel to the object and return. [ENG] *See* optical rangefinder.

rank [MATH] **1.** The rank of a matrix is its maximum number of linearly independent rows. **2.** The rank of a system of homogeneous linear equations equals the rank of the matrix of its coefficients. **3.** A tensor in an n-dimensional space is of rank r if it has n^r components. **4.** The rank of a group G is the number of elements in the basis of the quotient group of G over the subgroup consisting of all elements of G having finite period. **5.** The rank of a place or valuation is equal to the number of proper prime ideals in its valuation ring. **6.** The rank of a prime ideal P is the largest number n for which there exists a sequence $P_0 = P, P_1, P_2, \ldots, P_n$ of prime ideals such that P_i is a subset of P_{i-1}.

rank correlation [STAT] A nonparametric test of statistical dependence for a random sample of paired observations.

RANKINE CYCLE

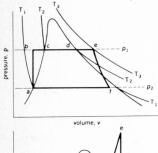

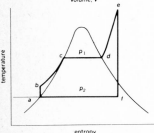

Rankine-cycle diagrams (pressure-volume and temperature-entropy) for a steam power plant using superheated steam. Typically, cycle has four phases: (1) heat addition b, c, d, e in a boiler at constant pressure p_1, changing water at b to superheated steam at e; (2) isentropic expansion e, f in a prime mover from initial pressure p_1 to back pressure p_2; (3) heat rejection f, a in a condenser at constant pressure p_2 with steam at f converted to a saturated liquid at a; (4) isentropic compression of a, b at water in feed pump from pressure p_2 to pressure p_1. T_1–T_3 are lines of constant temperature.

Rankine body [FL MECH] A fluid flow pattern formed by combining a uniform stream with a source and a sink of equal strengths, with the line joining the source and sink along the stream direction.

Rankine cycle [THERMO] An ideal thermodynamic cycle consisting of heat addition at constant pressure, isentropic expansion, heat rejection at constant pressure, and isentropic compression; used as an ideal standard for the performance of heat-engine and heat-pump installations operating with a condensable vapor as the working fluid, such as a steam power plant. Also known as steam cycle.

Rankine formula [MECH] An empirical formula,

$$F = \sigma A / [1 + \sigma l^2 / \pi^2 k^2 E],$$

giving the compression force F that causes a column of length l, cross-sectional area A, safe compressive stress σ, radius of gyration k, and Young's modulus E to collapse.

Rankine-Hugoniot equations [THERMO] Equations, derived from the laws of conservation of mass, momentum, and energy, which relate the velocity of a shock wave and the pressure, density, and enthalpy of the transmitting fluid before and after the shock wave passes.

Rankine temperature scale [THERMO] A scale of absolute temperature; the temperature in degrees Rankine (°R) is equal to $\frac{9}{5}$ of the temperature in kelvins and to the temperature in degrees Fahrenheit plus 459.67.

Rankine vortex [FL MECH] A vortex with a vertical axis and circular motion, in which the motion is that of a rotating solid cylinder inside some fixed radius, and the circulation is constant outside this radius.

rank of an observation [STAT] The number assigned to an observation if a collection of observations is ordered from smallest to largest and each observation is given the number corresponding to its place in the order.

rank-order statistics [STAT] Statistics computed from rankings of the observations rather than from the observations themselves.

rank tests [STAT] Tests which use the ranks of observations with respect to one another rather than the observations themselves.

Raoult's law [PHYS CHEM] The law that the vapor pressure of a solution equals the product of the vapor pressure of the pure solvent and the mole fraction of solvent.

rare-earth chelate laser *See* chelate laser.

rare-earth element [CHEM] The name given to any of the group of chemical elements with atomic numbers 58 to 71; the name is a misnomer since they are neither rare nor earths; examples are cerium, erbium, and gadolinium.

rare-earth magnet [ELECTROMAG] Any of several types of magnets made with rare-earth elements, such as rare-earth-cobalt magnets, which have coercive forces up to ten times that of ordinary magnets; used for computers and signaling devices.

rarefaction [ACOUS] The instantaneous, local reduction in density of a gas resulting from passage of a sound wave, or the region in which the density is reduced at some instant.

rarefaction wave [FL MECH] A pressure wave or rush of air or water induced by rarefaction; it travels in the opposite direction to that of a shock wave directly following an explosion. Also known as suction wave.

rarefied gas [FL MECH] A gas whose pressure is much less than atmospheric pressure.

Rarita-Schwinger equation [QUANT MECH] A partial differential equation, similar in form to the Dirac equation, relating the spatial and time dependence of a 16-component wave function describing a free relativistic particle with intrinsic spin $\frac{3}{2}$, and its antiparticle.

rate [SCI TECH] The amount of change of some quantity during a time interval divided by the length of the time interval.

rate action *See* derivative action.

Rateau formula [FL MECH] A formula, $m = A_2 p_1 (16.367 - 0.96 \log p_1)/1000$, for determining the discharge m of saturated steam in pounds per second through a well-rounded convergent orifice; A_2 is the area of the orifice in square inches, and p_1 the reservoir pressure in pounds per square inch.

rate effect [ELECTR] The phenomenon of a *pnpn* device switching to a high-conduction mode when anode voltage is applied suddenly or when high-frequency transients exist.

rate feedback [ELECTR] The return of a signal, proportional to the rate of change of the output of a device, from the output to the input.

rate-grown transistor [ELECTR] A junction transistor in which both impurities (such as gallium and antimony) are placed in the melt at the same time and the temperature is suddenly raised and lowered to produce the alternate *p*-type and *n*-type layers of rate-grown junctions. Also known as graded-junction transistor.

rate gyroscope [MECH ENG] A gyroscope that is suspended in just one gimbal whose bearings form its output axis and which is restrained by a spring; rotation of the gyroscope frame about an axis perpendicular to both spin and output axes produces precession of the gimbal within the bearings proportional to the rate of rotation.

rate integrating gyroscope [MECH ENG] A single-degree-of-freedom gyro having primarily viscous restraint of its spin axis about the output axis; an output signal is produced by gimbal angular displacement, relative to the base, which is proportional to the integral of the angular rate of the base about the input axis.

rate meter *See* counting rate meter.

rate of change *See* derivative.

rate of flow *See* flow rate.

rate process [PHYS] Any process in which the derivatives with respect to time of one or more variables, evaluated at any given time t_0, depend on the values of the variables at time t_0 and possibly at times earlier than t_0.

rate response [ENG] Quantitative expression of the output rate of a control system as a function of its input signal.

rate servomechanism *See* velocity servomechanism.

ratio [MATH] A ratio of two quantities or mathematical objects A and B is their quotient or fraction A/B.

ratio arm circuit [ELEC] Two adjacent arms of a Wheatstone bridge, designed so they can be set to provide a variety of indicated resistance ratios.

ratio control system [CONT SYS] Control system in which two process variables are kept at a fixed ratio, regardless of the variation of either of the variables, as when flow rates in two separate fluid conduits are held at a fixed ratio.

ratio estimator [STAT] A ratio of two random variables that is used as an estimator.

ratio meter [ENG] A meter that measures the quotient of two electrical quantities; the deflection of the meter pointer is proportional to the ratio of the currents flowing through two coils.

rational function [MATH] A function which is a quotient of polynomials.

rational horizon *See* celestial horizon.

rationalized units [ELEC] A system of electrical units, such as occurs in the International System, in which the factor of 4π is removed from the field equations and appears instead in the explicit expressions for the fields of a point charge and current element.

rational number [MATH] A number which is the quotient of two integers.

ratio of specific heats [PHYS CHEM] The ratio of specific heat at constant pressure to specific heat at constant volume, $\gamma = C_p/C_v$.

ratio of transformer [ELEC] Ratio of the number of turns in one winding of a transformer to the number of turns in the other, unless otherwise specified.

ratio resistor [ELEC] One of the resistors in a Wheatstone or Kelvin bridge whose resistances appear in a pair of ratios which are equal in a balanced bridge.

ratio scale [STAT] A rule or system for assigning numbers to objects which has all the properties of an interval scale and, in addition, has a natural origin, so that ratios of numbers assigned to different objects have meaning.

raw score [STAT] Data as originally obtained from a test before any transformations are made.

ray [ASTRON] One of the broad streaks that radiate from some craters on the moon, especially Copernicus and Tycho; they consist of material of high reflectivity and are seen from earth best at full moon. [MATH] A straight-line segment emanating from a point. Also known as half line. [OPTICS] A curve whose tangent at any point lies in the direction of propagation of a light wave. [PHYS] A moving particle or photon of ionizing radiation.

ray acoustics [ACOUS] The study of the behavior of sound under the assumption that sound traversing a homogeneous medium travels along straight lines or rays. Also known as geometrical acoustics.

Raychuraduri equation [RELAT] An equation of general relativity theory, useful in proving singularity theorems, that relates the expansion, convergence, and shear of a congruence of timelike or null curves to the amount of matter present.

ray ellipsoid *See* Fresnel ellipsoid.

rayl [ACOUS] A unit of specific acoustical impedance, equal to a sound pressure of 1 dyne per square centimeter divided by a sound particle velocity of 1 centimeter per second. Also known as specific acoustical ohm (Ω_s); unit-area acoustical ohm.

rayleigh [OPTICS] A unit of brightness, used to measure the brightness of the night sky and aurorae, equal to $10^{10}/4\pi$ quanta per square meter per second per steradian.

Rayleigh balance [ELECTROMAG] An apparatus for assigning the value of the ampere in which the force exerted on a movable circular coil by larger circular coils above and below, but coaxial with, the movable coil is compared with the gravitational force on a known mass.

Rayleigh criterion [OPTICS] A criterion for the resolving power of an optical instrument which states that the images of two point objects are resolved when the principal maximum of the diffraction pattern of one falls exactly on the first minimum of the diffraction pattern of the other.

Rayleigh cycle [ELECTROMAG] A cycle of magnetization that does not extend beyond the initial portion of the magnetization curve, between zero and the upward bend.

Rayleigh disk [ACOUS] An acoustic radiometer, used to measure particle velocity, consisting of a thin disk set at an angle of 45° to a sound beam; the particle velocity is calculated from the resulting torque on the disk.

Rayleigh distribution [STAT] A normal distribution of two uncorrelated variates with the same variance.

Rayleigh flow [FL MECH] An idealized type of gas flow in which heat transfer may occur, satisfying the assumptions that the flow takes place in constant-area cross section and is frictionless and steady, that the gas is perfect and has constant specific heat, that the composition of the gas does not change, and that there are no devices in the system which deliver or receive mechanical work.

Rayleigh function [MATH] One of the functions $\sigma_{2n}(\nu)$, $n = 1, 2, 3, \ldots$, with value equal to the sum over $m = 1, 2, 3, \ldots$ of $j_{\nu,m}^{-2n}$, where $j_{\nu,m}$ is the mth positive zero of J_ν, the Bessel function of the first kind of order ν.

Rayleigh interferometer [OPTICS] An optical interferometer in which two rays of light, emanating from a single slit, are collimated by a lens, pass through separate slits and cells, and are brought to focus by a second lens so that interference fringes become visible. Also known as Rayleigh refractometer.

Rayleigh-Jansen method [FL MECH] A method for solving equations for compressible fluid flow past a body, in which the velocity potential for the difference between the fluid velocity and the velocity distant from the body (V) is expressed as a power series in the square of the Mach number corresponding to V.

Rayleigh-Jeans law [STAT MECH] A law giving the intensity of radiation emitted by a blackbody within a narrow band of wavelengths; it states that this intensity is proportional to the temperature divided by the fourth power of the wavelength; it is a good approximation to the experimentally verified Planck radiation formula only at long wavelengths.

Rayleigh law [ELECTROMAG] **1.** For small values of the magnetic field strength H, the normal permeability of a material is approximated by $a + bH$, where a is the initial permeability and b is a constant. **2.** In a magnetic material subject to cyclic magnetization, with maximum magnetic field strength small compared with the coercive force, the hysteresis loss per cycle is proportional to the cube of the maximum value of the magnetic induction. [OPTICS] In Rayleigh scattering, the intensity of light scattered in a direction making an angle θ with the incident direction is proportional to $1 + \cos^2 \theta$ and inversely proportional to the fourth power of the wavelength of the incident radiation.

Rayleigh line [MECH] A straight line connecting points corresponding to the initial and final states on a graph of pressure versus specific volume for a substance subjected to a shock wave. [SPECT] Spectrum line in scattered radiation which has the same frequency as the corresponding incident radiation.

Rayleigh loop [ELECTROMAG] A parabolic approximation to a magnetic hysteresis loop.

Rayleigh number 1 [FL MECH] A dimensionless number used in studying the breakup of liquid jets, equal to Weber number 2. Symbolized N_{Ra_1}.

Rayleigh number 2 [THERMO] A dimensionless number used in studying free convection, equal to the product of the Grashof number and the Prandtl number. Symbolized R'_2.

Rayleigh number 3 [THERMO] A dimensionless number used in the study of combined free and forced convection in vertical tubes, equal to Rayleigh number 2 times the Nusselt number times the tube diameter divided by its entry length. Symbolized Ra_3.

Rayleigh polynomials [MATH] The polynomials $\phi_{2n}(\nu)$, $n = 1, 2, 3, \ldots$, given by the product over k from 1 to n of

RAYLEIGH INTERFEROMETER

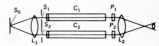

Schematic drawing of the Rayleigh interferometer. S_1, S_2 are slits illuminated with parallel light from lens L_1 whose focus is a single narrow slit S_0. Second lens L_2 brings the two beams to a focus, where interference fringes become visible. Cells (C_1, C_2) are placed in front of each slit; material of interest is placed in one cell and reference material in other. One of the plates of glass (P_1, P_2) is tilted until effect of cells on interference fringes is cancelled.

RAYLEIGH PRISM

Rayleigh prism system; the arrow shows the path of light through the prism.

$4^n(\nu + k)^{[n/k]}\sigma_{2n}(\nu)$, where $[n/k]$ is the greatest integer less than or equal to n/k, and $\sigma_{2n}(\nu)$ is a Rayleigh function.

Rayleigh prism   [OPTICS] A system of prisms used to produce greater dispersion of light than would be produced by a single prism.

Rayleigh quotient   [MATH] For a matrix **A** and a vector **x**, the quantity $x^\dagger A x / x^\dagger x$, where $x^\dagger$ is the Hermitian conjugate of **x**.

Rayleigh ratio   [OPTICS] Light-scattering relationship defined by the ratio of intensities of incident and scattered light at a specified distance; used in photometric and refractometric analyses.

Rayleigh reciprocity theorem   [ELECTROMAG] Reciprocal relationship for an antenna when it is transmitting or receiving; the effective heights, radiation resistance, and the radiation pattern are alike, whether the antenna is transmitting or receiving.

Rayleigh refractometer   *See* Rayleigh interferometer.

Rayleigh-Ritz method   [MATH] An approximation method for finding solutions of functional equations in terms of finite systems of equations.

Rayleigh scattering   [ELECTROMAG] Scattering of electromagnetic radiation by independent particles which are much smaller than the wavelength of the radiation.

Rayleigh's dissipation function   [MECH] A function which enters into the equations of motion of a system undergoing small oscillations and represents frictional forces which are proportional to velocities; given by a positive definite quadratic form in the time derivatives of the coordinates. Also known as dissipation function.

Rayleigh's equation for group velocity   [PHYS] The equation $c' = c - \lambda\, dc/d\lambda$, expressing the group velocity c' of a wave packet in terms of the phase velocity c and the wavelength λ.

Rayleigh's principle   [MATH] A principle useful in calculating eigenvalues of matrices, which states that the Rayleigh quotient of a Hermitian matrix **A** for a vector **x** is stationary when **x** is an eigenvector of **A**.

Rayleigh-Taylor instability   [FL MECH] The instability of the interface separating two fluids having different densities when the lighter fluid is accelerated toward the heavier fluid.

Rayleigh wave   [GEOPHYS] In seismology, a surface wave with a retrograde, elliptical motion at the free surface. Also known as R wave.   [MECH] A wave which propagates on the surface of a solid; particle trajectories are ellipses in planes normal to the surface and parallel to the direction of propagation. Also known as surface wave.

ray surface   [OPTICS] The locus of points reached in a unit time in an anisotropic medium by an electromagnetic disturbance that starts from the origin.

ray tracing   [OPTICS] Calculation of the paths followed by rays of light through an optical system, using Snell's law and trigonometrical formulas.

Rb   *See* rubidium.

RBE   *See* relative biological effectiveness.

R-C amplifier   *See* resistance-capacitance coupled amplifer.

R-C circuit   *See* resistance-capacitance circuit.

R-C constant   *See* resistance-capacitance constant.

R-C coupled amplifier   *See* resistance-capacitance coupled amplifier.

R center   [SOLID STATE] A color center whose absorption band lies between the F band and M band, and which is produced by prolonged irradiation with light in the F band or prolonged x-ray exposure at room temperature. Also known as D center; E center.

R-C network   *See* resistance-capacitance network.

R-C oscillator   *See* resistance-capacitance oscillator.

rd   *See* rutherford.

Re   *See* rhenium.

reactance   [ELEC] The imaginary part of the impedance of an alternating-current circuit.

reactance amplifier   *See* parametric amplifier.

reactance drop   [ELEC] The component of the phasor representing the voltage drop across a component or conductor of an alternating-current circuit which is perpendicular to the current.

reaction   [CONT SYS] *See* positive feedback.   [MECH] The equal and opposite force which results when a force is exerted on a body, according to Newton's third law of motion.   [NUC PHYS] *See* nuclear reaction.

reactive   [ELEC] Pertaining to either inductive or capacitive reactance; a reactive circuit has a high value of reactance in comparison with resistance.

reactive component   [ELEC] In the phasor representation of quantities in an alternating-current circuit, the component of current, voltage, or apparent power which does not contribute power, and which results from inductive or capacitive reactance in the circuit, namely, the reactive current, reactive voltage, or reactive power. Also known as idle component; quadrature component; wattless component.

reactive current   [ELEC] In the phasor representation of alternating current, the component of the current perpendicular to the voltage, which contributes no power but increases the power losses of the system. Also known as idle current; quadrature current; wattless current.

reactive factor   [ELEC] The ratio of reactive power to apparent power.

reactive load   [ELEC] A load having inductive or capacitive reactance.

reactive power   [ELEC] The power value obtained by multiplying together the effective value of current in amperes, the effective value of voltage in volts, and the sine of the angular phase difference between current and voltage. Also known as wattless power.

reactive voltage   [ELEC] In the phasor representation of alternating current, the voltage component that is perpendicular to the current.

reactive volt-ampere hour   *See* var hour.

reactivity   [NUCLEO] A measure of the deviation of a nuclear reactor from the critical state at any instant of time such that positive and negative values correspond to reactors above and below critical, respectively; measured in percent k, millikays, dollars, or inhours.

reactor   [ELEC] A device that introduces either inductive or capacitive reactance into a circuit, such as a coil or capacitor. Also known as electric reactor.   [NUC PHYS] *See* nuclear reactor.

reactor fuel   *See* nuclear fuel.

reactor fuel cycle   [NUCLEO] The processes of preparing fuel elements and assemblies for use in a reactor, using these elements in reactor operation, recovering radioactive by-products from spent fuel, and reprocessing remaining fissionable material into new fuel elements. Also known as fuel cycle; nuclear fuel cycle.

reactor fuel element   *See* nuclear fuel element.

reactor physics   [NUCLEO] The science of the interaction of the elementary particles and radiations characteristic of nuclear reactors with matter in bulk.

reactor vessel   [NUCLEO] A large tanklike structure built to prevent radioactive materials from escaping from the reactor and associated equipment.

Read diode   [ELECTR] A high-frequency semiconductor di-

REACTIVE POWER

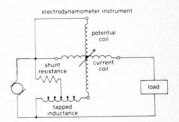

Diagram of a circuit for single-phase reactive power measurement which uses a tapped inductance. This inductance and shunt resistance are adjusted until currents in potential coil and current coil are in quadrature.

ode consisting of an avalanching *pn* junction, biased to fields of several hundred thousand volts per centimeter, at one end of a high-resistance carrier serving as a drift space for the charge carriers.

real axis [MATH] The horizontal axis of the cartesian coordinate system for the euclidean or complex plane.

real closed field [MATH] A real field which has no algebraic extensions other than itself.

real closure [MATH] A real closure of a real field F is a real closed field which is an algebraic extension of F.

real crystal [CRYSTAL] A crystal for which the finite extent of the crystal and its various imperfections and defects are taken into account.

real field [MATH] A field F such that -1 is not equal to a sum of squares of elements in F.

real fluid flow [FL MECH] The flow in which effects of tangential or shearing forces are taken into account; these forces give rise to fluid friction, because they oppose the sliding of one particle past another.

real gas [THERMO] A gas, as considered from the viewpoint in which deviations from the ideal gas law, resulting from interactions of gas molecules, are taken into account. Also known as imperfect gas.

real image [OPTICS] An optical image such that all the light from a point on an object that passes through an optical system actually passes close to or through a point on the image.

realizability [CONT SYS] Property of a transfer function that can be realized by a network that has only resistances, capacitances, inductances, and ideal transformers.

realization of a stochastic process [STAT] A probability space whose points are sample paths of the stochastic process and whose probability is obtained from the joint probability distributions of the random variables in the process.

real line [MATH] The set of all real numbers.

real number [MATH] Any member of the unique (to within isomorphism) complete ordered field.

real object [OPTICS] A collection of points which actually serves as a source of light rays in an optical system.

real orthogonal group [MATH] The group composed of orthogonal matrices having real number entries.

real part [MATH] The real part of a complex number $z = x + iy$ is the real number x.

real power [ELEC] The component of apparent power that represents true work; expressed in watts, it is equal to voltamperes multiplied by the power factor.

real source [ACOUS] A source of sound consisting of a macroscopic body that is composed of materials different from those of the medium in which the sound propagates and has sharply delineated physical extent, and which generates sound by executing complex motions while immersed in the medium.

real unimodular group [MATH] The group of all square $n \times n$ matrices with real number entries and of determinant 1.

real variable [MATH] A variable that assumes real numbers for its values.

rearrangement [MATH] A rearrangement of the series

$$\sum_{i=1}^{\infty} a_i$$

is a series

$$\sum_{i=1}^{\infty} a_{\sigma(i)}$$

where σ is a permutation of the set of positive integers.

REAL IMAGE

Formation of real image. Rays leaving object point Q and passing through the refracting surface separating media n and n' are brought to a focus at the image point Q'. (*Modified from F. A. Jenkins and H. E. White, Fundamentals of Optics, 3d ed., McGraw-Hill, 1957*)

rearrangement reaction [NUC PHYS] A nuclear reaction in which nucleons are exchanged between nuclei.

Réaumur temperature scale [THERMO] Temperature scale where water freezes at 0°R and boils at 80°R.

receiver [ELECTR] The complete equipment required for receiving modulated radio waves and converting them into the original intelligence, such as into sounds or pictures, or converting to desired useful information as in a radar receiver.

receiver radiation [ELECTROMAG] Radiation of interfering electromagnetic fields by the oscillator of a receiver.

receiving area [ELECTROMAG] The factor by which the power density must be multiplied to obtain the received power of an antenna, equal to the gain of the antenna times the square of the wavelength divided by 4π.

receiving set *See* radio receiver.

recession of galaxies [ASTROPHYS] The increase in the velocity of recession (red shift) of galaxies with distance from an observer on earth.

rechargeable battery *See* storage battery.

reciprocal [MATH] The reciprocal of a number A is the number $1/A$.

reciprocal differences [MATH] An interpolation technique using successive quotients of a function with its values so as to obtain a continued fraction expansion approximating the given function by a rational function.

reciprocal ellipsoid *See* index ellipsoid.

reciprocal ferrite switch [ELECTROMAG] A ferrite switch that can be inserted in a waveguide to switch an input signal to either of two output waveguides; switching is done by a Faraday rotator when acted on by an external magnetic field.

reciprocal impedance [ELEC] Two impedances Z_1 and Z_2 are said to be reciprocal impedances with respect to an impedance Z (invariably a resistance) if they are so related as to satisfy the equation $Z_1Z_2 = Z^2$.

reciprocal junction [ELECTROMAG] A waveguide junction in which the transmission coefficient from the ith port to the jth port is the same as that from the jth port to the ith port; that is, the S matrix is symmetrical.

reciprocal lattice [CRYSTAL] A lattice array of points formed by drawing perpendiculars to each plane (hkl) in a crystal lattice through a common point as origin; the distance from each point to the origin is inversely proportional to spacing of the specific lattice planes; the axes of the reciprocal lattice are perpendicular to those of the crystal lattice.

reciprocal ohm *See* siemens.

reciprocal ohm centimeter *See* roc.

reciprocal ohm meter *See* rom.

reciprocal polar figures [MATH] Two plane figures consisting of lines and their points of intersection such that the points of each of them are the poles of the lines of the other with respect to a given conic.

reciprocal space *See* wave-vector space.

reciprocal spiral *See* hyperbolic spiral.

reciprocal vectors [CRYSTAL] For a set of three vectors forming the primitive translations of a lattice, the vectors that form the primitive translations of the reciprocal lattice.

reciprocal velocity region [NUC PHYS] The energy region in which the capture cross section for neutrons by a given element is inversely proportional to neutron velocity.

reciprocal wavelength *See* wave number.

reciprocity theorem Also known as principle of reciprocity. [ACOUS] The theorem that, in an acoustic system consisting of a fluid medium with boundary surfaces and subject to no impressed body forces, if p_1 and p_2 are the pressure fields produced respectively by the components of the fluid velocities V_1 and V_2 normal to the boundary surfaces, then the integral over the boundary surfaces of $p_1V_2 - p_2V_1$ vanishes. [ELEC] **1.** The electric potentials V_1 and V_2 produced at some arbitrary point, due to charge distributions having total charges of q_1 and q_2 respectively, are such that $q_1V_2 = q_2V_1$. **2.** In an electric network consisting of linear passive impedances, the ratio of the electromotive force introduced in any branch to the current in any other branch is equal in magnitude and phase to the ratio that results if the positions of electromotive force and current are exchanged. [ELECTROMAG] Given two loop antennas, a and b, then $I_{ab}/V_a = I_{ba}/V_b$, where I_{ab} denotes the current received in b when a is used as transmitter, and V_a denotes the voltage applied in a; I_{ba} and V_b are the corresponding quantities when b is the transmitter, a the receiver; it is assumed that the frequency and impedances remain unchanged. [PHYS] In general, any theorem that expresses various reciprocal relations for the behavior of some physical system in which input and output can be interchanged without altering the response of the system to a given excitation.

recoil electron [PHYS] An electron that has been set into motion by a collision.

recoil particle [PHYS] A particle that has been set into motion by a collision or by a process involving the ejection of another particle.

recombination [PHYS] The combination and resultant neutralization of particles or objects having unlike charges, such as a hole and an electron or a positive ion and a negative ion.

recombination coefficient [ELECTR] The rate of recombination of positive ions with electrons or negative ions in a gas, per unit volume, divided by the product of the number of positive ions per unit volume and the number of electrons or negative ions per unit volume.

recombination electroluminescence *See* injection electroluminescence.

recombination energy [PHYS] The energy released when two oppositely charged portions of an atom or molecule rejoin to form a neutral atom or molecule.

recombination radiation [SOLID STATE] The radiation emitted in semiconductors when electrons in the conduction band recombine with holes in the valence band.

recombination velocity [ELECTR] On a semiconductor surface, the ratio of the normal component of the electron (or hole) current density at the surface to the excess electron (or hole) charge density at the surface.

recontrol time *See* deionization time.

recording thermometer *See* thermograph.

recoupling [QUANT MECH] A transformation between eigenfunctions of total angular momentum resulting from coupling eigenfunctions of three or more angular momenta in some order, and eigenfunctions of total angular momentum resulting from coupling of the same eigenfunctions in a different order.

recoverable shear [FL MECH] Measure of the elastic content of a fluid, related to elastic recovery (mechanicallike property of elastic recoil); found in unvulcanized, unfilled natural rubber, and certain polymer solutions, soap gels, and biological fluids.

recovery [MECH] The return of a body to its original dimensions after it has been stressed, possibly over a considerable period of time.

recovery temperature *See* adiabatic recovery temperature.

recovery time [ELECTR] **1.** The time required for the control electrode of a gas tube to regain control after anode-current interruption. **2.** The interval required, after a sudden decrease in input signal amplitude to a system or component, to attain a specified percentage (usually 63%) of the ultimate change in amplification or attenuation due to this decrease. [NUCLEO] The minimum time from the start of a counted pulse to the instant a succeeding pulse can attain a specific percentage of the maximum value of the counted pulse in a Geiger counter.

rectangle [MATH] A plane quadrilateral having four interior right angles and opposite sides of equal length.

rectangular cartesian coordinate system *See* cartesian coordinate system.

rectangular cavity [ELECTROMAG] A resonant cavity having the shape of a rectangular parallelepiped.

rectangular distribution *See* uniform distribution.

rectangular game *See* matrix game.

rectangular hyperbola [MATH] A hyperbola whose major and minor axes are equal.

rectangular parallelepiped [MATH] A parallelepiped with bases as rectangles all perpendicular to its lateral faces. Also known as rectangular solid.

rectangular pulse [ELECTR] A pulse in which the wave amplitude suddenly changes from zero to another value at which it remains constant for a short period of time, and then suddenly changes back to zero.

rectangular solid *See* rectangular parallelepiped.

rectangular wave [ELECTR] A periodic wave that alternately and suddenly changes from one to the other of two fixed values. Also known as rectangular wave train.

rectangular waveguide [ELECTROMAG] A waveguide having a rectangular cross section.

rectangular wave train *See* rectangular wave.

rectifiable curve [MATH] A curve whose length can be computed and is finite.

rectification [ELEC] The process of converting an alternating current to a unidirectional current.

rectification factor [ELECTR] Quotient of the change in average current of an electrode by the change in amplitude of the alternating sinusoidal voltage applied to the same electrode, the direct voltages of this and other electrodes being maintained constant.

rectified value of an alternating quantity [ELEC] Average of all the positive (or negative) values of the quantity during an integral number of periods.

rectifier [ELEC] A nonlinear circuit component that allows more current to flow in one direction than the other; ideally, it allows current to flow in one direction unimpeded but allows no current to flow in the other direction.

rectifier instrument [ENG] Combination of an instrument sensitive to direct current and a rectifying device whereby alternating current (or voltages) may be rectified for measurement.

rectifier rating [ELECTR] A performance rating for a semiconductor rectifier, usually on the basis of the root-mean-square value of sinusoidal voltage that it can withstand in the reverse direction and the average current density that it will pass in the forward direction.

rectifier stack [ELECTR] A dry-disk rectifier made up of layers or stacks of disks of individual rectifiers, as in a selenium rectifier or copper-oxide rectifier.

rectifying plane [MATH] The plane that contains the tangent and binormal to a curve at a given point on the curve.

rectilinear [MATH] Consisting of or bounded by lines.

rectilinear generators [MATH] Straight lines which generate ruled surfaces.

RECTANGLE

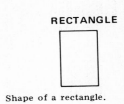

Shape of a rectangle.

RECTILINEAR MOTION

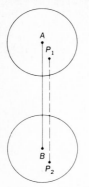

Diagram of rectilinear motion. Motion of center of mass of a body from A to B is along the straight line connecting these points. Path of any particle in the body is a straight line parallel to or coinciding with AB, such as the straight line connecting P_1 and P_2.

rectilinear lens [OPTICS] A lens that is free from distortion, imaging straight lines onto straight lines regardless of their orientation.

rectilinear motion [MECH] A continuous change of position of a body so that every particle of the body follows a straight-line path. Also known as linear motion.

rectilinear scanning [ELECTR] Process of scanning an area in a predetermined sequence of narrow parallel strips.

rectilinear system *See* orthoscopic system.

recurrence formula methods [MATH] Methods of calculating numerical solutions of differential equations in which the equation is written in the form of a recurrence relation between values of the solution function at successive points by replacing the derivatives with corresponding finite difference expressions.

recurrent transformation [MATH] **1.** A measurable function from a measure space T to itself such that for every measurable set A in the space and every point x in A there is a positive integer n such that $T^n(x)$ is also in A. **2.** A continuous function from a topological space T to itself such that for every open set A in the space and every point x in A there is a positive integer n such that $T^n(x)$ is also in A.

recurring decimal *See* repeating decimal.

recursion formula [MATH] An algorithm allowing computation of a succession of quantities. Also known as recursion relation.

recursion relation *See* recursion formula.

recursive functions [MATH] Functions that can be obtained by a finite number of operations, computations, or algorithms.

recycling [ELECTR] Returning to an original condition, as to 0 or 1 in a counting circuit. [NUCLEO] Reuse of fissionable nuclear reactor fuel by chemical processing, reenriching, and refabricating into new fuel elements.

red [OPTICS] The hue evoked in an average observer by monochromatic radiation having a wavelength in the approximate range from 622 to 770 nanometers; however, the same sensation can be produced in a variety of other ways.

red dwarf star [ASTRON] A red star of low luminosity, so designated by E. Hertzsprung; dwarf stars are commonly those main-sequence stars fainter than an absolute magnitude of +1, and red dwarfs are the faintest and coolest of the dwarfs.

red giant star [ASTRON] A star whose evolution has progressed to the point where hydrogen core burning has been completed, the helium core has become denser and hotter than originally, and the envelope has expanded to perhaps 100 times its initial size.

redox cell [ELEC] Cell designed to convert the energy of reactants to electrical energy; an intermediate reductant, in the form of liquid electrolyte, reacts at the anode in a conventional manner; it is then regenerated by reaction with a primary fuel.

red shift [ASTROPHYS] A systematic displacement toward longer wavelengths of lines in the spectra of distant galaxies and also of the continuous portion of the spectrum; increases with distance from the observer. Also known as Hubble effect.

Red Spot [ASTRON] A semipermanent marking of the planet Jupiter; some fluctuations in visibility exist; it does not rotate uniformly with the planet, indicating that it is a disturbance of Jupiter's atmosphere.

reduced Compton wavelength [QUANT MECH] The Compton wavelength of a particle divided by 2π.

RED SPOT

Appearance of Red Spot on Jupiter in 1933. *(After L. Rudaux)*

reduced distance [OPTICS] A distance in a medium divided by the medium's index of refraction.

reduced equation of state [PHYS] An equation relating the reduced pressure, reduced volume, and reduced temperature of a substance.

reduced frequency *See* Strouhal number.

reduced inspection [IND ENG] The decrease in the number of items inspected from that specified in the original sampling plan because the quality of the item has consistently improved.

reduced mass [MECH] For a system of two particles with masses m_1 and m_2 exerting equal and opposite forces on each other and subject to no external forces, the reduced mass is the mass m such that the motion of either particle, with respect to the other as origin, is the same as the motion with respect to a fixed origin of a single particle with mass m acted on by the same force; it is given by $m = m_1m_2/(m_1 + m_2)$.

reduced pressure [THERMO] The ratio of the pressure of a substance to its critical pressure.

reduced property *See* reduced value.

reduced temperature [THERMO] The ratio of the temperature of a substance to its critical temperature.

reduced value [THERMO] The actual value of a quantity divided by the value of that quantity at the critical point. Also known as reduced property.

reduced volume [THERMO] The ratio of the specific volume of a substance to its critical volume.

reducible polynomial [MATH] A polynomial relative to some field which can be written as the product of two polynomials of degree at least 1.

reducing glass [OPTICS] A double-concave lens that reduces the apparent size of objects viewed through it; used by illustrators and painters to create an artificial sense of distance from their work.

reduction of observations [SCI TECH] The mathematical analysis of data from observations to obtain the desired information.

reduction potential [PHYS CHEM] The potential drop involved in the reduction of a positively charged ion to a neutral form or to a less highly charged ion, or of a neutral atom to a negatively charged ion.

reduction to the sun [ASTRON] In the spectroscopic determination of a star's radial motion referred to the sun, the correction that is needed to be applied to the observed radial velocity of the star to compensate for the motion of the earth with respect to the sun.

redundancy [MATH] A repetitive statement. [MECH] A statically indeterminate structure.

red white dwarf star [ASTRON] A star type that is considered an anomaly; these are objects 10,000 times fainter than the sun, with surface temperature below 4000 K so that surface radiation has cooled the star at an unexpectedly rapid rate.

reed [ENG] A thin bar of metal, wood, or cane that is clamped at one end and set into transverse elastic vibration, usually by wind pressure; used to generate sound in musical instruments, and as a frequency standard, as in a vibrating-reed frequency meter.

reference acoustic pressure [ACOUS] Magnitude of any complex sound that will produce a sound-level meter reading equal to that produced by a sound pressure of 0.0002 dyne per square centimeter at 1000 hertz. Also known as reference sound level.

reference frame *See* frame of reference.

reference noise [ELECTR] The power level used as a basis of comparison when designating noise power expressed in decibels above reference noise (dBrn); the reference usually used is 10^{-12} watt (-90 decibels above 1 milliwatt; dBm) at 1000 hertz.

reference sound level *See* reference acoustic pressure.

reference volume [ACOUS] The audio volume level that gives a reading of 0 VU (volume units) on a standard volume indicator; the sensitivity of the volume indicator is adjusted so reference volume or 0 is read when the instrument is connected across a 600-ohm resistance to which is delivered a power of 1 milliwatt at 1000 hertz.

refinement of a tower [MATH] A tower that can be obtained by inserting a finite number of subsets in the given tower.

reflectance *See* reflectivity.

reflectance spectrophotometry [SPECT] Measurement of the ratio of spectral radiant flux reflected from a light-diffusing specimen to that reflected from a light-diffusing standard substituted for the specimen.

reflected pressure [PHYS] The pressure from an explosion (especially an airburst bomb), which is reflected back from a solid object or surface, rather than dissipated in the air.

reflected ray [PHYS] A ray extending outward from a point of reflection.

reflected ultraviolet method [GRAPHICS] A method of ultraviolet photography in which an ultraviolet source is used and the camera is provided with a filter which permits only reflected ultraviolet light to reach the film.

reflected wave [PHYS] A wave reflected from a surface, discontinuity, or junction of two different media, such as the sky wave in radio, the echo wave from a target in radar, or the wave that travels back to the source end of a mismatched transmission line.

reflecting antenna [ELECTROMAG] An antenna used to achieve greater directivity or desired radiation patterns, in which a dipole, slot, or horn radiates toward a larger reflector which shapes the radiated wave to produce the desired pattern; the reflector may consist of one or two plane sheets, a parabolic or paraboloidal sheet, or a paraboloidal horn.

reflecting galvanometer *See* mirror galvanometer.

reflecting microscope [OPTICS] A microscope whose objective is composed of two mirrors, one convex and the other concave; its imaging properties are independent of the wavelength of light, allowing it to be used even for infrared and ultraviolet radiation.

REFLECTING PRISM

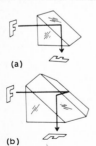

(a)

(b)

Examples of reflecting prism.
(a) Right-angle. *(b)* Amici roof prism.

reflecting prism [OPTICS] A prism used in place of a mirror for deviating light, usually designed so that there is no dispersion of light; the light undergoes at least one internal reflection.

reflecting spectrograph [OPTICS] A solar spectrograph in which the collimator and camera element are long-focus concave mirrors.

reflecting telescope [OPTICS] A telescope in which a concave parabolic mirror gathers light and forms a real image of an object. Also known as reflector telescope.

reflection [PHYS] The return of waves or particles from surfaces on which they are incident.

reflection angle *See* angle of reflection.

reflection coefficient [PHYS] The ratio of the amplitude of a wave reflected from a surface to the amplitude of the incident wave. Also known as coefficient of reflection.

reflection density [OPTICS] The common logarithm of the ratio of the luminance of a nonabsorbing perfect diffuser to that of the surface under consideration, when both are illuminated at an angle of 45° to the normal and the direction of measurement is perpendicular to the surface.

reflection diffraction [PHYS] Type of electron diffraction analysis in which the electron beam grazes the sample surface.

reflection law [PHYS] When a wave, such as electromagnetic radiation or sound, is reflected from a surface in a sharply defined direction, the reflected and incident waves travel in directions that make the same angle with a perpendicular to the surface and lie in a common plane with it. Also known as law of reflection.

reflection lobes [ELECTROMAG] Three-dimensional sections of the radiation pattern of a directional antenna, such as a radar antenna, which results from reflection of radiation from the earth's surface.

reflection nebula [ASTRON] A type of bright diffuse nebula composed mainly of cosmic dust; it is visible because of starlight from nearby stars or nebula stars that is scattered by the dust particles.

reflection plane *See* plane of mirror symmetry; plane of reflection.

reflection rainbow [OPTICS] A rainbow formed by light rays which have been reflected from an extended water surface; not to be confused with a reflected rainbow whose image may be seen in a still body of water.

reflection x-ray microscopy [ENG] A technique for producing enlarged images in which a beam of x-rays is successively reflected at grazing incidence, from two crossed cylindrical surfaces; resolution is about 0.5–1 micrometer.

reflective coating *See* mirror coating.

reflectivity [PHYS] The ratio of the energy carried by a wave which is reflected from a surface to the energy carried by the wave which is incident on the surface. Also known as reflectance.

reflectometer [ELECTROMAG] *See* microwave reflectometer. [ENG] A photoelectric instrument for measuring the optical reflectance of a reflecting surface.

reflector [ELECTR] *See* repeller. [ELECTROMAG] **1.** A single rod, system of rods, metal screen, or metal sheet used behind an antenna to increase its directivity. **2.** A metal sheet or screen used as a mirror to change the direction of a microwave radio beam. [NUCLEO] A layer of water, graphite, beryllium, or other scattering material placed around the core of a nuclear reactor to reduce the loss of neutrons. Also known as tamper.

reflector telescope *See* reflecting telescope.

reflex angle [MATH] An angle greater than 180° and less then 360°.

reflexive relation [MATH] A relation among the elements of a set such that every element stands in that relation to itself.

reflex klystron [ELECTR] A single-cavity klystron in which the electron beam is reflected back through the cavity resonator by a repelling electrode having a negative voltage; used as a microwave oscillator. Also known as reflex oscillator.

reflex oscillator *See* reflex klystron.

reflex sight [OPTICS] An optical or computing sight that reflects a reticle image or images onto a reflector plate for superimposition on the target by the eye.

refracted ray [PHYS] A ray extending onward from the point of refraction.

refracted wave [PHYS] That portion of an incident wave which travels from one medium into a second medium. Also known as transmitted wave.

refracting angle *See* apical angle.

refracting edge [OPTICS] The intersection of the two refracting faces of a prism.

refracting sphere [OPTICS] A sphere made of a transparent

REFLECTION NEBULA

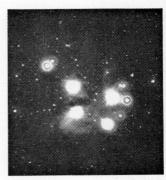

Reflection nebula associated with the Pleiades star cluster. Clouds with brushlike pattern are composed of small solid particles about 1 micrometer in diameter. Radiation pressure is probably responsible for streaked appearance of the clouds. Photographed with the Curtis-Schmidt telescope, University of Michigan.

REFLECTION X-RAY MICROSCOPY

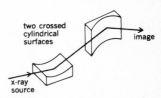

Principle for reflection x-ray microscopy.

material whose index of refraction differs from the medium surrounding it, so that it refracts light passing through it.

refracting telescope [OPTICS] A telescope in which a lens gathers light and forms a real image of an object. Also known as refractor telescope.

refraction [PHYS] The change of direction of propagation of any wave, such as an electromagnetic or sound wave, when it passes from one medium to another in which the wave velocity is different, or when there is a spatial variation in a medium's wave velocity.

refraction loss [ELECTROMAG] Portion of the transmission loss that is due to refraction resulting from nonuniformity of the medium.

refraction of lines of force [ELECTROMAG] The change in direction of lines of force of an electric or magnetic field at a boundary between media with different permittivities or permeabilities.

refractivity [ELECTROMAG] **1.** Some quantitative measure of refraction, usually a measure of the index of refraction. **2.** The index of refraction minus 1.

refractometer [ENG] An instrument used to measure the index of refraction of a substance in any one of several ways, such as measurement of the refraction produced by a prism, measurement of the critical angle, observation of an interference pattern produced by passing light through the substance, and measurement of the substance's dielectric constant.

refractometry [OPTICS] The measurement of the index of refraction of a substance; it is an important tool in analytical chemistry.

refractor telescope *See* refracting telescope.

refracture index *See* index of refraction.

refrangible [PHYS] Capable of being refracted.

refrigeration cycle [THERMO] A sequence of thermodynamic processes whereby heat is withdrawn from a cold body and expelled to a hot body.

RE galaxy [ASTRON] A type of ring galaxy that consists of a single, relatively empty, ringlike structure, without any prominent condensation or nucleus.

regelation [THERMO] Phenomenon in which ice (or any substance which expands upon freezing) melts under intense pressure and freezes again when this pressure is removed; accounts for phenomena such as the slippery nature of ice and the motion of glaciers.

regenerate [ELECTR] To restore pulses to their original shape.

regeneration [CONT SYS] *See* positive feedback. [NUCLEO] Restoration of contaminated nuclear fuel to a usable condition.

regenerative cooling [ENG] A method of cooling gases in which compressed gas is cooled by allowing it to expand through a nozzle, and the cooled expanded gas then passes through a heat exchanger where it further cools the incoming compressed gas.

regenerative cycle [THERMO] An engine cycle in which low-grade heat that would ordinarily be lost is used to improve the cyclic efficiency.

regenerative feedback *See* positive feedback.

regenerative reactor [NUCLEO] A nuclear reactor that produces fissionable material in addition to energy; when loaded with fissionable uranium-235, and nonfissionable uranium-238 or thorium, it converts the uranium-238 or thorium into fissionable materials which are then used as fuel in the core of the reactor.

Regge calculus [RELAT] A transcription of the equations of general relativity in terms of the properties of simplices approximating a given space-time.

Reggeism [PARTIC PHYS] An attempt to account for and correlate hadron resonances and the asymptotic behavior of scattering amplitudes of hadrons at high energies in terms of Regge poles.

Regge pole [PARTIC PHYS] A pole singularity of a scattering amplitude in the complex angular momentum plane; the scattering amplitude is formed by continuing partial wave amplitudes from positive integer values of the angular momentum to the complex plane.

Regge recurrence [PARTIC PHYS] One of a sequence of hadrons, with successive hadrons increasing by one in spin and also increasing in mass, but with the same values of other quantum numbers, except for parity, charge parity and G parity, which alternate in sign; it is believed that they are rotationally excited states of a particle, and that they alternate between two Regge trajectories.

Regge trajectory [PARTIC PHYS] **1.** The path followed by a Regge pole in the complex angular momentum plane as the center-of-mass energy is varied. **2.** The relationship between the spin and mass of a sequence of hadrons, with successive hadrons increasing by 2 in spin and also increasing in mass, but with the same values of other quantum numbers; the hadrons are thought to correspond to energies at which a Regge pole passes near positive integers (or half integers).

region *See* domain.

register ton *See* ton.

regression [STAT] Given two stochastically dependent random variables, regression functions measure the mean expectation of one relative to the other.

regression analysis [STAT] The description of the nature of the relationship between two or more variables; it is concerned with the problem of describing or estimating the value of the dependent variable on the basis of one or more independent variables.

regression coefficient [STAT] The coefficient of the independent variables in a regression equation.

regression curve [STAT] A plot of a regression equation; for two variables, the independent variable is plotted as the abscissa and the dependent variable as the ordinate; for three variables, a solid model can be constructed or the representation can be reduced by an isometric chart or stereogram.

regression estimate [STAT] An estimate of one variable obtained by substituting the known value of another variable in a regression equation calculated on sample values of the two variables.

regression line [STAT] A linear regression equation with two or more variables.

regression of nodes [ASTRON] The westward movement of the nodes of the moon's orbit; one cycle is completed in about 18.6 years.

regret criterion *See* Savage principle.

regula falsi [MATH] A method of calculating an unknown quantity by first making an estimate and then using this and the properties of the unknown to obtain it. Also known as rule of false position.

regular [ELECTROMAG] In a definite direction; not diffused or scattered, when applied to reflection, refraction, or transmission.

regular Baire measure [MATH] A Baire measure such that the measure of any Baire set E is equal to both the greatest lower bound of measures of open Baire sets containing E, and to the least upper bound of closed, compact sets contained in E.

regular Borel measure [MATH] A Borel measure such that the measure of any Borel set E is equal to both the greatest lower bound of measures of open Borel sets containing E, and to the least upper bound of measures of compact sets contained in E. Also known as Radon measure.

regular dodecahedron [CRYSTAL] *See* pyritohedron. [MATH] A regular polyhedron of 12 faces.

regular extension [MATH] An extension field K of a field F such that F is algebraically closed in K and K is separable over F; equivalently, an extension field K of a field F such that K and $\bar{F}$ are linearly disjoint over F, where $\bar{F}$ is the algebraic closure of F.

regular function [MATH] An analytic function of one or more complex variables.

regular icosohedron [MATH] A 20-sided regular polyhedron, having five equilateral triangles meeting at each face.

regularization [QUANT MECH] A formal procedure used to eliminate ambiguities which arise in evaluating certain integrals in a quantized field theory; corresponds to adding extra fields whose masses are allowed to approach infinity.

regular map *See* normal map.

regular octahedron [MATH] A regular polyhedron of eight faces.

regular polygon [MATH] A polygon with congruent sides and congruent interior angles.

regular polyhedron [MATH] A polyhedron all of whose faces are regular polygons, and whose polyhedral angles are congruent.

regular reflection *See* specular reflection.

regular reflector *See* specular reflector.

regular representation [MATH] A regular representation of a finite group is an isomorphism of it with a group of permutations.

regular singular point [MATH] A regular singular point of a differential equation is a singular point of the equation at which none of the solutions has an essential singularity.

regular tetrahedron [MATH] A regular polyhedron of four faces.

regular topological space [MATH] A topological space where any point and a closed set not containing it can be enclosed in disjoint open sets.

regulating rod [NUCLEO] A control rod intended to accomplish rapid, fine, and sometimes continuous adjustment of the reactivity of a nuclear reactor; it usually can move much more rapidly than a shim rod but makes a smaller change in reactivity.

regulating system *See* automatic control system.

regulation [CONT SYS] The process of holding constant a quantity such as speed, temperature, voltage, or position by means of an electronic or other system that automatically corrects errors by feeding back into the system the condition being regulated; regulation thus is based on feedback, whereas control is not. [ELEC] The change in output voltage that occurs between no load and full load in a transformer, generator, or other source. [ELECTR] The difference between the maximum and minimum tube voltage drops within a specified range of anode current in a gas tube.

regulator [CONT SYS] A device that maintains a desired quantity at a predetermined value or varies it according to a predetermined plan.

Rehbock weir formula [FL MECH] Probably the most accurate formula for the rate of flow of water over a rectangular suppressed weir; it includes a correction for the velocity of approach for normal, or fairly uniform, velocity distribution in the upstream channel; the formula is $Q = [3.234\ +$

$5.347/(320h - 3) + 0.428h/d_0]lh^{3/2}$, where Q is the flow rate in cubic feet per second, l is the width of the weir in feet, h is the head of water above the crest of the weir in feet, and d_0 is the height of weir or depth of water at zero head in feet.

reheating [THERMO] A process in which the gas or steam is reheated after a partial isentropic expansion to reduce moisture content. Also known as resuperheating.

Reissner-Nordstrom solution [RELAT] The unique solution of general relativity theory describing a nonrotating, charged black hole.

rejection band Also known as stop band. [ELECTROMAG] The band of frequencies below the cutoff frequency in a uniconductor waveguide. [PHYS] A frequency band within which electrical or electromagnetic signals are reduced or eliminated.

rejection number [IND ENG] A predetermined number of defective items in a batch which, if not exceeded, requires acceptance of the batch.

rejector *See* trap.

rejector impedance *See* dynamic impedance.

rel [ELECTROMAG] Unit of reluctance equal to 1 ampere-turn per magnetic line of force.

relation [MATH] A set of ordered pairs.

relative biological effectiveness [NUCLEO] A factor used to compare the biological effectiveness of different types of ionizing radiation; it is the inverse ratio of the amount of absorbed radiation required to produce a given effect to a standard (or reference) radiation required to produce the same effect. Abbreviated RBE.

relative coordinate system [PHYS] Any coordinate system which is moving with respect to an inertial coordinate system.

relative damping ratio *See* damping ratio.

relative density *See* specific gravity.

relative-density bottle *See* specific-gravity bottle.

relative dielectric constant *See* dielectric constant.

relative efficiency [STAT] **1.** Of an estimator, the comparative efficiency of the two estimators of the same parameter. **2.** For experimental design, the number of replications each design requires to reach the same precision.

relative frequency [STAT] The ratio of the number of occurrences of a given type of event or the number of members of a population in a given class to the total number of events or the total number of members of the population.

relative frequency distribution *See* percentage distribution.

relative frequency table *See* percentage distribution.

relative gain of an antenna [ELECTROMAG] Gain of an antenna in a given direction when the reference antenna is a half-wave, loss-free dipole isolated in space whose equatorial plane contains the given direction.

relative ionospheric opacity meter *See* riometer.

relative luminosity factor *See* luminosity function.

relatively closed set [MATH] A subset of a topological space is relatively closed if it is a closed set in some relative topology of a subset.

relatively compact set *See* conditionally compact set.

relatively open set [MATH] A subset of a topological space is relatively open if it is an open set in some relative topology of a subset.

relatively prime [MATH] Integers m and n are relatively prime if there are integers p and q so that $pm + qn = 1$; equivalently, if they have no common factors other than 1.

relative Mach number *See* Mach number.

relative magnetometer [ENG] Any magnetometer which must be calibrated by measuring the intensity of a field whose strength is accurately determined by other means; opposed to absolute magnetometer.

relative maximum [MATH] A value of a function at a point x_0 which is equal to or greater than the values of the function at all points in some neighborhood of x_0.

relative minimum [MATH] A value of a function at a point x_0 which is equal to or less than the values of the function at all points in some neighborhood of x_0.

relative momentum [MECH] The momentum of a body in a reference frame in which another specified body is fixed.

relative motion [MECH] The continuous change of position of a body with respect to a second body or to a reference point that is fixed. Also known as apparent motion.

relative permeability [ELECTROMAG] The ratio of the permeability of a substance to the permeability of a vacuum at the same magnetic field strength.

relative permittivity *See* dielectric constant.

relative primes [MATH] Two positive integers with no common positive divisor other than 1.

relative resistance [ELEC] The ratio of the resistance of a piece of a material to the resistance of a piece of specified material, such as annealed copper, having the same dimensions and temperature.

relative roughness factor [FL MECH] Roughness of pipe-wall interior (distance from peaks to valleys) divided by pipe internal diameter; used to modify Reynolds number calculations for fluid flow through pipes.

relative scatter intensity [OPTICS] For scattering of radiation under any given set of physical conditions, the ratio of the radiant intensity scattered in any given direction to the radiant intensity scattered in the direction of the incident beam.

relative sunspot number [ASTRON] A measure of sunspot activity, computed from the formula $R = k(10g + f)$, where R is the relative sunspot number, f the number of individual spots, g the number of groups of spots, and k a factor that varies with the observer (his personal equation), the seeing, and the observatory (location and instrumentation). Also known as sunspot number; sunspot relative number; Wolf number; Wolf-Wolfer number; Zurich number.

relative tensor [MATH] A relative tensor of weight w is an object whose transformation law differs from that of a tensor by the presence of the Jacobian $|\partial x/\partial \bar{x}|$ raised to the wth power in the expression of the components in the $\bar{x}_1, \bar{x}_2, \dots, \bar{x}_n$ coordinate system in terms of those in the $x_1, x_2, \dots, x_n$ coordinate system.

relative topology [MATH] In a topological space X any subset A has a topology on it relative to the given one by intersecting the open sets of X with A to obtain open sets in A.

relative velocity [MECH] The velocity of a body with respect to a second body; that is, its velocity in a reference frame where the second body is fixed.

relativistic beam [RELAT] A beam of particles traveling at a speed comparable with the speed of light.

relativistic electrodynamics [ELECTROMAG] The study of the interaction between charged particles and electric and magnetic fields when the velocities of the particles are comparable with that of light.

relativistic kinematics [RELAT] A description of the motion of particles compatible with the special theory of relativity, without reference to the causes of motion.

relativistic mass [RELAT] The mass of a particle moving at a velocity exceeding about one-tenth the velocity of light; it is significantly larger than the rest mass.

relativistic mechanics [RELAT] **1.** Any form of mechanics compatible with either the special or the general theory of relativity. **2.** The nonquantum mechanics of a system of particles or of a fluid interacting with an electromagnetic field, in the case when some of the velocities are comparable with the speed of light.

relativistic particle [RELAT] A particle moving at a speed comparable with the speed of light.

relativistic quantum theory [QUANT MECH] The quantum theory of particles which is consistent with the special theory of relativity, and thus can describe particles moving arbitrarily close to the speed of light.

relativistic theory [PHYS] Any theory which is consistent with the special or general theory of relativity.

relativity [PHYS] Theory of physics which recognizes the universal character of the propagation speed of light and the consequent dependence of space, time, and other mechanical measurements on the motion of the observer performing the measurements; it has two main divisions, the special theory and the general theory.

relaxation [MATH] *See* relaxation method. [MECH] **1.** Relief of stress in a strained material due to creep. **2.** The lessening of elastic resistance in an elastic medium under an applied stress resulting in permanent deformation. [PHYS] A process in which a physical system approaches a steady state after conditions affecting it have been suddenly changed, and in which the presence of dissipative agents prevents the system from overshooting and then oscillating about this state.

relaxation circuit [ELECTR] Circuit arrangement, usually of vacuum tubes, reactances, and resistances, which has two states or conditions, one, both, or neither of which may be stable; the transient voltage produced by passing from one to the other, or the voltage in a state of rest, can be used in other circuits.

relaxation method [MATH] A successive approximation method for solving systems of equations where the errors from an initial approximation are viewed as constraints to be minimized or relaxed within a toleration limit. Also known as relaxation.

relaxation oscillations [PHYS] Oscillations having a sawtooth waveform in which the displacement increases to a certain value and then drops back to zero, after which the cycle is repeated.

relaxation oscillator [ELECTR] An oscillator whose fundamental frequency is determined by the time of charging or discharging a capacitor or coil through a resistor, producing waveforms that may be rectangular or sawtooth.

relaxation time [MECH] The ratio η/k of a Maxwell liquid whose deformation γ is related to the shearing stress S acting on it by the equation $d\gamma/dt = (dS/dt)/k + S/\eta$. [PHYS] For many physical systems undergoing relaxation, a time τ such that the displacement of a quantity from its equilibrium value at any instant of time t is the exponential of $-t/\tau$. [SOLID STATE] The travel time of an electron in a metal before it is scattered and loses its momentum.

relaxed peak process *See* quasi-fission.

relay [ELEC] A device that is operated by a variation in the conditions in one electric circuit and serves to make or break one or more connections in the same or another electric circuit. Also known as electric relay.

relay control system [CONT SYS] A control system in which the error signal must reach a certain value before the controller reacts to it, so that the control action is discontinuous in amplitude.

reliability [STAT] **1.** The amount of credence placed in a result. **2.** The precision of a measurement, as measured by the variance of repeated measurements of the same object.

reluctance [ELECTROMAG] A measure of the opposition presented to magnetic flux in a magnetic circuit, analogous to resistance in an electric circuit; it is equal to magnetomotive force divided by magnetic flux. Also known as magnetic reluctance.

reluctivity [PHYS] The reciprocal of magnetic permeability; the reluctivity of empty space is unity. Also known as magnetic reluctivity; specific reluctance.

rem [NUCLEO] A unit of ionizing radiation, equal to the amount that produces the same damage to humans as 1 roentgen of high-voltage x-rays. Derived from roentgen equivalent man.

remainder [MATH] **1.** The remaining integer when a division of an integer by another is performed; if $l = m \cdot p + r$, where $l, m, p,$ and r are integers and r is less than p, then r is the remainder when l is divided by p. **2.** The remaining polynomial when division of a polynomial is performed; if $l = m \cdot p + r$, where $l, m, p,$ and r are polynomials, and the degree of r is less than that of p, then r is the remainder when l is divided by p. **3.** The remaining part of a convergent infinite series after a computation, for some n, of the first n terms.

remainder formula [MATH] A formula by which the remainder resulting from an approximation of a function by a partial sum of a power series can be computed or analyzed.

remainder theorem [MATH] Dividing a polynomial $p(x)$ by $(x - a)$ gives a remainder equaling the number $p(a)$.

remanence [ELECTROMAG] The magnetic flux density that remains in a magnetic circuit after the removal of an applied magnetomotive force; if the magnetic circuit has an air gap, the remanence will be less than the residual flux density.

remanent magnetization [GEOPHYS] That component of a rock's magnetization whose direction is fixed relative to the rock and which is independent of moderate, applied magnetic fields.

remote-cutoff tube *See* variable-mu tube.

removable discontinuity [MATH] A point where a function is discontinuous, but it is possible to redefine the function at this point so that it will be continuous there.

renormalizability [QUANT MECH] The property of some quantum field theories whereby all infinite quantities can be absorbed into a renormalization of physical parameters such as mass and charge.

renormalization [QUANT MECH] In certain quantum field theories, a procedure in which nonphysical bare values of certain quantities such as mass and charge are eliminated and the corresponding physically observable quantities are introduced.

renormalization group methods [STAT MECH] Methods for treating the behavior of substances near critical points, in which the canonical ensemble is generalized by dividing a substance into cells of arbitrary size and forming an ensemble consisting of all microscopic configurations consistent with specified values of the thermodynamic variables in each of these cells.

rep [NUCLEO] A unit of ionizing radiation, equal to the amount that causes absorption of 93 ergs per gram of soft tissue. Derived from roentgen equivalent physical. Also known as parker; tissue roentgen.

repeated load [MECH] A force applied repeatedly, causing variation in the magnitude and sometimes in the sense, of the internal forces.

repeated measurements model [STAT] A product model in which each factor is the same.

repeated root   *See* multiple root.

repeat glass   [OPTICS] Used by textile and wallpaper design-ers, a device consisting of four lenses formed in one piece of glass; when a single drawing or pattern is viewed, the subject matter is repeated four times.

repeating decimal   [MATH] A decimal that is either finite or infinite with a finite block of digits repeating indefinitely. Also known as recurring decimal.

repeller   [ELECTR] An electrode whose primary function is to reverse the direction of an electron stream in an electron tube. Also known as reflector.

replica grating   [OPTICS] A diffraction grating made by flow-ing a plastic solution over an original grating, evaporating the solvent, and removing the resulting film, which has the lines of the original grating impressed on it.

replication   [STAT] In experimental design, the repetition of an experiment or parts of an experiment to secure more data as an aid to determining the experimental error and to arrive at better estimates of the effects of various treatments with smaller standard errors.

representation of groups   [MATH] A representation of a group is given by a homomorphism of it onto some group either of matrices or unitary operators of a Hilbert space.

representation theory   [MATH] **1.** The study of groups by the use of their representations. **2.** The determination of repre-sentations of specific groups.   [QUANT MECH] Quantum-me-chanical device in which one selects the common eigenfunc-tions of a complete set of quantum-mechanical operators as a basis of vectors in a Hilbert space, and expresses wave functions and operators in terms of column matrices and square matrices, respectively, which correspond to this basis.

representative sample   [STAT] A sample whose characteris-tics reflect those of the population from which it is drawn.

repulsion   [MECH] A force which tends to increase the dis-tance between two bodies having like electric charges, or the force between atoms or molecules at very short distances which keeps them apart. Also known as repulsive force.

repulsive force   *See* repulsion.

research reactor   [NUCLEO] A reactor primarily designed to supply neutrons or other ionizing radiation for experimental purposes; it may also be used for training, materials testing, and production of radioisotopes. Also known as teaching reactor.

reset action   [CONT SYS] Floating action in which the final control element is moved at a speed proportional to the extent of proportional-position action.

reset rate   [ENG] The number of times per minute that the effect of the proportional-position action upon the final control element is repeated by the proportional-speed float-ing action.

residence time   [NUCLEO] The time during which radioactive material remains in the atmosphere following the detonation of a nuclear explosive; it is usually expressed as a half-time, since the time for all material to leave the atmosphere is not well known.

residual charge   [ELEC] The charge remaining on the plates of a capacitor after initial discharge.

residual current   [ELECTR] Current flowing through a thermi-onic diode when there is no anode voltage, due to the velocity of the electrons emitted by the heated cathode.

residual error ratio   [PHYS] The difference between an opti-mum result derived from experience or experiment and a supposedly exact result derived from theory.

residual field   [ELECTROMAG] The magnetic field left in an iron core after excitation has been removed.

residual flux density [ELECTROMAG] The magnetic flux density at which the magnetizing force is zero when the material is in a symmetrically and cyclically magnetized condition. Also known as residual induction; residual magnetic induction; residual magnetism.

residual induction *See* residual flux density.

residual ionization [PHYS] Ionization of air or other gas in a closed chamber, not accounted for by recognizable neighboring agencies; now attributed to cosmic rays.

residual magnetic induction *See* residual flux density.

residual magnetism *See* residual flux density.

residual radiation [NUCLEO] Nuclear radiation emitted by radioactive material deposited after an atomic burst, including fission products, unfissioned nuclear material, and material in which radioactivity may have been induced by neutron bombardment. Also known as reststrahlen. [OPTICS] The nearly monochromatic radiation resulting from several reflections of light or other radiation from polished surfaces of certain substances such as quartz and rock salt, due to high reflectivity of these substances in certain bands of wavelengths.

residual resistance [SOLID STATE] The value to which the electrical resistance of a metal drops as the temperature is lowered to near absolute zero, caused by imperfections and impurities in the metal rather than by lattice vibrations.

residual set [MATH] In a topological space, the complement of a set which is a countable union of nowhere dense sets.

residual spectrum [MATH] Those members λ of the spectrum of a linear operator A on a Banach space X for which $(A - \lambda I)^{-1}$, I the identity operator, is unbounded with domain not dense in X.

residual stress *See* internal stress.

residual sum of squares *See* error sum of squares.

residual variance [STAT] In analysis of variance and regression analysis, that part of the variance which cannot be attributed to specific causes.

residual vibration *See* zero-point vibration.

residual voltage [ELEC] Vector sum of the voltages to ground of the several phase wires of an electric supply circuit.

residue [MATH] **1.** The residue of a complex function $f(z)$ at an isolated singularity z_0 is given by $(1/2\pi i) \int f(z) \, dz$ along a simple closed curve interior to an annulus about z_0; equivalently, the coefficient of the term $(z - z_0)^{-1}$ in the Laurent series expansion of $f(z)$ about z_0. **2.** In general, a coset of an ideal in a ring.

residue class [MATH] A set of numbers satisfying a congruency relation.

residue class ring *See* quotient ring.

residue theorem [MATH] The value of the integral of a complex function, taken along a simple closed curve enclosing at most a finite number of isolated singularities, is given by $2\pi i$ times the sum of the residues of the function at each of the singularities.

resilience [MECH] **1.** Ability of a strained body, by virtue of high yield strength and low elastic modulus, to recover its size and form following deformation. **2.** The work done in deforming a body to some predetermined limit, such as its elastic limit or breaking point, divided by the body's volume.

resistance [ELEC] **1.** The opposition that a device or material offers to the flow of direct current, equal to the voltage drop across the element divided by the current through the element. Also known as electrical resistance. **2.** In an alternating-current circuit, the real part of the complex impedance. [MECH] In damped harmonic motion, the ratio of the frictional resistive force to the speed. Also known as damping coefficient; damping constant; mechanical resistance.

resistance box [ELEC] A box containing a number of precision resistors connected to panel terminals or contacts so that a desired resistance value can be obtained by withdrawing plugs (as in a post-office bridge) or by setting multicontact switches.

resistance-capacitance circuit [ELEC] A circuit which has a resistance and a capacitance in series, and in which inductance is negligible. Abbreviated *R-C* circuit.

resistance-capacitance constant [ELEC] Time constant of a resistive-capacitive circuit; equal in seconds to the resistance value in ohms multiplied by the capacitance value in farads. Abbreviated *R-C* constant.

resistance-capacitance coupled amplifier [ELECTR] An amplifier in which a capacitor provides a path for signal currents from one stage to the next, with resistors connected from each side of the capacitor to the power supply or to ground; it can amplify alternating-current signals but cannot handle small changes in direct currents. Also known as *R-C* amplifier; *R-C* coupled amplifier; resistance-coupled amplifier.

resistance-capacitance network [ELEC] Circuit containing resistances and capacitances arranged in a particular manner to perform a specific function. Abbreviated *R-C* network.

resistance-capacitance oscillator [ELECTR] Oscillator in which the frequency is determined by resistance and capacitance elements. Abbreviated *R-C* oscillator.

resistance coefficient 1 [FL MECH] A dimensionless number used in the study of flow resistance, equal to the resistance force in flow divided by ½ the product of fluid density, the square of fluid velocity, and the square of a characteristic length. Symbolized c_f.

resistance coefficient 2 *See* Darcy number 1.

resistance-coupled amplifier *See* resistance-capacitance coupled amplifier.

resistance drop [ELEC] The voltage drop occurring between two points on a conductor due to the flow of current through the resistance of the conductor; multiplying the resistance in ohms by the current in amperes gives the voltage drop in volts. Also known as *IR* drop.

resistance gage [ENG] An instrument for determining high pressures from the change in the electrical resistance of manganin or mercury produced by these pressures.

resistance loss [ELEC] Power loss due to current flowing through resistance; its value in watts is equal to the resistance in ohms multiplied by the square of the current in amperes.

resistance magnetometer [ENG] A magnetometer that depends for its operation on variations in the electrical resistance of a material immersed in the magnetic field to be measured.

resistance measurement [ELEC] The quantitative determination of that property of an electrically conductive material, component, or circuit called electrical resistance.

resistance meter [ENG] Any instrument which measures electrical resistance. Also known as electrical resistance meter.

resistance noise *See* thermal noise.

resistance pyrometer *See* resistance thermometer.

resistance thermometer [ENG] A thermometer in which the sensing element is a resistor whose resistance is an accurately known function of temperature. Also known as electrical resistance thermometer; resistance pyrometer.

resisting moment [MECH] A moment produced by internal tensile and compressive forces that balances the external bending moment on a beam.

resistivity *See* electrical resistivity.

RESISTANCE THERMOMETER

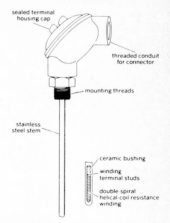

sealed terminal housing cap

threaded conduit for connector

mounting threads

stainless steel stem

ceramic bushing

winding terminal studs

double-spiral helical-coil resistance winding

Industrial-type resistance thermometer. *(From D. M. Considine, ed., Process Instruments and Controls Handbook, McGraw-Hill, 1957)*

RESISTOR NETWORK

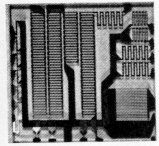

A thin-film resistor network.
(*TRW Inc.*)

RESISTOR-TRANSISTOR LOGIC

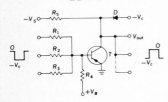

Resistor-transistor logic circuit:
R_1, R_2, R_3 are resistors coupling
the logic circuit to preceding
logic circuits. With resistor R_4
and positive supply voltage $+V_B$
they form the gate circuit. D
is the diode and T the transistor.
V_S = input signal voltage; V_c =
collector voltage; V_{out} = output
voltage.

resistor [ELEC] A device designed to have a definite amount of resistance; used in circuits to limit current flow or to provide a voltage drop. Also known as electrical resistor.

resistor-capacitor-transistor logic [ELECTR] A resistor-transistor logic with the addition of capacitors that are used to enhance switching speed.

resistor color code [ELEC] Code adopted by the Electronic Industries Association to mark the values of resistance on resistors in a readily recognizable manner; the first color represents the first significant figure of the resistor value, the second color the second significant figure, and the third color represents the number of zeros following the first two figures; a fourth color is sometimes added to indicate the tolerance of the resistor.

resistor core [ELEC] Insulating support on which a resistor element is wound or otherwise placed.

resistor element [ELEC] That portion of a resistor which possesses the property of electric resistance.

resistor network [ELEC] An electrical network consisting entirely of resistances.

resistor-transistor logic [ELECTR] One of the simplest logic circuits, having several resistors, a transistor, and a diode. Abbreviated RTL.

resolution [OPTICS] *See* resolving power. [PHYS] **1.** For a measurement of energy or momentum of a collection of particles, the difference between the highest and lowest energies at which the response of an instrument to a beam of monoenergetic particles is at least half its maximum value, divided by the energy of the particles. **2.** The procedure of breaking up a vectorial quantity into its components. [SPECT] *See* resolving power.

resolution chart [OPTICS] A device to test resolving power; usually alternate black and white lines of equal width arranged in groups of decreasing line width, identified as the number of line pairs per millimeter.

resolution of a vector [MATH] The determination of vectors parallel to specified (usually perpendicular) axes such that their sum equals a given vector.

resolution of the identity [MATH] A family of linear projection operators on a Banach space used in studying the spectra of linear operators.

resolution reading [OPTICS] A number indicating how many lines per millimeter are contained in the finest group which can be distinguished on a resolution chart.

resolvable balanced incomplete block design [MATH] A balanced incomplete block design such that the blocks themselves are partitioned into r families of v/k blocks, such that every element occurs in exactly one block of each of these families.

resolvent kernel [MATH] A function appearing as an integrand in an integral representation for a solution of a linear integral equation which often completely determines the solutions.

resolvent of an operator [MATH] The function, defined on the complement of the spectrum of a linear operator T of Banach spaces, given by $(T - \lambda I)^{-1}$ for each λ in this complement, where I is the identity operator; this enables a study of T relative to its eigenvalues.

resolvent set [MATH] Those scalars λ for which the operator $T - \lambda I$ has a bounded inverse, where T is some linear operator on a Banach space, and I is the identity operator.

resolving power [ELECTROMAG] **1.** The reciprocal of the beam width of a unidirectional antenna, measured in degrees. **2.** *See* resolution. [OPTICS] A quantitative measure of the ability of an optical instrument to produce separable images of different points on an object; usually, the smallest angular or linear separation of two object points for which they may be resolved according to the Rayleigh criterion. Also known as resolution. [PHYS] A measure of the ability of a mass spectroscope to separate particles of different masses, equal to the ratio of the average mass of two particles whose mass spectrum lines can just be completely separated, to the difference in their masses. [SPECT] A measure of the ability of a spectroscope or interferometer to separate spectral lines of nearly equal wavelength, equal to the average wavelength of two equally strong spectral lines whose images can barely be separated, divided by the difference in wavelengths; for spectroscopes, the lines must be resolved according to the Rayleigh criterion; for interferometers, the wavelengths at which the lines have half of maximum intensity must be equal. Also known as resolution.

resolving time [ENG] Minimum time interval, between events, that can be detected; resolving time may refer to an electronic circuit, to a mechanical recording device, or to a counter tube.

resonance [ELEC] A phenomenon exhibited by an alternating-current circuit in which there are relatively large currents near certain frequencies, and a relatively unimpeded oscillation of energy from a potential to a kinetic form; a special case of the physics definition. [PHYS] A phenomenon exhibited by a physical system acted upon by an external periodic driving force, in which the resulting amplitude of oscillation of the system becomes large when the frequency of the driving force approaches a natural free oscillation frequency of the system. [QUANT MECH] *See* resonance absorption; resonance level.

resonance absorption [NUCLEO] The absorption of neutrons having a narrow range of energies corresponding to a nuclear resonance level of the absorber in a nuclear reactor. [QUANT MECH] The absorption of electromagnetic radiation by a quantum-mechanical system at a characteristic frequency satisfying the Bohr frequency condition. Also known as resonance.

resonance bridge [ELEC] A four-arm alternating-current bridge used to measure inductance, capacitance, or frequency; the inductor and the capacitor, which may be either in series or in parallel, are tuned to resonance at the frequency of the source before the bridge is balanced.

resonance capture [NUC PHYS] The combination of an incident particle and a nucleus in a resonance level of the resulting compound nucleus, characterized by having a large cross section at and very near the corresponding resonance energy.

resonance curve [ELEC] Graphical representation illustrating the manner in which a tuned circuit responds to the various frequencies in and near the resonant frequency.

resonance fluorescence [ATOM PHYS] *See* resonance radiation. [NUC PHYS] Resonant scattering from an atomic nucleus.

resonance frequency [PHYS] A frequency at which some measure of the response of a physical system to an external periodic driving force is a maximum; three types are defined, namely, phase resonance, amplitude resonance, and natural resonance, but they are nearly equal when dissipative effects are small. Also known as resonant frequency. [QUANT MECH] A characteristic frequency, satisfying the Bohr frequency condition, at which a quantum-mechanical system absorbs radiation.

resonance lamp [ATOM PHYS] An evacuated quartz bulb containing mercury, which acts as a source of radiation at the wavelength of the pure resonance line of mercury when irradiated by a mercury-arc lamp.

resonance level [QUANT MECH] An unstable state of a compound system capable of being formed in a collision between two particles, and associated with a peak in a graph of cross section versus energy for the scattering of the particles. Also known as resonance.

resonance luminescence *See* resonance radiation.

resonance method [ELEC] A method of determining the impedance of a circuit element, in which resonance frequency of a resonant circuit containing the element is measured.

resonance radiation [ATOM PHYS] The emission of radiation by a gas or vapor as a result of excitation of atoms to higher energy levels by incident photons at the resonance frequency of the gas or vapor; the radiation is characteristic of the particular gas or vapor atom but is not necessarily the same frequency as the absorbed radiation. Also known as resonance fluorescence; resonance luminescence.

resonance scattering [NUC PHYS] A peak in the cross section of a nucleus for elastic scattering of neutrons at energies near a resonance level, accompanied by an anomalous phase shift in the scattered neutrons.

resonance spectrum [SPECT] An emission spectrum resulting from illumination of a substance (usually a molecular gas) by radiation of a definite frequency or definite frequencies.

resonance transformer [ELEC] A high-voltage transformer in which the secondary circuit is tuned to the frequency of the power supply. [ELECTR] An electrostatic particle accelerator, used principally for acceleration of electrons, in which the high-voltage terminal oscillates between voltages which are equal in magnitude and opposite in sign.

resonance vibration [MECH] Forced vibration in which the frequency of the disturbing force is very close to the natural frequency of the system, so that the amplitude of vibration is very large.

resonant antenna [ELECTROMAG] An antenna for which there is a sharp peak in the power radiated or intercepted by the antenna at a certain frequency, at which electric currents in the antenna form a standing-wave pattern.

resonant capacitor [ELEC] A tubular capacitor that is wound to have inductance in series with its capacitance.

resonant cavity *See* cavity resonator.

resonant-cavity maser [PHYS] A maser in which the paramagnetic active material is placed in a cavity resonator.

resonant chamber *See* cavity resonator.

resonant circuit [ELEC] A circuit that contains inductance, capacitance, and resistance of such values as to give resonance at an operating frequency.

resonant coupling [ELEC] Coupling between two circuits that reaches a sharp peak at a certain frequency.

resonant detector [PHYS] A detector of electromagnetic radiation which is responsive to radiation only at certain frequencies at which resonance is created in the detector.

resonant diaphragm [ELECTROMAG] Diaphragm, in waveguide technique, so proportioned as to introduce no reactive impedance at the design frequency.

resonant element *See* cavity resonator.

resonant frequency *See* resonance frequency.

resonant gate transistor [ELECTR] Surface field-effect transistor incorporating a cantilevered beam which resonates at a specific frequency to provide high-Q-frequency discrimination.

resonant iris [ELECTROMAG] A resonant window in a circular waveguide; it resembles an optical iris.

resonant scattering [QUANT MECH] Scattering of a photon by

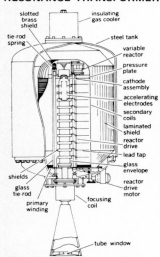

RESONANCE TRANSFORMER

slotted brass shield
insulating gas cooler
tie-rod spring
steel tank
variable reactor
pressure plate
cathode assembly
accelerating electrodes
secondary coils
laminated shield
reactor drive
lead tap
glass envelope
reactor drive motor
shields
glass tie-rod
focusing coil
primary winding
tube window

Schematic diagram of 1-MeV resonance transformer that is used as an electrostatic particle accelerator. (*General Electric Co.*)

a quantum-mechanical system (usually an atom or nucleus) in which the system first absorbs the photon by undergoing a transition from one of its energy states to one of higher energy, and subsequently reemits the photon by the exact inverse transition.

resonant voltage step-up [ELEC] Ability of an inductor and a capacitor in a series resonant circuit to deliver a voltage several times greater than the input voltage of the circuit.

resonant window [ELECTROMAG] A parallel combination of inductive and capacitive diaphragms, used in a waveguide structure to provide transmission at the resonant frequency and reflection at other frequencies.

resonate [ELEC] To bring to resonance, as by tuning.

resonating cavity [ELECTROMAG] Short piece of waveguide of adjustable length, terminated at either or both ends by a metal piston, an iris diaphragm, or some other wave-reflecting device; it is used as a filter, as a means of coupling between guides of different diameters, and as impedance networks corresponding to those used in radio circuits.

resonator [PHYS] A device that exhibits resonance at a particular frequency, such as an acoustic resonator or cavity resonator.

resonator grid [ELECTR] Grid that is attached to a cavity resonator in velocity-modulated tubes to provide coupling between the resonator and the electron beam.

resonator wavemeter [ELECTROMAG] Any resonant circuit used to determine wavelength, such as a cavity-resonator frequency meter.

response [CONT SYS] A quantitative expression of the output of a device or system as a function of the input. Also known as system response. [STAT] The value of some measurable quantity after a treatment has been applied.

response characteristic [CONT SYS] The response as a function of an independent variable, such as direction or frequency, often presented in graphical form.

response time [CONT SYS] The time required for the output of a control system or element to reach a specified fraction of its new value after application of a step input or disturbance. [ELEC] The time it takes for the pointer of an electrical or electronic instrument to come to rest at a new value, after the quantity it measures has been abruptly changed.

rest density [RELAT] The density of a small portion of a fluid in a Lorentz frame in which that portion of the fluid is at rest.

rest energy [RELAT] The energy equivalent to the rest mass m_0 of a particle or body; that is, the quantity of m_0c^2, where c is the speed of light; often expressed in electron volts.

rest frame [RELAT] The Lorentz frame in which the total momentum of a system equals zero; for an accelerated system, the rest frame varies from instant to instant.

rest mass [RELAT] The mass of a particle in a Lorentz reference frame in which it is at rest.

restricted internal rotation [PHYS CHEM] Restrictions on the rotational motion of molecules or parts of molecules in some substances, such as solid methane, at certain temperatures.

reststrahlen *See* residual radiation.

resultant *See* vector sum.

resultant of forces [MECH] A system of at most a single force and a single couple whose external effects on a rigid body are identical with the effects of the several actual forces that act on that body.

resuperheating *See* reheating.

retardation [OPTICS] In interference microscopy, the difference in optical path between the light passing through the specimen and the light bypassing the specimen. Also known as optical-path difference.

retardation sheet *See* wave plate.

retardation theory [OPTICS] General methods of calculating the effect of one or more wave plates on light which is normally incident on the plates and which is initially polarized in some fashion.

retardation time [MECH] The ratio η/k of a Kelvin body whose shearing stress S is related to its deformation γ by the equation $S = k\gamma + \eta \, d\gamma/dt$, where η and k are constants and t represents time.

retarded field [ELECTROMAG] An electric or magnetic field strength as found from the retarded potentials.

retarded potentials [ELECTROMAG] The electromagnetic potentials at an instant in time t and a point in space r as a function of the charges and currents that existed at earlier times at points on the past light cone of the event r,t.

retarding-field oscillator [ELECTR] An oscillator employing an electron tube in which the electrons oscillate back and forth through a grid that is maintained positive with respect to both the cathode and anode; the field in the region of the grid exerts a retarding effect through the grid in either direction. Also known as positive-grid oscillator.

retarding potential [PHYS] A potential which causes the speed of a moving particle to be reduced.

retentivity [ELECTROMAG] The residual flux density corresponding to the saturation induction of a magnetic material.

Retgers' law [SOLID STATE] The law that the properties of crystalline mixtures of isomorphous substances are continuous functions of the percentage composition.

reticle [OPTICS] A series of intersecting fine lines, wires, or the like which are placed in the focus of the objective of an optical instrument to aid in measurement of angles or distances.

reticular density [MATH] The number of points per unit area in a two-dimensional lattice, such as the plane of a crystal lattice.

retinal illuminance [OPTICS] A psychophysiological quantity which is a measure of the brightness of a visual sensation; it is measured in trolands.

retract [MATH] A subset R of a topological space X is a retract of X if there is a continuous map f from X to R, with $f(r) = r$ for all points r of R.

retroaction *See* positive feedback.

retrodirective mirror [OPTICS] **1.** An optical system consisting of two mutually perpendicular plane mirrors; it reflects any beam of light which lies in a plane perpendicular to the mirrors into a direction antiparallel to its original direction. **2.** An optical system consisting of three mutually perpendicular plane mirrors; it reflects any beam of light into a direction antiparallel to its original direction.

retrograde motion [ASTRON] **1.** An apparent backward motion of a planet among the stars resulting from the observation of the planet from the planet earth which is also revolving about the sun at a different velocity. Also known as retrogression. **2.** *See* retrograde orbit.

retrograde orbit [ASTRON] Motion in an orbit opposite to the usual orbital direction of celestial bodies within a given system; specifically, of a satellite, motion in a direction opposite to the direction of rotation of the primary. Also known as retrograde motion.

retrogression *See* retrograde motion.

retroreflection [PHYS] Reflection wherein the reflected rays of radiation return along paths parallel to those of their corresponding incident rays.

retroreflector [PHYS] Any instrument used to cause reflected radiation to return along paths parallel to those of their corresponding incident rays; one type, the corner reflector, is an efficient radar target.

return *See* echo.

return streamer [GEOPHYS] The intensely luminous streamer which propagates upward from earth to cloud base in the last phase of each lightning stroke of a cloud-to-ground discharge. Also known as main stroke; return stroke.

return stroke *See* return streamer.

reverberation [ACOUS] The prolongation of sound at a given point after direct reception from the source has ceased, due to such causes as reflections from bounding surfaces, scattering from inhomogeneities in a medium, and vibrations excited by the original sound.

reverberation chamber [ACOUS] An enclosure with heavy surfaces which randomly reflect as great an amount of sound as possible; used in acoustic measurements. Also known as random diffusion chamber.

reverberation time [ACOUS] The time in seconds required for the average sound-energy density at a given frequency to reduce to one-millionth of its initial steady-state value after the sound source has been stopped; this corresponds to a decrease of 60 decibels.

reversal spectrum [SPECT] A spectrum which may be observed in intense white light which has traversed luminous gas, in which there are dark lines where there were bright lines in the emission spectrum of the gas.

reversal temperature [SPECT] The temperature of a blackbody source such that, when light from this source is passed through a luminous gas and analyzed in a spectroscope, a given spectral line of the gas disappears, whereas it appears as a bright line at lower blackbody temperatures, and a dark line at higher temperatures.

reverse bias [ELECTR] A bias voltage applied to a diode or a semiconductor junction with polarity such that little or no current flows; the opposite of forward bias.

reverse-blocking tetrode thyristor *See* silicon controlled switch.

reverse-blocking triode thyristor *See* silicon controlled rectifier.

reverse Brayton cycle [THERMO] A refrigeration cycle using air as the refrigerant but with all system pressures above the ambient. Also known as dense-air refrigeration cycle.

reverse Carnot cycle [THERMO] An ideal thermodynamic cycle consisting of the processes of the Carnot cycle reversed and in reverse order, namely, isentropic expansion, isothermal expansion, isentropic compression, and isothermal compression.

reverse current [ELECTR] Small value of direct current that flows when a semiconductor diode has reverse bias.

reverse curve [MATH] An S-shaped curve, that is, one having two arcs with their centers on opposite sides of the curve. Also known as S curve.

reversed image *See* inverted image.

reverse feedback *See* negative feedback.

reverse voltage [ELEC] In the case of two opposing voltages, voltage of that polarity which produces the smaller current.

reversibility principle [OPTICS] The principle that if a beam of light is reflected back on itself, it will traverse the same path or paths as it did before reversal. [STAT MECH] *See* microscopic reversibility.

reversible capacitance [ELECTR] Limit, as the amplitude of an applied sinusoidal capacitor voltage approaches 0, of the ratio of the amplitude of the resulting in-phase fundamental-frequency component of transferred charge to the amplitude

REVERSE BRAYTON CYCLE

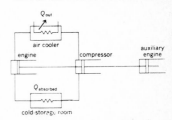

Schematic arrangement of the reverse Brayton cycle; air undergoes isentropic compression followed by reversible constant-pressure cooling; high-pressure air next expands reversibly in engine and exhausts at low temperature; cooled air passes through the cold storage chamber, picks up heat at constant pressure, and then returns to suction side of compressor. Q = heat.

of the applied voltage, for a given constant bias voltage superimposed on the sinusoidal voltage.

reversible engine [THERMO] An ideal engine which carries out a cycle of reversible processes.

reversible path [THERMO] A path followed by a thermodynamic system such that its direction of motion can be reversed at any point by an infinitesimal change in external conditions; thus the system can be considered to be at equilibrium at all points along the path.

reversible process [THERMO] An ideal thermodynamic process which can be exactly reversed by making an indefinitely small change in the external conditions. Also known as quasistatic process.

reversible transducer [ELECTR] Transducer whose loss is independent of transmission direction.

reversing layer [ASTROPHYS] A layer of relatively cool gas forming the lower part of the sun's chromosphere, just above the photosphere, that gives rise to absorption lines in the sun's spectrum.

reversing thermometer [ENG] A mercury-in-glass thermometer which records temperature upon being inverted and thereafter retains its reading until returned to the first position.

revolution [MECH] The motion of a body around a closed orbit.

revolution per minute [MECH] A unit of angular velocity equal to the uniform angular velocity of a body which rotates through an angle of 360° (2π radians), so that every point in the body returns to its original position, in 1 minute. Abbreviated rpm.

revolution per second [MECH] A unit of angular velocity equal to the uniform angular velocity of a body which rotates through an angle of 360° (2π radians), so that every point in the body returns to its original position, in 1 second. Abbreviated rps.

reyn [FL MECH] A unit of dynamic viscosity equal to the dynamic viscosity of a fluid in which there is a tangential force of 1 poundal per square foot resisting the flow of two parallel fluid layers past each other when their differential velocity is 1 foot per second per foot of separation; equal to approximately 14.8816 poise.

Reynolds analogy [CHEM ENG] Relationship showing the similarity between the transfer of mass, heat, and momentum.

Reynolds criterion [FL MECH] The principle that the type of fluid motion, that is, laminar flow or turbulent flow, in geometrically similar flow systems depends only on the Reynolds number; for example, in a pipe, laminar flow exists at Reynolds numbers less than 2000, turbulent flow at numbers above about 3000.

Reynolds equation [FL MECH] A form of the Navier-Stokes equation which is $\rho\partial u/\partial t = (\partial/\partial x)\,(p_{xx} - \rho u^2) + (\partial/\partial y)(p_{xy} - \rho uv) + (\partial/\partial z)(p_{xz} - \rho uw)$, where ρ is the fluid density, u, v, and w are the components of the fluid velocity, and p_{xx}, p_{xy}, and p_{xz} are normal and shearing stresses.

Reynolds law [FL MECH] The pressure head needed to maintain a liquid flow at constant speed v through a pipe of length l and radius r is equal to klv^p/r^q, where k, p, and q are constants, with p equal to approximately 1 and q to approximately 2.

Reynolds number [FL MECH] A dimensionless number which is significant in the design of a model of any system in which the effect of viscosity is important in controlling the velocities or the flow pattern of a fluid; equal to the density of a fluid, times its velocity, times a characteristic length, divided by the fluid viscosity. Symbolized N_{Re}. Also known as Damköhler number V (DaV).

Reynolds stress [FL MECH] The net transfer of momentum across a surface in a turbulent fluid because of fluctuations in fluid velocity. Also known as eddy stress.

Reynolds stress tensor [FL MECH] A tensor whose components are the components of the Reynolds stress across three mutually perpendicular surfaces.

rf *See* radio frequency.

Rh *See* rhodium.

rhe [FL MECH] **1.** A unit of dynamic fluidity, equal to the dynamic fluidity of a fluid whose dynamic viscosity is 1 centipoise. **2.** A unit of kinematic fluidity, equal to the kinematic fluidity of a fluid whose kinematic viscosity is 1 centistoke.

Rheinberg illumination [OPTICS] An illumination technique used in optical microscopes that is a modification of the dark-field method; the central disk is transparent and colored; an annulus of a complementary color fills the remaining condenser aperture; the specimen is seen in the color of the annulus against the background of the central disk. Also known as optical staining.

rhenium [CHEM] A metallic element, symbol Re, atomic number 75, atomic weight 186.2; a transition element.

rheogoniometry [MECH] Rheological tests to determine the various stress and shear actions on Newtonian and non-Newtonian fluids.

rheology [MECH] The study of the deformation and flow of matter, especially non-Newtonian flow of liquids and plastic flow of solids.

rheometer [ENG] An instrument for determining flow properties of solids by measuring relationships between stress, strain, and time.

rheopectic fluid [FL MECH] A fluid for which the structure builds up on shearing; this phenomenon is regarded as the reverse of thixotropy.

rheopexy [PHYS CHEM] A property of certain sols, having particles shaped like rods or plates, which set to gel form more quickly when mechanical means are used to hasten the orientation of the particles.

rheostat [ELEC] A resistor constructed so that its resistance value may be changed without interrupting the circuit to which it is connected. Also known as variable resistor.

rheostriction *See* pinch effect.

rheotron *See* betatron.

RHM *See* roentgen-per-hour-at-one-meter.

rhodium [CHEM] A chemical element, symbol Rh, atomic number 45, atomic weight 102.905.

rhomb *See* rhombohedron.

rhombic antenna [ELECTROMAG] A horizontal antenna having four conductors forming a diamond or rhombus; usually fed at one apex and terminated with a resistance or impedance at the opposite apex. Also known as diamond antenna.

rhombic lattice *See* orthorhombic lattice.

rhombic system *See* orthorhombic system.

rhombohedral lattice [CRYSTAL] A crystal lattice in which the three axes of a unit cell are of equal length, and the three angles between axes are the same, and are not right angles. Also known as trigonal lattice.

rhombohedral system [CRYSTAL] A division of the trigonal crystal system in which the rhombohedron is the basic unit cell.

rhombohedron [CRYSTAL] A trigonal crystal form that is a parallelepiped, the six identical faces being rhombs. Also

RHENIUM

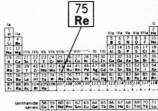

Periodic table of the chemical elements showing the position of rhenium.

RHODIUM

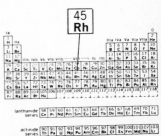

Periodic table of the chemical elements showing the position of rhodium.

known as rhomb. [MATH] A prism with six parallelogram faces.

rhomboid [MATH] A parallelogram whose adjacent sides are not equal.

rhomboidal prism [OPTICS] A prism with four parallel sides and two slanting, or oblique, parallel ends; it will divert the path of light entering its ends without changing the form of the light.

rhombus [MATH] A parallelogram with all sides equal.

rho meson [PARTIC PHYS] Collective name for vector meson resonances belonging to a charge multiplet with total isospin 1, hypercharge 0, negative charge conjugation parity, positive G parity, mass of about 770 MeV, and width of about 146 MeV. Designated $\rho(770)$.

rhumbatron *See* cavity resonator.

ribbon [MATH] The plane figure generated by a straight line which moves so that it is always perpendicular to the path traced by its middle point.

ribbon lightning [GEOPHYS] Ordinary streak lightning that appears to be spread horizontally into a ribbon of parallel luminous streaks when a very strong wind is blowing at right angles to the observer's line of sight; successive strokes of the lightning flash are then displaced by small angular amounts and may appear to the eye or camera as distinct paths. Also known as band lightning; fillet lightning.

Riblet coupler *See* three-decibel coupler.

Riccati-Bessel functions [MATH] Solutions of a second-order differential equation in a complex variable which have the form $zf(z)$, where $f(z)$ is a function in terms of polynomials and cos (z), sin (z).

Riccati equation [MATH] **1.** A first-order differential equation having the form $y' = A_0(x) + A_1(x)y + A_2(x)y^2$; every second-order linear differential equation can be transformed into an equation of this form. **2.** A matrix equation of the form $dP(t)/dt + P(t)F(t) + F^T(t)P(t) - P(t)G(t)R^{-1}(t)G^T(t)P(t) + Q(t) \neq 0$, which frequently arises in control and estimation theory.

Ricci equations [MATH] Equations relating the components of the Ricci tensor, the curvature tensor, and an arbitrary tensor of a Riemann space. Also known as Ricci identities.

Ricci identities *See* Ricci equations.

Ricci scalar *See* scalar curvature.

Ricci tensor *See* contracted curvature tensor.

Ricci theorem [MATH] The covariant derivative vanishes for either of the fundamental tensors of a Riemann space.

rice grains [ASTRON] Bright patches that stand out against the darker background of the surface of the sun; they are short-lived, and the pattern changes in a matter of minutes.

Richardson-Dushman equation [ELECTR] An equation for the current density of electrons that leave a heated conductor in thermionic emission. Also known as Dushman equation.

Richardson number [FL MECH] A dimensionless number used in studying the stratified flow of multilayer systems; equal to the acceleration of gravity times the density gradient of a fluid, divided by the product of the fluid's density and the square of its velocity gradient at a wall. Symbolized N_{Ri}.

Richardson plot [ELECTR] A graph of log (J/T^2) against $1/T$, where J is the current density of electrons leaving a heated conductor in thermionic emission, and T is the temperature of the conductor; according to the Richardson-Dushman equation, this is a straight line.

ridge waveguide [ELECTROMAG] A circular or rectangular waveguide having one or more longitudinal internal ridges that serve primarily to increase transmission bandwidth by lowering the cutoff frequency.

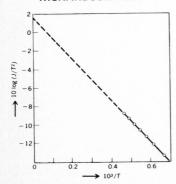

RICHARDSON PLOT

Richardson plot for tungsten, an important thermionic emitter. *(After G. Herrmann and S. Wagener, The Oxide-Coated Cathode, vol. 2, Chapman and Hall, 1951)*

Rieke diagram [ELECTR] A chart showing contours of constant power output and constant frequency for a microwave oscillator, drawn on a Smith chart or other polar diagram whose coordinates represent the components of the complex reflection coefficient at the oscillator load.

Riemann-Christoffel tensor [MATH] The basic tensor used for the study of curvature of a Riemann space; it is a fourth-rank tensor, formed from Christoffel symbols and their derivatives, and its vanishing is a necessary condition for the space to be flat. Also known as curvature tensor.

Riemann equations [FL MECH] Equations for the propagation of sound waves in one-dimensional, compressible fluids, indicating that the quantity $\sigma + v$ is propagated in the positive direction with speed $c + v$, while the quantity $\sigma - v$ is propagated in the negative direction with velocity $c - v$, where v is the fluid velocity, c is the speed of sound for small-amplitude waves, and σ is a function of density ρ given by the integral of $(1/\rho)(\partial P/\partial \rho)^{1/2} \, d\rho$, where P is the pressure.

Riemann function [MATH] A type of Green's function used in solving the Cauchy problem for a real hyperbolic partial differential equation.

Riemann hypothesis [MATH] The conjecture that the only zeros of the Riemann zeta function with positive real part must have their real part equal to $\frac{1}{2}$.

Riemannian curvature [MATH] A general notion of space curvature at a point of a Riemann space which is directly obtained from orthonormal tangent vectors there.

Riemannian geometry *See* elliptic geometry.

Riemannian manifold [MATH] A differentiable manifold where the tangent vectors about each point have an inner product so defined as to allow a generalized study of distance and orthogonality.

Riemannian space-time [RELAT] The space-time of general relativity, having the mathematical structure of a four-dimensional Riemann space.

Riemann integral [MATH] The Riemann integral of a real function $f(x)$ on an interval (a,b) is the unique limit (when it exists) of the sum of $f(a_i)(x_i - x_{i-1})$, $i = 1, \ldots, n$, taken over all partitions of (a,b), $a = x_0 < a_1 < x_1 < \cdots < a_n < x_n = b$, as the maximum distance between x_i and x_{i-1} tends to zero.

Riemann-Lebesgue lemma [MATH] If the absolute value of a function is integrable over the interval where it has a Fourier expansion, then its Fourier coefficients a_n tend to zero as n goes to infinity.

Riemann-Liouville fractional integral [MATH] A generalization of repeated integration of a function f from 0 to a positive number x, given by the integral from 0 to x of $[\Gamma(\alpha)]^{-1}(x - t)^{\alpha-1}f(t) \, dt$, where the real part of α is strictly positive. Also known as Euler transform of the first kind.

Riemann mapping theorem [MATH] Any simply connected domain in the plane with boundary containing more than one point can be conformally mapped onto the interior of the unit disk.

Riemann method [MATH] A method of solving the Cauchy problem for hyperbolic partial differential equations.

Riemann P function [MATH] A scheme for exhibiting the singular points of a second-order ordinary differential equation, and the orders at these points of solutions of the equation.

Riemann scalar *See* scalar curvature.

Riemann space [MATH] A Riemannian manifold or subset of a euclidean space where tensors can be defined to allow a general study of distance, angle, and curvature.

Riemann sphere [MATH] The two-sphere whose points are

identified with all complex numbers by a stereographic projection. Also known as complex sphere.

Riemann standard forms [MATH] The integrals of dt/y, $(1-\lambda t)\,dt/y$, and $dt/(t-a)y$, where $y = \sqrt{t(1-t)(1-\lambda t)}$; they are obtained from the Legendre normal elliptic integrals by making the substitutions $t = x^2$, $\lambda = k^2$, and $a = -1/n^2$.

Riemann-Stieltjes integral [MATH] Let f and g be real-valued functions defined on an interval $[a,b]$; suppose there is a number A such that for each positive ε there is a positive δ for which

$$\left| \sum_{i=1}^{n} f(c_i)[g(x_i) - g(x_{i-1})] - A \right| < \varepsilon$$

for every sequence $\{x_i\}$ with $a = x_0 \leq x_1 \leq \cdots \leq x_n = b$ and max $(x_i - x_{i-1}) < \delta$, and for every sequence $\{c_i\}$ with $x_{i-1} \leq c_i \leq x_i$ for $i = 1, 2, \ldots, n$; the number A is called the Riemann-Stieltjes integral of f with respect to g, denoted

$$A = \int_a^b f\,dg.$$

Riemann surfaces [MATH] Sheets or surfaces obtained by analyzing multiple-valued complex functions and the various choices of principal branches.

Riemann tensors [MATH] Various types of tensors used in the study of curvature for a Riemann space.

Riemann zeta function [MATH] The complex function $\zeta(z)$ defined by an infinite series with nth term $e^{-z \log n}$. Also known as zeta function.

Riesz-Fischer theorem [MATH] The vector space of all real- or complex-valued functions whose absolute value squared has a finite integral constitutes a complete inner product space.

Riesz index [MATH] The smallest integer n such that the null space of $(L-\lambda I)^n$ is the same as the null space of $(L-\lambda I)^{n+1}$, where L is a specified linear transformation, λ is an eigenvalue of L, and I is the identity operator.

Righi experiment [OPTICS] An experiment in which a rotating Nicol prism, a Fresnel mirror, a quarter-wave plate, and a fixed Nicol prism are used to produce effects in light beams similar to beats between sounds with slightly different frequencies.

Righi-Leduc effect [PHYS] The phenomenon wherein, if a magnetic field is applied at right angles to the direction of a temperature gradient in a conductor, a new temperature gradient is produced perpendicular to both the direction of the original temperature gradient and to the magnetic field. Also known as Leduc effect.

right angle [MATH] An angle of 90°.

right-angle prism [OPTICS] A type of prism used to turn a beam of light through a right angle (90°); it will invert (turn upside-down) or will revert (turn right for left), according to the position of the prism, any light reflected by it.

right ascension [ASTRON] A celestial coordinate; the angular distance taken along the celestial equator from the vernal equinox eastward to the hour circle of a given celestial body.

right circular cylinder [MATH] A solid bounded by two parallel planes and by a cylindrical surface consisting of the straight lines perpendicular to the planes and passing through a circle in one of them.

right coset [MATH] A right coset of a subgroup H of a group G is a subset of G consisting of all elements of the form ha, where a is a fixed element of G and h is any element of H.

right-handed coordinate system [MATH] **1.** A three-dimensional rectangular coordinate system such that when the thumb of the right hand extends in the positive direction of the first (or x) axis the fingers fold in the direction in which the second (or y) axis could be rotated about the first axis to coincide with the third (or z) axis. **2.** A Riemann space which has negative scalar density function.

right-hand helicity [QUANT MECH] Property of a particle whose spin is parallel to its momentum.

right-hand limit *See* limit on the right.

right-hand polarization [ELECTROMAG] In elementary particle discussions, circular or elliptical polarization of an electromagnetic wave in which the electric field vector at a fixed point in space rotates in the right-hand sense about the direction of propagation; in optics, the opposite convention is used; in facing the source of the beam, the electric vector is observed to rotate clockwise.

right-hand rule [ELECTROMAG] **1.** For a current-carrying wire, the rule that if the fingers of the right hand are placed around the wire so that the thumb points in the direction of current flow, the fingers will be pointing in the direction of the magnetic field produced by the wire. **2.** For a moving wire in a magnetic field, such as the wire on the armature of a generator, if the thumb, first, and second fingers of the right hand are extended at right angles to one another, with the first finger representing the direction of magnetic lines of force and the second finger representing the direction of current flow induced by the wire's motion, the thumb will be pointing in the direction of motion of the wire. Also known as Fleming's rule.

right-hand taper [ELEC] Taper in which there is greater resistance in the clockwise half of the operating range of a rheostat or potentiometer (looking from the shaft end) than in the counterclockwise half.

right identity [MATH] In a set on which a binary operation ∘ is defined, an element e with the property that $a ∘ e = a$ for every element a in the set.

righting lever [FL MECH] The horizontal distance from the center of mass of a floating body, slightly displaced from the equilibrium position, to a vertical line passing through the center of buoyancy.

right section [MATH] A right section of a surface is the plane section produced by a perpendicular plane.

right triangle [MATH] A triangle one of whose angles is a right angle.

rigid body [MECH] An idealized extended solid whose size and shape are definitely fixed and remain unaltered when forces are applied.

rigid-body dynamics [MECH] The study of the motions of a rigid body under the influence of forces and torques.

rill [ASTRON] A crooked, narrow crack on the moon's surface; may be a kilometer or more in width and a few to several hundred kilometers in length. Also spelled rille.

rille *See* rill.

ring [MATH] **1.** An algebraic system with two operations called multiplication and addition; the system is a commutative group relative to addition, and multiplication is associative, and is distributive with respect to addition. **2.** A ring of sets is a collection of sets where the union and difference of any two members is also a member.

ring circuit [ELECTROMAG] In waveguide practice, a hybrid T junction having the physical configuration of a ring with radial branches.

ring counter [ELECTR] A loop of binary scalers or other bistable units so connected that only one scaler is in a specified state at any given time; as input signals are counted, the position of the one specified state moves in an ordered sequence around the loop.

ring current [GEOPHYS] A westward electric current which is believed to circle the earth at an altitude of several earth radii during the main phase of geomagnetic storms, resulting in a large worldwide decrease in the geomagnetic field horizontal component at low latitudes.

ring discharge [ELECTR] A ring-shaped discharge generated by a high-frequency oscillating electromagnetic field produced by an external coil. Also known as toroidal discharge.

ring galaxy [ASTRON] A class of galaxy whose ringlike structure has clumps of ionized hydrogen clouds on its periphery, may have a nucleus of stars, and is usually accompanied by a small galaxy; probably formed when a small galaxy crashes through the disk of a spiral galaxy.

ringing [CONT SYS] An oscillatory transient occurring in the output of a system as a result of a sudden change in input.

ringing circuit [ELECTR] A circuit which has a capacitance in parallel with a resistance and inductance, with the whole in parallel with a second resistance; it is highly underdamped and is supplied with a step or pulse input.

ring isomorphism [MATH] An isomorphism between rings.

ring micrometer [OPTICS] A flat, thin ring in the focal plane of a telescope; used to measure differences in right ascension and declination.

ring of fractions *See* quotient ring.

ring of operators *See* von Neumann algebra.

rings and brushes [OPTICS] An interference pattern produced by ordinary and extraordinary rays when a uniaxial crystal is placed between two polarizers.

ring theory [MATH] The study of the structure of rings in algebra.

ring vortex *See* vortex ring.

riometer [ENG] An instrument that measures changes in ionospheric absorption of electromagnetic waves by determining and recording the level of extraterrestrial cosmic radio noise. Derived from relative ionospheric opacity meter.

ripple [ELEC] The alternating-current component in the output of a direct-current power supply, arising within the power supply from incomplete filtering or from commutator action in a dc generator. [FL MECH] *See* capillary wave.

ripple filter [ELECTR] A low-pass filter designed to reduce ripple while freely passing the direct current obtained from a rectifier or direct-current generator. Also known as smoothing circuit; smoothing filter.

ripple quantity [PHYS] Alternating component of a pulsating quantity when this component is small relative to the constant component.

ripple tank [PHYS] A shallow tray containing a liquid and equipped with means for generating surface waves; used to illustrate several types of wave phenomena, such as interference and diffraction.

ripple voltage [ELEC] The alternating component of the unidirectional voltage from a rectifier or generator used as a source of direct-current power.

rise time [CONT SYS] The time it takes for the output of a system to change from a specified small percentage (usually 5 or 10%) of its steady-state increment to a specified large percentage (usually 90 or 95%). [ELEC] The time for the pointer of an electrical instrument to make 90% of the change to its final value when electric power suddenly is applied from a source whose impedance is high enough that it does not affect damping.

Risley prism system [OPTICS] A type of dispersing prism used to test ocular convergence in ophthalmology; consists of

RISLEY PRISM SYSTEM

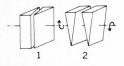

Diagram of Risley prism system. In orientation 1, combined deviation of prisms is zero; when both have been rotated 90° as in orientation 2, combined deviation is at a maximum.

two thin prisms mounted so that they can be rotated simultaneously in opposite directions.

Ritchey-Chrétien optics [OPTICS] A modification of the Cassegrain optical system used in large optical telescopes; it has a hyperbolic image-forming primary mirror, no spherical aberration, and no coma; it has a larger usable field than either Newtonian or Cassegrain optical systems.

Ritchie's experiment [THERMO] An experiment that uses a Leslie cube and a differential air thermometer to demonstrate that the emissivity of a surface is proportional to its absorptivity.

Ritchie wedge [OPTICS] A photometer in which a test source and a standard source of light illuminate two perpendicular white, diffusing surfaces which intersect in a movable wedge, and these surfaces are viewed from a direction perpendicular to a line connecting the sources.

Ritz formula [ATOM PHYS] A particular expansion of an equation used in studying the spectra of atoms.

Ritz method [MATH] A method of solving boundary value problems based upon reformulating the given problem as a minimization problem.

Ritz's combination principle [SPECT] The empirical rule that sums and differences of the frequencies of spectral lines often equal other observed frequencies. Also known as combination principle.

RK galaxy [ASTRON] A type of ring galaxy which consists of a ringlike structure with a large, bright condensation or knot within the ring itself.

R meson [PARTIC PHYS] A meson resonance observed in missing mass experiments involving scattering of negative pions by protons; it has been resolved into three peaks labeled R_1, R_2, and R_3, having masses of 1640, 1700, and 1750 MeV respectively, and is believed to have a spin of 3.

R meter [NUCLEO] An ionization instrument calibrated to indicate the intensity of γ-rays, x-rays, and other ionizing radiation in roentgens.

rms value See root-mean-square value.

Rn See radon.

RN galaxy [ASTRON] A type of ring galaxy which consists of a ringlike structure with a nucleus somewhere within it, the nucleus being somewhat like those seen in ordinary spiral galaxies but typically lying off the center of the ring.

Robertson-Walker solutions [RELAT] A class of general relativistic models for the universe; divided into three classes depending on whether spatial sections are open, closed, or flat; conventionally accepted as describing the real universe.

Robin law [PHYS] The law that an increase in pressure on a system in chemical or physical equilibrium favors the system formed with a decrease in volume, and conversely, a change in pressure does not affect a system formed with no change in volume.

Robin's problem [MATH] The determination of a harmonic function f within a finite region of three-dimensional space enclosed by a closed surface, when the function satisfies the equation $uf + v(\partial f/\partial n) = w$ on the surface, where u, v, and w are specified continuous functions on the surface, and $\partial f/\partial n$ is the normal derivative of f with respect to the surface.

roc [ELEC] A unit of electrical conductivity equal to the conductivity of a material in which an electric field of 1 volt per centimeter gives rise to a current density of 1 ampere per square centimeter. Derived from reciprocal ohm centimeter.

Roche lobes [MECH] **1.** Regions of space surrounding two massive bodies revolving around each other under their mutual gravitational attraction, such that the gravitational attraction of each body dominates the lobe surrounding it. **2.**

R MESON

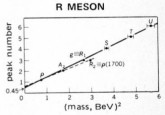

(Mass)² plot for the prominent mesonic (MM) peaks observed in the reaction $\pi^- p \rightarrow p$ (MM)⁻ with the Missing Mass spectrometer at CERN (Geneva), against their order in this sequence. The straight-line fit has generally been interpreted as a Regge trajectory, the ordinate being identified with the spin J of the resonance state. The R-meson peak has been resolved into several peaks, and spin 3 is associated at least with the peak R_2 at 1700 MeV: a Regge trajectory associated with the ρ-meson (spin 1) would then be expected to pass through the point R_2, as shown by the dashed line.

In particular, the effective potential energy (referred to a system of coordinates rotating with the bodies) is equal to a constant V_0 over the surface of the lobes, and if a particle is inside one of the lobes and if the sum of its effective potential energy and its kinetic energy is less than V_0, it will remain inside the lobe.

Roche's limit [ASTROPHYS] The limiting distance below which a satellite orbiting a celestial body would be disrupted by the tidal forces generated by the gravitational attraction of the primary; the distance depends on the relative densities of the bodies and on the orbit of the satellite; it is computed by $R = 2.45(Lr)$, where L is a factor that depends on the relative densities of the satellite and the body, R is the radius of the satellite's orbit measured from the center of the primary body, and r is the radius of the primary body; if the satellite and the body have the same density, the relationship is $R = 2.45r$.

Rochon polarizing prism [OPTICS] A device for producing linearly polarized beams of light, consisting of two adjacent quartz wedges, the first of which has its optic axis parallel to the beam, while the second has its optic axis perpendicular to the beam; one of the beams (the ordinary ray) is undeviated, and is therefore not spread into a spectrum.

rocket astronomy [ASTRON] The discipline comprising measurements of the electromagnetic radiation from the sun, planets, stars, and other bodies, of wavelengths that are almost completely absorbed below the 250-kilometer level, by using a rocket to carry instruments above 250 kilometers to measure these phenomena.

rocky point effect [ELECTR] Transient but violent discharges between electrodes in high-voltage transmitting tubes.

rod [MECH] *See* perch. [NUCLEO] A relatively long and slender body of material used in, or in conjunction with, a nuclear reactor; may contain fuel, absorber, or fertile material or other material in which activation or transmutation is desired.

Rodrigues formula [MATH] **1.** The equation giving the nth function in a class of special functions in terms of the nth derivatives of some polynomial. **2.** The formula $d\mathbf{n} + k\,d\mathbf{r} = 0$, expressing the difference $d\mathbf{n}$ in the unit normals to a surface at two neighboring points on a line of curvature, in terms of the difference $d\mathbf{r}$ in the position vectors of the two points and the principal curvature k. **3.** The formula for a matrix that is used to transform the cartesian coordinates of a vector in three-space under a rotation through a specified angle about an axis with specified direction cosines.

rod thermistor [ELECTR] A type of thermistor that has high resistance, long time constant, and moderate power dissipation; it is extruded as a long vertical rod 0.250–2.0 inches (0.63–5.1 centimeters) long and 0.050–0.110 inch (0.13–0.28 centimeter) in diameter, of oxide-binder mix and sintered; ends are coated with conducting paste and leads are wrapped on the coated area.

roentgen [NUCLEO] An exposure dose of x- or γ-radiation such that the electrons and positrons liberated by this radiation produce, in air, when stopped completely, ions carrying positive and negative charges of 2.58×10^{-4} coulomb per kilogram of air. Abbreviated R (formerly r). Also spelled röntgen.

roentgen diffractometry *See* x-ray crystallography.

roentgen equivalent man *See* rem.

roentgen equivalent physical *See* rep.

roentgenography [PHYS] Radiography by means of x-rays.

roentgenoluminescence [PHYS] Luminescence which can be produced by x-rays.

roentgen optics *See* x-ray optics.

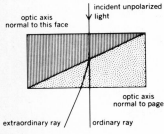

ROCHON POLARIZING PRISM

incident unpolarized light

optic axis normal to this face

optic axis normal to page

extraordinary ray

ordinary ray

Rochan prism.

roentgen-per-hour-at-one-meter [NUCLEO] A unit of γ-ray source strength, corresponding to a dose rate of 1 roentgen per hour at a distance of 1 meter in air. Symbolized RHM.

roentgen rays *See* x-rays.

Roget's spiral [ELEC] A spiral wire, suspended vertically with the lower end in mercury, that is made to go through a cycle in which an electric current passing through the wire produces mutual attraction between the coils, causing the wire to lift out of the mercury and breaking the current; the spiral then expands under its own weight, so that the lower end drops back into the mercury and the current is reestablished.

roll [MECH] Rotational or oscillatory movement of an aircraft or similar body about a longitudinal axis through the body; it is called roll for any degree of such rotation.

roll acceleration [MECH] The angular acceleration of an aircraft or missile about its longitudinal or X axis.

roll axis [MECH] A longitudinal axis through an aircraft, rocket, or similar body, about which the body rolls.

Rolle's theorem [MATH] If a function $f(x)$ is continuous on the closed interval $[a,b]$ and differentiable on the open interval (a,b) and if $f(a) = f(b)$, then there exists x_0, $a < x_0 < b$, such that $f'(x_0) = 0$.

Rollin film *See* helium film.

rolling [MECH] Motion of a body across a surface combined with rotational motion of the body so that the point on the body in contact with the surface is instantaneously at rest.

rolling contact [MECH] Contact between bodies such that the relative velocity of the two contacting surfaces at the point of contact is zero.

rolling friction [MECH] A force which opposes the motion of any body which is rolling over the surface of another.

roll-off [ELECTR] Gradually increasing loss or attenuation with increase or decrease of frequency beyond the substantially flat portion of the amplitude-frequency response characteristic of a system or transducer.

rom [ELEC] A unit of electrical conductivity, equal to the conductivity of a material in which an electric field of 1 volt per meter gives rise to a current density of 1 ampere per square meter. Derived from reciprocal ohm meter.

Römer method [OPTICS] A method of measuring the speed of light, in which apparent changes in the periods of satellites of another planet, such as Jupiter, whose distance from the earth is known, are observed throughout the year.

Ronchi test [OPTICS] An improvement on the Foucault knife-edge test for testing curved mirrors, in which the knife edge is replaced with a transmission grating with 15–80 lines per centimeter, and the pinhole source is replaced with a slit or a section of the same grating.

röntgen *See* roentgen.

rood [MECH] A unit of area, equal to ¼ acre, or 10,890 square feet, or 1011.7141056 square meters.

room acoustics [ACOUS] The study of the behavior of sound waves in an enclosed room.

root [MATH] **1.** A root of a given real or complex number is a number which when raised to some exponent equals that number. Also known as radix. **2.** A root of a polynomial $p(x)$ is a number a such that $p(a) = 0$.

root locus plot [CONT SYS] A plot in the complex plane of values at which the loop transfer function of a feedback control system is a negative number.

root-mean-square current *See* effective current.

root-mean-square deviation [STAT] The square root of the sum of squared deviations from the mean divided by the number of observations in the sample.

root-mean-square error [STAT] The square root of the second moment corresponding to the frequency function of a random variable.

root-mean-square sound pressure *See* effective sound pressure.

root-mean-square value Abbreviated rms value. [PHYS] The square root of the time average of the square of a quantity; for a periodic quantity the average is taken over one complete cycle. Also known as effective value. [STAT] The square root of the average of the squares of a series of related values.

root of an equation [MATH] A number or quantity which satisfies a given equation.

root of unity [MATH] A root of unity in a field F is an element a in F such that $a^n = 1$ for some positive integer n.

root-squaring methods [MATH] Methods of solving algebraic equations which involve calculating the coefficients in a sequence of equations, each of which has roots which are the squares of the roots in the previous equation.

root-sum-square value [PHYS] The square root of the sum of the squares of a series of related values; commonly used to express total harmonic distortion.

root test [MATH] An infinite series of nonnegative terms a_n converges if, after some term, the ith root of a_i is less than a fixed number smaller than 1.

Rosa and Dorsey method [ELECTROMAG] A method of measuring the speed of light by comparing the capacitance of a capacitor in electromagnetic units, as measured experimentally, with values of currents determined from a current balance, to the capacitance of the same capacitor in electrostatic units, as determined from its geometrical dimensions.

rose [MATH] A graph consisting of loops shaped like rose petals arising from the equations in polar coordinates $r = a \sin n\theta$ or $r = a \cos n\theta$.

Rossby diagram [THERMO] A thermodynamic diagram, named after its designer, with mixing ratio as abscissa and potential temperature as ordinate; lines of constant equivalent potential temperature are added.

Rossby number [FL MECH] The nondimensional ratio of the inertial force to the Coriolis force for a given flow of a rotating fluid, given as $R_0 = U/fL$, where U is a characteristic velocity, f the Coriolis parameter (or, if the system is cylindrical rather than spherical, twice the system's rotation rate), and L a characteristic length.

Rossby parameter [FL MECH] The northward variation of the Coriolis parameter, arising from the sphericity of the earth. Also known as Rossby term.

Rossby regime [FL MECH] A type of flow pattern in a rotating fluid with differential radial heating in which the major radial transport of shear and momentum is effected by horizontal eddies of low wave-number; this regime occurs for low values of the Rossby number (of the order of 0.1).

Rossby term *See* Rossby parameter.

Ross objective [OPTICS] A type of wide-field lens objective in cameras used for astrometric work.

rot *See* curl.

rotary coupler *See* rotating joint.

rotary dispersion [OPTICS] The change in the angle through which an optically active substance rotates the plane of polarization of plane polarized light as the wavelength of the light is varied. Also known as rotatory dispersion.

rotary joint *See* rotating joint.

rotary polarization *See* optical activity.

rotary voltmeter [ENG] Type of electrostatic voltmeter used for measuring high voltages.

ROSS OBJECTIVE

The four component lenses of the Ross objective.

rotating coordinate system [MECH] A coordinate system whose axes as seen in an inertial coordinate system are rotating.

rotating crystal method [SOLID STATE] Any method of studying crystalline structures by x-ray or neutron diffraction in which a monochromatic, collimated beam of x-rays or neutrons falls on a single crystal that is rotated about an axis perpendicular to the beam.

rotating-cylinder method [FL MECH] A method of measuring the viscosity of a fluid in which the fluid fills the space between two concentric cylinders, and the torque on the stationary inner cylinder is measured when the outer cylinder is rotated at constant speed.

rotating joint [ELECTROMAG] A joint that permits one section of a transmission line or waveguide to rotate continuously with respect to another while passing radio-frequency energy. Also known as rotary coupler; rotary joint.

rotating Reynolds number [FL MECH] A nondimensional number arising in problems of a rotating viscous fluid and, in particular, in problems involving the agitation of such a fluid by an impeller, equal to the product of the square of the impeller's diameter and its angular velocity divided by the kinematic viscosity of the fluid. Symbolized Re_r.

rotating viscometer vacuum gage [ENG] Vacuum (reduced-pressure) measurement device in which the torque on a spinning armature is proportional to the viscosity (and the pressure) of the rarefied gas being measured; sensitive for absolute pressures of 1 millimeter of mercury (133.32 newtons per square meter), down to a few tens of micrometers.

rotating wedge [OPTICS] A circular optical wedge mounted to be rotated in the path of light to divert the line of sight to a limited degree.

rotation [MATH] *See* curl. [MECH] Also known as rotational motion. **1.** Motion of a rigid body in which either one point is fixed, or all the points on a straight line are fixed. **2.** Angular displacement of a rigid body. **3.** The motion of a particle about a fixed point.

rotational constant [PHYS CHEM] That constant inversely proportioned to the moment of inertia of a linear molecule; used in calculations of microwave spectroscopy quantums.

rotational energy [MECH] The kinetic energy of a rigid body due to rotation. [PHYS CHEM] For a diatomic molecule, the difference between the energy of the actual molecule and that of an idealized molecule which is obtained by the hypothetical process of gradually stopping the relative rotation of the nuclei without placing any new constraint on their vibration, or on motions of electrons.

rotational field [PHYS] A vector field whose curl does not vanish. Also known as circuital field; vortical field.

rotational flow [FL MECH] Flow of a fluid in which the curl of the fluid velocity is not zero, so that each minute particle of fluid rotates about its own axis. Also known as rotational motion.

rotational impedance [MECH] A complex quantity, equal to the phasor representing the alternating torque acting on a system divided by the phasor representing the resulting angular velocity in the direction of the torque at its point of application. Also known as mechanical rotational impedance.

rotational inertia *See* moment of inertia.

rotational level [PHYS CHEM] An energy level of a diatomic or polyatomic molecule characterized by a particular value of the rotational energy and of the angular momentum associated with the motion of the nuclei.

rotational motion *See* rotation; rotational flow.

rotational quantum number [PHYS CHEM] A quantum number J characterizing the angular momentum associated with the motion of the nuclei of a molecule; the angular momentum is $(h/2\pi)\sqrt{J(J+1)}$ and the largest component is $(h/2\pi)J$, where h is Planck's constant.

rotational reactance [MECH] The imaginary part of the rotational impedance. Also known as mechanical rotational reactance.

rotational resistance [MECH] The real part of rotational impedance; it is responsible for dissipation of energy. Also known as mechanical rotational resistance.

rotational spectrum [SPECT] The molecular spectrum resulting from transitions between rotational levels of a molecule which behaves as the quantum-mechanical analog of a rotating rigid body.

rotational stability [MECH] Property of a body for which a small angular displacement sets up a restoring torque that tends to return the body to its original position.

rotational sum rule [SPECT] The rule that, for a molecule which behaves as a symmetric top, the sum of the line strengths corresponding to transitions to or from a given rotational level is proportional to the statistical weight of that level, that is, to $2J+1$, where J is the total angular momentum quantum number of the level.

rotational transform [PL PHYS] Property possessed by a magnetic field, in a system used to confine a plasma, in which magnetic lines of force do not close in on themselves after making a circuit around the system, but are rotationally displaced.

rotational viscometer *See* Couette viscometer.

rotation axis [CRYSTAL] A symmetry element of certain crystals in which the crystal can be brought into a position physically indistinguishable from its original position by a rotation through an angle of $360°/n$ about the axis, where n is the multiplicity of the axis, equal to 2, 3, 4, or 6. Also known as symmetry axis.

rotation camera [SOLID STATE] An instrument for studying crystalline structure by x-ray or neutron diffraction, in which a monochromatic, collimated beam of x-rays or neutrons falls on a single crystal which is rotated about an axis perpendicular to the beam and parallel to one of the crystal axes, and the various diffracted beams are registered on a cylindrical film concentric with the axis of rotation.

rotation coefficients [MECH] Factors employed in computing the effects on range and deflection which are caused by the rotation of the earth; they are published only in firing tables involving comparatively long ranges.

rotation group [MATH] The group consisting of all orthogonal matrices or linear transformations having determinant 1.

rotation-inversion axis [CRYSTAL] A symmetry element of certain crystals in which a crystal can be brought into a position physically indistinguishable from its original position by a rotation through an angle of $360°/n$ about the axis followed by an inversion, where n is the multiplicity of the axis, equal to 1, 2, 3, 4, or 6. Also known as inversion axis.

rotation moment *See* torque.

rotation-reflection axis [CRYSTAL] A symmetry element of certain crystals in which a crystal can be brought into a position physically indistinguishable from its original position by a rotation through an angle of $360°/n$ about the axis followed by a reflection in the plane perpendicular to the axis, where n is the multiplicity of the axis, equal to 1, 2, 3, 4, or 6.

rotation Reynolds number *See* rotating Reynolds number.

rotation spectrum [PHYS CHEM] Absorption-spectrum (absorbed electromagnetic energy) wavelengths produced if only

ROTATION CAMERA

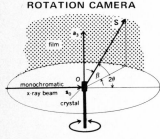

Schematic of rotation camera. Single crystal is located at O. 2θ represents angle between primary beam s_0 and diffracted beam S. β represents angle that S makes with plane perpendicular to crystal axis a_3 about which crystal is rotated.

the rotational energy of a molecule is affected during excitation.

rotation twin [CRYSTAL] A twin crystal in which the parts will coincide if one part is rotated 180° (sometimes 30, 60, or 120°).

rotation-vibration spectrum [PHYS CHEM] Absorption-spectrum (absorbed electromagnetic energy) wavelengths produced when both the energy of vibration and energy of rotation of a molecule are affected by excitation.

rotator [ELECTROMAG] A device that rotates the plane of polarization of a plane-polarized electromagnetic wave, such as a twist in a waveguide. [MECH] A rotating rigid body. [QUANT MECH] A molecule or other quantum-mechanical system which behaves as the quantum-mechanical analog of a rotating rigid body. Also known as top.

rotatory [OPTICS] Having the capability to rotate the plane of polarization of polarized electromagnetic radiation.

rotatory dispersion *See* rotary dispersion.

rotatory power [OPTICS] A substance's capability to rotate the plane of polarization of polarized electromagnetic radiation.

roton [PHYS] A quantum of rotational motion in a liquid, such as superfluid helium.

rotor [ELEC] The rotating member of an electrical machine or device, such as the rotating armature of a motor or generator, or the rotating plates of a variable capacitor. [MECH ENG] *See* impeller.

rotor plate [ELEC] One of the rotating plates of a variable capacitor, usually directly connected to the metal frame.

Rouche's theorem [MATH] If analytic functions $f(z)$ and $g(z)$ in a simply connected domain satisfy on the boundary $|g(z)| < |f(z)|$, then $f(z)$ and $f(z) + g(z)$ have the same number of zeros in the domain.

roughness [FL MECH] Distance from peaks to valleys in pipe-wall irregularities; used to modify Reynolds number calculations for fluid flow through pipes.

roughness factor [FL MECH] A correction factor used in fluid-flow calculations to allow for flow resistance caused by the roughness of the surface over which the fluid must flow.

rounding [MATH] Dropping or neglecting decimals after some significant place. Also known as truncation.

rounding error [MATH] The computational error due to always rounding numbers in a calculation. Also known as round-off error.

round off [MATH] To truncate the least significant digit or digits of a numeral, and adjust the remaining numeral to be as close as possible to the original number.

round-off error *See* rounding error.

Rousseau diagram [OPTICS] A geometric construction used to determine the total luminous flux of a lamp from a number of polar diagrams which give the effective luminous intensity of the lamp in various directions.

Routh's procedure [MECH] A procedure for modifying the Lagrangian of a system so that the modified function satisfies a modified form of Lagrange's equations in which ignorable coordinates are eliminated.

Routh's rule [MATH] A rule that the number of roots with positive real parts of an algebraic equation is equal to the number of changes of algebraic sign of a sequence whose terms are formed from coefficients of the equation in a specified manner. Also known as Routh test.

Routh's rule of inertia [MECH] The rule that the moment of inertia of a body about an axis of symmetry equals $M(a^2 + b^2)/n$, where M is the body's mass, a and b are the lengths of the body's two other perpendicular semiaxes, and n equals 3,

ROWLAND CIRCLE

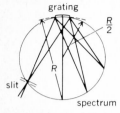

R represents a radius of the circle of which the curved surface of the grating forms a part. *R*/2 represents a radius of the Rowland circle. The camera may be located anywhere on this circle.

RUBIDIUM

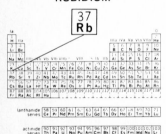

Periodic table of the chemical elements showing the position of rubidium.

4, or 5 depending on whether the body is a rectangular parallelepiped, elliptic cylinder, or ellipsoid, respectively.

Routh test *See* Routh's rule.

rowland [SPECT] A unit of length, formerly used in spectroscopy, equal to 999.81/999.94 angstrom, or approximately 0.99987×10^{-10} meter.

Rowland circle [SPECT] A circle drawn tangent to the face of a concave diffraction grating at its midpoint, having a diameter equal to the radius of curvature of a grating surface; the slit and camera for the grating should lie on this circle.

Rowland ghost [SPECT] A false spectral line produced by a diffraction grating, arising from periodic errors in groove position.

Rowland grating *See* concave grating.

Rowland mounting [SPECT] A mounting for a concave grating spectrograph in which camera and grating are connected by a bar forming a diameter of the Rowland circle, and the two run on perpendicular tracks with the slit placed at their junction.

Rowland ring [ELECTROMAG] A ring-shaped sample of magnetic material, generally surrounded by a coil of wire carrying a current.

rpm *See* revolution per minute.

rps *See* revolution per second.

RR Lyrae stars [ASTRON] Pulsating variable stars with a period of 0.05–1.2 days in the halo population of the Milky Way Galaxy; color is white, and they are mostly stars of spectral class A. Also known as cluster cepheids; cluster variables.

RTG *See* radioisotope thermoelectric generator.

RTL *See* resistor-transistor logic.

Ru *See* ruthenium.

rubidium [CHEM] A chemical element, symbol Rb, atomic number 37, atomic weight 85.47; a reactive alkali metal; salts of the metal may be used in glass and ceramic manufacture.

rubidium magnetometer *See* rubidium-vapor magnetometer.

rubidium-vapor frequency standard [PHYS] An atomic frequency standard in which the frequency is established by a gas cell containing rubidium vapor and a neutral buffer gas.

rubidium-vapor magnetometer [ENG] A highly sensitive magnetometer in which the spin precession principle is combined with optical pumping and monitoring for detecting and recording variations as small as 0.01 gamma (0.1 micro-oersted) in the total magnetic field intensity of the earth. Also known as rubidium magnetometer.

ruby laser [OPTICS] An optically pumped solid-state laser that uses a ruby crystal to produce an intense and extremely narrow beam of coherent red light.

ruby maser [PHYS] A maser that uses a ruby crystal in the cavity resonator.

ruled surface [MATH] A surface that can be generated by the motion of a straight line.

rule of false cofactors [MATH] This states that if A is an $n \times n$ matrix with elements a_{ij} and cofactors A_{ij}, then the sum of $a_{ik}A_{jk}$ and the sum of $a_{ki}A_{kj}$ over $k = 1, 2, \ldots, n$ are both equal to 0 when i does not equal j.

rule of false position *See* regula falsi.

ruling engine [SPECT] A machine operated by a long micrometer screw which rules equally spaced lines on an optical diffraction grating.

run [STAT] The occurrence of the same characteristic in a series of observations; can be used to test whether or not two random samples come from populations having the same frequency distribution.

runaway electron [ELECTR] An electron, in an ionized gas to which an electric field is applied, that gains energy from the field faster than it loses energy by colliding with other particles in the gas.

Runge-Kutta method [MATH] A numerical approximation technique for solving differential equations.

Runge vector [MECH] A vector which describes certain unchanging features of a nonrelativistic two-body interaction obeying an inverse-square law, either in classical or quantum mechanics; its constancy is a reflection of the symmetry inherent in the inverse-square interaction.

R unit *See* German R unit; Solomon R unit.

Russell-Saunders coupling [PHYS] A process for building many-electron single-particle eigenfunctions of orbital angular momentum and spin; the orbital functions are combined to make an eigenfunction of the total orbital angular momentum, the spin functions are combined to make an eigenfunction of the total spin angular momentum, and then the results are combined into eigenfunctions of the total angular momentum of the system. Also known as LS coupling.

Russell's paradox [MATH] The paradox concerning the concept of all sets which are not members of themselves which forces distinctions in set theory between sets and classes.

ruthenium [CHEM] A chemical element, symbol Ru, atomic number 44, atomic weight 101.07.

rutherford [NUCLEO] Abbreviated rd. **1.** A unit used to express the decay rate of radioactive material, equal to 10^6 disintegrating atoms per second. **2.** That amount of a substance which is undergoing 10^6 disintegrations per second.

Rutherford nuclear atom [ATOM PHYS] A theory of atomic structure in which nearly all the mass is concentrated in a small nucleus, electrons surrounding the nucleus fill nearly all the atom's volume, the number of these electrons equals the atomic number, and the positive charge on the nucleus is equal in magnitude to the negative charge of the electrons.

Rutherford scattering [ATOM PHYS] Scattering of heavy charged particles by the Coulomb field of an atomic nucleus.

RV Tauri stars [ASTRON] A class of stars; they are long-period pulsating variable types with periods from about 50 to 150 days; otherwise they are like the shorter-period W Virginis stars; they are found in both the Milky Way Galaxy and the globular clusters.

RW Aurigae stars [ASTRON] A class of stars that are variable, and whose light variations are rapid and irregular.

R wave *See* Rayleigh wave.

ry *See* rydberg.

rydberg [ATOM PHYS] A unit of energy used in atomic physics, equal to the square of the charge of the electron divided by twice the Bohr radius; equal to 13.60583 ± 0.00004 electron volts. Symbolized ry. [SPECT] *See* kayser.

Rydberg constant [ATOM PHYS] **1.** An atomic constant which enters into the formulas for wave numbers of atomic spectra, equal to $2\pi^2me^4/ch^3$, where m and e are the rest mass and charge of the electron, c is the speed of light, and h is Planck's constant; equal to $109,737.31 \pm 0.01$ inverse centimeters. Symbolized R_∞. **2.** For any atom, the Rydberg constant (first definition) divided by $1 + m/M$, where m and M are the masses of an electron and of the nucleus.

Rydberg correction [ATOM PHYS] A term inserted into a formula for the energy of a single electron in the outermost shell of an atom to take into account the failure of the inner electron shells to screen the nuclear charge completely.

Rydberg series formula [SPECT] An empirical formula for the wave numbers of various lines of certain spectral series such as neutral hydrogen and alkali metals; it states that the wave number of the nth member of the series is λ_∞ —

RUTHENIUM

Periodic table of the chemical elements showing the position of ruthenium.

$R/(n+a)^2$, where λ_∞ is the series limit, R is the Rydberg constant of the atom, and a is an empirical constant.

Rydberg spectrum [SPECT] An ultraviolet absorption spectrum produced by transitions of atoms of a given element from the ground state to states in which a single electron occupies an orbital farther from the nucleus.

s *See* second; strange quark.

S *See* secondary winding; siemens; stoke; sulfur.

Saalschütz theorem [MATH] **1.** The equation

$$\Gamma(z) = \int_0^\infty t^{z-1} (e^{-t} - 1 + t - t^2/2! + \cdots$$
$$+ (-1)^{k+1} t^k/k!) \, dt$$

is satisfied for negative values of the real part of z, $\mathrm{Re}(z)$, where k is the integer such that $k < -\mathrm{Re}(z) < k + 1$. **2.** The equation

$${}_3F_2(a, b, -n; c, 1 + a + b - c - n; 1) = \frac{(c-a)_n(c-b)_n}{(c)_n(c-a-b)_n}$$

is valid for $n = 1, 2, 3, \ldots$, where ${}_3F_2$ is a generalized hypergeometric function and $(a)_n$ is Pochhammer's symbol.

Sabattier effect [GRAPHICS] Formation of a positive image on a developed photographic plate or film exposed to diffuse light; the image is formed by exposure of silver halide in undeveloped regions of the emulsion. Also known as solarization.

sabin [ACOUS] A unit of sound absorption for a surface, equivalent to 1 square foot (0.09290304 square meter) of perfectly absorbing surface. Also known as absorption unit; open window unit (OW unit); square-foot unit of absorption.

Sabine formula [ACOUS] An empirical equation for the reverberation time of sound in a room; its form is identical to that of the Franklin equation.

Sackur-Tetrode equation [STAT MECH] An equation for the translational entropy of an ideal gas made up of free fermions.

saddle point [MATH] A point where all the first partial derivatives of a function vanish but which is not a local maximum or minimum.

saddle point method *See* steepest descent method.

saddle point theory [MATH] The study of differentiable functions and their derivatives from the viewpoint of saddle points, especially applicable to the calculus of variations.

safety button [NUCLEO] A device worn by workers exposed to nuclear radiation to warn of excessive exposure.

safety factor *See* factor of safety.

safety rod [NUCLEO] A control rod capable of shutting down a reactor quickly in case of failure of the ordinary control system using regulating rods and shim rods; a safety rod may be suspended above the core by a magnetic coupling and allowed to fall in when power reaches a predetermined level. Also known as scram rod.

sagitta [MATH] The distance between the midpoint of an arc and the midpoint of its chord.

sagittal coma [OPTICS] The radius of the circle formed in the focal plane by rays from an off-axis point passing near the edge of a lens that displays coma.

sagittal focus   *See* secondary focus.

sagittal plane   [OPTICS] A plane that is perpendicular to the meridional plane of an optical system and contains a specified ray. Also known as equatorial plane.

sagittal surface   [OPTICS] A surface containing the secondary foci of points in a plane perpendicular to the optical axis of an astigmatic system.

Sagittarius star cloud   [ASTRON] A large star cloud within the Milky Way; its extension is about 1500 to 6000 light-years (1.42×10^{19} to 5.68×10^{19} meters) from the sun.

Saha ionization   [STAT MECH] The ionization of a gas which exists when the gas is in thermal equilibrium at a given temperature, in the absence of external influences; it increases with increasing temperature. Also known as thermal ionization.

Saha's equation   [STAT MECH] An equation for Saha ionization of a monatomic gas in terms of the temperature and pressure of the gas, the ionization potential, and statistical weights of ion, electron, and atom.

Saint Elmo's fire   [ELEC] A visible electric discharge, sometimes seen on the mast of a ship, on metal towers, and on projecting parts of aircraft, due to concentration of the atmospheric electric field at such projecting parts.

Saint Venant's principle   [MECH] The strains that result from application, to a small part of a body's surface, of a system of forces that are statically equivalent to zero force and zero torque become negligible at distances which are large compared with the dimensions of the part.

Sakata-Taketani equation   [QUANT MECH] A relativistic wave equation for a particle with spin 1 whose form resembles that of the nonrelativistic Schrödinger equation.

Salam-Weinberg theory   *See* Weinberg-Salam theory.

salt screen   [GRAPHICS] An intensifying screen consisting of a chemical salt which fluoresces when bombarded by x-rays.

saltus   *See* oscillation of a function.

samarium   [CHEM] Group III rare-earth metal, atomic number 62, symbol Sm; melts at 1350°C, tarnishes in air, ignites at 200–400°C.

samarium-cobalt magnet   [ELECTROMAG] A rare-earth permanent magnet that is more efficient, has lower leakage and greater resistance to demagnetization, and can be magnetized to higher levels than conventional permanent magnets.

sample   [STAT] A selection of a certain collection from a larger collection.

sample correlation coefficient   [STAT] The ratio of the sample covariance of x and y to the standard deviation of x times the standard deviation of y. Also known as product-moment coefficient.

sampled data   [STAT] Data that are obtained at discrete rather than continuous intervals.

sampled-data control system   [CONT SYS] A form of control system in which the signal appears at one or more points in the system as a sequence of pulses or numbers usually equally spaced in time.

sample design   [STAT] A procedure or plan drawn up before any data are collected to obtain a sample from a given population. Also known as sampling plan; survey design.

sample function   [STAT] A function or procedure which, when applied repeatedly to a given population, produces a collection of samples.

sample path   [MATH] If $\{X_t: t \text{ in } T\}$ is a stochastic process, a sample path for the process is the function on T to the range of the process which assigns to each t the value $X_t(w)$, where w is a previously given fixed point in the domain of the process.

SAMARIUM

Periodic table of the chemical elements showing the position of samarium.

sampler [CONT SYS] A device, used in sampled-data control systems, whose output is a series of impulses at regular intervals in time; the height of each impulse equals the value of the continuous input signal at the instant of the impulse.

sample size [STAT] The number of objects in the sample.

sample space [STAT] A concept in probability theory which considers all possible outcomes of an experiment, game, and so on, as points in a space.

sample survey [STAT] A survey of a population made by using only a portion of the population.

sampling [SCI TECH] The obtaining of small representative quantities of materials (gas, liquid, solid) for the purpose of analysis. [STAT] A drawing of a collection from a given population.

sampling distribution [STAT] A distribution of the estimates that can be made by each of all possible samples of a fixed size that could be taken from a universe.

sampling error [STAT] That portion of the difference between the value of a statistic derived from observations and the value that it is supposed to estimate; attributed to the fact that samples represent only a portion of a population.

sampling fraction [STAT] The ratio of the sample size to the population size.

sampling interval [CONT SYS] The time between successive sampling pulses in a sampled-data control system.

sampling plan *See* sample design.

sampling process [ENG] The process of obtaining a sequence of instantaneous values of some quantity that varies continuously with time.

sampling spark chamber [NUCLEO] A spark chamber that yields as output data the coordinates of a single point on the track (or tracks) in each gap; all narrow-gap chambers are of this type.

sampling techniques [STAT] The methods used in drawing samples from a population usually in such a manner that the sample will facilitate determination of some hypothesis concerning the population.

sampling theory [STAT] The mathematical study of sampling techniques.

sampling voltmeter [ENG] A special type of voltmeter that detects the instantaneous value of an input signal at prescribed times by means of an electronic switch connecting the signal to a memory capacitor; it is particularly effective in detecting high-frequency signals (up to 12 gigahertz) or signals mixed with noise.

sand heap analogy *See* sand hill analogy.

sand hill analogy [MECH] A formal identity between the differential equation and boundary conditions for a stress function for torsion of a perfectly plastic prismatic bar, and those for the height of the surface of a granular material, such as dry sand, which has a constant angle of rest. Also known as sand heap analogy.

Sargent curve [NUC PHYS] A graph of logarithms of decay constants of radioisotopes subject to beta-decay against logarithms of the corresponding maximum beta-particle energies; most of the points fall on two straight lines.

Sargent cycle [THERMO] An ideal thermodynamic cycle consisting of four reversible processes: adiabatic compression, heating at constant volume, adiabatic expansion, and isobaric cooling.

saros [ASTRON] A cycle of time after which the centers of the sun and moon and the nodes of the moon's orbit return to the same relative position; this period is 18 years $11\frac{1}{3}$ days, or 18 years $10\frac{1}{3}$ days if 5 rather than 4 leap years are included.

Sa spiral [ASTRON] A class of spiral galaxy, including those galaxies that have the largest center sections and closely wound galactic arms.

satellite infrared spectrometer [SPECT] A spectrometer carried aboard satellites in the Nimbus series which measures the radiation from carbon dioxide in the atmosphere at several different wavelengths in the infrared region, giving the vertical temperature structure of the atmosphere over a large part of the earth. Abbreviated SIRS.

saturable-core magnetometer [ENG] A magnetometer that depends for its operation on the changes in permeability of a ferromagnetic core as a function of the magnetic field to be measured.

saturable-core reactor *See* saturable reactor.

saturable reactor [ELECTROMAG] An iron-core reactor having an additional control winding that carries direct current whose value is adjusted to change the degree of saturation of the core, thereby changing the reactance that the alternating-current winding offers to the flow of alternating current; with appropriate external circuits, a saturable reactor can serve as a magnetic amplifier. Also known as saturable-core reactor; transductor.

saturated activity [NUCLEO] The maximum activity obtainable by activation in a definite flux in a nuclear reactor.

saturated air [METEOROL] Moist air in a state of equilibrium with a plane surface of pure water or ice at the same temperature and pressure; that is, air whose vapor pressure is the saturation vapor pressure and whose relative humidity is 100%.

saturated vapor [THERMO] A vapor whose temperature equals the temperature of boiling at the pressure existing on it.

saturation [ELECTR] 1. The condition that occurs when a transistor is driven so that it becomes biased in the forward direction (the collector becomes positive with respect to the base, for example, in a *pnp* type of transistor). 2. *See* anode saturation. 3. *See* temperature saturation. [ELECTROMAG] *See* magnetic saturation. [NUCLEO] 1. The condition in which the decay rate of a given radionuclide is equal to its rate of production in an induced nuclear reaction. 2. The condition in which the voltage applied to an ionization chamber is high enough to collect all the ions formed by radiation but not high enough to produce ionization by collision. [OPTICS] *See* color saturation. [PHYS] The condition in which a further increase in some cause produces no further increase in the resultant effect.

saturation current [ELECTR] 1. In general, the maximum current which can be obtained under certain conditions. 2. In a vacuum tube, the space-charge-limited current, such that further increase in filament temperature produces no specific increase in anode current. 3. In a vacuum tube, the temperature-limited current, such that a further increase in anode-cathode potential difference produces only a relatively small increase in current. 4. In a gaseous-discharge device, the maximum current which can be obtained for a given mode of discharge. 5. In a semiconductor, the maximum current which just precedes a change in conduction mode. [NUCLEO] The ionization current in a gas tube when the applied potential is large enough to collect all ions produced by ionizing radiation.

saturation flux density *See* saturation induction.

saturation induction [ELECTROMAG] The maximum intrinsic induction possible in a material. Also known as saturation flux density.

saturation of forces [PHYS] Property exhibited by forces between particles wherein each particle can interact strongly with only a limited number of other particles, as in the forces between atoms in a molecule, and between nucleons in a nucleus.

saturation scale [OPTICS] A series of colors which appear to have equal differences in color saturation.

saturation specific humidity [THERMO] A thermodynamic function of state; the value of the specific humidity of saturated air at the given temperature and pressure.

saturation spectroscopy [SPECT] A branch of spectroscopy in which the intense, monochromatic beam produced by a laser is used to alter the energy-level populations of a resonant medium over a narrow range of particle velocities, giving rise to extremely narrow spectral lines that are free from Doppler broadening; used to study atomic, molecular, and nuclear structure, and to establish accurate values for fundamental physical constants.

saturation vapor pressure [THERMO] The vapor pressure of a thermodynamic system, at a given temperature, wherein the vapor of a substance is in equilibrium with a plane surface of that substance's pure liquid or solid phase.

Saturn [ASTRON] The second largest planet in the solar system (mass is 95.3 compared to earth's 1) and the sixth in the order of distance to the sun; it is visible to the naked eye as a yellowish first-magnitude star except during short periods near its conjunction with the sun; it is surrounded by a series of rings.

sausage instability *See* kink instability.

Savage principle [MATH] A technique used in decision theory; a criterion is used to construct a regret matrix in which each outcome entry represents a regret defined as the difference between best possible outcome and the given outcome; the matrix is then used as in decision making under risk with expected regret as the decision-determining quality. Also known as regret criterion.

savart [ACOUS] A unit of pitch interval, such that the interval between two frequencies measured in savarts is equal to 1000 times the common logarithm of the ratio of the frequencies; one octave equals approximately 301.030 savarts.

Savart plate [OPTICS] A device consisting of a pair of calcite plates having the same thickness, cut along the natural cleavage faces, and mounted with corresponding faces perpendicular to each other; used to detect polarization of light by means of interference fringes.

Savart polariscope [OPTICS] A polariscope consisting of a specially constructed double-plate polarizer and a tourmaline plate analyzer; polarized light passing through the instrument is indicated by the presence of parallel colored fringes, while unpolarized light results in a uniform field.

SAW *See* surface acoustic wave.

sawtooth generator [ELECTR] A generator whose output voltage has a sawtooth waveform; used to produce sweep voltages for cathode-ray tubes.

sawtooth pulse [ELECTR] An electric pulse having a linear rise and a virtually instantaneous fall, or conversely, a virtually instantaneous rise and a linear fall.

sawtooth waveform [ELECTR] A waveform characterized by a slow rise time and a sharp fall, resembling a tooth of a saw.

Saybolt Furol viscosimeter [ENG] An instrument for measuring viscosity of very thick fluids, for example, heavy oils; similar to the Saybolt Universal viscosimeter, but with a larger-diameter tube so that the efflux time is about one-tenth that of the Universal instrument.

Saybolt Universal viscosimeter [ENG] An instrument for measuring viscosity by the time it takes a fluid to flow through

SAWTOOTH WAVEFORM

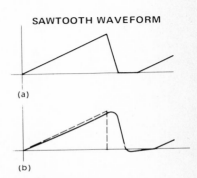

(a)

(b)

Sawtooth waveforms.
(a) An ideal linear sawtooth.
(b) Approximate sawtooth generated by actual circuits.

a calibrated tube; used for the lighter petroleum products and lubricating oils.

sb *See* stilb.

Sb *See* antimony.

Sb spiral galaxy [ASTRON] A class of spiral galaxy characterized by smaller central bodies and more open, larger arms.

Sc *See* scandium.

scalar [MATH] One of the algebraic quantities which form a field, usually the real or complex numbers, by which the vectors of a vector space are multiplied.   [PHYS] **1.** A quantity which has magnitude only and no direction, in contrast to a vector. **2.** A quantity which has magnitude only, and has the same value in every coordinate system. Also known as scalar invariant.

scalar curvature [MATH] A scalar function on a Riemannian space, equal to the sum over indices i and j of $g_{ij}R^{ij}$, where the g_{ij} are components of the metric tensor and the R^{ij} are the components of the contracted curvature tensor. Also known as curvature scalar; Ricci scalar; Riemann scalar.

scalar density [MATH] A quantity defined on an n-dimensional space which transforms under a change of coordinate system according to the law

$$\bar{s}(\bar{x}_1, \bar{x}_2, \ldots, \bar{x}_n) = |\partial x / \partial \bar{x}| s(x_1, x_2, \ldots, x_n)$$

where $|\partial x / \partial \bar{x}|$ is the Jacobian of the coordinate transformation.

scalar field [MATH] **1.** The field consisting of the scalars of a vector space. **2.** A function on a vector space into the scalars of the vector space.   [PHYS] A field which is characterized by a function of position and time whose value at each point is a scalar.

scalar function [MATH] A function from a vector space to its scalar field.   [PHYS] A function of position and time whose value at each point is a scalar.

scalar gradient [MATH] The gradient of a function.

scalar invariant *See* scalar.

scalar meson [PARTIC PHYS] A meson which has spin 0 and positive parity, and may be described by a scalar field.

scalar multiplication [MATH] The multiplication of a vector from a vector space by a scalar from the associated field; this usually contracts or expands the length of a vector.

scalar polynomial curvature singularity [RELAT] A singularity in space-time at which a scalar, formed as a polynomial in the curvature tensor, diverges.

scalar potential [PHYS] A scalar function whose negative gradient is equal to some vector field, at least when this field is time-independent; for example, the potential energy of a particle in a conservative force field, and the electrostatic potential.

scalar product [MATH] **1.** A symmetric, alternating, or Hermitian form. **2.** *See* inner product.

scalar triple product [MATH] The scalar triple product of vectors v_1, v_2, and v_3 from euclidean three-dimensional space determines the volume of the parallelepiped with these vectors as edges; it is given by the determinant of the 3×3 matrix whose rows are the components of v_1, v_2, and v_3.

scale [ACOUS] A series of musical notes arranged from low to high by a specified scheme of intervals suitable for musical purposes.   [PHYS] **1.** A one-to-one correspondence between numbers and the value of some physical quantity, such as the centigrade or Kelvin temperature scales on the API or Baumé scales of specific gravity. **2.** To determine a quantity at some order of magnitude by using data or relationships which are known to be valid at other (usually lower) orders of magnitude.

scale effect [FL MECH] An effect in fluid flow that results from changing the scale, but not the shape, of a body around which the flow passes; this effect is relevant to wind tunnel experiments.

scale factor [ENG] The factor by which the reading of an instrument or the solution of a problem should be multiplied to give the true final value when a corresponding scale factor is used initially to bring the magnitude within the range of the instrument or computer.

scale height [GEOPHYS] A measure of the thickness of an ionized layer in the upper atmosphere, using the equation $H = kT/mg = R^*T/Mg$, where k is the Boltzmann constant equal to 1.3804×10^{-16} erg/degree, T the absolute temperature, m and M the mean molecular mass and mean gram-molecular weight respectively of the layer, g the acceleration of gravity, and R^* the universal gas constant.

scalene triangle [MATH] A triangle where no two angles are equal.

scale-of-ten circuit *See* decade scaler.

scale-of-two circuit *See* binary scaler.

scaler [ELECTR] A circuit that produces an output pulse when a prescribed number of input pulses is received. Also known as counter; scaling circuit.

scaling [MECH] Expressing the terms in an equation of motion in powers of nondimensional quantities (such as a Reynolds number), so that terms of significant magnitude under conditions specified in the problem can be identified, and terms of insignificant magnitude can be dropped.

scaling circuit *See* scaler.

scaling factor [ELECTR] The number of input pulses per output pulse of a scaling circuit. Also known as scaling ratio. [PHYS] A constant of proportionality which appears in a scaling law.

scaling law [PHYS] A law, stating that two quantities are proportional, which is known to be valid at certain orders of magnitude and is used to calculate the value of one of the quantities at another order of magnitude.

scaling ratio *See* scaling factor.

scan [ENG] **1.** To examine an area, a region in space, or a portion of the radio spectrum point by point in an ordered sequence; for example, conversion of a scene or image to an electric signal or use of radar to monitor an airspace for detection, navigation, or traffic control purposes. **2.** One complete circular, up-and-down, or left-to-right sweep of the radar, light, or other beam or device used in making a scan.

scandium [CHEM] A metallic group III element, symbol Sc, atomic number 21; melts at 1200°C; found associated with rare-earth elements.

scanner [ENG] **1.** Any device that examines an area or region point by point in a continuous systematic manner, repeatedly sweeping across until the entire area or region is covered; for example, a flying-spot scanner. **2.** A device that automatically samples, measures, or checks a number of quantities or conditions in sequence, as in process control.

scanning circuit *See* sweep circuit.

scanning electron microscope [ELECTR] A type of electron microscope in which a beam of electrons, a few hundred angstroms in diameter, systematically sweeps over the specimen; the intensity of secondary electrons generated at the point of impact of the beam on the specimen is measured, and the resulting signal is fed into a cathode-ray-tube display which is scanned in synchronism with the scanning of the specimen.

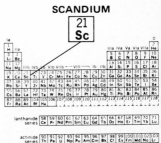

SCANDIUM

Periodic table of the chemical elements showing the position of scandium.

scanning transmission electron microscope [ELECTR] A type of electron microscope which scans with an extremely narrow beam that is transmitted through the sample; the detection apparatus produces an image whose brightness depends on atomic number of the sample. Abbreviated STEM.

scanning yoke *See* deflection yoke.

scatter angle *See* scattering angle.

scatter diagram [STAT] A plot of the pairs of values of two variates in rectangular coordinates.

scattering [ELECTROMAG] Diffusion of electromagnetic waves in a random manner by air masses in the upper atmosphere, permitting long-range reception, as in scatter propagation. Also known as radio scattering. [PHYS] **1.** The change in direction of a particle or photon because of a collision with another particle or a system. **2.** Diffusion of acoustic or electromagnetic waves caused by inhomogeneity or anisotropy of the transmitting medium. **3.** In general, causing a collection of entities to assume a less orderly arrangement.

scattering amplitude [QUANT MECH] A quantity, depending in general on the energy and scattering angle, which specifies the wave function of particles scattered in a collision, and whose squared modulus is proportional to the number of particles scattered in a given direction.

scattering angle [PHYS] The angle between the initial and final directions of motion of a scattered particle. Also known as scatter angle.

scattering coefficient [ELECTROMAG] One of the elements of the scattering matrix of a waveguide junction; that is, a transmission or reflection coefficient of the junction. [PHYS] The fractional decrease in intensity of a beam of electromagnetic radiation or particles per unit distance traversed, which results from scattering rather than absorption. Also known as dissipation coefficient.

scattering cross section [ELECTROMAG] The power of electromagnetic radiation scattered by an antenna divided by the incident power. [PHYS] The sum of the cross sections for elastic and inelastic scattering.

scattering function [ELECTROMAG] The intensity of scattered radiation in a given direction per lumen of flux incident upon the scattering material.

scattering length [NUC PHYS] A parameter used in analyzing nuclear scattering at low energies; as the energy of the bombarding particle becomes very small, the scattering cross section approaches that of an impenetrable sphere whose radius equals this length. Also known as scattering power.

scattering loss [ELECTROMAG] The portion of the transmission loss that is due to scattering within the medium or roughness of the reflecting surface.

scattering matrix [ELECTROMAG] A square array of complex numbers consisting of the transmission and reflection coefficients of a waveguide junction. [QUANT MECH] A matrix which expresses the initial state in a scattering experiment in terms of the possible final states. Also known as collision matrix; S matrix.

scattering operator [PHYS] An operator which acts in the vector space of solutions of a wave equation, transforming solutions representing incoming waves in a domain exterior to a bounded obstacle into solutions representing outgoing waves.

scattering power *See* scattering length.

scatter propagation [ELECTROMAG] Transmission of radio waves far beyond line-of-sight distances by using high power and a large transmitting antenna to beam the signal upward into the atmosphere and by using a similar large receiving antenna to pick up the small portion of the signal that is scattered by the atmosphere. Also known as beyond-the-horizon communication; forward-scatter propagation; over-the-horizon propagation.

scfh [FL MECH] Cubic feet per hour of gas flow at specified standard conditions of temperature and pressure.

scfm [FL MECH] Cubic feet per minute of gas flow at specified standard conditions of temperature and pressure.

Schauder's fixed-point theorem [MATH] A continuous mapping from a closed, compact, convex set in a Banach space into itself has at least one fixed point.

schedule [STAT] A group or list of questions used by an interviewer to obtain information directly from a subject.

Schering bridge [ELEC] A four-arm alternating-current bridge used to measure capacitance and dissipation factor; bridge balance is independent of frequency.

schiller *See* play of color.

schillerization [OPTICS] Development of schiller in crystals due to the pattern of inclusions.

Schläfli's integral formula [MATH] **1.** The formula

$$P_n(z) = \frac{1}{2\pi i} \int_c \frac{(t^2 - 1)^n \, dt}{2^n (t - z)^{n+1}}$$

for the Legendre polynomials P_n, where c is a contour that circles the point z once in a counterclockwise direction. **2.** The formula

$$J_\nu(z) = \frac{(\frac{1}{2}z)^\nu}{2\pi i} \int_c t^{-\nu-1} \exp\left(t - \frac{z^2}{4t}\right) dt$$

where J_ν is the Bessel function of the first kind of order ν, and c is a contour that starts at $-\infty$ on the real axis, circles the origin once in a positive direction, and returns to its starting point.

Schleiermacher's method [THERMO] A method of determining the thermal conductivity of a gas, in which the gas is placed in a cylinder with an electrically heated wire along its axis, and the electric energy supplied to the wire and the temperatures of wire and cylinder are measured.

schlieren [OPTICS] In atmospheric optics, parcels or strata of air having densities sufficiently different from that of their surroundings so that they may be discerned by means of refraction anomalies in transmitted light.

schlieren method [OPTICS] An optical technique that detects density gradients occurring in a fluid flow; in its simplest form, light from a slit is collimated by a lens and focused onto a knife-edge by a second lens, the flow pattern is placed between these two lenses, and the diffraction pattern that results on a screen or photographic film placed behind the knife-edge is observed.

Schlömilch series [MATH] An infinite series whose nth term is of the form $a_n J_\nu(nx)$, where J_ν is the Bessel function of the first kind of order ν, and the a_n are constants.

Schlömilch's form of remainder [MATH] An expression for the difference R_n between the value of a function $f(x)$ and the sum of the first $n + 1$ terms of its Taylor series about a point a; it may be written $R_n = h^{n+1}(1 - \theta)^{n+1-p} f^{(n+1)}(a + \theta h)/n! p$, where p is a positive number, $h = x - a$, and θ is some number between 0 and 1.

Schlömilch's integral equation [MATH] The equation

$$f(x) = \frac{2}{\pi} \int_0^{\pi/2} g(x \sin \theta) \, d\theta$$

where $f(x)$ is a known function which is continuous for $-\pi \leq x \leq \pi$, and $g(x)$ is to be determined.

Schmidt camera *See* Schmidt system.

SCHERING BRIDGE

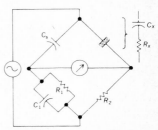

Circuit diagram of Schering bridge used to measure capacitance C_X and resistance R_X of equivalent series-circuit representation of capacitor. Standard capacitor C_S is assumed free from loss. Bridge is balanced when $C_X R_1 = C_S R_1$; $R_X C_S = R_2 C_1$.

SCHLIEREN METHOD

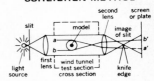

Optical system used in the schlieren method. Light rays from a and b are bent by density gradients in test section. Ray from a is interrupted by knife-edge so that a dark spot occurs on the screen at a'; ray from b escapes knife-edge so that light spot occurs on screen at b'.

Schmidt-Cassegrain telescope [OPTICS] A variant of the Schmidt system which uses a Schmidt corrector plate together with a pair of spheroidal or slightly aspherical mirrors arranged as in a Cassegrain telescope.

Schmidt correction plate [OPTICS] In the Schmidt system, a glass plate with one face a plane and the other aspherical and deviating from a plane in such a way that it bends light, traveling to the system's spherical mirror, so as to correct for spherical aberration and coma. Also known as Schmidt lens.

Schmidt lens *See* Schmidt correction plate.

Schmidt lines [NUC PHYS] Two lines, on a graph of nuclear magnetic moment versus nuclear spin, on which points describing all nuclides should lie, according to the independent particle model; experimentally, however, points describing nuclides are scattered between the lines.

Schmidt number 1 *See* Prandtl number.

Schmidt number 2 *See* Semenov number 1.

Schmidt number 3 [PHYS CHEM] A dimensionless number used in electrochemistry, equal to the product of the dielectric susceptibility and the dynamic viscosity of a fluid divided by the product of the fluid density, electrical conductivity, and the square of a characteristic length. Symbolized Sc_3.

Schmidt objective *See* Schmidt system.

Schmidt optics *See* Schmidt system.

Schmidt reflector [OPTICS] A telescope employing the Schmidt system.

Schmidt system [OPTICS] An optical system designed to eliminate spherical aberration and coma, which, in its original form, consists of a spherical mirror, a Schmidt correction plate near the focus of the mirror, and usually a curved reflecting plate at the focus of the mirror; used in astronomical telescopes with unusually wide fields of view and in spectroscopes, and to project a television image from a cathode-ray tube onto a screen. Also known as projection optics; Schmidt camera; Schmidt objective; Schmidt optics.

Schmitt circuit [ELECTR] A bistable pulse generator in which an output pulse of constant amplitude exists only as long as the input voltage exceeds a certain value. Also known as Schmitt limiter; Schmitt trigger.

Schmitt limiter *See* Schmitt circuit.

Schmitt trigger *See* Schmitt circuit.

Schönflies crystal symbols [CRYSTAL] Symbols denoting the 32 crystal point groups or symmetry classes; capital letters indicate the general type of class, and subscripts the multiplicity of rotation axes and the existence of additional symmetries.

Schottky barrier [ELECTR] A transition region formed within a semiconductor surface to serve as a rectifying barrier at a junction with a layer of metal.

Schottky barrier diode [ELECTR] A semiconductor diode formed by contact between a semiconductor layer and a metal coating; it has a nonlinear rectifying characteristic; hot carriers (electrons for *n*-type material or holes for *p*-type material) are emitted from the Schottky barrier of the semiconductor and move to the metal coating that is the diode base; since majority carriers predominate, there is essentially no injection or storage of minority carriers to limit switching speeds. Also known as hot-carrier diode; Schottky diode.

Schottky defect [SOLID STATE] **1.** A defect in an ionic crystal in which a single ion is removed from its interior lattice site and relocated in a lattice site at the surface of the crystal. **2.** A defect in an ionic crystal consisting of the smallest number of positive-ion vacancies and negative-ion vacancies which leave the crystal electrically neutral.

Schottky diode *See* Schottky barrier diode.

SCHMIDT SYSTEM

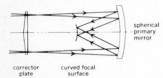

corrector plate curved focal surface

Optics of the Schmidt system showing axial and the extra-axial light paths.

Schottky effect [SOLID STATE] The enhancement of the thermionic emission of a conductor resulting from an electric field at the conductor surface.

Schottky line [SOLID STATE] A graph of the logarithm of the saturation current from a thermionic cathode as a function of the square root of anode voltage; it is a straight line according to the Schottky theory.

Schottky noise *See* shot noise.

Schottky theory [SOLID STATE] A theory describing the rectification properties of the junction between a semiconductor and a metal that result from formation of a depletion layer at the surface of contact.

Schrödinger equation [QUANT MECH] A partial differential equation governing the Schrödinger wave function ψ of a system of one or more nonrelativistic particles; $\hbar(\partial \psi / \partial t) = H\psi$, where H is a linear operator, the Hamiltonian, which depends on the dynamics of the system, and $\hbar$ is Planck's constant divided by 2π.

Schrödinger-Klein-Gordon equation *See* Klein-Gordon equation.

Schrödinger operator [MATH] A linear operator defined on differentiable functions of three variables, x_1, x_2, and x_3, which associates to each such function $f(x_1,x_2,x_3)$ the function $[-\Delta + V(x_1,x_2,x_3)]f(x_1,x_2,x_3)$, where Δ represents the Laplace operator and $V(x_1,x_2,x_3)$ is a specified function.

Schrödinger-Pauli equation [QUANT MECH] A modification of the Schrödinger equation to describe a particle with spin of $\frac{1}{2}\hbar$, where $\hbar$ is Planck's constant divided by 2π; the wave function has two components, corresponding to the particle's spin pointing in either of two opposite directions.

Schrödinger picture [QUANT MECH] A mode of description of a quantum-mechanical system in which dynamical states are represented by vectors which evolve in the course of time, and physical quantities are represented by stationary operators, in contrast to the Heisenberg picture.

Schrödinger representation *See* position representation.

Schrödinger's wave mechanics [QUANT MECH] The version of nonrelativistic quantum mechanics in which a system is characterized by a wave function which is a function of the coordinates of all the particles of the system and time, and obeys a differential equation, the Schrödinger equation; physical quantities are represented by differential operators which may act on the wave function, and expectation values of measurements are equal to integrals involving the corresponding operator and the wave function. Also known as wave mechanics.

Schrödinger wave function [QUANT MECH] A function of the coordinates of the particles of a system and of time which is a solution of the Schrödinger equation and which determines the average result of every conceivable experiment on the system. Also known as probability amplitude; psi function; wave function.

Schroeder-Bernstein theorem [MATH] If a set A has at least as many elements as another set B and B has at least as many elements as A, then A and B have the same number of elements.

Schuler pendulum [MECH] Any apparatus which swings, because of gravity, with a natural period of 84.4 minutes, that is, with the same period as a hypothetical simple pendulum whose length is the earth's radius; the pendulum arm remains vertical despite any motion of its pivot, and the apparatus is therefore useful in navigation.

Schuler tuning [ENG] The designing of gyroscopic devices so that their periods of oscillation will be about 84.4 minutes.

Schumann region [OPTICS] The most extreme ultraviolet

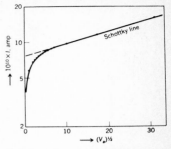

SCHOTTKY LINE

Graph of the logarithm of thermionic emission current I of tungsten as function of square root of anode voltage V_a. (*After W. B. Nottingham, Phys. Rev., 58:927–928, 1940*)

portion of the electromagnetic spectrum that will affect a photographic plate.

Schur's lemma [MATH] For certain types of modules M, the ring consisting of all homomorphisms of M to itself will be a division ring.

Schur's theorem [MATH] If the components of the Riemann-Christoffel tensor R_{hijk} are equal to $K(g_{hj}g_{ik} - g_{hk}g_{ij})$, where g is the metric tensor and K is a function of the coordinates, then K is a constant.

Schuster method [SPECT] A method for focusing a prism spectroscope without using a distant object or a Gauss eyepiece.

Schwartz's theory of distributions [MATH] A theory that treats distributions as continuous linear functionals on a vector space of continuous functions which have continuous derivatives of all orders and vanish appropriately at infinity.

Schwarz-Christoffel transformations [MATH] Those complex transformations which conformally map the interior of a given polygon onto the portion of the complex plane above the real axis.

Schwarzian derivative [MATH] The Schwarzian derivative of a dependent variable y with respect to an independent variable x is $(y''/2y')' - (y''/2y')^2$, where primes denote differentiation with respect to x.

Schwarz reflection principle [MATH] To obtain the analytic continuation of a given function $f(z)$ analytic in a region R, whose boundary contains a segment of the real axis, into a region reflected from R through this segment, one takes the complex conjugate function $f(\bar{z})$.

Schwarzschild anastigmat [OPTICS] A Gregorian telescope whose surfaces are altered to reduce astigmatism.

Schwarzschild radius [RELAT] Conventionally taken to be twice the black hole mass appearing in the general relativistic Schwarzschild solution times the gravitational constant divided by the square of the speed of light; the event horizon in a Schwarzschild solution is at the Schwarzschild radius.

Schwarzschild singularity [RELAT] The coordinate singularity at the event horizon that exists in a certain coordinate system describing a nonrotating black hole in empty space.

Schwarzschild solution [RELAT] The unique solution of general relativity theory describing a nonrotating black hole in empty space.

Schwarz's inequality *See* Cauchy-Schwarz inequality.

Schwarz's lemma [MATH] If an analytic function of the unit disk to itself sends the origin to the origin then it must be distance-decreasing.

Schwinger's variational principle [PHYS] A method used in electromagnetic theory, or similar disciplines, to calculate an approximate value or a linear of quadratic functional, such as a scattering amplitude or reflection coefficient, when the function for which the functional is evaluated is the solution of an integral equation.

scintillation [NUCLEO] A flash of light produced in a phosphor by an ionizing particle or photon. [OPTICS] Rapid changes of brightness of stars or other distant, celestial objects caused by variations in the density of the air through which the light passes.

scintillation camera [NUCLEO] A camera that gives a complete image of radionuclide distribution in a particular area of the human body in one exposure, in contrast to line-scanning techniques.

scintillation counter [NUCLEO] A device in which the scintillations produced in a fluorescent material by an ionizing radiation are detected and counted by a multiplier phototube and associated circuits; used in medical and nuclear research

SCINTILLATION COUNTER

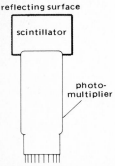

Diagram of a scintillation counter.

and in prospecting for radioactive ores. Also known as scintillation detector; scintillometer.

scintillation counter crystal [NUCLEO] A substance (a fluor, such as thallium-activated calcium tungstate) that emits a flash of light (scintillation) when contacted by a high-energy particle, for example, an alpha, beta, or gamma ray.

scintillation detector *See* scintillation counter.

scintillation spectrometer [NUCLEO] A scintillation counter adapted to measuring the energy and intensity of gamma rays from radioactive elements.

scintillator [NUCLEO] A material that emits optical photons in response to ionizing radiation.

scintillometer *See* scintillation counter.

scopometer [OPTICS] An instrument used to measure the absorption or scattering of light in a solution containing solid particles by measuring the contrast between an illuminated line placed behind the solution and a field of constant brightness.

scotoscope [ELECTR] A telescope which employs an image intensifier to see in the dark.

SCR *See* silicon controlled rectifier.

scram [NUCLEO] **1.** A sudden shutting down of a nuclear reactor, usually by dropping safety rods, when a predetermined neutron flux or other dangerous condition occurs. **2.** To close down a reactor by bringing about a scram.

scram rod *See* safety rod.

scram system [NUCLEO] A system which causes a scram in a nuclear reactor when dangerous conditions arise.

screen [ELECTR] **1.** The surface on which a television, radar, x-ray, or cathode-ray oscilloscope image is made visible for viewing; it may be a fluorescent screen with a phosphor layer that converts the energy of an electron beam to visible light, or a translucent or opaque screen on which the optical image is projected. Also known as viewing screen. **2.** *See* screen grid. [ELECTROMAG] Metal partition or shield which isolates a device from external magnetic or electric fields.

screen dissipation [ELECTR] Power dissipated in the form of heat on the screen grid as the result of bombardment by the electron stream.

screen grid [ELECTR] A grid placed between a control grid and an anode of an electron tube, and usually maintained at a fixed positive potential, to reduce the electrostatic influence of the anode in the space between the screen grid and the cathode. Also known as screen.

screening [ATOM PHYS] The reduction of the electric field about a nucleus by the space charge of the surrounding electrons. [ELECTROMAG] *See* electric shielding.

screening constant [ATOM PHYS] The difference between the atomic number of an element and the apparent atomic number for a given process; this difference results from screening.

screening inspection *See* total inspection.

screw axis [CRYSTAL] A symmetry element of some crystal lattices, in which the lattice is unaltered by a rotation about the axis combined with a translation parallel to the axis and equal to a fraction of the unit lattice distance in this direction.

screw dislocation [CRYSTAL] A dislocation in which atomic planes form a spiral ramp winding around the line of the dislocation.

screw displacement [MECH] A rotation of a rigid body about an axis accompanied by a translation of the body along the same axis.

Sc spiral galaxy [ASTRON] A class of spiral galaxy characterized by spirals with the largest and most loosely coiled arms and the smallest central portion.

S curve   *See* reverse curve.

Se   *See* selenium.

search coil   *See* exploring coil.

searchlight   [OPTICS] A type of light projector designed to produce a beam of high intensity and minimum divergence; it usually employs a specular paraboloidal reflector to produce parallel rays from a light source located at the focus.

seasonal variation   [GEOPHYS] The variation in the ionization of different parts of the ionosphere, and the resulting variation in transmission of radio signals over large distances, with the season.

sec   *See* secant; second; secondary winding.

SEC   *See* secondary electron conduction.

secant   [MATH] **1.** The function given by the reciprocal of the cosine function. Abreviated sec. **2.** The secant of an angle A is $1/\cos A$.

sech   *See* hyperbolic secant.

secohm   *See* international henry.

second   [MATH] A unit of plane angle, equal to 1/60 minute, or 1/3,600 degree, or $\pi/648,000$ radian. [MECH] The fundamental unit of time equal to 9,192,631,770 periods of the radiation corresponding to the transition between the two hyperfine levels of the ground state of an atom of cesium-133. Abbreviated s: sec.

secondary   *See* secondary winding.

secondary battery   *See* storage battery.

secondary bow   *See* secondary rainbow.

secondary cell   *See* storage cell.

secondary cosmic rays   [GEOPHYS] Radiation produced when primary cosmic rays enter the atmosphere and collide with atomic nuclei and electrons.

secondary creep   [MECH] The change in shape of a substance under a minimum and almost constant differential stress, with the strain-time relationship a constant. Also known as steady-state creep.

secondary electron   [ELECTR] **1.** An electron emitted as a result of bombardment of a material by an incident electron. **2.** An electron whose motion is due to a transfer of momentum from primary radiation.

secondary electron conduction   [ELECTR] Transport of charge by secondary electrons moving through the interstices of a porous material under the influence of an externally applied electric field. Abbreviated SEC.

secondary emission   [ELECTR] The emission of electrons from the surface of a solid or liquid into a vacuum as a result of bombardment by electrons or other charged particles.

secondary extinction   [PHYS] Increased absorption or decreased diffraction of x-rays by a crystal lattice, due to previous reflection of the x-rays from suitably placed crystal planes.

secondary flow   [FL MECH] A field of fluid motion which can be considered as superposed on a primary field of motion through the action of friction usually in the vicinity of solid boundaries. Also known as frictional secondary flow.

secondary focus   [OPTICS] In an astigmatic system, a line at which some of the bundle of rays from an off-axis point meet; this line lies in a plane which contains the point and the optical axis, and has a greater image distance than the primary focus. Also known as sagittal focus.

secondary grid emission   [ELECTR] Electron emission from a grid resulting directly from bombardment of its surface by electrons or other charged particles.

secondary ionization coefficient   *See* Townsend second ionization coefficient.

secondary ion mass analyzer   [ENG] A type of secondary ion mass spectrometer that provides general surface analysis and depth-profiling capabilities.

secondary ion mass spectrometer [ENG] An instrument for microscopic chemical analysis, in which a beam of primary ions with an energy in the range 5–20 keV bombards a small spot on the surface of a sample, and positive and negative secondary ions sputtered from the surface are analyzed in a mass spectrometer. Abbreviated SIMS. Also known as ion microprobe; ion probe.

secondary lobe *See* minor lobe.

secondary radiation [PHYS] Particles or photons produced by the action of primary radiation on matter, such as Compton recoil electrons, delta rays, secondary cosmic rays, and secondary electrons.

secondary rainbow [OPTICS] A faint rainbow of angular radius about 50° which may appear outside the primary rainbow of 42° radius, and which has its colors in reverse order to those of the primary. Also known as secondary bow.

secondary standard [PHYS] **1.** A unit, as of length, capacitance, or weight, used as a standard of comparison in individual countries or localities, but checked against the one primary standard in existence somewhere. **2.** A unit defined as a specified multiple or submultiple of a primary standard, such as the centimeter.

secondary voltage [ELECTROMAG] The voltage across the secondary winding of a transformer.

secondary wave [GEOPHYS] *See* S wave. [OPTICS] One of the waves that radiate from each point on a wavefront, according to Huygens' principle.

secondary winding [ELECTROMAG] A transformer winding that receives energy by electromagnetic induction from the primary winding; a transformer may have several secondary windings, and they may provide alternating-current voltages that are higher, lower, or the same as that applied to the primary winding. Abbreviated sec. Also known as secondary.

second breakdown [ELECTR] Destructive breakdown in a transistor, wherein structural imperfections cause localized current concentrations and uncontrollable generation and multiplication of current carriers; reaction occurs so suddenly that the thermal time constant of the collector regions is exceeded, and the transistor is irreversibly damaged.

second fundamental form [MATH] For a surface with curvilinear coordinates u^i, $i = 1, 2$, this is the quadratic differential form

$$\sum_{i,j} b_{ij}\, du^i\, du^j$$

where the b_{ij} are fundamental magnitudes of second order; it is related to the normal curvature of the surface along a line element from u^i to $u^i + du^i$. Also known as second ground form.

second ground form *See* second fundamental form.

second law of motion *See* Newton's second law.

second law of the mean for integrals [MATH] **1.** If $f(x)$ and $g(x)$ are continuous functions on a closed interval $[a,b]$ and $g(x)$ does not change sign on $[a,b]$, then there exists x_0, $a < x_0 < b$, such that

$$\int_a^b f(x)g(x)\, dx = f(x_0) \int_a^b g(x)\, dx.$$

2. If $f(x)$ and $g(x)$ are continuous functions on $[a,b]$ and $g(x)$ is a monotone function on $[a,b]$, then there exists x_0, $a < x_0 < b$, such that

$$\int_a^b f(x)g(x)\,dx = g(a) \int_a^{x_0} f(x)\,dx + g(b) \int_{x_0}^b f(x)\,dx.$$

Also known as Du Bois-Reymond's mean value theorem. **3.** If $f(x)$ and $g(x)$ are continuous functions on $[a,b]$ and $g(x)$ is a nonnegative, monotone function on $[a,b]$, then there exits x_0, $a < x_0 < b$, such that

$$\int_a^b f(x)g(x)\,dx = g(a) \int_a^{x_0} f(x)\,dx$$

if $g(x)$ is nonincreasing, or

$$\int_a^b f(x)g(x) = g(b) \int_{x_0}^b f(x)\,dx$$

if $g(x)$ is nondecreasing. Also known as Bonnet's mean value theorem.

second law of thermodynamics [THERMO] A general statement of the idea that there is a preferred direction for any process; there are many equivalent statements of the law, the best known being those of Clausius and of Kelvin.

second moment of area *See* geometric moment of inertia.

second-order equation [MATH] A differential equation where some term includes the second derivative of the unknown function and no derivative of higher order is present.

second-order transition [THERMO] A change of state through which the free energy of a substance and its first derivatives are continuous functions of temperature and pressure, or other corresponding variables.

second quantization [QUANT MECH] A procedure in which the dependent variables of a classical field or a quantum-mechanical wave function are regarded as operators on which commutation rules are imposed; this produces a formalism in which particles may be created and destroyed.

second radiation constant [STAT MECH] A constant appearing in the Planck radiation formula, equal to the speed of light times Planck's constant divided by Boltzmann's constant, or approximately 1.4388 degree-centimeters. Symbolized c_2; C_2.

second sound [CRYO] A type of wave propagated in the superfluid phase of liquid helium (helium II), in which temperature and entropy variations propagate with no appreciable variation in density or pressure.

seconds pendulum [HOROL] A pendulum measuring 99.353 centimeters between its center of suspension and center of oscillation, and requiring exactly 1 second to swing from one side to another at sea level and 45° latitude.

sectional radiography [ELECTR] The technique of making radiographs of plane sections of a body or an object; its purpose is to show detail in a predetermined plane of the body, while blurring the images of structures in other planes. Also known as laminography; planigraphy; tomography.

section modulus [MECH] The ratio of the moment of inertia of the cross section of a beam undergoing flexure to the greatest distance of an element of the beam from the neutral axis.

sector [ELECTROMAG] Coverage of a radar as measured in azimuth. [MATH] A portion of a circle bounded by two radii and an arc joining their end points.

sectoral harmonic [MATH] A spherical harmonic which is 0 on a set of equally spaced meridians of a sphere with center at the origin of spherical coordinates, dividing the sphere into sectors.

sector boundary [ASTROPHYS] The rapid transition from one polarity to another in the interplanetary magnetic field.

sector disk [PHYS] A device used to reduce the intensity of a beam of light or other electromagnetic radiation by an

SECOND SOUND

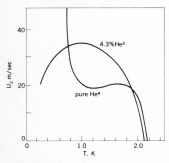

Velocity of second sound U_2 as a function of temperature (kelvins) for pure He^4 and a 4.3% He^3 mixture. *(From J. C. King and H. A. Fairbanks, Second sound in He^3–He^4 mixtures below 1°K, Phys. Rev., 93:21, 1954)*

accurately known amount; in its simplest form, it consists of a circular, opaque disk with one or more sectors cut out of it, rapidly rotating in the path of the beam.

sector structure [ASTROPHYS] The polarity pattern of the interplanetary magnetic field observed during a solar rotation.

secular determinant [MATH] For a square matrix A, the determinant of the matrix whose off-diagonal components are equal to those of A, and whose diagonal components are equal to the difference between those of A and a parameter λ; it is equal to the characteristic polynomial in λ of the linear transformation represented by A.

secular equilibrium [NUCLEO] Radioactive equilibrium in which the parent has such a small decay constant that there has been no appreciable change in the quantity of parent present by the time the decay products have reached radioactive equilibrium.

secular parallax [ASTRON] An apparent angular displacement of a star, resulting from the sun's motion.

secular perturbations [ASTROPHYS] Changes in the orbit of a planet, or of a satellite, that operates in extremely long cycles.

secular trend [STAT] A concept in time series analysis that refers to a movement or trend in a series over very long periods of time. Also known as long-time trend.

secular variable [ASTRON] A star whose brightness appears to have slowly lessened or increased over a time period of centuries.

secular variation [ASTRON] A perturbation of the moon's motion caused by variations in the effect of the sun's gravitational attraction on the earth and moon as their relative distances from the sun vary during the synodic month. [GEOPHYS] The changes, measured in hundreds of years, in the magnetic field of the earth. Also known as geomagnetic secular variation.

SED *See* skin erythema dose.

Seebeck coefficient [ELECTR] The ratio of the open-circuit voltage to the temperature difference between the hot and cold junctions of a circuit exhibiting the Seebeck effect.

Seebeck effect [ELECTR] The development of a voltage due to differences in temperature between two junctions of dissimilar metals in the same circuit.

seed core [NUCLEO] A reactor core that includes a relatively small volume of highly enriched uranium, surrounded by a much larger volume of natural uranium or thorium in a blanket; as a result of fission in the seed uranium, neutrons are supplied to the blanket, which is made to furnish a substantial fraction of the core power. Also known as spiked core.

seeing [ASTRON] The clarity and steadiness of an image of a star in a telescope. [ELECTR] The introduction of atoms with a low ionization potential into a hot gas to increase electrical conductivity.

seepage [FL MECH] The slow movement of water or other fluid through a porous medium.

segment [MATH] **1.** A segment of a line or curve is any connected piece. **2.** A segment of a circle is a portion of the circle bounded by a chord and an arc subtended by the chord. **3.** A segment of a totally ordered Abelian group G is a subset D of G such that if a is in D then so are all elements b satisfying $-a \leq b \leq a$.

Segrè chart [NUC PHYS] A chart of the nuclides which is laid off in squares, each square displaying data about a nuclide; each column contains nuclides with a given neutron number and each row contains nuclides with a given atomic number; successive columns and rows represent successively higher numbers of neutrons and protons.

seiche [FL MECH] An oscillation of a fluid body in response to the disturbing force having the same frequency as the natural frequency of the fluid system.

Seidel aberrations [OPTICS] The five types of aberration of monochromatic light deduced from the Seidel theory, namely, spherical aberration, coma, astigmatism, curvature of field, and distortion.

Seidel method [MATH] A basic iterative procedure for solving a system of linear equations by reducing it to triangular form. Also known as Gauss-Seidel method.

Seidel theory [OPTICS] A theory of aberrations in which the sine of the angle which a ray makes with the optical axis is approximated by the first two terms in the sine's Taylor expansion (the first- and third-order terms), rather than the first term alone, as in a first-order theory.

seignette-electric *See* ferroelectric.

seismic activity *See* seismicity.

seismic detector [ENG] An instrument that receives seismic impulses.

seismic discontinuity [GEOPHYS] **1.** A surface at which velocities of seismic waves change abruptly. **2.** A boundary between seismic layers of the earth. Also known as interface; velocity discontinuity.

seismicity [GEOPHYS] The phenomena of earth movements. Also known as seismic activity.

seismochronograph [ENG] A chronograph for determining the time at which an earthquake shock appears.

seismograph [ENG] An instrument that records vibrations in the earth, especially earthquakes.

seismology [GEOPHYS] **1.** The study of earthquakes. **2.** The science of strain-wave propagation in the earth.

seismometer [ENG] An instrument that detects movements in the earth.

seismoscope [ENG] An instrument for recording only the occurrence or time of occurrence (not the magnitude) of an earthquake.

selected areas *See* Kapteyn selected areas.

selection bias [STAT] A bias built into an experiment by the method used to select the subjects which are to undergo treatment.

selection rules [PHYS] Rules summarizing the changes that must take place in the quantum numbers of a quantum-mechanical system for a transition between two states to take place with appreciable probability; transitions that do not agree with the selection rules are called forbidden and have considerably lower probability.

selective absorption [ELECTROMAG] A greater absorption of electromagnetic radiation at some wavelengths (or frequencies) than at others.

selective permeability [PHYS] The property of a membrane or other material that allows some substances to pass through it more easily than others.

selective radiator [PHYS] An object that emits electromagnetic radiation whose spectral energy distribution differs from that of a blackbody with the same temperature.

selective reflection [ELECTROMAG] Reflection of electromagnetic radiation more strongly at some wavelengths (or frequencies) than at others.

selective scattering [ELECTROMAG] Scattering of electromagnetic radiation more strongly at some wavelengths than at others.

s-electron [ATOM PHYS] An atomic electron that is described by a wave function with orbital angular momentum quantum number of zero in the independent particle approximation.

selenium [CHEM] A highly toxic, nonmetallic element in

SELENIUM

Periodic table of the chemical elements showing the position of selenium.

group VI, symbol Se, atomic number 34; steel-gray color; soluble in carbon disulfide, insoluble in water and alcohol; melts at 217°C; and boils at 690°C; used in analytical chemistry, metallurgy, and photoelectric cells, and as a lube-oil stabilizer and chemicals intermediate.

selenium cell [ELECTR] A photoconductive cell in which a thin film of selenium is used between suitable electrodes; the resistance of the cell decreases when the illumination is increased.

selenium diode [ELECTR] A small-area selenium rectifier which has characteristics similar to those of selenium rectifiers used in power systems.

selenium rectifier [ELECTR] A metallic rectifier in which a thin layer of selenium is deposited on one side of an aluminum plate and a conductive metal coating is deposited on the selenium.

selenocentric [ASTRON] Pertaining to the moon's center.

selenodesy [ASTRON] The branch of applied mathematics that determines, by observation and measurement, the exact positions of points on the moon's surface, as well as the shape and size of the moon.

selenography [ASTRON] Studies pertaining to the physical geography of the moon; specifically, referring to positions on the moon measured in latitude from the moon's equator and in longitude from a reference meridian.

selenology [ASTRON] A branch of astronomy that treats of the moon, including such attributes as magnitude, motion and constitution. Also known as lunar geology.

selenomorphology [ASTRON] The study of landforms on the moon, including their origin, evolution, and distribution.

self-absorption [NUCLEO] Absorption of ionizing radiation by the material that emits the radiation, thus reducing the radiation level against which further shielding must be provided. Also known as self-shielding. [SPECT] Reduction of the intensity of the center of an emission line caused by selective absorption by the cooler portions of the source of radiation. Also known as self-reduction; self-reversal.

self-adapting system [SYS ENG] A system which has the ability to modify itself in response to changes in its environment.

self-adjoint operator [MATH] A linear operator which is identical with its adjoint operator.

self-bias [ELECTR] A grid bias provided automatically by the resistor in the cathode or grid circuit of an electron tube; the resulting voltage drop across the resistor serves as the grid bias. Also known as automatic C bias; automatic grid bias.

self-bias transistor circuit [ELECTR] A transistor with a resistance in the emitter lead that gives rise to a voltage drop which is in the direction to reverse-bias the emitter junction; the circuit can be used even if there is zero direct-current resistance in series with the collector terminal.

self-charge [QUANT MECH] A contribution to a particle's electric charge arising from the vacuum polarization in the neighborhood of the bare charge.

self-conjugate particle [PARTIC PHYS] An elementary particle which is identical to its antiparticle; it must have zero charge, lepton number, baryon number, and hypercharge.

self-consistent field method *See* Hartree method.

self-diffusion [SOLID STATE] The spontaneous movement of an atom to a new site in a crystal of its own species.

self-energy [PHYS] **1.** Classically, the contribution to the energy of a particle that arises from the interaction between different parts of the particle. **2.** In a quantized field theory, the contribution to the energy of a particle due to virtual emission and absorption of other particles, in particular, mesons and photons.

self-excited [ELEC] Operating without an external source of alternating-current power.

self-excited vibration *See* self-induced vibration.

self-fields [ELECTROMAG] The electric and magnetic fields generated by an intense beam of charged particles, which act on the beam itself; they limit the beam intensities which can be achieved in storage rings.

self-focusing fiber [OPTICS] A type of optical fiber in which the refractive index decreases continuously along the radius, but progressively more rapidly with distance from the radius, so that light rays which travel longer distances are speeded up, and nearly all light rays travel with the same net axial velocity.

self-healing dielectric breakdown [ELECTR] A dielectric breakdown in which the breakdown process itself causes the material to become insulating again.

self-impedance *See* mesh impedance.

self-induced transparency [OPTICS] A phenomenon in which a pulse of coherent light, with a certain frequency, amplitude, and duration, is transmitted by a normally opaque medium; energy absorbed from the first half of the pulse, whose frequency is at or near the average resonance peak of a band of coherent atomic two-quantum-level optical oscillators, is returned to the last half of the pulse by the medium in the form of coherently emitted light. Abbreviated SIT.

self-induced vibration [MECH] The vibration of a mechanical system resulting from conversion, within the system, of nonoscillatory excitation to oscillatory excitation. Also known as self-excited vibration.

self-inductance [ELECTROMAG] **1.** The property of an electric circuit whereby an electromotive force is produced in the circuit by a change of current in the circuit itself. **2.** Quantitatively, the ratio of the electromotive force produced to the rate of change of current in the circuit.

self-induction [ELECTROMAG] The production of a voltage in a circuit by a varying current in that same circuit.

self-organizing system [SYS ENG] A system that is able to affect or determine its own internal structure.

self-reduction *See* self-absorption.

self-reversal *See* self-absorption.

self-selection bias [STAT] Bias introduced into an experiment by having the subjects decide themselves whether or not they will receive treatment.

self-shielding [NUCLEO] **1.** Shielding of the inner portion of the fuel in a nuclear reactor by the outer portion of the fuel. **2.** *See* self-absorption.

Sellmeier's equation [ELECTROMAG] An equation for the index of refraction of electromagnetic radiation as a function of wavelength in a medium whose molecules have oscillators of different frequencies.

Semenov number 1 [PHYS CHEM] A dimensionless number used in reaction kinetics, equal to a mass transfer constant divided by a reaction rate constant. Symbolized S_m. Formerly known as Schmidt number 2.

Semenov number 2 [PHYS] The reciprocal of the Lewis number.

semianechoic room [ACOUS] A room having surfaces which reduce the reflection of sound to less than normal, although not to as great an extent as an anechoic room.

semiconducting compound [SOLID STATE] A compound which is a semiconductor, such as copper oxide, mercury indium telluride, zinc sulfide, cadmium selenide, and magnesium iodide.

semiconducting crystal [SOLID STATE] A crystal of a semiconductor, such as silicon, germanium, or gray tin.

semiconductor [SOLID STATE] A solid crystalline material whose electrical conductivity is intermediate between that of a metal and an insulator, ranging from about 10^5 mhos to 10^{-7} mho per meter, and is usually strongly temperature-dependent.

semiconductor detector [NUCLEO] A particle detector which detects ionization produced by energetic charged particles in the depletion layer of a reverse-biased *pn* junction in a semiconductor, usually a very pure single crystal of silicon or germanium.

semiconductor device [ELECTR] Electronic device in which the characteristic distinguishing electronic conduction takes place within a semiconductor.

semiconductor diode [ELECTR] Also known as crystal diode; crystal rectifier; diode. **1.** A two-electrode semiconductor device that utilizes the rectifying properties of a *pn* junction or a point contact. **2.** More generally, any two-terminal electronic device that utilizes the properties of the semiconductor from which it is constructed.

semiconductor doping *See* doping.

semiconductor intrinsic properties [SOLID STATE] Properties of a semiconductor that are characteristic of the ideal crystal.

semiconductor junction [ELECTR] Region of transition between semiconducting regions of different electrical properties, usually between *p*-type and *n*-type material.

semiconductor laser [OPTICS] A laser in which stimulated emission of coherent light occurs at a *pn* junction when electrons and holes are driven into the junction by carrier injection, electron-beam excitation, impact ionization, optical excitation, or other means. Also known as diode laser; laser diode.

semiconductor rectifier *See* metallic rectifier.

semiconductor thermocouple [ELECTR] A thermocouple made of a semiconductor, which offers the prospect of operation with high-temperature gradients, because semiconductors are good electrical conductors but poor heat conductors.

semiconductor trap *See* trap.

semicubical parabola [MATH] A plane curve whose equation in cartesian coordinates x and y is $y^2 = x^3$.

semidiameter [ASTRON] Measured at the observer, half the angle subtended by the visible disk of a celestial body.

semidiurnal [ASTRON] Having a period of, occurring in, or related to approximately half a day.

semigroup [MATH] A set which is closed with respect to a given associative binary operation.

semigroup theory [MATH] The formal algebraic study of the structure of semigroups.

semi-interquartile range *See* quartile deviation.

semi-invariants *See* cumulants.

semilinear map *See* antilinear map.

semilogarithmic coordinate paper [MATH] Paper ruled with two sets of mutually perpendicular, parallel lines, one set being spaced according to the logarithms of consecutive numbers, and the other set uniformly spaced.

semimajor axis [MATH] Either of the equal line segments into which the major axis of an ellipse is divided by the center of symmetry.

semimetric [MATH] A real valued function $d(x,y)$ on pairs of points from a topological space which has all the same properties as a metric save that $d(x,y)$ may be zero even if x and y are distinct points. Also known as pseudometric.

semiminor axis [MATH] Either of the equal line segments

into which the minor axis of an ellipse is divided by the center of symmetry.

seminorm [MATH] A scalar-valued function on a real or complex vector space satisfying the axioms of a norm, except that the seminorm of a nonzero vector may equal zero.

semipermeable membrane [PHYS] A membrane which allows a solvent to pass through it, but not certain dissolved or colloidal substances.

semiregular variables [ASTRON] Red giant stars with absolute magnitude of about 0 or − 1 that are variable and have quasi periods of from about 40 to 150 days.

semisimple module [MATH] A module which is the sum of a family of simple modules.

semisimple representation of a group *See* completely reducible representation of a group.

semisimple ring [MATH] A ring in which 1 does not equal 0, and which is semisimple as a left module over itself.

semistable elementary particle *See* quasi-stable elementary particle.

semitone [ACOUS] The interval between two sounds whose frequencies have a ratio approximately equal to the twelfth root of 2. Also known as half step.

semitransparent photocathode [ELECTR] Photocathode in which radiant flux incident on one side produces photoelectric emission from the opposite side.

sending-end impedance [ELEC] Ratio of an applied potential difference to the resultant current at the point where the potential difference is applied; the sending-end impedance of a line is synonymous with the driving-point impedance of the line.

sensation unit [ACOUS] A unit of loudness, no longer in use; the loudness of a sound is $20 \log_{10}(p/p_0)$ sensation units above threshold, where p is the pressure level of the sound, and p_0 is the pressure of a sound which can just be detected by the ear.

sensibility [PHYS] The ability of a magnetic compass card to align itself with the magnetic meridian after deflection.

sensible heat *See* enthalpy.

sensible heat factor [THERMO] The ratio of space sensible heat to space total heat; used for air-conditioning calculations. Abbreviated SHF.

sensible heat flow [THERMO] The heat given up or absorbed by a body upon being cooled or heated, as the result of the body's ability to hold heat; excludes latent heats of fusion and vaporization.

sensing element *See* sensor.

sensitive flame [ACOUS] A gas flame for which the gas pressure or size of orifice is adjusted so that the shape or height of the flame changes when sound waves fall on it.

sensitive time [NUCLEO] The duration of supersaturation, sufficient for track formation, following expansion in a cloud chamber.

sensitive volume [NUCLEO] The portion of a radiation-counter tube that responds to a specific radiation.

sensitivity [ELECTR] **1.** The minimum input signal required to produce a specified output signal, for a radio receiver or similar device. **2.** Of a camera tube, the signal current developed per unit incident radiation, that is, per watt per unit area. [SCI TECH] **1.** The ability of the output of a device, system, or organism to respond to an input stimulus. **2.** Mathematically, the ratio of the response or change induced in the output to a stimulus or change in the input.

sensitivity function [CONT SYS] The ratio of the fractional change in the system response of a feedback-compensated feedback control system to the fractional change in an open-loop parameter, for some specified parameter variation.

sensitization *See* activation.

sensitometer [ENG] An instrument for measuring the sensitivity of light-sensitive materials.

sensor [ENG] The generic name for a device that senses either the absolute value or a change in a physical quantity such as temperature, pressure, flow rate, or pH, or the intensity of light, sound, or radio waves and converts that change into a useful input signal for an information-gathering system; a television camera is therefore a sensor, and a transducer is a special type of sensor. Also known as primary detector; sensing element.

sentential calculus *See* propositional calculus.

sentential connectives *See* propositional connectives.

separable degree [MATH] Let E be an algebraic extension of a field F, and let f be any embedding of F in a field L such that L is the algebraic closure of the image of F under f; the separable degree of E over F is the number of distinct embeddings of E in L which are extensions of f.

separable element [MATH] An element a is said to be separable over a field F if it is algebraic over F and if the extension field of F generated by a is a separable extension of F.

separable extension [MATH] **1.** An algebraic field extension K of a field F such that every element of K is the root of a separable polynomial whose coefficients are elements of F. **2.** More generally, a field extension K of a field F (which may or may not be algebraic) such that every subfield of K containing F and finitely generated over F is separably generated.

separable polynomial [MATH] A polynomial with no multiple roots.

separable space [MATH] A topological space which has a countable subset that is dense.

separably generated extension [MATH] A separably generated extension E of a field F is an extension field E of F for which there exists a finite transcendence base of E over F such that E is algebraic and separable over the field generated by F and the transcendence base.

separated-function synchrotron [NUCLEO] A proton synchrotron in which separate groups of magnets are used to focus and deflect the beam so that it follows a circular path.

separated sets [MATH] Sets A and B in a topological space are separated if both the closure of A intersected with B and the closure of B intersected with A are disjoint.

separating calorimeter [PHYS] A device for measuring the moisture content of steam.

separating transcendence base [MATH] A transcendence base of a field E over a field F such that E is algebraic and separable over the field generated by F and the transcendence base.

separation axioms [MATH] Properties of topological spaces such as Hausdorff, regular, and normal which reflect how points and closed sets may be enclosed in disjoint neighborhoods.

separation energy [NUC PHYS] The energy needed to remove a proton, neutron, or alpha particle from a nucleus.

separation factor [NUCLEO] The abundance ratio of two isotopes after processing, divided by their abundance ratio before processing.

separation of variables [MATH] **1.** A technique where certain differential equations are rewritten in the form $f(x)\,dx = g(y)\,dy$ which is then solvable by integrating both sides of the equation. **2.** A method of solving partial differential equations in which the solution is written in the form of a product of functions, each of which depends on only one of the independent variables; the equation is then arranged so that each of the terms involves only one of the variables and its corresponding function, and each of these terms is then set equal to a constant, resulting in ordinary differential equations. Also known as product-solution method.

separation theorem [CONT SYS] A theorem in optimal control theory which states that the solution to the linear quadratic Gaussian problem separates into the optimal deterministic controller (that is, the optimal controller for the corresponding problem without noise) in which the state used is obtained as the output of an optimal state estimator.

separative work unit [NUC PHYS] A fundamental measure of work required to separate a quantity of isotopic mixture into two component parts, one having a higher percentage of concentration of the desired isotope and one having a lower percentage.

septate coaxial cavity [ELECTROMAG] Coaxial cavity having a vane or septum, added between the inner and outer conductors, so that it acts as a cavity of a rectangular cross section bent transversely.

septillion [MATH] **1.** The number 10^{24}. **2.** In British and German usage, the number 10^{42}.

septinary number [MATH] A number in which the quantity represented by each figure is based on a radix of 7.

septum [ELECTROMAG] A metal plate placed across a waveguide and attached to the walls by highly conducting joints; the plate usually has one or more windows, or irises, designed to give inductive, capacitive, or resistive characteristics.

sequence [MATH] A listing of mathematical entities $x_1, x_2,$ ... which is indexed by the positive integers; more precisely, a function whose domain is an infinite subset of the positive integers. Also known as infinite sequence.

sequential analysis [STAT] The continuous analysis of data, obtained via sampling, performed as the amount of sampling increases.

sequential compactness [MATH] A topological space is sequentially compact if every sequence formed from its points has a convergent sequence contained in it.

sequential logic element [ELECTR] A circuit element having at least one input channel, at least one output channel, and at least one internal state variable, so designed and constructed that the output signals depend on the past and present states of the inputs.

sequential trials [STAT] The outcome of each trial is known before the next trial is performed.

Serber potential [NUC PHYS] A potential between nucleons, equal to $\frac{1}{2}(1 + M)V(r)$, where $V(r)$ is a function of the distance between the nucleons, and M is an operator which exchanges the spatial coordinates of the particles but not their spins (corresponding to the Majorana force).

serial correlation [STAT] The correlation between values of events in a time series and those values ahead or behind by a fixed amount in time or space or between parts of two different time series.

serially ordered set *See* linearly ordered set.

series [ELEC] An arrangement of circuit components end to end to form a single path for current. [MATH] An expression of the form $x_1 + x_2 + x_3 + \cdots$, where x_i are real or complex numbers.

series circuit [ELEC] A circuit in which all parts are connected end to end to provide a single path for current.

series compensation *See* cascade compensation.

series connection [ELEC] A connection that forms a series circuit.

series decay *See* radioactive series.

series disintegration [NUC PHYS] The successive radioactive transformations in a radioactive series. Also known as chain decay; chain disintegration.

series feed [ELECTR] Application of the direct-current voltage to the plate or grid of a vacuum tube through the same impedance in which the alternating-current flows.

series of lines [SPECT] A collection of spectral lines of an atom or ion for a set of transitions, with the same selection rules, to a single final state; often the frequencies have a general formula of the form $[R/(a + c_1)^2] - [R/(n + c_2)^2]$, where R is the Rydberg constant for the atom, a and c_1 and c_2 are constants, and n takes on the values of the integers greater than a for the various lines in the series.

series-parallel circuit [ELEC] A circuit in which some of the components or elements are connected in parallel, and one or more of these parallel combinations are in series with other components of the circuit.

series resonance [ELEC] Resonance in a series resonant circuit, wherein the inductive and capacitive reactances are equal at the frequency of the applied voltage; the reactances then cancel each other, reducing the impedance of the circuit to a minimum, purely resistive value.

series resonant circuit [ELEC] A resonant circuit in which the capacitor and coil are in series with the applied alternating-current voltage.

series-shunt network *See* ladder network.

series T junction *See* E-plane T junction.

series transistor regulator [ELECTR] A voltage regulator whose circuit has a transistor in series with the output voltage, a Zener diode, and a resistor chosen so that the Zener diode is approximately in the middle of its operating range.

series-tuned circuit [ELEC] A simple resonant circuit consisting of an inductance and a capacitance connected in series.

serioscopy [NUCLEO] A radiographic technique enabling three-dimensional exploration by moving two of the three components of the system (tube, subject, film) in order to register the radiographic image of one plane in the object while images outside this slice have a continuous relative displacement and are blurred.

serpentine curve [MATH] The curve given by the equation $x^2y + b^2y - a^2x = 0$, passing through and having symmetry about the origin while being asymptotic to the x axis in both directions.

Serret-Frenet formulas *See* Frenet-Serret formulas.

servo *See* servomotor.

servolink [CONT SYS] A power amplifier, usually mechanical, by which signals at a low power level are made to operate control surfaces requiring relatively large power inputs, for example, a relay and motor-driven actuator.

servo loop *See* single-loop servomechanism.

servomechanism [CONT SYS] An automatic feedback control system for mechanical motion; it applies only to those systems in which the controlled quantity or output is mechanical position or one of its derivatives (velocity, acceleration, and so on). Also known as servo system.

servomotor [CONT SYS] The electric, hydraulic, or other type of motor that serves as the final control element in a servomechanism; it receives power from the amplifier element and drives the load with a linear or rotary motion. Also known as servo.

servo system *See* servomechanism.

sesquilinear form [MATH] A mapping $f(x,y)$ from $E \times F$ into R, where R is a commutative ring with an automorphism with period 2 and $E \times F$ is the cartesian product of two modules E and F over R, such that for each x in E the function which

SERIES-PARALLEL CIRCUIT

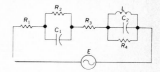

Circuit diagram of a typical series-parallel circuit. E = source of electromotive force; C_1, C_2 = capacitors; L = inductor; R_1, R_2, R_3 = resistors.

SERIES TRANSISTOR REGULATOR

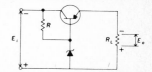

Circuit diagram of a series transistor regulator. E_O = output voltage; E_i = input voltage; R = resistor; R_L = load resistance.

takes y into $f(x,y)$ is antilinear, and for each y in F the function which takes x into $f(x,y)$ is linear.

sessile bubble method [FL MECH] A method of measuring the surface tension of a liquid that involves measuring the dimensions of a bubble in the liquid resting under a plane or concave-downward surface.

sessile drop method [FL MECH] A method of measuring surface tension in which the depth and mass of a drop resting on a surface that it does not wet are measured; from this, the shape of the drop and, in turn, the surface tension are determined.

set [MATH] A collection of objects which has the property that, given any thing, it can be determined whether or not the thing is in the collection.

set analyzer *See* analyzer.

set point [CONT SYS] The value selected to be maintained by an automatic controller.

set theory [MATH] The study of the structure and size of sets from the viewpoint of the axioms imposed.

settling time *See* correction time.

sexadecimal *See* hexadecimal.

sexadecimal number system *See* hexadecimal number system.

sexagesimal counting table [MATH] A table for converting numbers using the 60 system into decimals, for example, minutes and seconds.

sexagesimal measure of angles [MATH] A system of angular units in which a complete revolution is divided into 360 degrees, a degree into 60 minutes, and a minute into 60 seconds.

sextant [MATH] A unit of plane angle, equal to 60° or $\pi/3$ radians.

sextic [MATH] Having the sixth degree or order.

sextile aspect [ASTRON] The position of two celestial bodies when they are 60° apart.

sextillion [MATH] **1.** The number 10^{21}. **2.** In British and German usage, the number 10^{36}.

Seyfert galaxy [ASTRON] A galaxy that has a small, bright nucleus from which violent explosions may occur.

sferics receiver [ELECTR] An instrument which measures, electronically, the direction of arrival, intensity, and rate of occurrence of atmospherics; in its simplest form, the instrument consists of two orthogonally crossed antennas, whose output signals are connected to an oscillograph so that one loop measures the north-south component while the other measures the east-west component; the signals are combined vertically to give the azimuth. Also known as lightning recorder.

shade glass [OPTICS] A darkened transparency that can be moved into the line of sight of an optical instrument, such as a sextant, to reduce the intensity of light reaching the eye.

shadow [OPTICS] A region of darkness caused by the presence of an opaque object interposed between such a region and a source of light. [PHYS] A region which some type of radiation, such as sound or x-rays, does not reach because of the presence of an object, which the radiation cannot penetrate, interposed between the region and the source of radiation.

shadow bands [ASTRON] Rippling bands of shadow that appear on every white surface of flat terrestrial objects a few minutes before the total eclipse of the sun.

shadow factor [ELECTROMAG] The ratio of the electric-field strength that would result from propagation of waves over a sphere to that which would result from propagation over a plane under comparable conditions. [OPTICS] A multiplication factor derived from the sun's declination, the latitude of the target and the time of photography, used in determining the heights of objects from shadow length. Also known as tan alt.

shadowgram [PHYS] A plot or display of a shadow.

shadowgraph [OPTICS] A simple method of making visible the disturbances that occur in fluid flow at high velocity, in which light passing through a flowing fluid is refracted by density gradients in the fluid, resulting in bright and dark areas on a screen placed behind the fluid.

shadow microscopy *See* projection microradiography.

shadow photometer [ENG] A simple photometer in which a rod is placed in front of a screen and two light sources to be compared are adjusted in position until their shadows touch and are equal in intensity.

shadow scattering [QUANT MECH] Scattering that results from the interference of the incident wave and scattered waves.

shadow zone [ACOUS] A region, usually under water or in the atmosphere, which sound waves will not reach, according to ray acoustics.

shaker [ELECTROMAG] An electromagnetic device capable of imparting known and usually controlled vibratory acceleration to a given object. Also known as electrodynamic shaker; shake table.

shake table *See* shaker.

shake wave *See* S wave.

Shannon formula [COMMUN] A theorem in information theory which states that the highest number of binary digits per second which can be transmitted with arbitrarily small frequency of error is equal to the product of the bandwidth and $\log_2 (1 + R)$, where R is the signal-to-noise ratio.

Shannon limit [COMMUN] Maximum signal-to-noise ratio improvement which can be achieved by the best modulation technique as implied by Shannon's theorem relating channel capacity to signal-to-noise ratio.

Shannon-McMillian-Breiman theorem [MATH] Given an ergodic measure preserving transformation T on a probability space and a finite partition ζ of that space the limit as $n \to \infty$ of $1/n$ times the information function of the common refinement of ζ, $T^{-1}\zeta$, . . . , $T^{-n+1}\zeta$ converges almost everywhere and in the L_1 metric to the entropy of T given ζ.

Shannon's theorems [MATH] These results are foundational to the mathematical study of information; mathematically they link the concept of entropy with the amount of efficient transmittal and reception of information.

shape factor [ELEC] *See* form factor. [FL MECH] The quotient of the area of a sphere equivalent to the volume of a solid particle divided by the actual surface of the particle; used in calculations of gas flow through beds of granular solids. [OPTICS] For a lens, the quantity $(R_2 + R_1)/(R_2 - R_1)$, where R_1 and R_2 are the radii of the first and second surface of the lens. Also known as Coddington shape factor.

shaping circuit *See* corrective network.

shaping network *See* corrective network.

sharp-cutoff tube [ELECTR] An electron tube in which the control-grid openings are uniformly spaced; the anode current then decreases linearly as the grid voltage is made more negative, and cuts off sharply at a particular grid voltage.

sharpness of resonance [ELEC] The narrowness of the frequency band around the resonance at which the response of an electric circuit exceeds an arbitrary fraction of its maximum response, often 70.7%.

sharp series [SPECT] A series occurring in the line spectra of many atoms and ions with one, two, or three electrons in the

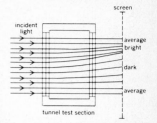

SHADOWGRAPH

Shadowgraph produced by light passing through fluid in test section indicates flow pattern.

outer shell, in which the total orbital angular momentum quantum number changes from 0 to 1.

sheaf [MATH] A fiber bundle with algebraic and topological structure usually associated to a differentiable manifold M which reflects the local behavior of differentiable functions on M.

sheaf of planes [MATH] All the planes passing through a given point.

shear *See* shear strain.

shear center *See* center of twist.

shear diagram [MECH] A diagram in which the shear at every point along a beam is plotted as an ordinate.

shear fracture [MECH] A fracture resulting from shear stress.

shear-gravity wave [GEOPHYS] A combination of gravity waves and a Helmholtz wave on a surface of discontinuity of density and velocity.

shearing field [PL PHYS] A special type of magnetic field, used to confine a plasma whose rotational transform angle changes with distance from the magnetic axis.

shearing instability *See* Helmholtz instability.

shearing stress [MECH] A stress in which the material on one side of a surface pushes on the material on the other side of the surface with a force which is parallel to the surface. Also known as shear stress; tangential stress.

shear modulus *See* modulus of elasticity in shear.

shear plane [MECH] A confined zone along which fracture occurs in metal cutting.

shear rate [FL MECH] The relative velocities in laminar flow of parallel adjacent layers of a fluid body under shear force.

shear strain [MECH] Also known as shear. **1.** A deformation of a solid body in which a plane in the body is displaced parallel to itself relative to parallel planes in the body; quantitatively, it is the displacement of any plane relative to a second plane, divided by the perpendicular distance between planes. **2.** The force causing such deformation.

shear strength [MECH] **1.** The maximum shear stress which a material can withstand without rupture. **2.** The ability of a material to withstand shear stress.

shear stress *See* shearing stress.

shear thickening [FL MECH] Viscosity increase of non-Newtonian fluids (for example, complex polymers, proteins, protoplasm) that undergo viscosity increases under conditions of shear stress (that is, viscometric flow).

shear thinning [FL MECH] Viscosity reduction of non-Newtonian fluids (for example, polymers and their solutions, most slurries and suspensions, lube oils with viscosity-index improvers) that undergo viscosity reductions under conditions of shear stress (that is, viscometric flow).

shear-viscosity function [FL MECH] The expression of the viscometric flow of a purely viscous, non-Newtonian fluid in terms of velocity gradient and shear stress of the flowing fluid.

shear wave [GEOPHYS] *See* S wave. [MECH] A wave that causes an element of an elastic medium to change its shape without changing its volume. Also known as rotational wave.

sheath [ELEC] A protective outside covering on a cable. [ELECTR] A space charge formed by ions near an electrode in a gas tube. [ELECTROMAG] The metal wall of a waveguide. [SCI TECH] A protective case or cover.

sheath-reshaping converter [ELECTROMAG] In a waveguide, a mode converter in which the change of wave pattern is achieved by gradual reshaping of the sheath of the waveguide and of conducting metal sheets mounted longitudinally in the guide.

shed [NUC PHYS] A unit of cross section, used in studying collisions of nuclei and particles, equal to 10^{-24} barn, or 10^{-48} square centimeter.

sheet cavitation [FL MECH] A type of cavitation in which cavities form on a solid boundary and remain attached as long as the conditions that led to their formation remain unaltered. Also known as steady-state cavitation.

sheet grating [ELECTROMAG] Three-dimensional grating consisting of thin, longitudinal, metal sheets extending along the inside of a waveguide for a distance of about a wavelength, and used to stop all waves except one predetermined wave that passes unimpeded.

sheet polarizer [OPTICS] A mechanism for obtaining linear polarized light; there are several types, one of which is a microcrystalline polarizer in which small crystals of a dichroic material (quinine iodosulfate), oriented parallel to each other in a plastic medium, absorb one polarization and transmit the other.

sheffer stroke *See* NAND.

shell model [NUC PHYS] A model of the nucleus in which the shell structure is either postulated or is a consequence of other postulates; especially the model in which the nucleons act as independent particles filling a preassigned set of energy levels as permitted by the quantum numbers and Pauli principle.

shell structure [NUC PHYS] Structure of the nucleus in which nucleons of each kind occupy quantum states which are in groups of approximately the same energy, called shells, the number of nucleons in each shell being limited by the Pauli exclusion principle.

Shenstone effect [ELECTR] An increase in photoelectric emission of certain metals following passage of an electric current.

Sheppard's corrections [STAT] A correction for moments computed for a frequency distribution of grouped data to adjust for the error that is introduced by the assumption that all the data within a class are at the midpoint of the class; the adjustment is made by subtracting one-twelfth of the square of a grouping unit from the estimated variance.

Sherwood number *See* Nusselt number.

Sherwood project [NUCLEO] The overall research program in the United States for producing useful power from nuclear fusion.

SHF *See* sensible-heat factor.

shield [NUCLEO] The material placed around a nuclear reactor, or other source of radiation, to reduce escaping radiation or particles to a permissible level. Also known as shielding.

shielded line [ELECTROMAG] Transmission line, the elements of which confine the propagated waves to an essentially finite space; the external conducting surface is called the sheath.

shield grid [ELECTR] A grid that shields the control grid of a gas tube from electrostatic fields, thermal radiation, and deposition of thermionic emissive material; it may also be used as an additional control electrode.

shielding [ELECTROMAG] *See* electric shielding. [NUCLEO] **1.** Reducing the ionizing radiation reaching one region of space from another region by using a shield or other device. **2.** *See* shield.

shielding distance *See* Debye shielding length.

shielding ratio [ELECTROMAG] The ratio of a field in a specified region when electrical shielding is in place to the field in that region when the shielding is removed.

shifting theorem [MATH] **1.** If the Fourier transform of $f(t)$ is $F(x)$, then the Fourier transform of $f(t-a)$ is $\exp(iax)F(x)$. **2.** If the Laplace transform of $f(x)$ is $F(y)$, then the Laplace transform of $f(x-a)$ is $\exp(-ay)F(y)$.

shift of spectral line [SPECT] A small change in the position of a spectral line that is due to a corresponding change in frequency which, in turn, results from one or more of several causes, such as the Doppler effect.

shim rod [NUCLEO] A control rod used for making occasional coarse adjustments in the reactivity of a nuclear reactor.

shipping ton *See* ton.

SHM *See* harmonic motion.

shock [MECH] A pulse or transient motion or force lasting thousandths to tenths of a second which is capable of exciting mechanical resonances; for example, a blast produced by explosives.

shock diamonds [PHYS] The shock waves that appear in the exhaust stream of a rocket; they are made visible by their luminosity and describe an approximate diamond configuration in side view.

shock front [PHYS] The outer side of a shock wave whose pressure rises from zero up to its peak value. Also known as pressure front.

shock heating [PHYS] The nonisentropic heating of a fluid which takes place when a shock wave passes through it.

Shockley diode [ELECTR] A *pnpn* silicon controlled switch having characteristics that permit operation as a unidirectional diode switch.

Shockley partial dislocation [SOLID STATE] A partial dislocation in which the Burger's vector lies in the fault plane, so that it is able to glide, in contrast to a Frank partial dislocation. Also known as glissile dislocation.

shock resistance [ENG] The property which prevents cracking or general rupture when impacted.

shock tube [FL MECH] A long tube divided into two parts by a diaphragm; the volume on one side of the diaphragm constitutes the compression chamber, the other side is the expansion chamber; a high pressure is developed by suitable means in the compression chamber, and the diaphragm ruptured; the shock wave produced in the expansion chamber can be used for the calibration of air blast gages, or the chamber can be instrumented for the study of the characteristics of the shock wave.

shock tunnel [ENG] A hypervelocity wind tunnel in which a shock wave generated in a shock tube ruptures a second diaphragm in the throat of a nozzle at the end of the tube, and gases emerge from the nozzle into a vacuum tank with Mach numbers of 6 to 25.

shock wave [PHYS] A fully developed compression wave of large amplitude, across which density, pressure, and particle velocity change drastically.

shock-wave lip [PHYS] The shock wave obtained from the lip of a free jet nozzle, because of failure to match the stream pressure and the ambient exhaust pressure.

shooting star [ASTRON] A small meteor that has the brief appearance of a darting, starlike object.

short *See* short circuit.

short antenna [ELECTROMAG] An antenna shorter than about one-tenth of a wavelength, so that the current may be assumed to have constant magnitude along its length, and the antenna may be treated as an elementary dipole.

short circuit [ELEC] A low-resistance connection across a voltage source or between both sides of a circuit or line, usually accidental and usually resulting in excessive current flow that may cause damage. Also known as short.

short-circuit impedance [ELEC] Of a line or four-terminal network, the driving point impedance when the far-end is short-circuited.

SHOCK TUBE

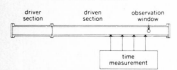

Schematic representation of a shock tube.

SHOCK WAVE

Schlieren photograph of supersonic flow over blunt object, showing approximately parabolic shock wave detached from object. (*Avco Everett Research Laboratory, Inc.*)

short hundredweight *See* hundredweight.

short-path principle *See* Hittorf principle.

short-pulse laser [OPTICS] A laser designed to generate a pulse of light lasting on the order of nanoseconds or less, and having very high power, such as by Q switching or mode-locking.

short-range force [PHYS] A force between two particles which is negligible when the distance between the particles is greater than a certain amount; in particular, nuclear forces whose range is several times 10^{-15} meter.

short-range order [PHYS] A regularity in the arrangement of atoms in a disordered solid or a liquid in which the probability of a given type of atom having neighbors of a given type is greater than one would expect on a purely random basis.

short-slot coupler *See* three-decibel coupler.

Shortt clock [HOROL] An accurate clock manufactured by the Shortt-Synchronome Corporation; it is a pendulum clock in which the master pendulum is enclosed in an airtight, nearly evacuated chamber; this pendulum receives a small impulse to maintain its swing. Also known as free-pendulum clock.

short ton *See* ton.

shortwave radiation [ELECTROMAG] A term used loosely to distinguish radiation in the visible and near-visible portions of the electromagnetic spectrum (roughly 0.4 to 1.0 micrometer in wavelength) from long-wave radiation (infrared radiation).

shot effect *See* shot noise.

shot noise [ELECTR] Noise voltage developed in a thermionic tube because of the random variations in the number and the velocity of electrons emitted by the heated cathode; the effect causes sputtering or popping sounds in radio receivers and snow effects in television pictures. Also known as Schottky noise; shot effect.

shower *See* cosmic-ray shower.

shower unit [NUCLEO] The mean path length required to reduce the energy of relativistic charged particles to half value as they pass through matter; one shower unit is equal to 0.693 radiation length.

Shubnikov–de Haas effect [SOLID STATE] Oscillations of the resistance or Hall coefficient of a metal or semiconductor as a function of a strong magnetic field, due to the quantization of the electron's energy.

Shubnikov groups [SOLID STATE] The point groups and space groups of crystals having magnetic moments. Also known as black-and-white groups; magnetic groups.

shunt [ELEC] **1.** A precision low-value resistor placed across the terminals of an ammeter to increase its range by allowing a known fraction of the circuit current to go around the meter. Also known as electric shunt. **2.** To place one part in parallel with another. **3.** *See* parallel. [ELECTROMAG] A piece of iron that provides a parallel path for magnetic flux around an air gap in a magnetic circuit.

shunt-excited [ELECTROMAG] Having field windings connected across the armature terminals, as in a direct-current generator.

shunt feed *See* parallel feed.

shunting [ELEC] The act of connecting one device to the terminals of another so that the current is divided between the two devices in proportion to their respective admittances.

shunt loading [ELEC] Loading in which reactances are applied in shunt across the conductors.

shunt peaking [ELECTR] The use of a peaking coil in a parallel circuit branch connecting the output load of one stage to the input load of the following stage, to compensate for high-frequency loss due to the distributed capacitances of the two stages.

shunt regulator [ELEC] A regulator that maintains a constant output voltage by controlling the current through a dropping resistance in series with the load.

shunt T junction *See* H-plane T junction.

shunt wound [ELEC] Having armature and field windings in parallel, as in a direct-current generator or motor.

shutter [NUCLEO] A movable plate of absorbing material used to cover a window or a beam hole in a reactor when radiation is not desired, or used to shut off a flow of slow neutrons. [OPTICS] A mechanical device that cuts off a beam of light by opening and closing at different rates of speed to expose film or plates; used in cameras and motion picture projectors.

shuttered image converter [ELECTR] An image tube whose photoelectrons can be rapidly switched off to allow a camera to record the image on its screen.

shuttle *See* rabbit.

Si *See* silicon.

SI *See* International System of Units.

Siacci method [MECH] An accurate and useful method for calculation of trajectories of high-velocity missiles with low quadrant angles of departure; basic assumptions are that the atmospheric density anywhere on the trajectory is approximately constant, and the angle of departure is less than about 15°.

SID *See* sudden ionospheric disturbance.

sideband [ELECTROMAG] **1.** The frequency band located either above or below the carrier frequency, within which fall the frequency components of the wave produced by the process of modulation. **2.** The wave components lying within such bands.

side direction [MECH] In stress analysis, the direction perpendicular to the plane of symmetry of an object.

side echo [ELECTROMAG] Echo due to a side lobe of an antenna.

side lobe *See* minor lobe.

sidereal [ASTRON] Referring to a quantity, such as time, to indicate that it is measured in relation to the apparent motion or position of the stars.

sidereal clock [HOROL] A clock set at 0 hours 0 minutes 0 seconds (midnight) as the vernal equinox crosses the meridian; it is the astronomical clock.

sidereal day [ASTRON] The time between two successive upper transits of the vernal equinox; this period measures one sidereal day.

sidereal hour angle [ASTRON] The angle along the celestial equator formed between the hour circle of a celestial body and the hour circle of the vernal equinox, measuring westward from the vernal equinox through 360°.

sidereal month [ASTRON] The time period of one revolution of the moon about the earth relative to the stars; this period varies because of perturbations, but it is a little less than 27 1/3 days.

sidereal noon [ASTRON] The instant in time that the vernal equinox is on the meridian.

sidereal period [ASTRON] The length of time required for one revolution of a celestial body about its primary, with respect to the stars.

sidereal time [ASTRON] Time based on diurnal motion of stars; it is used by astronomers but is not convenient for ordinary purposes.

sidereal year [ASTRON] The time period relative to the stars of one revolution of the earth around the sun; it is about 365.2564 mean solar days.

siderostat [OPTICS] A more precise model of a heliostat; the siderostat uses a modified mirror mounting so that the image of a star is kept steady while the rest of the field is in rotation about the center.

siegbahn [SPECT] A unit of length, formerly used to express wavelengths of x-rays, equal to exactly 1/3029.45 of the spacing of the (200) planes of calcite at 18°C, or to $(1.00202 \pm 0.00003) \times 10^{-13}$ meter. Also known as x-ray unit; X unit. Symbolized X; XU.

siemens [ELEC] A unit of conductance, admittance, and susceptance, equal to the conductance between two points of a conductor such that a potential difference of 1 volt between these points produces a current of 1 ampere; the conductance of a conductor in siemens is the reciprocal of its resistance in ohms. Formerly known as mho ($\mho$); reciprocal ohm. Symbolized S.

sieve of Eratosthenes [MATH] An iterative procedure which determines all the primes less than a given number.

sievert [NUCLEO] A unit of radiation dose, equal to the dose delivered by a point source of 1 milligram of radium, enclosed in a platinum container with walls $\frac{1}{2}$ millimeter thick, to a sample at a distance of 1 centimeter over a period of 1 hour; equal to approximately 8.38 roentgens. Also known as milli-curie-of-intensity-hour.

sight distance [OPTICS] The distance from which an object at eye level remains visible to an observer.

sighting [OPTICS] **1.** The act or procedure of aiming with the aid of a sight. **2.** The action of bringing something into view; the action of seeing something.

sighting tube [ENG] A tube, usually ceramic, inserted into a hot chamber whose temperature is to be measured; an optical pyrometer is sighted into the tube to observe the interior end of the tube to give a temperature reading.

sight unit [OPTICS] A compact sighting device composed of an elbow or panoramic telescope, mount, or adapter, usually used for pointing a weapon for direct or indirect fire; it may be attached to, or used in conjunction with, a weapon rocket launcher, or the like.

σ^2 *See* error variance.

sigma algebra [MATH] A collection of subsets of a given set which contains the empty set and is closed under countable union and complementation of sets. Also known as sigma field.

sigma field *See* sigma algebra.

sigma finite [MATH] A measure is sigma finite on a space X if X is a countable disjoint union of sets each of which is measurable and has finite measure.

sigma function [THERMO] A property of a mixture of air and water vapor, equal to the difference between the enthalpy and the product of the specific humidity and the enthalpy of water (liquid) at the thermodynamic wet-bulb temperature; it is constant for constant barometric pressure and thermodynamic wet-bulb temperature.

sigma hyperon [PARTIC PHYS] **1.** The collective name for three semistable baryons with charges of $+1, 0$, and -1 times the proton charge, designated $\Sigma^+, \Sigma^0, \Sigma^-$, all having masses of approximately 1193 MeV, spin of $\frac{1}{2}$, and positive parity; they form an isotopic spin multiplet with a total isotopic spin of 1 and a hypercharge of 0. **2.** Any baryon belonging to an isotopic spin multiplet having a total isotopic spin of 1 and a hypercharge of 0; designated $\Sigma_{J}P(m)$, where m is the mass of the baryon in MeV, and J and P are its spin and parity; the $\Sigma_{3/2+}$ (1385) is sometimes designated Σ^*.

sigma-minus hyperonic atom [ATOM PHYS] An atom consisting of a negatively charged sigma hyperon orbiting around an ordinary nucleus. Designated Σ^--hyperonic atom.

sigma pile [NUCLEO] An assembly of moderating material containing a neutron source, used to study the absorption cross sections and other neutron properties of the material.

sigma ring [MATH] A ring of sets where any countable union of its members is also a member.

sigmatron [NUCLEO] A cyclotron and betatron operating in tandem to produce billion-volt x-rays.

sign [MATH] **1.** A symbol which indicates whether a quantity is greater than zero or less than zero; the signs are often the marks + and − respectively, but other arbitrarily selected symbols are used, especially in automatic data processing. **2.** A unit of plane angle, equal to 30° or $\pi/6$ radians.

signal carrier *See* carrier.

signal-flow graph [SYS ENG] An abbreviated block diagram in which small circles, called nodes, represent variables of the system, and the nodes are connected by lines, called branches, which represent one-way signal multipliers; an arrow on the line indicates direction of signal flow, and a letter near the arrow indicates the multiplication factor. Also known as flow graph.

signal-to-noise ratio [ELECTR] The ratio of the amplitude of a desired signal at any point to the amplitude of noise signals at that same point; often expressed in decibels; the peak value is usually used for pulse noise, while the root-mean-square (rms) value is used for random noise. Abbreviated S/N; SNR.

signal voltage [ELEC] Effective (root-mean-square) voltage value of a signal.

sign convention [OPTICS] A convention as to which quantities, such as angles, distances, and radii of curvature, are positive and which are negative in computations involving a lens or a mirror.

sign count function [MATH] An integer-valued function equal to the number of changes in sign between successive members of the Sturm sequence of a specified polynomial.

signed measure [MATH] An extended real-valued function m defined on a sigma algebra of subsets of a set S such that (1) the value of m on the empty set is 0, (2) the value of m on a countable union of disjoint sets is the sum of its values on each set, and (3) m assumes at most one of the values $+\infty$ and $-\infty$.

significance [MATH] The arbitrary rank, priority, or order of relative magnitude assigned to a given position in a number.

significance level *See* level of significance of a test.

significance probability [STAT] The probability of observing a value of a test statistic as significant as, or even more significant than, the value actually observed.

significant digit *See* significant figure.

significant figure [MATH] A prescribed decimal place which determines the amount of rounding off to be done; this is usually based upon the degree of accuracy in measurement. Also known as significant digit.

sign test [STAT] A test which can be used whenever an experiment is conducted to compare a treatment with a control on a number of matched pairs, provided the two treatments are assigned to the members of each pair at random.

signum [MATH] The real function sgn (x) defined for all x different from 0, where sgn $(x) = 1$ if $x > 0$ and sgn $(x) = -1$ if $x < 0$.

silent discharge [ELECTR] An inaudible electric discharge in air that occurs at high voltage and consumes a relatively large amount of energy.

silicide resistor [ELECTR] A thin-film resistor that uses a silicide of molybdenum or chromium, deposited by direct-current sputtering in an integrated circuit when radiation hardness or high resistance values are required.

silicon [CHEM] A group IV nonmetallic element, symbol Si, with atomic number 14, atomic weight 28.086; dark-brown crystals that burn in air when ignited; soluble in hydrofluoric acid and alkalies; melts at 1410°C; used to make silicon-containing alloys, as an intermediate for silicon-containing compounds, and in rectifiers and transistors.

silicon capacitor [ELECTR] A capacitor in which a pure silicon-crystal slab serves as the dielectric; when the crystal is grown to have a *p* zone, a depletion zone, and an *n* zone, the capacitance varies with the externally applied bias voltage, as in a varactor.

silicon controlled rectifier [ELECTR] A semiconductor rectifier that can be controlled; it is a *pnpn* four-layer semiconductor device that normally acts as an open circuit, but switches rapidly to a conducting state when an appropriate gate signal is applied to the gate terminal. Abbreviated SCR. Also known as reverse-blocking triode thyristor.

silicon controlled switch [ELECTR] A four-terminal switching device having four semiconductor layers, all of which are accessible; it can be used as a silicon controlled rectifier, gate-turnoff switch, complementary silicon controlled rectifier, or conventional silicon transistor. Abbreviated SCS. Also known as reverse-blocking tetrode thyristor.

silicon detector *See* silicon diode.

silicon diode [ELECTR] A crystal diode that uses silicon as a semiconductor; used as a detector in ultra-high- and super-high-frequency circuits. Also known as silicon detector.

silicon-on-sapphire [ELECTR] A semiconductor manufacturing technology in which metal oxide semiconductor devices are constructed in a thin single-crystal silicon film grown on an electrically insulating synthetic sapphire substrate. Abbreviated SOS.

silicon rectifier [ELECTR] A metallic rectifier in which rectifying action is provided by an alloy junction formed in a high-purity silicon slab.

silicon resistor [ELECTR] A resistor using silicon semiconductor material as a resistance element, to obtain a positive temperature coefficient of resistance that does not appreciably change with temperature; used as a temperature-sensing element.

silicon solar cell [ELECTR] A solar cell consisting of *p* and *n* silicon layers placed one above the other to form a *pn* junction at which radiant energy is converted into electricity.

silicon transistor [ELECTR] A transistor in which silicon is used as the semiconducting material.

Silsbee effect [CRYO] The ability of an electric current to destroy superconductivity by means of the magnetic field that it generates, without raising the cryogenic temperature.

silver [CHEM] A white metallic element in group I, symbol Ag, with atomic number 47; soluble in acids and alkalies, insoluble in water; melts at 961°C, boils at 2212°C; used in photographic chemicals, alloys, conductors, and plating.

silver-cadmium storage battery [ELEC] A storage battery that combines the excellent space and weight characteristics of silver-zinc batteries with long shelf life and other desirable properties of nickel-cadmium batteries.

silvered mica capacitor [ELECTR] A mica capacitor in which a coating of silver is deposited directly on the mica sheets to serve in place of conducting metal foil.

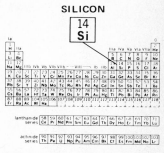

SILICON

Periodic table of the chemical elements showing the position of silicon.

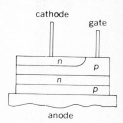

SILICON CONTROLLED RECTIFIER

Diagrammatic view of typical silicon controlled rectifier showing four alternate layers of *p*-type and *n*-type material.

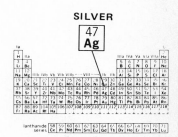

SILVER

Periodic table of the chemical elements showing the position of silver.

silver-zinc storage battery [ELEC] A storage battery that gives higher watt current output and greater watt-hour capacity per unit of weight and volume than most other types, even at high discharge rates; used in missiles and torpedoes, where its high cost can be tolerated.

similar figures [MATH] Two figures or bodies that are identical except for size; similar figures can be placed in perspective, so that straight lines joining corresponding parts of the two figures will pass through a common point.

similarity principle *See* principle of dynamical similarity.

similarity transformation [MATH] **1.** A transformation of a euclidean space obtained from such transformations as translations, rotations, and those which either shrink or expand the length of vectors. **2.** A mapping that associates with each linear transformation P on a vector space the linear transformation $R^{-1}PR$ that results when the coordinates of the space are subjected to a nonsingular linear transformation R. **3.** A mapping that associates with each square matrix P the matrix $Q = R^{-1}PR$, where R is a nonsingular matrix and R^{-1} is the inverse matrix of R; if P is the matrix representation of a linear transformation, then this definition is equivalent to the second definition.

similar matrices [MATH] Two square matrices A and B related by the transformation $B = SAT$, where S and T are nonsingular matrices and T is the inverse matrix of S.

similitude [PHYS] The use in scientific studies and engineering designs of the corresponding behavior between large and small objects or systems which are of similar nature and, more precisely, have geometrical, kinematic, and dynamical similarity.

Simon liquefier [CRYO] A device for liquefying helium in which helium is first cooled at high pressure by liquid or solid hydrogen and is then liquefied by a single adiabatic expansion.

simple aggregative index [STAT] A statistic computed for a collection of items by taking the ratio of the sum of their given-year values or amounts to the sum of their base-year values or amounts and usually multiplying by 100 to express the figure as a percentage.

simple alternative [STAT] An alternative to the null hypothesis which completely specifies the distribution of the observed random variables.

simple arc [MATH] The image of a closed interval under a continuous, injective mapping from the interval into a plane. Also known as Jordan arc.

simple balance [ENG] An instrument for measuring weight in which a beam can rotate about a knife-edge or other point of support, the unknown weight is placed in one of two pans suspended from the ends of the beam and the known weights are placed in the other pan, and a small weight is slid along the beam until the beam is horizontal.

simple character [MATH] The character of an irreducible representation of a group.

simple closed curve [MATH] The image of the closed interval $[0,1]$ under some continuous function f from the interval into the plane, where $f(0) = f(1)$, and $x < y$, $f(x) = f(y)$, implies $x = 0$, $y = 1$. Also known as Jordan curve.

simple cubic lattice [CRYSTAL] A crystal lattice whose unit cell is a cube, and whose lattice points are located at the vertices of the cube.

simple function *See* step function.

simple group [MATH] A group G that is nontrivial and contains no normal subgroups other than the identity element and G itself.

simple harmonic current [ELEC] Alternating current, the instantaneous value of which is equal to the product of a constant, and the cosine of an angle varying linearly with time. Also known as sinusoidal current.

simple harmonic electromotive force [ELEC] An alternating electromotive force which is equal to the product of a constant and the cosine or sine of an angle which varies linearly with time.

simple harmonic motion *See* harmonic motion.

simple hypothesis [STAT] A hypothesis which completely specifies the distribution of the observed random variables.

simple lens [OPTICS] A lens consisting of a single element. Also known as single lens.

simple microscope [OPTICS] A diverging lens system, which can form an enlarged image of a small object. Also known as hand lens; magnifier; magnifying glass.

simple module [MATH] A module M that is nontrivial and contains no submodules other than 0 and M itself.

simple order *See* linear order.

simple oscillator *See* harmonic oscillator.

simple pendulum [MECH] A device consisting of a small, massive body suspended by an inextensible object of negligible mass from a fixed horizontal axis about which the body and suspension are free to rotate.

simple random samples [STAT] Samples in which every possible sample of size n, that is, every combination of n items from the number in the population, is equally likely to be part of the sample.

simple results [STAT] Results of observations such that on each trial of an experiment one and only one of these results will occur.

simple ring [MATH] A semisimple ring R such that for any two left ideals in R there is an isomorphism of R which maps one onto the other.

simple root [MATH] A polynomial $f(x)$ has c as a simple root if $(x - c)$ is a factor but $(x - c)^2$ is not.

simple tone [ACOUS] Also known as pure tone. **1.** A sound wave whose instantaneous sound pressure is a simple sinusoidal function of time. **2.** A sound sensation characterized by singleness of pitch.

simplex [MATH] An n-dimensional simplex in a euclidean space consists of $n + 1$ linearly independent points $p_0, p_1, \ldots, p_n$ together with all line segments $a_0p_0 + a_1p_1 + \cdots + a_np_n$ where the $a_i \geq 0$ and $a_0 + a_1 + \cdots + a_n = 1$; a triangle with its interior and a tetrahedron with its interior are examples.

simplex method [MATH] A finite iterative algorithm used in linear programming whereby successive solutions are obtained and tested for optimality.

simplicial complex [MATH] A set consisting of finitely many simplices where either two simplices are disjoint or intersect in a simplex which is a face common to each.

simplicial homology [MATH] A homology for a topological space where the nth group reflects how the space may be filled out by n-dimensional simplicial complexes and detects the presence of analogs of n-dimensional holes.

simplicial subdivision [MATH] A decomposition of the simplices composing a simplicial complex which results in a simplicial complex with a larger number of simplices.

simply connected region [MATH] A region having no holes; all closed curves can be shrunk to a point without passing through points in the complement of the region.

simply connected space [MATH] A topological space whose fundamental group consists of only one element; equivalently, all closed curves can be shrunk to a point.

simply ordered set *See* linearly ordered set.

SIMPLE MICROSCOPE

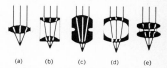

(a) (b) (c) (d) (e)

Types of lenses used in simple microscope. *(a)* Double convex. *(b)* Doublet. *(c)* Coddington. *(d)* Hastings triplet. *(e)* Achromat. *(From F. A. Jenkins and H. E. White, Fundamentals of Optics, 3d ed., McGraw-Hill, 1957)*

simply periodic function [MATH] A periodic function $f(x)$ for which there is a period a such that every period of $f(x)$ is an integral multiple of a. Also known as singly periodic function.

Simpson's rule [MATH] Also known as parabolic rule. **1.** A basic approximation formula for definite integrals which states that the integral of a real-valued function f on an interval $[a,b]$ is approximated by $h[f(a) + 4f(a + h) + f(b)]/3$, where $h = (b - a)/2$; this is the area under a parabola which coincides with the graph of f at the abscissas a, $a + h$, and b. **2.** A method of approximating a definite integral over an interval which is equivalent to dividing the interval into equal subintervals and applying the formula in the first definition to each subinterval.

SIMS *See* secondary ion mass spectrometer.

simson *See* Simson line.

Simson line [MATH] The Simson line of a point P on the circumcircle of a triangle ABC is the line passing through the collinear points L, M, and N, where L, M, and N are the projections of P upon the sides BC, CA, and AB, respectively. Also known as simson.

simulate [ENG] To mimic some or all of the behavior of one system with a different, dissimilar system, particularly with computers, models, or other equipment.

simulator [ENG] A computer or other piece of equipment that simulates a desired system or condition and shows the effects of various applied changes, such as a flight simulator.

simultaneity [MECH] Two events have simultaneity, relative to an observer, if they take place at the same time according to a clock which is fixed relative to the observer.

simultaneous equations [MATH] A collection of equations considered to be a set of joint conditions imposed on the variables involved.

sin A *See* sine.

Sincov's equation [MATH] The functional equation $f(x,z) = f(x,y) + f(y,z)$.

sine [MATH] The sine of an angle A in a right triangle with hypotenuse of length c is given by the ratio a/c, where a is the length of the side opposite A; more generally, the sine function assigns to any real number A the ordinate of the point on the unit circle obtained by moving from $(1,0)$ counterclockwise A units along the circle, or clockwise $|A|$ units if A is less than 0. Denoted sin A.

sine galvanometer [ENG] A type of magnetometer in which a small magnet is suspended in the center of a pair of Helmholtz coils, and the rest position of the magnet is measured when various known currents are sent through the coils.

sine integral [MATH] A function whose value at x is equal to the integral from 0 to x of $[(\sin t)/t] dt$. Denoted si.

sine wave [PHYS] A wave whose amplitude varies as the sine of a linear function of time. Also known as sinusoidal wave.

sine-wave response *See* frequency response.

singing [CONT SYS] An undesired, self-sustained oscillation in a system or component, at a frequency in or above the passband of the system or component; generally due to excessive positive feedback.

singing arc *See* Duddle arc.

singing flame [ACOUS] A jet of hydrogen or carbon dioxide burning in an open tube under such conditions as to produce a musical sound with relatively few overtones.

singing margin [CONT SYS] The difference in level, usually expressed in decibels, between the singing point and the operating gain of a system or component.

singing point [CONT SYS] The minimum value of gain of a system or component that will result in singing.

SINE WAVE

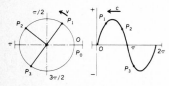

Relation of sine wave, simple harmonic motion, and uniform circular motion. Point P moves around the circle at constant tangential speed v; projection of P on vertical line moves up and down, executing simple harmonic motion. Plane on which the projection is plotted travels to the left at speed c, generating a sine wave.

single axis gyroscope [ENG] A gyroscope suspended in just one gimbal whose bearings form its output axis; an example is a rate gyroscope.

single-carrier theory [SOLID STATE] A theory of the behavior of a rectifying barrier which assumes that conduction is due to the motion of carriers of only one type; it can be applied to the contact between a metal and a semiconductor.

single crystal [CRYSTAL] A crystal, usually grown artificially, in which all parts have the same crystallographic orientation.

single cusp of the first kind *See* keratoid cusp.

single cusp of the second kind *See* ramphoid cusp.

single-degree-of-freedom gyro [MECH] A gyro the spin reference axis of which is free to rotate about only one of the orthogonal axes, such as the input or output axis.

single-layer solenoid [ELECTROMAG] A solenoid which has only one layer of wire, wound in a cylindrical helix.

single lens *See* simple lens.

single-loop feedback [CONT SYS] A system in which feedback may occur through only one electrical path.

single-loop servomechanism [CONT SYS] A servomechanism which has only one feedback loop. Also known as servo loop.

single-phase [ELEC] Energized by a single alternating voltage.

single-phase circuit [ELEC] Either an alternating-current circuit which has only two points of entry, or one which, having more than two points of entry, is intended to be so energized that the potential differences between all pairs of points of entry are either in phase or differ in phase by 180°.

single-pole double-throw [ELEC] A three-terminal switch or relay contact arrangement that connects one terminal to either of two other terminals. Abbreviated SPDT.

single-pole single-throw [ELEC] A two-terminal switch or relay contact arrangement that opens or closes one circuit. Abbreviated SPST.

single scattering [PHYS] A change in direction of a particle or photon because of a single collision.

single-shot multivibrator *See* monostable multivibrator.

single-throw switch [ELEC] A switch in which the same pair of contacts is always opened or closed.

single tuned circuit [ELEC] A circuit whose behavior is the same as that of a circuit with a single inductance and a single capacitance, together with associated resistances.

single-unit semiconductor device [ELECTR] Semiconductor device having one set of electrodes associated with a single carrier stream.

singly periodic function *See* simply periodic function.

singular arc [CONT SYS] In an optimal control problem, that portion of the optimal trajectory in which the Hamiltonian is not an explicit function of the control inputs, requiring higher-order necessary conditions to be applied in the process of solution.

singular integral equation [MATH] An integral equation where the integral appearing either has infinite limits of integration or the kernel function has points where it is infinite.

singular integral of a differential equation *See* singular solution of a differential equation.

singularity [MATH] A point where a function of real or complex variables is not differentiable or analytic. Also known as singular point of a function. [RELAT] A singularity exists in a space-time when it is not possible to extend all timelike and null geodesics to arbitrary values of their affine parameter.

singularity theorems [RELAT] Theorems proving that singularities must develop in certain space-times, such as the universe, given only broad conditions, such as causality, and the existence of a trapped surface.

singular matrix [MATH] A matrix which has no inverse; equivalently, its determinant is zero.

singular point of a curve [MATH] A point on a curve at which the curve possesses no smoothly turning tangent, or crosses or touches itself, or has a cusp or isolated point.

singular point of a differential equation [MATH] A point which is a singularity for at least one of the known functions appearing in the equation.

singular point of a function *See* singularity.

singular solution of a differential equation [MATH] A solution which is not generic, that is, not obtainable from the general solution. Also known as singular integral of a differential equation.

singular transformation [MATH] A linear transformation which has no corresponding inverse transformation.

singular values of a matrix [MATH] For a matrix A these are the positive square roots of the eigenvalues of A^*A, where A^* denotes the adjoint matrix of A.

sinh *See* hyperbolic sine.

sink [ELECTROMAG] The region of a Rieke diagram where the rate of change of frequency with respect to phase of the reflection coefficient is maximum for an oscillator; operation in this region may lead to unsatisfactory performance by reason of cessation or instability of oscillations. [PHYS] A device or system where some extensive entity is absorbed, such as a heat sink, a sink flow, a load in an electrical circuit, or a region in a nuclear reactor where neutrons are strongly absorbed.

sink flow [FL MECH] **1.** In three-dimensional flow, a point into which fluid flows uniformly from all directions. **2.** In two-dimensional flow, a straight line into which fluid flows uniformly from all directions at right angles to the line.

sinking [OPTICS] In atmospheric optics, a refraction phenomenon, the opposite of looming, in which an object on, or slightly above, the geographic horizon apparently sinks below it.

sinusoidal angular modulation *See* angle modulation.

sinusoidal current *See* simple harmonic current.

sinusoidal function [MATH] The real or complex function $\sin(u)$ or any function with analogous continuous periodic behavior.

sinusoidal wave *See* sine wave.

siriometer [ASTRON] A unit of length, formerly used in astronomical measurement, equal to 10^6 astronomical units, or 1.496×10^{17} meters.

SIRS *See* satellite infrared spectrometer.

SIT *See* self-induced transparency.

six-j-symbol [QUANT MECH] A coefficient that appears in the transformation between various modes of coupling eigenfunctions of three angular momenta; it is equal to the Racah coefficient, except perhaps in sign, and has greater symmetry than the Racah coefficient.

Six's thermometer [ENG] A combination maximum thermometer and minimum thermometer; the tube is shaped in the form of a U with a bulb at either end; one bulb is filled with creosote which expands or contracts with temperature variation, forcing before it a short column of mercury having iron indexes at either end; the indexes remain at the extreme positions reached by the mercury column, thus indicating the maximum and minimum temperatures; the indexes can be reset with the aid of a magnet.

sixty degrees Fahrenheit British thermal unit *See* British thermal unit.

six-vector [RELAT] An antisymmetrical, second-rank tensor in Minkowski space; that is, a tensor whose components, $T_{\mu\nu}$, with $\mu,\nu = 1, 2, 3, 4$ satisfy $T_{\mu\nu} = -T_{\nu\mu}$; it has six independent components.

size of a critical region [STAT] For statistical hypotheses, the probability of committing a type I error, that is, rejecting the hypothesis tested when it is true.

skeleton crystal [CRYSTAL] A crystal formed in microscopic outline with incomplete filling in of the faces.

skewed density function [STAT] A density function which is not symmetrical, and which depends not only on the magnitude of the difference between the average value and the variate, but also on the sign of this difference.

skew field [MATH] A ring whose nonzero elements form a non-Abelian group with respect to the multiplicative operation.

skew Hermitian matrix [MATH] A square matrix which equals the negative of its adjoint.

skew lines [MATH] Lines which do not lie in the same plane in euclidean three-dimensional space.

skew product [MATH] A multiplicative operation or structure induced upon a cartesian product of sets, where each has some algebraic structure.

skew surface [MATH] A ruled surface whose total curvature vanishes everywhere.

skew-symmetric matrix *See* antisymmetric matrix.

skew-symmetric tensor [MATH] A tensor where interchanging two indices will only change the sign of the corresponding component.

skiascope [OPTICS] An instrument used to study optical refraction within the eye.

skin depth [ELECTROMAG] The depth beneath the surface of a conductor, which is carrying current at a given frequency due to electromagnetic waves incident on its surface, at which the current density drops to one neper below the current density at the surface.

skin effect [ELEC] The tendency of alternating currents to flow near the surface of a conductor thus being restricted to a small part of the total sectional area and producing the effect of increasing the resistance. Also known as conductor skin effect; Kelvin skin effect.

skin erythema dose [NUCLEO] A unit of radioactive dose resulting from exposure to electromagnetic radiation, equal to the dose that slightly reddens or browns the skin of 80% of all persons within 3 weeks after exposure; it is approximately 1000 roentgens for gamma rays, 600 roentgens for x-rays. Abbreviated SED.

skin friction [FL MECH] A type of friction force which exists at the surface of a solid body immersed in a much larger volume of fluid which is in motion relative to the body.

skin resistance [ELEC] For alternating current of a given frequency, the direct-current resistance of a layer at the surface of a conductor whose thickness equals the skin depth.

skiograph [ELECTR] An instrument used to measure the intensity of x-rays.

skip zone [ACOUS] A region in the air surrounding a source of sound in which no sound is heard, although the sound becomes audible at greater distances. Also known as zone of silence. [COMMUN] The area between the outer limit of reception of radio high-frequency ground waves and the inner limit of reception of sky waves, where no signal is received.

skot [OPTICS] A unit of luminance, used particularly to

SKIN FRICTION

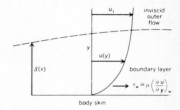

The thin boundary layer, in which fluid flow is distorted by the passage of the body, fills the region between the body skin and the dashed line. $\delta(x)$ is the thickness of the boundary layer; u_1 is the velocity of fluid relative to the body outside the boundary layer; $u(y)$, the fluid velocity within the boundary at a distance y from the body skin, is graphed by the solid curved line. The friction force per unit surface area of the body, τ_w, equals the rate of deformation of an adjacent fluid element, $(\partial u/\partial y)_w$, times the dynamic viscosity μ.

measure low-level luminance, equal to 10^{-3} apostilb, or $10^{-3}/\pi$ nit.

skylight   *See* diffuse sky radiation.

sky radiation   *See* diffuse sky radiation.

sky wave   [ELECTROMAG] A radio wave that travels upward into space and may or may not be returned to earth by reflection from the ionosphere. Also known as ionospheric wave.

slant visibility   *See* oblique visual range.

Slater determinant   [QUANT MECH] A quantum-mechanical wave function for n fermions, which is an $n \times n$ determinant whose entries are n different one-particle wave functions depending on the coordinates of each of the particles in the system.

slave clock   [HOROL] The part of a Shortt clock which consists of a complete clock with pendulum and indicating mechanism, and which receives trigger pulses from a master clock as synchronizing signals.

sleeve antenna   [ELECTROMAG] A single vertical half-wave radiator, the lower half of which is a metallic sleeve through which the concentric feed line runs; the upper radiating portion, one quarter-wavelength long, connects to the center of the line.

slender-body theory   [FL MECH] The theory of compressible inviscid fluid flow past bodies which have pointed noses and bases, or flat bases in supersonic flow only, and which satisfy the following conditions: (1) the ratio r of the maximum thickness to the length of the body must be small compared with unity, (2) the angle between the tangent plane to the body and the direction of motion must be small and of order r, and (3) the smoothness conditions.

slew rate   [ELECTR] The maximum rate at which the output voltage of an operational amplifier changes for a square-wave or step-signal input; usually specified in volts per microsecond.

slicer   *See* amplitude gate.

slicer amplifier   *See* amplitude gate.

slide-back voltmeter   [ELECTR] An electronic voltmeter in which an unknown voltage is measured indirectly by adjusting a calibrated voltage source until its voltage equals the unknown voltage.

slide rule   [MATH] A mechanical device, composed of a ruler with sliding insert, marked with various number scales, which facilitates such calculations as division, multiplication, finding roots, and finding logarithms.

slide-wire bridge   [ELEC] A bridge circuit in which the resistance in one or more branches is controlled by the position of a sliding contact on a length of resistance wire stretched along a linear scale.

slide-wire potentiometer   [ELEC] A potentiometer (variable resistor) which employs a movable sliding connection on a length of resistance wire.

sliding friction   [MECH] Rubbing of bodies in sliding contact.

sliding vector   [MECH] A vector whose direction and line of application are prescribed, but whose point of application is not prescribed.

slip   [CRYSTAL] The movement of one atomic plane over another in a crystal; it is one of the ways that plastic deformation occurs in a solid. Also known as glide. [FL MECH] The difference between the velocity of a solid surface and the mean velocity of a fluid at a point just outside the surface.

slipband   [CRYSTAL] One of the microscopic parallel lines (Lüders' lines) on the surface of a crystalline material stretched beyond its elastic limit, located at the intersection of

SLIDE RULE

Left end of slide rule, showing position of sliding insert for multiplication of 1.5 by 2.

the surface with intracrystalline slip planes in the grains of the material. Also known as slip line.

slip direction [CRYSTAL] The crystallographic direction in which the translation of slip occurs.

slip flow [FL MECH] A situation in which the mean free path of a gas is between 1 and 65% of the channel diameter; the gas layer next to the channel wall assumes a velocity of slip past the liquid, known as slip flow.

slip line *See* slipband.

slip plane *See* glide plane.

slip velocity [FL MECH] The difference in velocities between liquids and solids (or gases and liquids) in the vertical flow of two-phase mixtures through a pipe because of the slip between the two phases.

slitless spectrograph [OPTICS] A type of astronomical spectrograph that does not use a slit, sufficient resolution being obtained from the small image sizes of individual stars, and through the use of an objective prism in front of the telescope.

slit spectrograph [OPTICS] A type of astronomical spectrograph that uses a slit to provide resolution.

slope [MATH] **1.** The slope of a line through the points (x_1, y_1) and (x_2, y_2) is the number $(y_2 - y_1)/(x_2 - x_1)$. **2.** The slope of a curve at a point p is the slope of the tangent line to the curve at p.

slope angle [MATH] The angle of inclination of a line in the plane, where this angle is measured from the positive x-axis to the line in the counterclockwise direction.

slope of fall [MECH] Ratio between the drop of a projectile and its horizontal movement; tangent of the angle of fall.

slot antenna [ELECTROMAG] An antenna formed by cutting one or more narrow slots in a large metal surface fed by a coaxial line or waveguide.

slot radiator [ELECTROMAG] Primary radiating element in the form of a slot cut in the walls of a metal waveguide or cavity resonator or in a metal plate.

slotted line *See* slotted section.

slotted section [ELECTROMAG] A section of waveguide or shielded transmission line in which the shield is slotted to permit the use of a traveling probe for examination of standing waves. Also known as slotted line; slotted wave guide.

slotted wave guide *See* slotted section.

slow death [ELECTR] The gradual change of transistor characteristics with time; this change is attributed to ions which collect on the surface of the transistor.

slowing-down [NUCLEO] A decrease in the energy of a particle as a result of collisions with nuclei.

slowing-down area [NUCLEO] One-sixth of the mean square distance from the source of a neutron in an infinite, homogeneous medium to the point at which the neutron reaches a given energy.

slowing-down density [NUCLEO] A measure of the rate at which neutrons lose energy in a nuclear reactor through collisions; equal to the number of neutrons that fall below a given energy per unit volume per unit time.

slowing-down kernel [NUCLEO] The probability per unit volume that a neutron will travel from one position in a homogeneous medium to another while slowing down through a specified energy range.

slowing of clocks [RELAT] According to the special theory of relativity, a clock appears to tick less rapidly to an observer moving relative to the clock than to an observer who is at rest with respect to the clock. Also known as time dilation effect.

slow neutron [NUC PHYS] **1.** A neutron having low-kinetic energy, up to about 100 electron volts. **2.** *See* thermal neutron.

slow reactor [NUCLEO] A nuclear reactor in which fission is induced primarily by slow neutrons, as in a thermal reactor.

slow-vibration direction [OPTICS] The direction of the electric field vector of the ray of light that travels with the smallest velocity in an anisotropic crystal and therefore corresponds to the maximum refractive index.

slow wave [ELECTROMAG] A wave having a phase velocity less than the velocity of light, as in a ridge wave guide.

slug [MECH] A unit of mass in the British gravitational system of units, equal to the mass which experiences an acceleration of 1 foot per second per second when a force of 1 pound acts on it; equal to approximately 32.1740 pounds mass or 14.5939 kilograms. Also known as geepound. [NUCLEO] A short fuel rod inserted in a hole or channel in the active lattice of a nuclear reactor.

slug tuner [ELECTROMAG] Waveguide tuner containing one or more longitudinally adjustable pieces of metal or dielectric.

Sm *See* samarium.

small-angle scattering [PHYS] Scattering of a beam of electromagnetic or acoustic radiation, or particles, at small angles by particles or cavities whose dimensions are many times as large as the wavelength of the radiation or the de Broglie wavelength of the scattered particles. Also known as low-angle scattering.

small calorie *See* calorie.

Small Magellanic Cloud [ASTRON] The smaller of the two star clouds near the south celestial pole; it is about 170,000 light-years (1.61×10^{21} meters) away and contains a wide assortment of giant and variable stars, star clusters, and nebulae. Also known as Nubecula Minor.

small perturbation [PHYS] A disturbance imposed on a system in steady state, with amplitude assumed small of the first order; that is, the square of the amplitude is negligible in comparison with the amplitude, and the derivatives of the perturbation are assumed to be of the same order of magnitude as the perturbation.

S matrix *See* scattering matrix.

S-matrix theory [PARTIC PHYS] A theory of elementary particles based on the scattering matrix, and on its properties such as unitarity and analyticity. Also known as scattering-matrix theory.

smectic phase [PHYS CHEM] A form of the mesomorphic (liquid crystal) state in which liquid motion is more of a gliding than a flowing nature, drops are often formed which exhibit a series of fine lines, especially under polarized light, and x-ray diffraction patterns can be obtained in only one direction.

Smith-Baker microscope [OPTICS] A type of interference microscope in which a beam of polarized light is split by a birefringent calcite plate cemented to the front lens of the condenser, and reunited by another such plate cemented to the objective.

Smith chart [ELECTROMAG] A special polar diagram containing constant-resistance circles, constant-reactance circles, circles of constant standing-wave ratio, and radius lines representing constant line-angle loci; used in solving transmission-line and waveguide problems.

Smith-Helmholtz law [OPTICS] For a single refracting surface of sufficiently small aperture, the product of the index of refraction, distance from the optical axis, and the angle which a light ray makes with the optical axis at the object point is equal to the corresponding product at the image point.

smooth [STAT] To modify a sequential set of numerical data items in a manner designed to reduce the differences in value between adjacent items.

smoothing [MATH] Approximating or perturbing a function by one which has a higher degree of differentiability. [STAT] A process that uses either freehand methods, moving averages, or fitting a curve by least squares method to remove fluctuations in the data in a time series.

smooth manifold [MATH] A differentiable manifold whose local coordinate systems depend upon those of euclidean space in an infinitely differentiable manner.

smooth map [MATH] **1.** An infinitely differentiable function. **2.** A map between smooth manifolds which is infinitely differentiable when referred to local coordinates.

smoothness conditions [FL MECH] Two conditions that must be satisfied by bodies studied in slender body theory: (1) the rate of change of the angle between the tangent plane to the body and the direction of the motion, evaluated along this direction, must be small and of the same order as the ratio of the maximum thickness of the body to its length; (2) the curvature of any section of the body in a plane normal to the direction of motion must be of the same order as the reciprocal of the maximum diameter of the section, at all points where the section is convex outward.

Sn *See* tin.

S/N *See* signal-to-noise ratio.

SNAP [NUCLEO] A small nuclear power plant in which heat from radioisotope decay in a fuel such as strontium-90 is converted into electric energy, to provide power for spacecraft instrumentation, telemetry, and other applications. Derived from systems for nuclear auxiliary power.

snap-off diode [ELECTR] Planar epitaxial passivated silicon diode that is processed so a charge is stored close to the junction when the diode is conducting; when reverse voltage is applied, the stored charge then forces the diode to snap off or switch rapidly to its blocking state.

Snell laws of refraction [OPTICS] When light travels from one medium into another the incident and refracted rays lie in one plane with the normal to the surface; are on opposite sides of the normal; and make angles with the normal whose sines have a constant ratio to one another. Also known as Descartes laws of refraction; laws of refraction.

snow point [PHYS CHEM] Referring to a gas mixture, the temperature at which the vapor pressure of the sublimable component is equal to the actual partial pressure of that component in the gas mixture; analogous to dew point.

SNR *See* signal-to-noise ratio.

SNU *See* solar neutrino unit.

Sobolev generalized derivative *See* weak derivative.

Sobolev generalized partial derivative *See* weak partial derivative.

Sobolev inequality [MATH] For any positive integer n and real number p, $1 \leq p < n$, there is a constant K such that

$$\left\{ \int \left| u(x_1, \ldots, x_n) \right|^q \, dx_1 \cdots dx_n \right\}^{1/q}$$

$$\leq K \left\{ \sum_{i=1}^{n} \int \left| \partial u(x_1, \ldots, x_n)/dx_i \right|^p \, dx_1 \cdots dx_n \right\}^{1/p}$$

where $q = np/(n - p)$, for any function $u(x_1, \ldots, x_n)$ for which the right-hand side of the inequality is defined.

Sobolev space [MATH] For a given number p satisfying $1 \leq p \leq \infty$, nonnegative integer m and open set S in n-dimensional euclidean space, the Sobolev space $W^{m,p}(S)$ consists of real-valued functions in $L^p(S)$ whose weak partial derivatives of order equal to or less than *m* exist and belong to $L^p(S)$. Also known as W space.

Soddy's displacement law [NUC PHYS] The atomic number of a nuclide decreases by 2 in alpha decay, increases by 1 in beta negatron decay, and decreases by 1 in beta positron decay and electron capture.

sodium [CHEM] A metallic element of group I, symbol Na, with atomic number 11, atomic weight 22.9898; silver-white, soft, and malleable; oxidizes in air; melts at 97.6°C; used as a chemical intermediate and in pharmaceuticals, petroleum refining, and metallurgy; the source of the symbol Na is natrium.

sodium-24 [NUC PHYS] A radioactive isotope of sodium, mass 24, half-life 15.5 hours; formed by deuteron bombardment of sodium; decomposes to magnesium with emission of beta rays.

sodium-cooled reactor [NUCLEO] A nuclear reactor in which sodium metal in liquid form is used as the coolant.

sodium-graphite reactor [NUCLEO] A nuclear reactor that uses slightly enriched uranium as fuel, graphite as moderator, and liquid sodium as the coolant.

sodium iodide scintillator [NUCLEO] A sodium iodide crystal activated with thallium; used especially in the measurement of the energy of gamma rays, by measuring the amplitude of pulses of light generated by electrons in the crystal which are excited by the gamma rays.

sodium-line reversal temperature measurement [PHYS] A method of measuring the temperature of a gas containing sodium vapor, in which the gas is placed in the path of a radiator of known temperature, and the temperature of the gas or the radiator is adjusted until the sodium D line disappears against the background of light from the radiator.

softening point [PHYS] For a substance which does not have a definite melting point, the temperature at which viscous flow changes to plastic flow.

soft-iron ammeter [ENG] An ammeter in which current in a coil causes two pieces of magnetic material within the coil, one fixed and one attached to a pointer, to become similarly magnetized and to repel each other, moving the pointer; used for alternating-current measurement.

soft magnetic material [ELECTROMAG] A magnetic material which is relatively easily magnetized or demagnetized.

soft radiation [PHYS] Radiation whose particles or photons have a low energy, and, as a result, do not penetrate any type of material readily.

soft shower [NUC PHYS] A cosmic-ray shower that cannot penetrate 15 to 20 centimeters of lead; consists mainly of electrons and positrons.

soft tube [ELECTR] **1.** An x-ray tube having a vacuum of about 0.000002 atmosphere (0.202650 newton per square meter), the remaining gas being left in intentionally to give less-penetrating rays than those of a more completely evacuated tube. **2.** *See* gassy tube.

soft x-ray [ELECTROMAG] An x-ray having a comparatively long wavelength and poor penetrating power.

soft x-ray absorption spectroscopy [SPECT] A spectroscopic technique which is used to get information about unoccupied states above the Fermi level in a metal or about empty conduction bands in an inoculator.

soft x-ray appearance potential spectroscopy [SPECT] A branch of electron spectroscopy in which a solid surface is bombarded with monochromatic electrons, and small but abrupt changes in the resulting total x-ray emission intensity are detected as the energy of the electrons is varied. Abbreviated SXAPS.

sogasoid [PHYS] A system of solid particles dispersed in a gas.

Sohncke's law [PHYS] The law that the stress per unit area normal to a crystallographic plane needed to produce a fracture in a crystal is a constant characteristic of a crystalline substance.

sol [CHEM] A colloidal solution consisting of a suitable dispersion medium, which may be gas, liquid, or solid, and the colloidal substance, the disperse phase, which is distributed throughout the dispersion medium.

Sol *See* sun.

solar activity [ASTRON] Disturbances on the surface of the sun; examples are sunspots, prominences, and solar flares.

solar apex [ASTRON] A point toward which the solar system is moving; it is about 10° southwest of the star Vega.

solar battery [ELECTR] An array of solar cells, usually connected in parallel and series.

solar burst [ASTROPHYS] A sudden increase in the radio-frequency energy radiated by the sun, generally associated with visible solar flares.

solar cell [ELECTR] A *pn*-junction device which converts the radiant energy of sunlight directly and efficiently into electrical energy.

solar corona [ASTRON] The upper, rarefied solar atmosphere which becomes visible around the darkened sun during a total solar eclipse. Also known as corona.

solar cosmic rays *See* energetic solar particles.

solar cycle [ASTRON] The periodic change in the number of sunspots; the cycle is taken as the interval between successive minima and is about 11.1 years.

solar day [ASTRON] A time measurement, the duration of one rotation of the earth on its axis with respect to the sun; this may be a mean solar day or an apparent solar day as the reference is to the mean sun or apparent sun.

solar eclipse [ASTRON] An eclipse that takes place when the new moon passes between the earth and the sun and the shadow formed reaches the earth; may be classified as total, partial, or annular.

solar-excited laser *See* sun-pumped laser.

solar faculae [ASTRON] Bright streaks or regions on the surface of the sun, especially near solar sunspots.

solar flare [ASTROPHYS] An abrupt increase in the intensity of the Hα and other emission near a sunspot region; the brightness may be many times that of the associated plage.

solarization [GRAPHICS] **1.** Reversal of a photographic image due to great overexposure. **2.** *See* Sabattier effect. [PHYS] Loss of transparency or coloration of glass exposed to sunlight or ultraviolet radiation.

solar magnetic field [ASTROPHYS] The magnetic field that pervades the ionized and highly conducting gas composing the sun.

solar magnetograph [ENG] An instrument that utilizes the Zeeman effect to directly measure the strength and polarity of the complex patterns of magnetic fields at the sun's surface; comprises a telescope, a differential analyzer, a spectrograph, and a photoelectric or photographic means of differencing and recording.

solar motion [ASTRON] The two main motions of the sun: relative motion with respect to the neighboring stars, or motion due to the rotation of the Milky Way of which the sun is a part.

solar neutrino unit [ASTROPHYS] A unit for measuring the capture rate of neutrinos emanating from the sun, equal to 10^{-36} per second per atom. Abbreviated SNU.

solar nutation [ASTRON] Nutation caused by the change in declination of the sun.

solar parallax [ASTRON] The sun's mean equatorial horizontal parallax p, which is the angle subtended by the equatorial radius r of the earth at mean distance a of the sun.

solar physics [ASTROPHYS] The scientific study of all physical phenomena connected with the sun; it overlaps with geophysics in the consideration of solar-terrestrial relationships, such as the connection between solar activity and auroras.

solar prominence [ASTRON] Sheets of luminous gas emanating from the sun's surface; they appear dark against the sun's disk but bright against the dark sky, and occur only in regions of horizontal magnetic fields.

solar radiation [ASTROPHYS] The electromagnetic radiation and particles (electrons, protons, and rarer heavy atomic nuclei) emitted by the sun.

solar radio emission [ASTROPHYS] Radio-frequency electromagnetic radiation emitted from the sun, and increasing greatly in intensity during sunspots and flares.

solar spectrum [ASTROPHYS] The spectrum of the sun's electromagnetic radiation extending over the whole electromagnetic spectrum, from wavelengths of 10^{-9} centimeter to 30 kilometers.

solar system [ASTRON] The sun and the celestial bodies moving about it; the bodies are planets, satellites of the planets, asteroids, comets, and meteor swarms.

solar telescope [OPTICS] An observational instrument of the solar astronomer; it is designed so that heating effects produced by the sun do not distort the images; the two classes consist of those designed for observations of the brilliant solar disk, and those designed for the study of the much fainter prominences and the still fainter corona through the relatively bright, scattered light of the sky.

solar-terrestrial phenomena [GEOPHYS] All observed physical effects that are caused by solar activity; the phenomena may be in the atmosphere or on the earth's surface; an example is the aurora borealis.

solar time [ASTRON] Time based on the rotation of the earth relative to the sun.

solar-type star [ASTRON] Any of the stars (yellow stars) of spectral type G, so called because the sun is in this class.

solar ultraviolet radiation [ASTROPHYS] That portion of the sun's electromagnetic radiation that has wavelengths from about 400 to about 4 nanometers; this radiation may sufficiently ionize the earth's atmosphere so that propagation of radio waves is affected.

solar units [ASTROPHYS] A set of units for measuring properties of stars, in which properties of the sun such as mass, diameter, density, and luminosity are set equal to unity.

solar wind [GEOPHYS] The supersonic flow of gas, composed of ionized hydrogen and helium, which continuously flows from the sun out through the solar system with velocities of 300 to 1000 kilometers per second; it carries magnetic fields from the sun.

soleil compensator [OPTICS] A compensator which resembles the Babinet compensator but is constructed so that the phase change is constant over the entire field.

solenoid [ELECTROMAG] Also known as electric solenoid. **1.** An electrically energized coil of insulated wire which produces a magnetic field within the coil. **2.** In particular, a coil that surrounds a movable iron core which is pulled to a central position with respect to the coil when the coil is energized by sending current through it.

solenoidal [MATH] A vector field has this property · in a

SOLENOID

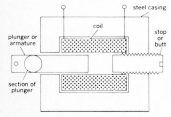

A cross-sectional view of a steel-clad solenoid showing coil and movable core known as plunger or armature.

region if its divergence vanishes at every point of the region.

solenoid group [MATH] A compact Abelian, topological group that is one-dimensional and connected.

solid [PHYS] **1.** A substance that has a definite volume and shape and resists forces that tend to alter its volume or shape. **2.** A crystalline material, that is, one in which the constituent atoms are arranged in a three-dimensional lattice, periodic in three independent directions.

solid angle [MATH] A surface formed by all rays joining a point to a closed curve.

solid-dielectric capacitor [ELEC] A capacitor whose dielectric is one of several solid materials such as ceramic, mica, glass, plastic film, or paper.

solid-electrolyte battery [ELEC] A primary battery whose electrolyte is either a solid crystalline salt, such as silver iodide or lead chloride, or an ion-exchange membrane; in either case, conductivity is almost entirely ionic.

solid electrolytic capacitor [ELEC] An electrolytic capacitor in which the dielectric is an anodized coating on one electrode, with a solid semiconductor material filling the rest of the space between the electrodes.

solid helium [CRYO] A certain state which is not attained by helium under its own vapor pressure down to absolute zero, but which requires an external pressure of 25 atmospheres (2.5 megapascals) at absolute zero.

solidification [PHYS] The change of a fluid (liquid or gas) into the solid state.

solid insulator [ELEC] An electric insulator made of a solid substance, such as sulfur, polystyrene, rubber, or porcelain.

solid ionization chamber [NUCLEO] A particle detector in which the gas filling a conventional ionization chamber is replaced by a large single crystal of suitably chosen material.

solid laser [OPTICS] A laser in which either a crystalline or amorphous solid material, usually in the form of a rod, is excited by optical pumping; the most common crystalline materials are ruby, neodymium-doped ruby, and neodymium-doped yttrium aluminum garnet.

solid moment of inertia [PHYS] The integral of the products of the mass of each of the infinitesimal elements of the solid with the square of their distance from a given axis.

solid Schmidt telescope [OPTICS] A type of Schmidt system which is constructed from a single block of glass, designed to operate at very small aperture ratios.

solid solution [PHYS] A homogeneous crystalline phase composed of several distinct chemical species, occupying the lattice points at random and existing in a range of concentrations.

solid state [PHYS] The condition of a substance in which it is a solid.

solid-state circuit [ELECTR] Complete circuit formed from a single block of semiconductor material.

solid-state component [ELECTR] A component whose operation depends on the control of electrical or magnetic phenomena in solids, such as a transistor, crystal diode, or ferrite device.

solid-state counter [NUCLEO] A radiation counter whose sensitive material is a crystalline solid; for example, a crystal counter or a scintillation counter.

solid-state device [ELECTR] A device, other than a conductor, which uses magnetic, electrical, and other properties of solid materials, as opposed to vacuum or gaseous devices.

solid-state lamp *See* light-emitting diode.

solid-state laser [OPTICS] A laser in which a semiconductor material produces the coherent output beam.

solid-state maser [PHYSICS] A maser in which a semicon-

SOLID-ELECTROLYTE BATTERY

lead (anode) silver chloride (cathode)

− ○———○ +

lead chloride (electrolyte) silver backing

Schematic diagram of typical solid-electrolyte cell with solid crystalline salt electrolyte.

ductor material produces the coherent output beam; two input waves are required: one wave, called the pumping source, induces upward energy transitions in the active material, and the second wave, of lower frequency, causes downward transitions and undergoes amplification as it absorbs photons from the active material.

solid-state physics [PHYS] The branch of physics centering about the physical properties of solid materials.

solid-state switch [ELECTR] A microwave switch in which a semiconductor material serves as the switching element; a zero or negative potential applied to the control electrode will reverse-bias the switch and turn it off, and a slight positive voltage will turn it on.

solid-state thyratron [ELECTR] A semiconductor device, such as a silicon controlled rectifier, that approximates the extremely fast switching speed and power-handling capability of a gaseous thyratron tube.

solid tantalum capacitor [ELEC] An electrolytic capacitor in which the anode is a porous pellet of tantalum; the dielectric is an extremely thin layer of tantalum pentoxide formed by anodization of the exterior and interior surfaces of the pellet; the cathode is a layer of semiconducting manganese dioxide that fills the pores of the anode over the dielectric.

solidus [PHYS CHEM] In a constitution or equilibrium diagram, the locus of points representing the temperature below which the various compositions finish freezing on cooling, or begin to melt on heating.

solidus curve [PHYS CHEM] A curve on the phase diagram of a system with two components which represents the equilibrium between the liquid phase and the solid phase.

solion [ELEC] An electrochemical device in which amplification is obtained by controlling and monitoring a reversible electrochemical reaction.

solitary wave [PHYS] A traveling wave in which a single disturbance is neither preceded by nor followed by other such disturbances, but which does not involve unusually large amplitudes or rapid changes in variables, in contrast to a shock wave.

soliton [MATH] A solution of a nonlinear differential equation that propagates with a characteristic constant shape.

Solomon R unit [NUCLEO] A unit of radiation dose rate due to x-rays, equal to 2100 roentgens per hour. Also known as R unit.

solstice [ASTRON] The two days (actually, instants) during the year when the earth is so located in its orbit that the inclination (about $23\frac{1}{2}°$) of the polar axis is toward the sun; the days are June 21 for the North Pole and December 22 for the South Pole; because of leap years, the dates vary a little.

soluble [CHEM] Capable of being dissolved.

solution [CHEM] A single, homogeneous liquid, solid, or gas phase that is a mixture in which the components (liquid, gas, solid, or combinations thereof) are uniformly distributed throughout the mixture.

solution poison [NUCLEO] A soluble nuclear poison, such as boric acid, added to the coolant of a nuclear reactor for purposes of reactivity control; generally used only during shutdown periods, and chemically removed from the coolant prior to resuming operation.

solvable extension [MATH] A finite extension E of a field F such that the Galois group of the smallest Galois extension of F containing E is a solvable group.

solvable group [MATH] A group G which has subgroups G_0, $G_1, \ldots, G_n$, where $G_0 = G$, G_n = the identity element alone, and each G_i is a normal subgroup of G_{i-1} with the quotient group G_{i-1}/G_i Abelian.

solvent [CHEM] That part of a solution that is present in the largest amount, or the compound that is normally liquid in the pure state (as for solutions of solids or gases in liquids).

solvent extraction [NUCLEO] A process for removing uranium fuel residue from used fuel elements of a reactor; it generally involves decay cooling under water for up to 6 months, removal of cladding, dissolution, separation of reusable fuel, decontamination, and disposal of radioactive wastes. Also known as liquid extraction.

solvmanifold [MATH] A homogeneous space obtained by factoring a connected solvable Lie group by a closed subgroup.

solvolysis [CHEM] A reaction in which a solvent reacts with the solute to form a new substance.

Sommerfeld equation *See* Sommerfeld formula.

Sommerfeld fine-structure constant *See* fine-structure constant.

Sommerfeld formula [ELECTROMAG] An approximate formula for the field strength of electromagnetic radiation generated by an antenna at distances small enough so that the curvature of the earth may be neglected, in terms of radiated power, distance from the antenna, and various constants and parameters. Also known as Sommerfeld equation.

Sommerfeld law for doublets [ATOM PHYS] According to the Bohr-Sommerfeld theory, the splitting in frequency of regular or relativistic doublets is $\alpha^2 R(Z - \sigma)^4/n^3(l + 1)$, where α is the fine structure constant, R is the Rydberg constant of the atom, Z is the atomic number, σ is a screening constant, n is the principal quantum number, and l is the orbital angular momentum quantum number.

Sommerfeld model *See* free-electron theory of metals.

Sommerfeld theory *See* free-electron theory of metals.

Sommerfeld-Watson transformation *See* Watson-Sommerfeld transformation.

sonar [ENG] A system that uses underwater sound, at sonic or ultrasonic frequencies, to detect and locate objects in the sea, or for communication; the commonest type is echo-ranging sonar, other versions are passive sonar, scanning sonar, and searchlight sonar. Derived from sound navigation and ranging.

sonde [ENG] An instrument used to obtain weather data during ascent and descent through the atmosphere, in a form suitable for telemetering to a ground station by radio, as in a radiosonde.

sone [ACOUS] A unit of loudness, equal to the loudness of a simple 1000-hertz tone with a sound pressure level 40 decibels above 0.0002 microbar; a sound that is judged by listeners to be n times as loud as this tone has a loudness of n sones.

sonic [ACOUS] **1.** Of or pertaining to the speed of sound. **2.** Pertaining to that which moves at acoustic velocity, as in sonic flow. **3.** Designed to operate or perform at the speed of sound, as in sonic leading edge.

sonic boom [ACOUS] A noise caused by a shock wave that emanates from an aircraft or other object traveling at or above sonic velocity.

sonic radiation [ACOUS] Acoustic radiation with a frequency between about 16 hertz and about 20,000 hertz.

sonics [ACOUS] The technology of sound, or elastic wave motion, as applied to problems of measurement, control, and processing.

sonic spark chamber [NUCLEO] A spark chamber in which the position of a spark is determined by measuring the time it takes for sound from the spark to arrive at each of two microphones.

sonic speed *See* speed of sound.

sonic thermometer [ENG] A thermometer based upon the

SONE

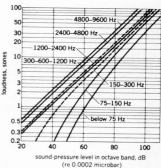

Functions relating loudness, in sones, to sound pressure, in decibels above 0.0002 microbar, for bands of noise one octave wide in a sound field.

principle that the velocity of a sound wave is a function of the temperature of the medium through which it passes.

sonic velocity   *See* speed of sound.

Sonine polynomials   [MATH] A sequence of orthogonal polynomials of the form

$$S_m^n(z) = \sum_{r=0}^{n} \frac{(r+m+1)(r+m+2)\cdots(m+n)(-z)^r}{r!(n-r)!}$$

where m is called the index, n is the order, and z is the argument.

sonograph   [ENG] **1.** An instrument for recording sound or seismic vibrations. **2.** An instrument for converting sounds into seismic vibrations.

sonoluminescence   [PHYS] Luminescence produced by high-frequency sound waves or by phonons.

sonotone   [ENG] In general, any device consisting of a thin metallic wire stretched over two bridges that are usually mounted on a soundboard; used to measure the vibration frequency, tension, density, or diameter of the wire, or to verify relations between these quantities. Also known as monochord.

soot luminŏsity   [OPTICS] The portion of the luminosity of a flame attributable to soot particles in the flame.

Soret coefficient   [PHYS] A tabulated value used in binary thermal diffusion calculations in gaseous systems; expressed as $D'/D = \alpha X_1 X_2$, where D' is the coefficient of thermal diffusion, D is the coefficient of ordinary diffusion, α is the thermal diffusion constant, X_1 is the mole fraction of the lower-molecular-weight component, and X_2 is the mole fraction of the higher-molecular-weight component.

Soret effect   [PHYS] Thermal diffusion in liquids.

sorption   [PHYS CHEM] A general term used to encompass the processes of adsorption, absorption, desorption, ion exchange, ion exclusion, ion retardation, chemisorption, and dialysis.

sorption pumping   [ENG] A technique used to reduce the pressure of gas in an atmosphere; the gas is adsorbed on a granular sorbent material such as a molecular sieve in a metal container; when this sorbent-filled container is immersed in liquid nitrogen, the gas is sorbed.

sound   [ACOUS] **1.** An alteration of properties of an elastic medium, such as pressure, particle displacement, or density, that propagates through the medium, or a superposition of such alterations; sound waves having frequencies above the audible (sonic) range are termed ultrasonic waves; those with frequencies below the sonic range are called infrasonic waves. Also known as acoustic wave; sound wave. **2.** The auditory sensation which is produced by these alterations. Also known as sound sensation.

sound absorption   [ACOUS] A process in which sound energy is reduced when sound waves pass through a medium or strike a surface. Also known as acoustic absorption.

sound absorption coefficient   [ACOUS] The ratio of sound energy absorbed to that arriving at a surface or medium. Also known as acoustic absorption coefficient; acoustic absorptivity.

sound analyzer   [ENG] An instrument which measures the amount of sound energy in various frequency bands; it generally consists of a set of fixed electrical filters or a tunable electrical filter, along with associated amplifiers and a meter which indicates the filter output.

sound attenuation   [ACOUS] Diminution of the intensity of sound energy propagating in a medium; caused by absorption, spreading, and scattering.

sound band pressure level   [ACOUS] The effective sound

pressure for the sound energy in a given frequency band.

sound channel [ACOUS] A layer of seawater extending from about 700 meters down to about 1500 meters, in which sound travels at about 450 meters per second, the slowest it can travel in seawater; below 1500 meters the speed of sound increases as a result of pressure. [ELECTR] The series of stages that handles only the sound signal in a television receiver.

sound detection [ACOUS] The discrimination of a sound from background noise, either by the ear or by an electronic instrument such as a volume indicator.

sound energy [ACOUS] The difference between the total energy and the energy which would exist if no sound waves were present. Also known as acoustic energy.

sound energy density [ACOUS] Sound energy per unit volume; the commonly used unit is the erg per cubic centimeter.

sound energy flux [ACOUS] Average over one period of the rate of flow of sound energy through any specified area; the unit is the erg per second.

sounding velocity [ACOUS] The vertical velocity of sound in water, usually assumed to be constant at 800 to 820 fathoms (1464 to 1501 meters) per second for sounding measurements.

sound intensity [ACOUS] For a specified direction and point in space, the average rate at which sound energy is transmitted through a unit area perpendicular to the specified direction.

sound irradiator [ACOUS] A device for focusing sound waves so that sound of high intensity is produced at the focus.

sound lag [ACOUS] Time necessary for a sound wave to travel from its source to the point of reception.

sound level [ACOUS] The sound pressure level (in decibels) at a point in a sound field, averaged over the audible frequency range and over a time interval, with a frequency weighting and the time interval as specified by the American National Standards Association.

sound-level meter [ENG] An instrument used to measure noise and sound levels in a specified manner; the meter may be calibrated in decibels or volume units and includes a microphone, an amplifier, an output meter, and frequency-weighting networks.

sound masking [ACOUS] The ability of one sound to make the ear incapable of perceiving another sound.

sound navigation and ranging *See* sonar.

sound power [ACOUS] The total sound energy radiated by a source per unit time, generally expressed in ergs per second or watts. Also known as acoustic power.

sound pressure *See* effective sound pressure.

sound pressure level [ACOUS] A value in decibels equal to 20 times the logarithm to the base 10 of the ratio of the pressure of the sound under consideration to a reference pressure; reference pressures in common use are 0.0002 microbar and 1 microbar. Abbreviated SPL.

sound-ray diagram [ACOUS] A plot of the paths taken by sound rays in an acoustical system; analogous to a light-ray diagram in optics.

sound reduction factor [ACOUS] A measure of the reduction in the intensity of sound when it crosses an interface, equal to 10 times the common logarithm of the reciprocal of the sound transmission coefficient of the surface.

sound reflection coefficient *See* acoustic reflectivity.

sound sensation *See* sound.

sound spectrum [ACOUS] A plot of the strength of a sound at various frequencies.

sound transducer *See* electroacoustic transducer.

SOUND-LEVEL METER

General Radio Type 1551-C sound-level meter. *(General Radio Co.)*

sound transmission [ACOUS] Passage of a sound wave through a medium or series of media.

sound transmission coefficient [ACOUS] The ratio of transmitted to incident sound energy at an interface in a sound medium; the value depends on the angle of incidence of the sound. Also known as acoustic transmission coefficient; acoustic transmissivity.

sound volume velocity [ACOUS] The rate at which a substance flows through a specified area as a result of a sound wave.

sound wave *See* sound.

sound-wave photography [PHYS] A method of studying propagation, reflection, and refraction of sound waves, in which a sound wave is generated by a spark and is illuminated a fraction of a second later by a second spark, causing the wave to cast a shadow on a photographic plate.

source [ELEC] The circuit or device that supplies signal power or electric energy or charge to a transducer or load circuit. [ELECTR] The terminal in a field-effect transistor from which majority carriers flow into the conducting channel in the semiconductor material. [NUCLEO] A radioactive material packaged so as to produce radiation for experimental or industrial use. [PHYS] **1.** In general, a device that supplies some extensive entity, such as energy, matter, particles, or electric charge. **2.** A point, line, or area at which mass or energy is added to a system, either instantaneously or continuously. **3.** A point at which lines of force in a vector field originate, such as a point in an electrostatic field where there is positive charge. [SPECT] The arc or spark that supplies light for a spectroscope. [THERMO] A device that supplies heat.

source flow [FL MECH] **1.** In three-dimensional flow, a point from which fluid issues at a uniform rate in all directions. **2.** In two-dimensional flow, a line normal to the planes of flow, from which fluid flows uniformly in all directions at right angles to the line.

source-follower amplifier *See* common-drain amplifier.

source impedance [ELEC] Impedance presented by a source of energy to the input terminals of a device.

source level [ACOUS] The sound intensity, in decibels above a reference level, at a point which is a unit distance from a source and on an axis of the source.

source material [NUCLEO] Material from which fissionable material can be extracted.

southern lights *See* aurora australis.

south geomagnetic pole [GEOPHYS] The geomagnetic pole in the Southern Hemisphere at approximately 78.5°S, longitude 111°E, 180° from the north geomagnetic pole. Also known as south pole.

south pole [ELECTROMAG] The pole of a magnet at which magnetic lines of force are assumed to enter. [GEOPHYS] *See* south geomagnetic pole.

space [ASTRON] **1.** Specifically, the part of the universe lying outside the limits of the earth's atmosphere. **2.** More generally, the volume in which all celestial bodies, including the earth, move. [MATH] In context, usually a set with a topology on it or some other type of structure.

space attenuation [ACOUS] Loss of energy, expressed in decibels, of a signal in free air; caused by such factors as absorption, reflection, scattering, and dispersion.

space centrode [MECH] The path traced by the instantaneous center of a rotating body relative to an inertial frame of reference.

space charge [ELEC] The net electric charge within a given volume.

space-charge debunching [ELECTR] A process in which the mutual interactions between electrons in a stream spread out the electrons of a bunch.

space-charge effect [ELECTR] Repulsion of electrons emitted from the cathode of a thermionic vacuum tube by electrons accumulated in the space charge near the cathode.

space-charge grid [ELECTR] Grid operated at a low positive potential and placed between the cathode and control grid of a vacuum tube to reduce the limiting effect of space charge on the current through the tube.

space-charge layer *See* depletion layer.

space-charge limitation [ELECTR] The current flowing through a vacuum between a cathode and an anode cannot exceed a certain maximum value, as a result of modification of the electric field near the cathode due to space charge in this region.

space-charge polarization [ELEC] Polarization of a dielectric which occurs when charge carriers are present which can migrate an appreciable distance through the dielectric but which become trapped or cannot discharge at an electrode. Also known as interfacial polarization.

space-charge region [ELECTR] Of a semiconductor device, a region in which the net charge density is significantly different from zero.

space cone [MECH] The cone in space that is swept out by the instantaneous axis of a rigid body during Poinsot motion. Also known as herpolhode cone.

space coordinates [MATH] A three-dimensional system of cartesian coordinates by which a point is located by three magnitudes indicating distance from three planes which intersect at a point.

space current [ELECTR] Total current flowing between the cathode and all other electrodes in a tube; this includes the plate current, grid current, screen grid current, and any other electrode current which may be present.

space curve [MATH] A curve in three-dimensional euclidean space; it may be a twisted curve or a plane curve.

space factor [ELECTROMAG] 1. The ratio of the space occupied by the conductors in a winding to the total cubic content or volume of the winding, or the similar ratio of cross sections. 2. The ratio of the space occupied by iron to the total cubic content of an iron core.

space-filling curve *See* Peano curve.

space group [CRYSTAL] A group of operations which leave the infinitely extended, regularly repeating pattern of a crystal unchanged; there are 230 such groups.

space group extinction [CRYSTAL] The absence of certain classes of reflections in the x-ray diffraction pattern of a crystal due to the existence of symmetry elements in the space group of the crystal which are not present in its point group.

space inversion *See* inversion.

space lattice *See* lattice.

spacelike surface [RELAT] A three-dimensional surface in a four-dimensional space-time which has the property that no event on the surface lies in the past or the future of any other event on the surface.

spacelike vector [RELAT] A four vector in Minkowski space whose space component has a magnitude which is greater than the magnitude of its time component multiplied by the speed of light.

space permeability [ELECTROMAG] Factor that expresses the ratio of magnetic induction to magnetizing force in a vacuum; in the centimeter-gram-second electromagnetic system of units, the permeability of a vacuum is arbitrarily taken as unity; in the meter-kilogram-second-ampere system, it is $4\pi \times 10^{-7}$.

space polar coordinates [MATH] A system of coordinates by which a point is located in space by its distance from a fixed point called the pole, the colatitude or angle between the polar axis (a reference line through the pole) and the radius vector (a straight line connecting the pole and the point), and the longitude or angle between a reference plane containing the polar axis and a plane through the radius vector and polar axis.

space quadrature [PHYS] A difference of a quarter-wavelength in the position of corresponding points of a wave in space.

space quantization [QUANT MECH] The quantization of the component of the angular momentum of a system in some specified direction.

space reflection symmetry *See* parity.

space-time [RELAT] A four-dimensional space used to represent the universe in the theory of relativity, with three dimensions corresponding to ordinary space and the fourth to time. Also known as space-time continuum.

space-time continuum *See* space-time.

space velocity [ASTRON] A star's true velocity with reference to the sun.

space wave [ELECTROMAG] The component of a ground wave that travels more or less directly through space from the transmitting antenna to the receiving antenna; one part of the space wave goes directly from one antenna to the other; another part is reflected off the earth between the antennas.

spacistor [ELECTR] A multiple-terminal solid-state device, similar to a transistor, that generates frequencies up to about 10,000 megahertz by injecting electrons or holes into a space-charge layer which rapidly forces these carriers to a collecting electrode.

spallation [NUC PHYS] A nuclear reaction in which the energy of each incident particle is so high that more than two or three particles are ejected from the target nucleus and both its mass number and atomic number are changed. Also known as nuclear spallation.

spallation reaction [NUC PHYS] A high-energy nuclear reaction which results in the release of large numbers of nucleons as reaction products.

span [MATH] The span of a set of vectors is the set of all possible linear combinations of those vectors.

spanning tree [MATH] A spanning tree of a graph G is a subgraph of G which is a tree and which includes all the vertices in G.

spark [ELEC] A short-duration electric discharge due to a sudden breakdown of air or some other dielectric material separating two terminals, accompanied by a momentary flash of light. Also known as electric spark; spark discharge; sparkover.

spark chamber [NUCLEO] A particle-detecting device in which the trajectory of a charged particle is made visible by a series of sparks that are triggered by the particle as it passes through an array of spark gaps.

spark counter [NUCLEO] A particle detector in which high-speed charged particles ionize a gas, consisting of argon mixed with an organic gas, triggering a spark between two plane parallel metal electrodes.

spark discharge *See* spark.

spark excitation [SPECT] The use of an electric spark (10,000 to 30,000 volts) to excite spectral line emissions from otherwise hard-to-excite samples; used in emission spectroscopy.

spark gap [ELEC] An arrangement of two electrodes between which a spark may occur; the insulation (usually air) between the electrodes is self-restoring after passage of the spark; used as a switching device, for example, to protect equipment against lightning or to switch a radar antenna from receiver to transmitter and vice versa.

spark-ignition combustion cycle   *See* Otto cycle.

sparking potential   *See* breakdown voltage.

sparking voltage   *See* breakdown voltage.

sparkover   *See* spark.

spark photography   [OPTICS] Any type of photography in which a spark provides illumination and determines the length of exposure.

spark recorder   [ENG] Recorder in which the recording paper passes through a spark gap formed by a metal plate underneath and a moving metal pointer above the paper; sparks from an induction coil pass through the paper periodically, burning small holes that form the record trace.

spark spectrum   [SPECT] The spectrum produced by a spark discharging through a gas or vapor; with metal electrodes, a spectrum of the metallic vapor is obtained.

Sparrow's criterion   [OPTICS] A criterion for the resolution of two light sources, according to which the light sources are resolved if there is some central dip in their combined diffraction pattern.

spatial filter   [OPTICS] An optical filter that consists of a very small aperture, such as a pinhole.

SPDT   *See* single-pole double-throw.

Spearman-Brown formula   [STAT] A formula to estimate the reliability of a test n times as long as one for which reliability is known; the tests must be comparable in all aspects other than size.

Spearman's rank correlation coefficient   [STAT] A statistic used as a measure of correlation in nonparametric statistics when the data are in ordinal form; a product moment correlation coeficient. Also known as Spearman's rho.

Spearman's rho   *See* Spearman's rank correlation coefficient.

special functions   [MATH] The various families of solution functions corresponding to cases of the hypergeometric equation or functions used in the equation's study, such as the gamma function.

special Jordan algebra   [MATH] A Jordan algebra that can be written as a symmetrized product over a matrix algebra.

special nuclear material   [NUCLEO] Fissionable and related material controlled directly by the Atomic Energy Commission, such as uranium enriched in the isotopes U^{235} and U^{233}, and plutonium.

special orthogonal group of dimension n   [MATH] The Lie group of special orthogonal transformations on an n-dimensional real inner product space. Symbolized SO_n; $SO(n)$.

special orthogonal transformation   [MATH] An orthogonal transformation whose matrix representation has determinant equal to 1.

special relativity   [RELAT] The division of relativity theory which relates the observations of observers moving with constant relative velocities and postulates that natural laws are the same for all such observers.

special unitary group of dimension n   [MATH] The Lie group of special unitary transformations on an n-dimensional inner product space over the complex numbers. Symbolized $SU(n)$.

special unitary transformation   [MATH] A unitary transformation whose matrix representation has determinant equal to 1.

species   *See* nuclide.

specific   [PHYS] Indicating the amount of a physical quantity per unit mass, weight, volume, or area, or the ratio of the quantity for the substance under consideration to the same quantity for a standard substance, such as water.

specific acoustical impedance [ACOUS] The ratio of the pressure phasor associated with a sound wave at any given point in a medium to the velocity phasor at that point.

specific acoustical ohm *See* rayl.

specific acoustical reactance [ACOUS] The magnitude of the imaginary part of the specific acoustical impedance.

specific acoustical resistance [ACOUS] The real part of the specific acoustical impedance.

specific activity [NUCLEO] **1.** The activity of a radioisotope of an element per unit weight of element present in the sample. **2.** The activity per unit mass of a pure radionuclide. **3.** The activity per unit weight of a sample of radioactive material. In these three cases, specific activity can be expressed in such units as millicuries per gram, disintegrations per second per milligram, or counts per minute per milligram.

specific charge [ELEC] The ratio of a particle's charge to its mass.

specific conductance *See* conductivity.

specific energy [THERMO] The internal energy of a substance per unit mass.

specific gravity [MECH] The ratio of the density of a material to the density of some standard material, such as water at a specified temperature, for example, 4°C or 60°F, or (for gases) air at standard conditions of pressure and temperature. Abbreviated sp gr. Also known as relative density.

specific gravity bottle [ENG] A small bottle or flask used to measure the specific gravities of liquids; the bottle is weighed when it is filled with the liquid whose specific gravity is to be determined, when filled with a reference liquid, and when empty. Also known as density bottle; relative density bottle.

specific gravity hydrometer [ENG] A hydrometer which indicates the specific gravity of a liquid, with reference to water at a particular temperature.

specific heat [THERMO] **1.** The ratio of the amount of heat required to raise a mass of material 1 degree in temperature to the amount of heat required to raise an equal mass of a reference substance, usually water, 1 degree in temperature; both measurements are made at a reference temperature, usually at constant pressure or constant volume. **2.** The quantity of heat required to raise a unit mass of homogeneous material one degree in temperature in a specified way; it is assumed that during the process no phase or chemical change occurs.

specific inductive capacity *See* dielectric constant.

specific insulation resistance *See* volume resistivity.

specific ionization [NUCLEO] The number of ion pairs formed per unit distance along the track of an ion passing through matter. Also known as total specific ionization.

specific power [NUCLEO] The power produced per unit mass of fuel present in a nuclear reactor.

specific reluctance *See* reluctivity.

specific resistance *See* electrical resistivity.

specific rotation [OPTICS] The calculated rotation of light passing through a solution as related to the solution volume and depth, the amount of solute, and the observed optical rotation at a given wavelength and temperature.

specific susceptibility *See* mass susceptibility.

specific viscosity [FL MECH] The specific viscosity of a polymer is the relative viscosity of a polymer solution of known concentration minus 1; usually determined at low concentration of the polymer; for example, 0.5 gram per 100 milliliters of solution, or less.

specific volume [MECH] The volume of a substance per unit mass; it is the reciprocal of the density. Abbreviated sp vol.

specific weight [MECH] The weight per unit volume of a substance.

speckle [OPTICS] A phenomenon in which the scattering of light from a highly coherent source, such as a laser, by a rough surface or inhomogeneous medium generates a random-intensity distribution of light that gives the surface or medium a granular appearance.

spectral bandwidth [SPECT] The minimum radiant-energy bandwidth to which a spectrophotometer is accurate; that is, 1–5 nanometers for better models.

spectral centroid [OPTICS] An average wavelength; specifically, for a light filter or other light-transmitting device, a weighted average of the spectral energy distribution of the incident light, the transmittance of the device, and the luminosity function.

spectral characteristic [OPTICS] The relation between wavelength and some other variable, such as between wavelength and emitted radiant power of a luminescent screen per unit wavelength interval.

spectral classification [ASTRON] A classification of stars by characteristics revealed by study of their spectra; the six classes B, A, F, G, K, and M include 99% of all known stars.

spectral color [OPTICS] **1.** A color corresponding to light of a pure frequency; the basic spectral colors are violet, blue-green, yellow, orange, and red. **2.** A color that is represented by a point on the chromaticity diagram that lies on a straight line between some point on the spectral color (first definition) locus and the achromatic points; purple, for example, is not a spectral color.

spectral density [ELECTROMAG] *See* spectral energy distribution. [MATH] The density function for the spectral measure of a linear transformation on a Hilbert space.

spectral emissivity [THERMO] Monochromatic emissivity considered as a function of wavelength.

spectral energy distribution [ELECTROMAG] The power carried by electromagnetic radiation within some small interval of wavelength (or frequency) of fixed amount as a function of wavelength (or frequency). Also known as spectral density.

spectral extinction [OPTICS] The selective absorption of different wavelengths of light as a function of depth in water.

spectral factorization [MATH] A process sometimes used in the study of control systems, in which a given rational function of the complex variable s is factored into the product of two functions, $F_R(s)$ and $F_L(s)$, each of which has all of its poles and zeros in the right and left half of the complex plane, respectively.

spectral function [MATH] In the theory of stationary stochastic processes, the function

$$F(y) = (2/\pi) \int_0^\infty \rho(x)(\sin xy/x)(dx), 0 \leq y \leq \infty$$

where $\rho(x)$ is the autocorrelation function of a stationary time series.

spectral irradiance [OPTICS] The density of the radiant flux that is incident on a surface per unit of wavelength.

spectral line [SPECT] A discrete value of a quantity, such as frequency, wavelength, energy, or mass, whose spectrum is being investigated; one may observe a finite spread of values resulting from such factors as level width, Doppler broadening, and instrument imperfections. Also known as spectrum line.

spectral locus *See* spectrum locus.

spectral luminous efficacy [OPTICS] The ratio of the luminous flux emitted by a monochromatic light source in lumens to its radiant flux in watts, as a function of the wavelength of the emitted light.

spectral luminous efficiency *See* luminosity function.

spectral measure [MATH] A measure on the spectrum of an operator on a Hilbert space whose values are projection operators there; spectral theorems concerning linear operators often give an integral representation of the operator in terms of these projection valued measures.

spectral photography [OPTICS] A technique used in airborne surveys for mineral deposits; narrow-band-pass filters and special film are used to accentuate minor color effects caused by mineralization and alteration which would be undetectable by broad-band photography.

spectral pyrometer *See* narrow-band pyrometer.

spectral radius [MATH] For the spectrum of an operator, this is the least upper bound of the set of all $|\lambda|$, where λ is in the spectrum.

spectral regions [SPECT] Arbitrary ranges of wavelength, some of them overlapping, into which the electromagnetic spectrum is divided, according to the types of sources that are required to produce and detect the various wavelengths, such as x-ray, ultraviolet, visible, infrared, or radio-frequency.

spectral response *See* spectral sensitivity.

spectral sensitivity [ELECTR] Radiant sensitivity, considered as a function of wavelength. [PHYS] The response of a device or material to monochromatic light as a function of wavelength. Also known as spectral response.

spectral sequence [MATH] An E^k spectral sequence E is a sequence $\{E^r, d^r\}$ for $r \geq k$ such that (1) E^r is a bigraded module and d^r is a differential of bidegree $(-r, r - 1)$ on E^r, and (2) for $r \geq k$, $H(E^r)$ is isomorphic to E^{r+1}.

spectral series [SPECT] Spectral lines or groups of lines that occur in sequence.

spectral-shift reactor [NUCLEO] A reactor in which, for control or other purposes, the neutron spectrum may be adjusted by varying the properties or amount of moderator.

spectral theorems [MATH] Spectral theorems enable detailed study of various types of operators on Banach spaces by giving an integral or series representation of the operator in terms of its spectrum, eigenspaces, and simple projectionlike operators.

spectral transmission [OPTICS] The radiant flux which passes through a filter divided by the radiant flux incident upon it, for monochromatic light of a specified wavelength.

spectral type [ASTRON] A label used to indicate the physical and chemical characteristics of a star as indicated by study of the star's spectra; for example, the stars in the spectral type known as class B are blue-white, and are referred to as helium stars because the dominant lines in their spectra are the lines in helium spectra.

Spectra Pritchard photometer [OPTICS] A photoelectric instrument for measuring the luminance of surfaces; it has a telescopic viewing system for imaging the bright surface to be measured on the cathode of a photoemissive tube, and a separate unit that combines the power supply with the controls and readout meter.

spectrobolometer [SPECT] An instrument that measures radiation from stars; measurement can be made in a narrow band of wavelengths in the electromagnetic spectrum; the instrument itself is a combination spectrometer and bolometer.

spectrofluorometer [SPECT] A device used in fluorescence spectroscopy to increase the selectivity of fluorometry by

SPECTRA PRITCHARD PHOTOMETER

SPECTRA PRITCHARD photometer. (a) The viewing system. (b) The readout meter. (Photo Research Corp.)

SPECTROFLUOROMETER

Turner 210 "Spectro," an advanced spectrofluorometer. (G. K. Turner Associates)

passing emitted fluorescent light through a monochromator to record the fluorescence emission spectrum.

spectrography [SPECT] The use of photography to record the electromagnetic spectrum displayed in a spectroscope.

spectroheliocinematograph [OPTICS] A camera used to make motion pictures of, for example, prominences of the sun; the camera utilizes monochromatic light; it is composed of a camera and a spectrohelioscope.

spectroheliograph [OPTICS] An instrument used to photograph the sun in one spectral band.

spectrohelioscope [OPTICS] An instrument based on the principle of the spectroheliograph but used for visual observation, and not for photography.

spectrometer [SPECT] **1.** A spectroscope that is provided with a calibrated scale either for measurement of wavelength or for measurement of refractive indices of transparent prism materials. **2.** A spectroscope equipped with a photoelectric photometer to measure radiant intensities at various wavelengths.

spectrometry [SPECT] The use of spectrographic techniques for deriving the physical constants of materials.

spectrophotometer [SPECT] An instrument that measures transmission or apparent reflectance of visible light as a function of wavelength, permitting accurate analysis of color or accurate comparison of luminous intensities of two sources or specific wavelengths.

spectrophotometry [SPECT] A procedure to measure photometrically the wavelength range of radiant energy absorbed by a sample under analysis; can be by visible light, ultraviolet light, or x-rays.

spectropolarimeter [OPTICS] A device used to measure optical rotation in solutions for different light wavelengths.

spectropyrheliometer [SPECT] An astronomical instrument used to measure distribution of radiant energy from the sun in the ultraviolet and visible wavelengths.

spectroscope [SPECT] An optical instrument consisting of a slit, collimator lens, prism or grating, and a telescope or objective lens which produces a spectrum for visual observation.

spectroscopic binary star [ASTRON] A binary star that may be distinguished from a single star only by noting the Doppler shift of the spectral lines of one or both stars as they revolve about their common center of mass.

spectroscopic displacement law [SPECT] The spectrum of an un-ionized atom resembles that of a singly ionized atom of the element one place higher in the periodic table, and that of a doubly ionized atom two places higher in the table, and so forth.

spectroscopic parallax [ASTRON] Parallax as determined from examination of a stellar spectrum; critical spectral lines indicate the star's absolute magnitude, from which the star's distance, or parallax, can be deduced.

spectroscopic splitting factor *See* Landé g factor.

spectroscopy [PHYS] The branch of physics concerned with the production, measurement, and interpretation of electromagnetic spectra arising from either emission or absorption of radiant energy by various substances.

spectrum [MATH] If T is a linear operator of a normed space X to itself and I is the identity transformation ($I(x) \equiv x$), the spectrum of T consists of all scalars λ for which either $T - \lambda I$ has no inverse or the range of $T - \lambda I$ is not dense in X. [PHYS] **1.** A display or plot of intensity of radiation (particles, photons, or acoustic radiation) as a function of mass, momentum, wavelength, frequency, or some related quantity. **2.** The set of frequencies, wavelengths, or related quantities, involved in some process; for example, each element has a characteristic discrete spectrum for emission and absorption of light. **3.** A range of frequencies within which radiation has some specified characteristic, such as audio-frequency spectrum, ultraviolet spectrum, or radio spectrum.

spectrum analysis [PHYS] The measurement of the amplitude of the components of a complex waveform throughout the frequency range of the waveform.

spectrum line *See* spectral line.

spectrum locus [OPTICS] The locus of points representing the chromaticities of spectrally pure stimuli in a chromaticity diagram. Also known as spectral locus.

spectrum of turbulence [ASTROPHYS] A relationship between the size of turbulent eddies in the sun's atmosphere and their average speed.

specular reflection [PHYS] Reflection of electromagnetic, acoustic, or water waves in which the reflected waves travel in a definite direction, and the directions of the incident and reflected waves make equal angles with a line perpendicular to the reflecting surface, and lie in the same plane with it. Also known as direct reflection; mirror reflection; regular reflection.

specular reflection factor [OPTICS] The ratio of the specularly reflected light to the incident light.

specular reflection model [PHYS] A model for the behavior of gas molecules striking the surface of a solid body, in which the molecules are reflected so that the component of velocity tangent to the surface is unchanged while the component of velocity perpendicular to the surface is reversed.

specular reflector [OPTICS] A reflecting surface (polished metal or silvered glass) that gives a direct image of the source, with the angle of reflection equal to the angle of incidence. Also known as regular reflector.

specular transmittance [ELECTROMAG] The ratio of the power carried by electromagnetic radiation which emerges from a body and is parallel to a beam entering the body, to the power carried by the beam entering the body.

speed [MECH] The time rate of change of position of a body without regard to direction; in other words, the magnitude of the velocity vector. [OPTICS] **1.** The light-gathering power of a lens, expressed as the reciprocal of the *f* number. **2.** The time that a camera shutter is open. [PHYS] In general, the rapidity with which a process takes place.

speed of light [ELECTROMAG] The speed of propagation of electromagnetic waves in a vacuum, which is a physical constant equal to $299{,}792.4580 \pm 0.0012$ kilometers per second. Also known as electromagnetic constant; velocity of light.

speed of response [PHYS] The time required for a system to react to some signal; for example, the delay time for a photon detector to react to a radiation pulse, or the time needed for a current or voltage in a circuit to reach a definite fraction of its final value as a result of an abrupt change in the electromotive force.

speed of sound [ACOUS] The phase velocity of a sound wave. Also known as sonic speed; sonic velocity; velocity of sound.

speed-power product [ELECTR] The product of the gate speed or propagation delay of an electronic circuit and its power dissipation.

spent fuel [NUCLEO] Nuclear reactor fuel that has been irradiated to the extent that it can no longer effectively sustain a chain reaction because its fissionable isotopes have been partially consumed and fission-product poisons have accumulated in it.

Sperner set [MATH] A set S of subsets of a given set such that

SPECULAR REFLECTION

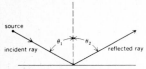

Specular reflection, such as from a polished surface. Angle θ_1, between incident ray and perpendicular to surface equals angle θ_2 between reflected ray and perpendicular.

if A and B are in S, and A does not equal B, then neither A nor B is a subset of the other.

Sperner's lemma [MATH] If S is a Sperner set of subsets of a set containing n elements, then the number of elements in S is equal to or less than

$$\binom{n}{r} = n!/r!\,(n-r)!$$

where $r = [\frac{1}{2}n]$ is the largest integer equal to or less than $\frac{1}{2}n$.

sp gr *See* specific gravity.

sphenoid [CRYSTAL] An open crystal, occurring in monoclinic crystals of the sphenoidal class, and characterized by two nonparallel faces symmetrical with an axis of twofold symmetry. [SCI TECH] Wedge-shaped.

spherator [PL PHYS] One of the class of low-β, low-density, quasi-steady-state closed devices (like Tokamak) used in studying production of electric power by fusion.

sphere [MATH] 1. The set of all points in a euclidean space which are a fixed common distance from some given point; in euclidean three-dimensional space the Riemann sphere consists of all points (x,y,z) which satisfy the equation $x^2 + y^2 + z^2 = 1$. 2. The set of points in a metric space whose distance from a fixed point is constant.

sphere gap [ELEC] A spark gap between two equal-diameter spherical electrodes.

sphere of attraction [PHYS CHEM] The distance within which the potential energy arising from mutual attraction of two molecules is not negligible with respect to the molecules' average thermal energy at room temperature.

sphere photometer *See* integrating-sphere photometer.

spherical aberration [OPTICS] Aberration arising from the fact that rays which are initially at different distances from the optical axis come to a focus at different distances along the axis when they are reflected from a spherical mirror or refracted by a lens with spherical surfaces.

spherical antenna [ELECTROMAG] An antenna having the shape of a sphere, used chiefly in theoretical studies.

spherical Bessel functions [MATH] Bessel functions whose order is half of an odd integer; they arise as the radial functions that result from solving Pockel's equation (or, equivalently, the time-independent Schrödinger equation for a free particle) by separation of variables in spherical coordinates.

spherical capacitor [ELEC] A capacitor made of two concentric metal spheres with a dielectric filling the space between the spheres.

spherical coordinates [MATH] A system of curvilinear coordinates in which the position of a point in space is designated by its distance r from the origin or pole, called the radius vector, the angle ϕ between the radius vector and a vertically directed polar axis, called the cone angle or colatitude, and the angle θ between the plane of ϕ and a fixed meridian plane through the polar axis, called the polar angle or longitude. Also known as spherical polar coordinates.

spherical cyclic curve *See* cyclic curve.

spherical degree [MATH] A solid angle equal to $\frac{1}{90}$ of a spherical right angle.

spherical distance [MATH] The length of a great circle arc between two points on a sphere.

spherical-earth factor [ELECTROMAG] The ratio of the electric field strength that would result from propagation over an imperfectly conducting spherical earth to that which would result from propagation over a perfectly conducting plane.

spherical excess [MATH] The sum of the angles of a spherical triangle, minus $180°$.

spherical harmonics [MATH] Solutions of Laplace's equation in spherical coordinates.

spherical indicatrix of binormal to a curve [MATH] All the end points of those radii from the sphere of radius one which are parallel to the positive direction of the binormal to a space curve.

spherical indicatrix of tangent to a curve [MATH] Those points on the unit sphere traced out by a radius moving from point to point always parallel with the tangent to the curve.

spherical lens [OPTICS] A lens whose surfaces form portions of spheres.

spherical mirror [OPTICS] A mirror, either convex or concave, whose surface forms part of a sphere.

spherical pendulum [MECH] A simple pendulum mounted on a pivot so that its motion is not confined to a plane; the bob moves over a spherical surface.

spherical polar coordinates *See* spherical coordinates.

spherical radius [MATH] For a circle on a sphere, the smaller of the spherical distances from one of the two poles of the circle to any point on the circle.

spherical sector [MATH] The cap and cone formed by the intersection of a plane with a sphere, the cone extending from the plane to the center of the sphere and the cap extending from the plane to the surface of the sphere.

spherical segment [MATH] A solid that is bounded by a sphere and two parallel planes which intersect the sphere or are tangent to it.

spherical stress [MECH] The portion of the total stress that corresponds to an isotropic hydrostatic pressure; its stress tensor is the unit tensor multiplied by one-third the trace of the total stress tensor.

spherical surface harmonics [MATH] Functions of the two angular coordinates of a spherical coordinate system which are solutions of the partial differential equation obtained by separation of variables of Laplace's equation in spherical coordinates. Also known as surface harmonics.

spherical triangle [MATH] A three-sided surface on a sphere the sides of which are arcs of great circles.

spherical trigonometry [MATH] The study of spherical triangles from the viewpoint of angle, length, and area.

spherical wave [PHYS] A wave whose equiphase surfaces form a family of concentric spheres; the direction of travel is always perpendicular to the surfaces of the spheres.

spherical wedge [MATH] The portion of a sphere bounded by two semicircles and a lune (the surface of the sphere between the semicircles).

spherocylindrical lens [OPTICS] A lens having one surface that is a portion of a sphere, while the other is a portion of a cylinder.

spheroid *See* ellipsoid of revolution.

spheroidal excess [MATH] The amount by which the sum of the three angles of a triangle on the surface of a spheroid exceeds 180°.

spheroidal galaxy *See* elliptical galaxy.

spheroidal group [CRYSTAL] A group in the tetragonal symmetry system; the sphenoid is the typical form.

spheroidal harmonics [MATH] Solutions to Laplace's equation when phrased in ellipsoidal coordinates.

spheroidal triangle [MATH] The figure formed by three geodesic lines joining three points on a spheroid. Also known as geodetic triangle.

spherotoric lens [OPTICS] A lens having one surface that is a portion of a sphere, while the other is a portion of a torus.

spicule [ASTRON] One of an irregular distribution of jets

SPICULE

Large-scale photograph of the chromosphere in H-α light, with the disk of the sun artificially eclipsed and the hairy spicules projecting about the continuous chromosphere. The length of this section is about 70,000 kilometers. *(Photography by R. B. Dunn, through 15-inch telescope, Sacramento Peak Observatory, operated by the Association of Universities for Research in Astronomy, Inc.)*

shooting up from the sun's chromosphere. Also known as solar spicule.

spike [PHYS] A short-duration transient whose amplitude considerably exceeds the average amplitude of the associated pulse or signal.

spike antenna *See* monopole antenna.

spiked core *See* seed core.

spill [NUCLEO] The accidental release of radioactive material.

spin [MECH] The rotation of a body about its axis. [QUANT MECH] The intrinsic angular momentum of an elementary particle or nucleus, which exists even when the particle is at rest, as distinguished from orbital angular momentum.

spin axis [PHYS] The axis of rotation of a gyroscope.

spin-decelerating moment [MECH] A couple about the axis of the projectile, which diminishes spin.

spin dependent force [PHYS] A force between two particles which depends in some way on the spin, possibly on the angle between their spin directions, or on the angles between their spin directions and a line joining the particles.

spin echo technique [NUC PHYS] A variation of the nuclear magnetic resonance technique in which the radio frequency field is applied in two pulses, separated by a time interval t, and a strong nuclear induction signal is observed at a time t after the second pulse.

spin-flip laser [OPTICS] A semiconductor laser in which the output wavelength is continuously tunable by a magnetic field; operation is based on exciting conduction-band electrons to a higher energy level by reversing the direction of the electrons as they spin about an axis in the direction of the magnetic field.

spin-flip scattering [QUANT MECH] Scattering of a particle with spin $\frac{1}{2}$ in which the direction of the particle's spin is reversed.

spin glass [SOLID STATE] A magnetic alloy in which the concentration of magnetic atoms is such that below a certain temperature their magnetic moments are no longer able to fluctuate thermally in time but are still directed at random, in loose analogy to the atoms of ordinary glass; if such an alloy is cooled in an external magnetic field from above to below this transition temperature and the external field is then removed, the magnetization decays slowly to 0, typically in a matter of hours.

spin-lattice relaxation [SOLID STATE] Magnetic relaxation in which the excess potential energy associated with electron spins in a magnetic field is transferred to the lattice.

spin magnetism [SOLID STATE] Paramagnetism or ferromagnetism that arises from polarization of electron spins in a substance.

Spinnbarkeit relaxation [FL MECH] A rheological effect illustrated by the pulling away of liquid threads when an object that has been immersed in a viscoelastic fluid is pulled out.

spinode *See* cusp.

spinor [MATH] **1.** A vector with two complex components, which undergoes a unitary unimodular transformation when the three-dimensional coordinate system is rotated; it can represent the spin state of a particle of spin $\frac{1}{2}$. **2.** More generally, a spinor of order (or rank) n is an object with 2^n components which transform as products of components of n spinors of rank one. **3.** A quantity with four complex components which transforms linearly under a Lorentz transformation in such a way that if it is a solution of the Dirac equation in the original Lorentz frame it remains a solution of the Dirac equation in the transformed frame; it is formed from two spinors (definition 1). Also known as Dirac spinor.

spin-orbit coupling [QUANT MECH] The interaction between a particle's spin and its orbital angular momentum.

spin-orbit multiplet [PHYS] A collection of atomic or nuclear states which differ in energy only on account of spin-orbit coupling; the total spin angular momentum quantum number S and total orbital angular momentum quantum number L are the same for all states; the energy levels are labeled by the total angular momentum quantum number J.

spin paramagnetism [SOLID STATE] Paramagnetism that arises from the electron spins in a substance.

spin-parity [PARTIC PHYS] A combined symbol J^P for an elementary particle's spin J, and its intrinsic parity P.

spin quantum number [QUANT MECH] The ratio of the maximum observable component of a system's spin to Planck's constant divided by 2π; it is an integer or a half-integer.

spin resonance *See* magnetic resonance.

spin space [MATH] The two-dimensional vector space over the complex numbers, whose unitary unimodular transformations are a two-dimensional double-valued representation of the three-dimensional rotation group; its vectors can represent the various spin states of a particle with spin ½, and its unitary unimodular transformations can represent rotations of this particle.

spin-spin energy [PHYS] An interaction energy proportional to the dot product of the spin angular momenta of two systems.

spin-spin relaxation [SOLID STATE] Magnetic relaxation, observed after application of weak magnetic fields, in which the excess potential energy associated with electron spins in a magnetic field is redistributed among the spins, resulting in heating of the spin system.

spin state [QUANT MECH] Condition of a particle in which its total spin, and the component of its spin along some specified axis, have definite values; more precisely, the particle's wave function is an eigenfunction of the operators corresponding to these quantities.

spin temperature [SOLID STATE] For a system of electron spins in a lattice, a temperature such that the population of the energy levels of the spin system is given by the Boltzmann distribution with this temperature.

spinthariscope [ELECTR] An instrument for viewing the scintillations of alpha particles on a luminescent screen, usually with the aid of a microscope.

spin wave [SOLID STATE] A sinusoidal variation, propagating through a crystal lattice, of that angular momentum which is associated with magnetism (mostly spin angular momentum of the electrons).

spiral [MATH] A simple curve in the plane which continuously winds about itself either into some point or out from some point.

spiral arms [ASTRON] The shape of sections of certain galaxies called spirals; these sections are two so-called arms composed of stars, dust, and gas extending from the center of the galaxy and coiled about it.

spiral gage *See* spiral pressure gage.

spiral galaxy [ASTRON] A type of galaxy classified on the basis of appearance of its photographic image; this type includes two main groups: normal spirals with circular symmetry of the nucleus and of the spiral arms, and barred spirals in which the dominant form is a luminous bar crossing the nucleus with spiral arms starting at the ends of the bar or tangent to a luminous rim on which the bar terminates.

spiral of Archimedes [MATH] The curve spiraling into the origin which in polar coordinates is given by the equation $r = a\theta$. Also known as Archimedes' spiral.

spiral pressure gage [ENG] A device for measurement of pressures; a hollow tube spiral receives the system pressure which deforms (unwinds) the spiral in direct relation to the pressure in the tube. Also known as spiral gage.

spiral thermometer [ENG] A temperature-measurement device consisting of a bimetal spiral that winds tighter or opens with changes in temperature.

spirit thermometer [ENG] A temperature-measurement device consisting of a closed capillary tube with a liquid (for example, alcohol) reservoir bulb at the bottom; as the bulb is heated, the liquid expands up into the capillary tubing, indicating the temperature of the bulb.

SPL *See* sound pressure level.

spline [MATH] A function used to approximate a specified function on an interval, consisting of pieces which are defined uniquely on a set of subintervals, usually as polynomials or some other simple form, and which match up with each other and the prescribed function at the end points of the subintervals with a sufficiently high degree of accuracy.

split cameras [OPTICS] An assembly of two cameras disposed at a fixed overlapping angle relative to each other.

split-half method [STAT] A method used to gage the reliability of a test; two sets of scores are obtained from the same test, one set from odd items and one set from even items, and the scores of the two sets are correlated.

split-lens interference [OPTICS] Interference produced by a Billet split lens.

split-plot design [STAT] An experimental design that enables an additional factor or treatment to be included at more than one level; each plot is split into two or more parts.

splitting field [MATH] **1.** A splitting field of a polynomial p with coefficients in a field k is an extension field K of k such that p splits into linear factors over K, that is, $p(x) = c(x - a_1) \cdots (x - a_n)$, where $c, a_1, \ldots, a_n$ are elements of K, and K is the smallest field containing k and $a_1, \ldots, a_n$. **2.** The splitting field of a family of polynomials with coefficients in a field k is an extension K of k such that each polynomial in the family splits into linear factors over K, and K is the smallest field containing k and the roots of all the polynomials in the family.

spoiler [ELECTROMAG] Rod grating mounted on a parabolic reflector to change the pencil-beam pattern of the reflector to a cosecant-squared pattern; rotating the reflector and grating 90° with respect to the feed antenna changes one pattern to the other.

spontaneous [PHYS] Occurring without application of an external agency, because of the inherent properties of an object.

spontaneous emission [ATOM PHYS] Emission of electromagnetic radiation by an atom or molecule that does not depend on any external agency.

spontaneous fission [NUC PHYS] Nuclear fission in which no particles or photons enter the nucleus from the outside.

spontaneous magnetization [ELECTROMAG] Magnetization which a substance possesses in the absence of an applied magnetic field.

spontaneous polarization [ELEC] Electric polarization that a substance possesses in the absence of an external electric field.

spontaneous process [THERMO] A thermodynamic process which takes place without the application of an external agency, because of the inherent properties of a system.

spontaneous symmetry breaking [PHYS] A situation in which the solution of a set of physical equations fails to exhibit a symmetry possessed by the equations themselves; an example is a magnet, in which the underlying equations

SPLIT-LENS INTERFERENCE

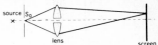

Billet split-lens interference. Light from source S_0 is transmitted through two separated parts of lens to screen, producing interference fringes on screen.

describing the metal do not distinguish any direction of space from any other, but the magnet certainly does, since it points in some definite direction.

sporadic E layer [GEOPHYS] A layer of intense ionization that occurs sporadically within the E layer; it is variable in time of occurrence, height, geographical distribution, penetration frequency, and ionization density.

sporadic simple group [MATH] A simple group which cannot be classified in any known infinite family of simple groups.

Spörer's law [ASTRON] A relationship to indicate the frequency of occurrence of sunspots and their progressive movement to lower latitudes on the sun.

spot [ELECTR] In a cathode-ray tube, the area instantaneously affected by the impact of an electron beam.

spot noise figure [ELECTR] Of a transducer at a selected frequency, the ratio of the output noise power per unit bandwidth to a portion thereof attributable to the thermal noise in the input termination per unit bandwidth, the noise temperature of the input termination being standard (290 K).

spread [STAT] The range within which the values of a variable quantity occur.

spreading coefficient [THERMO] The work done in spreading one liquid over a unit area of another, equal to the surface tension of the stationary liquid, minus the surface tension of the spreading liquid, minus the interfacial tension between the liquids.

spring [ASTRON] The period extending from the vernal equinox to the summer solstice; comprises the transition period from winter to summer.

spring balance [ENG] An instrument which measures force by determining the extension of a helical spring.

spring equinox *See* vernal equinox.

spring modulus [MECH] The additional force necessary to deflect a spring an additional unit distance; if a certain spring has a modulus of 100 newtons per centimeter, a 100-newton weight will compress it 1 centimeter, a 200-newton weight 2 centimeters, and so on.

s-process [NUC PHYS] The synthesis of elements, predominantly in the iron group, over long periods of time through the capture of slow neutrons which are produced mainly by the reactions of α-particles with carbon-13 and neon-21.

SPST *See* single-pole single-throw.

spur [MATH] *See* trace. [PHYS] A cluster of ionized molecules near the path of an energetic charged particle, consisting of the molecule ionized directly by the charged particle, and secondary ionizations produced by electrons released in the primary ionization; it usually forms a side track from the path of the particle.

spurious correlation [STAT] The value of the coefficient of correlation when it is computed correctly but its relationship implications are nonsensical or unreasonable.

spurious disk [OPTICS] The nearly round image of perceptible diameter of a star as seen through a telescope, due to diffraction of light in the telescope.

sputtering [ELECTR] Also known as cathode sputtering. **1.** The ejection of atoms or groups of atoms from the surface of the cathode of a vacuum tube as the result of heavy-ion impact. **2.** The use of this process to deposit a thin layer of metal on a glass, plastic, metal, or other surface in vacuum.

sputter-ion pump *See* getter-ion pump.

sp vol *See* specific volume.

sq *See* square.

square [MATH] **1.** The square of a number r is the number r^2, that is, r times r. **2.** The plane figure with four equal sides and four interior right angles. [MECH] Denotes a unit of area; if x is a unit of length, a square x is the area of a square whose sides have a length of $1x$: for example, a square meter, or a meter squared, is the area of a square whose sides have a length of 1 meter. Abbreviated sq.

square degree [MATH] A unit of a solid angle equal to $(\pi/180)^2$ steradian, or approximately 3.04617×10^{-4} steradian.

square-foot unit of absorption *See* sabin.

square grade [MATH] A unit of solid angle equal to $(\pi/200)^2$ steradian, or approximately 2.46740×10^{-4} steradian.

square matrix [MATH] A matrix with the same number of rows and columns.

squareness ratio [ELECTROMAG] **1.** The magnetic induction at zero magnetizing force divided by the maximum magnetic induction, in a symmetric cyclic magnetization of a material. **2.** The magnetic induction when the magnetizing force has changed half-way from zero toward its negative limiting value divided by the maximum magnetic induction in a symmetric cyclic magnetization of a material.

square-on *See* center.

square root [MATH] A square root of a real or complex number s is a number t for which $t^2 = s$.

square-root law [STAT] The standard deviation of the ratio of the number of successes to number of trials is inversely proportional to the square root of the number of trials.

square-root transformation [STAT] A conversion or transformation of data having a Poisson distribution where sample means are approximately proportional to the variances of the respective samples; replacing each measurement by its square root will often result in homogeneous variances.

square wave [ELEC] An oscillation the amplitude of which shows periodic discontinuities between two values, remaining constant between jumps.

squaring circuit [ELECTR] **1.** A circuit that reshapes a sine or other wave into a square wave. **2.** A circuit that contains nonlinear elements proportional to the square of the input voltage.

squegger *See* blocking oscillator.

squegging oscillator *See* blocking oscillator.

SQUID *See* superconducting quantum interference device.

sr *See* steradian.

Sr *See* strontium.

SS Cygni stars *See* U Geminorum stars.

S star [ASTRON] A spectral classification of stars, comprising red stars with surface temperature of about 2200 K; prominent in the spectra is zirconium oxide.

s-state [QUANT MECH] A single-particle state whose orbital angular momentum quantum number is zero.

SSU *See* Saybolt Seconds Universal.

St *See* stoke.

stability [CONT SYS] The property of a system for which any bounded input signal results in a bounded output signal. [FL MECH] The resistance to overturning or mixing in the water column, resulting from the presence of a positive (increasing downward) density gradient. [MECH] *See* dynamic stability. [PHYS] **1.** The property of a system which does not undergo any change without the application of an external agency. **2.** The property of a system in which any departure from an equilibrium state gives rise to forces or influences which tend to return the system to equilibrium. Also known as static stability. [PL PHYS] The property of a plasma which maintains its shape against externally applied forces (usually pressure of magnetic fields) and whose constituents can pass through confining fields only by diffusion of individual particles.

stability criterion [CONT SYS] A condition which is necessary and sufficient for a system to be stable, such as the Nyquist criterion, or the condition that poles of the system's overall transmittance lie in the left half of the complex-frequency plane.

stability exchange principle [CONT SYS] In a linear system, which is either dynamically stable or unstable depending on the value of a parameter, the complex frequency varies with the parameter in such a way that its real and imaginary parts pass through zero simultaneously; the principle is often violated.

stability factor [ELECTR] A measure of a transistor amplifier's bias stability, equal to the rate of change of collector current with respect to reverse saturation current.

stability matrix *See* stiffness matrix.

stability of a system [MATH] Stability theory of systems of differential equations deals with those solution functions possessing some particular property which still maintain the property after a perturbation.

stability subgroup *See* stabilizer.

stabilization [CONT SYS] *See* compensation. [ELECTR] Feedback introduced into vacuum tube or transistor amplifier stages to reduce distortion by making the amplification substantially independent of electrode voltages and tube constants. [ELECTROMAG] Treatment of a magnetic material to improve the stability of its magnetic properties.

stabilized feedback *See* negative feedback.

stabilizer [MATH] The stabilizer of a point x in a Riemann surface X, relative to a group G of conformal mappings of X onto itself, is the subgroup G_x of G consisting of elements g such that $g(x) = x$. Also known as stability subgroup.

stabilizing magnetic field [PL PHYS] A magnetic field which is added to a device confining a plasma in order to increase the plasma's stability.

stable [PHYS] Not subject to any change without the application of an external agency, such as radiation; said of a molecule, atom, nucleus, or elementary particle.

stable bundle [MATH] The bundle E^s of a hyperbolic structure.

stable equilibrium [SCI TECH] Equilibrium in which any departure from the equilibrium state gives rise to forces or influences which tend to return the system to equilibrium.

stable graph [MATH] A graph from which an edge can be deleted to produce a subgraph whose group of automorphisms is a subgroup of the group of automorphisms of the original graph.

stable homeomorphism conjecture [MATH] For dimension n, the assertion that each orientation-preserving homeomorphism of the real n space, R^n, into itself can be expressed as a composition of homeomorphisms, each of which is the identity on some nonempty open set in R^n.

stable isobar [NUC PHYS] One of two or more stable nuclides which have the same mass number but differ in atomic number.

stable isotope [NUC PHYS] An isotope which does not spontaneously undergo radioactive decay.

stable nucleus [NUC PHYS] A nucleus which does not spontaneously undergo radioactive decay.

stable orbit *See* equilibrium orbit.

stacking [ELECTROMAG] The placing of antennas one above the other, connecting them in phase to increase the gain.

stacking fault [CRYSTAL] A defect in a face-centered cubic or hexagonal close-packed crystal in which there is a change from the regular sequence of positions of atomic planes.

stactometer   *See* stalagmometer.

stage   [ELECTR] A circuit containing a single section of an electron tube or equivalent device or two or more similar sections connected in parallel, push-pull, or push-push; it includes all parts connected between the control-grid input terminal of the device and the input terminal of the next adjacent stage.

stagnation point   [FL MECH] A point in a field of flow about a body where the fluid particles have zero velocity with respect to the body.

stagnation pressure   *See* dynamic pressure.

stagnation temperature   *See* adiabatic recovery temperature.

stalagmometer   [ENG] An instrument for measuring the size of drops suspended from a capillary tube, used in the drop-weight method. Also known as stactometer.

stalogometer   [ENG] A device for measuring surface tension; a drop is suspended from a tube of known radius, and the radius of the drop is measured at the instant the drop detaches itself from the tube.

standard   [PHYS] An accepted reference sample used for establishing a unit for the measurement of a physical quantity.

standard atmosphere   *See* atmosphere.

standard candle   *See* international candle.

standard capacitor   [ELEC] A capacitor constructed in such a manner that its capacitance value is not likely to vary with temperature and is known to a high degree of accuracy. Also known as capacitance standard.

standard cell   [ELEC] A primary cell whose voltage is accurately known and remains sufficiently constant for instrument calibration purposes; the Weston standard cell has a voltage of 1.018636 volts at 20°C.

standard conditions   [PHYS] **1.** A temperature of 0°C and a pressure of 1 atmosphere (101,325 newtons per square meter). Also known as normal temperature and pressure (NTP); standard temperature and pressure (STP). **2.** According to the American Gas Association, a temperature of 60°F (15-5/9°C) and a pressure of 762 millimeters (30 inches) of mercury. **3.** According to the Compressed Gas Institute, a temperature of 20°C (68°F) and a pressure of 1 atmosphere. [SOLID STATE] The allotropic form in which a substance most commonly occurs.

standard deviation   [STAT] The positive square root of the expected value of the square of the difference between a random variable and its mean.

standard error   [STAT] A measure of the variability any statistical constant would be expected to show in taking repeated random samples of a given size from the same universe of observations.

standard error of the estimate   [STAT] Standard deviation of observed values about the regression line; computed by dividing the unexplained variation or the error sum of squares by its degrees of freedom.

standard error of the forecast   [STAT] Standard deviation of the estimate (point or interval) of a dependent variable for a given value of an independent variable.

standard error of the regression coefficient   [STAT] The standard deviation of an estimated regression coefficient; depends on sample size and model assumptions.

standard free energy increase   [THERMO] The increase in Gibbs free energy in a chemical reaction, when both the reactants and the products of the reaction are in their standard states.

standard gravity   [MECH] A value of the acceleration of gravity equal to 9.80665 meters per second per second.

standard heat of formation [THERMO] The heat needed to produce one mole of a compound from its elements in their standard state.

standard illuminants [OPTICS] Three standard sources of light, designated A, B, and C, used in specifying the light used when colors are matched; A is light from a filament at a color temperature of 2575°C, and B and C, representing noon sunlight and normal daylight respectively, are obtained by modifying A with rigorously specified filters.

standard inductor [ELECTROMAG] An inductor (coil) having high stability of inductance value, with little variation of inductance with current or frequency and with a low temperature coefficient; it may have an air core or an iron core; used as a primary standard in laboratories and as a precise working standard for impedance measurements.

standardized test statistic [STAT] A test statistic which has been reduced to standardized units.

standardized units [STAT] A random variable Z has been reduced to standardized units when it has zero expected value and standard deviation 1; this is accomplished by dividing the difference of Z and the expected value of Z by the standard deviation of Z.

standard lens [OPTICS] Usually the lens provided with a camera as standard equipment; in still cameras, the standard lens is one whose focal length is about equal to the length of the diagonal of the negative area normally provided by the camera; the normal field of view of a standard lens is about 53°.

standard measure *See* standard score.

standard noise temperature [ELECTR] The standard reference temperature for noise measurements, equal to 290 K.

standard noon [ASTRON] Twelve o'clock standard time, or the instant the mean sun is over the upper branch of the standard meridian.

standard pitch [ACOUS] A musical pitch based on 440 hertz for tone A; with this standard, the frequency of middle C is 261 hertz.

standard plane [CRYSTAL] The crystal plane whose Miller indices are (111), that is, whose intercepts on the crystal axes are proportional to the corresponding sides of a unit cell.

standard pressure [PHYS] A pressure of 1 atmosphere (101,325 newtons per square meter), to which measurements of quantities dependent on pressure, such as the volume of a gas, are often referred. Also known as normal pressure.

standard propagation [ELECTROMAG] Propagation of radio waves over a smooth spherical earth of specified dielectric constant and conductivity, under conditions of standard refraction in the atmosphere.

standard refraction [ELECTROMAG] Refraction which would occur in an idealized atmosphere in which the index of refraction decreases uniformly with height at a rate of 39×10^{-6} per kilometer; standard refraction may be included in ground wave calculations by use of an effective earth radius of 8.5×10^6 meters, or $\frac{4}{3}$ the geometrical radius of the earth.

standard score [STAT] A test score converted or transformed into a common scale, such as standard units, to effect a more reasonable scale of measurement in order to make comparisons between different tests. Also known as standard measure; z score.

standard star [ASTRON] A star whose position or other data are precisely known so that it is used as a reference to calculate positions of other celestial bodies, or of objects on earth.

standard state [PHYS] The stable and pure form of a substance at standard pressure and ordinary temperature.

standard time [ASTRON] The mean solar time, based on the transit of the sun over a specified meridian, called the time meridian, and adopted for use over an area that is called a time zone.

standard ton *See* ton.

standard trajectory [MECH] Path through the air that it is calculated a projectile will follow under given conditions of weather, position, and material, including the particular fuse, projectile, and propelling charge that are used; firing tables are based on standard trajectories.

standard volume [PHYS] The volume of 1 mole of a gas at a pressure of 1 atmosphere and a temperature of 0°C. Also known as normal volume.

standing wave [PHYS] A wave in which the ratio of an instantaneous value at one point to that at any other point does not vary with time. Also known as stationary wave.

standing-wave detector [ELECTROMAG] An electric indicating instrument used for detecting a standing electromagnetic wave along a transmission line or in a waveguide and measuring the resulting standing-wave ratio; it can also be used to measure the wavelength, and hence the frequency, of the wave. Also known as standing-wave indicator; standing-wave meter; standing-wave-ratio meter.

standing-wave indicator *See* standing-wave detector.

standing-wave loss factor [ELECTROMAG] The ratio of the transmission loss in an unmatched waveguide to that in the same waveguide when matched.

standing-wave meter *See* standing-wave detector.

standing-wave method [ELECTROMAG] Any method of measuring the wavelength of electromagnetic waves that involves measuring the distance between successive nodes or antinodes of standing waves.

standing-wave producer [ELECTROMAG] A movable probe inserted in a slotted waveguide to produce a desired standing-wave pattern, generally for test purposes.

standing-wave ratio [PHYS] **1.** The ratio of the maximum amplitude to the minimum amplitude of corresponding components of a wave in a transmission line or waveguide. **2.** The reciprocal of this ratio.

standing-wave-ratio meter *See* standing-wave detector.

Stanhope lens [OPTICS] A thick biconvex lens with front and back surfaces having radii of curvature of two-thirds and one-third the lens thickness; used as a magnifier by placing the object to be viewed in contact with the front surface.

Stanton diagram [FL MECH] The plot of the airflow friction coefficient against the Reynolds number.

Stanton number [THERMO] A dimensionless number used in the study of forced convection, equal to the heat-transfer coefficient of a fluid divided by the product of the specific heat at constant pressure, the fluid density, and the fluid velocity. Symbolized N_{St}. Also known as Margoulis number (M).

star [ASTRON] A celestial body consisting of a large, self-luminous mass of hot gas held together by its own gravity; the sun is a typical star. [NUCLEO] A star-shaped group of tracks made by ionizing particles originating at a common point in a nuclear emulsion, cloud chamber, or bubble chamber; some stars are produced by successive disintegrations of an atom in a radioactive series, and others by nuclear reactions of the spallation type, such as those due to cosmic-ray particles. Also known as nuclear star. [OPTICS] A light source that subtends a very small angle at the entrance pupil of an optical instrument and is used to test the instrument.

star algebra [MATH] A real or complex algebra on which an involution is defined.

star cloud [ASTRON] An aggregation of thousands or of

millions of stars spread over hundreds or thousands of light-years.

star cluster [ASTRON] A group of stars held together by gravitational attraction; the two chief types are open clusters (composed of from 12 to hundreds of stars) and globular clusters (composed of thousands to hundreds of thousands of stars).

star color [ASTRON] The color of a star as a function of its radiation and related to its temperature; colors range from blue-white to deep red.

star day [ASTRON] The time period between two successive passages of a star across the meridian.

star density [ASTRON] The average number of stars in a unit volume of space.

star drift [ASTRON] A description of two star groups in the Milky Way traveling through each other in opposite directions; individual stars have movements that are relative to each other. Also known as star stream.

star globe *See* celestial globe.

star group [ASTRON] A number of stars that move in the same general direction at the same time.

Stark effect [SPECT] The effect on spectrum lines of an electric field which is either externally applied or is an internal field caused by the presence of neighboring ions or atoms in a gas, liquid, or solid. Also known as electric field effect.

Stark-Einstein law *See* Einstein photochemical equivalence law.

Stark-Lunelund effect [ELECTROMAG] The polarization of light emitted from a beam of moving atoms in a region where there are no electric or magnetic fields.

Stark number *See* Stefan number.

star motions [ASTRON] For the Milky Way, these include a systematic rotation of the Galaxy, which is described with respect to an external frame of reference; superposed on this systematic rotation are the individual motions of a star; each star moves in a somewhat elliptical orbit, with respect to the local standard of rest, the standard moving in a circular orbit around the galactic center.

star stream *See* star drift.

star subalgebra [MATH] A subalgebra of a star algebra which is mapped onto itself by the involution operation.

starter [ELECTR] An auxiliary control electrode used in a gas tube to establish sufficient ionization to reduce the anode breakdown voltage. Also known as trigger electrode.

star test [OPTICS] A procedure in which a telescope is directed at a bright star and the in-focus and out-of-focus images and diffraction patterns of the star are examined to detect aberrations and abnormalities in the optical system.

starting friction *See* static friction.

start-stop multivibrator *See* monostable multivibrator.

stat [NUCLEO] A unit of radioactive disintegration rate equal to the disintegration rate of that quantity of radon that gives rise to a charge of 1 statcoulomb in 1 second when situated in air.

statΩ *See* statohm.

stat℧ *See* statmho.

stat- [ELEC] A prefix indicating an electrical unit in the electrostatic centimeter-gram-second system of units; it is attached to the corresponding SI unit.

statA *See* statampere.

statampere [ELEC] The unit of electric current in the electrostatic centimeter-gram-second system of units, equal to a flow of charge of 1 statcoulomb per second; equal to approximately 3.3356×10^{-10} ampere. Abbreviated statA.

STARK EFFECT

An example of the linear Stark effect in the 4144- and 4169-angstrom lines of helium showing large symmetrical pattern. Electric field strengths range continuously from 0 to 85,000 volts per centimeter. Symbols π and σ refer respectively to light polarized parallel and perpendicular to the electric field.

statC *See* statcoulomb.

statcoulomb [ELEC] The unit of charge in the electrostatic centimeter-gram-second system of units, equal to the charge which exerts a force of 1 dyne on an equal charge at a distance of 1 centimeter in a vacuum; equal to approximately 3.3356×10^{-10} coulomb. Abbreviated statC. Also known as franklin (Fr); unit charge.

state [CONT SYS] A minimum set of numbers which contain enough information about a system's history to enable its future behavior to be computed. [PHYS] The condition of a system which is specified as completely as possible by observations of a specified nature, for example, thermodynamic state, energy state. [QUANT MECH] The condition in which a system exists; the state may be pure and describable by a wave function or mixed and describable by a density matrix.

state equations [CONT SYS] Equations which express the state of a system and the output of a system at any time as a single valued function of the system's input at the same time and the state of the system at some fixed initial time.

state of matter [PHYS] One of three fundamental conditions of matter: the solid, liquid, and gaseous states.

state of strain [MECH] A complete description, including the six components of strain, of the deformation within a homogeneously deformed volume.

state of stress [MECH] A complete description, including the six components of stress, of a homogeneously stressed volume.

state parameter *See* thermodynamic function of state.

state space [CONT SYS] The set of all possible values of the state vector of a system.

state transition equation [CONT SYS] The equation satisfied by the $n \times n$ state transition matrix $\Phi(t,t_0)$: $\partial\Phi(t,t_0)/\partial t = A(t)\Phi(t,t_0)$, $\Phi(t_0,t_0) = I$; here I is the unit $n \times n$ matrix, and $A(t)$ is the $n \times n$ matrix which appears in the vector differential equation $dx(t)/dt = A(t)x(t)$ for the n-component state vector $x(t)$.

state transition matrix [CONT SYS] A matrix $\Phi(t,t_0)$ whose product with the state vector x at an initial time t_0 gives the state vector at a later time t; that is, $x(t) = \Phi(t,t_0)x(t_0)$.

state variable [CONT SYS] One of a minimum set of numbers which contain enough information about a system's history to enable computation of its future behavior. [THERMO] *See* thermodynamic function of state.

state vector [CONT SYS] A column vector whose components are the state variables of a system. [QUANT MECH] A vector in Hilbert space which corresponds to the state of a quantum-mechanical system.

statF *See* statfarad.

statfarad [ELEC] Unit of capacitance in the electrostatic centimeter-gram-second system of units, equal to the capacitance of a capacitor having a charge of 1 statcoulomb, across the plates of which the charge is 1 statvolt; equal to approximately 1.1126×10^{-12} farad. Abbreviated statF. Also known as centimeter.

statH *See* stathenry.

stathenry [ELECTROMAG] The unit of inductance in the electrostatic centimeter-gram-second system of units, equal to the self-inductance of a circuit or the mutual inductance between two circuits if there is an induced electromotive force of 1 statvolt when the current is changing at a rate of 1 statampere per second; equal to approximately 8.9876×10^{11} henries. Abbreviated statH.

static [PHYS] Without motion or change.

statically admissible loads [MECH] Any set of external loads and internal forces which fulfills conditions necessary to maintain the equilibrium of a mechanical system.

static breeze *See* convective discharge.

static characteristic [ELECTR] A relation between a pair of variables, such as electrode voltage and electrode current, with all other operating voltages for an electron tube, transistor, or other amplifying device maintained constant.

static charge [ELEC] An electric charge accumulated on an object.

static electricity [ELEC] **1.** The study of the effects of macroscopic charges, including the transfer of a static charge from one object to another by actual contact or by means of a spark that bridges an air gap between the objects. **2.** *See* electrostatics.

static equilibrium *See* equilibrium.

static error [STAT] Error independent of the time-varying nature of a variable.

static fluid column [FL MECH] An unchanging height of fluid in a vertical pipe, well bore, process vessel, or tank.

static friction [MECH] **1.** The force that resists the initiation of sliding motion of one body over the other with which it is in contact. **2.** The force required to move one of the bodies when they are at rest. Also known as starting friction.

static head [FL MECH] Pressure of a fluid due to the head of fluid above some reference point.

static level [FL MECH] Elevation of the water level or a pressure surface at rest.

static load [MECH] A nonvarying load; the basal pressure exerted by the weight of a mass at rest, such as the load imposed on a drill bit by the weight of the drill-stem equipment or the pressure exerted on the rocks around an underground opening by the weight of the superimposed rocks. Also known as dead load.

static machine [ELEC] A machine for generating electric charges, usually by electric induction, sometimes used to build up high voltages for research purposes.

static moment [MECH] **1.** A scalar quantity (such as area or mass) multiplied by the perpendicular distance from a point connected with the quantity (such as the centroid of the area or the center of mass) to a reference axis. **2.** The magnitude of some vector (such as force, momentum, or a directed line segment) multiplied by the length of a perpendicular dropped from the line of action of the vector to a reference point.

static pressure [ACOUS] The pressure that would exist at a point in a medium if no sound waves were present. [FL MECH] **1.** The normal component of stress, the force per unit area, exerted across a surface moving with a fluid, especially across a surface which lies in the direction of fluid flow. **2.** The average of the normal components of stress exerted across three mutually perpendicular surfaces moving with a fluid.

static-pressure tube [ENG] A smooth tube with a rounded nose that has radial holes in the portion behind the nose and is used to measure the static pressure within the flow of a fluid.

static reaction [MECH] The force exerted on a body by other bodies which are keeping it in equilibrium.

statics [MECH] The branch of mechanics which treats of force and force systems abstracted from matter, and of forces which act on bodies in equilibrium.

static sensitivity [ELECTR] In phototubes, quotient of the direct anode current divided by the incident radiant flux of constant value.

static stability *See* stability.

static tube [ENG] A device used to measure the static (not kinetic or total) pressure in a stream of fluid; consists of a perforated, tapered tube that is placed parallel to the flow, and has a branch tube that is connected to a manometer.

static universe [ASTRON] A postulated universe that has a finite static volume and is closed.

stationary distribution [PHYS] A time-independent distribution of a scalar quantity.

stationary ergodic noise [ELECTR] A stationary noise for which the probability that the noise voltage lies within any given interval at any time is nearly equal to the fraction of time that the noise voltage lies within this interval if a sufficiently long observation interval is recorded.

stationary field [PHYS] Field which does not change during the time interval under consideration. Also known as constant field.

stationary noise [ELECTR] A random noise for which the probability that the noise voltage lies within any given interval does not change with time.

stationary phase [MATH] A method used to find approximations to the integral of a rapidly oscillating function, based on the principle that this integral depends chiefly on that part of the range of integration near points at which the derivative of the trigonometric function involved vanishes.

stationary point [MATH] **1.** A point on a curve at which the tangent is horizontal. **2.** For a function of several variables, a point at which all partial derivatives are 0.

stationary state *See* energy state.

stationary stochastic process [MATH] A stochastic process $x(t)$ is stationary if each of the joint probability distributions is unaffected by a change in the time parameter t.

stationary time principle *See* Fermat's principle.

stationary time series [STAT] A time series which as a stochastic process is unchanged by a uniform increment in the time parameter defining it.

stationary value [MATH] **1.** The value $f(x_0)$ of a function f at a point x_0 in its domain at which the first derivative of the function $f'(x_0)$ is equal to 0. **2.** A stationary value of an integral

$$I = \int_a^b f(x, y, dy / dx) \, dx$$

is the value of I for a function $y(x)$ such that the first variation of I equals 0 for any variation $\phi(x)$ of $y(x)$ satisfying certain continuity conditions and the equation $\phi(a) = \phi(b) = 0$.

stationary wave *See* standing wave.

statistic [STAT] An estimate or piece of data, concerning some parameter, obtained from a sampling.

statistical analysis [STAT] The body of techniques used in statistical inference concerning a population.

statistical distribution *See* distribution of a random variable.

statistical hypothesis [STAT] A statement about the way a random variable is distributed.

statistical independence [STAT] Two events are statistically independent if the probability of their occurring jointly equals the product of their respective probabilities.

statistical inference [STAT] The process of reaching conclusions concerning a population upon the basis of random samplings.

statistical mechanics [PHYS] That branch of physics which endeavors to explain and predict the macroscopic properties and behavior of a system on the basis of the known characteristics and interactions of the microscopic constituents of the system, usually when the number of such constituents is very large. Also known as statistical thermodynamics.

statistical thermodynamics *See* statistical mechanics.

statistics [MATH] A discipline dealing with methods of obtaining data, analyzing and summarizing it, and drawing inferences from data samples by the use of probability theory.

statmho [ELEC] The unit of conductance, admittance, and susceptance in the electrostatic centimeter-gram-second system of units, equal to the conductance between two points of a conductor when a constant potential difference of 1 statvolt applied between the points produces in this conductor a current of 1 statampere, the conductor not being the source of any electromotive force; equal to approximately 1.1126 × 10^{-12} mho. Abbreviated stat℧. Also known as statsiemens (statS).

statohm [ELEC] The unit of resistance, reactance, and impedance in the electrostatic centimeter-gram-second system of units, equal to the resistance between two points of a conductor when a constant potential difference of 1 statvolt between these points produces a current of 1 statampere; equal to approximately 8.9876 × 10^{11} ohm. Abbreviated statΩ.

stator [ELEC] The portion of a rotating machine that contains the stationary parts of the magnetic circuit and their associated windings.

stator plate [ELEC] One of the fixed plates in a variable capacitor; stator plates are generally insulated from the frame of the capacitor.

statoscope [ENG] **1.** A barometer that records small variations in atmospheric pressure. **2.** An instrument that indicates small changes in an aircraft's altitude.

statS *See* statmho.

statsiemens *See* statmho.

statT *See* stattesla.

stattesla [ELECTROMAG] The unit of magnetic flux density in the electrostatic centimeter-gram-second system of units, equal to one statweber per square centimeter; equal to approximately 2.9979 × 10^6 tesla. Abbreviated statT.

statute mile *See* mile.

statV *See* statvolt.

statvolt [ELEC] The unit of electric potential and electromotive force in the electrostatic centimeter-gram-second system of units, equal to the potential difference between two points such that the work required to transport 1 statcoulomb of electric charge from one to the other is equal to 1 erg; equal to approximately 2.9979 × 10^2 volts. Abbreviated statV.

statWb *See* statweber.

statweber [ELECTROMAG] The unit of magnetic flux in the electrostatic centimeter-gram-second system of units, equal to the magnetic flux which, linking a circuit of one turn, produces in it an electromotive force of 1 statvolt as it is reduced to zero at a uniform rate in 1 second; equal to approximately 2.9979 × 10^2 weber. Abbreviated statWb.

steady flow [FL MECH] Fluid flow in which all the conditions at any one point are constant with respect to time.

steady state [PHYS] The condition of a body or system in which the conditions at each point do not change with time, that is after initial transients or fluctuations have disappeared.

steady-state cavitation *See* sheet cavitation.

steady-state conduction [THERMO] Heat conduction in which the temperature and heat flow at each point does not change with time.

steady-state creep *See* secondary creep.

steady-state current [ELEC] An electric current that does not change with time.

steady-state error [CONT SYS] The error that remains after transient conditions have disappeared in a control system.

steady-state reactor [NUCLEO] A reactor in which conditions such as temperature, reaction rate, and neutron flux do not change appreciably with time.

steady-state theory [ASTRON] A cosmological theory which holds that the average density of matter does not vary with space or time in spite of the expansion of the universe; this requires that matter be continuously created.

steady-state wave motion [PHYS] Wave motion in which the wave quantities at each point in the region through which the wave is passing repeat themselves periodically.

steam [PHYS] Water vapor, or water in its gaseous state; the most widely used working fluid in external combustion engine cycles.

steam calorimeter [ENG] **1.** A calorimeter, such as the Joly or differential steam calorimeter, in which the mass of steam condensed on a body is used to calculate the amount of heat supplied. **2.** *See* throttling calorimeter.

steam cycle *See* Rankine cycle.

steam line [THERMO] A graph of the boiling point of water as a function of pressure.

steam point [THERMO] The boiling point of pure water whose isotopic composition is the same as that of sea water at standard atmospheric pressure; it is assigned a value of 100°C on the International Practical Temperature Scale of 1968.

Steenrod algebra [MATH] The cohomology groups of a topological space have additive operations on them, which can be added and multiplied so as to form the Steenrod algebra.

Steenrod squares [MATH] Operations which associate elements from different cohomology groups of a topological space and produce an element in another of the groups; these operations can be so added and multiplied as to produce the Steenrod algebra.

steepest descent method [MATH] Certain functions can be approximated for large values by an asymptotic formula derived from a Taylor series expansion about a saddle point. Also known as saddle point method.

steerable antenna [ELECTROMAG] A directional antenna whose major lobe can be readily shifted in direction.

Stefan-Boltzmann constant [STAT MECH] The energy radiated by a blackbody per unit area per unit time divided by the fourth power of the body's temperature; equal to $(5.6696 \pm 0.0010) \times 10^{-8}$ (watts)(meter)$^{-2}$(kelvins)$^{-4}$.

Stefan-Boltzmann law [STAT MECH] The total energy radiated from a blackbody is proportional to the fourth power of the temperature of the body. Also known as fourth-power law; Stefan's law of radiation.

Stefan number [THERMO] A dimensionless number used in the study of radiant heat transfer, equal to the Stefan-Boltzmann constant times the cube of the temperature times the thickness of a layer divided by the layer's thermal conductivity. Symbolized $\overline{Sf}$. Also known as Stark number (Sk).

Stefan's law of radiation *See* Stefan-Boltzmann law.

Steiner's hypocycloid *See* deltoid.

Steiner's theorem *See* parallel axis theorem.

Steiner triple system [MATH] A balanced incomplete block design in which the number k of distinct elements in each block equals 3, and the number λ of blocks in which each combination of elements occurs together equals 1.

Steinheil lens [OPTICS] A type of magnifier lens in which a biconvex crown lens is cemented between a pair of flint lenses.

Steinmetz coefficient [ELECTROMAG] The constant of proportionality in Steinmetz's law.

Steinmetz's law [ELECTROMAG] The energy converted into

STEINHEIL LENS

Schematic of Steinheil aplanatic magnifier showing the crown lens within the flint lenses.

heat per unit volume per cycle during a cyclic change of magnetization is proportional to the maximum magnetic induction raised to the 1.6 power, the constant of proportionality depending only on the material.

stellar [ASTRON] Relating to or consisting of stars.

stellar association [ASTRON] A loose grouping of stars which may have had a common origin.

stellar atmosphere [ASTRON] The envelope of gas and plasma surrounding a star; consists of about 90% hydrogen atoms and 9% helium atoms, by number of atoms.

stellarator [PL PHYS] A device for confining a high-temperature plasma, consisting of a tube, which closes in on itself in a figure-eight or race-track configuration, and external coils which generate magnetic fields whose lines of force run parallel to the walls of the tube and prevent the plasma from touching the walls.

stellar crystal *See* plane-dendritic crystal.

stellar evolution [ASTROPHYS] The changes in spectrum and luminosity that take place in the life of a star.

stellar flare [ASTRON] Ejection of material from a star in an eruption that may last from a few minutes to an hour or more.

stellar interferometer [OPTICS] An optical interferometer for measuring angular diameters of stars; it is attached to a telescope and measures interference rings at the telescope's focus.

stellar luminosity [ASTRON] A star's brightness; it is measured either in ergs per second or in units of solar luminosity or in absolute magnitude.

stellar magnetic field [ASTROPHYS] A magnetic field, generally stronger than the earth's magnetic field, possessed by many stars.

stellar magnitude *See* magnitude.

stellar mass [ASTROPHYS] The mass of a star, usually expressed in terms of the sun's mass.

stellar model [ASTROPHYS] A mathematical characterization of the internal properties of a star. Also known as star model.

stellar parallax [ASTRON] The subtended angle at a star formed by the mean radius of the earth's orbit; it indicates distance to the star.

stellar photometry [ASTRON] The measurement of the brightness of stars.

stellar population [ASTRON] Either of two classes of stars, termed population I and population II; population I are relatively young stars, found in the arms of spiral galaxies, especially the blue stars of high luminosity; population II stars are the much older, more evolved stars of lower metallic content; many high luminosity red giants and many variable stars are members of population II.

stellar pulsation [ASTROPHYS] Expansion of a star followed by contraction so that its surface temperature and intrinsic brightness undergo periodic variation.

stellar rotation [ASTRON] Axial rotation of stars; surface rotational equatorial velocities of stars range from a few to 500 kilometers per second.

stellar scintillation *See* astronomical scintillation.

stellar spectroscopy [ASTRON] The techniques of obtaining spectra of stars and their study.

stellar spectrum [ASTRON] The spectrum of a star normally obtained with a slit spectrograph by black-and-white photography; the spectrum of a star in a large majority of cases shows absorption lines superposed on a continuous background.

stellar structure [ASTROPHYS] The mathematical study of a rotating, chemically homogeneous mass of gas held together by its own gravitation; a representative model of the observ-

STELLARATOR

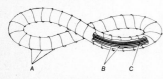

Diagram of a stellarator. External, closely spaced, current-carrying windings *A* produce a magnetic field whose lines of force *B* run parallel to the walls, confining plasma *C* in tube's center.

able properties of a star; thermonuclear reactions are postulated to be the main source of stellar energy.

stellar system [ASTRON] A gravitational system of stars.

stellar temperature [ASTROPHYS] Any temperature above several million degrees, such as occurs naturally in the interior of the sun and other stars.

STEM *See* scanning transmission electron microscope.

stem correction [THERMO] A correction which must be made in reading a thermometer in which part of the stem, and the thermometric fluid within it, is at a temperature which differs from the temperature being measured.

stenometer [ENG] An instrument for measuring distances; employs a telescope in which two target images a known distance apart are superimposed by turning a micrometer screw.

step-and-repeat camera [OPTICS] A type of camera providing a gridlike pattern of latent image frames in a given sequence.

step-by-step system [CONT SYS] A control system in which the drive motor moves in discrete steps when the input element is moved continuously.

step-down transformer [ELEC] A transformer in which the alternating-current voltages of the secondary windings are lower than those applied to the primary winding.

step function [MATH] **1.** A function f defined on an interval $[a,b]$ so that $[a,b]$ can be partitioned into a finite number of subintervals on each of which f is a constant. Also known as simple function. **2.** More generally, a real function with finite range.

stepped leader [GEOPHYS] The initial streamer of a lightning discharge; an intermittently advancing column of high ion density which established the channel for subsequent return streamers and dart leaders.

stepping *See* zoning.

step-recovery diode [ELECTR] A varactor in which forward voltage injects carriers across the junction, but before the carriers can combine, voltage reverses and carriers return to their origin in a group; the result is abrupt cessation of reverse current and a harmonic-rich waveform.

step response [CONT SYS] The behavior of a system when its input signal is zero before a certain time and is equal to a constant nonzero value after this time.

step-up transformer [ELEC] Transformer in which the energy transfer is from a low-voltage winding to a high-voltage winding or windings.

sterad *See* steradian.

steradian [MATH] The unit of measurement for solid angles; it is equal to the solid angle subtended at the center of a sphere by a portion of the surface of the sphere whose area equals the square of the sphere's radius. Abbreviated sr; sterad.

steradiancy *See* radiance.

stère [MECH] A unit of volume equal to 1 cubic meter; it is used mainly in France, and in measuring timber volumes.

steregon [MATH] The entire solid angle bounded by a sphere; equal to 4π steradians.

stereo- [PHYS] A prefix used to designate a three-dimensional characteristic.

stereo camera *See* stereoscopic camera.

stereochemistry [PHYS CHEM] The study of the spatial arrangement of atoms in molecules and the chemical and physical consequences of such arrangement.

stereocomparagraph [OPTICS] A projection device in which two-dimensional aerial photographs taken at slightly different angles are combined so as to give the appearance of tridimensionality.

STEREOGNOMOGRAM

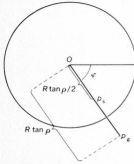

Stereognomogram of pole defined by azimuthal angle φ and colatitude ρ. Both projection planes are superimposed; P_s and P_g are on a radius making an angle φ with the origin, and at distances respectively equal to $R \tan \rho/2$ and $R \tan \rho$.

STEREOSCOPIC RANGEFINDER

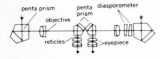

Lens system of a typical stereoscopic rangefinder. Objective forms images in plane of respective reticles. Diasporometer is used to vary angle between beams reaching eyes from target until it equals angle between beams reaching eyes from two reticles.

stereo comparator [OPTICS] An instrument that may be used to view two photographs taken of the stars in the same section of sky at different times; viewing the images stereoscopically may reveal stars that have moved between exposures or stars of varying brightness.

stereo effect [ACOUS] Reproduction of sound in such a manner that the listener receives the sensation that individual sounds are coming from different locations, just as did the original sounds reaching the stereo microphone system.

stereofluoroscopy [ELECTR] A fluoroscopic technique that gives three-dimensional images.

stereognomogram [CRYSTAL] The projection resulting from the superposition of the projection planes of a stereogram and a gnomogram.

stereographic projection [CRYSTAL] A method of displaying the positions of the poles of a crystal in which poles are projected through the equatorial plane of the reference sphere by lines joining them with the south pole for poles in the upper hemisphere, and with the north pole for poles in the lower hemisphere. [MATH] The projection of the Riemann sphere onto the euclidean plane performed by emanating rays from the north pole of the sphere through a point on the sphere.

stereomicrography [OPTICS] The taking of two microphotographs of the same field at different angles (a stereo pair), then viewing them simultaneously with a stereo viewer.

stereomicrometer [ENG] An instrument attached to an optical instrument (such as a telescope) to measure small angles.

stereophonics [ENG ACOUS] The study of reproducing or reinforcing sound in such a way as to produce the sensation that the sound is coming from sources whose spatial distribution is similar to that of the original sound sources.

stereo rangefinder *See* stereoscopic rangefinder.

stereoscope [OPTICS] An optical instrument in which each eye views one of two photographs taken with the camera or object of study displaced, or simultaneously with two cameras, or with a stereoscopic camera, so that a sensation of depth is produced.

stereoscopic camera [OPTICS] A camera which takes photographs simultaneously with two similar lenses a few inches apart, for use in a stereoscope or other optical system which gives a sensation of depth to the viewer. Also known as stereo camera.

stereoscopic heightfinder *See* stereoscopic rangefinder.

stereoscopic microscope [OPTICS] A microscope with two eyepieces and two objectives, giving the viewer a sensation of depth.

stereoscopic photography [OPTICS] A technique that simulates stereoscopic vision, in which two photographs are made with the camera or object of study displaced, or simultaneously with two cameras, or with a stereoscopic camera, and each of the photographs is viewed by one eye, using a stereoscope or other optical system.

stereoscopic power [OPTICS] The magnifying power of a binoculars or other stereo system multiplied by the ratio of the distance between the objective axes to the distance between the eyepiece axes; it is a measure of the stereoscopic radius. Also known as total relief.

stereoscopic rangefinder [OPTICS] An optical rangefinder which utilizes stereoscopic vision; it is essentially a large binoculars fitted with special reticles which allow a skilled user to superimpose the stereoscopic image formed by the pair of reticles over the images of the target seen in the eyepieces, so that the correct range is obtained when the reticle marks appear to be suspended over the target and at the same apparent distance. Also known as stereo rangefinder; stereoscopic heightfinder.

stereoscopic system [OPTICS] An optical system such as a binoculars or stereoscope that produces two images of the same object viewed from slightly different positions, so that a sensation of depth is created when one image is presented to each eye.

Stern-Gerlach effect [ATOM PHYS] The splitting of a beam of atoms passing through a strong, inhomogeneous magnetic field into several beams.

stern layer [PHYS CHEM] One of two electrically charged layers of electrolyte ions, the layer of ions immediately adjacent to the surface, in the neighborhood of a negatively charged surface.

Stern-Zartman experiment [STAT MECH] An experiment in which the distribution in speed of atoms or molecules in a beam emitted from an opening in an oven is measured by having the beam impinge on a rotating cylindrical drum, with a slit cut parallel to the drum axis, and measuring the density of atoms or molecules deposited on the inner surface of the drum, roughly opposite the slit, as a function of distance from a point directly opposite the slit; it is used to test the Maxwell-Boltzmann distribution law.

sthène [MECH] The force which, when applied to a body whose mass is 1 metric ton, results in an acceleration of 1 meter per second per second; equal to 1000 newtons. Formerly known as funal.

stibium [CHEM] The Latin name for antimony, thus the symbol Sb for the element.

stick-slip friction [MECH] Friction between two surfaces that are alternately at rest and in motion with respect to each other.

stiction [MECH] Friction that tends to prevent relative motion between two movable parts at their null position.

Stieltjes integral [MATH] The Stieltjes integral of a real function $f(x)$ relative to a real function $g(x)$ of bounded variation on an interval $[a,b]$ is defined, analogously to the Riemann integral, as a limit of a sum of terms $f(a_i) [g(x_i) - g(x_{i-1})]$ taken as partitions of the interval shrink. Denoted

$$\int_a^b f(x) \, dg(x).$$

Stieltjes transform [MATH] A form of the Laplace transform of a function where the usual Riemann integral is replaced by a Stieltjes integral.

stiffness [ACOUS] *See* acoustic stiffness [MECH] The ratio of a steady force acting on a deformable elastic medium to the resulting displacement.

stiffness coefficient [MECH] The ratio of the force acting on a linear mechanical system, such as a spring, to its displacement from equilibrium.

stiffness constant [MECH] Any one of the coefficients of the relations in the generalized Hooke's law used to express stress components as linear functions of the strain components. Also known as elastic constant.

stiffness matrix [MECH] A matrix $\mathbf{K}$ used to express the potential energy V of a mechanical system during small displacements from an equilibrium position, by means of the equation $V = \frac{1}{2}\mathbf{q}^T \mathbf{Kq}$, where $\mathbf{q}$ is the vector whose components are the generalized components of the system with respect to time and $\mathbf{q}^T$ is the transpose of $\mathbf{q}$. Also known as stability matrix.

stiffness reactance *See* acoustic stiffness reactance.

stigma [MECH] A unit of length used mainly in nuclear measurements, equal to 10^{-12} meter. Also known as bicron.

stigmatic [OPTICS] **1.** Property of an optical system whose focal power is the same in all meridians. **2.** *See* homocentric.

stigmatic concave grating [OPTICS] An optical element with many parallel grooves on a concave optical surface; combines the two functions of light dispersion and focusing in one dispersive element; used in space optics, in food and metal analysis, and as a dispersive element of spectrophotometers and spectrographs.

stilb [OPTICS] A unit of luminance, equal to 1 candela per square centimeter. Abbreviated sb.

stimulated emission [ATOM PHYS] Emission of electromagnetic radiation by an atom or molecule as a result of its interaction with incident radiation of the same frequency. Also known as induced emission.

stimulated emission device [ELECTR] A device that uses the principle of amplification of electromagnetic waves by stimulated emission, namely, a maser or a laser.

stimulus [CONT SYS] A signal that affects the controlled variable in a control system.

Stirling cycle [THERMO] A regenerative thermodynamic power cycle using two isothermal and two constant volume phases.

Stirling numbers [MATH] The coefficients which occur in the Stirling interpolation formula for a difference operator.

Stirling numbers of the second kind [MATH] The numbers $S(n,r)$ giving the numbers of ways that n elements can be distributed among r indistinguishable cells so that no cell remains empty.

Stirling's formula [MATH] The expression $(n/e)^n\sqrt{2\pi n}$ is asymptotic to factorial n; that is, the limit as n goes to ∞ of their ratio is 1.

Stirling's interpolation formula [MATH] A formula for estimating the value of a smooth function at an intermediate value of the independent variable, when its values are known at a sequence of equidistant points (such as those that appear in a table), in terms of the central differences of the function; if $y_i = f(x_0 + ih)$, with i running over the integers, are the known values of the function, the formula states that $f(x_0 + uh)$, where $|u| < \frac{1}{2}$, is approximated by a series whose kth term is $S_k\delta^k y_0$ for even k, and

$$S_k[\delta^k f(x - h/2) + \delta^k f(x + h/2)]$$

for odd k, where δ is the central difference operator and the S_k are polynomial functions of u.

stirring [PHYS] A turbulent process in which molecular diffusion and molecular heat conduction are speeded up.

stirring effect [ELECTROMAG] The circulation in a molten metal carrying electric current as a result of the combined forces of the pinch and motor effects.

stochastic [MATH] Pertaining to random variables.

stochastic calculus [MATH] The mathematical theory of stochastic integrals and differentials, and its application to the study of stochastic processes.

stochastic chain rule [MATH] A generalization of the ordinary chain rule to stochastic processes; it states that the process $U_t = u(X_t^1, X_t^2, \ldots, X_t^n)$ satisfies

$$dU = \sum_i \partial_i u\, dX^i + \frac{1}{2}\sum_{i,j} \partial_i \partial_j u\, dX^i\, dX^j$$

with the conventions $(dt)^2 = 0$ and $dW^\alpha\, dW^\beta = \partial_{\alpha\beta}\, dt$, where the X^i are processes satisfying

$$dX_i = a_i\, dt + \sum_{\alpha=1}^{m} b_i{}^\alpha\, dW^\alpha, \quad i = 1, 2, \ldots, n;$$

$$\{W_t^\alpha, t \geq 0\}, \alpha = 1, 2, \ldots, m$$

are independent Wiener processes; the dW_t^α are the corresponding random disturbances occurring in the infinitesimal time interval dt; the a_t^i and $b_t^{i\alpha}$ are independent of future disturbances, and $u(x_1, x_2, \ldots, x_n)$ is a function whose derivatives $\partial_i u$ and $\partial_i \partial_j u$ are continuous. Also known as Itô's formula.

stochastic differential [MATH] An expression representing the random disturbances occurring in an infinitesimal time interval; it has the form dW_t, where $\{ W_t, t \geq 0 \}$ is a Wiener process.

stochastic integral [MATH] An integral used to construct the sample functions of a general diffusion process from those of a Wiener process; it has the form

$$\int_{W_0}^{W_s} a_t \, dW_t$$

where $\{ W_t, t \geq 0 \}$ is a Wiener process, dW_t represents the random disturbances occurring in an infinitesimal time interval dt, and a_t is independent of future disturbances. Also known as Itô's integral.

stochastic matrix [MATH] A square matrix with nonnegative real entries such that the sum of the entries of each row is equal to 1.

stochastic process [MATH] A family of random variables, dependent upon a parameter which usually denotes time. Also known as random process.

stochastic variable *See* random variable.

Stodola method [MECH] A method of calculating the deflection of a uniform or nonuniform beam in free transverse vibration at a specified frequency, as a function of distance along the beam, in which one calculates a sequence of deflection curves each of which is the deflection resulting from the loading corresponding to the previous deflection, and these deflections converge to the solution.

stoke [FL MECH] A unit of kinematic viscosity, equal to the kinematic viscosity of a fluid with a dynamic viscosity of 1 poise and a density of 1 gram per cubic centimeter. Symbolized St (formerly S). Also known as lentor (deprecated usage); stokes.

stokes *See* stoke.

Stokes drift [FL MECH] The drift of particles in a gravity wave, which arises from the fact that particle velocities are periodic with a mean which is not zero.

Stokes frequencies [OPTICS] Scattered (secondary) light in the Raman effect (when a high-intensity light beam passes through a transparent medium) that occurs at frequencies smaller than the frequency of the primary beam.

Stokes' integral theorem [MATH] The analog of Green's theorem in n-dimensional euclidean space; that is, a line integral of $F_1(x_1, x_2, \ldots, x_n) \, dx_1 + \cdots + F_n(x_1, x_2, \ldots, x_n) \, dx_n$ over a closed curve equals an integral of an expression containing various partial derivatives of $F_1, \ldots, F_n$ over a surface bounded by the curve.

Stokes' law [FL MECH] At low velocities, the frictional force on a spherical body moving through a fluid at constant velocity is equal to 6π times the product of the velocity, the fluid viscosity, and the radius of the sphere. [SPECT] The wavelength of luminescence excited by radiation is always greater than that of the exciting radiation.

Stokes' lens [OPTICS] A variable-power compound lens made up of cylindrical lenses mounted so that the angle between their axes can be varied.

Stokes line [MATH] Any of the lines in the complex plane at which changes in the asymptotic behavior of the Airy function $Ai(z)$ occur, located at $\arg z = 0$, $\arg z = \frac{2}{3}\pi$, and arg $z = -\frac{2}{3}\pi$. [SPECT] A spectrum line in luminescent radiation whose wavelength is greater than that of the radiation which excited the luminescence, and thus obeys Stokes' law.

Stokes number 1 [FL MECH] A dimensionless number used in the study of the dynamics of a particle in a fluid, equal to the product of the dynamic viscosity of the fluid and the particle's vibration time, divided by the product of the fluid density and a characteristic length. Symbol St.

Stokes number 2 [ENG] A dimensionless number used in the calibration of rotameters, equal to $1.042\, m_f g\rho\, (1 - \rho/\rho_f)R^3/\mu^2$, where ρ and μ are the density and dynamic viscosity of the fluid respectively, m_f and ρ_f are the mass and density of the float respectively, and R is the ratio of the radius of the tube to the radius of the float. Symbol St_2.

Stokes phenomenon [MATH] A change in the asymptotic representation of certain analytic functions that occurs in passing from one section of the complex plane to another.

Stokes shift [SPECT] The displacement of spectral lines or bands of luminescent radiation toward longer wavelengths than those of the absorption lines or bands.

Stokes stream function [FL MECH] A one-component vector potential function used in analyzing and describing a steady, axially symmetric fluid flow; at any point it is equal to $\frac{1}{2}\pi$ times the mass rate of flow inside the surface generated by rotating the streamline on which the point is located about the axis of symmetry.

stone [MECH] A unit of mass in common use in the United Kingdom, equal to 14 pounds or 6.35029318 kilograms.

Stone-Čech compactification [MATH] The Stone-Čech compactification of a completely regular space X is a compact Hausdorff space $\beta(X)$ such that X is a dense subset of $\beta(X)$ and for any continuous function f from X to a compact space Y there is a unique continuous function from $\beta(X)$ to Y which is an extension of f.

Stone's representation theorem [MATH] This theorem determines the nature of all unitary representations of locally compact Abelian groups.

Stone's theorem [MATH] Every Boolean ring is isomorphic to a ring of subsets of some set.

Stone-Weierstrass theorem [MATH] If S is a collection of continuous real-valued functions on a compact space E, which contains the constant functions, and if for any pair of distinct points x and y in E there is a function f in S such that $f(x)$ is not equal to $f(y)$, then for any continuous real-valued function g on E there is a sequence of functions, each of which can be expressed as a polynomial in the functions of S with real coefficients, that converges uniformly to g.

stop [OPTICS] The aperture or useful opening of a lens, usually adjustable by means of a diaphragm.

stop band *See* rejection band.

stop down [OPTICS] To reduce the size of a lens aperture.

stop number *See* f number.

stopping [NUCLEO] The decrease in kinetic energy of an ionizing particle as a result of energy losses along its path through matter.

stopping capacitor *See* coupling capacitor.

stopping number [NUCLEO] An expression appearing in the Bethe-Bloch formula for linear stopping power of a material, equal to $Z[\ln (2mc^2\beta^2/I) - \ln (1 - \beta^2) - \beta^2 - C/Z - \delta]$, where Z is the atomic number of the material, m the electron mass, c the speed of light, β the ratio of the incident particle speed to the speed of light, I the average excitation energy of the material, C a correction factor for the effect of inner electron shells, and δ a density correction that is important at high energies.

stopping potential [ELECTR] Voltage required to stop the outward movement of electrons emitted by photoelectric or thermionic action.

stopping power [NUCLEO] The energy lost by a charged particle passing through a substance per unit length of path; related concepts are mass, atomic, molecular, and relative stopping power. Also known as linear energy transfer (LET); linear stopping power.

stopping rule [STAT] A rule which specifies when observation is to be discontinued in sequential trials.

storage battery [ELEC] A connected group of two or more storage cells or a single storage cell. Also known as accumulator; accumulator battery; rechargeable battery; secondary battery.

storage cell [ELEC] An electrolytic cell for generating electric energy, in which the cell after being discharged may be restored to a charged condition by sending a current through it in a direction opposite to that of the discharging current. Also known as secondary cell.

storage factor *See* Q.

storage oscilloscope [ELECTR] An oscilloscope that can retain an image for a period of time ranging from minutes to days, or until deliberately erased to make room for a new image.

storage rings [NUCLEO] Annular vacuum chambers in which charged particles can be stored, without acceleration, by a magnetic field of suitable focusing properties; they are used to stretch effectively the duty cycle of a particle accelerator or to produce colliding beams of particles, resulting in a greater possible center of mass energy.

straggling [PHYS] Random variations in some property associated with the passage of ions through matter.

strain [MECH] Change in length of an object in some direction per unit undistorted length in some direction, not necessarily the same; the nine possible strains form a second-rank tensor.

strain axis *See* principal axis of strain.

strain ellipsoid [MECH] A mathematical representation of the strain of a homogeneous body by a strain that is the same at all points or of unequal stress at a particular point. Also known as deformation ellipsoid.

strain energy [MECH] The potential energy stored in a body by virtue of an elastic deformation, equal to the work that must be done to produce this deformation.

strain figure [PHYS] A series of markings, such as Lüder's lines, that may appear on the surface of a body subjected to stress, indicating its state of deformation. Also known as flow figure.

strain gage [ENG] A device which uses the change of electrical resistance of a wire under strain to measure pressure.

strain rosette [MECH] A pattern of intersecting lines on a surface along which linear strains are measured to find stresses at a point.

strain shadow *See* undulatory extinction.

stranded conductor *See* stranded wire.

strangeness conservation [PARTIC PHYS] The principle that the sum of the strangeness numbers of the hadrons in an isolated system is constant; it is violated by the weak interactions.

strangeness number [PARTIC PHYS] A quantum number carried by hadrons, equal to the hypercharge minus the baryon number. Symbol S.

strange particle [PARTIC PHYS] A hadron whose strangeness

STRAIN ROSETTE

(a)

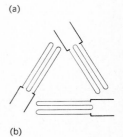

(b)

Strain rosettes. *(a)* 45° type; *(b)* 60° type.

number is not 0, for example, a *K* meson or a Σ-hyperon.

strange quark [PARTIC PHYS] A quark with an electric charge of $-\frac{1}{3}$, baryon number of $\frac{1}{3}$, strangeness of -1, and 0 charm. Symbolized *s*.

strategy [MATH] In game theory a strategy is a specified collection of moves, which cover all possible situations, for the complete play of a given game.

strategy vector [MATH] A vector characterizing a mixed strategy, whose components are the probability weights of the strategy.

stratified fluid [FL MECH] A fluid having density variation along the axis of gravity, usually implying upward decrease of density, that is, a stratification characterized by static stability.

stratified sampling [STAT] A random sample of specified size is drawn from each stratum of a population.

stratoscope [OPTICS] A balloon-borne astronomical telescope for taking solar or other celestial photographs at high altitudes; subsequently, the photos are transmitted to a ground receiving station.

Stratton pseudoscope [OPTICS] A type of Wheatstone stereoscope in which the mirrors transpose the right- and left-eye views, producing a reversed stereoscopic effect.

stray emission [PHYS] Emission of radiation that serves no useful purpose.

stray field [ELECTROMAG] Leakage of magnetic flux that spreads outward from a coil and does no useful work.

streak camera [OPTICS] A special type of high-speed motion picture camera that records an image as a continuous spread-out picture rather than as a sequence of separate frames; special viewing equipment must be used to analyze the image and reconstitute individual pictures.

streak line [FL MECH] A line within a fluid which, at a given instant, is formed by those fluid particles which at some previous instant have passed through a specified fixed point in the fluid; an example is the line of color in a flow into which a dye is continuously introduced through a small tube, all dyed fluid particles having passed the tube's end.

streamer [GEOPHYS] A sinuous channel of very high ion-density which propagates itself through a gas by continual establishment of an electron avalanche just ahead of its advancing tip; in lightning discharges, the stepped leader, and return streamer all constitute special types of streamers.

stream function *See* Lagrange stream function.

streaming current [ELEC] The electric current which is produced when a liquid is forced to flow through a diaphragm, capillary, or porous solid.

streaming potential [ELEC] The difference in electric potential between a diaphragm, capillary, or porous solid and a liquid that is forced to flow through it.

streamline [FL MECH] A line which is every where parallel to the direction of fluid flow at a given instant.

streamline flow [FL MECH] Flow of a fluid in which there is no turbulence: particles of the fluid follow well-defined continuous paths, and the flow velocity at a fixed point either remains constant or varies in a regular fashion with time.

stream tube [FL MECH] In fluid flow, an imaginary tube whose wall is generated by streamlines passing through a closed curve.

strength [ACOUS] The maximum instantaneous rate of volume displacement produced by a sound source when emitting a wave with sinusoidal time variation. [MECH] The stress at which material ruptures or fails.

stress [MECH] The force acting across a unit area in a solid

STREAMLINE FLOW

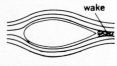

Flow indicated about a streamlined body which is traveling at subsonic speed.

material in resisting the separation, compacting, or sliding that tends to be induced by external forces.

stress analysis [PHYS] The determination of the stresses produced in a solid body when subjected to various external forces.

stress axis *See* principal axis of stress.

stress birefringence *See* mechanical birefringence.

stress concentration [MECH] A condition in which a stress distribution has high localized stresses; usually induced by an abrupt change in the shape of a member; in the vicinity of notches, holes, changes in diameter of a shaft, and so forth, maximum stress is several times greater than where there is no geometrical discontinuity.

stress concentration factor [MECH] A theoretical factor K_t expressing the ratio of the greatest stress in the region of stress concentration to the corresponding nominal stress.

stress crack [MECH] An external or internal crack in a solid body (metal or plastic) caused by tensile, compressive, or shear forces.

stress difference [MECH] The difference between the greatest and the least of the three principal stresses.

stress ellipsoid [MECH] A mathematical representation of the state of stress at a point that is defined by the minimum, intermediate, and maximum stresses and their intensities.

stress function [MECH] A single function, such as the Airy stress function, or one of two or more functions, such as Maxwell's or Morera's stress functions, that uniquely define the stresses in an elastic body as a function of position.

stress lines *See* isostatics.

stress-optic law [OPTICS] In a transparent, isotropic plate subjected to a biaxial stress field, the relative retardation R_t between the two components produced by temporary double refraction is equal to $Ct(p - q)$, which in turn is equal to $n\lambda$; C is the stress-optic coefficient, t the plate thickness, p and q the principal stresses, n the number of fringes which have passed the point during application of the load, and λ the wavelength of the light.

stress-strain curve *See* deformation curve.

stress tensor [MECH] A second-rank tensor whose components are stresses exerted across surfaces perpendicular to the coordinate directions.

stress trajectories *See* isostatics.

striation [ELECTR] A succession of alternately luminous and dark regions sometimes observed in the positive column of a glow-discharge tube near the anode.

striation technique [ACOUS] A technique for making sound waves visible by using their ability to refract light waves.

strich *See* millimeter.

strike note [ACOUS] The note which is the loudest heard when a bell is struck, and whose pitch is generally assigned to the bell.

striking potential [ELECTR] **1.** Voltage required to start an electric arc. **2.** Smallest grid-cathode potential value at which plate current begins flowing in a gas-filled triode.

striking velocity *See* impact velocity.

string electrometer [ENG] An electrometer in which a conducting fiber is stretched midway between two oppositely charged metal plates; the electrostatic field between the plates displaces the fiber laterally in proportion to the voltage between the plates.

string galvanometer [ENG] A galvanometer consisting of a silver-plated quartz fiber under tension in a magnetic field, used to measure oscillating currents. Also known as Einthoven galvanometer.

stripped atom [ATOM PHYS] An ionized atom which has

STRESS CONCENTRATION

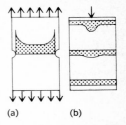

(a)   (b)

Stress concentrations. *(a)* Tensile stress distribution in a plate reduced by circular notches is shown qualitatively; stress at root of notch is about three times stress at end of plate. *(b)* Bar under concentrated end load; load concentrated at end of bar produces nonuniformly distributed normal stresses on adjacent sections with variation decreasing at more remote sections.

appreciably fewer electrons than it has protons in the nucleus.

stripping reaction [NUC PHYS] A nuclear reaction in which part of the incident nucleus combines with the target nucleus, and the other part proceeds with most of its original momentum in practically its original direction; especially the reaction in which the incident nucleus is a deuteron and only a proton emerges from the target.

strip transmission line [ELECTROMAG] A microwave transmission line consisting of a thin, narrow, rectangular metal strip that is supported above a ground-plane conductor or between two wide ground-plane conductors and is usually separated from them by a dielectric material.

strobe circuit [ELECTR] A circuit that produces an output pulse only at certain times or under certain conditions, such as a gating circuit or a coincidence circuit.

strobe pulse [ELECTR] Pulse of duration less than the time period of a recurrent phenomenon used for making a close investigation of that phenomenon; the frequency of the strobe pulse bears a simple relation to that of the phenomenon, and the relative timing is usually adjustable.

stroboscope [ENG] An instrument for making moving bodies visible intermittently, either by illuminating the object with brilliant flashes of light or by imposing an intermittent shutter between the viewer and the object; a high-speed vibration can be made visible by adjusting the strobe frequency close to the vibration frequency.

stroboscopic lamp *See* flash lamp.

strong causality condition [RELAT] The strong causality condition is said to hold at a point p in a space-time if every neighborhood of p contains a neighborhood of p which no timelike or null curve intersects more than once.

strong derivative [MATH] The derivative $f'(x_0)$ of a function f at a point x_0 is said to be a strong derivative of f at x_0 if for every $\varepsilon > 0$ there exists a $\delta > 0$ such that for any numbers x_1, x_2 the inequalities $|x_1 - x_0| < \delta$, $|x_2 - x_0| < \delta$ imply that $|f(x_2) - f(x_1) - (x_2 - x_1)f'(x_0)| \leq \varepsilon |x_2 - x_1|$.

strong interaction [PARTIC PHYS] One of the fundamental interactions of elementary particles, primarily responsible for nuclear forces and other interactions among hadrons.

strongly continuous semigroup [MATH] A semigroup of bounded linear operators on a Banach space B, together with a bijective mapping T from the positive real numbers onto the semigroup, such that $T(0)$ is the identity operator on B, $T(s + t) = T(s)T(t)$ for any two positive numbers s and t and, for each element x of B, $T(t)x$ is a continuous function of t.

strongly damped collision *See* quasi-fission.

strongly future asymptotically predictable space-time [RELAT] A future asymptotically predictable space-time such that a neighborhood of the event horizon is also predictable.

strong topology [MATH] The topology on a normed space obtained from the given norm; the basic open neighborhoods of a vector x are sets consisting of all those vectors y where the norm of $x - y$ is less than some number.

strontium [CHEM] A metallic element in group IIA, symbol Sr, with atomic number 38, atomic weight 87.62; flammable, soft, pale-yellow solid; soluble in alcohol and acids, decomposes in water; melts at 770°C; boils at 1380°C; chemistry is similar to that of calcium; used as electron-tube getter.

strontium-90 [NUC PHYS] A poisonous, radioactive isotope of strontium; 28-year half life with β radiation; derived from reactor-fuel fission products; used in thickness gages, medical treatment, phosphor activation, and atomic batteries.

strontium unit [NUCLEO] A unit of concentration of strontium-90 in a medium relative to the concentration of calcium, equal to 10^{-12} curie of strontium per gram of calcium.

STRIP TRANSMISSION LINE

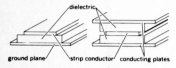

Components of strip transmission lines.

STRONTIUM

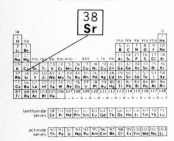

Periodic table of the chemical elements showing the position of strontium.

Abbreviated SU. Also known as sunshine unit (deprecated usage).

strophoid [MATH] The curve traced in the plane by a point P moving along a varying line L passing through a fixed point, where the distance of P to L's intersection with the y axis always is equal to the y-intercept value.

Strouhal number [MECH] A dimensionless number used in studying the vibrations of a body past which a fluid is flowing; it is equal to a characteristic dimension of the body times the frequency of vibrations divided by the fluid velocity relative to the body; for a taut wire perpendicular to the fluid flow, with the characteristic dimension taken as the diameter of the wire, it has a value between 0.185 and 0.2 Symbolized S_r. Also known as reduced frequency.

structural stability [MATH] Property of a differentiable flow on a compact manifold whose orbit structure is insensitive to small perturbations in the equations governing the flow or in the vector field generating the flow.

structure amplitude [SOLID STATE] The absolute value of a structure factor.

structure cell *See* unit cell.

structure constants [MATH] A set of numbers that serve as coefficients in expressing the commutators of the elements of a Lie algebra.

structure factor [SOLID STATE] A factor which determines the amplitude of the beam reflected from a given atomic plane in the diffraction of an x-ray beam by a crystal, and is equal to the sum of the atomic scattering factors of the atoms in a unit cell, each multiplied by an appropriate phase factor.

structure-sensitive property [SOLID STATE] A property of a substance that depends on impurities and the imperfections of the crystal structure.

structure type [CRYSTAL] The structural arrangement of a crystal, regardless of the atomic elements present; it corresponds to the crystal's space group.

Struve function [MATH] A solution of a generalization of the Bessel equation,

$$x^2 y'' + xy' + (x^2 - \nu^2)y = 4(x/2)^{\nu + 1}/\sqrt{\pi}\,\Gamma(\nu + \frac{1}{2}),$$

where ν is a parameter.

stub [ELECTROMAG] **1.** A short section of transmission line, open or shorted at the far end, connected in parallel with a transmission line to match the impedance of the line to that of an antenna or transmitter. **2.** A solid projection one-quarter-wavelength long, used as an insulating support in a waveguide or cavity.

stub matching [ELECTROMAG] Use of a stub to match a transmission line to an antenna or load; matching depends on the spacing between the two wires of the stub, the position of the shorting bar, and the point at which the transmission line is connected to the stub.

stub-supported line [ELECTROMAG] A transmission line that is supported by short-circuited quarter-wave sections of coaxial line; a stub exactly a quarter-wavelength long acts as an insulator because it has infinite reactance.

stub tuner [ELECTROMAG] Stub which is terminated by movable short-circuiting means and used for matching impedance in the line to which it is joined as a branch.

studentization [STAT] A method of taking care of the complications caused by the presence of a nuisance parameter by computing a statistic whose sampling distribution is independent of the parameter.

Student's distribution [STAT] The probability distribution used to test the hypothesis that a random sample of n

observations comes from a normal population with a given mean.

Student's t-statistic [STAT] A one-sample test statistic computed by $T = \sqrt{n}(\bar{X} - \mu_H)/S$, where $\bar{X}$ is the mean of a collection of n observations, S is the square root of the mean square deviation, and μ_H is the hypothesized mean.

Student's t-test [STAT] A test in a one-sample problem which uses Student's t-statistic.

Sturges rule [STAT] A rule for determining the desirable number of groups into which a distribution of observations should be classified; the number of groups or classes is 1 + 3.3 log n, where n is the number of observations.

Sturm-Liouville equation [MATH] A second-order homogeneous linear differential equation of the form

$$\frac{d}{dx}\left[p(x)\frac{du}{dx}\right] + \left[\lambda\rho(x) - q(x)\right]u = 0$$

where λ is a parameter; p, ρ, and q are real-valued functions of x; and the functions p and ρ are positive.

Sturm-Liouville expansion [MATH] The expression of a continuous function $f(x)$ defined on a finite interval as a series with general term $c_n\phi(x,\lambda_n)$, where the c_n are constants and the $\phi(x,\lambda_n)$ are normalized solutions of a Sturm-Liouville problem with eigenvalue λ_n.

Sturm-Liouville problem [MATH] The general problem of solving a given linear differential equation of order $2n$ together with $2n$-boundary conditions. Also known as eigenvalue problem.

Sturm-Liouville system [MATH] A given differential equation together with its boundary conditions having Sturm-Liouville problem form.

Sturm sequence [MATH] For a polynomial $p(x)$, this is the sequence of functions $f_0(x), f_1(x), \ldots$, where $f_0(x) = p(x), f_1(x) = p'(x)$, and $f_n(x)$ is the negative remainder that occurs by finding the greatest common divisor of $f_{n-2}(x)$ and $f_{n-1}(x)$ via the euclidean algorithm.

Sturm's theorem [MATH] This gives a method to determine the number of real roots of a polynomial $p(x)$ which lie between two given values of x; the Sturm sequence of $p(x)$ provides the necessary information.

SU *See* strontium unit.

subadditive function [MATH] A function F is subadditive if $f(x + y)$ is less than or equal to $f(x) + f(y)$ for all x and y in its domain.

subalgebra [MATH] **1.** A subset of an algebra which itself forms an algebra relative to the same operations. **2.** A subalgebra (of sets) is any algebra (of sets) contained in some given algebra.

subatomic particle [PHYS] A particle which is smaller than an atom, namely, an elementary particle or an atomic nucleus.

subbase for a topology [MATH] A family S of subsets of a topological space X where by taking all finite intersections of sets from S and all unions of such intersections the entire topology of open sets of X is obtained.

subcritical [NUCLEO] Having an effective multiplication constant less than one, so that a self-supporting chain reaction cannot be maintained in a nuclear reactor.

subcritical assembly *See* subcritical reactor.

subcritical flow *See* subsonic flow.

subcritical mass [NUCLEO] A piece of fissionable material having an effective multiplication constant of less than one, so that it does not give rise to a self-supporting chain reaction.

subcritical reactor [NUCLEO] A reactor having an effective multiplication constant of less than one, so that a self-supporting chain reaction cannot be maintained. Also known as subcritical assembly; teaching reactor.

subdivided capacitor [ELEC] Capacitor in which several capacitors known as sections are mounted so that they may be used individually or in combination.

subdivision graph [MATH] A graph which can be obtained from a given graph by breaking up each edge into one or more segments by inserting intermediate vertices between its two ends.

subdwarf star [ASTRON] An intermediate star type; luminosity is between that of main sequence stars and the white dwarf stars on the Hertzsprung-Russell diagram; spectral classes F, G, and K are most numerous.

subfield [MATH] **1.** A subset of a field which itself forms a field relative to the same operations. **2.** A subfield (of sets) is any field (of sets) contained in some given field of sets.

subgiant star [ASTRON] A member of the family of stars whose luminosity is intermediate between giants and the main sequence in the Hertzsprung-Russell diagram; spectral classes G and K are most frequent.

subgraph [MATH] A graph contained in a given graph which has as its vertices some subset of the vertices of the original.

subgroup [MATH] A subset N of a group G which is itself a group relative to the same operation.

subharmonic [PHYS] A sinusoidal quantity having a frequency that is an integral submultiple of the frequency of some other sinusoidal quantity to which it is referred; a third subharmonic would be one-third the fundamental or reference frequency.

subharmonic function [MATH] A continuous function is subharmonic in a region R of the plane if its value at any point z_0 of R is less than or equal to its integral along a circle centered at z_0.

subjective probability *See* personal probability.

sublevel *See* subshell.

sublimation [THERMO] The process by which solids are transformed directly to the vapor state or vice versa without passing through the liquid phase.

sublimation cooling [THERMO] Cooling caused by the extraction of energy to produce sublimation.

sublimation curve [THERMO] A graph of the vapor pressure of a solid as a function of temperature.

sublimation energy [THERMO] The increase in internal energy when a unit mass, or 1 mole, of a solid is converted into a gas, at constant pressure and temperature.

sublimation point [THERMO] The temperature at which the vapor pressure of the solid phase of a compound is equal to the total pressure of the gas phase in contact with it; analogous to the boiling point of a liquid.

sublimation pressure [THERMO] The vapor pressure of a solid.

sublime [THERMO] To change from the solid to the gaseous state without passing through the liquid phase.

submetallic [OPTICS] Referring to a luster intermediate between metallic and nonmetallic, such as exhibited by the mineral chromite.

submillimeter wave [ELECTROMAG] An electromagnetic wave whose wavelength is less than 1 millimeter, corresponding to frequencies above 300 gigahertz.

submodule [MATH] A subset N of a module M over a ring R such that, if x and y are in N and a is in R, then $x + y$ and ax are in N, so that N is also a module over R.

submultiple resonance [PHYS] Resonance at a frequency that is a submultiple of the frequency of the exciting impulses.

subnormal operator [MATH] An operator A on a Hilbert

space **H** is said to be subnormal if there exists a normal operator B on a Hilbert space **K** such that **H** is a subspace of **K**, the subspace **H** is invariant under the operator B, and the restriction of B to **H** coincides with A.

suboptimization [SYS ENG] The process of fulfilling or optimizing some chosen objective which is an integral part of a broader objective; usually the broad objective and lower-level objective are different.

subpopulation [STAT] A subset of population. Also known as stratum.

subring [MATH] **1.** A subset I of a ring R where I is also a ring relative to the operations of R. **2.** A subring (of sets) is any ring (of sets) contained in some given ring (of sets).

subsampling [STAT] Taking samples from a sample of a population.

subscriber set *See* subset.

subscript [SCI TECH] A letter or symbol written below, and usually to the right, of another symbol for any of various purposes, such as to identify a particular element or elements of a set, to denote a constant value of a variable, or, in a chemical formula, to indicate the number of atoms of a particular kind in a molecule.

subsequence [MATH] A subsequence of a given sequence is any sequence all of whose entries appear in the original sequence and in the same manner of succession.

subset [MATH] A subset A of a set B is a set all of whose elements are included in B.

subshell [ATOM PHYS] Electrons of an atom within the same shell (energy level) and having the same azimuthal quantum numbers. Also known as sublevel.

subsonic [ACOUS] *See* infrasonic. [PHYS] Of, pertaining to, or dealing with speeds less than acoustic velocity, as in subsonic aerodynamics.

subsonic flow [FL MECH] Flow of a fluid at a speed less than that of the speed of sound in the fluid. Also known as subcritical flow.

subsonic speed [FL MECH] A speed relative to surrounding fluid less than that of the speed of sound in the same fluid.

subspace [MATH] A subset of a space which, in the appropriate context, is a space in its own right.

substance [PHYS] Tangible material, occurring in macroscopic amounts.

substitution method [PHYS] Any method of measurement, such as substitution weighing, in which a quantity is determined by substituting for it a known quantity which produces the same effect.

substitution weighing [MECH] A method of weighing to allow for differences in lengths of the balance arms, in which the object to be weighed is first balanced against a counterpoise, and the known weights needed to balance the same counterpoise are then determined. Also known as counterpoise method.

substrate [ELECTR] The physical material on which a microcircuit is fabricated; used primarily for mechanical support and insulating purposes, as with ceramic, plastic, and glass substrates; however, semiconductor and ferrite substrates may also provide useful electrical functions.

subtend [MATH] A line segment or an arc of a circle subtends an angle with vertex at a specified point if the end points of the line segment or arc lie on the sides of the angle.

subtraction [MATH] The addition of one quantity with the negative of another; in a system with an additive operation this is formally the sum of one element with the additive inverse of another.

subtractive primaries [OPTICS] The three colors, usually yellow, magenta, and cyan (greenish-blue), which are mixed together in a subtractive process.

subtractive process [OPTICS] The process of producing colors by mixing absorbing media or filters of subtractive primary colors.

subtrahend [MATH] A quantity which is to be subtracted from another given quantity.

successive approximations [MATH] Any method of solving a problem in which an approximate solution is first calculated, this solution is then used in computing an improved approximation, and the process is repeated as many times as desired.

Sucksmith ring balance [ENG] A magnetic balance in which the specimen is rigidly suspended from a phosphor bronze ring carrying two mirrors that convert small deflections of the specimen in a nonuniform magnetic field into large deflections of a light beam; used chiefly to measure paramagnetic susceptibility.

suction wave *See* rarefaction wave.

sudden commencement [GEOPHYS] Magnetic storms which start suddenly (within a few seconds) and simultaneously all over the earth.

sudden ionospheric disturbance [GEOPHYS] A complex combination of sudden changes in the condition of the ionosphere following the appearance of solar flares, and the effects of these changes. Abbreviated SID.

sufficiency [STAT] Condition of an estimator that uses all the information about the population parameter contained in the sample observations.

Suhl amplifier [SOLID STATE] A parametric microwave amplifier which utilizes the instability of certain spin waves in a ferromagnetic material subjected to intense microwave fields.

Suhl effect [ELECTR] When a strong transverse magnetic field is applied to an n-type semiconducting filament, holes injected into the filament are deflected to the surface, where they may recombine rapidly with electrons or be withdrawn by a probe.

sulfur [CHEM] A nonmetallic element in group VIa, symbol S, atomic number 16, atomic weight 32.064, existing in a crystalline or amorphous form and in four stable isotopes; used as a chemical intermediate and fungicide, and in rubber vulcanization.

sulfur-35 [NUC PHYS] Radioactive sulfur with mass number 35; radiotoxic, with 87.1-day half-life, β-radiation; derived from pile irradiation; used as a tracer to study chemical reactions, engine wear, and protein metabolism.

sum [MATH] **1.** The addition of numbers or mathematical objects in context. **2.** The sum of an infinite series is the limit of the sequence consisting of all partial sums of the series.

summability method [MATH] A method, such as Hölder summation or Cesaro summation, of attributing a sum to a divergent series by using some process to average the terms in the series.

summation convention [MATH] An abbreviated notation used particularly in tensor analysis and relativity theory, in which a product of tensors is to be summed over all possible values of any index which appears twice in the expression.

summation tone [ACOUS] Combination tone, heard under certain circumstances, whose pitch corresponds to a frequency equal to the sum of the frequencies of the two components.

summer solstice [ASTRON] **1.** The sun's position on the ecliptic when it reaches its greatest northern declination. Also known as first point of Cancer. **2.** The date, about June 21, on which the sun has its greatest northern declination.

SULFUR

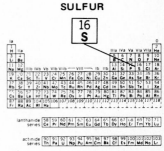

Periodic table of the chemical elements showing the position of sulfur.

summing amplifier [ELECTR] An amplifier that delivers an output voltage which is proportional to the sum of two or more input voltages or currents.

summing network [ELEC] A passive electric network whose output voltage is proportional to the sum of two or more input voltages. Also known as summation network.

sum of matrices [MATH] The sum $A + B$ of two matrices A and B, with the same number of rows and columns, is the matrix whose element c_{ij} in row i and column j is the sum of corresponding elements a_{ij} in A and b_{ij} in B.

sum of states *See* partition function.

sum over states *See* partition function.

sum rule [QUANT MECH] A formula which sets some quantity equal to the sum, over all the states of a system, of another quantity, usually involving the square of the magnitude of the matrix element of some operator between a given state and the state being summed over.

sun [ASTRON] The star about which the earth revolves; it is a globe of gas 1.4×10^6 kilometers in diameter, held together by its own gravity; thermonuclear reactions take place in the deep interior of the sun converting hydrogen into helium releasing energy which streams out. Also known as Sol.

sun-pumped laser [OPTICS] A continuous-wave laser in which pumping is achieved by concentrating the energy of the sun on the laser crystal with a parabolic mirror. Also known as solar-excited laser.

sunshine unit *See* strontium unit.

sunspot [ASTRON] A dark area in the photosphere of the sun caused by a lowered surface temperature.

sunspot cycle [ASTRON] Variation of the size and number of sunspots in an 11-year cycle which is shared by all other forms of solar activity.

sunspot maximum [ASTRON] The time in the solar cycle when the number of sunspots reaches a maximum value.

sunspot number *See* relative sunspot number.

sunspot relative number *See* relative sunspot number.

sun's way [ASTRON] The path of the solar system through space.

sup *See* least upper bound.

superaerodynamics [FL MECH] That branch of gas dynamics dealing with the flow of gases at such low density that the molecular mean free path is not negligibly small; under these conditions the gas no longer behaves as a continuous fluid.

supercompressibility factor *See* compressibility factor.

superconducting circuit [CRYO] An electric circuit having elements which are in a superconducting state at least part of the time, such as a cryotron.

superconducting device *See* cryogenic device.

superconducting magnet [CRYO] An electromagnet whose coils are made of a type II superconductor with a high transition temperature and extremely high critical field, such as niobium tin, Nb_3Sn; it is capable of generating magnetic fields of 100,000 oersteds and more with no steady power dissipation.

superconducting material *See* superconductor.

superconducting quantum interference device [ELECTR] A superconducting ring that couples with one or two Josephson junctions; applications include high-sensitivity magnetometers, near-magnetic-field antennas, and measurement of very small currents or voltages. Abbreviated SQUID.

superconducting thin film [CRYO] A thin film of indium, tin, or other superconducting element, used as a cryogenic switching or storage device, as in a thin-film cryotron.

superconductivity [SOLID STATE] A property of many metals, alloys, and chemical compounds at temperatures near absolute zero by virtue of which their electrical resistivity vanishes and they become strongly diamagnetic.

superconductor [SOLID STATE] Any material capable of exhibiting superconductivity; examples include iridium, lead, mercury, niobium, tin, tantalum, vanadium, and many alloys. Also known as cryogenic conductor; superconducting material.

supercooling [THERMO] Cooling of a substance below the temperature at which a change of state would ordinarily take place without such a change of state occurring, for example, the cooling of a liquid below its freezing point without freezing taking place; this results in a metastable state.

supercritical [NUCLEO] Having an effective multiplication constant greater than 1, so that the rate of reaction rises rapidly in a nuclear reactor.

supercritical flow *See* supersonic flow.

supercritical reactor [NUCLEO] A nuclear reactor in which the effective multiplication constant is greater than 1 and consequently a reactor that is increasing its power level; if uncontrolled, a supercritical reactor will undergo a sudden and dangerous rise in power level.

supercurrent [SOLID STATE] In the two-fluid model of superconductivity, the current arising from motion of superconducting electrons, in contrast to the normal current.

superelastic collision [ATOM PHYS] A collision between an atom in an excited state and an electron in which the atom returns to the ground state and nearly all of its excitation energy is transferred to the electron.

superexchange [SOLID STATE] A phenomenon in which two electrons from a double negative ion (such as oxygen) in a solid go to different positive ions and couple with their spins, giving rise to a strong antiferromagnetic coupling between the positive ions, which are too far apart to have a direct exchange interaction.

superfluid [PHYS] A collection of particles which obey Bose-Einstein statistics and are all in the lowest energy state allowed by quantum mechanics, having zero entropy and zero resistance to motion; examples are a fraction of the atoms in liquid helium II and a fraction of the pairs of electrons in a superconductor.

superfluidity [CRYO] The frictionless flow of liquid helium at temperatures very close to absolute zero through holes as small as 10^{-7} centimeter in diameter, and for particle velocities below a few centimeters per second.

supergalaxy [ASTRON] A hypothetical very large group of galaxies which together fill an ellipsoidal space.

supergiant star [ASTRON] A member of the family containing the intrinsically brightest stars, populating the top of the Hertzsprung-Russell diagram; supergiant stars occur at all temperatures from 30,000 to 3000 K and have luminosities from 10^4 to 10^6 times that of the sun; the star Betelgeuse is an example.

supergravity [PHYS] A supersymmetry which is used to unify general relativity and quantum theory; it is formed by adding to the Poincaré group, as a symmetry of space-time, four new generators that behave as spinors and vary as the square root of the translations.

superharmonic function [MATH] A continuous complex function f whose value at a point z_0 exceeds its average values computed by the integral of f around a circle centered at z_0.

superheated vapor [THERMO] A vapor that has been heated above its boiling point.

superheating [THERMO] Heating of a substance above the temperature at which a change of state would ordinarily take place without such a change of state occurring, for example, the heating of a liquid above its boiling point without boiling taking place; this results in a metastable state.

superionic conductor [SOLID STATE] An ionic solid whose ionic conductivity is extremely high, on the order of 100 times that normally observed.

superior conjunction [ASTRON] A conjunction when an astronomical body is opposite the earth on the other side of the sun.

superior mirage [OPTICS] A spurious image of an object formed above the object's position by abnormal refraction conditions; opposite to an inferior mirage.

superior planet [ASTRON] Any of the planets that are farther than the earth from the sun; includes Mars, Jupiter, Saturn, Uranus, Neptune, and Pluto.

superior transit *See* upper transit.

superlattice [ELECTR] A structure consisting of alternating layers of two different semiconductor materials, each several nanometers thick. [SOLID STATE] An ordered arrangement of atoms in a solid solution which forms a lattice superimposed on the normal solid solution lattice. Also known as superstructure.

superleak *See* lambda leak.

supermassive star [ASTRON] A star with a mass exceeding about 50 times that of the sun.

supermode laser [OPTICS] Frequency-modulated laser, the output of which is passed through a second phase modulator driven 180° out of phase and with the same modulation index as the first modulator; all of the energy of the previously existing laser modes is compressed into a single frequency with nearly the full power of the laser concentrated in that signal.

supermultiplet [QUANT MECH] A set of quantum-mechanical states each of which has the same value of some fundamental quantum numbers and differs from the other members of the set by other quantum numbers, which take values from a range of numbers dictated by the fundamental quantum numbers.

supernova [ASTRON] A star that suddenly bursts into very great brilliance as a result of its blowing up; it is orders of magnitude brighter than a nova.

supernumerary rainbow [OPTICS] One of a set of weakly colored rainbow arcs sometimes discernible inside a primary rainbow; they are of smaller angular width and fade toward the common center.

superparamagnetism *See* collective paramagnetism.

superposition [MATH] The principle of superposition states that any given geometric figure in a euclidean space can be so moved about as not to change its size or shape. [PHYS] Addition of phenomena when the sum of two physically realizable disturbances is also physically realizable; for example, sound waves are superposable in this sense, but shock waves are not.

superposition integral [CONT SYS] An integral which expresses the response of a linear system to some input in terms of the impulse response or step response of the system; it may be thought of as the summation of the responses to impulses or step functions occurring at various times.

superposition principle *See* principle of superposition.

superposition theorem *See* principle of superposition.

supersaturation [PHYS CHEM] The condition existing in a solution when it contains more solute than is needed to cause saturation. Also known as supersolubility.

super-Schmidt telescope [OPTICS] A type of Schmidt system that has a compound corrector plate consisting of a pair of opposing meniscus lenses and an achromatic doublet.

supersolubility *See* supersaturation.

supersolvable group [MATH] A group G which has subgroups $G_0, G_1, \ldots, G_n$ where $G_0 = G$, G_n is the identity element alone, each G_i is a normal subgroup of G_{i-1}, and the quotient group G_{i-1}/G_i is a cyclic group of prime order.

supersonic [ACOUS] *See* ultrasonic. [PHYS] Of, pertaining to, or dealing with speeds greater than the speed of sound.

supersonic aerodynamics [FL MECH] The study of aerodynamics of supersonic speeds.

supersonic flow [FL MECH] Flow of a fluid over a body at speeds greater than the speed of sound in the fluid, and in which the shock waves start at the surface of the body. Also known as supercritical flow.

superspace [PHYS] The space of all three-geometries on a three-manifold used in discussions of quantum gravity.

superstructure *See* superlattice.

supersymmetry [PARTIC PHYS] A generalization of previously known symmetries of elementary particles to new kinds of supermultiplets that include both bosons and fermions; it is based on graded Lie algebras rather than on Lie algebras.

superturbulent flow [FL MECH] The flow of water in which the energy loss by friction is so great that Reynolds criterion for the transition of laminar to turbulent flow does not apply.

supervoltage [ELEC] A voltage in the range of 500 to 2000 kilovolts, used for some x-ray tubes.

supplementary angle [MATH] One angle is supplementary to another angle if their sum is 180°.

supplementary condition [QUANT MECH] In a quantized field theory, an auxiliary condition required of a state vector to make it correspond to an actual state.

support [MATH] The support of a real-valued function f on a topological space is the closure of the set of points where f is not zero.

supported end [MECH] An end of a structure, such as a beam, whose position is fixed but whose orientation may vary; for example, an end supported on a knife-edge.

suppressor *See* suppressor grid.

suppressor grid [ELECTR] A grid placed between two positive electrodes in an electron tube primarily to reduce the flow of secondary electrons from one electrode to the other; it is usually used between the screen grid and the anode. Also known as suppressor.

supremum *See* least upper bound.

surface [MATH] A subset of three-space consisting of those points whose cartesian coordinates x, y, and z satisfy equations of the form $x = f(u,v)$, $y = g(u,v)$, $z = h(u,v)$, where f, g, and h are differentiable real-valued functions of two parameters u and v which take real values and vary freely in some domain.

surface acoustic wave [ACOUS] A sound wave that propagates along and is bound to the surface of a solid; ordinarily it contains both compressional and shear components. Abbreviated SAW.

surface acoustic wave device [ELECTR] Any device, such as a filter, resonator, or oscillator, which employs surface acoustic waves with frequencies in the range 10^7–10^9 hertz, traveling on the optically polished surface of a piezoelectric substrate, to process electronic signals.

surface acoustic wave filter [ELECTR] An electric filter consisting of a piezoelectric bar with a polished surface along which surface acoustic waves can propagate, and on which are deposited metallic transducers, one of which is connected, via thermocompression-bonded leads, to the electric source, while the other drives the load.

surface barrier [ELECTR] A potential barrier formed at a

SURFACE BARRIER

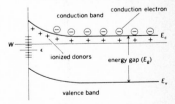

Energy diagram of a surface barrier as employed in an n-type semiconductor. W = Fermi level; ϵ = highest energy level of surface state filled by electrons when surface is electrically neutral; E_c = energy of bottom of conduction band; E_v = energy of top of valence band.

surface of a semiconductor by the trapping of carriers at the surface.

surface-barrier diode [ELECTR] A diode utilizing thin-surface layers, formed either by deposition of metal films or by surface diffusion, to serve as a rectifying junction.

surface-barrier transistor [ELECTR] A transistor in which the emitter and collector are formed on opposite sides of a semiconductor wafer, usually made of n-type germanium, by training two jets of electrolyte against its opposite surfaces to etch and then electroplate the surfaces.

surface-charge transistor [ELECTR] An integrated-circuit transistor element based on controlling the transfer of stored electric charges along the surface of a semiconductor.

surface-coated mirror [OPTICS] A mirror produced by depositing a thin film of highly reflective material on a glass surface.

surface color [OPTICS] The color of light reflected from the surface of a body; in contrast to the color of light that is reflected after penetrating some distance into the body.

surface density [PHYS] The quantity of anything distributed over a surface per unit area of surface.

surface drag [FL MECH] That portion of drag which is caused by skin friction.

surface duct [GEOPHYS] Atmospheric duct for which the lower boundary is the surface of the earth.

surface energy [FL MECH] The energy per unit area of an exposed surface of a liquid; generally greater than the surface tension, which equals the free energy per unit surface.

surface force [MECH] An external force which acts only on the surface of a body; an example is the force exerted by another object with which the body is in contact.

surface friction [GEOPHYS] The drag or skin friction of the earth on the atmosphere, usually expressed in terms of the shearing stress of the wind on the earth's surface.

surface harmonics *See* spherical surface harmonics.

surface integral [MATH] The integral of a function of several variables with respect to surface area over a surface in the domain of the function.

surface leakage [ELEC] The passage of current over the surface of an insulator.

surface magnetic wave [ELECTROMAG] A magnetostatic wave that can be propagated on the surface of a magnetic material, as on a slab of yttrium iron garnet.

surface of discontinuity [FL MECH] A surface within a fluid across which there is a discontinuity in fluid velocity; often generated in the wake of a body moving relative to the fluid.

surface of revolution [MATH] A surface realized by rotating a planar curve about some axis in its plane.

surface passivation [ELECTR] A method of coating the surface of a p-type wafer for a diffused junction transistor with an oxide compound, such as silicon oxide, to prevent penetration of the impurity in undesired regions.

surface pressure [FL MECH] The reduction in the surface tension of a liquid resulting from the presence of a surface film. [PHYS] *See* film pressure.

surface recombination rate [SOLID STATE] The rate at which free electrons and holes at the surface of a semiconductor recombine, thus neutralizing each other.

surface recombination velocity [SOLID STATE] A measure of the rate of recombination between electrons and holes at the surface of a semiconductor, equal to the component of the electron or hole current density normal to the surface divided by the excess electron or hole volume charge density close to the surface.

surface resistivity [ELEC] The electric resistance of the sur-

SURFACE-BARRIER TRANSISTOR

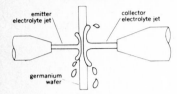

emitter electrolyte jet

collector electrolyte jet

germanium wafer

Technique for making surface-barrier transistor.

face of an insulator, measured between the opposite sides of a square on the surface; the value in ohms is independent of the size of the square and the thickness of the surface film.

surface state [SOLID STATE] An electron state in a semiconductor whose wave function is restricted to a layer near the surface.

surface tension [FL MECH] The force acting on the surface of a liquid, tending to minimize the area of the surface; quantitatively, the force that appears to act across a line of unit length on the surface. Also known as interfacial force; interfacial tension; surface tensity.

surface tension number [FL MECH] A dimensionless number used in studying mass transfer in packed columns equal to the square of the dynamic viscosity of a fluid times the length of the perimeter of a packing element, divided by the product of the surface area of the packing element, the surface tension, and the density of the liquid. Symbol T_s.

surface tensity *See* surface tension.

surface thermometer [ENG] A thermometer, mounted in a bucket, used to measure the temperature of the sea surface.

surface wave [ELECTROMAG] A wave that can travel along an interface between two different mediums without radiation; the interface must be essentially straight in the direction of propagation; the commonest interface used is that between air and the surface of a circular wire. [FL MECH] A wave that distorts the free surface that separates two fluid phases, usually a liquid and a gas. [MECH] *See* Rayleigh wave.

surge [ELEC] A momentary large increase in the current or voltage in an electric circuit. [FL MECH] A wave at the free surface of a liquid generated by the motion of a vertical wall, having a change in the height of the surface across the wavefront and violent eddy motion at the wavefront.

surge admittance [ELEC] Reciprocal of surge impedance.

surge generator [ELEC] A device for producing high-voltage pulses, usually by charging capacitors in parallel and discharging them in series.

surge impedance *See* characteristic impedance.

surjection [MATH] A mapping f from a set A to a set B such that for every element b of B there is an element a of A such that $f(a) = b$. Also known as surjective mapping.

surjective mapping *See* surjection.

survey [NUCLEO] Measurement of radiation in the vicinity of a nuclear reactor or other source.

survey design *See* sample design.

survey foot [MECH] A unit of length, used by the U.S. Coast and Geodetic Survey, equal to 12/39.37 meter, or approximately 1.000002 feet.

survey instrument [NUCLEO] A portable instrument used to detect and measure radiation. Also known as survey meter.

survey meter *See* survey instrument.

survival curve [NUCLEO] The curve obtained by plotting the number or percentage of organisms surviving at a given time against the dose of radiation, or the number surviving at different intervals after a particular dose of radiation.

susceptance [ELEC] The imaginary component of admittance.

susceptibility *See* electric susceptibility; magnetic susceptibility.

susceptometer [ENG] An instrument that measures paramagnetic, diamagnetic, or ferromagnetic susceptibility.

suspended solids *See* suspension.

suspended transformation [THERMO] The cessation of change before true equilibrium is reached, or the failure of a system to change immediately after a change in conditions,

SURGE

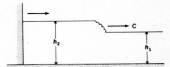

Sketch of a surge wave (fl mech). C = speed of propagation; h_1 = height of water before passage of surge; h_2 = height of water after passage of surge.

such as in supercooling and other forms of metastable equilibrium.

sustained oscillation [CONT SYS] Continued oscillation due to insufficient attenuation in the feedback path. [PHYS] Oscillation in which forces outside the system, but controlled by the system, maintain a periodic oscillation of the system at a period or frequency that is nearly the natural period of the system.

Sutherland's formula [STAT MECH] **1.** The absolute viscosity of a gas is proportional to $T^{3/2}/(C + T)$, where T is the absolute temperature and C is a constant for a given gas. **2.** The mean free path of a molecule in a gas is proportional to $1/nd\sqrt{1 + C/T}$, where n is the number of molecules per unit volume, d is the diameter of a molecule, T is the absolute temperature, and C is a constant.

svedberg [PHYS CHEM] A unit of sedimentation coefficient, equal to 10^{-13} second.

Svedberg equation [STAT MECH] An equation which states that the amplitude of vibration of a particle which exhibits Brownian motion is proportional to its period.

sverdrup [FL MECH] A unit of volume transport equal to 1,000,000 cubic meters per second.

SW *See* switch.

Swann bands [ASTROPHYS] Particular bands seen in the visible spectra of comets; they arise from the presence of dimeric carbon (C_2) in the comet's tail.

swastika [MATH] A plane curve whose equation in cartesian coordinates x and y is $y^4 - x^4 = xy$.

S wave [GEOPHYS] A seismic body wave propagated in the crust or mantle of the earth by a shearing motion of material; speed is 3.0–4.0 kilometers per second in the crust and 4.4–4.6 in the mantle. Also known as distortional wave; equivoluminal wave; rotational wave; secondary wave; shake wave; shear wave; tangential wave; transverse wave.

sweep [ELECTR] **1.** The steady movement of the electron beam across the screen of a cathode-ray tube, producing a steady bright line when no signal is present; the line is straight for a linear sweep and circular for a circular sweep. **2.** The steady change in the output frequency of a signal generator from one limit of its range to the other.

sweep amplifier [ELECTR] An amplifier used with a cathode-ray tube, such as in a television receiver or cathode-ray oscilloscope, to amplify the sawtooth output voltage of the sweep oscillator, to shape the waveform for the deflection circuits of a television picture tube, or to provide balanced signals to the deflection plates.

sweep circuit [ELECTR] The sweep oscillator, sweep amplifier, and any other stage used to produce the deflection voltage or current for a cathode-ray tube. Also known as scanning circuit.

sweep generator Also known as sweep oscillator. [ELECTR] **1.** An electronic circuit that generates a voltage or current, usually recurrent, as a prescribed function of time; the resulting waveform is used as a time base to be applied to the deflection system of an electron-beam device, such as a cathode-ray tube. Also known as time-base generator; timing-axis oscillator. **2.** A test instrument that generates a radio-frequency voltage whose frequency varies back and forth through a given frequency range at a rapid constant rate; used to produce an input signal for circuits or devices whose frequency response is to be observed on an oscilloscope.

sweep oscillator *See* sweep generator.

sweep voltage [ELECTR] Periodically varying voltage applied to the deflection plates of a cathode-ray tube to give a beam displacement that is a function of time, frequency, or other data base.

swimming-pool reactor   *See* pool reactor.

swing   [ELEC] Variation in frequency or amplitude of an electrical quantity.

switch   [ELEC] A manual or mechanically actuated device for making, breaking, or changing the connections in an electric circuit. Also known as electric switch. Symbolized SW.

switching   [ELEC] Making, breaking, or changing the connections in an electrical circuit.

switching diode   [ELECTR] A crystal diode that provides essentially the same function as a switch; below a specified applied voltage it has high resistance corresponding to an open switch, while above that voltage it suddenly changes to the low resistance of a closed switch.

switching gate   [ELECTR] An electronic circuit in which an output having constant amplitude is registered if a particular combination of input signals exists; examples are the OR, AND, NOT, and INHIBIT circuits. Also known as logical gate.

switching surface   [CONT SYS] In feedback control systems employing bang-bang control laws, the surface in state space which separates a region of maximum control effort from one of minimum control effort.

switching transistor   [ELECTR] A transistor designed for on/off switching operation.

SXAPS   *See* soft x-ray appearance potential spectroscopy.

Sylow subgroup   [MATH] A subgroup H of a given group G such that the order of H is p^n, where p is a prime and n is an integer, and p^n is the highest power of p dividing the order of G.

Sylvester's theorem   [MATH] **1.** If A is a matrix with distinct eigenvalues $\lambda_1, \ldots, \lambda_n$, then any analytic function $f(A)$ can be realized from the $\lambda_i, f(\lambda_i)$, and the matrices $A - \lambda_i I$, where I is the identity matrix. **2.** If k is an ordered field and E is an n-dimensional vector space over k with a nondegenerate symmetric form g, then there exists a nonnegative integer r such that for any orthogonal basis $(v_1, \ldots, v_n)$ of E precisely r among the n elements $g(v_1,v_1), \ldots, g(v_n,v_n)$ are greater than 0, and $n - r$ among these elements are less than 0.

symbiotic objects   [ASTRON] Stars whose spectra have characteristics of two disparate spectral classes.

symbolic age of neutrons   *See* Fermi age.

symbolic logic   [MATH] The formal study of symbolism and its use in the foundations of mathematical logic.

symmetrical achromat lens   [OPTICS] An older type of camera lens in which two positive achromatic meniscus lenses are symmetrically arranged about the stop.

symmetrical alternating quantity   [PHYS] Alternating quantity of which all values separated by a half period have the same magnitude but opposite sign.

symmetrical avalanche rectifier   [ELECTR] Avalanche rectifier that can be triggered in either direction, after which it has a low impedance in the triggered direction.

symmetrical band-pass filter   [ELECTR] A band-pass filter whose attenuation as a function of frequency is symmetrical about a frequency at the center of the pass band.

symmetrical deflection   [ELECTR] A type of electrostatic deflection in which voltages that are equal in magnitude and opposite in sign are applied to the two deflector plates.

symmetrical distribution   [STAT] A distribution in which observations equidistant from the central maximum have the same frequency.

symmetrical inductive diaphragm   [ELECTROMAG] A waveguide diaphragm which consists of two plates that leave a

SYMMETRICAL ACHROMAT LENS

Geometry of symmetrical achromat lens.

space at the center of the waveguide, and which introduces an inductance in the waveguide.

symmetrical lens [OPTICS] A lens system consisting of two parts, each of which is the mirror image of the other.

symmetrical transducer [ELECTR] A transducer is symmetrical with respect to a specified pair of terminations when the interchange of that pair of terminations will not affect the transmission.

symmetric connection [MATH] An affine connection whose components satisfy the equation $\Gamma^i_{jk} = \Gamma^i_{kj}$, where the components may be defined by the equation

$$\delta A^i = -\sum \Gamma^i_{jk} A^j \, dx^k,$$

relating the vector A^i at a point with coordinates x^i and the associated vector $A^i + \delta A^i$ at a point with coordinates $x^i + dx^i$.

symmetric design [MATH] A balanced incomplete block design in which the number b of blocks equals the number v of elements arranged among the blocks.

symmetric difference [MATH] The symmetric difference of two sets consists of all points in one or the other of the sets but not in both.

symmetric form [MATH] A bilinear form f which is unchanged under interchange of its independent variables; that is, $f(x,y) = f(y,x)$ for all values of the independent variables x and y.

symmetric function [MATH] A function whose value is unchanged for any permutation of its variables.

symmetric group [MATH] The group consisting of all permutations of a finite set of symbols.

symmetric matrix [MATH] A matrix which equals its transpose.

symmetric relation [MATH] The property of a relation on a set that requires y to be related to x whenever x is related to y.

symmetric space [MATH] A differentiable manifold which has a differentiable multiplication operation that behaves similarly to the multiplication of a complex number and its conjugate.

symmetric top molecule [PHYS CHEM] A nonlinear molecule which has one and only one axis of threefold or higher symmetry.

symmetry [MATH] A geometric object G has this property relative to some configuration S of its points if S determines two pieces of G which can be reflected onto each other through S. [PHYS] *See* invariance.

symmetry axis *See* axis of symmetry; rotation axis.

symmetry center *See* center of symmetry.

symmetry class *See* crystal class.

symmetry element [CRYSTAL] 1. Some combination of rotations and reflections and translations which brings a crystal into a position that cannot be distinguished from its original position. Also known as symmetry operation; symmetry transformation. 2. The rotational axes, mirror planes, and center of symmetry characteristic of a given crystal.

symmetry function *See* symmetry transformation.

symmetry group [MATH] A group composed of all rigid motions or similarity transformations of some geometric object onto itself.

symmetry law *See* invariance principle.

symmetry operation *See* symmetry element.

symmetry operation of the second kind [CRYSTAL] A combination of rotations, reflections, and translations that brings a crystal into a position that is a mirror image of its original position.

symmetry plane   *See* plane of mirror symmetry.

symmetry principle   [MATH] The centroid of a geometrical figure (line, area, or volume) is at a point on a line or plane of symmetry of the figure. [PHYS] *See* invariance principle.

symmetry transformation   [CRYSTAL] *See* symmetry element. [MATH] A rigid motion sending a geometric object onto itself; examples are rotations and, for the case of a polygon, permutations of the vertices. Also known as symmetry function.

sympathetic vibration   [PHYS] The driving of a mechanical or acoustical system at its resonant frequency by energy from an adjacent system vibrating at the same frequency.

symplectic group of dimension n   [MATH] The Lie group of symplectic transformations on an n-dimensional vector space over the quaternions. Symbolized Sp(n).

symplectic matrix   [MATH] A $2n \times 2n$ matrix A which satisfies the equation $A^T J A = J$, where A^T is the transpose of A, J, is the matrix

$$\begin{pmatrix} O_n & I_n \\ I_n & O_n \end{pmatrix},$$

I_n is the unit $n \times n$ matrix, and O_n is the $n \times n$ matrix each of whose elements is 0.

symplectic transformation   [MATH] A linear transformation of a vector space over the quaternions that leaves the lengths of vectors unchanged.

synchrocyclotron   [NUCLEO] A circular particle accelerator for accelerating protons, deuterons, or alpha particles, in which the frequency of the accelerating voltage is modulated to maintain synchronism with the frequency of the particles which spiral outward and attain energies at which the relativistic mass increase becomes significant. Also known as frequency-modulated cyclotron; synchrophasotron.

synchronism   [PHYS] Condition of two periodic quantities which have the same frequency, and whose phase difference is either constant or varies around a constant average value.

synchronized blocking oscillator   [ELECTR] A blocking oscillator which is synchronized with pulses occurring at a rate slightly faster than its own natural frequency.

synchronous detector   *See* cross-correlator.

synchrophasotron   *See* synchrocyclotron.

synchroscope   [ELECTR] A cathode-ray oscilloscope designed to show a short-duration pulse by using a fast sweep that is synchronized with the pulse signal to be observed. [ENG] An instrument for indicating whether two periodic quantities are synchronous; the indicator may be a rotating-pointer device or a cathode-ray oscilloscope providing a rotating pattern; the position of the rotating pointer is a measure of the instantaneous phase difference between the quantities.

synchrotron   [NUCLEO] A device for accelerating electrons or protons in closed orbits in which the frequency of the accelerating voltage is varied (or held constant in the case of electrons) and the strength of the magnetic field is varied so as to keep the orbit radius constant.

synchrotron process   [ELECTROMAG] The emission of electromagnetic radiation by relativistic electrons orbiting in a magnetic field.

synchrotron radiation   [ELECTROMAG] Electromagnetic radiation generated by the acceleration of charged relativistic particles, usually electrons, in a magnetic field.

synclastic   [MATH] Property of a surface or portion of a surface for which the centers of curvature of the principal sections at each point lie on the same side of the surface.

SYNCHROCYCLOTRON

View of the vacuum tank and upper magnet coil of the 184-inch (4.67-meter) synchrocyclotron at the University of California. Pumps are in the foreground; target-handling equipment is at the lower left and the radio-frequency oscillator housing is at the right-hand edge. Note the magnet return yoke and the shielding at the top of the picture. *(University of California Lawrence Berkeley Laboratory)*

synodic [ASTRON] Referring to conjunction of celestial bodies.

synodic month [ASTRON] A month based on the moon's phases.

synodic period [ASTRON] The time period between two successive astronomical conjunctions of the same celestial objects.

synoptic correlation *See* Eulerian correlation.

synthetic division [MATH] A long division process for dividing a polynomial $p(x)$ by a polynomial $(x - a)$ where only the coefficients of these polynomials are used.

syntony [ELEC] Condition in which two oscillating circuits have the same resonant frequency.

system analysis [CONT SYS] The use of mathematics to determine how a set of interconnected components whose individual characteristics are known will behave in response to a given input or set of inputs.

systematic distortion [ELEC] Periodic or constant distortion, such as bias or characteristic distortion; the direct opposite of fortuitous distortion.

systematic error [STAT] An error which results from some bias in the measurement process and is not due to chance, in contrast to random error.

systematic sampling [SCI TECH] A sampling procedure in which a random starting point is selected and then every kth element after it becomes an item in the sample.

system bandwidth [CONT SYS] The difference between the frequencies at which the gain of a system is $\sqrt{2}/2$ (that is, 0.707) times its peak value.

system design [CONT SYS] A technique of constructing a system that performs in a specified manner, making use of available components. Also known as synthesis.

Système International d'Unités *See* International System of Units.

system engineering *See* systems engineering.

system of distinct representatives [MATH] A family of subsets S_i of a given finite set S such that the family has as many members as there are elements in S, and such that it is possible to assign each element x_i of S to a distinct subset S_i with x_i in S_i.

system response *See* response.

systems analysis [ENG] The analysis of an activity, procedure, method, technique, or business to determine what must be accomplished and how the necessary operations may best be accomplished.

systems engineering [ENG] The design of a complex interrelation of many elements (a system) to maximize an agreed-upon measure of system performance, taking into consideration all of the elements related in any way to the system, including utilization of manpower as well as the characteristics of each of the system's components. Also known as system engineering.

systems for nuclear auxiliary power *See* SNAP.

syzygy [ASTRON] **1.** One of the two points in a celestial object's orbit where it is in conjunction with or opposition to the sun. **2.** Those points in the moon's orbit where the moon, earth, and sun are in a straight line.

Szilard-Chalmers effect [NUCLEO] The breaking of a chemical bond between a radioactive atom formed in a nuclear reaction and the molecule to which the atom belonged; used in the separation of isotopes by chemical means.

T

t *See* troy system.

T *See* tera-; tesla.

Ta *See* tantalum.

table [MATH] An array or listing of computed quantities.

tablespoonful [MECH] A unit of volume used particularly in cookery, equal to 4 fluid drams or ½ fluid ounce; in the United States this is equal to approximately 14.7868 cubic centimeters, in the United Kingdom to approximately 14.2065 cubic centimeters. Abbreviated tbsp.

tabular crystal [CRYSTAL] A crystal that appears broad and flat due to two prominent parallel faces.

tabular interpolation [MATH] Method of finding from a table the values of the dependent variable for intermediate values of the independent variable.

tachyon [PARTIC PHYS] A hypothetical particle that travels faster than light, consistent with the theory of relativity.

tacnode *See* double cusp; point of osculation.

tag *See* isotopic tracer.

tail event [MATH] An element of a tail sigma field.

tail of a stochastic process [MATH] A tail of a stochastic process represented by $x(t_1)$, $x(t_2)$, . . . is the process obtained by deleting the first n terms, for some n.

tail sigma field [MATH] For a sequence of random variables $f_1, f_2, . . .$, the intersection of the nonincreasing sequence of sigma fields $\{S_n\}$, where S_n is the sigma field consisting of inverse images of Borel sets under the random sequence $(f_n(x), f_{n+1}(x), . . .)$.

talbot [OPTICS] A unit of luminous energy equal to the luminous energy carried by a luminous flux of 1 lumen during a period of 1 second.

Talbot's bands [OPTICS] A series of dark bands that appear in the spectrum of white light when a glass plate of the proper thickness is placed across one half of the aperture of a spectroscope on the side of the blue end of the spectrum.

Talbot's curve [MATH] The negative pedal of an ellipse, with eccentricity greater than $\sqrt{2}/2$, with respect to its center.

Talbot's law [OPTICS] The law that apparent brightness of an object flashing at a frequency greater than about 10 hertz is equal to its actual brightness times the ratio of the exposure time to the total time.

Tamm-Dancoff method [QUANT MECH] A method of forming an approximate wave function of a system of interacting particles, particularly nucleons and mesons, by describing it as an algebraic sum of several possible states, the number of such states determining the order of the approximation.

tamper *See* reflector.

tan *See* tangent.

tandem [ELEC] Two-terminal pair networks are in tandem when the output terminals of one network are directly connected to the input terminals of the other network.

tandem accelerator [NUCLEO] An electrostatic accelerator in which negative hydrogen ions generated in a special ion source are accelerated as they pass from ground potential up to a high-voltage terminal, both electrons are then stripped from the negative ion by passage through a very thin foil or gas cell, and the proton is again accelerated as it passes to ground potential.

tandem compensation *See* cascade compensation.

tandem connection *See* cascade connection.

tangent [MATH] **1.** A line is tangent to a curve at a fixed point P if it is the limiting position of a line passing through P and a variable point on the curve Q, as Q approaches P. **2.** The function which is the quotient of the sine function by the cosine function. Abbreviated tan. **3.** The tangent of an angle is the ratio of its sine and cosine. Abbreviated tan.

tangent bundle [MATH] The fiber bundle $T(M)$ associated to a differentiable manifold M which is composed of the points of M together with all their tangent vectors. Also known as tangent space.

tangent galvanometer [ENG] A galvanometer in which a small compass is mounted horizontally in the center of a large vertical coil of wire; the current through the coil is proportional to the tangent of the angle of deflection of the compass needle from its normal position parallel to the magnetic field of the earth.

tangential acceleration [MECH] The component of linear acceleration tangent to the path of a particle moving in a circular path.

tangential coma [OPTICS] For a lens that displays coma, the length of a tangent from the vertex of the patch of light formed in the focal plane by rays from an off-axis point to the comatic circle of rays from this point that pass near the edge of the lens.

tangential component [MATH] A component of a given vector acting at right angles to a given radius of a given circle.

tangential curvature *See* geodesic curvature.

tangential developable *See* tangent surface.

tangential focus *See* primary focus.

tangential plane *See* meridional plane.

tangential polar equation [MATH] An equation of a curve expressed in terms of the distance of a point P on the curve from a reference point O and the perpendicular distance from O to the tangent to the curve at P.

tangential stress *See* shearing stress.

tangential surface [OPTICS] A surface containing the primary foci of points in a plane perpendicular to the optical axis of an astigmatic system.

tangential velocity [MECH] **1.** The instantaneous linear velocity of a body moving in a circular path; its direction is tangential to the circular path at the point in question. **2.** The component of the velocity of a body that is perpendicular to a line from an observer or reference point to the body.

tangential wave *See* S wave.

tangent plane to a surface [MATH] The tangent plane to a surface at a point is the plane having every line in it tangent to some curve on the surface at that point.

tangent space [MATH] **1.** The vector space of all tangent vectors at a given point of a differentiable manifold. **2.** *See* tangent bundle.

tangent surface [MATH] The ruled surface generated by the tangents to a specified space curve. Also known as tangential developable.

tangent vector [MATH] A tangent vector to a point p of a differential manifold is a linear functional L on the real-valued smooth functions on the manifold satisfying the equation $L(fg) = f(p)L(g) + g(p)L(f)$ for all smooth f and g; in local coordinates, it is a linear combination, with numerical coefficients, of first partial derivatives evaluated at p.

tanh   *See* hyperbolic tangent.

tank   *See* tank circuit.

tank circuit   [ELECTR] A circuit which exhibits resonance at one or more frequencies, and which is capable of storing electric energy over a band of frequencies continuously distributed about the resonant frequency, such as a coil and capacitor in parallel. Also known as electrical resonator; tank.

tank reactor   [NUCLEO] A nuclear reactor in which the core is suspended in a closed tank, as distinct from an open-pool reactor.

tantalum   [CHEM] Metallic element in group V, symbol Ta, atomic number 73, atomic weight 180.948; black powder or steel-blue solid soluble in fused alkalies, insoluble in acids (except hydrofluoric and fuming sulfuric); melts at about 3000°C.

tantalum capacitor   [ELEC] An electrolytic capacitor in which the anode is some form of tantalum; examples include solid tantalum, tantalum-foil electrolytic, and tantalum-slug electrolytic capacitors.

tantalum-foil electrolytic capacitor   [ELEC] An electrolytic capacitor that uses plain or etched tantalum foil for both electrodes, with a weak acid electrolyte.

tantalum-slug electrolytic capacitor   [ELEC] An electrolytic capacitor that uses a sintered slug of tantalum as the anode, in a highly conductive acid electrolyte.

tap   [ELEC] A connection made at some point other than the ends of a resistor or coil.

tap crystal   [ELECTR] Compound semiconductor that stores current when stimulated by light and then gives up energy as flashes of light when it is physically tapped.

tapered transmission line   *See* tapered waveguide.

tapered waveguide   [ELECTROMAG] A waveguide in which a physical or electrical characteristic changes continuously with distance along the axis of the waveguide. Also known as tapered transmission line.

tape-wound core   [ELECTROMAG] A length of ferromagnetic material in tape form, wound in such a way that each turn falls directly over the preceding turn.

tapped resistor   [ELEC] A wire-wound fixed resistor having one or more additional terminals along its length, generally for voltage-divider applications.

tare   [MECH] The weight of an empty vehicle or container; subtracted from gross weight to ascertain net weight.

tare effect   [FL MECH] In wind tunnel testing, the forces and moments due to support assembly and mutual interference between support assembly and model.

target   [ATOM PHYS] The atom or nucleus in an atomic or nuclear reaction which is initially stationary.   [ELECTR] In an x-ray tube, the anode or anticathode which emits x-rays when bombarded with electrons.   [PHYS] An object or substance subjected to bombardment or irradiation by particles or electromagnetic radiation.

target strength   [ACOUS] A measure of the reflecting power of a sonar target, which is expressed in decibels by the equation $E + 2L - S$, where E is the echo level, L is the total transmission loss, and S is the source level.

Tauberian theorems   [MATH] A class of theorems, such as Tauber's first theorem, Tauber's second theorem, or Wiener's Tauberian theorem, which are converses of Abelian theorems with some additional condition attached.

Tauber's first theorem   [MATH] If the series Σa_n is Abel-

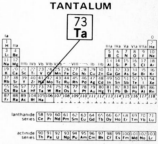

TANTALUM

Periodic table of the chemical elements showing the position of tantalum.

summable to s, and if a_n/n approaches 0 as n approaches infinity, then Σa_n coverges to s.

Tauber's second theorem [MATH] If the series Σa_n is Abel-summable to s, then a necessary and sufficient condition for its convergence to s is that the quantity $(a_1 + 2a_2 + \cdots + na_n)/n$ approaches 0 as n approaches infinity.

Taub NUT space [RELAT] An exact homogeneous, aniso-tropic solution of Einstein's equations of general relativity that contains many of the pathologies (such as closed timelike lines) possible in a space-time.

tau meson [PARTIC PHYS] Former name for the K meson, especially one which decays into three pions.

tau particle [PARTIC PHYS] A heavy, charged lepton with a mass of approximately 1800 MeV, believed to have been observed as a resonance in electron-positron collisions.

Taurus cluster [ASTRON] A cluster of stars observed in the region of the constellation Taurus; it is about 130 light-years $(1.23 \times 10^{18}$ meters) from the sun, and 58 light-years $(5.49 \times 10^{17}$ meters) in diameter.

Taylor effect [FL MECH] A phenomenon in which the relative motion of a homogeneous rotating liquid tends to be the same in all planes perpendicular to the axis of rotation.

Taylor number [FL MECH] A nondimensional number arising in problems of a rotating viscous fluid, written as $T = (f^2h^4)/\nu^2$, where f is the Coriolis parameter (or, for a cylindri-cal system, twice the rate of rotation of the system), h the depth of the fluid, and ν the kinematic viscosity; the square root of the Taylor number is a rotating Reynolds number, and the fourth root is proportional to the ratio of the depth h to the depth of the Ekman layer.

Taylor-Orowan dislocation *See* edge dislocation.

Taylor series [MATH] The Taylor series corresponding to a function $f(x)$ at a point x_0 is the infinite series whose nth term is $(1/n!) \cdot f^{(n)}(x_0)(x - x_0)^n$, where $f^{(n)}(x)$ denotes the nth deriva-tive of $f(x)$.

Taylor's theorem [MATH] The theorem that under certain conditions a real or complex function can be represented, in a neighborhood of a point where it is infinitely differentiable, as a power series whose coefficients involve the various order derivatives evaluated at that point.

Tb *See* terbium.

TBE *See* binding energy.

tbsp *See* tablespoonful.

Tc *See* technetium.

Tchebycheff *See* Chebyshev.

Tchuprow-Neymann allocation [STAT] A technique of strati-fied sampling in which the size of each strata sample is proportional to the size of the strata population and the variance of the strata.

T circulator [ELECTROMAG] A circulator in which three iden-tical rectangular waveguides are joined asymmetrically to form a T-shaped structure, with a ferrite post or wedge at its center; power entering any waveguide emerges from only one adjacent waveguide.

t distribution [STAT] A distribution used to test a hypothesis about a population mean when the population standard deviation is not known, the sample size is small, and the normal distribution is assumed for the sample mean.

Te *See* tellurium.

TEA *See* transferred-electron amplifier.

teaching reactor *See* research reactor; subcritical reactor.

TEA laser [OPTICS] A gas laser in which a glow discharge is maintained without arc formation at atmospheric pressure (which is relatively high for a gas laser) by using a discharge which is transverse rather than parallel to the optic axis.

Derived from transversely excited atmospheric pressure laser.

tear strength [MECH] The force needed to initiate or to continue tearing a sheet or fabric.

teaspoonful [MECH] A unit of volume used particularly in cookery and pharmacy, equal to 1⅓ fluid drams, or ⅓ tablespoonful; in the United States this is equal to approximately 4.9289 cubic centimeters, in the United Kingdom to approximately 4.7355 cubic centimeters. Abbreviated tsp; tspn.

technetium [CHEM] A member of group VII, symbol Tc, atomic number 43; derived from uranium and plutonium fission products; chemically similar to rhenium and manganese; isotope Tc⁹⁹ has a half-life of 2×10^5 years; used to absorb slow neutrons in reactor technology.

technetron [ELECTR] High-power multichannel field-effect transistor.

technical atmosphere [MECH] A unit of pressure in the metric technical system equal to one kilogram-force per square centimeter. Abbreviated at.

tectonophysics [GEOPHYS] A branch of geophysics dealing with the physical processes involved in forming geological structures.

telecentric system [OPTICS] A telescopic system whose aperture stop is located at one of the foci of the objective lens.

telegrapher's equation [MATH] The partial differential equation $(\partial^2 f/\partial x^2) = a^2(\partial^2 f/\partial y^2) + b(\partial f/\partial y) + cf$, where a, b, and c are constants; appears in the study of atomic phenomena.

telemeter [ENG] **1.** The complete measuring, transmitting, and receiving apparatus for indicating or recording the value of a quantity at a distance. **2.** To transmit the value of a measured quantity to a remote point.

teleoperator [CONT SYS] A general-purpose, remotely controlled, cybernetic, dexterous person-machine system.

telephoto lens [OPTICS] A lens for photographing distant objects; it is designed in a compact manner so that the distance from the front of the lens to the film plane is less than the focal length of the lens.

telephotometer [ENG] A photometer that measures the received intensity of a distant light source.

telephotometry [OPTICS] The body of principles and techniques concerned with measuring atmospheric extinction by using various types of telephotometers.

telescope [OPTICS] An optical instrument which, in order to obtain a better resolution, increases the angle under which a distant object, terrestrial or astronomical, is seen.

telescopic comet [ASTRON] A comet in which only the coma is observed.

teleseismology [GEOPHYS] The aspect of seismology dealing with records made at a distance from the source of the impulse.

Teller-Redlich rule [PHYS CHEM] For two isotopic molecules, the product of the frequency ratio values of all vibrations of a given symmetry type depends only on the geometrical structure of the molecule and the masses of the atoms, and not on the potential constants.

telluric current *See* earth current.

telluric line [SPECT] Any of the spectral bands and lines in the spectrum of the sun and stars produced by the absorption of their light in the atmosphere of the earth.

tellurium [CHEM] A member of group VI, symbol Te, atomic number 52, atomic weight 127.60; dark-gray crystals, insoluble in water, soluble in nitric and sulfuric acids and potassium hydroxide; melts at 452°C, boils at 1390°C; used in alloys (with lead or steel), glass, and ceramics.

TECHNETIUM

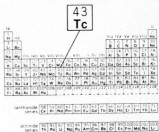

Periodic table of the chemical elements showing the position of technetium.

TELLURIUM

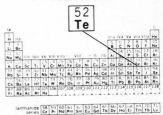

Periodic table of the chemical elements showing the position of tellurium.

TEM mode *See* transverse electromagnetic mode.

TE mode *See* transverse electric mode.

temperament [ACOUS] The adjustment of the pitch of the notes of a keyboard instrument so that the diatonic scale in all keys approximates just tuning; this permits modulation to any key.

temperature [THERMO] A property of an object which determines the direction of heat flow when the object is placed in thermal contact with another object: heat flows from a region of higher temperature to one of lower temperature; it is measured either by an empirical temperature scale, based on some convenient property of a material or instrument, or by a scale of absolute temperature, for example, the Kelvin scale.

temperature bath [THERMO] A relatively large volume of a homogeneous substance held at constant temperature, so that an object placed in thermal contact with it is maintained at the same temperature.

temperature coefficient [PHYS] The rate of change of some physical quantity (such as resistance of a conductor or voltage drop across a vacuum tube) with respect to temperature.

temperature coefficient of reactivity [NUCLEO] The change in reactor reactivity (per degree of temperature) occurring when the operating temperature of the reactor changes.

temperature color scale [THERMO] The relation between an incandescent substance's temperature and the color of the light it emits.

temperature compensation [ELECTR] The process of making some characteristic of a circuit or device independent of changes in ambient temperature.

temperature gradient [THERMO] For a given point, a vector whose direction is perpendicular to an isothermal surface at the point, and whose magnitude equals the rate of change of temperature in this direction.

temperature resistance coefficient [ELEC] The ratio of the change of electrical resistance in a wire caused by a change in its temperature of 1°C as related to its resistance at 0°C.

temperature saturation [ELECTR] The condition in which the anode current of a thermionic vacuum tube cannot be further increased by increasing the cathode temperature at a given value of anode voltage; the effect is due to the space charge formed near the cathode. Also known as filament saturation; saturation.

temperature scale [THERMO] An assignment of numbers to temperatures in a continuous manner, such that the resulting function is single valued; it is either an empirical temperature scale, based on some convenient property of a substance or object, or it measures the absolute temperature.

temperature wave [CRYO] A disturbance in which a variation in temperature propagates through a medium; the chief example of this is second sound. Also known as thermal wave.

TEM wave *See* transverse electromagnetic wave.

tenebrescence [PHYS] Darkening and bleaching under suitable irradiation; materials having this property are called scotophors; darkening may be produced by primary x-rays or cathode rays, while bleaching may be produced by heat or by photons of appropriate wavelength.

ten's complement [MATH] In decimal arithmetic, the unique numeral that can be added to a given N-digit numeral to form a sum equal to 10^N (that is, a one followed by N zeros).

tensile modulus [MECH] The tangent or secant modulus of elasticity of a material in tension.

tensile strength [MECH] The maximum stress a material subjected to a stretching load can withstand without tearing. Also known as hot strength.

tensile stress [MECH] Stress developed by a material bearing a tensile load.

tensimeter [ENG] A device for measuring differences in the vapor pressure of two liquids in which the liquids are placed in sealed, evacuated bulbs connected by a differential manometer.

tensiometer method [FL MECH] A method of determining the surface tension of a liquid that involves measuring the force necessary to remove a ring of known radius from the liquid surface, usually by means of a torsion balance.

tension [MECH] **1.** The condition of a string, wire, or rod that is stretched between two points. **2.** The force exerted by the stretched object on a support.

tensor [MATH] **1.** An object relative to a locally euclidean space which possesses a specified system of components for every coordinate system and which changes under a transformation of coordinates. **2.** A multilinear function on the cartesian product of several copies of a vector space and the dual of the vector space to the field of scalars on the vector space.

tensor analysis [MATH] The abstract study of mathematical objects having components which express properties similar to those of a geometric tensor; this study is fundamental to Riemannian geometry and the structure of euclidean spaces. Also known as tensor calculus.

tensor calculus *See* tensor analysis.

tensor contraction [MATH] For a tensor having an upper and a lower index, summation over the components in which these indexes have the same value, in order to obtain a new tensor two lower in rank.

tensor density [MATH] A relative tensor of weight 1.

tensor differentiation [MATH] An operation on a tensor in which a term involving a Christoffel symbol is subtracted from the ordinary derivative, to obtain another tensor of one higher rank.

tensor field [MATH] A tensor or collection of tensors defined in some open subset of a Riemann space.

tensor force [NUC PHYS] A spin-dependent force between nucleons, having the same form as the interaction between magnetic dipoles; it is introduced to account for the observed values of the magnetic dipole moment and electric quadrupole moment of the deuteron.

tensorial set [MATH] Any collection of quantities that are associated with a system of spatial coordinates and which undergo a linear transformation when this system rotates; examples are the components of a tensor and the eigenfunctions of a quantum mechanical operator.

tensor product [MATH] **1.** The product of two tensors is the tensor whose components are obtained by multiplying those of the given tensors. **2.** In algebra, a multiplicative operation performed between modules.

tensor quantity [MATH] A quantity mathematically represented by a tensor or possessing properties analogous to a tensor.

tensor space [MATH] A fiber bundle composed of the points of a Riemannian manifold and tensor fields.

tenthmeter *See* angstrom.

tera- [MATH] A prefix representing 10^{12}, which is 1,000,000,000,000 or a million million. Abbreviated T.

terahertz [PHYS] A unit of frequency, equal to 10^{12} hertz, or 1,000,000 megahertz. Abbreviated THz.

teraohm [ELEC] A unit of electrical resistance, equal to 10^{12} ohms, or 1,000,000 megohms. Abbreviated TΩ.

teraohmmeter [ENG] An ohmmeter having a teraohm range for measuring extremely high insulation resistance values.

TERBIUM

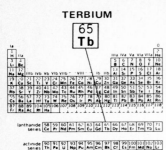

Periodic table of the chemical elements showing the position of terbium.

terawatt [PHYS] A unit of power, equal to 10^{12} watts, or 1,000,000 megawatts. Abbreviated TW.

terbium [CHEM] A rare-earth element, symbol Tb, in the yttrium subgroup of group III, atomic number 65, atomic weight 158.924.

tercentesimal thermometric scale *See* approximate absolute temperature.

term [SPECT] A set of $(2S+1)(2L+1)$ atomic states belonging to a definite configuration and to definite spin and orbital angular momentum quantum numbers S and L.

terminal [ELEC] **1.** A screw, soldering lug, or other point to which electric connections can be made. Also known as electric terminal. **2.** One of the electric input or output points of a circuit or component.

terminal indecomposable future [RELAT] A means of attaching a causal boundary to a space-time; it is the future of a past inextendible timelike curve. Abbreviated TIF.

terminal indecomposable past [RELAT] A means of attaching a causal boundary to a space-time; it is the past of a future inextendible timelike curve. Abbreviated TIP.

terminal pair [ELEC] An associated pair of accessible terminals, such as the input or output terminals of a device or network.

terminal velocity [FL MECH] The velocity with which a body moves relative to a fluid when the resultant force acting on it (due to friction, gravity, and so forth) is zero. [PHYS] The maximum velocity attainable, especially by a freely falling body, under given conditions.

terminator [ASTRON] The line of demarcation between the dark and light portions of the moon or planets.

term splitting [QUANT MECH] The separation of the energies of the states in a term; in the Russell-Saunders case this is produced by the spin-orbit interaction.

ternary expansion [MATH] The numerical representation of a real number relative to the base 3, the digits determined by how the given number can be written in terms of powers of 3.

ternary notation [MATH] A system of notation using the base of 3 and the characters 0, 1, and 2.

ternary ring [MATH] A system, useful in representing projective planes, consisting of a pair (R,t), where R is a set and t is a mapping from the cartesian product $R \times R \times R$ into R such that (1) there exist elements 0 and 1 in R such that $0 \neq 1$, $t(0,a,b) = t(a, 0, b) = b$ and $t(1, a, 0) = t(a,1,0) = a$ for all a and b in R; (2) for any $a, b, c,$ and d in R such that $a \neq c$, there exists a unique x in R such that $t(x,a,b) = t(x,c,d)$; (3) for any $a, b,$ and c in R there exists a unique x in R such that $t(a,b,x) = c$; and (4) for any $a, b, c,$ and d in R such that $a \neq c$, there exists a unique pair (x,y) in $R \times R$ such that $t(a,x,y) = b$ and $t(c,x,y) = d$.

terrestrial magnetism *See* geomagnetism.

terrestrial planet [ASTRON] One of the four small planets near the sun (Earth, Mercury, Venus, and Mars).

terrestrial radiation [GEOPHYS] Electromagnetic radiation originating from the earth and its atmosphere at wavelengths determined by their temperature. Also known as earth radiation; eradiation.

terrestrial refraction [OPTICS] Any refraction phenomenon observed in the light originating from a source lying within the earth's atmosphere; this is applied only to refraction caused by inhomogeneities of the atmosphere itself, not, for example, to that caused by ice crystals suspended in the atmosphere.

terrestrial scintillation [OPTICS] A generic term for scintillation phenomena observed in light that reaches the eye from sources lying within the earth's atmosphere. Also known as atmospheric boil; atmospheric shimmer; optical haze.

terrestrial telescope [OPTICS] Any telescope which produces an erect image.

tertiary pyroelectricity [SOLID STATE] The polarization due to temperature and gradients and corresponding nonuniform stresses and strains when the crystal is heated nonuniformly; found in pyroelectric and nonpyroelectric crystals, that is, crystals which have no polar directions. Also known as false pyroelectricity.

Tesaar lens [OPTICS] An anastigmatic lens made up of a negative lens at the aperture stop with two positive lenses, one in front and the other in back; the last positive lens is a cemented doublet.

tesla [ELECTROMAG] The System International unit of magnetic flux density, equal to one weber per square meter. Symbolized T.

Tesla coil [ELECTROMAG] An air-core transformer used with a spark gap and capacitor to produce a high voltage at a high frequency.

tesselation [MATH] A covering of a plane without gaps or overlappings by polygons, all of which have the same size and shape.

tesseral harmonic [MATH] A spherical harmonic which is 0 on both a set of equally spaced meridians and a set of parallels of latitude of a sphere with center at the origin of spherical coordinates, dividing the sphere into rectangular and triangular regions.

test function [MATH] An infinitely differentiable function of several real variables used in studying solutions of partial differential equations.

test of hypothesis [STAT] A rule for rejecting or accepting a hypothesis concerning a population which is based upon a given sample of data.

test of significance [STAT] A test of a hypothetical population property against a sample property where an acceptance interval is used as the rule for rejection.

test reactor [NUCLEO] A nuclear reactor designed to test the behavior of materials and components under the neutron and gamma fluxes and temperature conditions of an operating reactor.

test statistic [STAT] A numerical value which summarizes the information contained in the sample data and which is a basis for testing a given hypothesis.

tetrachoric correlation [STAT] A measure of the association between two continuous, normally distributed, linearly related variables; the information is obtained from a 2×2 table or dichotomy of the distribution of the two variates; it is not a product moment correlation coefficient.

tetrad axis [CRYSTAL] A rotation axis whose multiplicity is equal to 4.

tetradic [MATH] An operator that transforms one dyadic into another.

tetragonal lattice [CRYSTAL] A crystal lattice in which the axes of a unit cell are perpendicular, and two of them are equal in length to each other, but not to the third axis.

tetragonal tristetrahedron See deltohedron.

tetrahedral symmetry [PHYS] Having the same rotation symmetries as a regular tetrahedron.

tetrahedron [CRYSTAL] An isometric crystal form in cubic crystals, in the shape of a four-faced polyhedron, each face of which is a triangle. [MATH] A four-sided polyhedron.

tetrahexahedron [CRYSTAL] A form of regular crystal system with four triangular isosceles faces on each side of a cube; there are altogether 24 congruent faces.

TESLA COIL

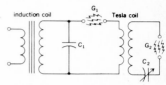

Circuit diagram of Tesla coil. G_1 and G_2 are spark gaps, C_1 is capacitor, C_2 is variable capacitor. The coils and C_1 act as a resonant circuit to produce high-frequency oscillation. Current in secondary (right-hand winding) of Tesla coil is both high-voltage and high-frequency.

tetratohedral crystal [CRYSTAL] A crystal which has one quarter of the maximum number of faces allowed by the crystal system to which the crystal belongs.

tetrode [ELECTR] A four-electrode electron tube containing an anode, a cathode, a control electrode, and one additional electrode that is ordinarily a grid.

tetrode junction transistor See double-base junction transistor.

texture [CRYSTAL] The nature of the orientation, shape, and size of the small crystals in a polycrystalline solid.

TE wave See transverse electric wave.

th See thermie.

Th See thorium.

thallium [CHEM] A metallic element in group III, symbol Tl, atomic number 81, atomic weight 204.37; insoluble in water, soluble in nitric and sulfuric acids, melts at 302°C, boils at 1457°C.

thallium-beam clock [HOROL] A device similar to a cesium-beam clock, but using a beam of thallium atoms instead of cesium; advantages over cesium are reduced sensitivity to magnetic fields and a higher frequency; the beam, however, is much harder to detect and deflect.

thallofide cell [ELECTR] A photoconductive cell in which the active light-sensitive material is thallium oxysulfide in a vacuum; it has maximum response at the red end of the visible spectrum and in the near infrared.

theodolite [OPTICS] An optical instrument used in surveying which consists of a sighting telescope mounted so that it is free to rotate around horizontal and vertical axes, and graduated scales so that the angles of rotation may be measured; the telescope is usually fitted with a right-angle prism so that the observer continues to look horizontally into the eyepiece, whatever the variation of the elevation angle; in meteorology, it is used principally to observe the motion of a pilot balloon.

theorem [MATH] A proven mathematical statement.

theorem of corresponding states [STAT MECH] A theorem stating that two substances which have the same reduced temperature and the same reduced pressure have the same reduced volume.

theoretical cutoff frequency [ELEC] Of an electric structure, a frequency at which, disregarding the effects of dissipation, the attenuation constant changes from zero to a positive value or vice versa.

theoretical frequency [STAT] A distributional frequency that would result if the data followed a theoretical distribution law rather than the actual observed frequencies.

theoretical physics [PHYS] The description of natural phenomena in mathematical form.

theory [MATH] The collection of theorems and principles associated with some mathematical object or concept. [SCI TECH] An attempt to explain a certain class of phenomena by deducing them as necessary consequences of other phenomena regarded as more primitive and less in need of explanation.

theory of equations [MATH] The study of polynomial equations from the viewpoint of solution methods, relations among roots, and connections between coefficients and roots.

therm [THERMO] A unit of heat energy, equal to 10^5 international table British thermal units, or approximately 1.055×10^8 joules.

thermal [THERMO] Of or concerning heat.

thermal agitation [SOLID STATE] Random movements of the free electrons in a conductor, producing noise signals that

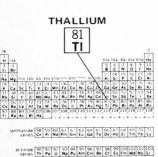

THALLIUM

Periodic table of the chemical elements showing the position of thallium.

may become noticeable when they occur at the input of a high-gain amplifier. Also known as thermal effect.

thermal ammeter *See* hot-wire ammeter.

thermal-arrest calorimeter [ENG] A vacuum device for measurement of heats of fusion; a sample is frozen under vacuum and allowed to melt as the calorimeter warms to room temperature.

thermal bond [NUCLEO] A thermally conductive bond, providing maximum transfer of heat, as between nuclear-reactor fuel and its protective cladding.

thermal breeder reactor [NUCLEO] A breeder reactor in which the fission chain reaction is sustained by thermal neutrons.

thermal bulb [ENG] A device for measurement of temperature; the liquid in a bulb expands with increasing temperature, pressuring a spiral Bourdon-type tube element and causing it to deform (unwind) in direct relation to the temperature in the bulb.

thermal capacitance [THERMO] The ratio of the entropy added to a body to the resulting rise in temperature.

thermal capacity *See* heat capacity.

thermal charge *See* entropy.

thermal column [NUCLEO] A column of moderating material extending through the shield into the reflector of a nuclear reactor, used to provide a source of thermal neutrons for research purposes.

thermal conductance [THERMO] The amount of heat transmitted by a material divided by the difference in temperature of the surfaces of the material. Also known as conductance.

thermal conductimetry [THERMO] Measurement of thermal conductivities.

thermal conductivity [THERMO] The heat flow across a surface per unit area per unit time, divided by the negative of the rate of change of temperature with distance in a direction perpendicular to the surface. Also known as heat conductivity.

thermal conductivity cell *See* katharometer.

thermal conductivity gage [ENG] A pressure measurement device for high-vacuum systems; an electrically heated wire is exposed to the gas under pressure, the thermal conductivity of which changes with changes in the system pressure.

thermal convection *See* heat convection.

thermal converter [ELECTR] A device that converts heat energy directly into electric energy by using the Seebeck effect; it is composed of at least two dissimilar materials, one junction of which is in contact with a heat source and the other junction of which is in contact with a heat sink. Also known as thermocouple converter; thermoelectric generator; thermoelectric power generator; thermoelement.

thermal coulomb [THERMO] A unit of entropy equal to 1 joule per kelvin.

thermal cross section [NUCLEO] The cross section of a thermal neutron.

thermal detector *See* bolometer.

thermal diffusion [PHYS CHEM] A phenomenon in which a temperature gradient in a mixture of fluids gives rise to a flow of one constituent relative to the mixture as a whole. Also known as thermodiffusion.

thermal effect *See* thermal agitation.

thermal efficiency *See* efficiency.

thermal effusion *See* thermal transpiration.

thermal electromotive force [PHYS] An electromotive force arising from a difference in temperature at two points along a circuit, as in the Seebeck effect.

thermal emissivity *See* emissivity.

THERMAL CONVERTER

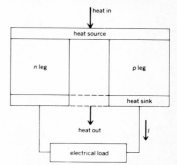

Simple thermal converter. One leg is *n*-type semiconductor material, the other is *p*-type material.

THERMAL HYSTERESIS

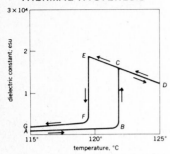

Plot of dielectric constant versus temperature for a single crystal of barium titanate, showing thermal hysteresis. On heating, dielectric constant was observed to follow path ABCD, and on cooling path DCEFG. (After M. E. Drougard and D. R. Young, Phys. Rev., 95:1152–1153, 1954)

THERMAL PULSE METHOD

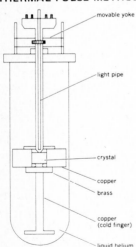

Apparatus for measuring heat pulses at various temperatures, showing components. Laser pulse propagates down light pipe onto absorbing film on face of crystal, and heat pulse propagates through crystal. (From R. J. von Gutfeld and A. H. Nethercot, in W. P. Mason, ed., Physical Acoustics, vol. 5, 1969)

thermal energy [NUCLEO] Energy which is characteristic for thermal neutrons at room temperature, about 0.025 electron volt.

thermal equilibrium [THERMO] Property of a system all parts of which have attained a uniform temperature which is the same as that of the system's surroundings.

thermal excitation [ATOM PHYS] The process in which atoms or molecules acquire internal energy in collisions with other particles.

thermal expansion [PHYS] The dimensional changes exhibited by solids, liquids, and gases for changes in temperature while pressure is held constant.

thermal expansion coefficient [PHYS] The fractional change in length or volume of a material for a unit change in temperature.

thermal farad [THERMO] A unit of thermal capacitance equal to the thermal capacitance of a body for which an increase in entropy of 1 joule per kelvin results in a temperature rise of 1 kelvin.

thermal flux See heat flux.

thermal henry [THERMO] A unit of thermal inductance equal to the product of a temperature difference of 1 kelvin and a time of 1 second divided by a rate of flow of entropy of 1 watt per kelvin.

thermal hysteresis [THERMO] A phenomenon sometimes observed in the behavior of a temperature-dependent property of a body; it is said to occur if the behavior of such a property is different when the body is heated through a given temperature range from when it is cooled through the same temperature range.

thermal imagery [ELECTR] Imagery produced by measuring and recording electronically the thermal radiation of objects.

thermal inductance [THERMO] The product of temperature difference and time divided by entropy flow.

thermal instability [FL MECH] The instability resulting in free convection in a fluid heated at a boundary.

thermal ionization See Saha ionization.

thermalize [NUC PHYS] To bring neutrons into thermal equilibrium with the surroundings.

thermal magnon [SOLID STATE] A magnon with a relatively short wavelength, on the order of 10^{-6} centimeter.

thermal neutron [NUCLEO] One of a collection of neutrons whose energy distribution is identical with or similar to the Maxwellian distribution in the material in which they are found; the average kinetic energy of such neutrons at room temperature is about 0.025 electron volt. Also known as slow neutron.

thermal noise [ELECTR] Electric noise produced by thermal agitation of electrons in conductors and semiconductors. Also known as Johnson noise; resistance noise.

thermal ohm [THERMO] A unit of thermal resistance equal to the thermal resistance for which a temperature difference of 1 kelvin produces a flow of entropy of 1 watt per kelvin. Also known as fourier.

thermal photography See thermography.

thermal potential difference [THERMO] The difference between the thermodynamic temperatures of two points.

thermal pulse method [SOLID STATE] A method of measuring properties of insulating and conducting crystals, in which a heat pulse of known duration is measured after propagating through a crystal; the pulse can be generated by directing a laser pulse at an absorbing film evaporated onto one face of the crystal, and detected by a thin-film circuit on the other face.

thermal radiation See heat radiation.

thermal reactor [NUCLEO] A nuclear reactor in which fission is induced primarily by neutrons of such low energy that they are in substantial thermal equilibrium with the material of the core.

thermal resistance [ELECTR] *See* effective thermal resistance. [THERMO] A measure of a body's ability to prevent heat from flowing through it, equal to the difference between the temperatures of opposite faces of the body divided by the rate of heat flow. Also known as heat resistance.

thermal resistivity [THERMO] The reciprocal of the thermal conductivity.

thermal resistor [ELEC] A resistor designed so its resistance varies in a known manner with changes in ambient temperature.

thermal Rossby number [FL MECH] The nondimensional ratio of the inertial force due to the thermal wind and the Coriolis force in the flow of a fluid which is heated from below.

thermal scattering [SOLID STATE] Scattering of electrons, neutrons, or x-rays passing through a solid due to thermal motion of the atoms in the crystal lattice.

thermal shield [NUCLEO] A high-density heat-conducting portion of a shield placed close to the reflector of a nuclear reactor to absorb thermal neutrons, gamma rays, beta rays, and x-rays, whose absorption in the outer shield would generate excessive heat.

thermal stress [MECH] Mechanical stress induced in a body when some or all of its parts are not free to expand or contract in response to changes in temperature.

thermal transducer [ENG] Any device which converts energy from some form other than heat energy into heat energy; an example is the absorbing film used in the thermal pulse method.

thermal transpiration [THERMO] The formation of a pressure gradient in gas inside a tube when there is a temperature gradient in the gas and when the mean free path of molecules in the gas is a significant fraction of the tube diameter. Also known as thermal effusion.

thermal utilization factor [NUCLEO] The probability that a thermal neutron which is absorbed is absorbed usefully, as in a fissionable material.

thermal value [THERMO] Heat produced by combustion, usually expressed in calories per gram or British thermal units per pound.

thermal volt *See* kelvin.

thermal wave [CRYO] *See* temperature wave. [SOLID STATE] A sound wave in a solid which has a short wavelength.

thermal x-rays [ELECTROMAG] The electromagnetic radiation, mainly in the soft (low-energy) x-ray region.

thermie [THERMO] A unit of heat energy equal to the heat energy needed to raise 1 metric ton of water from 14.5 to 15.5°C at a constant pressure of 1 standard atmosphere (101,325 newtons per square meter); equal to 10^6 fifteen-degrees calories or $(4.1855 \pm 0.0005) \times 10^6$ joules. Abbreviated th.

thermion [ELECTR] A charged particle, either negative or positive, emitted by a heated body, as by the hot cathode of a thermionic tube.

thermionic [ELECTR] Pertaining to the emission of electrons as a result of heat.

thermionic arc [ELECTR] An electric arc in which the arc current itself heats the cathode.

thermionic cathode *See* hot cathode.

thermionic converter [ELECTR] A device in which heat energy is directly converted to electric energy; it has two electrodes, one of which is raised to a sufficiently high temperature to become a thermionic electron emitter, while the other, serving as an electron collector, is operated at a significantly lower temperature. Also known as thermionic generator; thermionic power generator; thermoelectric engine.

thermionic current [ELECTR] Current due to directed movements of thermions, such as the flow of emitted electrons from the cathode to the plate in a thermionic vacuum tube.

thermionic diode [ELECTR] A diode electron tube having a heated cathode.

thermionic emission [ELECTR] **1.** The outflow of electrons into vacuum from a heated electric conductor. Also known as Edison effect; Richardson effect. **2.** More broadly, the liberation of electrons or ions from a substance as a result of heat.

thermionic fuel cell [ELECTR] A thermionic converter in which the space between the electrodes is filled with cesium or other gas, which lowers the work functions of the electrodes, and creates an ionized atmosphere, controlling the electron space charge.

thermionic generator *See* thermionic converter.

thermionic power generator *See* thermionic converter.

thermionics [ELECTR] The study and applications of thermionic emission.

thermionic triode [ELECTR] A three-electrode thermionic tube, containing an anode, a cathode, and a control electrode.

thermionic tube [ELECTR] An electron tube that relies upon thermally emitted electrons from a heated cathode for tube current. Also known as hot-cathode tube.

thermionic work function [ELECTR] Energy required to transfer an electron from the fermi energy in a given metal through the surface to the vacuum just outside the metal.

thermistor [ELECTR] A resistive circuit component, having a high negative temperature coefficient of resistance, so that its resistance decreases as the temperature increases; it is a stable, compact, and rugged two-terminal ceramiclike semiconductor bead, rod, or disk. Derived from thermal resistor.

thermoacoustic array [ACOUS] A sound source consisting of a light beam (usually a laser beam) modulated at the frequency of the sound to be generated; the resulting sound has its maximum value in a direction perpendicular to the axis of the light beam, and its directivity pattern has no side lobes.

thermoacoustic effect *See* optoacoustic effect.

thermoammeter [ENG] An ammeter that is actuated by the voltage generated in a thermocouple through which is sent the current to be measured; used chiefly for measuring radio-frequency currents. Also known as electrothermal ammeter; thermocouple ammeter.

thermochemical calorie *See* calorie.

thermochemistry [PHYS CHEM] The measurement, interpretation, and analysis of heat changes accompanying chemical reactions and changes in state.

thermocouple [ENG] A device consisting basically of two dissimilar conductors joined together at their ends; the thermoelectric voltage developed between the two junctions is proportional to the temperature difference between the junctions, so the device can be used to measure the temperature of one of the junctions when the other is held at a fixed, known temperature, or to convert radiant energy into electric energy.

thermocouple ammeter *See* thermoammeter.

thermocouple converter *See* thermal converter.

thermocouple pyrometer *See* thermoelectric pyrometer.

thermodiffusion *See* thermal diffusion.

THERMOCOUPLE

measuring junction   reference junction

metal A

metal B   metal B

M

Thermocouple circuit for measuring metal temperatures. Voltmeter *M* measures voltage developed between the junctions.

thermodynamic cycle [THERMO] A procedure or arrangement in which some material goes through a cyclic process and one form of energy, such as heat at an elevated temperature from combustion of a fuel, is in part converted to another form, such as mechanical energy of a shaft, the remainder being rejected to a lower temperature sink. Also known as heat cycle.

thermodynamic equation of state [THERMO] An equation that relates the reversible change in energy of a thermodynamic system to the pressure, volume, and temperature.

thermodynamic equilibrium [THERMO] Property of a system which is in mechanical, chemical, and thermal equilibrium.

thermodynamic function of state [THERMO] Any of the quantities defining the thermodynamic state of a substance in thermodynamic equilibrium; for a perfect gas, the pressure, temperature, and density are the fundamental thermodynamic variables, any two of which are, by the equation of state, sufficient to specify the state. Also known as state parameter; state variable; thermodynamic variable.

thermodynamic potential [THERMO] One of several extensive quantities which are determined by the instantaneous state of a thermodynamic system, independent of its previous history, and which are at a minimum when the system is in thermodynamic equilibrium under specified conditions.

thermodynamic potential at constant volume *See* free energy.

thermodynamic principles [THERMO] Laws governing the conversion of energy from one form to another.

thermodynamic probability [THERMO] Under specified conditions, the number of equally likely states in which a substance may exist; the thermodynamic probability Ω is related to the entropy S by $S = k \ln \Omega$, where k is Boltzmann's constant.

thermodynamic process [THERMO] A change of any property of an aggregation of matter and energy, accompanied by thermal effects.

thermodynamic property [THERMO] A quantity which is either an attribute of an entire system or is a function of position which is continuous and does not vary rapidly over microscopic distances, except possibly for abrupt changes at boundaries between phases of the system; examples are temperature, pressure, volume, concentration, surface tension, and viscosity. Also known as macroscopic property.

thermodynamics [PHYS] The branch of physics which seeks to derive, from a few basic postulates, relationships between properties of matter, especially those which are affected by changes in temperature, and a description of the conversion of energy from one form to another.

thermodynamic system [THERMO] A part of the physical world as described by its thermodynamic properties.

thermodynamic temperature scale [THERMO] Any temperature scale in which the ratio of the temperatures of two reservoirs is equal to the ratio of the amount of heat absorbed from one of them by a heat engine operating in a Carnot cycle to the amount of heat rejected by this engine to the other reservoir; the Kelvin scale and the Rankine scale are examples of this type.

thermodynamic variable *See* thermodynamic function of state.

thermoelasticity [PHYS] Dependence of the stress distribution of an elastic solid on its thermal state, or of its thermal conductivity on the stress distribution.

thermoelectric converter [ELECTR] A converter that changes solar or other heat energy to electric energy; used as a power source on spacecraft.

thermoelectric cooling [ENG] Cooling of a chamber based on the Peltier effect; an electric current is sent through a thermocouple whose cold junction is thermally coupled to the cooled chamber, while the hot junction dissipates heat to the surroundings. Also known as thermoelectric refrigeration.

thermoelectric effect *See* thermoelectricity.

thermoelectric engine *See* thermionic converter.

thermoelectric generator *See* thermal converter.

thermoelectric heating [ENG] Heating based on the Peltier effect, involving a device which is in principle the same as that used in thermoelectric cooling except that the current is reversed.

thermoelectricity [PHYS] The direct conversion of heat into electrical energy, or the reverse; it encompasses the Seebeck, Peltier, and Thomson effects but, by convention, excludes other electrothermal phenomena, such as thermionic emission. Also known as thermoelectric effect.

thermoelectric laws [ENG] Basic relationships used in the design and application of thermocouples for temperature measurement; for example, the law of the homogeneous circuit, the law of intermediate metals, and the law of successive or intermediate temperatures.

thermoelectric material [ELECTR] A material that can be used to convert thermal energy into electric energy or provide refrigeration directly from electric energy; good thermoelectric materials include lead telluride, germanium telluride, bismuth telluride, and cesium sulfide.

thermoelectric nuclear battery [NUCLEO] A low-voltage battery in which a heat source, consisting of a radioactive isotope such as polonium-210, is hermetically sealed in a strong, dense capsule, and a series of thermocouples are alternately connected thermally, but not electrically, to the heat source and the outer surface of the capsule.

thermoelectric power generator *See* thermal converter.

thermoelectric properties [PHYS] Properties of materials associated with thermoelectricity, namely, the electromotive force generated in the Seebeck effect, the heat generated or absorbed in the Peltier and Thomson effects, and the influence of magnetic fields upon these quantities.

thermoelectric pyrometer [ENG] An instrument which uses one or more thermocouples to measure high temperatures, usually in the range between 800 and 2400°F (425 and 1315°C). Also known as thermocouple pyrometer.

thermoelectric refrigeration *See* thermoelectric cooling.

thermoelectric series [MET] A series of metals arranged in order of their thermoelectric voltage-generating ratings with respect to some reference metal, such as lead.

thermoelectric solar cell [ELECTR] A solar cell in which the sun's energy is first converted into heat by a sheet of metal, and the heat is converted into electricity by a semiconductor material sandwiched between the first metal sheet and a metal collector sheet.

thermoelectric thermometer [ENG] A type of electrical thermometer consisting of two thermocouples which are series-connected with a potentiometer and a constant-temperature bath; one couple, called the reference junction, is placed in a constant-temperature bath, while the other is used as the measuring junction.

thermoelectromotive force [ELEC] Voltage developed due to differences in temperature between parts of a circuit containing two or more different metals.

thermoelectron [ELECTR] An electron liberated by heat, as from a heated filament. Also known as negative thermion.

thermoelement *See* thermal converter.

**THERMOELECTRIC
NUCLEAR BATTERY**

polonium-210 source

thermocouples (electrically insulated from source and outer container)

aluminum outer container

thermal insulation

Diagram of thermoelectric nuclear battery.

thermogalvanometer [ENG] Instrument for measuring small high-frequency currents by their heating effect, generally consisting of a direct-current galvanometer connected to a thermocouple that is heated by a filament carrying the current to be measured.

thermograph [ENG] An instrument that senses, measures, and records the temperature of the atmosphere. Also known as recording thermometer. [OPTICS] A far-infrared image-forming device that provides a thermal photograph by scanning a far-infrared image of an object or scene.

thermography [ENG] A method of measuring surface temperature by using luminescent materials: the two main types are contact thermography and projection thermography.

thermojunction [ELECTR] One of the surfaces of contact between the two conductors of a thermocouple. Also known as thermoelectric junction.

thermojunction battery [ELEC] Nuclear-type battery which converts heat into electrical energy directly by the thermoelectric or Seebeck effect.

thermoluminescence [ATOM PHYS] 1. Broadly, any luminescence appearing in a material due to application of heat. 2. Specifically, the luminescence appearing as the temperature of a material is steadily increased; it is usually caused by a process in which electrons receiving increasing amounts of thermal energy escape from a center in a solid where they have been trapped and go over to a luminescent center, giving it energy and causing it to luminesce.

thermomagnetic effect [PHYS] An electrical or thermal phenomenon occurring when a conductor or semiconductor is placed simultaneously in a temperature gradient and a magnetic field; examples are the Ettingshausen-Nernst effect and the Righi-Leduc effect.

thermomechanical effect *See* fountain effect.

thermometal [MET] A bimetallic strip which, on temperature change, deflects because of differences in the coefficients of expansion of the two bonded metals.

thermometer [ENG] An instrument that measures temperature.

thermometric conductivity *See* diffusivity.

thermometric fluid [THERMO] A fluid that has properties, such as a large and uniform thermal expansion coefficient, good thermal conductivity, and chemical stability, that make it suitable for use in a thermometer.

thermometry [THERMO] The science and technology of measuring temperature, and the establishment of standards of temperature measurement.

thermomigration [ELECTR] A technique for doping semiconductors in which exact amounts of known impurities are made to migrate from the cool side of a wafer of pure semiconductor material to the hotter side when the wafer is heated in an oven.

thermonuclear [NUCLEO] Referring to any process in which a very high temperature is used to bring about the fusion of light nuclei, with the accompanying liberation of energy.

thermonuclear device [NUCLEO] A fusion bomb used for peaceful purposes, tests, or experiments.

thermonuclear reaction [NUC PHYS] A nuclear fusion reaction which occurs between various nuclei of light elements when they are constituents of a gas at very high temperature.

thermopile [ENG] An array of thermocouples connected either in series to give higher voltage output or in parallel to give higher current output, used for measuring temperature or radiant energy or for converting radiant energy into electric power.

THERMOLUMINESCENCE

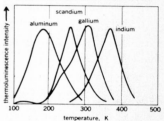

Plot of thermoluminescence intensity as a function of temperature in several zinc sulfide phosphors, each of which contains traces of copper, giving rise to luminescent centers, and different trivalent ions (as indicated), giving rise to traps. Curves reach a maximum and then decrease as traps are emptied.

THERMOSTAT

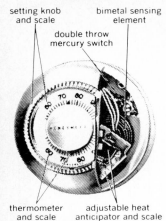

setting knob and scale

bimetal sensing element

double throw mercury switch

thermometer and scale

adjustable heat anticipator and scale

Typical thermostat uses bimetal sensing element to control room temperature by switching a heating or cooling device on and off. Heat anticipator switches device off prematurely to prevent excessive temperature swings. (Honeywell Inc.)

THERMOVOLTMETER

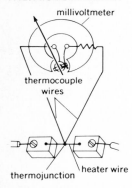

millivoltmeter

thermocouple wires

thermojunction

heater wire

Elementary thermovoltmeter circuit. (General Electric Co.)

THÉVENIN'S THEOREM

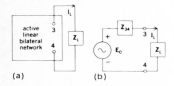

active linear bilateral network

(a)

(b)

The theorem states that circuit to left of terminals 3 and 4 in (a) may be replaced by that in (b). Impedance $\mathbf{Z}_L$ and current $\mathbf{I}_L$ are same in both cases. $\mathbf{E}_0$ is voltage of voltage generator. $\mathbf{Z}_{34}$ is series impedance.

thermopower [ELEC] A measure of the temperature-induced voltage in a conductor.

thermoregulator [ENG] A high-accuracy or high-sensitivity thermostat; one type consists of a mercury-in-glass thermometer with sealed-in electrodes, in which the rising and falling column of mercury makes and breaks an electric circuit.

thermorelay *See* thermostat.

thermostat [ENG] An instrument which measures changes in temperature and directly or indirectly controls sources of heating and cooling to maintain a desired temperature. Also known as thermorelay.

thermovoltmeter [ENG] A voltmeter in which a current from the voltage source is passed through a resistor and a fine vacuum-enclosed platinum heater wire; a thermocouple, attached to the midpoint of the heater, generates a voltage of a few millivolts, and this voltage is measured by a direct-current millivoltmeter.

theta functions [MATH] Complex functions used in the study of Riemann surfaces and of elliptic functions and elliptic integrals; they are:

$$\theta_1(z) = 2\sum_{n=0}^{\infty} (-1)^n q^{(n+\frac{1}{2})^2} \sin(2n+1)z$$

$$\theta_2(z) = 2\sum_{n=0}^{\infty} q^{(n+\frac{1}{2})^2} \cos(2n+1)z$$

$$\theta_3(z) = 1 + 2\sum_{n=1}^{\infty} q^{n^2} \cos 2z$$

$$\theta_4(z) = 1 + 2\sum_{n=1}^{\infty} (-1)^n q^{n^2} \cos 2nz$$

where $q = \exp \pi i \tau$, and τ is a constant complex number with positive imaginary part.

thetagram [THERMO] A thermodynamic diagram with coordinates of pressure and temperature, both on a linear scale.

theta pinch [PL PHYS] A device for producing a controlled nuclear fusion reaction, in which plasma in a long torus or skinny tube is confined by a magnetic field produced by current-carrying coils, and is shock-heated and compressed by pulses in this field to produce the high temperatures at which fusion reactions take place; the magnetic field is then sustained in order to maintain the plasma confinement.

theta polarization [ELECTROMAG] State of a wave in which the E vector is tangential to the meridian lines of some given spherical frame of reference.

Thévenin generator [ELEC] The voltage generator in the equivalent circuit of Thévenin's theorem.

Thévenin's theorem [ELEC] A valuable theorem in network problems which allows calculation of the performance of a device from its terminal properties only: the theorem states that at any given frequency the current flowing in any impedance, connected to two terminals of a linear bilateral network containing generators of the same frequency, is equal to the current flowing in the same impedance when it is connected to a voltage generator whose generated voltage is the voltage at the terminals in question with the impedance removed, and whose series impedance is the impedance of the network looking back from the terminals into the network with all generators replaced by their internal impedances. Also known as Helmholtz's theorem.

thick-film capacitor [ELEC] A capacitor in a thick-film circuit, made by successive screen-printing and firing processes.

thick-film circuit [ELECTR] A microcircuit in which passive components, of a ceramic-metal composition, are formed on a ceramic substrate by successive screen-printing and firing processes, and discrete active elements are attached separately.

thick-film resistor [ELEC] Fixed resistor whose resistance element is a film well over one-thousandth of an inch (over 25 micrometers) thick.

thick lens [OPTICS] A lens in which the separation between the two surfaces is too great to be ignored in calculations of such quantities as focal length and magnification.

thimble ionization chamber [NUCLEO] A small cylindrical or spherical ionization chamber, usually with walls made of organic material or air walls.

thin film [ELECTR] A film a few molecules thick deposited on a glass, ceramic, or semiconductor substrate to form a capacitor, resistor, coil, cryotron, or other circuit component.

thin-film capacitor [ELEC] A capacitor that can be constructed by evaporation of conductor and dielectric films in sequence on a substrate; silicon monoxide is generally used as the dielectric.

thin-film circuit [ELECTR] A circuit in which the passive components and conductors are produced as films on a substrate by evaporation or sputtering; active components may be similarly produced or mounted separately.

thin-film cryotron [ELECTR] A cryotron in which the transition from superconducting to normal resistivity of a thin film of tin or indium, serving as a gate, is controlled by current in a film of lead that crosses and is insulated from the gate.

thin-film ferrite coil [ELECTROMAG] An inductor made by depositing a thin flat spiral of gold or other conducting metal on a ferrite substrate.

thin-film integrated circuit [ELECTR] An integrated circuit consisting entirely of thin films deposited in a patterned relationship on a substrate.

thin-film material [ELECTR] A material that can be deposited as a thin film in a desired pattern by a variety of chemical, mechanical, or high-vacuum evaporation techniques.

thin-film resistor [ELEC] A fixed resistor whose resistance element is a metal, alloy, carbon, or other film having a thickness of about one-millionth inch (about 25 nanometers).

thin-film semiconductor [ELECTR] Semiconductor produced by the deposition of an appropriate single-crystal layer on a suitable insulator.

thin-film solar cell [ELECTR] A solar cell in which a thin film of gallium arsenide, cadmium sulfide, or other semiconductor material is evaporated on a thin, flexible metal or plastic substrate; the rather low efficiency (about 2%) is compensated by the flexibility and light weight, making these cells attractive as power sources for spacecraft.

thin-film transducer [SOLID STATE] A film a few molecules thick, usually consisting of cadmium sulfide, evaporated on a crystal substrate, used to convert microwave radiation into hypersonic sound waves in the crystal.

thin-film transistor [ELECTR] A field-effect transistor constructed by thin-film techniques, for use in thin-film circuits.

thin lens [OPTICS] A lens whose thickness is small enough to be neglected in calculations of such quantities as object distance, image distance, and magnification.

third fundamental form [MATH] For a surface with curvilinear coordinates u^i, $i = 1, 2$, this is the quadratic differential form

$$\sum_{i,j} c_{ij} \, du^i \, du^j$$

where the c_{ij} are fundamental magnitudes of third order. Also known as third ground form.

third ground form *See* third fundamental form.

third harmonic [PHYS] A sine-wave component having three times the fundamental frequency of a complex signal.

third law of motion *See* Newton's third law.

third law of thermodynamics [THERMO] The entropy of all perfect crystalline solids is zero at absolute zero temperature.

third sound [CRYO] A type of wave propagated in thin films of superfluid helium (helium II), consisting of variations in film thickness and temperature.

thixotropy [PHYS CHEM] Property of certain gels which liquefy when subjected to vibratory forces, such as ultrasonic waves or even simple shaking, and then solidify again when left standing.

Thomas cyclotron [NUCLEO] A circular particle accelerator which operates like an ordinary cyclotron but employs a magnetic field that is variable in azimuth in such a way that cyclotron resonance at a fixed orbital frequency and radial and axial focusing can be maintained simultaneously.

Thomas-Fermi atom model [ATOM PHYS] A method of approximating the electrostatic potential and the electron density in an atom in its ground state, in which these two quantities are related by the Poisson equation on the one hand, and on the other hand by a semiclassical formula for the density of quantum states in phase space.

Thomas-Fermi equation [ATOM PHYS] The differential equation $x^{1/2}(d^2y/dx^2) = y^{3/2}$ that arises in calculating the potential in the Thomas-Fermi atom model; the physically meaningful solution satisfies the boundary conditions $y(0) = 1$ and $y(\infty) = 0$.

Thomas precession [RELAT] The precession of a vector in an accelerated system, relative to an observer for whom the system has a given velocity and acceleration, when this vector appears to be constant to an observer attached to the system; this precession is the kinematical basis of one type of spin-orbit coupling.

Thomas-Reiche-Kuhn sum rule *See* f-sum rule.

Thomson bridge *See* Kelvin bridge.

Thomson coefficient [PHYS] The ratio of the voltage existing between two points on a metallic conductor to the difference in temperature of those points.

Thomson cross section [ELECTROMAG] The total scattering cross section for Thomson scattering, equal to $(8/3)\pi(e^2/mc^2)^2$, where e and m are the charge (in electrostatic units) and mass of the scattering particle, and c is the speed of light.

Thomson effect [PHYS] A thermoelectric effect in which heat flows into or out of a homogeneous conductor when an electric current flows between two points in the conductor at different temperatures, the direction of heat flow depending upon whether the current flows from colder to warmer metal or from warmer to colder.

Thomson formula [ELECTROMAG] **1.** The formula for the intensity of scattered electromagnetic radiation in Thomson scattering as a function of the scattering angle ϕ; the intensity is proportional to $1 + \cos^2 \phi$. **2.** A formula for the period of oscillation of a current when a capacitor is discharged. Also known as Kelvin's formula.

Thomson functions *See* Kelvin functions.

Thomson heat [PHYS] The heat generated or absorbed in the Thomson effect in a reversible manner when a current passes through a conductor in which there is a temperature gradient; it is proportional to the product of the current and the temperature gradient.

Thomson parabolas [ELECTROMAG] A pattern of parabolas which appear on a photographic plate exposed to a beam of ions of an element which has passed through electric and magnetic fields applied in the same direction normal to the path of the ions; each parabola corresponds to a different charge-to-mass ratio, and thus to a different isotope.

Thomson relations [PHYS] Equations in the study of thermoelectricity, relating the Peltier coefficient and the Thomson coefficient to the Seebeck voltage; they are derived by thermodynamics. Also known as Kelvin relations.

Thomson scattering [ELECTROMAG] Scattering of electromagnetic radiation by free (or very loosely bound) charged particles, computed according to a classical nonrelativistic theory: energy is taken away from the primary radiation as the charged particles accelerated by the transverse electric field of the radiation, radiate in all directions.

Thomson voltage [PHYS] The voltage that exists between two points that are at different temperatures in a conductor.

Thoraeus filter [NUCLEO] A primary radiological filter of tin, combined with a secondary filter of copper to absorb the characteristic radiation of the tin and a third filter of aluminum to absorb the characteristic radiation of the copper; in the range of 200 to 400 kilovolts, such a filter hardens x-rays more efficiently than the usual combination of copper and aluminum.

thorium [CHEM] An element of the actinium series, symbol Th, atomic number 90, atomic weight 232; soft, radioactive, insoluble in water and alkalies, soluble in acids, melts at 1750°C, boils at 4500°C.

thorium reactor [NUCLEO] A nuclear reactor in which thorium surrounds the central enriched uranium core to give breeder operation.

thorium series [NUCLEO] The series of nuclides resulting from the decay of thorium-232.

thoron [NUC PHYS] The conventional name for radon-220. Symbolized Tn.

thou *See* mil.

thousandth mass unit [PHYS] A unit of energy equal to the energy equivalent of a mass of 10^{-3} atomic mass unit according to the Einstein mass-energy relation, that is, to the product of 10^{-3} atomic mass unit and the square of the speed of light; equal to approximately 1.49176×10^{-13} joule.

three-body problem [MECH] The problem of predicting the motions of three objects obeying Newton's laws of motion and attracting each other according to Newton's law of gravitation.

three-decibel coupler [ELECTROMAG] Junction of two waveguides having a common H wall; the two guides are coupled together by H-type aperture coupling; the coupling is such that 50% of the power from either channel will be fed into the other. Also known as Riblet coupler; short-slot coupler.

three-decision problem [STAT] A problem in which a choice must be made among three possible courses of action.

three-dimensional [SCI TECH] Giving the illusion of depth, in three dimensions.

three-dimensional flow [FL MECH] Any fluid flow which is not a two-dimensional flow.

three-eighths rule [MATH] **1.** An approximation formula for definite integrals which states that the integral of a real-valued function f on an interval $[a,b]$ is approximated by $(3/8)h[f(a) + 3f(a + h) + 3f(a + 2h) + f(b)]$, where $h = (b - a)/3$; this is the integral of a third-degree polynomial whose value equals that of f at a, $a + h$, $a + 2h$, and b. **2.** A method of approximating a definite integral over an interval which is equivalent to dividing the interval into equal subintervals and applying the formula in the first definition to each subinterval.

three-j number [QUANT MECH] A coefficient used in coupling eigenfunctions of two commuting angular momenta to form

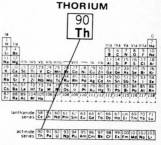

THORIUM

Periodic table of the chemical elements showing the position of thorium.

eigenfunctions of the total angular momentum; closely related to the Clebsch-Gordan coefficients. Also known as Wigner three-j symbol.

three-junction transistor [ELECTR] A *pnpn* transistor having three junctions and four regions of alternating conductivity; the emitter connection may be made to the *p* region at the left, the base connection to the adjacent *n* region, and the collector connection to the *n* region at the right, while the remaining *p* region is allowed to float.

three-layer diode [ELECTR] A junction diode with three conductivity regions.

three-level laser [OPTICS] A laser involving three energy levels, one of which is the ground state; laser action usually occurs between the intermediate and ground states.

three-level maser [PHYS] A solid-state maser in which three energy levels are used; successful operation has been obtained with crystals of gadolinium ethyl sulfate and crystals of potassium chromecyanide at the temperature of liquid helium.

three-phase system [PHYS] Any physical system in which three distinct phases coexist; phases can be liquid, solid, vapor (gas), or three mutually insoluble liquids, or any combination thereof.

three-space [MATH] A vector space over the real numbers whose basis has three vectors.

three-way switch [ELEC] An electric switch with three terminals used to control a circuit from two different points.

threshold [MATH] A logic operator such that, if $P, Q, R, S, \ldots$ are statements, then the threshold will be true if at least N statements are true, false otherwise. [PHYS] The minimum level of some input quantity needed for some process to take place, such as a threshold energy for a reaction, or the minimum level of pumping at which a laser can go into self-excited oscillation.

threshold contrast [OPTICS] The smallest contrast of luminance (or brightness) that is perceptible to the human eye under specified conditions of adaptation luminance and target visual angle. Also known as contrast sensitivity; contrast threshold; liminal contrast.

threshold detector [NUC PHYS] An element or isotope in which radioactivity is induced only by the capture of neutrons having energies in excess of a certain characteristic threshold value; used to determine the neutron spectrum from a nuclear explosion.

threshold dose [NUCLEO] The minimum radiation dose that will produce a detectable specified effect.

threshold frequency [ELECTR] The frequency of incident radiant energy below which there is no photoemissive effect.

threshold illuminance [OPTICS] The lowest value of illuminance which the eye is capable of detecting under specified conditions of background luminance and degree of dark adaptation of the eye. Also known as flux density threshold.

threshold of reaction [PHYS] The minimum energy, for an incident particle or photon, below which a particular reaction does not occur.

threshold value [CONT SYS] The minimum input that produces a corrective action in an automatic control system.

throat velocity *See* critical velocity.

throttled flow [FL MECH] Flow which is forced to pass through a restricted area, where the velocity must increase.

throttling [CONT SYS] Control by means of intermediate steps between full on and full off. [THERMO] An adiabatic, irreversible process in which a gas expands by passing from one chamber to another chamber which is at a lower pressure than the first chamber.

throttling calorimeter [ENG] An instrument utilizing the principle of constant enthalpy expansion for the measurement of the moisture content of steam; steam drawn from a steampipe through sampling nozzles enters the calorimeter through a throttling orifice and moves into a well-insulated expansion chamber in which its temperature is measured. Also known as steam calorimeter.

thrust [MECH] The force exerted in any direction by a fluid jet or by a powered screw.

thulium [CHEM] A rare-earth element, symbol Tm, group IIIB, of the lanthanide group, atomic number 69, atomic weight 168.934; reacts slowly with water, soluble in dilute acids, melts at 1550°C, boils at 1727°C; the dust is a fire hazard; used as x-ray source and to make ferrites.

thulium-170 [NUC PHYS] The radioactive isotope of thulium, with mass number 170; used as a portable x-ray source.

thyratron [ELECTR] A hot-cathode gas tube in which one or more control electrodes initiate but do not limit the anode current except under certain operating conditions. Also known as hot-cathode gas-filled tube.

thyristor [ELECTR] A transistor having a thyratronlike characteristic; as collector current is increased to a critical value, the alpha of the unit rises above unity to give high-speed triggering action.

THz *See* terahertz.

Ti *See* titanium.

tidal correction [GEOPHYS] A correction made in gravity observations to remove the effect of the earth's tides.

tide-producing force [GEOPHYS] The slight local difference between the gravitational attraction of two astronomical bodies and the centrifugal force that holds them apart.

tied rank [STAT] If two distinct observations have the same value, thus being given the same rank, they are said to be tied; this presents difficulties in the Wilcoxon two-sample test, the sign test, and the Fisher-Irwin test.

tie line [PHYS CHEM] A line on a phase diagram joining the two points which represent the composition of systems in equilibrium. Also known as conode.

Tietze extension theorem [MATH] A topological space X is normal if and only if every continuous function of a closed subset to [0,1] has a continuous extension to all of X.

TIF *See* terminal indecomposable future.

tight binding approximation [SOLID STATE] A method of calculating energy states and wave functions of electrons in a solid in which the wave function is assumed to be a sum of pure atomic wave functions centered about each of the atoms in the lattice, each multiplied by a phase factor; it is suitable for deep-lying energy levels.

tight coupling *See* close coupling.

tilt boundary [SOLID STATE] A boundary between two crystals that differ in orientation by only a few degrees, consisting of a series of edge dislocations; it is formed during polygonization. Also known as bend plane; polygon wall.

timbre [ACOUS] That attribute of auditory sensation in terms of which a listener can judge that two sounds similarly presented and having the same loudness and pitch are dissimilar. Also known as musical quality.

time [PHYS] **1.** The dimension of the physical universe which, at a given place, orders the sequence of events. **2.** A designated instant in this sequence, as the time of day. Also known as epoch.

time base [ELECTR] A device which moves the fluorescent spot rhythmically across the screen of the cathode-ray tube.

time-base generator *See* sweep generator.

time constant [PHYS] **1.** The time required for a physical

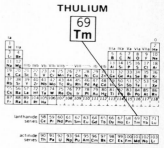

THULIUM

Periodic table of the chemical elements showing the position of thulium.

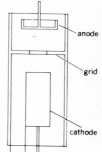

THYRATRON

Diagram of construction of a negative-grid thyratron.

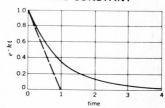

TIME CONSTANT

Graph of exponentially decreasing variable $e-kt$ (def. 2) as a function of time in time constants. Dotted line indicates variable that continues to decrease at same rate that exponential variable decreases at time $t = 0$.

quantity to rise from zero to $1 - 1/e$ (that is, 63.2%) of its final steady value when it varies with time t as $1 - e^{-kt}$. **2.** The time required for a physical quantity to fall to $1/e$ (that is, 36.8%) of its initial value when it varies with time t as e^{-kt}. **3.** Generally, the time required for an instrument to indicate a given percentage of the final reading resulting from an input signal. Also known as lag coefficient.

time delay [PHYS] The time required for a signal to travel between two points in a circuit or for a wave to travel between two points in space.

time-delay circuit [ELECTR] A circuit in which the output signal is delayed by a specified time interval with respect to the input signal. Also known as delay circuit.

time dilation effect *See* slowing of clocks.

time-interval measurement [HOROL] A process that consists either in calculating the duration between two known epochs, or in counting the repetitions of a recurring phenomenon from an arbitrary starting point, as with an electronic digital-reading counter, which counts the cycles of an oscillator.

time-invariant system [CONT SYS] A system in which all quantities governing the system's behavior remain constant with time, so that the system's response to a given input does not depend on the time it is applied.

timelike surface [RELAT] A surface in space-time whose normal vector is everywhere spacelike.

timelike vector [RELAT] A four vector in Minkowski space whose space component has a magnitude which is less than the magnitude of its time component multiplied by the speed of light.

time-mark generator [ELECTR] A signal generator that produces highly accurate clock pulses which can be superimposed as pips on a cathode-ray screen for timing the events shown on the display.

time measurement [HOROL] A process that consists in counting the repetitions of any recurring phenomenon and, if the interval between successive recurrences is sensible, in subdividing it.

time meridian [ASTRON] Any meridian used as a reference for reckoning time, particularly a zone or standard meridian.

time of flight [MECH] Elapsed time in seconds from the instant a projectile or other missile leaves a gun or launcher until the instant it strikes or bursts. [PHYS] The elapsed time from the instant a particle leaves a source to the instant it reaches a detector.

time-of-flight mass spectrometer [SPECT] A mass spectrometer in which all the positive ions of the material being analyzed are ejected into the drift region of the spectrometer tube with essentially the same energies, and spread out in accordance with their masses as they reach the cathode of a magnetic electron multiplier at the other end of the tube.

time-phase [PHYS] Two disturbances are in time phase if they reach corresponding peak values at the same instants of time, though not necessarily at the same points in space.

time quadrature [PHYS] **1.** Differing by a time interval corresponding to one-fourth the time of one cycle of the frequency in question. **2.** An integration over time.

time reversal [PHYS] The replacement of the time coordinate t by its negative $-t$ in the equations of motion of a dynamical system; the time reversal operator, a symmetry operator for a quantum-mechanical system, contains also the complex conjugation operator and a matrix operating on the spin coordinate.

time-reversal test [STAT] A test used with index numbers that is satisfied when the new index is the reciprocal of the original index if the functions of the base period and given

---

TIME-DELAY CIRCUIT

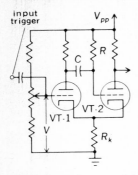

Circuit diagram of a monostable multivibrator used as a time-delay circuit. A small input trigger pulse gives rise to output signals at the plates of vacuum tubes VT-1 and VT-2 whose duration is proportional to resistance R when product of R and capacitance C is held fixed. R_k = resistance of cathode lead; V_{pp} = plate supply voltage.

period are interchanged; the advantage of index numbers meeting the criteria of the test is that a symmetric comparison of the two periods is obtained and the results are consistent whether one or the other period is used as a base.

time series [STAT] A statistical process analogous to the taking of data at intervals of time.

time series analysis [MATH] The general study of mathematical systems or processes analogous to that of data taken at time intervals.

time standard [HOROL] A recurring phenomenon, used as a reference for establishing a unit of time; the presently accepted standard is the second, defined to be 9,192,631,770 transitions between two specified hyperfine levels of the atom of cesium-133.

time-varying system [CONT SYS] A system in which certain quantities governing the system's behavior change with time, so that the system will respond differently to the same input at different times.

time zone [ASTRON] To avoid the inconvenience of the continuous change of mean solar time with longitude, the earth is divided into 24 time zones, each about 15° wide and centered on standard longitudes, 0°, 15°, 30°, and so on; within each zone the time kept is the mean solar time of the standard meridian.

timing-axis oscillator *See* sweep generator.

tin [CHEM] Metallic element in group IV, symbol Sn, atomic number 50, atomic weight 118.69; insoluble in water, soluble in acids and hot potassium hydroxide solution; melts at 232°C, boils at 2260°C.

tint [OPTICS] The mixture of a pure color with white.

tint of passage [OPTICS] The color produced when a plate which is colorless, but which rotates the plane of polarization of polarized light passing through it by an amount which depends on the wavelength of the light, is placed between crossed polarizers.

tintometer [OPTICS] A device used to estimate the intensity of a colored solution by comparing it with standard solutions or colored glass slides, as with the Lovibond tintometer.

TIP *See* terminal indecomposable past.

tissue dose [NUCLEO] The dose received by a tissue in the region of interest, expressed in roentgens for x-rays and gamma rays.

tissue roentgen *See* rep.

titanium [CHEM] A metallic element in group IV, symbol Ti, atomic number 22, atomic weight 47.90; ninth most abundant element in the earth's crust; insoluble in water, melts at 1660°C, boils above 3000°C.

Titchmarsh's theorem [MATH] If $f(x)$ and $g(x)$ are continuous functions on the positive real numbers and are not identically equal to 0, then their convolution is not identically 0.

Titius-Bode law *See* Bode's law.

T junction [ELECTR] A network of waveguides with three waveguide terminals arranged in the form of a letter T; in a rectangular waveguide a symmetrical T junction is arranged by having either all three broadsides in one plane or two broadsides in one plane and the third in a perpendicular plane.

Tl *See* thallium.

T²L *See* transistor-transistor logic.

Tm *See* thulium.

TME *See* metric-technical unit of mass.

TM mode *See* transverse magnetic mode.

TM wave *See* transverse magnetic wave.

Tn *See* thoron.

TIN

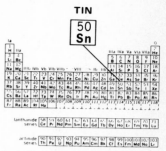

Periodic table of the chemical elements showing the position of tin.

TITANIUM

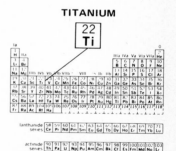

Periodic table of the chemical elements showing the position of titanium.

T JUNCTION

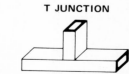

Drawing showing one type of waveguide T junction.

T network [ELEC] A network composed of three branches, with one end of each branch connected to a common junction point, and with the three remaining ends connected to an input terminal, an output terminal, and a common input and output terminal, respectively.

TNT equivalent [NUCLEO] A measure of the energy released in the detonation of a nuclear weapon, expressed in terms of the weight of TNT that would release the same amount of energy when exploded; usually expressed in kilotons or megatons of TNT; based on the release of 10^9 calories (approximately 4.18×10^9 joules) of energy by 1 ton of TNT.

Toepler-Holtz machine [ELEC] An early type of machine for continuously producing electrical charges at high voltage by electrostatic induction, superseded by the Wimhurst machine. Also known as Holtz machine.

tokamak [PL PHYS] A device for confining a plasma within a toroidal chamber, which produces plasma temperatures, densities, and confinement times greater than that of any other such device; confinement is effected by a very strong externally applied toroidal field, plus a weaker poloidal field produced by a toroidally directed plasma current, and this current causes ohmic heating of the plasma.

tolerance dose *See* permissible dose.

Tolman and Stewart effect [ELEC] The development of negative charge at the forward end of a metal rod which is suddenly stopped after rapid longitudinal motion.

tomography *See* sectional radiography.

ton [MECH] **1.** A unit of weight in common use in the United States, equal to 2000 pounds or 907.18474 kilograms-force. Also known as just ton; net ton; short ton. **2.** A unit of mass in common use in the United Kingdom equal to 2240 pounds, or to 1016.0469088 kilograms-force. Also known as gross ton; long ton. **3.** A unit of weight in troy measure, equal to 2000 troy pounds, or to 746.4834432 kilograms-force. **4.** *See* tonne. [NUCLEO] The energy released by one metric ton of chemical high explosives calculated at the rate of 1000 calories per gram; equal to 4.18×10^9 joules; used principally in expressing the energy released by a nuclear bomb.

tondal [MECH] A unit of force equal to the force which will impart an acceleration of 1 foot per second to a mass of 1 long ton; equal to approximately 309.6911 newtons.

tone [ACOUS] **1.** A sound oscillation capable of exciting an auditory sensation having pitch. **2.** An auditory sensation having pitch.

tonne [MECH] A unit of mass in the metric system, equal to 1000 kilograms or to approximately 2204.62 pounds-mass. Also known as metric ton; millier; ton; tonneau.

tonneau *See* tonne.

top [MECH] A rigid body, one point of which is held fixed in an inertial reference frame, and which usually has an axis of symmetry passing through this point; its motion is usually studied when it is spinning rapidly about the axis of symmetry. [QUANT MECH] *See* rotator.

Topogon lens [OPTICS] A periscopic lens with supplementary thick menisci to permit the correction of aperture aberrations for a moderate aperture and a large field; one or two plane-parallel plates are sometimes added to correct distortion.

topological dynamics [MATH] The study and application of transformations, or groups of such transformations (particularly topological transformation groups), defined on a topological space (usually compact), with particular regard to properties of interest in the qualitative theory of differential equations.

topological groups [MATH] Groups which also have a topol-

TOPOGON LENS

Components of a Topogon lens.

ogy with the property that the group operation and the inverse operation determine continuous functions.

topological linear space   *See* topological vector space.

topologically closed set   *See* closed set.

topological mapping   *See* homeomorphism.

topological product   [MATH] The topological space obtained from taking the cartesian product of topological spaces.

topological property   [MATH] A property that holds true for any topological space homeomorphic to one possessing the property.

topological space   [MATH] A set endowed with a topology.

topological transformation group   [MATH] A topological group of continuous transformations on a topological space; more precisely, a triple (G,X,π), where G is a topological group, X is a topological space, and π is a continuous mapping of $G \times X$ onto X satisfying the equation $\pi(gh,x) = \pi[g, \pi(h,x)]$, for all g and h in G and x in X. Also known as transformation group.

topological vector space   [MATH] A vector space which has a topology with the property that vector addition and scalar multiplication are continuous functions. Also known as linear topological space; topological linear space.

topology   [MATH] **1.** A collection of subsets of a set X, which includes X and the empty set, and has the property that any union or finite intersection of its members is also a member. **2.** The generalized study of properties of spaces invariant under deformations and stretchings.

topology of circuits   [ELEC] The study of electric networks in terms of the geometry of their connections only; used in finding such properties of circuits as equivalence and duality, and in analyzing and synthesizing complex circuits.

toric lens   [OPTICS] A lens whose surfaces form portions of toric surfaces. Also known as toroidal lens.

toric surface   [MATH] A surface generated by rotating an arc of a circle about a line that lies in the plane of the circle but does not pass through its center. Also known as toroidal surface.

toroid   *See* doughnut; toroidal magnetic circuit.

toroidal   [SCI TECH] Shaped like a doughnut.

toroidal coil   *See* toroidal magnetic circuit.

toroidal coordinate system   [MATH] A three-dimensional coordinate system whose coordinate surfaces are the toruses and spheres generated by rotating the families of circles defining a two-dimensional bipolar coordinate system about the perpendicular bisector of the line joining the common points of intersection of one of the families, together with the planes passing through the axis of rotation.

toroidal core   [ELECTROMAG] The doughnut-shaped piece of magnetic material in a toroidal magnetic circuit.

toroidal discharge   *See* ring discharge.

toroidal lens   *See* toric lens.

toroidal magnetic circuit   [ELECTROMAG] Doughnut-shaped piece of magnetic material, together with one or more coils of current-carrying wire wound about the doughnut, with the permeability of the magnetic material high enough so that the magnetic flux is almost completely confined within it. Also known as toroid; toroidal coil.

toroidal surface   *See* toric surface.

torque   [MECH] **1.** For a single force, the cross product of a vector from some reference point to the point of application of the force with the force itself. Also known as moment of force; rotation moment. **2.** For several forces, the vector sum of the torques (first definition) associated with each of the forces.

torque-coil magnetometer   [ENG] A magnetometer that de-

**TOROIDAL
MAGNETIC CIRCUIT**

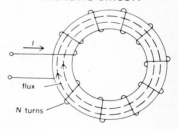

Diagram of toroidal magnetic circuit. I represents current. (From A. E. Fitzgerald, D. E. Higginbotham, and A. Grabel, Basic Electrical Engineering, McGraw-Hill, 1967)

pends for its operation on the torque developed by a known current in a coil that can turn in the field to be measured.

torquemeter [ENG] An instrument to measure torque.

torque-type viscometer [ENG] A device that measures liquid viscosity by the torque needed to rotate a vertical paddle submerged in the liquid; used for both Newtonian and non-Newtonian liquids and for suspensions.

torr [MECH] A unit of pressure, equal to 1/760 atmosphere; it differs from 1 millimeter of mercury by less than one part in seven million; approximately equal to 133.3224 pascals.

Torricellian barometer *See* mercury barometer.

Torricellian vacuum [FL MECH] The space enclosed above a column of mercury when a tube, closed at one end, is filled with mercury and then placed, open end downward, in a well of mercury; this space is evacuated except for mercury vapor.

Torricelli's law of efflux [FL MECH] The velocity of efflux of liquid from an orifice in a container is equal to that which would be attained by a body falling freely from rest at the free surface of the liquid to the orifice.

torsion [MECH] A twisting deformation of a solid body about an axis in which lines that were initially parallel to the axis become helices.

torsional angle [MECH] The total relative rotation of the ends of a straight cylindrical bar when subjected to a torque.

torsional compliance [MECH] The reciprocal of the torsional rigidity.

torsional hysteresis [MECH] Dependence of the torques in a twisted wire or rod not only on the present torsion of the object but on its previous history of torsion.

torsional modulus [MECH] The ratio of the torsional rigidity of a bar to its length. Also known as modulus of torsion.

torsional pendulum [MECH] A device consisting of a disk or other body of large moment of inertia mounted on one end of a torsionally flexible elastic rod whose other end is held fixed; if the disk is twisted and released, it will undergo simple harmonic motion, provided the torque in the rod is proportional to the angle of twist.

torsional rigidity [MECH] The ratio of the torque applied about the centroidal axis of a bar at one end of the bar to the resulting torsional angle, when the other end is held fixed.

torsional vibration [MECH] A periodic motion of a shaft in which the shaft is twisted about its axis first in one direction and then in the other; this motion may be superimposed on rotational or other motion.

torsional wave [MECH] A wave resulting from torsional vibrations in one or more parts of a substance.

torsion balance [ENG] An instrument, consisting essentially of a straight vertical torsion wire whose upper end is fixed while a horizontal beam is suspended from the lower end; used to measure minute gravitational, electrostatic, or magnetic forces.

torsion element [MATH] 1. A torsion element of an Abelian group G is an element of G with finite period. 2. A torsion element of a module M over an entire, principal ring R is an element x in M for which there exists an element a in R such that $a \neq 0$ and $ax = 0$.

torsion-free group [MATH] A group whose only torsion element is the unit element.

torsion function [MECH] A harmonic function, $\phi(x,y) = w/\tau$, expressing the warping of a cylinder undergoing torsion, where the $x, y,$ and z coordinates are chosen so that the axis of torsion lies along the z axis, w is the z component of the displacement, and τ is the torsion angle. Also known as warping function.

torsion galvanometer [ENG] A galvanometer in which the

TORSION

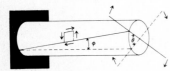

Cylindrical bar in torsion, showing deformation of a small element of the bar; θ represents torsional angle; ϕ represents helical angle.

TORSIONAL PENDULUM

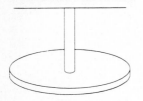

Diagram of a torsional pendulum.

force between the fixed and moving systems is measured by the angle through which the supporting head of the moving system must be rotated to bring the moving system back to its zero position.

torsion group [MATH] **1.** A group whose elements all have finite period. **2.** For a topological space X, one of a sequence of finite groups $G_n(X)$ such that the homology group $H_n(X)$ is the direct sum of $G_n(X)$ and a number of infinite cyclic groups.

torsion hygrometer [ENG] A hygrometer in which the rotation of the hygrometric element is a function of the humidity; such hygrometers are constructed by taking a substance whose length is a function of the humidity and twisting or spiraling it under tension in such a manner that a change in length will cause a further rotation of the element.

torsion module [MATH] A module M over an entire principal ring R is said to be a torsion module if for any element x in M there exists an element a in R such that $a \neq 0$ and $ax = 0$.

torsion of a curve [MATH] For a given space curve, the rate at which the curve turns out of its osculating plane relative to arc length; it is defined in terms of the binormals.

torsion-string galvanometer [ENG] A sensitive galvanometer in which the moving system is suspended by two parallel fibers that tend to twist around each other.

torsion subgroup [MATH] The torsion subgroup of an Abelian group G is the subset of all torsion elements of G.

torsion submodule [MATH] The torsion submodule of a module E over an entire principal ring is the submodule consisting of all torsion elements of E.

torus [MATH] **1.** The surface of a doughnut. **2.** The topological space obtained by identifying the opposite sides of a rectangle. **3.** The group which is the product of two circles.

total binding energy *See* binding energy.

total curvature *See* Gaussian curvature.

total curvature of a lens [OPTICS] The difference between the reciprocals of the radii of curvature of the two surfaces of a lens.

total differential [MATH] The total differential of a function of several variables, $f(x_1, x_2, \ldots, x_n)$, is the function given by the sum of terms $(\partial f/\partial x_i)\, dx_i$ as i runs from 1 to n. Also known as differential.

total eclipse [ASTRON] An eclipse that obscures the entire surface of the moon or sun.

total head [FL MECH] The sum of the velocity head and the pressure head corresponding to the static pressure.

total heat *See* enthalpy.

total heat of dilution *See* heat of dilution.

total heat of solution *See* heat of solution.

total internal reflection [OPTICS] A phenomenon in which electromagnetic radiation in a given medium which is incident on the boundary with a less-dense medium (one having a lower index of refraction) at an angle less than the critical angle is completely reflected from the boundary.

totally bounded set *See* precompact set.

totally disconnected [MATH] A topological space has this property if the largest connected subset containing any given point is only the point itself.

totally imaginary field [MATH] An extension field F of the field of rational numbers such that no embedding of F in the complex numbers is contained in the real numbers.

total order [MATH] **1.** The total order of an analytic function in a domain D is the algebraic sum of its orders at all poles and zeros in D. **2.** *See* linear order.

total pressure [FL MECH] *See* dynamic pressure. [MECH] The gross load applied on a given surface.

total radiation pyrometer [ENG] A pyrometer which focuses

TOTAL RADIATION PYROMETER

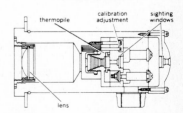

Diagram of total radiation pyrometer. Lens focuses heat radiation onto thermopile. (*Honeywell Inc.*)

heat radiation emitted by a hot object on a detector (usually a thermopile or other thermal type detector), and which responds to a broad band of radiation, limited only by absorption of the focusing lens, or window and mirror.

total relief *See* stereoscopic power.

total space of a bundle [MATH] The topological space E in the bundle (E,p,B).

total specific ionization *See* specific ionization.

total variation [MATH] **1.** For a real function defined on an interval, the least upper bound of the function's variation relative to all possible partitions of the interval. **2.** The total variation of a signed measure m is a set function $|m|$ defined for every measurable set E by $|m|(E) = m^+(E) + m^-(E)$, where m^+ and m^- are the upper and lower variations of m.

total vorticity [FL MECH] Usually, the magnitude of the vorticity vector, all components included, as opposed to the vertical (component of the) vorticity.

toughness [MECH] A property of a material capable of absorbing energy by plastic deformation; intermediate between softness and brittleness.

tower [MATH] For a set S with a given algebraic structure, this is a set of subsets, $S_0 = S, S_1, S_2, \ldots, S_n$, such that S_{i+1} is a subset of S_i, $i = 1, 2, \ldots, n-1$, and each S_i is closed under all possible operations in the algebraic structure of S.

Townsend avalanche *See* avalanche.

Townsend characteristic [ELECTR] Current-voltage characteristic curve for a phototube at constant illumination and at voltages below that at which a glow discharge occurs.

Townsend coefficient [ELECTR] The number of ionizing collisions by an electron per centimeter of path length in the direction of the applied electric field in a radiation counter.

Townsend discharge [ELECTR] A discharge which occurs at voltages too low for it to be maintained by the electric field alone, and which must be initiated and sustained by ionization produced by other agents; it occurs at moderate pressures, above about 0.1 torr, and is free of space charges.

Townsend ionization *See* avalanche.

Townsend second ionization coefficient [NUCLEO] The number of electrons released from the cathode of an ionization chamber per initial ionizing collision in the gas. Also known as secondary ionization coefficient.

Tr *See* trace.

trace [ELECTR] The visible path of a moving spot on the screen of a cathode-ray tube. Also known as line. [MATH] **1.** The trace of a matrix is the sum of the entries along its principal diagonal. Designated Tr. Also known as spur. **2.** The trace of a linear transformation on a finite-dimensional vector space is the trace (in the sense of the first definition) of the matrix associated with it. **3.** Let E be a finite extension of a field F, and let a be an element of E; the trace of a from E to F is the element

$$q \sum_{j=1}^{p} f_j(a)$$

where q is the inseparable degree of E over F, p is the separable degree of E over F, and f_j, $j = 1, \ldots, p$, are the distinct embeddings of E in $\overline{F}$ over F, where $\overline{F}$ is the algebraic closure of F.

tracer [CHEM] A foreign substance, usually radioactive, that is mixed with or attached to a given substance so the distribution or location of the latter can later be determined; used to trace chemical behavior of a natural element in an organism. Also known as tracer element.

tracer element   *See* tracer.

trace sensitivity   [ELECTR] The ability of an oscilloscope to produce a visible trace on the scope face for a specified input voltage.

track   [NUCLEO] **1.** The visible path of an ionizing particle in a particle detector, such as a cloud chamber, bubble chamber, spark chamber, or nuclear photographic emulsion. **2.** *See* race track.

tracking problem   [CONT SYS] The problem of determining a control law which when applied to a dynamical system causes its output to track a given function; the performance index is in many cases taken to be of the integral square error variety.

tracking telescope   [OPTICS] A long-focal-length telescope mounted to track missiles in flight precisely while collecting missile performance data.

traction   [MECH] Pulling friction of a moving body on the surface on which it moves.

tractional force   [FL MECH] The force exerted on particles under flowing water by the current; it is proportional to the square of the velocity.

tractrix   [MATH] A curve in the plane where every tangent to it has the same length.

trajectory   [MECH] The curve described by an object moving through space, as of a meteor through the atmosphere, a planet around the sun, a projectile fired from a gun, or a rocket in flight.

trammel   [ENG] A device consisting of a bar, each of whose ends is constrained to move along one of two perpendicular lines; used in drawing ellipses and in the Rowland mounting.

transadmittance   [ELECTR] A specific measure of transfer admittance under a given set of conditions, as in forward transadmittance, interelectrode transadmittance, short-circuit transadmittance, small-signal forward transadmittance, and transadmittance compression ratio.

transcendence base   [MATH] A transcendence base of a field E over a subfield F is a subset S of E which is algebraically independent over F and is not a proper subset of any other subset S' which is algebraically independent over F.

transcendence degree   [MATH] The transcendence degree of a field E of a subfield F is the number of elements in a transcendence base of E over F. Also known as transcendence dimension.

transcendence dimension   *See* transcendence degree.

transcendental curve   [MATH] The graph of a transcendental function.

transcendental element   [MATH] An element of a field K is transcendental relative to a subfield F if it satisfies no polynomial whose coefficients come from F.

transcendental field extension   [MATH] A field extension K of F where the elements of K not in F are all transcendental relative to F.

transcendental functions   [MATH] Functions which cannot be given by any algebraic expression involving only their variables and constants.

transcendental number   [MATH] An irrational number that is the root of no polynomial with rational-number coefficients.

transconductance   [ELECTR] An electron-tube rating, equal to the change in plate current divided by the change in control-grid voltage that causes it, when the plate voltage and all other voltages are maintained constant. Symbolized G_m ; g_m. Also known as grid-anode transconductance; grid-plate transconductance; mutual conductance.

transducer   [ENG] Any device or element which converts an input signal into an output signal of a different form; examples include the microphone, phonograph pickup, loudspeaker, barometer, photoelectric cell, automobile horn, doorbell, and underwater sound transducer.

transductor   *See* magnetic amplifier; saturable reactor.

transfer admittance   [ELECTR] An admittance rating for electron tubes and other transducers or networks; it is equal to the complex alternating component of current flowing to one terminal from its external termination, divided by the complex alternating component of the voltage applied to the adjacent terminal on the cathode or reference side; all other terminals have arbitrary external terminations.

transfer characteristic   [ELECTR] **1.** Relation, usually shown by a graph, between the voltage of one electrode and the current to another electrode, with all other electrode voltages being maintained constant. **2.** Function which, multiplied by an input magnitude, will give a resulting output magnitude. **3.** Relation between the illumination on a camera tube and the corresponding output-signal current, under specified conditions of illumination.

transfer constant   [ENG] A transducer rating, equal to one-half the natural logarithm of the complex ratio of the product of the voltage and current entering a transducer to that leaving the transducer when the latter is terminated in its image impedance; alternatively, the product may be that of force and velocity or pressure and volume velocity; the real part of the transfer constant is the image attenuation constant, and the imaginary part is the image phase constant. Also known as transfer factor.

transfer factor   *See* transfer constant.

transfer function   [CONT SYS] The mathematical relationship between the output of a control system and its input: for a linear system, it is the Laplace transform of the output divided by the Laplace transform of the input under conditions of zero initial-energy storage.

transfer impedance   [ELEC] The ratio of the voltage applied at one pair of terminals of a network to the resultant current at another pair of terminals, all terminals being terminated in a specified manner.

transfer matrix method   [MECH] A method of analyzing vibrations of complex systems, in which the system is approximated by a finite number of elements connected in a chainlike manner, and matrices are constructed which can be used to determine the configuration and forces acting on one element in terms of those on another.

transfer ratio   [ENG] From one point to another in a transducer at a specified frequency, the complex ratio of the generalized force or velocity at the second point to the generalized force or velocity applied at the first point; the generalized force or velocity includes not only mechanical quantities, but also other analogous quantities such as acoustical and electrical; the electrical quantities are usually electromotive force and current.

transfer reaction   [NUC PHYS] A nuclear reaction in which one or more nucleons are exchanged between the target nucleus and an incident projectile.

transferred-electron amplifier   [ELECTR] A diode amplifier, which generally uses a transferred-electron diode made from doped n-type gallium arsenide, that provides amplification in the gigahertz range to well over 50 gigahertz at power outputs typically below 1 watt continuous-wave. Abbreviated TEA.

transferred-electron device   [ELECTR] A semiconductor device, usually a diode, that depends on internal negative resistance caused by transferred electrons in gallium arsenide or indium phosphide at high electric fields; transit time is minimized, permitting oscillation at frequencies up to several hundred megahertz.

transferred-electron effect [SOLID STATE] The variation in the effective drift mobility of charge carriers in a semiconductor when significant numbers of electrons are transferred from a low-mobility valley of the conduction band in a zone to a high-mobility valley, or vice versa.

transfinite induction [MATH] A reasoning process by which if a theorem holds true for the first element of a well-ordered set N and is true for an element n whenever it holds for all predecessors of n, then the theorem is true for all members of N.

transfinite number [MATH] Any ordinal or cardinal number equal to or exceeding aleph null.

transform [MATH] **1.** A conjugate of an element of a group. **2.** An expression, commonly used in harmonic analysis, formed from a given function f by taking an integral of $f \cdot g$, where g is a member of an orthogonal family of functions. **3.** The value of a transformation at some point.

transformation [CRYSTAL] *See* inversion. [ELEC] For two networks which are equivalent as far as conditions at the terminals are concerned, a set of equations which give the admittances or impedances of the branches of one circuit in terms of the admittances or impedances of the other. [MATH] A function, usually between vector spaces.

transformation constant *See* decay constant.

transformation group [MATH] **1.** A collection of transformations which forms a group with composition as the operation. **2.** *See* topological transformation group.

transformation matrix [ELECTROMAG] A two-by-two matrix which relates the amplitudes of the traveling waves on one side of a waveguide junction to those on the other.

transformation methods [MATH] A category of numerical methods for finding the eigenvalues of a matrix, in which a series of orthogonal transformations are used to reduce the matrix to some simpler matrix, usually a triple-diagonal one, before an attempt is made to find the eigenvalues.

transformation series *See* radioactive series.

transformation theory [QUANT MECH] The study of coordinate and other transformations in quantum mechanics, especially those which leave some properties of the system invariant.

transformation twin [CRYSTAL] A crystal twin developed by a growth transformation from a higher to a lower symmetry.

transformer [ELECTROMAG] An electrical component consisting of two or more multiturn coils of wire placed in close proximity to cause the magnetic field of one to link the other; used to transfer electric energy from one or more alternating-current circuits to one or more other circuits by magnetic induction.

transformer coupling [ELEC] *See* inductive coupling. [ELECTR] Interconnection between stages of an amplifier which employs a transformer for connecting the plate circuit of one stage to the grid circuit of the following stage; a special case of inductive coupling.

transient [PHYS] A pulse, damped oscillation, or other temporary phenomenon occurring in a system prior to reaching a steady-state condition.

transient analyzer [ELECTR] An analyzer that generates transients in the form of a succession of equal electric surges of small amplitude and adjustable waveform, applies these transients to a circuit or device under test, and shows the resulting output waveforms on the screen of an oscilloscope.

transient equilibrium [NUCLEO] Radioactive equilibrium in which the lifetime of the parent is sufficiently short that the quantity present decreases appreciably in the period under consideration.

transient motion [PHYS] An oscillatory or other irregular motion occurring while a quantity is changing to a new steady-state value.

transient overshoot [PHYS] The maximum value of the overshoot of a quantity as a result of a sudden change in conditions.

transient phenomena [ELEC] Rapidly changing actions occurring in a circuit during the interval between closing of a switch and settling to a steady-state condition, or any other temporary actions occurring after some change in a circuit.

transient problem *See* initial-value problem.

transient response [PHYS] The behavior of a system following a sudden change in its input.

transistance [ELECTR] The characteristic that makes possible the control of voltages or currents so as to accomplish gain or switching action in a circuit; examples of transistance occur in transistors, diodes, and saturable reactors.

transistor [ELECTR] An active component of an electronic circuit consisting of a small block of semiconducting material to which at least three electrical contacts are made, usually two closely spaced rectifying contacts and one ohmic (nonrectifying) contact; it may be used as an amplifier, detector, or switch.

transistor biasing [ELECTR] Maintaining a direct-current voltage between the base and some other element of a transistor.

transistor characteristics [ELECTR] The values of the impedances and gains of a transistor.

transistor chip [ELECTR] An unencapsulated transistor of very small size used in microcircuits.

transistor input resistance [ELECTR] The resistance across the input terminals of a transistor stage. Also known as input resistance.

transistor-transistor logic [ELECTR] A logic circuit containing two transistors, for driving large output capacitances at high speed. Abbreviated T^2L; TTL.

transit [ASTRON] **1.** A celestial body's movement across the meridian of a place. Also known as meridian transit. **2.** Passage of a smaller celestial body across a larger one. **3.** Passage of a satellite's shadow across the disk of its primary.

transit circle [ENG] A type of astronomical transit instrument having a micrometer eyepiece that has an extra pair of moving wires perpendicular to the vertical set to measure the zenith distance or declination of the celestial object in conjunction with readings taken from a large, accurately calibrated circle attached to the horizontal axis. Also known as meridian circle; meridian transit.

transit instrument *See* transit telescope.

transition [QUANT MECH] The change of a quantum-mechanical system from one energy state to another. [THERMO] A change of a substance from one of the three states of matter to another.

transitional flow [FL MECH] A flow in which the viscous and Reynolds stresses are of approximately equal magnitude; it is transitional between laminar flow and turbulent flow.

transition element [CHEM] One of a group of metallic elements in which the members have the filling of the outermost shell to 8 electrons interrupted to bring the penultimate shell from 8 to 18 or 32 electrons; includes elements 21 through 29 (scandium through copper), 39 through 47 (yttrium through silver), 57 through 79 (lanthanum through gold), and all known elements from 89 (actinium) on. [ELECTROMAG] An element used to couple one type of transmission system to another, as for coupling a coaxial line to a waveguide.

transition frequency [QUANT MECH] The characteristic fre-

TRANSIT CIRCLE

Six-inch (15-centimeter) transit circle, U.S. Naval Observatory. *(Official U.S. Naval Observatory photograph)*

quency of radiation emitted or absorbed by a quantum-mechanical system as it changes from one energy state to another; equal to the energy difference between the states divided by Planck's constant.

transition moment [QUANT MECH] Any type of multipole moment which determines radiative transitions between states; it consists of an integral of the product of the conjugate of the final state wave function, a multipole moment operator, and the initial state wave function.

transition point [ELECTROMAG] A point at which the constants of a circuit change in such a way as to cause reflection of a wave being propagated along the circuit. [THERMO] Either the temperature at which a substance changes from one state of aggregation to another (a first-order transition), or the temperature of culmination of a gradual change, such as the lambda point, or Curie point (a second-order transition). Also known as transition temperature.

transition probability [MATH] Conditional probability concerning a discrete Markov chain giving the probabilities of change from one state to another. [QUANT MECH] The probability per unit time that a quantum-mechanical system will make a transition from a given initial state to a given final state.

transition region [SOLID STATE] The region between two homogeneous semiconductors in which the impurity concentration changes.

transition temperature *See* transition point.

transition zone [FL MECH] Those conditions of fluid flow in which the nature of the flow is changing from laminar to turbulent.

transitive group [MATH] A group of permutations of a finite set such that for any two elements in the set there exists an element of the group which takes one into the other.

transitive relation [MATH] A relation $<$ on a set such that if $a < b$ and $b < c$, then $a < c$.

transit telescope [OPTICS] A telescopic instrument adapted to the observation of the passage, or transit, of an astronomical object across the meridian of an observer; consists of a telescope mounted on a single fixed horizontal axis of rotation which has a central hollow cube (sometimes a sphere) and two conical semiaxes ending in cylindrical pivots; the objective and eyepiece halves of the instrument are also fastened to the cube of the instrument, perpendicular to the horizontal axis. Also known as transit instrument.

transit time [ELECTR] The time required for an electron or other charge carrier to travel between two electrodes in an electron tube or transistor.

transit-time microwave diode [ELECTR] A solid-state microwave diode in which the transit time of charge carriers is short enough to permit operation in microwave bands.

transit-time mode [ELECTR] One of the three operating modes of a transferred-electron diode, in which space-charge domains are formed at the cathode and travel across the drift region to the anode.

translation [MATH] **1.** A function changing the coordinates of a point in a euclidean space into new coordinates relative to axes parallel to the original. **2.** A function on a group to itself given by operating on each element by some one fixed element. **3.** Let E be a finitely generated extension of a field k, F be an extension of k, and both E and F be contained in a common field; the translation of E to F is the extension EF of F, where EF is the compositum of E and F. Also known as lifting. [MECH] The linear movement of a point in space without any rotation.

translational motion [MECH] Motion of a rigid body in such

Large transit telescope, Pulkovo Observatory, Soviet Union, in use since 1838. *(Courtesy of B. L. Klock)*

a way that any line which is imagined rigidly attached to the body remains parallel to its original direction.

translation gliding *See* crystal gliding.

translation group [CRYSTAL] The collection of all translation operations which carry a crystal lattice into itself.

translation operation [PHYS] The process of moving an object along a straight line in such a way that any line which is fixed with respect to the object remains parallel to its original direction.

translation operator [MATH] An operator T on the space C of all continuous bounded functions on the positive real numbers, defined by the formula $(Tf)(x) = f(x + a)$, where f is any element of C, x is any positive number, and a is some positive constant.

translucent medium [OPTICS] A medium which transmits rays of light so diffused that objects cannot be seen distinctly; examples are various forms of glass which admit considerable light but impede vision.

transmission *See* transmittance.

transmission band [ELECTROMAG] Frequency range above the cutoff frequency in a waveguide, or the comparable useful frequency range for any other transmission line, system, or device.

transmission coefficient [PHYS] **1.** The value of some quantity associated with the resultant field produced by incident and reflected waves at a given point in a transmission medium divided by the corresponding quantity in the incident wave. **2.** The ratio of transmitted to incident energy flux or flux of some other quantity at a discontinuity in a transmission medium; for sound waves, it is called the sound transmission coefficient. **3.** The ratio of the transmitted flux of some quantity to the incident flux for a substance of unit thickness. [QUANT MECH] *See* penetration probability.

transmission electron microscope [ELECTR] A type of electron microscope in which the specimen transmits an electron beam focused on it, image contrasts are formed by the scattering of electrons out of the beam, and various magnetic lenses perform functions analogous to those of ordinary lenses in a light microscope.

transmission electron radiography [ELECTR] A technique used in microradiography to obtain radiographic images of very thin specimens; the photographic plate is in close contact with the specimen, over which is placed a lead foil and then a light-tight covering; hardened x-rays shoot through the light-tight covering.

transmission factor [PHYS] The ratio of the flux of some quantity transmitted through a body to the incident flux.

transmission function [GEOPHYS] A mathematical formulation of relationships between infrared transmission in the atmosphere, the path length, and the concentration of absorbing gases.

transmission gain *See* gain.

transmission grating [OPTICS] A diffraction grating produced on a transparent base so radiation is transmitted through the grating instead of being reflected from it.

transmission line [ELEC] A system of conductors, such as wires, waveguides, or coaxial cables, suitable for conducting electric power or signals efficiently between two or more terminals.

transmission mode *See* mode.

transmission plane [OPTICS] The plane of vibration of polarized light that will pass through a Nicol prism or other polarizer.

transmission range *See* night visual range.

transmissivity [ELECTROMAG] The ratio of the transmitted radiation to the radiation arriving perpendicular to the boundary between two mediums.

transmissometer [ENG] An instrument for measuring the extinction coefficient of the atmosphere and for the determination of visual range. Also known as hazemeter; transmittance meter.

transmissometry [OPTICS] The technique of determining the extinction characteristics of a medium by measuring the transmission of a light beam of known initial intensity directed into that medium.

transmittance [ELECTROMAG] The radiant power transmitted by a body divided by the total radiant power incident upon the body. Also known as transmission.

transmittance meter *See* transmissometer.

transmittancy [ELECTROMAG] The transmittance of a solution divided by that of the pure solvent of the same thickness.

transmitted wave *See* refracted wave.

transmittivity [ELECTROMAG] The internal transmittance of a piece of nondiffusing substance of unit thickness.

transmutation [NUC PHYS] A nuclear process in which one nuclide is transformed into the nuclide of a different element. Also known as nuclear transformation.

transonic [PHYS] That which occurs or is occurring within the range of speed in which flow patterns change from subsonic to supersonic (or vice versa), about Mach 0.8 to 1.2, as in transonic flight or transonic flutter.

transonic speed [FL MECH] The speed of a body relative to the surrounding fluid at which the flow is in some places on the body subsonic and in other places supersonic.

transparent [PHYS] Permitting passage of radiation or particles.

transparent medium [OPTICS] **1.** A medium which has the property of transmitting rays of light in such a way that the human eye may see through the medium distinctly. **2.** A medium transparent to other regions of the electromagnetic spectrum, such as x-rays and microwaves.

transplutonium element [INORG CHEM] An element having an atomic number greater than that of plutonium (94).

transport cross section [PHYS] The product of the total scattering cross section and the average value of $1 - \cos \theta$, where θ is the laboratory scattering angle.

transport mean free path [NUCLEO] **1.** A path length equal to three times the diffusion coefficient of neutron flux in a nuclear reactor when Fick's law is applicable. **2.** A modification of the mean free path to take into account anisotropy of scattering and the persistence of velocities.

transport properties [PHYS] Properties of a compound or material associated with mass or heat transport; for example, viscosity and thermal conductivity of liquids, gases, or solids.

transpose of a matrix [MATH] The matrix obtained from the original matrix by interchanging its rows and columns.

transposition [MATH] A permutation of a set of symbols which exchanges exactly two while leaving all others unaffected.

transrectifier [ELECTR] Device, ordinarily a vacuum tube, in which rectification occurs in one electrode circuit when an alternating voltage is applied to another electrode.

transuranic elements [CHEM] Elements that have atomic numbers greater than 92; all are radioactive, are products of artificial nuclear changes, and are members of the actinide group. Also known as transuranium elements.

transuranium elements *See* transuranic elements.

transversal [MATH] **1.** A line intersecting a given family of lines. **2.** A curve orthogonal to a hypersurface. **3.** If π is a given map of a set X onto a set Y, a transversal for π is a subset

T of X with the property that T contains exactly one point of $\pi^{-1}(y)$ for each $y \, \varepsilon \, Y$.

transversality conditions [MATH] Additional equations which must be satisfied by the independent variable x and the solution functions $y_1, y_2, \ldots, y_n$ of a problem of minimizing the integral of $f(y_1, y_2, \ldots, y_n, y_1', \ldots, y_n', x)$, when $y_i' = dy_i/dx$, with one of the limits of the integral given implicitly by $m \le n$ relations among x and the y_i.

transverse cylindrical orthomorphic projection *See* transverse Mercator projection.

transverse Doppler effect [ELECTROMAG] An aspect of the optical Doppler effect, occurring when the direction of motion of the source relative to an observer is perpendicular to the direction of the light received by the observer; the observed frequency is smaller than the source frequency by the factor $[1 - (v/c)^2]^{1/2}$, where v is the speed of the source and c is the speed of light.

transverse electric mode [ELECTROMAG] A mode in which a particular transverse electric wave is propagated in a waveguide or cavity. Abbreviated TE mode. Also known as H mode (British usage).

transverse electric wave [ELECTROMAG] An electromagnetic wave in which the electric field vector is everywhere perpendicular to the direction of propagation. Abbreviated TE wave. Also known as H wave (British usage).

transverse electromagnetic mode [ELECTROMAG] A mode in which a particular transverse electromagnetic wave is propagated in a waveguide or cavity. Abbreviated TEM mode.

transverse electromagnetic wave [ELECTROMAG] An electromagnetic wave in which both the electric and magnetic field vectors are everywhere perpendicular to the direction of propagation. Abbreviated TEM wave.

transversely excited atmospheric pressure laser *See* TEA laser.

transverse magnetic mode [ELECTROMAG] A mode in which a particular transverse magnetic wave is propagated in a waveguide or cavity. Abbreviated TM mode. Also known as E mode (British usage).

transverse magnetic wave [ELECTROMAG] An electromagnetic wave in which the magnetic field vector is everywhere perpendicular to the direction of propagation. Abbreviated TM wave. Also known as E wave (British usage).

transverse magnetoresistance [ELECTROMAG] One of the galvanomagnetic effects, in which a magnetic field perpendicular to an electric current gives rise to an electrical potential change in the direction of the current.

transverse mass [RELAT] The ratio of a force acting on a relativistic particle in a direction perpendicular to its velocity to the resulting acceleration; equal to $m_0 (1 - v^2/c^2)^{-1/2}$, where m_0 is the particle's rest mass, v is its speed, and c is the speed of light.

transverse piezoelectric effect [SOLID STATE] The manifestation of the piezoelectric effect in which the applied stress is perpendicular to the direction of the resultant electric field, or in which the applied electric field is perpendicular to the direction of the resultant stress.

transverse stability [ENG] The ability of a ship or aircraft to recover an upright position after waves or wind roll it to one side.

transverse vibration [MECH] Vibration of a rod in which elements of the rod move at right angles to the axis of the rod.

transverse wave [GEOPHYS] *See* S wave. [PHYS] A wave in which the direction of the disturbance at each point of the medium is perpendicular to the wave vector and parallel to surfaces of constant phase.

TRANSVERSE VIBRATION

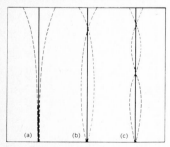

Transverse vibrations of a long circular rod, rigidly clamped at one end and free at the other. *(a)* The fundamental mode. *(b)* The first-overtone mode. *(c)* The second-overtone mode.

trap [SOLID STATE] Any irregularity, such as a vacancy, in a semiconductor at which an electron or hole in the conduction band can be caught and trapped until released by thermal agitation. Also known as semiconductor trap.

TRAPATT diode [ELECTR] A *pn* junction diode, similar to the IMPATT diode, but characterized by the formation of a trapped space-charge plasma within the junction region; used in the generation and amplification of microwave power. Derived from trapped plasma avalanche transit time diode.

trapezium [MATH] A quadrilateral where no sides are parallel.

trapezium distortion [ELECTR] A defect in a cathode-ray tube in which the trace is confined within a trapezium rather than a rectangle, usually as a result of interaction between the two pairs of deflection plates.

trapezohedron [CRYSTAL] An isometric crystal form of 24 faces, each face of which is an irregular four-sided figure. Also known as icositetrahedron; leucitohedron; tetragonal trisoctahedron.

trapezoid [MATH] A quadrilateral having two parallel sides.

trapezoidal integration [MATH] A numerical approximation of an integral by means of the trapezoidal rule.

trapezoidal pulse [ELECTR] An electrical pulse in which the voltage rises linearly to some value, remains constant at this value for some time, and then drops linearly to the original value.

trapezoidal rule [MATH] The integral from a to b of a real function $f(x)$ is approximated by

$$\frac{b-a}{2n}\left[f(a) + \sum_{j=1}^{n-1} 2f(x_j) + f(b)\right]$$

where $x_0 = a$, $x_j = x_{j-1} + (b-a)/n$ for $j = 1, 2, \ldots, n-1$.

trapezoidal wave [ELECTR] A wave consisting of a series of trapezoidal pulses.

trapped plasma avalanche transit time diode *See* TRAPATT diode.

traveling microscope [OPTICS] A low-power microscope equipped with a graticule and rails enabling it to move horizontally or vertically, used to make accurate length determinations.

traveling salesman problem [MATH] The problem of performing successively a number of tasks, represented by vertices of a graph, with the least expenditure on transitions from one task to another, represented by edges of the graph with journey costs attached.

traveling wave [PHYS] A wave in which energy is transported from one part of a medium to another, in contrast to a standing wave.

traveling-wave antenna [ELECTROMAG] An antenna in which the current distributions are produced by waves of charges propagated in only one direction in the conductors. Also known as progressive-wave antenna.

traveling-wave maser [PHYS] A ruby maser used with a comblike slow-wave structure and a number of yttrium iron garnet isolators to give L-band amplification (390 to 1550 megahertz); operation is at the temperature of liquid helium (4.2 K).

traveling-wave phototube [ELECTR] A traveling-wave tube having a photocathode and an appropriate window to admit a modulated laser beam; the modulated laser beam causes emission of a current-modulated photoelectron beam, which in turn is accelerated by an electron gun and directed into the helical slow-wave structure of the tube.

traveling-wave tube [ELECTR] An electron tube in which a stream of electrons interacts continuously or repeatedly with a guided electromagnetic wave moving substantially in synchronism with it, in such a way that there is a net transfer of energy from the stream to the wave; the tube is used as an amplifier or oscillator at frequencies in the microwave region.

treble [ACOUS] High audio frequencies, such as those handled by a tweeter in a sound system.

tree [ELECTR] A set of connected circuit branches that includes no meshes; responds uniquely to each of the possible combinations of a number of simultaneous inputs. Also known as decoder. [MATH] A connected graph contained in a given connected graph having all the vertices of the original but without any closed circuit.

trend [STAT] The general drift, tendency, or bent of a set of statistical data as related to time or another related set of statistical data.

Tresca criterion [MECH] The assumption that plastic deformation of a material begins when the difference between the maximum and minimum principal stresses equals twice the yield stress in shear.

Trevelyan rocker [PHYS] A prismatic metal block having one edge grooved to form two ridges; it vibrates when heated and placed on the grooved edge, providing a simple example of heat-maintained vibrations.

triad axis [CRYSTAL] A rotation axis whose multiplicity is equal to 3.

trial [STAT] One of a series of duplicate experiments.

triangle [MATH] The figure realized by connecting three noncollinear points by line segments.

triangle inequality [MATH] For real or complex numbers or vectors in a normed space x and y, the absolute value or norm of $x + y$ is less than or equal to the sum of the absolute values or norms of x and y.

triangle of forces [MECH] A triangle, two of whose sides represent forces acting on a particle, while the third represents the combined effect of these forces.

triangle of vectors [MATH] A triangle, two of whose sides represent vectors to be added, while the third represents the sum of these two vectors.

triangular matrix [MATH] A matrix where either all entries above or all entries below the principal diagonal are zero.

triangular pulse [ELECTR] An electrical pulse in which the voltage rises linearly to some value, and immediately falls linearly to the original value.

triangular wave [ELECTR] A wave consisting of a series of triangular pulses.

triangulation [MATH] A decomposition of a topological manifold into subsets homeomorphic with a polyhedron in some euclidean space.

triangulation problem [MATH] The problem of whether each topological n manifold admits a piecewise linear structure.

triaxial pinch [PL PHYS] A device for heating a confined plasma, in which a discharge in an annular space between two concentric cylindrical conductors forms a cylindrical sheet of plasma, and this plasma is then confined and compressed by magnetic fields produced by currents flowing in the axial direction in the discharge itself and in the two conductors.

tribo- [PHYS] A prefix meaning pertaining to or resulting from friction.

triboelectricity *See* frictional electricity.

triboelectric series [ELEC] A list of materials that produce an electrostatic charge when rubbed together, arranged in such an order that a material has a positive charge when rubbed with a material below it in the list, and has a negative charge when rubbed with a material above it in the list.

triboelectrification [ELEC] The production of electrostatic charges by friction.

tribology [PHYS] The study of the phenomena and mechanisms of friction, lubrication, and wear of surfaces in relative motion.

triboluminescence [ATOM PHYS] Luminescence produced by friction between two materials.

tribometer [ENG] A device for measuring coefficients of friction, consisting of a loaded sled subject to a measurable force.

trichroism [OPTICS] Phenomenon exhibited by certain optically anisotropic transparent crystals when subjected to white light, in which a cube of the material is found to transmit a different color through each of the three pairs of parallel faces.

trichromatic theory [OPTICS] A theory of color vision which states that three primary colors may be chosen in such a way that, combined in various proportions, they can match any color.

triclinic crystal [CRYSTAL] A crystal whose unit cell has axes which are not at right angles, and are unequal. Also known as anorthic crystal.

triclinic system [CRYSTAL] The most general and least symmetric crystal system, referred to by three axes of different length which are not at right angles to one another.

Tricomi's equation [MATH] The partial differential equation $y\, \partial^2 u/\partial x^2 + \partial^2 u/\partial y^2 = 0$.

Tricomi's function [MATH] The confluent hypergeometric function

$$\Psi(\alpha,\rho,z) = \frac{\Gamma(1-\rho)}{\Gamma(\alpha-\rho+1)} \,_1F_1(\alpha;\rho;z)$$

$$+ \frac{\Gamma(\rho-1)}{\Gamma(\alpha)} z^{1-\rho} \,_1F_1(\alpha-\rho+1; 2-\rho; z)$$

where $_1F_1$ is Kummer's function.

trident of Newton [MATH] The curve in the plane given by the equation $xy = ax^3 + bx^2 + cx + d$, where $a \neq 0$; this cuts the x axis in one or three points and is asymptotic to the y axis if $d \neq 0$.

trigger [ELECTR] 1. To initiate an action, which then continues for a period of time, as by applying a pulse to a trigger circuit. 2. The pulse used to initiate the action of a trigger circuit. 3. *See* trigger circuit.

trigger circuit [ELECTR] 1. A circuit or network in which the output changes abruptly with an infinitesimal change in input at a predetermined operating point. Also known as trigger. 2. A circuit in which an action is initiated by an input pulse, as in a radar modulator. 3. *See* bistable multivibrator.

trigger electrode *See* starter.

trigistor [ELECTR] A *pnpn* device with a gating control acting as a fast-acting switch similar in nature to a thyratron.

trigonal lattice *See* rhombohedral lattice.

trigonal system [CRYSTAL] A crystal system which is characterized by threefold symmetry, and which is usually considered as part of the hexagonal system since the lattice may be either hexagonal or rhombohedral.

trigonometric functions [MATH] The real-valued functions such as $\sin(x)$, $\tan(x)$, and $\cos(x)$ obtained from studying certain ratios of the sides of a right triangle. Also known as circular functions.

trigonometric parallax [ASTRON] A parallax that may be determined for the nearest stars (less than 300 light-years or 28.38×10^{17} m) by a direct method utilizing trigonometry.

trigonometric polynomial [MATH] A finite series of functions of the form $a_n \cos nx + b_n \sin nx$; occasionally used synonymously with trigonometric series.

trigonometric series [MATH] An infinite series of functions with nth term of the form $a_n \cos nx + b_n \sin nx$.

trigonometry [MATH] The study of triangles and the trigonometric functions.

trihedral [MATH] Any figure obtained from three noncoplanar lines intersecting in a common point.

trillion [MATH] **1.** The number 10^{12}. **2.** In British and German usage, the number 10^{18}.

trim [ELECTR] Fine adjustment of capacitance, inductance, or resistance of a component during manufacture or after installation in a circuit.

trimmer capacitor [ELEC] A relatively small variable capacitor used in parallel with a larger variable or fixed capacitor to permit exact adjustment of the capacitance of the parallel combination.

trimmer potentiometer [ELEC] A potentiometer which is used to provide a small-percentage adjustment and is often used with a coarse control.

trimuon event [PARTIC PHYS] An inelastic collision of a neutrino or antineutrino with a nucleus in which there are three muons among the products of the collision.

trinomial [MATH] A polynomial having three terms.

trinomial distribution [STAT] A multinomial distribution in which there are three distinct outcomes.

triode [ELECTR] A three-electrode electron tube containing an anode, a cathode, and a control electrode.

triode laser [ELECTR] Gas laser whose light output may be modulated by signal voltages applied to an integral grid.

triode transistor [ELECTR] A transistor that has three terminals.

triple collision [PHYS] A process in which three particles collide simultaneously.

triple-diagonal matrix *See* continuant matrix.

triple harmonic [PHYS] A harmonic whose frequency is three times the fundamental frequency.

triple point [PHYS CHEM] A particular temperature and pressure at which three different phases of one substance can coexist in equilibrium.

triple scalar product *See* scalar triple product.

triplet [OPTICS] A compound lens made up of three components.

triplet state [ATOM PHYS] Electronic state of an atom or molecule whose total spin angular momentum quantum number is equal to 1. [QUANT MECH] Any multiplet having three states.

triple vector product [MATH] The triple vector product of vectors **a**, **b**, and **c** is the cross product of **a** with the cross product of **b** and **c**; written $\mathbf{a} \times (\mathbf{b} \times \mathbf{c})$.

trisectrix [MATH] The planar curve given by $x^3 + xy^2 + ay^2 - 3ax^2 = 0$ which is symmetric about the x axis and asymptotic to the line $x = -a$; this is useful in studying the trisection of an angle problem. Also known as trisectrix of Maclaurin.

trisectrix of Catalan *See* Tschirnhausen's cubic.

trisectrix of Maclaurin *See* trisectrix.

trisistor [ELECTR] Fast-switching semiconductor consisting of an alloyed junction *pnp* device in which the collector is capable of electron injection into the base; characteristics resemble those of a thyratron electron tube, and switching time is in the nanosecond range.

tristate logic [ELECTR] A form of transistor-transistor logic in which the output stages or input and output stages can assume three states; two are the normal low-impedance 1 and 0 states, and the third is a high-impedance state that allows many tristate devices to time-share bus lines.

tristimulus colorimeter [OPTICS] A colorimeter that measures a color stimulus in terms of tristimulus values.

tristimulus values [OPTICS] The magnitudes of three standard stimuli needed to match a given sample of light.

tritium [NUC PHYS] The hydrogen isotope having mass number 3; it is one form of heavy hydrogen, the other being deuterium. Symbolized H^3; T.

triton [NUC PHYS] The nucleus of tritium.

trochoid [MATH] The path in the plane obtained from a point on the radius of a circle as the circle rolls along a fixed straight line.

trochoidal mass analyzer [PHYS] A mass spectrometer in which the ion beams traverse trochoidal paths within mutually perpendicular electric and magnetic fields.

Trojan planet [ASTRON] One of a group of asteroids whose periods of revolution are about equal to that of Jupiter, or about 12 years; these bodies move close to one or the other of two positions called Lagrangian points, 60° ahead of or 60° behind Jupiter; the asteroids near these positions are known as Greeks and Pure Trojans respectively.

troland [OPTICS] A unit of retinal illuminance, equal to the retinal illuminance produced by a surface whose luminance is one nit when the apparent area of the entrance pupil of the eye is 1 square millimeter. Also known as luxon; photon.

tropical month [ASTRON] The average period of the revolution of the moon about the earth with respect to the vernal equinox, a period of 27 days 7 hours 43 minutes 4.7 seconds, or approximately $27\frac{1}{3}$ days.

tropical year [ASTRON] A unit of time equal to the period of one revolution of the earth about the sun measured between successive vernal equinoxes; it is 365.2422 mean solar days or 365 days 5 hours 48 minutes 46 seconds.

tropospheric superrefraction [GEOPHYS] Phenomenon occurring in the troposphere whereby radio waves are bent sufficiently to be returned to the earth.

Trouton-Noble experiment [ELECTROMAG] An experiment to detect ether drift by measuring the deflection of a charged parallel plate capacitor which is suspended so that it is free to turn.

Trouton's rule [THERMO] The rule that, for a nonassociated liquid, the latent heat of vaporization in calories is equal to approximately 22 times the normal boiling point on the Kelvin scale.

troy ounce *See* ounce.

troy pound *See* pound.

troy system [MECH] A system of mass units used primarily to measure gold and silver; the ounce is the same as that in the apothecaries' system, being equal to 480 grains or 31.1034768 grams. Abbreviated t. Also known as troy weight.

troy weight *See* troy system.

true anomaly *See* anomaly.

true complement *See* radix complement.

true condensing point *See* critical condensation temperature.

true freezing point [PHYS CHEM] The temperature at which the liquid and solid forms of a substance exist in equilibrium at a given pressure (usually 1 standard atmosphere, or 101,325 newtons per square meter).

true horizon [OPTICS] The boundary of a horizontal plane passing through a point of vision, or in photogrammetry, the perspective center of a lens system.

true solar day *See* apparent solar day.

TROJAN PLANET

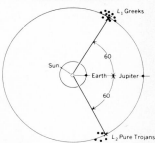

Diagram showing position of Trojan planets. Crosses indicate Lagrangian points L_1 and L_2, near which are clustered Greeks and Pure Trojans respectively.

true solar time   *See* apparent solar time.

true sun   *See* apparent sun.

truncate   [MATH] **1.** To drop digits at the end of a numerical value; the number 3.14159265 is truncated to five figures in 3.1415, whereas it would be 3.1416 if rounded off to five figures. **2.** To approximate the sum of an infinite series by the sum of a finite number of its terms. **3.** To terminate an infinite sequence of successively better approximations of a quantity after a finite number of such approximations.

truncated distribution   [STAT] A distribution fashioned from another distribution by deleting that part of the distribution to the right or left of a random variable value.

truncated paraboloid   [ELECTROMAG] Paraboloid antenna in which a portion of the top and bottom have been cut away to broaden the main lobe in the vertical plane.

truncation   [MATH] **1.** Approximating the sum of an infinite series by the sum of a finite number of its terms. **2.** *See* rounding.

truncation error   [MATH] **1.** The computation error resulting from use of only a finite number of terms of an infinite series. **2.** The error resulting from the approximation of a derivative or differential by a finite difference.

truth table   [MATH] A table listing statements concerning an event and their respective truth values.

Tschirnhausen's cubic   [MATH] A plane curve consisting of the envelope of the line through a variable point P on a parabola which is perpendicular to the line from the focus of the parabola to P. Also known as l'Hôpital's cubic; trisectrix of Catalan.

T score   [STAT] A score utilized in setting up norms for standardized tests; obtained by linearly transforming normalized standard scores.

T-section filter   [ELEC] T network used as an electric filter.

tsi   [MECH] A unit of force equal to 1 ton-force per square inch; equal to approximately 1.54444×10^7 pascals.

tsp   *See* teaspoonful.

T_1 space   [MATH] A topological space where, for each pair of distinct points, each one has a neighborhood not containing the other.

T_2 space   *See* Hausdorff space.

T_3 space   *See* regular topological space.

T_4 space   *See* normal space.

T_6 space   [MATH] A topological space where, for each pair of distinct points, at least one has a neighborhood not containing the other.

tspn   *See* teaspoonful.

T system   *See* Chebyshev system.

T Tauri star   [ASTRON] A star classification; an extrinsic variable star whose variation in luminosity may be stimulated by the associated nebulosity. Also known as nebular variable.

t-test   [STAT] A statistical test involving means of normal populations with unknown standard deviations; small samples are used, based on a variable t equal to the difference between the mean of the sample and the mean of the population divided by a result obtained by dividing the standard deviation of the sample by the square root of the number of individuals in the sample.

TTL   *See* transistor-transistor logic.

tube   *See* electron tube.

tube coefficient   [ELECTR] Any of the constants that describe the characteristics of a thermionic vacuum tube, such as amplification factor, mutual conductance, or alternating-current plate resistance.

tube noise [ELECTR] Noise originating in a vacuum tube, such as that due to shot effect and thermal agitation.

tube of flux *See* tube of force.

tube of force [ELEC] A region of space bounded by a tubular surface consisting of the lines of force which pass through a given closed curve. Also known as tube of flux.

tube voltage drop [ELECTR] In a gas tube, the anode voltage during the conducting period.

tubular capacitor [ELEC] A paper or electrolytic capacitor having the form of a cylinder, with leads usually projecting axially from the ends; the capacitor plates are long strips of metal foil separated by insulating strips, rolled into a compact tubular shape.

tunable filter [ELECTR] An electric filter in which the frequency of the passband or rejection band can be varied by adjusting its components.

tunable laser [OPTICS] A laser in which the frequency of the output radiation can be tuned over part or all of the ultraviolet, visible, and infrared regions of the spectrum.

tune [ELECTR] To adjust for resonance at a desired frequency.

tuned amplifier [ELECTR] An amplifier in which the load is a tuned circuit; load impedance and amplifier gain then vary with frequency.

tuned cavity *See* cavity resonator.

tuned circuit [ELECTR] A circuit whose components can be adjusted to make the circuit responsive to a particular frequency in a tuning range. Also known as tuning circuit.

tuned resonating cavity [ELECTROMAG] Resonating cavity half a wavelength long or some multiple of a half wavelength, used in connection with a waveguide to produce a resultant wave with the amplitude in the cavity greatly exceeding that of the wave in the waveguide.

tuner [ELECTR] The portion of a receiver that contains circuits which can be tuned to accept the carrier frequency of the alternating current supplied to the primary, thereby causing the secondary voltage to build up to higher values than would otherwise be obtained.

tungsten [CHEM] Also known as wolfram. A metallic element in group VI, symbol W, atomic number 74, atomic weight 183.85; soluble in mixed nitric and hydrofluoric acids; melts at 3400°C.

tungsten filament [ELEC] A filament used in incandescent lamps, and as an incandescent cathode in many types of electron tubes, such as thermionic vacuum tubes.

tungsten-halogen lamp [ELECTR] A lamp containing a halogen, usually iodine or bromine, which combines with tungsten evaporated from the filament.

tuning [ELECTR] The process of adjusting the inductance or the capacitance or both in a tuned circuit, for example, in a radio, television, or radar receiver or transmitter, so as to obtain optimum performance at a selected frequency.

tuning capacitor [ELEC] A variable capacitor used for tuning purposes.

tuning circuit *See* tuned circuit.

tuning fork [ENG] A U-shaped bar of hard steel, fused quartz, or other elastic material that vibrates at a definite natural frequency when struck or when set in motion by electromagnetic means; used as a frequency standard.

tunnel diode [ELECTR] A heavily doped junction diode that has a negative resistance at very low voltage in the forward bias direction, due to quantum-mechanical tunneling, and a short circuit in the negative bias direction. Also known as Esaki tunnel diode.

tunnel effect [QUANT MECH] The ability of a particle to pass

TUNGSTEN

Periodic table of the chemical elements showing the position of tungsten.

TUNING FORK

A tuning fork vibrating at its fundamental frequency.

through a region of finite extent in which the particle's potential energy is greater than its total energy; this is a quantum-mechanical phenomenon which would be impossible according to classical mechanics.

tunnel rectifier [ELECTR] Tunnel diode having a relatively low peak-current rating as compared with other tunnel diodes used in memory-circuit applications.

tunnel resistor [ELECTR] Resistor in which a thin layer of metal is plated across a tunneling junction, to give the combined characteristics of a tunnel diode and an ordinary resistor.

tunnel triode [ELECTR] Transistorlike device in which the emitter-base junction is a tunnel diode and the collector-base junction is a conventional diode.

turbidimeter [OPTICS] A device that measures the loss in intensity of a light beam as it passes through a solution with particles large enough to scatter the light.

turbidity [ANALY CHEM] **1.** Measure of the clarity (using APHA or colorimetric scales) of an otherwise clear liquid. **2.** Cloudy or hazy appearance in a naturally clear liquid caused by a suspension of colloidal liquid droplets or fine solids.

turbidity coefficient [OPTICS] A factor in the absorption (light) law equation that describes the extinction of the incident light beam.

turbidity factor [GEOPHYS] A measure of the atmospheric transmission of incident solar radiation; if I_0 is the flux density of the solar beam just outside the earth's atmosphere, I the flux density measured at the earth's surface with the sun at a zenith distance which implies an optical air mass m, and $I_{m,w}$ the intensity which would be observed at the earth's surface for a pure atmosphere containing 1 centimeter of precipitable water viewed through the given optical air mass, then turbidity factor θ is given by $\theta = (\ln I_0 - \ln I)/(\ln I_0 - \ln I_{m,w})$.

turbulence _See_ turbulent flow.

turbulence energy _See_ eddy kinetic energy.

turbulent boundary layer [FL MECH] The layer in which the Reynolds stresses are much larger than the viscous stresses.

turbulent diffusion _See_ eddy diffusion.

turbulent flow [FL MECH] Motion of fluids in which local velocities and pressures fluctuate irregularly, in a random manner. Also known as turbulence.

turbulent flux _See_ eddy flux.

turbulent Lewis number [PHYS] A dimensionless number used in the study of combined turbulent heat and mass transfer, equal to the ratio of the eddy mass diffusivity to the eddy thermal diffusivity. Symbolized Le_T.

turbulent Prandtl number [PHYS] A dimensionless number used in the study of heat transfer in turbulent flow, equal to the ratio of the eddy viscosity to the eddy thermal diffusivity. Symbolized Pr_T.

turbulent Schmidt number [FL MECH] A dimensionless number used in the study of mass transfer in turbulent flow, equal to the ratio of the eddy viscosity to the eddy mass diffusivity. Symbolized Sc_T.

turbulent shear force [FL MECH] A shear force in a fluid which arises from turbulent flow.

Turing machine [ADP] A mathematical idealization of a computing automation similar in some ways to real computing machines; used by mathematicians to define the concept of computability.

turn [ELEC] One complete loop of wire. [MATH] _See_ circle.

turning value [MATH] A relative maximum or relative minimum of a function.

turnover frequency _See_ transition frequency.

turns ratio [ELEC] The ratio of the number of turns in a secondary winding of a transformer to the number of turns in the primary winding.

TW *See* terawatt.

Twaddell scale [ENG] A scale for specific gravity of solutions that is the first two digits to the right of the decimal point multiplied by two; for example, a specific gravity of 1.4202 is equal to 84.04°Tw.

twilight arch *See* bright segment.

twin *See* twin crystal.

twin crystal [CRYSTAL] A compound crystal which has one or more parts whose lattice structure is the mirror image of that in the other parts of the crystal. Also known as twin.

twin law [CRYSTAL] A statement relating two or more individuals of a twin to one another in terms of their crystallography (twin plane, twin axis, and so on).

twinning [CRYSTAL] The development of a twin crystal by growth, translation, or gliding.

twinning plane *See* twin plane.

twin paradox *See* clock paradox.

twin plane [CRYSTAL] The plane common to and across which the individual crystals or components of a crystal twin are symmetrically arranged or reflected. Also known as twinning plane.

twin-T filter [ELEC] An electric filter consisting of a parallel-T network with values of network elements chosen in such a way that the outputs due to each of the paths precisely cancel at a specified frequency.

twin-T network *See* parallel-T network.

twist [ELECTROMAG] A waveguide section in which there is a progressive rotation of the cross section about the longitudinal axis of the waveguide.

twist boundary [SOLID STATE] A boundary between two crystals that differ in orientation by only a few degrees, consisting of a series of screw dislocations.

twisted curve [MATH] A curve that does not lie wholly in any one plane.

two-beam interference [PHYS] Interference between two waves.

two-body force [PHYS] A force between two particles which is not affected by the existence of other particles in the vicinity, such as a gravitational force or a Coulomb force between charged particles.

two-body problem [MECH] The problem of predicting the motions of two objects obeying Newton's laws of motion and exerting forces on each other according to some specified law such as Newton's law of gravitation, given their masses and their positions and velocities at some initial time.

two-carrier theory [SOLID STATE] A theory of the conduction properties of a material in bulk or in a rectifying barrier which takes into account the motion of both electrons and holes.

two-component neutrino theory [PARTIC PHYS] A theory according to which the neutrino and antineutrino have exactly zero rest mass, and the neutrino spin is always antiparallel to its motion, while the antineutrino spin is parallel to its motion.

two-decision problem [STAT] The problem of deciding, using statistical information, between two actions or decisions.

two-degrees-of-freedom gyro [MECH] A gyro whose spin axis is free to rotate about two orthogonal axes, not counting the spin axis.

two-dimensional flow [FL MECH] Fluid flow in which all flow occurs in a set of parallel planes with no flow normal to them, and the flow is identical in each of these parallel planes.

TWIN CRYSTAL

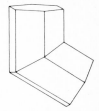

Example of twin crystal, consisting of hexagonal prism twinned on a pyramid face.

TWIST

A 90° twist for a rectangular waveguide.

two-fluid model [CRYO] A theoretical model of helium II which assumes that it consists of two interpenetrating components, a normal fluid and a superfluid with zero entropy, viscosity, and thermal conductivity.

two-part experiment [STAT] Experiments in which two operations or actions are performed; for example, throwing two dice, drawing two marbles from a box, throwing a die and then drawing a marble from a box.

two-person game [MATH] A game consisting of exactly two players with competing interests.

two-phase [PHYS] Having a phase difference of one quarter-cycle or 90°. Also known as quarter-phase.

two-phase flow [CRYO] Flow of helium II, or of electrons in a superconductor, thought of as consisting of two interpenetrating, noninteracting fluids, a superfluid component which exhibits no resistance to flow and is responsible for superconducting properties, and a normal component, which behaves as does an ordinary fluid or as conduction electrons in a nonsuperconducting metal. [FL MECH] Cocurrent movement of two phases (for example, gas and liquid) through a closed conduit or duct (for example, a pipe).

two-port junction [ELECTROMAG] A waveguide junction with two openings; it can consist either of a discontinuity or obstacle in a waveguide, or of two essentially different waveguides connected together.

two-port system [CONT SYS] A system which has only one input or excitation and only one response or output.

two's complement [MATH] A number derived from a given n-bit number by requiring the two numbers to sum to a value of 2^n.

two-sided ideal [MATH] A two-sided ideal I is a subring of a ring R where the products xy and yx are always in I for every x in R and y in I.

two-sided test [STAT] A test which rejects the null hypothesis when the test statistic T is either less than or equal to c or greater than or equal to d, where c and d are critical values.

two-sphere [MATH] The surface of a ball; the two-dimensional sphere in euclidean three-dimensional space obtained from all points whose distance from the origin is one.

two-stage design [STAT] The design of an experiment which employs a pilot study in order to decide how to design the main experiment.

two-stage experiment [STAT] An experiment in two parts, the outcome of the first part deciding the procedure for the second.

two-stage sampling [STAT] Sampling from a population whose members are themselves sets of objects and then sampling from the sets selected in the first sampling; for example, to first draw a sample of states and then to draw a sample of representatives to Congress from each state selected.

two-valued variable [MATH] A variable which assumes values in a set containing exactly two elements, often symbolized as 0 and 1.

Twyman-Green interferometer [OPTICS] An interferometer similar to the Michelson interferometer except that it is illuminated with a point source of light instead of an extended source.

Twystron [ELECTR] Very-high-power, hybrid microwave tube, combining the input section of a high-power klystron with the output section of a traveling wave tube, characterized by high operating efficiency and wide bandwidths.

Tychonoff space *See* completely regular space.

Tychonoff theorem [MATH] A product of topological spaces is compact if and only if each individual space is compact.

Tyndall cone [OPTICS] The luminous path of a beam of light resulting from the Tyndall effect.

Tyndall effect [OPTICS] Visible scattering of light along the path of a beam of light as it passes through a system containing discontinuities, such as the surfaces of colloidal particles in a colloidal solution

Type II Cepheids *See* W Virginis stars.

type I error [STAT] One of two types of errors in testing hypotheses: incorrectly rejecting the hypothesis tested when it is true. Also known as error of the first kind.

type II error [STAT] One of two types of error in testing hypotheses: incorrectly accepting the hypothesis tested when an alternate hypothesis is true. Also known as error of the second kind.

type I superconductor [CRYO] A superconductor for which there is a single critical magnetic field; magnetic flux is completely excluded from the interior of the material at field strengths below this critical field, while at field strengths above the critical field, magnetic flux penetrates the superconductor completely and it reverts to the normal state.

type II superconductor [CRYO] A superconductor for which there are two critical magnetic fields; magnetic flux is completely excluded from the interior of the material only at field strengths below the smaller critical field, and at field strengths between the two critical fields the magnetic flux consists of flux vortices in the form of filaments embedded in the superconducting material. Also known as high-field superconductor (HFS).

TYNDALL CONE

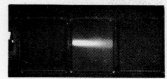

The luminous light path known as the Tyndall cone. *(H. Steeves and R. G. Babcock)*

u *See* up quark.

U *See* uranium.

u-band [OPTICS] The absorption band in the ultraviolet resulting from a U-center type of point lattice defect.

UBV photometry [ASTRON] A system of three-color photometry used to obtain specific stellar magnitudes; the system is based on the comparison of stars' magnitudes with a standard sequence of about 400 stars.

U center [CRYSTAL] The color-center type of point lattice defect in ionic crystals created by the incorporation of an impurity such as hydrogen into alkali halides.

U coefficient [QUANT MECH] A coefficient that appears in the transformation between modes of coupling eigenfunctions of three angular momenta; it is equal to the product of the Racah coefficient and $[(2j_{12} + 1) (2j_{23} + 1)]^{1/2}$, where j_{12} and j_{23} are the intermediate angular momenta in the respective modes.

Uda antenna *See* Yagi-Uda antenna.

U figure *See* U index.

U Geminorum stars [ASTRON] A class of variable stars known as dwarf novae; their light curves resemble those of novae, with range brightness variations of about 4 magnitudes; examples are U Gemini and SS Cygni. Also known as SS Cygni stars.

Uhlbricht sphere [OPTICS] A sphere whose inside surface has a diffusely reflecting white finish, used in an integrating sphere photometer.

U index [GEOPHYS] The difference between consecutive daily mean values of the horizontal component of the geomagnetic field. Also known as U figure.

UJT *See* unijunction transistor.

ultimate lines [ASTRON] Special spectral lines that can be used to indicate the existence of an element in the sun or other star.

ultimate strength [MECH] The tensile stress, per unit of the original surface area, at which a body will fracture, or continue to deform under a decreasing load.

ultracentrifuge [ENG] A laboratory centrifuge that develops centrifugal fields of more than 100,000 times gravity.

ultra-cold neutron [PHYS] A neutron whose energy is of the order of 10^{-7} electron volt or less, so that it is totally reflected from various materials and suitably constructed magnetic fields, regardless of the angle of incidence, and can be stored in suitably constructed bottles.

ultrafilter [MATH] A filter on a set S which is not contained in any other filter on S.

ultra-high vacuum [PHYS] A vacuum in which the pressure is of the order of 10^{-10} millimeter of mercury (1.33×10^{-8} pascal) or less.

ultramicrobalance [ENG] A differential weighing device with accuracies better than 1 microgram; used for analytical weighings in microanalysis.

ultramicroscope [OPTICS] An instrument for investigating particles of submicroscopic dimensions: it consists of a high-intensity illumination system for producing a Tyndall cone in a colloidal system, coupled with a compound microscope to examine the points of light scattered from the individual particles.

ultraphotic rays [ELECTROMAG] Rays outside the visible part of the spectrum, including infrared and ultraviolet rays.

ultrasonic [ACOUS] Pertaining to signals, equipment, or phenomena involving frequencies just above the range of human hearing, hence above about 20,000 hertz. Also known as supersonic (deprecated usage).

ultrasonic camera [ELECTR] A device which produces a picture display of ultrasonic waves sent through a sample to be inspected or through live tissue; a piezoelectric crystal is used to convert the ultrasonic waves to voltage differences, and the voltage pattern on the crystal modulates the intensity of an electronic beam scanning the crystal; this beam in turn controls the intensity of a beam in a television tube.

ultrasonic coagulation [PHYS] The bonding of small particles into large aggregates by the action of ultrasonic waves.

ultrasonic echo [ACOUS] An ultrasonic wave that has been reflected or otherwise returned with sufficient magnitude and time delay to be perceived in some manner as a wave distinct from that directly transmitted.

ultrasonic generator [ENG ACOUS] A generator consisting of an oscillator driving an electroacoustic transducer, used to produce acoustic waves above about 20 kilohertz.

ultrasonic imaging *See* acoustic imaging.

ultrasonic light modulator [OPTICS] Device containing a fluid which, by action of ultrasonic waves passing through the fluid, modulates a beam of light passed transversely through the fluid.

ultrasonic microscope [OPTICS] A special type of microscope which employs ultrasonic radiation.

ultrasonic radiation [ACOUS] Ultrasonic waves propagating through a solid, liquid, or gaseous medium.

ultrasonics [ACOUS] The science of ultrasonic sound waves.

ultrasonic transducer [ENG ACOUS] A transducer that converts alternating-current energy above 20 kilohertz to mechanical vibrations of the same frequency; it is generally either magnetostrictive or piezoelectric.

ultrasonic wave [ACOUS] A sound wave that has a frequency above about 20,000 hertz.

ultraspherical polynomials *See* Gegenbauer polynomials.

ultraviolet [PHYS] Pertaining to ultraviolet radiation. Abbreviated UV.

ultraviolet absorber [OPTICS] Any substance that absorbs ultraviolet radiant energy, then dissipates the energy in a harmless form; used in plastics and rubbers to decrease light sensitivity.

ultraviolet absorption [OPTICS] Absorption of specific ultraviolet radiation wavelengths by a material; for example, by a sample solution during spectroscopic analysis.

ultraviolet absorption spectrophotometry [SPECT] The study of the spectra produced by the absorption of ultraviolet radiant energy during the transformation of an electron from the ground state to an excited state as a function of the wavelength causing the transformation.

ultraviolet catastrophe [STAT MECH] The prediction of the Rayleigh-Jeans law that the energy radiated by a blackbody at extremely short wavelengths is extremely large, and the total energy radiated is infinite, whereas in reality it must be finite.

ultraviolet imagery [ELECTROMAG] That imagery produced as a result of sensing ultraviolet radiations reflected from a given target surface.

ultraviolet lamp [ELECTR] A lamp providing a high proportion of ultraviolet radiation, such as various forms of mercury-vapor lamps.

ultraviolet light See ultraviolet radiation.

ultraviolet microscope [OPTICS] A special type of microscope which uses electromagnetic radiation in the range 180–400 nanometers; it requires reflecting optics or special quartz and crystal objectives.

ultraviolet radiation [ELECTROMAG] Electromagnetic radiation in the wavelength range 4–400 nanometers; this range begins at the short-wavelength limit of visible light and overlaps the wavelengths of long x-rays (some scientists place the lower limit at higher values, up to 40 nanometers). Also known as ultraviolet light.

ultraviolet spectrometer [SPECT] A device which produces a spectrum of ultraviolet light and is provided with a calibrated scale for measurement of wavelength.

ultraviolet spectrophotometry [SPECT] Determination of the spectra of ultraviolet absorption by specific molecules in gases or liquids (for example, Cl_2, SO_2, NO_2, CS_2, ozone, mercury vapor, and various unsaturated compounds).

ultraviolet spectroscopy [SPECT] Absorption spectroscopy involving electromagnetic wavelengths in the range 4–400 nanometers.

ultraviolet spectrum [ELECTROMAG] **1.** The range of wavelengths of ultraviolet radiation, covering 4–400 nanometers. **2.** A display or graph of the intensity of ultraviolet radiation emitted or absorbed by a material as a function of wavelength or some related parameter.

umbilic See umbilical point.

umbilical point [MATH] A point on a surface at which the normal curvature is the same in all directions. Also known as navel point; umbilic.

umbra [OPTICS] That portion of a shadow which is screened from light rays emanating from any part of an extended source.

Umkehr effect [OPTICS] An anomaly of the relative zenith intensities of scattered sunlight at certain wavelengths in the ultraviolet as the sun approaches the horizon; it is due to the presence of the ozone layer.

Umklapp process [SOLID STATE] The interaction of three or more waves in a solid, such as lattice waves or electron waves, in which the sum of the wave vectors is not equal to zero but, rather, is equal to a vector in the reciprocal lattice. Also known as flip-over process.

unary operation [MATH] An operation in which only a single operand is required to produce a unique result; some examples are negation, complementation, square root, transpose, inverse, and conjugate.

unavailable energy [THERMO] That part of the energy which, when an irreversible process takes place, is initially in a form completely available for work and is converted to a form completely unavailable for work.

unbalanced line [ELEC] A transmission line in which the voltages on the two conductors are not equal with respect to ground; a coaxial line is an example.

unbiased estimate [STAT] An estimate for a parameter θ whose expected value is θ.

unblanking pulse [ELECTR] Voltage applied to a cathode-ray tube to overcome bias and cause trace to be visible.

ULTRAVIOLET LAMP

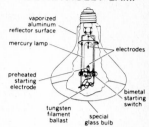

Features of a sunlamp, a type of ultraviolet lamp that produces radiation in the middle ultraviolet region (280–320 nanometers) together with infrared radiation and light from the filament.

unbounded manifold [MATH] A manifold with no boundary.

unbounded set of real numbers [MATH] A set with the property that if R is any positive real number, there is a number in the set which is smaller than $-R$ or a number larger than R.

unbounded wave [PHYS] A wave which propagates through a nondissipative, homogeneous medium which is infinite in extent, without any boundaries.

uncertainty principle [QUANT MECH] The precept that the accurate measurement of an observable quantity necessarily produces uncertainties in one's knowledge of the values of other observables. Also known as Heisenberg uncertainty principle; indeterminacy principle.

uncertainty relation [QUANT MECH] The relation whereby, if one simultaneously measures values of two canonically conjugate variables, such as position and momentum, the product of the uncertainties of their measured values cannot be less than approximately Planck's constant divided by 2π. Also known as Heisenberg uncertainty relation.

unconditional convergence [MATH] A convergent series converges unconditionally if every series obtained by rearranging its terms also converges; equivalent to absolute convergence.

unconditional inequality [MATH] An inequality which holds true for all values of the variables involved, or which contains no variables; for example, $y + 2 > y$, or $4 > 3$. Also known as absolute inequality.

uncorrelated random variables [STAT] Two random variables whose correlation coefficient is zero.

uncountable set [MATH] An infinite set which can not be put in 1 to 1 correspondence with the set of integers; for example, the set of real numbers.

uncoupling phenomena [SPECT] Deviations of observed spectra from those predicted in a diatomic molecule as the magnitude of the angular momentum increases, caused by interactions which could be neglected at low angular momenta.

undamped wave [PHYS] A continuous wave produced by oscillations having constant amplitude.

underbunching [ELECTR] In velocity-modulated electron streams, a condition representing less than the optimum bunching.

underdamping [PHYS] Condition of a system when the amount of damping is sufficiently small so that, when the system is subjected to a single disturbance, either constant or instantaneous, one or more oscillations are executed by the system.

undersaturated fluid [PHYS CHEM] Any fluid (liquid or gas) capable of holding additional vapor or liquid components in solution at specified conditions of pressure and temperature.

underspin [MECH] Property of a projectile having insufficient rate of spin to give proper stabilization.

underwater acoustics [ACOUS] Study of the propagation of sound waves in water, especially in the oceans, and of phenomena produced by these sound waves. Also known as hydroacoustics.

underwater sound [ACOUS] The production, transmission, and reception of sounds in the ocean; used for locating submarines and other submerged objects, and to determine the physical structure of the ocean and its bottom, and to study organisms found in the sea.

underwater transducer [ENG ACOUS] A device used for the generation or reception of underwater sounds.

undetermined multipliers *See* Lagrangian multipliers.

UNDERWATER SOUND

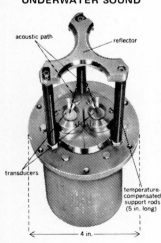

acoustic path

reflector

transducers

temperature-compensated support rods (5 in. long)

4 in.

A typical commercial velocimeter with protecting cage removed, for direct measurement of the velocity of underwater sound (1 inch = 2.54 centimeters). *(NUS Corp.)*

undisturbed motion [PHYS] The steady state of a system before perturbations are introduced.

undulatory extinction [OPTICS] Extinction that occurs successively in adjacent areas as the microscope stage is turned. Also known as oscillatory extinction; strain shadow; wavy extinction.

ungula [MATH] A solid bounded by a portion of a circular cylindrical surface and portions of two planes, one of which is perpendicular to the generators of the cylindrical surface.

uniaxial crystal [OPTICS] A doubly refracting crystal which has a single axis along which light can propagate without exhibiting double refraction.

uniaxial stress [MECH] A state of stress in which two of the three principal stresses are zero.

unidirectional [PHYS] **1.** Flowing in only one direction, such as direct current. **2.** Radiating in only one direction.

unidirectional antenna [ELECTROMAG] An antenna that has a single well-defined direction of maximum gain.

unidirectional log-periodic antenna [ELECTROMAG] Broadband antenna in which the cut-out portions of a log-periodic antenna are mounted at an angle to each other, to give a unidirectional radiation pattern in which the major radiation is in the backward direction, off the apex of the antenna; impedance is essentially constant for all frequencies, as is the radiation pattern.

unified field theory [RELAT] Any theory which attempts to express gravitational theory and electromagnetic theory within a single unified framework; usually, an attempt to generalize Einstein's general theory of relativity from a theory of gravitation alone to a theory of gravitation and classical electromagnetism.

unifilar suspension [ENG] The suspension of a body from a single thread, wire, or strip.

uniform bound [MATH] A number M such that $|f_n(x)| < M$ for every x and for every function in a given sequence of functions $\{f_n(x)\}$.

uniform boundedness principle [MATH] A family of pointwise bounded, real-valued continuous functions on a complete metric space X is uniformly bounded on some open subset of X.

uniform circular motion [MECH] Circular motion in which the angular velocity remains constant.

uniform continuity [MATH] A property of a function f on a set, namely: given any $\varepsilon > 0$ there is a $\delta > 0$ such that $|f(x_1) - f(x_2)| < \varepsilon$ provided $|x_1 - x_2| < \delta$ for any pair x_1, x_2 in the set.

uniform convergence [MATH] A sequence of functions $\{f_n(x)\}$ converges uniformly on E to $f(x)$ if given $\varepsilon > 0$ there is an N such that $|f_n(x) - f(x)| < \varepsilon$ for all x in E provided $n > N$.

uniform distribution [STAT] The distribution of a random variable in which each value has the same probability of occurrence. Also known as rectangular distribution.

uniform field [PHYS] A field which, at the instant under consideration, has the same value at every point in the region under consideration.

uniform line [ELEC] Line which has substantially identical electrical properties through its length.

uniform load [MECH] A load distributed uniformly over a portion or over the entire length of a beam; measured in pounds per foot.

uniform luminance [OPTICS] Property of a surface for which the luminous intensity of any area of the surface is proportional to the area.

uniformly most powerful test [STAT] A test which is simultaneously most powerful for all alternatives of interest in an experiment.

uniform plane wave [ELECTROMAG] Plane wave in which the electric and magnetic intensities have constant amplitude over the equiphase surfaces; such a wave can only be found in free space at an infinite distance from the source.

uniform space [MATH] A topological space X whose topology is derived from a family of subsets of $X \times X$, called a uniformity; intuitively, this gives a notion of "nearness" which is uniform throughout the space.

unijunction transistor [ELECTR] An n-type bar of semiconductor with a p-type alloy region on one side; connections are made to base contacts at either end of the bar and to the p region. Abbreviated UJT. Formely known as double-base diode; double-base junction diode.

unilateral conductivity [ELECTR] Conductivity in only one direction, as in a perfect rectifier.

unilateralization [ELECTR] Use of an external feedback circuit in a high-frequency transistor amplifier to prevent undesired oscillation by canceling both the resistive and reactive changes produced in the input circuit by internal voltage feedback; with neutralization, only the reactive changes are canceled.

unilateral surface [MATH] A one-sided surface; equivalently, any nonorientable two-dimensional manifold such as the Möbius strip and the Klein bottle.

unimodal [STAT] Referring to a distribution with only one mode.

unimodular matrix [MATH] A unimodulus matrix with integer entries.

unimodulus matrix [MATH] A square matrix whose determinant is 1.

union [MATH] A union of a given family of sets is a set consisting of those elements which are members of at least one set in the family.

union of Boolean matrices [MATH] The union of two Boolean matrices A and B, with the same number of rows and columns, is the Boolean matrix whose element c_{ij} in row i and column j is the union of corresponding elements a_{ij} in A and b_{ij} in B.

union of sets [MATH] A set consisting of those elements which are members of at least one set in a given family of sets.

union rule of probability [STAT] The probability that the union of two events E_i and E_j equals the sum total of the probability of the sample points in either E_i or E_j minus the probability of being in both E_i and E_j.

unipolar [ELEC] Having but one pole, polarity, or direction; when applied to amplifiers or power supplies, it means that the output can vary in only one polarity from zero and, therefore, must always contain a direct-current component.

unipolar transistor [ELECTR] A transistor that utilizes charge carriers of only one polarity, such as a field-effect transistor.

unipole [ELECTROMAG] A hypothetical antenna that radiates or receives signals equally well in all directions. Also known as isotropic antenna.

unipotential cathode *See* indirectly heated cathode.

unipotential electrostatic lens [ELECTR] An electrostatic lens in which the focusing is produced by application of a single potential difference; in its simplest form it consists of three apertures of which the outer two are at a common potential, and the central aperture is at a different, generally lower, potential.

unique factorization domain [MATH] An integral domain in which every nonzero element has a unique factorization into

UNIJUNCTION TRANSISTOR

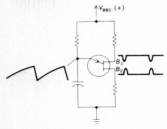

Circuit diagram of a unijunction transistor used as a relaxation oscillator. Sawtooth waveform on the left and output trigger waveform on the right. V_{BB2} = supply voltage of base 2 with respect to emitter; B_1, B_2 = base 1, base 2.

UNIPOTENTIAL ELECTROSTATIC LENS

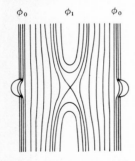

A unipotential electrostatic lens, a type of axially symmetric electrostatic lens. ϕ_0 = common potential for outer two apertures; ϕ_1 = lower potential for central aperture. *(From E. G. Ramberg and G. A. Morton, J. Appl. Phys., vol. 10, 1939, and V. K. Zworykin et al., Electron Optics and the Electron Microscope, Wiley, 1945)*

irreducible elements. Also known as factorial ring; unique factorization ring.

unique factorization into irreducible elements [MATH] An element $a \neq 0$ in a commutative ring with identity R is said to have unique factorization into irreducible elements if (1) it has a factorization into irreducible elements and (2) for any two such factorizations, $a = u p_1 \cdots p_n$ and $a = u' q_1 \cdots q_m$, with units u and u' and irreducible elements $p_1, \ldots, p_n, q_1, \ldots, q_m$, one has $n = m$, and there is a permutation of indices i such that p_i and q_i are associates for $i = 1, \ldots, n$.

unique factorization ring *See* unique factorization domain.

unique factorization theorem [MATH] A positive integer may be expressed in precisely one way as a product of prime numbers.

unit [ENG] An assembly or device capable of independent operation, such as a radio receiver, cathode-ray oscilloscope, or computer subassembly that performs some inclusive operation or function. [MATH] An element of a ring with identity that has both a left inverse and a right inverse. [PHYS] A quantity adopted as a standard of measurement.

unit-area acoustical ohm *See* rayl.

unitary decuplet [PARTIC PHYS] A collection of 10 hadrons whose isospin and hypercharge values form a symmetrical pattern, and which are related by unitary symmetry operations.

unitary group [MATH] The group of unitary transformations on a k-dimensional complex vector space. Usually denoted $U(k)$.

unitary matrix [MATH] A matrix whose inverse is identical with its conjugate transpose.

unitary octet [PARTIC PHYS] A collection of eight hadrons whose isospin and hypercharge values form a symmetrical pattern, and which are related by unitary symmetry operations.

unitary space *See* inner product space.

unitary spin [PARTIC PHYS] A quantum number associated with SU_3 symmetry and which determines the SU_3 supermultiplet to which a particle belongs, such as singlet, octet, or decuplet.

unitary symmetry [PARTIC PHYS] An approximate symmetry law obeyed by the strong interactions of elementary particles; it may be described as the equivalence of three fundamental particles, termed quarks, out of which all hadrons could be assumed to be composed. Also known as SU_3 symmetry.

unitary transformation [MATH] A linear transformation on a vector space which preserves inner products and norms; alternatively, a linear operator whose adjoint is equal to its inverse.

unit ball [MATH] The set of all points in euclidean n-space whose distance from the origin is at most 1.

unit binormal [MATH] A unit vector in the same direction as the binormal to a point on a surface or space curve.

unit cell [CRYSTAL] A parallelepiped which will fill all space under the action of translations which leave the crystal lattice unchanged. Also known as structure cell.

unit charge *See* statcoulomb.

unit circle [MATH] The locus of points in the plane which are precisely one unit from the origin.

unit element [MATH] An element in a ring which acts as a multiplicative identity.

unit magnetic pole [ELECTROMAG] Two equal magnetic poles of the same sign have unit value when they repel each other with a force of 1 dyne if placed 1 centimeter apart in a vacuum.

unit normal [MATH] A unit vector in the direction of the principal normal to a surface or space curve.

unit operator [MATH] The identity operator.

unit sphere [MATH] The set of points in three-space (more generally n-space) which are precisely one unit distance from the origin.

unit strain [MECH] **1.** For tensile strain, the elongation per unit length. **2.** For compressive strain, the shortening per unit length. **3.** For shear strain, the change in angle between two lines originally perpendicular to each other.

unit stress [MECH] The load per unit of area.

unit systems [OPTICS] Optical systems that have a lateral magnification of $+1$ or -1. [PHYS] Groups of units suitable for use in measurement of physical quantities and in the convenient statement of physical laws relating physical quantities.

unit tangent [MATH] A unit vector in the tangent plane at a point of a surface.

unit vector [MATH] A vector whose length is one unit.

unity coupling [ELECTROMAG] Perfect magnetic coupling between two coils, so that all magnetic flux produced by the primary winding passes through the entire secondary winding.

unity gain bandwidth [ELECTR] Measure of the gain-frequency product of an amplifier; unity gain bandwidth is the frequency at which the open-loop gain becomes unity, based on 6 decibels per octave crossing.

unity power factor [ELEC] Power factor of 1.0, obtained when current and voltage are in phase, as in a circuit containing only resistance or in a reactive circuit at resonance.

univalent function *See* injection.

univariant system [THERMO] A system which has only one degree of freedom according to the phase rule.

univariate distribution [STAT] A frequency distribution of only one variate.

universal algebra [MATH] The study of algebraic systems such as groups, rings, modules, and fields and the examination of what families of theorems are analogous in each system.

universal element [MATH] An element of a Boolean algebra that includes every element of the algebra.

universality [STAT MECH] The hypothesis that the critical exponents of a substance are the same within broad classes of substances of widely varying characteristics, and depend only on the microscopic symmetry properties of the substance.

universality class [STAT MECH] A class of substances which have the same critical exponents according to the universality hypothesis.

universally attracting object [MATH] An object O in a category C such that there exists a unique morphism of each object of C into O.

universally repelling object [MATH] An object O of a category C such that there exists a unique morphism of O into each object of C.

universal object [MATH] An object which is universally attracting or universally repelling.

universal quantifier [MATH] A logical relation, often symbolized $\forall$, that may be expressed by the phrase "for all" or "for every"; if P is a predicate, the statement $(\forall x)P(x)$ is true if $P(x)$ is true for all values of x in the domain of P, and is false otherwise.

universal resonance curve [ELEC] A plot of Y/Y_0 against $Q_0\delta$ for a series-resonant circuit, or of Z/Z_0 against $Q_0\delta$ for a parallel-resonant circuit, where Y and Z are the admittance and impedance of a circuit, Y_0 and Z_0 are the values of these quantities at resonance, Q_0 is the Q value of the circuit at resonance, and δ is the deviation of the frequency from resonance divided by the resonant frequency; it can be applied to all resonant circuits.

universal shunt   *See* Ayrton shunt.

universal stage   [OPTICS] A stage attached to the rotating stage of a polarizing microscope that has three, four, or five axes and thin sections of low-symmetry minerals to be tilted about two mutually perpendicular horizontal axes. Also known as Fedorov stage; U stage.

universal time   *See* Greenwich mean time.

universal transmission function   [GEOPHYS] A mathematical relationship that attempts to describe quantitatively the complex infrared propagation (including absorption and reradiation) in the atmosphere.

universal wavelength function   [OPTICS] One of four functions which enable one to compute easily, with reasonable accuracy, the refractive index of glass or other transparent material when this index is known for four standard wavelengths.

universe   [ASTRON] The totality of astronomical things, events, relations, and energies capable of being described objectively.

univibrator   *See* monostable multivibrator.

unloaded Q   [ELECTR] The Q of a system when there is no external coupling to it.

unloading amplifer   [ELECTR] Amplifier that is capable of reproducing or amplifying a given voltage signal while drawing negligible current from the voltage source.

unpaired electron   [ATOM PHYS] an orbital electron for which there is no other electron in the same atom with the same energy but opposite spin.

unpitched sound   [ACOUS] Sound that includes a wide range of frequencies and thus does not have a definite pitch.

unpolarized light   [OPTICS] Light in which the electric vector is oriented in a random, unpredictable fashion.

unpolarized particle beam   [PHYS] A beam of particles with spin in which the directions of the spins are random.

unrelated frequencies   [STAT] The long-run frequency of any result in one part of an experiment is approximately equal to the long-run conditional frequency of that result, given that any specified result has occurred in the other part of the experiment.

unsaturated standard cell   [ELEC] One of two types of Weston standard cells (batteries); used for voltage calibration work not requiring an accuracy greater than 0.01%.

unstable   [PHYS] Capable of undergoing spontaneous change, as in a radioactive nuclide or an excited nuclear system.

unstable bundle   [MATH] The bundle E^u of a hyperbolic structure.

unstable equilibrium   [PHYS] An equilibrium state of a system in which any departure of the system from equilibrium gives rise to forces or tendencies moving the system further away from equilibrium; for example, mechanical equilibrium in which the potential energy is a maximum, as a sphere sitting on top of a hill.

unstable graph   [MATH] A graph from which it is not possible to delete an edge to produce a subgraph whose group of automorphisms is a subgroup of the group of automorphisms of the original graph.

unstable isotope   *See* radioisotope.

unstable particle   [PARTIC PHYS] **1.** Any elementary particle that spontaneously decays into other particles. **2.** An elementary particle that can decay through the strong interactions, as opposed to a semistable particle; it has a lifetime on the order of 10^{-23} second.

unstable wave [PHYS] A wave motion whose amplitude increases with time or whose total energy increases at the expense of its environment.

unsteady flow [FL MECH] Fluid flow in which properties of the flow change with respect to time.

unsteady-state flow [FL MECH] A condition of fluid flow in which the volumetric ratios of two or more phases (liquid-gas, liquid-liquid, and so on) vary along the course of flow; can be the result of changes in temperature, pressure, or composition.

untuned [ELEC] Not resonant at any of the frequencies being handled.

Unwin coefficients [FL MECH] The constants k, p, and q which appear in Reynolds' law.

up-and-down method [STAT] A technique which uses a unit sequential method of testing; the level of variable at each experiment is raised or lowered depending on the outcome of the previous test.

upper bound [MATH] If S is a subset of an ordered set A, an upper bound b for S in A is an element b of A such that $x \leq b$ for all x belonging to A.

upper consolute temperature *See* consolute temperature.

upper control limit [IND ENG] A horizontal line on a control chart at a specified distance above the central line; if all the plotted points fall between the upper and lower control lines, the process is said to be in control.

upper critical solution temperature *See* consolute temperature.

upper culmination *See* upper transit.

upper Darboux integral [MATH] For a bounded function $f(x)$ defined on a closed interval $[a,b]$, the limit specified by Darboux's theorem for the sum S derived from least upper bounds of $f(x)$ on subintervals of $[a,b]$; written

$$\overline{\int_a^b} f(x)\ dx.$$

upper derivative [MATH] The upper derivative of a real-valued function of a real variable $f(x)$ at a point x_0 is the limit superior of $[f(x)-f(x_0)]/(x - x_0)$ as x approaches x_0.

upper half-power frequency [ELECTR] The frequency on an amplifier response curve which is greater than the frequency for peak response and at which the output voltage is $1/\sqrt{2}$ (that is, 0.707) of its midband or other reference value.

upper integral [MATH] The upper Riemann integral for a real-valued function $f(x)$ on an interval is computed to be the infimum of all finite sums over all partitions of the interval, the sums having terms given by $(x_i - x_{i-1})y_i$, where the x_i are from a partition, and y_i is the largest value of $f(x)$ over the interval from x_{i-1} to x_i.

upper limb [ASTRON] That half of the outer edge of a celestial body having the greatest altitude.

upper metric density [MATH] The upper metric density of a measurable subset S of the real numbers at a point x is the limit as n approaches infinity of $\bar{\phi}_n(x)$, where $\bar{\phi}_n(x)$ is the least upper bound, over all intervals I containing x and having length $l(I)$ less than $1/n$, of $m(S \cap I)/l(I)$, where $m(S \cap I)$ is the measure of the intersection of S with I.

upper semicontinuous decomposition [MATH] A partition of a topological space with the property that for every member D of the partition and for every open set U containing D there is an open set V containing D which is contained in U and is the union of members of the partition.

upper semicontinuous function [MATH] A real-valued function $f(x)$ is upper semicontinuous at a point x_0 if for any small positive ε, $f(x)$ always is less than $f(x_0) + \varepsilon$ for all x in some neighborhood of x_0.

upper transit [ASTRON] The movement of a celestial body across a celestial meridian's upper branch. Also known as superior transit; upper culmination.

upper variation [MATH] The upper variation of a signed measure m is the set function m^+ defined for every measurable set E by $m^+(E) = m(E \cap A)$, where A is the member of the Hahn decompostion which is positive with respect to m.

up quark [PARTIC PHYS] A quark with an electric charge of $+\frac{2}{3}$, baryon number of $\frac{1}{3}$, and 0 strangeness and charm. Symbolized u.

upward bias [STAT] The overestimation or overstatement by a statistical measure of the event it is attempting to describe.

uranium [CHEM] A metallic element in the actinide series, symbol U, atomic number 92, atomic weight 238.03; highly toxic and radioactive; ignites spontaneously in air and reacts with nearly all nonmetals; melts at 1132°C, boils at 3818°C; used in nuclear fuel and as the source of U^{235} and plutonium.

uranium decay series *See* uranium series.

uranium enrichment [NUCLEO] A process carried out on uranium, in which the ratio of the abundance of the isotope uranium-235 to that of the isotope uranium-238 is increased above that found in natural uranium.

uranium-radium series *See* uranium series.

uranium reactor [NUCLEO] A nuclear reactor in which the principal fuel is uranium; the uranium may be natural, with the naturally occurring ratio of 1 atom of uranium-235 to about 139 atoms of uranium-238, or may be enriched to have a higher proportion of fissile uranium-233 or uranium-235 atoms.

uranium series [NUC PHYS] The series of nuclides resulting from the decay of uranium-238, including uranium I, II, X_1, X_2, Y, and Z, and radium A, B, C, C', C'', D, E, E'', F, and G. Also known as uranium decay series; uranium-radium series.

Uranus [ASTRON] A planet, seventh in the order of distance from the sun; it has five known satellites, and its equatorial diameter is about four times that of the earth.

Ursa Major cluster [ASTRON] A group of about 126 stars, including 5 stars 'of the constellation Ursa Major; the sun is passing through this cluster which occupies a spherical space of about 450 light-years (4.26×10^{18} meters) in diameter.

Urysohn lemma [MATH] If A and B are disjoint, closed sets in a normal space X, there is a real-valued function f such that $0 \leq f(x) \leq 1$ for all $x \, \varepsilon \, X$, and $f(A) = 0$ and $f(B) = 1$.

U-shaped distribution [STAT] A frequency distribution whose shape approximates that of the letter U.

U stage *See* universal stage.

U-tube manometer [ENG] A manometer consisting of a U-shaped glass tube partly filled with a liquid of known specific gravity; when the legs of the manometer are connected to separate sources of pressure, the liquid rises in one leg and drops in the other; the difference between the levels is proportional to the difference in pressures and inversely proportional to the liquid's specific gravity. Also known as liquid-column gage.

UV Ceti stars [ASTRON] A class of stars that have brief outbursts of energy over their surface areas; they may have an increase of about 1 magnitude for periods of 1 hour; the type star is UV Ceti. Also known as flare stars.

URANIUM

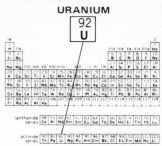

Periodic table of the chemical elements showing the position of uranium.

URANUS

Telescopic appearance of Uranus, when the earth is in the equatorial plane of the globe.

U-TUBE MANOMETER

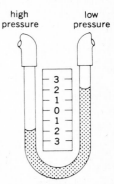

Components of U-tube manometer for measuring liquid flow.

V *See* vanadium; volt.

VA *See* volt-ampere.

vac *See* millibar.

vacancy [SOLID STATE] A defect in the form of an unoccupied lattice position in a crystal.

vacuum [PHYS] **1.** Theoretically, a space in which there is no matter. **2.** Practically, a space in which the pressure is far below normal atmospheric pressure so that the remaining gases do not affect processes being carried on in the space.

vacuum capacitor [ELEC] A capacitor with separated metal plates or cylinders mounted in an evacuated glass envelope to obtain a high breakdown voltage rating.

vacuum correction [PHYS] The correction to the reading of a mercury barometer required by the imperfections in the vacuum above the mercury column, due to the presence of water vapor and air; this correction is a function of both temperature and pressure.

vacuum polarization [QUANT MECH] A process in which an electromagnetic field gives rise to virtual electron-positron pairs that effectively alter the distribution of charges and currents that generated the original electromagnetic field.

vacuum pump [MECH ENG] A compressor for exhausting air and noncondensable gases from a space that is to be maintained at subatmospheric pressure.

vacuum tube [ELECTR] An electron tube evacuated to such a degree that its electrical characteristics are essentially unaffected by the presence of residual gas or vapor.

vacuum-tube electrometer [ELECTR] An electrometer in which the ionization current in an ionization chamber is amplified by a special vacuum triode having an input resistance above 10,000 megohms.

vacuum-tube rectifier [ELECTR] A rectifier in which rectification is accomplished by the unidirectional passage of electrons from a heated electrode to one or more other electrodes within an evacuated space.

vacuum-tube voltmeter [ENG] Any of several types of instrument in which vacuum tubes, acting as amplifiers or rectifiers, are used in circuits for the measurement of alternating-current or direct-current voltage. Abbreviated VTVM. Also known as tube voltmeter.

vacuum ultraviolet spectroscopy [SPECT] Absorption spectroscopy involving electromagnetic wavelengths shorter than 200 nanometers; so called because the interference of the high absorption of most gases necessitates work with evacuated equipment.

Vaisala comparator [OPTICS] An interferometer measuring distances on the order of 100 meters with accuracies on the order of 1 part in 10^7

valence [CHEM] A positive number that characterizes the combining power of an element for other elements, as measured by the number of bonds to other atoms which one atom of the given element forms upon chemical combination; hydrogen is assigned valence 1, and the valence is the number of hydrogen atoms, or their equivalent, with which an atom of the given element combines.

valence angle   *See* bond angle.

valence band   [SOLID STATE] The highest electronic energy band in a semiconductor or insulator which can be filled with electrons.

valence-bond method   [PHYS CHEM] A method of calculating binding energies and other parameters of molecules by taking linear combinations of electronic wave functions, some of which represent covalent structures, others ionic structures; the coefficients in the linear combination are calculated by the variational method. Also known as valence-bond resonance method.

valence-bond resonance method   *See* valence-bond method.

valence-bond theory   [CHEM] A theory of the structure of chemical compounds according to which the principal requirements for the formation of a covalent bond are a pair of electrons and suitably oriented electron orbitals on each of the atoms being bonded; the geometry of the atoms in the resulting coordination polyhedron is coordinated with the orientation of the orbitals on the central atom.

valence crystal   *See* covalent crystal.

valence electron   [ATOM PHYS] An electron that belongs to the outermost shell of an atom. [SOLID STATE] *See* conduction electron.

valence number   [CHEM] A number that is equal to the valence of an atom or ion multiplied by $+1$ or -1, depending on whether the ion is positive or negative, or equivalently on whether the atom in the molecule under consideration has lost or gained electrons from its free state.

valence shell   [ATOM PHYS] The electrons that form the outermost shell of an atom.

validity   [MATH] Correctness; especially the degree of closeness by which iterated results approach the correct result.

valley attenuation   [ELECTR] For an electric filter with an equal ripple characteristic, the maximum attenuation occurring at a frequency between two frequencies where the attenuation reaches a minimum value.

valuation   [MATH] A mapping v on a field K to an ordered field which has properties similar to an absolute value; specifically, (1) $v(x) \geq 0$ for all x in K and $v(x) = 0$ if and only if $x = 0$, (2) for all x and y in K, $v(xy) = v(x)v(y)$, and (3) for all x and y in K, $v(x + y) \leq v(x) + v(y)$.

valuation ring   [MATH] **1.** A ring R contained in a field K such that for every x in K either x is in R or x^{-1} is in R. **2.** The valuation ring of a place (second definition) is the ring K_p that forms the domain of the place. **3.** The valuation ring of a discrete valuation v on a field K consists of the elements x in K such that $v(x) \geq 0$.

value   [SCI TECH] The magnitude of a quantity.

value group   [MATH] For a discrete valuation v on a field K, this is the group formed by the elements $v(x)$ corresponding to nonzero elements x in K.

value index   [STAT] An index member which is the ratio of the value of all items in a given period to the value of all items in the base period.

value of a function   [MATH] The value of a function f at an element x is the element y which f associates with x; that is, it is $y = f(x)$.

value of a matrix game   [MATH] The expected payoff of a matrix game when each player follows an optimal strategy.

valve   *See* electron tube.

vanadium [CHEM] A metal in group Vb, symbol V, atomic number 23; soluble in strong acids and alkalies; melts at 1900°C, boils about 3000°C; used as a catalyst.

Van Allen radiation belt [GEOPHYS] One of the belts of intense ionizing radiation in space about the earth formed by high-energy charged particles which are trapped by the geomagnetic field.

Van de Graaff accelerator [ELECTR] A Van de Graaff generator equipped with an evacuated tube through which charged particles may be accelerated.

Van de Graaff generator [ELECTR] A high-voltage electrostatic generator in which electrical charge is carried from ground to a high-voltage terminal by means of an insulating belt and is discharged onto a large, hollow metal electrode.

Vandermonde determinant [MATH] The determinant of the $n \times n$ matrix whose ith row appears as $1, x_i, x_i{}^2, \ldots, x_i{}^{n-1}$ where the $x_i{}^k$ appear as variables in a given polynomial equation; this provides information about the roots.

Vandermonde's theorem [MATH] Represents a binomial $(x + y)^a$, where a is an exponent involving the variables x and y, in terms of a sum of expressions $x^c y^d$ where the exponents c and d involve the variables x and y also.

Van der Pol oscillator [ELECTR] A type of relaxation oscillator which has a single pentode tube and an external circuit with a capacitance that causes the device to switch between two values of the screen voltage. [PHYS] A vibrating system governed by an equation of the form $\ddot{x} + \varepsilon(-\dot{x} + \tfrac{1}{3}\dot{x}^3) + x = 0$.

van der Waals adsorption [PHYS CHEM] Adsorption in which the cohesion between gas and solid arises from van der Waals forces.

van der Waals attraction *See* van der Waals force.

van der Waals covolume [PHYS CHEM] The constant b in the van der Waals equation, which is approximately four times the volume of an atom of the gas in question multiplied by Avogadro's number.

van der Waals equation [PHYS CHEM] An empirical equation of state which takes into account the finite size of the molecules and the attractive forces between them: $p = [RT/(v - b)] - (a/v^2)$, where p is the pressure, v is the volume per mole, T is the absolute temperature, R is the gas constant, and a and b are constants.

van der Waals force [PHYS CHEM] An attractive force between two atoms or nonpolar molecules, which arises because a fluctuating dipole moment in one molecule induces a dipole moment in the other, and the two dipole moments then interact. Also known as dispersion force; London dispersion force; van der Waals attraction.

van der Waals–London interactions [PHYS CHEM] The interaction associated with the van der Waals force.

van der Waals molecule [PHYS CHEM] A molecule that is held together by van der Waals forces.

van der Waals structure [CRYSTAL] The structure of a molecular crystal.

van der Waals surface tension formula [THERMO] An empirical formula for the dependence of the surface tension on temperature: $\gamma = K p_c{}^{2/3} T_c{}^{1/3} (1 - T/T_c)^n$, where γ is the surface tension, T is the temperature, T_c and p_c are the critical temperature and pressure, K is a constant, and n is a constant equal to approximately 1.23.

vane attenuator *See* flap attenuator.

V antenna [ELECTROMAG] An antenna having a V-shaped arrangement of conductors fed by a balanced line at the apex; the included angle, length, and elevation of the conductors are

VANADIUM

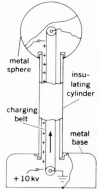

Periodic table of the chemical elements showing the position of vanadium.

VAN DE GRAAFF GENERATOR

metal sphere
insulating cylinder
charging belt
metal base
+ 10 kv

Schematic drawing of Van de Graaff generator for operation in air at atmospheric pressure.

VAN DER POL OSCILLATOR

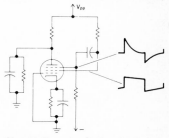

Circuit diagram of Van der Pol oscillator showing waveform. V_{pp} = plate supply voltage.

proportioned to give the desired directivity. Also spelled vee antenna.

Van Vleck equation [QUANT MECH] An equation based on quantum theory for the molar paramagnetism of a magnetically susceptible material from magnetic moment, absolute temperature, and various constants.

Van Vleck paramagnetism [QUANT MECH] The paramagnetism of a collection of atoms, ions, or molecules, as computed by quantum theory; the atoms, ions, or molecules in a magnetic field are distributed among the various allowed energy levels according to a Boltzmann distribution, and the magnetization of the system is computed by finding the average component of angular momentum parallel to the field.

vapor [THERMO] A gas at a temperature below the critical temperature, so that it can be liquefied by compression, without lowering the temperature.

vapor cycle [THERMO] A thermodynamic cycle, operating as a heat engine or a heat pump, during which the working substance is in, or passes through, the vapor state.

vapor-filled thermometer [ENG] A gas- or vapor-filled temperature measurement device that moves or distorts in response to temperature-induced pressure changes from the expansion or contraction of the sealed, vapor-containing chamber.

vaporization *See* volatilization.

vaporization coefficient [THERMO] The ratio of the rate of vaporization of a solid or liquid at a given temperature and corresponding vapor pressure to the rate of vaporization that would be necessary to produce the same vapor pressure at this temperature if every vapor molecule striking the solid or liquid were absorbed there.

vapor lamp *See* discharge lamp.

vapor pressure [THERMO] For a liquid or solid, the pressure of the vapor in equilibrium with the liquid or solid.

vapor-pressure thermometer [ENG] A thermometer in which the vapor pressure of a homogeneous substance is measured and from which the temperature can be determined; used mostly for low-temperature measurements.

varactor [ELECTR] A semiconductor device characterized by a voltage-sensitive capacitance that resides in the space-charge region at the surface of a semiconductor bounded by an insulating layer. Also known as varactor diode; variable-capacitance diode; varicap; voltage-variable capacitor.

varactor diode *See* varactor.

var hour [ELEC] A unit of the integral of reactive power over time, equal to a reactive power of 1 var integrated over 1 hour; equal in magnitude to 3600 joules. Also known as reactive volt-ampere hour; volt-ampere-hour reactive.

variable-bandwidth filter [ELECTR] An electric filter whose upper and lower cutoff frequencies may be independently selected, so that almost any bandwidth may be obtained; it usually consists of several stages of R-C filters, each separated by buffer amplifiers; tuning is accomplished by varying the resistance and capacitance values.

variable-capacitance diode *See* varactor.

variable capacitor [ELEC] A capacitor whose capacitance can be varied continuously by moving one set of metal plates with respect to another.

variable coupling [ELEC] Inductive coupling that can be varied by moving one coil with respect to another.

variable diode function generator [ELECTR] An improvement of a diode function generator in which fully adjustable potentiometers are used for breakpoint and slope resistances, permitting the programming of analytic, arbitrary, and empirical functions, including inflections. Abbreviated VDFG.

variable field [PHYS] Field which changes during the time under consideration.

variable flow [FL MECH] Fluid flow in which the velocity changes both with time and from point to point.

variable-focal-length lens *See* zoom lens.

variable-focus condenser [OPTICS] A condenser that is used to obtain a large illuminated field area, and has two lenses, the first of which can be adjusted to bring light to a focus between the lenses.

variable-mu tube [ELECTR] An electron tube in which the amplification factor varies in a predetermined manner with control-grid voltage; this characteristic is achieved by making the spacing of the grid wires vary regularly along the length of the grid, so that a very large negative grid bias is required to block anode current completely. Also known as remote-cutoff tube.

variable nebula [ASTRON] A nebula whose shape and brightness vary; an example is in the constellation Monoceros.

variable resistor *See* rheostat.

variable star [ASTRON] A star that has a detectable change in its intensity which is often accompanied by other physical changes; changes in brightness may be a few thousandths of a magnitude to 20 magnitudes or even more.

variable transformer [ELEC] An iron-core transformer having provisions for varying its output voltage over a limited range or continuously from zero to maximum output voltage, generally by means of a contact arm moving along exposed turns of the secondary winding. Also known as adjustable transformer; continuously adjustable transformer.

variance [STAT] The square of the standard deviation.

variance ratio test [STAT] A technique for comparing the spreads or variabilities of two sets of figures to determine whether the two sets of figures were drawn from the same population. Also known as F test.

variate *See* random variable.

variate difference method [STAT] A technique for estimating the correlation between the random parts of two given time series.

variation [GEOPHYS] *See* declination. [MATH] **1.** A variation $\delta y(x)$ of a function $y(x)$ is any function which is added to $y(x)$ to produce a new function $y(x) + \delta y(x)$. **2.** For a real function f defined on an interval $[a,b]$, relative to a partition $a = x_0 < x_1 < x_2 < \cdots < x_n = b$, the sum over $k = 1, 2, \ldots, n$ of $|f(x_k) - f(x_{k-1})|$.

variational method [QUANT MECH] A method of calculating an upper bound on the lowest energy level of a quantum-mechanical system and an approximation for the corresponding wave function; in the integral representing the expectation value of the Hamiltonian operator, one substitutes a trial function for the true wave function, and varies parameters in the trial function to minimize the integral.

variational principle [MATH] A technique for solving boundary value problems that is applicable when the given problem can be rephrased as a minimization problem.

varicap *See* varactor.

varifocal lens *See* zoom lens.

Varignon's theorem [MECH] The theorem that the moment of a force is the algebraic sum of the moments of its vector components acting at a common point on the line of action of the force.

varindor [ELECTROMAG] Inductor in which the inductance varies markedly with the current in the winding.

variometer [ELECTROMAG] A variable inductance having two coils in series, one mounted inside the other, with provisions

VARIOMETER

Variometer for measuring horizontal intensity or declination of terrestrial magnetic field, equipped with Helmholtz coil for calibration. *(U.S. Coast and Geodetic Survey)*

for rotating the inner coil in order to vary the total inductance of the unit over a wide range. [ENG] A geomagnetic device for detecting and indicating changes in one of the components of the terrestrial magnetic field vector, usually magnetic declination, the horizontal intensity component, or the vertical intensity component.

varistor [ELECTR] A two-electrode semiconductor device having a voltage-dependent nonlinear resistance; its resistance drops as the applied voltage is increased. Also known as voltage-dependent resistor.

var measurement [ELEC] The measurement of reactive power in a circuit.

varmeter [ENG] An instrument for measuring reactive power in vars. Also known as reactive volt-ampere meter.

V coefficient [QUANT MECH] Either of two coefficients used in the coupling of eigenfunctions of two angular momenta, differing from the Wigner 3-j symbol by at most a sign. Symbolized V and $\bar{V}$.

VDFG *See* variable diode function generator.

vector [MATH] An element of a vector space. [PHYS] A quantity which has both magnitude and direction, and whose components transform from one coordinate system to another in the same manner as the components of a displacement. Also known as polar vector.

vector analysis [MATH] The formal study of vectors.

vector bundle [MATH] A locally trivial bundle whose fibers are isomorphic vector spaces.

vector coupling coefficient [QUANT MECH] One of the coefficients used to express an eigenfunction of the sum of two angular momenta in terms of sums of products of eigenfunctions of the original two angular momenta. Also known as Clebsch-Gordan coefficient; Wigner coefficient.

vector current [PARTIC PHYS] A current which behaves as a vector under Lorentz transformations, rather than as an axial vector.

vector density [MATH] A quantity which differs from a vector by the presence of the Jacobian $|\partial x/\partial \bar{x}|$ in the transformation law; that is, the contravariant components transform according to the equation $\bar{v}^i = |\partial x/\partial \bar{x}|(\partial \bar{x}^i/\partial x^j)v^j$, and the covariant components transform according to the equation $\bar{v}_i = |\partial x/\partial \bar{x}|(\partial x^j/\partial \bar{x}^i)v_j$.

vector equation [MATH] An equation involving vectors.

vector field [MATH] **1.** The field of vectors arising from considering a system of differential equations on a differentiable manifold. **2.** A function whose range is in a vector space. [PHYS] A field which is characterized by a vector function.

vector function [PHYS] A function of position and time whose value at each point is a vector. Also known as vector point function.

vector impedance meter [ENG] An instrument that not only determines the ratio between voltage and current, to give the magnitude of impedance, but also determines the phase difference between these quantities, to give the phase angle of impedance.

vector meson [PARTIC PHYS] A meson which has spin quantum number 1 and negative parity, and may be described by a vector field; examples include the ω, ρ, ϕ, and K^* mesons.

vector model of atomic structure [ATOM PHYS] A model of atomic structure in which spin and orbital angular momenta of the electrons are represented by vectors, with special rules for their addition imposed by underlying quantum-mechanical considerations.

vector momentum *See* momentum.

vector point function *See* vector function.

vector potential [ELECTROMAG] A vector function whose curl is equal to the magnetic induction. Symbolized **A**. Also known as magnetic vector potential. [PHYS] Any vector function whose curl is equal to some solenoidal vector field.

vector power [ELEC] Vector quantity equal in magnitude to the square root of the sum of the squares of the active power and the reactive power.

vector-power factor [ELEC] Ratio of the active power to the vector power; it is the same as power factor in the case of simple sinusoidal quantities.

vector product See cross product.

vector space [MATH] A system of mathematical objects which have an additive operation producing a group structure and which can be multiplied by elements from a field in a manner similar to contraction or magnification of directed line segments in euclidean space. Also known as linear space.

vector sum [MATH] For a set of located vectors in euclidean space, $v_1, v_2, \ldots, v_n$, this is the vector whose initial point is the initial point of v_1 and whose terminal point is the terminal point of v_n, when the vectors are laid end to end so that the terminal point of one vector v_i is the initial point of the next vector v_{i+1}. Also known as resultant.

vector voltmeter [ENG] A two-channel high-frequency sampling voltmeter that measures phase as well as voltage of two input signals of the same frequency.

vee antenna See V antenna.

veiling glare [OPTICS] The reduction in contrast of an optical image caused by superposition of scattered light.

Veil Nebula [ASTRON] A nebula in the constellation Cygnus; speculation is that the nebula is the remnant of a gigantic supernova which occurred 30,000 years ago.

Vela pulsar [ASTRON] A pulsar with a period of 80 milliseconds, about 1500 light-years (1.4×10^{19} meters) away in the constellation Vela, whose variation has been detected at radio, gamma-ray, and optical wavelengths.

velocity [MECH] **1.** The time rate of change of position of a body; it is a vector quantity having direction as well as magnitude. Also known as linear velocity. **2.** The speed at which the detonating wave passes through a column of explosives, expressed in meters or feet per second.

velocity coefficient [FL MECH] The ratio of the actual velocity of gas emerging from a nozzle to the velocity calculated under ideal conditions; it is less than 1 because of friction losses. Also known as coefficient of velocity.

velocity constant [CONT SYS] The ratio of the rate of change of the input command signal to the steady-state error, in a control system where these two quantities are proportional.

velocity discontinuity See seismic discontinuity.

velocity-distance relation [ASTRON] The relation wherein all the exterior galaxies are moving away from the galaxy that the sun is part of, with velocities that are greater with increasing distance of the galaxy.

velocity distribution [STAT MECH] For the molecules of a gas, a function of velocity whose value at any velocity v is proportional to the number of molecules with velocities in an infinitesimal range about v, per unit velocity range.

velocity-focusing mass spectrograph See velocity spectrograph.

velocity gradient [FL MECH] The rate of change of velocity of propagation with distance normal to the direction of flow.

velocity head [FL MECH] The square of the speed of flow of a fluid divided by twice the acceleration of gravity; it is equal to the static pressure head corresponding to a pressure equal to the kinetic energy of the fluid per unit volume.

velocity level [ACOUS] A sound rating in decibels, equal to 20 times the logarithm to the base 10 of the ratio of the particle velocity of the sound to a specified reference particle velocity.

velocity of light *See* speed of light.

velocity of sound *See* speed of sound.

velocity potential [FL MECH] For a fluid flow, a scalar function whose gradient is equal to the velocity of the fluid.

velocity profile [FL MECH] A graph of the speed of a fluid flow as a function of distance perpendicular to the direction of flow.

velocity resonance *See* phase resonance.

velocity servomechanism [CONT SYS] A servomechanism in which the feedback-measuring device generates a signal representing a measured value of the velocity of the output shaft. Also known as rate servomechanism.

velocity spectrograph [PHYS] A mass spectrograph in which only positive ions having a certain velocity pass through all three slits and enter a chamber where they are deflected by a magnetic field in proportion to their charge-to-mass ratio. Also known as velocity-focusing mass spectrograph.

vena contracta [FL MECH] The contraction of a jet of liquid which comes out of an opening in a container to a cross section smaller than the opening.

Venn diagram [MATH] A pictorial representation of set theoretic operations such as union, intersection, and complementation of sets.

venturi meter [ENG] An instrument for efficiently measuring fluid flow rate in a piping system; a nozzle section increases velocity and is followed by an expanding section for recovery of kinetic energy.

venturi tube [ENG] A constriction that is placed in a pipe and causes a drop in pressure as fluid flows through it, consisting essentially of a short straight pipe section or throat between two tapered sections; it can be used to measure fluid flow rate (a venturi meter), or to draw fuel into the main flow stream, as in a carburetor.

Venus [ASTRON] The planet second in distance from the sun; the linear diameter, about 12,200 kilometers, includes the top of a cloud layer; the diameter of the solid globe is about 50 kilometers less; the mass is about 0.815 (earth = 1).

Verdet constant [OPTICS] A constant of proportionality in the equation of the Faraday effect; it is equal to the angle of rotation of plane-polarized light in a magnetized substance divided by the product of the length of the light path in the substance and the strength of the magnetic field.

vernal equinox [ASTRON] The sun's position on the celestial sphere about March 21; at this time the sun's path on the ecliptic crosses the celestial equator. Also known as first point of Aries; March equinox; spring equinox.

vernier [ENG] A short, auxiliary scale which slides along the main instrument scale to permit accurate fractional reading of the least main division of the main scale.

vernier capacitor [ELEC] Variable capacitor placed in parallel with a larger tuning capacitor to provide a finer adjustment after the larger unit has been set approximately to the desired position.

vers *See* versed sine.

versed cosine *See* coversed sine.

versed sine [MATH] The versed sine of A is $1 - \text{cosine } A$. Denoted vers. Also known as versine.

versine *See* versed sine.

vertex [MATH] For a polygon or polyhedron, any of those finitely many points which together with line segments or plane pieces determine the figure or solid. [OPTICS] One of

VERNIER

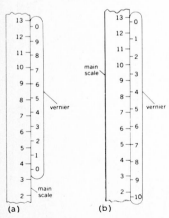

Two types of vernier scales.
(a) Direct (reading 3.6).
(b) Retrograde (reading 12.7).

the points where the surface of a lens intersects the optical axis.

vertex power [OPTICS] The reciprocal of the back focal length of a lens.

vertical angles [MATH] The two angles produced by a pair of intersecting lines and lying on opposite sides of the point of intersection.

vertical component effect *See* antenna effect.

vertical illuminator [OPTICS] A microscope designed for observing surfaces of opaque substances such as metals, which has a mechanism for passing light down through the objective lens in order to illuminate the surface to be observed with a beam perpendicular to the surface.

vertical-incidence transmission [ELECTROMAG] Transmission of a radio wave vertically to the ionosphere and back.

vertical intensity [GEOPHYS] The magnetic intensity of the vertical component of the earth's magnetic field, reckoned positive if downward, negative if upward.

vertical intensity variometer [ENG] A variometer employing a large permanent magnet and equipped with very fine steel knife-edges or pivots resting on agate planes or saddles and balanced so that its magnetic axis is horizontal. Also known as Z variometer.

vertical metal oxide semiconductor technology [ELECTR] For semiconductor devices, a technology that involves essentially the formation of four diffused layers in silicon and etching of a V-shaped groove to a precisely controlled depth in the layers, followed by deposition of metal over silicon dioxide in the groove to form the gate electrode. Abbreviated VMOS technology.

very-long-baseline interferometry [ELECTR] A method of improving angular resolution in the observation of radio sources; these are simultaneously observed by two radio telescopes which are very far apart, and the signals are recorded on magnetic tapes which are combined electronically or on a computer. Abbreviated VLBI.

vibrating capacitor [ELEC] A capacitor whose capacitance is varied in a cyclic manner to produce an alternating voltage proportional to the charge on the capacitor; used in a vibrating-reed electrometer.

vibrating-reed electrometer [ENG] An instrument using a vibrating capacitor to measure a small charge, often in combination with an ionization chamber.

vibrating-reed magnetometer [ENG] An instrument that measures magnetic fields by noting their effect on the vibration of reeds excited by an alternating magnetic field.

vibration [MECH] A continuing periodic change in a displacement with respect to a fixed reference.

vibrational energy [PHYS CHEM] For a diatomic molecule, the difference between the energy of the molecule idealized by setting the rotational energy equal to zero, and that of a further idealized molecule which is obtained by gradually stopping the vibration of the nuclei without placing any new constraint on the motions of electrons.

vibrational level [PHYS CHEM] An energy level of a diatomic or polyatomic molecule characterized by a particular value of the vibrational energy.

vibrational quantum number [PHYS CHEM] A quantum number v characterizing the vibrational motion of nuclei in a molecule; in the approximation that the molecule behaves as a quantum-mechanical harmonic oscillator, the vibrational energy is $h(v + \frac{1}{2})f$, where h is Planck's constant and f is the vibration frequency.

vibrational spectrum [SPECT] The molecular spectrum resulting from transitions between vibrational levels of a molecule which behaves like the quantum-mechanical harmonic oscillator.

vibrational sum rule [SPECT] **1.** The rule that the sums of the band strengths of all emission bands with the same upper state is proportional to the number of molecules in the upper state, where the band strength is the emission intensity divided by the fourth power of the frequency. **2.** The sums of the band strengths of all absorption bands with the same lower state is proportional to the number of molecules in the lower state, where the band strength is the absorption intensity divided by the frequency.

vibration galvanometer [ENG] An alternating-current galvanometer in which the natural oscillation frequency of the moving element is equal to the frequency of the current being measured.

vibration meter *See* vibrometer.

vibration pickup [ELEC] An electromechanical transducer capable of converting mechanical vibrations into electrical voltages.

vibrograph [ENG] An instrument that provides a complete oscillographic record of a mechanical vibration; in one form a moving stylus records the motion being measured on a moving paper or film.

vibrometer [ENG] An instrument designed to measure the amplitude of a vibration. Also known as vibration meter.

vicinal faces [CRYSTAL] Macroscopic crystal faces which are inclined only a few minutes of arc to crystal faces with low Miller indices, and which therefore must have high Miller indices themselves.

Vieta's sequence [MATH] An expression for π:

$$2/\pi = \sqrt{\tfrac{1}{2}}\ \sqrt{\tfrac{1}{2}+\tfrac{1}{2}\ \sqrt{\tfrac{1}{2}+\cdots}}\ \cdots.$$

view camera [OPTICS] A camera that can be focused at both front and back, with adjustments for tilts, swings, shifts, and rise and fall, to control the shape of the subject in the image; it has a groundglass on the back which enables the photographer to view the image to be recorded.

viewing screen *See* screen.

vignetting [OPTICS] Reduction in intensity of illumination near the edges of an optical instrument's field of view caused by obstruction of light rays by the edge of the aperture.

Villari effect [PHYS] A change of magnetic induction within a ferromagnetic substance in a magnetic field when the substance is subjected to mechanical stress.

Villari reversal [PHYS] A change in the sign of the Villari effect which occurs with some ferromagnetic materials when the magnetic field strength reaches a certain value.

violet [OPTICS] The hue evoked in an average observer by monochromatic radiation having a wavelength in the approximate range from 390 to 455 nanometers; however, the same sensation can be produced in a variety of other ways.

violle [OPTICS] A unit of luminous intensity, equal to the luminous intensity of 1 square centimeter of platinum at its temperature of solidification; it is found experimentally to be equal to 20.17 candelas.

virgin neutron [NUCLEO] A neutron from any source, before it makes a collision.

Virgo A [ASTRON] A radio galaxy; it is associated with the galaxy M 87 (NGC 4486).

Virgo cluster [ASTRON] A cluster of galaxies which is the nearest to the galaxy that includes the sun; the cluster is centered in the constellation Virgo and is about 19 megaparsecs (5.9×10^{23} meters) from earth.

virial coefficients [THERMO] For a given temperature T, one

VIBROMETER

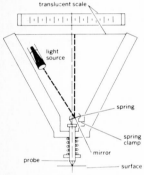

Mechanooptical vibrometer. The motion given to the probe by the vibrating surface is used to rock a mirror and thereby actuate an optical lever arm. A light beam reflected from the mirror and focused onto the scale provides an indication of the vibration amplitude. *(General Electric Co.)*

of the coefficients in the expansion of P/RT in inverse powers of the molar volume, where P is the pressure and R is the gas constant.

virial of a system [STAT MECH] The average over a long period of time of $-\frac{1}{2}$ the sum over the particles in the system of the scalar product of the total force acting on the particle and its radius vector.

virial theorem *See* Clausius virial theorem.

virtual cathode [ELECTR] The locus of a space-charge-potential minimum such that only some of the electrons approaching it are transmitted, the remainder being reflected back to the electron-emitting cathode.

virtual displacement [MECH] **1.** Any change in the positions of the particles forming a mechanical system. **2.** An infinitesimal change in the positions of the particles forming a mechanical system, which is consistent with the geometrical constraints on the system.

virtual entropy [THERMO] The entropy of a system, excluding that due to nuclear spin. Also known as practical entropy.

virtual image [OPTICS] An optical image from which rays of light only appear to diverge, without actually being focused there.

virtual leak [ENG] An apparent leak in a vacuum system caused by the escape of vapors that were condensed on cold surfaces.

virtual level [NUC PHYS] The energy of a virtual state.

virtual object [OPTICS] A collection of points which may be regarded as a source of light rays for a portion of an optical system but which does not actually have this function.

virtual particle *See* virtual quantum.

virtual process [QUANT MECH] A process which contributes in a stage of a theoretical model but is not, by itself, physically realizable.

virtual quantum [QUANT MECH] A photon or other particle in an intermediate state which appears in matrix elements connecting initial and final states in second- and higher-order perturbation theory; energy is not conserved in the transitions to or from the intermediate state. Also known as virtual particle.

virtual source [ACOUS] A source of sound which is composed of the same material as that in which the sound propagates and which does not have sharply delineated boundaries; such sources include thermal sources (such as lightning and thermoacoustic arrays), turbulence (as in rocket and jet engines), and sound itself (as in a parametric acoustic array).

virtual state [NUC PHYS] An unstable state of a compound nucleus which has a lifetime many times longer than the time it takes a nucleon, with the same energy as it has in the virtual state, to cross the nucleus.

virtual work [MECH] The work done on a system during any displacement which is consistent with the constraints on the system.

virtual work principle *See* principle of virtual work.

viscoelastic fluid [FL MECH] A fluid that displays viscoelasticity.

viscoelasticity [MECH] Property of a material which is viscous but which also exhibits certain elastic properties such as the ability to store energy of deformation, and in which the application of a stress gives rise to a strain that approaches its equilibrium value slowly.

viscoelastic theory [MECH] The theory which attempts to specify the relationship between stress and strain in a material displaying viscoelasticity.

VISCOMETRY

Picture of a Laray viscometer; used to measure viscosity and other rheological properties of ink.

viscometer gage [ENG] A vacuum gage in which the gas pressure is determined from the viscosity of the gas.

viscometric analysis [FL MECH] Measurement of the flow properties of substances by viscometry.

viscometry [ENG] A branch of rheology; the study of the behavior of fluids under conditions of internal shear; the technology of measuring viscosities of fluids.

viscosity [FL MECH] Energy dissipation and generation of stresses in a fluid by the distortion of fluid elements; quantitatively, when otherwise qualified, the absolute viscosity. Also known as flow resistance; internal friction.

viscosity coefficient [FL MECH] An empirical number used in equations of fluid mechanics to account for the effects of viscosity.

viscosity curve [FL MECH] A graph showing the viscosity of a liquid or gaseous material as a function of temperature.

viscosity gage *See* molecular gage.

viscosity manometer *See* molecular gage.

viscous dissipation function [FL MECH] A quadratic function of spatial derivatives of components of fluid velocity which gives the rate at which mechanical energy is converted into heat in a viscous fluid per unit volume. Also known as dissipation function.

viscous drag [FL MECH] That part of the rearward force on an aircraft that results from the aircraft carrying air forward with it through viscous adherence.

viscous-drag gas-density meter [ENG] A device to measure gas-mixture densities; driven impellers in sample and standard chambers create measurable turbulences (drags) against respective nonrotating impellers.

viscous flow [FL MECH] 1. The flow of a viscous fluid. 2. The flow of a fluid through a duct under conditions such that the mean free path is small in comparison with the smallest, transverse section of the duct.

viscous fluid [FL MECH] A fluid whose viscosity is sufficiently large to make the viscous forces a significant part of the total force field in the fluid.

viscous force [FL MECH] The force per unit volume or per unit mass arising from viscous effects in fluid flow.

visibility factor [ELECTR] The ratio of the minimum signal input detectable by ideal instruments connected to the output of a receiver, to the minimum signal power detectable by a human operator through a display connected to the same receiver. Also known as display loss.

visibility function *See* luminosity function.

visibility meter [OPTICS] A type of photometer that operates on the principle of artificially reducing the visibility of objects to threshold values (borderline of seeing and not seeing) and measuring the amount of the reduction on an appropriate scale.

visible absorption spectrophotometry [SPECT] Study of the spectra produced by the absorption of visible-light energy during the transformation of an electron from the ground state to an excited state as a function of the wavelength causing the transformation.

visible radiation *See* light.

visible spectrophotometry [SPECT] In spectrophotometric analysis, the use of a spectrophotometer with a tungsten lamp that has an electromagnetic spectrum of 380–780 nanometers as a light source, glass or quartz prisms or gratings in the monochromator, and a photomultiplier cell as a detector.

visible spectrum [SPECT] 1. The range of wavelengths of visible radiation. 2. A display or graph of the intensity of visible radiation emitted or absorbed by a material as a function of wavelength or some related parameter.

visual achromatism [OPTICS] In an optical system, the removal of chromatic aberration or chromatic differences of magnification between light at the wavelength of the Fraunhofer C line at 656.3 nanometers and the F line at 486.1 nanometers in order to minimize these defects at wavelengths at which the human eye is most sensitive. Also known as optical achromatism.

visual angle [OPTICS] The angle which an object subtends at the nodal point of the eye of an observer.

visual binaries [ASTRON] Binary stars that to the naked eye seem to be single stars, but when viewed through the telescope, are separated into pairs. Also known as visual doubles.

visual colorimetry [ANALY CHEM] A procedure for the determination of the color of an unknown solution by visual comparison to color standards (solutions or color-tinted disks).

visual magnitude [ASTRON] A celestial body's magnitude as seen by the eye of the observer.

visual photometer [OPTICS] A photometer in which the luminance of two surfaces is compared by human vision; it usually utilizes the Lummer-Brodhun sightbox or some adaptation of its principles.

Vitali covering [MATH] A family J of sets in euclidean n-space is said to be a Vitali covering of a set S, or to cover S in the sense of Vitali, if for every point x in S there is a positive number $a(x)$ and a sequence of sets $U_1, U_2, \ldots$ in J, each containing x, whose diameters approach 0 and each of which is contained in a cube C_n whose measure is less than $a(x)$ times the measure of U_n.

Vitali covering theorem [MATH] A Vitali covering J of a set S in euclidean n-space contains a finite or denumerable collection of disjoint sets whose union contains all of S except for a set of measure zero.

Vitali-Hahn-Saks theorem [MATH] **1.** Let m be a signed measure on a sigma field S, and f_n a sequence of vector- or scalar-valued additive set functions on S such that

$$\lim_{|m|(E)\to 0} f_n(E) = 0$$

for each n, where $|m|$ denotes the total variation of m; if $f_n(E)$ approaches a limit as $n \to \infty$ for each E in S, then

$$\lim_{|m|(E)\to 0} f_n(E) = 0$$

uniformly over n. **2.** If μ_n is a sequence of finite measures on a sigma field S and

$$\mu(E) = \lim_{n\to\infty} \mu_n(E)$$

exists for each $E \, \varepsilon \, S$, then μ is a measure on S.

vitreous luster [OPTICS] A type of luster resembling that of glass.

vitreous state [SOLID STATE] A solid state in which the atoms or molecules are not arranged in any regular order, as in a crystal, and which crystallizes only after an extremely long time. Also known as glassy state.

Vlasov equation [PL PHYS] A modification of the Boltzmann transport equation for the study of a plasma, in which particles interact only through the mutually induced space-charge field, and collisions are assumed to be negligible. Also known as collisionless Boltzmann equation.

VLBI *See* very-long-baseline interferometry.

VMOS technology *See* vertical metal oxide semiconductor technology.

void coefficient [NUCLEO] A rate of change in the reactivity of a water reactor system resulting from a formation of steam bubbles as the power level and temperature increase.

void swelling [NUCLEO] An increase in the external dimension of solid materials after irradiation.

Voigt body *See* Kelvin body.

Voigt effect [OPTICS] Double refraction of light passing through a substance that is placed in a magnetic field perpendicular to the direction of light propagation.

Voigt notation [MECH] A notation employed in the theory of elasticity in which elastic constants and elastic moduli are labeled by replacing the pairs of letters *xx, yy, zz, yz, zx,* and *xy* by the number 1, 2, 3, 4, 5, and 6 respectively.

vol *See* volume.

volatility [THERMO] The quality of having a low boiling point or subliming temperature at ordinary pressure or, equivalently, of having a high vapor pressure at ordinary temperatures.

volatilization [THERMO] The conversion of a chemical substance from a liquid or solid state to a gaseous or vapor state by the application of heat, by reducing pressure, or by a combination of these processes. Also known as vaporization.

volt [ELEC] The unit of potential difference or electromotive force in the meter-kilogram-second system, equal to the potential difference between two points for which 1 coulomb of electricity will do 1 joule of work in going from one point to the other. Symbolized V.

Volta effect *See* contact potential difference.

voltage [ELEC] Potential difference or electromotive force measured in volts.

voltage amplification [ELECTR] The ratio of the magnitude of the voltage across a specified load impedance to the magnitude of the input voltage of the amplifier or other transducer feeding that load; often expressed in decibels by multiplying the common logarithm of the ratio by 20.

voltage amplifier [ELECTR] An amplifier designed primarily to build up the voltage of a signal, without supplying appreciable power.

voltage coefficient [ELEC] For a resistor whose resistance varies with voltage, the ratio of the fractional change in resistance to the change in voltage.

voltage-current dual [ELEC] A pair of circuits in which the elements of one circuit are replaced by their dual elements in the other circuit according to the duality principle; for example, currents are replaced by voltages, capacitances by resistances.

voltage-dependent resistor *See* varistor.

voltage divider [ELEC] A tapped resistor, adjustable resistor, potentiometer, or a series arrangement of two or more fixed resistors connected across a voltage source; a desired fraction of the total voltage is obtained from the intermediate tap, movable contact, or resistor junction. Also known as potential divider.

voltage doubler [ELECTR] A transformerless rectifier circuit that gives approximately double the output voltage of a conventional half-wave vacuum-tube rectifier by charging a capacitor during the normally wasted half-cycle and discharging it in series with the output voltage during the next half-cycle. Also known as doubler.

voltage drop [ELEC] The voltage developed across a component or conductor by the flow of current through the resistance or impedance of that component or conductor.

voltage gain [ELECTR] The difference between the output signal voltage level in decibels and the input signal voltage level in decibels; this value is equal to 20 times the common logarithm of the ratio of the output voltage to the input voltage.

voltage generator [ELECTR] A two-terminal circuit element in which the terminal voltage is independent of the current through the element.

voltage gradient [ELEC] The voltage per unit length along a resistor or other conductive path.

voltage multiplier [ELEC] *See* instrument multiplier. [ELECTR] A rectifier circuit capable of supplying a direct-current output voltage that is two or more times the peak value of the alternating-current voltage.

voltage node [ELECTROMAG] Point having zero voltage in a stationary wave system, as in an antenna or transmission line; for example, a voltage node exists at the center of a half-wave antenna.

voltage phasor [ELEC] A line whose length represents the magnitude of a sinusoidally varying voltage and whose angle with the positive x-axis represents its phase.

voltage quadrupler [ELECTR] A rectifier circuit, containing four diodes, which supplies a direct-current output voltage which is four times the peak value of the alternating-current input voltage.

voltage-range multiplier *See* instrument multiplier.

voltage ratio [ELEC] The root-mean-square primary terminal voltage of a transformer divided by the root-mean-square secondary terminal voltage under a specified load.

voltage regulation [ELEC] The ratio of the difference between no-load and full-load output voltage of a device to the full-load output voltage, expressed as a percentage.

voltage regulator [ELECTR] A device that maintains the terminal voltage of a generator or other voltage source within required limits despite variations in input voltage or load. Also known as automatic voltage regulator; voltage stabilizer.

voltage-regulator diode [ELECTR] A diode that maintains an essentially constant direct voltage in a circuit despite changes in line voltage or load.

voltage-regulator tube [ELECTR] A glow-discharge tube in which the tube voltage drop is approximately constant over the operating range of current; used to maintain an essentially constant direct voltage in a circuit despite changes in line voltage or load. Also known as VR tube.

voltage saturation *See* anode saturation.

voltage stabilizer *See* voltage regulator.

voltage transformer [ELEC] An instrument transformer whose primary winding is connected in parallel with a circuit in which the voltage is to be measured or controlled. Also known as potential transformer.

voltage-variable capacitor *See* varactor.

voltaic cell [ELEC] A primary cell consisting of two dissimilar metal electrodes in a solution that acts chemically on one or both of them to produce a voltage.

voltammeter [ELEC] An instrument that may be used either as a voltmeter or ammeter.

volt-ampere [ELEC] The unit of apparent power in the International System; it is equal to the apparent power in a circuit when the product of the root-mean-square value of the voltage, expressed in volts, and the root-mean-square value of the current, expressed in amperes, equals 1. Abbreviated VA.

volt-ampere hour [ELEC] A unit for expressing the integral of apparent power over time, equal to the product of 1 volt-ampere and 1 hour, or to 3600 joules.

volt-ampere-hour reactive *See* var hour.

volt-ampere reactive [ELEC] The unit of reactive power in the International System; it is equal to the reactive power in a circuit carrying a sinusoidal current when the product of the

VOLTAGE REGULATOR

Circuit diagram of Zener-diode voltage regulator. Output load voltage V_L across load resistance R_L is maintained constant despite variation in input E_i. Current through series resistance R_S is sum of diode current i_1 and i_2.

VOLTAGE-REGULATOR TUBE

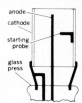

Construction of a typical voltage-regulator tube. Cathode is large cylinder; anode is thin wire.

root-mean-square value of the voltage, expressed in volts, by the root-mean-square value of the current, expressed in amperes, and by the sine of the phase angle between the voltage and the current, equals 1. Abbreviated var. Also known as reactive volt-ampere.

volt box [ELEC] A series of resistors arranged so that a desired fraction of a voltage can be measured, and the voltage thereby computed.

Volterra dislocation [SOLID STATE] A model of a dislocation which is formed in a ring of crystalline material by cutting the ring, moving the cut surfaces over each other, and then rejoining them.

Volterra equations [MATH] Given functions $f(x)$ and $K(x,y)$, these are two types of equations with unknown function y:

$$f(x) = \int_a^x K(x,t)y(t)\ dt$$

$$y(x) = f(x) + \lambda \int_a^x K(x,t)y(t)\ dt.$$

Volterra-Lotka equations *See* predator-prey equations.

Volterra's population equation [MATH] The integral equation

$$dN(t)/dt = aN(t) - b[N(t)]^2 - N(t)\int_c^t f(t-t')N(t')\ dt',$$

used in studying rate processes, where a and b are positive numbers, c is equal to 0 or $-\infty$, and f is a known function.

voltmeter [ENG] An instrument for the measurement of potential difference between two points, in volts or in related smaller or larger units.

voltmeter-ammeter [ENG] A voltmeter and an ammeter combined in a single case but having separate terminals.

voltmeter-ammeter method [ELEC] A method of measuring resistance in which simultaneous readings of the voltmeter and ammeter are taken, and the unknown resistance is calculated from Ohm's law.

voltmeter sensitivity [ELEC] Ratio of the total resistance of the voltmeter to its full scale reading in volts, expressed in ohms per volt.

volt-ohm-milliammeter [ENG] A test instrument having a number of different ranges for measuring voltage, current, and resistance. Also known as circuit analyzer; multimeter; multiple-purpose tester.

volume [ACOUS] The intensity of a sound. [MATH] A measure of the size of a body or definite region in three-dimensional space; it is equal to the least upper bound of the sum of the volumes of nonoverlapping cubes that can be fitted inside the body or region, where the volume of a cube is the cube of the length of one of its sides. Abbreviated vol.

volume acoustic wave *See* bulk acoustic wave.

volume dose *See* integral dose.

volume flow rate [FL MECH] The volume of the fluid that passes through a given surface in a unit time.

volume integral [MATH] An integral of a function of several variables with respect to volume measure taken over a three-dimensional subset of the domain of the function.

volume lifetime [SOLID STATE] Average time interval between the generation and recombination of minority carriers in a homogeneous semiconductor.

volume meter [ENG] Any flowmeter in which the actual flow is determined by the measurement of a phenomenon associated with the flow.

volumenometer [ENG] An instrument for determining the volume of a body by measuring the pressure in a closed air space when the specimen is present and when it is absent.

volume recombination rate [SOLID STATE] The rate at which free electrons and holes within the volume of a semiconductor recombine and thus neutralize each other.

volume resistivity [ELEC] Electrical resistance between opposite faces of a 1-centimeter cube of insulating material, commonly expressed in ohm-centimeters. Also known as specific insulation resistance.

volume susceptibility [PHYS CHEM] The magnetic susceptibility of a specified volume (for example, 1 cubic centimeter) of a magnetically susceptible material.

volumeter [ENG] Any instrument for measuring volumes of gases, liquids, or solids.

volumetric strain [MECH] One measure of deformation; the change of volume per unit of volume.

volume velocity [ACOUS] The rate of flow of a medium through a specified area due to a sound wave.

von Mises yield criterion [MECH] The assumption that plastic deformation of a material begins when the sum of the squares of the principal components of the deviatoric stress reaches a certain critical value.

von Mises transformation [FL MECH] A generalization of the Joukowski transformation which assigns to each complex number z the number $w = z + (a_1/z) + (a_2/z^2) + \cdots + (a_n/z^n)$.

von Neumann algebra [MATH] A subalgebra A of the algebra $B(H)$ of bounded linear operators on a complex Hilbert space, such that the adjoint operator of any operator in A is also in A, and A is closed in the strong operator topology in $B(H)$. Also known as ring of operators; W* algebra.

vortex [FL MECH] **1.** Any flow possessing vorticity; for example, an eddy, whirlpool, or other rotary motion. **2.** A flow with closed streamlines, such as a free vortex or line vortex. **3.** *See* vortex tube.

vortex distribution method [FL MECH] An analytic method used in ideal aerodynamics which ignores the thickness of the profile of the aerodynamic figure being studied.

vortex filament [FL MECH] The line of concentrated vorticity in a line vortex. Also known as vortex line.

vortex line [FL MECH] **1.** A line drawn through a fluid such that it is everywhere tangent to the vorticity. **2.** *See* vortex filament.

vortex ring [FL MECH] A line vortex in which the line of concentrated vorticity is a closed curve. Also known as collar vortex; ring vortex.

vortex shedding [FL MECH] In the flow of fluids past objects, the shedding of fluid vortices periodically downstream from the restricting object (for example, smokestacks, pipelines, or orifices).

vortex sheet [FL MECH] A surface across which there is a discontinuity in fluid velocity, such as in slippage of one layer of fluid over another; the surface may be regarded as being composed of vortex filaments.

vortex street [FL MECH] A series of vortices which are systematically shed from the downstream side of a body around which fluid is flowing rapidly. Also known as vortex trail; vortex train.

vortex trail *See* vortex street.

vortex train *See* vortex street.

vortex tube [FL MECH] A tubular surface consisting of the collection of vortex lines which pass through a small closed curve. Also known as vortex.

vortical field *See* rotational field.

vorticity [FL MECH] For a fluid flow, a vector equal to the curl of the velocity of flow.

vorticity equation [FL MECH] An equation of fluid mechanics describing horizontal circulation in the motion of particles around a vertical axis: $(d/dt)\,(S + f) = -\,(S + f)\,\mathrm{div}_h c$, where $(S + f)$ is the absolute vorticity and $\mathrm{div}_h c$ is the horizontal divergence of the fluid velocity.

vorticity-transport hypothesis [FL MECH] The hypothesis that, owing to the existence of pressure fluctuations, vorticity, and not momentum, is conservative in turbulent eddy flux.

V particle [PARTIC PHYS] The name first used for the unstable particles whose decay is responsible for the production of characteristic V-shaped tracks observed in cloud chambers exposed to cosmic radiation; they are neutral semistable particles such as neutral K mesons or Λ hyperons.

VR tube *See* voltage-regulator tube.

W *See* tungsten; watt.

Wadsworth mounting [OPTICS] **1.** A device in which light passes through a prism and is then reflected from a plane mirror; it has the effect of a constant-deviation prism. **2.** A mounting for a diffraction grating in which the slit is placed at the principal focus of a concave mirror, so that the light falling on the grating is in a parallel beam; it greatly reduces astigmatism.

wafer [ELECTR] A thin semiconductor slice on which matrices of microcircuits can be fabricated, or which can be cut into individual dice for fabricating single transistors and diodes.

Wagner earth connection *See* Wagner ground.

Wagner ground [ELEC] A ground connection used with an alternating-current bridge to minimize stray capacitance errors when measuring high impedances; a potentiometer is connected across the bridge supply oscillator, with its movable tap grounded. Also known as Wagner earth connection.

Waidner-Burgess standard [OPTICS] A unit of luminous intensity equal to the luminous intensity of 1 square centimeter of a blackbody at the melting point of platinum, or to 60 candelas.

wake [FL MECH] The region behind a body moving relative to a fluid in which the effects of the body on the fluid's motion are concentrated.

wake flow [FL MECH] Turbulent eddying flow that occurs downstream from bluff bodies.

Wald-Wolfowitz run test [STAT] A procedure used in non-parametric statistics to determine whether the means of two independently drawn samples were taken from the same population.

W* algebra *See* von Neumann algebra.

wall effect [ELECTR] The contribution to the ionization in an ionization chamber by electrons liberated from the walls.

wall energy [SOLID STATE] The energy per unit area of the boundary between two ferromagnetic domains which are oriented in different directions.

wall friction [FL MECH] The drag created in the flow of a liquid or gas because of contact with the wall surfaces of its conductor, such as the inside surfaces of a pipe.

Wallis formulas [MATH] Formulas that determine the values of the definite integrals from 0 to $\pi/2$ of the functions $\sin^n (x)$, $\cos^n (x)$, and $\cos^m (x) \sin^n (x)$ for positive integers m and n. Also known as Wallis theorem.

Wallis product [MATH] An infinite product representation of $\pi/2$, namely,

$$\frac{\pi}{2} = \frac{2}{1}\frac{2}{3}\frac{4}{3}\frac{4}{5} \cdots \frac{2n}{2n-1}\frac{2n}{2n+1} \cdots$$

WAKE

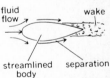

Wake formed downstream from a streamlined body.

Walsh functions [MATH] A complete, orthonormal system of functions $\{w_n(x)\}$ on the interval [0,1], where $w_n(x)$ is the product over $k = 1, 2, 3, \ldots$ of $[r_k(x)]^{a_k(n)}$, where $a_k(n)$ is the kth place to the left of the decimal point in the binary notation for n and the $r_k(x)$ are Rademacher functions.

Wanner optical pyrometer [ENG] A type of polarizing pyrometer in which beams from the source under investigation and a comparison lamp are polarized at right angles and then passed through a Nicol prism and a red filter; the source temperature is determined from the angle through which the Nicol prism must be rotated in order to equalize the intensities of the resulting patches of light.

Wannier function [SOLID STATE] The Fourier transform of a Bloch function defined for an entire band, regarded as a function of the wave vector.

warble tone [ACOUS] A tone whose frequency varies periodically several times per second over a small range; used to prevent standing-wave patterns from forming in reverberation chambers.

Ward-Leonard speed-control system [CONT SYS] A system for controlling the speed of a direct-current motor in which the armature voltage of a separately excited direct-current motor is controlled by a motor-generator set.

Waring problem [MATH] The problem of whether, for each fixed integral exponent k, there exists an integer s such that the equation $x_1^k + x_2^k + \cdots + x_s^k = n$ has integer solutions for every positive integer n.

warping function *See* torsion function.

washer thermistor [ELECTR] A thermistor in the shape of a washer, which may be as large as 0.75 inch (1.9 centimeters) in diameter and 0.50 inch (1.3 centimeters) thick; it is formed by pressing and sintering an oxide-binder mixture.

water-activated battery [ELEC] A primary battery that contains the electrolyte but requires the addition of or immersion in water before it is usable.

water boiler *See* water-boiler reactor.

water-boiler reactor [NUCLEO] A homogeneous reactor that uses enriched uranium as fuel and ordinary water as moderator; the fuel is uranyl sulfate dissolved in water. Also known as water boiler.

water calorimeter [ENG] A calorimeter that measures radiofrequency power in terms of the rise in temperature of water in which the rf energy is absorbed.

water-cooled reactor [NUCLEO] A nuclear reactor in which water is used as a primary coolant.

water-cooled tube [ELECTR] An electron tube that is cooled by circulating water through or around the anode structure.

water dropper [ELEC] A simple electrostatic generator in which each of two series of water drops falls through cylindrical metal cans into lower cans with funnels, and the cans are electrically connected in such a way that charge accumulates on them, energy being supplied by the gravitational force on the water drops.

water hammer [FL MECH] Pressure rise in a pipeline caused by a sudden change in the rate of flow or stoppage of flow in the line.

water-moderated reactor [NUCLEO] A nuclear reactor in which water is the principal moderator.

water rheostat *See* electrolytic rheostat.

water vapor [PHYS] Water in the form of a vapor, especially when below the boiling point and diffused.

water-vapor laser [OPTICS] A laser whose active substance is water vapor, and which emits infrared radiation at wavelengths of 27.97, 47.7, 78.46, and 118.6 micrometers.

Watson-Sommerfeld transformation [MATH] A procedure

WATER-ACTIVATED BATTERY

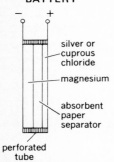

silver or cuprous chloride

magnesium

absorbent paper separator

perforated tube

Schematic diagram of cell of water-activated battery.

for transforming a series whose *l*th term is the product of the *l*th Legendre polynomial and a coefficient, a_l, having certain properties, into the sum of a contour integral of $a(l)$ and terms involving poles of $a(l)$, where $a(l)$ is a meromorphic function such that $a(l)$ equals a_l at integral values of *l*; used in studying rainbows, propagation of radio waves around the earth, scattering from various potentials, and scattering of elementary particles. Also known as Sommerfeld-Watson transformation.

watt [PHYS] The unit of power in the meter-kilogram-second system of units, equal to 1 joule per second. Symbolized W.

wattage rating [ELEC] A rating expressing the maximum power that a device can safely handle continuously.

watt-hour [ELEC] A unit of energy used in electrical measurements, equal to the energy converted or consumed at a rate of 1 watt during a period of 1 hour, or to 3600 joules. Abbreviated whr.

wattless current *See* reactive current.

wattless power *See* reactive power.

wattmeter [ENG] An instrument that measures electric power in watts ordinarily.

Watt's curve [MATH] The curve traced out by the midpoint of a line segment whose end points move along two circles of equal radius.

watt-second [PHYS] Amount of energy corresponding to 1 watt acting for 1 second; 1 watt-second is equal to 1 joule.

Watt's law [THERMO] A law which states that the sum of the latent heat of steam at any temperature of generation and the heat required to raise water from 0°C to that temperature is constant; it has been shown to be substantially in error.

wave [FL MECH] A disturbance which moves through or over the surface of a liquid, as of a sea. [PHYS] A disturbance which propagates from one point in a medium to other points without giving the medium as a whole any permanent displacement.

wave acoustics [ACOUS] The study of the propagation of sound based on its wave properties.

wave amplitude [PHYS] The magnitude of the greatest departure from equilibrium of the wave disturbance.

wave antenna [ELECTROMAG] Directional antenna composed of a system of parallel, horizontal conductors, varying from a half to several wavelengths long, terminated to ground at the far end in its characteristic impedance.

wave celerity *See* phase velocity.

wave converter [ELECTROMAG] Device for changing a wave of a given pattern into a wave of another pattern, for example, baffle-plate converters, grating converters, and sheath-reshaping converters for waveguides.

wave-corpuscle duality *See* wave-particle duality.

wave crest [PHYS] The position at which the disturbance of a progressive wave attains its maximum positive value.

wave duct [ELECTROMAG] **1.** Waveguide, with tubular boundaries, capable of concentrating the propagation of waves within its boundaries. **2.** Natural duct, formed in air by atmospheric conditions, through which waves of certain frequencies travel with more than average efficiency.

wave equation [PHYS] **1.** In classical physics, a special equation governing waves that suffer no dissipative attenuation; it states that the second partial derivative with respect to time of the function characterizing the wave is equal to the square of the wave velocity times the Laplacian of this function. Also known as classical wave equation; d'Alembert's wave equation. **2.** Any of several equations which relate the spatial and time dependence of a function characterizing some

WATTMETER

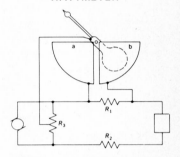

Electrostatic wattmeter circuit diagram showing two quadrants *a* and *b*.

WAVEFORM

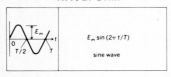

$E_m \sin(2\pi t/T)$

sine wave

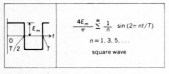

$$\frac{4E_m}{\pi} \sum_{n=1,3,5,\ldots}^{\infty} \frac{1}{n} \sin(2\pi nt/T)$$

square wave

Waveforms of sine wave and square wave. Fourier series for waveforms are given at right as functions of time t. E_m is maximum value of wave. T is period of wave.

physical entity which can propagate as a wave, including quantum-wave equations for particles.

wave filter [ELEC] A transducer for separating waves on the basis of their frequency; it introduces relatively small insertion loss to waves in one or more frequency bands and relatively large insertion loss to waves of other frequencies.

waveform [PHYS] The pictorial representation of the form or shape of a wave, obtained by plotting the displacement of the wave as a function of time, at a fixed point in space.

waveform analysis [PHYS] The determination of the amplitude and phase of the components of a complex waveform, either mathematically or by means of electronic instruments.

wavefront [PHYS] **1.** A surface of constant phase. **2.** The portion of a wave envelope that is between the beginning zero point and the point at which the wave reaches its crest value, as measured either in time or distance.

wavefront splitting [OPTICS] Any method of producing interference in which light from a single source is split into two parts which can then be recombined; examples include Young's two-slit experiment, the Fresnel double mirror, and the Fresnel biprism.

wave function *See* Schrödinger wave function.

wave group [PHYS] A series of waves in which the wave direction, length, and height vary only slightly.

waveguide [ELECTROMAG] **1.** Broadly, a device which constrains or guides the propagation of electromagnetic waves along a path defined by the physical construction of the waveguide; includes ducts, a pair of parallel wires, and a coaxial cable. Also known as microwave waveguide. **2.** More specifically, a metallic tube which can confine and guide the propagation of electromagnetic waves in the lengthwise direction of the tube.

waveguide bend [ELECTROMAG] A section of waveguide in which the direction of the longitudinal axis is changed; an **E**-plane bend in a rectangular waveguide is bent along the narrow dimension, while an **H**-plane bend is bent along the wide dimension. Also known as waveguide elbow.

waveguide cavity [ELECTROMAG] A cavity resonator formed by enclosing a section of waveguide between a pair of waveguide windows which form shunt susceptances.

waveguide connector [ELECTROMAG] A mechanical device for electrically joining and locking together separable mating parts of a waveguide system. Also known as waveguide coupler.

waveguide coupler *See* waveguide connector.

waveguide critical dimension [ELECTROMAG] Dimension of waveguide cross section which determines the cutoff frequency.

waveguide cutoff frequency [ELECTROMAG] Frequency limit of propagation along a waveguide for waves of a given field configuration.

waveguide elbow *See* waveguide bend.

waveguide filter [ELECTROMAG] A filter made up of waveguide components, used to change the amplitude-frequency response characteristic of a waveguide system.

waveguide hybrid [ELECTROMAG] A waveguide circuit that has four arms so arranged that a signal entering through one arm will divide and emerge from the two adjacent arms, but will be unable to reach the opposite arm.

waveguide junction *See* junction.

waveguide plunger *See* piston.

waveguide probe *See* probe.

waveguide resonator *See* cavity resonator.

waveguide shim [ELECTROMAG] Thin resilient metal sheet

inserted between waveguide components to ensure electrical contact.

waveguide slot [ELECTROMAG] A slot in a waveguide wall, either for coupling with a coaxial cable or another waveguide, or to permit the insertion of a traveling probe for examination of standing waves.

waveguide switch [ELECTROMAG] A switch designed for mechanically positioning a waveguide section so as to couple it to one of several other sections in a waveguide system.

waveguide window *See* iris.

wave height [PHYS] Twice the wave amplitude.

wave impedance [ELECTROMAG] The ratio, at every point in a specified plane of a waveguide, of the transverse component of the electric field to the transverse component of the magnetic field.

wave intensity [PHYS] The average amount of energy transported by a wave in the direction of wave propagation, per unit area per unit time.

wave interference *See* interference.

wavelength [PHYS] The distance between two points having the same phase in two consecutive cycles of a periodic wave, along a line in the direction of propagation.

wavelength constant *See* phase constant.

wavelength shifter [ELECTR] A photofluorescent compound used with a scintillator material to increase the wavelengths of the optical photons emitted by the scintillator, thereby permitting more efficient use of the photons by the phototube or photocell.

wavelength standards [SPECT] Accurately measured lengths of waves emitted by specified light sources for the purpose of obtaining the wavelengths in other spectra by interpolating between the standards.

wave-making resistance *See* wave resistance.

wave mechanics *See* Schrödinger's wave mechanics.

wavemeter [ENG] A device for measuring the geometrical spacing between successive surfaces of equal phase in an electromagnetic wave.

wave motion [PHYS] The process by which a disturbance at one point is propagated to another point more remote from the source with no net transport of the material of the medium itself; examples include the motion of electromagnetic waves, sound waves, hydrodynamic waves in liquids, and vibration waves in solids. Also known as propagation; wave propagation.

wave normal [PHYS] **1.** A unit vector which is perpendicular to an equiphase surface of a wave, and has its positive direction on the same side of the surface as the direction of propagation. **2.** One of a family of curves which are everywhere perpendicular to the equiphase surfaces of a wave.

wave number [PHYS] The reciprocal of the wavelength of a wave, or sometimes 2π divided by the wavelength. Also known as reciprocal wavelength.

wave optics [OPTICS] The branch of optics which treats of light (or electromagnetic radiation in general) with explicit recognition of its wave nature.

wave packet [PHYS] In wave phenomena, a superposition of waves of differing lengths, so phased that the resultant amplitude is negligibly small except in a limited portion of space whose dimensions are the dimensions of the packet.

wave-particle duality [QUANT MECH] The principle that both matter and electromagnetic radiation exhibit phenomena in which they behave as waves and other phenomena in which they behave as particles, the two aspects being associated by the de Broglie relations. Also known as duality principle; wave-corpuscle duality.

WAVE MOTION

Relation between frequency f, wavelength λ, and velocity c in wave motion.

wave period [PHYS] The time between the attainment of successive maxima, at a fixed point, of a quantity characterizing a wave.

wave plate [OPTICS] A plate of material which is linearly birefringent. Also known as retardation sheet.

wave propagation *See* wave motion.

wave refraction [PHYS] The process by which the direction of a wave train moving in shallow water at an angle to the contours is changed.

wave resistance [FL MECH] The portion of fluid resistance to a body moving on the surface of a liquid that results from energy dissipation in the formation of waves on the liquid surface. Also known as wave-making resistance.

wave-shaping circuit [ELECTR] An electronic circuit used to create or modify a specified time-varying electrical quantity, usually voltage or current, using combinations of electronic devices, such as vacuum tubes or transistors, and circuit elements, including resistors, capacitors, and inductors.

wave speed *See* phase velocity.

wave tail [ELECTR] Part of a signal-wave envelope (in time or distance) between the steady-state value (or crest) and the end of the envelope.

wave theory of light [OPTICS] A theory which assumes that light is a wave motion, rather than a stream of particles.

wave train [PHYS] A series of waves produced by the same disturbance.

wave trough [PHYS] The lowest part of a wave form between successive wave crests.

wave vector [PHYS] A vector whose direction is the direction of phase propagation of a wave at each point in space, and whose magnitude is sometimes set at $2\pi/\lambda$ and sometimes at $1/\lambda$, where λ is the wavelength.

wave-vector space [SOLID STATE] The space of the wave vectors of the state functions of some system; this would be used, for example, for electron wave functions in a crystal and thermal vibrations of a lattice. Also known as **k**-space; reciprocal space.

wave velocity *See* phase velocity.

wavy extinction *See* undulatory extinction.

wax-block photometer *See* Joly photometer.

waxy-electrolyte battery [ELEC] A primary battery in which the electrolyte is a waxy material, such as polyethylene glycol, in which is dissolved a small amount of a salt, such as zinc chloride; the electrodes are frequently made of zinc and manganese dioxide, and the electrolyte is melted and painted on a paper sheet to form the separator.

Wb *See* weber.

W boson [PARTIC PHYS] Collective name for two of the hypothetical intermediate vector bosons which carry electric charge; it is believed that charged current interactions are produced by the exhange of these bosons.

W coefficient *See* Racah coefficient.

weak convergence [MATH] A sequence of elements $x_1, x_2, \ldots$ from a topological vector space X converges weakly if the sequence $f(x_1), f(x_2), \ldots$ converges for every continuous linear functional f on X.

weak coupling [PARTIC PHYS] The coupling of four fermion fields in the weak interaction, having a strength many orders of magntiude weaker than that of the strong or electromagnetic interactions.

weak derivative [MATH] For a function $f(x)$ which is locally integrable on the real numbers, this is a locally integrable function $g(x)$ such that, for any infinitely differentiable function $t(x)$ vanishing outside some interval $[a,b]$, the inte-

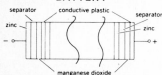

WAXY-ELECTROLYTE BATTERY

separator   conductive plastic   separator
zinc   zinc
− ○   ○ +
manganese dioxide

Diagram of battery stack of cells using a waxy electrolyte.

gral from a to b of $f(x)t'(x) + g(x)t(x)$ equals 0. Also known as distributional derivative; Sobolev generalized derivative.

weak energy condition [RELAT] The condition in general relativity theory that all observers see a nonnegative energy density.

weak interaction [PARTIC PHYS] One of the fundamental interactions among elementary particles, responsible for beta decay of nuclei, and for the decay of elementary particles with lifetimes greater than about 10^{-10} second, such as muons, K mesons, and lambda hyperons; it is several orders of magnitude weaker than the strong and electromagnetic interactions, and fails to conserve strangeness or parity. Also known as beta interaction.

weak partial derivative [MATH] Let $f(x)$ be a locally integrable function on an open set S in n-dimensional euclidean space, and let D^α denote the partial derivative

$$\partial^{\alpha_1 + \cdots + \alpha_n} / \partial x_1^{\alpha_1} \partial x_2^{\alpha_2} \cdots \partial x_n^{\alpha_n};$$

the weak partial derivative D^α of $f(x)$ is a locally integrable function $g(x)$ such that, for any infinitely differentiable function $t(x)$ vanishing outside a compact subset of S, the integral over S of $f(x)D^\alpha t(x) - (-1)^{|\alpha|}g(x)t(x)$ equals 0, where $|\alpha| = \alpha_1 + \cdots + \alpha_n$. Also known as distributional partial derivative; Sobolev generalized partial derivative.

weak topology [MATH] A topology on a topological vector space X whose open neighborhoods around a point x are obtained from those points y of X for which every $f_i(x)$ is close to $f_i(y)$, f_i appearing in a finite list of linear functionals.

weak* topology [MATH] A topology on the dual space T^* of a topological vector space T whose open neighborhoods around an element f are unions of sets consisting of elements g of T^* for which every $g(x_i)$ is close to $f(x_i)$ for x_i appearing in a finite list of elements of T. Also known as w* topology.

Webb-Kapteyn theory [MATH] A method of studying Neumann series (second definition) based on the theory of functions of a real variable.

weber [ELECTROMAG] The unit of magnetic flux in the meter-kilogram-second system, equal to the magnetic flux which, linking a circuit of one turn, produces in it an electromotive force of 1 volt as it is reduced to zero at a uniform rate in 1 second. Symbolized Wb.

Weber differential equation [MATH] A special case of the confluent hypergeometric equation that has as solution a confluent hypergeometric series. Also known as Weber-Hermite equation.

Weber function [MATH] A function $E_\nu(z)$ related to the Anger function and Struve functions; it is equal to the integral from 0 to π of $\pi^{-1} \sin(\nu\theta - z \sin\theta) \, d\theta$.

Weber-Hermite equation *See* Weber differential equation.

Weber number 1 [FL MECH] A dimensionless number used in the study of surface tension waves and bubble formation, equal to the product of the square of the velocity of the wave or the fluid velocity, the density of the fluid, and a characteristic length, divided by the surface tension. Symbolized N_{We_1}, We.

Weber number 2 [FL MECH] A dimensionless number, equal to the square root of Weber number 1. Symbolized N_{We_2}.

Weber number 3 [CHEM ENG] A dimensionless number used in interfacial area determination in distillation equipment, equal to the surface tension divided by the product of the liquid density, the acceleration of gravity, and the depth of liquid on the tray under consideration. Symbolized N_{We_3}.

Weddle's rule [MATH] The integral of a real-valued function

f on an interval $[a,b]$ is approximated by $(3/10)h\,[f(a) + 5f(a+h) + f(a+2h) + 6f(a+3h) + f(a+4h) + 5f(a+5h) + f(b)]$, where $h = (b-a)/6$.

WEDGE

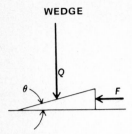

The shape of a wedge and a diagram of forces acting on it. F is smaller applied force, Q is larger force to be exerted, θ is angle between surfaces of wedge.

wedge [ELECTROMAG] A waveguide termination consisting of a tapered length of dissipative material introduced into the guide, such as carbon. [MATH] A polyhedron whose base is a rectangle and whose lateral faces consist of two equilateral triangles and two trapezoids. [OPTICS] An optical filter in which the transmission decreases continuously or in steps from one end to the other.

wedge filter [NUCLEO] A radiation filter so constructed that its thickness or transmission characteristics vary continuously or in steps from one edge to the other; used to increase the uniformity of radiation in certain types of treatment.

wedge photometer [ENG] A photometer in which the luminous flux density of light from two sources is made equal by pushing into the beam from the brighter source a wedge of absorbing material; the wedge has a scale indicating how much it reduces the flux density, so that the luminous intensities of the sources may be compared.

wedge product [MATH] A product defined on forms such that a wedge product of a p-form and a q-form results in a $p + q$ form.

wedge spectrograph [SPECT] A spectrograph in which the intensity of the radiation passing through the entrance slit is varied by moving an optical wedge.

Wehnelt cathode *See* oxide-coated cathode.

Weibull distribution [STAT] A distribution that describes lifetime characteristics of parts and components.

Weierstrass' approximation theorem [MATH] A continuous real-valued function on a closed interval can be uniformly approximated by polynomials.

Weierstrass functions [MATH] Used in the calculus of variations, these determine functions satisfying the Euler-Lagrange equation and Jacobi's condition while maximizing a given definite integral.

Weierstrassian elliptic function [MATH] A function that plays a central role in the theory of elliptic functions; for z, g_2 and g_3 real or complex numbers, let y be that number such that

$$z = \int_y^\infty \frac{dt}{\sqrt{4t^3 - g_2 t - g_3}} \; ;$$

the Weierstrassian elliptic function of z with parameters g_2 and g_3 is $p(z;g_2,g_3) = y$.

Weierstrass M test [MATH] An infinite series of numbers will converge or functions will converge uniformly if each term is dominated in absolute value by a nonnegative constant M_n, where these M_n form a convergent series. Also known as Weierstrass' test for convergence.

Weierstrass normal elliptic integrals [MATH] Three standard elliptic integrals having the property that any elliptic integral is a linear combination of these integrals and elementary functions; they are integrals of dx/y, $x\,dx/y$, and $dx/(x-c)y$, where y is the radicial in Weierstrass normal form and c is a real or complex number.

Weierstrass normal form [MATH] **1.** The standard form, $y = \sqrt{4x^3 - g_2 x - g_3}$, to which the radical in an elliptic integral can be reduced by a transformation on the integration variable x, where g_2 and g_3 are real or complex numbers. **2.** The form of the Weierstrass normal elliptic integrals.

Weierstrass' test for convergence *See* Weierstrass M test.

Weierstrass transform [MATH] This transform of a real function $f(y)$ is the function given by the integral from $-\infty$ to ∞

of $(4\pi t)^{-1/2} \exp[-(x-y)^2/4] f(y) \, dy$; this is used in studying the heat equation.

weight [MECH] **1.** The gravitational force with which the earth attracts a body. **2.** By extension, the gravitational force with which a star, planet, or satellite attracts a nearby body.

weight density [PHYS] The weight of a body or portion of a body divided by its volume.

weighted aggregate index [STAT] A statistic for a collection of items weighted so as to reflect the relative importance of the items with regard to the overall phenomenon which the index is designed to describe; a price index is an example.

weighted average [STAT] The number obtained by adding the product of α_i times the ith number in a set of N numbers for $i = 1, 2, \ldots, N$, where α_i are numbers (weights) such that $\alpha_1 + \alpha_2 + \cdots + \alpha_N = 1$.

weighted moving average [STAT] A method used for smoothing data in a time series in which each observation being averaged is given a weight which reflects its relative importance in calculating the average.

weight factor [STAT MECH] The number of microstates that correspond to a given macrostate.

weight function [MATH] Two real valued functions f and g are orthogonal relative to a weight function σ on an interval if the integral over the interval of $f \cdot g \cdot \sigma$ vanishes.

weighting [ENG] The artificial adjustment of measurements to account for factors that, in the normal use of the device, would otherwise be different from conditions during the measurements.

weightlessness [MECH] A condition in which no acceleration, whether of gravity or other force, can be detected by an observer within the system in question. Also known as zero gravity.

weight thermometer [ENG] A glass vessel for determining the thermal expansion coefficient of a liquid by measuring the mass of liquid needed to fill the vessel at two different temperatures.

Weinberg-Salam theory [PARTIC PHYS] A gage theory in which the electromagnetic and weak nuclear interactions are described by a single unifying framework in which both have a characteristic coupling parameter equal to the fine-structure constant; it predicts the existence of intermediate vector bosons and neutral current interactions. Also known as Salam-Weinberg theory.

Weingarten formulas [MATH] Equations concerning the normals to a surface at a point.

Weissenberg method [SOLID STATE] A method of studying crystal structure by x-ray diffraction in which the crystal is rotated in a beam of x-rays, and a photographic film is moved parallel to the axis of rotation; the crystal is surrounded by a sleeve which has a slot that passes only diffraction spots from a single layer of the reciprocal lattice, permitting positive identification of each spot in the pattern.

Weiss magneton [ATOM PHYS] A unit of magnetic moment, equal to 1.853×10^{-21} erg/oersted, about one-fifth of the Bohr magneton; it is experimentally derived, the magnetic moments of certain molecules being close to integral multiples of this quantity.

Weiss molecular field [SOLID STATE] The effective magnetic field postulated in the Weiss theory of ferromagnetism, which acts on atomic magnetic moments within a domain, tending to align them, and is in turn generated by these magnetic moments.

Weiss theory [SOLID STATE] A theory of ferromagnetism based on the hypotheses that below the Curie point a ferromagnetic substance is composed of small, spontaneously

WEIGHTLESSNESS

Film magazine (arrow) floats in weightless environment of space cabin. *(NASA)*

magnetized regions called domains, and that each domain is spontaneously magnetized because a strong molecular magnetic field tends to align the individual atomic magnetic moments within the domain. Also known as molecular field theory.

Weizäcker-Williams method [QUANT MECH] A method of calculating the bremsstrahlung emitted when two particles, whose relative kinetic energies are much larger than their rest energies, collide; in the rest frame of one of the particles, the field of the other is equivalent to a set of virtual photons, and Compton scattering of these photons by the particle at rest is computed.

Weizsaecker's theory [ASTRON] A theory of the origin of the solar system; it hypothesizes primeval turbulent eddies which become permanent and self-gravitating; Weizsaecker does not discuss the origin of the gas clouds.

well-ordered set [MATH] A linearly ordered set where every subset has a least element.

well-ordering principle [MATH] The proposition that every set can be endowed with an order so that it becomes a well-ordered set; this is equivalent to the axiom of choice.

well-type manometer [ENG] A type of double-leg, glass-tube manometer; one leg has a relatively small diameter, and the second leg is a reservoir; the level of the liquid in the reservoir does not change appreciably with change of pressure; a mercury barometer is a common example.

Wentzel-Kramers-Brillouin method [QUANT MECH] Method of approximating quantum-mechanical wave functions and energy levels, in which the logarithm of the wave function is expanded in powers of Planck's constant, and all except the first two terms are neglected. Also known as phase integral method; WKB method.

Werner band [SPECT] A band in the ultraviolet spectrum of molecular hydrogen extending from 116 to 125 nanometers.

Werner complex *See* coordination compound.

Wertheim effect *See* Wiedemann effect.

Westergaard formulas [MECH] Special cases of the Kolosov-Muskhelishvili formulas for plane stress which satisfy the boundary condition that the stress component σ_{xy} equals 0 along the line $y = 0$, or the condition that $\sigma_{yy} = 0$ on $y = 0$.

Weston standard cell [ELEC] A standard cell used as a highly accurate voltage source for calibrating purposes; the positive electrode is mercury, the negative electrode is cadmium, and the electrolyte is a saturated cadmium sulfate solution; the Weston standard cell has a voltage of 1.018636 volts at 20°C.

Westphal balance [ENG] A direct-reading instrument for determining the densities of solids and liquids; a plummet of known mass and volume is immersed in the liquid whose density is to be measured or, alternatively, a sample of the solid whose density is to be measured is immersed in a liquid of known density, and the loss in weight is measured, using a balance with movable weights.

wet [PHYS] A liquid is said to wet a solid if the contact angle between the solid and the liquid, measured through the liquid, lies between 0 and 90°, and not to wet the solid if the contact angle lies between 90 and 180°.

wet and dry bulb thermometer *See* psychrometer.

wet-bulb thermometer [ENG] A thermometer having the bulb covered with a cloth, usually muslin or cambric, saturated with water.

wet cell [ELEC] A primary cell in which there is a substantial amount of free electrolyte in liquid form.

wet criticality [NUCLEO] Reactor criticality achieved with the coolant present.

WESTON STANDARD CELL

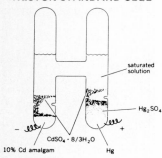

saturated solution

Hg_2SO_4

$CdSO_4 \cdot 8/3H_2O$

10% Cd amalgam — Hg

Cross-sectional diagram of Weston standard cell.

WET

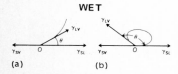

(a)   (b)

Contact angle between the solid and the liquid is indicated by θ. (a) Liquid wets solid, (b) liquid does not wet solid. Arrows represent interfacial tensions γ_{SL}, γ_{SV}, and γ_{LV}, where S, L, and V refer to solid, liquid and vapor.

wet electrolytic capacitor [ELEC] An electrolytic capacitor employing a liquid electrolyte.

wettability [CHEM] The ability of any solid surface to be wetted when in contact with a liquid; that is, the surface tension of the liquid is reduced so that the liquid spreads over the surface.

wetting angle [FL MECH] A contact angle which lies between 0 and 90°.

Weyl fractional integral [MATH] A generalization of repeated integration of a function f from a number x to infinity, given by the integral from x to infinity of $[\Gamma(\alpha)]^{-1}(t-x)^{\alpha-1}f(t)\,dt$, where the real part of α is positive. Also known as Euler transform of the second kind.

Weyl tensor [RELAT] A tensor with the symmetries of the curvature tensor such that all contractions on its indices vanish; the curvature tensor is decomposable in terms of the metric, the scalar curvature, and the Weyl tensor.

Wheatstone bridge [ELEC] A four-arm bridge circuit, all arms of which are predominantly resistive; used to measure the electrical resistance of an unknown resistor by comparing it with a known standard resistance. Also known as resistance bridge; Wheatstone network.

Wheatstone network *See* Wheatstone bridge.

Wheatstone stereoscope [OPTICS] A type of stereoscope that uses plane mirrors to enable the eyes to form a fused image of two pictures whose separation is greater than the interocular distance.

Wheeler-Feynman theory [RELAT] A relativistic action-at-a-distance theory in which it is assumed that there are enough absorbers in the universe to serve as sinks for all actions that emanate from any charged particle; radiation damping is a consequence of the theory.

whispering gallery [ACOUS] A domed gallery in which weak sounds can be heard at great distances.

whistler [GEOPHYS] An effect that occurs when a plasma disturbance, caused by a lightning discharge, travels out along lines of magnetic force of the earth's field and is reflected back to its origin from a magnetically conjugate point on the earth's surface; the disturbance may be picked up electromagnetically and converted directly to sound; the characteristic drawn-out descending pitch of the whistler is a dispersion effect due to the greater velocity of the higher-frequency components of the disturbance.

whistler wave *See* electron cyclotron wave.

white body [PHYS] A hypothetical substance whose surface absorbs no electromagnetic radiation of any wavelength, that is, one which exhibits zero absorptivity for all wavelengths.

white dwarf star [ASTRON] An intrinsically faint star of very small radius and high density; the mass is about 0.6 that of the sun and the average radius is about 8000 kilometers; it is one final stage of stellar evolution with thermonuclear energy sources extinct.

white light [OPTICS] Any radiation producing the same color sensation as average noon sunlight.

whitening filter [ELECTR] An electrical filter which converts a given signal to white noise. Also known as prewhitening filter.

white noise [PHYS] Random noise that has a constant energy per unit bandwidth at every frequency in the range of interest.

white object [OPTICS] An object that reflects all wavelengths of light with substantially equal high efficiencies and with considerable diffusion.

white radiation *See* continuous radiation.

white rainbow *See* fogbow.

Whitney sum [MATH] A tangent bundle TX over a differentia-

WHEATSTONE BRIDGE

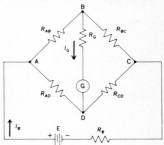

Circuit diagram of Wheatstone bridge, used to measure resistance R_{CD} in terms of known resistances R_{AB}, R_{BC}, and R_{AD}; the latter are adjusted until current I_G through detector G equals zero. R_G = internal resistance of detector. Current I_B flows through battery with open-circuit voltage E and internal resistance R_B.

ble manifold X is a Whitney sum of continuous bundles A and B over X if for each x the fibers of A and B at x are complementary subspaces of the tangent space at x.

Whittaker differential equation [MATH] A special form of Gauss' hypergeometric equation with solutions as special cases of the confluent hypergeometric series.

Whittaker functions [MATH] The functions

$$W_{k,m}(z) = e^{-z/2} z^{c/2} \Psi (a,c;z)$$

$$M_{k,m}(z) = e^{-z/2} z^{c/2} {}_1F_1 (a,c;z)$$

where Ψ is Tricomi's function, ${}_1F_1$ is Kummer's function, $a = \frac{1}{2} - k + m$, and $c = 2m + 1$.

whole-body counter [NUCLEO] A radiation counter that directly measures radioactivity in the entire human body.

whole step *See* whole tone.

whole tone [ACOUS] The interval between two sounds whose basic frequency ratio is approximately equal to the sixth root of 2. Also known as whole step.

whr *See* watt-hour.

wide-angle lens [OPTICS] An optical lens having a large angular field, generally greater than 80°.

wide band [ELECTR] Property of a tuner, amplifier, or other device that can pass a broad range of frequencies.

Wiedemann effect [ELECTROMAG] The twist produced in a current-carrying wire when placed in a longitudinal magnetic field. Also known as Wertheim effect.

Wiedemann-Franz law [SOLID STATE] The law that the ratio of the thermal conductivity of a metal to its electrical conductivity is a constant, independent of the metal, times the absolute temperature. Also known as Lorentz relation.

Wiedemann's additivity law [PHYS CHEM] The law that the mass (or specific) magnetic susceptibility of a mixture or solution of components is the sum of the proportionate (by weight fraction) susceptibilities of each component in the mixture.

Wien bridge oscillator [ELECTR] A phase-shift feedback oscillator that uses a Wien bridge as the frequency-determining element.

Wien capacitance bridge [ELEC] A four-arm alternating-current bridge used to measure capacitance in terms of resistance and frequency; two adjacent arms contain capacitors respectively in parallel and in series with resistors, while the other two arms are nonreactive resistors; bridge balance depends on frequency.

Wien constant [STAT MECH] The product of the temperature and the wavelength at which the intensity of radiation from a blackbody reaches its maximum; it is equal to approximately 2898 micrometer-kelvins.

Wien-DeSauty bridge *See* DeSauty's bridge.

Wien effect [PHYS CHEM] An increase in the conductance of an electrolyte at very high potential gradients.

Wiener experiment [OPTICS] An experiment in which a front-faced mirror is covered with a thick photographic emulsion which is then exposed to light incident perpendicular to the surface; upon development, it is found that standing waves are set up in the emulsion whose nodes coincide with those of the electric vector, rather than those of the magnetic vector.

Wiener-Hopf equations [MATH] Integral equations arising in the study of random walks and harmonic analysis; they are

$$g(x) = \int_0^\infty K(|x - t|) f(t) \, dt$$

$$f(x) = \int_0^\infty K(|x + t|) f(t) \, dt + g(x)$$

where g and K are known functions on the positive real numbers and f is the unknown function.

Wiener-Hopf technique [MATH] A method used in solving certain integral equations, boundary-value problems, and other problems, which involves writing a function that is holomorphic in a vertical strip of the complex z plane as the product of two functions, one of which is holomorphic both in the strip and everywhere to the right of the strip, while the other is holomorphic in the strip and everywhere to the left of the strip.

Wiener-Khintchine theorem [MATH] The theorem that determines the form of the correlation function of a given stationary stochastic process.

Wiener process [MATH] A stochastic process with normal density at each stage, arising from the study of Brownian motion, which represents the limit of a sequence of experiments. Also known as Gaussian noise.

Wiener's Tauberian theorem [MATH] If $g(x)$ is a Lebesgue integrable function whose Fourier transform

$$G(t) = \int_{-\infty}^{\infty} \exp(-ixt)g(x)\,dx$$

is not equal to 0 for any real number t, and if $f(x)$ is bounded, and if

$$\int_{-\infty}^{\infty} g(x-y)f(y)\,dy$$

approaches 0 as y approaches infinity, then for any integrable function $h(x)$,

$$\int_{-\infty}^{\infty} h(x-y)f(y)\,dy$$

approaches 0 as y approaches infinity.

Wien frequency bridge [ELEC] A modification of the Wien capacitance bridge, used to measure frequencies.

Wien inductance bridge [ELEC] A four-arm alternating-current bridge used to measure inductance in terms of resistance and frequency; two adjacent arms contain inductors respectively in parallel and in series with resistors, while the other two arms are nonreactive resistors; bridge balance depends on frequency.

Wien-Maxwell bridge *See* Maxwell bridge.

Wien's displacement law [STAT MECH] A law for blackbody radiation which states that the wavelength at which the maximum amount of radiation occurs is a constant equal to approximately 2898 times the product of 1 micrometer and 1 kelvin. Also known as displacement law; Wien's radiation law.

Wien's distribution law [STAT MECH] A formula for the spectral distribution of radiation from a blackbody, which is a good approximation to the Planck radiation formula at sufficiently low temperatures or wavelengths, for example, in the visible region of the spectrum below 3000 K. Also known as Wien's radiation law.

Wien's radiation law [STAT MECH] **1.** The law that the intensity of radiation emitted by a blackbody per unit wavelength, at that wavelength at which this intensity reaches a maximum, is proportional to the fifth power of the temperature. **2.** *See* Wien's displacement law. **3.** *See* Wien's distribution law.

Wierl equation [ELECTR] A formula for the intensity of an electron beam scattered through a specified angle by diffraction from the molecules in a gas.

Wigner coefficient *See* vector coupling coefficient.

Wigner-Eckart theorem [QUANT MECH] A theorem in the quantum theory of angular momentum which states that the matrix elements of a tensor operator can be factored into two

WIEN FREQUENCY BRIDGE

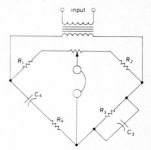

Circuit diagram of Wien frequency bridge. The frequency of the input can be determined from the resistances and capacitances when the latter are adjusted so that current through detector (the circles) is zero.

quantities, the first of which is a vector-coupling coefficient, and the second of which contains the information about the physical properties of the particular states and operator, and is completely independent of the magnetic quantum numbers.

Wigner effect *See* discomposition effect.

Wigner force [NUCLEO] A short-range nonexchange force between nucleons, postulated to explain various phenomena.

Wigner nuclides [NUC PHYS] The most important class of mirror nuclides, comprising pairs of odd-mass-number isobars for which the atomic number and the neutron number differ by 1.

Wigner-Seitz cell [CRYSTAL] A polyhedron about an atom in a face-centered cubic structure, made by drawing planes which perpendicularly bisect the lines to the nearest neighbors; in a body-centered cubic structure, bisecting planes of lines to nearest neighbors and next-nearest neighbors are used; such polyhedra fill space.

Wigner-Seitz method [SOLID STATE] A method of approximating the band structure of a solid: Wigner-Seitz cells surrounding atoms in the solid are approximated by spheres, and band solutions of the Schrödinger equation for one electron are estimated by using the assumption that an electronic wave function is the product of a plane wave function and a function whose gradient has a vanishing radial component at the sphere's surface.

Wigner's theorem [QUANT MECH] **1.** The theorem that, if ψ is an eigenfunction of the Hamiltonian operator and R is a symmetry element of the Hamiltonian, then $R\psi$ is an eigenfunction of the Hamiltonian having the same eigenvalue as ψ. **2.** Angular momentum of the electron spin is conserved in a collision of the second kind.

Wigner supermultiplet [NUC PHYS] A set of quantum-mechanical states of a collection of nucleons which form the basis of a representation of SU(4), especially appropriate when spin and isospin dependence of the nuclear interaction may be disregarded; several combinations of spin and isospin multiplets may occur in a supermultiplet.

Wigner three-j symbol *See* three-*j* number.

Wilcoxon one-sample test [STAT] A rank test for testing the hypothesis $\mu = \mu_H$ against the alternative $\mu > \mu_H$ under the assumption that observations are symmetrically distributed about μ_H; here μ_H is a given number and μ is the (unknown) mean of a random variable.

Wilcoxon paired comparison distribution [STAT] The distribution of the rank sum V_- (or V_+) of the negative differences (or positive differences) of observations in paired comparisons.

Wilcoxon paired comparison test [STAT] The test based upon the rank sum V_- (or V_+) of the negative differences (or positive differences) of observations in paired comparisons.

Wilcoxon two-sample distribution [STAT] The distribution of the Wilcoxon two-sample test statistic; it consists of the rank sums of treated subjects.

Wilcoxon two-sample test [STAT] The test based upon the rank sum of treated (or untreated) subjects.

Williams-Hazen formula [FL MECH] In a liquid-flow system, a method for calculation of head loss due to the friction in a pipeline.

Williams refractometer [OPTICS] A refractometer in which light from a single slit is divided into two beams by a pentagonal prism.

Wilson cloud chamber [NUCLEO] A cloud chamber containing air supersaturated with water vapor by sudden expansion, in which rapidly moving nuclear particles such as alpha or beta rays produced ionization tracks by condensation of vapor on the ions produced by the rays.

Wilson electroscope [ELEC] An electroscope that has a single gold leaf which, when charged, is attracted to a grounded metal plate inclined at an angle that maximizes the instrument's sensitivity.

Wilson experiment [ELECTROMAG] An experiment that tests the validity of electromagnetic theory; a hollow cylinder of dielectric material, having layers of metal on its outer and inner cylindrical surfaces, is rotated about its axis in a magnetic field parallel to the axis; a sensitive electrometer, connected to the metal layers, indicates a charge that has the magnitude and sign predicted by theory.

Wilson's theorem [MATH] The number $(n-1)! + 1$ is divisible by n if and only if n is a prime.

Wimshurst machine [ELEC] An electrostatic generator consisting of two glass disks rotating in opposite directions, having sectors of tinfoil and collecting combs so arranged that static electricity is produced for charging Leyden jars or discharging across a gap.

wind drift [ACOUS] Shift in the apparent position of a sound source or target observed by sound apparatus; it is caused by the effect of wind on sound waves, which changes their direction and increases or decreases sound lag.

winding [ELEC] **1.** One or more turns of wire forming a continuous coil for a transformer, relay, rotating machine, or other electric device. **2.** A conductive path, usually of wire, that is inductively coupled to a magnetic storage core or cell.

winding number [MATH] The number of times a given closed curve winds in the counterclockwise direction about a designated point in the plane.

window [ELECTR] A material having minimum absorption and minimum reflection of radiant energy, sealed into the vacuum envelope of a microwave or other electron tube to permit passage of the desired radiation through the envelope to the output device. [ELECTROMAG] A hole in a partition between two cavities or waveguides, used for coupling. [GEOPHYS] Any range of wavelengths in the electromagnetic spectrum to which the atmosphere is transparent. [NUCLEO] **1.** An aperture for the passage of particles or radiation in a nuclear reactor. **2.** An energy range of relatively high transparency in the total neutron cross section of a material; such windows arise from interference between potential and resonance scattering in elements of intermediate atomic weight, and can be of importance in neutron shielding.

wind tunnel [ENG] A duct in which the effects of airflow past objects can be determined.

winter [ASTRON] The period from the winter solstice, about December 22, to the vernal equinox, about March 21; popularly and for most meteorological purposes, winter is taken to include December, January, and February in the Northern Hemisphere, and June, July, and August in the Southern Hemisphere.

winter solstice [ASTRON] **1.** The sun's position on the ecliptic (about December 22). Also known as first point of Capricorn. **2.** The date (December 22) when the greatest southern declination of the sun occurs.

wire [ELEC] A single bare or insulated metallic conductor having solid, stranded, or tinsel construction, designed to carry current in an electric circuit. Also known as electric wire. [OPTICS] A filament, usually consisting of a stretched strand of spider's web or a fine metal wire, mounted in the field of view of a telescope eyepiece to serve as a reference or for measurements. Also known as web.

wiregrating [ELECTROMAG] A series of wires placed in a

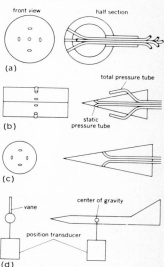

WIND TUNNEL

front view · half section

(a)

total pressure tube

(b) static pressure tube

(c)

center of gravity

vane

position transducer

(d)

Instruments used to measure direction of flow in a wind tunnel. First three instruments measure pressure distribution over surface of body. (a) Yaw sphere. (b) Wedge. (c) Cone. (d) Vane.

waveguide that allow one or more types of waves to pass and block all others.

wire-wound cryotron [CRYO] A cryotron that consists of a central insulated wire surrounded by a control coil; it is designed so that a relatively small current passed through the control coil produces a magnetic field which makes the gate resistive.

wire-wound potentiometer [ELEC] A potentiometer which is similar to a slide-wire potentiometer, except that the resistance wire is wound on a form and contact is made by a slider which moves along an edge from turn to turn.

wire-wound resistor [ELEC] A resistor employing as the resistance element a length of high-resistance wire or ribbon, usually Nichrome, wound on an insulating form.

wire-wound rheostat [ELEC] A rheostat in which a sliding or rolling contact moves over resistance wire that has been wound on an insulating core.

witch of Agnesi [MATH] The curve, symmetric about the y axis and asymptotic in both directions to the x axis, given by $x^2y = 4a^2(2a - y)$.

Witt group [MATH] The group of isometry classes of symmetric forms on vector spaces over a given field, where the product of two such forms is given by their orthogonal sum.

Witt-Grothendieck group [MATH] The Grothendieck group of the monoid consisting of isometry classes of nondegenerate symmetric forms on vector spaces over a given field, where the product of two such forms is given by their orthogonal sum.

Witt's theorem [MATH] If F and F' are subspaces of a vector space E with a nondegenerate, symmetric form g, then an isometry of g from F onto F' can be extended to an isometry of g from E onto itself.

WKB method *See* Wentzel-Kramers-Brillouin method.

Wobbe index [THERMO] A measure of the amount of heat released by a gas burner with a constant orifice, equal to the gross calorific thermal value of the gas in British thermal units per cubic foot at standard temperature and pressure divided by the square root of the specific gravity of the gas.

wobbulator [ELECTR] A signal generator in which a motor-driven variable capacitor is used to vary the output frequency periodically between two known limits, as required for displaying a frequency-response curve on the screen of a cathode-ray oscilloscope.

wolf [ACOUS] A dissonant interval which appears when the meantone scale is extended to include chromatic notes.

Wolf number *See* relative sunspot number.

wolfram *See* tungsten.

Wolf-Rayet star [ASTRON] A member of a class of very hot stars (100,000–35,000 K) which characteristically show broad bright emission lines in their spectra; luminosities are high, probably in the range 10^4–10^5 times that of the sun; these stars are probably very young and represent an early short-lived stage in stellar evolution.

Wolf-Wolfer number *See* relative sunspot number.

Wollaston polarizing prism [OPTICS] A device for producing linearly polarized beams of light, consisting of two adjacent quartz wedges with their optic axes perpendicular to each other and to the direction of incident light.

Wollaston wire [ENG] An extremely fine platinum wire, produced by enclosing a platinum wire in a silver sheath, drawing them together, and using acid to dissolve away the silver; used in electroscopes, microfuses, and hot-wire instruments.

Wood effect [OPTICS] Transparence of alkali metals to ultraviolet light.

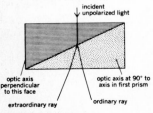

WOLLASTON POLARIZING PRISM

incident unpolarized light

optic axis perpendicular to this face

optic axis at 90° to axis in first prism

extraordinary ray

ordinary ray

Diagram of Wollaston polarizing prism.

Wood's glass [MATER] A type of glass that has a high transmission factor for ultraviolet radiation but is relatively opaque to visible radiation.

work [ELEC] *See* load. [MECH] The transference of energy that occurs when a force is applied to a body that is moving in such a way that the force has a component in the direction of the body's motion; it is equal to the line integral of the force over the path taken by the body.

work function [SOLID STATE] The minimum energy needed to remove an electron from the Fermi level of a metal to infinity; usually expressed in electron volts. [THERMO] *See* free energy.

working Q *See* loaded Q.

work-kinetic energy theorem [MECH] The theorem that the change in the kinetic energy of a particle during a displacement is equal to the work done by the resultant force on the particle during this displacement.

work of adhesion *See* adhesional work.

world [RELAT] Pertaining to Lorentz transformations and four-dimensional space-time, rather than rotations and three-dimensional space, as in world scalar, world vector, world line.

wrench [MECH] The combination of a couple and a force which is parallel to the torque exerted by the couple.

Wright telescope [OPTICS] A modification of the Schmidt system in which the spherical primary mirror is replaced by an ellipsoidal mirror, and the corrector plate is modified accordingly.

Wronskian [MATH] An $n \times n$ matrix whose ith row is a list of the $(i - 1)$st derivatives of a set of functions $f_1, \ldots, f_n$; ordinarily used to determine linear independence of solutions of linear homogeneous differential equations.

W space *See* Sobolev space.

W stars [ASTRON] Stars of the W spectral class; their spectra contain an abundance of highly ionized elements such as He, C, N, and O, and they are intensely hot with surface temperatures of about 50,000 to 100,000 K.

w* topology *See* weak* topology.

Wulf electrometer [ENG] **1.** A variant of the string electrometer in which charged metal plates are replaced by charged knife-edges. **2.** An electrometer in which two conducting fibers are placed side by side, and their separation upon charging is measured.

W Ursae Majoris stars [ASTRON] Eclipsing variable stars whose brightness is continuously varying in periods of a few hours; they are composed of two close stars that have a common gaseous envelope.

W Virginis stars [ASTRON] Periodic variable stars with periods of about 10 to 30 days; they exhibit two surges of activity from the same star so that there is a doubling of their spectral lines. Also known as Type II Cepheids.

wye [ELEC] Polyphase circuit whose phase differences are 120° and which when drawn resembles the letter Y.

X _See_ siegbahn.

x axis [CRYSTAL] A reference axis in a quartz crystal. [MATH] **1.** A horizontal axis in a system of rectangular coordinates. **2.** That line on which distances to the right or left (east or west) of the reference line are marked, especially on a map, chart, or graph.

X coefficient _See_ nine-_j_ symbol.

x coordinate [MATH] One of the coordinates of a point in a two- or three-dimensional cartesian coordinate system, equal to the directed distance of a point from the _y_ axis in a two-dimensional system, or from the plane of the _y_ and _z_ axes in a three-dimensional system, measured along a line parallel to the _x_ axis.

X cut [CRYSTAL] A quartz-crystal cut made in such a manner that the _x_ axis is perpendicular to the faces of the resulting slab.

Xe _See_ xenon.

xenon [CHEM] An element, symbol Xe, member of the noble gas family, group O, atomic number 54, atomic weight 131.30; colorless, boiling point −108°C (1 atmosphere, or 101,325 newtons per square meter), noncombustible, nontoxic, and nonreactive; used in photographic flash lamps, luminescent tubes, and lasers, and as an anesthetic.

xenon-135 [NUC PHYS] A radioactive isotope of xenon produced in nuclear reactors; readily absorbs neutrons; half-life is 9.2 hours.

xenon override [NUCLEO] In a nuclear reactor, the excess reactivity provided to compensate for the poisoning effect of xenon buildup.

xenon poisoning [NUCLEO] The accumulation in a nuclear reactor of xenon-135, formed by beta decay of iodine-135; xenon-135 has the highest cross section for thermal neutron capture of any known reactor poison.

xi hyperon [PARTIC PHYS] Also known as xi particle. **1.** Collective name for the xi-minus and xi-zero particles, which form an isotopic-spin multiplet of quasi-stable baryons, designated Ξ, having a hypercharge of −1, a total isotopic spin of $\frac{1}{2}$, a spin of $\frac{1}{2}$, positive parity, and an average mass of approximately 1318 MeV. Also known as cascade hyperon; cascade particle. **2.** A baryon belonging to any isotopic-spin multiplet having a hypercharge of −1 and a total isotopic spin of $\frac{1}{2}$; designated by $\Xi_{JP(m)}$, where m is the mass of the baryon in MeV, and J and P are its spin and parity (if known); the $\Xi_{3/2}+$ (1530) is sometimes designated Ξ^*.

xi-minus particle [PARTIC PHYS] A negatively charged xi hyperon, designated Ξ^-. Also known as cascade particle.

xi particle _See_ xi hyperon.

xi-zero particle [PARTIC PHYS] An uncharged xi hyperon, designated Ξ^0.

XPS _See_ x-ray photoelectron spectroscopy.

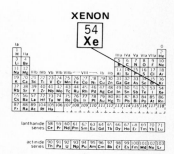

Periodic table of the chemical elements showing the position of xenon.

x-radiation *See* x-rays.

x-ray absorption [ELECTROMAG] The taking up of energy from an x-ray beam by a medium through which the beam is passing.

x-ray analysis [PHYS] The use of x-ray radiations to detect heavy elements in the presence of lighter ones, to give critical-edge absorption to identify elemental composition, and to identify crystal structures by diffraction patterns.

x-ray astronomy [ASTRON] The study of x-rays mainly from sources outside the solar system; it includes the study of novae and supernovae in the Milky Way Galaxy, together with extragalactic radio sources.

x-ray burster [ASTRON] One of a class of celestial x-ray sources which produce bursts of x-rays in the 1- to 20-kilo-electron-volt range and which are characterized by rise times of less than a few seconds and decay times of a few seconds to a few minutes; the peak luminosity is of the order of 10^{38} ergs/per second (10^{31} watts) and the sources have an average equivalent temperature of 10^8 K.

x-ray crystallography [CRYSTAL] The study of crystal structure by x-ray diffraction techniques. Also known as roentgen diffractometry.

x-ray crystal spectrometer [SPECT] An instrument designed to produce an x-ray spectrum and measure the wavelengths of its components, by diffracting x-rays from a crystal with known lattice spacing.

x-ray diffraction [PHYS] The scattering of x-rays by matter, especially crystals, with accompanying variation in intensity due to interference effects. Also known as x-ray microdiffraction.

x-ray diffraction analysis [CRYSTAL] Analysis of the crystal structure of materials by passing x-rays through them and registering the diffraction (scattering) image of the rays.

x-ray diffractometer [ENG] An instrument used in x-ray analysis to measure the intensities of the diffracted beams at different angles.

x-ray emission *See* x-ray fluorescence.

x-ray film [GRAPHICS] A film base coated, usually on both sides, with an emulsion designed for use with x-rays.

x-ray fluorescence [ATOM PHYS] Emission by a substance of its characteristic x-ray line spectrum upon exposure to x-rays. Also known as x-ray emission.

x-ray fluorescence analysis [ANALY CHEM] A nondestructive physical method used for chemical analyses of solids and liquids; the specimen is irradiated by an intense x-ray beam and the lines in the spectrum of the resulting x-ray fluorescence are diffracted at various angles by a crystal with known lattice spacing; the elements in the specimen are identified by the wavelengths of their spectral lines, and their concentrations are determined by the intensities of these lines. Also known as x-ray fluorimetry.

x-ray fluorescent emission spectrometer [SPECT] An x-ray crystal spectrometer used to measure wavelengths of x-ray fluorescence; in order to concentrate beams of low intensity, it has bent reflecting or transmitting crystals arranged so that the theoretical curvature required can be varied with the diffraction angle of a spectrum line.

x-ray fluorimetry *See* x-ray fluorescence analysis.

x-ray generator [ELECTR] A metal from whose surface large amounts of x-rays are emitted when it is bombarded with high-velocity electrons; metals with high atomic weight are the most efficient generators.

x-ray goniometer [ENG] A scale designed to measure the angle between the incident and refracted beams in x-ray diffraction analysis.

x-ray hardness [ELECTROMAG] The penetrating ability of x-rays; it is an inverse function of the wavelength.

x-ray image spectrography [SPECT] A modification of x-ray fluorescence analysis in which x-rays irradiate a cylindrically bent crystal, and Bragg diffraction of the resulting emissions produces a slightly enlarged image with a resolution of about 50 micrometers.

x-ray irradiation [PHYS] Subjection of a material, object, or patient to x-rays.

x-ray microdiffraction *See* x-ray diffraction.

x-ray microscope [ENG] **1.** A device in which an ultra-fine-focus x-ray tube or electron gun produces an electron beam focused to an extremely small image on a transmission-type x-ray target that serves as a vacuum seal; the magnification is by projection; specimens being examined can thus be in air, as also can the photographic film that records the magnified image. **2.** Any of several instruments which utilize x-radiation for chemical analysis and for magnification of 100–1000 diameters; it is based on contact or projection microradiography, reflection x-ray microscopy, or x-ray image spectrography.

x-ray monochromator [ENG] An instrument in which x-rays are diffracted from a crystal to produce a beam having a narrow range of wavelengths.

x-ray nebulae [ASTRON] The remnant of an ancient supernova that has been identified as a source of x-rays; an example is the Crab Nebula.

x-ray optics [ELECTROMAG] A title-by-analogy of those phases of x-ray physics in which x-rays demonstrate properties similar to those of light waves. Also known as roentgen optics.

x-ray photoelectron spectroscopy [SPECT] A form of electron spectroscopy in which a sample is irradiated with a beam of monochromatic x-rays and the energies of the resulting photoelectrons are measured. Abbreviated XPS. Also known as electron spectroscopy for chemical analysis (ESCA).

x-ray powder diffractometer *See* powder diffraction camera.

x-ray powder method *See* powder method.

x-ray projection microscopy *See* projection microradiography.

x-rays [PHYS] A penetrating electromagnetic radiation, usually generated by accelerating electrons to high velocity and suddenly stopping them by collision with a solid body, or by inner-shell transitions of atoms with atomic number greater than 10; their wavelengths range from about 10^{-5} angstrom to 10^3 angstroms (1 femtometer to 100 nanometers), the average wavelength used in research being about 1 angstrom (0.1 nanometer). Also known as roentgen rays; x-radiation.

x-ray spectrograph [SPECT] An x-ray spectrometer equipped with photographic or other recording apparatus; one application is fluorescence analysis.

x-ray spectrometer [SPECT] An instrument for producing the x-ray spectrum of a material and measuring the wavelengths of the various components.

x-ray spectrometry [SPECT] The measure of wavelengths of x-rays by observing their diffraction by crystals of known lattice spacing. Also known as roentgen spectrometry; x-ray spectroscopy.

x-ray spectroscopy *See* x-ray spectrometry.

x-ray spectrum [SPECT] A display or graph of the intensity of x-rays, produced when electrons strike a solid object, as a function of wavelengths or some related parameter; it consists of a continuous bremsstrahlung spectrum on which are

X-RAY IMAGE SPECTROGRAPHY

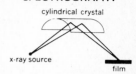

Arrangement of x-ray source, cylindrical crystal, and photographic film in x-ray image spectrography.

X-RAY SPECTROGRAPH

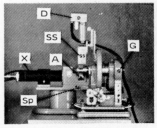

Modern x-ray spectrograph for fluorescence analysis. X is x-ray tube; Sp, specimen; A, crystal analyzer; SS, Soller (parallel) slits; D, counter tube detector; G, goniometer. *(Philips Electronic Instruments, Inc.)*

superimposed groups of sharp lines characteristic of the elements in the target.

x-ray star [ASTROPHYS] A source of x-rays from outside the solar system; examples are the point x-ray sources Scorpius X-1, Cygnus X-2, and the Crab x-ray source.

x-ray target [ELECTR] The metal body with which high-velocity electrons collide, in a vacuum tube designed to produce x-rays.

x-ray telescope [ENG] An instrument designed to detect x-rays emanating from a source outside the earth's atmosphere and to resolve the x-rays into an image; they are carried to high altitudes by balloons, rockets, or space vehicles; although several types of x-ray detector, involving gas counters, scintillation counters, and collimators, have been used, only one, making use of the phenomenon of total external reflection of x-rays from a surface at grazing incidence, is strictly an x-ray telescope.

x-ray tube [ELECTR] A vacuum tube designed to produce x-rays by accelerating electrons to a high velocity by means of an electrostatic field, then suddenly stopping them by collision with a target.

x-ray unit *See* siegbahn.

X test [STAT] A one-sample test which rejects the hypothesis $\mu = \mu_H$ in favor of the alternative $\mu > \mu_H$ if $X - \mu_H \geq c$ where c is an appropriate critical value, X is the arithmetic mean of observations, μ_H is a given number, and μ is the (unknown) expected value of the random variable X.

XU *See* siegbahn.

X unit *See* siegbahn.

X wave *See* extraordinary wave.

X-RAY TELESCOPE

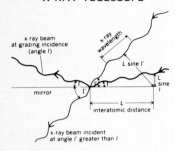

Diagram illustrating phenomenon of total external reflection of x-rays at grazing incidence, on which x-ray telescope is based. Beam incident at angle *I* is totally reflected, because *L* sine *I* is much smaller than x-ray wavelength, while beam incident at angle *I'* greater than *I* is not reflected.

Y *See* yttrium.

Yagi antenna *See* Yagi-Uda antenna.

Yagi-Uda antenna [ELECTROMAG] An end-fire antenna array having maximum radiation in the direction of the array line; it has one dipole connected to the transmission line and a number of equally spaced unconnected dipoles mounted parallel to the first in the same horizontal plane to serve as directors and reflectors. Also known as Uda antenna; Yagi antenna.

yag laser *See* yttrium-aluminum-garnet laser.

yard [MECH] A unit of length in common use in the United States and United Kingdom, equal to 0.9144 meter, or 3 feet. Abbreviated yd.

yaw [MECH] **1.** The rotational or oscillatory movement of a ship, aircraft, rocket, or the like about a vertical axis. Also known as yawing. **2.** The amount of this movement, that is, the angle of yaw. **3.** To rotate or oscillate about a vertical axis.

yaw acceleration [MECH] The angular acceleration of an aircraft or missile about its normal or Z axis.

yaw axis [MECH] A vertical axis through an aircraft, rocket, or similar body, about which the body yaws; it may be a body, wind, or stability axis. Also known as yawing axis.

yawing *See* yaw.

yawing axis *See* yaw axis.

y axis [CRYSTAL] A line perpendicular to two opposite parallel faces of a quartz crystal. [MATH] **1.** A vertical axis in a system of rectangular coordinates. **2.** That line on which distances above or below (north or south) the reference line are marked, especially on a map, chart, or graph.

Yb *See* ytterbium.

Y circulator [ELECTROMAG] Circulator in which three identical retangular waveguides are joined to form a symmetrical Y-shaped configuration, with a ferrite post or wedge at its center; power entering any waveguide will emerge from only one adjacent waveguide.

Y connection *See* Y network.

y coordinate [MATH] One of the coordinates of a point in a two- or three-dimensional coordinate system, equal to the directed distance of a point from the x axis in a two-dimensional system, or from the plane of the x and z axes in a three-dimensional coordinate system, measured along a line parallel to the y axis.

Y cut [CRYSTAL] A quartz-crystal cut such that the y axis is perpendicular to the faces of the resulting slab.

yd *See* yard.

Y-delta transformation [ELEC] One of two electrically equivalent networks with three terminals, one being connected internally by a Y configuration and the other being connected

YAGI-UDA ANTENNA

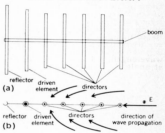

Yagi-Uda antenna. *(a)* View from above. *(b)* Side view; arrows show energy flow close to parasitic elements.

internally by a delta transformation. Also known as delta-Y transformation; pi-T transformation.

year [ASTRON] Any of several units of time based on the revolution of the earth about the sun; the tropical year to which the calendar is adjusted is the period required for the sun's longitude to increase 360°; it is about 365.24220 mean solar days. Abbreviated yr.

yellow [OPTICS] The hue evoked in an average observer by monochromatic radiation having a wavelength in the approximate range from 577 to 597 nanometers; however, the same sensation can be produced in a variety of other ways.

yield [MECH] That stress in a material at which plastic deformation occurs.

yield point [MECH] The lowest stress at which strain increases without increase in stress.

yield strength [MECH] The stress at which a material exhibits a specified deviation from proportionality of stress and strain.

yield stress [FL MECH] The minimum stress needed to cause a Bingham plastic to flow. [MECH] The lowest stress at which extension of the tensile test piece increases without increase in load.

yig device [ELECTR] A filter, oscillator, parametric amplifier, or other device that uses an yttrium-iron-garnet crystal in combination with a variable magnetic field to achieve wideband tuning in microwave circuits. Derived from yttrium-iron-garnet device.

yig filter [ELECTR] A filter consisting of an yttrium-iron-garnet crystal positioned in a magnetic field provided by a permanent magnet and a solenoid; tuning is achieved by varying the amount of direct current through the solenoid; the bias magnet serves to tune the filter to the center of the band, thus minimizing the solenoid power required to tune over wide bandwidths.

Y junction [ELECTROMAG] A waveguide in which the longitudinal axes of the waveguide form a Y.

ylem [ASTROPHYS] The primordial matter which according to the big bang theory existed just prior to the formation of the chemical elements.

Y network [ELEC] A star network having three branches. Also known as Y connection.

yoke [ELECTR] *See* deflection yoke. [ELECTROMAG] Piece of ferromagnetic material without windings, which permanently connects two or more magnet cores.

Yonden square [STAT] An experimental design that is an incomplete block design derived from a Latin square by dropping one or more rows and by treating columns as blocks. Also known as incomplete Latin square.

Young construction [OPTICS] A graphical procedure for tracing a light ray through a boundary between two media having different refractive indices.

Young-Helmholtz laws [MECH] Two laws describing the motion of bowed strings; the first states that no overtone with a node at the point of excitation can be present; the second states that when the string is bowed at a distance of $1/n$ times the string's length from one of the ends, where n is an integer, the string moves back and forth with two constant velocities, one of which has the same direction as that of the bow and is equal to it, while the other has the opposite direction and is $n - 1$ times as large.

Young's modulus [MECH] The ratio of a simple tension stress applied to a material to the resulting strain parallel to the tension.

Young's two-slit interference [OPTICS] Interference of light from two parallel slits which are illuminated by light from a single slit, which in turn is illuminated by a source; the interference can be seen by letting the light fall on a screen, which then shows a series of parallel fringes.

y parameter [ELECTR] One of a set of four transistor equivalent-circuit parameters, used especially with field-effect transistors, that conveniently specify performance for small voltage and current in an equivalent circuit; the equivalent circuit is a current source with shunt impedance at both input and output.

yr *See* year.

yrast state [NUC PHYS] An energy state of a nucleus whose energy is less than that of any other state with the same spin.

ytterbium [CHEM] A rare-earth metal of the yttrium subgroup, symbol Yb, atomic number 70, atomic weight 173.04; lustrous, malleable, soluble in dilute acids and liquid ammonia, reacts slowly with water; melts at 824°C, boils at 1427°C; used in chemical research, lasers, garnet doping, and x-ray tubes.

yttrium [CHEM] A rare-earth metal, symbol Y, atomic number 39, atomic weight 88.905; dark-gray, flammable (as powder), soluble in dilute acids and potassium hydroxide solution, and decomposes in water; melts at 1500°C, boils at 2927°C; used in alloys and nuclear technology and as a metal deoxidizer.

yttrium-aluminum-garnet laser [OPTICS] A four-level infrared laser in which the active material is neodymium ions in an yttrium-aluminum-garnet crystal; it can provide a continuous output power of several watts. Abbreviated yag laser.

yttrium-iron-garnet device *See* yig device.

Yukawa force [NUC PHYS] The strong, short-range force between nucleons, as calculated on the assumption that this force is due to the exchange of a particle of finite mass (Yukawa meson), just as electrostatic forces are interpreted in quantum electrodynamics as being due to the exchange of photons.

Yukawa meson [PARTIC PHYS] A particle, having a finite rest mass, whose exchange between nucleons is postulated to account for the strong, short-range forces between them; such a contributor is the pi meson.

Yukawa potential [NUC PHYS] The potential function that is associated with the Yukawa force, with the form $V(r) = -V_0(b/r) \exp(-r/b)$, where r is the distance between the nucleons and V_0 and b are constants, giving measures of the strength and range of the force respectively.

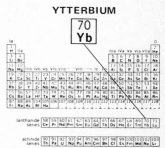

YTTERBIUM

Periodic table of the chemical elements showing the position of ytterbium.

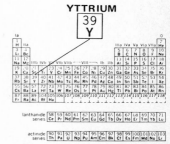

YTTRIUM

Periodic table of the chemical elements showing the position of yttrium.

Z

ZAA spectrometry *See* Zeeman-effect atomic absorption spectrometry.

z axis [CRYSTAL] The optical axis of a quartz crystal, perpendicular to both the x and y axes. [MATH] One of the three axes in a three-dimensional cartesian coordinate system; in a rectangular coordinate system it is perpendicular to the x and y axes.

Z^0 boson [PARTIC PHYS] A hypothetical intermediate vector boson which carries no electric charge; it is believed that neutral current interactions are produced by exchange of these bosons.

Z cam [ASTRON] A representative type of variable star; it is eruptive with a cycle of about 10–600 days; magnitude ranges from 2 to 6.

Z coefficient [QUANT MECH] A coefficient used in the transformation between modes of coupling eigenfunctions of three angular momenta, and especially in calculating matrix elements in beta decay and similar problems.

z coordinate [MATH] One of the coordinates of a point in a three-dimensional coordinate system, equal to the directed distance of a point from the plane of the x and y axes, measured along a line parallel to the z axis.

Zeeman displacement [SPECT] The separation, in wave numbers, of adjacent spectral lines in the normal Zeeman effect in a unit magnetic field, equal (in centimeter-gram-second Gaussian units) to $e/4\pi mc^2$, where e and m are the charge and mass of the electron, or to approximately 4.67×10^{-5} (centimeter)$^{-1}$(gauss)$^{-1}$.

Zeeman effect [SPECT] A splitting of spectral lines in the radiation emitted by atoms or molecules in a static magnetic field.

Zeeman-effect atomic absorption spectrometry [SPECT] A type of atomic absorption spectrometry in which either the light source or the sample is placed in a magnetic field, splitting the spectral lines under observation into polarized components, and a rotating polarizer is placed between the source and the sample, enabling the absorption caused by the element under analysis to be separated from background absorption. Abbreviated ZAA spectrometry.

Zeeman energy [ATOM PHYS] The energy of interaction between an atomic or molecular magnetic moment and an applied magnetic field.

Zener breakdown [ELECTR] Nondestructive breakdown in a semiconductor, occurring when the electric field across the barrier region becomes high enough to produce a form of field emission that suddenly increases the number of carriers in this region. Also known as Zener effect.

Zener diode [ELECTR] A semiconductor breakdown diode, usually constructed of silicon, in which reverse-voltage breakdown is based on the Zener effect.

Zener diode voltage regulator   *See* diode voltage regulator.

Zener effect   *See* Zener breakdown.

Zener voltage   *See* breakdown voltage.

Zeno's paradox   [MATH] An erroneous group of paradoxes dealing with motion; the most famous one concerns two objects, one chasing the other which has a given head start, where the chasing one moves faster yet seemingly never catches the other.

Zepp antenna   [ELECTROMAG] Horizontal antenna which is a multiple of a half-wavelength long and is fed at one end by one lead of a two-wire transmission line that is some multiple of a quarter-wavelength long.

Zernike polynomials   *See* circle polynomials.

zero   [MATH] The additive identity element of an algebraic system.

zero bias   [ELECTR] The condition in which the control grid and cathode of an electron tube are at the same direct-current voltage.

zero branch   [SPECT] A spectral band whose Fortrat parabola lies between two other Fortrat parabolas, with its vertex almost on the wave number axis.

zero divisor   *See* divisor of zero.

zero-field emission   *See* field-free emission current.

zero geodesic   *See* null geodesic.

zero gravity   *See* weightlessness.

zero level   [ENG ACOUS] Reference level used for comparing sound or signal intensities; in audio-frequency work, a power of 0.006 watt is generally used as zero level; in sound, the threshold of hearing is generally assumed as the zero level.

zero method   *See* null method.

zero of a function   [MATH] Any point where a function assumes the value zero.

zero of a set of polynomials   [MATH] A zero of a set S of polynomials in n variables with coefficients in a field F, in an extension field L of F, is an n-tuple of elements $(c_1, \ldots, c_n)$ in L such that $p(c_1, \ldots, c_n) = 0$ for every polynomial p in S.

zero-order hold   [CONT SYS] A device which converts a sampled output into an output which is held constant between samples at the last sampled value.

zero-point energy   [STAT MECH] The kinetic energy retained by the molecules of a substance at a temperature of absolute zero.

zero-point entropy   [STAT MECH] The entropy that a substance such as glass, which is not in thermodynamic equilibrium, retains at a temperature of absolute zero.

zero-point vibration   [STAT MECH] The vibrational motion which molecules in a crystal lattice, or particles in any oscillator potential, retain at a temperature of absolute zero; it is quantum-mechanical in origin. Also known as residual vibration.

zero potential   [ELEC] Expression usually applied to the potential of the earth, as a convenient reference for comparison.

zero-power reactor   [NUCLEO] An experimental nuclear reactor operated at low neutron flux and at a power level so low that no forced cooling is required; fission product activity in the fuel is then sufficiently low to permit handling the fuel after use.

zero-sum game   [MATH] A two-person game where the sum of the payoffs to the two players is zero for each move.

zeroth law of thermodynamics   [THERMO] If two systems are separately in thermal equilibrium with a third, they are in equilibrium with each other.

zeta function   *See* Riemann zeta function.

Zeta Geminorum stars   [ASTRON] A subgroup of classical Cepheid variable stars whose variation of magnitude with time for one complete cycle produces a quasi-bell-shaped curve.

zeta potential [PHYS] The electrical potential that exists across the interface of all solids and liquids. Also known as electrokinetic potential.

zinc [CHEM] A metal of group IIb, symbol Zn, atomic number 30, atomic weight 65.37; explosive as powder; soluble in acids and alkalies, insoluble in water; strongly electropositive; melts at 419°C, boils at 907°C.

zinc-65 [NUC PHYS] A radioactive isotope of zinc, which has a 250-day half-life with beta and gamma radiation; used in alloy-wear tracer studies and body metabolism studies.

zirconium [CHEM] A metallic element of group IVb, symbol Zr, atomic number 40, atomic weight 91.22; occurs as crystals, flammable as powder; insoluble in water, soluble in hot, concentrated acids; melts at 1850°C, boils at 4377°C.

zirconium-95 [NUC PHYS] A radioactive isotope of zirconium; half-life of 63 days with beta and gamma radiation; used to trace petroleum-pipeline flows and in the circulation of a catalyst in a cracking plant.

zitterbewegung [QUANT MECH] An oscillatory motion of an electron suggested in some interpretations of the Dirac electron theory, having a frequency greater than $4\pi mc^2/h$, where m is the electron's mass, c is the speed of light, and h is Planck's constant, or approximately 1.5×10^{21} hertz.

Zn *See* zinc.

zodiac [ASTRON] A band of the sky extending 8° on each side of the ecliptic, within which the moon and principal planets remain.

zodiacal cone *See* zodiacal pyramid.

zodiacal counterglow *See* gegenschein.

zodiacal light [GEOPHYS] A diffuse band of luminosity occasionally visible on the ecliptic; it is sunlight diffracted and reflected by dust particles in the solar system within and beyond the orbit of the earth.

zodiacal pyramid [GEOPHYS] The pattern formed by the zodiacal light. Also known as zodiacal cone.

zonal harmonics [MATH] Spherical harmonics which do not depend on the azimuthal angle; they are proportional to Legendre polynomials of cos θ, where θ is the colatitude.

zone [CRYSTAL] A set of crystal faces which intersect (or would intersect, if extended) along edges which are all parallel. [MATH] The portion of a sphere lying between two parallel planes that intersect the sphere.

zone axis [CRYSTAL] A line through the center of a crystal which is parallel to all the faces of a zone.

zone indices [CRYSTAL] Three integers identifying a zone of a crystal; they are the crystallographic coordinates of a point joined to the origin by a line parallel to the zone axis.

zone melting crystallization [CHEM ENG] A method for purification of crystalline solids; the sample, packed in a narrow column, is heated so that a molten zone passes down through the sample, carrying impurities with it.

zone of avoidance [ASTRON] An irregularly shaped area in the Milky Way Galaxy in which no extragalactic nebulae are observed because of the presence of interstellar matter.

zone of silence *See* skip zone.

zone plate [OPTICS] A plate with alternate transparent and opaque rings, designed to block off every other Fresnel half-period zone; light from a point source passing through the plate produces an intense point image much like that produced by a lens.

zoning [ELECTROMAG] The displacement of various portions of the lens or surface of a microwave reflector so the resulting

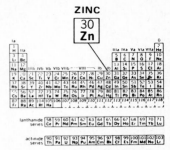

ZINC

Periodic table of the chemical elements showing the position of zinc.

ZIRCONIUM

Periodic table of the chemical elements showing the position of zirconium.

phase front in the near field remains unchanged. Also known as stepping.

zoom lens [OPTICS] A system of lenses in which two or more parts are moved with respect to each other to obtain a continuously variable focal length and hence magnification, while the image is kept in the same image plane. Also known as variable-focal-length lens; varifocal lens.

Zorn's lemma [MATH] If every linearly ordered subset of a partially ordered set has a maximal element in the set, then the set has a maximal element.

Zr *See* zirconium.

z score *See* standard score.

Z time *See* Greenwich mean time.

z-transfer function *See* pulsed transfer function.

z-transform [MATH] The z-transform of a sequence whose general term is f_n is the sum of a series whose general term is $f_n z^{-n}$, where z is a complex variable; n runs over the positive integers for a one-sided transform, over all the integers for a two-sided transform.

Zurich number *See* relative sunspot number.

Z variometer *See* vertical intensity variometer.

Appendix

U.S. Customary System and the metric system

Scientists and engineers have been using two major systems of units in measurement. They are commonly called the U.S. Customary System (inherited from the British Imperial System) and the metric system.

In the U.S. Customary System the units yard and pound with their divisions, such as the inch, and multiples, such as the ton, are basic. The metric system was evolved during the 18th century and has been adopted for general use by most countries. Nearly everywhere it is used for precise measurements in science. The meter and kilogram with their multiples, such as the kilometer, and fractions, such as the gram, are basic to the metric system.

In the U.S. Customary System, units of the same kind are related almost at random. For example, there are the units of length, the inch, yard, and mile. In the metric system the relationships between units of the same kind are strictly decimal (millimeter, meter, and kilometer).

However, to complicate matters in scientific writing, there is no uniformity within each of these two systems as to the choice of units for the same quantities. For example, the hour or the second, the foot or the inch, and the centimeter or the millimeter could be chosen by a scientist as the unit of measurement for the quantities time and length.

The International System, or SI

To simplify matters and to make communication more understandable, an internationally accepted system of units is coming into use. This is termed the International System of Units, which is abbreviated SI in all languages.

Fundamentally the system is metric with the base units derived from scientific formulas or natural constants. For example, the meter in the SI is defined as the length equal to 1 650 763.73 wavelengths in vacuum of the radiation corresponding to the transition between the electronic energy levels $2p_{10}$ and $5d_5$ of the krypton-86 atom. Previously, in the metric system, the meter was defined as the distance between two marks on a specific metal bar.

In a similar way the second in the SI is defined as the duration of 9 192 631 770 periods of the radiation corresponding to the transition between two hyperfine levels of the ground state of the cesium-133 atom.

Interestingly, the kilogram, the SI unit of mass, is still the mass of the kilogram kept at Sèvres, France. However, it is possible that eventually the unit will be redefined in terms of atomic mass.

Although the SI is increasing in usage by scientists and engineers, there are some units in everyday use which will probably remain, for example, minute, hour, day, degree (angle), and liter. The point should be made, however, that these terms will not be employed in a scientific context if the SI is fully adopted.

Because of their extremely common use among scientists, several units are still permitted in conjunction with SI units, for example, the electron volt, rad, roentgen, barn, and curie. In time their usage might be phased out.

One further point is that in October, 1967, the Thirteenth General Conference of Weights and Measures decided to name the SI unit of thermodynamic temperature "kelvin" (symbol K) instead of "degree Kelvin" (symbol °K). For example, the notation is 273 K and not 273°K.

The base units and derived units of the SI are shown in **Table 1** and **Table 2.** Some common units defined in terms of SI units are given in **Table 3** (the definitions in the fourth column are exact).

In the SI the prefixes differ from a unit in steps of 10^3. A list of prefix terms, symbols, and their factors is given in **Table 4.** Prefixes are used as follows:

$$1000 \text{ m} = 1 \text{ kilometer} \quad = 1 \text{ km}$$
$$1000 \text{ V} = 1 \text{ kilovolt} \quad = 1 \text{ kV}$$
$$1\,000\,000 \text{ } \Omega = 1 \text{ megohm} \quad = 1 \text{ M}\Omega$$
$$0.000\,000\,001 \text{ s} = 1 \text{ nanosecond} = 1 \text{ ns}$$

Only one prefix is to be employed for a unit. For example:

$$1000 \text{ kg} = 1 \text{ Mg} \qquad \text{not } 1 \text{ kkg}$$
$$10^{-9} \text{ s} = 1 \text{ ns} \qquad \text{not } 1 \text{ m}\mu\text{s}$$
$$1\,000\,000 \text{ m} = 1 \text{ Mm} \qquad \text{not } 1 \text{ kkm}$$

Also, when a unit is raised to a power, the power applies to the whole unit including the prefix. For example:

$$\text{km}^2 = (\text{km})^2 = (1000 \text{ m})^2 = 10^6 \text{ m}^2 \qquad \text{not } 1000 \text{ m}^2$$

Table 1. Base units of the International System

Quantity	Name of unit	Unit symbol
length	meter	m
mass	kilogram	kg
time	second	s
electric current	ampere	A
temperature	kelvin	K
luminous intensity	candela	cd
amount of substance	mole	mol

Table 2. Derived units of the International System

Quantity	Name of unit	Unit symbol, where differing from basic form	Unit expressed in terms of base or supplementary units*
area	square meter		m^2
volume	cubic meter		m^3
frequency	hertz	Hz	s^{-1}
density	kilogram per cubic meter		kg/m^3
velocity	meter per second		m/s
angular velocity	radian per second		rad/s
acceleration	meter per second squared		m/s^2
angular acceleration	radian per second squared		rad/s^2
volumetric flow rate	cubic meter per second		m^3/s
force	newton	N	$kg \cdot m/s^2$
surface tension	newton per meter, joule per square meter	N/m, J/m^2	kg/s^2
pressure	newton per square meter, pascal	N/m^2, Pa	$kg/m \cdot s^2$
viscosity, dynamic	newton-second per square meter, pascal-second	$N \cdot s/m^2$, Pa·s	$kg/m \cdot s$
viscosity, kinematic	meter squared per second		m^2/s
work, torque, energy, quantity of heat	joule, newton-meter, watt-second	J, N·m, W·s	$kg \cdot m^2/s^2$
power, heat flux	watt, joule per second	W, J/s	$kg \cdot m^2/s^3$
heat flux density	watt per square meter	W/m^2	kg/s^3
volumetric heat release rate	watt per cubic meter	W/m^3	$kg/m \cdot s^3$

Table 2. Derived units of the International System (cont.)

Quantity	Name of unit	Unit symbol, where differing from basic form	Unit expressed in terms of base or supplementary units*
heat transfer coefficient	watt per square meter kelvin	$W/m^2 \cdot K$	$kg/s^3 \cdot K$
heat capacity (specific)	joule per kilogram kelvin	$J/kg \cdot K$	$m^2/s^2 \cdot K$
capacity rate	watt per kelvin	W/K	$kg \cdot m^2/s^3 \cdot K$
thermal conductivity	watt per meter kelvin	$W/m \cdot K, \dfrac{J \cdot m}{s \cdot m^2 \cdot K}$	$kg \cdot m/s^3 \cdot K$
quantity of electricity	coulomb	C	$A \cdot s$
electromotive force	volt	$V, W/A$	$kg \cdot m^2/A \cdot s^3$
electric field strength	volt per meter	V/m	V/m
electric resistance	ohm	$\Omega, V/A$	$kg \cdot m^2/A^2 \cdot s^3$
electric conductivity	ampere per volt meter	$A/V \cdot m$	$A^2 \cdot s^3/kg \cdot m^3$
electric capacitance	farad	$F, A \cdot s/V$	$A^3 \cdot s^4/kg \cdot m^2$
magnetic flux	weber	$Wb, V \cdot s$	$kg \cdot m^2/A \cdot s^2$
inductance	henry	$H, V \cdot s/A$	$kg \cdot m^2/A^2 \cdot s^2$
magnetic permeability	henry per meter	H/m	$kg \cdot m/A^2 \cdot s^2$
magnetic flux density	tesla, weber per square meter	$T, Wb/m^2$	$kg/A \cdot s^2$
magnetic field strength	ampere per meter		A/m
magnetomotive force	ampere		A
luminous flux	lumen	lm	$cd \cdot sr$
luminance	candela per square meter		cd/m^2
illumination	lux, lumen per square meter	$lx, lm/m^2$	$cd \cdot sr/m^2$
activity (of radionuclides)	becquerel	Bq	s^{-1}
absorbed dose	gray	Gy	J/kg

*Supplementary units are: plane angle, radian (rad), solid angle, steradian (sr).

Table 3. Some common units defined in terms of SI units

Quantity	Name of unit	Unit symbol	Definition of unit
length	inch	in.	2.54×10^{-2} m
mass	pound (avoirdupois)	lb	0.45359237 kg
force	kilogram-force	kgf	9.80665 N
pressure	atmosphere	atm	101325 N·m^{-2}
pressure	torr	torr	$(101325/760)$ N·m^{-2}
pressure	conventional millimeter of mercury*	mmHg	$13.5951 \times 980.665 \times 10^{-2}$ N·m^{-2}
energy	kilowatt-hour	kWh	3.6×10^6 J
energy	thermochemical calorie	cal	4.184 J
energy	international steam table calorie	cal$_{IT}$	4.1868 J
thermodynamic temperature (T)	degree Rankine	°R	$(5/9)$ K
customary temperature (t)	degree Celsius	°C	$t(°C) = T(K) - 273.15$
customary temperature (t)	degree Fahrenheit	°F	$t(°F) = T(°R) - 459.67$
radioactivity	curie	Ci	3.7×10^{10} s^{-1}
energy†	electron volt	eV	eV $\approx 1.60219 \times 10^{-19}$ J
mass†	unified atomic mass unit	u	u $\approx 1.66057 \times 10^{-27}$ kg

*The conventional millimeter of mercury, symbol mmHg (not mm Hg), is the pressure exerted by a column exactly 1 mm high of a fluid of density exactly 13.5951 g· cm^{-3} in a place where the gravitational acceleration is exactly 980.665 cm· s^{-2}. The mmHg differs from the torr by less than 2×10^{-7} torr.

†These units defined in terms of the best available experimental values of certain physical constants may be converted to SI units. The factors for conversion of these units are subject to change in the light of new experimental measurements of the constants involved.

Table 4. Prefixes for units in the International System

Prefix	Symbol	Power	Example
tera	T	10^{12}	
giga	G	10^{9}	
mega	M	10^{6}	megahertz (MHz)
kilo	k	10^{3}	kilometer (km)
hecto	h	10^{2}	
deka	da	10^{1}	
deci	d	10^{-1}	
centi	c	10^{-2}	
milli	m	10^{-3}	milligram (mg)
micro	μ	10^{-6}	microgram (μg)
nano	n	10^{-9}	nanosecond (ns)
pico	p	10^{-12}	picofarad (pf)
femto	f	10^{-15}	
atto	a	10^{-18}	

Conversion factors for the measurement systems

Because it will take some years for all scientists and engineers to convert to the SI, this dictionary has often retained the U.S. Customary units, followed by parenthetical SI equivalents. Conversion factors between the three measurement systems are given in **Table 5** for some prevalent units; in each of the subtables the user proceeds as follows:

To convert a quantity expressed in a unit in the left-hand column to the equivalent in a unit in the top row of a subtable, multiply the quantity by the factor common to both units.

The factors have been carried out to seven significant figures, as derived from the fundamental constants and the definitions of the units. However, this does not mean that the factors are always known to that accuracy. Numbers followed by ellipses are to be continued indefinitely with repetition of the same pattern of digits. Factors written with fewer than seven significant digits are exact values.

Table 5. Conversion factors for the U.S. Customary System, metric system, and International System

A. UNITS OF LENGTH

Units	cm	m	in.	ft	yd	mile
1 cm =	1	0.01*	0.3937008	0.03280840	0.01093613	6.213712×10^{-6}
1 m =	100.	1	39.37008	3.280840	1.093613	6.213712×10^{-4}
1 in. =	2.54*	0.0254	1	0.08333333...	0.02777777....	1.578283×10^{-5}
1 ft =	30.48	0.3048	12.*	1	0.3333333...	$1.893939... \times 10^{-4}$
1 yd =	91.44	0.9144	36.	3.*	1	$5.681818... \times 10^{-4}$
1 mile =	1.609344×10^{5}	1.609344×10^{3}	6.336×10^{4}	5280.*	1760.	1

B. UNITS OF AREA

Units	cm²	m²	in.²	ft²	yd²	mile²
1 cm² =	1	10^{-4}*	0.1550003	1.076391×10^{-3}	1.195990×10^{-4}	3.861022×10^{-11}
1 m² =	10^{4}	1	1550.003	10.76391	1.195990	3.861022×10^{-7}
1 in.² =	6.4516*	6.4516×10^{-4}	1	$6.944444 \times 10^{-3}...$	7.716049×10^{-4}	2.490977×10^{-10}
1 ft² =	929.0304	0.09290304	144.*	1	0.1111111...	3.587007×10^{-8}
1 yd² =	8361.273	0.8361273	1296.	9.*	1	3.228306×10^{-7}
1 mile² =	2.589988×10^{10}	2.589988×10^{6}	4.014490×10^{9}	2.78784×10^{7}*	3.0976×10^{6}	1

Table 5. Conversion factors for the U.S. Customary System, metric system, and International System (cont.)

C. UNITS OF VOLUME

Units	m³	cm³	liter	in.³	ft³	qt	gal
1 m³	= 1	10^6	10^3	6.102374×10^4	35.31467	1.056688×10^3	264.1721
1 cm³	$= 10^{-6}$	1	10^{-3}	0.06102374	3.531467×10^{-5}	1.056688×10^{-3}	2.641721×10^{-4}
1 liter	$= 10^{-3}$	1000.*	1	61.02374	0.03531467	1.056688	0.2641721
1 in.³	$= 1.638706 \times 10^{-5}$	16.38706*	0.01638706	1	5.787037×10^{-4}	0.01731602	4.329004×10^{-3}
1 ft³	$= 2.831685 \times 10^{-2}$	28316.85	28.31685	1728.*	1	2.992208	7.480520
1 qt	$= 9.46353 \times 10^{-4}$	946.353	0.946353	57.75	0.0342014	1	0.25
1 gal (U.S.)	$= 3.785412 \times 10^{-3}$	3785.412	3.785412	231.*	0.1336806	4.*	1

D. UNITS OF MASS

Units	g	kg	oz	lb	metric ton	ton
1 g	= 1	10^{-3}	0.03527396	2.204623×10^{-3}	10^{-6}	1.102311×10^{-6}
1 kg	= 1000.	1	35.27396	2.204623	10^{-3}	1.102311×10^{-3}
1 oz (avdp)	= 28.34952	0.02834952	1	0.0625	2.834952×10^{-5}	$5. \times 10^{-4}$
1 lb (avdp)	= 453.5924	0.4535924	16.*	1	4.535924×10^{-4}	0.0005
1 metric ton	= 10^6	1000.*	35273.96	2204.623	1	1.102311
1 ton	= 907184.7	907.1847	32000.	2000.*	0.9071847	1

*For footnotes, see page A11.

continued

Table 5. Conversion factors for the U.S. Customary System, metric system, and International System (cont.)

E. UNITS OF DENSITY

Units	$g \cdot cm^{-3}$	$g \cdot l^{-1}, kg \cdot m^{-3}$	$oz \cdot in.^{-3}$	$lb \cdot in.^{-3}$	$lb \cdot ft^{-3}$	$lb \cdot gal^{-1}$
$1\ g \cdot cm^{-3}$	$= 1$	1000.	0.5780365	0.03612728	62.42795	8.345403
$1\ g \cdot l^{-1}, kg \cdot m^{-3}$	$= 10^{-3}$	1	5.780365×10^{-4}	3.612728×10^{-5}	0.06242795	8.345403×10^{-3}
$1\ oz \cdot in.^{-3}$	$= 1.729994$	1729.994	1	0.0625	108.	14.4375
$1\ lb \cdot in.^{-3}$	$= 27.67991$	27679.91	16.	1	1728.	231.
$1\ lb \cdot ft^{-3}$	$= 0.01601847$	16.01847	9.259259×10^{-3}	5.7870370×10^{-4}	1	0.1336806
$1\ lb \cdot gal^{-1}$	$= 0.1198264$	119.8264	4.749536×10^{-3}	4.3290043×10^{-3}	7.480519	1

F. UNITS OF PRESSURE

Units	$Pa, N \cdot m^{-2}$	$dyn \cdot cm^{-2}$	bar	atm	$kg(wt) \cdot cm^{-2}$	mmHg (torr)	in. Hg	$lb(wt) \cdot in.^{-2}$
$1\ Pa, 1\ N \cdot m^{-2}$	$= 1$	10	10^{-5}	9.869233×10^{-6}	1.019716×10^{-5}	7.500617×10^{-3}	2.952999×10^{-4}	1.450377×10^{-4}
$1\ dyn \cdot cm^{-2}$	$= 0.1$	1	10^{-6}	9.869233×10^{-7}	1.019716×10^{-6}	7.500617×10^{-4}	2.952999×10^{-5}	1.450377×10^{-5}
$1\ bar$	$= 10^{6*}$	10^6	1	0.9869233	1.019716	750.0617	29.52999	14.50377
$1\ atm$	$= 101325.0^*$	1013250.	1.013250	1	1.033227	760.	29.92126	14.69595
$1\ kg(wt) \cdot cm^{-2}$	$= 98066.5$	980665.	0.980665	0.9678411	1	735.5592	28.95903	14.22334
$1\ mmHg(torr)$	$= 133.3224$	1333.224	1.333224×10^{-3}	1.3157895×10^{-3}	1.3595099×10^{-3}	1	0.03937008	0.01933678
$1\ in.\ Hg$	$= 3386.388$	33863.88	0.03386388	0.03342105	0.03453155	25.4	1	0.4911541
$1\ lb(wt) \cdot in.^{-2}$	$= 6894.757$	68947.57	0.06894757	0.06804596	0.07030696	51.71493	2.036021	1

Table 5. Conversion factors for the U.S. Customary System, metric system, and International System (cont.)

G. UNITS OF ENERGY†

Units	g mass‡	J	int J	cal	cal$_{IT}$	Btu$_{IT}$	k Wh	hp hr	ft-lb (wt)	cu ft-lb (wt) in.²	atm
1 g mass‡ =1	1	8.987552×10^{13}	8.986069×10^{13}	2.148076×10^{13}	2.146640×10^{13}	8.518555×10^{10}	2.496542×10^{7}	3.347918×10^{7}	6.628878×10^{13}	4.603388×10^{11}	8.870024×10^{11}
1 J	$=1.112650 \times 10^{-14}$	1	0.999835	0.2390057	0.2388459	9.478172×10^{-4}	$2.777777... \times 10^{-7}$	3.725062×10^{-7}	0.7375622	5.121960×10^{-3}	9.869233×10^{-3}
1 int J	$=1.112834 \times 10^{-14}$	1.000165	1	0.2390452	0.2388853	9.479735×10^{-4}	2.778236×10^{-7}	3.725676×10^{-7}	0.7376839	5.122805×10^{-3}	9.870862×10^{-3}
1 cal	$=4.655328 \times 10^{-14}$	4.184*	4.183310	1	0.9993312	3.965667×10^{-3}	$1.1622222... \times 10^{-6}$	1.558562×10^{-6}	3.085960	2.143028×10^{-2}	0.04129287
1 cal$_{IT}$	$=4.658443 \times 10^{-14}$	4.1868*	4.186109	1.000669	1	3.968321×10^{-3}	1.16300×10^{-6}	1.559609×10^{-6}	3.088025	2.144462×10^{-2}	0.04132050
1 Btu$_{IT}$	$=1.173908 \times 10^{-11}$	1055.056	1054.882	252.1644	251.9958*	1	2.930711×10^{-4}	3.930148×10^{-4}	778.1693	5.403953	10.41259
1 k Wh	$=4.005540 \times 10^{-8}$	3600000.*	3599406.	860420.7	859845.2	3412.142	1	1.341022	2655224.	18439.06	35529.24
1 hp hr	$=2.986931 \times 10^{-8}$	2684519.	2684077.	641615.6	641186.5	2544.33	0.7456998	1	1980000.*	13750.	26494.15
1 ft-lb (wt)	$=1.508551 \times 10^{-14}$	1.355818	1.355594	0.3240483	0.3238315	1.285067×10^{-3}	3.766161×10^{-7}	$5.050505... \times 10^{-7}$	1	$6.944444... \times 10^{-3}$	0.01338088
1 cu ft-lb (wt) in.²	$=2.172313 \times 10^{-12}$	195.2378	195.2056	46.66295	46.63174	0.1850497	5.423272×10^{-5}	$7.272727... \times 10^{-5}$	144.*	1	1.926847
1 liter-atm	$=1.127393 \times 10^{-12}$	101.3250	101.3083	24.21726	24.20106	0.09603757	2.814583×10^{-5}	3.774419×10^{-5}	74.73349	0.5189825	1

*Numbers followed by an asterisk are definitions of the relation between the two units. †The electrical units are those in terms of which certification of standard cells, standard resistances, and so forth, is made by the National Bureau of Standards. Unless otherwise indicated, all electrical units are absolute. ‡Energy equivalent.

Fundamental constants

Fundamental constants

Quantity	Symbol	Numerical Value *	Uncert. (ppm)	SI †	← Units →	cgs ‡
Speed of light in vacuum	c	299792458(1.2)	0.004	$m \cdot s^{-1}$		$10^2 \ cm \cdot s^{-1}$
Permeability of vacuum	μ_0	4π		$10^{-7} \ H \cdot m^{-1}$		
		$=12.5663706144$		$10^{-7} \ H \cdot m^{-1}$		
Permittivity of vacuum, $1/\mu_0 c^2$	ϵ_0	8.854187818(71)	0.008	$10^{-12} \ F \cdot m^{-1}$		
Fine-structure constant, $[\mu_0 c^2/4\pi](e^2/\hbar c)$	α α^{-1}	7.2973506(60) 137.03604(11)	0.82 0.82	10^{-3}		10^{-3}
Elementary charge	e	1.6021892(46) 4.803242(14)	2.9 2.9	$10^{-19} \ C$		10^{-20} emu 10^{-10} esu
Planck constant	h $\hbar = h/2\pi$	6.626176(36) 1.0545887(57)	5.4 5.4	$10^{-34} \ J \cdot s$ $10^{-34} \ J \cdot s$		10^{-27} erg·s 10^{-27} erg·s
Avogadro constant	N_A	6.022045(31)	5.1	$10^{23} \ mol^{-1}$		$10^{23} \ mol^{-1}$
Atomic mass unit, $10^{-3} kg \cdot mol^{-1} N_A^{-1}$	u	1.6605655(86)	5.1	$10^{-27} \ kg$		$10^{-24} \ g$
Electron rest mass	m_e	9.109534(47) 5.4858026(21)	5.1 0.38	$10^{-31} \ kg$ $10^{-4} \ u$		$10^{-28} \ g$ $10^{-4} \ u$
Proton rest mass	m_p	1.6726485(86) 1.007276470(11)	5.1 0.011	$10^{-27} \ kg$ u		$10^{-24} \ g$ u
Ratio of proton mass to electron mass	m_p/m_e	1836.15152(70)	0.38			
Neutron rest mass	m_n	1.6749543(86) 1.008665012(37)	5.1 0.037	$10^{-27} \ kg$ u		$10^{-24} \ g$ u
Electron charge to mass ratio	e/m_e	1.7588047(49) 5.272764(15)	2.8 2.8	$10^{11} \ C \cdot kg^{-1}$		$10^7 \ emu \cdot g^{-1}$ $10^{11} \ esu \cdot g^{-1}$
Magnetic flux quantum, $[c]^{-1}(hc/2e)$	Φ_0 h/e	2.0678506(54) 4.135701(11) 1.3795215(36)	2.6 2.6 2.6	$10^{-15} \ Wb$ $10^{-15} \ J \cdot s \cdot C^{-1}$		$10^{-7} \ G \cdot cm^2$ $10^{-7} \ erg \cdot s \cdot emu^{-1}$ $10^{-17} \ erg \cdot s \cdot esu^{-1}$
Josephson frequency-voltage ratio	$2e/h$	4.835939(13)	2.6	$10^{14} \ Hz \cdot V^{-1}$		
Quantum of circulation	$h/2m_e$ h/m_e	3.6369455(60) 7.273891(12)	1.6 1.6	$10^{-4} \ J \cdot s \cdot kg^{-1}$ $10^{-4} \ J \cdot s \cdot kg^{-1}$		$erg \cdot s \cdot g^{-1}$ $erg \cdot s \cdot g^{-1}$
Faraday constant, $N_A e$	F	9.648456(27) 2.8925342(82)	2.8 2.8	$10^4 \ C \cdot mol^{-1}$		$10^3 \ emu \cdot mol^{-1}$ $10^{14} \ esu \cdot mol^{-1}$
Rydberg constant, $[\mu_0 c^2/4\pi]^2(m_e e^4/4\pi \hbar^3 c)$	R_∞	1.097373177(83)	0.075	$10^7 \ m^{-1}$		$10^5 \ cm^{-1}$
Bohr radius, $[\mu_0 c^2/4\pi]^{-1}(\hbar^2/m_e e^2) = \alpha/4\pi R_\infty$	a_0	5.2917706(44)	0.82	$10^{-11} \ m$		$10^{-9} \ cm$
Classical electron radius, $[\mu_0 c^2/4\pi](e^2/m_e c^2) = \alpha^3/4\pi R_\infty$	$r_e = \alpha \lambda_C$	2.8179380(70)	2.5	$10^{-15} \ m$		$10^{-13} \ cm$
Thomson cross section, $(8/3)\pi r_e^2$	σ_e	0.6652448(33)	4.9	$10^{-28} \ m^2$		$10^{-24} \ cm^2$
Free electron g-factor, or electron magnetic moment in Bohr magnetons	$g_e/2 = \mu_e/\mu_B$	1.0011596567(35)	0.0035			
Free muon g-factor, or muon magnetic moment in units of $[c](e\hbar/2m_\mu c)$	$g_\mu/2$	1.00116616(31)	0.31			
Bohr magneton, $[c](e\hbar/2m_e c)$	μ_B	9.274078(36)	3.9	$10^{-24} \ J \cdot T^{-1}$		$10^{-21} \ erg \cdot G^{-1}$
Electron magnetic moment	μ_e	9.284832(36)	3.9	$10^{-24} \ J \cdot T^{-1}$		$10^{-21} \ erg \cdot G^{-1}$

*For footnotes, see page A14.

Fundamental constants (cont.)

Quantity	Symbol	Numerical Value *	Uncert. (ppm)	SI †	← Units →	cgs ‡
Gyromagnetic ratio of protons in H_2O	γ'_p	2.6751301(75)	2.8	10^8 $s^{-1} \cdot T^{-1}$		10^4 $s^{-1} \cdot G^{-1}$
	$\gamma'_p/2\pi$	4.257602(12)	2.8	10^7 $Hz \cdot T^{-1}$		10^3 $Hz \cdot G^{-1}$
γ'_p corrected for diamagnetism of H_2O	γ_p	2.6751987(75)	2.8	10^8 $s^{-1} \cdot T^{-1}$		10^4 $s^{-1} \cdot G^{-1}$
	$\gamma_p/2\pi$	4.257711(12)	2.8	10^7 $Hz \cdot T^{-1}$		10^3 $Hz \cdot G^{-1}$
Magnetic moment of protons in H_2O in Bohr magnetons	μ'_p/μ_B	1.52099322(10)	0.066	10^{-3}		10^{-3}
Proton magnetic moment in Bohr magnetons	μ_p/μ_B	1.521032209(16)	0.011	10^{-3}		10^{-3}
Ratio of electron and proton magnetic moments	μ_e/μ_p	658.2106880(66)	0.010			
Proton magnetic moment	μ_p	1.4106171(55)	3.9	10^{-26} $J \cdot T^{-1}$		10^{-23} $erg \cdot G^{-1}$
Magnetic moment of protons in H_2O in nuclear magnetons	μ'_p/μ_N	2.7927740(11)	0.38			
μ'_p/μ_N corrected for diamagnetism of H_2O	μ_p/μ_N	2.7928456(11)	0.38			
Nuclear magneton, $[c](e\hbar/2m_pc)$	μ_N	5.050824(20)	3.9	10^{-27} $J \cdot T^{-1}$		10^{-24} $erg \cdot G^{-1}$
Ratio of muon and proton magnetic moments	μ_μ/μ_p	3.1833402(72)	2.3			
Muon magnetic moment	μ_μ	4.490474(18)	3.9	10^{-26} $J \cdot T^{-1}$		10^{-23} $erg \cdot G^{-1}$
Ratio of muon mass to electron mass	m_μ/m_e	206.76865(47)	2.3			
Muon rest mass	m_μ	1.883566(11)	5.6	10^{-28} kg		10^{-25} g
		0.11342920(26)	2.3	u		u
Compton wavelength of the electron, $h/m_ec=\alpha^2/2R_\infty$	λ_C	2.4263089(40)	1.6	10^{-12} m		10^{-10} cm
	$\lambdabar_C=\lambda_C/2\pi=\alpha a_0$	3.8615905(64)	1.6	10^{-13} m		10^{-11} cm
Compton wavelength of the proton, h/m_pc	$\lambda_{C,p}$	1.3214099(22)	1.7	10^{-15} m		10^{-13} cm
	$\lambdabar_{C,p}=\lambda_{C,p}/2\pi$	2.1030892(36)	1.7	10^{-16} m		10^{-14} cm
Compton wavelength of the neutron, h/m_nc	$\lambda_{C,n}$	1.3195909(22)	1.7	10^{-15} m		10^{-13} cm
	$\lambdabar_{C,n}=\lambda_{C,n}/2\pi$	2.1001941(35)	1.7	10^{-16} m		10^{-14} cm
Molar volume of ideal gas at s.t.p.	V_m	22.41383(70)	31	10^{-3} $m^3 \cdot mol^{-1}$		10^3 $cm^3 \cdot mol^{-1}$
Molar gas constant, V_mp_0/T_0	R	8.31441(26)	31	$J \cdot mol^{-1} \cdot K^{-1}$		10^7 $erg \cdot mol^{-1} \cdot K^{-1}$
($T_0 \equiv 273.15$ K; $p_0 \equiv 101325$ Pa$\equiv$1atm)		8.20568(26)	31	10^{-5} $m^3 \cdot atm \cdot mol^{-1} \cdot K^{-1}$		10 $cm^3 \cdot atm \cdot mol^{-1} \cdot K^{-1}$
Boltzmann constant, R/N_A	k	1.380662(44)	32	10^{-23} $J \cdot K^{-1}$		10^{-16} $erg \cdot K^{-1}$
Stefan-Boltzmann constant, $\pi^2k^4/60\hbar^3c^2$	σ	5.67032(71)	125	10^{-8} $W \cdot m^{-2} \cdot K^{-4}$		10^{-5} $erg \cdot s^{-1} \cdot cm^{-2} \cdot K^{-4}$
First radiation constant, $2\pi hc^2$	c_1	3.741832(20)	5.4	10^{-16} $W \cdot m^2$		10^{-5} $erg \cdot cm^2 \cdot s^{-1}$
Second radiation constant, hc/k	c_2	1.438786(45)	31	10^{-2} $m \cdot K$		$cm \cdot K$
Gravitational constant	G	6.6720(41)	615	10^{-11} $m^3 \cdot s^{-2} \cdot kg^{-1}$		10^{-8} $cm^3 \cdot s^{-2} \cdot g^{-1}$
Ratio, kx-unit to ångström, $\Lambda=\lambda(\text{Å})/\lambda(\text{kxu})$; $\lambda(CuK\alpha_1) \equiv 1.537400$ kxu	Λ	1.0020772(54)	5.3			
Ratio, Å* to ångström, $\Lambda^*=\lambda(\text{Å})/\lambda(\text{Å}^*)$; $\lambda(WK\alpha_1) \equiv 0.2090100$ Å*	Λ^*	1.0000205(56)	5.6			

continued

Fundamental constants (cont.): Energy conversion factors and equivalents

Quantity	Symbol	Numerical Value *	Units	Uncert. (ppm)
1 kilogram ($kg \cdot c^2$)		8.987551786(72)	10^{16} J	0.008
		5.609545(16)	10^{29} MeV	2.9
1 Atomic mass unit ($u \cdot c^2$)		1.4924418(77)	10^{-10} J	5.1
		931.5016(26)	MeV	2.8
1 Electron mass $m_e \cdot c^2$)		8.187241(42)	10^{-14} J	5.1
		0.5110034(14)	MeV	2.8
1 Muon mass ($m_\mu \cdot c^2$)		1.6928648(96)	10^{-11} J	5.6
		105.65948(35)	MeV	3.3
1 Proton mass ($m_p \cdot c^2$)		1.5033015(77)	10^{-10} J	5.1
		938.2796(27)	MeV	2.8
1 Neutron mass ($m_n \cdot c^2$)		1.5053738(78)	10^{-10} J	5.1
		939.5731(27)	MeV	2.8
1 Electron volt		1.6021892(46)	10^{-19} J	2.9
			10^{-12} erg	2.9
	1 eV/h	2.4179696(63)	10^{14} Hz	2.6
	1 eV/hc	8.065479(21)	10^5 m^{-1}	2.6
			10^3 cm^{-1}	2.6
	1 eV/k	1.160450(36)	10^4 K	31
Voltage-wavelength conversion, hc		1.986478(11)	10^{-25} $J \cdot m$	5.4
		1.2398520(32)	10^{-6} $eV \cdot m$	2.6
			10^{-4} $eV \cdot cm$	2.6
Rydberg constant	$R_\infty hc$	2.179907(12)	10^{-18} J	5.4
			10^{-11} erg	5.4
		13.605804(36)	eV	2.6
	$R_\infty c$	3.28984200(25)	10^{15} Hz	0.075
	$R_\infty hc/k$	1.578885(49)	10^5 K	31
Bohr magneton	μ_B	9.274078(36)	10^{-24} $J \cdot T^{-1}$	3.9
		5.7883785(95)	10^{-5} $eV \cdot T^{-1}$	1.6
	μ_B/h	1.3996123(39)	10^{10} $Hz \cdot T^{-1}$	2.8
	μ_B/hc	46.68604(13)	$m^{-1} \cdot T^{-1}$	2.8
			10^{-2} $cm^{-1} \cdot T^{-1}$	2.8
	μ_B/k	0.671712(21)	$K \cdot T^{-1}$	31
Nuclear magneton	μ_N	5.505824(20)	10^{-27} $J \cdot T^{-1}$	3.9
		3.1524515(53)	10^{-8} $eV \cdot T^{-1}$	1.7
	μ_N/h	7.622532(22)	10^6 $Hz \cdot T^{-1}$	2.8
	μ_N/hc	2.5426030(72)	10^{-2} $m^{-1} \cdot T^{-1}$	2.8
			10^{-4} $cm^{-1} \cdot T^{-1}$	2.8
	μ_N/k	3.65826(12)	10^{-4} $K \cdot T^{-1}$	31

* Note that the numbers in parentheses are the one standard-deviation uncertainties in the last digits of the quoted value computed on the basis of internal consistency, that the unified atomic mass scale $^{12}C \triangleq 12$ has been used throughout, that u=atomic mass unit, C=coulomb, F=farad, G=gauss, H=henry, Hz=hertz=cycle/s, J=joule, K=kelvin (degree Kelvin), Pa=pascal=N•m⁻², T=tesla (10^4 G), V=volt, Wb=weber= T•m², and W=watt. In cases where formulas for constants are given (e.g., R_∞), the relations are written as the product of two factors. The second factor, in parentheses, is the expression to be used when all quantities are expressed in cgs units, with the electron charge in electrostatic units. The first factor, in brackets, is to be included only if all quantities are expressed in SI units. We remind the reader that with the exception of the auxiliary constants which have been taken to be exact, the uncertainties of these constants are correlated, and therefore the general law of error propagation must be used in calculating additional quantities requiring two or more of these constants.

† Quantities given in u and atm are for the convenience of the reader; these units are not part of the International System of Units (SI).

‡ In order to avoid separate columns for "electromagnetic" and "electrostatic" units, both are given under the single heading "cgs Units." When using these units, the elementary charge e in the second column should be understood to be replaced by e_m or e_s, respectively.

SOURCE: Compiled by E. R. Cohen and B. N. Taylor under the auspices of the CODATA Task Group on Fundamental Constants. This set has been officially adopted by CODATA and is taken from *J. Phys. Chem. Ref. Data*, 2(4):663, 1973, and CODATA Bulletin No. 11, December 1973.

Special mathematical constants

π $= 3.14159\ 26535\ 89793\ 23846\ 2643\ldots$

e $= 2.71828\ 18284\ 59045\ 23536\ 0287\ldots = \lim\limits_{n\to\infty}\left(1+\dfrac{1}{n}\right)^{n}$
$=$ natural base of logarithms

$\sqrt{2}$ $= 1.41421\ 35623\ 73095\ 0488\ldots$

$\sqrt{3}$ $= 1.73205\ 08075\ 68877\ 2935\ldots$

$\sqrt{5}$ $= 2.23606\ 79774\ 99789\ 6964\ldots$

$\sqrt[3]{2}$ $= 1.25992\ 1050\ldots$

$\sqrt[3]{3}$ $= 1.44224\ 9570\ldots$

$\sqrt[5]{2}$ $= 1.14869\ 8355\ldots$

$\sqrt[5]{3}$ $= 1.24573\ 0940\ldots$

e^{π} $= 23.14069\ 26327\ 79269\ 006\ldots$

π^{e} $= 22.45915\ 77183\ 61045\ 47342\ 715\ldots$

e^{e} $= 15.15426\ 22414\ 79264\ 190\ldots$

$\log_{10} 2$ $= 0.30102\ 99956\ 63981\ 19521\ 37389\ldots$

$\log_{10} 3$ $= 0.47712\ 12547\ 19662\ 43729\ 50279\ldots$

$\log_{10} e$ $= 0.43429\ 44819\ 03251\ 82765\ldots$

$\log_{10} \pi$ $= 0.49714\ 98726\ 94133\ 85435\ 12683\ldots$

$\log_{e} 10$ $= \ln 10 = 2.30258\ 50929\ 94045\ 68401\ 7991\ldots$

$\log_{e} 2$ $= \ln 2 = 0.69314\ 71805\ 59945\ 30941\ 7232\ldots$

$\log_{e} 3$ $= \ln 3 = 1.09861\ 22886\ 68109\ 69139\ 5245\ldots$

γ $= 0.57721\ 56649\ 01532\ 86060\ 6512\ldots =$ Euler's constant
$= \lim\limits_{n\to\infty}\left(1+\dfrac{1}{2}+\dfrac{1}{3}+\cdots+\dfrac{1}{n}-\ln n\right)$

e^{γ} $= 1.78107\ 24179\ 90197\ 9852\ldots$

$\sqrt{e}$ $= 1.64872\ 12707\ 00128\ 1468\ldots$

$\sqrt{\pi}$ $= \Gamma(\tfrac{1}{2}) = 1.77245\ 38509\ 05516\ 02729\ 8167\ldots$
where Γ is the gamma function

$\Gamma(\tfrac{1}{3})$ $= 2.67893\ 85347\ 07748\ldots$

$\Gamma(\tfrac{1}{4})$ $= 3.62560\ 99082\ 21908\ldots$

1 radian $= 180°/\pi = 57.29577\ 95130\ 8232\ldots°$

$1°$ $= \pi/180$ radians $= 0.01745\ 32925\ 19943\ 2957\ldots$ radians

Schematic electronic symbols*

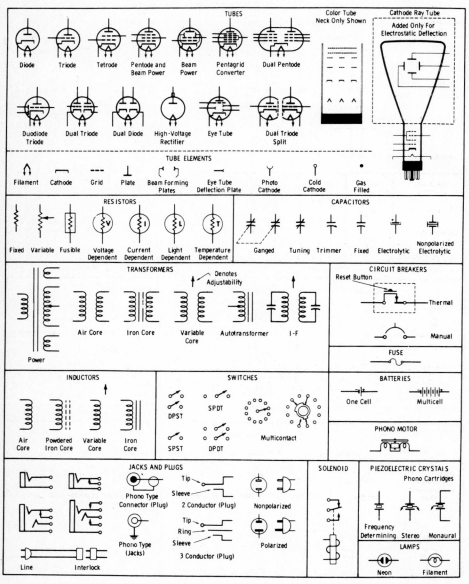

*From R. F. Graf, *Modern Dictionary of Electronics*, 4th ed., Indianapolis: Howard Sams, 1972.

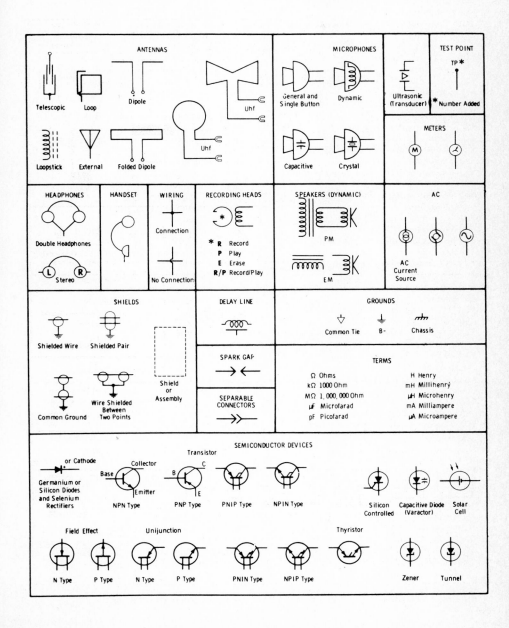

ANTENNAS

Telescopic Loop Dipole Uhf

Loopstick External Folded Dipole Uhf

MICROPHONES

General and Single Button Dynamic

Capacitive Crystal

TEST POINT

TP *

Ultrasonic (Transducer) * Number Added

METERS

HEADPHONES

Double Headphones

Stereo

HANDSET

WIRING

Connection

No Connection

RECORDING HEADS

* R Record
P Play
E Erase
R/P Record/Play

SPEAKERS (DYNAMIC)

PM

EM

AC

AC Current Source

SHIELDS

Shielded Wire Shielded Pair

Shield or Assembly

Common Ground Wire Shielded Between Two Points

DELAY LINE

SPARK GAP

SEPARABLE CONNECTORS

GROUNDS

Common Tie B- Chassis

TERMS

Ω Ohms
kΩ 1000 Ohm
MΩ 1,000,000 Ohm
µF Microfarad
pF Picofarad

H Henry
mH Millihenry
µH Microhenry
mA Milliampere
µA Microampere

SEMICONDUCTOR DEVICES

Transistor

or Cathode

Germanium or Silicon Diodes and Selenium Rectifiers

Base Collector C
Emitter
NPN Type

B C
E
PNP Type

PNIP Type

NPIN Type

Silicon Controlled Capacitive Diode (Varactor) Solar Cell

Field Effect

N Type P Type

Unijunction

N Type P Type

PNIN Type NPIP Type

Thyristor

Zener Tunnel

Semiconductor symbols and abbreviations*

A, a — Anode

B, b — Base

b_{fs} — Common-source small-signal forward transfer susceptance

b_{is} — Common-source small-signal input susceptance

b_{os} — Common-source small-signal output susceptance

b_{rs} — Common-source small-signal reverse transfer susceptance

C, c — Collector

C_{cb} — Collector-base interterminal capacitance

C_{ce} — Collector-emitter interterminal capacitance

C_{ds} — Drain-source capacitance

C_{du} — Drain-substrate capacitance

C_{eb} — Emitter-base interterminal capacitance

C_{ibo} — Common-base open-circuit input capacitance

C_{ibs} — Common-base short-circuit input capacitance

C_{ieo} — Common-emitter open-circuit input capacitance

C_{ies} — Common-emitter short-circuit input capacitance

C_{iss} — Common-source short-circuit input capacitance

C_{obo} — Common-base open-circuit output capacitance

C_{obs} — Common-base short-circuit output capacitance

C_{oeo} — Common-emitter open-circuit output capacitance

C_{oes} — Common-emitter short-circuit output capacitance

C_{oss} — Common-source short-circuit output capacitance

C_{rbs} — Common-base short-circuit reverse transfer capacitance

C_{rcs} — Common-collector short-circuit reverse transfer capacitance

C_{res} — Common-emitter short-circuit reverse transfer capacitance

C_{rss} — Common-source short-circuit reverse transfer capacitance

C_{tc} — Collector depletion-layer capacitance

C_{te} — Emitter depletion-layer capacitance

D, d — Drain

E, e — Emitter

η — Intrinsic standoff ratio

f_{hfb} — Common-base small-signal short-circuit forward current transfer ratio

cutoff frequency

f_{hfc} — Common-collector small-signal short-circuit forward current transfer ratio cutoff frequency

f_{hfe} — Common-emitter small-signal short-circuit forward current transfer ratio cutoff frequency

f_{max} — Maximum frequency of oscillation

f_T — Transition frequency (frequency at which common-emitter small-signal forward current transfer ratio extrapolates to unity)

G, g — Gate

g_{fs} — Common-source small-signal forward transfer conductance

g_{is} — Common-source small-signal input conductance

g_{MB} — Common-base static transconductance

g_{MC} — Common-collector static transconductance

g_{ME} — Common-emitter static transconductance

g_{os} — Common-source small-signal output conductance

G_{PB} — Common-base large-signal insertion power gain

G_{pb} — Common-base small-signal insertion power gain

G_{PC} — Common-collector large-signal insertion power gain

G_{pc} — Common-collector small-signal insertion power gain

G_{PE} — Common-emitter large-signal insertion power gain

G_{pe} — Common-emitter small-signal insertion power gain

G_{pg} — Common-gate small-signal insertion power gain

G_{ps} — Common-source small-signal insertion power gain

g_{rs} — Common-source small-signal reverse transfer conductance

G_{TB} — Common-base large-signal transducer power gain

G_{tb} — Common-base small-signal transducer power gain

G_{TC} — Common-collector large-signal transducer power gain

G_{tc} — Common-collector small-signal transducer power gain

G_{TE} — Common-emitter large-signal transducer power gain

*Recommended by the Joint Electron Device Engineering Council of the Electronic Industries Association and the National Electrical Manufacturers Assocation for use in semiconductor device data sheets and specifications.
From R. F. Graf, *Modern Dictionary of Electronics*, 4th ed., Indianapolis: Howard Sams, 1972.

$\mathbf{G_{te}}$—Common-emitter small-signal transducer power gain

$\mathbf{G_{tg}}$—Common-gate small-signal transducer power gain

$\mathbf{G_{ts}}$—Common-source small-signal transducer power gain

$\mathbf{h_{FB}}$—Common-base static forward current transfer ratio

$\mathbf{h_{fb}}$—Common-base small-signal short-circuit forward current transfer ratio

$\mathbf{h_{FC}}$—Common-collector static forward current transfer ratio

$\mathbf{h_{fc}}$—Common-collector small-signal short-circuit forward current transfer ratio

$\mathbf{h_{FE}}$—Common-emitter static forward current transfer ratio

$\mathbf{h_{fe}}$—Common-emitter small-signal short-circuit forward current transfer ratio

$\mathbf{h_{FEL}}$—Inherent large-signal forward current transfer ratio

$\mathbf{h_{IB}}$—Common-base static input resistance

$\mathbf{h_{ib}}$—Common-base small-signal short-circuit input impedance

$\mathbf{h_{IC}}$—Common-collector statis input resistance

$\mathbf{h_{ic}}$—Common-collector small-signal short-circuit input impedance

$\mathbf{h_{IE}}$—Common-emitter static input resistance

$\mathbf{h_{ie}}$—Common-emitter small-signal short-circuit input impedance

$\mathbf{h_{ie(imag)}}$—Imaginary part of common-emitter small-signal short-circuit input impedance

$\mathbf{h_{ie(real)}}$—Real part of common-emitter small-signal short-circuit input impedance

$\mathbf{h_{ob}}$—Common-base small-signal open-circuit output admittance

$\mathbf{h_{oc}}$—Common-collector small-signal open-circuit output admittance

$\mathbf{h_{oe}}$—Common-emitter small-signal open-circuit output admittance

$\mathbf{h_{oe(imag)}}$—Imaginary part of common-emitter small-signal open-circuit output admittance

$\mathbf{h_{oe(real)}}$—Real part of common-emitter small-signal open-circuit output admittance

$\mathbf{h_{rb}}$—Common-base small-signal open-circuit reverse voltage transfer ratio

$\mathbf{h_{rc}}$—Common-collector small-signal open-circuit reverse voltage transfer ratio

$\mathbf{h_{ro}}$—Common-emitter small-signal open-circuit reverse voltage transfer ratio

$\mathbf{I_B}$—Base-terminal dc current

$\mathbf{I_b}$—Alternating component (rms value) of base-terminal current

$\mathbf{i_B}$—Instantaneous total value of base-terminal current

$\mathbf{I_{BEV}}$—Base cutoff current, dc

$\mathbf{I_{B2(mod)}}$—Interbase modulated current

$\mathbf{I_C}$—Collector-terminal dc current

$\mathbf{I_c}$—Alternating component (rms value) of collector-terminal current

$\mathbf{i_C}$—Instantaneous total value of collector-terminal current

$\mathbf{I_{CBO}}$—Collector cutoff current (dc), emitter open

$\mathbf{I_{CEO}}$—Collector cutoff current (dc), base open

$\mathbf{I_{CER}}$—Collector cutoff current (dc), specified resistance between base and emitter

$\mathbf{I_{CES}}$—Collector cutoff current (dc), base shorted to emitter

$\mathbf{I_{CEV}}$—Collector cutoff current (dc), specified voltage between base and emitter

$\mathbf{I_{CEX}}$—Collector cutoff current (dc), specified circuit between base and emitter

$\mathbf{I_D}$—Drain current; dc

$\mathbf{I_{D(off)}}$—Drain cutoff current

$\mathbf{I_{D(on)}}$—On-state drain current

$\mathbf{I_{DSS}}$—Zero-gate-voltage drain current

$\mathbf{I_E}$—Emitter-terminal dc current

$\mathbf{I_e}$—Alternating component (rms value) of emitter-terminal current

$\mathbf{i_E}$—Instantaneous total value of emitter-terminal current

$\mathbf{I_{EBO}}$—Emitter cutoff current (dc), collector open

$\mathbf{I_{EB20}}$—Emitter reverse current

$\mathbf{I_{EC(ofs)}}$—Emitter-collector offset current

$\mathbf{I_{ECS}}$—Emitter cutoff current (dc), base short-circuited to collector

$\mathbf{I_{EIE2(off)}}$—Emitter cutoff current

$\mathbf{I_F}$—For voltage-regulator and voltage-reference diodes: dc forward current. For signal diodes and rectifier diodes: dc forward current (no alternating component)

$\mathbf{I_f}$—Alternating component of forward current (rms value)

$\mathbf{i_F}$—Instantaneous total forward current

$\mathbf{I_{F(AV)}}$—Forward current, dc (with alternating component)

$\mathbf{I_{FM}}$—Maximum (peak) total forward current

$\mathbf{I_{F(OV)}}$—Forward current, overload

$\mathbf{I_{FRM}}$—Maximum (peak) forward current, repetitive

$\mathbf{I_{F(RMS)}}$—Total rms forward current

$\mathbf{I_{FSM}}$—Maximum (peak) forward current, surge

$\mathbf{I_G}$—Gate current, dc

$\mathbf{I_{GF}}$—Forward gate current

$\mathbf{I_{GR}}$—Reverse gate current

$\mathbf{I_{GSS}}$—Reverse gate current, drain short-

continued

I_{GSSF} — Forward gate current, drain short-circuited to source

I_{GSSR} — Reverse gate current, drain short-circuited to source

I_I — Inflection-point current

$Im(h_{oe})$ — Imaginary part of common-emitter small-signal short-circuit input impedance

$Im(h_{oe})$ — Imaginary part of common-emitter small-signal open-circuit output admittance

I_O — Average forward current, 180° conduction angle, 60-Hz half sine wave

I_P — Peak-point current

I_R — For voltage-regulator and voltage-reference diodes: dc reverse current. For signal diodes and rectifier diodes: dc reverse current (no alternating component)

I_r — Alternating component of reverse current (rms value)

i_R — Instantaneous total reverse current

$I_{R(AV)}$ — Reverse current, dc (with alternating component)

I_{RM} — Maximum (peak) total reverse current

I_{RRM} — Maximum (peak) reverse current, repetitive

$I_{R(RMS)}$ — Total rms reverse current

I_{RSM} — Maximum (peak) surge reverse current

I_S — Source current, dc

I_{SDS} — Zero-gate-voltage source current

$I_{S(off)}$ — Source cutoff current

I_V — Valley-point current

I_Z — Regulator current, reference current (dc)

I_{ZK} — Regulator current, reference current (dc near breakdown knee)

I_{ZM} — Regulator current, reference current (dc maximum rated current)

K, k — Cathode

L_c — Conversion loss

M — Figure of merit

NF_o — Overall noise figure

NR_o — Output noise ratio

P_{BE} — Power input (dc) to base, common emitter

p_{BE} — Instantaneous total power input to base, common emitter

P_{CB} — Power input (dc) to collector, common base

p_{CB} — Instantaneous total power input to collector, common base

P_{CE} — Power input (dc) to collector, common emitter

p_{CE} — Instantaneous total power input to collector, common emitter

P_{EB} — Power input (dc) to emitter, common base

p_{EB} — Instantaneous total power input to emitter, common base

P_F — Forward power dissipation, dc (no alternating component)

p_F — Instantaneous total forward power dissipation

$P_{F(AV)}$ — Forward power dissipation, dc (with alternating component)

P_{FM} — Maximum (peak) total forward power dissipation

P_{IB} — Common-base large-signal input power

p_{ib} — Common-base small-signal input power

P_{IC} — Common-collector large-signal input power

p_{ic} — Common collector small-signal input power

P_{IE} — Common-emitter large-signal input power

p_{ie} — Common-emitter small-signal input power

P_{OB} — Common-base large-signal output power

p_{ob} — Common-base small-signal output power

P_{OC} — Common-collector large-signal output power

p_{oc} — Common-collector small-signal output power

P_{OE} — Common-emitter large-signal output power

p_{oe} — Common-emitter small-signal output power

P_R — Reverse power dissipation, dc (no alternating component)

p_R — Instantaneous total reverse power dissipation

$P_{R(AV)}$ — Reverse power dissipation, dc (with alternating component)

P_{RM} — Maximum (peak) total reverse power dissipation

P_T — Total nonreactive power input to all terminals

p_T — Nonreactive power input, instantaneous total, to all terminals

Q_S — Stored charge

r_{BB} — Interbase resistance

$r_b'C_c$ — Collector-base time constant

$r_{CE(sat)}$ — Saturation resistance, collector-to-emitter

$r_{DS(on)}$ — Static drain-source on-state resistance

$r_{ds(on)}$ — Small-signal drain-source on-state resistance

$Re(h_{ie})$ — Real part of common-emitter small-signal short-circuit input impedance

$Re(h_{oe})$ — Real part of common-emitter small-signal open-circuit output admittance

$r_{ele2(on)}$ — Small-signal emitter-emitter on-state resistance

r_i — Dynamic resistance at inflection point

R_θ — Thermal resistance

$R_{\theta CA}$ — Thermal resistance, case to ambient

$R_{\theta JA}$ — Thermal resistance, junction to ambient

$R_{\theta JC}$ — Thermal resistance, junction to case

S, s — Source

T_A — Ambient temperature or free-air temperature

T_C — Case temperature

t_d — Delay time

$t_{d(off)}$ — Turn-off delay time

$t_{d(on)}$ — Turn-on delay time

t_f — Fall time

t_{fr} — Forward recovery time

T_j — Junction temperature

t_{off} — Turn-off time

t_{on} — Turn-on time

t_p — Pulse time

t_r — Rise time

t_{rr} — Reverse recovery time

t_s — Storage time

TSS — Tangential signal sensitivity

T_{stg} — Storage temperature

t_w — Pulse average time

U, u — Bulk (substrate)

V_{BB} — Base supply voltage (dc)

V_{BC} — Average or dc voltage, base to collector

v_{bc} — Instantaneous value of alternating component of base-collector voltage

V_{BE} — Average or dc voltage, base to emitter

v_{be} — Instantaneous value of alternating component of base-emitter voltage

$V_{(BR)}$ — Breakdown voltage (dc)

$v_{(BR)}$ — Breakdown voltage (instantaneous total)

$V_{(BR)CBO}$ — Collector-base breakdown voltage, emitter open

$V_{(BR)CEO}$ — Collector-emitter breakdown voltage, base open

$V_{(BR)CER}$ — Collector-emitter breakdown voltage, resistance between base and emitter

$V_{(BR)CES}$ — Collector-emitter breakdown voltage, base shorted to emitter

$V_{(BR)CEV}$ — Collector-emitter breakdown voltage, specified voltage between base and emitter

$V_{(BR)CEX}$ — Collector-emitter breakdown voltage, specified circuit between base and emitter

$V_{(BR)EBO}$ — Emitter-base breakdown voltage, collector open

$V_{(BR)ECO}$ — Emitter-collector breakdown voltage, base open

$V_{(BR)E1E2}$ — Emitter-emitter breakdown voltage

$V_{(BR)GSS}$ — Gate-source breakdown voltage

$V_{(BR)GSSF}$ — Forward gate-source breakdown voltage

$V_{(BR)GSSR}$ — Reverse gate-source breakdown voltage

V_{B2B1} — Interbase voltage

V_{CB} — Average or dc voltage, collector to base

v_{cb} — Instantaneous value of alternating component of collector-base voltage

$V_{CB(fl)}$ — Collector-base dc open-circuit voltage (floating potential)

V_{CBO} — Collector-base voltage, dc, emitter open

V_{CC} — Collector supply voltage (dc)

V_{CE} — Average or dc voltage, collector to emitter

v_{ce} — Instantaneous value of alternating component of collector-emitter voltage

$V_{CE(fl)}$ — Collector-emitter dc open-circuit voltage (floating potential)

V_{CEO} — Collector-emitter voltage (dc), base open

$V_{CE(ofs)}$ — Collector-emitter offset voltage

V_{CER} — Collector-emitter voltage (dc), resistance between base and emitter

V_{CES} — Collector-emitter voltage (dc), base shorted to emitter

$V_{CE(sat)}$ — Collector-emitter dc saturation voltage

V_{CEV} — Collector-emitter voltage (dc), specified voltage between base and emitter

V_{CEX} — Collector-emitter voltage (dc), specified circuit between base and emitter

V_{DD} — Drain supply voltage (dc)

V_{DG} — Drain-gate voltage

V_{DS} — Drain-source voltage

$V_{DS(on)}$ — Drain-source on-state voltage

V_{DU} — Drain-substrate voltage

V_{EB} — Average or dc voltage, emitter to base

v_{eb} — Instantaneous value of alternating component of emitter-base voltage

$V_{EB(fl)}$ — Emitter-base dc open-circuit voltage (floating potential)

V_{EBO} — Emitter-base voltage (dc), collector open

$V_{EBl(sat)}$ — Emitter saturation voltage

V_{EC} — Average or dc voltage, emitter to collector

v_{ec} — Instantaneous value of alternating component of emitter-collector voltage

$V_{EC(fl)}$ — Emitter-collector dc open-circuit voltage (floating potential)

continued

$V_{EC(ofs)}$ — Emitter-collector offset voltage

V_{EE} — Emitter supply voltage (dc)

V_F — For voltage-regulator and voltage-reference diodes: dc forward voltage. For signal diodes and rectifier diodes: dc forward voltage (no alternating component)

V_f — Alternating component of forward voltage (rms value)

v_F — Instantaneous total forward voltage

$V_{F(AV)}$ — Forward voltage, dc (with alternating component)

V_{FM} — Maximum (peak) total forward voltage

$V_{F(RMS)}$ — Total rms forward voltage

V_{GG} — Gate supply voltage (dc)

V_{GS} — Gate-source voltage

V_{GSF} — Forward gate-source voltage

$V_{GS(off)}$ — Gate-source cutoff voltage

V_{GSR} — Reverse gate-source voltage

$V_{GS(th)}$ — Gate-source threshold voltage

V_{GU} — Gate-substrate voltage

V_I — Inflection-point voltage

V_{OB1} — Base-1 peak voltage

V_P — Peak-point voltage

V_{PP} — Projected peak-point voltage

V_R — For voltage-regulator and voltage-reference diodes: dc reverse voltage. For signal diodes and rectifier diodes: dc reverse voltage (no alternating component)

V_r — Alternating component of reverse voltage (rms value)

v_R — Instantaneous total reverse voltage

$V_{R(AV)}$ — Reverse voltage, dc (with alternating component)

V_{RM} — Maximum (peak) total reverse voltage

V_{RRM} — Repetitive peak reverse voltage

$V_{R(RMS)}$ — Total rms reverse voltage

V_{RSM} — Nonrepetitive peak reverse voltage

V_{RT} — Reach-through voltage

V_{RWM} — Working peak reverse voltage

V_{SS} — Source supply voltage (dc)

V_{SU} — Source-substrate voltage

$V_{(TO)}$ — Threshold voltage

V_V — Valley-point voltage

V_Z — Regulator voltage, reference voltage (dc)

V_{ZM} — Regulator voltage, reference voltage (dc at maximum rated current)

y_{fb} — Common base small-signal short-circuit forward transfer admittance

y_{fc} — Common-collector small-signal short-circuit forward transfer admittance

y_{fe} — Common-emitter small-signal short-circuit forward transfer admittance

y_{fs} — Common-source small-signal short-circuit forward transfer admittance

$y_{fs(imag)}$ — Common-source small-signal forward transfer susceptance

$y_{fs(real)}$ — Common-source small-signal forward transfer conductance

y_{ib} — Common-base small-signal short-circuit input admittance

y_{ic} — Common-collector small-signal short-circuit input admittance

y_{ie} — Common-emitter small-signal short-circuit input admittance

$y_{ie(imag)}$ — Imaginary part of small-signal short-circuit input admittance (common-emitter)

$y_{ie(real)}$ — Real part of small-signal short-circuit input admittance (common-emitter)

y_{is} — Common-source small-signal short-circuit input admittance

$y_{is(imag)}$ — Common-source small-signal input susceptance

$y_{is(real)}$ — Common-source small-signal input conductance

y_{ob} — Common-base small-signal short-circuit output admittance

y_{oc} — Common-collector small-signal short-circuit output admittance

y_{oe} — Common-emitter small-signal short-circuit output admittance

$y_{oe(imag)}$ — Imaginary part of small-signal short-circuit output admittance (common-emitter)

$y_{oe(real)}$ — Real part of small-signal short-circuit output admittance (common-emitter)

y_{os} — Common-source small-signal short-circuit output admittance

$y_{os(imag)}$ — Common-source small-signal output susceptance

$Y_{os(real)}$ — Common-source small-signal output conductance

y_{rb} — Common-base small-signal short-circuit reverse transfer admittance

y_{rc} — Common-collector small-signal short-circuit reverse transfer admittance

y_{re} — Common-emitter small-signal short-circuit reverse transfer admittance

y_{rs} — Common-source small-signal short-circuit reverse transfer admittance

$y_{rs(imag)}$ — Common-source small-signal reverse transfer susceptance

$y_{rs(real)}$ — Common-source small-signal reverse transfer conductance

z_{if} — Intermediate-frequency impedance

z_m — Modulator-frequency load impedance

z_{rf} — Radio-frequency impedance

$Z_{\theta JA(t)}$ — Junction-to-ambient transient thermal impedance

$Z_{\theta JC(t)}$ — Junction-to-case transient thermal impedance

$Z_{\theta(t)}$ — Transient thermal impedance

z_v — Video impedance

z_z — Regulator impedance, reference impedance (small-signal at I_z)

z_{zk} — Regulator impedance, reference impedance (small-signal at I_{ZK})

Z_{zm} — Regulator impedance, reference impedance (small-signal at I_{ZM})

Elementary particles

Table 1. The more stable elementary particles (mean lives longer than 10^{-20} sec)*

Symbol and name	J^P (and C, if self-conjugate)	Mass, MeV	Mean life, sec	Principal decay modes and branching ratios, %
Classons				
(graviton)	$2^+(C=+)$	0	Stable	
γ(photon)	$1^-(C=-)$	0	Stable	
Leptons				
ν_e(e neutrino) $\}$	1/2	0	Stable	
$\bar{\nu}_e$(e antineutrino) $\}$				
e^-(electron) $\}$	1/2	0.511	Stable	
e^+(positron) $\}$				
ν_μ(μ neutrino) $\}$	1/2	0	Stable	
$\bar{\nu}_\mu$(μ antineutrino) $\}$				
$\left.\begin{matrix}\mu^-\\ \mu^+\end{matrix}\right\}$ (muon)	1/2	105.7	2.20×10^{-6}	$\begin{cases}e^-\bar{\nu}_e\nu_\mu & 100\\ e^+\nu_e\bar{\nu}_\mu & 100\end{cases}$

*For footnote, see page A25.

continued

Table 1. The more stable elementary particles (mean lives longer than 10^{-20} sec) (cont.)*

Symbol and name	J^P (and C, if self-conjugate) Hadronic quantum numbers	Mass, MeV	Mean life, sec	Principal decay modes and branching ratios, %
Hadrons				
Mesons				
π (pion)	$\begin{bmatrix} Y=0 \\ I_G=1 \end{bmatrix} 0^-$			
	$I_3=+1:\pi^+$	139.6	2.60×10^{-8}	$\mu^+\nu_\mu$ 100
	$I_3=0:\pi^0\,(C=+)$	135.0	0.8×10^{-16}	$\gamma\gamma$ 99
				γe^+e^- 1
	$I_3=-1:\pi^-\,(=\overline{\pi^+})$	139.6	2.60×10^{-8}	$\mu^-\bar{\nu}_\mu$ 100
	$I_3=+1/2:K^+$	493.7	1.24×10^{-8}	$\mu^+\nu_\mu$ 64
				$\pi^+\pi^0$ 21
				$\pi^+\pi^-\pi$ 7
				$\pi^0e^+\nu_e$ 5
				$\pi^0\mu^+\nu_\mu$ 3
$\left.\begin{array}{l}K \\ \overline{K}\end{array}\right\}$ (kaon)	$\begin{bmatrix} Y=\left\{\begin{array}{l}+1 \\ -1\end{array}\right. \\ I=1/2 \end{bmatrix} 0^-$			
	$Y=+1\left\{\begin{array}{l} I_3=-1/2:K^0 \\ I_3=+1/2:\overline{K^0}\end{array}\right.$ 498 $\left\{\begin{array}{l}K_S\,(CP\approx+) \\ K_L\,(CP\approx-)\end{array}\right.$		0.89×10^{-10}	$\pi^+\pi^-$ 69
				$\pi^0\pi^0$ 31
			0.52×10^{-7}	$\pi^0\pi^0\pi^0$ 21
				$\pi^+\pi^-\pi^0$ 12
				$\pi e\nu$ 39
				$\pi\mu\nu$ 27
				$\pi e\nu\gamma\approx$ 1
				$\pi^+\pi^-$ 0.2
	$Y=-1\left\{\ I_3=-1/2:K^-(=\overline{K^+},\text{ see }K^+)\right.$			$\pi^0\pi^0$ 0.1
η (eta meson)	$\begin{bmatrix} Y=0 \\ I_G=0_+ \end{bmatrix} 0^-\,(C=+)$	549	$\approx 0.8\times10^{-18}$ (Full width $= 0.85\pm0.12$ keV)	$\gamma\gamma$ 38
				$\pi^0\pi^0\pi^0$ 30
				$\pi^+\pi^-\pi^0$ 24
				$\pi^+\pi^-\gamma$ 5
				$\pi^0\gamma\gamma$ 3(?)

Table 1. The more stable elementary particles (mean lives longer than 10^{-20} sec) (cont.)*

Symbol and name	J^P (and C, if self-conjugate) Hadronic quantum numbers	Mass, MeV	Mean life, sec	Principal decay modes and branching ratios, %
Hadrons (cont).				
J/ψ	$\begin{bmatrix} Y=0 \\ I_G=0_- \end{bmatrix}$ $1^-(C=-)$	3098 ± 3	$\approx 1.0 \times 10^{-20}$ (Full width $= 67 \pm 12$ keV)	mesons 86; e^+e^- 7; $\mu^+\mu^-$ 7
Baryons (all have distinct antiparticles)				
N (nucleon)	$\begin{bmatrix} Y=+1 \\ I=1/2 \end{bmatrix}$ $1/2^+$ $\begin{cases} I_3=+1/2:p \text{ (proton)} \\ I_3=-1/2:n \text{ (neutron)} \end{cases}$	938.3 939.6	Stable 0.9×10^3	$pe^-\bar{\nu}_e$ 100
Λ (lambda hyperon)	$\begin{bmatrix} Y=0 \\ I_G=0 \end{bmatrix}$ $1/2^+$ Λ^0	1115.6	2.6×10^{-10}	$p\pi^-$ 64; $n\pi^0$ 36
Σ (sigma hyperon)	$\begin{bmatrix} Y=0 \\ I=1 \end{bmatrix}$ $1/2^+$ $\begin{cases} I_3=+1:\Sigma^+ \\ I_3=0:\Sigma^0 \\ I_3=-1:\Sigma^- \end{cases}$	1189.4 1192.5 1197.3	0.80×10^{-10} $\approx 6 \times 10^{-20}$ 1.5×10^{-10}	$p\pi^0$ 52; $n\pi^+$ 48; $\Lambda\gamma$ 100; $n\pi^-$ 100
Ξ (xi or cascade hyperon)	$\begin{bmatrix} Y=-1 \\ I=1/2 \end{bmatrix}$ $1/2^+$ $\begin{cases} I_3=+1/2:\Xi^0 \\ I_3=-1/2:\Xi^- \end{cases}$	1315 1321	3×10^{-10} 1.7×10^{-10}	$\Lambda\pi^0$ 100; $\Lambda\pi^-$ 100
Ω (omega hyperon)	$\begin{bmatrix} Y=-2 \\ I=0 \end{bmatrix}$ $(3/2^+)$ Ω^-	1672	$\approx 10^{-10}$	$\Xi\pi$?; ΛK ?

*This table does not include the τ (tau) lepton and certain charmed particles (included in Table 2) whose mean lives are expected to be longer than 10^{-20} sec but have not been experimentally established.

Table 2. The hadrons (strongly interacting elementary particles)

Hypercharge, strangeness, charm, and i-spin	Symbol (mass, MeV)	J_C^P†	Width, MeV	Decay products and branching ratios, %
Mesons				
$I=1$	$\pi_A(138) = \pi$	0_+^-	Hadronically stable	
	$\rho_N(770) = \rho$	1_-^-	150	$\pi\pi 100$
	$\pi_N(970) = \delta$	0_+^+	50	$\eta\pi, K\bar{K}$
	$\pi_A(1100) = A_1$	1_+^+	300	$\rho\pi 100?$
	$\rho_A(1230) = B$	1_+^+	125	$\omega\pi 100$
	$\pi_N(1310) = A_2$	2_+^+	100	$\rho\pi 70, \eta\pi 15, \omega\pi\pi 10, K\bar{K} 5$
	$_A(1540) = F_1$	?	40	$K^*\bar{K} + \bar{K}^*K, ?$
	$\rho_N(1600) = \rho'$	1_-^-	400	$\pi\pi\pi\pi$
	$\pi_A(1640) = A_3$	2_+^-	300	$f\pi 100?$
	$\rho_N(1680) = g$	3_-^-	180	$\pi\pi 25, \omega\pi, A_2\pi, \rho\rho, \ldots$
$Y=0$ $S=0$ $C=0$ $I=0$	$\eta_A(549) = \eta$	0_+^-	Hadronically stable	
	$\omega_N(783) = \omega$	1_-^-	10	$\pi\pi\pi 90, \pi\gamma 9, \pi\pi 1$
	$\eta_A'(958) = \eta'/X_0$	0_+^-	1	$\eta\pi\pi 68, \rho^0\gamma 30, \gamma\gamma 2$
	$\eta_N'(993) = S^*$	0_+^+	40	$K\bar{K}, \pi\pi$
	$\phi_N(1020) = \phi$	1_-^-	4	$K\bar{K} 80, \rho\pi 15, \eta\gamma 2$
	$\eta_N(1200) = \epsilon$	0_+^+	600	$\pi\pi 100$
	$\eta_N(1270) = f$	2_+^+	180	$\pi\pi 80, K\bar{K} 3, \pi\pi\pi\pi 3$
	$\eta_A(1285) = D$	$1_+^-?$	30	$\delta\pi?$
	$\eta_A'(1415) = E$	$0_-^-/1_+^+$	60	$\delta\pi 80, K^*\bar{K} + \bar{K}^*K\ 20$
	$\eta_N'(1515) = f'$	2_+^+	40	$K\bar{K}\ 100?$
	$\omega_N(1675)$	$3_-^-?$	150	$\rho\pi, ?$
	$\eta_N(2040) = h$	4_+^+	200	$\pi\pi, K\bar{K}$
	$\eta''(2830) = \eta_c$	0_+^-		$\gamma\gamma, p\bar{p}, \text{mesons}?$
	$\psi(3100) = \psi/J$	1_-^-	0.07	$\text{mesons } 86, \mu\bar{\mu}7, e\bar{e}7, \eta_c\gamma$
	$\chi(3415)$	$0_+^+?$		$\pi\pi, K\bar{K}, \pi\pi\pi\pi, \ldots, J\gamma?$
	$\chi(3510) = P_c$	$1_+^+?$		$J\gamma, \pi\pi\pi\pi, \ldots$
	$\chi(3550)$	$2_+^+?$		$\pi\pi, KK, \pi\pi\pi\pi, \ldots, J\gamma$
	$\psi(3685) = \psi'$	1_-^-	0.25	$J\pi\pi 50, \chi(3415)\gamma 7, \chi(3510)\gamma 7,$ $\chi(3550)\gamma 7, J\eta 4, \mu\bar{\mu}1, e\bar{e}1$
	$\psi(3770) = \psi''$	$1_-^-?$	30	$D\bar{D}100?$
	$\psi(4000-4500)$	1_-^-		‡
	$\Upsilon(9400) = \Upsilon$	$1_-^-?$		$\mu\bar{\mu}, ?$
	$\Upsilon(10,000) = \Upsilon'$	$1_-^-?$		$\mu\bar{\mu}, ?$
$Y=+1, I=1/2$ $S=+1, C=0$	$K_A(495) = K$	0^-	Hadronically stable	
	$K_N(892) = K^*$	1^-	50	$K\pi 100$
	$K_N(\approx 1250) = \kappa$	0^+	≈ 450	$K\pi$
	$K_A(1240) \Big\} = Q$	$1^+?$	125	$K\pi\pi = (K^*\pi, K\rho, K\epsilon?)$
	$K_A(1350)$	$1^+?$	200?	
	$K_N(1420)$	2^+	110	$K\pi 55, K^*\pi 30, K\rho 7, K\omega 5$
	$K_A(1770) = L$	$2^-?$	150	$K\pi\pi, K\pi\pi\pi$
	$K_N(1850)$	$3^-?$	300	$K\pi, ?$
$Y=+1, I=1/2$ $S=0, C=+1$	$D(1866)$	$0^-?$	<2§	$Ke\nu_e$ or $K^*e\nu_e 7, K\pi\pi, K\pi\pi\pi, K\pi, ?$
	$D(2007) = D^*$	$1^-?$	<2	$D\pi, D\gamma$
$Y=+2, I=0$ $S=+1, C=+1$	$F(2030) = F$	$0^-?$	§	$\eta\pi, ?$
	$F(2140) = F^*$	$1^-?$		$F\gamma$

†For footnotes, see page A28.

Table 2. The hadrons (strongly interacting elementary particles) (cont.)

Hypercharge, strangeness, charm, and i-spin	Symbol (mass, MeV)	$J_c^{P\dagger}$	Width, MeV	Decay products and branching ratios, %			
Baryons							
	$N(939) = N$	$1/2^+8$	Hadronically stable				
				$N\pi$	$\Delta\pi$	$N\rho$	$N\epsilon$
$I = 1/2$	$N(1450) = N'$	$1/2^+$	200	50,	25,	10,	10
	$N(1520)$	$3/2^-9$	150	60,	25,	10,	5?
	$N(1510)$	$1/2^-9$	100	25,	5?,	5?,	5?, $N\eta$60
	$N(1660)$	$5/2^-8$	150	45, 55			
	$N(1680)$	$5/2^+8$	125	60,	10,	15,	15
	$N(1680)$	$1/2^-$	150	50,	5,	10,	10
	$N(1800)$	$3/2^+8?$	200	20,		75	
	$N(2190)$	$7/2^-9$	250	25			
	$N(2220)$	$9/2^+8?$	300	20			
$I = 3/2$	$\Delta(1232) = \Delta$	$3/2^+10$	115	100			
	$\Delta(1650)$	$1/2^-$	150	40,	60		
	$\Delta(1700)$	$3/2^-$	200	20,	35,	45	
	$\Delta(1850)$	$5/2^+$	250	20,	20,	60	
	$\Delta(1900)$	$1/2^+$	200	25			
	$\Delta(1900)$	$3/2^+$	200	20,	70		
	$\Delta(1900)$	$7/2^+10$	220	40,	20,	20	
	$\Delta(2400)$	$11/2^+?10?$	300	10			

$Y = +1$
$S = 0$
$C = 0$

i-spin	Symbol (mass, MeV)	$J_c^{P\dagger}$	Width, MeV			
	$\Lambda(1116) = \Lambda$	$1/2^+8$	Hadronically stable			
				$N\overline{K}$	$\Sigma\pi$	
$I = 0$	$\Lambda'(1405)$	$1/2^-9$	40		100,	
	$\Lambda'(1520)$	$3/2^-9$	15	45,	40, $\Lambda\pi\pi$10	
	$\Lambda(1670)$	$1/2^-9$	40	30	40, $\Lambda\eta$20	
	$\Lambda(1690)$	$3/2^-9$	60	25,	30, $\Lambda\pi\pi$25, $\Sigma\pi\pi$20	
	$\Lambda(1815)$	$5/2^+8$	85	60,	10, $\Sigma^*\pi$20	
	$\Lambda(1830)$	$5/2^-8$	100	<10,	35 − 75	
	$\Lambda(1860)$	$3/2^+$	80	25,	5 − 10	
	$\Lambda'(2100)$	$7/2^-9$	250	30,	5	
	$\Lambda(2350)$	$7/2^-?9?$	120	20		
	$\Sigma(1193) = \Sigma$	$1/2^+8$	Hadronically stable			

$Y = 0$
$S = -1$
$C = 0$

i-spin	Symbol (mass, MeV)	$J_c^{P\dagger}$	Width, MeV			
				$N\overline{K}$	$\Lambda\pi$	$\Sigma\pi$
$I = 1$	$\Sigma(1385) = \Sigma^*$	$3/2^+10$	35		90,	10
	$\Sigma(1670)$	$3/2^-9$	50	10,	10,	50
	$\Sigma(1750)$	$1/2^-9$	75	≈30,	≈10,	≈5, $\Sigma\eta \approx 20$
	$\Sigma(1765)$	$5/2^-8$	130	40,	10	$\Lambda(1520)\pi$15, $\Sigma^*\pi$10
	$\Sigma(1915)$	$5/2^+8$	100	10,	20,	5
	$\Sigma(1940)$	$3/2^-$	220	<20,	5,	5
	$\Sigma(2030)$	$7/2^+10$	180	20,	20,	5
	$\Sigma(2250)$	$7/2^-?9?$	150	10,	15,	75

continued

Table 2. The hadrons (strongly interacting elementary particles) (cont.)

Hypercharge, strangeness, charm, and i-spin	Symbol (mass, MeV)	J_c^{P}†	Width, MeV	Decay products and branching ratios, %
Baryons (cont.)				
$Y=-1, I=1/2$ $S=-2, C=0$	$\Xi(1318)=\Xi$	$1/2^+$ 8	Hadronically stable	
	$\Xi(1530)$	$3/2^+$10	10	$\Xi\pi$100
	$\Xi(1820)$	$3/2^-$?9?	50	
	$\Xi(1940)$	$5/2^-$?8?	100	
$Y=-2, I=0$ $S=-3, C=0$	$\Omega^-(1672)$	$3/2^+$?10?	Hadronically stable	
$Y=+2$ $S=0$ $C=+1$ $I=0$	$\Lambda_c^+(2260)$	$1/2^+$?	$<$75§	$\Lambda\pi\pi\pi$
$I=1$	$\Sigma_c(2430)$	$1/2^+$?		$\Lambda_c^+\pi$

†Numbers following J_c^P in baryon entries give multiplicity of SU(3) supermultiplet.
‡Two or more mesons in this region, not yet clearly resolved.
§Expected to be hadronically stable.

Chemical elements

Table 1. Natural isotopic abundances of the elements

Element		Mass no.	At. %	Element		Mass no.	At. %
1	H*	1	99.985	23	V	50	0.25
		2	0.015			51	99.75
		3	0.00013	24	Cr	50	4.35
2	He*	4	≈100.			52	83.79
3	Li*	6	7.5			53	9.50
		7	92.5			54	2.36
4	Be	9	100.	25	Mn	55	100.
5	B*	10	19.8	26	Fe	54	5.85
		11	80.2			56	91.7
6	C*	12	98.89			57	2.14
		13	1.11			58	0.31
7	N*	14	99.64	27	Co	59	100.
		15	0.36	28	Ni	58	68.3
8	O*	16	99.756			60	26.1
		17	0.039			61	1.1
		18	0.205			62	3.6
9	F	19	100.			64	0.9
10	Ne*†	20	90.51	29	Cu	63	69.2
		21	0.27			65	30.8
		22	9.22	30	Zn	64	48.9
11	Na	23	100.			66	27.8
12	Mg	24	78.99			67	4.1
		25	10.00			68	18.6
		26	11.01			70	0.6
13	Al	27	100.	31	Ga	69	60.0
14	Si	28	92.2			71	40.0
		29	4.7	32	Ge	70	20.7
		30	3.1			72	27.5
15	P	31	100.			73	7.7
16	S*	32	95.00			74	36.4
		33	0.76			76	7.7
		34	4.22	33	As	75	100.
		36	0.02	34	Se	74	0.9
17	Cl	35	75.77			76	9.0
		37	24.23			77	7.6
18	Ar†	36	0.34			78	23.5
		38	0.07			80	49.8
		40	99.59			82	9.2
19	K	39	93.26	35	Br	79	50.69
		40	0.01			81	49.31
		41	6.73	36	Kr†	78	0.35
20	Ca	40	96.937			80	2.25
		42	0.65			82	11.6
		43	0.14			83	11.5
		44	2.08			84	57.0
		46	0.003			86	17.3
		48	0.19	37	Rb	85	72.17
21	Se	45	100.			87	27.83
22	Ti	46	8.0	38	Sr†	84	0.56
		47	7.5			86	9.84
		48	73.7			87	7.0
		49	5.5			88	82.6
		50	5.3	39	Y	89	100.

*For footnotes, see page A31.

continued

Table 1. Natural isotopic abundances of the elements (cont.)

Element		Mass no.	At. %	Element		Mass no.	At. %
40	Zr	90	51.4	52	Te	120	0.1
		91	11.2			122	2.5
		92	17.1			123	0.9
		94	17.5			124	4.6
		96	2.8			125	7.0
41	Nb	93	100.			126	18.7
42	Mo	92	14.8			128	31.7
		94	9.1			130	34.5
		95	15.9	53	I	127	100.
		96	16.7	54	Xe†	124	0.1
		97	9.5			126	0.1
		98	24.4			128	1.9
		100	9.6			129	26.4
44	Ru	96	5.5			130	4.1
		98	1.9			131	21.2
		99	12.7			132	26.9
		100	12.6			134	10.4
		101	17.1			136	8.9
		102	31.6	55	Cs	133	100.
		104	18.6	56	Ba	130	0.1
45	Rh	103	100.			132	0.1
46	Pd	102	1.0			134	2.4
		104	11.0			135	6.6
		105	22.2			136	7.9
		106	27.3			137	11.2
		108	26.7			138	71.7
		110	11.8	57	La	138	0.09
47	Ag	107	51.83			139	99.91
		109	48.17	58	Ce	136	0.2
48	Cd	106	1.2			138	0.3
		108	0.9			140	88.4
		110	12.4			142	11.1
		111	12.8	59	Pr	141	100.
		112	24.0	60	Nd	142	27.1
		113	12.3			143	12.2
		114	28.8			144	23.9
		116	7.6			145	8.3
49	In	113	4.3			146	17.2
		115	95.7			148	5.7
50	Sn	112	1.0			150	5.6
		114	0.7	62	Sm	144	3.1
		115	0.4			147	15.0
		116	14.7			148	11.2
		117	7.7			149	13.8
		118	24.3			150	7.4
		119	8.6			152	26.7
		120	32.4			154	22.8
		122	4.6	63	Eu	151	47.8
		124	5.6			153	52.2
51	Sb	121	57.3	64	Gd	152	0.2
		123	42.7			154	2.2
						155	14.9
						156	20.6
						157	15.7
						158	24.7
						160	21.7
				65	Tb	159	100.

Table 1. Natural isotopic abundances of the elements (cont.)

Element		Mass no.	At. %	Element		Mass no.	At. %
66	Dy	156	0.06	75	Re	185	37.40
		158	0.1			187	62.60
		160	2.34	76	Os†	184	0.02
		161	18.9			186	1.58
		162	25.5			187	1.6
		163	24.9			188	13.3
		164	28.2			189	16.1
67	Ho	165	100.			190	26.4
68	Er	162	0.1			192	41.0
		164	1.6	77	Ir	191	37.4
		166	33.4			193	62.6
		167	22.9	78	Pt	190	0.01
		168	27.0			192	0.79
		170	15.0			194	32.9
69	Tm	169	100.			195	33.8
70	Yb	168	0.1			196	25.3
		170	3.1			198	7.2
		171	14.3	79	Au	197	100.
		172	21.9	80	Hg	196	0.2
		173	16.2			198	10.1
		174	31.7			199	16.9
		176	12.7			200	23.1
71	Lu	175	97.4			201	13.2
		176	2.6			202	29.7
72	Hf	174	0.2			204	6.8
		176	5.2	81	Tl	203	29.5
		177	18.5			205	70.5
		178	27.1	82	Pb†	204	1.4
		179	13.8			206	24.1
		180	35.2			207	22.1
73	Ta	180	0.012			208	52.4
		181	99.988	83	Bi	209	100.
74	W	180	0.1	90	Th	232	100.
		182	26.3	92	U*	234	0.0054
		183	14.3			235	0.7200
		184	30.7			238	99.2746
		186	28.6				

*Isotopic composition of the element may be somewhat variable with specific geological or biological origin of the sample. Commercial chemicals may have, in some cases, quite anomalous composition as the result of processes of isotope separation.

†The element may vary in isotopic composition in some samples because one or more of the isotopes result from radioactive decay, or from nuclear processes in nature, such as spontaneous fission of uranium or α,n reactions on light elements.

Table 2. Properties of the elements[a,h]

Name	Symbol	Atomic number	Atomic weight[a]	Density, g/cm³ at 20°C	Melting point, °C	Boiling point, °C	Valence	Electronic configuration	Ground term	Ionization potential, eV
Actinium	Ac	89	(227)[f]		1050	(3330)	3	[Rn]$6d7s^2$	$^2D_{3/2}$	6.9
Aluminum	Al	13	26.98154	2.6984	660.2	2447	3	[Ne]$3s^23p$	$^2P^{\circ}_{1/2}$	5.986
Americium	Am	95	(243)[f]	13.67	>800	(2600)	3	[Rn]$5f^77s^2$	$^8S^{\circ}_{7/2}$	6.0
Antimony	Sb	51	121.75*	6.684	630.5	1640	3, 5	[Kr]$4d^{10}5s^25p^3$	$^4S^{\circ}_{3/2}$	8.641
Argon	Ar	18	39.948*[b,e]	0.0017824	−189.38	−185.87	0	[Ne]$3s^23p^6$	$^1S_0$	15.759
Arsenic	As	33	74.9216	5.72 gray 2.026 yellow 4.7 black	817 (28 atm)	613 subl	5, ±3	[Ar]$3d^{10}4s^24p^3$	$^4S^{\circ}_{3/2}$	9.81
Astatine	At	85	(210)[f]		302	334	. . .	[Xe]$4f^{14}5d^{10}6s^26p^5$	$^2P^{\circ}_{3/2}$	. . .
Barium	Ba	56	137.33	3.59	850	1537	2	[Xe]$6s^2$	$^1S_0$	5.211
Berkelium	Bk	97	(247)[f]				4, 3	[Rn]$5f^86d7s^2$	. . .	. . .
Beryllium	Be	4	9.01218	1.86	1285	2970	2	[He]$2s^2$	$^1S_0$	9.322
Bismuth	Bi	83	208.9804	9.80	271.3	1560	3, 5	[Xe]$4f^{14}5d^{10}6s^26p^3$	$^4S^{\circ}_{3/2}$	7.289
Boron	B	5	10.81[b,c]	2.46	2074	3675	3	[He]$2s^22p$	$^2P^{\circ}_{1/2}$	8.298
Bromine	Br	35	79.904	3.119 l	−7.08	58.76	±1, 5	[Ar]$3d^{10}4s^24p^5$	$^2P^{\circ}_{3/2}$	11.84
Cadmium	Cd	48	112.41	8.642	320.9	767	2	[Kr]$4d^{10}5s^2$	$^1S_0$	8.993
Calcium	Ca	20	40.08*	1.55	851	1487	2	[Ar]$4s^2$	$^1S_0$	6.113
Californium	Cf	98	(251)[f]				3	[Rn]$5f^{10}7s^2$	$^1S_0$	. . .
Carbon	C	6	12.011[b]	2.267 graphite 3.515 diamond	4000 (63 atm)	3850 subl	±4, 2	[He]$2s^22p^2$	$^3P_0$	11.259
Cerium	Ce	58	140.12	6.771	795	3470	3, 4	[Xe]$4f5d6s^2$	$^1G_4$	5.47
Cesium	Cs	55	132.9054	1.8785[15]	28.6	670	1	[Xe]$6s$	$^2S_{1/2}$	3.894
Chlorine	Cl	17	35.453	0.00295 g	−101.0	−34.05	±1, 7, 5	[Ne]$3s^23p^5$	$^2P^{\circ}_{3/2}$	13.01
Chromium	Cr	24	51.996	7.20	1900	2640	6, 3, 2	[Ar]$3d^54s$	$^7S_3$	6.766
Cobalt	Co	27	58.9332	8.9	1495	3550	3, 2	[Ar]$3d^74s^2$	$^4F_{9/2}$	7.86
Copper	Cu	29	63.546*[b]	8.92	1083	2582	2, 1	[Ar]$3d^{10}4s$	$^2S_{1/2}$	7.726
Curium	Cm	96	(247)[f]	13.51			3	[Rn]$5f^76d7s^2$	$^9D^{\circ}_2$	. . .
Dysprosium	Dy	66	162.50*	8.536	1407	2600	3	[Xe]$4f^{10}6s^2$	$^5I_8$	5.93
Einsteinium	Es	99	(254)[f]				. . .	[Rn]$5f^{11}7s^2$	. . .	. . .
Erbium	Er	68	167.26*	9.051	1495	2900	3	[Xe]$4f^{12}6s^2$	$^3H_6$	6.10
Europium	Eu	63	151.96	5.259	826	1440	3, 2	[Xe]$4f^76s^2$	$^8S^{\circ}_{7/2}$	5.68
Fermium	Fm	100	(257)[f]				. . .	[Rn]$5f^{12}7s^2$	. . .	. . .

Table 2. Properties of the elements (cont.)[g,h]

Name	Symbol	Atomic number	Atomic weight[a]	Density, g/cm³ at 20°C	Melting point, °C	Boiling point, °C	Valence	Electronic configuration	Ground term	Ionization potential, eV
Fluorine	F	9	18.998403	0.001580 g	−219.62	−188.14	−1	[He]$2s^2 2p^5$	$^2P^o_{3/2}$	17.422
Francium	Fr	87	(223)[f]		(27)		1	[Rn]$7s$	...	...
Gadolinium	Gd	64	157.25*	7.895	1312	3000	3	[Xe]$4f^7 5d 6s^2$	$^9D^o_2$	6.16
Gallium	Ga	31	69.72	5.907	29.75	1980	3	[Ar]$3d^{10} 4s^2 4p$	$^2P^o_{1/2}$	6.00
Germanium	Ge	32	72.59*	5.323	937	2830	4	[Ar]$3d^{10} 4s^2 4p^2$	$^3P_0$	7.88
Gold	Au	79	196.9665	19.3	1063	2707	3, 1	[Xe]$4f^{14} 5d^{10} 6s$	$^2S_{1/2}$	9.22
Hafnium	Hf	72	178.49	13.31	2225	~5200	4	[Xe]$4f^{14} 5d^2 6s^2$	$^3F_2$	6.8
Helium	He	2	4.00260	0.17847 g (STP)	−272.2 (25 atm)	−268.935	0	$1s^2$	$^1S_0$	24.5868
Holmium	Ho	67	164.9304	8.803	1461	2600	3	[Xe]$4f^{11} 6s^2$	$^4I^o_{15/2}$	6.02
Hydrogen	H	1	1.0079[b]	0.8987 × 10⁻⁴ (STP)	−259.20	−252.77	1	$1s$	$^2S_{1/2}$	13.5981
Indium	In	49	114.82	7.28	156.4	2050	3	[Kr]$4d^{10} 5s^2 5p$	$^2P^o_{1/2}$	5.786
Iodine	I	53	126.9045	4.660 s	113.6	184.4	−1, 5, 7	[Kr]$4d^{10} 5s^2 5p^5$	$^2P^o_{3/2}$	10.457
Iridium	Ir	77	192.22*	22.65	2448	4500	3, 4, 6	[Xe]$4f^{14} 5d^7 6s^2$	$^4F^o_{9/2}$	9.1
Iron	Fe	26	55.847*	7.86	1530	3000	3, 2	[Ar]$3d^6 4s^2$	$^5D_4$	7.87
Krypton	Kr	36	83.80[c]	0.003736	−157.2	−153.4	0	[Ar]$3d^{10} 4s^2 4p^6$	$^1S_0$	14.000
Lanthanum	La	57	138.9055*	6.174	920	3470	3	[Xe]$5d 6s^2$	$^2D_{3/2}$	5.577
Lawrencium	Lr	103	(260)[f]				...	[Rn]$5f^{14} 6d 7s^2$?	...	...
Lead	Pb	82	207.2[b,e]	11.34	327.4	1751	2, 4	[Xe]$4f^{14} 5d^{10} 6s^2 6p^2$	$^3P_0$	7.417
Lithium	Li	3	6.941*[b,c,e]	0.535	179	1336	1	$1s^2 2s$	$^2S_{1/2}$	5.3916
Lutetium	Lu	71	174.97	9.842	1652	3330	3	[Xe]$4f^{14} 5d 6s^2$	$^2D_{3/2}$	6.15
Magnesium	Mg	12	24.305[e]	1.74	650	1117	2	[Ne]$3s^2$	$^1S_0$	7.646
Manganese	Mn	25	54.9380	7.30	1244	2120	7, 4, 2, 6, 3	[Ar]$3d^5 4s^2$	$^6S_{5/2}$	7.434
Mendelevium	Md	101	(258)[f]				...	[Rn]$5f^{13} 7s^2$	...	...
Mercury	Hg	80	200.59*	13.5939	−38.87	356.58	2, 1	[Xe]$4f^{14} 5d^{10} 6s^2$	$^1S_0$	10.43
Molybdenum	Mo	42	95.94	10.2	2625	4800	6, 3, 5	[Kr]$4d^5 5s$	$^7S_3$	7.10
Neodymium	Nd	60	144.24*	7.004	1024	3030	3	[Xe]$4f^4 6s^2$	$^5I_4$	5.49
Neon	Ne	10	20.179*[c]	1.207 l b.p.	−248.6	−246.1	0	$1s^2 2s^2 2p^6$	$^1S_0$	21.565
Neptunium	Np	93	237.0482[d]	20.45	630		6, 5, 4, 3	[Rn]$5f^4 6d 7s^2$	$^6L_{11/2}$	...
Nickel	Ni	28	58.70	8.90	1455	2840	2, 3	[Ar]$3d^8 4s^2$	$^3F_4$	7.635
Niobium	Nb	41	92.9064	8.57	2468	5127	5, 3	[Kr]$4d^5 5s$	$^6D_{1/2}$	6.88
Nitrogen	N	7	14.0067	0.001165 g	−209.97	−195.798	−3, 5, 2	$1s^2 2s^2 2p^3$	$^4S^o_{3/2}$	14.53
Nobelium	No	102	(255)[f]				...	[Rn]$5f^{14} 7s^2$?	...	...

[a]For footnotes, see page A35.

continued

Table 2. Properties of the elements (cont.)[g,h]

Name	Symbol	Atomic number	Atomic weight[a]	Density, g/cm³ at 20°C	Melting point, °C	Boiling point, °C	Valence	Electronic configuration	Ground term	Ionization potential, eV
Osmium	Os	76	190.2[c]	22.61	2727	(4100)	4, 6, 8	[Xe]$4f^{14}5d^6 6s^2$	$^5D_4$	8.5
Oxygen	O	8	15.9994*[b]	0.001331 g	−218.787	−182.98	−2	$1s^2 2s^2 2p^4$	$^3P_2$	13.617
Palladium	Pd	46	106.4	12.023	1552	2870	2, 4	[Kr]$4d^{10}$	$^1S_0$	8.33
Phosphorus	P	15	30.97376	1.828 white 2.34 red 2.699 black	44.2 597 610	280.3 431 subl 453 subl	5, ±3	[Ne]$3s^2 3p^3$	$^4S^o_{3/2}$	10.487
Platinum	Pt	78	195.09*	21.45	1774	~3800	4, 2	[Xe]$4f^{14}5d^9 6s$	$^3D_3$	9.0
Plutonium	Pu	94	(244)[f]	19.82	638	3235	6, 5, 4, 3	[Rn]$5f^6 7s^2$	$^7F_0$	. . .
Polonium	Po	84	(209)[f]	9.20 cub.	254	962	2, 4	[Xe]$4f^{14}5d^{10}6s^2 6p^4$	$^3P_2$	8.43
Potassium	K	19	39.0983*	0.87	63.5	758	1	[Ar]$4s$	$^2S_{1/2}$	4.340
Praseodymium	Pr	59	140.9077	6.782	935	3130	3	[Xe]$4f^3 6s^2$	$^4I^o_{9/2}$	5.42
Promethium	Pm	61	(145)[f]		(1027)	(2727)	3	[Xe]$4f^5 6s^2$	$^6H^o_{5/2}$	5.55
Protactinium	Pa	91	231.0359[d]	15.37	(1227)	(4027)	5	[Rn]$5f^2 6d 7s^2$	$^4K_{11/2}$	. . .
Radium	Ra	88	226.02544[e]	ca. 6	ca. 700	(1525)	2	[Rn]$7s^2$	$^1S_0$	5.278
Radon	Rn	86	(222)[f]	4.4 l.b.p.	−71	−62	0	[Xe]$4f^{14}5d^{10}6s^2 6p^6$	$^1S_0$	10.749
Rhenium	Re	75	186.207	21.04	3180	5885	7, 4, −1	[Xe]$4f^{14}5d^5 6s^2$	$^6S_{5/2}$	7.87
Rhodium	Rh	45	102.9055	12.41	1966	(3700)	3, 4	[Kr]$4d^8 5s$	$^4F_{9/2}$	7.46
Rubidium	Rb	37	85.4678*	1.53	39.0	700	1	[Kr]$5s$	$^2S_{1/2}$	4.177
Ruthenium	Ru	44	101.07*	12.45	2430	3700	3, 4, 6, 8	[Kr]$4d^7 5s$	$^5F_5$	7.366
Samarium	Sm	62	150.4	7.536	1052	1900	3	[Xe]$4f^6 6s^2$	$^7F_0$	5.63
Scandium	Sc	21	44.9559	2.992	1397	2730	3	[Ar]$3d 4s^2$	$^2D_{3/2}$	6.54
Selenium	Se	34	78.96*	4.792 hex-rhb 4.48 red, mn	217 170	685	6, 4, −2	[Ar]$3d^{10}4s^2 4p^4$	$^3P_2$	9.75
Silicon	Si	14	28.0855*[b]	2.33	1415	2680	4	[Ne]$3s^2 3p^2$	$^3P_0$	8.151
Silver	Ag	47	107.868	10.50	960.15	2177	1	[Kr]$4d^{10}5s$	$^2S_{1/2}$	7.576
Sodium	Na	11	22.98977	0.97	97.8	883	1	[Ne]$3s$	$^2S_{1/2}$	5.139
Strontium	Sr	38	87.62*	2.60	774	1366	2	[Kr]$5s^2$	$^1S_0$	5.693
Sulfur	S	16	32.06[b]	2.08 α 1.96 β 1.92 γ	112.8 114.6 106.8	444.60	6, 4, −2	[Ne]$3s^2 3p^4$	$^3P_2$	10.360
Tantalum	Ta	73	180.9479*	16.60	2980	5425	5	[Xe]$4f^{14}5d^3 6s^2$	$^4F_{3/2}$	7.88
Technetium	Tc	43	(97)[f]	11.487	2200	(4700)	7	[Kr]$4d^5 5s^2$	$^6S_{5/2}$	7.28

Table 2. Properties of the elements (cont.)^{g,h}

Name	Symbol	Atomic number	Atomic weight^a	Density, g/cm³ at 20°C	Melting point, °C	Boiling point, °C	Valence	Electronic configuration	Ground term	Ionization potential, eV
Tellurium	Te	52	127.60*	6.24	450	994	4, 6, −2	[Kr]$4d^{10}5s^25p^4$	$^3P_2$	9.01
Terbium	Tb	65	158.9254	8.272	1356	2800	3	[Xe]$4f^96s^2$	$(^8G_{13/2})$	5.85
Thallium	Tl	81	204.37*	11.80	304	1470	1, 3	[Xe]$4f^{14}5d^{10}6s^26p$	$^3P_0$	6.108
Thorium	Th	90	232.0381^{d,e}	11.71	1800	~4200	4	[Rn]$6d^27s^2$	$^3F_2$	$\cdots$
Thullium	Tm	69	168.9342	9.332	1545	1730	3	[Xe]$4f^{13}6s^2$	$^2F^0_{7/2}$	6.18
Tin	Sn	50	118.69*	7.28 white	231.89	2687	4, 2	[Kr]$4d^{10}5s^25p2$	$^3P_0$	7.344
Titanium	Ti	22	47.90*	4.507 α 4.32 β	1672	3260	4, 3	[Ar]$3d^24s^2$	$^3F_2$	6.82
Tungsten (Wolfram)	W	74	183.85*	19.35	3415	5000	6	[Xe]$4f^{14}5d^46s^2$	$^5D_0$	7.98
Uranium	U	92	238.029^{c,e}	19.05	1132	3818	6, 5, 4, 3	[Rn]$5f^36d^17s^2$	$^5L_6$	~6.0
Vanadium	V	23	50.9414*	6.1	1919	3400	5, 4, 2	[Ar]$3d^34s^2$	$^4F_{3/2}$	6.74
Wolfram (see Tungsten)										
Xenon	Xe	54	131.30^f	0.0058971⁰	−111.8	−108.1	0	[Kr]$4d^{10}5s^25p^6$	$^1S_0$	12.130
Ytterbium	Yb	70	173.04*	6.977	824	1430	3, 2	[Xe]$4f^{14}6s^2$	$^1S_0$	6.25
Yttrium	Y	39	88.9059	4.478	1509	2930	3	[Kr]$4d5s^2$	$^2D_{3/2}$	6.38
Zinc	Zn	30	65.38	7.14	419.47	907	2	[Ar]$3d^{10}4s^2$	$^1S_0$	9.393
Zirconium	Zr	40	91.22	6.52³⁰	1855	4375	4	[Kr]$4d^25s^2$	$^3F_2$	6.84

^{a}Scaled to the relative atomic mass. $A_r(^{12}C) = 12$. Values apply to elements as they exist naturally on Earth and to certain artificial elements. They are considered reliable to ±1 in the last digit or ±3 when followed by an asterisk.

Data from Inorganic Chemistry Division, Commission on Atomic Weights, International Union of Pure and Applied Chemistry, incorporating changes approved at 28th IUPAC Conference, Madrid, 1975. Published in *Pure Appl. Chem.*, 37(4):591–603, 1974.

^{b}Known variations in isotopic composition in terrestrial material prevent a more precise atomic weight from being given.

^{c}Substantial variations from the value given can occur in commercially available material because of inadvertent or undisclosed change of isotopic composition.

^{d}Atomic weight of most commonly available long-lived isotope.

^{e}Geological specimens are known in which element has anomalous isotopic composition.

^{f}Numbers in parentheses denote atomic mass number of isotope of longest known half-life.

^{g}This table does not include elements 104, 105, 106, and 107, whose official names and symbols have not yet been assigned, and whose properties are not yet well established.

^{h}The following abbreviations are used: atm = atmosphere; b.p. = boiling point; cub. = cubic; g = gas; hex-rhb = hexagonal-rhombic; l = liquid; mn = monoclinic; rh = rhombohedral; s = solid; STP = standard temperature and pressure; subl = sublimes.

Crystal structure

Crystal lattice types are as follows:*

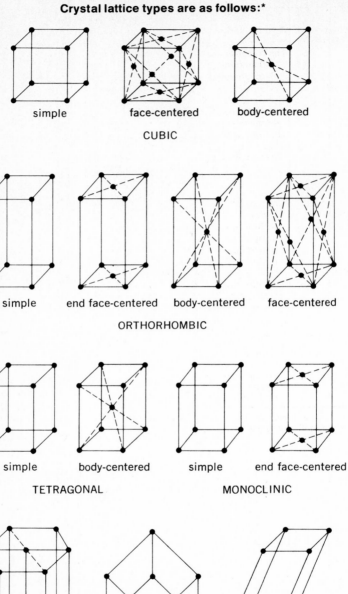

simple · face-centered · body-centered

CUBIC

simple · end face-centered · body-centered · face-centered

ORTHORHOMBIC

simple · body-centered · simple · end face-centered

TETRAGONAL · MONOCLINIC

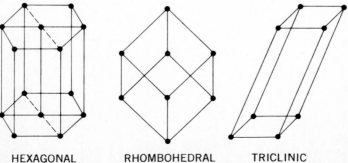

HEXAGONAL · RHOMBOHEDRAL · TRICLINIC

*From J. A. Dean, ed., *Lange's Handbook of Chemistry*, 11th ed., copyright 1973 by McGraw-Hill Book Company; used with permission.

Characteristics and properties of crystal systems*

System	Characteristics	Essential symmetry	Axes in unit cell	Angles in unit cell
Cubic	Three axes equal and mutually perpendicular	Four threefold axes	$a = b = c$	$\alpha = \beta = \gamma = 90°$
Tetragonal	Two equal axes and one unequal axis mutually perpendicular	One fourfold axis	$a = b \neq c$	$\alpha = \beta = \gamma = 90°$
Orthorhombic (or rhombic)	Three unequal axes mutually perpendicular	Three mutually perpendicular twofold axes, or two planes intersecting in a twofold axis	$a \neq b \neq c$	$\alpha = \beta = \gamma = 90°$
Hexagonal or trigonal	Three equal axes inclined at 120° with a fourth axis unequal and perpendicular to the other three	One sixfold axis or one threefold axis	$a = b \neq c$ $a = b = c$	$\alpha = \beta = 90°;$ $\gamma = 120°$ $\alpha = \beta = \gamma \neq 90°$
Monoclinic	Two axes at an oblique angle with a third perpendicular to the other two	One twofold axis or one plane	$a \neq b \neq c$	$\alpha = \beta = 90°;$ $\gamma \neq 90°$
Triclinic	Three unequal axes intersecting obliquely	No planes or axes of symmetry	$a \neq b \neq c$	$\alpha \neq \beta \neq \gamma \neq 90°$
Rhombohedral	Two equal axes making equal angle with each other			

*From J. A. Dean, ed., *Lange's Handbook of Chemistry*, 11th ed., copyright 1973 by McGraw-Hill Book Company; used with permission.

Mathematical signs and symbols

Symbol	Definition	Symbol	Definition
$+$	plus (sign of addition)	π	(pi), $= 3.14159 +$
$+$	positive	$°$	degrees
$-$	minus (sign of subtraction)	$'$	minutes
$-$	negative	$''$	seconds
$\pm$ ($\mp$)	plus or minus (minus or plus)	$\angle$	angle
$\times$	times, by (multiplication sign)	dx	differential of x
$\cdot$	multiplied by	Δ	(delta) difference
$\div$	sign of division	Δx	increment of x
$/$	divided by	$\partial u/\partial x$	partial derivative of u with respect to x
$:$	ratio sign, divided by, is to		
$::$	equals, as (proportion)	$\int$	integral of
$<$	less than	$\int_b^a$	integral of, between limits a and b
$>$	greater than		
$\ll$	much less than	$\oint$	line integral around a closed path
$\gg$	much greater than	Σ	(sigma) summation of
$=$	equals	$f(x), F(x)$	functions of x
$\equiv$	identical with	$\exp x = e^x$	($e =$ naperian log base) (abbreviation for e^x)
$\sim$	similar to		
$\approx$	approximately equals	∇	del or nabla, vector differential operator
$\cong$	approximately equals, congruent		
$\leq$	equal to or less than	∇^2	Laplacian operator
$\geq$	equal to or greater than	$\pounds$	Laplace operational symbol
$\neq$ $\neq$	not equal to	$4!$	factorial $4 = 1 \times 2 \times 3 \times 4$
$\rightarrow \doteq$	approaches	$\|x\|$	absolute value of x
$\propto$	varies as	$\dot{x}$	first derivative of x with respect to time
∞	infinity		
$\sqrt{}$	square root of	$\ddot{x}$	second derivative of x with respect to time
$\sqrt[3]{}$	cube root of		
$\therefore$	therefore	$\mathbf{A} \times \mathbf{B}$	vector product; magnitude of $\mathbf{A}$ times magnitude of $\mathbf{B}$ times sine of the angle from $\mathbf{A}$ to $\mathbf{B}$; $AB \sin \overline{AB}$
$\parallel$	parallel to		
$()[]\{\}$	parentheses, brackets and braces; quantities enclosed by them to be taken together in multiplying, dividing, etc.	$\mathbf{A} \cdot \mathbf{B}$	scalar product of $\mathbf{A}$ and $\mathbf{B}$; magnitude of $\mathbf{A}$ times magnitude of $\mathbf{B}$ times cosine of the angle from $\mathbf{A}$ to $\mathbf{B}$; $AB \cos \overline{AB}$
$\overline{AB}$	length of line from A to B		

Mathematical notation

Mathematical logic.

$p, q, P(x)$	Sentences, propositional functions, propositions
$\neg p, \sim p, \text{non } p, Np$	Negation, read "not p" ($\neq$: read "not equal")
$p \vee q, p + q, Apq$	Disjunction, read "p or q," "p, q," or both
$p \wedge q, p \cdot q, p \& q, Kpq$	Conjunction, read "p and q"
$p \rightarrow q, p \supset q, p \Rightarrow q,$ Cpq	Implication, read "p implies q" or "if p then q"
$p \leftrightarrow q, p \equiv q, p \Longleftrightarrow q,$ $Epq, p \text{ iff } q$	Equivalence, read "p is equivalent to q" or "p if and only if q"
n.a.s.c.	Read "necessary and sufficient condition"
$(), [], \{\}, \ldots, \cdot\cdot$	Parentheses
$V, \forall, \Sigma$	Universal quantifier, read "for all" or "for every"
$\exists, \exists, \Pi$	Existential quantifier, read "there is a" or "there exists"
$\vdash$	Assertion sign ($p \vdash q$: read "q follows from p"; $\vdash p$: read "p is or follows from an axiom," or "p is a tautology"
$0, 1$	Truth, falsity (values)
$=$	Identity
$\overset{\text{Df}}{=}, \overset{\text{df}}{=}, =, \equiv$ $\underset{\text{df}}{=}$	Definitional identity
$\blacksquare$	"End of proof"; "QED"

Set theory, relations, functions.

X, Y	Sets
$x \in X$	x is a member of the set X
$x \notin X$	x is not a member of X
$A \subset X, A \subseteq X$	Set A is contained in set X
$A \not\subset X, A \not\subseteq X$	A is not contained in X
$X \cup Y, X + Y$	Union of sets X and Y
$X \cap Y, X \cdot Y$	Intersection of sets X and Y
$+, \dot{+}, \bigcirc$	Symmetric difference of sets
$\cup X_i, \Sigma X_i$	Union of all the sets X_i
$\cap X_i, \Pi X_i$	Intersection of all the sets X_i
$\varnothing, 0, \Lambda$	Null set, empty set
$X', \mathbf{C}X, CX$	Complement of the set X
$X - Y, X \backslash Y$	Difference of sets X and Y
$\hat{x}(P(x)), \{x | P(x)\},$ $\{x : P(x)\}$	The set of all x with the property P

$(x,y,z), \langle x,y,z \rangle$	Ordered set of elements x, y, and z; to be distinguished from (x,z,y), for example
$\{x,y,z\}$	Unordered set, the set whose elements are x, y, z, and no others
$\{a_1, a_2, \ldots, a_n\},$ $\{a_i\}_{i=1,2,\ldots,n}, \{a_i\}_{i=1}^n$	The set whose members are a_i, where i is any whole number from 1 to n
$\{a_1, a_2, \ldots\},$ $\{a_i\}_{i=1,2,\ldots}, \{a_i\}_{i=1}^\infty$	The set whose members are a_i, where i is any positive whole number
$X \times Y$	Cartesian product, set of all (x,y) such that $x \in X, y \in Y$
$\{a_i\}_{i \in I}$	The set whose elements are a_i, where $i \in I$
$xRy, R\{x,y\}$	Relation
$\equiv, \cong, \sim, \simeq$	Equivalence relations, for example, congruence
$\geqq, \geq, >, \succcurlyeq, \gg, \leqq,$ $\leq, <$	Transitive relations, for example, numerical order
$f : X \rightarrow Y, X \overset{f}{\rightarrow} Y,$ $X \rightarrow Y, f \in Y^X$	Function, mapping, transformation
$f^{-1}, \overset{-1}{f}, X \overset{f^{-1}}{\longleftarrow} Y$	Inverse mapping
$g \circ f$	Composite functions: $(g \circ f)(x) = g(f(x))$
$f(X)$	Image of X by f
$f^{-1}(X)$	Inverse-image set, counter image
1-1, one-one	Read "one-to-one correspondence"
$\begin{matrix} X & \overset{f}{\rightarrow} & Y \\ \phi \downarrow & & \downarrow \psi \\ W & \overset{g}{\rightarrow} & Z \end{matrix}$	Diagram: the diagram is commutative in case $\psi \circ f = g \circ \phi$
$f|A$	Partial mapping, restriction of function f to set A
$X, \text{card } X, |X|$	Cardinal of the set A
$\aleph_0, d$	Denumerable infinity
$\mathfrak{c}, c, 2^{\aleph_0}$	Power of continuum
ω	Order type of the set of positive integers
σ-	Read "countably"

Number, numerical functions.

1.4; 1,4; 1·4	Read "one and four-tenths"
1(1)20(10)100	Read "from 1 to 20 in intervals of 1, and from 20 to 100 in intervals of 10"
const	Constant

$A \geqq 0$ — The number A is nonnegative, or, the matrix A is positive definite, or, the matrix A has non-negative entries

$x|y$ — Read "x divides y"

$x \equiv y \bmod p$ — Read "x congruent to y modulo p"

$a_0 + \dfrac{1}{a_1 +} \dfrac{1}{a_2 +} \cdots ,$

$a_0 + \dfrac{1|}{|a_1} + \cdots$ — Continued fractions

$[a,b]$ — Closed interval

$[a,b),\ [a,b[$ — Half-open interval (open at the right)

$(a,b),\ ]a,b[$ — Open interval

$[a,\infty),\ [a,\rightarrow[$ — Interval closed at the left, infinite to the right

$(-\infty, \infty),\ ]\leftarrow,\rightarrow[$ — Set of all real numbers

$\max_{x \epsilon X} f(x),$
$\max \{f(x)|x \epsilon X\}$ — Maximum of $f(x)$ when x is in the set X

min — Minimum

sup, l.u.b. — Supremum, least upper bound

inf, g.l.b. — Infimum, greatest lower bound

$\lim_{x \to a} f(x) = b,$
$\lim_{x = a} f(x) = b,$
$f(x) \to b$ as $x \to a$ — b is the limit of $f(x)$ as x approaches a

$\lim_{x \to a-} f(x),$
$\lim_{x = a-0} f(x), f(a-)$ — Limit of $f(x)$ as x approaches a from the left

$\limsup, \overline{\lim}$ — Limit superior

$\liminf, \underline{\lim}$ — Limit inferior

l.i.m. — Limit in the mean

$z = x + iy = re^{i\theta},$
$\zeta = \xi + i\eta,$
$w = u + iv = \rho e^{i\phi}$ — Complex variables

$\bar{z}\ z^*$ — Complex conjugate

Re, $\Re$ — Real part

Im, $\Im$ — Imaginary part

arg — Argument

$\dfrac{\partial(u,v)}{\partial(x,y)},$
$\dfrac{D(u,v)}{D(x,y)}$ — Jacobian, functional determinant

$\displaystyle\int_E f(x)\, d\mu(x)$ — Integral (for example, Lebesgue integral) of function f over set E with respect to measure μ

$f(n) \sim \log n$ as $n \to \infty$ — $f(n)/\log n$ approaches 1 as $n \to \infty$

$f(n) = O(\log n)$ as $n \to \infty$ — $f(n)/\log n$ is bounded as $n \to \infty$

$f(n) = o(\log n)$ — $f(n)/\log n$ approaches zero

$f(x) \nearrow b, f(x) \uparrow b$ — $f(x)$ increases, approaching the limit b

$f(x) \downarrow b, f(x) \searrow b$ — $f(x)$ decreases, approaching the limit b

a.e., p.p. — Almost everywhere

ess sup — Essential supremum

$C^0, C^0(X), C(X)$ — Space of continuous functions

$C^k, C^k[a,b]$ — The class of functions having continuous kth derivative (on $[a,b]$)

C' — Same as C^1

$\text{Lip}_\alpha, \text{Lip}\,\alpha$ — Lipschitz class of functions

$L^p, L_p, L^p[a,b]$ — Space of functions having integrable absolute pth power (on $[a,b]$)

L' — Same as L^1

$(C,\alpha), (C,p)$ — Cesàro summability

Special functions.

$[x]$ — The integral part of x

$\binom{n}{k}, {}^nC_k, {}_nC_k$ — Binomial coefficient $n!/k!(n-k)!$

$\left(\dfrac{n}{p}\right)$ — Legendre symbol

$e^x, \exp x$ — Exponential function

$\sinh x, \cosh x, \tanh x$ — Hyperbolic functions

$\operatorname{sn} x, \operatorname{cn} x, \operatorname{dn} x$ — Jacobi elliptic functions

$\wp(x)$ — Weierstrass elliptic function

$\Gamma(x)$ — Gamma function

$J_\nu(x)$ — Bessel function

$\chi_X(x)$ — Characteristic function of the set X: $\chi_X(x) = 1$ in case $x \epsilon X$, otherwise $\chi_X(x) = 0$

$\operatorname{sgn} x$ — Signum: $\operatorname{sgn} 0 = 0$, while $\operatorname{sgn} x = x/|x|$ for $x \neq 0$

$\delta(x)$ — Dirac delta function

Algebra, tensors, operators.

$+, \cdot, \times, \circ, \top, \tau$ — Laws of composition in algebraic systems

$e, 0$ — Identity, unit, neutral element (of an additive system)

$e, 1, I$ — Identity, unit, neutral element (of a general algebraic system)

e, ϵ, E, P — Idempotent

a^{-1} — Inverse of a

$\operatorname{Hom}(M,N)$ — Group of all homomorphisms of M into N

G/H — Factor group, group of cosets

$[K:k]$ — Dimension of K over k

$\oplus, +$ — Direct sum

$\otimes$ — Tensor product, Kronecker product

$\wedge$ — Exterior product, Grassmann product

$\vec{x}, \mathbf{x}, \mathfrak{x}, \underline{x}$ — Vector

$\vec{x} \cdot \vec{y}$, $\mathbf{x} \cdot \mathbf{y}$, $(\mathfrak{x},\mathfrak{h})$	Inner product, scalar product, dot product
$\mathbf{x} \times \mathbf{y}$, $[\mathfrak{x},\mathfrak{h}]$, $\mathbf{x} \wedge \mathbf{y}$	Outer product, vector product, cross product
$\|x\|$, $|x|$, $\|x\|$, $\|x\|_p$	Norm of the vector x
Ax, xA	The image of x under the transformation A
δ_{ij}	Kronecker delta: $\delta_{ii}=1$, while $\delta_{ij}=0$ for $i \neq j$
A', tA, A^t, ${}^tA$	Transpose of the matrix A
A^*, $\tilde{A}$	Adjoint, Hermitian conjugate of A
$\operatorname{tr} A$, $\operatorname{Sp} A$	Trace of the matrix A
$\det A$, $|A|$	Determinant of the matrix A
$\Delta^n f(x)$, $\Delta_h^n f$, $\underset{h}{\Delta^n} f(x)$	Finite differences
$[x_0,x_1]$, $[x_0,x_1,x_2]$, $\underset{x_1}{\Delta}u_{x_0}$, $[x_0,x_1]_f$	Divided differences
∇f, $\operatorname{grad} f$	Read "gradient of f"
$\nabla \cdot \mathbf{v}$, $\operatorname{div} \mathbf{v}$	Read "divergence of $\mathbf{v}$"
$\nabla \times \mathbf{v}$, $\operatorname{curl} \mathbf{v}$, $\operatorname{rot} \mathbf{v}$	Read "curl of $\mathbf{v}$"
∇^2, Δ, $\operatorname{div} \operatorname{grad}$	Laplacian
$[X,Y]$	Poisson bracket, or commutator, or Lie product
$\mathrm{GL}(n,R)$	Full linear group of degree n over field R
$\mathrm{O}(n,R)$	Full orthogonal group
$\mathrm{SO}(n,R)$, $\mathrm{O}^+(n,R)$	Special orthogonal group

Topology.

E^n	Euclidean n-space
S^n	n-sphere
$A \times B$	Cartesian product of set A and set B
$\rho(p,q)$, $d(p,q)$	Metric, distance (between points p and q)
$\overline{X}$, X^-, $\operatorname{cl} X$, X^c	Closure of the set X
$\operatorname{Fr}X$, $\operatorname{fr}X$, ∂X, $\operatorname{bdry} X$, $X^{\cdot}$	Frontier, boundary of X
$\operatorname{int} X$, $\mathring{X}$	Interior of X
T_2 space	Hausdorff space
F_σ	Union of countably many closed sets
G_δ	Intersection of countably many open sets
$\dim X$	Dimensionality, dimension of X
$\pi_1(X)$	Fundamental group of the space X
$\pi_n(X)$, $\pi_n(X,A)$	Homotopy groups
$H_n(X)$, $H_n(X,A;G)$, $H_*(X)$	Homology groups
$H^n(X)$, $H^n(X,A;G)$, $H^*(X)$	Cohomology groups

Probability and statistics.

X, Y	Random variables
$P(X \leqq 2)$, $\Pr\{X \leqq 2\}$	Probability that $X \leqq 2$
$P(X \leqq 2|Y \geqq 1)$	Conditional probability
$E(X)$, $\mathrm{E}(X)$	Expectation of X
$E(X|Y \geqq 1)$	Conditional expectation
c.d.f.	Cumulative distribution function
p.d.f.	Probability density function
c.f.	Characteristic function
$\bar{x}$	Mean (especially, sample mean)
σ, s.d.	Standard deviation
σ^2, Var, var	Variance
μ_1, μ_2, μ_3, μ_i, μ_{ij}	Moments of a distribution
ρ	Coefficient of correlation
$\rho_{12 \cdot 34}$	Partial correlation coefficient

Differential geometry.

R_{abcd}	Riemann-Christoffel tensor, curvature tensor
R_{ab}	Ricci tensor, contracted curvature tensor
R	Ricci scalar, scalar curvature
C_{abcd}	Weyl tensor

Causal structure.

I^+	Chronological future
I^-	Chronological past
J^+	Causal future
J^-	Causal past
D^+	Future Cauchy development
D^-	Past Cauchy development
H^+	Future Cauchy horizon
H^-	Past Cauchy horizon
$\mathscr{I}^+$	Boundary of the past of future null infinity, event horizon

Boundary of space-time.

Δ	C-boundary of a space-time
$\mathscr{M}^*$	Union of a space-time $\mathscr{M}$ with its C-boundary
∂	b-boundary of a space-time
$\mathscr{M}^+$	Union of a space-time $\mathscr{M}$ with its b-boundary

C-boundaries of asymptotically simple and empty spaces:

$\mathscr{I}^+$	Future null infinity (symbol pronounced "scri plus")
$\mathscr{I}^-$	Past null infinity (symbol pronounced "scri minus")
i^+	Future timelike infinity
i^-	Past timelike infinity
i^0	Spatial infinity

Planets

Table 1. Elements of planetary orbits

Planet	Symbol	Mean distance (semimajor axis)		Sidereal period of revolution		Synodic period, days	Mean velocity, km/sec	Eccentricity	Inclination
		AU	10^6 km	Years	Days				
Mercury	☿	0.387	57.9	0.241	87.97	115.88	47.90	0.206	7°00′
Venus	♀	0.723	108.2	0.615	224.70	583.92	35.05	0.007	3°24′
Earth	⊕	1.000	149.6	1.000	365.26		29.80	0.017	0°00′
Mars	♂	1.524	227.9	1.881	686.98	779.94	24.14	0.093	1°51′
Jupiter	♃	5.203	778.3	11.862	4332.59	398.88	13.06	0.048	1°18′
Saturn	♄	9.546	1428.	29.458	10759.	378.09	9.65	0.056	2°30′
Uranus	♂	19.20	2872.	84.018	30687.	369.66	6.80	0.047	0°46′
Neptune	♆	30.09	4498.	164.78	60184.	367.49	5.43	0.009	1°47′
Pluto	♇	39.5	5910.	248.4	90700.	366.74	4.74	0.247	17°09′

Table 2. Physical elements of planets

Planet	Equatorial radius, r_e (♂ = 1) km		Ellipticity	Volume (♂ = 1)	Mass (♂ = 1)	Density, g/cm³	Escape velocity, km/sec	Rotation period	Inclination of axis*
Mercury	0.38	2439	0.000	0.055	0.053	5.3	4.2	58.65d	0
Venus	0.95	6052	0.000	0.87	0.815	5.2	10.4	243d	177
Earth	1.00	6378	0.0034	1.00	1.000	5.52	11.2	23h56m4.1s	23.45
Mars	0.53	3397	0.004	0.150	0.107	3.95	5.0	24h37m22.7s	24.0
Jupiter	11.20	71400	0.062	1317.	318.00	1.33	61.	9h50m30s†	3.1
Saturn	9.47	60100	0.096	762.	95.22	0.69	37.	10h14m‡	26.75
Uranus	4.06	25900	0.06	67.	14.55	1.20	21.	10h49m	97.9
Neptune	3.87	24700	0.02	58.	17.23	1.64	24.	15h40m	29
Pluto	0.4?	2700?	?		0.06?	0.1?	5 ?	5?	6.3d	?

*To perpendicular to orbit, in arc degrees.
†Latitude < 12° (system I); 9h55m40.6s, latitude > 12° (system II).
‡Near Equator; 10h38m at intermediate latitudes.

Satellites

Table 3. Properties of satellites

Planet	Satellite	Mean distance of planet from satellite, 10^6 km	Sidereal period, days	Diameter, km	Visual magnitude at mean opposition
Earth	Moon	0.384	27.322	3476	−12.7
Mars	Phobos	0.009	0.319	$20 \times 23 \times 28$*	+11.5
	Deimos	0.023	1.262	$10 \times 12 \times 16$*	+12.6
Jupiter	I Io	0.422	1.769	3640	+ 4.9
	II Europa	0.671	3.551	3050	+ 5.3
	III Ganymede	1.070	7.155	5270	+ 4.6
	IV Callisto	1.883	16.689	5000	+ 5.6
	V Amalthea	0.181	0.498	160	+13
	VI Himalia	11.476	250.566	100	+14.2
	VII Elara	11.737	259.65	24	+17
	VIII Pasiphae	23.5	739	20	+18
	IX Sinope	23.6	758	18	+18.6
	X Lysithea	11.7	259.22	16	+18.8
	XI Carme	22.6	692	18	+18.6
	XII Ananke	21.2	630	16	+18.7
	XIII Leda	11.1	~240	8	+20†
	XIV	−	−	−	+21
Saturn	I Mimas	0.186	0.942	400	+12.1
	II Enceladus	0.238	1.370	500	+11.8
	III Tethys	0.295	1.888	1000	+10.3
	IV Dione	0.377	2.737	800	+10.4
	V Rhea	0.527	4.518	1500	+ 9.7
	VI Titan	1.222	15.945	5800	+ 8.4
	VII Hyperion	1.481	21.277	200	+14.2
	VIII Iapetus	3.560	79.331	1500	+10.2 − 11.9
	IX Phoebe	12.930	550.45	300	+16.5
	X Janus	0.159	0.750	220	+14
Uranus	I Ariel	0.192	1.282	700	+14.3
	II Umbriel	0.267	1.786	500	+15.1
	III Titania	0.438	2.930	1000	+13.9
	IV Oberon	0.586	3.919	900	+14.1
	V Miranda	0.130	0.872	240	+16.8
Neptune	Triton	0.355	5.877	3800	+13.6
	Nereid	5.562	359.88	240	+19.1

*Dimensions.
†Photographic magnitude.

Stars

Table 4. The 26 nearest stars (from P. van de Kamp)[a]

Name	Parallax, seconds of arc	Distance, light-years	Annual proper motion, seconds of arc	Radial velocity, km/sec	Transverse velocity, km/sec	Apparent magnitude and spectrum	Absolute magnitude
Sun						−26.7 G2	+4.8
α Centauri[b]	0.760	4.3	3.68	−25	23	+ 0.3 G2(1.7 K5)	4.7(6.1)[e]
Barnard's star	.545	6.0	10.30	−108	90	9.5 M5	13.2
Wolf 359	.421	7.7	4.84	+13	54	13.5 M6e, v[f]	16.6
Luyten 726−8	.410	7.9	3.35	+29	38	12.5 M6e (13.0 M6e)v	15.6(16.1)
Lalande 21185	.398	8.2	4.78	−86	57	7.5 M2	10.5
Sirius[c]	.375	8.7	1.32	−8	16	−1.5 A0(8.7 DA)	1.4(11.6)
Ross 154	.351	9.3	0.67	−4	9	10.6 M5e	13.3
Ross 248	.316	10.3	1.58	−81	23	12.2 M6e	14.7
ε Eri	.303	10.8	0.97	+15	15	3.8 K2	6.2
Ross 128	.298	10.9	1.40	−13	22	11.1 M5	13.5
61 Cyg	.293	11.1	5.22	−64	84	5.6 K6 (6.3 M0)	7.9(8.6)
Luyten 789−6	.292	11.2	3.27	−60	53	12.2 M6	14.5
Procyon	.288	11.3	1.25	−3	20	0.5 F5 (10.8 DA?)	2.8(13.1)
ε Indi	.285	11.4	4.67	−40	77	4.7 K5	7.0
Σ 2398	.280	11.6	2.29	+1	38	8.9 M4 (9.7 M4)	11.1(11.9)
Groombr. 34	.278	11.7	2.91	+14	49	8.1 M2e (10.9 M4e), v	10.3(13.1)
τ Cet	.275	11.8	1.92	−16	33	3.6 G4	5.8
Lacaille 9352	.273	11.9	6.87	+10	118	7.2 M2	9.4
+5° 1668	.263	12.4	3.73	+26	67	9.8 M4	12.0
Lacaille 8760	.255	12.8	3.46	+23	64	6.6 M1	8.6
Kapetyn's star	.251	13.0	8.79	+242	166	9.2 M0	11.2
Ross 614	.251	13.1	0.97	+24	18	11.1 M5e(14.8)	13.1(16.8)
Kruger 60	.249	13.1	0.87	−24	16	9.9 M4 (11.4 M5e)	11.9(13.4)
−12° 4523	.244	13.4	1.24	−13	24	10.0 M5	11.9
vMa 2[d]	.236	13.8	2.98	+70	59	12.3 DG	14.2

[a]In 26 listings there are 36 individual stars, and some have still-undiscovered faint companions.
[b]α Centauri has a second companion (dM5e) of absolute magnitude + 15.4.
[c]Sirius and Procyon each have a white dwarf companion.
[d]vMa 2 is a white dwarf.
[e]Parentheses indicate uncertain value.
[f]Here v= variable flare star which may emit bursts of light and even radio noise.

Table 5. The 25 brightest stars (from H. L. Johnson)*

Star	Name	Spectrum	Absolute visual magnitude, M_v	Visual brightness, V	Color index, B − V	Remarks
α CMa	Sirius	A1 V	+1.4	−1.43	0.00	
α Car	Canopus	F0 Ia	−4.5	−0.73	+0.15	
α Cen		G2 V	+4.7	−0.27	+0.66	Double
α Boo	Arcturus	K2 IIIp	−0.1	−0.06	+1.23	
α Lyr	Vega	A0 V	+0.5	+0.04	0.00	
α Aur	Capella	G0 IIIp	−0.6	+0.09	+0.80	Spectroscopic binary, double
β Ori	Rigel	B8 Ia	−7	+0.15	−0.04	Double
α CMi	Procyon	F5 IV-V	+2.7	+0.37	+0.41	
α Eri	Achernar	B3 V	−2	+0.53	−0.16	
β Cen		B0.5 V	−4	+0.66	−0.21	Double
α Ori	Betelgeuse	M2 Iab	−5	+0.7	+1.87	Variable
α Aql	Altair	A7 IV-V	+2.2	+0.80	+0.22	
α Tau	Aldebaran	K5 III	−0.7	+0.85	+1.52	Variable, double
α Cru		B0.5 V	−4	+0.87	−0.24	Double
α Sco	Antares	M1 Ib	−4	+0.98	+1.80	Double, variable
α Vir	Spica	B1 V	−3	+1.00	−0.23	Spectroscopic binary
α PsA	Fomalhaut	A3 V	+1.9	+1.16	+0.09	
β Gem	Pollux	K0 III	+1.0	+1.16	+1.01	
α Cyg	Deneb	A2 Ia	−7	+1.26	+0.09	
β Cru		B0.5 IV	−4	+1.31	−0.23	
α Leo	Regulus	B7 V	−0.7	+1.36	−0.11	Double
ε CMa	Adhara	B2 II	−5	+1.49	−0.17	
α Gem	Castor	A0	+0.9	+1.59	+0.05	Double, spectroscopic binary
λ Sco	Shaula	B2 IV	−3	+1.62	−0.23	
γ Ori	Bellatrix	B2 III	−4	+1.64	−0.23	

*The spectra are from H. L. Johnson and W. W. Morgan; colors and magnitudes are photoelectric, V being the equivalent of a visual brightness and B − V a blue minus visual color index. The absolute visual magnitudes M_v are based on measured parallaxes; when only one significant figure is given, however, they are only estimates.

Table 6. Approximate physical parameters of the stars

Properties of the sun:

G2, $L_\odot = 3.9 \times 10^{33}$ ergs/sec, $M_\odot = 2 \times 10^{33}$ g, $T = 5750$ K, $M_v = +4.64$, $R_\odot = 6.96 \times 10^{10}$ cm, $\log g = +4.44$, $\log \rho = +0.16$

Type	Color B − V	M_v	$\log L/L_\odot$	$T/1000°$	$\log R/R_\odot$	$\log M/M_\odot$	Remarks
Main sequence							
O8	−0.3 :*	− 5	+5.05	35	+0.96	+1.25	Uncertain, wide range
B0	−0.32:	− 4.3	+4.66	25	+1.05	+1.15	Uncertain
B1	−0.28	− 3.5	+4.06	22	+0.85	+1.05	
B2	− .24	− 2.8	+3.72	20	+0.76	+0.95	
B5	− .16	− 1.3	+2.96	15	+ .62	+ .78	
A0	0.00	+ 0.8	+1.73	11	+ .31	+ .45	
A5	+ .19	+ 1.9	+1.21	8.7	+ .24	+ .25	
F0	+ .37	+ 2.5	+0.85	7.6	+ .18	+ .14	
F5	+ .47	+ 3.5	+ .49	6.6	+ .12	+ .08	
G0	+ .60	+ 4.2	+ .21	6.0	+ .06	+ .04	
G5	+ .70	+ 5.2	− .19	5.5	− .06	− .17	
K0	+ .86	+ 6.1	− .55	5.1	− .18	− .22	
K5	+1.24	+ 7.5	− .84	4.4	− .18	− .25	
M0	+1.45:	+ 9.0	−1.23	3.6	− .22	− .30	
M2	+1.5 :	+10.0	−1.48	3.2	− .24	− .40	
M4	+1.6 :	+12.0	−1.91	3.1	− .42	− .55	Uncertain, because of
M6	+1.8 :	+15.0	−2.71	2.9	− .76	−0.90	steepness of main sequence
Giant stars							
G0		+ 0.7	+1.65	5.3	+0.90		Masses, to 4 M$_\odot$
K0		+ 0.2	+2.04	4.2	+1.30		Masses, to 4 M$_\odot$
M0		− 0.4	+2.72	3.3	+1.84		Masses, to 4 M$_\odot$
Supergiant stars							
B0		− 7	+5.66	25	+1.55		All data have wide range
A0		− 7	+4.77	10	+1.90		All data have wide range
K0		− 7	+5.26	3.6	+3.03		All data have wide range
M0		− 7	+5.66	3.0	+3.39		All data have wide range

*Colons indicate discordant determinations.